Functional Plant Ecology

Second Edition

BOOKS IN SOILS, PLANTS, AND THE ENVIRONMENT

Soil Biochemistry, Volume 1, edited by A. D. McLaren and G. H. Peterson

Soil Biochemistry, Volume 2, edited by A. D. McLaren and J. Skujins

Soil Biochemistry, Volume 3, edited by E. A. Paul and A. D. McLaren

Soil Biochemistry, Volume 4, edited by E. A. Paul and A. D. McLaren

Soil Biochemistry, Volume 5, edited by E. A. Paul and J. N. Ladd

Soil Biochemistry, Volume 6, edited by Jean-Marc Bollag and G. Stotzky

Soil Biochemistry, Volume 7, edited by G. Stotzky and Jean-Marc Bollag

Soil Biochemistry, Volume 8, edited by Jean-Marc Bollag and G. Stotzky

Soil Biochemistry, Volume 9, edited by G. Stotzky and Jean-Marc Bollag

Organic Chemicals in the Soil Environment, Volumes 1 and 2, edited by C. A. I. Goring and J. W. Hamaker

Humic Substances in the Environment, M. Schnitzer and S. U. Khan

Microbial Life in the Soil: An Introduction, T. Hattori

Principles of Soil Chemistry, Kim H. Tan

Soil Analysis: Instrumental Techniques and Related Procedures, edited by Keith A. Smith

Soil Reclamation Processes: Microbiological Analyses and Applications, edited by Robert L. Tate III and Donald A. Klein

Symbiotic Nitrogen Fixation Technology, edited by Gerald H. Elkan

Soil--Water Interactions: Mechanisms and Applications, Shingo Iwata and Toshio Tabuchi with Benno P. Warkentin

Soil Analysis: Modern Instrumental Techniques, Second Edition, edited by Keith A. Smith

Soil Analysis: Physical Methods, edited by Keith A. Smith and Chris E. Mullins

Growth and Mineral Nutrition of Field Crops, N. K. Fageria, V. C. Baligar, and Charles Allan Jones

Semiarid Lands and Deserts: Soil Resource and Reclamation, edited by J. Skujins

Plant Roots: The Hidden Half, edited by Yoav Waisel, Amram Eshel, and Uzi Kafkafi

Plant Biochemical Regulators, edited by Harold W. Gausman

Maximizing Crop Yields, N. K. Fageria

Transgenic Plants: Fundamentals and Applications, edited by Andrew Hiatt

Soil Microbial Ecology: Applications in Agricultural and Environmental Management, edited by F. Blaine Metting, Jr.

Principles of Soil Chemistry: Second Edition, Kim H. Tan

Water Flow in Soils, edited by Tsuyoshi Miyazaki

Handbook of Plant and Crop Stress, edited by Mohammad Pessarakli

Genetic Improvement of Field Crops, edited by Gustavo A. Slafer

Agricultural Field Experiments: Design and Analysis, Roger G. Petersen

Environmental Soil Science, Kim H. Tan

Mechanisms of Plant Growth and Improved Productivity: Modern Approaches, edited by Amarjit S. Basra

Selenium in the Environment, edited by W. T. Frankenberger, Jr. and Sally Benson

Plant–Environment Interactions, edited by Robert E. Wilkinson

Handbook of Plant and Crop Physiology, edited by Mohammad Pessarakli

Handbook of Phytoalexin Metabolism and Action, edited by M. Daniel and R. P. Purkayastha

Soil–Water Interactions: Mechanisms and Applications, Second Edition, Revised and Expanded, Shingo Iwata, Toshio Tabuchi, and Benno P. Warkentin

Stored-Grain Ecosystems, edited by Digvir S. Jayas, Noel D. G. White, and William E. Muir

Agrochemicals from Natural Products, edited by C. R. A. Godfrey

Seed Development and Germination, edited by Jaime Kigel and Gad Galili

Nitrogen Fertilization in the Environment, edited by Peter Edward Bacon

Phytohormones in Soils: Microbial Production and Function, William T. Frankenberger, Jr., and Muhammad Arshad

Handbook of Weed Management Systems, edited by Albert E. Smith

Soil Sampling, Preparation, and Analysis, Kim H. Tan

Soil Erosion, Conservation, and Rehabilitation, edited by Menachem Agassi

Plant Roots: The Hidden Half, Second Edition, Revised and Expanded, edited by Yoav Waisel, Amram Eshel, and Uzi Kafkafi

Photoassimilate Distribution in Plants and Crops: Source–Sink Relationships, edited by Eli Zamski and Arthur A. Schaffer

Mass Spectrometry of Soils, edited by Thomas W. Boutton and Shinichi Yamasaki

Handbook of Photosynthesis, edited by Mohammad Pessarakli

Chemical and Isotopic Groundwater Hydrology: The Applied Approach, Second Edition, Revised and Expanded, Emanuel Mazor

Fauna in Soil Ecosystems: Recycling Processes, Nutrient Fluxes, and Agricultural Production, edited by Gero Benckiser

Soil and Plant Analysis in Sustainable Agriculture and Environment, edited by Teresa Hood and J. Benton Jones, Jr.

Seeds Handbook: Biology, Production, Processing, and Storage, B. B. Desai, P. M. Kotecha, and D. K. Salunkhe

Modern Soil Microbiology, edited by J. D. van Elsas, J. T. Trevors, and E. M. H. Wellington

Growth and Mineral Nutrition of Field Crops: Second Edition, N. K. Fageria, V. C. Baligar, and Charles Allan Jones

Fungal Pathogenesis in Plants and Crops: Molecular Biology and Host Defense Mechanisms, P. Vidhyasekaran

Plant Pathogen Detection and Disease Diagnosis, P. Narayanasamy

Agricultural Systems Modeling and Simulation, edited by Robert M. Peart and R. Bruce Curry

Agricultural Biotechnology, edited by Arie Altman

Plant–Microbe Interactions and Biological Control, edited by Greg J. Boland and L. David Kuykendall

Handbook of Soil Conditioners: Substances That Enhance the Physical Properties of Soil, edited by Arthur Wallace and Richard E. Terry

Environmental Chemistry of Selenium, edited by William T. Frankenberger, Jr., and Richard A. Engberg

Principles of Soil Chemistry: Third Edition, Revised and Expanded, Kim H. Tan

Sulfur in the Environment, edited by Douglas G. Maynard

Soil–Machine Interactions: A Finite Element Perspective, edited by Jie Shen and Radhey Lal Kushwaha

Mycotoxins in Agriculture and Food Safety, edited by Kaushal K. Sinha and Deepak Bhatnagar

Plant Amino Acids: Biochemistry and Biotechnology, edited by Bijay K. Singh

Handbook of Functional Plant Ecology, edited by Francisco I. Pugnaire and Fernando Valladares

Handbook of Plant and Crop Stress: Second Edition, Revised and Expanded, edited by Mohammad Pessarakli

Plant Responses to Environmental Stresses: From Phytohormones to Genome Reorganization, edited by H. R. Lerner

Handbook of Pest Management, edited by John R. Ruberson

Environmental Soil Science: Second Edition, Revised and Expanded, Kim H. Tan

Microbial Endophytes, edited by Charles W. Bacon and James F. White, Jr.

Plant–Environment Interactions: Second Edition, edited by Robert E. Wilkinson

Microbial Pest Control, Sushil K. Khetan

Soil and Environmental Analysis: Physical Methods, Second Edition, Revised and Expanded, edited by Keith A. Smith and Chris E. Mullins

The Rhizosphere: Biochemistry and Organic Substances at the Soil–Plant Interface, edited by Roberto Pinton, Zeno Varanini, and Paolo Nannipieri

Woody Plants and Woody Plant Management: Ecology, Safety, and Environmental Impact, Rodney W. Bovey

Metals in the Environment, M. N. V. Prasad

Plant Pathogen Detection and Disease Diagnosis: Second Edition, Revised and Expanded, P. Narayanasamy

Handbook of Plant and Crop Physiology: Second Edition, Revised and Expanded, edited by Mohammad Pessarakli

Environmental Chemistry of Arsenic, edited by William T. Frankenberger, Jr.

Enzymes in the Environment: Activity, Ecology, and Applications, edited by Richard G. Burns and Richard P. Dick

Plant Roots: The Hidden Half, Third Edition, Revised and Expanded, edited by Yoav Waisel, Amram Eshel, and Uzi Kafkafi

Handbook of Plant Growth: pH as the Master Variable, edited by Zdenko Rengel

Biological Control of Major Crop Plant Diseases edited by Samuel S. Gnanamanickam

Pesticides in Agriculture and the Environment, edited by Willis B. Wheeler

Mathematical Models of Crop Growth and Yield, , Allen R. Overman and Richard Scholtz

Plant Biotechnology and Transgenic Plants, edited by Kirsi-Marja Oksman Caldentey and Wolfgang Barz

Handbook of Postharvest Technology: Cereals, Fruits, Vegetables, Tea, and Spices, edited by Amalendu Chakraverty, Arun S. Mujumdar, G. S. Vijaya Raghavan, and Hosahalli S. Ramaswamy

Handbook of Soil Acidity, edited by Zdenko Rengel

Humic Matter in Soil and the Environment: Principles and Controversies, edited by Kim H. Tan

Molecular Host Plant Resistance to Pests, edited by S. Sadasivam and B. Thayumanayan

Soil and Environmental Analysis: Modern Instrumental Techniques, Third Edition, edited by Keith A. Smith and Malcolm S. Cresser

Chemical and Isotopic Groundwater Hydrology, Third Edition, edited by Emanuel Mazor

Agricultural Systems Management: Optimizing Efficiency and Performance, edited by Robert M. Peart and W. David Shoup

Physiology and Biotechnology Integration for Plant Breeding, edited by Henry T. Nguyen and Abraham Blum

Global Water Dynamics: Shallow and Deep Groundwater: Petroleum Hydrology: Hydrothermal Fluids, and Landscaping, , edited by Emanuel Mazor

Principles of Soil Physics, edited by Rattan Lal

Seeds Handbook: Biology, Production, Processing, and Storage, Second Edition, Babasaheb B. Desai

Field Sampling: Principles and Practices in Environmental Analysis, edited by Alfred R. Conklin

Sustainable Agriculture and the International Rice-Wheat System, edited by Rattan Lal, Peter R. Hobbs, Norman Uphoff, and David O. Hansen

Plant Toxicology, Fourth Edition, edited by Bertold Hock and Erich F. Elstner

Drought and Water Crises: Science, Technology, and Management Issues, edited by Donald A. Wilhite

Soil Sampling, Preparation, and Analysis, Second Edition, Kim H. Tan

Climate Change and Global Food Security, edited by Rattan Lal, Norman Uphoff, B. A. Stewart, and David O. Hansen

Handbook of Photosynthesis, Second Edition, edited by Mohammad Pessarakli

Environmental Soil-Landscape Modeling: Geographic Information Technologies and Pedometrics, edited by Sabine Grunwald

Water Flow In Soils, Second Edition, Tsuyoshi Miyazaki

Biological Approaches to Sustainable Soil Systems, edited by Norman Uphoff, Andrew S. Ball, Erick Fernandes, Hans Herren, Olivier Husson, Mark Laing, Cheryl Palm, Jules Pretty, Pedro Sanchez, Nteranya Sanginga, and Janice Thies

Plant–Environment Interactions, Third Edition, edited by Bingru Huang

Biodiversity In Agricultural Production Systems, edited by Gero Benckiser and Sylvia Schnell

Organic Production and Use of Alternative Crops, Franc Bavec and Martina Bavec

Handbook of Plant Nutrition, edited by Allen V. Barker and David J. Pilbeam

Modern Soil Microbiology, Second Edition, edited by Jan Dirk van Elsas, Janet K. Jansson, and Jack T. Trevors

The Rhizosphere: Biochemistry and Organic Substances at the Soil-Plant Interface, Second Edition, edited by Roberto Pinton, Zeno Varanini, and Paolo Nannipieri

Functional Plant Ecology, Second Edition, edited by Francisco I. Pugnaire and Fernando Valladares

Functional Plant Ecology

Second Edition

edited by

Francisco I. Pugnaire
Fernando Valladares

CRC Press is an imprint of the
Taylor & Francis Group, an **informa** business

CRC Press
Taylor & Francis Group
6000 Broken Sound Parkway NW, Suite 300
Boca Raton, FL 33487-2742

Printed in the United States of America on acid-free paper
10 9 8 7 6 5 4 3 2

International Standard Book Number-10: 0-8493-7488-X (Hardcover)
International Standard Book Number-13: 978-0-8493-7488-3 (Hardcover)

Library of Congress Cataloging-in-Publication Data

Functional plant ecology / edited by Francisco Pugnaire and Fernando Valladares. -- 2nd ed.
p. cm. -- (Books in soils, plants, and the environment ; 120)
Rev. ed. of: Handbook of functional plant ecology / edited by Francisco I. Pugnaire, Fernando Valladares. c1999.
Includes bibliographical references and index.
ISBN 978-0-8493-7488-3
1. Plant ecology. 2. Plant ecophysiology. I. Pugnaire, Francisco I., 1957- II. Valladares, Fernando, 1965- III. Handbook of functional plant ecology. IV. Title. V. Series.

QK901.H295 2007
581.7--dc22 2007000591

Visit the Taylor & Francis Web site at
http://www.taylorandfrancis.com

and the CRC Press Web site at
http://www.crcpress.com

Table of Contents

Preface

Diversity of plant form and life history and their distribution onto different habitats suggest that plant functions should underlie this diversity, providing tools to successfully and differentially thrive in every habitat. The knowledge of these functions is then the key to understand community and ecosystem structure and functioning, something that attracted the interest and effort of many plant ecologists trying to establish patterns of adaptive specialization in plants.

This volume on *Functional Plant Ecology* is an updated version of a successful first edition in which we tried to put together chapters from all areas of plant ecology to provide readers the broadest view of functional approaches to plant ecology. Our aim was to gather original reviews with an attractive presentation, giving a comprehensive overview of the topic with a historical perspective when needed. The book is intended for a broad audience, from plant ecologists to students, with characteristics of both a textbook and an essay book. We were not interested in presentation of new experimental data, novel theoretical interpretations, or hypotheses, but rather asked the authors to provide easy-to-read, up-to-date, and suggestive introductions to each topic.

Deciding the book composition was not an easy task, as many attractive, substantial topics emerged at first glimpse. Finally, only a short number made their way into the book, and we are aware that many important questions have been left out, but practical and technical reasons limited the extent of the volume. The book follows a bottom-up approach, from the more specific, detailed studies focusing on plant organs to the broadest ecosystem approaches, each gathering chapters on the most outstanding aspects.

The history, aims, and potentials of functional approaches are established in the first chapter, which also sets the limits of functional plant ecology, a science centered in the study of whole plants and that attempts to predict responses in plant functioning caused by environmental clues, emphasizing plant influence on ecosystem functions, services, and products, and aiming to extract patterns and functional laws from comparative analyses. The search for these patterns is likely to be most effective if driven by specific hypotheses tested on the basis of comparative analyses at the broadest possible scale. Functional laws thus developed may hold predictive power irrespective of whether they represent direct cause–effect relationships. Yet, the nested nature of the control of functional responses implies uncertainties when scaling functional laws, either toward lower or higher levels of organization.

We would like to express our sincere thanks to the authors who contributed to this volume for their efforts in updating their chapters and for meeting the deadlines over already busy timetables. Finally, we want to thank John Sulzycki for his support throughout and Pat Roberson for her help and patience. All of them made possible and greatly improved the quality of this work.

Editors

Francisco I. Pugnaire is a research scientist at the Arid Zones Experimental Station in Almería, Spain, an Institute of the Spanish National Council for Scientific Research. Author of numerous publications, his research interests focus on physiological ecology and community dynamics and on the effects of current environmental changes on biodiversity. He is actively involved in several EU and national initiatives in the science–policy interface. Centered on semiarid environments, he has also worked in desert ecosystems, tropical forests, and high mountain environments. Member of the editorial board of several international journals, he is also a member of the ecological societies of Spain, UK, and USA and vice president of the European Ecological Federation.

Fernando Valladares is currently a senior scientist of the Spanish Research Council (CSIC) at the Centre of Environmental Sciences and associate professor at the Rey Juan Carlos University of Madrid. He has published more than 150 scientific articles, and is the author or coauthor of 5 books. Dr. Valladares is a member of the editorial board of three of the most prestigious international journals of plant ecology and physiology, and he is an active member of the main ecological societies of Spain, UK, and USA. Dr. Valladares is involved in several panels and expert committees addressing global change issues and transferring ecological knowledge to society and policy makers at both national and international levels. His research on functional plant ecology and evolution is focused on phenotypic plasticity and on the physiological and morphological strategies of photosynthetic organisms to cope with changing and adverse environmental conditions. The research has been carried out in a range of ecosystems, spanning from Antarctica to the tropical rain forest, although his current research objectives are centered on low-productivity Mediterranean ecosystems.

Contributors

Cristina Armas
Department of Biology
Duke University
Durham, North Carolina

Luis Balaguer
Department of Plant Biology
Complutense University
Madrid, Spain

Elena Baraza
Laboratory of Ecology of Communities
Institute of Ecology
UNAM, Mexico D.F.

Jeremy Barnes
Institute for Research on the Enviornment and Sustainability
Newcastle University
Newcastle upon Tyne, England

Frank Berendse
Centre for Ecosystem Studies
Wageningen University
Wageningen, The Netherlands

Wolfram Beyschlag
Department of Experimental and Systems Ecology
University of Bielefeld
Bielefeld, Germany

Timothy M. Bleby
Department of Biology
Duke University
Durham, North Carolina

Wim G. Braakhekke
Centre for Ecosystem Studies
Wageningen University
Wageningen, The Netherlands

Ragan M. Callaway
Division of Biological Sciences
University of Montana
Missoula, Montana

Sofia R. Costa
Department of Botany
University of Coimbra
Coimbra, Portugal

Claire Damesin
Plant Ecophysiology
Paris-Sud University
Orsay, France

Alan Davison
School of Biology and Psychology: Division of Biology
Newcastle University
Newcastle upon Tyne, England

Carlos M. Duarte
Mediterranean Institute for Advanced Studies
Spanish Council for Scientific Research
Blanes, Spain

Helena Freitas
Department of Botany
University of Coimbra
Coimbra, Portugal

John A. Gamon
Department of Biology and Microbiology
California State University
Los Angeles, California

Eric Garnier
Centre for Functional and Evolutionary Ecology
CNRS
Montpellier, France

José M. Gómez
Department of Ecology
University of Granada
Granada, Spain

T.G. Allan Green
Department of Biological Sciences
Waikato University
Hamilton, New Zealand

Sarah E. Hobbie
Department of Ecology, Evolution, and Behavior
University of Minnesota
St. Paul, Minnesota

José A. Hódar
Terrestrial Ecology Group
Department of Ecology
University of Granada
Granada, Spain

William A. Hoffmann
Department of Plant Biology
North Carolina State University
Raleigh, North Carolina

Robert B. Jackson
Department of Biology
Duke University
Durham, North Carolina

Anna Jakobsson
Mediterranean Institute for Advanced Studies
Esporles, Mallorca
Balearic Islands, Spain

Richard Joffre
Centre for Functional and Evolutionary Ecology
CNRS
Montpellier, France

Ludger Kappen
Department of Botany
University of Kiel
Kiel, Germany

Kaoru Kitajima
Department of Botany
University of Florida
Gainesville, Florida

Hans de Kroon
Department of Experimental Plant Ecology
Institute of Water and Wetland Research
Radboud University
Nijmegen, The Netherlands

Esteban Manrique-Reol
Department of Agricultural and Environmental Science
University of Newcastle
Newcastle upon Tyne, England

Ernesto Medina
Center of Ecology
Venezuelan Institute for Scientific Investigations
Caracas, Venezuela

Ülo Niinemets
Institute of Environment and Agriculture
Estonian University of Life Sciences
Tartu, Estonia

Robert W. Pearcy
Division of Biological Sciences
University of California
Davis, California

William T. Pockman
Department of Biology
University of New Mexico
Albuquerque, New Mexico

Hendrik Poorter
Institute of Environmental Biology
Utrecht University
Utrecht, The Netherlands

Francisco I. Pugnaire
Arid Zones Research Station
Spanish Council for Scientific Research
Almeria, Spain

Hong-Lie Qiu
Department of Geography and Urban Analysis
California State University
Los Angeles, California

Tara K. Rajaniemi
Biology Department
University of Massachusetts
Dartmouth, Massachusetts

Serge Rambal
Centre for Functional and Evolutionary Ecology
CNRS
Montpellier, France

Heather L. Reynolds
Department of Biology
Indiana University
Bloomington, Indiana

Susana Rodríguez-Echeverría
Department of Botany
University of Coimbra
Coimbra, Portugal

Ronald J. Ryel
Department of Wildland Resources and the Ecology Center
Utah State University
Logan, Utah

Arturo Sanchez-Azofeifa
Department of Earth and Atmospheric Sciences
University of Alberta
Edmonton, Alberta, Canada

Leopoldo G. Sancho
Department Vegetal II
Complutense University
Madrid, Spain

Burkhard Schroeter
Botanical Institute
University of Kiel
Kiel, Germany

and

IPN – Leibniz Institute for Science Education
University of Kiel
Kiel, Germany

Anna Traveset
Mediterranean Institute for Advanced Studies
Esporles, Mallorca
Balearic Islands, Spain

Melvin T. Tyree
Aiken Forestry Sciences Laboratory
Burlington, Vermont

Fernando Valladares
Centre for Environmental Studies
Spanish Council for Scientific Research
Madrid, Spain

Mark Westoby
Department of Biological Sciences
Macquarie University
Sydney, Australia

S. Joseph Wright
Smithsonian Tropical Research Institute
Balboa, Republic of Panama

Regino Zamora
Department of Ecology
University of Granada
Granada, Spain

1 Methods in Comparative Functional Ecology

Carlos M. Duarte

CONTENTS

DEVELOPMENT OF FUNCTIONAL PLANT ECOLOGY

The quest to describe the diversity of extant plants and the identification of the basic mechanisms that allow them to occupy different environments have shifted scientists' attention from ancient Greece to the present. This interest was prompted by two fundamental aims: (1) a pressing need to understand the basic functions and growth requirements of plants because they provide direct and indirect services to human kind and (2) the widespread belief that the distribution of organisms was not random, for there was essential order in nature, and that there ought to be a fundamental link between differences in the functions of these organisms and their dominance in contrasting habitats. The notion that differences in plant functions are essential components of their fitness, accounting for their relative dominance in differential habitats, was, therefore, deeply rooted in the minds of early philosophers and, later on, naturalists. While animal functions were relatively easy to embrace from a simple parallel with our own basic functions, those of plants appeared more inaccessible to our ancestors, and the concepts of "plant" and "plant functions" have unfolded through the history of biology.

The examination of plant functions in modern science has largely followed a reductionistic path aimed at the explanation of plant functions in terms of the principles of physics and chemistry (Salisbury and Ross 1992). This reductionistic path is linked to the parallel transformation of traditional agricultural science into plant science and the technical developments needed to evolve from the examination of the coarser, integrative functions to those occurring at the molecular level. While this reductionistic path has led us toward a thorough catalog and understanding of plant functions, its limited usefulness to explain and predict the distribution of plants in nature has been a source of frustration. This is largely because of the multiple interactions that are expected to be involved in the responses of plants to a changing environment (Chapin et al. 1987). Yet, the need to achieve this predictive power has now transcended the academic arena to be a critical component of our ability to forecast the large-scale changes expected from on-going climatic change. For instance, increased CO_2 concentrations are expected to affect the water and nutrient requirements of plants, but resource availability is itself believed to be influenced by rising temperatures. Such feedback effects cannot be appropriately predicted from knowledge of the controls that individual factors exert on specific functions. Moreover, the changes expected to occur from climate change are likely to derive mostly from changes in vegetation and dominant plant types rather than from altered physiological responses of extant plants to the new conditions (Betts et al. 1997).

Failure of plant physiology and plant science to provide reliable predictions of the response of vegetation to changes in their environment likely derives from the hierarchical nature of plants. The response of higher organizational levels is not predictable from the dynamics of those at smaller scales, although these set constraints on the larger-scale responses of hierarchical systems. Component functions do not exist in isolation, as the dominant molecular approaches in modern plant physiology investigate them. Rather, these individual functions are integrated within the plants, which can modulate the responses expected from particular functions, leading to synergism, whether amplifying the responses through multiplicative effects or maintaining homeostasis against external forcing.

Recognition of the limitations of modern physiology to provide the needed predictions at the ecological scale led to the advent of plant ecophysiology, which tried to produce more relevant knowledge by the introduction of larger plant components, such as plant organs (instead of cells or organelles), as the units of analysis. Plant ecophysiology represented, therefore, an effort toward approaching the relevant scale of organization, by examining the functions of plant organs. Most often, however, practitioners of the discipline laid somewhere between the molecular approaches dominant in plant physiology and the more integrative approaches championed by plant physiological ecology. Because of the strong roots in the tradition of plant physiology, the suite of plant functions addressed by plant ecophysiology still targeted basic functions (e.g., photosynthesis, respiration, etc.) that can be studied through chemical and physical laws (Salisbury and Ross 1992). As a consequence, plant ecophysiology failed to consider more integrative plant functions, such as plant growth, which do not have a single physiological basis, but which are possibly the most relevant function for the prediction of plant performance in nature (cf. Chapter 3).

The efforts of plant ecophysiology proved, therefore, to be insufficient to achieve the prediction of how plant function allows the prediction of plant distribution and changes in plant abundance in a changing environment. Realization that the knowledge required to effectively address this question would be best achieved through a more integrative approach led to the advent of a new approach, hereafter referred to as "Functional Plant Ecology," which is emerging as a coherent research program (cf. Duarte et al. 1995). Functional plant ecology is centered on whole plants as the units of analysis, the responses of which to external forcing are examined in nature or under field conditions. Functional plant ecology, therefore, attempts to bypass the major uncertainties derived from the extrapolation of responses to nature (tested in isolated plant organs maintained under carefully controlled laboratory conditions) and to incorporate the integrated responses to multiple stresses displayed by plants onto the research program.

Although centered in whole plants, functional plant ecology encompasses lower and higher scales of organization, including studies at the organ or cellular level (e.g., Chapter 8), as well as the effect of changes in plant architecture or functions (e.g., Chapters 4 and 5), and the importance of life history traits (e.g., Chapters 15 and 16), interactions with neighbors (e.g., Chapters 17 and 18), and those with other components of the ecosystem (e.g., Chapter 19). In fact, this research program is also based on a much broader conception of plant functions than hitherto formulated. The plant functions that represent the core of present efforts in functional plant ecology are those by which plants influence ecosystem functions, particularly those that influence the services and products provided by ecosystems (Costanza et al. 1997). Hence, studies at lower levels of organization are conducted with the aim of being subsequently scaled up to the ecosystem level (e.g., Chapter 10).

Because of the emphasis on the prediction of the consequences of changes in vegetation structure and distribution for the ecosystem, functional plant ecology strives to encompass the broadest possible range of functional responses encountered within the biosphere. Yet, the elucidation of the range of possible functional responses of plants is not possible with the use of model organisms that characterize most of plant (and animal) physiology. Functional plant ecology arises, therefore, as an essentially comparative science concerned with the elucidation of the range of variations in functional properties among plants and the search for patterns and functional laws accounting for this variation (Duarte et al. 1995). While practitioners of functional plant ecology share the emphasis on the comparative analysis of plant function, the approaches used to achieve these comparisons range broadly. These differences rely largely on the breadth of the comparison and the description of the subject organisms in the analysis. The implications of these choices have not, however, been subject of explicit discussions despite their considerable epistemological implications and their impact on the power of the approach.

SCREENING, BROAD-SCALE COMPARISONS, AND THE DEVELOPMENT OF FUNCTIONAL LAWS

The success and the limitations of comparative functional plant ecology depend on the choices of approach made, involving the aims and scope of the comparison, as well as the methods to achieve them. The aims of the comparisons range widely, from the compilation of a "functional taxonomy" of particular sets of species or floras to efforts to uncover patterns of functional properties that may help formulate predictions or identify possible controlling factors. Many available floras incorporate considerable knowledge, albeit rarely quantitative, on the ecology of the species, particularly as to habitat requirements. An outstanding example is the Biological Flora of the British Isles (cf. *Journal of Ecology*), which incorporates some functional properties of the plants (e.g., Aksoy et al. 1998). The likely reason why "functional" floras are still few is the absence of standardized protocols to examine these properties while ensuring comparability of the results obtained. A step toward solving this bottleneck was provided by Hendry and Grime (1993), who described a series of protocols to obtain estimates of selected basic functional traits of plants in a comparable manner. Unfortunately, while exemplary, those protocols were specifically designed for use within the screening program of the British flora conducted by those investigators (Grime et al. 1988), rendering them of limited applicability in broader comparisons or comparisons of other vegetation types.

The screening approach may, if pursued further, generate an encyclopedic catalog of details on functional properties of different plants. Some ecologists may hold the hope that, once completed, such catalogs will reveal by themselves a fundamental order in the functional diversity of the plants investigated, conforming to a predictive sample similar to a "periodic table" of plant functional traits. While I do not dispute here that this goal may be eventually achieved, the resources required to produce such catalogs are likely to be overwhelming, since, by definition, such a screening procedure is of an exploratory nature, where the search for pattern is made a posteriori. Provided the number of elements to be screened and the potentially large number of traits to be tested, the cost-effectiveness of the approach is likely to prove suboptimal. A screening approach to functional plant ecology is, therefore, unlikely to improve our predictive power or to uncover basic patterns unless driven by specific hypotheses. Moreover, a hypothesis-driven search for pattern is likely to be most effective if based on a comparative approach, encompassing the broadest possible relevant range of plants. It is not necessary to test every single plant species to generate and test such general laws.

The comparisons attempted may differ greatly in scope, from comparisons of variability within species to broad-scale comparisons encompassing the broadest possible range of phototrophic organisms, from the smallest unicells to trees (e.g., Agustí et al. 1994, Nielsen et al. 1996). Experience shows, however, that the patterns obtained at one level of analysis may differ greatly from those observed at a broader level (Duarte 1990), without necessarily involving a conflict (Reich 1993). The scope of the comparison depends on the question that is posed. However, whenever possible, progress in comparative functional plant ecology should evolve from the general to the particular, thereby evolving from comparisons at the broadest possible scales to comparisons within species or closely related species. In doing so, we shall first draw the overall patterns, which yield the functional laws that help identify the constraints of possible functional responses in organisms.

The simplest possible comparison involves only two subjects, which are commonly enunciated under the euphemism of "contrasting" plant types. Such simple comparisons between one or a few subject plants are very common in the literature. These simple comparisons are, however, deceiving, for they cannot possibly be conclusive as to the nature of the differences or similarities identified. The implicit suggestion in these contrasts is that the trait on which the contrast is based (e.g., stress resistance vs. stress tolerance) is the cause underlying any observed differences in functional traits. This is fallacious and at odds with the simplest principles of method in science. Hence, contrasts are unlikely to be an effective approach to uncover regular patterns in plant function, since the degrees of freedom involved are clearly insufficient to venture any strong inferences on the outcome of the comparison.

Broad-scale comparisons involving functional responses across widely different species are, therefore, the approach of choice when the description of general laws is sought. The formulation of the comparative analysis of plant functions at the broadest possible level has been strongly advocated (Duarte et al. 1995), on the grounds that it will be most likely to disclose the basic rules that govern functional differences among plants. Broad-scale comparisons are most effective when encompassing the most diverse range of plant types possible (e.g., Agustí et al. 1994, Niklas 1994). In addition, they are most powerful when the functional properties are examined in concert with quantification of plant traits believed to influence the functions examined, for comparisons based on qualitative or nominal plant traits cannot be readily falsified and remain, therefore, unreliable tools for prediction. Hence, the development of broad-scale comparisons requires that both the functional property examined and the plant traits, which account for the differences in functional properties among the plants, are to be tested and carefully selected.

Broad-scale comparisons must be driven by a sound hypothesis or questions. Yet, this approach is of a statistical nature, often involving allometric relationships (e.g., Niklas 1994), so that observation of robust patterns is no guaranty of underlying cause and effect relationships, which must be tested experimentally. Nevertheless, the functional laws developed through broad-scale comparative analysis may hold predictive power, irrespective of whether they represent direct cause–effect relationships. This use requires, however, that the independent, predictor variable be simpler than the functional trait examined, if the law is to have practical application. Examples of such functional laws are many (e.g., Niklas 1994, Agustí et al. 1994, Duarte et al. 1995, Enríquez et al. 1996, Nielsen et al. 1996) and have been generally derived from the compilation of literature data and the use of plant cultures in phytotrons or the use of the functional diversity found, for instance, in botanical gardens (e.g., Nielsen et al. 1998). This choice of subject organisms is appropriate whenever the emphasis is on the functional significance of intrinsic properties. However, the effect of environment conditions can hardly be approached in this manner, and functional ecologists must transport the research to the field, which is the ultimate framework of relevance for this research program.

The comparative approach is also a powerful tool to examine the effect of environmental conditions in situ. Gradient analysis, where functional responses are examined along a clearly defined environmental gradient, has proven a powerful approach to investigate the relationship between plant function and environmental conditions (Vitousek and Matson 1990). Gradient analysis is particularly prone to spurious relationships where the relationship between the gradient property and the functional response reflects a functional relationship to a hidden factor covarying with that nominally defining the gradient. Inferences from gradient analysis are, therefore, also statistical in nature and have to be confirmed experimentally to elucidate the nature of underlying relationships.

Broad-scale comparisons often entail substantial uncertainty—typically in the order-of-magnitude range—in their predictions, which is a result of the breadth—typically four or more orders of magnitude—in the functions examined. This imprecision limits the applicability of these functional laws and renders their value greatest in the description of general, large-scale patterns, over which the effect of less-general functional regulatory factors, both intrinsic and extrinsic, is superimposed. Hence, multiple factors that constrain the functional responses of plants are nested in a descending rank of generality, whereby the total number of traits involved in the control is very large and only a few of them are general across a broad spectrum of plants.

The nested nature of the control of functional responses implies uncertainties when scaling functional laws, either toward lower or higher levels of organization (Duarte 1990). There is, therefore, no guaranty that the patterns observed at the broad-scale level will apply when focusing on particular functional types. Changes in the nature of the patterns when shifting across scales have prompted unnecessary disagreement in the past (Reich 1993). A thorough investigation of functional properties of plants should include, whenever possible, a nested research program, whereby the hypotheses on functional controls examined are first investigated at the broadest possible scale, to focus subsequently on particular subsets of species or functional groups, along environmental gradients.

The chapters in this volume provide a clear guide to functional ecology with examples, emphasizing the nested nature of the research program both within the chapters and in the manner in which they have been linked into different parts. The chapters also provide an overview of the entire suite of approaches available to address the goals of functional ecology, providing, therefore, a most useful tool box for prospective practitioners of the research program. The resulting set provides, therefore, a heuristic description of functional ecology, which should serve the dual role of providing a factual account of the achievements of functional ecology while endowing the reader with the tools to design research within this important research program.

REFERENCES

Agustí, S., S. Enríquez, H. Christensen, K. Sand-Jensen, and C.M. Duarte, 1994. Light harvesting among photosynthetic organisms. Functional Ecology 8: 273–279.

Aksoy, A., J.M. Dixon, and W.H.G. Hale, 1998. Capsella bursa-pastoris (L.) Medikus (Thlapsi bursa-pastoris L., Bursa bursa-pastoris (L.). Shull, Bursa pastoris (L.) Weber). Journal of Ecology 86: 171–186.

Betts, R.A., P.M. Cox, S.E. Lee, and F.I. Woodward, 1997. Contrasting physiological and structural vegetation feedbacks in climate change simulations. Nature 387: 796–799.

Chapin III, F.S., A.J. Bloom, C.B. Field, and R.H. Waring, 1987. Plant responses to multiple environmental factors. Bioscience 37: 49–57.

Costanza R., R. d'Arge, R. de Groo, S. Farber, M. Grasso, B. Hannon, K. Limburg, S. Naeem, R.V. O'Neill, J. Paruelo, R.G. Raskin, P. Sutton, and M. van der Belt, 1997. The value of the world's ecosystem services and natural capital. Nature 387: 253–260.

Duarte, C.M., K. Sand-Jensen, S.L. Nielsen, S. Enríquez, and S. Agustí, 1995. Comparative functional plant ecology: Rationale and potentials. Trends in Ecology and Evolution 10: 418–421.

Enríquez, S., S.L. Nielsen, C.M. Duarte, and K. Sand-Jensen, 1996. Broad-scale comparison of photosynthetic rates across phototrophic organisms. Oecologia (Berlin) 108: 197–206.

Grime, G.P., J.G. Hodgson, and R. Hunt, 1988. Comparative plant ecology. Unwin Hyman, Boston, MA.

Hendry, G.A.F. and J.P. Grime, 1993. Methods in Comparative Plant Ecology. A Laboratory Manual. Chapman and Hall, London.

Nielsen, S.L., S. Enríquez, and C.M. Duarte, 1998. Control of PAR-saturated CO_2 exchange rate in some C_3 and CAM plants. Biologia Plantarum 40: 91–101.

Nielsen, S.L., S. Enríquez, C.M. Duarte, and K. Sand-Jensen, 1996. Scaling of maximum growth rates across photosynthetic organisms. Functional Ecology 10: 167–175.

Niklas, K.J., 1994. Plant Allometry. The Scaling of Form and Process. The University of Chicago Press, Chicago, IL.

Salisbury, F.B. and C.W. Ross, 1992. Plant Physiology, 4th edn. Wadsworth, Belmont, CA.

Vitousek, P.M. and P.A. Matson, 1990. Gradient analysis of ecosystems. In: J.J. Cole, G. Lovett, and S. Findlay, eds. Comparative Ecology of Ecosystems: Patterns, Mechanisms, and Theories. Springer-Verlag, NY, pp. 287–298.

2 Opportunistic Growth and Desiccation Tolerance: The Ecological Success of Poikilohydrous Autotrophs

Ludger Kappen and Fernando Valladares

CONTENTS

POIKILOHYDROUS WAY OF LIFE

Poikilohydry, or the lack of control of water relations, has typically been a subject studied by lichenologists and bryologists. For many years, much was unknown about poikilohydrous vascular plants, and evidence for their abilities was mostly anecdotal. A small number of these plants were studied by a few physiologists and ecologists who were fascinated by the capability of these "resurrection plants" to quickly switch from an anabiotic to a biotic state and vice versa (Pessin 1924, Heil 1925, Walter 1931, Oppenheimer and Halevy 1962, Kappen 1966, Vieweg and Ziegler 1969). Recently, a practical demand has released an unprecedented interest in poikilohydrous plants. The increasing importance of developing and improving technologies for preserving living material in the dry state for breeding and medical purposes has induced tremendous research activity aimed at uncovering the molecular and biochemical basis of desiccation tolerance. Poikilohydrous plants have proven to be very suitable for exploring the basis of this tolerance with the target of genetic engineering (Stewart 1989, Oliver and Bewley 1997, Yang et al. 2003, Bernacchia and Furini 2004, Alpert 2006). Consequently, much of the current literature discusses poikilohydrous plants mainly as a means of explaining basic mechanisms of desiccation tolerance (Hartung et al. 1998, Scott 2000, Bartels and Salamini 2001, Rascio and Rocca 2005) instead of exploring their origin, life history, and ecology (Raven 1999, Porembski and Barthlott 2000, Belnap and Lange 2001, Ibisch et al. 2001, Proctor and Tuba 2002, Heilmeier et al. 2005).

Many new resurrection plants have been discovered during the last 25 years, especially in the Tropics and the Southern Hemisphere (Gaff 1989, Kubitzki 1998, Proctor and Tuba 2002). This has provided new insights into the biology of these organisms. In this chapter, structural and physiological features of poikilohydrous autotrophs and the different strategies in different ecological situations are discussed. As desiccation tolerance itself is the most—but not only—striking feature, our goal is to assess in addition the life style and the ecological success of poikilohydrous autotrophs. We give attention to the productivity of poikilohydrous autotrophs, how they manage to live in extreme environments, the advantage of their opportunistic growth, and what happens to structure and physiology during desiccation and resurrection.

POIKILOHYDROUS CONSTITUTION VERSUS POIKILOHYDROUS PERFORMANCE: TOWARD A DEFINITION OF POIKILOHYDRY

According to Walter (1931), poikilohydry in plants can be understood as analogous to poikilothermy in animals. The latter show variations of their body temperature as a function of ambient temperature, whereas poikilohydrous autotrophs (chlorophyll-containing organisms) exhibit variations of their hydration levels as a function of ambient water status (Walter and Kreeb 1970). The term autotroph is used here to comprise an extensive and heterogenous list of autotrophic unicellular and multicellular organisms (cyanobacteria, algae, bryophytes, and vascular plants), including the lichen symbiosis. Poikilohydrous performance (from the Greek words *poikilos*, changing or varying, and *hydor*, water) is applied to organisms that passively change their water content in response to water availability ("hydrolabil"; Stalfelt 1939), eventually reaching a hydric equilibrium with the environment. This fact does not necessarily imply that the organism tolerates complete desiccation (Table 2.1). There is no general consensus on the definition of poikilohydrous autotrophs. The Greek word poikilos also means malicious, which, figuratively speaking, may apply to the difficulty of comprising the outstanding structural and functional heterogeneity of this group of organisms.

It is difficult to be precise about the vast number of poikilohydrous nonvascular taxa, comprising 2000 Cyanophyta, *c.*23,000 Phycophyta, *c.*16,000 Lichenes, and *c.*25,000 Bryophyta. The number of poikilohydrous vascular plant species could be almost 1500 if the

TABLE 2.1
Desiccation Tolerance of Isolated Chloroplasts and of Poikilohydrous Autotrophs[a]

Species	Degree of Desiccation Survived	Time of Drought Survived	Reference
Isolated chloroplasts (*Beta*)	15% RWC	Until equilibrium	Santarius 1967
Nostoc and *Chlorococcum*	Air-dry	73 years	Shields and Durrell 1964
Lichens			
Lichens (50 species)	2%–9% d.wt.	38–78 week air-dry	Lange 1953
		24–56 week P_2O_5	
Pseudocyphellaria dissimilis	45%–65% rh	8–10 h until equilibrium	Green et al. 1991
Bryophytes			
Ctenidium molluscum	Varying between <10% rh and >60% rh	40 h	Dircksen 1964
Dicranum scoparium	<10% rh in winter, >50% in summer	40 h	Dircksen 1964
Epiphytic mosses (14 species)	0%–30% rh	Until equilibrium	Hosokawa and Kubota 1957
Fissidens cristatus	Varying between <10% rh and >60% rh	40 h	Dircksen 1964
Mnium punctatum	<25% rh in winter and >80% rh in summer	40 h	Dircksen 1964
Rhacomitrium lanuginosum	32% rh	239 days	Dilks and Proctor 1974
Sphagnum sp.	Air-dry	<5 days	Wagner and Titus 1984
Exormetheca holstii	Air-dry	8 months	Dinter 1921 (S. Hellwege et al. 1994)
Riccia canescens (bulbils)	Air-dry	7 year	Jovet-Ast 1969
Pteridophytes			
Selaginella lepidophylla	Air-dry	About 1 year	Eickmeier 1979
Isoetes australis	0% rh	Until equilibrium	Gaff and Latz 1978
Asplenium ruta-muraria	5%–10% RWC	1–3 days H_2SO_4 (winter)	Kappen 1964
Asplenium septentrionale	7%–20% RWC	1–3 days H_2SO_4 (winter)	Kappen 1964
Asplenium trichomanes	7%–15% RWC	1–3 d H_2SO_4 (winter)	Kappen 1964
Camptosorus rhizophyllus (gametophyte)		H_2SO_4	Picket 1914
Cheilanthes (8 species)	2%–20% rh	Until equilibrium	Gaff and Latz 1978
Hymenophyllum tunbridgense	43%	15 days	Proctor 2003
Hymenophyllum wilsonii	20%	30 days	Proctor 2003
Notholaena maranthae	6% RWC		Iljin 1931
Paraceterach sp.	15% rh	Until equilibrium	Gaff and Latz 1978
Pellea (2 species)	2%–30% rh	Until equilibrium	Gaff and Latz 1978
Pleurosorus rutifolius	2% rh	Until equilibrium	Gaff and Latz 1978
Polypodium polypodioides	3% RWC	50 h	Stuart 1968
Polypodium vulgare	3%–10% RWC	10 D P_2O_5 (winter)	Kappen 1964
Polystichum lobatum	7%–10% RWC	24 h (air) winter	Kappen 1964
Gametophytes of ferns (5 species)	20–65 rh	36 h	Kappen 1965

(*continued*)

TABLE 2.1 (continued)
Desiccation Tolerance of Isolated Chloroplasts and of Poikilohydrous Autotrophs[a]

Species	Degree of Desiccation Survived	Time of Drought Survived	Reference
Angiosperms			
Dicotyledons			
Ramonda serbica	Air-dry	6–12 months	Markowska et al. 1994
Ramonda myconi	2% RWC	2 days	Kappen 1966
Haberlea rhodopensis	Air dry	6–12 months	Markowska et al. 1994
Boea hygroscopia	0% rh	Until equilibrium	Gaff and Latz 1978
Chamaegigas intrepidus	Dry desert soil	10 months	Heil 1925
C. intrepidus, floating leaves	96% rh	Until equilibrium	Gaff 1971
C. intrepidus, submersed leaves	5% rh	4.5 months	Gaff 1971
Craterostigma (2 species)	0%–15% rh	Until equilibrium	Gaff 1971
Limosella grandiflora (corms)	5% rh	4.5 months (but both species decayed at 100% rh)	Gaff and Giess 1986
Myrothamnus flabellifolius	Air-dry, 0%	Leaves (several year)	Ziegler and Vieweg 1969, Gaff 1971
Blossfeldia liliputana	18% of initial weight	33 months	Barthlott and Porembski 1996
Monocotyledons			
Poaceae from India (10 species)	0–2 (11)% rh	3 months	Gaff and Bole 1986
Southern African Poaceae (11 species)	0%–5% rh	2–7 months	Gaff and Ellis 1974
Poaceae from Africa and Kenya (5 species)	0%–15% rh	Until equilibrium	Gaff and Latz 1978
Trilepis pilosa (African Inselberg)	8% RWC	Up to 1 year	Hambler 1961
Coleochloa setifera	Air-dry	5 years	Gaff 1977
Oropetium sp.	0%–15% rh	Until equilibrium	Gaff 1971
Australian Poaceae (6 species)	0%–15% rh	Until equilibrium	Gaff and Latz 1978
Australian Cyperaceae (4 species)	0%–2% rh	Until equilibrium	Gaff and Latz 1978
Southern African Cyperaceae (4 species)	0%–5% rh	27 months	Gaff and Ellis 1975
Cyperaceae from Africa and Kenya (3 species)	Tissues: 0%–5% rh		Gaff 1986a
Australian Liliaceae (2 species)	0%–5% rh	Until equilibrium	Gaff and Latz 1978
Borya nitida	Air dry	>4 year	Gaff and Churchill 1976
Xerophyta squarrosa	Air-dry	5 year	Gaff 1971
Xerophyta scabrida	Air-dry	5 year	Csintalan et al. 1996
Xerophyta (5 species)	0%–15% rh	Until equilibrium	Gaff 1971
Velloziaceae from Africa and Kenya (2 species)	Mature leaf tissues: 5%–30% rh		Gaff 1986a

Abbreviation: RWC, relative water content; d.wt., dry weight; rh, relative humidity; until equilibrium, until equilibrium between moisture content and ambient air relative humidity.

[a] The list in this table is not exhaustive.

hydrophytes (*c.*940 spp.) are included. Among the land plants, we know nearly 90 species of pteridophytes and approximately 350 of the angiosperms (Gaff 1989, Proctor and Tuba 2002). Within the angiosperms only 10 families have to be taken into account, the Myrothamnaceae, Cactaceae, Acanthaceae, Gesneriaceae, Scrophulariaceae, and Lamiaceae (contributing in total only 35 dicotyledonous species), and the Cyperaceae, Boryaceae (*sensu* Lazerides 1992), Poaceae, and Velloziaceae (together 300 monocotyledonous species). Solely the latter, old and isolated family comprises 8 genera with nearly 260 species (Kubitzki 1998), all most likely desiccation tolerant, and more Velloziaceae species may be discovered in the future (Ibisch et al. 2001). Gaff (1989) suggests an early specialization of the poikilohydrous taxa within their small and often isolated genera.

Nonvascular autotrophs (cyanobacteria, algae, bryophytes, and lichens) are considered constitutively poikilohydrous because they lack the means of controlling water relations (Stocker and Holtheide 1938, Biebl 1962, Walter and Kreeb 1970). This is in contrast with vascular plants, which in general have constitutively homoiohydrous "sporophytes," and keep their hydration state within certain limits by such means of roots, conducting tissues, epidermis, cuticles, and stomata. The poikilohydrous performance of vascular plants is to be taken as an acquired ("secondary": Raven 1999) trait and is realized in phylogenetically unrelated plant species, genera, or families (Oliver et al. 2000). Because poikilohydry is constitutional in nonvascular autotrophs and rare among vascular plants, it is tempting to consider it a primitive property and to suggest that evolutionarily early terrestrial, photosynthetic organisms based their survival on tolerance (Raven 1999) instead of avoidance mechanisms. However, poikilohydry is not an indicator of an early evolutionary stage among vascular plants. Although several recent pteridophytes are poikilohydrous, there is no known poikilohydrous recent gymnosperm, and poikilohydry is frequent only in highly derived angiosperm families (Oliver and Bewley 1997, Oliver et al. 2000). Therefore, poikilohydrous performance by vascular plants can be interpreted evolutionarily as an adaptive response to climates and habitats with infrequent moist periods (see also Proctor and Tuba 2002).

The term *resurrection* has been commonly used for some species and, in general, matches the capability of poikilohydrous plants to quickly reactivate after falling into a period of anabiosis caused by dehydration. It is very appropriate for spikemosses (*Selaginella*) and certain bryophytes and lichens that curl strongly with water loss and unfold conspicuously upon rehydration. Similar performance can be observed in the dead remnants of plants in deserts and steppes. In addition, in fact, the annual homoiohydrous species *Anastatica hierochuntica* was called a resurrection plant by some investigators (Wellburn and Wellburn 1976) because of the dramatic change between a curled and shriveled stage in the dry season and the spreading of the dead branches in the rainy season to release the seeds. Consequently, *resurrection*, in a broad, intuitive sense, could also be applied to certain homoiohydrous desert perennials (e.g., *Aloe*, Mesembryanthemaceae, and certain cacti). On the other hand, the shape and appearance of some constitutively poikilohydrous autotrophs, such as terrestrial unicellular algae and crustose lichens, do not visibly change. To add to the confusion, water loss can be dramatic in some homoiohydrous desert plants, whereas it can be minor in constitutively poikilohydrous plants such as *Hymenophyllum tunbridgense* or bryophytes and lichens from moist environments. Therefore, the resurrection phenomenon (visible changes in shape and aspect with hydration) is only part of the poikilohydrous performance and it is not exhibited to the same extent by all poikilohydrous autotrophs.

Ferns are dual because they produce constitutively poikilohydrous gametophytes and a cormophytic sporophyte with the full anatomy of a homoiohydrous plant. Knowledge about gametophytes is scant. They are usually found in humid, sheltered habitats where hygric and mesic bryophytes also grow. Previous literature reports on extremely desiccation-tolerant prothallia of the North American *Camptosorus rhizophyllus*, and of *Asplenium platyneuron* and *Ceterach officinarum* (= *Asplenium ceterach*) (Walter and Kreeb 1970). The desiccation tolerance of prothallia of some European fern species varied with species and season (Kappen 1965). They usually overwinter, and their ability to survive low temperatures and freezing is based on increased desiccation tolerance. Prothallia of rock-colonizing species (*Asplenium* species, *Polypodium vulgare*) could withstand 36 h drying in 40% relative humidity; some species were partly damaged but could regenerate from surviving tissue. Prothallia of other ferns from European forests were more sensitive to desiccation (Table 2.1).

The poikilohydrous nature of a terrestrial vascular plant is frequently defined by the combination of a passive response to ambient water relations and a tolerance to desiccation (Gaff 1989), but the emphasis on the different functional aspects involved and the actual limits of poikilohydry are matters of debate. Some poikilohydrous species cannot even tolerate a water loss greater than 80% of their maximal water content (Gaff and Loveys 1984), and others can be shown to gain their tolerance only by a preconditioning procedure. Boundaries between poikilohydrous and homoiohydrous plants can be rather blurry, especially if we include examples of xerophytes that can survive extremely low water potentials (Kappen et al. 1972). Surviving at very low relative humidities is not a useful indicator because a limit of 0%–10% relative humidity excludes many nonvascular plants that are undoubtedly poikilohydrous. Considering the photosynthetic performance and low tolerance to desiccation of certain forest lichens (Green et al. 1991) and the fact that, in particular, endohydric bryophytes depend on moist environments, Green and Lange (1994) concluded that the passive response to ambient moisture conditions of poikilohydrous autotrophs varies in a species- and environment-specific manner.

The conflict between ecologically based and physiologically or morphologically based criteria cannot be easily solved. However, a compromise can be reached by distinguishing between stenopoikilohydrous (narrow range of water contents) and eurypoikilohydrous (broad range of water contents) autotrophs. This distinction is especially useful for nonvascular, that is, for constitutively poikilohydrous autotrophs. For instance, microfungi that spend all their active lifetime within a narrow range of air humidity are stenopoikilohydrous. As xeric species they grow in equilibrium with relative humidities as low as 60% (Pitt and Christian 1968, Zimmermann and Butin 1973). Aquatic algae and cyanobacteria are also typically stenopoikilohydrous. The so-called hygric and mesic bryophytes and filmy ferns that are not able to survive drying to less than 60% water content or less than 95% relative humidity also belong to the stenopoikilohydrous type. The same is the case with some wet forest lichens that have low desiccation tolerance (Green et al. 1991). A stenopoikilohydrous performance is also apparent in those ephemeral bryophytes that germinate after heavy rain and then quickly develop gametophytes and sporogons. Some examples with this drought evasion strategy are the genera *Riella, Riccia*, and species of Sphaerocarpales, Pottiaceae, and Bryobatramiaceae. These annual shuttle species are characteristic of seepage areas and pond margins where the soil remains wet for a few weeks (Volk 1984). The many vascular plant species growing permanently submersed in water have also a stenopoikilohydrous life style (see Raven 1999).

All nonvascular and vascular species that are extremely tolerant to desiccation and typically perform as resurrection plants (Gaff 1972, 1977, Proctor 1990) belong to the eurypoikilohydrous group. Because many of these species grow in dry or desert environments, poikilohydry was often associated with xerophytism (Hickel 1967, Patterson 1964, Gaff 1977). However, seasonal changes in the tolerance to desiccation can confound this distinction between stenopoikilohydrous and eurypoikilohydrous organisms. These changes have been found in bryophytes (Dilks and Proctor 1976b) and ferns (Kappen 1964) and are very likely to occur in angiosperms. As suggested by Kappen (1964), such plants may be considered as temporarily poikilohydrous. Hence, the number of eurypoikilohydrous bryophyte species cannot be fixed until temporal changes of desiccation tolerance are better studied in mesic species (Proctor 1990, Davey 1997, Proctor and Tuba 2002). Most of the available information and, consequently, most of what follows, involves eurypoikilohydrous autotrophs. The different groups of poikilohydrous autotrophs that can be identified according to the range of water contents experienced in nature or tolerated are summarized later.

Ecology and Distribution of Poikilohydrous Autotrophs

Nowhere else in the world are poikilohydrous autotrophs more conspicuous than in arid and climatically extreme regions (e.g., Namib desert, Antarctica). It is somewhat paradoxical that precisely in habitats with extreme water deficits the dominant organisms are the least protected against water loss. Additionally, the poikilohydrous angiosperms show in general no typical features against water loss, but they can compete well with extremely specialized taxa of homoiohydrous plants. However, again the distinction between stenopoikilohydrous and eurypoikilohydrous plants becomes important, because stenopoikilohydrous autotrophs can be very abundant in moist habitats (e.g., cloud forests: Gradstein 2006). In the moist and misty climate of San Miguel, Azores, even *Sphagnum* species are able to grow as epiphytes on small trees. However, eurypoikilohydrous autotrophs, which are capable of enduring prolonged drought and extreme temperatures, represent the most interesting group because they have more specifically exploited the ecological advantages of their opportunistic strategy. The remainder of this chapter presents examples of poikilohydrous autotrophs living under very limiting ecological conditions in many different regions of the Earth.

In temperate climates, poikilohydrous autotrophs are mainly represented by aerophytic algae, bryophytes, and lichens. Depending on their habitat, bryophytes can be eurypoikilohydrous or stenopoikilohydrous. Among the temperate vascular plants, poikilohydrous performance is realized in some mainly rock-colonizing fern genera such as *Asplenium, Ceterach, Cheilanthes, Hymenophyllum, Notholaena*, and *Polypodium* and the phanerogamous genera *Haberlea* and *Ramonda*.

From the arctic region, no poikilohydrous vascular plants are known, and most parts of Antarctica are inhabited solely by algae, bryophytes, lichens, and fungi, which are mainly eurypoikilohydrous. In the polar regions and in hot, extremely arid deserts, nonvascular autotrophs may be restricted to clefts and rock fissures or even grow inside the rock as endolithic organisms or hypolithic on the underside of more or less translucent rock particles and stones (Friedmann and Galun 1974, Scott 1982, Danin 1983, Kappen 1988, 1993b, Nienow and Friedmann 1993).

In subtropical regions bryophytes, algae, and lichens are well known as crust-forming elements on open soils (Belnap and Lange 2001). The coastal Namib desert, with extremely scattered rainfall, consists of wide areas where no vascular plants can be found, but a large cover of mainly lichens forms a prominent vegetation. In rocky places of the Near East, southern Africa, arid northwest North America, coastal southwest North America, and the South American westcoast, lichens and bryophytes coexist with xeromorphic or succulent plants. They also occupy rock surfaces and places where vascular plants do not find enough soil, or they grow as epiphytes on shrubs and cacti. Under such extreme conditions, lichens and bryophytes share the habitat with poikilohydrous vascular plants as for instance *Borya nitida* on temporarily wet granitic outcrops (Figure 2.1) with shallow soil cover in southern and western Australia (Gaff and Churchill 1976).

In Africa, subfruticose poikilohydrous plants such as *Lindernia crassifolia* and *Lindernia acicularis* grow in sheltered rock niches (Fischer 1992). The same is true for the fruticose poikilohydrous species, *Myrothamnus flabellifolius*, occurring in southern Africa and Madagascar from Namibia (Child 1960, Puff 1978, Sherwin et al. 1998), which is frequently associated with other resurrection plants (e.g., *Pellaea viridis, Pellaea calomelanos*). In the wet season, these plants benefit from run off water that floods the shallow ground (Child 1960). Particularly remarkable are poikilohydrous aquatic *Lindernia* species (*L. linearifolia,*

FIGURE 2.1 Two very different examples of poikilohydrous autotrophs co-occurring on a shallow depression of a granite outcrop near Armadale, western Australia: the monocotyledonous plant *B. nitida* (*left*), mosses, and the whitish fruticose lichen *Siphula* sp. (Photograph from Kappen, L.)

L. monrio, L. conferta) and *Chamaegigas (Lindernia) intrepidus* (Heil 1925, Hickel 1967, Gaff and Giess 1986, Heilmeier et al. 2005), which grow in small temporarily water-filled basins of granitic outcrops in Africa (Angola, Zaire, Zimbabwe, South Africa, Namibia). Many of the poikilohydrous grass species (less than 20 cm high) and sedges (30–50 cm high) are pioneering perennial plants colonizing shallow soil pans in southern Africa (Gaff and Ellis 1974). In Kenya and West Africa, the resurrection grasses, sedges, and Vellociaceae (in Africa 30 species, Ibisch et al. 2001) are confined to rocky areas, except *Sporobolus fimbriatus* and *Sporobolus pellucidatus*. *Eragrostis invalida* is the tallest poikilohydrous grass species known with a foliage up to 60 cm (Gaff 1986). *Vellozia schnitzleinia* is a primary mat former following algae and lichens on shallow soils of African inselbergs, persisting during the dry season with brown, purple-tinged rolled leaves that turn green in the wet season (Owoseye and Sandford 1972).

The resurrection flora of North America is represented mainly by pteridophytes. Most of the poikilohydrous fern species so far known are preferentially found in rock cervices, gullies, or sheltered in shady rocky habitats (Nobel 1978, Gildner and Larson 1992). By contrast, the most famous resurrection plant *Selaginella lepidophylla* colonizes open plains in Texas (Eickmeier 1979, 1983). In Middle and South America, 220 species of the Velloziaceae form the dominant part of the poikilohydrous flora. They grow in various habitats and even in alpine regions. The endemic *Vellozia andina* seems to be an opportunistic species as it takes benefit from degraded formerly forested sites (Ibisch et al. 2001). Fire resistance is typical of many Velloziaceae species (Kubitzki 1998). Gaff (1987) has enumerated 12 fern species for South America. *Pleopeltis mexicana* and *Trichomanes bucinatum* may also be candidates (Hietz and Briones 1998). One of the most remarkable poikilohydrous vascular plants could be *Blossfeldia liliputana*, a tiny cactus that grows in shaded rock crevices of the eastern Andean chain (Bolivia to northern Argentine) at altitudes between 1200 and 2000 m (Barthlott and Porembski 1996). This plant is unable to maintain growth and shape during periods of drought, and it persists in the dry state (18% of initial weight) for 12–14 months, looking like a piece of paper. When water is again available, it can rehydrate and resume CO_2 assimilation within 2 weeks; it is the only known example of a succulent poikilohydrous plant.

From a plant–geographical perspective, inselberg regions in Africa, Madagascar, tropical South America, and Western Australia have the largest diversity of poikilohydrous vascular plants in the World. Porembski and Barthlott (2000) state that 90% of the known vascular poikilohydrous plant species occur on tropical inselbergs. The presence of almost all known genera with poikilohydrous plants could be recorded from such sites. Despite the existence of similar potential habitats for poikilohydrous vascular plants in Australia, species are less numerous there than in southern Africa. Lazarides (1992) suggested that this biogeographical difference between Australia and southern Africa is due to the fact that the Australian arid flora has been exposed to alternating arid and pluvial cycles for a shorter geological period of time than the arid flora of southern Africa. The former has experienced these alternations since the Tertiary, whereas the latter has been exposed to dry–wet cycles since the Cretaceous. Ferns, represented by a relatively large number of species [14], and most of the poikilohydrous grasses found in Australia [10] grow in xeric rocky sites (Lazarides 1992). We have very few records about poikilohydrous vascular plants from Asia, although such a type of plant must exist there as well. Gaff and Bole (1986) recorded 10 poikilohydrous Poaceae (genera *Eragrostidella, Oropetium, Tripogon*) for India. The Gesneriaceae *Boea hygrometrica*, closely related to the Australian *Boea hygroscopica*, is a poikilohydrous representative in China (see Yang et al. 2003).

Most of the resurrection plants are confined to lowland and up to 2000 m a. s. l. However, a few Velloziaceae species such as *Xerophyta splendens* reach altitudes of 2800 m in Malawi (Porembski 1996) and *Barbaceniopsis boliviensis* reach 2900 m in the Andes (Ibisch et al. 2001), the latter staying in anabiosis with reddish-brown leaves for half a year. In such high altitudes, they are exposed to frost periods.

Does Poikilohydry Rely on Specific Morphological Features?

Poikilohydrous performance cannot be typified by any one given set of morphological and anatomical features because of the heterogeneity of this functional group of photosynthetic organisms. Poikilohydry can be found in autotrophs ranging from those with the most primitive unicellular or thallose organization to those with the most highly derived vascular anatomy. In angiosperms, desiccation tolerance is, in general, inversely related to anatomical complexity. It seems that plants can operate either by avoidance or tolerance mechanisms at all levels of organization if they are adapted to temporarily dry habitats. Gaff (1977) called resurrection plants "true xerophytic" just because they live in xeric environments. However, poikilohydrous angiosperms do not necessarily have xeromorphic traits. Xeromorphic features such as small and leathery leaves are typical for *Myrothamnus*; xeromorphic narrow or needle-like leaves for many Velloziaceae, Cyperaceae, and the genus *Borya* (see Figure 2.1); and massive sclerenchymatic elements, for example, several Velloziaceae and *Borya* (Gaff and Churchill 1976, Lazarides 1992, Kubitzki 1998). Hairs on leaves (e.g., Velloziaceae, Gesneriaceae) are mostly small, and scales (e.g., *Ceterach*) or succulence (*Blossfeldia*) are the exception rather than the rule in poikilohydrous vascular plants. Xeromorphic structures would also counteract the potential of rehydration during the wet period. However, curling and uncurling of leaves, frequently enabled by contraction mechanism, is a widespread phenomenon in poikilohydrous vascular plants.

Poikilohydrous vascular plants are mainly perennials represented by various types of hemicryptophytic and chamaephytic life forms but no trees. Lignification of stems is not rare, and the two existing *Myrothamnus* species are true shrubs reaching approximately 1.5 m height. Within the monocotyledons, a tree-like habit is achieved either by an enhanced primary growth of the main axis or by secondary thickening, and trunks may reach up to 4 m length. Such pseudostems are realized in the genus *Borya* (secondary growth) and by some Cyperaceae and Velloziaceae (Gaff 1997, Kubitzki 1998, Porembski and Barthlott 2000). For instance, a sample of *Vellozia kolbekii* was looking with its stem (covered by roots and leaf sheaths) like a tree fern, was 3 m tall, and was estimated to be about 500 years old (Alves 1994).

As most of the phanerogamous resurrection plants do not show peculiar or uniform anatomical features, it is hard to decide whether a particular plant is poikilohydrous just from herbarium material or from short-term observations in the field (Gaff and Latz 1978). It is still uncertain, for instance, whether more members of the Lindernieae can be identified as poikilohydrous in studies such as that by Fischer (1992) and Proctor (2003). Many species that grow in shady habitats or that colonize temporarily inundated habitats exhibit a hygromorphic tendency (Volk 1984, Fischer 1992, Markowska et al. 1994). For instance, *Chamaegigas intrepidus* has, like other aquatic plants, aerenchyma and two types of leaves, floating and submerged. *Blossfeldia liliputana*, the only known poikilohydrous Cactaceae, combines a succulent habit with a typically hygromorphic anatomy: very thin cuticle, no thickened outer cell walls, absence of hypodermal layers, and extremely low stomatal density (Barthlott and Porembski 1996). Poikilohydrous vascular plants exhibit, in general, very low stomatal control of transpiration (Gebauer 1986, Sherwin et al. 1998, Proctor 2003). The leaves of *Satureja gilliesii* even have protruding stomata on the underside (Montenegro et al. 1979).

The secondarily poikilohydrous nature of aquatic vascular plants has rarely been acknowledged (Raven 1999). Most of them have reduced xylem structure and no sustaining function. Roots merely act to fix to the substratum, and nutrients are taken up by the leaves. Cuticles are thin and stomata are scattered and frequently nonfunctional (*Isoetes, Litorella, Elodea, Vallisneria*, Potamogetonaceae, etc). Living in streams and underwater rapids in the Tropics, the Podostemaceae are very remarkable examples with a drastic reduction of their homoiohydrous architecture. With their thallus-like shoots they resemble foliose liverworts.

Small size is recognized frequently as typical of the shape of the poikilohydrous autotrophs. Indeed, only a few vascular species are fruticose and reach more than 50 cm height. Alpert (2006) discusses whether there is a trade-off between low growth and desiccation tolerance in the sense of a disadvantage, because the plant has to invest in protection mechanisms instead of extension growth as most of the homoiohydrous plants do. Proctor and Tuba (2002) on the other hand, refer to poikilohydry as an advantage particularly for living in temporarily dry environments. High desiccation tolerance is the ultimate drought-evading mechanism. The resurrection strategy is ecologically as successful as that of homoiohydrous plants with CAM or the adaptation to live on heavy-metal soils or in raised bogs. In addition, the slow growth and small size of poikilohydrous plants is not only a function of changing water status but also of nutrient deficiency, which is obvious from most of their natural habitats. *C. intrepidus*, for instance, has to use urea as nitrogen source by means of free urease in the sediments of rock pools (Heilmeier et al. 2000) and free amino acids (Schiller 1998).

Living under water, the nonvascular autotrophs are able to develop a size (*Macrocystis* spp.: 60 m) comparable to that of tall trees, and the vascular plant species *Elodea canadensis* may produce up to 6 m long shoots (see Raven 1999). Endohydrous mosses such as the Dawsoniaceae and Polytrichaceae may reach a height of 1 m in the damp atmosphere of rain forests. This demonstrates that the small size of autotrophs which are eurypoikilohydrous is an adaptive trait to respond flexibly to drought events rather than remaining principally handicapped with respect to growth and productivity.

EXPLOITING AN ERRATIC RESOURCE

Water is evasive in many terrestrial habitats, and plants in general have to deal with the changing availability of this crucial resource. This is especially true for poikilohydrous autotrophs, which have successfully explored many different strategies within their general tolerance to water scarcity. However, some of the features that make tolerance to desiccation possible are irreconcilable with those that enhance water use. Poikilohydrous autotrophs, therefore, have had to trade-off between surviving desiccation against uptake, transport, and storage of water. Some adaptive conflicts appear, for instance, when a particular feature retards water loss. The important functional problems that arise when the plant has to resume water transport after desiccation might have limited the range of growth forms and plant sizes compatible with poikilohydry.

Different Modes of Water Uptake and Transport

Plants must be efficient in acquiring water, particularly in arid regions where rainfall is scarce and sometimes the only available water comes from dew, mist, or fog. Poikilohydrous plants can outcompete their homoiohydric counterparts in dry habitats if they can rehydrate efficiently. The following section describes the different possibilities for water capture exhibited by poikilohydrous autotrophs, with emphasis on the role of the growth form and of the morphology and anatomy of the structures involved.

Aerophytic algae and lichens with green-algal photobionts can take up enough water from humid atmospheres to become metabolically active (Lange 1969b, Blum 1973, Lange and Kilian 1985, Lange et al. 1990a, Bertsch 1996a,b). Rehydration in lichens from humid atmospheres may take 1–4 days until equilibrium, whereas mist and dewfall yield water saturation within hours (Kappen et al. 1979, Lange and Redon 1983, Lange et al. 1991). Even water vapor over ice and snow serves as an effective water source for the activation of lichens in polar regions (Kappen 2000, Pannewitz et al. 2006; see Chapter 14). As a consequence, lichens in deserts can survive well with sporadic or even no rainfall (Kappen 1988, Lange et al. 1990c, 1991). Anatomical structures such as long cilia, rhizines, branching, or a reticulate thallus structure are characteristic of lichens from fog deserts (e.g., *Ramalina melanothrix, Teloschistes capensis, Ramalina menziesii*), suggesting that these structures are means for increased water absorption (Rundel 1982a). In lichens, liquid water is absorbed by the entire body (thallus), usually within a few minutes (Blum 1973, Rundel 1982a, 1988). The thallus swells and can unfold lobes or branches. However, there is little evidence of a water transport system in these organisms (Green and Lange 1994). Nevertheless, not all lichens have the same capacity for exploiting the various forms of water from the environment. For example, lichens with cyanobacteria as photobiont cannot exist without liquid water (Lange et al. 1988). For the Australian erratic green-algal *Chondropsis semiviridis*, rainwater is necessary to allow photosynthetic production because the curled lobes must be unfolded (Rogers and Lange 1971, Lange et al. 1990a).

The kinetics of water uptake seems to be similar in lichens and mosses, and the larger the surface area to weight ratio, the more rapid the water uptake (Larson 1981). Rundel (1982a) suggested that thin cortical layers of coastal Roccellaceae in desert regions may be a morphological adaptation to increase rates of water uptake. However, textural features of the upper cortex seem to be more important for water uptake than just thickness (Larson 1984, Valladares 1994a). Valladares (1994a) found that species of Umbilicariaceae that possess the most porous and hygroscopic upper cortex (equal to filter paper) are adapted to live mainly from water vapor (aero-hygrophytic), whereas species that have an almost impervious cortex were more frequently exploiting liquid water from the substratum (substrate-hygrophytic; Sancho and Kappen 1989).

Most bryophytes need a humid environment or externally adhered water to keep a level of hydration high enough for metabolic functions. Many species form cushions, turfs, or mats that aid to keep capillary water around the single shoots (Gimingham and Smith 1971, Giordano et al. 1993). At full saturation, the water content of mosses (excluding external water) can vary between 140% and 250% dry weight (d.wt.) (Dilks and Proctor 1979), which is similar to that of macrolichens. Thallose hygrophytic liverworts require higher levels of hydration, and their maximal water content can be more than 800% d.wt. In shaded or sheltered habitats, hygric and some mesic bryophytes are able to keep their water content relatively constant throughout the year, which is characteristic for a stenopoikilohydrous lifestyle (Green and Lange 1994). In more open and exposed sites, the fluctuations in water content are very large (Dilks and Proctor 1979).

The more complex and differentiated morphology and anatomy of bryophytes, in comparison with lichens, allow for more varied modes of water uptake (Proctor 1982, 1990, Rundel 1982b). Bryophytes can take up water vapor to limited extent and reach only low (less than 30% of maximum water content) relative values (Rundel and Lange 1980, Dhindsa 1985, Lange et al. 1986). Dew uptake was recorded for *Tortula ruralis* (Tuba et al. 1996a) and for 10 sand-dune mosses (Scott 1982). Leaves of certain desert mosses (e.g., Pottiaceae) act as focus for condensation of water vapor and mist by means of their recurved margins, papillose surfaces, and hair points (Scott 1982). However, the presence of lamellae, filaments, and other outcrops on the adaxial surface of the leaves, which is common in arid zone mosses, may act more as sun shelter rather than as means to enhance water uptake. The role of scales and hyaline structures on the midrib of desert liverworts (e.g., *Riccia, Exormotheca,* and *Grimaldia*), which is inverted and exposed to the open when the thallus is dry, is not clear, but they start absorbing rainwater and swelling to turn down rapidly and may help in storing water (Rundel and Lange 1980). Mosses of the family Polytrichaceae have so-called rhizomes or root-like structures, which are not very efficient for water uptake (Hebant 1977). In general, water uptake of mosses from the soil is poor and needs to be supplemented by external water absorption.

Two main groups of bryophytes have been described according to the mode of water transport. Ectohydrous species resemble lichens because they take up water over all or most of their thallus surface and have no internal water transport system, whereas endohydrous species have various water-proofed surfaces (cuticles), often well developed near to the gas exchange pores (stomata on the sporophytes), and have a significant water-transport pathway (Proctor 1984, Green and Lange 1994). These properties of the latter are similar to those of homoiohydrous plants (Hebant 1977). However, they differ from vascular plants in that their conductive structures are not lignified, and all these properties are functional only in moist environments. Therefore desert mosses are typically ectohydrous (Longton 1988a), and the water transport in eurypoikilohydrous bryophytes growing in dry environments is predominantly external. However, some eurypoikilohydrous mosses (Fabronianaceae, Orthotrichiaceae) have large masses of stereom tissue (usually a supporting tissue), that is considered to be an alternative route for the conduction of water (Zamski and Trachtenberg 1976).

Proctor (1982) summarized four different pathways or modes by which water moves in a bryophyte: (1) inside elongated conductive cells (hydroids), forming a central strand in the stems of mosses and some liverworts; (2) by the cell walls, which are frequently thickened (in fact, bryophyte cell walls have higher water conductivity than those of vascular plants); (3) through intervening walls and membranes; and (4) by extracellular capillary spaces. The highest internal conduction for water in Polytrichaceae at 70% relative humidity was 67% of the total conduction (Hebant 1977).

Water uptake in poikilohydrous vascular plants can be very complex because of interactions between different organs. For instance, in the fern *Cheilanthes fragrans*, water uptake through the leaf surface from a water vapor-saturated atmosphere allows it to reach 80% of its maximal water content within 50 h (Figure 2.2a). Petiolar water uptake was also efficient, but only if the leaves were in high air humidity (Figure 2.2b). Stuart (1968) found that the fern *Polypodium polypodioides* was not able to rehydrate by soil moistening if the air was dry, and the leaves reached only 50% of their maximal water content within 2–3 days, even in a water

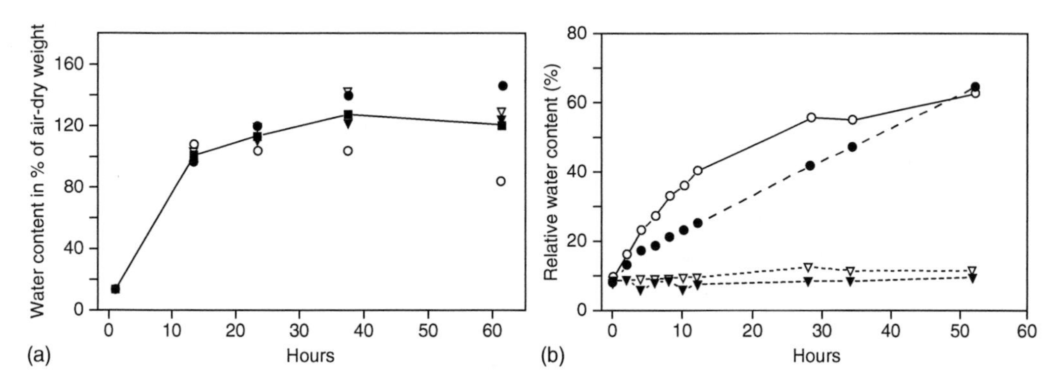

FIGURE 2.2 (a) Water-vapor uptake of leaves of the fern *Cheilanthes fragrans* with sealed petioles in a moist chamber. The different symbols stand for four replicates (L. Kappen, unpublished results). (b) Water uptake of leaves of *C. fragrans* placed on filter paper in a moist chamber (open circles); with petiole in a vessel with water and standing in a moist chamber (closed circles), and (open and closed triangles) with petiole in water in a room (approximately 60% rh) (L. Kappen, unpublished results).

vapor-saturated atmosphere (Stuart 1968, confirming the results of Pessin 1924). Fronds of the highly desiccation-tolerant *Polypodium virginianum* were, however, not able to absorb water from air as was shown with deuterium-labeled water (Matthes-Sears et al. 1993). Thus, the capacity of the leaves to take up water vapor varies significantly among species and seems not to be associated with the tolerance to desiccation. In contrast, liquid-water uptake by leaves has been shown to be a common feature in poikilohydrous vascular plants. Detached leaves of *P. polypodioides* regained full saturation within 20–30 min if submersed in liquid water (Stuart 1968). However, leaves attached to the rhizome needed 10 times longer for saturation than detached leaves. Stuart explained this by alluding to anaerobic conditions that impede rapid water uptake. Rapid water uptake by leaves was also shown in *Selaginella lepidophylla* (Eickmeier 1979). It seems that, in pteridophytes, water uptake through leaves is an important mechanism for reestablishing water relations of the whole plant and for resuming xylem function. Similarly, rehydration of the whole plant solely by watering the soil in dry air is also incomplete in poikilohydrous angiosperms (Gaff 1977).

Water uptake from mist or from saturated atmospheres is insignificant in poikilohydrous angiosperms (Vieweg and Ziegler 1969), as has been shown for isolated leaves of *Ramonda myconi* (Gebauer et al. 1987). In addition, exposure to dewfall could only raise the relative water content to less than 13% in *Craterostigma wilmsii* (Gaff 1977). Foliar water uptake by desert plants has been investigated, particularly with respect to dew uptake (Barthlott and Capesius 1974), but it seems to be insignificant in homoiohydrous plants except in the genus *Tillandsia* (Rundel 1982b). In contrast, foliar water uptake from rain by poikilohydrous vascular plants may be important to resume functioning of the hydraulic system, as Gaff (1977) found that leaves of resurrection plants in contact with liquid water can rehydrate within 1–14 h, depending on the species. The quickest uptake was measured in *Chamaegigas intrepidus* (Hickel 1967). The cuticle of vascular plants is generally considered an efficient protection against water loss. However, the cuticle of poikilohydrous vascular plants may also enhance water uptake by leaves (e.g., *Borya*; Gaff 1977). The permeability of the cuticle to water was assumed for *C. interpidus* (Hickel 1967). Barthlott and Capesius (1974) suggested that the cuticle of some of these plants seems to be more permeable to water from outside than from inside the leaf. However, this is not clear as some studies attribute permeability to the state of the cuticular layer rather than to the cuticle itself (Schönherr 1982). According to

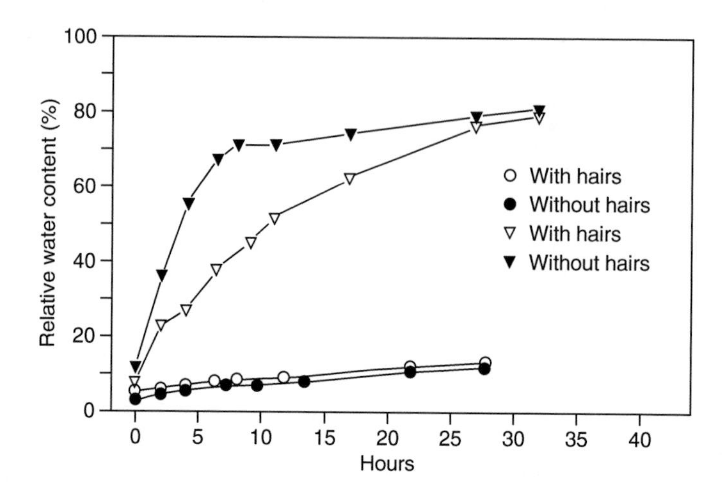

FIGURE 2.3 Water uptake of leaves of *Ramonda myconi* (Gesneriaceae) with sealed petioles. Leaves with hairs (open triangles) and after removing the hairs (closed triangles) soaking from sprayed water; and leaves with hairs (open circles) and without hairs (closed circles) in a moist chamber. (From Gebauer, R., Lösch, R., and Kappen, L., *Verh. Ges. Ökologie*., XVI, 231, 1987.)

Kerstiens (1996), water uptake through the cuticle is most likely, but evidence needs to be shown.

Hairs and scales can function as auxiliary structures for water uptake because they absorb water more easily than the leaf epidermis. The lower surface of the curled and folded leaflets of ferns like *Ceterach officinarum*, densely covered with scales and trichomes, should enhance water capture (Oppenheimer and Halevy 1962). The so-called hydathodes on the leaves of *Myrothamnus* may actually function as water-absorbing trichomes (Rundel 1982b). However, Sherwin and Farrant (1996) do not believe in any water uptake by leaves of this species. In addition, scales of *P. polypodioides* did not facilitate water uptake, but allowed the water to spread homogeneously on the leaf surface (Pessin 1924, Stuart 1968), and the scales on the leaves of several species of *Ceterach* and *Cheilanthes* retarded water uptake for several hours because of the air that was trapped between the scales (Oppenheimer and Halevy 1962, Gaff 1977, Gebauer 1986). The hairs of the leaves of *R. myconi* (and of other *Ramonda* species) had the same effect (Figure 2.3). Spraying of the detached hairy leaves resulted in less water uptake than immersion in water or spraying hairless leaves. The retarding effect of scales and hairs suggests that a very rapid water uptake after desiccation could be injurious to the leaf cells.

Problems of Resuming Water Transport

Poikilohydrous plants that possess an internal system for water transport (endohydrous bryophytes and vascular plants) are exposed to cavitation (blockade of a vessel by air bubbles) during desiccation, which compromise the functioning of the conducting tissues upon rehydration. This was particularly investigated in trees (Sperry and Tyree 1988, Tyree and Sperry 1988, Hargrave et al. 1994, Lewis et al. 1994, Kolb et al. 1996, Tyree, Chapter 6, this volume). Emboli in a fraction of the conductive elements confines water transport into a diminished number of vessels, which requires an increased tension and further increased the risk of embolism. Embolized conduits can become functional again through bubble dissolution or expulsion, which requires a positive pressurization (Zimmermann and Milburn 1982). Poikilohydrous plants face the dilemma of restoring water transport through their old, embolized tissues or investing in new conducting tissues, which reduces the resources available to be allocated elsewhere in the plant. The fact that most of the poikilohydrous vascular plants are herbaceous and smaller than 50 cm might be explained by the difficulties of restoration of conductivity of the xylem. The same mechanical difficulties of resuming water conductivity of embolized tissues might also be behind the remarkable lack of poikilohydrous species among the gymnosperms, which consist only of trees and shrubs. The low flexibility due to the xylem anatomy of gymnosperms was demonstrated by Ingrouille (1995).

Poikilohydrous vascular plants that would be able to resume water transport only if their shoot tissues have been hydrated by external water uptake perform like the endohydrous bryophytes, where water conduction in the hydroids is supported by lateral and apoplastic water transport. The imbibition of the cell walls of leaf and stem tissues generates the necessary pressure to induce dissolution of emboli in the tracheary tissues. Capillary forces in poikilohydrous plants at 40% relative humidity and under laboratory conditions could move water to a height of 2–12 cm (Galace 1974, cited in Gaff 1977). These forces are sufficient to rehydrate many of the small herbaceous poikilohydrous species. Other mechanisms to eliminate emboli are temperature-associated osmosis at the plant apex (Pickard 1989) and generation of a root pressure that is able to dissolve gas bubbles in the conduits of small herbs and grasses (Zimmermann and Milburn 1982). The latter was shown to be able to restore full liquid continuity and was assumed to be important in larger poikilohydrous species like *Xerophyta eglandulosa* (Gaff 1977). Reversal of almost complete embolism in stems of the homoiohydrous *Salvia multiflora* was related to the presence of narrow vessels and tracheids, which were better to refill than wider conduits (Hargrave et al. 1994). Very narrow vessels (approximately 14 μm) are also true for *Myrothamnus flabellifolius*. In addition, it has reticular perforation plates and knob-like protuberances on the outer walls of the vessels and tracheids, obviously to provide stability when the tissues swell and shrink. The hydraulic conductivity is, however, low and the shrub needs approximately 70 h to regain turgor (Sherwin et al. 1998). Water rise in the axes is substantially aided by root pressure (which develops 3–4 h after watering the plants and ceases after 4–5 days) and additionally, mechanisms that disintegrate lipid films on the lumen walls of the xylem elements, such as radial water flow along the xylem parenchyma, the phloem, and cortical cells (Schneider et al. 2000). This demonstrates that a woody poikilohydrous plant has to take great efforts for its rehydration. As soon as a leaf is reached by the waterfront, it is unfolded and hydrated to 65% RWC within 2 h and soon photosynthetically functional (Sherwin and Farrant 1996).

The slow recovery upon rehydration of whole plants when compared with detached leaves (Stuart 1968, Gebauer 1986) might be because of the time required to form new roots. This topic has been better explored in homoiohydrous plants from arid environments. For instance, the hydraulic conductivity of roots of *Agave deserti* and *A. acanthodes*, which decreased dramatically after several days of drought, rapidly recovered when water was again available, not by formation of new roots but by refilling the extant tissues, which were made up of flexible and unlignified vessels (North and Nobel 1991, Ewers et al. 1992). In fact, a flexible structure in the conductive system was identified in the poikilohydrous *Blossfeldia liliputana* (Barthlott Porembski 1996). Contractive tracheids evidently enable the dry, contracted, and submersed leaves of *Chamaegigas intrepidus* to swell by 800%–900% in water (Schiller et al. 1999). Another means of rapid water uptake is provided in this poikilohydrous species (Heilmeier et al. 2000) and in the genus *Borya*, Cyperaceae, and *Xerophyta pinifolia* (Porembski and Barthlott 1995, 2000) by a *Velamen radicum*, which is otherwise typical for epiphytic orchids. Earlier literature has recorded the formation of new adventitious roots in poikilohydrous plants subsequent to rehydration (Walter and Kreeb 1970). This was confirmed recently for Velloziaceae (e.g., *Xerophyta scabrida*), *Borya sphaerocephala*, and *Craterostigma plantagineum*, where such roots appeared after the regreening of the plant, replacing the drought-killed original roots (Tuba et al. 1993a, Porembski and Barthlott 2000, Norwood et al. 2003).

Retarding Water Loss

Poikilohydrous autotrophs can maintain their water content at a constant level only to a limited extent, but they can extend hydration into the dry period by certain, mostly structural, mechanisms. By retarding water loss, such organisms can enhance their exploitation of the transient periods of water availability. Retarding water loss could, however, counteract some advantageous aspects of the poikilohydrous strategy. For instance, extending hydration sometimes reduces water capture. Poikilohydry also provides a remarkable tolerance for desiccated autotrophs to other stresses that usually occur with drought, such as heat and excessive light (see Section "Preventing Damage and Tolerating Stresses"). Thus, if metabolic activity is extended into these harmful periods, it could not only reduce overall productivity but also compromise survival.

One way of retarding water loss in lichens is by increasing the water that can be stored within the plectenchyma (Valladares 1994b, Valladares et al. 1998). Anatomical characteristics, such as porous and thick medulla layers and rhizinae, have been suggested as means of increasing water storage in lichens (Snelgar and Green 1981, Valladares et al. 1993, Valladares and Sancho 1995). However, because water and CO_2 share the pathway in lichens, enhanced water storage can hamper CO_2 diffusion and consequently reduce photosynthetic carbon uptake (Green et al. 1985, Lange et al. 1996, Maguas et al. 1997). Thus, again, these plants must reach a compromise between two opposing situations. Large foliose lichens can possibly separate photosynthesis and water storage in space somewhat, as young, growing zones of the thallus optimize gas diffusion, whereas old, thick regions act primarily as water reservoirs, sacrificing gas diffusion and carbon gain (Green et al. 1985, Valladares et al. 1994). This trade-off between gas diffusion and water storage seems to be flexible, and lichens have been shown to exhibit a remarkable phenotypic plasticity in their water storage and retention capacities in response to habitat conditions (Larson 1979, 1981, Pintado et al. 1997). Dry habitats induce increased water storage (Tretiach and Brown 1995), but there are complex interactions with light availability for photosynthesis. In shaded sites without access to liquid water, the Antarctic lichen *Catillaria corymbosa* enhanced both water storage and photosynthesis via increased light harvesting by chlorophylls (Sojo et al. 1997), whereas in exposed sites (dry and receiving high irradiance), the lichen *Ramalina capitata* enhanced photosynthetic utilization of brief periods of activity via improved gas diffusion at the expense of reducing water storage capacity (Pintado et al. 1997). These problems are not faced by bryophytes because most of them have rather complex photosynthetic tissues, where the CO_2 exchange surface is separated from water storage volumes (Green and Lange 1994). The capacity of bryophytes to keep high rates of photosynthesis at high water contents is a very likely explanation for the dominance of these organisms in wet habitats (Green and Lange 1994). Zotz et al. (2000) demonstrated experimentally with *Grimmia pulvinata* a positive relationship between cushion size and water retention capacity and also an increasing CO_2 gain up to an optimum water content. However, despite mechanistic differences between mosses and lichens, in certain cases overall performance can be similar.

Discussing the role of morphological properties for lichens, Rundel (1982a) concluded that evaporative water loss can be reduced by a decrease in surface/volume ratio, but such a decrease reduces uptake of water vapor similarly. This seems not to be the case for liquid water, since structures on the lower side of the thallus such as the rhizinomorphs of the lichen family Umbilicariaceae enhance capture and storage of water from run-offs without significantly increasing water loss by evaporation (Larson 1981, Valladares 1994b, Valladares et al. 1998). Reduction of evaporative water loss by structural traits such as a tomentum on the upper surface (Snelgar and Green 1981), a thick cortex (Larson 1979, Büdel 1990), or a decreased surface/volume ratio is very limited in open habitats, but can be significant in sheltered or humid places or in areas with frequently overcast skies (Kappen 1988) or under the influence of drip water from trees or from antennae on roofs. Bryophytes can also retard water loss by an increased boundary layer resistance and due to growth forms of low surface-to-volume ratios (Gimingham and Smith 1971, Proctor 1982, Giordano et al. 1993).

Comparing water retention of a terricolous moss and a terricolous lichen, Klepper (1968) found that both remained hydrated for the same period of time after the rainfall although the moss stored initially more water. Then, regardless of the initial water content, they dried out quickly when the ambient water vanished and the atmosphere became dry. Measurements on desert lichens have clearly shown that the thalli start drying as soon as the sun is rising and their period of hydration depends solely on evaporative conditions (Kappen et al. 1979, Kappen 1988).

As has been repeatedly demonstrated, the water content of vascular resurrection plants varies with the soil moisture and, like their nonvascular counterparts, they dehydrate within a few hours or days after soil water supply has declined (Gaff and Churchill 1976, Gaff 1977). In their shaded habitats some resurrection plants, especially the ferns, take profit from less evaporative stress than in the open. Water-storing tissues of the succulent *Blossfeldia liliputana* and those within the leaves of some Velloziaceae (e.g., *V. tubiflora, V. luteola, Barbacenia reflexa*; Kubitzki 1998) may retard water loss. Since stomatal conductance of resurrection vascular plants is, in general, rather high (Tuba et al. 1994), water loss must be retarded by structural features such as scales, as was shown for leaves of *Cetrerach officinarum* (Oppenheimer and Halevy 1962). However, scales were almost ineffective in *Cheilanthes maranthae* (Gebauer 1986, Schwab et al. 1989). A retarding effect of hairs in leaves of *Ramonda* (Figure 2.4a) has long been known (Bewley and Krochko 1982) and the efficiency of leaf pubescence in retarding water loss is increased when the leaves shrink (Gebauer et al. 1987, Figure 2.4b).

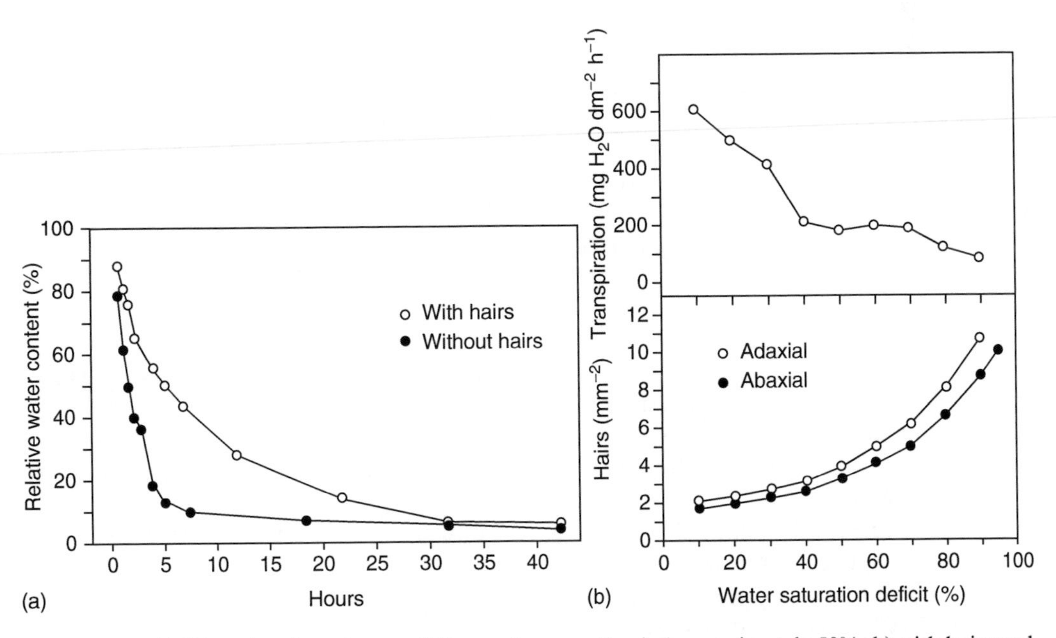

FIGURE 2.4 (a) Water loss from leaves of *Ramonda myconi* in air (approximately 50% rh) with hairs and after hairs were removed before the experiment. (After Gebauer, R., Lösch, R., and Kappen, L., *Verh Ges Ökologie.*, XVI, 231, 1987.) (b) Transpiration rates of leaves of *R. myconi* with increasing water saturation deficit. Hair density on the upper (adaxial) and the lower (abaxial) axial leaf surface increases as the leaf shrinks with increasing water loss. At approximately 40% saturation deficit, stomatal closure becomes effective. (After Gebauer, R., *Wasserabgabe und Wasseraufnahme poikilohydrer höherer Pflanzen im Hinblick auf ihre physiologische Aktivität.* Diploma thesis, Universität zu Kiel., 97 pp., 1986.)

The effect of hair density in reducing the transpiration rate strongly increased when the leaf water-saturation deficit went beyond 40%. Reduction of the exposed surface is also a typical mechanism to reduce transpiration. This is the case with poikilohydrous ferns (Pessin 1924); with *Selaginella lepidophylla*, which curls the whole shoot rosette (Lebkuecher und Eickmeier 1993); with *Myrothamnus flabellifolius*, which regularly pleats its fan-like leaves (Vieweg and Ziegler 1969, Puff 1978); and with Velloziaceae, which fold or curl their leaves; and with other xeromorphous structures in *Sporobolus stapfianus* (Gaff 1977, Kubitzki 1998, Vecchia et al. 1998). Leaves of *Craterostigma plantagineum* can reduce surface area to 15% (Sherwin and Farrant 1998) and those of *Xerophyta scabrida* to 30% (Tuba et al. 1997b) of the original size. Leaf or shoot movements of most of these plants are due to differential imbibition of the tissues involved, rather than to osmotic phenomena, because they still operate in dead plants. Slow drying over periods of several days may typically occur in monocotyledonous plants such as *X. scabrida* (Tuba et al. 1997b). Its ecological significance for hardening and conditioning is discussed later.

PREVENTING DAMAGE AND TOLERATING STRESSES

Desiccation Tolerance

Recent literature focuses on the phenomenal desiccation tolerance (although this is only relevant for the eurypoikilohydrous organisms) and wonders whether it is a primitive strategy (Oliver et al. 2005) or how it can be genetically traced down (Illing et al. 2005) or whether it constrains growth and competetivity (Alpert 2006). Today, 74 pteridophyte and 145 angiosperm poikilohydrous species have been investigated with respect to desiccation tolerance (see Proctor and Pence 2002).

The first, but not the only, stress during desiccation is the lack of water itself, which imposes dramatic structural and physiological changes on the tissues of poikilohydrous organisms. Some poikilohydrous plants exhibit a remarkable tolerance to intense desiccation. The moss *T. ruralis* survived a water content as low as 0.008%, which is equivalent to −6000 bars (Schonbeck and Bewley 1981a). Desiccation tolerance of eurypoikilohydrous autotrophs can be defined as the capacity to withstand equilibrium with a relative humidity less than 20% (Lange 1953, Biebl 1962, Bertsch 1966b, Gaff 1986). The lowest relative water contents tolerated by vascular plants were reported to be between 1.4% and 8.4%, which is equivalent to 4% and 15% of their dry weight (Kaiser et al. 1985) (see Table 2.1 for a detailed list). Detached leaves of resurrection plants proved to be much less tolerant to severe water loss (Gaff 1980). Table 2.1 also shows tolerance to extended periods of anabiosis. *Nostoc* may tolerate 5 years of desiccation, but other cyanobacteria do not survive desiccation at all (Biebl 1962, Scherer et al. 1986). Most lichens, bryophytes (Lange 1953, Biebl 1962, Proctor 1982), and poikilohydrous vascular plants (Hallam and Gaff 1978, Lazarides 1992) survive dry periods that last for a few months. Although Marchantiales were reported to be exposed to 6–8 months of drought in their habitat in Namibia (Volk 1979), such extremely long periods of desiccation are usually rare for this kind of bryophyte. Several vascular resurrection plants have been shown to tolerate air-dry periods of 2–5 years (Hickel 1967, Gaff and Ellis 1974, Gaff 1977), and some lichens can become photosynthetically active again after a period of 10 years frozen in the dry state (Larson 1988).

Bryophytes comprise taxa that have an intrinsic capacity of desiccation tolerance, as well as taxa that need acclimation (Proctor and Tuba 2002). Although the so-called xerophytic moss species such as *Syntrichia ruralis, Rhacomitrium canescens, Neckera crispa*, and others always immediately survive extreme desiccation, mesophytic and even hygrophytic species such as *Bryum caespititium, Plagiothecium platyphyllum, Pohlia elongata*, or *Mnium seligeri* become extremely desiccation tolerant only if they were pretreated at a relative humidity of

96% for 24 h (Abel 1956, Biebl 1962). Most remarkable is that the water moss *Fontinalis squamosa* is as tolerant as a xerophytic moss, whereas *Fontinalis antipyretica* is drought-sensitive even when pretreated at 96% relative humidity. The intrinsic desiccation tolerance of eurypoikilohydrous bryophytes can be modified by the rapidity of desiccation processes (Gaff 1980). Desiccation tolerance was observed to vary seasonally in many bryophytes (Dircksen 1964, Dilks and Proctor 1976a,b), in ferns (Kappen 1964), and also in *Borya nitida* (Gaff 1980) as they can acclimate to frost desiccation in winter or to summer drought. Several bryophytes (*Dicranum scoparium* and *Mnium punctatum*) and ferns (*Asplenium spp.* and *Polypodium vulgare*) gained an extremely high desiccation tolerance in winter (Dircksen 1964, Kappen 1964). Like seeds, air-dry lichens (Lange 1953), bryophytes (Hosokawa and Kubota 1957, Gaff 1980, Proctor 1990), and vascular resurrection plants can persist longer if stored at humidities lower than 30% relative humidity (Leopold 1990). This was explained by the fact that intermediate-to-low water contents allow some enzyme activity and lead to respiratory carbon loss, destructive processes, and infections (Gaff and Churchill 1976 [*B. nitida*], Proctor 1982). Most of our current knowledge of the effects of environmental conditions before and during desiccation on the tolerance to desiccation and related stresses comes from experiments under controlled conditions.

More important for eurypoikilohydrous autotrophs is their capacity to withstand repeated changes between dry and moist states. Lichens in deserts and the Mediterranean regions, for instance, oscillate regularly between periods of a few hours of activity and anabiosis for the rest of the day (Kappen et al. 1979, Redon and Lange 1983, Kappen 1988, Lange et al. 1991, Sancho et al. 1997). Similarly, it is typical of several poikilohydrous phanerogams such as *Chamaegigas intrepidus* living in ephemeral rock pools (Woitke et al. 2004) and changing repeatedly between hydrated and dried state within one season. The capacity to tolerate several changes between dry and wet states was tested experimentally for different moss species by Dilks and Proctor (1976a, 1979). *Tortula ruralis*, as an eurypoikilohydrous species, performed well during up to 63 changes within a period of 18 months, whereas *Rhytidiadelphus loreus* was killed when continually dry for 18 months or when the oscillation phase was 1 day wet/1 day dry, but it retained 50% of its normal net photosynthetic rate if the wet periods were longer (6 or 7 days) or the dry period was shorter. Mosses such as *T. ruralis* and the angiosperm *B. nitida* (Schonbeck and Bewley 1981b) were actually able to increase their desiccation tolerance if drying and rehydration were repeated. However, cultivation under moist conditions for 2 weeks can decrease the desiccation tolerance of most eurypoikilohydrous autotrophs (algae and lichens: Kappen 1973, Farrar 1976b; bryophytes: Schonbeck and Bewley 1981b, Hellwege et al. 1994; vascular plants: Gaff 1977, 1980). In contrast, continuous hydration over several days decreased the tolerance to desiccation in *T. ruralis* (Schonbeck and Bewley 1981b), a result also found for some lichens (Ahmadjian 1973). Apparently, in the absence of contrasting oscillations of moisture content, the algal partner of lichens grows excessively, altering its symbiotic relation with the mycobiont (Farrar 1976b). Repeated drying seems to be essential for the internal metabolic balance of lichens (McFarlane and Kershaw 1982), and also in poikilohydrous vascular plants, as indicated by the fact that cultivation of *Myrothamnus flabellifolius* is only successful if the plants dry occasionally (Puff 1978). For several desert cyanobacteria, hydration is a very rare event. Pleurococcoid green algae (e.g., *Apatococcus lobatus*) and many epilithic lichens are water-repellent (Bertsch 1966a). Species of the genera *Chrysothrix, Lepraria*, and *Psilolechia* growing under overhanging rocks never receive liquid water during their lifetime (Wirth 1987).

Cellular and Physiological Changes during Desiccation

Great attention has been given in the last 25 years to the investigation of the ultrastructural changes and the biochemical processes that take place during dehydration and rehydration of poikilohydrous autotrophs. Rather than providing a detailed account here, we refer the interested reader to some reviews (Gaff 1980, 1989, Bewley and Krochko 1982, Stewart 1989, Leopold 1990, Proctor 1990, Bewley and Oliver 1992, Ingram and Bartels 1996, Hartung et al. 1998, Bartels 2005, Rascio and Rocca 2005). Two main mechanisms are involved, one that downregulates the processes and structures, leading to desiccation tolerance, and another that contributes to full metabolic and structural recovery (Bernacchia et al. 1996).

Nonvascular autotrophs are desiccation tolerant if they can retain cellular integrity and limit cellular damage during drying. To accomplish this with *bryophytes* we summarize some salient facts from reviews by Bewley and Oliver (1992), Oliver and Bewley (1997), and Oliver et al. (2000, 2005). The capability of xeric bryophytes like *Tortula ruralis* to recover quickly even from extremely rapid desiccation indicates that their desiccation tolerance is intrinsic. Membrane structure does not suffer from drying, however, protein synthesis ceases rapidly during drying (Bewley and Krochko 1982). Specific protective substances or mechanisms are not apparent (Bewley and Oliver 1992, Oliver and Bewley 1997), as sugar content does not change significantly, and no membrane protective mechanism is detectable. Upon rehydration a transient leakage indicates membrane phase transitions, and cells regain normal shape and structure within 24 h. During the first 2 h of rehydration an extensive alteration in gene expression indicates synthesis of proteins, the *rehydrins*. The rehydrins are late embryo abundant (LEA) or LEA-like proteins (see Section "Synthesis of Proteins and Protective Substances"), which may stabilize membranes lipids or help to enable quick lipid transport for reconstitution of eventually damaged membranes (Oliver 2005). They may be more important for the mesic, less-tolerant mosses and liverworts, where according to Bewley (1979) changes in membranes and metabolism during dehydration were observed. However, mesic bryopytes may increase tolerance by previous drying treatment or by addition of ABA. The presence of ABA was found in moss (e.g., *Funaria hygrometrica*: Werner et al. 1991) and in hepatic species (e.g., the europoikilohydric *Exormotheca holstii*: Hellwege et al. 1994, Hartung personal communication). Accumulation of ABA during drying and disappearance in wet culture and the capacity of inducing hardiness when applied to dehardened gametopytes reveal its role in the drought hardening process, except in *T. ruralis* where ABA was not detected (Reynolds and Bewley 1993a).

The mechanism of lichens for dealing with desiccation remains broadly unknown but is strongly related to their high content of polyols (Farrar 1976a). Membrane leakage in lichens as a consequence of repeated drying and wetting (Farrar 1976a) is harmless. The role of ABA, recently detected in lichens, is unclear, as this hormone is produced by the fungal biont and, as opposed to its activity in plants, increases as a response to water uptake (Dietz and Hartung 1998).

Pteridophytes have a performance, intermediate between bryophytes and angiosperms. For instance, as was shown with *Polypodium virginianum* (Oliver et al. 2000), they carry out synthesis of proteins upon dehydration (dydrins) as well as subsequent to rehydration (rehydrins). Interestingly, the proteins synthesized during drying disappear rapidly upon rehydration, but later (after 24 h) some of these proteins increase again. In addition, immediately after rehydration novel polypeptides are synthesized, which are not exclusively related to the desiccation regime, and another set of new proteins is produced after 24 h. ABA is present in *Polypodium* but notably decreases during drying. Nonetheless, ABA application can induce synthesis of proteins similar to dehydrins and survival of otherwise letal-rapid desiccation (Reynolds and Bewley 1993a,b). From dehydrated microphylls of *S. lepidophylla*, an expressed gene sequence tag (EST) database was generated and compared with that of other, not poikilohydrous *Selaginella* species. The percentage of functional categories, which were disease/defense-related comprising induction of secondary metabolism, molecular chaperons, and LEA proteins, was significantly higher in the poikilohydrous *Selaginella* species (Iturriaga et al. 2006).

In *S. lepidophylla*, 74% of the activity of 9 enzymes of the carbohydrate metabolism was conserved during drying. The conservation of photosynthetic enzymes was lower than that for respiratory enzymes (Eickmeier 1986).

Based on the fact that most monocotyledonous poikilohydrous species show a dramatic color change when desiccated and most dicots do not, we can distinguish between poikilochlorophyllous and homoiochlorophyllous angiosperms. Comparative, mainly electron-microscopical studies illustrate these differences (Sherwin and Farrant 1996, 1998, Farrant 2000).

In the poikilochlorophyllous species, structural changes are considerable (see Gaff 1980, Hetherington and Smillie 1982, Tuba et al. 1996b, 1997), except in the nucleus (Barthley and Hallam 1979), containing a dense mass of chromatin (Hethrington and Smillie 1982). Ultrastructural changes in poikilochlorophyllous plants may be even greater than in desiccation-sensitive homoiohydrous species at a comparable dehydration level (Gaff 1989). Investigations with *Xerophyta villosa* and *X. scabrida* show the following: The polysome content rises significantly with water loss (*X. villosa*; Gaff 1989). Virtually all thylacoids and most of the carotenoids content are lost. Obviously, the destruction of the chlorophyll is structured, as grana are retained (Bartley and Hallam 1979). However, as most of the chlorophyll can be preserved in *X. scabrida*, if it is desiccated in darkness, Tuba et al. (1997) suggest that under natural conditions chlorophyll loss is due to photooxidation rather than metabolic destruction. Although most of their cristae disappear, mitochondria remain functional (see Hallam and Capicchianano 1974). A continuation of respiration still measurable below −3.2 MPa (Tuba et al. 1996b) in desiccated plants of *X. scabrida* suggests that energy is required for dismantling the thylacoids and the formation of the so-called desiccoplasts. Thus, rather than being deleterious, these organelle changes involve an organized remobilization of cell resources in these resurrection plants.

Structural changes by desiccation are usually small in the homoiochlorophyllous species, and thylacoid membranes and associated chlorophyll complexes apparently remain widely preserved (Owoseye and Sandford 1972, Hallam and Cappicchiano 1974, Gaff and McGregor 1979, Gaff 1989). However, changes in the chloroplast structure and loss of chlorophyll by 20% in *Myrothamnus flabellifolius* (Wellburn and Wellburn 1976, Farrant 2000) and by 40%–50% in *Craterostigma wilmsii* (Sherwin and Farrant 1998) were observed, which recovered within 24–45 h subsequent to rehydration. Kaiser and Heber (1981) and Schwab and Heber (1984) state that the lens shape of the chloroplasts permits dehydration without greater surface area reduction, and in vivo rupture of chloroplasts during desiccation is rarely observed. In dry *Talbotia elegans*, the mitochondria are reduced to membrane-bound sacks (Hallam and Gaff 1978). Transient membrane leakage has been reported for several species. Vacuoles, fragmented into numerous vesicles, become filled with a nonaqueous substance, obviously generating a backpressure for the desiccating cell (Farrant 2000) and lysosomes seem to be maintained intact (Gaff 1980, 1989, Hartung et al. 1998). As a consequence, repair processes may last for 2 days in small herbaceous species (*Ramonda* and *Haberlea* species: Gaff 1989, Markowska et al. 1995), and the extended time for recovery of the large fruticose *M. flabellifolius* is due to the restoration of its hydraulic system.

As a rule, the processes of desiccation and recovery take more time in the poikilochlorophyllous than in the homoiochlorophyllous species. Tuba et al. (1997) state that under field conditions *X. scabrida* may take more than 2 weeks until the plants are dry and inactive. Correspondingly, also repair processes are extended over periods of several days (*X. viscosa* 92–120 h: Sherwin and Farrant 1996, 1998). Poikilochlorophyllous performance is obviously a highly derived adaptation. It allows avoidance of stress from free radicals. In ecological terms, such species are adapted to sites where drought periods are extended over weeks or months, whereas homoiochorophyllous species easily manage in sites with frequent oscillations between dry and wet stages in shorter periods of time.

Synthesis of Proteins and Protective Substances

Obviously, a network of genes with presumably different functions is activated by water stress. Hartung et al. (1998) estimated that 800–3000 genes could be involved in the response of plants to desiccation. Poikilohydrous plants exhibit a great variety of down- and upregulation of cellular processes, which can be retained at very low water potentials (Leopold 1990). Particularly, genes that code for enzymes relevant to photosynthesis, both in vascular plants and in mosses (Ingram and Bartels 1996, Bernacchia et al. 1996, Oliver and Bewley 1997) were downregulated. In general, the decline of total protein is smaller than in drought-sensitive plants. Loss of water-insoluble proteins is common in resurrection plants, especially in the poikilochlorophyllous species, probably because of degradation of the lipoproteins of the membrane (Gaff 1980). The preservation of polysomes and of RNA may enable protein synthesis after drought (Bewley 1973). Many novel proteins (*dehydrins*) are synthesized during desiccation, most of which were considered specific to extremely desiccation-tolerant plants (Hallam and Luff 1980, Eickmeier 1988, Bartels et al. 1990, Piatkowski et al. 1990, Bartels et al. 1993, Kuang et al. 1995). Nevertheless, certain polypeptides, such as those found in desiccated *Polypodium virginianum*, are not exclusive to the desiccation regime (Reynolds and Bewley 1993b). The majority of the dehydrins belongs to LEA proteins, they are hydrophilic and resistant to denaturation, and typical of orthodox seeds. They are believed to protect desiccation-sensitive enzymes and to stabilize membranes during dehydration (Schneider et al. 1993, Bernacchia et al. 1996, Ingram and Bartels 1996, Bartels 2005). Proteins are necessary also during the rehydration phase. They can be gained by translation of already existing transcripts, as was shown for poikilohydrous species (Dace et al. 1998). Bartels and Salamini (2001) suggest that desiccation tolerance (of *Craterostigma plantagineum*) is in most cases not due to structural genes, unique to resurrection plants and could be present as well in desiccation-sensitive homoiohydrous plants. However, the latter may have less amounts of LEA proteins and the expression pattern may be different. Only one, a LEA-6 protein, was identified as typical exclusively for *Xerophyta humilis* and seeds (Illing et al. 2005).

In most resurrection plants, including the aquatic species *Chamaegigas intrepidus*, abscisic acid (ABA) is strongly accumulated and is involved in attaining desiccation tolerance and in stimulating the synthesis of dehydrins (Gaff 1980, 1989, Gaff and Loveys 1984, Reynolds and Bewley 1993a, Hellwege et al. 1994, Schiller et al. 1997). As was hypothesized by Bartels et al. (1990), Nelson et al. (1994), and Oliver and Bewley (1997), there is evidence now in vascular plants that ABA is necessary to induce the genes for desiccation tolerance. With experiments of mutants of *C. plantagineum*, the so-called CDT-1/2 gene family was shown to function by ABA signal transduction (Smith-Espinoza et al. 2005). Leaves of *Myrothamnus flabellifolius* and *Borya nitida* did not survive dehydration if they were dried so rapidly that ABA could not be accumulated (Gaff and Loveys 1984). Abscisic acid accumulation obviously can occur only in leaves attached to the whole plant (Hartung et al. 1998).

A common phenomenon in drought stress is the accumulation of organic compatible solutes because they stabilize proteins and membranes (Levitt 1980, Crowe and Crowe 1992). Lichens are permanently rich in sugar alcohols, which are assumed to be the basis of their remarkable desiccation tolerance (Kappen 1988). Cowan et al. (1979) have demonstrated that the synthesis of amino acids and sugar alcohols was active in lichens in equilibrium with humidities as low as 50%. In contrast, desiccation-tolerant bryophytes contain a low amount of sugars, mainly sucrose, and show no or very little increase in sugar content during drying (Bewley and Pacey 1978, Santarius 1994). Strong sugar accumulation, mainly sucrose, during desiccation has been demonstrated in seeds and many resurrection grasses, species of *Ramonda*, *Haberlea*, and *Boea*, and *X. villosa* (Kaiser et al. 1985, Scott 2000, Zirkovic et al. 2005). Other resurrection plants for example, of the genera *Ceterach* and *Craterostigma* already contain comparatively high amounts of sugar in the leaves when turgid

(Schwab and Gaff 1986). In these and other species (e.g., *M. flabellifolius*), sugar composition was observed to be changed during dehydration (Bianchi et al. 1991, Hartung et al. 1998). Unusual sugars such as stachyose that appear in the turgid leaves and roots are storage products, but they are converted into sucrose during the drying process (Bianchi et al. 1991, 1993, Albini et al. 1994, Heilmeier and Hartung 2001, Norwood et al. 2003). For instance, 2-octulose is typical for hydrated leaves of *C. plantagineum* and is converted into sucrose upon dehydration (Bartels and Salamini 2001). Thus, sucrose accumulation during desiccation is generally recruited from metabolizing storage carbohydrates rather than directly from photosynthesis.

Inositol, present in *Xerophyta viscosa*, may be an effective osmoprotectant (Mayee et al. 2005). Proline concentration in many plant species associated with water stress was comparatively high in *Ceterach* and *Craterostigma*, but did not significantly increase during dehydration (Schwab and Heber 1984). Application of proline had no effect on detached leaves of *B. nitida* and *M. flabellifolius* (Gaff 1980). In the latter species, polyphenols have been identified (Moore et al. 2005) that might be relevant to desiccation tolerance, as provenances from Namibia subjected to greater drought stress were genetically different from those in South Africa and contained more and different polyphenols (e.g., 3,4,5-tri-*O*-galloquinic acid).

One of the internal hazards of desiccation is the increase in oxidative processes, which occurs in plants exposed to a wide range of environmental stresses (Smirnoff 1995). An increase in or a high level of defense enzymes of the ascorbate/glutathione cycle was associated with the protection of the membrane lipids in *Sporobolus stapfianus* during drying (Sgherri et al. 1994a). Oxidized glutathione was much lower in slowly dried (unimpaired) samples than in rapidly (injured) dried samples of *Boea hygroscopica* (Sgherri et al. 1994b), and glutathione was shown to play the primary role in maintaining the sulfhydryl groups of thylacoid proteins in reduced state during desiccation (Navarri-Izzo et al. 1997). The reversible decrease in phenolic acids far below the level in the hydrated state, which is joined by a decrease in the enzyme ascorbate peroxidase (AP) while antioxidants were accumulated, indicates that *Ramonda serbica* leaves are able to keep up an antioxidative status when subjected to desiccation (Sgherri et al. 2004). Kranner and Grill (1997) postulate that glutathione reductase (GR) and glucose-6-phosphate dehydrogenase are needed for the reduction of desiccation-induced oxidized glutathione. It is suggested that this pathway provides the NADPH during the critical rehydration phase when photosynthesis is still inactive. Accordingly, when photosynthesis is recovered a decrease in antioxidants and production of reactive oxygen species was observed in a lichen species (Weissman et al. 2005). Antioxidants such as AP, GR, and SOD (superoxide dismutase) were increased but to various extents in subsequent phases of water loss in *Craterostigma wilmsii*, *M. flabellifolius* (mainly GR), and *X. viscosa* (Sherwin and Farrant 1998) and went down to normal level when the tissues were rehydrated. Anthocyanins recognized as antioxidants (Smirnoff 1993) were observed to increase in drying leaves of poikilochlorophyllous species such as *Eragrostis nindensis* (Van der Willigen et al. 2001) and particularly in *Xerophyta humilis* (Farrant 2000). In *C. plantagineum*, lipoxygenase, which catalyzes lipid peroxidation at membranes, becomes increasingly inhibited during drying (Smirnoff 1993). Similar processes may also operate in desiccation-tolerant bryophytes (Dhindsa and Matowe 1981, Seel et al. 1992a) in which lipid peroxidation during drought is low. Oxidative processes can take place both in the presence and the absence of light, and light can exacerbate oxidation. This situation—oxidative stress caused or accentuated by light—is discussed in the following section.

Photoprotection of the Photosynthetic Units

If plants absorb more light than required during photosynthesis, they are exposed to the risk of photooxidative destruction of their photosynthetic apparatus (Long et al. 1994).

Photooxidative stress can therefore be an important limiting factor for poikilohydrous autotrophs. Bryophytes and algae that are restricted to shady habitats were shown to have limited photoprotective capacities (Öquist and Fork 1982b). Negative effects of strong light were observed in hydrated lichens in the tropical, temperate (Coxson 1987a,b), and Mediterranean region (Manrique et al. 1993, Valladares et al. 1995). Nevertheless, tolerance to strong light can be enhanced by acclimation. Cyanobacterial mats taken from exposed habitats proved to be highly tolerant to high irradiance, whereas cyanobacteria from shaded sites were very sensitive (Lüttge et al. 1995). Field studies have revealed, for instance, that the cyanobacterial lichen *Peltigera rufescens* was at least photoinhibited under certain conditions in winter (Leisner et al. 1996). On the other hand, cryptogam species in Antarctica such as *Umbilicaria aprina*, *Leptogium puberulum*, *Xanthoria mawsonii*, and *Hennediella heimii* were very resistant to the combination of low temperatures and high irradiance while the thallus was photosynthetically active (Schlensog et al. 1997, Kappen et al. 1998a, Pannewitz et al. 2003, 2006).

In hydrated autotrophs, photosynthetic productivity is maintained because only that part of the light energy that is in excess to that used for energy conservation is thermally dissipated by a mechanism that requires zeaxanthin, a carotenoid of the xanthophyll cycle, and the protonation of a special thylacoid protein (Niyogi 1999, Heber et al. 2006). Thermal energy dissipation should be in equilibrium in hydrated autotrophs with ongoing photosynthesis. This means that energy dissipation is in equilibrium with energy conservation based on charge separation, the production of a strong oxidant and a reductant in the reaction centers of PS II. If energy dissipation caused is speeded up (photostress), it would inhibit photosynthesis (Wiltens et al. 1978, Öquist and Fork 1982a, Demmig-Adams et al. 1990b). Downregulation of photosynthetic processes and the so-called dynamic or recoverable photoinhibition (i.e., inhibition of photosynthesis by light, but no damage) has been observed in a number of poikilohydrous plants, bryophytes, and lichens (e.g., Seel et al. 1992, Leisner et al. 1996, Ekmekci et al. 2005), and as a result, avoidance of photooxidation (Eickmeier et al. 1993, Valladares et al. 1995, Calatayud et al. 1997, Heber et al. 2000, 2001, Bukhov et al. 2001).

A prevention of photooxidative damage by drying may be apparent from the fact that isolated *Trebouxia* as well as green-algal lichens resisted photostress in the field by quick desiccation under high irradiances (Öquist and Fork 1982b, Leisner et al. 1996). This would resemble in effect the strategy of poikilochlorophyllous plants that radically destruct the photosynthetic apparatus during desiccation (Smirnoff 1993). It was hypothesized that the photosynthetic apparatus of homoiochlorophyllous autotrophs cannot be affected by strong irradiance because it undergoes a functional dissociation between light harvesting complexes and photosystem II during desiccation (Bilger et al. 1989, Smirnoff 1993). However, water content has been proved to influence both dynamic and chronic photoinhibition of lichens (Valladares et al. 1995, Calatayud et al. 1997). Some air-dried lichens typical of shady habitats exhibited even damage after exposure to high light (Valladares et al. 1995, Gauslaa and Solhaug 1996, 1999, 2000, Gauslaa et al. 2001). In addition, stenopoikilohydrous mosses were more damaged by drying at high irradiance than at low irradiance (Seel et al. 1992a).

According to recent findings since Shuvalov and Heber (2003), it has become apparent that reaction centers are capable of charge separation even in the absence of water (Heber et al. 2006a). Thus, functional reaction centers would cause damaging oxidative reactions. A revised and more comprehensive approach to understanding photoprotection in desiccated autotrophs has recently come from Heber and coauthors (Heber et al. 2000, 2001, Heber and Shuvalov 2005, Kopecky et al. 2005, Heber et al. 2006a,b) who have demonstrated that more than one photoprotective mechanism of energy dissipation is active in lichens and bryophytes. Available evidence suggests that zeaxanthin-dependent energy dissipation remains active upon desiccation (Eickmeier et al. 1993, Kopecky et al. 2005, Georgieva et al. 2005), but it is not clear whether the zeaxanthin-dependent energy dissipation is fast enough to prevent charge separation in functional reaction centers particularly in lichens and xeric bryophytes.

A second protective mechanism was evident from the finding that functions of the reaction center of PS II can change upon desiccation (Heber et al. 2006a,b). In desiccated samples of the moss *Rhytidiadelphus squarrosus*, energy dissipation has been shown to occur in PS II reaction centers. In this case, a photoreaction is responsible for the formation of a quencher of fluorescence in the reaction center (Heber et al. 2006b). A third protective mechanism was apparent from the observation that light was not even necessary for the formation of a quencher during desiccation. After lichen thalli were carefully predarkened to avoid light activation of the mechanism of energy dissipation, fluorescence was quenched after desiccation took place in darkness. This reveals the activation of the third mechanism of thermal energy dissipation (Heber et al. 2006b, Heber personal communication). The latter two protective mechanisms were operating only under desiccation and ceased by rehydration. Lichens with cyanobacteria as photobionts lack the zeaxanthin-dependent thermal energy dissipation mechanism; however, other carotenoids may play a role at least in the hydrated state (Demmig-Adams et al. 1990b, Leisner et al. 1994, Lange et al. 1999). In the desiccated state, the desiccation-induced thermal energy dissipation mechanism (mechanism 3) may be operating (Heber, personal communication). As soon as water becomes available to the chloroplasts of homoiochlorophyllous autotrophs, photosynthetic water oxidation is resumed (Kopecky et al. 2005, Heber et al. 2006a).

In vascular plants, only the zeaxanthin-dependent dissipation mechanism is known to be protective also in the desiccated state. It also operates in homoiohydrous xerophytic plants (e.g., *Nerium oleander*: Demmig et al. 1988; *Clusia* spp. with CAM: Winter and Königer 1989). In resurrection plants, it was shown with *Selaginella lepidophylla* (Casper et al. 1993, Eickmeier et al. 1993). Accumulation of zeaxanthin upon drying was reported for the poikilohydrous species *Craterostigma Plantagineum* (Amalillo and Bartels 2001) and *Boea hygrometric* (see Yang et al. 2003). In *R. serbica*, the photoprotection appeared to be achieved, when dry, by the zeaxanthin-dependent dissipation as well as by ascorbate and glutathione (Augusti et al. 2001). In the poikilochlorophyllous *X. scabrida*, 22% of the carotenoids were still preserved in the dry leaves when the photosynthetic apparatus was dismantled, but the carotenoids seemed to be protective or essential when the chloroplasts reorganized during rehydration (Tuba et al. 1993b).

Poikilohydrous autotrophs exhibit various photoprotective mechanisms in addition to the thermal energy dissipation via carotenoids. In the case of lichens, filtering or screening effects of the upper cortex formed by the mycobiont and of certain secondary compounds such as parietin have been shown to be potentially important in reducing the risk of photodamage of the photosynthetic units (Büdel 1987, Solhaug and Gauslaa 1996, Kappen et al. 1998a, Gauslaa and Ustvedt 2003, Gauslaa and McEvoy 2005). Anthocyanins, which reflect photosynthetically active light, may prevent excessive light–chlorophyll interaction. An accumulation of anthocyanins was observed in sun-exposed leaves of several resurrection plants, mainly on the abaxial leaf face, which is everted when the leaves are curled or folded (Farrant 2000). Leaf and shoot curling during drought, despite not effective in certain bryophytes (Seel et al. 1992a), can confer photoprotection simply by shadowing. For instance, leaf curling by everting the reflectant abaxial leaf surface was effective in *Polypodium polypodioides*, a species sensitive to strong light under water stress (Muslin and Homann 1992, Helseth and Fischer 2005). Additionally, photorespiration and light-activated photosynthetic enzyme activity were not affected by intense radiation in *S. lepidophylla* due to leaf curling (Lebkuecher and Eickmeier 1991, 1993). The protective role of curling is important for plants growing in open and exposed habitats and, although not yet thoroughly explored, it may be relevant also for lichens such as *Parmelia convoluta* and *Chondropsis semiviridis*, or for thallose desert liverworts all everting a whitish underside when dry (Lange et al. 1990a).

Another radiative stressor is UV B, particularly in polar and alpine regions; however, experiments proved that it never harmed lichens and mosses in open habitats (Lud et al. 2003,

Nybakken et al. 2004). Moreover, UV B (280–320 nm) was shown to be an essential requisite for the synthesis of sun-screening pigments (parietin, melanin) in lichens (Solhaug et al. 2003).

Desiccation Tolerance: An Old Heritage

Taking desiccation as a fundamental heritage, geneticists and molecular biologists were challenged to trace down this capability of plants to the beginning of plant evolution and to ask whether this is unique to poikilohydrous organisms and why it is absent in the vegetative parts of the majority of vascular plants and certain bryophytes, and, whether it is mono- or polyphyletically evolved, and, on which genes desiccation tolerance is located.

According to Oliver et al. (2000) recent synthetic phylogenetic studies confirm the idea that vegetative desiccation tolerance is primitively present in the bryophytes but was then lost in the evolution of vascular plants. It is hypothesized that desiccation tolerance was crucial for ancestral fresh-water autotrophs to live on land. The tolerance might have been lost on the way of a more complex (homoiohydrous) organization of the vascular plants, and independent evolution or re-evolution of desiccation tolerance has happened in the Selaginellales, leptosporangiate ferns, and the 10 families of angiosperms with poikilohydrous species. Evolutionary progress may be evident from the fact that pteridophytes, in common with bryophytes, are able to synthesize rehydrins but can also synthesize dehydrins, the only capability of angiosperms. Tracing possible pathways by identifying possible gene orthologs of LEA-like proteins that are synthesized upon dehydration provided a network of land plant phylogeny. Oliver et al. (2005) proved their earlier theory and extended it by discussing the fact that several genera of bryophytes (*Haplomitrium, Sphagnum, Takakia, Tetraphis*) lack the particular genes and are not desiccation tolerant either constitutively or due to loss of phenotype, which involves the possibility that precursors of these bryophytes also did not realize desiccation tolerance in the vegetative state. There was also negative evidence that ancestors of the hornworts and of all vascular plants were desiccation tolerant. Keeping up the idea that desiccation tolerance of spores may elucidate the problem, Illing et al. (2005) demonstrated by molecular responses a crucial role of orthodox seeds as the carriers of genes for desiccation tolerance, which may, or not, be expressed in the vegetative part of vascular plants. Analyzing sequences of genes of *Arabidopsis* seeds and a set of vegetative homoio- and poikilohydrous plant species with respect to certain LEA proteins, RNA transcripts for antioxidants that are expressed during desiccation, and in addition, sugar accumulation revealed the following results: Among the antioxidant activities only that with 1-cys peroxidase appeared to be a mechanism specific to desiccation tolerance of resurrection plants and seeds. Most of the antioxidants are activated also as a response to various other abiotic stresses. However, genes belonging to the LEA six superfamily are uniquely associated with dehydration stress in resurrection plants and in seeds. Moreover, sucrose accumulation upon drying seems to be a desiccation-tolerance specific mechanism of resurrection plants and orthodox seeds, albeit it was also observed in some recalcitrant seeds (Kermode and Finch-Savage 2002). As a general result, desiccation tolerance is not considered the result of parallel origins, and the loss of desiccation tolerance can be interpreted as the result of suppression of latent genes in the vegetative state of the plant. However, single gene loss in certain instances (Dickie and Pritchard 2002) cannot be excluded.

Generally speaking of desiccation tolerance suggests that it relates to a fundamental uniform mechanism; however, a differentiated view is obvious now. On the one hand, it has to be realized that the most primitive autotrophs, such as Cyanobacteria, are extremely desiccation tolerant, whether they live in water or in the porous space of rock. This indicates that desiccation tolerance is a very early trait in evolution. Moreover, the fact that organelles such as chloroplasts of spinach (Table 2.1) are desiccation tolerant and that vegetative tissues can be lyophilized without harm suggests that it is the way of withdrawing water that can be damaging but not the dehydration itself. Thus, it can be concluded that the capability of surviving dehydration must be basic and ubiquitous. On the other hand, phenotypic tolerance must have been acquired at various advanced stages of organismic complexity by evolving metabolic or structural protective mechanisms. These must be very specific, either because of a distinct combination or quantitatively selected ubiquitous "house-keeping" mechanisms (Illing et al. 2005), or because of inventing genetic structures that act species-specific, as most likely the CDT-1 gene in *C. plantagineum* (Furini et al. 1997, Bartels and Salamini 2001). So far it is reasonable to consider desiccation tolerance of vascular plants (and highly derived bryophytes) as modified (see Oliver and Bewley 1997) or secondary trait (Raven 1999).

Tolerance to Extreme Temperatures: A Property Linked to Poikilohydry

Plants capable of surviving desiccation may be tolerant also to other environmental stresses such as extreme temperatures. Some lichen species, for example, are extremely resistant to freezing both in hydrated and dehydrated states. The freezing tolerance of hydrated lichens can exceed by far the temperature stresses that occur in winter or in polar environments (Kappen 1973). Bryophytes are less freezing-tolerant, but many species are well adapted to live and persist in cold environments (Richardson 1981, Proctor 1982, Longton 1988b). Fern gametophytes and vascular resurrection plants from the temperate zone have a moderate (−9°C to −18°C) freezing tolerance in winter (Kappen 1964, 1965). Most poikilohydrous vascular plants come from subtropical climates and are sensitive to freezing. For example, *Borya nitida* can only survive temperatures between −1°C and −2°C in winter (Gaff and Churchill 1976), and poikilohydrous grass species do not survive temperatures below 0°C (Lazarides 1992). However, desiccated leaves, fronds, or other vegetative parts become highly tolerant to cold depending on the remaining water content. For instance, experiments with *Ramonda myconi, Polypodium vulgare, Myrothamnus flabellifolius*, and gametophytes of several fern species revealed that the tissues resisted −196°C if they were desiccated to a relative water content of, for example, 6% (Kappen 1966, Vieweg and Ziegler 1969, Pence 2000). Thus, the tolerance of these vascular poikilohydrous plants to low temperatures was not different from that of dry algae, lichens, and bryophytes (Levitt 1980), which also indicates water content as a crucial factor in the freezing tolerance of organisms in general.

Heat tolerance of plants can also be increased if they are desiccated (Kappen 1981). This fact becomes ecologically relevant if we consider that dry plants can easily heat up to over 50°C in their natural habitats. Thallus temperatures in dry moss turfs and dry lichens can reach 60°C–70°C under field conditions (Lange 1953, 1955, Richardson 1981, Proctor 1982, Kershaw 1985). A similarly high temperature was reported for desert soil around *Selaginella lepidophylla* (Eickmeier 1986). The temperature in the rock pools where *Chamaegigas intrepidus* lives can reach 41°C (Gaff and Giess 1986). Although some lichens such as *Peltigera praetextata* and *Cladonia rangiferina* did not survive temperatures higher than 35°C, even in the desiccated state (Tegler and Kershaw 1981,Gauslaa and Solhaug 1999, 2001), tolerance to 70°C and even to 115°C was recorded for many other dry lichen and bryophyte species (Lange 1953, 1955, Proctor 1982, Meyer and Santarius 1998). Heat tolerance of bryophytes and vascular plant species varies with season. The maximal temperature tolerated in the turgescent state by the temporarily poikilohydrous fern *P. vulgare* and by *R. myconi* was highest in winter (approximately 48°C), and by decreasing water content, the heat tolerance could be increased to approximately 55°C (Kappen 1966, Figure 2.5). *C. intrepidus* was able to thrive at 60°C temperatures (Heil 1925), and dry leaves of *M. flabellifolius* were reported to have resisted at 80°C (Vieweg and Ziegler 1969). Eickmeier (1986) could demonstrate an increase in photosynthetic repair capacity of *S. lepidophylla* if dry plants were subjected to 45°C and 65°C, but he found also that desiccation tolerance decreased with increasing temperatures (25°C, 45°C, 65°C). However,

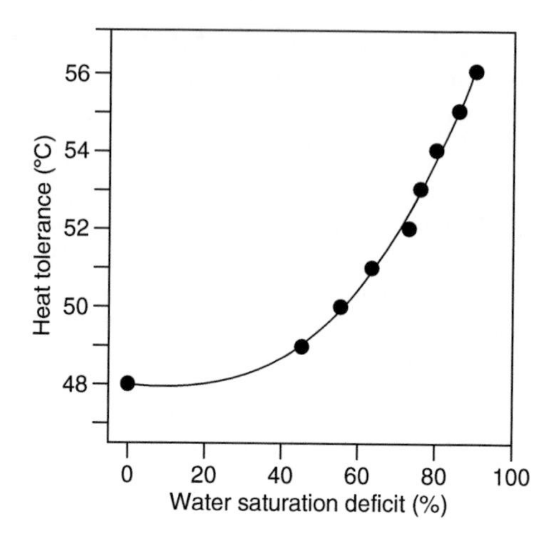

FIGURE 2.5 Shift of heat tolerance limit with increasing water saturation deficit in leaves of *Ramonda myconi.* (After Kappen, L., *Flora*, 156, 427, 1966.)

since the water content of these dry plants was not defined, it still remains unclear whether metabolic disturbance at the higher temperature was weakening the plants.

LIMITS AND SUCCESS OF POIKILOHYDRY

Poikilohydrous land plants are typically small and slow growing like many other ecological specialists, obviously as a response to environmental constraints. The carbon economy of poikilohydrous autotrophs is influenced by the frequency and duration of the periods of metabolic activity and by repair and other processes that delay maximal net photosynthesis once the plant enters the biotic state by rehydration. In fact, poikilohydrous species exhibit significant differences in their speed of recovery from dehydration and in their photosynthetic performance while they are active.

Photosynthesis

While cyanobacteria, algae, lichens, and bryophytes can reach maximal photosynthetic rates within 10–30 min after rehydration, poikilohydrous vascular plants usually require much longer (Table 2.2). Generally, the threshold for respiratory activity is lower than that for net photosynthesis (Lange and Redon 1983). The capacity of chloroplasts and mitochondria to function at low water potentials is remarkable in aerophytic green algae and green-algal lichens (−50 and −38 MPa, respectively; Nash et al. 1990, Bertsch 1996a) and also in homoiohydrous vascular plants (−19 MPa in the mesophyll of *Valerianella locusta*, Bertsch 1967; beet chloroplasts, Santarius 1967). However, each particular poikilohydrous organism has a suite of morphological and functional traits that exert specific influences on photosynthetic yield and overall performance during periods of water availability. These organisms are covered in the following sections.

Lichens and Bryophytes

Lichens can be photosynthetically active at water contents as low as 20% d.wt. (Lange 1969b, Lange et al. 1990b), under high saline stress (Nash et al. 1990), and at temperatures as low as −20°C (Kappen 1989, 1993a, Schroeter et al. 1994; see also Chapter 14). Schroeter and Scheidegger (1995) demonstrated that net photosynthesis was still positive when the algal

TABLE 2.2
Time Required by Different Poikilohydrous Autotrophs to Resume Photosynthetic Activity (Net Photosynthesis) after Rehydration with Water Vapor (>90% Relative Humidity) or with Liquid Water

Species	Water Source	Time	Reference
Parmelia hypoleucina	Water vapor	<1 h	Lange and Kilian 1985
Cladonia portentosa	Water vapor	6 h	Lange and Kilian 1985
Caloplaca regalis	Water vapor	24 h	Lange and Kilian 1985
Lobaria pulmonaria	Liquid	<30 s[a]	Scheidegger et al. 1997
Hypnum cupressiforme	Water vapor	12 h	Lange 1969
Tortula ruralis	Liquid	90 min	Scott and Oliver 1994
Exormotheca holstii	Liquid	2 h[a]	Hellwege et al. 1994
Polypodium virginianum	Liquid	6 h[b]	Reynolds and Bewley 1993
Craterostigma plantagineum	Liquid	15 h[b]	Bernachia et al. 1969
Selaginella lepidophylla	Liquid	24 h[b]	Eickmeier 1979
Ramonda serbica	Liquid	24 h	Markowska et al. 1997
Myrothamnus flabellifolius	Liquid	*c.* 40 h	Sherwin and Farrar 1996
Xerophyta scabrida	Liquid	72 h	Tuba et al. 1993b

[a] Chlorophyll a fluorescence positive.
[b] Chloroplasts reconstituted.

cells were shriveled and the fungal hyphae cavitated because of the low water content of the thallus. At a given low water potential, different lichen species exhibited different fractions of collapsed photobiont cells and different photosynthetic rates (Figure 2.6). It has therefore

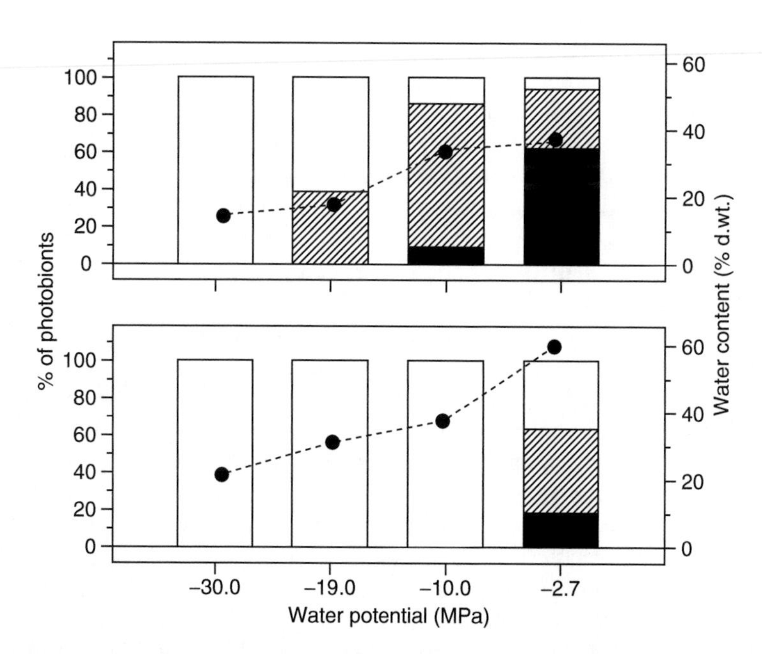

FIGURE 2.6 Amount of photobionts (open bars, heavily collapsed; hatched bars, partially folded; solid bars, globular) and water content (solid circles, % dry weight) of thalli of the lichens *Ramalina maciformis* (*top*) and *Pseudevernia furfuracea* (*bottom*) at an equilibrium with water potentials between −30 and −2.7 MPa. (After Scheidegger, C., Schroeter, B., and Frey, B., *Planta*, 197, 399, 1995.)

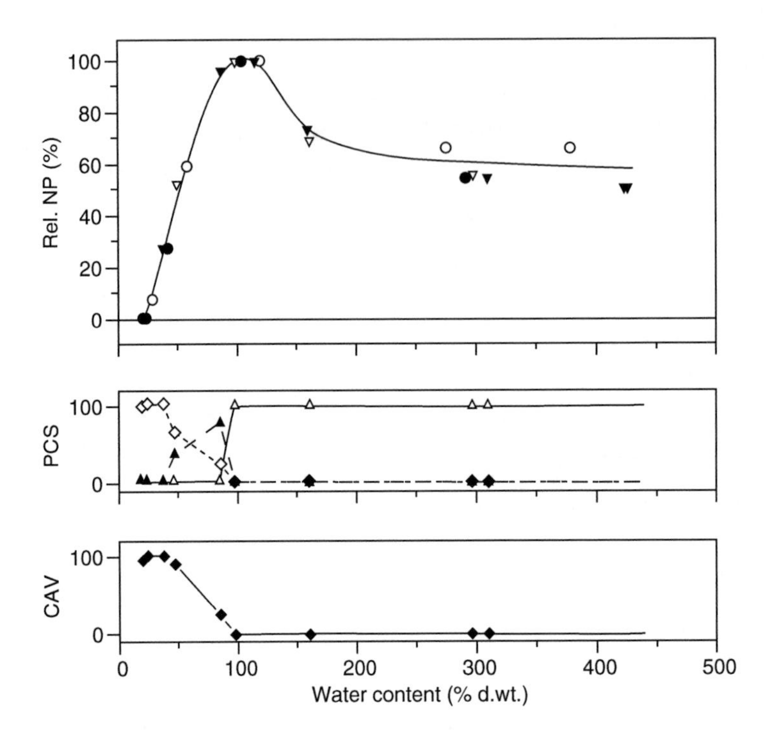

FIGURE 2.7 Relative net photosynthetic rates (rel. NP as % of the maximum rate) in the lichen *Lobaria pulmonaria*, changes of photobiont cell shape (PCS) given as percentage of globular (▵), slightly indented (▴), and heavily collapsed cells (▪), and percentage of cavitated hyphae (CAV) in the upper cortex in relation to water content (% dry weight). (After Scheidegger, C., Schroeter, B., and Frey, B., *Planta*, 197, 399, 1995.)

been suggested that an increase in net photosynthesis is, in general, proportional to an increase in the amount of functional photobiont cells (Figure 2.7). However, the levels of hydration at which algal cells reach maximal and minimal photosynthetic rates are unknown. The fungal part of the lichens proved to be totally impotent in aiding the algal cells to absorb water vapor, as was shown by comparing lichens and isolated photobionts (Lange et al. 1990b, Kappen 1994). Most importantly, these lichens are able to start photosynthesis by water-vapor uptake at low water potentials, as was observed in the laboratory (Lange and Bertsch 1965, Lange and Kilian 1985, Scheidegger et al. 1995) as well as in the field (Kappen 1993a, Schroeter et al. 1994).

Lichens with cyanobacteria as photobionts need liquid water for activation of their photobionts (Figure 2.8). Their water content for starting photosynthesis (85%–100% d.wt.) is 3–6 times higher than that of green-algal lichens (15%–30% d.wt.; Lange et al. 1988). The deficiency of cyanobacteria to become photosynthetically activated by water vapor cannot be compensated by the symbiosis with the fungus (Kappen 1994) nor in photosymbiodeme lichens, where only the green-algae fraction of the thallus becomes activated (Green et al. 2002). The explanation for this inability to resume turgidity by water-vapor uptake remains unclear, but the possibility that diffusion resistance by the gelatinose sheath of cyanobacteria could impede water vapor absorption has been excluded (Büdel and Lange 1991).

Quick recovery of CO_2 exchange during the daily wetting and drying cycles has been demonstrated in lichens from various habitats (Kappen et al. 1979, Lange and Redon 1983, Lange et al. 1990c, 1991, Bruns-Strenge and Lange 1991). However, in some cases, CO_2 exchange of lichens and bryophytes revealed only respiratory CO_2 release, frequently at high rates (lichens 1–3 times of normal), during periods of 15 min up to 7 h (lichens) after soaking

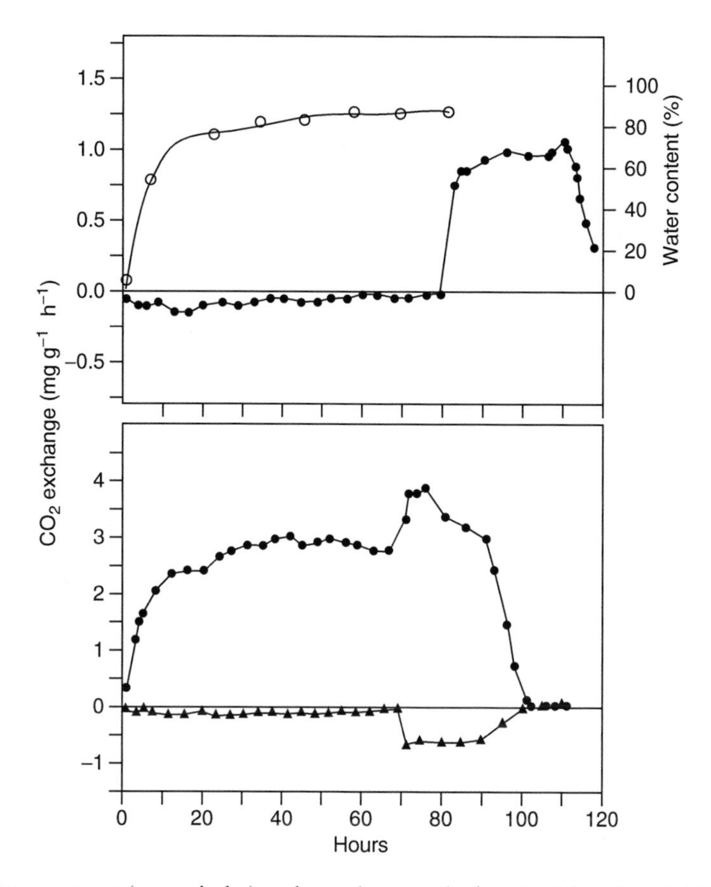

FIGURE 2.8 Water content (open circles) and net photosynthetic rates (closed circles) following water-vapor and then liquid-water uptake of the lichen species *Nephroma resupinatum* with cyanobacterial photobionts (*top*) and *Ramalina menziesii* with green-algal photobionts (*bottom*). In *N. resupinatum*, water-vapor uptake (open circles) resulted in maximum thallus water content of 85% d.wt. but induced only a weak respiratory response in light, whereas spraying activated photosynthetic CO_2 uptake. In *R. menziesii*, both dark respiration (closed triangles) and net photosynthesis (closed circles) were activated within the first 30 min of water-vapor uptake. (After Lange, O.L., Kilian, E., and Ziegler, H., *Oecologia*, 71, 104, 1986.)

with water. This resaturation respiration indicates recovery processes (Ried 1960, Hinshiri and Proctor 1971, Smith and Molesworth 1973, Dilks and Proctor 1974, 1976a, Farrar 1976b, Sundberg et al. 1999), which cause a significant carbon loss (Lechowicz 1992). Slow recovery was shown by species of moist habitats (stenopoikilohydrous species), whereas quick recovery was found in species subjected to frequent and severe drying (eurypoikilohydrous species).

Most bryophytes require liquid water for activation of photosynthesis, although some species (e.g., *Hypnum cupressiforme*) are able to activate photosynthesis by uptake of water vapor (Lange 1969a). Thalli of the desert moss *Barbula aurea* showed high respiration rates in a water-saturated atmosphere, but could achieve an almost negligible net photosynthetic rate when irradiated (Rundel and Lange 1980). In bryophytes as well as in lichens, net photosynthetic rate can become maximal within 15–30 min subsequent to moistening with liquid water (see Proctor and Tuba 2002). Photosynthetic rates of mosses decreased with increasing length of the drought period before hydration. Repeated dry–wet cycles caused a greater decrease in net photosynthesis in mesic stenopoikilohydrous mosses than in xeric species (Davey 1997).

As mentioned in the Section "Exploiting an Erratic Resource," photosynthetic CO_2 uptake can be severely impeded by excessive water contents. This has been intensively studied in lichens (Lange 1980, Lange and Tenhunen 1981, Lange et al. 1993, Green et al. 1994), where the major resistance to CO_2 diffusion seems to result from the soaking of the cortex (Cowan et al. 1992), but the effect is specific to species and can also be absent. According to Lange et al. (1993), four different types of net photosynthetic response to water content can be discerned in lichens (Figure 2.9): (1) net photosynthesis follows a saturation curve and does not decrease at high water content; (2) photosynthesis is maximal over a wider range of water contents but deceases slightly at high water content; (3) similar to (2) but net photosynthesis becomes zero or even negative at high water contents; and (4) the range of water content for high rates of net photosynthesis is narrow, with a clear optimum followed by a sharp decrease at high water contents where a low but positive net photosynthesis occurs. Surprisingly, these different photosynthetic responses to hydration could not be related with the environmental conditions under which each species was found, but species from xeric habitats have their maximal photosynthetic rates at relatively low water contents (60%–120% d.wt.) (Lange 1980, Lange and Matthes 1981, Kappen 1983).

In bryophytes (mainly endohydrous species; Green and Lange 1994), a similar effect can be identified (Dilks and Proctor 1979, Rundel and Lange 1980, Tuba et al. 1996a). However, the effect of diffusion resistance is diminished in ectohydrous bryophyte species as their growth form and anatomy allow a separation of the water-storing sites from the photosynthetic tissues and as they have ventilated structures (Proctor 1990, Green and Lange 1994). It is remarkable that photosynthetic response curves of intertidal algae show also a depression at high water contents (Quadir et al. 1979, Bewley and Krochko 1982). This indicates that increased CO_2 resistance with hydration is not only due to free water blocking gaseous pathways within the thallus, but also due to tissue swelling, which seems to be a widespread phenomenon among nonvascular plants.

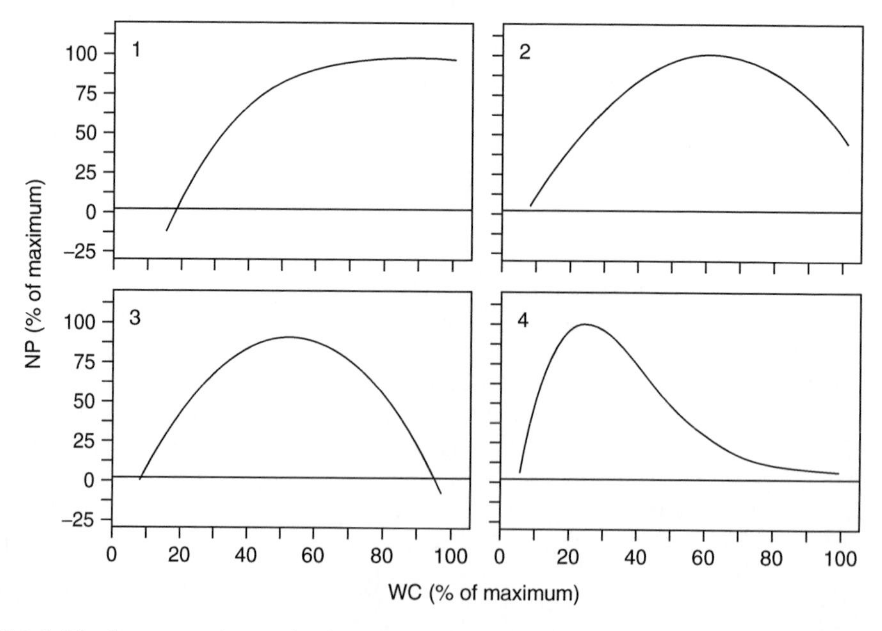

FIGURE 2.9 The four types (see text) of response of net photosynthesis (NP) to thallus water content (WC). Schematic curves based on measurements on 11 different lichen species from forests in New Zealand. (After Green, T.G.A., Lange, O.L., and Cowan, I.R., *Crypt. Bot.*, 4, 166, 1994.)

Maximal net photosynthetic rates of lichens are generally low, although there are rare exceptions (Table 2.3). The most productive lichens are epiphytes from shady forests (Green and Lange 1994). The photobionts of lichens from open habitats are frequently screened by a pigmented or thick upper cortex (Büdel 1987, Kappen et al. 1998a). Thus, although the chlorophyll content of lichens is usually enough to harvest more than 80% of the incident light (Valladares et al. 1996), the photosynthetic units of lichens experience light limitations even in high-light environments (Büdel and Lange 1994). This fact, together with some additional structural and physiological features (thylakoid structure, low-light compensation point, and large amounts of light-harvesting protein complexes), has led to the suggestion that lichens generally resemble shade plants (Green and Lange 1994).

Vascular Plants

Although C_3 photosynthesis is by far the most common photosynthetic pathway among poikilohydrous vascular plants, all of the other pathways have also been found in these plants. In many resurrection grasses of the genera *Eragrostidella*, *Sporobolus*, and *Tripogon* from Australia, Kranz-type anatomy indicated that they were C_4 plants (Lazarides 1992). Certain poikilohydrous vascular plants such as *Ramonda serbica* and *Haberlea rhodopensis*, with slightly succulent leaves when soaked, exhibited a crassulacean acid metabolism (CAM) under water stress (Kimenov et al. 1989, Markowska et al. 1994, 1997, Markowska 1999). This was confirmed by the diel CO_2 exchange pattern and by phosphoenol pyruvate (PEP) carboxylase activity. CAM activity was interpreted as a mechanism to delay periods of metabolic inactivity and to recycle CO_2. As this enhances reductive capacity, photo- and drought-induced oxidative processes can be ameliorated. The only known truly succulent resurrection plant, *Blossfeldia liliputana*, is a CAM plant according to its carbon isotopic and carbon discrimination values (Barthlott and Porembski 1996). The presence of CAM activity in poikilohydrous plants demonstrates that water conservation mechanisms can be combined with a desiccation–resurrection strategy.

Resurrection plants strictly depend on liquid water to recover their photosynthetic activity after desiccation, and they need longer periods of time than nonvascular autotrophs to regain full activity (Table 2.2). Ferns and allies require 4.5–48 h to gain full or reduced photosynthetic rates (Oppenheimer and Halevy 1962, Stuart 1968, Proctor and Tuba 2002), whereas in other cases, up to 4 days of hydration are necessary to recover net photosynthesis (Hoffmann 1968, Vieweg and Ziegler 1969). *Craterostigma plantagineum* recovered chlorophyll-a-fluorescence totally after 18 h of rehydration and *Myrothamnus flabellifolius* after 60 h, and full photosynthetic capacity of *C. wilmsii* was even reached before the chlorophyll content was fully recovered (Sherwin and Farrant 1996). The compensation point of net photosynthesis can be surpassed at 40%–50% RWC (Farrant 2000). In contrast, dark respiration of most poikilohydrous plants is active still at lower RWC (>10%: *Craterostigma wilmsii, M. flabellifolius*: Farrant 2000), and therefore can be long lasting if desiccation and rehydration phases are extended. The deleterious role of extended respiration is most likely the reason why *Hymenophyllum wilsonii* dies when subjected to 74% rh over 15 days (Proctor 2003). Plants of *Selaginella lepidophylla* that were dry during a whole year recovered full photosynthetic capacity within 20 h whether the plants were rehydrated in light or in darkness (Eickmeier 1979). During the first 6 h, the plants exhibited only respiration, and the photosynthetic recovery was linked to the regaining of ribulose bis-phosphate carboxylase activity, which was conserved by 60% during the long abiotic period (Eickmeier 1979). *Selaginella* species differed in the rapidity of recovering photosynthesis and the length of the initial period of respiration (Eickmeier 1980).

Long periods of drought, too-frequent cycles of drying and wetting, and too-rapid drying can decrease the productivity of poikilohydrous plants by affecting the maximal rates of net photosynthesis. In *Selaginella*, rates of net photosynthesis after hydration decreased with increasing

TABLE 2.3
Maximal Rate of Net Photosynthesis, Photosynthetic Light Compensation, and Saturation Points, and Minimum Water Content Allowing Photosynthetic Activity in Different Poikilohydrous Autotrophs

Organism	Maximum Rate of Net Photosynthesis (Different Units)	Light Compensation Point (μmol Photons m^{-2} s^{-1})	Light Saturation Point (μmol Photons m^{-2} s^{-1})	Minimum Water Content for Net Photosynthesis	Source
Fucus distichus (high interdial zone)	6.3 mg CO_2 g^{-1} h^{-1}			80% desiccation	Quadir et al. 1979
Lichens (in general)	0.2–5.0 mg CO_2 g d.wt.$^{-1}$ h^{-1}				Kallio and Kärenlampi 1975, Green et al. 1984
Collema tenax	7.0 μmol CO_2 m^{-2} s^{-1}				Lange et al. 1998
Lecidella crystallina	5.9 μmol CO_2 m^{-2} s^{-1}	28–43		0.13 mg m^{-2}	Lange et al. 1994
Barbula aurea	1.4 mg CO_2 g d.wt.$^{-1}$ h^{-1}	20	200	23% d.wt.	Rundel and Lange 1983
Tortula ruralis	4.8 mg CO_2 g d.wt.$^{-1}$ h^{-1}			10% d.wt.	Tuba 1987
Polypodium virginianum	3.5 μmol CO_2 m^{-2} s^{-1}	20–30	100–200		Gildner and Larson 1992
Cheilanthes maranthae	3.48 mg CO_2 g d.wt.$^{-1}$ h^{-1}	5–50	500–750		Gebauer 1986
Cheilanthes maderensis	4.04 mg CO_2 g d.wt.$^{-1}$ h^{-1}	5–25	100–500		Gebauer 1986
Notholaena parryi	5.5 μmol CO_2 m^{-2} s^{-1}		150		Nobel 1978
Selaginella lepidophylla	2.44 mg CO_2 g d.wt.$^{-1}$ h^{-1}		>2000		Eickmeier 1979
Ramonda myconi	2.39 mg CO_2 g d.wt.$^{-1}$ h^{-1}	30	400–500	20% RWC	Gebauer 1986
Xerophyta scabrida	4.0 μmol CO_2 m^{-2} s^{-1}				Tuba et al. 1994
Homoiohydrous sclerophylls of dry regions	7.9–23.8 μmol CO_2 m^{-2} s^{-1}				Larcher 1980
Homoiohydrous sun plants	10–20 μmol CO_2 m^{-2} s^{1}				Lüttge et al. 1988

Abbreviation: d.wt., dry weight; RWC, relative water content.

length of the period of desiccation (Eickmeier 1979). In *Ramonda myconi*, net photosynthesis decreased to one-third after the fifth cycle of drying and wetting (Gebauer 1986). In *S. lepidophylla*, intermediate drying speeds (52–94 h until complete curling of the plant was reached) led to maximal recovery, whereas either rapid (5.5 h) or very slow (175 h) drying was associated with significantly reduced photosynthetic rates (Eickmeier 1983). Rapid drying implied increased membrane dysfunction, whereas slow drying caused retarded de novo protein synthesis.

Poikilochlorophyllous plants typically undergo a long hiatus in photosynthetic activity during the periods of drying and rehydration because of the destruction of the photosynthetic apparatus and resynthesis of the desiccoplasts. Under natural conditions, the desiccation process may take several days or weeks (Hetherington and Smillie 1982, Tuba et al. 1997). In slowly drying leaves of *Xyrophyta scabrida*, net photosynthesis became negative after 3 days when leaf RWC was 54%, and respiration ceased after 15 days at 8% leaf RWC. A similar value limiting respiratory activity (less than 10%) was reported for *X. humilis* and also the homoihydrous *Craterostigma wilmsii* and *Myrothamnus flabellifolius* (Farrant 2000). Accordingly, respiration of *X. scabrida* was activated within 20 min of rehydration and reached full rates within 6 h, even before turgor was restored in the cells (Tuba et al. 1994). However, chlorophyll resynthesis started only after 12 h of rehydration and was not complete until 36 h had elapsed. The time required to fully restore photosynthetic capacity upon rehydration was not different whether the plants were exposed to air with 350 or 700 ppm of CO_2, but downregulation of photosynthetic rates was found at 700 ppm (Tuba et al. 1996b). During regreening upon rehydration, photosystem I activity appeared to recover faster than that of photosystem II (Gaff and Hallam 1974, Hetherington and Smillie 1982).

Available studies on poikilohydrous plants have mainly focused on the cellular and molecular mechanisms underlying inactivation and reactivation of photosynthesis and respiration. With a few exceptions, seemingly relevant aspects of the recovery of plant CO_2 exchange such as stomatal responses (Lichtenthaler and Rinderle 1988, Schwab et al. 1989, Tuba et al. 1994), leaf performance (Stuart 1968, Matthes-Sears et al. 1993), and whole plant behavior (Eickmeier et al. 1992) have been widely neglected. Field studies of the CO_2 exchange during the dehydration/rehydration cycles, such as those conducted with lichens (see Section "Opportunistic Metabolic Activity *In Situ*"), are necessary to reach a better ecological understanding of the carbon economy of poikilohydrous plants.

Different Strategies

Poikilohydrous life style does not obligatively mean that the organism is a perennial, as several of the constitutively poikilohydrous bryophytes and a few lichen taxa are ephemeral (Longton 1988a). Compared with the strategy of homoiohydrous plants, it can be characterized by the hydration range and rapidity with which the poikilohydrous plant enters into the anabiotic state during desiccation and with which it recovers normal activity during rehydration (Figure 2.10). According to the features discussed throughout this chapter, we can discern four kinds of strategies among the eurypoikilohydrous organisms.

The *Ready type* is exhibited by constitutively poikilohydrous nonvascular species (cyanobacteria, algae, lichens, and some bryophytes), which can easily lose and absorb water, switching their metabolism on and off very quickly. Their cellular structures are highly strengthened and do not need extensive, time-consuming repair processes, which allows them to oscillate frequently (daily or even within a few hours) between anabiosis and active state.

Certain bryophytes and vascular plants need variable periods of time (hours in the case of the former, days in the case of the latter) to recover completely from desiccation and therefore represent a *Repair type*. Preventing rapid water uptake may protect from deleterious effects during rehydration. The most drastic example of this strategy is poikilochlorophyllous plants. The partial destruction of organelles and photosynthetic pigments during desiccation seems

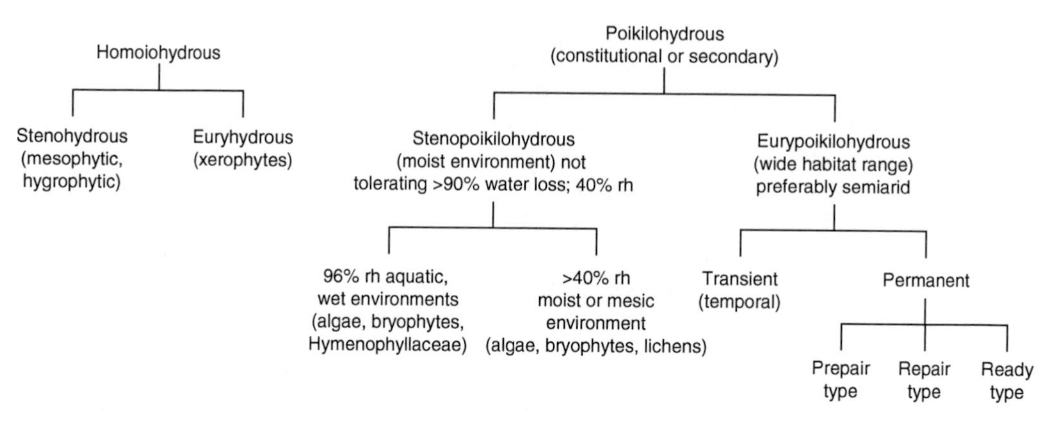

FIGURE 2.10 Water status-related plant performance.

to be disadvantageous, since it reduces the period of activity and limits photosynthetic carbon gain. However, this response type prevents oxidative membrane deterioration, particularly in plants that grow in open-exposed habitats. Such plants need rather long and continuous periods of activity and oscillate between dry and wet states only once or a few times per year (Gaff and Gies 1986). Homoiochlorophyllous plants manage better with repeated changes in hydration.

Certain bryophytes and vascular plants are not capable of tolerating extreme desiccation without a previous acclimation or preconditioning and are therefore typical of the *Prepare type*. Tolerance to desiccation is increased if they are exposed to a slow water loss, or if water loss occurs under a low vapor pressure deficit (e.g., *Bryum caespititium* and *Pohlia elongata*). Such plants either make use of structural features that retard water loss or grow in sheltered habitats where the evaporative potential is low (rock colonizing ferns, *Borya nitida*, some Velloziaceae).

The *Transient type* includes certain bryophytes and ferns that acquire a eurypoikilohydrous character only temporarily by hardening as their fronds become extremely desiccation-tolerant during the winter or the dry season (mesic bryophytes, *Polypodium vulgare, Asplenium* species).

A type with *mixed strategy* is realized with plants that act poikilohydrously only with a major part of the individual. By shifting between functioning with larger "dolichoblasts" in the rainy season and small, extremely desiccation-tolerant "brachyblasts" in the dry season, the small shrub *Satureja gillesii* can reduce its transpiring leaf surface (Montenegro et al. 1979). Brachyblasts have a mesophytic anatomy and are covered by filamentous trichomes. Among the plants that live in ephemeral rock pools, those species (*Aponogetum desertorum* and *Limosella* species) that are preserved by desiccation-tolerant rhizomes or corms (Gaff and Gies 1986) may also reveal a mixed strategy type. However, the heterophyllous *Chamaegigas intrepidus* typically does. It survives with dry rhizome and contracted conic basal leaves most of the year. On flooding the pool the basal submersed leaves do expend (it is not clear whether they carry out significant photosynthesis). The floating leaves can be produced within a few days and are the productive part of the plant. They perform according to the repair-type, as they can pass periods of desiccation repeatedly (up to 20 times) during one season and are able to regain full photosynthetic capacity within 18 h (Woitke et al. 2004).

Opportunistic Metabolic Activity *in situ*

The photosynthetic activity of lichens has been monitored in situ in various climatic regions (Kappen 1988). Figure 2.11 depicts diurnal courses for lichens in two xeric habitats, the

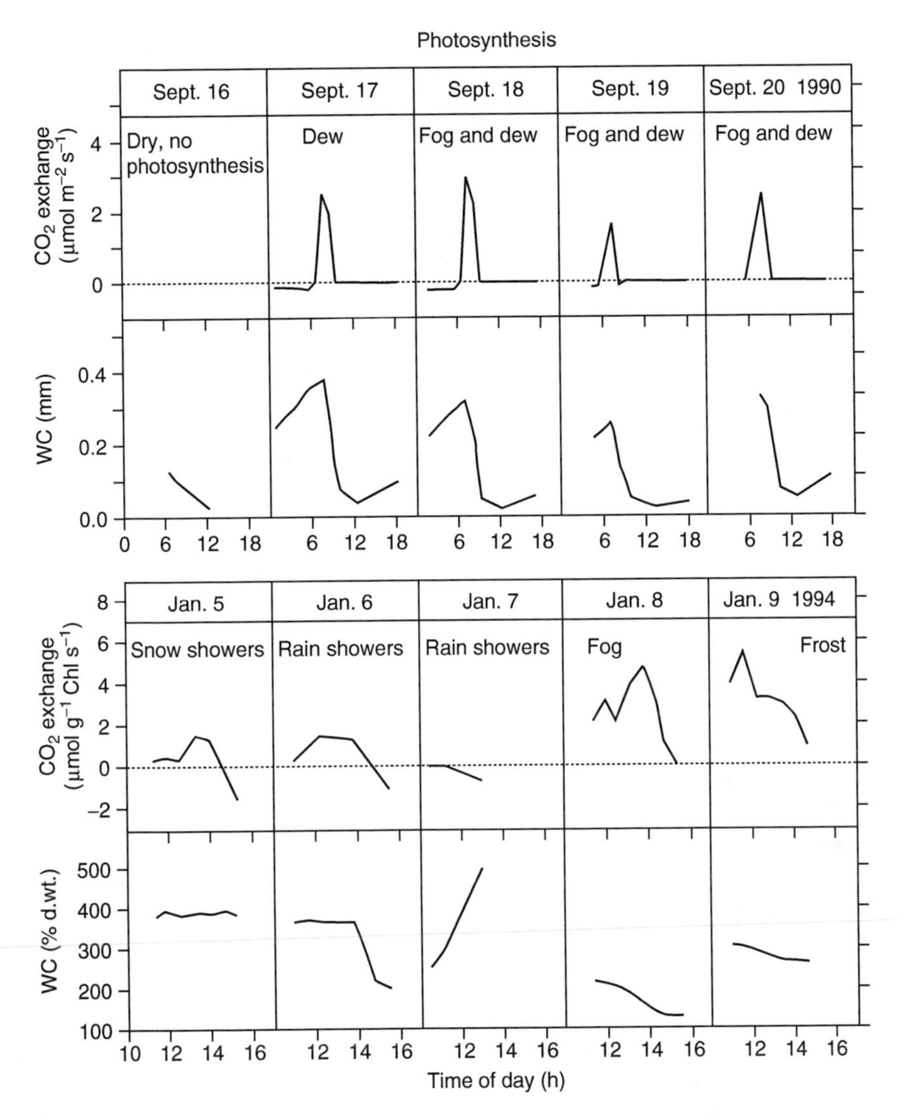

FIGURE 2.11 Series of diurnal courses of lichen water content (WC) and CO_2 exchange of *Lecidella crystallina* (on an area basis) from southern Africa in spring (September 16–20, 1990) and of *Umbilicaria spodochroa* (on chlorophyll basis) from Norway in winter (January 5–9, 1994). (After Lange, O.L., Meyer, A., Zellner, H., and Heber, U., *Funct. Ecol.*, 8, 253, 1994; Kappen, L., Schroeter, B., Hestmark, G., and Winkler, J.B., *Botanica Acta*, 109, 292, 1996.)

Namib desert in spring (Lange et al. 1990c, 1994) and south-exposed rocks of southern Norway in winter (Kappen et al. 1996). It may be noted here that frozen water is able to maintain CO_2 exchange in winter if irradiance is sufficient.

Different types of diurnal courses of gas exchange of lichens result from the existence of several possible sources of moisture (dew, fog, and rainfall, each one alone or in combination). Lösch et al. (1997) have proposed a simplified scheme for the seasonal variation of the periods of photosynthetic activity of nonvascular autotrophs with regard to the climatic conditions. In subpolar and polar regions, the most productive and extended periods of activity occur in spring, summer, and fall. In temperate regions, the periods of activity are regular but rather short and occur during all seasons, although they are somewhat limited in

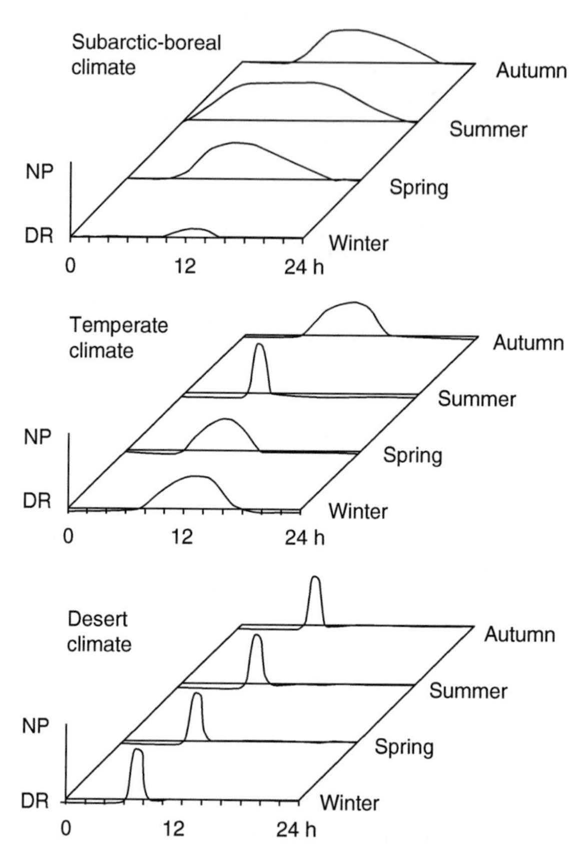

FIGURE 2.12 Schematic description of typical diurnal photosynthetic activity in the seasonal course for bryophytes and lichens in different climatic regions. NP, net photosynthesis; DR, dark respiration. (After Lösch, R., Pause, R., and Mies, B., *Bibl. Lichenol.*, 67, 145, 1997.)

summer. In hot arid regions, the periods of activity are brief but frequent, since they potentially occur throughout the year or every day during the wet season, depending on the regions (Figure 2.12). Examples of field studies in different environments are those by Hahn et al. (1989, 1993), Bruns-Strenge and Lange (1991), Sancho et al. (1997), and in the Antarctic (see Chapter 14 of this volume). The daily periods of moistening may last from midnight to up to 3 h in sunlight if dewfall is involved. In arid regions, the length of the period of lichen hydration is directly influenced by the exposure and compass direction of the site. In warm and temperate climates, shaded habitats allow the longest periods of hydration, but in frigid climates, hydration is combined with insolation of the habitat (Kappen et al. 1980, Kappen 1982, 1988,1998b, Nash and Moser 1982, Pintado et al. 1997).

A long-term investigation of the lichen *Ramalina maciformis* in the Negev desert revealed that thalli were active most days of the year: Dewfall caused 306 days of metabolic activity (of which carbon balance was positive on 218 days and negative on 88 days), whereas rainfall produced activity on only 29 days per year (Kappen et al. 1979). In high mountains and polar regions, melting snow can extend the productive period of lichens and bryophytes over several days or weeks (Kappen et al. 1995, 1998b). In Antarctica, the lichen *Usnea aurantiaco-atra* was active for a total of 3359 h within 268 days of 1 year in the wet maritime region (Schroeter et al. 1997), whereas the period of activity was reduced to one-fifth of this value within 120 days in the Antarctic dry valleys for cryptoendolithic microorganisms (see Table 2.4)

TABLE 2.4
Number of Days with Metabolic Activity of Lichens within 1 Year in Various Regions

	Active	Positive CO_2 Balance	Negative CO_2 Balance	Source
Negev 1971/1972, *Ramalina maciformis*	306	218	88	Kappen et al. 1979
Maritime Antarctic 1992/1993, *Usnea aurantiaco-atra*	268	150	118	Schroeter et al. 1997
Continental Antarctic 1985/1988, *Cryptoendolithic* lichen community	120	120(?)	0(?)	Friedmann et al. 1993

(Friedmann et al. 1993, Nienow and Friedmann 1993). The latter agrees with the findings on lichens in Botany Bay, continental Antarctica, where the productive period of the lichens was restricted to only the melt period in spring and some moist summer days since it was too dark and too cold under the snow cover in the cold season and frequently too dry in the warm season (Pannewitz et al. 2003).

Diurnal cycles of field-measured photosynthetic activity of bryophytes have been recorded for temperate (Lösch et al. 1997), subarctic, and polar habitats (Hicklenton and Oechel 1977, Oechel and Sveinbjörnson 1978), demonstrating the dependency of the periods of activity on weather conditions and habitat factors, as shown in lichens. In polar regions and high mountains, the photosynthetic activity of typically shade-adapted bryophytes may be depressed by temporarily high insolation combined with snow melt (Kappen et al. 1989, Valanne 1994).

For poikilohydrous vascular plants, we can only refer to one diurnal course measured in the fern *Notholaena parryi* of the western Colorado desert, showing full light-driven CO_2 exchange (Nobel 1978), and two studies on *Chamaegigas intrepidus* (Gaff and Gies 1986, Woitke et al. 2004). In one case the authors found 11 events of activity, each lasting for 2–23 days over a period of 1.5 years (Figure 2.13), and in the other 20 events per year were counted.

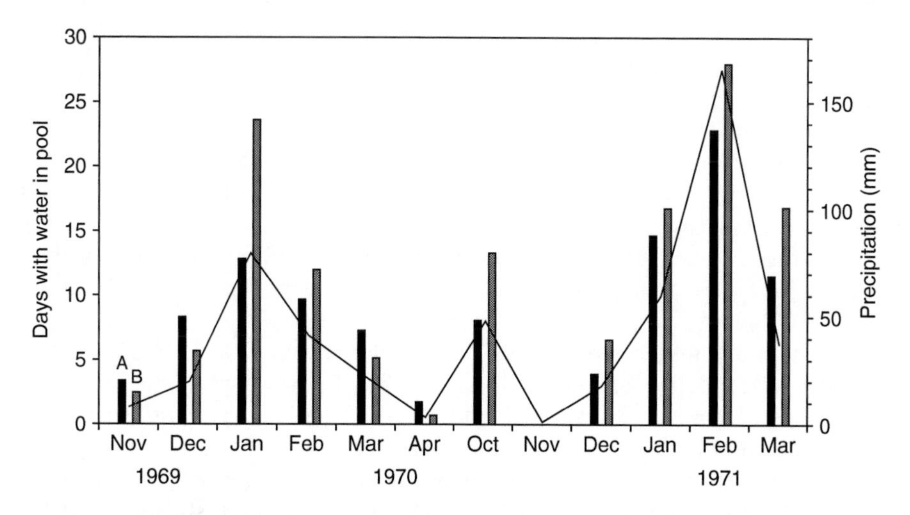

FIGURE 2.13 Amount of rainfall (line) and periods of filling two rockpools (columns) in Fritz Gaerdes Reserve, Okahandja, Namibia, where *Chamaegigas intrepidus* (Scrophulariaceae) grows. (After Gaff, D.F. and Giess, W., *Dinteria*, 18, 17, 1986.)

Place in Plant Communities

Their remarkable tolerance to climatically extreme conditions, their capacity to colonize both the exterior and the interior of rocks, and their relative success as epiphytes allow many poikilohydrous nonvascular autotrophs to survive and grow in the absence of strong competition by homoiohydrous plants. This is evident with the numerous lichen and bryophyte communities formed in polar regions, high mountains, and on rock outcrops, everywhere where vascular plants fail to grow. In real desert lands (Shields et al. 1957, Scott 1982), and also the Mediterranean and steppe regions, deserted land in southern Australia (Rogers and Lange 1971) and on dunes, bryophytes, lichens, and soil algae cover the soil as crusts with characteristic community structure such as for instance of many others like the famous Fulgensietum fulgentis in Europe or the Crossidio crassinervis-Tortuletum obtusatae in the Irano-Turanian territory (Galun and Garty 2001). These types of vegetation, especially the crusts, are ecologically crucial because they reduce erosion and contribute to the preservation of hydric, chemical, and physical properties of the soil (Danin and Garty 1983, Belnap and Lange 2001).

A particular phenomenon in semiarid regions is the occurrence of erratic (vagrant) lichens (Kappen 1988, Rosentreter 1993, Perez 1997) and bryophytes (Scott 1982). With their curled lobes or shoots, they are blown by the wind to shallow depressions and await a flush of water or just a bit of rain to unfold their thalli and to become productive. Without roots and not dependent on any soil formation, lichens and bryophytes typically form pioneer communities on young volcanic material or areas around receding glaciers as long as these sites are prepared for the establishment of homoiohydrous plants (Kappen 1988). In many habitats, mosaics are formed between vascular plant and cryptogam communities like in the boreal region, or lichens and bryophytes together with poikilohydrous vascular plants (see Figure 2.1) form communities in temporarily moist habitats with shallow soil cover (Gaff 1977, Volk 1984, Müller 1985, Lazarides 1992, Belnap and Lange 2001).

Communities with dominant poikilohydrous angiosperms were recognized since Lebrun (1947) who described a Craterostigmetum nanolanceolati, comprising *Craterostigma lanceolatum*, *C. plantagineum*, *C. hirsutum*, *Lindernia philcoxii* together with *Riccia* and Cyperaceae on lateritic crusts in Zaire. Typical are communities between resurrection plants and bryophytes and elsewhere lichens that have exploited temporal ponds or run-offs on Inselbergs in subtropical and tropical regions, such as the Xerophytetum humilis in Namibia (Volk 1984) where *Xerophyta* (= *Barbacenia humilis*) is associated with *Riccia*, *Bryum*, and cyanobacteria and locally with *C. plantagineum*, some succulents, and grasses. Similar communities with dominating poikilohydrous Scrophulariaceae were described for the African savannah region (Fischer 1992). Volk (1984) observed that communities of the Nanocyperion teneriffae are typical of pioneering vegetation that combines the strategies of therophytes and resurrection plants. In these places, the resurrection plants can outcompete therophytes, as they are able to defend their place against invasion of seeds and seedlings. Forming stable herds, turfs, or mats, their dominant position on inselbergs was demonstrated by Porembski and Barthlott (1997).

A transition from dominance to exclusion of poikilohydrous vascular plants was found in southern Africa as a function of soil depth (Gaff 1977). Initial stages are characteristic with poikilohydrous water plants (*Chamaegigas intrepidus* in Namibia or *Craterostigma monroi* in Zimbabwe) and lead to grass–herb communities over several steps to, finally, a pluvio-therophytic grass vegetation or associations between *Xerophyta* species and perennial succulent life forms. Although they grow slowly, they are also very successful in cracks and rock ledges with changing water supply once they are established. Species-rich communities of nonvascular autotrophs are frequently formed on tree trunks, branches, and even leaves in wet forests, particularly on mountain slopes.

Algae, lichens, and bryophytes grow potentially everywhere because of their small and easily transportable vegetative and generative propagules (Kappen 1995). The evolution of a great variety of asexual means of reproduction has made nonvascular plants extremely successful in the colonization of remote and difficult sites. We have only limited knowledge about the reproductive strategies and success of poikilohydrous vascular plants. *Myrothamnus flabellifolius* forms perianthless inconspicuous monoecic flowers, which are most likely anemogamous (Puff 1978, Child 1960). According to Puff (1978), pollen tetrades may increase the fertility on a receptaculum (success of mating) in this species. The seeds are extremely small and can be dispersed over long distances by the wind, but we have no records about periods and conditions of flowering. The flowers of poikilohydrous Scrophulariaceae do not differ much from those of the homoiohydrous members of the family (Heil 1925, Hickel 1967, Smook 1969, Gaff 1977). However, the rapidity of producing flowers is remarkable in species such as *Chamaegigas intrepidus*, as flower buds appeared simultaneously with the floating levels. As expected, the reproductive phase occurred predominantly during the wet season, and a general requirement was a period of photosynthetic activity before reproduction, as was shown for *Vellozia schnitzleinia* (Nigeria: Owoseye at Sandford 1972) and poikilohydrous grass species in Australia (Lazarides 1972). Nevertheless, a few grass species can reproduce during the dry period. *Chamaegigas intrepidus* has a high genetic variability within and between populations on one site. Although gene flow over long distances is low, there is no evidence of genetic isolation of populations by distance (Heilmeier et al. 2005). Describing the reproductive biology and ecology of the Velloziaceae (Kubitzki 1998, Ibisch et al. 2001) helps to interpret the ecological strategy of poikilohydrous plants and to compare their relative success with that of homoiohydrous co-occurring species.

Primary Production of Poikilohydrous Autotrophs

A measure of the success of a given plant is its rate of biomass production and its primary production integrated throughout a growth period. As previously discussed, photosynthetic and growth rates of poikilohydrous plants are low in comparison with homoiohydrous species (Table 2.3). Thus, the time period during which the plant is hydrated becomes a crucial factor for the primary production of poikilohydrous plants in relation to that of the potentially more productive homoiohydrous plants. To our knowledge, no data are available regarding the primary production of poikilohydrous vascular plants. Studies with lichens and bryophytes have revealed the existence of different diel cycles of activity depending on the season, the habitat, and the weather (Figure 2.12), each combination resulting in very different carbon balance. In the Namib desert, fog resulted in an annual carbon gain of 20% of the standing carbon mass for the lichens studied (Lange et al. 1990c). Annual carbon gain was 20%–30% in the maritime Antarctic and northern Europe (Schroeter et al. 1995, Kallio and Kärenlampi 1975) and went up to 79% in a coastal habitat with high precipitation in northern Germany (Bruns-Strenge and Lange 1992). In contrast, the estimated annual carbon gain was only 3.8% in the harsh conditions of the continental Antarctic (Kappen 1985).

The biomass of lichens can be surprisingly high in tundra ecosystems, especially when compared with that of vascular plants. For instance, in the Alaskan tundra (Atkasook), lichen biomass was 76 g dry matter m^{-2}, which represented 26% of the total aboveground biomass of the system, and it reached 1372 g m^{-2} near Anatuvuk Pass in the Brooks Range (Lechowicz 1981). In Scotland, the annual increase in dry biomass of the lichen *Cladonia portentosa* alone was 47 g m^{-2} (Prince, quoted by Bruns-Strenge and Lange 1992). Lichen primary production is also surprising in deserts, where they locally exhibit values similar to or higher than those of homoiohydrous vascular plants. A field study of the desert lichen *Ramalina maciformis* found an annual carbon gain of 60–195 mg CO_2 g^{-1} d.wt., which is equivalent to a net carbon gain of 4.5%–15.0% (Kappen et al. 1980). On the poor soils of the Namib desert, the lichen

Teloschistes capensis forms dense turfs of a biomass of 250 g m^{-2} (Kappen 1988). Similar values were found in the Negev desert, where the lichen biomass was within the range of that of the vascular plant vegetation (145–323 g m^{-2}; Kappen et al. 1980). The annual carbon production of soil crusts with dominating lichens can be comparatively high with 12–37 g C m^{-2} $year^{-1}$ in Arizona and Utah and in the Namib fog desert approximately 32 g C m^{-2} $year^{-1}$ (Evans and Lange 2001).

Bryophytes are frequent in deserts, but they are not very productive in these arid environments. They rarely surpass a 0.3%–8% cover and a biomass of 2 g m^{-2} (Nash et al. 1977). However, in tundra ecosystems they can be as productive and abundant as lichens and vascular plants. For example, in the Alaskan Arctic, bryophytes exhibit an average cover of almost 60%, and they contribute 36% of the net aboveground production of the ecosystem (Webber 1978). In certain areas of the maritime Antarctic, where cryptogamic phytomass reaches values of 1900 g m^{-2}, bryophytes are dominant (Kappen 1993b). However, most of them are stenopoikilohydrous moss species that, in warmer and moist habitats, form up to 4750 g m^{-2} measured in the uppermost 10 cm of the profile (Longton 1988b). *Sphagnum* species, for instance, may cover 1%, and lichens were estimated to dominate in total approximately 8% of the land surface of the Globe (Clymo 1970, Larson 1987, Ahmadjian 1995).

ACKNOWLEDGMENTS

The authors wish to thank Professors Carmen Ascaso, T.G. Allan Green, Ulrich Heber, Wolfram Hartung, Otto. L. Lange, and Rainer Lösch for kindly providing information and discussion on poikilohydrous plants. Allan Green also helped to edit the manuscript. Stefan Pannewitz (Kiel, Germany) helped to acquire the literature and organize the reference list.

REFERENCES

Abel, W.O., 1956. Die Austrocknungsresistenz der Laubmoose. Sitzungsberichte der Wiener Akademie der Wissenschaften. Mathematisch-Naturwissenschaftliche Klasse, Abteilung I. 165: 619–707.

Ahmadjian, V., 1973. Resynthesis of lichens. In: V. Ahmadjian and M.E. Hale, eds. The Lichens. Academic Press, New York, pp. 565–580.

Ahmadjian, V., 1995. Lichens are more important than you think. BioScience 45: 124.

Alamillo, J.M. and D. Bartels, 2001. Effects of desiccation on photosynthesis pigments and the ELIP-like dsp-22 protein complexes in the resurrection plant *Craterostigma plantagineum*. Plant Science 160: 1161–1170.

Albini, F.M., C. Murelli, G. Patritti, M. Rovati, P. Zienna, and P.V. Finzi, 1994. Low-molecular weight substances from the resurrection plant *Sporobolus stapfianus*. Phytochemistry 37: 137–142.

Alpert, P., 2006. Constraints of tolerance: why are desiccation-tolerant organisms so small and rare. Journal of Experimental Biology 209: 1575–1584.

Alves, R.J.V., 1994. Morphological age determination and longevity in some *Vellozia* populations in Brazil. Folia Geobotanica and Phytotaxonomica 29: 55–59.

Augusti, A., A. Scartazza, F. Navarri-Izzo, C.L.M. Sgherri, B. Stefanovic, and E. Brugnoli, 2001. Photosystem II photochemical efficiency, zeaxanthin and antioxidant contents in the poikilohydric *Ramonda serbica* during dehydration and rehydration. Photosynthesis Research 67: 79–88.

Balaguer, L., F.I. Pugnaire, E. Martinez-Ferri, C. Armas, C., F. Valladares, and E. Manrique, 2000. Ecophysiological significance of chlorophyll loss and reduced photochemical efficiency under extreme aridity in *Stipa tenacissima* L. Plant and Soil 240: 343–352.

Bartels, D., 2005. Desiccation tolerance studied in the resurrection plant *Craterostigma plantagineum*. Integrative and Comparative Biology 45: 696–701.

Bartels, D. and F. Salamini, 2001. Desiccation tolerance in the resurrection plant *Craterostigma plantagineum*. A contribution to the study of drought tolerance at the molecular level. Plant Physiology 127: 1346–1353.

Bartels, D., K. Schneider, G. Terstappen, D. Piatkowski, and F. Salamini, 1990. Molecular cloning of abscisic acid-modulated genes which are induced during desiccation of the resurrection plant *Craterostigma plantagineum*. Planta 181: 27–34.

Bartels, D., R. Alexander, K. Schneider, R. Elster, R. Velasco, J. Alamillo, G. Bianchi, D. Nelson, and F. Salamini, 1993. Dessication-related gene products analysed in a resurrection plant and barley embryos. In: T.J. Close and E.A. Bray, eds. Plant Responses to Cellular Dehydration During Environmental Stress. Current Topics in Plant Physiology, Vol. 10. American Society of Plant Physiology, New York, pp. 199–127.

Barthlott, W. and I. Capesius, 1974. Wasserabsorption durch Blatt- und Sprossorgane einiger Xerophyten. Zeitschrift für Pflanzenphysiologie 72: 443–455.

Barthlott, W. and S. Porembski, 1996. Ecology and morphology of *Blossfeldia liliputana* (Cactaceae): A poikilohydric and almost astomate succulent. Botanica Acta 109: 161–166.

Bartley, M. and N.D. Hallam, 1979. Changes in the fine structure of the desiccation tolerant sedge *Coleochloa setifera* (Ridley) Gilly under water stress. Australian Journal of Botany 27: 531–545.

Beckett, R.P., 1995. Some aspects of the water relations of lichens from habitats of contrasting water status studied using thermocouple psychrometry. Annals of Botany 76: 211–217.

Belnap, J. and O.L. Lange, 2001. Biological Soil Crusts: Structure, Function, and Management. Ecological Studies 150, Springer, Heidelberg, New York.

Bernacchia, G. and A. Furini, 2004. Biochemical and molecular responses to water stress in resurrection plants. Physiologia Plantarum 121: 175–181.

Bernacchia, G.S., F. Salamini, and D. Bartels, 1996. Molecular characterization of the rehydration process in the resurrection plant *Craterostigma plantagineum*. Plant Physiology 111: 1043–1050.

Bertsch, A., 1966a. Über den CO_2-Gaswechsel einiger Flechten nach Wasserdampfaufnahme. Planta 68: 157–166.

Bertsch, A., 1966b. CO_2-Gaswechsel und Wasserhaushalt der aerophilen Grünalge *Apatococcus lobatus*. Planta 70: 46–72.

Bertsch, A., 1967. CO_2-Gaswechsel, Wasserpotential und Sättigungsdefizit bei der Austrocknung epidermisfreier Blattscheiben von *Valerianella*. Naturwissenschaften 8: 204.

Bewley, J.D., 1973. Polyribosomes conserved during desiccation of the moss *Tortula ruralis* are active. Plant Physiology 51: 285–288.

Bewley, J.D., 1979. Physiological aspects of desiccation tolerance. Annual Review of Plant Physiology 30: 195–238.

Bewley, J.D., 1995. Physiological aspects of desiccation tolerance—a retrospect. International Journal of Plant Science 156: 393–403.

Bewley, J.D. and J.E. Krochko, 1982. Desiccation tolerance. In: A. Pirson and M.H. Zimmermann, eds. Encyclopedia of Plant Physiology. New Series Vol. 12b. Springer-Verlag, Berlin, Heidelberg, pp. 325–378.

Bewley, J.D. and M.J. Oliver, 1992. Desiccation-tolerance in vegetative plant tissues and seeds: protein synthesis in relation to desiccation and a potential role for protection and repair mechanisms. In: C.B. Osmond and G. Somero, eds. Water and Life: A Comparative Analysis of Water Relationship at the Organismic, Cellular and Molecular Levels. Springer-Verlag, Berlin, pp. 141–60.

Bewley, J.D. and J. Pacey, 1978. Desiccation-induced ultrastructural changes in droughtsensitive and drought-tolerant plants. In: J.W. Crowe and J.S. Clegg, eds. Academic Press, New York, pp. 53–73.

Bianchi, G., A. Gamba, C. Murelli, F. Salamini, and D. Bartels, 1991. Novel carbohydrate metabolism in the resurrection plant *Craterostigma plantagineum*. Plant Journal 1: 355–359.

Bianchi, G., A. Gamba, C.R. Limiroli, N. Pozzi, R. Elster, F. Salamini, and D. Bartels, 1993. The unusual sugar composition in leaves of the resurrection plant *Myrothamnus flabellifolia*. Physiologia Plantarum 87: 223–226.

Biebl, R. 1962. Protoplasmatische Ökologie der Pflanzen, Wasser und Temperature. Protoplasmatologia XII, 1, Springer-Verlag, Wien, 344 pp.

Bilger, W., S. Rimke, U. Schreiber, and O.L. Lange, 1989. Inhibition of energy-transfer to photosystem II in lichens by dehydration: different properties of reversibility with green and blue-green phycobionts. Journal of Plant Physiology 134: 261–268.

Blum, O.B., 1973. Water relations. In: V. Ahamadjian and M.E. Hale, eds. The Lichens. Academic Press, New York, pp. 381–400.

Bruns-Strenge, S. and O.L. Lange, 1991. Photosynthetische Primärproduktion der Flechte *Cladonia portentosa* an einem Dünenstandort auf der Nordseeinsel Baltrum I. Freilandmessungen von Mikroklima, Wassergehalt und CO_2-Gaswechsel. Flora 185: 73–97.

Bruns-Strenge, S. and O.L. Lange, 1992. Photosynthetische Primärproduktion der Flechte *Cladonia portentosa* an einem Dünenstandort auf der Nordseeinsel Baltrum III. Anwendung des Photosynthesemodells zur simulation von Tagesabläufen des CO_2-Gaswechsel und zur Abschätzung der Jahresproduktion. Flora 186: 127–140.

Büdel, B., 1987. Zur Biologie und Systematik der Flechtengattung *Heppia* und *Peltula* im südlichen Afrika. Bibliotheca Lichenologica 23: 1–105.

Büdel, B., 1990. Anatomical adaptions to the semiarid/arid environment in the lichen genus *Peltula*. Bibliotheca Lichenologica 38: 47–61.

Büdel, B. and O.L. Lange, 1991. Water status of green and blue-green phycobionts in lichen thalli after hydration by water vapor uptake: Do they become turgid? Botanica Acta 104: 345–404.

Büdel, B. and O.L. Lange, 1994. The role of the cortical and epineral layers in the lichen genus *Peltula*. Cryptogamic Botany 4: 262–269.

Büdel, B. and D.C.J. Wessels, 1991. Rock-inhabiting blue-green algae (cyanobacteria) from hot arid regions. Algological Studies 64: 385–398.

Bukhov, N.G., J. Kopecky, E.E. Pfündel, C. Klughammer, and U. Heber, 2001. A few molecules of zexanthin per reaction centre of photosystem II permit effective thermal dissipation of light energy in photosystem II of a poikilohydric moss. Planta 212: 739–748.

Calatayud, A., V.I. Deltoro, E. Barreno, and S. del Valle-Tascón, 1997. Changes in *in vivo* chlorophyll fluorescence quenching in lichen thalli as a function of water content and suggestion of zeaxanthin-associated photoprotection. Physiologia Plantarum 101: 93–102.

Casper, C., W.G. Eickmeier, and C.B. Osmond, 1993. Changes of fluorescence and xanthophylls pigments during dyhydration in the resurrection plant *Selaginella lepidophylla* in low and medium light intensities. Oecologia 94: 528–533.

Child, G.F., 1960. Brief notes on the ecology of the resurrection plant *Myrothamnus flabellifolia* with mention of its water-absorbing abilities. Journal of South African Botany 26: 1–8.

Clymo, R.S., 1970. The growth of *Sphagnum:* Methods and measurements. Journal of Ecology 58: 13–49.

Cowan, D.A., T.G.A. Green, and A.T. Wilson, 1979. Lichen metabolism I. The use of tritium-labeled water in studies of anhydrobiotic metabolism in *Ramalina celestri* and *Peltigera polydactyla*. New Phytologist 82: 489–503.

Cowan, I.R., O.L. Lange, and T.G.A. Green, 1992. Carbon-dioxide exchange in lichens: determination of transport and carboxylation characteristics. Planta 187: 282–294.

Coxson, D.S., 1987a. Photoinhibition of net photosynthesis in *Stereocaulon virgatum* and *S. tomentosum*, a tropical-temperate comparison. Canadian Journal of Botany 65: 1707–1715.

Coxson, D.S., 1987b. The temperature dependence of photoinhibition in the tropical basidiomycete lichen *Cora pavonia* E. Fries. Oecologia 73: 447–453.

Crowe, J.H. and L.M. Crowe, 1992. Membrane integrity in anhydrobiotic organisms: towards a mechanism for stabilizing dry cells. In: G.N. Somero, C.B. Osmond, and C.L. Bolin, eds. Water and Life. A Comparative Analysis of Water Relationships at the Organismic, Cellular and Molecular levels. Springer-Verlag, Berlin, pp. 87–113.

Csintalan, Z., Z. Tuba, H. Lichtenthaler, and J. Grace, 1996. Reconstitution of photosynthesis upon rehydration in the desiccated leaves of the poikilochlorophyllus shrub *Xerophyta scabrida* at elevated CO_2. Journal of Plant Physiology 148: 345–350.

Dace, H., H.W. Sherwin, N. Illing, and J.M. Farrant, 1998. Use of metabolic inhibitors to elucidate mechanisms of recovery from desiccation stress in the resurrection plant *Xerophyta humilis*. Plant Growth Regulation 24: 159–166.

Danin, A. and J. Garty, 1983. Distribution of Cyanobacteria and lichens on hillsides of the Negev Highlands and their impact on biogenic weathering. Zeitschrift für Geomorphologie N.F. 27: 423–444.

Davey, M.C., 1997. Effects of continuous and repeated dehydration on carbon fixation by bryophytes from the maritime Antarctic. Oecologia 110: 25–31.

Demmig, B., K. Winter, K. Krüger, and F.-C. Czygan, 1988. Photoinhibition and the heat dissipation of excess light energy in *Nerium oleander* exposed to a combination of high light and water stress. Plant Physiology 87: 17–24.

Demmig-Adams, B. and W.W. Adams, 1992. Photoprotection and other responses of plants to high light stress. Annual Review of Plant Physiology and Plant Molecular Biology 43: 599–626.

Demmig-Adams, B., C. Máguas, W.W.I. Adams, A. Meyer, E. Kilian, and O.L. Lange, 1990a. Effect of light on the efficiency of photochemical energy conversion in a variety of lichen species with green and blue-green phycobionts. Planta 180: 400–409.

Demmig-Adams, B., W.W. Adams III, F.-C. Czygan, U. Schreiber, and O.L. Lange, 1990b. Differences in the capacity for radiationless energy dissipation in the photochemical apparatus of green and blue-green algal lichens associated with differences in carotenoid composition. Planta 180: 582–589.

Dickie, J.B. and H.W. Pritchard, 2002. Systematic and evolutionary aspects of desiccation tolerance in seeds. In: M. Black and H.W. Pritchard, eds. Desiccation and Survival of Plants—Drying without Dying. CABI Publishing, Wallingford, pp. 239–259.

Dhindsa, R.S., 1985. Non-autotrophic CO_2 fixation and drought tolerance in mosses. Journal of experimental Botany 36: 980–988.

Dhindsa, R.S., 1991. Drought stress, enzymes of glutathion metabolism, oxidation injury, and protein synthesis in *Tortula ruralis*. Plant Physiology 95: 648–651.

Dhindsa, R.S. and W. Matowe, 1981. Drought tolerance in two mosses: Correlated with enzymatic defence against lipid peroxidation. Journal of Experimental Botany 32: 79–91.

Dietz, S. and W. Hartung, 1998. Abscisic acid in lichens: Variation, water relations and metabolism. New Phytologist 138: 99–106.

Dilks, T.J.K. and M.C.F. Proctor, 1974. The pattern of recovery of bryophytes after desiccation. Journal of Bryology 8: 97–115.

Dilks, T.J.K. and M.C.F. Proctor, 1976a. Effects of intermittent desiccation on bryophytes. Journal of Bryology 9: 249–264.

Dilks, T.J.K. and M.C.F. Proctor, 1976b. Seasonal variation in desiccation tolerance in some British bryophytes. Journal of Bryology 9: 239–247.

Dilks, T.J.K. and M.C.F. Proctor, 1979. Photosynthesis, respiration and water content in bryophytes. New Phytologist 82: 97–114.

Dircksen, A., 1964. Vergleichende Untersuchungen zur Frost-, Hitze- und Austrocknungsresistenz einheimischer Laub- und Lebermoose unter besonderer Berücksichtigung jahreszeitlicher Veränderungen. Doctor Thesis, Universität Göttingen.

Drazig, G., N. Mihailovic, and B. Stefanovic, 1999. Chlorophyll metabolism in leaves of higher poikilohydric plants *Ramonda serbica* Panc. and *Ramonda nathaliae* Panc. et Petrov. during dehydration and rehydration. Journal of Plant Physiology 154: 379–384.

Eickmeier, W.G., 1979. Photosynthetic recovery in the resurrection plant *Selaginella lepidophylla* after wetting. Oecologia 39: 93–106.

Eickmeier, W.G., 1980. Photosynthetic recovery of resurrection spike mosses from different hydration regimes. Oecologia 46: 380–385.

Eickmeier, W.G., 1982. Protein synthesis and photosynthetic recovery in the resurrection plant *Selaginella lepidophylla*. Plant Physiology 69: 135–138.

Eickmeier, W.G., 1983. Photosynthetic recovery of the resurrection plant *Selaginella lepidophylla*: Effects of prior desiccation rate and mechanisms of desiccation damage. Oecologia 58: 115–120.

Eickmeier, W.G., 1986. The correlation between high-temperature and desiccation tolerances in a poikilohydric desert plant. Canadian Journal of Botany 64: 611–617.

Eickmeier, W.G., 1988. The effects of desiccation rate on enzyme and protein—synthesis dynamics in the desiccation-tolerant pteridophyte *Selaginella lepidophylla*. Canadian Journal of Botany 66: 2574–2580.

Eickmeier, W.G., J.G. Lebkuecher, and C.B. Osmond, 1992. Photosynthetic water oxidation and water stress in plants. In: G.N. Somero, G.N. Osmond, and C.L. Bolis, eds. Water and Life: Comparitive Analysis of Water Relationships at the Organismic, Cellular, and Molecular Levels. Springer-Verlag, Berlin, Heidelberg, New York, pp. 223–239.

Eickmeier, W.G., C. Casper, and C.B. Osmond, 1993. Chlorophyll fluorescence in the resurrection plant *Selaginella lepidophylla* (Hook & Grev) spring during high-light and desiccation stress, and evidence for zeaxanthin-associated photoprotection. Planta 189: 30–38.

Ekmekci, Y., A. Bohms, J.A. Thomson, and S.G. Mundree, 2005. Photochemical and antioxidant responses in the leaves of *Xerophyta viscosa* Baker and *Digitaria sanguinalis* L. under water deficit. Zeitschrift für Naturforschung/Jounal of Biosciences 60: 435–443.

Evans, R.D. and O.L. Lange, 2001. Biological soil crusts and ecosystem nitrogen and carbon dynamics. In: J. Belnap and O.L. Lange, eds. Biological Soil Crusts: Structure, Function, and Management. Ecological Studies 150. Springer, Berlin, Heidelberg, New York, pp. 263–279.

Ewers, F.W., G.B. North, and P.S. Nobel, 1992. Root-stem junctions of a desert monocotyledon and a dicotyledon: Hydraulic consequences under wet conditions and during drought. New Phytologist 121: 377–385.

Farrant, J.M., 2000. A comparation of mechanisms of desiccation tolerance among three angiosperm resurrection plant species. Plant Ecology 151: 29–39.

Farrar, J.F., 1976a. The lichen as an ecosystem: Observation and experiment. In: D.H. Brown, D.L. Hawksworth, and R.H. Bailey, eds. Lichenology: Progress and Problems. Academic Press, New York, pp. 385–406.

Farrar, J.F., 1976b. Ecological physiology of the lichen *Hypogymnia physodes* I. Some effects of constant water saturation. New Phytologist 77: 93–103.

Fischer, E., 1992. Systematik der afrikanischen Lindernieae (Scrophulariaceae). Franz Steiner Verlag, Stuttgart.

Friedmann, E.I. and M. Galun, 1974. Desert algae, lichens and fungi. In: G.W.J. Brown, ed. Desert Biology. Academic Press, New York, pp. 166–212.

Friedmann, E.I., L. Kappen, M.A. Meyer, and J.A. Nienow, 1993. Long-term productivity in the cryptoendolithic microbial community of the Ross desert, Antarctica. Microbial Ecology 25: 51–69.

Furini, A., C. Konez, F. Salamini, and D. Bartels, 1997. High level transcription of a member of a repeated gene family confers dehydration tolerance to callus tissue of *Craterostigma plantagineum*. EMBO Journal 16: 3599–3608.

Gaff, D.F., 1971. Desiccation-tolerant flowering plants in southern Africa. Science 174: 1033–1034.

Gaff, D.F., 1972. Drought resistance in *Welwitschia mirabilis* Hook. fil. Dinteria 7: 3–19.

Gaff, D.F., 1977. Desiccation tolerant vascular plants of southern Africa. Oecologia 31: 95–109.

Gaff, D.F., 1980. Protoplasmic tolerance of extreme water stress. In: N.C. Turner and P.J. Kramer, eds. Adaptation of Plants to Water and High Temperature Stress. Wiley, New York, pp. 207–230.

Gaff, D.F., 1986. Desiccation tolerant "resurrection" grasses from Kenya and west Africa. Oecologia 70: 118–120.

Gaff, D.F., 1987. Desiccation tolerant plants in South America. Oecologia 74: 133–136.

Gaff, D.F., 1989. Responses of Desiccation-Tolerant "Resurrection" Plants to Water Stress. SPB Academic Publishing, The Hague, The Netherlands.

Gaff, D.F. and P.V. Bole, 1986. Resurrection grasses in India. Oecologia 71: 159–160.

Gaff, D.F. and D.M. Churchill, 1976. *Borya nitida* Labill.—an Australian species in the Liliaceae with desiccation-tolerant leaves. Australian Journal of Botany 24: 209–224.

Gaff, D.F. and R.P. Ellis, 1974. Southern African grasses with foliage that revives after dehydration. Bothalia 11: 305–308.

Gaff, D.F. and W. Giess, 1986. Drought resistance in water plants in rock pools of southern Africa. Dinteria 18: 17–36.

Gaff, D.F. and N.D. Hallam, 1974. Resurrecting desiccated plants. In: R.L. Bieleski, A.R. Ferguson, and M.M. Cresswell, eds. The Royal Society of New Zealand, Wellington, pp. 389–393.

Gaff, D.F. and P.K. Latz, 1978. The occurrence of resurrection plants in the Australian flora. Australian Journal of Botany 26: 485–492.

Gaff, D.F. and B.R. Loveys, 1984. Abscisic-acid content and effects during dehydration of detached leaves of desiccation tolerant plants. Journal of Experimental Botany 35: 1350–1358.

Gaff, D.F. and G.R. McGregor, 1979. The effect of dehydration and rehydration on the nitrogen content of various fractions from resurrection plants. Biologia Plantarum (Praha) 21: 92–99.

Gaff, D.F., S.-Y. Zee, and T.P. O'Brien, 1976. The fine structure of dehydrated and reviving leaves of *Borya nitida* Labill.—a desiccation-tolerant plant. Australian Journal of Botany 24: 225–236.

Galun, M. and J. Garty, 2001. Biological soil crusts of the Middle East. In: J. Belnap and O.L. Lange, eds. Biological Soil Crusts: Structure, Function and Management. Ecological Studies 150. Springer, Heidelberg, pp. 95–106.

Gauslaa, Y. and E. McEvoy, 2005. Seasonal changes in solar radiation drive acclimation of the sun-screening compound parietin in the lichen *Xanthoria parietina*. Basic and Applied Ecology 6: 75–82.

Gauslaa, Y. and K.A. Solhaug, 1996. Differences in the susceptibility to light stress between epiphytic lichens of ancient and young boreal forest stands. Functional Ecology 10: 344–354.

Gauslaa, Y. and K.A. Solhaug, 1999. High-light damage in air-dry thalli of the old forest lichen *Lobaria pulmonaria*—interactions of irradiance, exposure duration and high temperature. Journal of Experimental Botany 50: 697–705.

Gauslaa, Y. and K.A. Solhaug, 2000. High-light-intensity damage to the foliose lichen *Lobaria pulmonaria* within a natural forest: The applicability of chlorophyll fluorescence methods. Lichenologist 32: 271–289.

Gauslaa, Y. and K.A. Solhaug, 2001. Fungal melanins as a sun screen for symbiotic green algae in the lichen *Lobaria pulmonaria*. Oecologia 126: 462–471.

Gauslaa, Y. and E.M. Ustvedt, 2003. Is parietin a UV-B or a blue–light screening pigment in the lichen *Xanthoria parietina*? Photochemical and Photobiological Sciences 2: 424–432.

Gauslaa, Y., M. Ohlson, K.A. Solhaug, W. Bilger, and L. Nybakken, 2001. Aspect-dependent high-irradiance damage in two transplanted foliose forest lichens, *Lobaria pulmonaria* and *Parmelia sulcata*. Canadian Journal of Forest Research 31: 1639–1649.

Gebauer, R., 1986. Wasserabgabe und Wasseraufnahme poikilohydrer höherer Pflanzen im Hinblick auf ihre physiologische Aktivität. Diploma thesis, Universität zu Kiel, 97 pp.

Gebauer, R., R. Lösch, and L. Kappen, 1987. Wassergehalt und CO_2-Gaswechsel des poikilohydren Kormophyten *Ramonda myconi* (L.) Schltz. Während der Austrocknung und Wiederaufsättigung. Verhandlungen Gesellschaft für Ökologie XVI: 231–236.

Georgieva, K., L. Maslenkova, V. Peeva, Y. Markovska, D. Stefanov, and Z. Tuba. 2005. Comparative study on the changes in photosynthetic activity of the homoiochlorophyllous desiccation-tolerant *Haberlea rhodopensis* and desiccation-sensitive spinach leaves during desiccation and rehydration. Photosynthesis Research 85: 191–203.

Gilnder, B.S. and D.W. Larson, 1992. Seasonal changes in photosynthesis in the desiccation-tolerant fern *Polypodium virginianum*. Oecologia 89: 383–389.

Gimingham, C.H. and R.I.L. Smith, 1971. Growth form and water relations of mosses in the maritime Antarctic. British Antarctic Survey Bulletin 25: 1–21.

Giordano, S., C. Colacino, V. Spagnuolo, A. Basile, A. Esposito, and R. Castaldo-Cobianchi, 1993. Morphological adaptation to water uptake and transport in the poikilohydric moss *Tortula ruralis*. Giornale Botanico Italiano 127: 1123–1132.

Gradstein, S.R., 2006. The lowland cloud forest of French Guayana—a liverwort hotspot. Cryptogamie Bryologie 27: 141–152.

Green, T.G.A. and O.L. Lange, 1994. Photosynthesis in poikilohydric plants: a comparison of lichens and bryophytes. In: E.-D. Schulze and M.M. Caldwell, eds. Ecophysiology of Photosynthesis. Ecological Studies 100. Springer-Verlag, Berlin, Heidelberg, New York, pp. 320–341.

Green, T.G.A., W.P. Snelgar, and A.L. Wilkins, 1985. Photosynthesis, water relations and thallus structure of Stictaceae lichens. In: D.H. Brown, ed. Lichen Physiology and Cell Biology. Plenum Press, New York, pp. 57–75.

Green, T.G.A., E. Kilian, and O.L. Lange, 1991. *Pseudocyphellaria dissimilis:* a desiccation-sensitive, highly shade-adapted lichen from New Zealand. Oecologia 85: 498–503.

Green, T.G.A., O.L. Lange, and I.R. Cowan, 1994. Ecophysiology of lichen photosynthesis: the role of water status and thallus diffusion resistances. Cryptogamic Botany 4: 166–178.

Green, T.G.A., M. Schlensog, L.G. Sancho, J.B. Winkler, F.D. Broom, and B. Schroeter, 2002. The photobiont determines the pattern of photosynthetic activity within a single lichen thallus containing cyanobacterial and green-algal sectors (photosymbiodeme). Oecologia 130: 191–198.

Hahn, S., D. Speer, A. Meyer, and O.L. Lange, 1989. Photosynthetic primary production of epigean lichens growing in local xerothermic steppe formations in Franconia. I. Diurnal time course of microclimate, water content and CO_2 exchange at different seasons. Flora 182: 313–339.

Hahn, S., D. Tenhunen, P.W. Popp, A. Meyer, and O.L. Lange, 1993. Upland tundra in the foothills of the Brooks Range, Alaska: Diurnal CO_2 exchange patterns of characteristic lichen species. Flora 188: 125–143.

Hallam, N.D. and P. Cappicchiano, 1974. Studies on desiccation tolerant plants using nonaqueous fixation methods. Proceeding Eighth International Congress on Electronic Microscopy, Canberra, Vol. 2, pp. 612–613.

Hallam, N.D. and D.F. Gaff, 1978. Regeneration of chloroplast structure in *Talbotia elegans*: a desiccation-tolerant plant. New Phytologist 81: 657–662.

Hallam, N.D. and S.E. Luff, 1980. Fine structure changes in the leaves of the desiccation-tolerant plant *Talbotia elegans* during extreme water stress. Botanical Gazette 141: 180–187.

Hambler, D.J., 1961. A poikilohydrous, poikilochlorophyllous angiosperm from Africa. Nature 191: 1415–1416.

Hargrave, K.R., K.J. Kolb, F.W. Ewers, and S.D. Davis, 1994. Conduit diameter and drought-induced embolism in *Salvia mellifera* Green (Labiatae). New Phytologist 126: 695–705.

Hartung, W., P. Schiller, and K.-J. Dietz, 1998. The physiology of poikilohydric plants. Progress in Botany 59: 299–327.

Hébant, C., 1977. The Conducting Tissues of Bryophytes. Gantner Verlag, Vaduz.

Heber, U. and V.A. Shuvalov, 2005. Photochemical reactions of chlorophyll in dehydrated photosystem II: Two chlorophyll forms (680 and 700 nm). Photosynthesis Reserch 84: 85–91.

Heber, U., W. Bilger, R. Bligny, and O.L. Lange, 2000. Phototolerance of lichens, mosses and higher plants in an alpine environment: Analysis of photoreactions. Planta 211: 770–780.

Heber, U., N.G. Bukhov, V.A., Shuvalov, Y. Kobayashi, and O.L. Lange, 2001. Protection of the photosynthetic apparatus against damage by excessive illumination in homoiohydric leaves and poikilohydric mosses and lichens. Journal of Experimental Botany 52: 1999–2006.

Heber, U., O.L. Lange, and V.A. Shuvalov, 2006a. Conservation and dissipation of light energy as complementary process: Homoiohydric and poikilohydric autotrophs. Journal of Experimental Botany 57: 1211–1223.

Heber, U., W. Bilger, V.A. Shuvalov, 2006b. Thermal energy dissipation in reaction centres of Photosystem II protects desiccated poikilohydric mosses against photo-oxidation. Journal of Experimental Botany, 57: 2993–3006.

Heil, H., 1925. *Chamaegigas intrepidus* Dtr., eine neue Auferstehungspflanze. Beiträge zum Botanischen Zentralblatt 41: 41–50.

Heilmeier, H. and W. Hartung, 2001. Survival strategies under extreme and complex environmental conditions: the aquatic resurrection plant *Chameagigas intrepidus*. Flora 196: 245–260.

Heilmeier, H., E.R.G. Ratcliffe, and W. Hartung, 2000. Urea: a nitrogen source for the aquatic resurrection plant *Chamaegigas intrepidus* Dinter. Oecologia 123: 9–14.

Heilmeier, H., W. Durka, M. Woitke, W. Hartung, 2005. Ephemeral pools as stressful and isolated habitats for the endemic aquatic resurrection plant *Chamaegigas intrepidus*. Phytocoenologia 35: 449–468.

Hellwege, E.M., K.J. Dietz, O.H. Volk, and W. Hartung, 1994. Abscisic acid and the induction of desiccation tolerance in the extremely xerophilic liverwort *Exormotheca holstii*. Planta 194: 525–531.

Helseth, L.E. and T.M. Fischer, 2005. Physical mechanisms of rehydration in *Polypodium polypodioides*, a resurrection plant. Physical Review E 71: 061903.

Hetherington, S.E. and R.M. Smillie, 1982. Humidity-sensitive degreening and regreening of leaves of *Borya nitida* Labill. as followed by changes in chlorophyll fluorescence. Australian Journal of Plant Physiology 9: 587–599.

Hickel, B., 1967. Zur Kenntnis einer xerophilen Wasserpflanze: *Chamaegigas intrepidus* Dtr. aus Südwestafrika. Hydrobiologie 52: 361–400.

Hicklenton, P.R. and W.C. Oechel, 1977. The influence of light intensity and temperature on the field carbon dioxide exchange of *Dicranum fuscescens* in the subarctic. Arctic and Alpine Research 9: 407–419.

Hietz, P. and Briones, O., 1998. Correlation between water relations and within—canopy distribution of epiphytic ferns in a Mexican cloud forest. Oecologia 114: 305–316.

Hinshiri, H.M. and M.C.F. Proctor, 1971. The effect of desiccation on subsequent assimilation and respiration of the bryophytes *Anomodon viticulosus* and *Porella platyphylla*. New Phytologist 70: 527–538.

Hoffmann, P., 1968. Pigmentgehalt und Gaswechsel von *Myrothamnus*-Blättern nach Austrocknung und Wiederaufsättigung. Photosynthetica 2: 245–252.

Hosokawa, T. and H. Kubota, 1957. On the osmotic pressure and resistance to desiccation of epiphytic mosses from a beech forest, South-West Japan. Journal of Ecology 45: 579–591.

Ibisch, P.L., C. Nowicki, R. Vasques, and K. Koch, 2001. taxonomy and biology of Andean Velloziaceae: *Vellozia andina* sp nov. and notes on *Barbaceniopsis* (including *Barbaceniopsis cartillonii* comb. nov.) Systematic Botany 26: 5–16.

Iljin, W.S., 1931. Austrocknungsresistenz des Farnes *Notholaena maranthae* R.Br. Protoplasma 10: 379–414.

Illing, N., K.J. Denby, H. Collett, A. Shen, and J.M. Farrant, 2005. The signature of seeds in resurrection plants: A molecular and physiological comparison of desiccation tolerance in seeds and vegetative tissues. Integrative and Comparative Biology 45: 771–789.

Ingram, J. and D. Bartels, 1996. The molecular basis of dehydration tolerance in plants. Annual Review of Plant Physiology and Plant Molecular Biology 47: 377–403.

Ingrouille, M., 1995. Diversity and Evolution of Land Plants. Chapman and Hall, London.

Iturriaga, G., M.A.F. Cushman, and J.C. Cushman, 2006. An EST catalogue from the resurrection plant *Selaginella lepidophylla* reveals abiotic stress-adaptive genes. Plant Science 170: 1173–1184.

Jovet-Ast, S., 1969. Le caryotype des Ricciaceae. Revue Bryologie Lichenologique 36: 573–689.

Kaiser, W.M., 1987. Effect of water deficit on photosynthetic activity. Physiologia Plantarum 71: 142–149.

Kaiser, W.M. and U. Heber, 1981. Photosynthesis under osmotic stress. Effect of high solute concentrations on the permeability properties of the chloroplast envelope and the activity of stoma enzymes. Planta 153: 429–432.

Kaiser, K., D.F. Gaff, and W.H. Outlaw Jr., 1985. Sugar contents of leaves of desiccation-tolerant plants. Naturwissenschaften 72: 608–609.

Kallio, P. and L. Kärenlampi, 1975. Photosynthesis in mosses and lichens. In: J.P. Cooper, ed. Photosynthesis and Productivity in Different Environments. Cambridge University Press, Cambridge, pp. 393–423.

Kappen, L., 1964. Untersuchungen über den Jahreslauf der Frost-, Hitze- und Austrocknungsresistenz von Sporophyten einheimischer Polypodiaceen (Filicinae). Flora 155: 123–166.

Kappen, L., 1965. Untersuchungen über die Widerstandsfähigkeit der Gametophyten einheimischer Polypodiaceen gegenüber Frost, Hitze and Trockenheit. Flora 156: 101–115.

Kappen, L., 1966. Der Einfluss des Wassergehaltes auf die Widerstandsfähigkeit von Pflanzen gegenüber hohen und tiefen Temperaturen, untersucht an Blättern einiger Farne und von *Ramonda myconi*. Flora 156: 427–445.

Kappen, L., 1973. Response to extreme environments. In: V. Ahmadjian, and M.E. Hale, eds. The Lichens. Academic Press, New York, pp. 311–380.

Kappen, L., 1981. Ecological significance of resistance to high temperature. In: O.L. Lange, P.S. Nobel, C.B. Osmond, and H. Ziegler, eds. Physiological Plants Ecology, Vol. I. Responses to the Physical Environment. Springer-Verlag, Berlin-Heidelberg, pp. 439–473.

Kappen, L., 1982. Lichen oases in hot and cold deserts. Journal of the Hattori Botanical Laboratory 53: 325–330.

Kappen, L., 1983. Ecology and physiology of the Antarctic fructicose lichen *Usnea sulphurea* (Koenig) Th. Fries. Polar Biology 1: 249–255.

Kappen, L., 1985. Water relations and net photosynthesis of *Usnea*. A comparison between *Usnea fasciata* (maritime Antarctic) and *Usnea sulphurea* (continental Antarctic). In: D.H. Brown, eds. Lichen Physiology and Cell Biology. Plenum Press, New York-London, pp. 41–56.

Kappen, L., 1988. Ecophysiological relationship in different climatic regions. In: M. Galun, ed. Handbook of Lichenology, Vol. 2. CRC Press, Boca Raton, FL, pp. 37–100.

Kappen, L., 1989. Field measurements of carbon dioxide exchange of the Antarctic lichen *Usnea sphacelata* in the frozen state. Antarctic Science 1: 31–34.

Kappen, L., 1993a. Plant activity under snow and ice, with particular reference to lichens. Arctic 46: 297–302.

Kappen, L., 1993b. Lichens in the Antarctic region. In: E.I. Friedmann, ed. Antarctic Microbiology. Wiley-Liss, New York, pp. 433–490.

Kappen, L., 1994. The lichen, a mutualistic system? Some mainly ecophysiological aspects. Cryptogamic Botany 4: 193–202.

Kappen, L., 1995. Terrestrische Mikroalgen und Flechten der Antarktis. In: K. Hausmann and B.P. Kremer, eds. Extremophile. Mikroorganismen in ausgefallenen Lebensräumen. VHC Verlagsgesellschaft mbH, Weinheim, New York Basel, Cambridge, Tokyo, pp. 3–25.

Kappen, L., 2000. Some aspects of the great success of lichens in Antarctica. Antarctic Science 12: 314–324.

Kappen, L., O.L. Lange, E.-D. Schulze, M. Evenari, and U. Buschbom, 1972. Extreme water stress and photosynthetic activity of the desert plant *Artemisia herba-alba* Asso. Oecologia 10: 177–182.

Kappen, L., O.L. Lange, E.-D. Schulze, M. Evenari, and U. Buschbom, 1979. Ecophysiological investigations on lichens of the Negev desert. VI. Annual course of photosynthetic production of *Ramalina maciformis* (Del). Bory. Flora 168: 85–108.

Kappen, L., O.L. Lange, E.-D Schulze, U. Buschbom, and M. Evenari, 1980. Ecophysiological investigations on lichens of the Negev desert. VII. The influence of the habitat exposure on dew imbibition and photosynthetic productivity. Flora 169: 216–229.

Kappen, L., R.I. Lewis Smith, and M. Meyer, 1989. Carbon dioxide exchange of two ecodemes of *Schistidium antarctici* in continental Antarctica. Polar Biology 9: 415–422.

Kappen, L., M. Sommerkorn, and B. Schroeter, 1995. Carbon acquisition and water relations of lichens in polar regions—potentials and limitations. Lichenologist 27: 531–545.

Kappen, L., B. Schroeter, G. Hestmark, and J.B. Winkler, 1996. Field measurements of photosynthesis of umbilicarious lichens in winter. Botanica Acta 109: 292–298.

Kappen, L., B. Schroeter, T.G.A. Green, and R.D. Seppelt, 1998a. Chlorophyll a fluorescence and CO_2 exchange of *Umbilicaria aprina* under extreme light stress in the cold. Oecologia 113: 325–331.

Kappen, L., B. Schroeter, T.G.A. Green, and R.D. Seppelt, 1998b. Microclimatic conditions, meltwater moistening, and the distributional pattern of *Buellia frigida* on rock in a southern continental Antarctic habitat. Polar Biology 19: 101–106.

Kermode, A.R. and B.E. Finch-Savage, 2002. Desiccation sensitivity in othodox and calcitrant seeds in relation to development. In: M. Black and H.W. Pritchard, eds. Desiccation and Survival of Plants—Drying without Dying. CABI Publishing, Wallingford, pp. 150–184.

Kershaw, K.A., 1985. Physiological Ecology of Lichens. Cambridge University Press, Cambridge.

Kerstiens, G., 1996. Cuticular water permeability and its physiological significance. Journal of Experimental Botany 47: 1813–1832.

Kimenov, G.P., Y.K. Markovska, and T.D. Tsonev, 1989. Photosynthesis and transpiration of *Haberlea rhodopensis* Friv. in dependence on water deficit. Photosynthetica 23: 368–371.

Klepper, B. 1968. A comparison of the water relations of a moss and a lichen. Nova Hedwigia 15: 13–20.

Kolb, K.J., J.S. Sperry, and B.B. Lamont, 1996. A method for measuring xylem hydraulic conductance embolism in entire root and shoot system. Journal of Experimental Botany 47: 1805–1810.

Kopecky, J., M. Azarkovich, E.-E. Pfundel, V.A. Shuvalov, and U. Heber, 2005. Thermal dissipation of light energy is regulated differently and by different mechanisms in lichens and higher plants. Plant Biology 7: 156–167.

Kranner, I. and D. Grill, 1997 Desiccation and the subsequent recovery of cryptogamics that are resistant to drought. Phyton 37: 139–150.

Kuang, J., D.F. Gaff, R.D. Gianello, C.K. Blomstedt, A.D. Neale, and J.D. Hamill, 1995. Changes in *in vivo* protein complements in drying leaves of the desiccation-tolerant grass *Sporobolus stapfianus* and the desiccation-sensitive grass *Sporobolus pyramidalis.* Australian Journal of Plant Physiology 22: 1027–1034.

Kubitzki, K., 1998.Velloziaceae. In: K. Kubitzki, ed. The Families and Genera of Vascular Plants, Vol. 3, Springer, Berlin.

Lange, O.L., 1953. Hitze- und Trockenresistenz der Flechten in Beziehung zu ihrer Verbreitung. Flora 140: 39–97.

Lange, O.L., 1955. Untersuchungen über die Hitzeresistenz der Moose in Beziehung zu ihrer Verbreitung. I. Die Resistenz stark ausgetrockneter Moose. Flora 142: 381–399.

Lange, O.L., 1969a. CO_2 Gaswechsel von Moosen nach Wasserdampfaufnahme aus dem Luftraum. Planta 89: 90–94.

Lange, O.L., 1969b. Experimentell-ökologische Untersuchungen an Flechten der Negev-Wüste. I. CO_2-Gaswechsel von *Ramalina maciformis* (Del.) Bory unter kontrollierten Bedingungen im Laboratorium. Flora 158: 324–359.

Lange, O.L., 1980. Moisture content and CO_2 exchange of lichens. I. Influence of temperature on moisture-dependent net photosynthesis and dark respiration in *Ramalina maciformis.* Oecologia 45: 82–87.

Lange, O.L. and A. Bertsch, 1965. Photosynthese der Wüstenflechte *Ramalina maciformis* nach Wasserdampfaufnahme aus dem Luftraum. Naturwissenschaften 52: 215–216.

Lange, O.L. and E. Kilian, 1985. Reaktivierung der Photosynthese trockener Flechten durch Wasserdampfaufnahme aus dem Luftraum: Artspezifisch unterschiedliches Verhalten. Flora 176: 7–23.

Lange, O.L. and U. Matthes, 1981. Moisture-dependent CO_2 exchange of lichens. Photosynthetica 15: 555–574.

Lange, O.L. and J. Redon, 1983. Epiphytische Flechten im Bereich einer chilenischen "Nebeloase" (Fray Jorge) II. Ökophysiologische Charakterisierung von CO_2-Gaswechsel und Wasserhaushalt. Flora 174: 245–284.

Lange, O.L. and J.D. Tenhunen, 1981. Moisture content and CO_2 exchange of lichens. II. Depression of net photosynthesis in *Ramalina maciformis* at high water content is caused by increased thallus carbon dioxide diffusion resistance. Oecologia 51: 426–429.

Lange, O.L., E. Kilian, and H. Ziegler, 1986. Water vapor uptake and photosynthesis of lichens: performance differences in species with green and blue-green algae as phycobionts. Oecologia 71: 104–110.

Lange, O.L., T.G.A. Green, and H. Ziegler, 1988. Water status related photosynthesis and carbon isotope discrimination in species of the lichen genus *Pseudocyphellaria* with green or blue-green phycobionts and in photosymbiodemes. Oecologia 75: 494–501.

Lange, O.L., E. Kilian, and H. Ziegler, 1990a. Photosynthese von Blattflechten mit hygroskopischen Thallusbewegungen bei Befeuchtung durch Wasserdampf oder mit flüssigem Wasser. Bibliotheca Lichenologica 38: 311–323.

Lange, O.L., H. Pfanz, E. Kilian, and A. Meyer, 1990b. Effect of low water potential on photosynthesis in intact lichens and their liberated algal components. Planta 182: 467–472.

Lange, O.L., A. Meyer, H. Zellner, I. Ullmann, and D.C.J. Wessels, 1990c. Eight days in the life of a desert lichen: Water relations and photosynthesis of *Teloschistes capensis* in the coastal fog zone of the Namib desert. Madoqua 17: 17–30.

Lange, O.L., A. Meyer, I. Ullmann, and H. Zellner, 1991. Mikroklima, Wassergehalt und Photosynthese von Flechten in der küstennahen Nebelzone der Namib-Wüste: Messungen während der herbstlichen Witterungsperiode. Flora 185: 233–266.

Lange, O.L., B. Büdel, U. Heber, A. Meyer, H. Zellner, and T.G.A. Green, 1993. Temperate rainforest lichens in New Zealand: High thallus water content can severely limit photosynthetic CO_2 exchange. Oecologia 95: 303–313.

Lange, O.L., A. Meyer, H. Zellner, and U. Heber, 1994. Photosynthesis and water relations of lichen soil crusts: Field measurements in the coastal fog zone of the Namib desert. Functional Ecology 8: 253–264.

Lange, O.L., T.G.A. Green, H. Reichenberger, and A. Meyer, 1996. Photosynthetic depression at high water contents in lichens: Concurrent use of gas exchange and fluorescence techniques with cyanobacterial and green algal *Peltigera* species. Botanica Acta 109: 43–50.
Lange, O.L., J. Belnap, H. Reichenberger, 1998. Photosynthesis of the cyanobacterial soil crust lichen *Collema tenax* from arid lands in southern Utah, USA: Role of water content on light and temperature response of CO_2 exchange. Functional Ecology 12: 195–202.
Lange, O.L., J.M.R. Leisner, and W. Bilger, 1999. Chlorophyll fluorescence characteristics of the cyanobacterial lichen Peltigera rufescens under field conditions II. Diel and annual distribution of metabolic activity and possible mechanisms to avoid photoinhibition. Flora: 413–430.
Larcher, W., 1980. Ökologie der Pflanzen. Verlag Eugen Ulmer, Stuttgart.
Larson, D.W., 1979. Lichen water relations under drying conditions. New Phytologist 82: 713–731.
Larson, D.W., 1981. Differential wetting in some lichens and mosses: The role of morphology. Bryologist 84: 1–15.
Larson, D.W., 1984. Habitat overlap/niche segregation in two *Umbilicaria* lichens: a possible mechanism. Oecologia 62: 118–115.
Larson, D.W., 1987. The absorption and release of water by lichens. Bibliotheca Lichenologica 25: 351–360.
Larson, D.W., 1988. The impact of ten years at −20°C on gas exchange in five lichen species. Oecologia 78: 87–92.
Lazarides, M., 1992. Resurrection grasses (*Poaceae*) in Australia. In: G.P. Chapman, ed. Desertified Grasslands: Their Biology and Management. Academic Press, London, pp. 213–234.
Lebkuecher, J.G. and W.G. Eickmeier, 1991. Reduced photoinhibition with stem curling in the resurrection plant *Selaginella lepidophylla*. Oecologia 88: 597–604.
Lebkuecher, J.G. and W.G. Eickmeier, 1993. Physiological benefits of stem curling for resurrection plants in the field. Ecology 74: 1073–1080.
Lebrun, J., 1947. La vegetation de la pleine alluviale au sud du Lac Edouard. Exploration du Parc National Albert. Mission J. Lebrun (1937–1938). Institut des Parces Nationaux du Congo Belge, Bruxelles, 100 pp.
Lechowicz, M.J., 1981. The effects of climatic pattern on lichen productivity: *Cetraria cucullata* (Bell). Ach. in the Arctic tundra of northern Alaska. Oecologia 50: 210–216.
Lechowicz, M.J., 1992. The niche at the organismal level: lichen photosynthetic response. In: G.C. Carroll and D.T. Wicklow, eds. The Fungal Community. Marcel Dekker, New York, pp. 29–42.
Leisner, J.M.R., W. Bilger, F.-C. Czygan, and O.L. Lange, 1994. Light exposure and the composition of lipophilous carotenoids in cyanobacterial lichens. Journal of Plant Physiology 143: 514–519.
Leisner, J.M.R., W. Bilger, and O.L. Lange, 1996. Chlorophyll fluorescence characteristics of the cyanobacterial lichen *Peltigera rufescens* under field conditions. Flora 191: 261–273.
Leopold, A.C., 1990. Coping with desiccation. In: R.G.Alscher, ed. Stress Responses in Plants: Adaptation and Acclimation Mechanisms.Wiley-Liss, New York, pp. 36–56.
Levitt, J., 1980. Responses of Plants to Environmental Stress. Chilling, Freezing, and High Temperature Stresses. Academic Press, New York.
Lewis, A.M., V.D. Harnden, and M.T. Tyree, 1994. Collapse of water-stress emboli in the tracheids of *Thuja occidentalis* L. Plant Physiology 106: 1639–1646.
Lichtenthaler, H.K. and U. Rinderle, 1988. The role of chlorophyll fluorescence in the detection of stress conditions in plants. CRC Critical Review Analytical Chemistry 19: 29–85.
Long, S.P., S. Humphries, and P.G. Falkowski, 1994. Photoinhibition of photosynthesis in nature. Annual Review of Plant Physiology and Plant Molecular Biology 45: 633–662.
Longton, R.E., 1988a. Life-history strategies among bryophytes of arid regions. Journal of the Hattori Botanical Laboratory 64: 15–28.
Longton, R.E., 1988b. Biology of Polar Bryophytes and Lichens. Cambridge University Press, Cambridge.
Lösch, R., R. Pause, and B. Mies, 1997. Poikilohydrie und räumlich-zeitliche Existenznische von Flechten und Moosen. Bibliotheca Lichenologica 67: 145–162.
Lud, D., M. Schlensog, B. Schroeter, and A.H.L. Huiskes, 2003. The influence of UV-B radiation on light-dependent photosynthetic performance in *Sanionia uncinata* (Hedw.) Loeske in Antarctica. Polar Biology 26: 225–232.

Lüttge, U., M. Kluge, and G. Bauer, 1988. Botanik, ein grundlegendes Lehrbuch. Verlage Chemie, Weinheim.

Lüttge, U., B. Büdel, E. Ball, F. Strobe, and P. Weber, 1995. Photosynthesis of terrestrial cyanobacteria under light and desiccation stress as expressed by chlorophyll fluorescence and gas exchange. Journal of Experimental Botany 46: 309–319.

MacFarlane, J.D. and K.A. Kershaw, 1982. Physiological-environmental interactions in lichens XIV. The environmental control of glucose movement from alga to fungus in *Peltigera polydactyla, P. rufescens* and *Collema furfuraceum*. New Phytologist 91: 93–101.

Maguas, C., F. Valladares, and E. Brugnoli, 1997. Effects of thallus size on morphology and physiology of foliose lichens: New findings with a new approach. Symbiosis 23: 149–164.

Majee, M. Batra, S.G. Mundree, and A.L. Majumder, 2005. Molecular cloning, bacterial overexpression and characterization of L-myo-inositol 1-phosphate synthase from a monocotyledonous resurrection plant, *Xerophyta viscosa*. Journal of Plant Biochemistry and Biotechnology 14: 95–9.

Manrique, E., L. Balaguer, J. Barnes, and A.W. Davison, 1993. Photoinhibition studies in lichens using chlorophyll fluorescence analysis. Bryologist 96: 443–449.

Markovska, Y.K., 1999. Gas exchange and malate accumulation in *Haberlea rhodopensis* grown under different irradiances. Biologia Plantarum 42: 559–565.

Markovska, Y.K., T.D. Tsonev, G.P. Kimenov, and A.A. Tutekova, 1994. Physiological changes in higher poikilohydric plants—*Haberlea rhodopensis* Friv. and *Ramonda serbica* Panc. during drought and rewatering at different light regimes. Journal of Plant Physiology 144: 100–108.

Markovska, Y.K., A.A. Tutekova, and G.P. Kimenov, 1995. Ultrastructure of chloroplasts of poikilohydric plants *Haberlea rhodopensis* Friv. and *Ramonda serbica* Panc. during recovery from desiccation. Photosynthetica 31: 613–620.

Markovska, Y., T. Tsonev, and G. Kimenov, 1997. Regulation of CAM and respiratory recycling by water supply in higher poikilohydric plants—*Haberlea rhodopensis* Friv. and *Ramonda serbica* Panc. at transition from biosis to anabiosis and *vice versa*. Botanica Acta 110: 18–24.

Masuch, G., 1993. Biologie der Flechten. UTB Quelle und Meyer, Heidelberg, Wiesbaden.

Matthes-Sears, U., P.E. Kelly, and D.W. Larson, 1993. Early spring gas exchange and uptake of deuterium-labelled water in the poikilohydric fern *Polypodium virginianum*. Oecologia 95: 9–13.

Meyer, H. and K.A. Santarius, 1998. Short-term thermal acclimation and heat tolerance of the gametophytes of mosses. Oecologia 115: 1–8.

Montenegro, G., A.J. Hoffmann, M.E. Aljaro, and A.E. Hoffmann, 1979. *Satureja gilliesii*: A poikilohydric shrub from the Chilean Mediterranean vegetation. Canadian Journal of Botany 57: 1206–1213.

Moore, J.P., J.M. Farrant, G.G. Lindsey, and W.F. Brandt, 2005. The South African and Namibian populations of the resurrection plant *Myrothamnus flabellifolius* are genetically distinct and display variation in their galloquinic acid composition. Journal of Chemical Ecology 31: 2823–2834.

Müller, M.A.N., 1985. Gräser Südwestafrika/Namibias. John Meinert (Pty.) Ltd., Windhoek, Südwestafrika/Namibia.

Muslin, E.H. and P.H. Homann, 1992. Light as a hazard for the desiccation-resistant "resurrection" fern *Polypodium polypodioides* L. Plant, Cell and Environment 15: 81–89.

Nash, T.H.I. and T.J. Moser, 1982. Vegetational and physiological patterns of lichens in North American deserts. Journal of the Hattori Botanical Laboratory 53: 331–336.

Nash, T.H.I., A. Reiner, B. Demmig-Adams, E. Kilian, W.M. Kaiser, and O.L. Lange, 1990. The effect of atmospheric desiccation and osmotic water stress on photosynthesis and dark respiration of lichens. New Phytologist 116: 269–276.

Nash, T.H.I., S.L. White, and J.E. Marsh, 1997. Lichen and moss distribution and biomass in hot desert ecosystems. Bryologist 80: 470–479.

Navari-Izzo, F., S. Meneguzzo, B. Loggini, C. Vazzana, and C.L.M. Sgherri, 1997. The role of the glutathione system in dehydration of *Boea hygroscopica*. Physiologia Plantarum 99: 23–30.

Nelson, D., F. Salamini, and D. Bartels, 1994. Abscisic acid promotes novel DNA-binding activity to a desiccation-related promoter of *Craterostigma plantagineum*. The Plant Journal 5: 451–458.

Nienow, J.A. and E.I. Friedmann, 1993. Terrestrial lithophytic (rock) communities. In: E.I. Friedmann, ed. Antarctic Microbiology. Wiley-Liss, New York, pp. 343–412.

Niyogi, K.K., 1999. Photoprotection revisited: genetic and molecular approaches. Annual Review of Plant Physiology and Plant Molecular Biology 50: 375–382.

Nobel, P.S., 1978. Microhabitat, water relations, and photosynthesis of a desert fern, *Notholaena parryi*. Oecologia 31: 293–309.
Nobel, P.S., D.J. Longstreth, and T.L. Hartsock, 1978. Effect of water stress on the temperature optima of net CO_2 exchange for the two desert species. Physiologia Plantarum 44: 97–101.
North, G.B. and P.S. Nobel, 1991. Changes in hydraulic conductivity and anatomy caused by drying and rewetting roots of Agave-deserti (Agavaceae). American Journal of Botany 78: 906–915.
Norwood, M., O. Toldi, A. Richter, and P. Scott, 2003. Investigation into the ability of roots of the poikilohydric plant *Craterostigma plantagineum* to survive dehydration stress. Journal of Experimental Botany 54: 2313–1321.
Nybakken, L.K.A. Solhaug, W. Bilger, and Y. Gauslaa, 2004. The lichens *Xanthoria parietina* and *Cetraria islandica* maintain a high protection against UV-B radiation in arctic habitats. Oecologia 140: 211–216.
Oechel, W.C. and B. Sveinbjörnsson, 1978. Primary production processes in the arctic bryophytes at Barrow, Alaska. In: L.L.Tieszen, ed. Vegetation and Production Ecology of an Alaskan Arctic Tundra. Ecological Studies 29. Springer-Verlag, New York, Heidelberg, Berlin, pp. 269–299.
Oliver, M.J. and J.D. Bewley, 1997. Desiccation-tolerance of plant tissues: A mechanistic overview. Horticultural Reviews 18: 171–213.
Oliver, M.J., Z. Tuba, and B.D. Mishler, 2000. The evolution of vegetative desiccation tolerance in land plants. Plant Ecology 15: 85–100.
Oliver, M.J., J. Velten, and B.D. Mishler, 2005. Desiccation tolerance in bryophytes: A reflection of the primitive strategy for plant survival in dehydrating habitats. Integrative and Comparative Biology 45: 788–799.
Oppenheimer, H.R. and A.H. Halevy, 1962. Anabiosis of *Ceterach officinarum*. Bulletin of the Research Council Israel 11: 127–164.
Öquist, G. and D.C. Fork, 1982a. Effects of desiccation on the exitation energy distribution from phycoerythrin to the two photosystems in the red alga *Porphyra perforata*. Physiologia Plantarum 56: 56–62.
Öquist, G. and D.C. Fork, 1982b. Effect of desiccation on the 77 Kelvin fluorescence properties of the liverwort *Porella navicularis* and the isolated lichen green alga *Trebouxia pyriformis*. Physiologia Plantarum 56: 63–68.
Owoseye, J.A. and W.W. Sanford, 1972. An ecological study of *Vellozia schnitzleinia*, a drought-enduring plant of northern Nigeria. Journal of Ecology 60: 807–817.
Pannewitz, S., M. Schlensog, T.G.A. Green, L.G. Sancho, and B. Schroeter, 2003. Are lichens active under snow in continental Antarctica? Oecologia 135: 30–38.
Pannewitz, S., T.G.A. Green, M. Schlensog, R. Seppelt, L.G. Sancho, and B. Schroeter, 2006. Photosynthetic performance of *Xanthoria mawsonii* C.W. Dodge in coastal habitats of the Ross Sea region, continental Antarctica. Lichenologist 38: 67–81.
Patterson, P.M., 1964. Problems presented by bryophytic xerophytism. Bryologist 67: 390–396.
Pence, V.C., 2000. Cryopreservation of *in vitro* grown fern gametophytes. American Fern Journal 90: 16–23.
Perez, F.L., 1997. Geoecology of erratic globular lichens of *Catapyrenium lachneum* in a high Andean Paramo. Flora 192: 241–259.
Pessin, L.J., 1924. A physiological and anatomical study of the leaves of *Polypodium polypodioides*. American Journal of Botany 11: 370–381.
Piatkowski, D., K. Schneider, F. Salamini, and D. Bartels, 1990. Characterization of five abscisic acid-responsive cDNA clones isolated from the desiccation-tolerant plant *Craterostigma plantagineum* and their relationship to other water-stress genes. Plant Physiology 94: 1682–1688.
Pickard, W.F., 1989. How might a tracheary element which is embolized by day be healed by night? Journal of Theoretical Biology 141: 259–279.
Picket, F.L., 1914. Ecological adaptations of certain fern prothallia. American Journal of Botany I: 477–498.
Pintado, A., F. Valladares, and L.G. Sancho, 1997. Exploring phenotypic plasticity in the lichen *Ramalina capitata*: morphology, water relations and chlorophyll content in North- and South-facing populations. Annals of Botany 80: 345–353.

Pitt, J.I. and J.H.B. Christian, 1968. Water relations of xerophilic fungi isolated from prunes. Applied Microbiology 16: 1853–1858.

Porembski, S., 1996. Notes on the vegetation of inselbergs in Malawi. Flora 191: 1–8.

Porembski, S. and W. Barthlott, 1995. On the occurrence of a *velamen radicum* in Cyperaceae and Velloziaceae. Nordic Journal of Botany 15: 625–629.

Porembski, S. and W. Barthlott, 1997. Seasonal dynamics of plant diversity on inselbergs in the Ivory Coast (West Africa). Botanica Acta: 466–472.

Porembski, S. and W. Barthlott, 2000. Granitic and gneissic outcrops (inselbergs) as centers of diversity for desiccation-tolerant vascular plants. Plant Ecology 151: 19–28.

Proctor, M.C.F., 1982. Physiological ecology: Water relations, light and temperature responses, carbon balance. In: A.J.E. Smith, ed. Bryophyte Ecology. Chapman & Hall, London, New York, pp. 333–381.

Proctor, M.C.F., 1984. Structure and ecological adaptation. In: A.F. Dyer and J.Q. Duckett, eds. The Experimental Biology of Bryophytes. Academic Press, London, pp. 9–37.

Proctor, M.C.F., 1990. The physiological basis of bryophyte production. Botanical Journal of the Linnean Society 104: 61–77.

Proctor, M.C.F., 2002. Desiccation tolerance of vegetative tissues. In: M. Black and H.W. Pritchard, eds. Desiccation and Survival of Plants—Drying without Dying. CABI Publishing, Wallingford, pp. 207–237.

Proctor, M.C.F., 2003. Comparative ecophysiological measurements on the light responses, water relations and desiccation tolerance of the filmy ferns *Hymenophyllum wilsonii* Hook. and *H. tunbridgense* (L.) Smith. Annals of Botany 91: 717–727.

Proctor, M.C.F. and Z. Tuba, 2002 Poikilohydry and homoiohydry: Anthithesis or spectrum of possibilities? New Phytologist 156: 327–348.

Puff, C., 1978. Zur Biologie von *Myrothamnus flabellifolius* Welw. (Myrothamnanceae). Dinteria 14: 1–20.

Quadir, A., P.J. Harrison, and R.E. Dewreede, 1979. The effects of emergence and subemergence on the photosynthesis and respiration of marine macrophytes. Phycologia 18: 83–88.

Rascio, N. and N. Rocca, 2005. Resurrection plants: the puzzle of surviving extreme vegetative desiccation. Critical Reviews in Plant Sciences 24: 209–225.

Raven, J.A., 1999. The size of cells and organisms in relation to the evolution of embryophytes. Plant Biology 1: 2–12.

Redon, J. and O.L. Lange, 1983. Epiphytische Flechten im Bereich einer chilenischen "Nebeloase" (Fray Jorge) I. Vegetationskundliche Gliederung and Standortsbedingungen. Flora 174: 213–243.

Reynolds, T.L. and J.D. Bewley, 1993a. Absisic acid enhances the ability of the desiccation tolerant fern *Polypodium virginianum* to withstand drying. Journal of Experimental Botany 44: 1771–1779.

Reynolds, T.L. and J.D. Bewley, 1993b. Characterization of protein synthetic changes in a desiccation tolerant fern, *Polypodium virginianum*. Comparison of the effects of drying and rehydration, and absicic acid. Journal of Experimental Botany 44: 921–928.

Richardson, D.H.S., 1981. The Biology of Mosses. Blackwell Scientific, Oxford.

Ried, A., 1960. Thallusbau und Assimilationshaushalt von Laub- und Krustenflechten. Biol. Zentralblatt 79: 129–151.

Rogers, R.W. and R.T. Lange, 1971. Lichen populations on arid soil crusts around sheep watering places in south Australia. Oikos 22: 93–100.

Rosentreter, R., 1993. Vagrant lichens in North America. Bryologist 96: 333–338.

Rundel, P.W., 1982a. The role of morphology in the water relations of desert lichens. Journal of the Hattori Botanical Laboratory 53: 315–320.

Rundel, P.W., 1982b. Water uptake by organs other than roots. In: O.L. Lange, P.S. Nobel, C.B. Osmond, and H. Ziegler, eds. Physiological Plant Ecology, Vol. II. Water Relations and Carbon Assimilation. Springer-Verlag, Berlin, Heidelberg, New York, pp. 111–134.

Rundel, P.W., 1988. Water relations. In: M. Galun, ed. Handbook of Lichenology, Vol. 2. CRC Press, Boca Raton, FL, pp. 17–36.

Rundel, P.W. and O.L. Lange, 1980. Water relations and photosynthetic response of a desert moss. Flora 169: 329–325.

Sancho, L.G. and L. Kappen, 1989. Photosynthesis and water relations and the role of anatomy in Umbilicariaceae (lichens) from central Spain. Oecologia 81: 473–480.

Sancho, L.G., B. Schroeter, and F. Valladares, 1997. Photosynthetic performance of two closely related *Umbilicaria* species in central Spain: Temperature as a key factor. Lichenologist 29: 67–82.

Santarius, K.A., 1967. Das Verhalten von CO_2-Assimilation, NADP- und PGS-Reduktion und ATP-Synthese intakter Blattzellen in Abhängigkeit vom Wassergehalt. Planta 73: 228–242.

Santarius, K.A., 1994. Apoplasmic water fractions and osmotic water potentials at full turgidity of some Bryidae. Planta 193: 32–37.

Scheidegger, C., B. Schroeter, and B. Frey, 1995. Structural and functional processes during water vapor uptake and desiccation in selected lichens with green algal photobionts. Planta 197: 399–409.

Scheidegger, C., B. Frey, and B. Schroeter, 1997. Cellular water uptake, translocation and PSII activation during activation during rehydration of desiccated *Lobaria pulmonaria* and *Nephroma bellum*. Bibliotheca Lichenologica 67: 105–117.

Scherer, S., T.-W. Chen, and P. Böger, 1986. Recovery of adenine-nucleotide pools in terrestrial blue-green algae after prolonged drought periods. Oecologia 68: 585–588.

Schiller, P., H. Heilmeier, and W. Hartung, 1997. Absisic acid (ABA) relations in the aquatic resurrection plant *Chamaegigas intrepidus* under naturally fluctuating environmental conditions. New Phytologist 136: 603–611.

Schiller, P., H. Heilmeier, and W. Hartung, 1998. Uptake of amino acids by the aquatic resurrection plant *Chamaegigas intrepidus* and its implication for N nutrition. Oecologia 117: 63–69.

Schiller, P.H., R. Wolf, and W. Hartung, 1999. A scanning electron microscopical study of hydrated and desiccated submerged leaves of the aquatic resurrection plant *Chamaegigas intrepidus*. Flora 194: 97–102.

Schlensog, M., B. Schroeter, L.G. Sancho, A. Pintado, and L. Kappen, 1997. Effect of strong irradiance and photosynthetic performance of the melt-water dependent cyanobacterial lichen *Leptogium puberulum* Hue (Collemataceae) from the maritime Antarctic. Bibliotheca Lichenologica 67: 235–246.

Schneider, K., B. Wells, E. Schmelzer, F. Salamini, and D. Bartels. 1993. Desiccation leads to the rapid accumulation of both cytosolic and chloroplastic proteins in the resurrection plant *Craterostigma plantagineum* Hochst. Planta 189: 120–131.

Schneider, H., N. Wituba, H.-J. Wagner, F. Thürmer, and U. Zimmermann, 2000. Water rise kinetics in refilling xylem after desiccation in a resurrection plant. New Phytologist 148: 221–238.

Schonbeck, M.W. and J.D. Bewley, 1981a. Responses of the moss *Tortula ruralis* to desiccation treatments. I. Effects on minimum water content and rates of dehydration and rehydration. Canadian Journal of Botany 59: 2698–2706.

Schonbeck, M.W. and J.D. Bewley, 1981b. Responses of the moss *Tortula ruralis* to desiccation treatments. II. Variations in desiccation tolerance. Canadian Journal of Botany 59: 2707–2712.

Schönherr, J., 1982. Resistance of plant surfaces to water loss: Transport properties of cutin, suberin, and associated lipids. In: O.L. Lange, P.S. Nobel, C.B. Osmond, and H. Zeigler, eds. Physiological Plant Ecology, Vol. II. Water Relations and Carbon Assimilation. Springer-Verlag, Berlin, Heidelberg, New York, pp. 153–180.

Schroeter, B., T.G.A. Green, L. Kappen, and R.D. Seppelt, 1994. Carbon dioxide exchange at subzero temperatures. Field measurements on *Umbilicaria aprina* in Antarctica. Cryptogamic Botany 4: 233–241.

Schroeter, B. and C. Scheidegger, 1995. Water relations in lichens at subzero temperatures; structural changes and carbon dioxide exchange in the lichen *Umbilicaria aprina* from continental Antarctic. New Phytologist 131: 273–285.

Schroeter, B., M. Olech, L. Kappen, and W. Heitland, 1995. Ecophysiological investigations of *Usnea antarctica* in the maritime Antarctic. I. Annual microclimatic conditions and potential primary production. Antarctic Science 7: 251–260.

Schroeter, B., L. Kappen, and F. Schulz, 1997. Long-term measurements of microclimatic conditions in the fruticose lichen *Usnea aurantiaco-atra* in the maritime Antarctic. In: J. Cacho and D. Serrat, eds. Actas del V Simposio de Estudios Antárcticos. CICYT, Madrid, pp. 63–69.

Schwab, K.B. and D.F. Gaff, 1986. Sugar and ion contents in leaf tissues of several drought tolerant plants under water stress. Journal of Plant Physiology 125: 257–265.

Schwab, K.B. and U. Heber, 1984. Thylakoid membrane stability in drought tolerant and drought sensitive plants. Planta 161: 37–45.

Schwab, K.B., U. Schreiber, and U. Heber, 1989. Response of photosynthesis and respiration of resurrection plants to desiccation and rehydration. Plant 177: 217–227.

Scott, G.A.M., 1982. Desert bryophytes. In: A.J.E. Smith, ed. Bryophyte Ecology. Chapman and Hall, London, New York, pp. 105–122.

Scott, P., 2000. Resurrection plants and the secrets of eternal leaf. Annals of Botany 85: 159–166.

Seel, W.E., G.A.F. Hendry, and J.A. Lee, 1992a. The combined effects of desiccation and irradiance on mosses from xeric and hydric habitats. Journal of Experimental Botany 43: 1023–1030.

Seel, W.E., N.R. Baker, and J.A. Lee, 1992b. Analysis of the decrease in photosynthesis on desiccation of mosses from xeric and hydric environments. Physiologia Plantarum 86: 451–458.

Sgherri, C.L.M., B. Loggini, S. Puliga, and F. Navari-Izzo, 1994a. Antioxidant system in *Sporobolus stapfianus*: changes in response to desiccation and rehydration. Phytochemistry 35: 561–565.

Sgherri, C.L.M., B. Loggini, A. Bochicchio, and F. Navari-Izzo, 1994b. Antioxidant system in *Boea hygroscopica*: Changes in response to desiccation and rehydration. Phytochemistry 37: 377–381.

Sgherri, G., B. Stevanivic, and F. Navarri-Izzo, 2004. Role of phenolics in the antioxidative status of the resurrection plant *Ramonda serbica* during dehydration and rehydration. Physiologia Plantarum 122: 478–485.

Sherwin, H.W. and J.M. Farrant, 1996. Differences in rehydration of three desiccation-tolerant angiosperm species. Annals of Botany 78: 703–710.

Sherwin, H.W. and J.M. Farrant, 1998. Protection mechanisms against excess light in the resurrection plants *Craterostigma wilmsii* and *Xerophyta viscosa*. Plant Growth Regulation 24: 203–210.

Sherwin, H.W., N. Pammenter, E. February, C. van der Willigen, and C. Farrant, 1998. Xylem hydraulic characteristics, water relations and wood anatomy of the resurrection plant *Myrothamnus flabellifolius* Welw. Annals of Botany 81: 567–575.

Shields, L.M., 1957. Algal and lichen floras in relation to nitrogen content of certain volcanic and arid range soils. Ecology 38: 661–663.

Shuvalov, V.A. and U. Heber, 2003. Photochemical reactions in dehydrated photosynthetic organisms, leaves, chloroplasts and photosystem II particles: Reversible reduction of pheophytin and chlorophyll and oxidation of ß-carotene. Chemical Physics 294: 227–237.

Smirnoff, N., 1993. Tansley review No. 52: The role of active oxygen in the response of plants to water deficit and desiccation. New Phytologist 125: 27–58.

Smirnoff, N., 1995. Antioxidant systems and plant response to the environment. In: Smirnoff, N., ed. Environment and Plant Metabolism, Flexibility and Acclimation. Bios Scientific Publishers, Oxford, pp. 217–244.

Smith, D.C. and S. Molesworth, 1973. Lichen physiology. XIII. Effects of rewetting dry lichens. New Phytologist 72: 525–533.

Smith-Espinoza, C.J., J.R. Phillips, F. Salamini, and D. Bartels, 2005. Identification of further *Craterostigma plantagineum* cdt mutants affected in abscisic acid mediated desiccation tolerance. Molecular Genetics and Genomics 274: 364–372.

Smook, L., 1969. Some observations on *Lindernia intrepidus* (Dinter) Oberm. (= *Chamaegigas intrepidas* Dinter). Dinteria 2: 13–21.

Snelgar, W.P. and T.G.A. Green, 1981. Ecologically-linked variation in morphology, acetylene reduction, and water relations in *Pseudocyphellaria dissimilis*. New Phytologist 87: 403–411.

Sojo, F., F. Valladares, and L.G. Sancho, 1997. Structural and physiological plasticity of the lichen *Catillaria corymbosa* in different microhabitats of the maritime Antarctic. Bryologist 100: 171–179.

Solhaug, K.A. and Y. Gauslaa, 1996. Parietin, a photoprotective secondary product of the lichen *Xanthoria parietina*. Oecologia 108: 412–418.

Solhaug, K.A., Y. Gauslaa, L. Nybakken, and W. Bilger, 2003. UV-induction of sun-screening pigments in lichens. New Phytologist 158: 91–100.

Sperry, J.S. and M.T. Tyree 1988. Mechanism of water-stress induced xylem embolism. Plant Physiology 88: 581–587.

Stålfelt, M.G., 1939. Vom System der Wasserversorgung abhängige Stoffwechselcharaktere. Botaniska Notiser 40: 176–192.

Stewart, G.R., 1989. Desiccation injury, anhydrobiosis and survival. In: H.G. Jones, F.J. Flowers, and M.B. Jones, eds. Plant Under Stress. Cambridge University Press, Cambridge, pp. 115–130.

Stocker, O. and W. Holdheide, 1938. Die Assimilation Helgoländer Gezeitenalgen während der Ebbezeit. Zeitschrift Botany 32: 1–59.

Stuart, T.S. 1968. Revival of respiration and photosynthesis in dried leaves of *Polypodium polypodioides*. Planta 83: 185–206.

Sundberg, B.A. Ekblad, T. Nasholm, and K. Palmqvist, 1999. Lichen respiration in relation to active time, temperature, nitrogen, and ergosterol concentrations. Functional Ecology 13: 119–125.

Tegler, B. and K.A. Kershaw, 1981. Physiological-environmental interactions in lichens. XII. The variation of the heat stress response of *Cladonia rangiferina*. New Phytologist 87: 395–401.

Tennant, J.R., 1954. Some preliminary observations on the water relations of a mesophytic moss, *Thamnium alopecurum* (Hedw.) B. & S. Transactions of the British Bryological Society 2: 439–445.

Trachtenberg, S. and E. Zamski, 1979. The apoplastic conduction of water in *Polytrichum juniperinum* Willed. gametophytes. New Phytologist 83: 49–52.

Tuba, Z., 1987. Light, temperature and desiccation responses of CO_2-exchange in the desiccation-tolerant moss, *Tortula ruralis*. Symposia Biologica Hungarica 35: 137–149.

Tuba, Z., H.K. Lichtenthaler, Z. Csintalan, and T. Pocs, 1993a. Regreening of desiccated leaves of the poikilochlorophyllous *Xerophyta scabrida* upon rehydration. Journal of Plant Physiology 142: 103–108.

Tuba, Z., H.K. Lichtenthaler, I. Maroti, and Z. Csintalan, 1993b. Resynthesis of thylakoids and chloroplast ultrastructure in the desiccated leaves of the poikilochlorophyllous plant *Xerophyta scabrida* upon rehydration. Journal of Plant Physiology 142: 742–748.

Tuba, Z., H.K. Lichtenthaler, Z. Csintalan, Z. Nagy, and K. Szente, 1994. Reconstitution of chlorophylls and photosynthetic CO_2 assimilation upon rehydration of the desiccated poikilochlorophyllous plant *Xerophyta scabrida* (Pax) Th. Dur. et Schinz. Planta 192: 414–420.

Tuba, Z., Z. Csintalan, and M.C.F. Proctor, 1996a. Photosynthetic responses of a moss, *Tortula ruralis*, ssp, and the lichens *Cladonia convoluta* and *C. furcata* to water deficit and short periods of desiccation, and their ecophysiological significance: a baseline study at present-day CO_2 concentration. New Phytologist 133: 353–361.

Tuba, Z., H.K. Lichtenthaler, Z. Csintalan, Z. Nagy, and K. Szente. 1996b. Loss of chlorophylls, cessation of photosynthetic CO_2 assimilation and respiration in the poikilochlorophyllous plant *Xerophyta scabrida* during desiccation. Physiologia Plantarum 96: 383–388.

Tuba, Z., N. Smirnoff, Z. Csintalan, K. Szente, and Z. Nagy, 1997. Respiration during slow desiccation of the poikilochlorophyllous desiccation tolerant plant *Xerophyta scabrida* at present-day CO_2 concentration. Plant Physiology and Biochemistry 35: 381–386.

Tymms, M.J., D.F. Gaff, and N.D. Hallam, 1982. Protein synthesis in the desiccation tolerant angiosperm *Xerophyta villosa* during dehydration. Journal of Experimental Botany 33: 332–343.

Tyree, M.T. and J.S. Sperry, 1988. Do woody plants operate near the point of catastrophic xylem dysfunction caused by dynamic water stress? Plant Physiology 88: 574–580.

Valanne, N., 1984, Photosynthesis and photosynthetic products in mosses. In: A.F. Dyer, and J.G. Duckett, eds. The Experimental Biology of Bryophytes. Academic Press, London, pp. 257–273.

Valladares, F., 1994a. Texture and hygroscopic features of the upper surface of the thallus in the lichen family Umbilicariaceae. Annals of Botany 73: 493–500.

Valladares, F., 1994b. Form-functional trends in Spanish Umbilicariaceae with special reference to water relations. Cryptogamie, Bryologie et Lichénologie 15: 117–127.

Valladares, F. and L.G. Sancho, 1995. Medullary structure of Umbilicariaceae. Lichenologist 27: 189–199.

Valladares, F., J. Wierzchos, and C. Ascaso, 1993. Porosimetric study of the lichen family Umbilicariaceae: anatomical interpretation and implications for water storage capacity of the thallus. American Journal of Botany 80: 263–272.

Valladares, F., C. Ascaso, and L.G. Sancho, 1994. Intrathalline variability of some structural and physical parameters in the lichen genus *Lasallia*. Canadian Journal of Botany 72: 415–428.

Valladares, F., A. Sanchez-Hoyos, and E. Manrique, 1995. Diurnal changes in photosynthetic efficiency and carotenoid composition of the lichen *Anaptychia ciliaris*: effects of hydration and light intensity. Bryologist 98: 375–382.

Valladares, F., L.G. Sancho, and C. Ascaso. 1996. Functional analysis of the intrathalline and intracellular chlorophyll concentration in the lichen family Umbilicariaceae. Annals of Botany 78: 471–477.

Valladares, F., L.G. Sancho, and C. Ascaso, 1998. Water storage in the lichen family Umbilicariaceae. Botanica Acta 111: 1–9.

Van der Willigen, C., C.N.W. Pammenter, S. Mundree, and J. Farrant, 2001. Some physiological comparisons between the resurrection grass *Eragrostis nindensis* and the related desiccation-sensitive *E. curvula*. Plant Growth Regulation 35: 121–129.

Vecchia, F.D., T. El Asmar, R. Calamassi, N. Rascio, and C. Vazzana, 1998. Morphological and ultrastructural aspects of dehydration and rehydration in leaves of *Sporobolus stapfianus*. Plant Growth Regulation 24: 219–228.

Vieweg, G.H. and H. Ziegler, 1969. Zur Physiology von *Myrothamnus flabellifolia*. Berichte der Deutschen Botanischen Gesellschaft 82: 29–36.

Volk, O.H., 1979. Beiträge zur Kenntnis der Lebermoose (Marchantiales) aus Südwestafrika (Namibia). I. Mitteilungen der Botanischen Staatssammlungen München 15: 223–242.

Volk, O.H., 1984. Pflanzenvergesellschaftungen mit *Riccia*-Arten in Südwestafrika (Namibia). Vegetatio 55: 57–64.

Volk, O.H. and H. Leippert, 1971. Vegetationsverhältnisse in Windhoeker Bergland, Südwestafrika. Journal der Südwestafrikanischen wissenschaftlichen Gesellschaft 25: 5–44.

Wagner, D.J. and J.E.Titus, 1984. Comparative desiccation tolerance of two *Sphagnum* mosses. Oecologia 62: 182–187.

Walter, H., 1931. Die Hydratur der Pflanze. Gustav Fischer-Verlag, Jena.

Walter, H. and K. Kreeb, 1970. Die Hydration und Hydratur des Protoplasmas der Pflanzen und ihre ökophysiologische Bedeutung. Springer-Verlag, Wien, New York.

Webber, P.J., 1978. Spatial and temporal variation of the vegetation and its production, Barrow, Alaska. In: L.L. Tieszen, ed. Vegetation and Production Ecology of and Alaskan Arctic Tundra. Ecological Studies 29. Springer-Verlag, New York, Heidelberg, Berlin, pp. 37–112.

Weissman, L., J. Garty, and A. Hochman, 2005. Rehydration of the lichen *Ramalina lacera* results in production of reactive oxygen species and nitric oxide and a decrease in antioxidants. Applied and Environmental Microbiology 71: 2121–2129.

Wellburn, F.A.M. and A.R. Wellburn, 1976. Novel chloroplasts and unusual cellular ultrastructure in the "resurrection" plant *Myrothamnus flabellifolia* Welw. (Myrothamnanceae). Botanical Journal of the Linnean Society 72: 51–54.

Winter, K. and M. Königer, 1989. Dithiothreitol, an inhibitor of violaxanthin de-epoxidation, increases the susceptibility of leaves of *Nerium oleander* L. to photoinhibition of photosynthesis. Planta 180: 24–31.

Wirth, V., 1987. Die Flechten Baden-Württembergs, Verbreitungsatlas. Eugen Ulmer Verlag, Stuttgart.

Woitke, M., W. Hartung, H. Gimmler, and H. Heilmeier, 2004. Chlorophyll fluorescence of submerged and floating leaves of the aquatic resurrection plant *Chamaegigas intrepidus*. Functional Plant Biology 31: 53–62.

Yang, W.L., Z.A. Hu, and T.Y. Kuang, 2003. Photosynthesis of resurrection angiosperms. Acta Botanica Sinica 45: 505–508.

Zamski, E. and S. Trachtenberg, 1976. Water movement through hydroids of a moss gametophyte. Israel Journal of Botany 25: 168–173.

Zimmermann, G. and H. Butin, 1990. Untersuchungen über die Hitze- und Trockenresistenz Holz bewohnender Pilze. Flora 162: 393–419.

Zimmermann, M.H. and J.A. Milburn, 1982. Transport and storage of water. In: O.L. Lange, P.S. Nobel, C.B. Osmond, and H. Ziegler, eds. Physiological Plant Ecology, Vol. II. Water Relations and Carbon Assimilation. Springer-Verlag, Berlin, Heidelberg, New York, pp. 135–152.

Zirkovic, T., M.F. Quartacci, B. Stevanovic, F. Marinone, and F. Navari-Izzo, 2005. Low molecular weight substances in the poikilohydric plant *Ramonda serbica* during dehydration and rehydration. Plant Science 168: 105–111.

Zotz, G., A. Schweikert, W. Jetz, and H. Westerman, 2000. Water relations and carbon gain are closely related to cushion size in the moss *Grimmia pulvinata*. New Phytologist 148: 59–67.

3 Ecological Significance of Inherent Variation in Relative Growth Rate and Its Components

Hendrik Poorter and Eric Garnier

CONTENTS

INTRODUCTION

An amazing number of higher plant species are present on earth, with estimations greater than 250,000 (e.g., Wilson 1992). These species are not randomly distributed, but often can be found in rather specific habitats. What is the reason that some species flourish in a desert and others in the tundra? Clearly, a certain degree of specialization must have taken place. It is the one major aim of functional ecology to explain the distribution of species (or genotypes within a species) from their functional attributes. These attributes can be related to the physiological, morphological, anatomical, and chemical characteristics of a plant species, but they could also depend on life history characteristics such as seed longevity, flowering time, life form, and so on. Since the question is why an individual of species A performs better in a given environment than an individual of species B, a comparative approach is needed (Bradshaw 1987). The strength of such an approach is that we not only gain insight into the processes that determine a plant's success or failure in a given habitat, but it also enables us to categorize the wide variety of species into a more limited number of functional groups. This may be an avenue toward simplification of a complex reality, with a great number of species in a given habitat. On the basis of functional groups we are probably better able to predict effects of environmental changes on vegetations (Hobbs 1997).

This chapter uses the comparative approach to analyze the characteristics and distribution of species varying in the maximum relative growth rate (RGR_{max}; see the next section for a definition and Table 3.1 for a listing of abbreviations and units used throughout this chapter) they can achieve. Plant species grown under uniform and more or less optimum conditions in the laboratory differ several-fold in RGR_{max} (100–400 mg g^{-1} day^{-1} for herbaceous species; 10–150 mg g^{-1} day^{-1} for woody species). Over the last four decades, evidence has accumulated that RGR_{max} is linked to the characteristics of the habitat from which the species originated (Parsons 1968, Chapin 1980, Lambers and Poorter 1992, discussed on pp. 73–76). This linkage is intriguing, and leads to a number of questions, which are addressed in this chapter. After introducing the different analytical concepts that are used and providing evidence of a relationship between RGR_{max} and habitat characteristics, the physiological, morphological, and anatomical attributes that lead to variation in RGR_{max} between

TABLE 3.1
Terms, Abbreviations, and Units Used in This Chapter

Abbreviation	Meaning	Preferred Units	Normal Range
LAR	Leaf area ratio	m^2 leaf kg^{-1} plant	5–50
LD	Leaf density	kg m^{-3}	100–400
LTh	Leaf thickness	μm	100–600
LMF	Leaf mass fraction	g leaf g^{-1} plant	0.15–0.60
LR_m	Rate of leaf respiration	nmol CO_2 g^{-1} leaf s^{-1}	5–50
MRT	Mean residence time	Year	0.2–20
NUR	Net nitrogen uptake rate	mmol N g^{-1} root day^{-1}	1–8
NP	Nitrogen productivity	g increase mol^{-1} N day^{-1}	20–100
NUE	Nitrogen use efficiency	kg increase mol^{-1} N taken up or lost	0.5–15
PCC	[C] in the plant	mmol C g^{-1} plant	25–33
PNC	[N] in the plant	mmol N g^{-1} plant	1–4
PS_a	Rate of photosynthesis	μmol CO_2 m^{-2} leaf s^{-1}	3–30
PS_m	Rate of photosynthesis	nmol CO_2 g^{-1} leaf s^{-1}	100–600
RGR	Relative growth rate	mg increase g^{-1} plant day^{-1}	40–400
RMF	Root mass fraction	g root g^{-1} plant	0.15–0.60
RR_m	Rate of root respiration	nmol CO_2 g^{-1} root s^{-1}	10–80
SLA	Specific leaf area	m^2 leaf kg^{-1} leaf	8–80
SMF	Stem mass fraction	g stem g^{-1} plant	0.05–0.30
SR_m	Rate of stem respiration	nmol CO_2 g^{-1} stem s^{-1}	?
ULR	Unit leaf rate	g increase m^{-2} leaf day^{-1}	2–20

Note: Normal ranges found in herbaceous species. All mass-based parameters are expressed per unit dry mass.

species are explored. We show that, whatever the evolutionary forces have been, fast- and slow-growing species grown under laboratory conditions show consistent suites of traits. To have an ecological meaning, these sets of traits should not only be found in the laboratory, but also in the field. After testing whether this is indeed the case, we analyze the ecological implications of interspecific differences in growth rate and of variation in the underlying parameters. This leads us to suggest that selection has acted on components of RGR, rather than on RGR itself.

ASSESSING THE GROWTH POTENTIAL OF A SPECIES

This section focuses on the analysis of inherent differences in RGR (Box 3.1). The concept of RGR was first introduced by Blackman (1919), who recognized that the increase in plant biomass over a given period of time was proportional to the biomass present at the beginning of this period. He saw a parallel with money in a bank account accumulating at compound interest. In the case of plants, newly formed biomass is immediately deployed to fix new carbon and take up extra nutrients and water, thus leading to an accelerated biomass increase. Borrowing from economic theory, he derived the following equation:

BOX 3.1
Relative Growth Rate

The problem of how to express the growth rate of a plant can best be illustrated by the following example. Suppose there are two plants, A and B, whose dry masses are 0.1 and 1.0 g, respectively. Given that both increase in biomass by 0.1 g in 24 h, can they be considered to grow at the same rate? One way to express growth is to consider the absolute growth rate (AGR), which is defined as the increase in plant mass M over a period of time t

$$\text{AGR} = \frac{\mathrm{d}M}{\mathrm{d}t}. \qquad (3.1)$$

Plant mass M_2 at time t_2 can be calculated for a given AGR when mass M_1 at time t_1 is known

$$M_2 = M_1\text{AGR}(t_2 - t_1). \qquad (3.2)$$

In the above, example plants A and B have the same AGR. However, A achieved this increase with far less starting material than B. To take this into account, the rate of biomass increase can be defined relative to the mass of the plant already present. This is the RGR

$$\text{RGR} = \frac{1}{M}\frac{\mathrm{d}M}{\mathrm{d}t}. \qquad (3.3)$$

For a given RGR and plant mass M_1 at time t_1, M_2 at time t_2 can be calculated by

$$M_2 = M_1 e^{\text{RGR}(t_2 - t_1)}. \qquad (3.4)$$

By taking the natural logarithm of both sides of Equation 3.4, and a little rearranging, we obtain a formula by which RGR can be calculated from experimental data

$$\text{RGR} = \frac{\ln M_2 - \ln M_1}{t_2 - t_1}. \qquad (3.5)$$

(*continued*)

BOX 3.1 (continued)
Relative Growth Rate

Note that RGR can be interpreted graphically as the slope of the line that connects the ln-transformed dry masses of plants at several harvesting times (Figure 3.1b). Because of the ln-transformation, the numerator has no dimension; therefore RGR has the unit 'day^{-1}'. As RGR values were based on plant masses, rather than on leaf area or another parameter of plant size, units could be expressed as g g^{-1} day^{-1} (gram increase per gram dry mass present and per unit of time). These numbers are generally low, and therefore we use 'mg g^{-1} day^{-1}' as the unit of expression.

In the previously given example, plant A has an RGR of 693 mg g^{-1} day^{-1}, whereas B has an RGR of 95 mg g^{-1} day^{-1}. We thus conclude that plant A is more efficient in terms of growth than B. (It may be surprising at first sight that plant A, that doubles its mass in 24 h, has an RGR less than 1000 mg g^{-1} day^{-1}. The reason for this is that the assumption underlying Equation 3.4 is that extra biomass, produced during the beginning of the day, is immediately deployed to fix new C and minerals, leading to compounding growth.)

There are two additional points to which we want to draw attention. First, the dry mass of the whole plant is generally taken as the basis for the RGR calculation. However, RGRs have also been calculated on the basis of fresh weight, shoot weight, leaf area, or leaf number. As long as plant growth is in steady state, that is, as long as there are no changes in the dry mass: fresh mass ratio, allocation, morphology, leaf size, and so on, the RGR values expressed in several ways should be equal. Second, plants with a low RGR can still achieve a large biomass when they start with a high seed mass or grow over prolonged periods of time (see Equation 3.4). For a discussion on various approaches for the experimental design in growth analysis and details on calculations see Causton and Venus (1981), Hunt (1982), Poorter and Garnier (1996), Hoffmann and Poorter (2002).

$$M_2 = M_1 e^{\mathrm{RGR}(t_2 - t_1)},$$

where M_1 and M_2 are the plant masses at time t_1 and t_2, respectively. The RGR in this equation indicates the dry mass increment per unit dry mass, which is already present in the plant per unit time. For a more detailed discussion on the background of RGR see Box 3.1.

Originally, Blackman thought of RGR as a physiological constant, which would be characteristic for a given species under given conditions. However, a constant RGR implies that plants grow exponentially throughout their life (Figure 3.1). In reality, plants hardly ever show a true exponential growth phase, as RGR changes continuously with ontogeny (Hunt and Lloyd 1987, Robinson 1991, Poorter and Pothmann 1992). During germination there is a gradual transition from growth dependent on seed reserves to complete autotrophy. When plants get older and larger, the upper leaves start to shade lower leaves. Moreover, larger plants have to allocate more resources away from the assimilating parts of leaves and roots and invest more in support tissue, especially in stems. Consequently, RGR decreases with size and time (Figure 3.1). Does this imply that the concept of RGR can only be used in the seedling stage, during what is often termed the "exponential growth phase?" Mathematically, there is no requirement for RGR to be constant, because it is a parameter that can be used as a quantification of growth at any point in time, even if growth is not strictly exponential. It can also be used as an average over a given time period (Evans 1972) or a given mass trajectory. Therefore, as long as one is convinced that the growth of the plants under study is somehow proportional to the plant biomass already present, RGR is the most appropriate parameter to use. However, it is not a parameter fully independent of plant size!

In the field, where plants experience a fluctuating environment, growth is restricted by a continuously changing array of abiotic factors (light, temperature, nutrients, and water) and affected by biotic interactions (competitors, herbivores, pathogens, but also

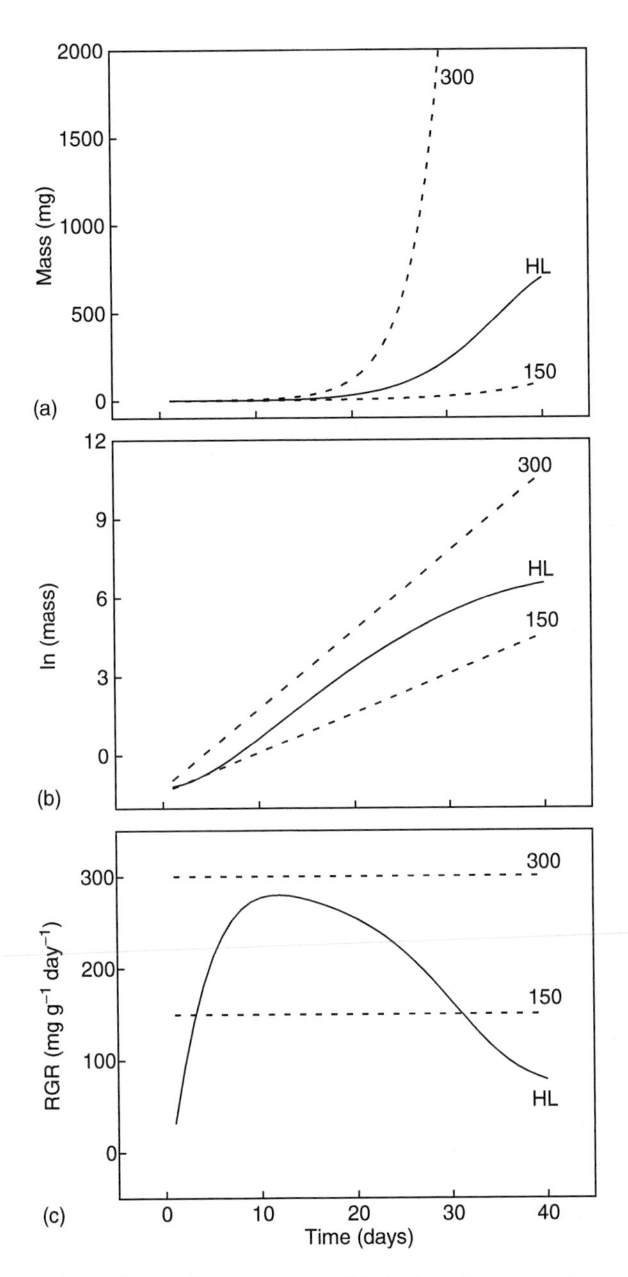

FIGURE 3.1 Time course in (a) total plant mass, (b) ln-transformed values of total plant mass, and (c) RGR of a theoretical plant population growing continuously with an RGR of 150 or 300 (dashed lines) mg g^{-1} day^{-1}, and experimental data on a population of *Holcus lanatus*. (Continuous line marked HL; adapted from Hunt, R. and Lloyd, P.S., *New Phytol.*, 106, 235, 1987.) All populations had a similar starting mass at day 0.

symbionts). In comparing species (or genotypes) it is of interest to know their genetic potential or growth achieved in the absence of constraining factors. Such a goal is difficult to achieve. It would require knowledge about the exact combination of factors that enables fastest growth for each species. Even if such a goal could be technically achieved, it would have the drawback of comparing species that had been grown in more or less different environments. The practical solution has been to choose a set of conditions that is close to optimal for growth of most species and technically achievable (Grime and Hunt 1975). Growth rate is then measured for relatively small and young plants over rather short time intervals (10–20 days), and without interference from other plants. The RGR value obtained in this way is considered to be RGR_{max}. These values are not absolute, because they depend on ontogeny as well as growth conditions. However, with the exception of very low nutrient levels (Shipley and Keddy 1988) or light levels (Mahmoud and Grime 1974), RGR ranking remains rather similar (see Poorter et al. 1995, Biere et al. 1996 for nutrients; Hunt and Cornelissen 1997, Poorter and Van der Werf 1998 for light). Therefore, ranking of species for RGR_{max} does not change strongly across experiments and can be used in a relative way to order species on the fast–slow continuum. However, there is variability in such relative rankings across experiments, with correlation coefficients approximately 0.6 (Table 3.2). Part of this variation is probably caused by imprecisions related to RGR determinations, especially in larger screening programs with a limited number of plants harvested per species (Poorter and Garnier 1996).

What do RGR_{max} values obtained in the laboratory tell about plant growth in the field? With respect to light, field-grown plants generally experience stronger fluctuations in instantaneous irradiance, and higher levels of total quantum input, when considered over the whole growing season (Garnier and Freijsen 1994). However, RGR does not strongly depend on the total daily quantum input above 20 mol m^{-2} day^{-1} (Poorter and Van der Werf 1998), a value quite often reached in growth chambers. Temperature is often lower in the field, especially during vegetative growth in temperate climates. With respect to nutrients, conditions are generally far more limiting in the field. Moreover, plants in the field encounter competition

TABLE 3.2
Correlation Coefficients

	G75	P89	P and O	H97	V98
G75	–	0.77	0.55	0.66	0.57
P89	8	–	0.98	–	–
P and O	23	7	–	0.73	0.76
H97	30	–	13	–	0.62
V98	33	–	13	13	–

Sources: Data are from Grime, J.P. and Hunt, R., *J. Ecol.*, 63, 393, 1975. (G75; 130 species), Poorter, H., *Causes and Consequences of Variation in Growth Rate and Productivity of Higher Plants*, H. Lambers, M.L. Cambridge, H. Konings, and T.L. Pons, eds, SPB Academic Publishing, The Hague, 1989 (P89; 9 species), Poorter, H. and Remkes, C., *Oecologia*, 83, 553, 1990 supplemented with data from Van der Werf, A., van Nuenen, M., Visser, A.J., and Lambers, H., *Physiol. Plant.*, 89, 563, 1993; Den Dubbelden, K.C. and Verburg, R.W., *Plant Soil*, 184, 341, 1996 and van Arendonk, J.J.C.M. (unpublished results), which were all grown under identical conditions (P&O; 47 species), Hunt, R. and Cornelissen, J.H.C., *New Phytol.*, 135, 395, 1997 (H97; 43 species), and from Van der Werf, A., Geerts, R.H.E.M., Jacobs, F.H.H., Korevaar, H., Oomes, M.J.M., and de Visser, W., *Inherent Variation in Plant Growth. Physiological Mechanisms and Ecological Consequences*, H. Lambers, H. Poorter, and M. van Vuuren, eds, Backhuys Publishers, Leiden, 1998 (V98; 71 species).

Note: RGR_{max} values of herbaceous species shared by some larger-scale comparative experiments (*upper right part*) and number of species in common on which the correlation coefficient is based (*lower left part*). Only those correlations are given when seven or more species were in common.

and other biotic interactions. Consequently, there is a large difference between the RGR_{max}, as measured in the lab, and RGR of plants in the field (reviewed by Garnier and Freijsen 1994, see also Villar 1998, 2005). Although we expect some relationship between the relative ranking in laboratory and field, data are too scarce for a proper evaluation. In this chapter, we treat RGR_{max} more as a representation of a complex of traits than as a predictor of growth rate under natural conditions. This aspect is discussed further in the sections on pages 73–76 and 84–90.

RGR_{MAX} AND PLANT ECOLOGY

RGR_{MAX} AND PLANT DISTRIBUTION

Bradshaw et al. (1964) were among the first to establish a relationship between the RGR_{max} of wild species, as measured under laboratory conditions, and the characteristics of the habitat they originated from. Others followed, but generally the number of species was rather low (less than 10) to infer strong conclusions. Grime and Hunt (1975) determined RGR_{max} of 130 species from England and classified them according to habitat. Fast-growing species were found relatively more often in fertile habitats, whereas species with a low potential growth rate tended to occupy infertile habitats. It is not always easy and straightforward to quantitatively classify a specific habitat along a fertility scale. A semiquantitative approach has been used by Ellenberg, who assigned so-called "*N*-numbers" to a wide range of species from Central Europe (Ellenberg 1988). The higher the value, the higher the fertility of the habitats in which such a species would generally occur. Plotting RGR_{max} data of herbaceous perennials against the *N*-number of Ellenberg generally yields positive relationships (Figure 3.2a). A positive relation between RGR_{max} and nutrient availability is also likely for woody species (Cornelissen et al. 1998). However, annuals seem to have a high RGR_{max} independent of soil fertility (Fichtner and Schulze 1992).

Is inherent variation in potential RGR of species also related to other environmental gradients? Evidence is less well-documented than in the case of nutrient availability. Alpine species have lower RGR_{max} under laboratory conditions than lowland species (Figure 3.2b). Dry (Rozijn and Van der Werf 1986, Figure 3.2c) or saline habitats (Figure 3.2d) harbor species that grow more slowly under optimal conditions, and the same relation between RGR_{max} and plant occurrence was found for sites with heavy metal pollution (Wilson 1988, Verkleij and Prast 1989). Disturbance regime may also be a source of variation in RGR_{max}: annuals that have to complete their life cycle in periodically disturbed habitats, display a higher RGR_{max} than congeneric perennials from more stable habitats (Garnier 1992), and species from early stages of secondary succession tend to have higher RGR_{max} than those from more advanced stages (Gleeson and Tilman 1994, Vile et al. 2006). The intensity of trampling has also been identified as selecting for species with different RGR_{max}: plants from trampled places have a lower RGR_{max} than those from nontrampled sites (Figure 3.2e). Finally, when determined at relatively high-light levels, species from strongly shaded habitats have a lower RGR than those from light-exposed environments. In most of the cases shown in Figure 3.2b through Figure 3.2f, however, it seems that differences in RGR_{max} between species from favorable and unfavorable habitats are not as clear as in the case of species adapted to habitats differing in fertility.

RGR_{MAX} AND ECOSYSTEM PRODUCTIVITY

Most of this chapter focuses on the comparison between plant species. However, in recent years there has been much attention for ecosystem functioning and a possible link with species composition. The fact that inherently fast-growing species are more often found in

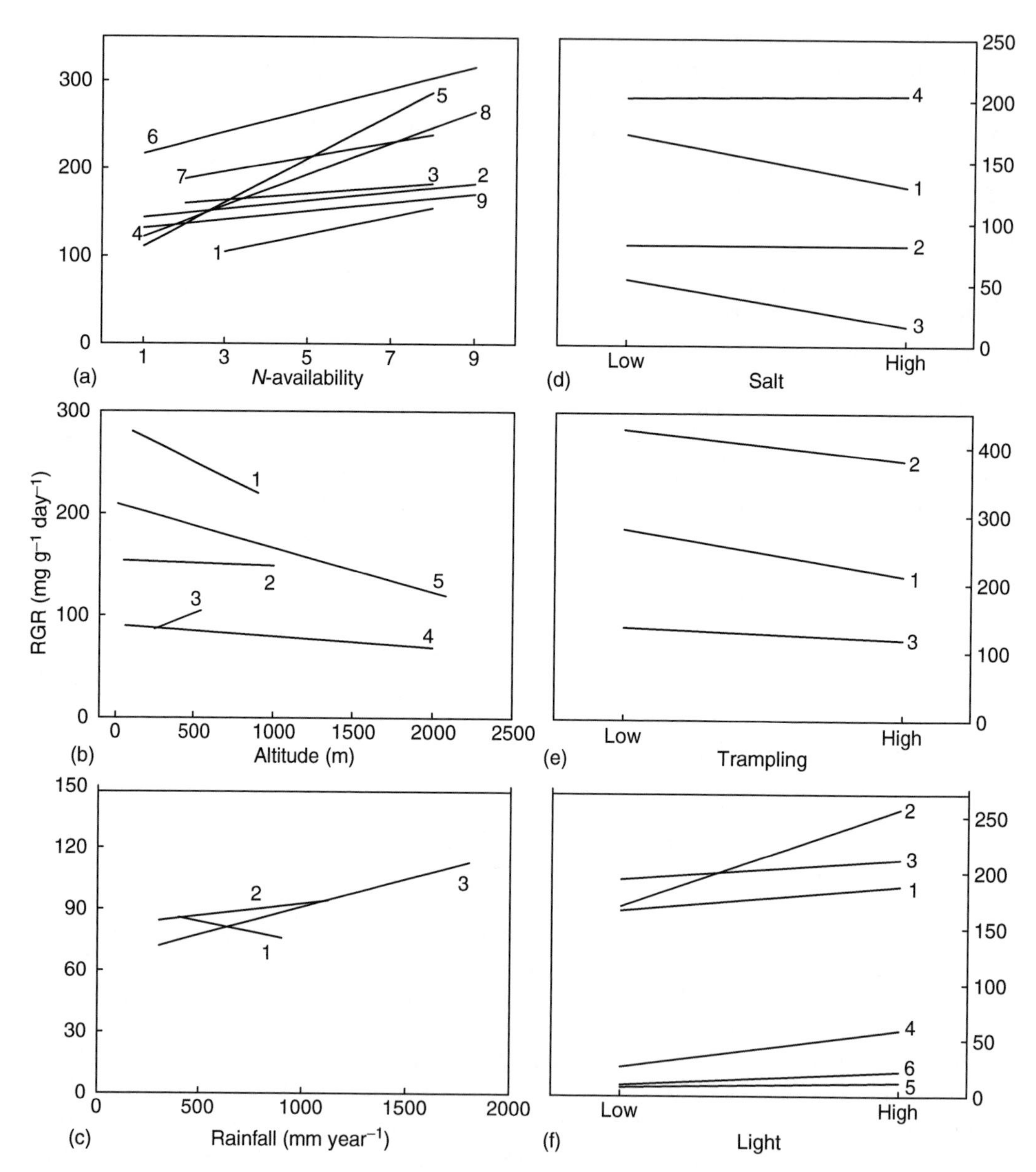

FIGURE 3.2 RGR_{max} of species originating from habitats differing in: (a) Nutrient availability, as indicated by the *N*-number of Ellenberg; (b) Altitude, in meters above sea level; (c) Rainfall; (d) Salt concentration; (e) Trampling intensity; and (f) Light intensity. Each study is represented by a regression line. Numbers refer to the following studies: A: data from (1) Rorison, I.H., *New Phytol.*, 67, 913, 1968; (2) Grime, J.P. and Hunt, R., *J. Ecol.*, 63, 393,1975; (3) Boorman, L.A., *J. Ecol.*, 70, 607, 1982; (4) Poorter, H., *Causes and Consequences of Variation in Growth Rate and Productivity of Higher Plants*, H. Lambers, M.L. Cambridge, H. Konings, and T.L. Pons, eds, SPB Academic Publishing, The Hague, 1989; (5) Poorter, H. and Remkes, C., *Oecologia*, 83, 553, 1990; (6) Reiling, K. and Davison, A.W., *New Phytol.*, 120, 29, 1992; (7) Stockey, A. and Hunt, R., *J. Appl. Ecol.*, 31, 543, 1994; (8) Van der Werf, A., Geerts, R.H.E.M., Jacobs, F.H.H., Korevaar, H., Oomes, M.J.M., and de Visser, W., *Inherent Variation in Plant Growth. Physiological Mechanisms and Ecological Consequences*, H. Lambers, H. Poorter, and M. van Vuuren, eds, Backhuys Publishers, Leiden, 1998; and (9) Schippers P. and Olff, H., *Plant Ecol.*, 149, 219, 2000; B: data from (1) Woodward, F.I., *New Phytol.*, 82, 385, 1979; (2) Woodward, F.I., *New Phytol.*, 96, 313, 1983; (3) Graves, J.D. and Taylor, K., *New Phytol.*, 104, 681, 1986; (4) Atkin, O.K. and Day, D.A., *Aust. J. Plant Physiol.*, 17, 517, 1990; and

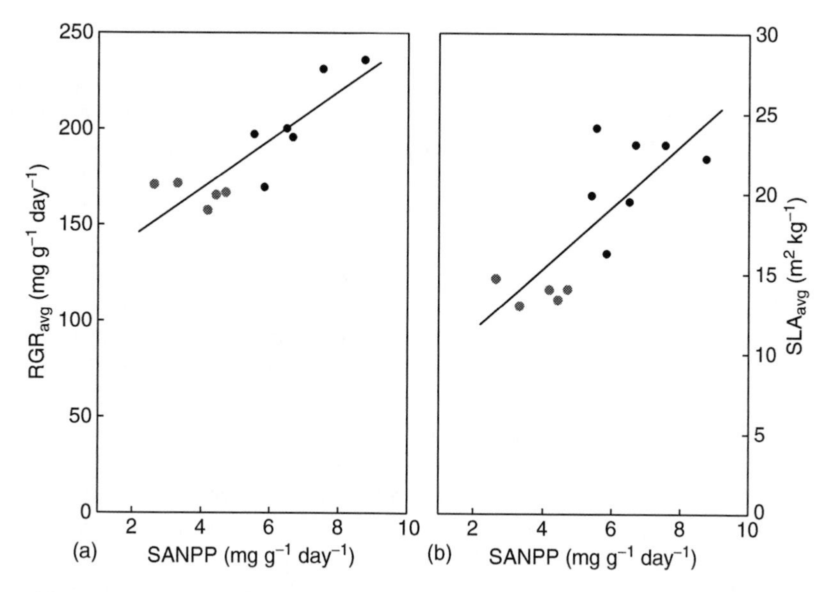

FIGURE 3.3 (a) Aggregated RGR_{max} and (b) aggregated SLA of plant species grown in the field plotted as a function of the specific aboveground net primary productivity (aboveground RGR) of 12 abandoned vineyards at different stages of secondary succession. Laboratory-derived data of the species present in the succession were averaged according to the proportion of biomass they represented in the field. (From Vile, D., Shipley, B., and Garnier, E., *Ecol. Lett.*, 9, 1061, 2006 and Garnier, E., Cortez, J., Billès, G., Navas, M.-L., Roumet, C., Debussche, M., Laurent, G., Blanchard, A., Aubry, D., Bellman, A., Neill, C., and Toussaint, J.-P., *Ecology*, 85, 2630, 2004.)

nitrogen-rich habitats (sensu Ellenberg 1988), and nitrogen-rich habitats often show a higher productivity, makes it likely that there is in this case a positive correlation between ecosystem behavior and the RGR of the composing species as determined in the laboratory. Vile et al. (2006) tested this correlation in a Mediterranean habitat, following secondary succession in abandoned vineyards. The aboveground net primary productivity, expressed per gram of biomass present at the beginning of the growing season, varied fourfold between sites and was negatively correlated with field age. The decrease in productivity was correlated with a change in species composition, such that species more abundantly present at later successional stages were those that were showing lowest RGR_{max} in growth room experiments (Figure 3.3a).

FIGURE 3.2 (continued) (5) Atkin, O.K., Botman, B., and Lambers, H., *Funct. Ecol.*, 10, 698, 1996a; C: data from (1) Mooney, H.A., Ferrar, P.J., and Slatyer, R.O., *Oecologia*, 36, 103, 1978; (2) Wright, F.I. and Westoby, M., *J. Ecol.*, 98, 85, 1999; and (3) Warren and Adams, *Oecologia*, 144, 373, 2005; D: data from (1) Ball, M.C., *Aust. J. Plant Physiol.*, 15, 447, 1988; (2) Van Diggelen, J., *A Comparative Study on the Ecophysiology of Salt Marsh Halophytes*, PhD Thesis, Free University, Amsterdam, 1988; (3) Schwarz and Gale, *J. Exp. Bot.*, 35, 193, 1984 and (4) Ishikawa S.I. and Kachi, N., *Ecol. Res.*, 15, 241, 2000; E: data from (1) Dijkstra, P. and Lambers, H., *New Phytol.*, 113, 283, 1989; (2) Meerts, P. and Garnier, E., *Oecologia*, 108, 438, 1996; and (3) Kobayashi, T., Ikeda, H., and Hori, Y., *Plant Biol.*, 1, 445, 1999; F: data from (1) Pons, T.L., *Acta Botanica Neerl.*, 26, 29, 1977; (2) Corré, W.J., *Acta Botanica Neerl.*, 32, 49, 1983a; (3) Corré, W.J., *Acta Botanica Neerl.*, 32, 185, 1983b; (4) Kitajama, K., *Oecologia*, 98, 419, 1994; (5) Osunkoya, O.O., Ash, J.E., Hopkins, M.S., and Graham, A.W., *J. Ecol.*, 82, 149, 1994; and (6) Poorter, L., *Funct. Ecol.*, 13, 396, 1999.

RGR_{MAX} AND PLANT STRATEGIES

We have shown earlier that in a variety of cases, there is a link between the potential growth rate of a species and its occurrence in a given habitat. As such, RGR_{max} forms one of the cornerstones in the plant strategy theory formulated by Grime (1979). According to this theory, plant strategies are shaped by the possible combinations of two factors experienced by plants: stress and disturbance. Stress in this sense is defined as the extent to which a combination of environmental variables retards growth (e.g., low nutrient availability, low or high temperature, low water availability). Disturbance is defined as the degree of physical disruption of the plant's biomass (e.g., grazing, trampling). Species from habitats with a high degree of stress and a low degree of disturbance are called "stress-tolerators." They are generally perennials with a low RGR_{max} (Figure 3.4). Species from sites with a high disturbance but with low stress are called "ruderals". They are mostly annuals that have a high RGR_{max}, enabling them to complete their life cycle quickly. This would ensure that seeds are produced before a disturbance event takes place, which kills the plant. Habitats in which both stress and disturbance are low are favorable for plant growth. According to the plant strategy theory, these are sites where a strong competition between plants is expected; consequently species that thrive here are called "competitors". They are generally perennials, and also have a high RGR_{max}. Sites with a high level of both stress and disturbance (volcanoes, nutrient-poor and strongly drifting sand dunes) do not bear plants, because there is no feasible strategy to cope with such an environment.

Given the importance that is generally attached to biomass gain and the relative fitness of a plant (discussed in McGraw and Garbutt 1990), it may be hypothesized that there has been a selection pressure in fertile (and favorable, even if only temporarily) habitats toward plant species with a high RGR, and toward a low potential RGR in places which are unfavorable (Grime 1979, Chapin 1980). However, RGR is a parameter that is the result of a combination of many physiological, morphological, anatomical, and biochemical traits. Alternatively, it could well be that it is one or more of these traits underlying RGR that has been the target of selection, rather than RGR itself (cf. Grime 1979, Coley 1983, Lambers and Dijkstra 1987). RGR would then merely be a by-product of selection. Before evaluating these contrasting hypotheses, we first have to investigate the components underlying RGR.

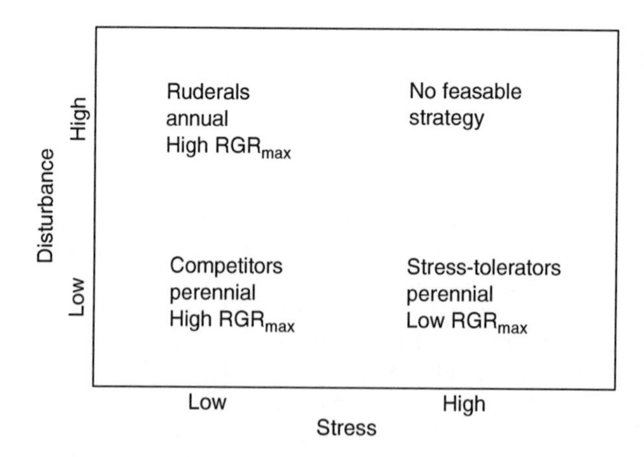

FIGURE 3.4 Relation between stress and disturbance and the growth strategy, life form, and RGR of the three types of strategies. (Adapted from Grime, J.P., *Plant Strategies and Vegetation Processes*, John Wiley & Sons, Chichester, 1979.)

COMPONENTS UNDERLYING RGR_{MAX}

Growth Parameters

Growth is more than photosynthesis. It is the balance between the carbon gain per unit leaf area (ULR) and the carbon losses in the plant (which depend on the respiration rate but also on the relative proportion of the assimilatory and nonassimilatory organs, and in the longer run on biomass turnover), corrected for the C-concentration of the newly formed biomass. Evaluating the relative importance of each of these factors requires a top-down approach, in which RGR is broken down into components. A common way to do so is factorizing RGR into the increase in mass per unit leaf area and time (ULR) and the leaf area per unit plant mass (LAR, leaf area ratio). LAR can be factorized further into the components leaf area: leaf mass (SLA, specific leaf area) and leaf mass: plant mass (LMF, leaf mass fraction). A definition of these components of RGR, as well as an explanation of the concept of growth parameters underlying RGR is given in Box 3.2.

BOX 3.2
Components of RGR

A simple framework to factorize RGR was developed at the beginning of the twentieth century (Blackman 1919, West et al. 1920). The basic assumption underlying this framework is that plant growth is dependent on photosynthesis and that leaf area is the plant variable driving total C-gain. RGR is then factorized into two components: ULR and LAR (Hunt 1982). ULR is defined as the increase in biomass per unit time and leaf area

$$\mathrm{ULR} = \frac{1}{A}\frac{\mathrm{d}M}{\mathrm{d}t}, \tag{3.6}$$

where A is the total leaf area of the plant, $\mathrm{d}M$ the increase in mass over period $\mathrm{d}t$. LAR is defined as the total leaf area per unit total plant mass

$$\mathrm{LAR} = \frac{A}{M} \tag{3.7}$$

and consequently

$$\mathrm{ULR.LAR} = \frac{1}{A}\frac{\mathrm{d}M}{\mathrm{d}t}\frac{A}{M} = \mathrm{RGR}. \tag{3.8}$$

By determining leaf mass and stem and root mass separately, one is also able to break down LAR into two components: specific leaf area (SLA) and leaf mass fraction (LMF). SLA is the amount of leaf area per unit leaf mass (M_L):

$$\mathrm{SLA} = \frac{A}{M_L} \tag{3.9}$$

and LMF is the fraction of total plant mass that is invested in the leaves

$$\mathrm{LMF} = \frac{M_L}{M} \tag{3.10}$$

and consequently

(*continued*)

BOX 3.2 (continued)
Components of RGR

$$\text{SLA.LMF} = \frac{A}{M_\text{L}} \frac{M_\text{L}}{M} = \text{LAR}. \tag{3.11}$$

The advantage of this approach is that it requires only data on progressions in leaf area and plant mass to obtain a good indication about the causes of variation in growth rate. This is because differences in ULR are often due to differences in the area-based rate of photosynthesis. However, ULR is, in fact, the net balance of total plant carbon gain and carbon losses, expressed per unit leaf area and corrected for the C-concentration of the newly formed biomass. If one really wants to obtain insight in the relation between the various C-fluxes and RGR, the following formula can be used

$$\text{RGR} = \frac{\text{PS}_\text{a}\,\text{FCI}}{\text{PCC}} \text{SLA LMF} \tag{3.12}$$

Poorter (2002), where PS_a is the total amount of C fixed per unit leaf area integrated over a 24 h period, FCI the fraction of that daily fixed carbon that is not respired in leaves, stems, or roots but invested in growth (so 1-RE/PS). The first four terms in the right-hand side of Equation 3.12 determine the net amount of C fixed per unit of biomass and per day. The denominator (PCC) converts this net amount of C into a biomass increase.

There are slightly different approaches, in which Equation 3.12 is written as the difference between carbon gain in photosynthesis and daily rate of respiration in leaves, stems, and roots. However, by using the form currently presented in Equation 3.12, with FCI (also termed carbon use efficiency), the equation is the product of five entities, which makes it amenable to the GRC analysis explained in the "Physiological Parameters" section on the next page. Note that in this equation C-losses due to exudation, or leaf and root turnover are considered to be negligible.

To what extent is inherent variation in RGR_{max} caused by variation in the components ULR and LAR? A wide variety of results has been published; some experiments found ULR to be the factor determining growth, others found LAR to be the cause of variation in growth, whereas others found intermediate results. Variation may be due to the choice of the species as well as growth conditions used and the experimental procedure followed (e.g., duration of the experiment and number of harvests). To enable a quantitative analysis of the cause of variation in RGR within a given experiment, we use the growth response coefficient (GRC). This coefficient can be calculated after determining the linear regression between the growth parameter X (which can be ULR, LAR, SLA, or LMF) as the dependent variable and RGR as the independent variable. GRC_X then is defined as the relative increase in growth parameter X divided by the relative increase in RGR

$$\text{GRC}_X = \frac{\text{d}X/X}{\text{dRGR/RGR}}$$

The sum of GRC_{ULR} and GRC_{LAR} for any experiment should be 1, and this is also the case for the sum of GRC_{ULR}, GRC_{SLA}, and GRC_{LMF}. A value of 1 for GRC_{ULR} indicates that species variation in RGR within an experiment fully scales with variation in ULR, whereas a value of 0 indicates no effect of this parameter at all.

What is the overall picture that emerges from the literature? Poorter and van der Werf (1998) analyzed a total of 111 experiments on herbaceous C_3 species, and calculated the

TABLE 3.3
Growth Response Coefficients for ULR, LAR, SLA, and LMF

	GRC
ULR	0.26
LAR	0.74
SLA	0.63
LMF	0.11

Note: Average values from a literature survey on causes of inherent variation in RGR. A total of 111 articles were compiled. Only those reports were considered where RGR differences between species or genotypes were at least 40 mg g^{-1} day^{-1}. More details are given in Poorter and van der Werf, 1998.

average GRC for the various growth parameters. They found that, on average, variation in SLA is by far the most dominant factor explaining variation in inherent RGR. ULR is second, and LMF is, on average, the quantitatively least important variable (Table 3.3). For a more elaborate analysis on GRC and the literature compilation see Poorter and van der Werf (1998).

It is not all that clear how the above-mentioned relationship between RGR of different species and the underlying growth parameters depend on environmental conditions or functional types. There have been suggestions that at high irradiance growth variation between species is due more to ULR than to SLA (Shipley 2006). Comparing variation in RGR between C3 and C4 species, or between sun and shade species, this seems indeed the case (Poorter 1989, Poorter 1999). However, for other species there is no indication that the relative importance of ULR and SLA shift with light environment (Poorter and van der Werf 1998). Growth temperature affects the relationships such that in cooler environments GRC_{ULR} gains in importance, in warmer environments GRC_{SLA} plays a dominant role (Loveys et al. 2002).

Physiological Parameters

The aforementioned technique of growth analysis has the advantage of being simple. Moreover, technical requirements to conduct such an experiment are low. A drawback is that ULR is a parameter that integrates various aspects of plant functioning (photosynthesis, respiration, chemical composition) and therefore cannot be related directly to a specific physiological process. One may assume that a considerable part of the variation in ULR is due to differences in the area-based rate of photosynthesis (Konings 1989). To obtain more insight into the physiological basis of variation in RGR, however, a more mechanistic approach has to be followed, in which growth is analyzed in terms of the plant's carbon (C) economy. This requires knowledge of the C-gain of the plant in photosynthesis, and C-losses in leaf, stem, and root respiration, all integrated over the day. The net increase in C over the day can be converted into a dry mass increase, if the C-concentration of the newly formed material is known. A formula to relate the C-fluxes to RGR is presented in the second part of Box 3.2.

Given that 85%–95% of plant dry matter is composed of carbon-based compounds (for a review see Poorter and Villar 1997), it is beyond doubt that almost all newly formed biomass is fixed during the photosynthetic process. However, that does not necessarily imply that variation in RGR must be due to variation in the rate of photosynthesis, measured per unit leaf area and per unit of time (PS_a). Only few attempts have been made to quantify both whole shoot photosynthesis at growth conditions (rather than determining photosynthetic capacity for the youngest expanded leaf) and RGR. In a number of cases no relationship at all has been observed (Dijkstra and Lambers 1989, Poorter et al. 1990, Van der Werf et al. 1993, Atkin et al. 1996a), but in others a positive relation was found (Garnier et al., unpublished). If indeed the growth parameter ULR is correlated well with the rate of photosynthesis per unit leaf area (Konings 1989, Poorter and van der Werf 1998), we might derive from the GRC_{ULR} value in Table 3.3 that, in the average experiment, there is a modestly positive relationship between PS_a and RGR.

It may well be that PS_a (and ULR) is not the best variable to consider if one seeks to understand the physiological basis of growth. Traditionally, in physiological research, the rate of photosynthesis is expressed per ULR. In this way the flux of C can be related to the flux of incoming photons and the efflux of water. In analyzing growth, however, it may be more relevant to consider how much C is fixed by 1 g of leaf biomass, which is the photosynthetic rate per unit leaf mass (PS_m). This would relate much better to RGR, which is the increase in biomass per unit of biomass and time. PS_m is the product of PS_a and SLA (Box 3.2), and this parameter is strongly associated with RGR (Poorter et al. 1990, Reich et al. 1992, Van der Werf et al. 1993, Walters et al. 1993, Atkin et al. 1996a, Garnier et al. unpublished results).

Both shoot and root respiration are generally positively correlated with RGR (Poorter et al. 1990, Van der Werf et al. 1993, Walters et al. 1993, see also Atkin et al. 1996a). This is partly a reflection of the higher rate of growth of the fast-growing species, which requires extra amounts of energy (ATP) and reducing power (NADH) per unit dry mass. As far as roots are concerned, the higher respiration rate is also a reflection of a much higher uptake rate of ions (Van der Werf et al. 1994).

An alternative approach to the carbon balance decomposition just described can be followed, focusing on the nitrogen economy of the plant. RGR can then be expressed as the product of the root mass fraction (RMF) and the nitrogen uptake rate divided by the mean plant nitrogen concentration (Garnier 1991, Lambers and Poorter 1992). Breaking down the RGR of fast- and slow-growing species in this way shows that faster growth is associated with a relatively lower allocation of biomass to roots, a higher plant nitrogen concentration, and a higher nitrogen uptake rate (NUR). In fact, NUR is the parameter that shows the widest variation and the strongest correlation with RGR in the fast-slow continuum (Garnier 1991).

Chemical and Anatomical Parameters

There is a wide range of other parameters for which fast- and slow-growing species differ under more or less optimal growth conditions. Most notably, fast-growing species have higher concentrations of reduced nitrogen in all organs, as well as higher amounts of minerals (Poorter and Bergkotte 1992, Reich et al. 1992, Van der Werf et al. 1993, Garnier and Vancaeyzeele 1994, Atkin et al. 1996b), observations that also hold for fast- and slow-growing tree species (Villar et al. 2006). They tend to have slightly lower concentrations of C (but see Garnier and Vancaeyzeele 1994 and Atkin et al. 1996a), but these differences are marginal for species within the same life form. Differences in C-concentration may become substantial when the growth differences between herbaceous and woody species are analyzed (Poorter 1989).

Given the importance of SLA in explaining variation in RGR_{max} (Table 3.3), it is appropriate to factorize this parameter further. SLA, or rather its inverse, 1/SLA (expressed in g m^{-2}), is the product of leaf thickness and leaf density (see Box 3.3). Generally, the lower SLA of slow-growing species is due more to a higher leaf density (LD) than that it is caused by a higher leaf thickness (Van Arendonk and Poorter 1994, Garnier and Laurent 1994, but see e.g., Körner and Diemer 1987 and Shipley 1995). The density of a leaf or root is strongly related to its water content per unit dry mass (Garnier and Laurent 1994). This can easily be understood by envisaging a cell as a box. A higher density is often caused by an extra deposition of cell wall material (lignin, cellulose). A doubling of the amount of cell wall material hardly affects the cell size or the amount of water in the cell (the volume of the box). However, the amount of water relative to the dry mass decreases, whereas the density, the dry

BOX 3.3
Components of Specific Leaf Area

SLA, leaf area per unit leaf mass, is a parameter that scales linearly and positively with RGR and is, in this respect, an easy parameter to use in growth analyses. However, if one would like to analyze what factors play a role in determining SLA, it is more appropriate to consider its inverse, 1/SLA, for which many other terms have been used (SLW, SLM, LMA). This is because components of the leaf, like leaf thickness, or anatomical and biochemical features, increase linearly with the inverse of SLA. Witkowski and Lamont (1991) factorized 1/SLA into two components: leaf thickness (LTh) and leaf density (LD). Leaf density is defined as the mass of a leaf per unit leaf volume:

$$\mathrm{LD} = \frac{M_\mathrm{L}}{\mathrm{LTh}\ A} \tag{3.13}$$

and therefore,

$$\frac{1}{\mathrm{SLA}} = \mathrm{LD\ LTh} = \frac{M_\mathrm{L}}{\mathrm{LTh}\ A}\mathrm{LTh} \tag{3.14}$$

A leaf can also be separated into its underlying anatomical tissues: epidermis, mesophyll, sclerenchyma, vascular tissue, and intercellular spaces. The 1/SLA is then the sum of the densities of the various tissues i, weighted by the volume per unit leaf area taken by each tissue (Garnier and Laurent 1994):

$$\frac{1}{\mathrm{SLA}} = \sum_{i=1}^{5} \frac{V_i}{A}\frac{M_i}{V_i} \tag{3.15}$$

Leaf biomass can also be separated into its various biochemical fractions. A simple grouping of the wide range of compounds is: Lipids, lignin, soluble phenolics, protein, structural carbohydrates (cellulose, hemicellulose, pectin), nonstructural carbohydrates (glucose, fructose, sucrose, starch), organic acids, and minerals (for a review see Poorter and Villar 1997). 1/SLA is then the sum of the masses of each of the eight classes of compounds expressed per unit area:

$$\frac{1}{\mathrm{SLA}} = \sum_{i=1}^{8} \frac{M_i}{A} \tag{3.16}$$

The three ways of breaking down 1/SLA are interrelated: a high leaf density, for example, can be the result of a high proportion of sclerenchyma, which shows up in the biochemical analysis as a high concentration of cell walls (lignin and structural carbohydrates).

mass per unit volume increases. It can therefore be expected that fast-growing species, with a low density of tissues, will have a high water content per unit biomass. This turns out to be the case and not only applies to leaves but also to stems and roots (Garnier 1992, Ryser 1996). Another factor that strongly affects the density is the relative volume of intercellular spaces.

RGR AND ITS COMPONENTS: A SYNTHESIS

It is quite clear from the previous section that inherent differences in RGR_{max} among species are associated with differences in a large array of plant traits (Lambers and Poorter 1992; Chapin et al. 1993). To what extent are these parameters related to each other? To answer this question, a principal component analysis (PCA) was conducted on two sets of data obtained from wild herbaceous species for which a large number of variables were measured. The first one (*Herbs*) concerns 11 grasses and 13 dicotyledonous species, the second (*Grasses*) is for 12 wild grasses. The results are shown in Figure 3.5.

The patterns confirm the differences between the fast- and slow-growing species as presented in the previous section: in both data sets a high RGR is strongly related to a high uptake capacity of both leaves (photosynthesis, PS_m) and roots (NUR) expressed per unit mass of leaves and roots. The correlation between RGR and PS_a (and ULR) is looser. Shoot respiration and root respiration (only available for Herbs) are also positively associated with RGR. Species with a high RGR also have a high SLA, a high nitrogen concentration in the leaves and a high concentration of minerals in their tissues (only available for Herbs). All these parameters can be found at the right-hand side of Figure 3.5. At the left-hand side are the variables associated with slow-growing species. They have a high density in leaves and roots and a high proportion of leaf biomass in cell walls. Moreover, a large fraction of the total C fixed daily in photosynthesis is spent in respiration (only available for Herbs). Similar results have been found for a range of *Hordeum spontaneum* genotypes differing in RGR_{max} (Poorter et al. 2005).

A PCA analysis can also be used to gain insight into how well the various other parameters of the plants are correlated with each other, and probably also mechanistically associated. For example, PS_a and ULR are strongly associated. The rates of shoot and root respiration (only available for Herbs) have a similar value in common for factor 1, but at the axis of factor 2 they are separated from each other. This may indicate that, although they are both highly correlated with RGR, the reason for correlation is—at least to some extent—different. Part of the respiration is necessary for growth and can be expected to be high in both organs for fast-growing species. Another part of the shoot respiration may have to do with a high transport of sugars to the sink, whereas part of the root respiration is related to nutrient uptake. It is this type of information we need if we want to understand the trade-offs that play a role in determining the functioning of the whole plant.

From the above analyses, suites of traits can be associated with fast and slow growth: high RGR_{max} is achieved in species exhibiting high rates of resource acquisition, made possible, among other things, through high-light interception per unit leaf mass (high SLA), high concentration of enzymatic machinery (reflected by a high nitrogen concentration) and low density of tissues (i.e., high water content). The latter two are an indication that fast-growing plants display a high concentration of protoplasmic elements (except for starch; see also Niemann et al. 1992). Opposite traits are found in slow-growing species, a low RGR_{max} associated with a high amount of cell walls and starch, which are metabolically inactive and lead to low acquisition rates per unit mass. In the section on pp. 90–93, we put forward the hypothesis that these contrasting suites of traits can be interpreted as a functional trade-off between high biomass productivity and efficient conservation of nutrients.

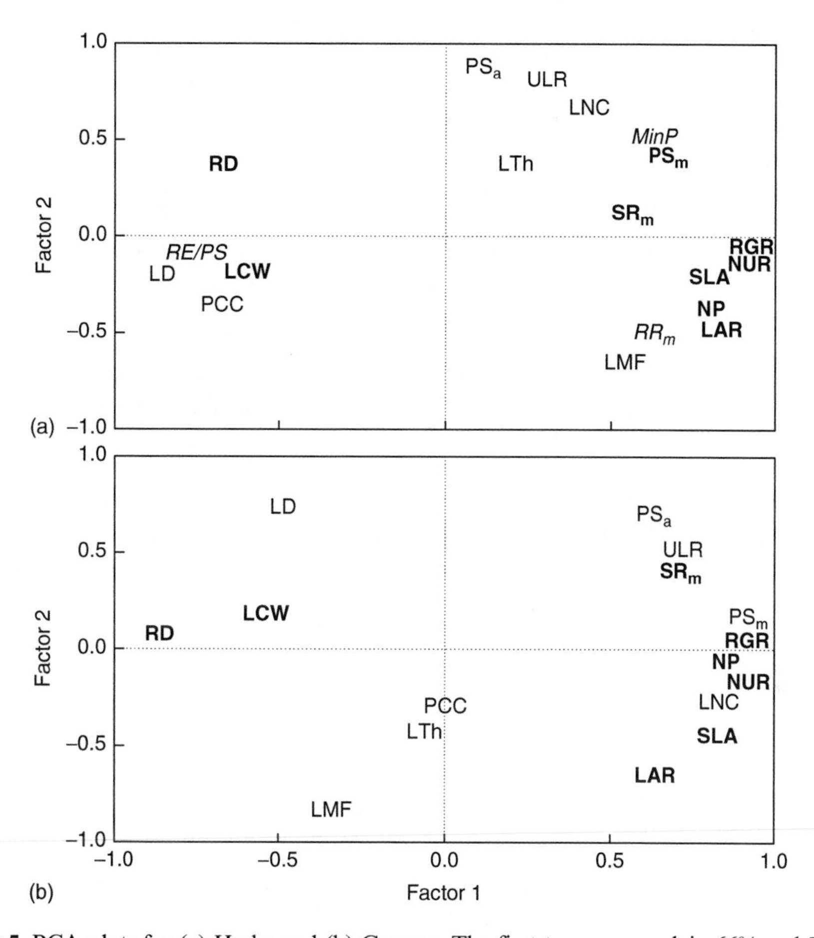

FIGURE 3.5 PCA plots for (a) Herbs and (b) Grasses. The first two axes explain 66% and 71% of the overall information for Herbs and Grasses, respectively. In both analyses, axis 1 can be interpreted as a biomass production axis; axis 2 appears to be mainly determined by gas exchange properties (photosynthesis and transpiration) per unit leaf area. Variables printed in bold are on similar places in the plane. Variables printed in italics were determined for Herbs only. Abbreviations not given in Table 3.1 are: LCW, proportion of cell walls in the leaves; LTh, leaf thickness; MinP, mineral concentration of the plant; RE/PS, the fraction of daily fixed photosynthesis that is respired again. In this analysis, two new variables (factor 1 and factor 2) are computed out of a combination of all original variables. For each of these variables it is calculated whether they contribute positively (close to 1.0), negatively (close to −1.0), or not (close to 0.0) to factor 1. The amount of variance thus explained is taken out of the data and the procedure is repeated with the remaining variance. The result is somewhat comparable to a two-dimensional electrophoresis. Variables that are close together (like RGR and NUR) are generally positively correlated, variables that are at opposite parts of the graph (like RGR and LCW) are negatively correlated, and variables that have values close to 1 or −1 for one factor and values close to 0 for the other axis [like RGR and PS_a in (a)] are generally not correlated at all. (Data for Herbs are from Poorter, H., Remkes, C., and Lambers, H., *Plant Physiol.*, 94, 621, 1990; Poorter, H. and Remkes, C., *Oecologia*, 83, 553, 1990; Poorter, H. and Bergkotte, M., *Plant, Cell Environ.*, 15, 221, 1992; and concern 11 grasses and 13 dicotyledonous species. Data for Grasses are from Garnier, E., *J. Ecol.*, 80, 665, 1992; Garnier, E. and Laurent, G., *New Phytol.*, 128, 725, 1994; Garnier, E. and Vancaeyzeele, S., *Plant, Cell Environ.*, 17, 399, 1994; and Garnier et al. (unpublished) and comprise 12 grass species.)

DO LABORATORY FINDINGS APPLY TO THE FIELD?

As stated earlier (see the second section on pp. 69–73), plants growing in the field certainly do not achieve their potential growth rate. However, to bear ecological significance, differences in the above-mentioned suite of traits between species, as found under laboratory conditions, should at least reflect to some extent the differences between the same species growing in their natural habitat. Is there evidence that this is the case? This question can be approached by comparing plant traits that have been measured for the same species both in the laboratory and in the field. This is done here for SLA and leaf nitrogen concentration on a leaf mass basis, which are positively related to RGR_{max}, and leaf density, which is negatively correlated with RGR_{max} for plants grown under optimum conditions in the lab (see the previous two sections, pp. 76–83). How are these attributes for plants growing naturally in the field, and is the ranking of species the same under both conditions?

Several data sets are available to address this question, comprising both herbaceous and woody plants. For the 86 species for which such data are available, SLA is on average 50% lower in the field than in the lab. Nonetheless, there is a good relationship between SLA measured in the field and in the laboratory for four of the five data sets taken individually (Figure 3.6a). Leaf nitrogen concentration is 34% lower in field grown plants, and the relationship between field and laboratory-grown plants is significant for three of the five data sets (Figure 3.6b). Finally, leaf density is approximately 40% higher in the field than in the laboratory, and the relationship between density measured in the field and in the laboratory is highly significant for the only data set with a substantial amount of data points (11 species: Figure 3.6c). Differences between laboratory and field-measured leaf traits are probably caused by a combination of a higher light intensity, higher wind speed, and lower nutrient availability in the latter environment.

Although there are differences in plant traits measured under field and lab conditions, these results show that the ranking of species for such important traits as SLA and leaf density are maintained under a wide range of growing conditions. This is further corroborated by the fact that the relationships between ecosystem productivity and community-averaged RGR_{max} on the one hand and ecosystem productivity and community-averaged SLA measured in the field on the other hand compare well (Figure 3.3). This may be less so for leaf nitrogen concentration. We may therefore expect that the suite of traits discussed in the previous section is likely to be maintained, at least partially, under field conditions.

SELECTION FOR RGR OR UNDERLYING COMPONENTS?

SELECTION FOR RGR

As discussed in the previous section, RGR is a parameter that is associated with a suite of traits. Based on the negative correlations between RGR_{max} and the harshness of the environment (Figure 3.2), it has been suggested that RGR was the target of selection (Grime 1977, Chapin 1988). Alternatively, it may have been one or more of the components of RGR that has been selected for (Grime 1977, Coley 1983, Dijkstra and Lambers 1987).

What are the arguments in favor of selection for a high RGR? For ruderals and other annual species, a fast completion of the life cycle seems of paramount importance (Grime 1979). A high RGR may be of help to reach the size required for high seed production (discussed by Benjamin and Hardwick 1986, McGraw and Garbutt 1990, and Garnier and Freijsen 1994). Competitive species, be it woody or herbaceous, seem to have an advantage by occupying quickly the available space within the vegetation. The acquisition of resources,

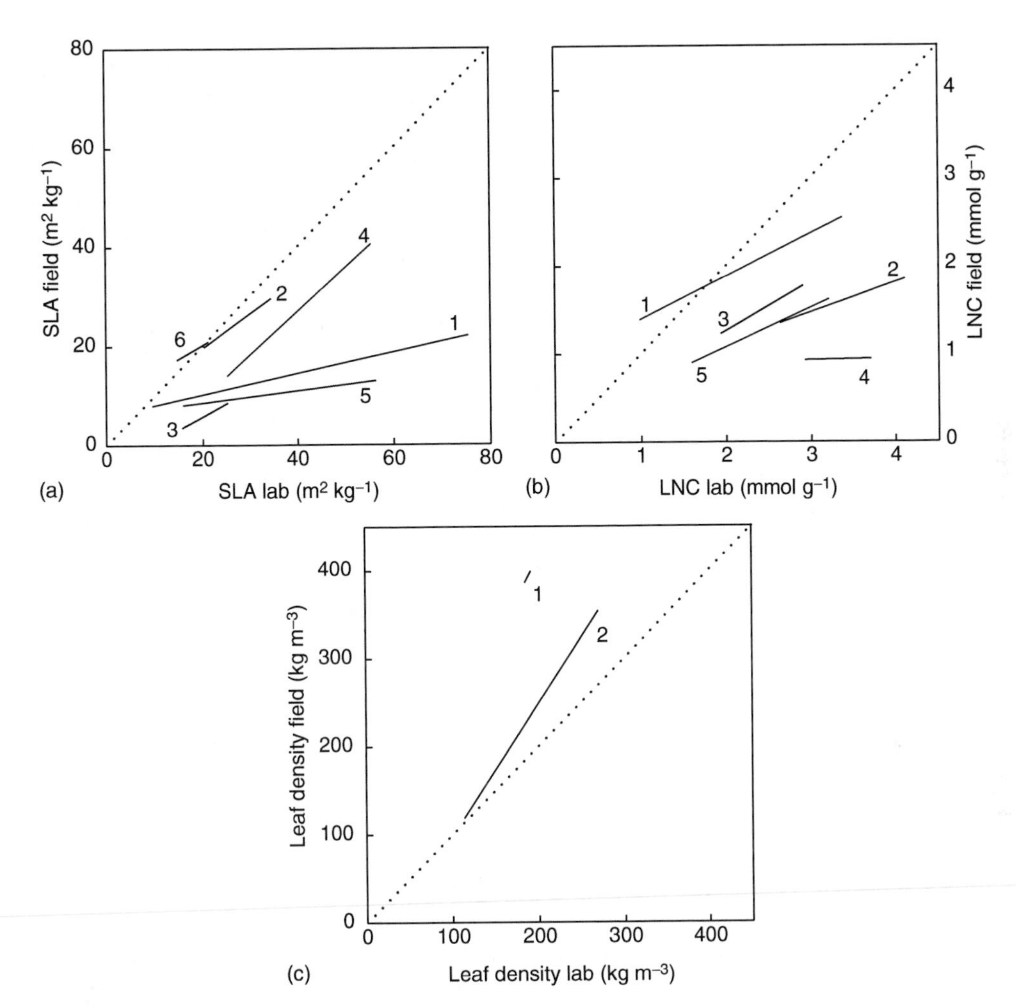

FIGURE 3.6 (a) SLA; (b) leaf nitrogen concentration, and (c) leaf density measured on the same species in field and laboratory conditions. Each study is represented by a regression line and contains several species. (Data from: a: (1) Cornelissen, J.H.C. et al. (unpublished, laboratory, and field); (2) Garnier, E. and Laurent, G., *New Phytol.*, 128, 725, 1994 (laboratory) and Garnier, E., Cordonnier, P., Guillerm, J.-L., and Sonié, L., *Oecologia*, 111, 490, 1997 (field); (3) Mooney, H.A., Ferrar, P.J., and Slatyer, R.O., *Oecologia*, 36, 103, 1978 (laboratory and field; (4) Poorter, H. and Remkes, C., *Oecologia*, 83, 553, 1990 (laboratory and Poorter, H. and de Jong, R., *New Phytol.*, 143, 163, 1999 (field); (5) Walters, M.B., Kruger, E.L., and Reich, P.B., *Oecologia*, 96, 219, 1993, and P.B. Reich (unpublished) (laboratory) and Reich, P.B., Walters, M.B., and Ellsworth, D.S., *Ecol. Monogr.*, 62, 365, 1992 (field); (6) Bloor J.M.G., *J. Trop. Ecol.*, 19, 163, 2003; b: (1) Cornelissen, J.H.C., Werger, M.J.A., Castro-Diez, P., van Rheenen, J.W.A., and Rowland, A.P., *Oecologia*, 111, 460, 1997; (2) Garnier, E. and Vancaeyzeele, S., *Plant, Cell Environ.*, 17, 399, 1994 (laboratory) and Garnier, E., Cordonnier, P., Guillerm, J.-L., and Sonié, L., *Oecologia*, 111, 490, 1997, (field); (3) Hull, J.C. and Mooney, H.A., *Acta Oecologica*, 11, 453, 1990 (laboratory and field); (4) Mooney, H.A., Ferrar, P.J., and Slatyer, R.O., *Oecologia*, 36, 103, 1978 (laboratory and field); (5) P.B. Reich (unpublished, laboratory) and Reich, P.B., Walters, M.B., and Ellsworth, D.S., *Ecol. Monogr.*, 62, 365, 1992 (field); C: Garnier, E. and Laurent, G., *New Phytol.*, 128, 725, 1994 (laboratory) and E. Garnier and J.-L. Cordonnier (unpublished); (2) Poorter, H. and Bergkotte, M., *Plant, Cell Environ.*, 15, 221, 1992 (laboratory) and Poorter, H. and de Jong, R., *New Phytol.*, 143, 163, 1999 (field).)

both above and belowground, depends on how much of the volume of soil and air is occupied by roots and leaves. A high RGR may enable this.

What would be the adaptive value of a low RGR? Some explanations have been offered, focusing on plant species from nutrient-poor habitats (Chapin 1980, 1988). However, these suggestions are questionable (Poorter 1989, Lambers and Poorter 1992):

1. *If plants grow slowly, they are less likely to deplete the available nutrient resources early in the season.* This does not seem to be an evolutionarily stable strategy. As soon as one species or genotype starts to take up as much nutrients as possible early in the season, the others miss out.
2. *Plants in nutrient-poor areas will never grow fast. If they have a low potential RGR, they will be closer to their optimum in the field.* For all plants it applies that their physiological optimum differs from the conditions they experience in the field (see the second section, pp. 69–73). However, a clear difference between physiological and ecological optimum does not necessarily imply a disadvantage. Fast- as well as slow-growing species have a large flexibility, morphological as well as physiological, to cope with varying conditions (Reynolds and d'Antonio 1996). At low nutrient supply, for example, it is found that species with a high RGR_{max} still grow faster or at similar rates than those with a low RGR_{max} (Shipley and Keddy 1988, Poorter et al. 1995).
3. *If plants grow slowly, they can accumulate sugars and nutrients during favorable times, so as to enable growth in later times when nutrients are less easily available.* As far as sugars are concerned this explanation does not hold. Plants experiencing nutrient stress are limited more in growth than in photosynthesis. They fix more C than can be used in growth, resulting in accumulation of nonstructural sugars (mainly starch; Poorter and Villar 1997). Therefore, it does not seem necessary to store sugars for times with a low-nutrient availability. For nutrients however, this reasoning could be valid. A prerequisite is that nutrients become available in flushes. Such flushes have been shown, for example during freeze-thaw events that lyse microbial cells or during drought–wet cycles (for discussion see Bilbrough and Caldwell 1997). However, although these processes have been shown to occur, there is at present not enough evidence to conclude that species found in sites of different nutrient availability (or those with contrasting RGR_{max}) differ in this respect. For plants grown hydroponically in the laboratory at a nonlimiting nutrient supply, we found fast-growing species to accumulate 4–5 times more NO_3^- than slow-growing species.
4. A last hypothesis does not explain why plants in low-resource environments do have a low RGR, but rather why they do not have a high potential RGR: *A high RGR cannot be realized in low-resource environments and therefore RGR is a selectively neutral trait.* Although a very high RGR would not be reached, a genotype with a slightly higher RGR could still occupy some extra space and consequently would be able to acquire some extra nutrients. Thus, in those cases RGR would not necessarily be a selectively neutral trait.

Selection for Components of RGR

Is there evidence that components of RGR have been selected for, rather than RGR itself? This is not a question that can be easily answered. If all of the components scale positively with RGR, it becomes impossible to separate between the two alternatives. Moreover, it is difficult to single out one component if the different traits have not been selected independently, or if they are functionally related. For example, a high rate of photosynthesis is achieved with a high concentration of protein. A high concentration of protein may imply a high rate of protein turnover, and therefore a high maintenance respiration. In addition, a high rate of photosynthesis may result in a high rate of export to the phloem, which also causes an increase in shoot respiration (cf. Figure 3.5). Given these mechanistic interrelations, it is not easy to break the correlation between photosynthesis and respiration.

Nevertheless, we believe that there are good reasons to think that selection in a low-resource environment has been for components related to a low SLA, rather than for RGR per se. Generally, SLA is the most important growth parameter associated with inherent variation in RGR_{max} (pp. 76–83). Moreover, differences in SLA observed in the laboratory are preserved in the field (p. 84), and in the case of the old-field succession described in the "RGR_{max} and Ecosystem Productivity," pp. 73–75, the correlation between the RGR of the vegetation and the aggregated SLA of the composing plant species, which was determined in the field, was almost as good as the correlation with the aggregated RGR derived from lab measurements (Figure 3.3b). What are the arguments in favor of this alternative hypothesis?

Selection for SLA-Related Traits in Adverse Environments

Plants from extremely nutrient-poor environments are often evergreens, with a high leaf longevity. As has been previously discussed, availability of photosynthates is not a growth-limiting factor in these environments, but the availability of nutrients is. A first problem of plants in a low-nutrient environment is to acquire nutrients; a second problem is to use them efficiently. Berendse and Aerts (1987) showed that an efficient use of, for example, N [nitrogen use efficiency (NUE); gram of growth per unit N taken up or lost by the plant] depends on two components: the biomass increase per unit N and per unit of time (NP, the mean annual nitrogen productivity) and the average time that a unit of N stays in the plant (MRT, mean residence time of nitrogen; see Box 3.4). They argue that there is a trade-off between these two components: a plant cannot achieve both a high NP and a high MRT (Figure 3.7). Such a trade-off has indeed been shown in several cases (e.g., Eckstein and Karlsson 1997; Figure 3.7), but is less evident in other experiments (reviewed by Garnier and Aronson 1998). The putative trade-off between NP and MRT is most likely due to the suite of characters discussed on pp. 82–83: a high nitrogen productivity is strongly correlated with both a high SLA and a low density of organs (characteristics found at the right-hand side in Figure 3.5). Plants with those characteristics are mainly directed toward attaining a high rate of resource capture (carbon, nutrients). This strategy is further discussed on p. 90. The other extreme is formed by the species with a high MRT. Species with such a strategy can be found in nutrient-poor environments. Why is this so? In nutrient-poor environments, it may be a disadvantage for an individual plant to lose acquired nutrients, as it is very questionable whether that individual is able to take them up again once they have entered the nutrient cycling process. Therefore, there is a premium to increase the residence time of nutrients. Theoretically, this can be achieved in two ways (Box 3.4): either a plant resorbs nutrients from senescing organs very efficiently, or it restricts the turnover of organs and thus the loss of biomass per unit of biomass present (Aerts 1990, Garnier and Aronson 1998). Although there is variation in resorption efficiency among species, it does not differ strongly among life-forms or species originating from habitats differing in fertility (reviewed by Aerts and Chapin 2000). Thus, the way species from nutrient-poor habitats achieve a high MRT is by a long life span of both leaves and roots (Aerts 1995, Ryser 1996, Eckstein and Karlsson 1997, Garnier and Aronson 1998, Hikosaka 2003).

BOX 3.4
Components of Nitrogen Use Efficiency

In Box 3.2, RGR was factorized into components based on the assumption that leaf area is the important plant variable driving photosynthesis, and thus growth. An alternative approach is to consider plant (organic) N as the driving variable, as proteins play a vital role in C-fixation, nutrient uptake, and in most other physiological processes of the plant. RGR is then factorized into the components NP (Nitrogen Productivity) and PNC (Plant Nitrogen Concentration; preferably restricted to organic nitrogen, as NH_4^+ and NO_3^- play less of a physiological role). NP is defined as the increase in biomass per unit time and plant nitrogen (Ingestad 1979):

$$\mathrm{NP} = \frac{1}{N}\frac{\mathrm{d}M}{\mathrm{d}t}, \tag{3.17}$$

where N is the total amount of nitrogen in the plant. PNC is the concentration of nitrogen in the plant

$$\mathrm{PNC} = \frac{N}{M} \tag{3.18}$$

and consequently

$$\mathrm{NP\ PNC} = \frac{1}{N}\frac{\mathrm{d}M}{\mathrm{d}t}\frac{N}{M} = \mathrm{RGR} \tag{3.19}$$

Considered over a short and fixed period of time, plants use their internal nitrogen efficiently if they have a high NP. Considered over a longer time scale, an alternative way to make efficient use of a unit of nitrogen taken up is to increase the time this unit remains in the plant. This time span is called the mean residence time (MRT, expressed in years). Under steady-state conditions (i.e., when the amount of nutrients taken up by the plant equals that lost by leaf and root shedding, herbivory, etc.), MRT is equal to the ratio between the average amount of nitrogen in the plant (N) and that taken up or lost over a given period of time ($\mathrm{d}N/\mathrm{d}t$; see, e.g., Frissel 1981). Berendse and Aerts (1987) have proposed that it is the product of NP and MRT that determines the nutrient use efficiency (NUE)

$$\mathrm{NP.MRT} = \frac{1}{N}\frac{\mathrm{d}M}{\mathrm{d}t}N\frac{\mathrm{d}t}{\mathrm{d}N} = \frac{\mathrm{d}M}{\mathrm{d}N} = \mathrm{NUE} \tag{3.20}$$

NUE is therefore the total amount of biomass produced per unit N taken up or lost. Note that in Equation 3.19, NP is generally determined over a short time scale (days to weeks), whereas in Equation 3.20 it is determined on an annual basis.

Under steady-state functioning MRT depends on two parameters in the following way (Garnier and Aronson 1998):

$$\mathrm{MRT} = \frac{1}{1 - R_{\mathrm{eff}}}\frac{1}{e}, \tag{3.21}$$

where R_{eff} is the nitrogen resorption efficiency during senescence (defined as the reduction in the amount of nutrient between mature and senesced organs, relative to the amount in mature organs: Aerts 1996, Killingbeck 1996), and e is the rate of biomass lost by the plant, which depends on the life span of organs. These two parameters are central to the definition of the nutrient conservation strategy of the various species.

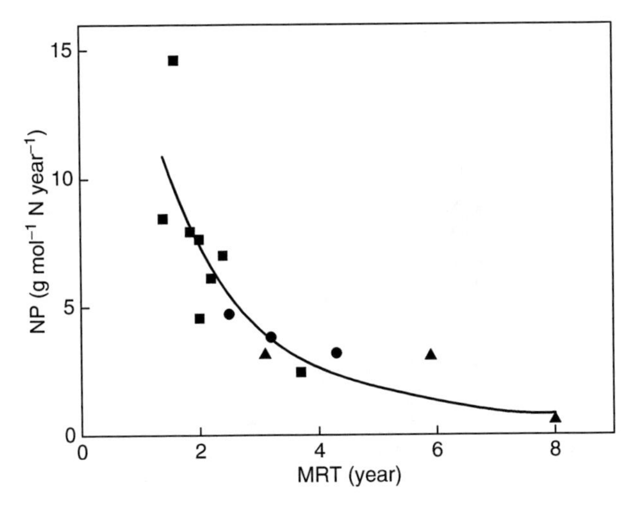

FIGURE 3.7 The relation between Nitrogen Productivity (NP; on an annual basis) and the MRT of nitrogen of 14 species grown in the field. Squares, herbaceous species; circles, deciduous shrubs; triangles, evergreen shrubs. (Data from Eckstein, R.L. and Karlsson, P.S., *Oikos*, 79, 311, 1997.)

What determines the life span of a leaf or root? For perennial species, there is an effect of the environment, with increases in leaf life span under relatively predictable low-resource conditions (low temperature, low irradiance). However, most of the variation in leaf life span is due to inherent differences between species (Reich 1998). Leaves of some species live for less than 2 months, whereas leaves of others have been reported to function for more than 20 years (Ewers and Schmid 1981). The physiological mechanism determining the life span of a leaf is unknown. However, it is evident that a leaf can only become long-lived if it can withstand adverse periods. Therefore, compared with a leaf with a short life span extra investments have to be made to survive periods of drought or coldness. It should also be less attractive to herbivores, and not be so frail that it is damaged in storms. Finally, in a nutrient-poor environment one could expect extra investments of plants to prevent nutrients leaching out of the leaf. Drought tolerance may be achieved by leaf hairs, a thick cuticle, an increase in lipid concentration and possibly small cell sizes (Jones 1983). Cold resistance may be acquired by increases in osmotic solutes. Herbivory may be counteracted by the accumulation of phenolics or other secondary compounds. In all these cases extra lignification may occur. Extra lignification and thicker cell walls can also be expected for plants that have adapted to a high degree of mechanical disturbance, such as trampling or strong winds. Nutrient leaching could be prevented by extra investment in wax layers.

Compared with a basic leaf with a short life span, all of these additional investments increase the biomass per unit leaf area and thus decrease SLA. Therefore, we might expect a negative relation between SLA (and the suite of traits positively associated with SLA) and leaf life span. This has indeed been shown in an analysis of plant species across a wide range of habitats (Reich et al. 1992, Wright et al. 2004, Figure 3.8). Data on the life span of roots are far less abundant (Eissenstat and Yani 1997). As far as data are available, they seem to indicate that species with a higher density of root tissue may have longer life spans (Ryser 1996). It is this connection between the nutrient economy (in the form of a high MRT through a long leaf or root life span) and the carbon economy (in the form of a low SLA) that may explain the success of species with a low RGR_{max} in nutrient-poor environments.

Up to now, we have focused discussion on RGR components of plants from habitats low in nutrients. Basically, similar considerations could work for plants in other environments

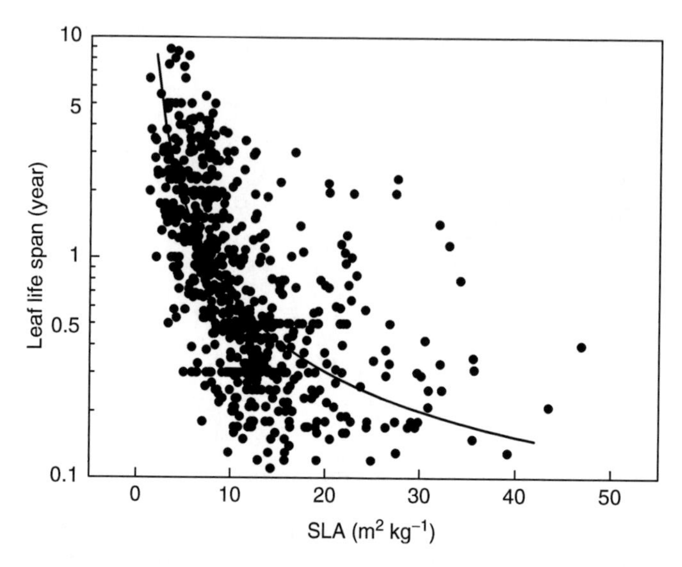

FIGURE 3.8 Leaf longevity as a function of SLA for a wide range of trees, shrubs, and herbaceous C_3 species found across a range of contrasting climates. (Wright, I.J., Reich, P.B., Westoby, M., Ackerly, D.D., Baruch, Z., Bongers, F., Cavender-Bares, J., +26 others, *Nature*, 428, 821, 2004.)

adverse to growth. In all cases where the production of biomass is difficult, one may expect a premium on maintaining existing biomass rather than replacing lost leaves or roots.

Selection for SLA-Related Traits in Favorable Environments

What about a possible selection for RGR components in an environment favorable for plant growth? Vegetation is dense there, with a fast leaf area development during the growing season. Consequently, there is a strong competition for light, necessitating a fast increase in stem height and an efficient light interception per unit leaf biomass. In simulations of competition under agricultural conditions, it has been shown how important SLA is during the period before canopy closure: high-SLA plants have a higher light-interception per unit leaf biomass and therefore faster growth than competing plants with a low SLA (Spitters and Aerts 1983, Gutschick 1988). In a closed canopy, maximum photosynthetic carbon gain of the vegetation as a whole is highest at lower values of SLA (Gutschick 1988). However, also under those conditions there may be a selection pressure towards an increase in SLA. Using game theory, Schieving and Poorter (1999) analyzed mathematically the total carbon gain of two putative genotypes, which were similar in all traits, except for SLA. Simulating competition in a dense stand on the basis of functions for the distribution of light, nitrogen and leaf area, they found that the genotype with the higher SLA always replaced the lower SLA genotypes.

EVIDENCE FOR THE IMPORTANCE OF SLA-RELATED TRAITS

As outlined in the previous section, it is difficult to really prove that selection has been for SLA-related traits rather than for RGR itself. At best we can show that SLA and the suite of leaf and root characteristics associated with it, vary along environmental gradients, or that high-SLA species that have been introduced in new continents displace low-SLA species in favorable habitats, but not in harsher environments. In analyzing these data, one should realize that SLA of a given species is not very sensitive to differences in nutrient availability, but rather sensitive to light.

Nutrient Availability

The most obvious difference in SLA is between woody deciduous and woody evergreen species. Evergreens have much lower SLAs (Villar and Merino 2001) and are generally found in nutrient-poor habitats (Monk 1966, Small 1972). Comparing productivity of various woody deciduous and evergreen species at the stand scale, a positive correlation has been found between SLA and annual productivity per unit leaf mass (Reich et al. 1997). Additionally, for herbaceous vegetations, there is a general trend that productive sites bear species with a higher SLA (Poorter and de Jong 1999). However, there is considerable variability around this trend, as only half of the total variation in SLA between species was explained by differences between sites, the other half due to differences in species within sites. This is in line with data of Van Andel and Biere (1989), who found a large variation in RGR_{max} and SLA for species co-occurring in the same habitat.

Species replacement has been observed in Venezuela, where two introduced C_4 grasses have outcompeted the original C_4 grass in most areas, but not in drier sites with low fertility (Baruch et al. 1985, Baruch 1996). Conforming to our expectations, the native species has a lower SLA, a higher proportion of sclerenchyma, and a lower concentration of N (Table 3.4).

A controlled experimental test to analyze the relation between growth parameters and field performance was carried out by Biere et al. (1996). They selected a range of families of *Lychnis flos-cuculi* and performed a growth analysis for each of these families in the glasshouse. At the same time, seeds of these families were sown in the field, along a fertility gradient. The aboveground biomass of all field plants at the end of the growing season, a good predictor of next year's fecundity, was estimated and correlated with the growth parameters determined in the glasshouse. At the poorest site, families with a high RGR_{max}, LAR, and SLA fared worst (Table 3.5). However, the higher the site fertility, the more these parameters gained importance, and in the most productive site families with a high SLA (and LAR) were those that attained the highest biomass. These results were confirmed in an experiment where early plant growth characteristics were correlated with fitness in *H. spontaneum* accessions (Verhoeven et al. 2004). At high, but not at low nutrient supply, high SLA accessions had the highest seed mass output. RGR itself was not correlated with fitness.

TABLE 3.4
Differences in Leaf Characteristics between Introduced and Native Species

	H. rufa Introduced	*T. plumosus* Native	*A. stolonifera* Introduced	*A. magellanica* Native
SLA ($m^2\ kg^{-1}$)	34	21	33	13
Leaf thickness (μm)	140	190	170	330
Leaf density ($kg\ m^{-3}$)	210	250	180	235
Volume (%)				
Epidermis	36	38	24	20
Mesophyll	53	39	72	62
Vascular	11	17	3	3
Sclerenchyma	1	5	2	16
Concentrations ($mg\ g^{-1}$)				
N	20	13	21	16

Note: The first comparison is a C_4 grass introduced in Venezuela (*Hyparrhenia rufa*) with a native C_4 grass species (*Trachypogon plumosus*; Data from Baruch, Z., Ludlow, M.M. and Davis, R., *Oecologia* 67, 388, 1985.) The second comparison is an *Agrostis* species introduced on sub Antarctic islands (*A. stolonifera*) with a native species from the same genus (*A. magellanica*; Data from Pammenter, N.W., Drennan, P.M., and Smith, V.R., *New Phytol.*, 102, 143, 1986.)

TABLE 3.5
Correlations between Plant Mass in the Field, Determined at Four Sites Differing in Productivity, and a Number of Growth-Related Parameters, Determined in the Glasshouse

Site Productivity	Seed Mass	Emergence	RGR	ULR	LAR	SLA	LMF
Low	**0.77**	0.07	**−0.70**	0.11	**−0.70**	**−0.76**	0.49
Low–intermed	0.25	−0.02	−0.06	0.15	−0.15	−0.11	−0.09
Intermed–high	0.12	−0.27	0.04	0.03	0.02	−0.02	0.12
High	−0.10	**−0.45**	0.21	−0.27	**0.38**	**0.41**	−0.02

Source: After Biere, A., *Plant Soil*, 182, 313, 1996, slightly simplified and reworked for this table.

Note: Data are family means ($n = 56$) of above-ground plant mass (in-transformed) of plants of *Lychnis flos-cuculi* sown in the field and seed mass, time to emergence and growth parameters as determined in the glass house. Values in bold are significant at $P < 0.05$.

Water Availability

Mooney et al. (1978) investigated *Eucalyptus* species along a rainfall gradient, and showed that, both for plants growing in the field and in the laboratory, the SLA of the low-rainfall species was lower than those of species growing at sites with higher rainfall. The high SLA species had the highest concentration of N per mass. In this case, the correlation between seedling RGR and rainfall was not very tight, whereas the correlation with SLA was tight. Similarly, the SLA of sun-lit leaves from *Eucalyptus* canopies, sampled along a climatic gradient in Australia was found to be inversely related to the potential evaporation at the various sites studied (Specht and Specht 1989). These results are not supported by Warren and Adams (2005) though, who showed a relationship between rainfall and biomass allocation, rather than with SLA for nine *Eucalypt* species. *Encelia farinosa* plants from dry places are reported to have lower SLA than those from wetter places, and lower SLA than *Encelia frutescens* plants growing close-by but tapping in on deeper water (Ehleringer and Cook 1984, Ehleringer 1988). The lower SLA of *E. farinosa* from the driest sites is completely due to a thick layer of leaf hairs, which reflect the sunlight and may take up 50% of the biomass of the leaf. Finally, Tsialtas et al. (2004) found low-SLA species to be more abundant in dry Mediterranean sites than high-SLA species.

Trampling and Wind Damage

A very clear case in which a low SLA is of survival value is trampling. Dijkstra and Lambers (1989) studied two subspecies of *Plantago major*. One is an annual, growing on occasionally flooded river banks, whereas the other is a perennial, occurring in frequently mown and trampled lawns. The leaves of the first subspecies have a high SLA and are erect. Those of the second are prostrate and have a low SLA, as well as a higher proportion of biomass in cell walls. An experiment showed that the subspecies with the low SLA survived trampling better (Table 3.6). Similarly, Meerts and Garnier (1996) found that *Polygonum aviculare* genotypes from trampled habitats show a lower SLA under laboratory conditions than those from nontrampled places, and so did Kobayashi et al. (1999) for trampling-resistant forbs and grasses.

Leaf destruction can also take place by strong winds. Pammenter et al. (1986) analyzed differences in leaf anatomy between two *Agrostis* species that occur at sub-Antarctic islands. *Agrostis magellanica* is native to these islands, *Agrostis stolonifera* has been introduced recently. The latter species has displaced *A. magellanica* in wind-sheltered areas, but not in

TABLE 3.6
Leaf Characteristics, Chemical Composition, and Trampling Survival, Determined for Lab-Grown Plants of Two Subspecies of *P. major*

	P. major River Bank	*P. major* Trampled Lawn
SLA (m^2 kg^{-1})	39	29
Leaf thickness (μm)	280	280
Leaf density (kg m^{-3})	90	125
Concentrations (mg g^{-1})		
N	42	42
Minerals	190	130
Cell walls	200	240
Trampling survival (%)	6	45

Source: Dijkstra, P. and Lambers, H., *New Phytol.*, 113, 283, 1989.

Note: One subspecies is an annual, occurring on irregularly flooded riverbanks; the other is perennial from a frequently mown and trampled lawn.

more open terrain. This is most likely explained by the fact that the leaves of the introduced species are relatively thin and fragile. *A. magellanica*, with an unusual high fraction of the leaf volume occupied by sclerenchyma, a high leaf thickness, and a much lower SLA (Table 3.4), seems better able to withstand wind damage.

Alpine species have been found to have lower SLA than lowland species, both in the field (Körner and Diemer 1987) and in the laboratory (Atkin et al. 1996a). Under both sets of conditions, this was at least partly due to thicker leaves. It has been suggested that the increased wind speed measured at higher elevations could play a role here as well. In an experiment with an upland, low-SLA species and a lowland high-SLA species grown in a wind tunnel at high wind speed, Woodward (1983) found much greater leaf damage in the high-SLA species.

Herbivory

Herbivory by insects and mammals may have a large impact on the vegetation and strongly damage individual plants. Species that have a low attractivity to herbivores are those with a low water content, a low organic nitrogen concentration, high concentrations of lignin and other cell wall components, a high concentration of secondary compounds like tannins and tough leaves in general (Scriber 1977, Grubb 1986, Coley and Barone 1996). These are all traits associated with a low SLA.

An interesting case where species with an inherently high SLA perform less is in the understorey of tropical forests. Plants generally acclimate to a low light environment by an increase in SLA, enabling to capture more light per unit leaf biomass. Therefore, one might expect at first to find high-SLA species in the understorey. Indeed, high-SLA pioneer species have a higher RGR at low light intensities than shade-tolerant species (Veneklaas and Poorter 1998). Notwithstanding their initially higher growth rate, these plants suffer from a high mortality rate under these conditions compared with the shade-tolerant seedlings with low SLA (Kitajima 1994). Susceptibility to herbivory may play a role here as well.

CONCLUSIONS

We have shown that there is a relationship between the potential RGR of a species, as measured under optimal conditions in the laboratory, and the distribution of species in the field. Fast-growing species are found in habitats favorable for plant growth, either on the short term (annuals, ruderals) or in the longer term (competitors according to the classification of Grime). Species found in harsh environments generally have a lower potential RGR. These interspecific differences in RGR_{max} are largely due to inherent differences in SLA. We have shown evidence that selection in the field may have acted, at least partly, on parameters related to SLA. An inherently low SLA (and the suite of traits associated with it) diminishes losses of nutrients or biomass due to grazing, trampling, or leaf turnover and may be advantageous for plants growing in harsh environments. If correct, this implies that a low RGR_{max} is merely a side effect of the low SLA.

Up to now, we have focused on the average fast- and slow-growing species, as representatives of two groups of functional types of species. However, it would be naive to think that any categorization into functional types can account for more than a modest amount of the variation that we encounter in the field. Moreover, it is evident that the plant traits discussed here are not the only ones that are shaped by evolution. To capture more of the variability, it would be desirable to include some other plant characteristics that are to a large extent independent of SLA and RGR. Westoby (1998) presented an interesting scheme, proposing to categorize species on the basis of SLA, seed mass, and maximum height achieved in the canopy. Seed mass is an important predictor of seed output per square meter of canopy cover, and a good indicator of seedling survival (Westoby 1998). The maximum height a plant can achieve is important for the type of vegetation in which it can survive. Westoby et al. (2002) further added leaf size and its relationship with twig size to this scheme. Such characterization across a wide range of species could be a promising avenue to further increase our understanding of the success or failure of species in a given habitat.

ACKNOWLEDGMENTS

We thank Arjen Biere, Hans Cornelissen, Lourens Poorter, Peter Reich, Jeroen van Arendonk, and Adrie van der Werf for generously providing published and unpublished data. Lourens Poorter, Owen Atkin, and an anonymous reviewer made helpful remarks on a previous version of this chapter.

REFERENCES

Aerts, R., 1990. Nitrogen use efficiency in evergreen and deciduous species from heathlands. Oecologia 84: 391–397.

Aerts, R., 1995. The advantages of being evergreen. Trends in Ecology and Evolution 10: 402–407.

Aerts, R. and F.S. Chapin, 2000. The mineral nutrition of wild plants revisited: a re-evaluation of processes and patterns. Advances in Ecological Research 30: 1–67.

Atkin, O.K. and D.A. Day, 1990. A comparison of the respiratory processes and growth rates of selected Australian alpine and related lowland plant species. Australian Journal of Plant Physiology 17: 517–526.

Atkin, O.K., B. Botman, and H. Lambers, 1996a. The causes of inherently slow growth in alpine plants: an analysis based on the underlying carbon economies of alpine and lowland *Poa* species. Functional Ecology 10: 698–707.

Atkin, O.K., B. Botman, and H. Lambers, 1996b. The relationship between the relative growth rate and nitrogen economy of alpine and lowland *Poa* species. Plant, Cell and Environment 19: 1324–1330.

Ball, M.C., 1988. Salinity tolerance in mangroves *Aegiceras corniculatum* and *Avicennia marina*. I. Water use in relation to growth, carbon partitioning, and salt balance. Australian Journal of Plant Physiology 15: 447–464.

Baruch, Z., 1996. Ecophysiological aspects of the invasion by African grasses and their impact on biodiversity and function of neotropical savannas. In: O.T. Solbrig, E. Medina, and J. Silva, eds. Biodiversity and Savanna Ecosystem Processes: A Global Perspective. Ecological Studies, Vol. 121, Springer-Verlag, Berlin, pp. 73–93.

Baruch, Z., M.M. Ludlow, and R. Davis, 1985. Photosynthetic responses of native and introduced C4 grasses from Venezuelan savannas. Oecologia 67: 388–393.

Benjamin, L.R. and R.C. Hardwick, 1986. Sources of variation and measures of variability in even-aged stands of plants. Annals of Botany 58: 757–778.

Berendse, F. and R. Aerts, 1987. Nitrogen-use-efficiency: A biologically meaningful definition? Functional Ecology 1: 293–296.

Biere, A., 1996. Intra-specific variation in relative growth rate: Impact on competitive ability and performance of *Lychnis flos-cuculi* in habitats differing in soil fertility. Plant and Soil 182: 313–327.

Bilbrough, C.J. and M.M. Caldwell, 1997. Exploitation of springtime ephemeral N pulses by six great basin plant species. Ecology 78: 231–243.

Blackman, V.H., 1919. The compound interest law and plant growth. Annals of Botany 33: 353–360.

Bloor, J.M.G., 2003. Light responses of shade-tolerant tropical tree species in north-east Queensland: a comparison of forest- and shadehouse-grown seedlings. Journal of Tropical Ecology 19: 163–170.

Boorman, L.A., 1982. Some plant growth patterns in relation to the sand dune habitat. Journal of Ecology 70: 607–614.

Bradshaw, A.D., 1987. Functional ecology: comparative ecology? Functional Ecology 1: 71.

Bradshaw, A.D., M.J. Chadwick, D. Jowett, and R.W. Snaydon, 1964. Experimental investigations into the mineral nutrition of several grass species. IV. Nitrogen level. Journal of Ecology 52: 665–676.

Causton, D.R. and J.C. Venus, 1981. The Biometry of Plant Growth. E. Arnold, London.

Chapin, F.S., 1980. The mineral nutrition of wild plants. Annual Review of Ecology and Systematics 11: 233–260.

Chapin, F.S., 1988. Ecological aspects of plant mineral nutrition. Advances in Mineral Nutrition 3: 161–191.

Chapin, F.S., K. Autumn, and F. Pugnaire, 1993. Evolution of suites of traits in response to environmental stress. American Naturalist 142: S78–S92.

Coley, P.D., 1983. Herbivory and defense characteristics of tree species in lowland tropical forest. Ecological Monographs 53: 209–233.

Coley, P.D. and J.A. Barone, 1996. Herbivory and plant defenses in tropical forests. Annual Review of Ecology and Systematics 27: 305–335.

Cornelissen, J.H.C., M.J.A. Werger, P. Castro-Diez, J.W.A. van Rheenen, and A.P. Rowland, 1997. Foliar nutrients in relation to growth, allocation and leaf traits in seedlings of a wide range of woody plant species and types. Oecologia 111: 460–469.

Cornelissen, J.H.C., P. Castro-Díez, and A.L. Carnelli, 1998. Variation in relative growth rate among woody species. In: H. Lambers, H. Poorter, and M. van Vuuren, eds. Inherent Variation in Plant Growth. Physiological Mechanisms and Ecological Consequences. Backhuys Publishers, Leiden, pp. 363–392.

Corré, W.J., 1983a. Growth and morphogenesis of sun and shade plants. I. The influence of light intensity. Acta Botanica Neerlandica 32: 49–62.

Corré, W.J., 1983b. Growth and morphogenesis of sun and shade plants. II. The influence of light quality. Acta Botanica Neerlandica 32: 185–202.

Den Dubbelden, K.C. and R.W. Verburg, 1996. Inherent allocation patterns and potential growth rates of herbaceous climbing plants. Plant and Soil 184: 341–347.

Dijkstra, P. and H. Lambers, 1989. Analysis of specific leaf area and photosynthesis of two inbred lines of *Plantago major* differing in relative growth rate. New Phytologist 113: 283–290.

Eckstein, R.L. and P.S. Karlsson, 1997. Above-ground growth and nutrient use by plants in a subarctic environment: effects of habitat, life-form and species. Oikos 79: 311–324.

Ehleringer, J.R., 1988. Comparative ecophysiology of *Encelia farinosa* and *Encelia frutescens*. I. Energy balance considerations. Oecologia 76: 553–561.

Ehleringer, J.R. and C.S. Cook, 1984. Photosynthesis in *Encelia farinosa* Gray in response to decreasing leaf water potential. Plant Physiology 75: 688–693.
Eissenstat, D.M. and R.D. Yanai, 1997. The ecology of root lifespan. Advances in Ecological Research 27: 1–60.
Ellenberg, H., 1988. Vegetation Ecology of Central Europe. Cambridge University Press, Cambridge.
Evans, G.C., 1972. The Quantitative Analysis of Plant Growth. Blackwell Scientific Publications, London.
Ewers, F.W. and R. Schmid, 1981. Longevity of needle fascicles of *Pinus longeava* (Bristlecone Pine) and other North American pines. Oecologia 51: 107–115.
Fichtner, K. and E.-D. Schulze, 1992. The effect of nitrogen nutrition on growth and biomass partitioning of annual plants originating from habitats of different nitrogen availability. Oecologia 92: 236–241.
Frissel, M.J., 1981. The definition of residence times in ecological models. In: F.E. Clark and T. Rosswall, eds. Terrestrial Nitrogen Cycles. Processes, Ecosystem Strategies and Management Impacts. Swedish Natural Science Research Council, Stockholm, pp. 117–122.
Garnier, E., 1991. Resource capture, biomass allocation and growth in herbaceous plants. Trends in Ecology and Evolution 6: 126–131.
Garnier, E., 1992. Growth analysis of congeneric annual and perennial grass species. Journal of Ecology 80: 665–675.
Garnier, E. and J. Aronson, 1998. Nitrogen-use-efficiency from leaf to stand level: clarifying the concept. In: H. Lambers, H. Poorter and M. van Vuuren, eds. Inherent Variation in Plant Growth. Physiological Mechanisms and Ecological Consequences. Backhuys Publishers, Leiden, pp. 525–538.
Garnier, E. and A.H.J. Freijsen, 1994. On ecological inference from laboratory experiments conducted under optimum conditions. In: J. Roy and E. Garnier, eds. A Whole Plant Perspective on Carbon-Nitrogen Interactions. SPB Academic Publishing, The Hague, pp. 267–292.
Garnier, E. and G. Laurent, 1994. Leaf anatomy, specific mass and water content in congeneric annual and perennial grass species. New Phytologist 128: 725–736.
Garnier, E. and S. Vancaeyzeele, 1994. Carbon and nitrogen content of congeneric annual and perennial species: relationships with growth. Plant, Cell and Environment 17: 399–407.
Garnier, E., P. Cordonnier, J.-L. Guillerm, and L. Sonié, 1997. Specific leaf area and leaf nitrogen concentration in annual and perennial grass species growing in Mediterranean old-fields. Oecologia 111: 490–498.
Garnier, E., J. Cortez, G. Billès, M.-L. Navas, C. Roumet, M. Debussche, G. Laurent, A. Blanchard, D. Aubry, A. Bellmann, C. Neill, and J.-P. Toussaint, 2004. Plant functional markers capture ecosystem properties during secondary succession. Ecology 85: 2630–2637.
Gleeson, S.K. and D. Tilman, 1994. Plant allocation, growth-rate and successional status. Functional Ecology 8: 543–550.
Graves, J.D. and K. Taylor, 1986. A comparative study of *Geum rivale* L. *and Geum urbanum* L. to determine those factors controlling their altitudinal distribution. I. Growth in controlled and natural environments. New Phytologist 104: 681–691.
Grime, J.P., 1977. Evidence for the existence of three primary strategies in plants and its relevance to ecological and evolutionary theory. American Naturalist 111: 1169–1194.
Grime, J.P., 1979. Plant Strategies and Vegetation Processes. John Wiley & Sons, Chichester.
Grime, J.P. and R. Hunt, 1975. Relative growth rate: Its range and adaptive significance in a local flora. Journal of Ecology 63: 393–422.
Grubb, P.J., 1986. Sclerophylls, pachyphylls and pycnophylls: The nature and significance of hard leaf surfaces. In: B. Juniper and R. Southwood, eds. Insects and Plant Surface. Edward Arnold, London, pp. 137–150.
Gutschick, V.P., 1988. Optimization of specific leaf mass, internal CO_2 concentration, and chlorophyll content in crop canopies. Plant Physiology and Biochemistry 26: 525–537.
Hikosaka, H., 2003. A model of dynamics of leaves and nitrogen in a plant canopy: An integration of canopy photosynthesis, leaf life span, and nitrogen use efficiency. American Naturalist 162: 149–164.
Hobbs, R.J., 1997. Can we use plant functional types to describe and predict responses to environmental changes? In: T.M. Smith, H.H. Shugart, and F.I. Woodward, eds. Plant Functional Types. Their Relevance to Ecosystem Properties and Global Change. Cambridge University Press, Cambridge, pp. 66–90.

Hoffmann, W.A. and H. Poorter, 2002. Avoiding bias in calculations of relative growth rate. Annals of Botany 80: 37–42.

Hull, J.C. and H.A. Mooney, 1990. Effects of nitrogen on photosynthesis and growth rates of four California annual species. Acta Oecologica 11: 453–468.

Hunt, R., 1982. Plant Growth Curves. The Functional Approach to Plant Growth Analysis. E. Arnold, London.

Hunt, R. and J.H.C. Cornelissen, 1997. Components of relative growth rate and their interrelations in 59 temperate plant species. New Phytologist 135: 395–417.

Hunt, R. and P.S. Lloyd, 1987. Growth and partitioning. New Phytologist 106: 235–249.

Ingestad, T., 1979. Nitrogen stress in birch seedlings. II. N, K, P, Ca and Mg nutrition. Physiologia Plantarum 45: 149–157.

Ishikawa, S.I. and N. Kachi, 2000. Differential salt tolerance of two Artemisia species growing in contrasting coastal habitats. Ecological Research 15: 241–247.

Jones, H.G., 1983. Plants and Microclimate. Cambridge University Press, Cambridge.

Killingbeck, K.T., 1996. Nutrients in senesced leaves: Keys to the search for potential resorption and resorption proficiency. Ecology 77: 1716–1727.

Kitajama, K., 1994. Relative importance of photosynthetic traits and allocation patterns as correlates of seedling shade tolerance of 13 tropical trees. Oecologia 98: 419–428.

Kobayashi, T., H. Ikeda, and Y. Hori, 1999. Growth analysis and reproductive allocation of Japanese forbs and grasses in relation to organ toughness under trampling. Plant Biology 1: 445–452.

Konings, H., 1989. Physiological and morphological differences between plants with a high NAR or a high LAR as related to environmental conditions. In: H. Lambers, M.L. Cambridge, H. Konings, and T.L. Pons, eds. Causes and consequences of variation in growth rate and productivity in higher plants. SPB Academic Publishing BV, The Hague, pp. 101–123.

Körner, C. and M. Diemer, 1987. *In situ* photosynthetic response to light, temperature and carbon dioxide in herbaceous plants from low and high altitude. Functional Ecology 1: 179–194.

Lambers, H. and P. Dijkstra, 1987. A physiological analysis of genotypic variation in relative growth rate: Can growth rate confer ecological advantage? In: J. van Andel, J.P. Bakker, and R.W. Snaydon, eds. Disturbance in Grasslands. Junk Publishers, Dordrecht, pp. 237–251.

Lambers, H. and H. Poorter, 1992. Inherent variation in growth rate between higher plants: a search for physiological causes and ecological consequences. Advances in Ecological Research 23: 187–261.

Loveys, B.R., I. Scheurwater, T.L. Pons, A.H. Fitter, and O.K. Atkin, 2002. Growth temperature influences the underlying components of relative growth rate: an investigation using inherently fast- and slow-growing species. Plant, Cell and Environment 25: 975–987.

Mahmoud, A. and J.P. Grime, 1974. An analysis of competitive ability in three perennial grasses. New Phytologist 77: 431–435.

McGraw, J.B. and K. Garbutt, 1990. The analysis of plant growth in ecological and evolutionary studies. Trends in Ecology and Evolution 5: 251–254.

Meerts, P. and E. Garnier, 1996. Variation in relative growth rate and its components in the annual *Polygonum aviculare* in relation to habitat disturbance and seed size. Oecologia 108: 438–445.

Monk, C.D., 1966. An ecological significance of evergreenness. Ecology 47: 504–505.

Mooney, H.A., P.J. Ferrar, and R.O. Slatyer, 1978. Photosynthetic capacity and carbon allocation patterns in diverse growth forms of *Eucalyptus*. Oecologia 36: 103–111.

Niemann, G.J., J.B.M. Pureveen, G.B. Eijkel, H. Poorter, and J.J. Boon, 1992. Differences in relative growth rate in 11 grasses correlate with differences in chemical composition as determined by pyrolysis mass spectrometry. Oecologia 89: 567–573.

Osunkoya, O.O., J.E. Ash, M.S. Hopkins, and A.W. Graham, 1994. Influence of seed size and ecological attributes on shade-tolerance of rainforest tree species in northern Queeensland. Journal of Ecology 82: 149–163.

Pammenter, N.W., P.M. Drennan, and V.R. Smith, 1986. Physiological and anatomical aspects of photosynthesis of two *Agrostis* species at a sub-Antarctic island. New Phytologist 102: 143–160.

Parsons, R.F., 1968. The significance of growth rate comparisons for plant ecology. American Naturalist 102: 595–597.

Pons, T.L., 1977. An ecophysiological study in the field layer of ash coppice. II. Experiments with *Geum urbanum* and *Cirsium palustre* in different light intensities. Acta Botanica Neerlandica 26: 29–42.

Poorter, H., 1989. Interspecific variation in relative growth rate: On ecological causes and physiological consequences. In: H. Lambers, M.L. Cambridge, H. Konings, and T.L. Pons, eds. Causes and Consequences of Variation in Growth Rate and Productivity of Higher Plants. SPB Academic Publishing, The Hague, pp. 45–68.

Poorter, H., 2002. Plant growth and carbon economy. In: Encyclopedia of Life Sciences, http://www.els.net, London: Nature Publishing Group.

Poorter, H. and M. Bergkotte, 1992. Chemical composition of 24 wild species differing in relative growth rate. Plant, Cell and Environment 15: 221–229.

Poorter, H. and R. de Jong, 1999. Specific leaf area, chemical composition and leaf construction costs of plant species from productive and unproductive habitats. New Phytologist 143: 163–176.

Poorter, H. and E. Garnier, 1996. Plant growth analysis: An evaluation of experimental design and computational methods. Journal of Experimental Botany 47: 1343–1351.

Poorter, H. and P. Pothmann, 1992. Growth and carbon economy of a fast-growing and a slow-growing grass species as dependent on ontogeny. New Phytologist 120: 159–166.

Poorter, H. and C. Remkes, 1990. Leaf area ratio and net assimilation rate of 24 wild species differing in relative growth rate. Oecologia 83: 553–559.

Poorter, H. and A.K. van der Werf, 1998. Is inherent variation in RGR determined by LAR at low irradiance and NAR at high irradiance? A review of herbaceous species. In: H. Lambers, H. Poorter, and M. van Vuuren, eds. Inherent Variation in Plant Growth. Physiological Mechanisms and Ecological Consequences. Backhuys Publishers, Leiden, pp. 309–336.

Poorter, H. and R. Villar, 1997. The fate of acquired carbon in plants: chemical composition and construction costs. In: F.A. Bazzaz and J. Grace, eds. Plant Resource Allocation. Academic Press, New York, pp. 39–72.

Poorter, H., C. Remkes, and H. Lambers, 1990. Carbon and nitrogen economy of 24 wild species differing in relative growth rate. Plant Physiology 94: 621–627.

Poorter, H., C.A.D.M. van de Vijver, R.G.A. Boot, and H. Lambers, 1995. Growth and carbon economy of a fast-growing and a slow-growing grass species as dependent on nitrate supply. Plant and Soil 171: 217–227.

Poorter, H., C.P.E. van Rijn, T.K. Vanhala, K.J.F. Verhoeven, Y.E.M. de Jong, P. Stam, and H. Lambers, 2005. A genetic analysis of Relative Growth Rate and underlying components in *Hordeum spontaneum*. Oecologia 142: 360–377.

Poorter, L. 1999. Growth responses of 15 rain-forest tree species to a light gradient: the relative importance of morphological and physiological traits. Functional Ecology 13: 396–410.

Reich, P.B., 1998.Variation among plant species in leaf turnover rates and associated traits: Implications for growth at all life stages. In: H. Lambers, H. Poorter, and M. van Vuuren, eds. Inherent Variation in Plant Growth. Physiological Mechanisms and Ecological Consequences. Backhuys Publishers, Leiden, pp. 467–487.

Reich, P.B., M.B. Walters, and D.S. Ellsworth, 1992. Leaf life-span in relation to leaf, plant, and stand characteristics among diverse ecosystems. Ecological Monographs 62: 365–392.

Reich, P.B., M.B. Walters, and D.S. Ellsworth, 1997. From tropics to tundra: Global convergence in plant functioning. Proceedings of the National Academy of Sciences (USA) 94: 13730–13734.

Reiling, K. and A.W. Davison, 1992. The response of native, herbaceous species to ozone: growth and fluorescence screening. New Phytologist 120: 29–37.

Reynolds, H.L. and C. d'Antonio, 1996. The ecological significance of plasticity in root weight ratio in response to nitrogen: Opinion. Plant and Soil 185: 75–97.

Robinson, D., 1991. Strategies for optimising growth in response to nutrient supply. In: J.R. Porter and D.W. Lawlor, eds. Plant Growth. Interactions with Nutrition and Environment. Cambridge University Press, Cambridge, pp. 177–199.

Rorison, I.H., 1968. The response to phosphorus of some ecologically distinct plant species. I. Growth rates and phosphorus absorption. New Phytologist 67: 913–923.

Rozijn, N.A.M.G. and D.C. van der Werf, 1986. Effect of drought during different stages in the life cycle on the growth and biomass of two *Aira* species. Journal of Ecology 74: 507–523.

Ryser, P., 1996. The importance of tissue density for growth and life span of leaves and roots: A comparison of five ecologically contrasting grasses. Functional Ecology 10: 717–723.

Schieving, F. and H. Poorter, 1999. Carbon gain in a multispecies canopy: the role of Specific Leaf Area and Photosynthetic Nitrogen-Use Efficiency in the tragedy of the commons. New Phytololgist 143: 201–211.

Schippers, P. and H. Olff, 2000. Biomass partitioning, architecture and turnover of six herbaceous species from habitats with different nutrient supply. Plant Ecology 149: 219–231.

Schwarz, M. and J. Gale, 1984. Growth response to salinity at high levels of carbon dioxide. Journal of Experimental Botany 35: 193–196.

Scriber, J.M., 1977. Limiting effects of leaf-water content on the nitrogen utilization, energy budget and larval growth of *Hyalophora cecropia* (Lepidoptera: Saturniidae). Oecologia 28: 269–287.

Shipley, B., 1995. Structures interspecific determinants of specific leaf area in 34 species of herbaceous angiosperms. Functional Ecology 9: 312–319.

Shipley, B., 2006. Net assimilation rate, specific leaf area and leaf mass ratio: which is most closely correlated with relative growth rate? A meta-analysis. Functional Ecology 20: 565–574.

Shipley, B. and P.A. Keddy, 1988. The relationship between relative growth rate and sensitivity to nutrient stress in twenty-eight species of emergent macrophytes. Journal of Ecology 76: 1101–1110.

Small, E., 1972. Photosynthetic rates in relation to nitrogen cycling as an adaptation to nutrient deficiency in peat bog plants. Canadian Journal of Botany 50: 2227–2233.

Specht, R.L. and A. Specht, 1989. Canopy structure in *Eucalyptus*-dominated communities in Australia along climatic gradients. Acta Oecologica, Oecologia Plantarum 10: 191–213.

Spitters, C.J.T. and R. Aerts, 1983. Simulation of competition for light and water in crop-weed associations. Aspects of Applied Biology 4: 467–483.

Stockey, A. and R. Hunt, 1994. Predicting secondary succession in wetland mesocosms on the basis of autecological information on seeds and seedlings. Journal of Applied Ecology 31: 543–559.

Tsialtas, J.T., T.S. Pritsa, and D.S. Veresoglou, 2004. Leaf physiological traits and their importance for species success in a Mediterranean grassland. Photosynthetica 42: 371–376.

Van Andel, J. and A. Biere, 1989. Ecological significance of variability in growth rate and plant productivity. In: H. Lambers, M.L. Cambridge, H. Konings, and T.L. Pons, eds. Causes and Consequences of Variation in Growth Rate and Productivity of Higher Plants. SPB Academic Publishing, The Hague, pp. 257–267.

Van Arendonk, J.J.C.M. and H. Poorter, 1994. The chemical composition and anatomical structure of leaves of grass species differing in relative growth rate. Plant, Cell and Environment 17: 963–970.

Van der Werf, A., M. van Nuenen, A.J. Visser, and H. Lambers, 1993. Effects of N-supply on the rates of photosynthesis and shoot and root respiration of inherently fast- and slow-growing monocotyledonous species. Physiologia Plantarum 89: 563–569.

Van der Werf, A., H. Poorter, and H. Lambers, 1994. Respiration as dependent on a species' inherent growth rate and on the nitrogen supply to the plant. In: J. Roy and E. Garnier, eds. A Whole Plant Perspective on Carbon-Nitrogen Interactions. SPB Academic Publishing, The Hague, pp. 91–110.

Van der Werf, A., R.H.E.M. Geerts, F.H.H. Jacobs, H. Korevaar, M.J.M. Oomes, and W. de Visser, 1998. The importance of relative growth rate and associated traits for competition between species. In: H. Lambers, H. Poorter, and M. van Vuuren, eds. Inherent Variation in Plant Growth. Physiological Mechanisms and Ecological Consequences. Backhuys Publishers, Leiden, pp. 489–502.

Van Diggelen, J., 1988. A Comparative Study on the Ecophysiology of Salt Marsh Halophytes. PhD Thesis, Free University, Amsterdam.

Veneklaas, E.J. and L. Poorter, 1998. Growth and carbon partitioning of tropical tree seedlings in contrasting light environments. In: H. Lambers, H. Poorter, and M. van Vuuren, eds. Inherent Variation in Plant Growth. Physiological Mechanisms and Ecological Consequences. Backhuys Publishers, Leiden, pp. 337–361.

Verhoeven, K.J.F., A. Biere, E. Nevo, and J.M.M. van Damme, 2004. Differential selection of growth rate-related traits in wild barley, Hordeum spontaneum, in contrasting greenhouse nutrient environments. Journal of Evolutionary Biology 17: 184–196.

Verkleij, J.A.C. and J.E. Prast, 1989. Cadmium tolerance and co-tolerance in *Silene vulgaris* (Moench.) Garcke (= *S. cucubalis* (L.) Wib.). New Phytologist 111: 637–645.

Vile, D., B. Shipley, and E. Garnier, 2006. Ecosystem productivity can be predicted from potential relative growth rate and species abundance. Ecology Letters 9: 1061–1067.

Villar, R. and J. Merino, 2001. Comparison of leaf construction costs in woody species with differing leaf life-spans in contrasting ecosystems. New Phytologist 151: 213–226.

Walters, M.B., E.L. Kruger, and P.B. Reich, 1993. Relative growth rate in relation to physiological and morphological traits for northern hardwood tree seedlings: species, light environment and ontogenetic considerations. Oecologia 96: 219–231.

Warren, C.R. and M.A. Adams, 2005. What determines interspecific variation in relative growth rate of Eucalyptus seedlings? Oecologia 144: 373–381.

West, C., G.E. Briggs, and F. Kidd, 1920. Methods and significant relations in the quantitative analysis of plant growth. New Phytologist 19: 200–207.

Westoby, M., 1998. A leaf-height-seed (LHS) plant ecology strategy scheme. Plant and Soil 199: 213–227.

Westoby, M., S.D. Falster, A.T. Moles, P.A. Vesk, and I.J. Wright, 2002. Plant ecological strategies: Some leading dimensions of variation between species. Annual Review of Ecology and Systematics 33: 125–159.

Wilson, E.O., 1992. The Diversity of Life. Harvard University Press, Harvard.

Wilson, J.B., 1988. The cost of heavy-metal tolerance. Evolution 42: 408–413.

Witkowski, E.T.F. and B.B. Lamont, 1991. Leaf specific mass confounds leaf density and thickness. Oecologia 88: 486–493.

Woodward, F.I., 1979. The differential temperature response of the growth of certain plant species from different altitudes. I. Growth analysis of *Phleum alpinum* L., *P. bertolonii* D.C., *Sesleria albicans* KIT. and *Dactylis glomerata* L. New Phytologist 82: 385–395.

Woodward, F.I., 1983. The significance of interspecific differences in specific leaf area to the growth of selected herbaceous species from different altitudes. New Phytologist 96: 313–323.

Wright, F.I. and M. Westoby, 1999. Differences in seedling growth behaviour among species: trait correlations across species, and trait shifts along nutrient compared to rainfall gradients. Journal of Ecology 98: 85–97.

Wright, I.J., P.B. Reich, M. Westoby, D.D. Ackerly, Z. Baruch, F. Bongers, J. Cavender-Bares, +26 others, 2004. The worldwide leaf economics spectrum. Nature 428: 821–827.

4 The Architecture of Plant Crowns: From Design Rules to Light Capture and Performance

Fernando Valladares and Ülo Niinemets

CONTENTS

INTRODUCTION

Plants exhibit a striking diversity of forms and structures, which are difficult to interpret. The functional approach to the study of plant form emerged as a separate discipline at the beginning of the twentieth century with the first classifications of growth forms in relation to climate and with tentative ecophysiological studies of plant responses to the environment (Waller 1986). Physiological ecology now makes detailed predictions on how physical and physiological characteristics affect plant photosynthesis, whereas plant population ecology translates patterns of growth into fitness of individuals and populations. And plant structure remains an essential tool for all these exercises of interpreting plant performance in natural habitats and for scaling from cellular and leaf-level to ecosystem processes (Ehleringer and FIeld 1993). Plant performance can be understood as the crucial link between its phenotype and its ecological success and the form becomes ecologically and evolutionary relevant when it affects performance (Koehl 1996). It is important to consider that misconceptions can arise from studies in which selective advantages of particular structures are not made with a mechanistic understanding of how the structural traits affect performance. Koehl (1996) showed that the relationship between morphology and performance can be nonlinear, context-dependent, and sometimes surprising. Remarkably, new functions and novel ecological consequences of morphological changes can arise simply as the result of changes in size or habitat.

While all would agree that structure is intrinsically coupled with function, the impetus is often stronger to investigate physiological mechanisms rather than the functional implications of plant form. This is not to say that functional plant architecture has been ignored. For instance, the role of canopy architecture in competition for light has been addressed in several works after the keystone study by Horn (1971). However, the architectural constraints of plant success, which is of plant persistence or expansion in the community, have not been explored extensively. Plant architecture involves the manner in which the foliage is positioned in different microenvironments and determines the flexibility of a shoot system to take advantage of unfilled gaps in the canopy, to allocate and utilize assimilates, and to recover from herbivory or mechanical damage (Caldwell et al. 1981, 1983, Küppers 1989). Except in particular or very extreme environments, plant physiology alone does not explain ecological success, since growth and competition have been clearly related to structural features (Küppers 1994). In agreement with Tomlinson (1987), the study of plant morphology is an integrative discipline rather than a subject restricted to the comparison of anatomical details of plant life cycles. We attempt to demonstrate that plant morphology in general and plant architecture in particular belongs more rightly within the fields of plant ecophysiology and plant population biology.

We analyze here plant shape from a functional point of view. The basic plant design and the many interpretations and implications of its modular nature are presented here as an indispensable starting point to enter into discussions on function and adaptive value of crown structural features. We further discuss the relationships between plant shape and light capture. Plants depend on the light energy that they capture by photosynthesis, and solar radiation is the major driving force affecting not only photosynthetic activity, but also leaf temperature, leaf water status, and many other physiological processes of the plant. Crown architecture is crucial for light capture and for the distribution of light to each particular photosynthetic unit of the crown, but must also serve several other functions. The architectural design of a given plant must provide safety margins to cope with gravity and wind; therefore, biomechanical constraints must be taken into account when assessing the influence of morphology and architecture on plant performance. The structural basis of light capture by plant crowns is explored here from the leaf level to the community level with special attention to leaf angle, phyllotaxis, branching patterns, and crown shape. Examples of plant architecture in extreme light environments are included, where the functional implications for light capture of a range of structural features can be better seen. In the analysis of plant shape at the community level, two main functional concepts involving plant interactions are discussed: the occupation of the space and the shading of neighbors. As there are important world-scale modifications in overall light availability and in the various components of solar radiation, in particular in the ratio of direct to diffuse irradiance (Roderick et al. 2001, Gu et al. 2002, Farquhar and Roderick 2003), understanding the fundamental relationships between plant architecture and efficiency of harvesting light is the precondition in grasping the global change effects on vegetation productivity.

PLANT DESIGN

The shape of a given plant is determined by the shape of the space that it fills, but most plants attain a characteristic shape when grown alone in the open due to an inherited developmental program (Horn 1971). This developmental program usually implies the reiterative addition of a series of structurally equivalent subunits (branches, axes, shoots, leaves), which confers plants a modular nature. This developmental program is the result of plant evolution under some general biomechanical constraints. For instance, the shape of the crown of a tree is constrained by the fact that the cost of horizontal branches is greater than that of vertical branches (Mattheck 1991). This section explores the functional implications of these two general aspects, the modular nature of plants and the biomechanical constraints of shape, which in addition to the environment where the plant grows determine plant architecture.

Basic Architecture of Terrestrial Plants

Terrestrial vascular plants must combine the structural requirements of water-conduction and gas-exchange systems with the problems of mechanical support of aerial structures and light capture by the photosynthetic surfaces. Many different solutions to these frequently opposing problems have been found during plant evolution (Niklas and Kerchner 1984, Speck and Vogellehner 1988, Niklas 1990, 1997). The diversification into trees, shrubs, and herbs occurred relatively rapidly (Raven 1986), and by the end of the Devonian, many alternative plant designs were successfully tested in most terrestrial systems. From primitive cylindrical or flat, two-dimensional photosynthetic surfaces restricted to liquid environments, terrestrial plants evolved complex three-dimensional arrangements of the photosynthetic units, which required stomata for control over water loss preventing embolism (Woodward 1998), lignified fibers for support, and a specialized root system for efficient competition for belowground resources (Jackson et al. 1999). However, because no one design dominates in all environments, specialization for efficiency in any given environment involved structural trade-offs that made the same plant less competitive in other environments (Waller 1986). Most of what follows in this chapter aims to explore the ecological implications and the trade-offs involved in the various and varying architectural designs of extant plants.

Modular Nature of Plants

In the crown of most vascular plants, it is easy to recognize a hierarchical series of subunits. The largest subunit is the branch, which is made up of modules (Porter 1983). *Module* is a general term that refers to a shoot with its leaves and buds, and the term can be applied to either determinate (structures whose apical meristem dies or produces a terminal inflorescence) or indeterminate shoot axes (Waller 1986). Modules are, in turn, made up of smaller subunits consisting of a leaf, its axillary buds, and the associated internode. These small subunits have been called metamers (White 1984). Since plants have many redundant modules or organs that have similar or identical functions (e.g., leaves or shoots transforming absorbed light into biomass), plants have been seen as metapopulations (White 1979). Such redundant modules are not fully dependent on one another, and, in fact, individual modules continue to function when neighbor organs are removed (Novoplansky et al. 1989). The existence of hundreds of redundant subunits within a single plant raises the question: To what degree do these structural subunits (e.g., shoots) respond independently to the environment?

Scaling Up and Down

The study of functional modularity of plants can be tackled at different scales. The smallest end is the so-called nutritional or physiological unit, comprising a unit of foliage, the axillary bud, and the corresponding portion of stem (Watson 1986). As pointed out by Sprugel et al. (1991), the opposite end of the spectrum would be the clonal herbs, in which each module (ramet) contains all of the structural parts necessary for independent existence. The branch is an intermediately scaled unit, which is very convenient because it is large enough to integrate most relevant physiological processed, but small enough to be used in ecophysiological experiments. For this reason, branches have been used extensively by ecologists and ecophysiologists to scale from leaf-level measurements to the whole plant or to the plant community (Gartner 1995).

All branches within a plant are structurally and physiologically connected to one another, but the mutual interactions are not always easy to elucidate. To make reliable scaling and generalization exercises, branch autonomy must be investigated thoroughly. Branch autonomy depends on the resource—carbon, water, or nutrients. The most clear aspect of branch autonomy is that related with carbon budget, since most branches fix all the carbon they need, and usually fix more, becoming exporters or sources of carbon in contrast with roots or reproductive structures, which are important carbon sinks (Geiger 1986). Although branches cannot be completely autonomous with respect to water and nutrients, which come from the roots via the stem, they exhibit different levels of uncoupling with the rest of the branches of the crown, that is, different levels of relative autonomy (Tyree 1999). In most species, branches are somewhat hydraulically isolated from the rest of the plant; thus, in words of Tyree (1988), branches can be treated as small, independent seedlings rooted in the main bole. Nevertheless, branches are imperfect substitutes for studies on whole plants, especially when exceptions to the general branch autonomy can be expected (Sprugel et al. 1991).

Ecology of Branch Autonomy

Branch autonomy has two major ecological advantages: (1) control of stress and damage, and (2) a more efficient exploitation of heterogeneous environments (Hardwick 1986). A compartmentalized plant may be less vulnerable to pathogens or herbivores than an integrated plant. It is well known that trees are capable of walling off injured or too-old branches, which provides an efficient protection against spreading of infections and against a net energy drain on the organism, respectively (Sprugel et al. 1991). A similar argument on the advantages of branch autonomy can be built for the prevention of runaway cavitation, for example, the formation of gas bubbles when transpiration rate exceeds water transport that block xylem vessels or tracheids (Tyree 1999, Zimmermann et al. 2000).

Because the different aerial parts of a plant (e.g., branches and leaves) are generally in different light environments, plants frequently face the problem of distributing limited resources in a way that would optimize the performance of units exposed to heterogeneous light conditions. Although plants do not forage in the classical sense of moving around to different prey locations, they do exhibit a foraging behavior (Hutchins and de Kroon 1994). Plants forage because they must spend energy producing the leaves and the associated supporting structures necessary to harvest light, and their fitness is increased if this energy is spent efficiently, that is, if leaves are arranged appropriately to maximize light capture. A plant that has new leaves in high-light areas of the crown has an advantage over one that remains symmetric and sets out leaves equally in all possible locations. Branch autonomy with respect to carbon budget enhances the efficiency of light foraging because branches exposed to high-light grows bigger, shaded branches stops growing, and no energy is wasted in producing leaves in shaded areas (Sprugel et al. 1991). However, this is the case only for woody plants with indeterminate or multiple flushing growth patterns, where photosynthate for new leaves at the top of a shoot has been shown to come primarily from the older leaves of the same shoot (Fujimori and Whitehead 1986). In woody plants with determinate, single-flush growth patterns, efficient light foraging is not achieved via branch autonomy but rather via increased bud production in high-light areas; the buds draws on reserves throughout the tree in the next growing season (Sprugel et al. 1991). Several evidences indicate that branches are interdependent so that a positive carbon budget by itself does not ensure branch survival, and a stressed branch on a tree where all other branches are also stressed does better than a similarly stressed branch on a tree where some branches are relatively unstressed. As stated by Sprugel (2002) although branch autonomy is an important and useful principle, it is not an absolute rule governing branch growth.

Modularity versus Integrity

Despite the ecological relevance and the functional evidence of a certain autonomy of the different modules of a given plant, many different studies suggest that a plant is more than just a population of redundant organs because it responds to the environment as an integrated individual and not as a simple colony with limited mutual aid (Sprugel et al. 1991, Sachs et al. 1993). The simplistic, albeit tempting, concept that a single plant is not a unit but a collection of independent subunitary parts became widespread during the nineteenth century and persisted until modern times (White 1979). It is reminiscent of the assumption that organismal structure and function can be understood by studying the cells, since cells have been considered the building blocks or organism form since the publication of the cell theory in 1938 (Kaplan and Hagemann 1991). In the advocacy of plant integrity, plants have been considered "metapopulations" (White 1979) in the context of the classical etymology of *meta-* as sharing. Therefore, plants are referred to as metapopulations when the shares elements that make up the morphological structure of an individual are emphasized. Under controlled conditions, plants have been shown to do more than respond locally to the degree to which they are damaged: interactions and mutual support between branches allowed treated plants for the comparison of available branches, and for the diversion of resources so as to increase the chances of greatest overall success (Sachs and Hassidim 1996, Sprugel 2002).

Nevertheless, two interesting lines of evidence support the notion that the modules of a plant are functionally independent of one another, at least to some extent: (1) independent patterns of phenology between branches, and (2) competitive interactions between modules for limited resources as a consequence of a eustelic arrangement. Each module undergoes a complete life cycle of birth, growth, maturation, senescence, and death; therefore, a plant can be studied as a dynamic population of modules with a distinct age structure following rigorous demographic analyses (Room et al. 1994). Individual leaves and foliage units are manifestly not all the same due to the two simple facts that they are not of the same age and that they are borne in different positions relative to each other (Harper 1989). In this sense, and considering the remarkable genetic variability of the different modules of a plant and the fitness differentials between modules, individual plants can be tackled as colonies of evolutionary individuals (Gill 1991).

Plant Biomechanics: Coping with Gravity and Wind

While plant architecture is an outcome of many selective pressures, the shapes of plant parts, their elasticity, and resistance to strain, are constrained by well-known mechanical principles (McMahon and Kronauer 1976, Niklas 1992). Because aerial plant parts face the obvious forces of gravity and wind, a fraction of the biomass must be devoted to support. As mechanical structures of similar shape become increasingly inefficient with increasing size, the fraction required to support plants increases rapidly with increasing plant size. For instance, the strength of a column (e.g., a branch or a stem) scales with the square of its diameter, whereas its mass increases with diameter squared times length (Gere and Timoshenko 1997). For any given plant, the mechanical costs associated with its crown geometry must be balanced with the photosynthetic benefits associated with its light-capture efficiency.

The height to which a plant should grow depends on the environment and on the height of the neighboring plants, that is, the goal is not to grow tall, but grow taller than the others (Waller 1986, King 1990). The taller a plant becomes in its competition for light, the more light it needs to support its preexisting biomass and to achieve growth. In fact, the maximum height of tree can be determined by the balance between maximal potential carbon gain that occurs in full sunlight and carbon required for construction and maintenance costs of crown and roots (Givnish 1988). Although small-statured plants have smaller growth maintenance requirements per unit of light-absorbing machinery than large plants, growing taller implies greater access to light. In general, the higher the plant, the more light it intercepts during the course of the day (Jahnke and Lawrence 1965, King 1981, 1990). Thus, there is a payback of investing in height that can be especially relevant under situations of strong competition for light. Of course, the reverse is also true: being tall requires on average higher irradiances due to extensive maintenance costs (Givnish 1988).

Mechanical stability imposes the minimum amount of tissue required to support the crown and its units. The most likely mode of stem failure is elastic toppling, rather than failure under the weight of the crown. Accordingly, stem diameter scales with height, with a safety margin that prevents elastic toppling but not compressive failure (McMahon 1973). For most plants, height varies with trunk diameter in such a way that there is a margin of safety against buckling (Niklas 1994). When trees grow in the open with little competition, their size and shape is conservative, being only one-quarter of their theoretical buckling height (McMahon 1975, McMahon and Kronauer 1976). However, when competition in a forest is strong, trees cannot afford large safety margins, especially when they have not reached the canopy. Based on this, Givnish (1995) predicted that shade-intolerant pioneer species should have lower mechanical safety margins than shade-tolerant species of similar stature. High wood density, usually reached in long-lived species with slow tissue turnover, provides resistance against mechanical failure and against attack by fungi and insects (King 1986). However, it adds extra weight for a given height or length of the stem or branch, so the biomechanical advantages of a stronger building material are frequently neutralized by the additional load. Structural costs are minimized by constructing stems of low-density wood, and for this reason softwoods can grow faster than hardwoods (Horn 1971). Hence, pioneer trees are expected to have light, energetically inexpensive wood, whereas late successional trees should have dense, highly lignified wood. Most studies in temperate and tropical forests confirm this trend (Horn 1971, Givnish 1995). The different biomechanics and associated costs of the crowns of hardwoods and softwoods can be crucial depending on the sign and intensity of factors such as frequency of storms, stability of the substrate, competition for light, or availability of water and nutrients.

Another important biomechanical aspect of the crown is the branching pattern. Branching angles should minimize both structural costs and leaf overlap to achieve optimal plant growth. However, these two features are mutually exclusive because branching patterns and leaf arrangements that reduce leaf overlap often require more investment in supporting tissues (Givnish 1995). Plants segregate in the cost and benefit trade-offs that their crown design entails in a given environment. In general, tree mechanisms concentrate on a good mechanical design only if light capture is sufficient (Mattheck 1991, 1995), but the biomechanical theory of crown design is still insufficient for integrated comparisons of the particular advantages of each crown architecture.

Although gravity leads to static loading of a plant based on the weight of individual parts, the dynamic loading caused by wind is often transitory (Grace 1977, King 1986, Speck et al. 1990). However, the wind exerts permanent modifications of the overall shape of plants and affects the anatomy and density of the wood, inducing biomechanical changes at architectural and anatomical levels (Coutts and Grace 1995, Ennos 1997). The greatest effects of strong winds on trees are seen near the tree line, where most species exhibit the so-called *krummholz* form (Ennos 1997). *Krummholz* refers to environmentally dwarfed trees, in which the crown is a prostrate cushion that extends leeward from the short trunk (Arno and Hammerly 1984). Despite the fact that light harvesting can be decreased by the *krummholz* habit, carbon gain is enhanced in comparison with upright trees in equivalent environmental conditions due to the increased photosynthetic rates exhibited by the leaves, which are deep in the boundary layer and warmed more by the sun (James et al. 1994). Another interesting, albeit little explored aspect of plant biomechanics and wind is the dynamic reconfiguration of crown shape while the wind is blowing. Branches and foliage bend away with the wind, which reduces drag. It has been suggested that drag reduction should lead to flexible twigs in windy environments (Vogel 1996), and also to pinnate or lobed leaves due to the great degree of reconfiguration of these leaves in comparison with that of simple leaves (Vogel 1989). Increasing evidence is pointing to the existence of two main strategies regarding the wind as an ecological factor: (1) pioneer trees in windy habitats with flexible branches and pinnate or lobed leaves to reduce aerodynamic drag; and (2) late-successional trees or species from sheltered sites with simple leaves and rigid branches to maintain optimal light interception (Vogel 1989, Ennos 1997). A similar reasoning was given for woody plants that dwell along shores of streams and torrents: flexible twigs and narrow, willow-like leaves should prove adaptive since they reduce pressure drag during flash floods (van Steenis 1981, Vogel 1996).

Unusual growth forms pose specific biomechanical problems, and precise studies are required to interpret certain plant designs. For instance, in most species of *Opuntia* (Cactaceae), shoots are formed as a sequence of short, flattened stem segments called cladodes. Cladodes have an elliptical base that supports the greatly enlarged upper portion and joins over only a small portion of their periphery so that there is considerable flexing at the cladode–cladode junctions (Nobel and Meyer 1991). Despite the fact that the contact between cladodes is only 20% of that occurring in a similar stem of constant width, the resulting shoot structure is rigid and resistant to typical wind and gravity loadings. The remarkable strength of this cladode–cladode junction cannot be fully explained from a biomechanical point of view (Nobel and Meyer 1991). Other interesting study cases are palm trees. Their lack of secondary thickening exposes them to a risk of toppling that increases with crown height. Mechanical safety of certain palm trees seems to be maintained by increasing the tissue density over time, and by proliferation of existing tissues that leads to an increase in actual cross-section of the stem (Rich 1986, 1987).

DEVELOPMENT OF A CROWN SHAPE

Despite the fact that most plants exhibit an indefinite growth, which produces a remarkable variability in their final size, they have a recognizable form. The many meristems of a plant are integrated into a galaxy of possible but not random morphologies. Understanding the mechanisms behind the production, arrangement, and turnover of plant modules led morphologists to group plants in a small number of architectural models.

CROWN ARCHITECTURE AND MODELS OF GROWTH

Plants exhibit an extraordinary variety of branching patterns and foliage arrangements. The luxuriance of structural details of a forest canopy or the diversity of morphologies displayed by the herbs of a subalpine meadow can be overwhelming. For this reason, botanists and plant ecologists have looked at the developmental organization (architecture) of plants with a reductionist approach, slimming the complexity of plant shape to a sequence of simpler processes, but retaining the holistic features that determine plant construction (Tomlinson 1987). The questions of how many possible ways there are to build a plant and how many architectural models are exhibited by real plants have led to several classifications of plant shape. One of the best-known detailed classifications of plant architecture is provided by Hallé et al. (1978). This review further provides an extensive, comparative study of the ontogenetic changes of the shape of tropical trees. In fact, most systematic descriptions and cataloging of architectural patterns have been based on trees. The most interesting features of these classifications are (1) a revival of the notion of modular construction and its importance in the generation of plant shape, and (2) an emphasis on understanding the mechanisms behind the dynamics of the arrangement, production, and turnover of plant modules and subunits (Porter 1989). This sort of information has made possible the realistic reconstruction of virtual plants, which is leading to in-depth understanding of plant growth in response to the environment and to promising orientations for plant breeding and pests and pathogens management, thanks to the potential of virtual experimentation (Room et al. 1996).

Branching: The Framework of a Crown

Branching complexity ranges from plants with a single axis to large trees with many orders of branching in three-dimensional space. However, the overall complex shape of a tree can be determined by surprisingly few parameters since a new branch is geometrically determined by just two parameters: branching angle and branch length (Honda et al. 1997). Repetition of the branching generates the distinctive complexity of plant crowns, and the relative simplicity of the process has resulted in the generation of numerous computer models that simulate branching and growth of plants with remarkable realism (Waller and Steingraeber 1985, Fisher 1992).

Although some trees have a single axis (e.g., most palms) and some have many similar branching axes, most species of trees have two or more types of axes that can be distinguished by their primary orientation, symmetry, or form. In general, leader axes are radially symmetrical, whereas lateral branch axes are dorsiventrally symmetrical (Fisher 1986). Differences in initial vigor of lateral branches results in a well-defined main axis, which is established commonly in a regular, alternating zigzag pattern (Fisher 1986). The branching and consequent growth of trees and shrubs can be characterized by vertical or longitudinal, and horizontal or lateral symmetries. Vertical symmetry is characterized by growth of branches at the top (acrotony) or at the base (basitony), whereas lateral symmetry is characterized by branch growth at the upper or lower side of the lateral branch (epitony and hypotony, respectively). Logically, shrubs exhibit a basitonic branching, whereas trees are characterized by acrotonic branching. Analogously, while typical trees exhibit a hypotonic branching, most shrubs and small trees exhibit epitonic branching. However, there are many exceptions to these rules. For instance, the pyramidal shape of the crown of many conifers is due to the combination of basitonic branching (typically a shrub pattern) with a monopodial growth of the bole. The dominance of branch development when branch originates from buds on the upper side of stems or main branches (epitonic shoots) appears to be important in shrub competition for space, since hypotonic branching confers the capacity of extending laterally but not overtopping an existing canopy (Schulze et al. 1986). More implications of branching patterns in the way shrubs and trees occupy and compete for space are discussed in Section "Structural Determinants of Light Capture".

Relative number of branches has been examined in trees using the Strahler ordering technique, which begins at the edge of the canopy (first-order branches) and works its way toward the trunk, incrementing the order of a branch each time it intersects the junction of two similarly ordered branches (Waller 1986). The bifurcation ratio, an index of the degree of branching from one order to the next, was initially related to the successional status of the tree (Whitney 1976). However, later studies have shown that it varies within a given species (Steingraeber 1982) and even within a given crown (Kellomäki and Väisänen 1995, Kull et al. 1999, Niinemets and Lukjanova 2003). The ratio between terminal and subterminal branches can be of ecological interest, but higher-order bifurcation ratios are difficult to interpret (Steingraeber 1982).

Plant form can be very complex due to the combination of regular and irregular pattern formation processes. While Euclidean geometry is very useful for studying linear, continuous, or regular structural properties of the objects, fractal geometry is a powerful tool to analyze nonlinear, discontinuous, or irregular structural properties, which are characteristic of plants (Hasting and Sugihara 1993). One of the properties of fractal objects is self-similarity, that is, the shape or geometry of the object does not change with the magnification or scale. The reiteration of a branching pattern in trees is a good example of this property, which was qualitatively described and used in the classification of architectural models of trees before fractals became popular (Hallé et al. 1978). Plant architecture has many fractal properties (see e.g., Prusinkiewicz and Lindenmayer 1990). A tree can be modeled as a fractal, and many functional aspects, such as efficiency of occupation of space by the leaves, total wood volume, stem surface area, and number of branch tips, can be calculated with more accuracy by using fractals rather than Euclidean geometry. However, forests, tree branches, plant crowns, or compound leaves, are most likely *multifractals*, because they are not strictly self-similar at every scale, that is, not exactly the same at all magnifications (Stewart 1988). This concept is clearly homologous to the *partial reiteration* concept that Hallé et al. (1978) and Hallé (1995) used in their classification of the architecture of trees.

Because symmetries and elegant geometric features of plants have always attracted mathematicians, models of plant shape and growth have received considerable attention. Models can be classified into two main groups: morphological and process-based models (Perttunen et al. 1996). However, the ideal model is a morphological model that deals with physiological processes or a process-based model that incorporates morphological information (Kurth 1994). Models vary greatly in scope and resolution, but very simple models can mimic response of real plants because the complex integrated growth patterns seem to be emergent properties of a simple system (Cheeseman 1993). Metamer dynamics have been simulated using the tools of population dynamics, which have rather simple mathematical formulation; however, this approach ignores structure and allows little scope for geometric analyses (Room et al. 1994). In geometric models, the spatial position and orientation of each structural component is considered, which allows the accurate simulation of interception of light by leaves (Pearcy and Yang 1996), of bending of branches due to gravity, and of collision between branches (Room et al. 1994). Geometric models also provide the information necessary to produce realistic images of plants (see Figure 4.10 and Figure 4.11), which has additional applications in education, entertainment, and art (Prusinkiewicz and Lindenmayer 1990, Prusinkiewicz 1998). For more examples of models and their applications, see reports by Kellomäki and Strandman (1995), Perttunen et al. (1996), and Küppers and List (1997).

From the many models available to simulate plant growth there are two systems that have the widest potential application for plant ecologists and physiologists: L-systems (initiated by Lindenmayer and further developed by Prusienkiewcz) and AMAP (Atelier pour la Modélisation de'Architecture des Plants) originated by de Reffye. AMAP uses stochastic mechanisms, and L-systems, although initially deterministic, can incorporate stochastic mechanisms as well. Despite the fact the AMAP has remarkable utility in agronomy by giving a central role to the structure of plants (Godin 2000), L-systems are inherently more versatile and hold greater promise (Room et al. 1994). Although an ideal growth model takes both internal and external factors into account, models to date focus on either one or the other. Room et al. (1994) revised all the internal and external parameters affecting metamer dynamics that should be considered in modeling plant growth. The advance of plant growth modeling is challenged by the difficulties of making virtual plants responsive to the environment and to neighboring plants in real time, and devising efficient methods of measuring plant structure, which is crucial information for the models that is usually hard to obtain.

Arrangement of Leaves

Fisher (1986) distinguished five different factors that determine the position of leaves. Among them, only two (phyllotaxis, which is addressed in this section, and secondary leaf reorientation by internode twisting, petiole bending, or pulvinus movement, which is addressed in the Section "Structural Determinants of Light Capture") apply to the leaves themselves. The other three concern the branching pattern and the position of the leaf-bearing branches. For instance, internode length affects the longitudinal distribution of leaves along the axis, or the existence of short and long shoots determines whether leaves would be produced every year or not, since, in general, only short shoots continue to produce leaves after one growing season. *Phyllotaxis*, as the sequence of origin of leaves on a stem (Figure 4.1), has a great impact not only on the shape of a crown (it affects the position of axillary buds or apical meristems and thus determines branching patterns), but also in many functional aspects of the crown since it affects the interception of light and the patterns of assimilate movement (Watson 1986). Phyllotaxis is responsible for the morphological contrast between plants with leaves along the sides of horizontal twigs, forming horizontal sprays of foliage, and those with leaves spiraling around erect twigs. With regard to leaves, there can be one per node (as in all monocotyledons and in some dicotyledons) or more than one per node (as in many dicotyledons). Leaves that lie directly above one another at different nodes form vertical ranks called orthostichies. When there is only one leaf per node, the phyllotaxis can be monostichous, distichous, tristichous, or spiral if the stem has one, two, three, or more than three orthostichies, respectively (Figure 4.1). Monostichous is a very rare phyllotaxis and is usually accompanied by a slight twist of the stem that arranges the leaves in a shallow helix; the corresponding phyllotaxis is called spiromonostichous (Bell 1993) (Figure 4.1 and also see Figure 4.10). In a distichous foliage, the two rows of leaves are 18° from each other, whereas in a tristichous foliage, leaves are in three rows with 120° between rows.

Spiral phyllotaxy results when each leaf is at a fixed angle from its predecessor in such a way that a line drawn through successive leaf bases forms a spiral (the genetic spiral) around the stem. This widespread phyllotaxis, also called disperse due to the apparent lack of geometrical pattern, can be mathematically described as a fraction in which the denominator is the number of leaves that develop before a direct vertical overlap between two leaves occurs, and the numerator is the number of turns around the stem before this happens (see Valladares 1999). This fraction times 360 is a measure of the angle around the stem between insertion of any two successive leaves (e.g., for a tristichous phyllotaxis, the fraction is 1/3, meaning that three leaves are developed before vertical overlap between two leaves, and this overlap happens in one turn around the stem, and the 120° between the orthostichies or between two successive leaves results from 1/3 times 360°). When the phyllotactic fraction of plants with spiral phyllotaxis was calculated and ordered, the following series was obtained: 1/2, 1/3, 2/5, 3/8, 5/13, 8/21, and so on. Interestingly, in this series both numerators and denominators form Fibonacci series since each number is the sum of the preceding two numbers. When multiplied by 360°, this series converges toward 137.5° (Fibonacci angle), which is the divergence angle between two successive leaves in most plants with spirally arranged leaves (Leigh 1972, Bell 1993). Other phyllotaxes can be observed when more than one leaf is present on each node. The simplest case is the opposite foliage, with two

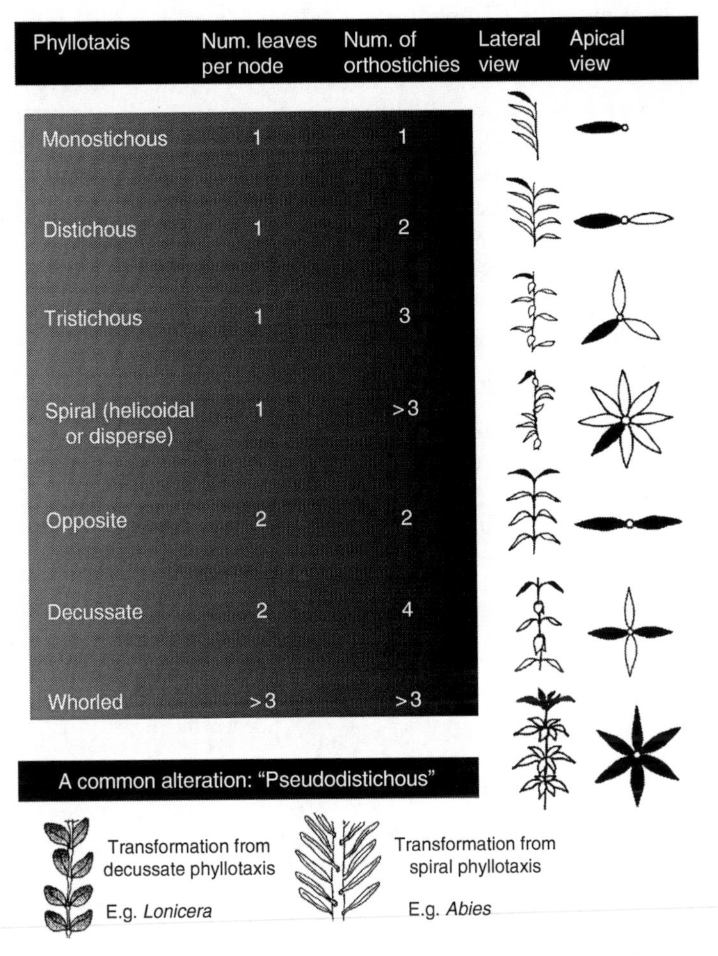

FIGURE 4.1 Main patterns of leaf arrangement (phyllotaxis) in plants. In the sketches, the black leaf (or leaves) represents the uppermost one in the shoot. A common alteration that results in a phyllotaxis that looks distichous (pseudodistichous), which has been generally interpreted as an adaptation to avoid self-shading, is shown in the lower part of the figure.

leaves 180° apart at each node, forming two orthostichies. A common variation is the decussate phyllotaxis, which has four orthostichies due to the fact that successive pairs of leaves are orientated 90° to each other (Figure 4.1). A more complex variation is the bijugate or spiral decussate phyllotaxis, where successive leaf pairs are less than 90° apart, leading to a double spiral (Bell 1993).

The ease with which the phyllotactic fraction is measured in a given plant is frequently confounded by internode twisting or leaf primordium displacement. It is relatively frequent that several phyllotaxes converge in an apparent distichous foliage. For example, the needles of some *Abies* are in two rows and look distichous, but their real phyllotaxis is spiral, as indicated by the petiole insertion (Figure 4.1). A similar case was found in the shade shoots of the chaparral shrub *Heteromeles arbutifolia*, which exhibited a pseudodistichous phyllotaxis instead of the characteristic spiral phyllotaxis of the species (Valladares and Pearcy 1998). Decussate phyllotaxis might also look distichous, as observed in horizontal shoots of *Lonicera* (Figure 4.1). Spiral and distichous leaf arrangements are also sometimes found in the same plant species. For instance, certain plants may first, as seedlings, set leaves spirally around an erect stem, and then, as mature individuals, develop a distichous foliage on the horizontal branches produced in the axes of the initial leaves. This seems to be the case for the tropical forest understory herb *Dichorisandra hexandra* (see Section "Structural Determinants of Light Capture" and Figure 4.10). Phyllotaxis is a clear case of phylogenetic constraint (Niklas 1988), but plants have solutions to compensate for the functional drawbacks of a given phyllotaxis. Spiral phyllotaxis can render contrasting shoot patterns with a simple change of 2° in the leaf divergence angle (Figure 4.2). However, despite the remarkable change in leaf overlap as seen from the top of the shoot, light capture is little affected by such a phyllotactic change (Valladares and Brites 2004). This negligible influence of the divergence angle on the light capture by spirally arranged shoots is in contrast with theoretical expectations: the intriguing trend of spiral phyllotaxis to converge in the golden angle, which allows for an infinite number of leaves to be arranged along a shoot without anyone fully blocking any other one, has been interpreted as a trend to maximize light capture efficiency (see Figure 4.2 and Valladares and Brites 2004). Significant differences in light capture efficiency are found, however, in comparisons of spiral versus opposite phyllotaxis, with a lower efficiency in the later (Figure 4.2). Nevertheless, the differences in light capture due to a given phyllotaxis can be easily compensated by an increased in either internode or petiole length (Pearcy and Yang 1998, Brites and Valladares 2005).

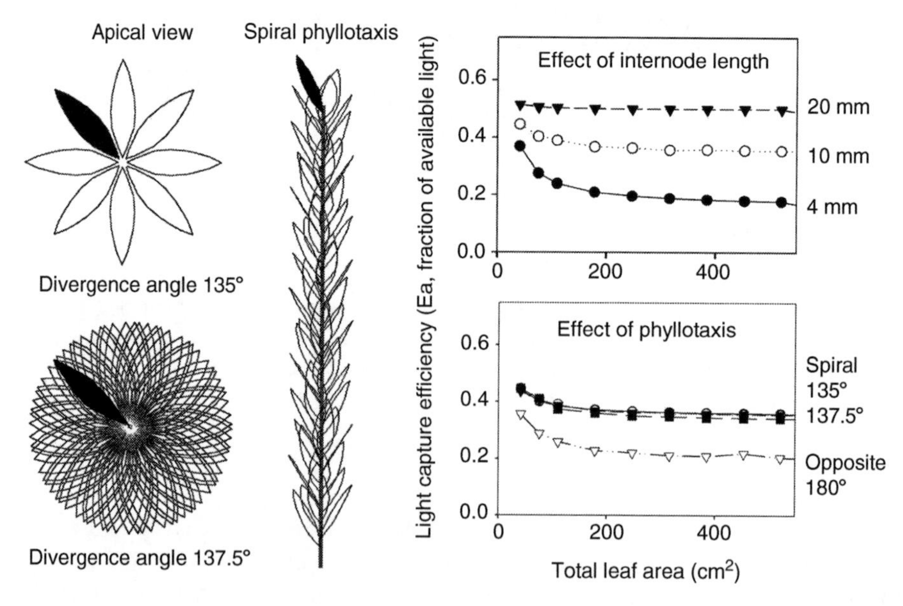

FIGURE 4.2 A vertical shoot such as that of *Heteromeles arbutifolia* (central drawing) generates contrasting views when seen from above (*left* images; the uppermost leaf is shown in black in the three figures). In a shoot with spirally arranged leaves such as the one of the figure, a mere 2.5° change in the divergence angle can dramatically change the number of leaves seen from above from only eight to all in the shoot (a golden angle of 137.5° generates an infinite number of ortostichies—see Figure 4.1). However, this contrasting arrangement had almost no effect in light capture efficiency (*right* graphs, note that values for 135° and 137.5° overlap), particularly when compared with simulations of the same shoot but with opposite phyllotaxis. By contrast, internode length (*upper right* graphs) had a very significant effect in light capture, so by modifying their internode plants can compensate phylogenetic constraints on light capture efficiency such as those imposed by an opposite phyllotaxis. Graphs on the right represent light capture efficiency versus total leaf area of the shoot. (Adapted from Valladares, F. and Brites, D., *Plant Ecol.*, 174, 11, 2004; Brites, D. and Valladares, F., *Trees: Struct. Funct.*, 19, 671, 2005.)

Classifying Crown Architectures

The first, and possibly the best known, classification of tree architecture was reported by Hallé et al. (1978). Basic features of this classification were dichotomic characteristics of the tree crown, such as monopodial or sympodial branching, basitonic or acrotonic branching, orthotropic or plagiotropic shoots, etc. (see Valladares 1999, for terms and for a key to these classic architectural models). From the practical point of view, this classification can be very difficult to use with certain species because the researcher must know the way by which the shape of the crown is achieved during the ontogeny of the tree from the seedling to sexual maturity, something that exceeds the time frame of most field studies dealing with long-lived plants. In addition, certain species exhibit architectural ambiguities, shifting from one model to another during their ontogeny or under different environmental conditions. Leigh (1990, 1998 #43993) modified Hallé et al. classification, simplifying it by merging some models that cannot be easily distinguished.

Architectural models are a convenient starting point for interpreting plant form, but there is a series of variations and exceptions to each program of development that complicates classification and suggests the search of additional descriptions of crown shapes. For instance, *Arbutus* sp. exhibit two different architectural patterns depending on the light environment, and *Acer pseudoplatanus*, as with many other woody plants, undergo significant changes of branching patterns during the ontogeny, switching from one model to another (Bell 1993). There are also many examples of metamorphosis (abrupt change from plagiotropic to orthotropic disposition of a branch) and intercalation of shoots infringing the rules of each model (Bell 1993). Nevertheless, architectural models are useful to predict the form that a plant assumes in the absence of unusual external forces or when affected by the common circumstance of losing a structural subunit (e.g., a branch) through injury. The modules that regrow when a tree loses a subunit usually mirror the architecture of the whole crown of the tree in a process called reiteration (Hallé et al. 1978, Hallé 1995). As the tree grows, the number of reiterated units tends to increase, but their size tends to diminish, and ultimately only parts of the architectural unit are reiterated in a so-called "partial reiteration" (Hallé 1995). This reiteration process that occurs during the growth of a large tree reinforces the idea that most trees are colonies, the elementary individual being not the bud, but the architectural unit. This idea of a plant as a colony (discussed in Section "Plant Design") dates back to eighteenth century: botanists such as de la Hire, Bradley and von Goethe (see references in White 1979), and Charles Darwin and his grandfather Erasmus Darwin thought that coloniality existed in trees. Although reiterated units have largely been considered as leafy branch systems, Hallé (1995) went one step beyond, posing the hypothesis that these units comprise their own root system, and thus the bole is made up of the aggregated root systems of all the reiterated units forming the tree crown. Needless to say, this hypothesis is controversial and may be somewhat heretical to certain readers, as acknowledged by Hallé himself (Hallé 1995 p. 41).

Functional Insights into Architectural Classifications

Since the shape of the crown influences important aspects of growth and survival of plants, such as light interception and competition for space, the adaptive significance of the architectural models of Hallé et al. (1978) has interested many ecologists dealing with plant form. While all investigators agree that crown shape is generally adaptive, there is no consensus regarding the ecological and evolutionary implications of these architectural models (Porter 1989). On the one hand, as observed by Porter (1989), fossil plants exhibit only three of 23 possible architectural models described by Hallé et al. (1978), mostly due to the remarkable lack of fossil examples of sympodial branching. This clumping of fossil trees among Hallé et al. models suggests that some plant forms may have paid an evolutionary penalty for their mode of whole plant development, that is, the limited number of ancestral architectures may have limited the number of architectural models that have survived. On the other hand, Ashton (1978) pointed out that in West Malaysia, certain models were very rare in shady habitats, whereas a very plastic type of organization (Troll's model) was very widespread. The relatively small number of models found in temperate deciduous forests (the conifer forests of the boreal regions have even fewer models) suggests that some models are selected against in some regions (Ingrouille 1995).

Three arguments have been given to support the notion that these architectural models are not adaptive: (1) all models only coexist in lowland tropical rainforests, so a single ecological region has not favored some models at the expense of others; (2) the same model exists at different levels in the forest canopy, despite the remarkable vertical gradients of light, predation, and nutrients; and (3) the same model exists in different growth forms from very tall trees to small herbs, which clearly do not share the same ecology (Fournier 1979). Actually, developmentally different models can produce functionally similar crown shapes (ecological convergence). And a single model shared by different plant species can produce functionally divergent crowns due to differences in factors such as the relative elongation of axes and the exact arrangements of leaves (Fisher and Hibbs 1982). Additionally, efficiency of leaf display, which is crucial in the ecological strategy of most species (see Section "Structural Determinants of Light Capture"), is not included in the parameters used to define the architectural models (Tomlinson 1987).

There is a wide plasticity allowable within one model of Hallé et al., so these models may lead to unequivocal ecological predictions only for the simplest crowns (Waller 1986). Because development plasticity is an intrinsic characteristic of plant form (see Section "Plasticity, Stress and Evolution"), any attempt to classify the architectural patterns of plants should include the structural response of each species to different environments or perturbations. And a response is a quantitative process, which would make the separation of species into discrete models very difficult. In many plants, and especially in long-lived trees, it is a challenge to distinguish the genetically determined structure from environmental damage and phenotypic plasticity (Fisher 1992). Consequently, searching for a single ecological classification of plant architecture seems a vain endeavor. The critical parameters for the classification of plant shape must vary depending on the problem at hand (Sachs and Novoplansky 1995).

The most widespread architectural classifications have been developed for trees, but they could be used with other plants if characteristics such as multiple stems are considered in detail. The multiple-stemmed characteristic results from the growth of buds from the below-ground level that escape apical dominance to form new stems or modules (Wilson 1995). Multiple-stemmed shrubs exhibit not only a different shape than single-stemmed shrubs, but also a different tolerance to perturbations (e.g., fire and pests). Multiple-stemmed shrubs can survive indefinitely as a clone by producing new stems, whereas single-stemmed shrubs die when the stem dies. The maintenance of an apical control; the tendency of the stems to bend toward the horizontal, producing vigorous vertical shoots in a series of arching segments; or the location of the underground buds of multiple-stemmed shrubs (on the basis of the shoot, along rhizomes, layered branches) are also important features to consider in the description of shrub architecture (Wilson 1995).

Real Crowns: Imperfect Architectures or Controlled Variability?

In contrast to human designs such as buildings, the final shape of the crown of a plant expresses a remarkable variability, which is evident even in comparisons of two halves of the very same individual (Sachs and Novoplansky 1995). However, there is a characteristic design or architectural pattern for each plant species. Therefore, the general shape of a crown is rather constant for a given species under a given environment, whereas many aspects of branch growth and survival do not follow a strict program, exhibiting an apparently stochastic behavior. At least the following three parameters have been shown to introduce variability in the shape of a plant: (1) the location and number of developing apices; (2) the developmental rates of individual apices; and (3) the shedding of branches (Sachs and Novoplansky 1995). Is this variability in the crown shape due to a malfunction of the genetic program that determines the development of the shape of a plant? How could the general form of a tree be more predictable than the individual events (e.g., production and shedding of branches) that lead to it? Variability in the final shape is not characteristic of primitive or maladapted plants, and it is not the result of errors in the developmental program. On the contrary, it has a crucial ecological role in changing and heterogeneous environments (see Section "Plasticity, Stress and Evolution"). On the other hand, predictable mature structures can result from selection of the most appropriate developmental events from an excess of possibilities that are genetically equivalent (epigenetic selection Sachs 1988). In this way, the final shape or pattern is genetically specified, but the development of the crown gravitates toward this final shape without a detailed genetic program. This tendency toward the final shape is accomplished by means of internal systems that control the variability in the aforementioned parameters, but allow for developmental plasticity. These control systems that constrain development variability include internal correlative interactions between branches, responses to local shading, and programed limitations of successful branches (Sachs and Novoplansky 1995). In conclusion, although the architecture of a plant limits its range of possible shapes, a plant's architectural model does not determine its final shape.

STRUCTURAL DETERMINANTS OF LIGHT CAPTURE

Canopy photosynthesis rate depends on the biochemical capacities of the foliage as well as on the distribution of light within the canopy (Wang and Jarvis 1990, Baldocchi and Harley 1995, Sinoquet et al. 2001). A major outcome of variation in crown architecture is modification of the overall light harvesting and the efficiency of light harvesting. The total leaf area supported by given crowns is the most basic structural property that affects the fraction of absorbed radiation. However, the distribution and arrangement of leaves within a crown can strongly modify the light harvesting efficiency of unit foliage area (Ross 1981, Cescatti and Niinemets 2004). As the three-dimensional arrangement of leaves in a crown is difficult to measure, light interception and canopy photosynthesis is often simulated assuming that foliage is randomly dispersed throughout the canopy volume (Beyschlag and Ryel 1999). However, recent development of three-dimensional ray-tracing models (Pearcy and Yang 1996, Sinoquet et al. 1998) as well as application of more advanced radiative transfer models combined with laborious harvesting of plant material (Baldocchi et al. 1984, Baldocchi and Collineau 1994, Niinemets et al. 2004a) has made it possible to resolve the effects of spatial clumping, foliage inclination angle, and foliage area density on distribution of solar energy in plant stands.

In most radiative transfer models, the sun is also considered as a point light source, and generally two classes of foliage—sunlit and shaded—are separated for any given situation (Wang and Leuning 1998). In reality, the radius of solar disk as seen from the earth is about 0.27 degrees. Due to finite size of solar disk, phytoelements can partially shade each other, resulting in intermediate situations between completely sunlit and shaded foliage, that is, in penumbral radiation. Recent advances in ray tracing approaches has made it possible to evaluate the importance of penumbral radiation for overall distribution of light in the canopy and on photosynthesis (Stenberg 1995, Cescatti and Niinemets 2004).

SHAPING THE FOLIAGE: THE SINGLE-CROWN LEVEL

Crown shape and the arrangement of foliage within the crown are the two most basic characteristics affecting the efficiency of light capture. From a photosynthetic perspective, the most efficient canopy is achieved when all of the leaves are evenly illuminated at quantum flux densities that saturate photosynthesis, that is, at intermediate quantum flux densities. Such ideal canopies are found in the nature rarely, if at all. Various crown shapes and different dispositions of leaves within the crown result in complex diurnal and seasonal patterns of light interception at both the single-leaf and the whole-crown levels. Leaves at the uppermost positions of the canopy are frequently exposed to high irradiances that are often in excess for photosynthesis. Lower leaves, in turn are often heavily shaded, and the light available for these depends not only on the amount of neighboring leaves, but is also affected by the general form of the crown and the angle and orientation of the surrounding units of the foliage (Niinemets and Valladares 2004). In addition, the level of incident photon irradiance can be regulated by diurnal movements of foliage units, that is, crowns can have their geometries changing over a short time interval.

Crown Size and Shape

The questions of whether there is a perfect crown shape that maximizes light interception in a given environment, and how far are the actual crown shapes of an optimal has attracted many researches (Jahnke and Lawrence 1965, Horn 1971, Terjung and Louie 1972, Oker-Blom and Kellomäki 1982, Kuuluvainen 1992, Chen et al. 1994). Probably most stimulating insight into the significance of variation in crown shape has been attained by studies investigating the role of different crown shapes in gradients of overall variation of available light during forest succession (Horn 1971), and in studies looking at the variation of solar radiation and average inclination of beam radiation with latitude (Kuuluvainen 1992).

The shape of the crown can be described by the absolute size, the ratio of height to width, and the convexity or shape of its contour. As the solar inclination angle decreases from equator to higher latitudes, crowns with differing height to width ratio have inherently varying efficiencies of light interception. Specifically, in high latitudes, light penetrates from high solar inclination angles, implying that beam path lengths become increasingly longer with increasing crown flatness. The beam path lengths are similar throughout the entire canopy for the narrow, vertically extended crowns that maximize the direct light interception of entire crown in high latitudes (Figure 4.3, Kuuluvainen 1992). In low latitudes, the beam path lengths are shortest for flat, horizontally extended crowns (Figure 4.3, Kuuluvainen 1992). The dominance of tall and thin conifers at high latitudes, and flat-topped Mediterranean conifers (*Pinus pinea* and *Pinus halepensis*) as well as acacia-like trees at low latitudes, partly confirms and supports the adaptive value of these two general crown shapes at different latitudes.

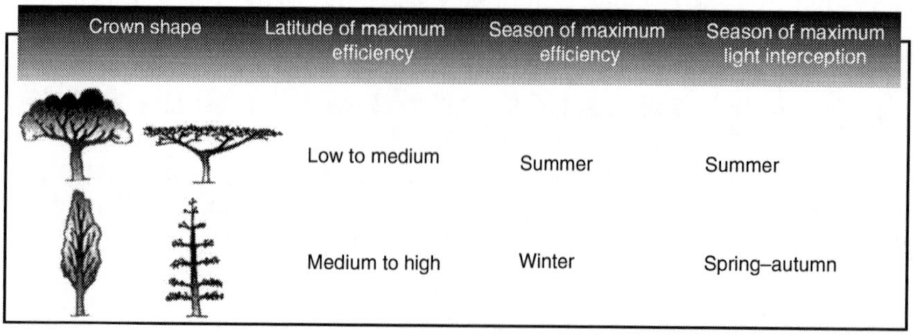

Crown shape	Latitude of maximum efficiency	Season of maximum efficiency	Season of maximum light interception
	Low to medium	Summer	Summer
	Medium to high	Winter	Spring–autumn

FIGURE 4.3 Latitude and season of maximum efficiency of light interception, and season of maximum light interception for two main types of crown shape: flat and broad versus thin and tall.

Contrary to these suggestions, Chen et al. (1994) discussed that the latitudinal variation of potential sunlight interception by different crown shapes does not match very well with the existing latitudinal gradients of crown shape. They suggested that this mismatch arises because (1) light is not the only factor affecting the crown shape variation along the latitudinal gradient, and that (2) in addition to crown shape, the geometry and distribution of the foliage alter crown light interception, partly compensating for differences in crown shape (Chen et al. 1994). As the result, crowns of different shapes can intercept a similar fraction of the available light.

It is further important that the crown shape can vary at any given height to width ratio. For low-solar inclination angles, the beam path length strongly increases with canopy depth for narrow ellipsoidal crowns. However, the beam path length is essentially the same for narrow conical crowns, in which the branches in lower canopy positions reach farther from the stem, implying that such crown can be very efficient at low latitudes. In general, the more extended the cone, the larger is the fraction of irradiance captured (Jahnke and Lawrence 1965). Simulations demonstrate that for a given latitude, either very small or very large values of the height-to-width ratio result in maximum direct light interception (Chen et al. 1994).

The crown height-to-width ratio must reach a balance between growth in height to reach the brighter areas of the canopy, and growth in width to intercept light and occupy enough space (Horn 1971, Givnish 1988, Küppers 1989). In addition, the greater the convexity of a crown, the greater the irradiance intercepted at most latitudes, but also the greater the amount of supporting and conductive tissues. Horn (1971) predicted that the optimal shape of trees varies in dependence of tree successional position and distinguished three different successional strategies: early successionals, late successionals, and early successionals in the mature forest (Figure 4.4). Because early succession is a race to form a canopy, fast-growing softwoods are favored over stronger hardwoods, and growth in height is favored over growth in width (Horn 1971, King 1991, 1994). For rapid height growth, stems of some early-successional species are even hollow (King 1994). Because of weak wood and relatively thin stems, early-successional trees cannot form extensive wide-reaching crowns.

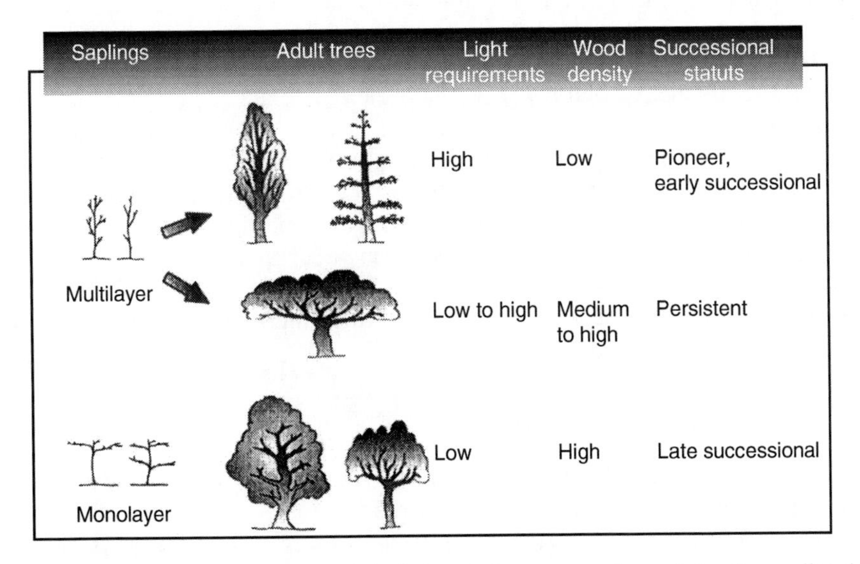

FIGURE 4.4 Crown shape as sapling and adult, light requirements, and wood density predicted for trees of different successional status. (Adapted from Horn, H.S., *The Adaptive Geometry of Trees*, Princeton University Press, Princeton, NJ, 1971.)

Horn (1971) predicted that crown shape of saplings of early-successional, shade-sensitive species is multilayered, consisting of short branches distributed over a long distance from top to bottom of the stem (Figure 4.4). Such a crown allows plants to expose a large leaf area in several independent layers to high irradiance. However, this crown shape is inefficient in low light because it results in extended self-shading within the crown. Late-successional, shade-tolerant species are predicted to be monolayered, distributing total sapling leaf area in a single layer by far-reaching extensive branch framework and thereby capturing more light in low irradiance (Horn 1971). Some multilayered trees persist by invading small openings in the forest and should have a mixed strategy. Because these species must initially race to the canopy, they must be tall, thin, multilayered, and made of softwood. Once they reach the canopy, they should spread out and dominate the forest gap. The height-to-width ratio decreases with age, and their wood should become harder to provide lateral support.

Although the predictions of crown shape variation during succession are based on only a single factor, light, and provide therefore an incomplete theory, as Horn himself acknowledged (Horn 1971 p. 121), these predictions provide an explicit list of testable assumptions. The experimental evidence of the successional sequence of crown shapes and foliage distribution has been scarce, but the available evidence from some temperate and tropical forests supports the gradual change from multilayer to monolayer species during succession (Horn 1971, Niinemets 1998, Sterck et al. 2001, 2003, Pearcy et al. 2005). Many observations reveal that plant species partition canopy light gradients through variation in adult stature and light demand, which has been well characterized in complex tropical forests (Poorter et al. 2005). Adult understory trees are typically shorter than similar-diameter juveniles of high-light species, since wide crowns allow intercepting light over a large area at the expense of a reduced height growth, whereas light-demanding species are characterized by orthotropic stems and branches, and large leaves (Poorter et al. 2005).

Functional analyses of the importance of crown shape often neglect the overall availability of light and the time of the year of maximum irradiance and light interception. Although the maximum efficiency of light interception by narrow-shaped trees is achieved during the winter, they intercept more light in spring and autumn. In broad-shaped trees, both light interception efficiency and the amount of light intercepted reach their maximum values during the summer. In addition, the fraction of diffuse radiation (radiation from all angles) in total irradiance importantly affects the efficiency of a crown light capture. In environments with frequent cloud cover as maritime temperate forests and mountain cloud forests, a large fraction of radiation is received as diffuse radiation, and as a result, the role of the crown shape less strongly affects the overall light interception. Understanding the relations between the latitudinal gradients in crown shape and the latitudinal variation of the light regime, requires both theoretical analyses of crown shape and light interception (like the one by Chen et al. 1994), and further case studies exploring the real light environment experienced by trees of different shapes at different latitudes. These studies should necessarily also investigate the modification of crown shape by other interfering factors and constraints, such as water, snow, gravity, and wind.

Geometry of Foliage Arrangement within the Crown

The amount of foliage supported by a given crown is measured by crown leaf area index, L ($m^2\ m^{-2}$), defined as total leaf area divided by the total ground area where it stands. The distribution, dispersion, and inclination of leaf area in space defines the probability for light beam penetration though a canopy gap to the lower leaves. Crowns with the same values of L can have widely differing efficiencies of light capture (Ross 1981, Baldocchi and Collineau 1994, Cescatti and Niinemets 2004).

Foliage Dispersion

Foliage dispersion is a major factor affecting the light-harvesting efficiency of unit foliage area. Simple light interception algorithms assume that plant canopies consist of randomly dispersed foliage elements. In real canopies, the foliage is often clumped to branches and shoots, resulting in greater fraction of canopy gaps and significantly larger light transmission relative to a clumped canopy (Figure 4.5, Ross 1981, Baldocchi and Collineau 1994, Cescatti and Niinemets 2004, Sinoquet et al. 2005). While clumped canopies intercept light less effectively, clumping allows the plants to expose larger leaf areas. Canopies with random dispersion intercept essentially all light above a L of 5 m^2 m^{-2}, whereas canopies with extensively aggregated foliage, as in some conifers, can support leaf area indices as high as 15 m^2 m^{-2} and more (Figure 4.5, Margolis et al. 1995, Van Pelt and Franklin 2000).

In addition to random and clumped foliage dispersions, which result in a relatively large canopy gap fraction, foliage can be arranged regularly. Arranging leaves side-by-side in a planar layer efficiently fills the gaps in the canopy and thereby results in greater light harvesting at a common L than either random or clumped dispersion (Figure 4.5). As regular dispersion is an extremely efficient strategy for light interception, it is favored in low-light environments and in late-successional mono-layer species (Horn 1971, Cescatti and Niinemets 2004).

It is possible to derive the estimates of whole canopy foliage aggregation structure from light transmission measurements that provide effective leaf area index (L_{eff}) and separately harvesting plants to estimate L (Kucharik et al. 1999 for a review). However, modification of foliage clumping can occur at the level of individual crowns, branching patterns and individual shoots (Oker-Blom 1986, Cescatti 1998). Crown-level clumping arises because crowns

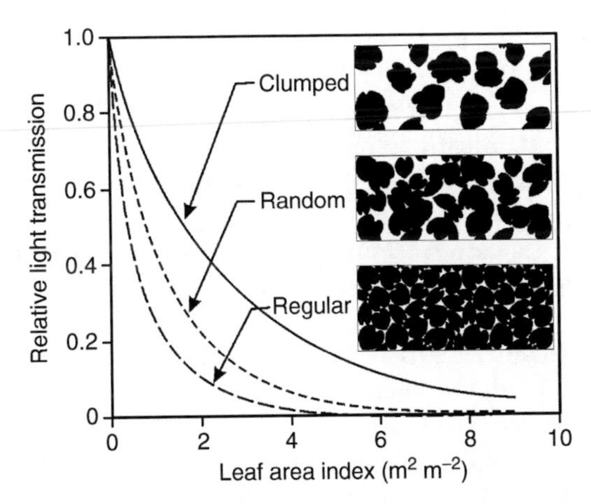

FIGURE 4.5 Light transmission relative to cumulative leaf area index for three hypothetical canopies. Relative to canopies with random foliage dispersion, canopies with clumped foliage intercept less light, and canopies with regular dispersion intercept more light. In these simulations, leaf angular distribution was assumed to be spherical, and light transmission was integrated over the entire sky hemisphere. Light transmission for nonrandom canopies was simulated using the theory of light penetration in nonrandom media (see Nilson 1971, Cescatti and Niinemets 2004 for details of light models). For the clumped canopy, we used a Markov model, using a clumping coefficient, $\lambda_0 = 0.5$, that corresponds to a moderately clumped canopy (λ_0 varies between 1 and 0, 1 corresponding to random dispersion and 0 to completely aggregated canopy) (Nilson 1971, Cescatti and Niinemets 2004). For the regular dispersion, we used a positive binomial model, with the parameter Δ_L (thickness of an independent leaf layer), set at 1.5 ($\Delta_L \rightarrow 0$ for a random dispersion, and the values increasing with the degree of regularity). In the boxes illustrating the concept of foliage dispersion, the number of leaves is equal for all dispersion types.

with different size and shapes result in different fraction of gaps in the canopy (Oker-Blom 1986, Cescatti 1998).

Branching modifies foliage dispersion via the frequency of branching (bifurcation ratio) and the branching angles. Modifying both of these characteristics can result in foliar displays that either minimizes the overlap among the leaf clusters on a horizontally spreading branch and results in regular foliage dispersion or results in strongly clumped foliage (Honda and Fisher 1978, Takenaka 1994b). As described earlier, there is a vast heterogeneity in the branch architectural models, but it is important to understand, that from a functional perspective, the branching architectural models mainly differ in the extent of foliage aggregation. For instance, the Aubréville's architectural model investigated in *Terminalia* results in arrangement of leaves side by side, minimizing the branch gap fraction and resulting in essentially regular leaf display (Honda and Fisher 1978, Fisher and Honda 1979a,b). In general, increases in the bifurcation ratio result in more clumped canopies, whereas lower bifurcation ratios result in random or regular canopies (Whitney 1976, Canham 1988). A branch system with a high bifurcation ratio allows plants to achieve a greater amount of foliar area for a given biomass investment in stem tissue, but such branch with enhanced clumping requires higher irradiance for full activity. Overall, the bifurcation ratio increases with increasing light availability (Kellomäki and Strandman 1995, Niinemets and Lukjanova 2003), demonstrating a general shift from highly divided branches with strong foliage clumping that require high light to less frequently bifurcating branches with more regular foliage display that require less light since they intercept it very efficiently.

At the shoot scale, light interception efficiency varies due to variations in the number of leaves per unit stem length, and differences in petiole length and leaf extension that modify the distance between the bulk of leaf area and shoot axis. Foliage is considered especially clumped in the shoots of conifers (Oker-Blom and Smolander 1988, Niinemets 1997, Stenberg et al. 2001), where the foliage in the shoots harvests light only with 10%–40% efficiency relative to the equivalent foliage area on an horizontal plane (Figure 4.6, Stenberg et al. 2001, Niinemets et al. 2002, Cescatti and Zorer 2003, Niinemets et al. 2006).

Increases in overall leaf extension and length of petioles strongly reduce shoot-level clumping, because these modifications reduce the shading by shoot axis as well as reduce the overlap of neighboring leaves (Figure 4.7, Takenaka 1994a, Figure 4.7, Pearcy and Yang 1998). At a global scale, there is a large variation in petiole length and foliage length. For instance, needle length varies between 2 and 35 cm among *Pinus* species (Figure 4.6). For the six conifer species depicted in Figure 4.6, which had contrasting foliage element length and shoot architecture, there was a uniform negative relationship between the degree of foliage clumping and foliage element length (Niinemets et al. 2006). However, an increase in the foliage element length and reduced clumping brings about lower foliage area density in the shoots with longer foliage elements (Figure 4.6). This implies that the canopies consisting of long-needled shoots do not cast deep shade and are open to invasion by competitors that can create denser foliage. Another disadvantage of increasing the length of foliage elements and petioles is the enhanced cost of support. In 17 clonal poplar stands, the whole canopy aggregation decreased with increased petiole length (Figure 4.7), but this resulted in overall greater fraction of foliage biomass invested in support (Niinemets et al. 2004a), implying a fundamental trade-off between efficiency of light harvesting and biomass investment in support.

In addition to the foliage and petiole lengths, shoot-level clumping can strongly vary with the distance between neighboring leaves on the shoot axis. Shorter distance between the leaves on shoot axis implies greater self-shading and aggregation within the shoot. Often, leaf number per unit shoot axis length increases in stressful environments due to stronger limitations on shoot length growth than on formation of leaves. Greater packing of needles on shoot axis explains greater aggregation and lower light harvesting efficiency of conifers on less

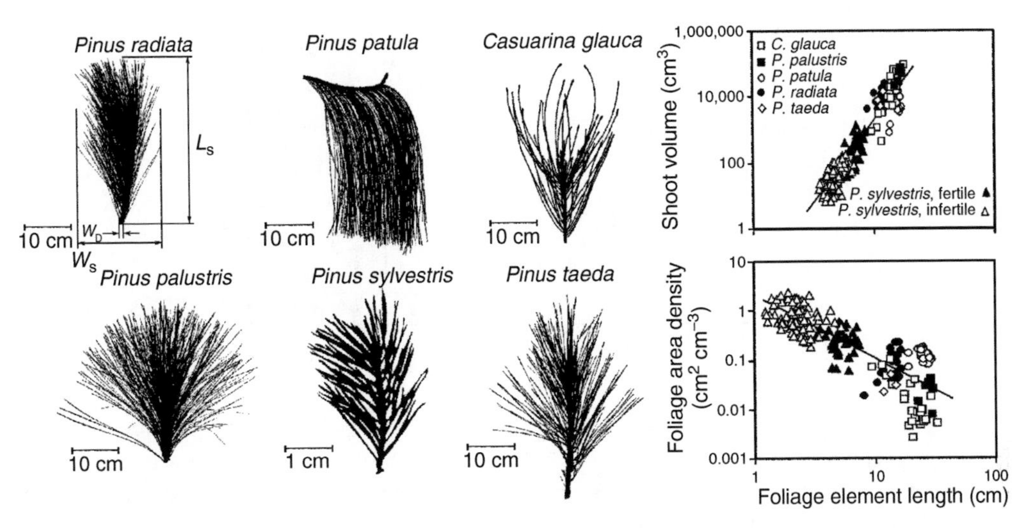

FIGURE 4.6 Illustration of shoot architecture in five *Pinus* species of contrasting needle length and in angiosperm conifer *Casuarina glauca*, and the relationships between shoot volume and foliage area density (ratio of half of the total foliage area to shoot volume) with the length of foliage elements (modified from Niinemets, Ü., Tobias, M., Cescatti, A., and Sparrrow, A.D., *Int. J. Plant Sci.*, 167, 19, 2006). Trees of *Pinus sylvestris* were sampled in two sites of contrasting fertility; needles were significantly shorter and shoots more clumped in the infertile site (Niinemets et al. 2002, 2006). Conifers have extensive clumping of foliage elements in the shoot. Light interception efficiency of unit leaf area, that is, the amount of light harvested by needles in their specific position in the shoot and with their specific cross-sectional geometry relative to the amount of light harvested by an equivalent flat surface, scales with the spherical average shoot silhouette area to total foliage area ratio $\overline{S_S}$ (Niinemets et al. 2002, Cescatti and Zorer 2003). The values of $\overline{S_S}$ (average ± SE for all shoots sampled per given species) were 0.2149 ± 0.0036 for *C. glauca*, 0.141 ± 0.008 for *Pinus palustris*, 0.1066 ± 0.0046 for *Pinus patula*, 0.0901 ± 0.0047 for *Pinus radiata*, 0.1562 ± 0.0047 for *P. sylvestris*, fertile site, 0.1147 ± 0.0019 for *P. sylvestris*, infertile site, and 0.147 ± 0.018 for *Pinus taeda*, demonstrating extreme inefficiency of light harvesting in these conifers. (Averages calculated from Niinemets, Ü., Tobias, M., Cescatti, A., and Sparrrow, A.D., *Int. J. Plant Sci.*, 167, 19, 2006.)

fertile sites (Niinemets et al. 2002, Palmroth et al. 2002). Analogously, greater clumping, and lower efficiency of shoot light harvesting in mature conifer trees relative to young trees is mainly associated with shorter and more densely leafed shoots in mature trees (Figure 4.8, Niinemets and Kull 1995, Niinemets et al. 2005). Several hypotheses have been advanced to explain the tree productivity decreases with tree age, mainly focusing on foliage physiological characteristics (Ryan et al. 1997), but there are important data demonstrating that the foliage clumping does increase in older stands (Brown and Parker 1994). Shoot-level observations suggest that enhanced foliage clumping due to arrested shoot growth may partly explain the curbed productivity in older trees.

Foliage Inclination and Orientation

Variation in vertical foliage angle and azimuthal orientation can generate large differences in diurnal patterns of light interception in canopies with similar degree of foliage aggregation. Steep leaves project a small fraction of their area to the sun during the central hours of the day, but the overall effect depends on foliage azimuth. Although steep leaf angle always reduces the light interception at individual leaf level, this reduction can vary from strongly limiting to negligibly affecting photosynthetic carbon fixation (Valladares and Pearcy 1999). Since crowns consist of large number of leaves that interact in determining the whole canopy

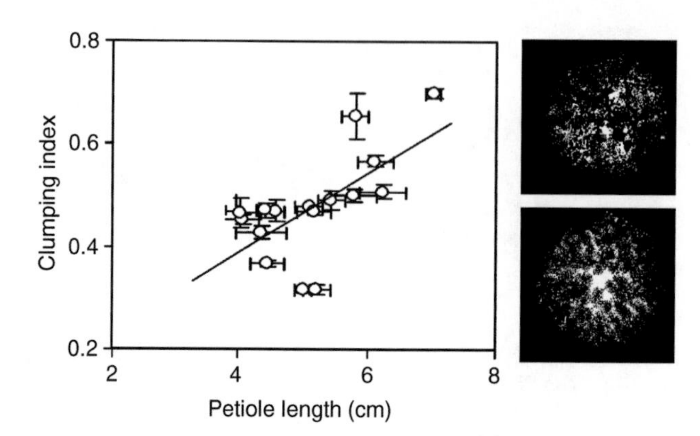

FIGURE 4.7 Relationship between the canopy clumping index (λ_0, Markov model of radiative transfer, Figure 4.5) and average petiole length for 17 different clonal stands of *Populus* (Niinemets et al. 2004a). The clumping index was derived from measurements of leaf area index by hemispherical photography (effective leaf area index L_{eff}) and actual measurements (L) and is given as $\lambda_0 = L_{eff}/L$. $L_{eff} = L$ ($\lambda_0 = 1$) for canopies with random dispersion, whereas L_{eff} becomes relatively smaller with increasing foliage aggregation. The hemispherical photographs illustrate two poplar canopies with similar effective leaf area index, but different total leaf area index and λ_0. Upper canopy photograph—*Populus deltoides* × *Populus nigra* 'Gibecq' ($\lambda_0 = 0.67$, $L_{eff} = 2.33$, $L = 3.74$ m^2 m^{-2}). Lower canopy photograph—*Populus nigra* 'Wolterson' ($\lambda_0 = 0.30$, $L_{eff} = 2.07$, $L = 6.84$ m^2 m^{-2}). (Modified from Niinemets, Ü., Al Afas, N., Cescatti, A., Pellis, A., and Ceulemans, R., *Tree Physiol.*, 24, 141, 2004a.)

light interception and light distribution, and single crowns can have leaves with differing angles, it is more appropriate to use leaf surface angle distributions to simulate the role of leaf angles in whole canopy light interception (Campbell and Norman 1989). To understand the effect of leaf angular distribution on whole canopy light harvesting, it is further important to integrate the light harvesting of the canopy over the entire day (Figure 4.7). Because solar position changes during the day, leaf angular distribution has generally a minor effect on total light interception and canopy photosynthesis for sparse canopies with a leaf area index (L) less than approximately 3 (Duncan 1971, Gutschick and Wiegel 1988). The effects of leaf angular distribution on canopy photosynthetic production are stronger for canopies with large leaf area (Duncan 1971, Gutschick and Wiegel 1988).

Horizontal leaves at the top of the crown exhibit their maximum light interception efficiency at times of the day and the year (midday and summer, respectively) when irradiance in sunny environments is well above the light saturation point for photosynthesis. Therefore, the superior light capture of horizontal leaves in high light usually translates into a negligible increase of potential carbon gain (Figure 4.9). For these reasons, erectophile crops have a marked yield advantage over those with horizontal leaves, especially at high values of L and at high solar elevations (Isebrands and Michael 1986). However, light interception by steep leaves themselves is poor, and if they represent a large fraction of the foliage or if their angle is too steep and if this is further combined with extensive clumping because the leaf blades are too close to each other (see computer images in Figure 4.9), light interception and potential carbon gain by the whole plant decrease. In a simulation of light interception and potential carbon gain by shoots of *H. arbutifolia* with leaves set at different angles, vertical foliages absorbed 20%–30% less photosynthetic photon flux density (PPFD) and had 30% lower daily carbon gain than normal shoots (average leaf angle = 71°) (Valladares and Pearcy 1998).

Leaf angular distribution is often considered constant in the canopy, but numerous observations demonstrate that leaves are more vertical in the upper canopy and become gradually horizontal in the lower canopy (e.g., Thomas and Winner 2000, Niinemets et al. 2004b, 2005).

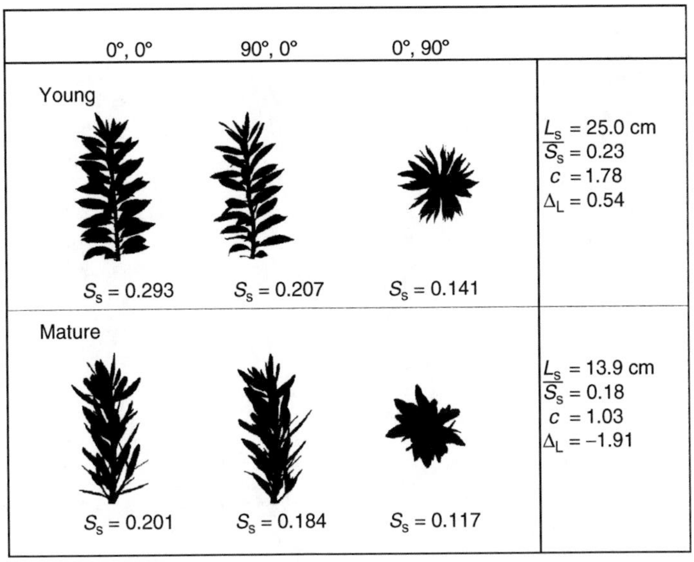

	0°, 0°	90°, 0°	0°, 90°	
Young	$S_S = 0.293$	$S_S = 0.207$	$S_S = 0.141$	$L_S = 25.0$ cm $\overline{S_S} = 0.23$ $c = 1.78$ $\Delta_L = 0.54$
Mature	$S_S = 0.201$	$S_S = 0.184$	$S_S = 0.117$	$L_S = 13.9$ cm $\overline{S_S} = 0.18$ $c = 1.03$ $\Delta_L = -1.91$

FIGURE 4.8 Representative shoot silhouettes for a young (tree height, $h = 4$ m) and a mature tree ($h = 18$ m) of temperate broadleaved conifer *Agathis australis* (data from Niinemets, Ü., Sparrow, A., and Cescatti, A., *Trees: Struct. Funct.*, 19, 177, 2005). The shoots were taken from similar high-light environments for both young (daily integrated seasonal average quantum flux density, $Q_{int} = 25.6$ mol m^{-2} day^{-1}) and mature ($Q_{int} = 26.8$ mol m^{-2} day^{-1}) tree. Shoots were photographed from various view directions. For the projection 0°, 0° (rotation, inclination angle) the upper part of the shoot is facing the view direction, for the 90°, 0° projection, the shoot is rotated 90° around its axis, and the projection 0°, 90° gives the shoot axial view. These and additional shoot projections were employed to derive the parameter of ellipsoidal distribution of leaf surface angles (c) and the degree of leaf clumping (Δ_L, defined in Figure 4.5) as described in detail in Niinemets et al. (2005). Ellipsoidal distribution of leaf angles assumes that the leaves are distributed parallel to an ellipsoid, and the parameter c is the ratio of ellipsoid major and minor semiaxes (Campbell 1986, Norman and Campbell 1989). $c = 1$ for a spherical distribution of leaf surface inclination angles, $c > 1$ for horizontal distributions, and $c < 1$ for vertical distributions. The clumping characteristic, $\Delta_L \to 0$ for a random dispersion, whereas positive values of Δ_L correspond to regular canopies (positive binomial model) and negative values (negative binomial model) to clumped canopies (Nilson 1971, Baldocchi and Collineau 1994). For every shoot projection, the ratio of silhouette to total surface area (S_S) and for every shoot, the spherical average of S_S ($\overline{S_S}$) and shoot length (L_S) are also provided.

As this pattern results in larger penetration of light to lower canopy layers, it results in a more uniform profile of light with the canopy than a distribution with constant leaf angles and maximizes whole-plant photosynthesis (Herbert 1991, Herbert and Nilson 1991). In a canopy with vertical inclination angles in the upper canopy and more horizontal leaves in the lower canopy, only a few leaves are light-saturated in the upper canopy, and the leaves at the base of the crown receive enough light for photosynthesis. Therefore, canopies with varying inclination angles can sustain greater foliage areas than canopies with constant inclination angles (Russell et al. 1989).

A little-explored aspect of leaf angle is how it interacts with leaf internal anatomical structure in modifying light harvesting and utilization at the chloroplast level. Leaf inclination affects the distribution of light between lower and upper surfaces, and depending on how efficiently foliage photosynthetic characteristics upper and lower surface of leaf acclimate to the long-term irradiance, modification of the fractional distribution of light interception between upper and lower surfaces of leaf can alter whole leaf photosynthesis (Poulson and DeLucia 1993, Valladares and Pearcy 1999). Large differences in mesophyll photosynthetic properties between the two sides of the leaves seem to depend on a complex interaction

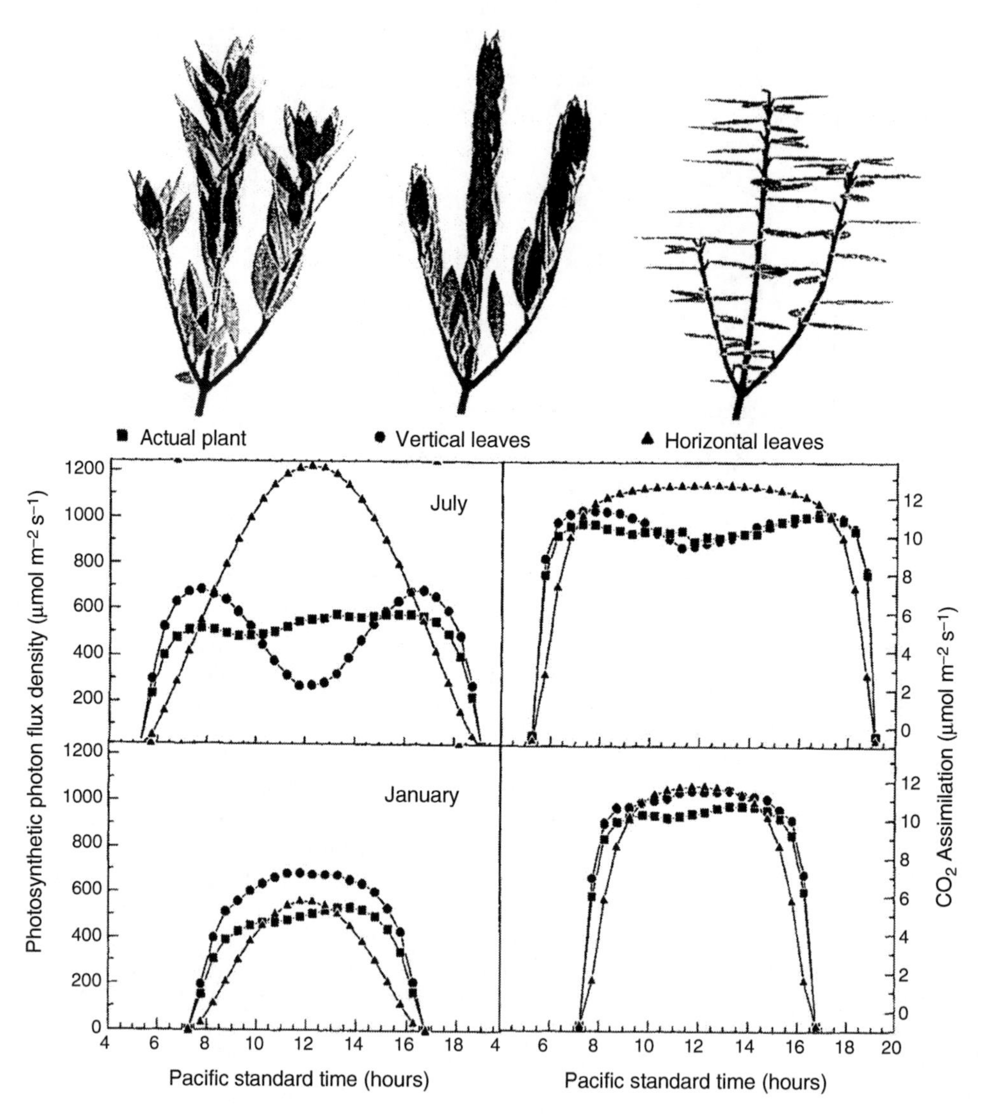

FIGURE 4.9 Diurnal course of interception of photosynthetically active radiation and CO_2 assimilation calculated for whole shoots of the chaparral shrub *H. arbutifolia* on a clear day of winter (*lower* graphs) and summer (*upper* graphs). Data were calculated for real shoots and for the same shoots with either vertical or horizontal leaves. Simulations were performed using the three-dimensional YPLANT model (Pearcy and Yang 1996). (Data from Valladares, F. and Pearcy, R.W., *Oecologia*, 121, 171, 1999.)

among light environment, leaf anatomy, and leaf angle (Myers et al. 1997). In addition, mesophyll cells (Smith et al. 1997) and bundle sheath extensions (Nikolopoulos et al. 2002) can function as optical fibres canalizing light into deeper leaf interior. However, to function as an optical fiber the leaf surface must be perpendicular to solar beams, implying that leaf inclination can modify the diurnal distribution of light penetration into the leaf.

Diffuse Light

We have so far considered the importance of leaf angular distribution for direct radiation interception, but diffuse light is an important component of incident radiation (Gutschick and Wiegel 1988, Herbert 1991). While the leaf angular distribution affects diffuse light transfer to a minor degree, foliage dispersion modifies diffuse light interception similarly to direct light interception (Cescatti and Zorer 2003). While the leaf angular distribution affects diffuse light transfer to a minor degree, foliage dispersion modifies diffuse light interception similarly to direct light interception (Cescatti and Zorer 2003). The geometry of the foliage, basically proximity of leaves and distribution of leaf angle throughout the canopy, affect the transport of diffuse light to lower layers. This can be relevant for whole-plant photosynthesis in both low- and high-light environments (Valladares and Pearcy 1998). Studies further demonstrate that the orientation of the crown and its leaves in the vicinities of forest gaps frequently respond to diffuse light rather than direct light (Ackerly and Bazzaz 1995, Clearwater and Gould 1995).

Penumbra

The majority of radiative transfer models assume that the sun is a point light source and separate only between shaded and sunlit foliage. As illustrated in Valladares (1999), this assumption can lead to significant errors in simulation of light interception and photosynthesis. The relevant parameter describing the relevance of considering that the solar disk has a finite size, is the ratio of canopy height to foliage element diameter (Cescatti and Niinemets 2004). The solar disk is completely blocked by a leaf (umbra) at a theoretical distance of 108 times the leaf diameter. An object farther than this distance is lit by at least part of the sun (penumbra). Empirically, this distance is approximately 50–70 times the leaf diameter with the sun at the zenith on a clear day (Horn 1971). Thus, long crowns with small leaves intercept a large fraction of light as penumbral radiation. In fact, in conifers, penumbral radiation can be more than 95% of total, whereas in broad-leaved herbs, penumbral radiation constitutes only a few percent (Cescatti and Niinemets 2004). Given the strong nonlinearity in photosynthetic light response, lack of consideration of penumbral radiation results in major underestimation of canopy productivity (Cescatti and Niinemets 2004). Overall, this discussion suggests that leaf size per se can play a major role in light harvesting by the whole crown and that future advancements of ray-tracing models can facilitate further insights into complex geometrical phenomena such as penumbra.

Changing Geometries: Leaf Movements and Rolling

Leaves from a number of species move during the day, keeping leaf blade either perpendicular (diaheliotropic movements) or parallel (paraheliotropic movements) to the direct rays of the sun. For the leaves tracking the sun, light interception can be enhanced by as much as 35% compared with a fixed leaf with a horizontal position (Ehleringer and Forseth 1980, Ehleringer and Werk 1986). For the leaves remaining parallel to the sunrays, can significantly reduce light interception and heat loads relative to a leaf remaining in a fixed position. Leaf solar tracking occurs in herbaceous species that do not form an extensive canopy and is most common in annuals (Ehleringer and Werk 1986). In drier sites, the frequency of leaf solar-tracking species seems to be inversely related to the length of the growing season, reaching values as high as 75% of the flora in the summer annuals of the Sonoran Desert (Ehleringer and Forseth 1980).

Leaf solar tracking poses a physiological dilemma when photosynthesis is impaired at midday by water or heat stress. Under these far-from-optimum conditions, diaheliotropic leaves can intercept excessive radiation that is potentially damaging to the photosynthetic apparatus. Certain species, such as the desert annual *Lupinus arizonicus*, avoid the dilemma, exhibiting either diaheliotropic or paraheliotropic leaf movements depending on the availability of water (Ehleringer and Werk 1986).

The overall effect of leaf movements on canopy productivity depends on stand leaf area index. When leaf area index is low, solar tracking enhances canopy productivity since leaves absorb photons that would otherwise pass through the sparse canopy. However, when leaf area index is greater than 4, leaf solar tracking reduces canopy productivity because the bulk of the canopy photosynthesis is restricted to the leaves of the upper parts of the crown (Ehleringer and Forseth 1989). In dense or very large crowns, leaf movements are restricted to the external layer of leaves, because the leaf movements require a high ratio of direct to diffuse components of the solar radiation (Ehleringer and Forseth 1989). By the same token, leaf movements are not expected to occur in habitats with a high incidence of overcast days or in understory habitats.

In addition to these short-term leaf movements, many species, in particular grasses, respond to drought by leaf rolling (Corlett et al. 1994, Turgut and Kadioglu 1998, Fernandez and Castrillo 1999). Leaves of these species have specific bulliform cells located near the vascular bundles. During drought, water is absorbed from these cells, resulting in inward rolling of the foliage (Moulia 1994).

Crown Architecture in Extreme Light Environments

Light can be a limiting resource in understories of dense stands or for plants subject to strong neighborhood competition, whereas light can be excessive and even harmful in open environments where plant metabolism is impaired by environmental stresses. Plant shape and size have been shown to change as a function of the light environment, and plants are capable of orienting their light-capturing surfaces in different ways to increase or decrease the leaf surface area projected in the direction of ambient light (Ellison and Niklas 1988, Stenberg et al. 1998, Cescatti and Niinemets 2004).

Plants exhibit a remarkable within-species and within-individual variability in their structural features. For instance, branching pattern of trees is not stationary, and it has been shown that the variation of branching pattern can be the result of developmental–phenotypic interaction (Steingraeber 1982). While in some cases it can be due to a malfunction of the genetic program, in most cases this variability is a plastic response to local conditions, and light is possibly the most spatially and temporally heterogeneous environmental factor affecting plant survival and growth. Structural plasticity of plants enables a fine-tuning with environmental changes so that the efficiency of the limiting processes at each stage is maximized. A common environmental change experienced by plants is the decreasing availability of light with the advance of succession. It has been shown for the succulent halophyte *Salicornia europaea* that morphological changes in the branching patterns during succession maximized light interception (Ellison and Niklas 1988). However, even phylogenetically close species differ in their capacity for a plastic response to the light environment (Valladares et al. 2000). Interestingly, certain species that exhibit an architecture suited to high irradiance conditions do not change significantly when grown in the shade. That was the case for mangroves in Malaysia: architecture and allometry of shaded mangroves were consistently more similar to those of exposed mangroves than to shaded, broad-leaved, evergreen, rainforest trees (Turner et al. 1995).

When Light Is Scarce

When plants grow in dense stands or in the understory, the resource of radiant energy becomes scarce, unpredictable, and patchy. In these environments, evolution has led to two principal approaches for survival: shade avoidance and shade tolerance. Angiosperms, in particular, have evolved an impressive capacity to avoid shade. The so-called shade-avoidance syndrome involves accelerated extension growth, strengthened apical dominance, and retarded leaf and chloroplast development, among other processes (Smith and Whitelam 1997, see Section "Plasticity, Stress and Evolution"). Here the focus is on the functional aspects of the crown of plants that tolerate shade and on the structural features that are relevant for such tolerance.

Tropical rainforests exhibit an outstanding diversity of plant species and growth forms (Medina 1999, Wright 1999). Despite the extremely low levels of irradiance experienced in the understory of late-successional rainforests, a relatively large number of shrubs, herbs, and seedlings can be found within a few hectares. These plants suffer shading not only from the forest canopy and neighboring plants, but also from the leaves of their own crowns. The efficiency of light capture of 24 understory species differing in their habit and growth form was compared, and the influence of phyllotaxis and leaf size and shape in the avoidance of self-shading was explored in a field study in a lowland tropical rainforest in Barro Colorado Island (Valladares et al. 2002c). The species studied included understory palms, saplings of canopy trees, shrubs, and a wide variety of monocots of contrasting architectures. Plant size and total leaf surface area also varied significantly among the species considered. Most of the phyllotaxes shown in Figure 4.1 were represented, and leaf size ranged from a few to several hundred square centimeters. Light harvesting efficiency was calculated with the three-dimensional plant architecture model YPLANT (Pearcy and Yang 1996). The most remarkable result of this study was the functional convergence of the different plant species co-occurring in the forest understory: most of the species intercepted between 80% and 90% of the available radiation, and mutual shading of the leaves during the brightest hours of the day was little, approximately 10% of the foliage area in most cases (Figure 4.10). Thus, the rare spiromonostichous phyllotaxis of *Costus pulverulentus* (Figure 4.10), apparently a unique solution to avoid self-shading, was no better for this purpose than the spiral phyllotaxis of the saplings of *Thevetia ahouai* or the pseudodistichous foliage of the shrub *Hybanthus prunifolius*. Nevertheless, significant differences among species were found when the fraction of the plant biomass invested in support was considered in the analysis of the efficiency of the different architectures. Monocots, with a lower investment in dry weight, generally reached a more favorable compromise in this simple cost–benefit analysis of plant architecture under limiting light conditions (Valladares et al. 2002c). The drawbacks of the monocot strategy are a reduced survival to mechanical damage, and in some cases, a shorter plant longevity and a limited capacity to reach the forest canopy.

There is evidence that both spiral and distichous phyllotaxis are more frequent in low-light environments, whereas opposite phyllotaxis are more frequent in open, high-light environments (Brites and Valladares 2005). Several plant species first set leaves spirally around an erect stem, and then produce horizontal branches bearing distichous leaves (Leigh 1998). This combination of two phyllotaxes has been interpreted as a way of minimizing leaf overlap. *Dichorisandra hexandra* exhibits this combination of spiral leaves around vertical stems and distichous leaves around horizontal branches (Figure 4.10), but leaf overlap is as reduced as in other understory species with different leaf arrangements. It seems more likely that this combination of two phyllotaxes is an efficient way of filling the space with leaves while growing in height.

Where there are many leaves in one spiral, long petioles in older leaves or narrow leaf bases in certain species can minimize leaf overlap (Leigh 1998). In the redwood forest understory plant *Adenocaulon bicolor*, which exhibits a spiral phyllotaxis with a mean divergence angle of 137° (phyllotactic fraction of 8/21), leaf overlap was reduced by particular combinations of leaf size and petiole length at successive nodes (both increasing initially and then decreasing). The petiole length observed in this plant corresponded to the optimal petiole length obtained in simulations of the dependence of light absorption efficiency on petiole length (Pearcy and Yang 1998).

In the search for light, the crown of certain plants becomes thin instead of broad and flat in the shade. Light interception is not favored by such transformation, which usually represent an escape strategy of shade-intolerant species (Peer et al. 1999). In some cases, the whole developmental sequence of the plant is changed in the shade. Shrubs such as *Arbutus* switch from a sympodial growth in the open (Leeuwemberg architectural model) to a monopodial

Costus pulverulentus *Dichorisandra hexandra*

0.07	Fraction of leaf area self-shaded during the central hours	0.06
0.84	Fraction of leaf area displayed during the central hours	0.85
0.49	Intercepted PPFD in a clear day of spring (mol m^{-2} day^{-1})	0.75
0.90	Intercepted PPFD in a clear day of spring (fraction of available)	0.93

Hybanthus prunifolius

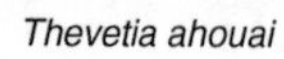

Thevetia ahouai

0.12	Fraction of leaf area self-shaded during the central hours	0.07
0.77	Fraction of leaf area displayed during the central hours	0.81
0.30	Intercepted PPFD in a clear day of spring (mol m^{-2} day^{-1})	0.26
0.86	Intercepted PPFD in a clear day of spring (fraction of available)	0.87

trunk (according to the model of Scarrone) in low-light environments (Bell 1993). Many plants accommodate their structure to the light environment, enhancing light interception efficiency under low-light conditions. This is the case of the chaparral shrub *H. arbutifolia*, which changes from orthotropic stems with spirally arranged leaves in the open to plagiotropic stems with pseudodistichous foliage when exposed to the moderate shade of a *Quercus* woodland (Valladares and Pearcy 1998). This structural change, in contrast to the escape strategy of more shade-intolerant species, significantly enhances light interception on a leaf area basis.

When Light Is Excessive

Plants in open environments are exposed to high irradiance, which frequently leads to a decline in the efficiency of photosynthesis (photoinhibition), particularly under adverse conditions (Horton et al. 1996, Osmond et al. 1999). Under these circumstances, plants exhibit remarkable physiological and architectural plasticity. Physiological adjustments result in protection of photosynthetic apparatus against light intensities in excess to those that can be used in photosynthesis (Osmond et al. 1999). Structural adjustments lead to the avoidance of excessive irradiance by structural features, overall reducing the total leaf area or the fraction of leaf area directly exposed to the sun.

Sun shoots of the chaparral shrub *H. arbutifolia* exhibited a remarkable structural photoprotection, and despite having seven times more photosynthetically active radiation available, they intercepted only four times more and had potential daily carbon gains only double of those of shade shoots (Valladares and Pearcy 1998). The resulting fraction of leaf area that was displayed during the central hours of a typical day of spring was only one-third of the total leaf area of the shoot (Figure 4.11). Leaf angle, the most plastic character in the response of *H. arbutifolia* shoot to high light, played a key role in achieving an efficient compromise between maximizing carbon gain while minimizing the time that the leaf surfaces were exposed to irradiance in excess of that required for light saturation of photosynthesis, and therefore potentially photoinhibitory (Figure 4.9). For relatively simple canopies, leaf angle and orientation are the main structural photoprotective features (Werk and Ehleringer 1984, Smith and Ullberg 1989), but mutual shading among leaves can be even more important in complex, multilayered canopies (Roberts and Miller 1977, Caldwell et al. 1986). In *H. arbutifolia*, 27% of the foliage was self-shaded during the central hours of a clear spring day (Figure 4.11), but this percentage was far higher for leaves of certain orientations, such as those facing south. A steeply oriented foliage and moderate self-shading that reduces the photosynthetic surface area displayed during the central hours of the day were also characteristic structural features of the crowns of two other plants from high-light environments: *Stipa tenacissima*, a tussock grass, ad *Retama sphaerocarpa*, a leguminuous, leafless shrub (Valladares and Pugnaire 1999). These two species exhibited similar leaf display and PPFD interception efficiencies to those of *H. arbutifolia* (Figure 4.11). The costs in terms of missed

FIGURE 4.10 Four plant species co-occurring in the understory of a tropical rainforest (Barro Colorado Island, Panama). *Costus pulverulentus* and *Dichorisandra hexandra* are monocot herbs, *Hybanthus prunifolius* is an understory shrub, and the individual of *Thevetia ahouai* (a canopy tree) presented is a 2-m-high sapling. Beneath each photograph, to computer images at dawn (*left*) and at noon (*right*) of a representative of each species are provided. A lighter gray in the computer images indicates overlap between two or more leaves as seen from the sunpath. For each species, the fraction of the total leaf area that is either self-shaded or displayed during the central hours of the day, and the PPFD intercepted in a clear day of spring (both as daily total and as a fraction of available) were calculated using the three-dimensional YPLANT model (Pearcy and Yang 1996). (Data from Valladares, F., *Handbook of Functional Plant Ecology*, F.I. Pugnaire and F. Valladares, eds, Marcel Dekker, Inc., New York, 1999.)

FIGURE 4.11 Three plant species from open, dry environments. *Heteromeles arbutifolia* is an evergreen sclerophyll of the California chaparral, *Stipa tenacissima* is a tussock grass frequent in the driest regions of the Iberian Peninsula, and *Retama sphaerocarpa* is a leguminous, leafless shrub also frequent in dry and warm areas of the Iberian Peninsula. Beneath each photograph, two computer images of a representative of each species are provided. A lighter gray in the computer images indicates overlap between two or more leaves as seen from the sunpath. For each species, the fraction of the total leaf area that is either self-shaded or displayed during the central hours of the day, and the photosynthetic photon flux density (PPFD) intercepted in a clear day of spring (both as daily total and as a fraction of available) were calculated using the three-dimensional YPLANT model (Pearcy and Yang 1996). (Data from Valladares, F. and Pearcy, R.W., *Oecologia*, 114, 1, 1998; Valladares, F. and Pugnaire, F.I., *Ann. Bot.*, 83, 459, 1999.)

opportunity for carbon gain (comparing plant crowns with equivalent horizontal photosynthetic surfaces) for these two species were similar to those imposed by the summer drought (approximately 50% of the potential carbon gain), the main limiting factor for plant survival in semiarid environments. This elevated cost of structural photoprotection emphasizes the ecological relevance of avoidance of high irradiance stress in these species.

Other stress factors occurring in high-light ecosystems, such as heat and water deficits, also favor increased inclination angles of leaves (Shackel and Hall 1979, Ehleringer and Forseth 1980, Comstock and Mahall 1985, Lovelock and Clough 1992) as well as greater degree of leaf rolling and folding (Fleck et al. 2003). Since different stresses co-occur at certain times of the day or during certain seasons (Niinemets and Valladares 2004), a protective strategy that is triggered by one type of stress (heat, water deficit, or excessive light) also increases protection or tolerance to other simultaneously occurring stresses is very adaptive. In addition to a series of structural features that clearly constitute an adaptive strategy helping to cope with multiple stress, a series of physiological responses such as down-regulation of photosynthesis and heat tolerance has been observed in plants from Mediterranean-type climates. As the structural adjustments, these physiological responses were found to be very efficient in protecting the high-light exposed plants to multiple stresses during the summer (Valladares and Pearcy 1997, Valladares et al. 2005).

Leaves of broad-leaved species are also often significantly rolled and curved, especially at high-upper canopy where high irradiances can be combined with water limitations (Farque et al. 2001, Fleck et al. 2003). Leaf rolling results in extreme reduction of leaf area, and therefore, in large decreases in radiation interception and transpiration (Figure 4.12). In fact, the leaves of broad-leaved species are never completely flat (Sinoquet et al. 1998). Leaf rolling in broad-leaved trees strongly reduces radiation interception of these leaves and can be a beneficial attribute in reducing photoinhibitory damage, heat stress, and transpiratory water loss.

It has been shown that structural avoidance of excessive irradiance by any of the means illustrated earlier can be crucial for survival under extreme conditions, even in plants capable of extensive physiological adjustment to stress (Valladares and Pearcy 1997). Stress itself can also direct influence crown architecture as it can modify the allocation patterns and the developmental processes of the plant. High irradiance can lead to high-leaf temperatures, especially in warm regions and when transpirational cooling is reduced due to water deficits, as in arid or Mediterranean-type environments (Figure 4.12 and Figure 4.13). The complex interplay between leaf size and shape, phyllotaxis, branching, and mutual shading among neighboring leaves can lead to contrasting temperatures in different plants and even in different leaves within the same plant as revealed by infrared thermographies (Figure 4.12 and Figure 4.13). Infrared thermography is a powerful tool to study the complex and heterogeneous pattern of leaf temperatures in plant crowns exposed to high light, which is the result of the combined effect of a large number of morphological, physiological, and environmental variables (Jones 2004).

Occupying Space and Casting Shade: The Community Level

Many analyses of adaptations in plant form have assumed that natural selection favors traits that tend to maximize the growth rate of a given plant (Givnish 1986). One important objection to this approach is the lack of consideration of the role of competitors. As Givnish (1986) has shown, certain features of plant architecture, such as leaf height, are examples of a trait in which the strategy that maximizes growth in the absence of competitors is the one that does most poorly in their presence. In addition, certain structural features are not very efficient for their primary function, but provide a competitive advantage. For instance, the generous amount of leaves within certain crowns, well above what could be strictly needed, appears to be competitively more effective by intercepting light that competitors might use than in providing photochemical energy (Margalef 1997). Thus, the evolution of plant form must be interpreted considering not only the immediate function of the organs and structures involved (e.g., light interception), but also its effect on the efficiency with which competitors exploit the resources.

Whenever plants are grown in close proximity there is competition for light and space. Branching poses an ecological dilemma to many plants, because a wide and low crown with a

FIGURE 4.12 Leaf size and shape, and their relative position in the shoot determine the exposure of leaf surfaces to sunlight, which in turn translates into different leaf temperatures under the same environmental conditions, particularly under clear skies. Left-hand side images (a, c, and e) are normal, visible photographs, whereas those at their right (b, d, and f) are the corresponding infrared thermal images. The curved leaves of *Ailanthus altissima* (a, b) generated within-leaf shading and this curling protected some leaf zones against overheating. The large leaves of *Ficus carica* (c, d) exhibited large within-leaf thermal ranges which were due to differential transpirational cooling and thermal properties over the leaf surface since leaves were almost flat; note the relatively high temperatures of major leaf veins. The multilayered crown of the climber *Hedera helix* (e, f) rendered contrasting leaf temperatures depending on the relative position of each leaf; note the 8.5°C difference between the exposed leaf and the shaded leaf immediately underneath. The scale at the right is in Celsius degrees. All images were taken with a Flir B50 thermal camera in the afternoon of a clear spring day in Madrid (Spain); air temperature during the measurements was 35°C, air relative humidity was 23%, and wind speed was 0.15 m s^{-1}.

lot of branches is highly efficient in terms of light capture versus construction costs, but is easily overtopped, whereas a tall crown with little branches represents the opposite trade-off. Certain plants have reached an interesting compromise by building compound leaves whose long rachises act as throw-away branches, extending the photosynthetic surface of the crown without investing in permanent and expensive support tissue (Givnish 1978). Probably one of the most extreme examples of this strategy is that of the Devil's walking stick *(Aralia spinosa)*, which avoids branching by producing long, light leaves, allowing for a fast growth of the

FIGURE 4.13 Phyllotaxis and shoot architecture influence the energy balance of individual leaves and of the whole plant. Steep and aggregated, spiral leaves of *Arbutus unedo* (a, b) reduce the exposure of leaf area leading to an important degree of self-shading; only some leaves with their laminas oriented to the south exhibit temperatures significantly above air temperature (up to 6°C higher). The erectophile foliage of *Rosmarinus oficinalis* (c, d), made up of thin and small leaves very close to each other (short internodes), is entirely at air temperature, in contrast with the surrounding soil, which is 20°C above air temperature. The complex and multi-branched crown of *Nerium oleander* (e, f) generates an intricate pattern of leaf temperatures. Left-hand side images (a, c, and e) are normal, visible photographs, whereas those at their right (b, d, and f) are the corresponding infrared thermal images. The scale at the right is in Celsius degrees. All images were taken as in Figure 4.12.

trunk. To support the same leaf area, a co-occurring tree, the flowering dogwood (*Cornus florida*), has to invest 7–15 times more in wood (White 1984).

Height growth is extended in the forest understory because light increases exponentially toward the canopy surface, whereas the costs of structural tissues escalate less rapidly with plant height (Givnish 1982 see Section "Plant Design" for further analyses of the economics of plant height). Competition between trees in its most general sense is competition to fill space, quickly in early succession, where r-selection seems to be dominant, but completely in late succession, where K-selection seems to dominate (Horn 1971). The dynamic nature of a forest canopy, with numerous gaps and clearings caused by treefalls, contributes to the coexistence of many more species of trees than could be expected in a theoretical analysis of crown architecture and monopolization of light; therefore, species of different successional status and crown shapes can coexist (Figure 4.4).

Forests are vertically stratified, from the towering emergent trees to the herbs on the forest floor, each strata comprising a distinct suite of plant species adapted for the conditions at each particular level, mainly light conditions. In the initial description of this idea of vertical strata in forests, an explicit mechanism that would account for stratification was not provided (see discussion in Terborgh 1992). One of the questions that remained unanswered in that description was why certain tree species cease their upward growth when they attain a given height, unlike canopy species that pursue an upward trajectory until they reach the open sky or die. Is there an optimum height for a midstory tree? Terborgh (1992) suggested an explanation, considering how direct sunlight passes through the holes of the forest canopy to the lower layers and eventually to the forest floor. As the sun progresses across the sky, sunlight penetrates into the forest over a wide range of angles. In a simplified and regular canopy, the sunlight passing into the forest interior through a single gap forms a triangular area on the way to the ground (Figure 4.14). At the upper parts of the triangle, the number of hours of sunlight is larger than at the lower parts. These triangular areas of direct sunlight

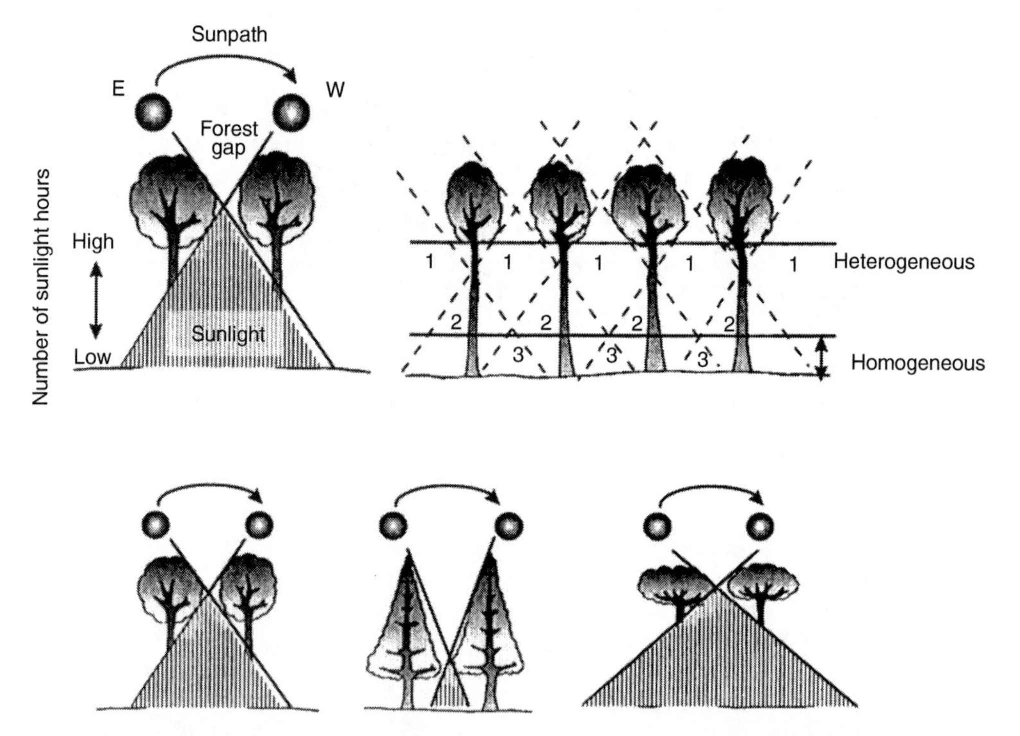

FIGURE 4.14 A gap in the forest canopy allows direct sunlight to reach the understory (*upper left*). The number of sunlight hours increases from the ground to the canopy. When the sunlight passing through more than one gap is considered, a more complex pattern is found (*upper right*), with understory areas affected by one, two, three, or more neighbor gaps (indicated by numbers). Where the cones of several gaps intersect, a spatially uniform light field is produced. Both the distance between the trees forming the limits of the gap, and the shape of the crown of these trees determine the duration of direct sunlight in the understory (lower graphs). Pyramidal crowns allow little sunlight to reach the understory, whereas the reverse is true for flat and broad crowns. (Adapted from Terborgh, J., *Diversity and the Tropical Forest*, Scientific American Library, New York, 1992.)

spread out below the canopy so that areas from adjacent gaps overlap. Some points receive direct sunlight twice a day (intersection of two areas), whereas others lower down in the forest receive sunlight from an increasing number of gaps, although for increasingly briefer periods of time (briefer "sunflecks" see Pearcy 1999). This generates a spatially uniform light field near the ground (Figure 4.14), and it was predicted that a midstory tree must grow as high as the higher limit of this field because above this point, at least part of the tree crown might not receive enough light to pay its costs, and the whole construction and maintenance costs of the tree would be increased at the expenses of other functions such as reproduction (Terborgh 1992). This prediction was found to be true (midstory trees were of the expected height) in a mature temperate forest in North America, but not in the more complex and irregular forests of the tropics. In addition to the fact that the canopy of tropical forests is uneven and complex, Terborgh (1992) pointed to the shape of the crown of the canopy trees as another factor to explain the lack of a predictable, uniform light field. The shape of their crown determines the size of the triangular area of sunlight beneath a gap (Figure 4.14). Crown shape tends to vary with latitude, with mushroom-like trees in the tropics and conical crowns in boreal regions (Figure 4.4), which allows for either generous shafts of direct sunlight or very little sunlight reaching the floor, respectively (Figure 4.14). Thus although the forest has plenty of understory plants in the tropics, it is nearly devoid of them in the boreal regions; temperate forests represent an intermediate situation, with their rather simple canopy structure being very suitable for Terborgh's theoretical description of vertical light gradients and for the corresponding predictions of optimal height of understory trees. Another prediction regarding crown architecture resulting from the thesis that forests are vertically stratified is that crown shape varies systematically with vertical position. This was found to be true in tropical forests with more than two plant strata, whereas emergent trees possessed crowns that were more broad than deep, those of trees immediately below were more deep than broad (for the rationale, see Terborgh 1992).

In their search for light, understory plants are not only exposed to the vertical gradient of light, but to other physical factors that interact and influence their architecture. If height growth in a low-light environment has the risk of too-expensive construction and maintenance costs, the situation becomes riskier or at least more complicated when the ground is not even, as is the case with hillsides. Since the lines of equal light intensity from the canopy to the ground run parallel to the ground (Horn 1971), the most efficient height growth occurs at right angles to the ground (Figure 4.15). However, to do this on a slope, trees should lean outward (Alexander 1997). Trunk inclination on slopes has been shown to be adaptive (Ishii and Higashi 1997), but the greater the angle of lean, the stronger the trunk of the tree needs to be for biomechanical reasons (Mattheck 1991, 1995), which entails additional costs. Under low-light conditions, leaning trees cannot grow a trunk as tall as it could if it were vertical, so their optimal angle on a slope is neither vertical nor perpendicular to the forest floor (Alexander 1997). Ishii and Higashi (1997) constructed a model to explore tree coexistence on a slope and to predict how tree survival is affected by trunk inclination. The predictions were that survival rate increases with slope angle more sharply under poorer light conditions. These predictions were supported by the understory tree *Rhododendron tashiroi*, which exhibited sharper trunk inclination and coexisted more successfully on steeper slopes with the dominant canopy trees (Ishii and Higashi 1997). Trunk inclination also seems to be affected by the shade tolerance of the species, with the relationship between slope and trunk inclination being more marked in shade-sensitive trees (Figure 4.15). This model provides an explanation based on optimizing processes of evolution by natural selection for the common observation that the trunks of trees on a slope often incline downward. This explanation is more complete and convincing than previous ones alluding to landslides or wind (King 1981, Del Tredici 1991, Mattheck 1991). Another way to enhance light capture on

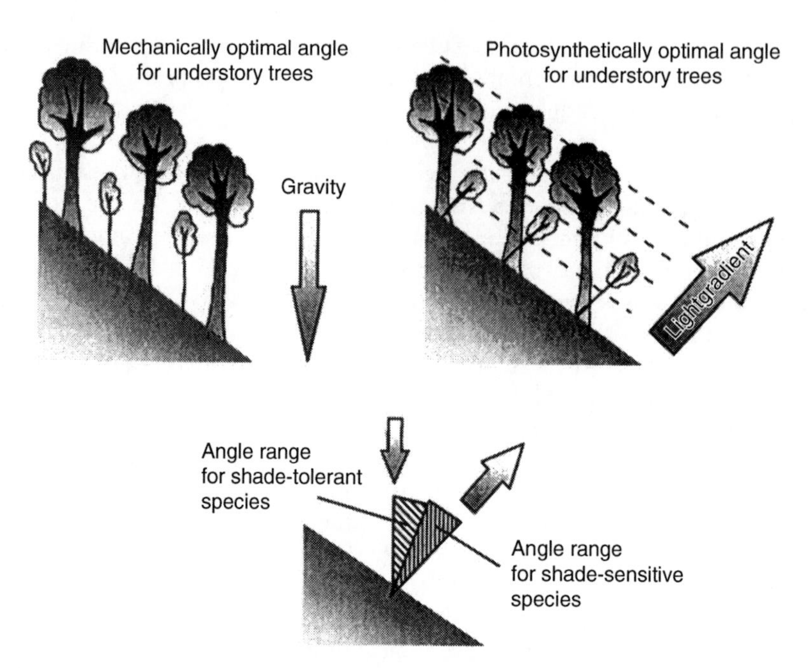

FIGURE 4.15 Understory trees that grow on slope are exposed to the dilemma of growing vertically, which is mechanically optimal, or with their trunks inclined downward, that is, parallel to the light gradient occurring from the ground to the upper canopy, which shortens the distance of their foliage from the canopy surface. Depending on their light requirement or shade tolerance, species are expected to exhibit two ranges of trunk angle, as shown in the lower figure. (Adapted from Alexander, R.M., *Nature*, 386, 327, 1997; Ishii, R. and Higashi, M., *Proc. R. Soc. Lond. Ser. B., Biol.*, 264, 133, 1997.)

slopes is with an asymmetrical crown, with more branches on the downhill side (Hallé et al. 1978, Alexander 1997). Although this architectural solution does not require trunk inclination, it must imply additional construction costs if the trees are not to snap or fall over.

Another interesting study regarding understory plants and slopes is that of leaning herbs that arrange their leaves in a distichous array along an arching stem (e.g., *Disporum, Polygonatum, Smilacina, Streptopus, Uvularia* in the Liliaceae, and *Renealmia* in the Zingiberaceae). Such crown architecture is mechanically less efficient than an umbrella-like arrangement that supports leaves at the same height on a vertical stem. Their competitive advantage derives from their tendency to orient strongly downslope (Givnish 1982, 1986). Above a critical slope inclination, leaning shoots become mechanically more efficient than other herb architectures and tend to supplant them. The correspondence between their observed and predicted distribution relative to slope inclination provides support for their competitive ability on slopes (Givnish 1986).

Hedgerows, linear arrangements of woody species that follow property boundaries between fields, are an interesting case of plant communities with strong competition for light and space. The interactions between carbon relationships, growth, and plant architecture have been thoroughly studied in these systems (Küppers 1994). An important conclusion from these studies was that plant architecture in combination with carbon input can lead to a better understanding of plant success since net primary production by itself is not sufficient to explain competitive relationships between woody species (Schulze et al. 1986). Branching and leaf exposure were principal. Mechanisms in competition: short internodes and thorns were important in early-successional species, whereas shading of neighbors with a minimum of self-shading (capacity for occupying new, higher aerial space coupled with maintenance of a closed leaf cover above the occupied space) provided a competitive advantage to late-successional species (Schulze et al. 1986). Particular branching patterns and lower costs of space occupation permitted late-successional species to grow a crown quickly and then outcompete shade-intolerant pioneers (Küppers 1989, 1994).

Using a stochastic model of plant growth, Ford (1987) reached some interesting conclusions on the implications of crown architecture for plant competition and spatial interference. Growth rates depended on branch probability and also on angle of branching in sympodial plants but not in monopodial plants. In sympodial plants, an optimal branching angle of approximately 30° was found when the ratio of the interference distance to the internode distance was 1.5:5. Sympodial plants with a branching angle of 30° outcompeted monopodial plants (Ford 1987). The light environment of a plant can vary due to the activity of nearby vegetation. Sensing their neighbors, perceiving the light opportunities, and responding in a timely fashion is crucial for plant survival at the community level (Ballaré 1994). An adaptive way of coping with competition is by a plastic response to the environmental changes caused by neighboring plants, an issue briefly addressed in the next section.

In plant communities, however, not everything is competition. In fact, facilitation is becoming better recognized and there is ample evidence pointing to facilitation as an important driver of community dynamics (Callaway and Walker 1997). In arid environments, plants can benefit from an attenuation of the stressful irradiance by being in the shade of others, which can act as nurse plants and improve early establishment and survival of seedlings (Gómez-Aparicio et al. 2004). However, to which extent this effect prevails over competition for water is uncertain, since there is conflicting evidence (Maestre et al. 2005). More studies on the net balance of plant–plant interactions in natural conditions are needed to solve this controversy, particularly in arid conditions where the increase of facilitation with drought stress is not always found (Lortie and Callaway 2006, Maestre et al. 2006).

PLASTICITY, STRESS, AND EVOLUTION

The shape of the crown is an adaptive compromise of conflicting strategies. Multiple functions of an architectural design and functional convergence of alternative architectural features make assessment of optimal design difficult. For instance, phyllotaxis is not strictly a developmental constraint because different phyllotaxes can be functionally equivalent (e.g., in terms of light interception efficiency (Niklas 1988)). The same argument can be built for branching patterns or for the general shape of the crown (Figure 4.3 and Figure 4.4), and these structural features may provide a paradigm for other features in plant evolution. This is the most likely reason why it has proved impossible to describe the many tree architectural models as adaptations. However, there are particular cases in which the several functions of a given structure do not appear to be a constraint in interpreting its functional optimization, such as the analysis of petiole length versus light capture in the understory herb *Adenocaulon bicolor* (Pearcy and Yang 1998). Certain genetic or developmental constraints can be overcome from a functional point of view by changes in other, more plastic structural features, such as petiole and leaf shape, size, and orientation. Consequently, evolution of architectural features cannot be interpreted without a minimum knowledge of the plastic response to the environment of the involved traits.

Shade avoidance, a feature that angiosperms have developed to a remarkable extent, is based on signals that anticipate that shade is going to change (via changes in the red-far red ratio). The so-called shade avoidance syndrome involves a highly plastic response in the shade with strong elongation of internodes and petioles, the production of less dry matter, larger and thinner leaves, a higher shoot–root ratio, and a series of remarkable physiological changes mediated by multiple phytochromes (Smith and Whitelam 1997). The variety of morphogenic programs triggered to move the photosynthetic area toward better-lit canopy regions, and all of the morphological and physiological adjustments to the light environment, correspond to the concept of foraging for light (Ballaré 1994, Ballaré et al. 1997).

Phenotypic plasticity in plants has three main functional roles: maintenance of homeostasis, foraging for resources, and defense. Although plant plasticity has been a commonplace observation, its ecological and evolutionary consequences are only beginning to be thoroughly explored. This is due to the fact that most ecological and ecophysiological studies of plant form to date have focused on adaptive, habitat-based specialization, and plasticity has been interpreted as a feature of generalists. The supposed superiority of specialized ecotypes or taxa over generalists has led biologists to focus on evolutionary specialization and neglect the ecological and evolutionary implications of plastic responses of the phenotype to the environment (Sultan 2004, 2005). Although the adaptive implications of plasticity for relative allocation of biomass or other fitness-related parameters have been widely recognized, plasticity has been traditionally viewed as an alternative to specialization. The evidence that plastic, and thus generalist, species are less able to compete with specialized species is weak at best (Niklas 1997). A different approach to this question has postulated that plasticity in some plant traits may, in fact, represent a product of specialization (Lortie and Aarssen 1996). The predictions for plasticity of specialized genotypes were proposed to depend on whether specialization is associated with the more or the less favorable end of an environmental gradient (Lortie and Aarssen 1996). Specialization to the more favorable end was proposed to increase plasticity, whereas specialization to the less favorable end was proposed to decrease plasticity. In a study of 16 species of the genus *Psychotria* occurring on Barro Colorado Island, Panama, we found that the mean phenotypic plasticity was significantly higher for the high-light species than for the low-light species (Valladares et al. 2000). Selection for greater plasticity may be stronger in the high-light species because forest gaps (high-light environments) exhibit a relatively predictable decrease in PPFD for which this plasticity could be adaptive. In contrast, the low-light species experience relatively unpredictable changes in light caused by infrequent gap formation. Under these conditions, phenotypic stability may have higher adaptive value. The results are consistent with the view that plasticity, rather than being an alternative to specialization, is indeed a specialization to the high-light end of the light gradient in tropical forests. Other studies have found greater photosynthetic plasticity in gap-dependent compared with shade-tolerant species (Bazzaz 1996). On the other hand, the relative stability exhibited by the low-light species is consistent with a stress-tolerant syndrome, with its low potential maximum growth rates, low maximum photosynthetic rates, and low leaf turnover (Valladares et al. 2000). We also found that plasticity was generally greater for the physiological than for the structural characteristics (Valladares et al. 2000). However, the different plasticity in morphological and physiological traits seems to be species-specific, with high-light species showing greater physiological plasticity and low-light species higher morphological plasticity (Valladares et al. 2002b, Niinemets and Valladares 2004).

In the analysis of plant plasticity, choice of which submits to count is essential, since within a plant there is clearly a hierarchy of plasticities, and not all structures exhibit the same degree of plastic response (White 1979). For instance, the range of variation of a phenotype (the norm of reaction) for vegetative structures tends to be broader than that for reproductive traits (Niklas 1997). The adaptive significance of the plasticity of a character becomes clear only when it is scaled up to the performance of the whole plant. Usually, if a plastic response in a feature improves performance, the result at the next level is an enhanced homeostasis (Pearcy 1999).

Since phenotypic plasticity is advantageous for sessile organisms due to the heterogeneous nature of most environments, the question of why all plants are not equally (and maximally) plastic is a very pertinent one. It is rare to find scenarios in which a plastic response to the environment could be maladaptive. Some examples have been pointed out for plants growing under extreme physiological conditions and usually in the absence of strong competition (Chapin 1991, Chapin et al. 1993). Such plants tend to show a conservative pattern involving slow, steady growth, even when conditions are temporarily favorable (Waller 1986). Under ideal growing conditions, these plants are more likely to store nutrients than to accelerate their growth (Chapin et al. 1986, Chapin et al. 1993) to avoid the production of a plant that is too large or structures that are too expensive to be sustained once conditions deteriorate. Specialization to a low-resource environment seems to start with a modification in a key growth-related character, which results in a cascade of effects that triggers the entire stress resistance syndrome (low rates of growth, photosynthesis, and nutrient absorption, high root: shoot ratios, low rates of tissue turnover, and high concentrations of secondary metabolites, Chapin et al. 1993). Low plasticity associated with stressful conditions has been found in several studies of Mediterranean woody plants (Valladares et al. 2002a, 2005).

Examples of proven adaptive plasticity in plants are scarce and most plastic responses actually may be "passive" rather than adaptive, suggesting that the evolution of adaptive plasticity is impeded by constraints (Weinig 2000, van Kleunen and Fischer 2005). And one ubiquitous constrain is stress. Stress dissipates energy that is otherwise available for homeostatic processes such as those controlling development and pattern formation (Parsons 1993). It has been shown that growth pattern and shape-related features are more sensitive to this energy dissipation caused by stress than just size (Alados et al. 1994, 1999). Stress introduces variation in structure, which is not adaptive, in contrast to plastic phenotypic responses to environmental changes. This kind of increased variability is known as developmental instability and the study of the so-called fluctuating asymmetry has been used to detect disruption of homeostasis (Alados et al. 1994, 1999). Stress also causes deviations in radial symmetry and in symmetry of scale, that is, in self-similarity at different spatial scales (Alados et al. 1994). Since both the distance between two successive leaves (internode length) and the internode diameter scale with node order, dispersion about the regression line between length or diameter and node order becomes a measure of the departure from perfect translational symmetry, which is another form of developmental instability (Escós et al. 1997). These are just some examples to illustrate the notions that there are many forms of asymmetry that can be associated with stress and that only a fraction of the phenotypic response to environment is adaptive.

ACKNOWLEDGMENTS

Tsvi Sachs has been most generous with his time and expertise in commenting on a first draft of this chapter. This chapter is based on research supported by the Spanish Ministry of Education and Science (grants RASINV, CGL2004-04884-C02-02/BOS, and PLASTOFOR, AGL2004-00536/FOR), by the Estonian Academy of Sciences, Consejo Superior de Investigaciones Científicas (CSIC, Spain) (grant for collaboration between scientific institutions in Estonia and the research institutes of CSIC), the Estonian Science Foundation, and the Estonian Ministry of Education and Science (grant SF1090065s07).

REFERENCES

Ackerly, D.D. and F.A. Bazzaz, 1995. Seedling crown orientation and interception of diffuse radiation in tropical forest gaps. Ecology 76: 1134–1146.

Alados, C.L., J.M. Escós, and J.M. Emlen, 1994. Scale asymmetry: A tool to detect developmental instability under the fractal geometry scope. In: M.M. Novak, ed. Fractals in the Natural and Applied Sciences. Elsevier, North-Holland, pp. 25–36.

Alados, C.L., T. Navarro, and B. Cabezudo, 1999. Tolerance assessment of Cistus ladanifer to serpentine soils by developmental stability analysis. Plant Ecology 143: 51–66.

Alexander, R.M., 1997. Leaning trees on a sloping ground. Nature 386: 327–329.

Arno, S.F. and R.P. Hammerly, 1984. Timberline: Mountain and Arctic Forest Frontiers. The Mountaineers, Seattle, WA.

Ashton, P.S., 1978. Crown characteristics of tropical trees. In: P.B. Tomlinson and M.H. Zimmermann, eds. Tropical Trees as Living Systems. The Proceedings of the Fourth Cabot Symposium held at Harvard Forest, Petersham Massachusetts on April 26–30, 1976. Cambridge University Press, Cambridge, New York, Melbourne, pp. 591–615.

Baldocchi, D. and S. Collineau, 1994. The physical nature of solar radiation in heterogeneous canopies: spatial and temporal attributes. In: M.M. Caldwell and R.W. Pearcy, eds. Exploitation of Environmental Heterogeneity by Plants. Ecophysiological Processes Above- and Belowground. Academic Press, San Diego, New York, Boston, London, Sydney, Tokyo, Toronto, pp. 21–71.

Baldocchi, D.D. and P.C. Harley, 1995. Scaling carbon dioxide and water vapour exchange from leaf to canopy in a deciduous forest. II. Model testing and application. Plant, Cell and Environment 18: 1157–1173.

Baldocchi, D.D., D.R. Matt, B.A. Hutchinson, and R.T. McMillen, 1984. Solar radiation within an oak-hickory forest: An evaluation of the extinction coefficients for several radiation components during fully-leafed and leafless periods. Agricultural and Forest Meteorology 32: 307–322.

Ballaré, C.L., 1994. Light gaps: Sensing the light opportunities in highly dynamic canopy environments. In: M.M. Caldwell and R.W. Pearcy, eds. Exploitation of Environmental Heterogeneity by Plants: Ecophysiological Processes Above- and Belowground. Academic Press, San Diego, CA, pp. 73–110.

Ballaré, C.L., A.L. Scopel, and R.A. Sánchez, 1997. Foraging for light: photosensory ecology and agricultural implications. Plant, Cell and Environment 20: 820–825.

Bazzaz, F.A., 1996. Plants in Changing Environments: Linking Physiological, Population, and Community Ecology. Cambridge University Press, Cambridge.

Bell, A.D., 1993. Plant Form. Oxford University Press, New York.

Beyschlag, W. and R.J. Ryel, 1999. Canopy photosynthesis modeling. In: F.I. Pugnaire and F. Valladares, eds. Handbook of Functional Plant Ecology. Marcel Dekker, Inc., New York, pp. 771–804.

Brites, D. and F. Valladares, 2005. Implications of opposite phyllotaxis for light interception efficiency of Mediterranean woody plants. Trees: Structure and Function 19: 671–679.

Brown, M.J. and G.G. Parker, 1994. Canopy light transmittance in a chronosequence of mixed-species deciduous forests. Canadian Journal of Forest Research 24: 1694–1703.

Caldwell, M.M., J.H. Richards, D.A. Johnson, R.S. Nowak, and R.S. Dzurec, 1981. Coping with herbivory: Photosynthetic capacity and resource allocation in two semiarid *Agropyron* bunchgrasses. Oecologia 50: 14–24.

Caldwell, M.M., T.J. Dean, R.S. Nowak, R.S. Dzurec, and J.H. Richards, 1983. Bunchgrass architecture, light interception, and water-use efficiency: Assessment by fiber optic point quadrats and gas exchange. Oecologia 59: 178–184.

Caldwell, M.M., H.P. Meister, J.D. Tenhunen, and O.L. Lange, 1986. Canopy structure, light microclimate and leaf gas exchange of *Quercus coccifera* L. in a Portugese macchia: Measurements in different canopy layers and simulations with a canopy model. Trees: Structure and Function 1: 25–41.

Callaway, R.M. and L.R. Walker, 1997. Competition and facilitation: a synthetic approach to interactions in plant communities. Ecology 78: 1958–1965.

Campbell, G.S., 1986. Extinction coefficients for radiation in plant canopies calculated using an ellipsoidal inclination angle distribution. Agricultural and Forest Meteorology 36: 317–321.

Campbell, G.S. and J.M. Norman, 1989. The description and measurement of plant canopy structure. In: G. Russell, B. Marshall, and P.G. Jarvis, eds. Plant Canopies: Their Growth, Form and Function. Cambridge University Press, Cambridge, New York, New Rochelle, Melbourne, Sydney, pp. 1–19.

Canham, C.D., 1988. Growth and canopy architecture of shade-tolerant trees: response to canopy gaps. Ecology 69: 786–795.

Cescatti, A., 1998. Effects of needle clumping in shoots and crowns on the radiative regime of a Norway spruce canopy. Annales des Sciences Forestieres 55: 89–102.

Cescatti, A. and Ü. Niinemets, 2004. Sunlight capture. Leaf to landscape. In: W.K. Smith, T.C. Vogelmann, and C. Chritchley, eds. Photosynthetic Adaptation. Chloroplast to Landscape. Springer-Verlag, Berlin, pp. 42–85.

Cescatti, A. and R. Zorer, 2003. Structural acclimation and radiation regime of silver fir (*Abies alba Mill.*) shoots along a light gradient. Plant, Cell and Environment 26: 429–442.

Clearwater, M.J. and K.S. Gould, 1995. Leaf orientation and light interception by juvenile *Pseudopanax crassifolius* (Cunn) C. Koch in a partially shaded forest environment. Oecologia 104: 363–371.

Comstock, J.P. and B.E. Mahall, 1985. Drought and changes in leaf orientation for two California chaparral shrubs: *Ceanothus megacarpus* and *Ceanothus crassifolius*. Oecologia 65: 531–535.

Corlett, J.E., H.G. Jones, A. Massacci, and J. Masojidek, 1994. Water deficit, leaf rolling and susceptibility to photoinhibition in field grown sorghum. Physiologia Plantarum 92: 423–430.

Coutts, M.P. and J. Grace, eds. 1995. Wind and Trees. Cambridge University Press, Cambridge.

Chapin, F., 1991. Integrated responses of plants to stress. BioScience 41: 29–36.

Chapin, F.S., III, P.M. Vitousek, and K. Van Cleve, 1986. The nature of nutrient limitation in plant communities. The American Naturalist 127: 48–58.

Chapin, F.S., III, K. Autumn, and F. Pugnaire, 1993. Evolution of suites of traits in response to environmental stress. The American Naturalist 142: S78–S92.

Cheeseman, J.M., 1993. Plant growth modelling without integrating mechanisms. Plant, Cell and Environment 16: 137–147.

Chen, S.G., R. Ceulemans, and I. Impens, 1994. Is there a light regime determined tree ideotype? Journal of Theoretical Biology 169: 153–161.

Del Tredici, P., 1991. Natural regeneration of *Ginkgo biloba* from downward growing cotyledonary buds (Basal Chichi). American Journal of Botany 79: 522–530.

Duncan, W.G., 1971. Leaf angles, leaf area and canopy photosynthesis. Crop Science 11: 482–485.

Ehleringer, J. and I. Forseth, 1980. Solar tracking by plants. Science 210: 1094–1098.

Ehleringer, J.R. and K.S. Werk, 1986. Modifications of solar-radiation absorption patterns and implications for carbon gain at the leaf level. In: T.J. Givnish, ed. On the Economy of Plant Form and Function. Proceedings of the Sixth Maria Moors Cabot Symposium, "Evolutionary constraints on primary productivity: adaptive patterns of energy capture in plants," Harvard Forest, August 1983. Cambridge University Press, Cambridge, London, New York, New Rochelle, Melbourne, Sydney, pp. 57–82.

Ehleringer, J.R. and I.N. Forseth, 1989. Diurnal leaf movements and productivity in canopies. In: G. Russell, B. Marshall, and P.G. Jarvis, eds. Plant Canopies: Their Growth, Form and Function. Cambridge University Press, Cambridge, New York, New Rochelle, Melbourne, Sydney, pp. 129–142. Ehleringer, J.R. and C.B. FIeld, eds. 1993. Scaling physiological processes. Academic Press, San Diego, CA.

Ellison, A.M. and K.J. Niklas, 1988. Branching patterns on *Salicornia europaea* (Chenopodiaceae) at different successional stages a comparison of theoretical and real plants. American Journal of Botany 75: 501–512.

Ennos, A.R., 1997. Wind as an ecological factor. Trends in Ecology and Evolution 12: 108–111.

Escós, J., C.L. Alados, and J.M. Emlen, 1997. The impact of grazing on plant fractal architecture and fitness of a Mediterranean shrub *Anthyllis cytisoides* L. Functional Ecology 11: 66–78.

Farque, L., H. Sinoquet, and F. Colin, 2001. Canopy structure and light interception in *Quercus petraea* seedlings in relation to light regime and plant density. Tree Physiology 21: 1257–1267.

Farquhar, G.D. and M.L. Roderick, 2003. Pinatubo, diffuse light, and the carbon cycle. Science 299: 1997–1998.

Fernandez, D. and M. Castrillo, 1999. Maize leaf rolling initiation. Photosynthetica 37: 493–497.

Fisher, J.B., 1986. Branching patterns and angles in trees. In: T.J.Givnish, ed. On the Economy of Plant Form and Function. Proceedings of the Sixth Maria Moors Cabot Symposium, "Evolutionary constraints on primary productivity: adaptive patterns of energy capture in plants," Harvard Forest, August 1983. Cambridge University Press, Cambridge, London, New York, New Rochelle, Melbourne, Sydney, pp. 493–523.

Fisher, J.B., 1992. How predictive are computer simulations of tree architecture? International Journal of Plant Sciences 153: S137–S146.
Fisher, J.B. and D.E. Hibbs, 1982. Plasticity of tree architecture: specific and ecological variations found in Aubrevilles's model. American Journal of Botany 69: 690–702.
Fisher, J.B. and H. Honda, 1979a. Branch geometry and effective leaf area: a study of *Terminalia*-branching pattern. I. Theoretical trees. American Journal of Botany 66: 633–644.
Fisher, J.B. and H. Honda, 1979b. Branch geometry and effective leaf area: a study of *Terminalia*-branching pattern. II. Survey of real trees. American Journal of Botany 66: 645–655.
Fleck, S., Ü. Niinemets, A. Cescatti, and J.D. Tenhunen, 2003. Three-dimensional lamina architecture alters light harvesting efficiency in *Fagus*: A leaf-scale analysis. Tree Physiology 23: 577–589.
Ford, H., 1987. Investigating the ecological and evolutionary significance of plant growth form using stochastic simulation. Annals of Botany 59: 487–494.
Fournier, A., 1979. Is architectural radiation adaptive? Université des Sciences et Techniques du Languedoc, Montpellier, France.
Fujimori, T. and D. Whitehead, 1986. Crown and Canopy Structure in Relation to Productivity. Forestry and Forestal Products Research Institute, Ibaraki, Japan.
Gartner, B.L., 1995. Plant Stems. Physiology and Functional Morphology. Academic Press, San Diego, CA.
Geiger, D.R., 1986. Processes affecting carbon allocation and partitioning among sinks. In: J. Cronshaw, W.J. Lucas, and R.T. Giaquinta, eds. Phloem Transport. Alan R. Liss, New York, pp. 375–388.
Gere, J.M. and S.P. Timoshenko, 1997. Mechanics of Materials, 4th edn. PWS Publishing Company, Boston, MA.
Gill, D.E., 1991. Individual plants as genetic mosaics: ecological organisms versus evolutionary individuals. In: M.J. Crawley, ed. Plant Ecology. Blackwell Scientific Publications, Wiltshire, pp. 321–344.
Givnish, T.J., 1978. On the adaptive significance of compound leaves, with special reference to tropical trees. In: P.B. Tomlinson and M.H. Zimmermann, eds. Tropical Trees as Living Systems. The Proceedings of the Fourth Cabot Symposium held at Harvard Forest, Petersham, Massachusetts on April 26–30, 1976. Cambridge University Press, Cambridge, New York, Melbourne, pp. 351–380.
Givnish, T.J., 1982. On the adaptive significance of leaf height in forest herbs. The American Naturalist 120: 353–381.
Givnish, T.J., 1986. Biomechanical constraints on self-thinning in plant populations. Journal of Theoretical Biology 119: 139–146.
Givnish, T.J., 1988. Adaptation to sun and shade: A whole-plant perspective. Australian Journal of Plant Physiology 15: 63–92.
Givnish, T.J., 1995. Plant stems: Biomechanical adaptation for energy capture and influence on species distributions. In: B.L. Gartner, ed. Plant Stems: Physiology and Functional Morphology. Academic Press, Inc, San Diego, CA, pp. 3–49.
Godin, C., 2000. Representing and encoding plant architecture: A review. Annals of Forest Science 57: 413–438.
Gómez-Aparicio, L., R. Zamora, J.M. Gómez, J.A. Hódar, J. Castro, and E. Baraza, 2004. Applying plant facilitation to reforestation: a meta-analysis of the use of shrubs as nurse plants. Ecological Applications 14: 1128–1138.
Grace, J., 1977. Plant Response to Wind. Academic Press, New York.
Gu, L., D. Baldocchi, S.B. Verma, T.A. Black, T. Vesala, E.M. Falge, and P.R. Dowty, 2002. Advantages of diffuse radiation for terrestrial ecosystem productivity. Journal of Geophysical Research 107: doi:10.1029/2001JD001242.
Gutschick, V.P. and F.W. Wiegel, 1988. Optimizing the canopy photosynthetic rate by patterns of investment in specific leaf mass. The American Naturalist 132: 67–86.
Hallé, F., 1995. Canopy architecture in tropical trees: a pictorial approach. In: M.D. Lowman and N.M. Nadkarni, eds. Forest Canopies. Academic Press, London, pp. 27–44.
Hallé, F., R.A.A. Oldeman, and P.B. Tomlinson, 1978. Tropical Trees and Forests: An Architectural Analysis. Springer-Verlag, Berlin, Heidelberg, New York.

Hardwick, R.C., 1986. Physiological consequences of modular growth in plants. Philosophical Transactions of the Royal Society of London, Biology 313: 161–173.

Harper, J.L., 1989. The value of a leaf. Oecologia 80: 53–58.

Hasting, H.M. and G. Sugihara, 1993. Fractals: A User's Guide for the Natural Sciences. Oxford University Press, Oxford.

Herbert, T.J., 1991. Variation in interception of the direct solar beam by top canopy layers. Ecology 72: 17–22.

Herbert, T.J. and T. Nilson, 1991. A model of variance of photosynthesis between leaves and maximization of whole plant photosynthesis. Photosynthetica 25: 597–606.

Honda, H. and J.B. Fisher, 1978. Tree branch angle: maximizing effective leaf area. Science 199: 888–890.

Honda, H., H. Hatta, and J.B. Fisher, 1997. Branch geometry in *Cornus kousa* (*Cornaceae*): computer simulations. American Journal of Botany 84: 745–755.

Horn, H.S., 1971. The Adaptive Geometry of Trees. Princeton University Press, Princeton, NJ.

Horton, P., A.V. Ruban, and R.G. Walters, 1996. Regulation of light harvesting in green plants. Annual Review of Plant Physiology and Plant Molecular Biology 47: 655–684.

Hutchins, M.J. and H. de Kroon, 1994. Foraging in plants: The role of morphological plasticity in resource acquisition. In: M. Begon and A.H. Fitter, eds. Advances in Ecological Research. Academic Press Ltd., London, pp. 159–238.

Ingrouille, M., 1995. Diversity and Evolution of Land Plants. Chapman & Hall, London.

Isebrands, J.G. and D.A. Michael, 1986. Effects of leaf morphology and orientation on solar radiation interception and photosynthesis in *Populus*. In: T. Fujimori and D. Whitehead, eds. Crown and Canopy Structure in Relation to Productivity. Forestry and Forest Products Research Institute, Ibaraki, pp. 359–381.

Ishii, R. and M. Higashi, 1997. Tree coexistence on a slope: an adaptive significance of trunk inclination. Proceedings of the Royal Society of London Series B, Biology 264: 133–140.

Jackson, R.B., W.T. Pockman, and W.A. Hoffman, 1999. The structure and function of root systems. In: F.I. Pugnaire and F. Valladares, eds. Handbook of Functional Plant Ecology. Marcel Dekker, Inc., New York, pp. 195–220.

Jahnke, L.S. and D.B. Lawrence, 1965. Influence of photosynthetic crown structure on potential productivity of vegetation, based primarily on mathematical models. Ecology 46: 319–326.

James, J.C., J. Grace, and S.P. Hoad, 1994. Growth and photosynthesis of *Pinus sylvestris* at its altitudinal limit in Scotland. Journal of Ecology 82: 297–306.

Jones, H.G., 2004. Application of thermal imaging and infrared sensing in plant physiology and ecophysiology. Advances in Botanical Research 41: 107–163.

Kellomäki, S. and H. Strandman, 1995. A model for the structural growth of young Scots pine crowns based on light interception by shoots. Ecological Modelling 80: 237–250.

Kellomäki, S. and H. Väisänen, 1995. Model computations on the impact of changing climate on natural regeneration of Scots pine in Finland. Canadian Journal of Forest Research 25: 929–942.

King, D.A., 1981. Tree dimensions: maximizing the rate of height growth in dense stands. Oecologia 51: 351–356.

King, D.A., 1986. Tree form, height growth, and susceptibility to wind damage in *Acer saccharum*. Ecology 67: 980–990.

King, D.A., 1990. The adaptive significance of tree height. The American Naturalist 135: 809–829.

King, D.A., 1991. Correlations between biomass allocation, relative growth rate and light environment in tropical forest saplings. Functional Ecology 5: 485–492.

King, D.A., 1994. Influence of light level on the growth and morphology of saplings in a Panamanian forest. American Journal of Botany 81: 948–957.

Koehl, M.A.R., 1996. When does morphology matter? Annual Review of Ecology and Systematics 27: 501–542.

Kucharik, C.J., J.M. Norman, and S.T. Gower, 1999. Characterization of radiation regimes in nonrandom forest canopies: Theory, measurements, and a simplified modeling approach. Tree Physiology 19: 695–706.

Kull, O., M. Broadmeadow, B. Kruijt, and P. Meir, 1999. Light distribution and foliage structure in an oak canopy. Trees: Structure and Function 14: 55–64.

Küppers, M., 1989. Ecological significance of aboveground architectural patterns in woody plants: A question of cost-benefit relationships. Trends in Ecology and Evolution 4: 375–379.

Küppers, M., 1994. Canopy gaps: competitive light interception and economic space filling—a matter of whole-plant allocation. In: M.M. Caldwell and R.W. Pearcy, eds. Exploitation of Environmental Heterogeneity by Plants. Ecophysiological Processes Above- and Belowground. Academic Press, San Diego, New York, Boston, London, Sydney, Tokyo, Toronto, pp. 111–144.

Küppers, M. and R. List, 1997. MADEIRA—a simulation of carbon gain, allocation, canopy architecture in competing woody plants. In: G. Jeremonidis and J.F.V. Vincent, eds. Plant Biomechanics: Conference Proceedings. Centre for Biomimetics, The University of Reading, Reading, pp. 321–329.

Kurth, W., 1994. Morphological models of plant growth: possibilities and ecological relevance. Ecological Modelling 75/76: 299–308.

Kuuluvainen, T., 1992. Tree architectures adapted to efficient light utilization—is there a basis for latitudinal gradients? Oikos 65: 275–284.

Leigh, E.G., 1972. The golden section and spiral leaf-arrangement. In: E.S. Deevey, ed. Growth by Intussusception. Archon Books, Hamden, CT, pp. 163–176.

Leigh, E.G., 1990. Tree shape and leaf arrangement: a quantitative comparison of montane forests, with emphasis on Malaysia and South India. In: J.C. Daniel and J.S. Serrao, eds. Conservation in Developing Countries: Problems and Prospects. Oxford University Press, Bombay, India, pp. 119–174.

Leigh, E.G., 1998. Tropical Forest Ecology. A View from Barro Colorado Island. Oxford University Press, Oxford.

Lortie, C.J. and L.W. Aarssen, 1996. The specialization hypothesis for phenotypic plasticity in plants. International Journal of Plant Sciences 157: 484–487.

Lortie, C.J. and R.M. Callaway, 2006. Re-analysis of meta-analysis: Support for the stress-gradient hypothesis. Journal of Ecology 94: 7–16.

Lovelock, C.E. and B.F. Clough, 1992. Influence of solar radiation and leaf angle on leaf xanthophyll concentrations in mangroves. Oecologia 91: 518–525.

Maestre, F.T., F. Valladares, and J.F. Reynolds, 2005. Is the change of plant–plant interactions with abiotic stress predictable? A meta-analysis of field results in arid environments. Journal of Ecology 93: 748–757.

Maestre, F.T., F. Valladares, and J.F. Reynolds, 2006. The stress-gradient hypothesis does not fit all relationships between plant–plant interactions and abiotic stress: further insights from arid environments. Journal of Ecology 94: 17–22.

Margalef, R., 1997. Our biosphere. Excellence in ecology 10. Ecology Institute of Germany, Wurzburg.

Margolis, H., R. Oren, D. Whitehead, and M.R. Kaufmann, 1995. Leaf area dynamics of conifer forests. In: W.K. Smith and T.M. Hinckley, eds. Ecophysiology of Coniferous Forests. Academic Press, San Diego, New York, Boston, London, Sydney, Tokyo, Toronto, pp. 181–223.

Mattheck, C., 1991. Trees: The Mechanical Design. Springer-Verlag, Berlin.

Mattheck, C., 1995. Biomechanical optimum in woody stems. In: B.L. Gartner, ed. Plant Stems: Physiology and Functional Morphology. Academic Press, San Diego, CA, pp. 3–49.

McMahon, T., 1973. Size and shape in biology. Elastic criteria impose limits on biological proportions, and consequently on metabolic rates. Science 179: 1201–1204.

McMahon, T., 1975. The mechanical design of trees. Scientific American 233: 92–102.

McMahon, T.A. and R.E. Kronauer, 1976. Tree structures: Deducing the principle of mechanical design. Journal of Theoretical Biology 59: 443–466.

Medina, E., 1999. Tropical forests: Diversity and function and dominant life-forms. In: F.I. Pugnaire and F. Valladares, eds. Handbook of Functional Plant Ecology. Marcel Dekker, Inc., New York, pp. 407–448.

Moulia, B., 1994. The biomechanics of leaf rolling. Biomimetics 2: 267–281.

Myers, D.A., D.N. Jordan, and T.C. Vogelmann, 1997. Inclination of sun and shade leaves influences chloroplast light harvesting and utilization. Physiologia Plantarum 99: 395–404.

Niinemets, Ü., 1997. Distribution patterns of foliar carbon and nitrogen as affected by tree dimensions and relative light conditions in the canopy of *Picea abies*. Trees: Structure and Function 11: 144–154.

Niinemets, Ü. 1998. Growth of young trees of *Acer platanoides* and *Quercus robur* along a gap—understory continuum: interrelationships between allometry, biomass partitioning, nitrogen, and shade-tolerance. International Journal of Plant Sciences 159: 318–330.

Niinemets, Ü. and O. Kull, 1995. Effects of light availability and tree size on the architecture of assimilative surface in the canopy of *Picea abies*: Variation in shoot structure. Tree Physiology 15: 791–798.

Niinemets, Ü. and A. Lukjanova, 2003. Total foliar area and average leaf age may be more strongly associated with branching frequency than with leaf longevity in temperate conifers. The New Phytologist 158: 75–89.

Niinemets, Ü. and F. Valladares, 2004. Photosynthetic acclimation to simultaneous and interacting environmental stresses along natural light gradients: optimality and constraints. Plant Biology 6: 254–268.

Niinemets, Ü., A. Cescatti, A. Lukjanova, M. Tobias, and L. Truus, 2002. Modification of light-acclimation of *Pinus sylvestris* shoot architecture by site fertility. Agricultural and Forest Meteorology 111: 121–140.

Niinemets, Ü., N. Al Afas, A. Cescatti, A. Pellis, and R. Ceulemans, 2004a. Determinants of clonal differences in light-interception efficiency in dense poplar plantations: petiole length and biomass allocation. Tree Physiology 24: 141–154.

Niinemets, Ü., A. Cescatti, and R. Christian, 2004b. Constraints on light interception efficiency due to shoot architecture in broad-leaved *Nothofagus* species. Tree Physiology 24: 617–630.

Niinemets, Ü., A. Sparrow, and A. Cescatti, 2005. Light capture efficiency decreases with increasing tree age and size in the southern hemisphere gymnosperm *Agathis australis*. Trees: Structure and Function 19: 177–190.

Niinemets, Ü., M. Tobias, A. Cescatti, and A.D. Sparrrow, 2006. Size-dependent variation in shoot light-harvesting efficiency in shade-intolerant conifers. International Journal of Plant Sciences 167: 19–32.

Niklas, K.J., 1988. The role of phyllotactic pattern as a "developmental constraint" on the interception of light by leaf surfaces. Evolution 42: 1–16.

Niklas, K.J., 1990. Biomechanics of *Psilotum nudum* and some early Paleozoic vascular sporophytes. American Journal of Botany 77: 590–606.

Niklas, K.J., 1992. Plant Biomechanics. An Engineering Approach to Plant Form and Function. The University of Chicago Press, Chicago, London.

Niklas, K.J., 1994. Interspecific allometries of critical buckling height and actual plant height. American Journal of Botany 81: 1275–1279.

Niklas, K.J., 1997. The Evolutionary Biology of Plants. The University of Chicago Press, Chicago, IL.

Niklas, K.J. and V. Kerchner, 1984. Mechanical and photosynthetic constraints on the evolution of plant shape. Paleobiology 10: 79–101.

Nikolopoulos, D., G. Liakopoulos, I. Drossopoulos, and G. Karabourniotis, 2002. The relationship between anatomy and photosynthetic performance of heterobaric leaves. Plant Physiology 129: 235–243.

Nilson, T., 1971. A theoretical analysis of the frequency of gaps in plant stands. Agricultural Meteorology 8: 25–38.

Nobel, P.S. and R.W. Meyer, 1991. Biomechanics of cladodes and cladode–cladode junctions for Opuntia ficus-indica (Cactaceae). American Journal of Botany 78: 1252–1259.

Norman, J.M. and G.S. Campbell, 1989. Canopy structure. In: R.W. Pearcy, J.R. Ehleringer, H.A. Mooney, and P.W. Rundel, eds. Plant Physiological Ecology. Field Methods and Instrumentation. Chapman & Hall, London, New York, pp. 301–325.

Novoplansky, A., D. Cohen, and T. Sachs, 1989. Ecological Implications of Correlative Inhibition Between Plant Shoots. Physiologia Plantarum 77: 136–140.

Oker-Blom, P., 1986. Photosynthetic radiation regime and canopy structure in modeled forest stands. Acta Forestalia Fennica 197: 1–44.

Oker-Blom, P. and S. Kellomäki, 1982. Theoretical computations on the role of crown shape in the absorption of light by forest trees. Mathematical Biosciences 59: 291–311.

Oker-Blom, P. and H. Smolander, 1988. The ratio of shoot silhouette to total needle area in Scots pine. Forest Science 34: 894–906.

Osmond, C.B., J.M. Anderson, M.C. Ball, and J.G. Egerton, 1999. Compromising efficiency: The molecular ecology of light-resource utilization in plants. In: M.C. Press, J.D. Scholes, and M.G. Barker, eds. Physiological Plant Ecology. The 39th Symposium of the British Ecological Society held at the University of York, 7–9 September 1998. Blackwell Science, Oxford, pp. 1–24.

Palmroth, S., P. Stenberg, S. Smolander, P. Voipio, and H. Smolander, 2002. Fertilization has little effect on light-interception efficiency of *Picea abies* shoots. Tree Physiology 22: 1185–1192.

Parsons, P.A., 1993. The importance and consequences of stress in living and fossil populations: From life-history variation to evolutionary change. American Naturalist 142: S5–S20.

Pearcy, R.W., 1999. Responses of plants to heterogeneous light environments. In: F.I. Pugnaire and F. Valladares, eds. Handbook of Functional Plant Ecology. Marcel Dekker, New York, pp. 269–314.

Pearcy, R.W. and W. Yang, 1996. A three-dimensional crown architecture model for assessment of light capture and carbon gain by understory plants. Oecologia 108: 1–12.

Pearcy, R.W. and W. Yang, 1998. The functional morphology of light capture and carbon gain in the redwood forest understorey plant, *Adenocaulon bicolor* Hook. Functional Ecology 12: 543–552.

Pearcy, R.W., H. Muraoka, and F. Valladares, 2005. Crown architecture in sun and shade environments: Assessing function and trade-offs with a three-dimensional simulation model. The New Phytologist 166: 791–800.

Peer, W.A., W.R. Briggs, and J.H. Langenheim, 1999. Shade-avoidance responses in two common coastal redwood forest species, *Sequoia sempervirens* (*Taxodiaceae*) and *Satureja douglasii* (*Lamiaceae*), occurring in various light quality environments. American Journal of Botany 86: 640–645.

Perttunen, J., R. Sievänen, E. Nikinmaa, H. Salminen, H. Saarenmaa, and J.Väkevä. 1996. LIGNUM: a tree model based on simple structural units. Annals of Botany 77: 87–98.

Poorter, L., L. Bongers, and F. Bongers, 2005. Architecture of 54 moist-forest tree species: traits, trade-offs, and functional groups. Ecology (Tempe) 87: 1289–1301.

Porter, J.R., 1983. A modular approach to analysis of plant growth. I. Theory and principles. New Phytologist 94: 183–190.

Porter, J.R., 1989. Modules, models and meristems in plant architecture. In: G. Russell, B. Marshall, and P.G. Jarvis, eds. Plant Canopies: Their Growth, Form and Function. Cambridge University Press, Cambridge, New York, New Rochelle, Melbourne, Sydney, pp. 143–159.

Poulson, M.E. and E.H. DeLucia, 1993. Photosynthetic and structural acclimation to light direction in vertical leaves of *Silphium terebinthinaceum*. Oecologia 95: 393–400.

Prusinkiewicz, P., 1998. Modeling of spatial structure and development of plants: A review. Scientia Horticulturae 74: 113–149.

Prusinkiewicz, P. and A. Lindenmayer, 1990. The Algorithmic Beauty of Plants. Springer-Verlag, New York.

Rich, P.M., 1986. Mechanical architecture of arborescent rainforest palms. Principles 30: 117–131.

Rich, P.M., 1987. Developmental anatomy of the stem of *Welfia georgii*, *Iriartea gigantea*, and other arborescent palms: Implications for mechanical support. American Journal of Botany 74: 792–802.

Roberts, S.W. and P.C. Miller, 1977. Interception of solar radiation as affected by canopy organization in two Mediterranean shrubs. Acta Oecologica/Oecologia Plantarum 12: 273–290.

Roderick, M.L., G.D. Farquhar, S.L. Berry, and I.R. Noble, 2001. On the direct effect of clouds and atmospheric particles on the productivity and structure of vegetation. Oecologia 129: 21–30.

Room, P., J. Hanan, and P. Prusinkiewicz, 1996. Virtual plants: new perspectives for ecologists, pathologists and agricultural scientists. Trends in Plant Science 1: 33–38.

Room, P.M., L. Maillette, and J.S. Hanan, 1994. Module and metamer dynamics and virtual plants. Advances in Ecological Research 25: 105–157.

Ross, J. 1981. The Radiation Regime and Architecture of Plant Stands. Dr. W. Junk, The Hague.

Russell, G., P.G. Jarvis, and J.L. Monteith, 1989. Absorption of radiation by canopies and stand growth. In: G. Russell, B. Marshall, and P.G. Jarvis, eds. Plant Canopies: Their Growth,

Form and Function. Cambridge University Press, Cambridge, New York, New Rochelle, Melbourne, Sydney, pp. 21–39.

Ryan, M.G., D. Binkley, and J.H. Fownes, 1997. Age-related decline in forest productivity: Pattern and process. Advances in Ecological Research 27: 213–262.

Sachs, T., 1988. Epigenetic selection: an alternative mechanism of pattern formation. Journal of Theoretical Biology 134: 547–559.

Sachs, T. and M. Hassidim, 1996. Mutual support and selection between branches of damaged plants. Vegetatio 127: 25–30.

Sachs, T. and A. Novoplansky, 1995. Tree form: Architectural models do not suffice. Israel Journal of Plant Sciences 43: 203–212.

Sachs, T., A. Novoplansky, and D. Cohen, 1993. Plants as competing populations of redundant organs. Plant, Cell and Environment 16: 765–770.

Schulze, E.D., M. Küppers, and R. Matyssek, 1986. The roles of carbon balance and branching pattern in the growth of woody species. In: T.J. Givnish, ed. On the Economy of Plant Form and Function. Proceedings of the Sixth Maria Moors Cabot Symposium, "Evolutionary constraints on primary productivity: adaptive patterns of energy capture in plants," Harvard Forest, August 1983. Cambridge University Press, Cambridge, London, New York, New Rochelle, Melbourne, Sydney, pp. 585–602.

Shackel, K.A. and A.E. Hall, 1979. Reversible leaflet movements in relation to drought adaptation of cowpeas, *Vigna unguiculata* (L.) Walp. Australian Journal of Plant Physiology 6: 265–276.

Sinoquet, H., S. Thanisawanyangkura, H. Mabrouk, and P. Kasemsap, 1998. Characterization of the light environment in canopies using 3D digitising and image processing. Annals of Botany 82: 203–212.

Sinoquet, H., X. Le Roux, B. Adam, T. Ameglio, and F.A. Daudet, 2001. RATP: A model for simulating the spatial distribution of radiation absorption, transpiration and photosynthesis within canopies: Application to an isolated tree crown. Plant, Cell and Environment 24: 395–406.

Sinoquet, H., G. Sonohat, J. Phattaralerphong, and C. Godin, 2005. Foliage randomness and light interception in 3D digitized trees: An analysis of 3D discretization of the canopy. Plant, Cell and Environment 29: 1158–1170.

Smith, H., and G.C. Whitelam, 1997. The shade avoidance syndrome: multiple responses mediated by multiple phytochromes. Plant, Cell and Environment 20: 840–844.

Smith, M. and D. Ullberg, 1989. Effect of leaf angle and orientation on photosynthesis and water relations in *Silphium terebinthinaceum*. American Journal of Botany 76: 1714–1719.

Smith, W.K., T.C. Vogelmann, E.H. DeLucia, D.T. Bell, and K.A. Shepherd, 1997. Leaf form and photosynthesis. Do leaf structure and orientation interact to regulate internal light and carbon dioxide? BioScience 47: 785–793.

Speck, T. and D. Vogellehner, 1988. Biophysical examinations of the bending stability of various stele types and the upright axes of early "vascular" land plants. Botanica Acta 101: 262–268.

Speck, T., H.C. Spatz, and D. Vogellehner, 1990. Contributes to the biomechanics of plants. I. Stabilities of plant stems with strengthening elements of differing cross-sections against weight and wind forces. Botanica Acta 103: 111–122.

Sprugel, D.G. 2002. When branch autonomy fails: Milton's Law of resource availability and allocation. Tree Physiology 22: 1119–1124.

Sprugel, D.G., T.M. Hinckley, and W. Schaap, 1991. The theory and practice of branch autonomy. Annual Review of Ecology and Systematics 22: 309–334.

Steingraeber, D.A., 1982. Phenotypic plasticity of branching pattern in sugar maple (*Acer saccharum*). American Journal of Botany 69: 638–640.

Stenberg, P., 1995. Penumbra in within-shoot and between-shoot shading in conifers and its significance for photosynthesis. Ecological Modelling 77: 215–231.

Stenberg, P., H. Smolander, D.G. Sprugel, and S. Smolander, 1998. Shoot structure, light interception, and distribution of nitrogen in an *Abies amabilis* canopy. Tree Physiology 18: 759–767.

Stenberg, P., S. Palmroth, B.J. Bond, D.G. Sprugel, and H. Smolander, 2001. Shoot structure and photosynthetic efficiency along the light gradient in a Scots pine canopy. Tree Physiology 21: 805–814.

Sterck, F.J., F. Bongers, and D.M. Newbery, 2001. Tree architecture in a Bornean lowland rain forest: intraspecific and interspecific patterns. Plant Ecology 153: 279–292.
Sterck, F., M. Martinez-Ramos, G. Dyer-Leal, J. Rodriguez-Velazquez, and L. Poorters, 2003. The consequences of crown traits for the growth and survival of tree saplings in a Mexican lowland rainforest. Functional Ecology 17: 194–200.
Stewart, I., 1988. A review of the science of fractal images. Nature 336: 289.
Strauss-Debenedetti, S., and F.A. Bazzaz, 1996. Photosynthetic characteristics of tropical trees along successional gradients. In: S.S. Mulkey, R.L. Chazdon, and A.P. Smith, eds. Tropical Forest Plant Ecophysiology. Chapman & Hall, New York, pp. 162–186.
Sultan, S.E., 2004. Promising directions in plant phenotypic plasticity. Perspectives in Plant Ecology Evolution and Systematics 6: 227–233.
Sultan, S.E., 2005. An emerging focus on plant ecological development. New Phytologist 166: 1–5.
Takenaka, A., 1994a. Effects of leaf blade narrowness and petiole length on the light capture efficiency of a shoot. Ecological Research 9: 109–114.
Takenaka, A., 1994b. A simulation model of tree architecture development based on growth response to local light environment. Journal of Plant Research 107: 321–330.
Terborgh, J., 1992. Diversity and the Tropical Forest. Scientific American Library, New York.
Terjung, W.H. and S.S.F. Louie, 1972. Potential solar radiation on plant shapes. International Journal of Biometeorology 16: 25–43.
Thomas, S.C. and W.E. Winner, 2000. A rotated ellipsoidal angle density function improves estimation of foliage inclination distributions in forest canopies. Agricultural and Forest Meteorology 100: 19–24.
Tomlinson, P.B., 1987. Architecture of tropical plants. Annual Review of Ecology and Systematics 18: 1–21.
Turgut, R., and A. Kadioglu, 1998. The effect of drought, temperature and irradiation on leaf rolling in *Ctenanthe setosa*. Biologia Plantarum 41: 629–633.
Turner, I.M., W.K. Gong, J.E. Ong, J.S. Bujang, and T. Kohyama, 1995. The architecture and allometry of mangrove saplings. Functional Ecology 9: 205–212.
Tyree, M.T., 1988. A dynamic model for water flow in a single tree: Evidence that models must account for hydraulic architecture. Tree Physiology 4: 195–217.
Tyree, M.T., 1999. Water relations and hydraulic architecture. In: F.I. Pugnaire and F. Valladares, eds. Handbook of Functional Plant Ecology. Marcel Dekker, New York, pp. 221–268.
Valladares, F., 1999. Architecture, ecology, and evolution of plant crowns. In: F.I. Pugnaire and F. Valladares, eds. Handbook of Functional Plant Ecology. Marcel Dekker, Inc., New York, pp. 121–194.
Valladares, F. and D. Brites, 2004. Leaf phyllotaxis: Does it really affect light capture? Plant Ecology 174: 11–17.
Valladares, F. and R.W. Pearcy, 1997. Interactions between water stress, sun-shade acclimation, heat tolerance and photoinhibition in the sclerophyll *Heteromeles arbutifolia*. Plant, Cell and Environment 20: 25–36.
Valladares, F. and R.W. Pearcy, 1998. The functional ecology of shoot architecture in sun and shade plants of *Heteromeles arbutifolia* M. Roem, a Californian chaparral shrub. Oecologia 114: 1–10.
Valladares, F. and R.W. Pearcy, 1999. The geometry of light interception by shoots of *Heteromeles arbutifolia*: Morphological and physiological consequences for individual leaves. Oecologia 121: 171–182.
Valladares, F. and F.I. Pugnaire, 1999. Tradeoffs between irradiance capture and avoidance in semi-arid environments assessed with a crown architecture model. Annals of Botany 83: 459–469.
Valladares, F., S.J. Wright, E. Lasso, K. Kitajima, and R.W. Pearcy, 2000. Plastic phenotypic response to light of 16 congeneric shrubs from a Panamanian rainforest. Ecology 81: 1925–1936.
Valladares, F., L. Balaguer, E. Martínez-Ferri, E. Perez-Corona, and E. Manrique, 2002a. Plasticity, instability and canalization: Is the phenotypic variation in seedlings of sclerophyll oaks consistent with the environmental unpredictability of Mediterranean ecosystems? The New Phytologist 156: 457–467.
Valladares, F., J.M. Chico, I. Aranda, L. Balaguer, P. Dizengremel, E. Manrique, and E. Dreyer, 2002b. The greater seedling high-light tolerance of *Quercus robur* over *Fagus sylvatica* is linked to a greater physiological plasticity. Trees: Structure and Function 16: 395–403.

Valladares, F., J.B. Skillman, and R.W. Pearcy, 2002c. Convergence in light capture efficiencies among tropical forest understory plants with contrasting crown architectures: A case of morphological compensation. American Journal of Botany 89: 1275–1284.
Valladares, F., I. Dobarro, D. Sánchez-Gómez, and R.W. Pearcy, 2005. Photoinhibition and drought in Mediterranean woody saplings: Scaling effects and interactions in sun and shade phenotypes. Journal of Experimental Botany 56: 483–494.
van Kleunen, M. and M. Fischer, 2005. Constraints on the evolution of adaptive phenotypic plasticity in plants. New Phytologist 166: 49–60.
Van Pelt, R. and J.F. Franklin, 2000. Influence of canopy structure on the understory environment in tall, old-growth, conifer forests. Canadian Journal of Forest Research 30: 1231–1245.
van Steenis, C.G.G.J., 1981. Rheophytes of the world: An account of flood-resistant flowering plants and ferns and the theory of autonomous evolution. Sijthoff and Noordhoof, Alphen aan den Rijn, The Netherlands.
Vogel, S., 1989. Drag and reconfiguration of broad leaves in high winds. Journal of Experimental Botany 40: 941–948.
Vogel, S., 1996. Life in moving fluids: The physical biology of flow. Princeton University Press, Princeton, NJ.
Waller, D.M., 1986. The dynamics of growth and form. In: M.J.Crawley, ed. Plant Ecology. Blackwell Scientific Publications, Oxford, London, Edinburgh, Boston, Palo Alto, Melbourne, pp. 291–320.
Waller, D.M. and D.A. Steingraeber, 1985. Branching and modular growth: Theoretical models and empirical patterns. In: J.B.C. Jackson, L.W. Buss, and R.E. Cook, eds. Population Biology and Evolution of Clonal Organisms. Yale University Press, New Haven, London, pp. 225–257.
Wang, Y.P. and P.G. Jarvis, 1990. Influence of crown structural properties on PAR absorption, photosynthesis, and transpiration in Sitka spruce: Application of a model (MAESTRO). Tree Physiology 7: 297–316.
Wang, Y.P. and R. Leuning, 1998. A two-leaf model for canopy conductance, photosynthesis and partitioning of available energy. I. Model description and comparison with a multi-layered model. Agricultural and Forest Meteorology 91: 89–111.
Watson, M.A., 1986. Integrated physiological units in plants. Trends in Ecology and Evolution 1: 119–123.
Weinig, C., 2000. Limits to adaptive plasticity: Temperature and photoperiod influence shade-avoidance responses. American Journal of Botany 87: 1660–1668.
Werk, K.S. and J.Ehleringer, 1984. Non-random leaf orientation in *Lactuca serriola* L. Plant, Cell and Environment 7: 81–87.
White, J., 1979. The plant as metapopulation. Annual Review of Ecology and Systematics 10: 109–145.
White, J., 1984. Plant metamerism. In: R. Dirzo and J. Sarukhán, eds. Perspectives on Plant Population Ecology. Sinauer, Sunderland, MA, pp. 176–185.
Whitney, G.G., 1976. The bifurcation ratio as an indicator of adaptive strategy in woody plant species. Bulletin of the Torrey Botanical Club 103: 67–72.
Wilson, B.F., 1995. Shrub stems: Form and function. In: B. Gartner, ed. Plant Stems. Physiology and Functional Morphology. Academic Press, San Diego, CA, pp. 91–103.
Woodward, F.I., 1998. Do plants really need stomata? Journal of Experimental Botany 49: 471–480.
Wright, S.J., 1999. Plant diversity in tropical forests. In: F.I. Pugnaire and F. Valladares, eds. Handbook of Functional Plant Ecology. Marcel Dekker, Inc., New York, pp. 449–471.
Zimmermann, U., H.-J. Wagner, H. Schneider, M. Rokitta, A. Haase, and F.-W. Bentrup, 2000. Water ascent in plants: The ongoing debate. Trends in Plant Science 5: 145–146.

5 Structure and Function of Root Systems

Robert B. Jackson, William T. Pockman, William A. Hoffmann, Timothy M. Bleby, and Cristina Armas

CONTENTS

INTRODUCTION

The study of root structure and functioning is centuries old (Hales 1727, reprinted 1961). While great progress has been made (e.g., Brouwer et al. 1981), our knowledge is limited by the difficulties in studying roots *in situ*. These limitations color our perception of plants. A typical layperson knows that forests can grow 50–100 m in height, but rarely recognizes that root systems can grow to similar depths (Canadell et al. 1996). The individual may also never consider the functional consequences of roots that typically spread well beyond the canopy line of most plants (e.g., Lyford and Wilson 1964). Just as there is a quiet bias in maps of the world that consistently present the northern hemisphere "on top," our perception of plants would change if they were drawn "upside-down"—roots on top and shoots underneath. A small shrub such as *Prosopis glandulosa* would suddenly appear as majestic as a tree, and many trees would suddenly seem shrubby. The bias of human perception shadows our view of the plant world.

Root and shoot functioning are often studied separately, in part because the techniques and equipment needed can differ substantially. In reality, roots and shoots are functionally integrated. This integration is evident in patterns of standing biomass and allocation. With the exception of forests, most natural systems have root:shoot ratios (the ratio of root to shoot biomass) between 1 and 7, including tundra, deserts, and grasslands (Jackson et al. 1996, 1997). Most forest systems typically have root:shoot ratios of approximately 0.2, with the majority of biomass stored as woody biomass in the boles of trees. Even for forests, half or more of annual primary production is usually allocated belowground (e.g., 60% for a *Liriodendron tulipfera* forest, Reichle et al. 1973). This is not to imply that roots are more important to the functioning of plants than are shoots, but just to demonstrate that they are no less important.

The purpose of this chapter is to provide an introduction to the structure and functioning of root systems. We begin by outlining the basics of root morphology and development. We next examine four broad categories of root functioning: anchoring, resource uptake, storage, and sensing the environment. Information on more specialized root functions such as reproduction and aeration is available elsewhere (e.g., Drew 1997). We discuss two important root symbioses, mycorrhizal associations, and symbiotic nitrogen fixation. We end by examining global patterns of root distributions for biomes and plant functional types. Interested readers will find references in each section that provide comprehensive detail on each topic.

ROOT MORPHOLOGY AND DEVELOPMENT

We begin by reviewing the generalized structure of primary roots, including the main tissue types within roots and changes that occur with secondary growth and tertiary root morphology. A more detailed discussion of these and other structural features can be found in complete botanical and anatomical texts (e.g., Esau 1977, Mauseth 1988).

PRIMARY ROOT ANATOMY

As in other plant organs, cell division during primary root growth occurs in the apical meristem, giving rise to the undifferentiated cells of the protoderm, ground meristem, and procambium (Figure 5.1a). Root length increases as these newly produced cells elongate and differentiate in the region immediately behind the root tip. As a result, root anatomy and functioning change with distance from the tip and all developmental stages may be present in a single root.

Epidermis

The epidermis is derived from the protoderm and generally consists of a single layer of cells forming the outermost root tissue (Figure 5.1b). A key feature of the epidermis is root hairs, elongated cells projecting into the surrounding soil (Hofer 1991). Root hair density is typically greatest in the most distal region of the primary root behind the root tip. This so-called root hair zone has been widely regarded as the location of most water and nutrient uptake (but see Section "Vascular Cylinder"). Although root hairs persist in some species, their distribution is generally restricted to the distal portion of the root because the oldest root hairs are lost and new ones are produced only near the root tip. Root hair density ranges from 20 to 2500 per cm^2 (Dittmer 1937, Kramer 1983) and can more than double the root surface area in contact with the soil, resulting in a greater accessible soil volume (Kramer 1969). In many but not all species, nutrient absorption increases in proportion to root hair density (Bole 1973, Itoh and Barber 1983).

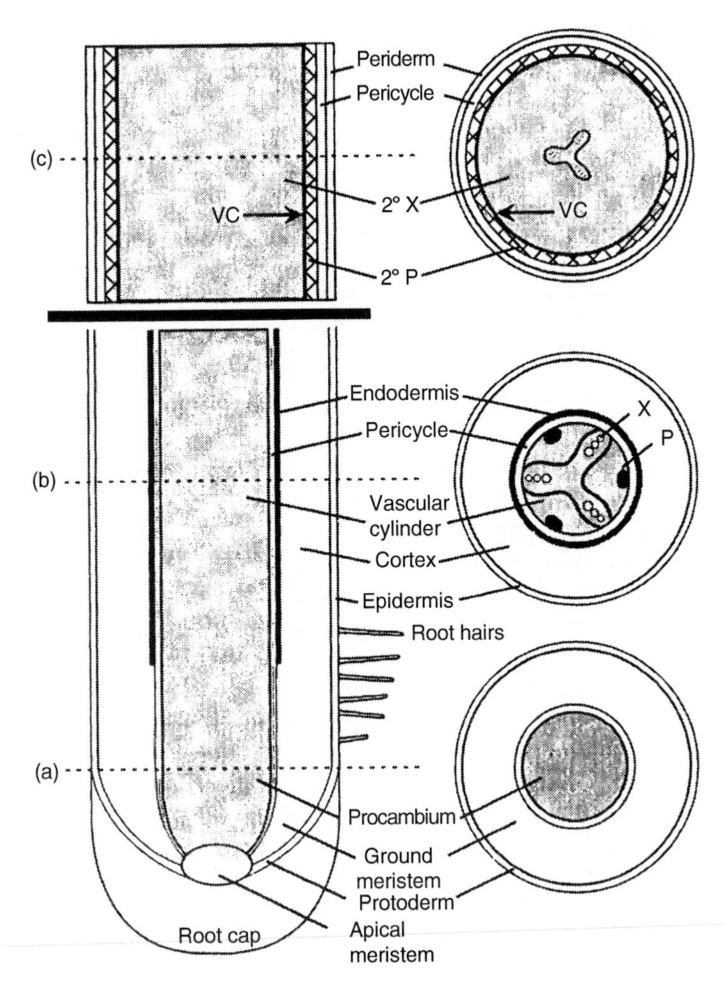

FIGURE 5.1 Longitudinal and cross-sectional schematics of generalized root anatomy. Cross sections are indicated by broken lines through the longitudinal section on the left. Drawing is not to scale to allow illustration of the key features of the root. (a) Undifferentiated cells of the protoderm, ground meristem, and procambium produced by divisions of the root apical meristem located immediately behind the root cap. (b) Differentiated primary tissues: epidermis (from protoderm), cortex and endodermis (from ground meristem) xylem (X), phloem (P), and pericycle (from procambium). (c) Root anatomy after substantial secondary growth. Differentiation of cells between xylem and phloem bundles (b) has produced the vascular cambium (VC). Divisions of the cambial initials give rise to secondary xylem (2° X) and phloem (2° P). The cortex, and with it the endodermis, has ruptured and sloughed off with growth of the vascular cylinder and is not visible at this stage. The suberized lignified periderm, ultimately derived from the pericycle, has developed and assumed the function of sealing the root from the surrounding soil. The primary xylem is visible at the center of the secondary xylem. (Adapted from Raven, P., Evert, R., and Eichhorn, S., *Biology of Plants*, 4th edition, Worth, New York, 1986. With permission.)

Cortex

The cortex is derived from the ground meristem, composed largely of parenchyma cells. It lies between the epidermis to the outside and the vascular cylinder at the center of the root (Figure 5.1b). The cortex may develop large air canals that increase oxygen availability to root cells, and it can be an important site for carbohydrate storage (Mauseth 1988). Perhaps the most studied feature of the cortex is the endodermis, a single cell layer defining the interior edge of the cortex. The central feature of the endodermis is the casparian band formed by the deposition of suberin in the primary cell wall and middle lamella of each adjoining endodermal cell. The result is a continuous suberized barrier preventing passage of soil solutes from the cortex into the vascular cylinder without crossing a cell membrane (Weatherley 1982, Clarkson 1993). Recent studies suggest that a more complex model of water and solute uptake may be appropriate but support the importance of the casparian band in providing control over solute flow into roots (Steudle 1994, Steudle and Meshcheryakov 1996). Many taxa also form an exodermis, an anatomically similar cell layer located at the outer edge of the cortex (Perumalla et al. 1990, Peterson and Perumalla 1990).

Vascular Cylinder

The vascular cylinder develops from the procambium in the center of the root and is delimited by a single layer of parenchyma cells called the pericycle (Figure 5.1b). In cross section, the primary xylem is arranged in finger-like projections from the center of the root toward the pericycle. Phloem bundles occur between these xylem projections. The smallest xylem conduits, the protoxylem, occur at the tips of these projections whereas the larger conduits of the metaxylem are located more centrally and mature later. Root uptake of water and nutrients requires axial transport through functional xylem conduits, which are dead at maturity. Although the root hair zone is often cited as the site of maximum uptake, studies of maize and soybean suggest that water absorption in the root hair zone may be restricted because the largest metaxylem conduits are still alive and nonconducting (McCully and Canny 1988). Future work addressing the relative timing of maturity of the xylem, endodermis, and root hairs across taxa will improve our understanding of the contribution of different developmental stages of the root to resource uptake (McCully 1995).

Lateral and Adventitious Roots

The production of lateral roots is an important feature determining the tertiary (three dimensional) structure of the root system and the distribution of surface area for resource uptake. Lateral roots occur in gymnosperms and dicots and arise from root primordia in the pericycle and, less commonly, the endodermis (Peterson and Peterson 1986). As a root primordium elongates, it passes through the cortex and epidermis. Vascular tissues differentiate within the developing root and are connected with the plant's vascular system at the base of the primordium. Many cortical cells are crushed during lateral root emergence, though anticlinal divisions of the endodermis (in which the cell plate forms perpendicular to the nearest tissue surface) minimize disruption of the casparian band around the emerging lateral root. Nevertheless, lateral root growth may provide a pathway for the flow of water in or out of the vascular cylinder that is not controlled by the casparian band (Kramer and Boyer 1995, Caldwell et al. 1998).

Adventitious roots originate from aerial or underground plant stems. They occur in most plant taxa and are particularly important in the monocotyledons, comprising most of the root system (for review see Davis and Haissig 1994). Adventitious roots develop from root primordial, which can arise in most tissues of plant stems. As in lateral roots, these primordia differentiate into cell types typical of a root, and vascular connections are formed with existing xylem and phloem at the root origin.

Secondary Growth

Secondary growth in roots is clearly important for the ability of plants to become "woody" and perennial. It occurs commonly among gymnosperms, to varying degrees among dicotyledonous plants, and is absent among the monocots (Mauseth 1988). The initiation of secondary growth is preceded by the formation of a vascular cambium derived from undifferentiated procambium and parts of the pericycle (Figure 5.1c). The vascular cambium forms a ring of meristematic cells between the phloem and xylem. These cambial initials give rise to xylem and phloem cells by periclinal divisions (the cell plate forms parallel to the nearest tissue surface) and accommodate increases in root diameter by occasional anticlinal divisions, which maintain the continuity of the cambium. Likewise, the pericycle undergoes both periclinal and anticlinal divisions giving rise to the phellogen, a meristematic tissue that produces the periderm. These changes are not accompanied by further divisions of the cortex, which becomes fractured and lost as the root increases in diameter. The periderm, which includes the suberized cells of the cork, becomes the outer surface of the root and assumes the protective function formerly provided by the epidermis and endodermis.

Tertiary Root Morphology

Root development among seed plants begins with the elongation of the taproot. For gymnosperms and dicots, the three-dimensional structure of the taproot and its branched lateral roots define the morphology of the root system (Figure 5.2). In contrast, the early demise of the taproot and the subsequent growth of adventitious roots in monocots result in a fibrous root system emanating from the base of the stem (limiting the rooting depth of monocots). This difference, combined with the lack of secondary growth in roots of monocots, defines an important functional difference between taxa that possess these very different root systems (Caldwell and Richards 1986). Although the absence of secondary growth limits the rooting depth of monocots, they often have very high root length densities in the soil volume they explore (Glinski and Lipiec 1990). In contrast, the taproot system of dicots and gymnosperms is often capable of exploring soil volume that extends both laterally and vertically beyond the reach of many monocots. Despite their different structural characteristics, both fibrous and taproot systems are capable of differential root proliferation in response to resource patches in the soil (e.g., Drew 1975, Bilbrough and Caldwell 1995). The exploitation of such resources is addressed later in this chapter.

FIGURE 5.2 Differences in rooting system morphology among prairie grasses (family Gramineae) and herbs: *h, Hieracium scouleri* (Compositae); *k, Koeleria cristata* (Gramineae); *b, Balsamina sagittata* (Balsaminaceae); *f, Festuca ovina ingrata* (Gramineae); *g, Geranium viscosissimum* (Geraniaceae); *p, Poa sandbergii* (Gramineae); *ho, Hoorebekia racemosa*; *po, Potentilla blaschkeana* (Rosaceae). (From Kramer, P. and Boyer, J., *Water Relations of Plants and Soils*, Academic Press, San Diego, CA, 1995; after Weaver, J.E., *The Ecological Relations of Roots*, Publication 286, Carnegie Institution of Washington, Washington D.C., 1919. With permission.)

ROOT FUNCTIONS

Root systems have at least five broad functions: anchoring plants, capturing resources, storing resources, and sensing and modifying the environment. Such distinctions are arbitrary, but provide a useful framework for examining root functioning.

ANCHORING

Probably the most fundamental root function is to hold plants in place. This need is most obvious in protecting trees from windthrow, but shrubs and herbaceous vegetation are also exposed to the vagaries of wind, trampling, and herbivores. Resistance to toppling has economic importance for crop species, too, whose root systems tend to be fairly shallow (Brady 1934).

In general, there are three kinds of mechanical failure in plants—uprooting, stem failure, and root failure. The biomechanics of root anchoring can be studied by uprooting plants mechanically and recording the resistance with a strain gauge (e.g., Somerville 1979, Mattheck 1991). Results from these and other experiments show that resistance to windthrow has two primary components: the resistance of leeward laterals to bending and the resistance of windward sinkers and taproots to uprooting. Bending tests on the leeward laterals of a deep-rooted larch species showed that they provided approximately 25% of tree anchorage support (Crook and Ennos 1996). Consequently, about three quarters of the stability in that system came from taproots and windward sinkers. Where there is a prevailing wind direction, there is often an asymmetrical development of structural roots—for example, greater root development on the leeward side than on the windward side of a tree (Nicoll and Ray 1996). The cross-sectional structure of individual roots can also differ depending on the location of the roots. Secondary growth above the center of a root can lead to a classic T-beam structure (swollen on top and thinner on the bottom). This type of thickening is more common in roots relatively close to the trunk (<1 m away), particularly on the leeward side of trees. Roots with an I-beam structure can be more prevalent at greater distances, especially on the windward side. Such roots resist vertical flexing (Nicoll and Ray 1996). Stokes et al. (1996) developed a theoretical model of anchoring, resistance to uprooting, and root branching patterns. Not surprisingly, deep roots were especially important.

RESOURCE UPTAKE

Leaves and roots play analogous roles in plants. Leaves are the structures primarily responsible for carbon and energy uptake and fine roots take up most of the water and nutrients acquired by plants. This dichotomy of structure and function is useful conceptually because above- and belowground resources are generally separated. In practice, however, it is difficult to disentangle aboveground and belowground processes (e.g., Donald 1958, Jackson and Caldwell 1992). Light availability powers the enzymes responsible for phosphate transport; adequate root surface area depends on the amount of CO_2 taken up by shoots. In turn, carbon and energy uptake requires soil water for the maintenance of turgor and stomatal conductance and nitrogen to build photosynthetic proteins such as RuBP Carboxylase. This interdependence has led to many perspectives on the "balance" of root and shoot processes (e.g., Brouwer 1963). In the following discussion of resource uptake by roots, we focus on the uptake of water, nitrogen, and phosphorus. A more detailed discussion can be found in Nye and Tinker (1977), Marschner (1995), and Casper and Jackson (1997), on which much of the following section is based. Additional perspectives can be found in Chapter 8, which describes the acquisition, use, and loss of nutrients.

Soil resources typically reach the surface of roots by three processes: root interception, mass flow of water and nutrients, and diffusion (Marschner 1995). Root interception occurs as a root grows through the soil, physically displacing soil particles and clay surfaces and acquiring water and nutrients. This process typically accounts for >10% of the resources taken up by roots. Mass flow, which is driven by plant transpiration, depends on the rate of H_2O movement to the root and the concentration of dissolved nutrients in the soil solution. Nutrient diffusion toward the root occurs when nutrient uptake by the root exceeds the supply by mass flow and a depletion zone around the root is created. The supply of nutrients by diffusion is especially important for those with large fractions bound to the soil matrix, such as K^+ and $H_2PO_4^-$. In nature, mass flow and diffusion work in concert to supply N, P, and K and are difficult to separate in the field (Nye and Tinker 1977).

Water moves into and out of roots passively based on the water potential gradient in the soil–plant system. In contrast, nutrient uptake is generally an enzymatic process that follows apparent Michaelis–Menten kinetics:

$$V = V_{max} C_l/(C_l + K_m),$$

where V is the flux of ion into the root per unit time, V_{max} is the maximum influx rate, C_l is the soil solution concentration at the root surface, and K_m is the soil solution concentration where influx is 50% of V_{max} (Nye and Tinker 1977). The equation sometimes includes a C_{min} term, the soil solution concentration at which net influx into the root is zero (Barber 1984). Because nutrient uptake is generally enzymatic, it is sensitive to reductions in photosynthesis by shoots (e.g., Jackson and Caldwell 1992).

The belowground competitive ability of plants is often directly proportional to the size of their root systems. This is in contrast to shoot systems, where a relatively small portion of leaves can overtop a canopy and acquire most of the available light. There are many examples where root systems with the highest densities and occupying the most space are the strongest competitors (Aerts et al. 1991, Casper and Jackson 1997). Consequently, a plant may grow higher root densities or extend the volume of soil explored to acquire more water and nutrients. Plants can also increase resource uptake by selective foraging. Plants frequently respond to enriched patches of soil water and nutrients by proliferating roots, selectively growing roots in the zone of enrichment (e.g., Duncan and Ohlrogge 1958, Berendse et al. 1999, chapter 8). Proliferated roots tend to be smaller in diameter and greater in density than those found in the background soil. A second, related factor that may increase resource uptake is a change in fine-root demography. In a Michigan hardwood forest, not only did roots proliferate in response to water and nitrogen patches, but the new roots lived significantly longer than new roots in control patches (Pregitzer and Hendrick 1993). Architectural adjustment (changes in root topology, length, or branching angles) is another type of morphological plasticity that can increase nutrient uptake. Fitter (1994) examined the architectural attributes of 11 herbaceous species and showed that roots in relatively high-nutrient patches typically had a more herringbone branching pattern than roots in low-nutrient patches, concentrating higher-order lateral roots in the patches and increasing the efficiency of nutrient uptake.

Physiological plasticity can selectively increase nutrient uptake by altering enzyme attributes or other physiological traits. A species with more enzymes per root surface area (greater V_{max}), a higher ion affinity of enzymes (smaller K_m), or a greater ability to draw nutrients down to a low level (smaller C_{min}) will be at a competitive advantage (ignoring other factors). Plants in the laboratory and in the field have been shown to increase V_{max} and decrease K_m in response to localized nutrients (e.g., Drew and Saker 1975, Jackson et al. 1990). For example, Jackson et al. (1990) showed that grass and shrub species in the field were able to selectively increase physiological rates of phosphate uptake in portions of their root system in fertilized soil patches (Figure 5.3). For water uptake, osmoregulation can lower cell water potential and maintain net uptake in the face of drying soils (Kramer and Boyer 1995). It is the suite of

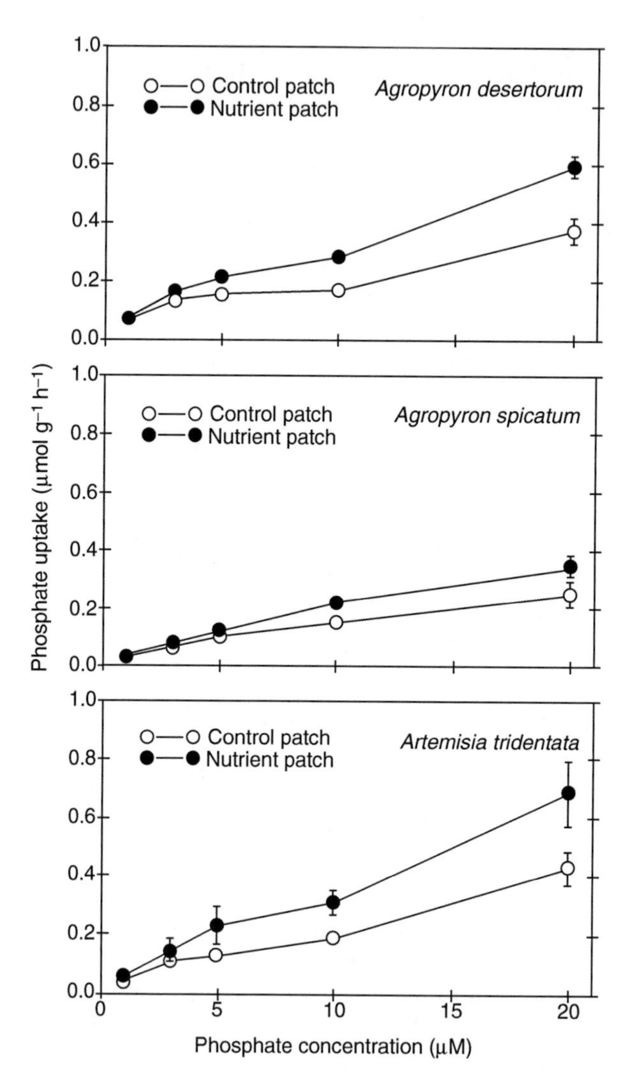

FIGURE 5.3 The rate of phosphate uptake for roots from enriched and control soil patches as a function of solution phosphate concentration (mean ± SEM; $n = 6$–8). Soil patches on opposite sides of plants in monoculture field plots were treated with 750 ml of nutrient solution or distilled water and samples of the soil patches were cored 1 week after treatment. Roots from each core were subsampled and immersed in ^{32}P solutions. Results were similar for 3-day experiments as for the week-long experiments shown here. (From Jackson, R.B., Manwaring, J., and Caldwell, M., *Nature*, 344, 58, 1990. With permission.)

morphological and physiological attributes that determine resource uptake by plants. While their genetic make-up plays a fundamental role in the type of root system plants possess, there is often great flexibility in how those genes are expressed based on environmental cues.

Extensive overviews of morphological and physiological plasticity and resource capture can be found in Hutchings and de Kroon (1994) and Robinson (1994). See also Chapter 8 in this volume.

Storage

Plants must cope with variable environments where the timing of resource availability and uptake may not coincide with demand by the plant. The storage of carbohydrates, nutrients, and, to a lesser extent, water when resources are abundant provides insurance for future periods of high demand. Although many plant organs are involved in storage, roots are the most important storage site for many species. Roots play a particularly important role in cases where complete regeneration of aerial biomass is necessary. Perennial and biennial herbaceous species enduring seasonal environments or intense herbivory must rely heavily on belowground reserves. Woody plants in fire-prone environments such as Mediterranean-type ecosystems and savannas also depend heavily on roots for nutrient and carbohydrate storage (Miyanishi and Kellman 1986, Bowen and Pate 1993, Bell et al. 1996). Even in environments not typically subjected to fire, woody roots are an important site for storage, with carbohydrate concentrations often exceeding those in stems (Loescher et al. 1990).

In many species, belowground storage occurs in modified stems such as rhizomes, bulbs, corms, stem tubers, lignotubers, and burls (de Kroon and Bobbink 1997). These organs are functionally similar to roots with respect to storage and are included in our discussion. Among true roots, storage can occur in all size classes of roots, but specialized, large-diameter roots such as root tubers and taproots often play the most important role.

Most of the mineral nutrients required by plants are stored in roots (Pate and Dixon 1982), but the majority of studies have focused on carbon, nitrogen, and phosphorus. Consequently, most of our discussion is devoted to these elements. A large number of chemical compounds are involved in their storage. In roots, starch is the most important form of carbon storage, though other polysaccharides can be important (Lewis 1984). In particular, fructan, a polymer of fructose, is common in many monocots and a few dicot families (Pollock 1986). Sucrose, monosaccharides, sugar alcohols, and lipids can also be prevalent (Glerum and Balatinecz 1980, Lewis 1984, Dickson 1991). Nitrogen is stored as specialized storage proteins, amino acids, amides, or nitrate (Tromp 1983, Staswick 1994). Phosphorus is stored primarily as phosphate, phytic acid, and polyphosphate (Bieleski 1973).

Within the cell, the vacuole is the most important site for the storage of sugars, phosphate, and nitrogen (Bieleski 1973, Willenbrink 1992), whereas starch storage occurs in plastids (Jenner 1992). In roots, storage occurs primarily in parenchyma cells (Bieleski 1973, Jenner 1992, Bell et al. 1996).

Plant storage of materials can be classified as accumulation or reserve storage (Chapin et al. 1990). Accumulation occurs when the uptake of a resource is greater than the plant's immediate capacity to use the resource. The plant would not be able to use the resource for other functions, so its storage does not compete with growth and maintenance. In contrast, reserve formation occurs at a time when the resource could otherwise be used for growth. Because reserve formation competes directly with growth and maintenance, there is a substantial cost in forming reserves.

Both accumulation and reserve formation are strongly influenced by resource availability. Shading reduces root carbohydrate concentrations (Jackson and Caldwell 1992, Bowen and Pate 1993), whereas elevated CO_2 can increase carbohydrate storage (Chomba et al. 1993). Nitrogen and phosphorus accumulation is typical under high nutrient availability (Chapin 1980), a response typically referred to as luxury consumption.

Low availability of one resource often increases the storage of other resources. For example, root carbohydrate storage has been found to be greater under water stress because tissue growth was more inhibited than photosynthesis (Busso et al. 1990). Low nutrient availability can also increase root carbohydrates (Jackson and Caldwell 1992) as has been shown for leaves (Waring et al. 1985, McDonald et al. 1991), presumably because low nutrient availability results in lower tissue production, reducing demand for photosynthate. These cases represent accumulation rather than reserve formation, because the storage results from low demand for the resource within the plant.

Demand for resources within the plant also largely determines the timing of storage. In herbaceous perennials and deciduous woody plants, root reserves are retranslocated at the beginning of the growing season to supply developing leaves, stems, and flowers. As a result, lowest reserve concentrations occur around the end of leaf flush (Woods et al. 1959, Daer and Willard 1981, Chapin et al. 1986, Keller and Loescher 1989, Wan and Sosebee 1990). After leaf maturation, the root undergoes a transition from carbohydrate source to carbohydrate sink, and replenishment of carbohydrate reserves begins. Root carbohydrate concentration reaches a peak around the time of leaf-fall (Woods et al. 1959, Daer and Willard 1981, Chapin et al. 1986, Keller and Loescher 1989, Wan and Sosebee 1990). In contrast to deciduous species, evergreen plants show lower seasonal fluctuations in the concentration of stored carbohydrates (Chapin et al. 1986, Dickson 1991).

Similar trends are observed in woody plants subjected to fire, but replenishment of the reserves often takes longer (Miyanishi and Kellman 1986, Bowen and Pate 1993). This slower replenishment may result from the greater amount of aerial biomass that must be replaced or may represent an adaptation to a less frequent form of disturbance. On the other hand, grasses tolerant to frequent grazing exhibit rapid recovery of carbohydrate reserves after grazing occurs (Owensby et al. 1970, Oesterheld and McNaughton 1988, Orodo and Trlica 1990).

If disturbance recurs before adequate replenishment is complete, further recovery will be compromised (Miyanishi and Kellman 1986, Kays and Canham 1991). Overall, the cost associated with rapid replenishment of reserves can be quite large if replenishment competes with plant growth (Chapin et al. 1990).

Producing Hormones and Sensing the Environment

In addition to anchoring the plant and taking up and storing resources, roots sense their environment (Mahall and Callaway 1991) and convey information on the balance between root and shoot functioning in the form of hormones. Plants produce a variety of hormones, including auxin, cytokinins, gibberrelins, ethylene, and abscisic acid (ABA). Depending on how broad the definition of a hormone is, other compounds such as jasmonates and polyamines may also be considered. Plant responses to hormones depend on changes in their concentration and in the sensitivity of tissues responding to the hormone (Trewavas 1981).

There are two groups of hormones where roots are the dominant sites of synthesis: cytokinins and ABA. Cytokinins, named for their role in stimulating cytokinesis, are produced primarily in roots and are then transported in the xylem to leaves where they retard senescence and maintain metabolic activity (Torrey 1976). Cytokinins affect numerous plant processes, including cell division and morphogenesis. They affect protein synthesis and stimulate chlorophyll development. Cytokinin production also helps to coordinate root and shoot activity in plants. Roots become active in the spring and produce cytokinins that are transported to the shoot and activate dormant buds (Mauseth 1995). Cytokinins help to balance total leaf area relative to the root system by affecting rates of leaf expansion. Roots that produce cytokinins delay the senescence of shoot tissue. Cytokinins also provide a link from N uptake and the N status of roots to the synthesis of proteins. There are many other examples of how cytokinins help integrate the functioning of roots and shoots.

Unlike most plant hormones, ABA is primarily a growth inhibitor. ABA and gibberellins are both synthesized from mevalonic acid, the pathway that produces carotenoids and the general class of compounds known as terpenes. First characterized in 1963, ABA was proposed to play a role in the abscision of cotton bolls (Ohkuma et al. 1963). Subsequent evidence suggested that ethylene rather than ABA is a more important controller of abscision. Instead ABA affects such plant processes as dormancy, senescence, stress responses (including water, freezing, and salt stress), water uptake, and stomatal regulation.

The role of roots in ABA synthesis is particularly important for stomatal closure and water stress. Changes in leaf cell turgor were originally thought to be the sole cause of stomatal closure in response to soil dehydration. However, when leaf cell turgor was maintained by pressurizing the root system, stomatal closure still occurred when part of the root system was in dry soil (Gollan et al. 1986). These data suggested that a chemical signal was produced in response to decreasing soil water status. Subsequent studies have identified ABA as the most important such signal (Davies and Zhang 1991). In leaves, ABA causes stomata to close at low concentrations ($<\mu$M) by inhibiting K^+ influx to guard cells and activating K^+ efflux; its concentration can increase 50-fold in leaves experiencing water stress (Salisbury and Ross 1985, Walton and Li 1995). Experimental evidence suggests that ABA production in the roots is stimulated by reduced cell turgor and ABA moves to the shoots in the xylem stream (Tardieu et al. 1992). Further evidence for the importance of ABA in regulating plant water status is that mutants unable to produce ABA wilt permanently (Taiz and Zeiger 1991).

The relative contributions of water potential and ABA signals in determining leaf responses to water stress are a matter of continuing research. Reports that stomata often begin to close before any damage or detectable change in leaf water status occurs (Davies and Zhang 1991) have been interpreted as evidence that ABA signals from the roots provide a feed-forward mechanism for avoiding water stress. Tardieu et al.(1991, 1992) showed that there was a strong correlation between [ABA] in the xylem and leaf stomatal conductance (g_s) in maize plants, but little relationship between g_s and leaf turgor or water potential. The time lag between ABA production in the root and its arrival at the leaf suggests that root signals may not be well suited for stomatal regulation over short timescales (Kramer 1988). Seasonal studies showed that ABA signals were correlated with maximum stomatal conductance during progressive drying over many days but not on a diurnal basis (Wartinger et al. 1990). Such data suggest that water potential and ABA root signals combine to affect stomatal responses to soil water status (Tardieu and Davies 1993).

A rich literature on the synthesis and effects of ABA and cytokinins is available for the interested reader (e.g., Addicott 1983, Davies and Jones 1991, Davies 1995). In addition, Davies (1995) provides an excellent physiological overview of the production and function of plant hormones.

Releasing Exudates and Modifying the Environment

Roots not only sense their environment, they have the capacity to modify it by secreting organic molecules and water. The secretion of organic molecules is an active process requiring metabolic energy, whereas the secretion of water is a passive process driven by natural gradients in water pressure, which exist between roots and soil. These secretions have a strong influence on the biochemical and physical properties of the rhizosphere and surrounding soil, and they play an important role in the establishment and maintenance of terrestrial plant communities (Walker et al. 2003).

Organic compounds secreted by roots are broadly referred to as root exudates. Root exudates are diverse in their composition, ranging from low-molecular mass carbon compounds (numbering upward of 100,000 compounds) to complex molecules such as polysaccharides and bioactive proteins, many of which are species-specific (Flores et al. 1999, Inderjit and Weston 2003). Overall, from 5% to 21% of all photosynthetically fixed carbon is transferred to the rhizosphere (Marschner 1995). In purely physical terms, root exudates play an important role in promoting healthy root growth by maintaining good root–soil contact, lubricating the root tip, protecting roots from desiccation, stabilizing soil microaggregates, and allowing selective adsorption and storage of ions (Griffin et al. 1976, Bengough and McKenzie 1997, Rougier 1981, Hawes et al. 2000). In biological terms, root exudates play an active role in rhizosphere processes, allowing individuals to regulate belowground interactions with neighboring plants and soil organisms, including bacteria, fungi, and insects (Bais et al. 2004).

The production of root exudates can lead to either positive or negative ecological interactions among plants and soil organisms. On the positive side, root exudates can be used to encourage beneficial symbioses. For example, the roots of some leguminous species secrete isoflavonoids and flavonoids, which promote the activation of *Rhizobium* genes responsible for the nodulation process, and which may also play a role in colonization by vesicular–arbuscular mycorrhiza (Peters et al. 1986, Trieu et al. 1997). On the negative side, parasitic plant species often use compounds secreted from host roots as chemical messengers to initiate the development of invasive organs (haustoria) required for heterotrophic growth (Estabrook and Yoder 1998, Keyes et al. 2000).

Many root exudates are designed to expose other organisms to allelopathic chemicals, which are used by a plant to prevent other organisms from growing too close to it. Exudates can possess antimicrobial or antifungal properties, providing the plant with defensive advantages, a means to limit herbivory, or phytoinhibitory properties to prevent the growth of competing plants (Inderjit and Weston 2003, Bais et al. 2004). For example, an invasive forb species in the western United States, the spotted knapweed (*Centaurea maculosa*), actively displaces native plant species by exuding a phytotoxin called catechin. This compound is secreted in two forms (actually mirror images of the same molecule): (−)-catechin, which has an allelochemical activity on native plant species; and (+)-catechin, which has antimicrobial properties. The secretion of these molecules accounts in part for the extremely invasive behavior of *C. maculosa* (Callaway and Aschehoug 2000, Bais et al. 2002, 2003). Interestingly, there is also evidence to suggest that allelopathic compounds may in fact help plants to direct root growth, by segregating roots from each other and serving as a root navigation tool to detect and circumvent physical obstacles (Falik et al. 2005).

The secretion of exudates is a remarkable metabolic feature of plant roots, but an equally remarkable physical feature is their ability to move water. Plants release water into the soil through a phenomenon known as hydraulic redistribution (Burgess et al. 1998), a passive mechanism whereby water moves from regions of wet soil to those of dry soil via roots. Quite simply, roots act as conduits for the movement of water across gradients in pressure, from regions of high water potential (wet soil) to regions of low water potential (dry soil). Water can be hydraulically "lifted" from deep, wet soil layers to shallow, dry soil layers (Dawson 1993), and in the same way, water can also be redistributed downward (e.g., Burgess et al. 2001) or laterally (e.g., Smart et al. 2005).

For redistribution to occur, however, conditions must be just right. Roots belonging to the same root system must simultaneously span regions of wet soil and dry soil, and contact between roots and soil must be close. Importantly, the water potential gradient between wet and dry soil must be strong enough to drive water movement toward dry soil rather than toward the atmosphere, which has a much lower water potential. For this reason, redistribution mostly occurs at night when stomata are closed and transpiration has ceased. Interestingly, this mechanism predicts that redistribution can occur in plants that are fully senesced (Leffler et al. 2005) or winter-dormant (Hultine et al. 2004) and that CAM plants can redistribute water during the day when their stomata are closed (Yoder and Nowak 1999).

Hydraulic redistribution is widespread. It has been documented in over 60 species covering a range of plant types, from grasses to trees, and in a range of environments, from deserts to rainforests, and we are beginning to understand that it has important ecological consequences (see reviews by Caldwell et al. 1998, Ryel 2004). By itself, hydraulic redistribution is a significant physical process affecting soil water dynamics because it allows water to move through the soil matrix in a manner that is not otherwise possible due to gravity, preferential flow, or "normal" infiltration.

Plants may use hydraulic redistribution to enhance survival and growth through benefits associated with increased soil moisture, particularly improved water status. During drought, for example, plants with dimorphic root systems (i.e., shallow lateral and deep taproots) can hydraulically lift water at night and then reabsorb it during the day. This can limit cavitation in fine roots and partially offset reductions in transpiration caused by a lack of soil moisture near the surface (Domec et al. 2004). The actual volume of water redistributed is often small, but for some tree species it can be relatively large, representing a substantial portion of the total amount of water transpired during the day, upward of 20% (Brooks et al. 2002). Redistributed water may also be useful for maintaining viable fine roots (Espeleta et al. 2004) and promoting root growth in soil layers that would otherwise be too dry to explore (Ryel et al. 2003), and redistribution may also benefit plants by increasing nutrient availability and microbial activity in dry soil (McCulley et al. 2004). Hydraulic redistribution can affect competition for water among neighboring plants, playing either a positive role, by increasing the amount of water available to be shared among competitors (Zou et al. 2005), or a negative role, by transferring water away from competitors (Hultine et al. 2003). Finally, at larger scales, there is growing evidence to suggest that hydraulic redistribution may play an important role in water, carbon, and nutrient cycling at the ecosystem scale (Jackson et al. 2000), and it may even affect the climate of densely vegetated ecosystems, such as the Amazon rainforests (Lee et al. 2005).

ROOT SYMBIOSES

There are numerous symbioses in nature. Arguably two of the most important for higher plants are mycorrhizae and symbiotic nitrogen fixation. Mycorrhizae (literally "fungus-root") are mutualistic associations between plant roots and soil fungi. In general, the fungus acts as a fine root matrix, exploring the soil and transporting nutrients and water back to the plant. In return, the plant typically supplies the fungus with a dependable carbon source. Fossil evidence indicates that mycorrhizae are as old as vascular land plants; they have been hypothesized as necessary for plant colonization of land in "soils" that would have been extremely nutrient poor (Pirozynski and Mallock 1975). Excellent introductory texts for the functioning and importance of mycorrhizae are Harley and Smith (1983) and Allen (1991).

Mycorrhizae are critical components of the rhizosphere (in all terrestrial biomes, and >90% of forest and grassland species are typically mycorrhizal [Brundrett 1991]). The direct and indirect benefits of mycorrhizae to plants include increased phosphorus status and, to a lesser extent, improved nitrogen and water uptake (e.g., Fitter 1989). Mycorrhizae have been shown to supply 80% of the P taken up by plants and 25% of the N (e.g., Marschner and Dell 1984). Increased resource uptake occurs primarily through the greater efficiency with which fungi explore the soil than plant roots, but there is also evidence that the fungi mobilize resources otherwise unavailable to the plant (see below). Such potential benefits come with pronounced carbon costs, estimated as 15% of net primary productivity for ectomycorrhizae (Vogt et al. 1982, for a Pacific fir forest) and 7%–17% of the energy translocated to roots for arbuscular mycorrhizae (e.g., Harris and Paul 1987).

There are four broad classes of mycorrhizae: arbuscular mycorrhizae (AM), ectomycorrhizae (ECM), ericoid mycorrhizae, and orchid mycorrhizae. Arbuscular mycorrhizae are a type of endomycorrhizae, with hyphae, arbuscules (exchange organs), and often storage vesicles produced within the cortex of root cells. Fungi in AM are limited to Zygomycetes. Only 150 or so fungal species have been shown to participate with vascular plants in AM, so there is little specificity between fungi and host plant (though this apparent lack of specificity may also reflect present limitations in fungal taxonomy). Ectomycorrhizae are characteristic of certain woody plants, particularly those in the pine, willow, and beech families. They are characterized by a mantle or hyphal sheath around highly branched roots and by a hyphal network called the Hartig net that grows between cortical root cells. In contrast to AM, ECM do not generally penetrate living root cells. There are thousands of fungal species that form ECM, usually Basidiomycetes, and there is much greater host-fungus specificity than for AM.

By far the majority of the world's mycorrhizal associations are AM or ECM. The absence of mycorrhizae is apparently limited primarily to early successional annuals of such relatively "advanced" families as the Brassicaceae, Chenopodiacea, Amaranthaceae, and Zygophyllaceae (Allen 1991). Two other less prevalent but important forms of mycorrhizae are those of plants in the Ericaceae and Orchidaceae. Ericaceous mycorrhizae are intermediate between AM and ECM. They sometimes form a sheath around the root but they also penetrate the outer root cortical cells with hyphal coils. Ericaceous plants often grow in acidic, nutrient-poor soils such as heathlands. The fungus can constitute 80% of their extensive mycorrhizal association by mass (Raven et al. 1986). These specialized, expensive mycorrhizae seem to be particularly beneficial for mobilizing and taking up organic nitrogen that would otherwise be unavailable to the plant (Read 1983). Orchids also have a unique mycorrhizal relationship. Their seeds will not germinate in nature in the absence of the mycorrhizal fungus, typically a Basidiomycete in such genera as *Rhizoctonia*. Orchids also have an early nonphotosynthetic stage of growth where carbon is obtained solely from the mycorrhizal fungus. Mature orchids receive more typical resources such as nutrients and water from mycorrhizae.

A second important root symbiosis is the fixation of nitrogen by plants and bacteria. Biological nitrogen fixation adds approximately 150×10^{12} g N to terrestrial ecosystems each year, roughly 1 g m^{-2} on average for all of the earth's land (e.g., Burns and Hardy 1975, Vitousek et al. 1997). Functionally the reciprocal benefits of symbiotic N fixation are similar to those for mycorrhizae; the plant supplies carbon and energy to the symbiont and the symbiont supplies a resource to the plant. The two symbioses differ in that biological fixation supplies only nitrogen and the symbiont is a bacterium rather than a fungus. In addition to sugars, the plant also supplies an environment conducive to N fixation in which bacterial enzymes are protected from atmospheric oxygen and light. A two-protein, enzyme complex called nitrogenase catalyzes the fixation of N through the following chemical reaction:

$$N_2 + 8H^+ + 6e^- + XATP \rightarrow 2NH_4^+ + XADP + XP_i,$$

where the value for the number of ATPs, ADPs, and P_is is approximately 15. The ammonium is converted to amino acids or ureides before it is transported to the plant.

The most important N-fixing bacteria are those in the genus *Rhizobium*, which colonize the roots of legumes. The bacterium enters the root through root hairs that curl in response to a chemical secreted by the bacterium. A bacterial infection thread then penetrates the root cortex and induces formation of the nodule where N fixation occurs. A second important group of symbionts are actinomycete bacteria; they fix N as in roots of such plant genera as *Alnus*, *Myrica*, and *Ceanothus*.

In addition to the symbioses of *Rhizobium*, actinomycetes, and higher plants, there are also rhizosphere bacteria such as *Azospirillum*, which fix nitrogen on the surface of roots of many plant species. It is debatable whether these relationships constitute a true symbiosis. There is clearly not the tight coupling that is seen with *Rhizobium* and legumes, and the quantities of N fixed are much smaller. Nevertheless, this source of N is important for a number of grasses and crop species.

GLOBAL PATTERNS OF ROOT DISTRIBUTIONS

Despite tremendous natural variation in the soil around plants and in the plasticity individual plants show to such variation (Snaydon 1962, Jackson and Caldwell 1993), there are broad patterns in the distributions of roots observed for biomes and plant functional types. Jackson et al. (1996, 1997) constructed a global database of climate, soil, and root attributes from more than 250 literature studies. The data were separated by terrestrial biomes and plant functional types and fitted to an asymptotic model of vertical root distribution:

$$Y = 1 - \beta^d,$$

where Y is the cumulative root fraction (a proportion between 0 and 1) from the soil surface to depth d (in cm) and β is the fitted extinction coefficient (Gale and Grigal 1987). β provides a simple numerical index of rooting distributions. Low β values (e.g., 0.90) correspond to proportionally more roots near the surface than do high β values (e.g., 0.98).

Examining the data for plant functional types, typical β values were 0.957 for grasses and 0.972 and 0.980 for trees and shrubs, respectively (Jackson et al. 1996). Cumulative root biomass in the top 30 cm of soil varied from 75% for a typical grass to 45% for an average shrub. Shrubs had 87% of their root biomass in the top meter of soil while grasses had almost 99%. These estimates for total root biomass are slightly deeper than those in Jackson et al. (1996) because only profiles to approximately 2 m soil depth or greater were used in this revised analysis. Terrestrial biomes also showed clear patterns for root distributions. Deserts, temperate coniferous forests, and savannas had some of the deepest distributions ($0.970 \leq \beta \leq 0.980$), while tundra, boreal forests, and temperate grasslands the shallowest profiles ($\beta = 0.913$, 0.943, and 0.943, respectively). Tundra typically had 60% of roots in the upper 10 cm of soil while deserts had approximately 20% of roots in the same depth increment.

The database was also used to examine global patterns of root biomass and annual belowground net primary production (NPP) (Jackson et al. 1996, 1997). Average root biomass ranged from <1 kg m^{-2} for deserts and croplands to 4–5 kg m^{-2} for most forest systems. When root biomass estimates were combined with the extent of each biome, global root biomass was estimated to be approximately 290×10^{15} g (or 140×10^{15} g C, equivalent to approximately 20% of atmospheric C). The standing biomass of live and dead fine roots (<2 mm diameter) was approximately 80×10^{15} g. Assuming conservatively that fine roots turn over once per year on average, they represent one-third of global NPP for plants, approximately 20×10^{15} g C $year^{-1}$ (Jackson et al. 1997). The average C:N:P ratio was 450:11:1 for fine roots and 850:11:1 for more coarse roots ($2 \leq x \leq 5$ mm).

The upper meter of soil contains the majority of root biomass in most systems. Nevertheless, what constitutes the functional rooting depth of an ecosystem is an important and more difficult question. For woody plants much of surface root biomass is in large-diameter roots that play a strong role in anchoring and transport but not in resource uptake. Furthermore, even where fine root biomass distributions are known with depth, root functioning is often not proportional to root biomass. As an example, Gregory et al. (1978) showed that winter wheat had only 3% of its root system by mass below 1 m soil depth; this small fraction of roots supplied almost 20% of the water transpired by the wheat canopy during midsummer. The importance of relatively deep roots may frequently be underestimated because few studies examine root abundance and functioning below 1 m soil depth (Jackson et al. 1996). Some of the root distributions estimated above are undoubtedly too shallow. There are a number of other uncertainties in this type of analysis, particularly seasonal and spatial dynamics that are masked by pooling data across space and time.

Although 2 or 3 m is "deep" for the typical ecological study, roots clearly grow much deeper. Of the 255 species examined for maximum rooting depth by Canadell et al. (1996), almost 10% grew roots below 10 m depth. At least eight woody species have been shown to grow roots below 40 m (Table 5.1). The functional significance of such roots can be profound. In the Brazilian cerrado and in Amazonian rainforests, roots have been found at least 18 m deep in the soil (Rawitscher 1948, Nepstad et al. 1994). More than 75% of transpired water in

TABLE 5.1
The Ten Deepest Records for Rooting Depth

Species	System	Maximum Rooting Depth (m)	Reference
Boscia albitrunca	Kalahari desert	68	Jennings (1974)
Juniperus monosperma	Colorado plateau	61	Cannon (1960)
Eucalyptus sp.	Australian forest	61	Jennings (1971)
Acacia erioloba	Kalahari desert	60	Jennings (1974)
Prosopis juliflora	Arizona desert	54	Phillips (1963)
Eucalyptus calophylla	Australian forest	45	Campion (1926)
Medicago sativa	Agricultural field	40	Meinzer (1927)
Eucalyptus marginata	Jarrah forest	40	Dell et al. (1983)
Acacia raddiana	Niger desert	35	Anonymous (1974)
Quercus douglassii	California woodland	24	Lewis and Burgy (1964)

Sources: Generated in part from Stone, E., Kalisz, P., *Forest Ecol. Manage.*, 46, 59, 1991. With permission; From Canadell, J., Jackson, R.B., Ehleringer, J.R., Mooney, H.A., Sala, O.E., Schulze, E.-D., *Oecologia*, 108, 583, 1996. With permission.

these systems can come from below 2 m soil depth, particularly during the dry season (Nepstad et al. 1994). Where resources, particularly water, are available at depth in the soil, deep roots can be disproportionately important for resource uptake. The depth to which roots grow and their distribution in the soil are also likely to be quite important for the maintenance of leaf area in evergreen systems and in determining the boundaries between evergreen and deciduous vegetation.

CONCLUDING REMARKS

Roots function in anchoring the plant, capturing soil resources, sensing and modifying the environment, and storing reserves. The physical difficulties in studying roots and the soil represent both a frustrating barrier and a tremendous opportunity for the creative scientist. There are numerous unanswered questions on root functioning in need of a novel approach (Jackson 1998). What is the relationship of root biomass to root functioning? In what systems are the phenologies of roots and shoots tightly coupled? Is the distribution of microbes and soil fauna in the soil coupled to plant rooting depth? How prevalent are deep roots and are they important only for water uptake? Do general patterns exist for such questions globally?

Fortunately, such tools as minirhizotrons and stable isotopes are improving our understanding of the turnover and functioning of root systems. Future progress will also likely be made by combining new approaches in ecological studies with the mechanistic insights of molecular biology. For example, Zhang and Forde (1998) recently showed that a nitrate-inducible gene in *Arabidopsis* (ANR1) is a key determinant of developmental plasticity in roots. Perhaps most of all, it is important to remember that the functioning of roots is intimately linked to that of shoots. This integration occurs through the mutual exchange of resources and plant signals. Attempts to understand root or shoot functioning must eventually take this integration into account.

ACKNOWLEDGMENTS

This chapter is reproduced unchanged from the first edition except for the addition of a new Section "Releasing exudates and modifying the environment."

We wish to thank the National Science Foundation, the Andrew W. Mellon Foundation, and the Department of Energy's National Institute for Climate Change Research for support of this work. C. Armas is supported by a Fulbright fellowship from the Spanish Government (FU2005-0282). L. Anderson, C. Bilbrough, H. de Kroon, and E. Jabbágy provided helpful comments on the manuscript.

REFERENCES

Aerts, R., R. Boot, and J. van der Aart, 1991. The relation between above- and belowground biomass allocation patterns and competitive ability. Oecologia 87: 551–559.

Allen, M., 1991. The Ecology of Mycorrhizae. Cambridge University Press, Cambridge.

Anonymous, 1974. L'Arbre du Ténéré est most. Bois et Forêts des Tropiques 153: 61–65.

Bais, H.P., T.S. Walker, F.R. Stermitz, R.A. Hufbauer, and J.M. Vivanco, 2002. Enantiomeric-dependent phytotoxic and antimicrobial activity of (±)-catechin; a rhizosecreted racemic mixture from *Centaurea maculosa* (spotted knapweed). Plant Physiology 128: 1173–1179.

Bais, H.P., R. Vepachedu, S. Gilroy, R.M. Callaway, and J.M. Vivanco, 2003. Allelopathy and exotic plant invasion: From molecules and genes to species interactions. Science 301: 1377–1380.

Bais, H.P., S.W. Park, T.L. Weir, R.M. Callaway, and J.M. Vivanco, 2004. How plants communicate using the underground information superhighway. Trends in Plant Science 9: 26–32.

Barber, S., 1984. Soil Nutrient Bioavailability. John Wiley, New York.

Bell, T., J. Pate, and K. Dixon, 1996. Relationship between fire response, morphology, root anatomy, and starch distribution in south-west Australian Epacridaceae. Annals of Botany 77: 357–364.

Bengough, A.G. and B.M. McKenzie, 1997. Sloughing of root cap cells decreases the frictional resistance to maize (*Zea mays* L) root growth. Journal of Experimental Botany 48: 885–893.

Berendse, F., H. de Kroon, and W.G. Braakhekke, 1999. Acquisition, use and loss of nutrients. In: F.I. Pugnaire and F. Valladares, eds. Handbook of Functional Plant Ecology. Marcel Dekker, New York, pp. 315–345.

Bieleski, R., 1973. Phosphate pools, phosphate transport, and phosphate availability. Annual Review of Plant Physiology 24: 225–252.

Bilbrough, C. and M. Caldwell, 1995. The effects of shading and N status on root proliferation in nutrient patches by the perennial grass *Agropyron desertorum* in the field. Oecologia 103: 10–16.

Bole, J., 1973. Influence of root hairs in supplying soil phosphorus to wheat. Canadian Journal of Soil Science 53: 169–175.

Bowen, B. and J. Pate, 1993. The significance of root starch in post-fire recovery of the resprouter *Stiringia latifolia* R.Br. (Proteaceae). Annals of Botany 72: 7–16.

Brady, J., 1934. Some factors influencing lodging in cereals. The Journal of Agricultural Science 72: 273–280.

Brooks, J.R., F.C. Meinzer, R. Coulombe, and J. Gregg, 2002. Hydraulic redistribution of soil water during summer drought in two contrasting Pacific Northwest coniferous forests. Tree Physiology 22: 1107–1117.

Brouwer, R., 1963. Some aspects of the equilibrium between overground and underground plant parts. Jaarb Int Bio Scheik Onderz Landbgewass 1963: 31–39.

Brouwer, R., O. Gasparíková, J. Kolek, and B. Loughman, 1981. Structure and Function of Plant Roots. Martinus Nijhoff, The Hague.

Brundrett, M., 1991. Mycorrhizas in natural ecosystems. Advances in Ecological Research 21: 171–313.

Burgess, S.S.O., M.A. Adams, N.C. Turner, and C.K. Ong, 1998. The redistribution of soil water by tree root systems. Oecologia 115: 306–311.

Burgess, S.S.O., M.A. Adams, N.C. Turner, D.A. White, and C.K. Ong, 2001. Tree roots: Conduits for deep recharge of soil water. Oecologia 126: 158–165.

Burns, R. and R. Hardy, 1975. Nitrogen Fixation in Bacteria and Higher Plants. Springer-Verlag, Berlin.

Busso, C., J. Richards, and N. Chatterton. 1990. Nonstructural carbohydrates and spring regrowth of two cool-season grasses: Interaction of drought and clipping. Journal of Range Management 43: 336–343.
Caldwell, M.M. and J.H. Richards, 1986. Competing root systems: morphology and models of absorption. In Givnish, T.J., ed. On the Economy of Plant Form and Function. Cambridge University Press, London, pp. 251–273.
Caldwell, M.M., T.E. Dawson, and J.H. Richards, 1998. Hydraulic lift: Consequences of water efflux from the roots of plants. Oecologia 113: 151–161.
Callaway, R.M. and E.T. Aschehoug, 2000. Invasive plants versus their new and old neighbors: A mechanism for exotic invasion. Science 290: 521–523.
Campion, W.E., 1926. The depth attained by roots. Australian Journal of Forestry 9: 128.
Canadell, J., R.B. Jackson, J.R. Ehleringer, H.A. Mooney, O.E. Sala, and E.-D. Schulze, 1996. Maximum rooting depth of vegetation types at the global scale. Oecologia 108: 583–595.
Cannon, H.L., 1960. The development of botanical methods of prospecting for uranium on the Colorado Plateau. U.S. Geological Survey, Bulletin 1085-A, Washington D.C., 50 pp.
Casper, B. and R. Jackson, 1997. Plant competition underground. Annual Review of Ecology and Systematics 28: 545–570.
Chapin, F.I., 1980. The mineral nutrition of wild plants. Annual Review of Ecology and Systematics 11: 233–260.
Chapin, F.I., J. Kendrick, and D. Johnson, 1986. Seasonal changes in carbon fractions in Alaskan tundra plants of differing growth forms: Implications for herbivory. Journal of Ecology 74: 707–731.
Chapin, F.S.I., E.D. Schulze, and H.A. Mooney, 1990. The ecology and economics of storage in plants. Annual Review of Ecology and Systematics 21: 423–447.
Chomba, B.M., R. Guy, and H. Werger, 1993. Carbohydrate reserve accumulation and depletion in Engelman spruce (*Picea engelmannii* Parry): Effects of cold storage and pre-storage CO_2 enrichment. Tree Physiology 13: 351–364.
Clarkson, D.T., 1993. Roots and the delivery of solutes to the xylem. Philosophical Transactions of the Royal Society London B 341: 5–17.
Crook, M. and A. Ennos, 1996. The anchorage mechanics of deep rooted larch, *Larix europea* x *L. japonica*. Journal of Experimental Botany 47: 1509–1517.
Daer, T. and E. Willard, 1981. Total nonstructural carbohydrate trends in bluebunch wheatgrass related to growth and physiology. Journal of Range Management 34: 377–379.
Davies, P., 1995. Plant Hormones: Physiology, Biochemistry and Molecular Biology. Kluwer, Dordrecht.
Davies, W. and H. Jones, 1991. Abscisic Acid: Physiology and Biochemistry. Bios Scientific, Oxford.
Davies, W. and J. Zhang, 1991. Root signals and the regulation of growth and development of plants in drying soil. Annual Review of Plant Physiology and Molecular Biology 42: 55–76.
Davis, T. and B. Haissig, 1994. Biology of Adventitious Root Formation. Plenum Press, New York.
Dawson, T.E., 1993. Hydraulic lift and water use by plants: Implications for water balance, performance and plant-plant interactions. Oecologia (Heidelberg) 95: 565–574.
de Kroon, H. and R. Bobbink, 1997. Clonal plant dominance under elevated nitrogen deposition, with special reference to *Brachypodium pinnatum* in chalk grassland. In: H. de Kroon and J. van Groenendael, eds. The Ecology and Evolution of Clonal Plants. Backhuys Publishers, Leiden, pp. 359–379.
Dell, B., J.R. Bartle, and W.H. Tracey, 1983. Root occupation and root channels of jarrah forest subsoils. Australian Journal of Botany 31: 615–627.
Dickson, R., 1991. Assimilate distribution and storage. In Raghavendra, A., ed. Physiology of Trees. John Wiley & Sons, New York, pp. 51–85.
Dittmer, H., 1937. A quantitative study of the roots and root hairs of a winter rye plant (*Secale cereale*). American Journal of Botany 24: 417–419.
Domec, J.C., J.M. Warren, F.C. Meinzer, J.R. Brooks, and R. Coulombe, 2004. Native root xylem embolism and stomatal closure in stands of Douglas-fir and ponderosa pine: Mitigation by hydraulic redistribution. Oecologia 141: 7–16.

Donald, C., 1958. The interaction of competition for light and nutrients. Australian Journal of Agricultural Research 9: 421–435.

Drew, M., 1975. Comparison of the effects of a localized supply of phosphate, nitrate, ammonium and potassium on the growth of the seminal root system. New Phytologist 75: 479–490.

Drew, M., 1997. Oxygen deficiency and root metabolism: injury and acclimation under hypoxia and anoxia. Annual Review of Plant Physiology and Plant Molecular Biology 48: 223–250.

Drew, M. and L. Saker, 1975. Nutrient supply and the growth of the seminal root system in barley. II. Localized, compensatory increases in lateral root growth and rates of nitrate uptake when nitrate supply is restricted to only part of the root system. Journal of Experimental Botany 26: 79–90.

Duncan, W. and A. Ohlrogge, 1958. Principles of nutrient uptake from fertilizer bands II. Root development in the band. Agricultural Journal 50: 605–608.

Esau, K., 1977. Anatomy of Seed Plants, 2 edn. John Wiley & Sons, New York.

Espeleta, J.F., J.B. West, and L.A. Donovan, 2004. Species-specific patterns of hydraulic lift in co-occurring adult trees and grasses in a sandhill community. Oecologia 138: 341–349.

Estabrook, E.M. and J.L. Yoder, 1998. Plant–plant communications: rhizosphere signaling between parasitic angiosperms and their hosts. Plant Physiology 116: 1–7.

Falik, O., P. Reides, M. Gersani, and A. Novoplansky, 2005. Root navigation by self inhibition. Plant, Cell and Environment 28: 562–569.

Fitter, A., 1989. The role and ecological significance of vesicular-arbuscular mycorrhizas in temperate ecosystems. Agriculture, Ecosystems and Environment 29: 137–151.

Fitter, A., 1994. Architecture and biomass allocation as components of the plastic response of root systems to soil heterogeneity. In: M. Caldwell, and R. Pearcy, eds. Exploitation of Environmental Heterogeneity by Plants. Academic Press, San Diego, CA, pp. 305–323.

Flores, H.E., J.M. Vivanco, and V.M. Loyola-Vargas, 1999. "Radicle" biochemistry: The biology of root-specific metabolism. Trends in Plant Science 4: 220–226.

Gale, M. and D. Grigal, 1987. Vertical root distributions of northern tree species in relation to successional status. Canadian Journal of Forest Research 17: 829–834.

Glerum, C. and J. Balatinecz, 1980. Formation and distribution of food reserves during autumn and their subsequent utilization in jack pine. Canadian Journal of Botany 58: 40–54.

Glinski, J. and J. Lipiec, 1990. Soil Physical Conditions and Plant Roots. CRC Press Inc., Boca Raton, FL.

Gollan, T., J.B. Passioura, and R. Munns, 1986. Soil water status affects the stomatal conductance of fully turgid wheat and sunflower leaves. Australian Journal of Plant Physiology 13: 459–464.

Gregory, P., M. McGowan, and P. Biscoe, 1978. Water relations of winter wheat. 2. Soil water relations. The Journal of Agricultural Science 91: 103–116.

Griffin, G.J., M.G. Hale, and F.J. Shay, 1976. Nature and quantity of sloughed organic matter produced by roots of axenic peanut plants. Soil Biology and Biochemistry 8: 29–32.

Hales, S., 1727. Vegetable Staticks. London Scientific Book Guild, London.

Harley, J. and S. Smith, 1983. Mycorrhizal Symbiosis. Academic Press, New York.

Harris, D. and E. Paul, 1987. Carbon requirements of vesicular-arbuscular mycorrhizae. In: G. Safir, ed. Ecophysiology of VA Mycorrhizal Plants. CRC Press, Boca Raton, FL, pp. 93–105.

Hawes, M.C., U. Gunawardena, S. Miyasaka, and X. Zhao, 2000. The role of root border cells in plant defense. Trends in Plant Science 5: 128–133.

Hofer, R.-M., 1991. Root Hairs. In: Y. Waisel, A. Eshel, and U. Kafkafi, eds. Plant Roots: The Hidden Half. Marcel Dekker Inc., New York, pp. 129–148.

Hultine, K.R., D.G. Williams, S.S.O. Burgess, and T.O. Keefer, 2003. Contrasting patterns of hydraulic redistribution in three desert phreatophytes. Oecologia 135: 167–175.

Hultine, K.R., R.L. Scott, W.L. Cable, D.C. Goodrich, and D.G. Williams, 2004. Hydraulic redistribution by a dominant, warm-desert phreatophyte: seasonal patterns and response to precipitation pulses. Functional Ecology 18: 530–538.

Hutchings, M. and H. de Kroon, 1994. Foraging in plants: the role of morphological plasticity in resource acquisition. Advances in Ecological Research 25: 159–238.

Inderjit and L.A. Weston, 2003. Root exudates: An Overview. In: H. de Kroon and E.J.W. Visser, eds. Root Ecology. Springer-Verlag, Berlin, pp. 235–255.

Itoh, S. and S. Barber, 1983. Phosphorus uptake by six plant species as related to root hairs. Agronomy Journal 75: 457–461.

Jackson, R., 1998. The importance of deep roots for hydrology, biogeochemistry, and ecosystem functioning. In: J. Tenhunen and P. Kabat, eds. Integrating Hydrology, Ecosystem Dynamics and Biogeochemistry in Complex Landscapes. John Wiley & Sons, Chichester, pp. 219–240.

Jackson, R. and M. Caldwell, 1992. Shading and the capture of localized soil nutrients: nutrient contents, carbohydrates, and root uptake kinetics of a perennial tussock grass. Oecologia 91: 457–462.

Jackson, R.B. and M.M. Caldwell, 1993. Geostatistical patterns of soil heterogeneity around individual perennial plants. Journal of Ecology 81: 683–692.

Jackson, R.B., J. Manwaring, and M. Caldwell, 1990. Rapid physiological adjustment of roots to localized soil enrichment. Nature 344: 58–60.

Jackson, R.B., J. Canadell, J.R. Ehleringer, H.A. Mooney, O.E. Sala, and E.-D. Schulze, 1996. A global analysis of root distributions for terrestrial biomes. Oecologia 108: 389–411.

Jackson, R., H. Mooney, and E.-D. Schulze, 1997. A global budget for fine root biomass, surface area, and nutrient contents. Proceedings of the National Academy of Sciences, USA 94, pp. 7362–7366.

Jackson, R.B., J.S. Sperry, and T.E. Dawson, 2000. Root water uptake and transport: Using physiological processes in global predictions. Trends in Plant Science 5: 482–488.

Jenner, C., 1992. Storage of Starch. In: F. Loewus and W. Tanner, eds. Plant Carbohydrates I: Intracellular Carbohydrates. Springer-Verlag, Berlin, pp. 700–747.

Jennings, C.M.H., 1974. The hydrology of Botswana. University of Natal, South Africa.

Jennings, J.N., 1971. Karst. MIT, Cambridge.

Kays, J. and C. Canham, 1991. Effects of time and frequency of cutting on hardwood root reserves and sprout growth. Forest Science 37: 524–539.

Keller, J. and H. Loescher, 1989. Nonstructural carbohydrate partitioning in perennial parts of sweet cherry. Journal of the American Society for Horticultural Science 114: 969–975.

Keyes, W.J., R.C. O'Malley, D. Kim, and D.G. Lynn, 2000. Signaling organogenesis in parasitic angiosperms: Xenognosin generation, perception, and response. Journal of Plant Growth Regulation 19: 217–231.

Kramer, P., 1969. Plant and Soil Water Relationships. McGraw-Hill, New York.

Kramer, P.J., 1983. Water relations of plants. Academic Press, New York.

Kramer, P.J., 1988. Changing concepts regarding plant water relations. Plant Cell and Environment 11: 565–568.

Kramer, P. and J. Boyer, 1995. Water relations of plants and soils. Academic Press, San Diego, CA.

Lee, J.E., R.S. Oliveira, T.E. Dawson, and I. Fung, 2005. Root functioning modifies seasonal climate. Proceedings of the National Academy of Sciences, USA 102, pp. 17576–17581.

Leffler, A.J., M.S. Peek, R.J. Ryel, C.Y. Ivans, and M.M. Caldwell. 2005. Hydraulic redistribution through the root systems of senesced plants. Ecology 86: 633–642.

Lewis, D.H., 1984. Occurrence and distribution of storage carbohydrates in vascular plants. In: D. Lewis, ed. Storage Carbohydrates in Vascular Plants. Cambridge University Press, Cambridge, pp. 1–52.

Lewis, D.C. and R.H. Burgy, 1964. The relationship between oak tree roots and groundwater in fractured rock as determined by tritium tracing. Journal of Geophysical Research 69: 2579–2588.

Loescher, W., T. McCamant, and J. Keller, 1990. Carbohydrate reserves, translocation, and storage in woody plant roots. Hortscience 25: 274–281.

Lyford, W. and B. Wilson, 1964. Development of the root system of *Acer rubrum* L. USDA Forest Service, Harvard Forest Paper 10, Patersham, MA.

Mahall, B.E. and R.M. Callaway, 1991. Root communication among desert shrubs. Proceedings of the National Academy of Science USA 88, pp. 874–876.

Marschner, H. 1995. Mineral Nutrition of Higher Plants, 2 edn. Academic Press, London.

Marschner, H. and B. Dell, 1984. Nutrient uptake in mycorrhizal symbiosis. Plant and Soil 159: 89–102.

Mattheck, C., 1991. Trees: The Mechanical Design. Springer-Verlag, Berlin.

Mauseth, J.D., 1988. Plant Anatomy. Benjamin/Cummings Publishing Co., Menlo Park.

Mauseth, J., 1995. Botany: An Introduction to Plant Biology. Saunders College, Philadelphia, PA.

McCully, M., 1995. How do real roots work? Some new views of root structure. Plant Physiology 109: 1–6.

McCully, M.E. and M.J. Canny, 1988. Pathways and processes of water transport and nutrient movement in roots. Plant and Soil 111: 159–170.

McCulley, R.L., E.G. Jobbagy, W.T. Pockman, and R.B. Jackson, 2004. Nutrient uptake as a contributing explanation for deep rooting in arid and semi-arid ecosystems. Oecologia 141: 620–628.

McDonald, A., T. Ericsson, and T. Ingestad, 1991. Growth and nutrition of tree seedlings. In: A. Raghavendra, ed. Physiology of Trees. John Wiley & Sons, New York, pp. 199–220.

Meinzer, O.E., 1927. Plants as indicators of ground water. U.S. Geological Survey, Water Supply Paper 577, Washington D.C., pp. 43–54.

Miyanishi, K. and M. Kellman, 1986. The role of root nutrient reserves in regrowth of two savanna shrubs. Canadian Journal of Botany 64: 1244–1248.

Nepstad, D.C., C.R. de Carvalho, E.A. Davidson, P.H. Jipp, P.A. Lefebvre, G.H. Negreiros, E.D. da Silva, T.A. Stone, S.E. Trumbore, and S. Vieira, 1994. The role of deep roots in the hydrological and carbon cycles of Amazonian forests and pastures. Nature 372: 666–669.

Nicoll, B. and D. Ray, 1996. Adaptive growth of tree root systems in response to wind action and site conditions. Tree Physiology 16: 891–898.

Nye, P. and P. Tinker, 1977. Solute Movement in the Soil-Root System. Blackwell, Berkeley, CA.

Oesterheld, M. and S. McNaughton, 1988. Intraspecific variation in the response of *Themeda triandra* to defoliation: the effect of time of recovery and growth rates on compensatory growth. Oecologia 77: 181–186.

Ohkuma, K., J. Lyon, F. Addicott, and O. Smith, 1963. Abscisin II, an abscission-accelerating substance from young cotton fruit. Science 142: 1592–1593.

Orodo, A. and M. Trlica, 1990. Clipping and long-term grazing effects on biomass and carbohydrates of Indian ricegras. Journal of Range Management 43: 52–57.

Owensby, C., G. Paulsen, and J. McKendrick, 1970. Effect of burning and clipping on big bluestem reserve carbohydrates. Journal of Range Management 23: 358–360.

Pate, J.S. and K.W. Dixon, 1982. Tuberous, Cormous and Bulbous Plants. University of Western Australia Press, Nedlands.

Perumalla, C., C. Peterson, and D. Enstone, 1990. A survey of angiosperm species to detect hypodermal Casparian bands. I. Roots with a uniseriate hypodermis and epidermis. Botanical Journal of the Linnean Society 103: 93–112.

Peters, N.K., J.W. Frost, and S.R. Long, 1986. A plant flavone, luteolin, induces expression of *Rhizobium meliloti* nodulation genes. Science 233: 977–980.

Peterson, C.A. and C.J. Perumalla, 1990. A survey of angiosperm species to detect hypodermal Casparian bands: II. Roots with a multiseriate hypodermis or epidermis. Botanical Journal of the Linnean Society 103: 113–125.

Peterson, R. and C. Peterson, 1986. Ontogeny and anatomy of lateral roots. In: M. Jackson, ed. New Root Formation in Plants and Cuttings. Martinus Nijhoff Publishers, Dordrecht, pp. 1–30.

Phillips, W.S., 1963, Depth of roots in soil. Ecology 44: 424.

Pirozynski, K. and D. Mallock, 1975. The origin of land plants: a matter of mycotrophism. Biosystems 6: 1533–1164.

Pollock, C., 1986. Fructans and the metabolism of sucrose in vascular plants. New Phytologist 104: 1–24.

Pregitzer, K. and R. Hendrick, 1993. The demography of fine roots in response to patches of water and nitrogen. New Phytologist 125: 575–580.

Raven, P., R. Evert, and S. Eichhorn, 1986. Biology of Plants, 4 edn. Worth, New York.

Rawitscher, F., 1948. The water economy of the vegetation of the "Campos Cerrados" in southern Brazil. Journal of Ecology 36: 237–268.

Read, D., 1983. The biology of mycorrhiza in the Ericales. Canadian Journal of Botany 61: 985–1004.

Reichle, D., B. Dinger, N. Edwards, W. Harris, and P. Sollins, 1973. Carbon flow and storage in a forest ecosystem. In: G. Woodwell and E. Pecan, eds. Carbon and the Biosphere. US Atomic Energy Commission, Washington D.C., pp. 345–365.

Robinson, D., 1994. The responses of plants to non-uniform supplies of nutrients. New Phytologist 127: 635–674.

Rougier, M., 1981. Secretory activity at the root cap. In: W. Tanner and F.A. Loews, eds. Encyclopedia of Plant Physiology. Springer Verlag, New York, pp. 542–574.

Ryel, R.J., 2004. Hydraulic Redistribution. In: Progress in Botany. Springer-Verlag, Berlin, pp. 413–435.

Ryel, R.J., M.M. Caldwell, A.J. Leffler, and C.K. Yoder, 2003. Rapid soil moisture recharge to depth by roots in a stand of *Artemisia tridentata*. Ecology 84: 757–764.

Salisbury, F. and C. Ross, 1985. Plant Physiology, 3rd edn. Wadsworth, Belmont.

Smart, D.R., E. Carlisle, M. Goebel, and B.A. Nunez, 2005. Transverse hydraulic redistribution by a grapevine. Plant Cell and Environment 28: 157–166.

Snaydon, R., 1962. Micro-distribution of *Trifolium repens* L and its relation to soil factors. Journal of Ecology 50: 133–143.

Somerville, A., 1979. Root anchorage and root morphology of *Pinus radiata* on a range of ripping treatments. New Zealand Journal of Forestry Science 9: 294–315.

Staswick, P., 1994. Storage proteins in vegetative plant tissues. Annual Review of Plant Physiology and Plant Molecular Biology 45: 303–322.

Steudle, E., 1994. Water transport across roots. Plant and Soil 167: 79–90.

Steudle, E. and A.B. Meshcheryakov, 1996. Hydraulic and osmotic properties of oak roots. Journal of Experimental Botany 47: 387–401.

Stokes, A., J. Ball, A. Fitter, P. Brain, and M. Coutts, 1996. An experimental investigation of the resistance of model root systems to uprooting. Annals of Botany 78: 415–421.

Stone, E. and P. Kalisz, 1991. On the maximum extent of tree roots. Forest Ecology and Management 46: 59–102.

Taiz, L. and E. Zeiger, 1991. Plant Physiology. Benjamin/Cummings, Redwood City.

Tardieu, F. and W.J. Davies, 1993. Integration of hydraulic and chemical signalling in the control of stomatal conductance and water status of droughted plants. Plant Cell and Environment 16: 341–349.

Tardieu, F., N. Katerji, O. Bethenod, J. Zhang, and W. Davies, 1991. Maize stomatal conductance in the field: its relationship with soil and plant water potentials, mechanical constraints and ABA concentration in the xylem sap. Plant Cell and Environment 14: 121–126.

Tardieu, F., J. Zhang, N. Katerji, O. Bethenod, S. Palmer, and W.J. Davies, 1992. Xylem ABA controls the stomatal conductance of field-grown maize subjected to soil compaction or soil drying. Plant Cell and Environment 15: 193–197.

Torrey, J., 1976. Root hormones and plant growth. Annual Review of Plant Physiology 27: 435–459.

Trewavas, A., 1981. How do plant growth substances work? Plant Cell and Environment 4: 203–228.

Trieu, A.T., M.L. van Buuren, and M.J. Harrison, 1997. Gene expression in mycorrhizal roots of Medicago truncatula. In: H.E. Flores, J.P. Lynch, and D. Eissentat, eds. Radical Biology: Advances and Perspectives on the Function of Plant Roots. American Society of Plant Physiologists, Rockville, MD, pp. 498–500.

Tromp, J., 1983. Nutrient reserves in roots of fruit trees, in particular carbohydrates and nitrogen. Plant and Soil 71: 401–413.

Vitousek, P., J. Aber, R. Howarth, G. Likens, P. Matson, D. Schindler, W. Schlesinger, and D. Tilman, 1997. Human alteration of the global nitrogen cycle: sources and consequences. Ecological Applications 7: 737–750.

Vogt, K., C. Grier, C. Meier, and R. Edmonds, 1982. Mycorrhizal role in net primary production and nutrient cycling in *Abies amabilis* ecosystems in western Washington. Ecology 63: 370–380.

Walker, T.S., H.P. Bais, E. Grotewold, and J.M. Vivanco, 2003. Root Exudation and Rhizosphere Biology. Plant Physiology 132: 44–51.

Walton, D. and Y. Li, 1995. Abscisic acid biosynthesis and metabolism. In: P. Davies, ed. Plant Hormones: Physiology, Biochemistry and Molecular Biology. Kluwer, Dordrecht, pp. 140–157.

Wan, C. and R.E. Sosebee, 1990. Relationship of photosynthetic rate and edaphic factors to root carbohydrate trends in honey mesquite. Journal of Range Management 43: 171–176.

Waring, R., A. McDonald, S. Laarson, T. Ericsson, E. Wiren, A. Arwidsson, A. Ericsson, and T. Lohammer, 1985. Differences in chemical composition of plants grown at constant relative growth rates with stable mineral nutrition. Oecologia 66: 157–160.

Wartinger, A., H. Heilmeier, W. Hartung, and E.-D. Schulze, 1990. Daily and seasonal courses of leaf conductance and abscisic acid in the xylem sap of almond trees (*Prunus dulcis*) under desert conditions. New Phytologist 116: 581–587.

Weatherley, P., 1982. Water uptake and flow in roots. In: O. Lange, P. Nobel, C. Osmond, and H. Ziegler, eds. Encyclopedia of Plant Physiology. Springer-Verlag, Berlin, pp. 79–109.

Weaver, J.E., 1919. The Ecological Relations of Roots. Publication 286, Carnegie Institution of Washington, Washington, D.C.

Willenbrink, J., 1992. Storage of sugars in higher plants. In: F.A. Loewus and W. Tanner, eds. Plant Carbohydrates I: Intracellular Carbohydrates. Springer Verlag, Berlin, pp. 684–699.

Woods, F., H. Harris, and R. Caldwell, 1959. Monthly variations of carbohydrates and nitrogen of sandhill oaks and wiregrass. Ecology 40: 292–295.

Yoder, C.K. and R.S. Nowak, 1999. Hydraulic lift among native plant species in the Mojave Desert. Plant and Soil 215: 93–102.

Zhang, H. and B.G. Forde, 1998. An *Arabidopsis* MADS Box gene that controls nutrient-induced changes in root architecture. Science 279: 407–409.

Zou, C.B., P.W. Barnes, S. Archer, and C.R. McMurtry, 2005. Soil moisture redistribution as a mechanism of facilitation in savanna tree–shrub clusters. Oecologia 145: 32–40.

6 Water Relations and Hydraulic Architecture

Melvin T. Tyree

CONTENTS

INTRODUCTION

Water relations of plants is a large and diverse subject. This chapter is confined to some basic concepts needed for a better understanding of the role of water relations in plant ecology. Readers seeking more details should consult Slatyer (1967), Kramer (1983), and Tyree and Zimmermann (2002).

Water movement in plants is purely passive. In contrast, plants are frequently involved in active transport of substances across membranes. Such transport involves membrane-bound proteins (enzymes) that impart metabolic energy to transport and increase the energy of the substance, which is transported. Although there have been claims of active water movement in the past, no claim of active water transport has ever been proved.

Passive movement of water (like passive movement of other substances or objects) still involves forces, but passive movement is defined as spontaneous movement in a system that is already out of equilibrium in such a way that the system tends toward equilibrium. Active movement, by contrast, requires the input of biological energy and moves the system further away from equilibrium or keeps it out of equilibrium despite continuous passive movement in the counterdirection.

The basic equation that describes passive movement is Newton's law of motion on earth where there is friction

$$v = (1/f)\ F, \tag{6.1}$$

where v is velocity of movement (m s^{-1}), F is the force causing the movement (N), and f is the coefficient of friction (N s m^{-1}). In the context of passive water or solute movement in plants, it is more convenient to measure moles moved per second per unit area, which is a unit of measure called a flux density (J). Fortunately, there is a simple relationship between J, v, and concentration (C, mol m^{-3}) of the substance moving: $J = Cv$. In addition, in a chemical/biological context, it is easier to measure the energy of a substance and how the energy changes as it moves than it is to measure the force acting on the substance. Passive movement of water or a substance occurs when it moves from a location where it has high energy to one where it has lower energy. The appropriate energy to measure is called the chemical potential, μ, and it has units of energy per mol (J mol^{-1}). The force acting on the water or solute is the rate of change of energy with distance, hence, $F = -(d\mu/dx)$, which has units of J m^{-1} mol^{-1} or N mol^{-1} (because $J =$ N m). Therefore, replacing F with $-(d\mu/dx)$ and J with v gives

$$J = -K(d\mu/dx), \tag{6.2}$$

where K is a constant equal to C/f. Equation 6.2 or some variation on it is used to describe water movement in soils and plants. The variations in Equation 6.2 generally involve measuring J in kg or m^3 of water rather than mol and measuring μ in pressure units rather than energy units. These changes can be accommodated by incorporating the conversion factor for unit changes into K in the variant equations as shown elsewhere in this chapter.

WATER RELATIONS OF PLANT CELLS

The water relations of plant cells can be described by the equation that gives the energy state of water in cells and how this energy state changes with water content, which can be understood through the Höfler diagram. First, the factors that determine the energy state of water in a cell is considered.

The energy content of water depends on temperature, height in the earth's gravitational field, pressure, and mole fraction of water (X_w) in a solution. For practical purposes, we evaluate the chemical potential of water in a plant cell in terms of how much it differs from pure water at ground level and at the same temperature as the water in the cell, that is, we measure

$$\Delta\mu = \mu - \mu_0, \tag{6.3}$$

where μ_0 is the chemical potential of water at ground level at the same temperature as the cell. It has become customary for plant physiologists to report $\Delta\mu$ (J mol^{-1}) in units of J m^{-3} of water because this has dimensions equal to a unit of pressure: P_a (= N m^{-2}), and this new quantity is called water potential (Ψ). The conversion involves dividing $\Delta\mu$ by the partial molal volume of water ($\overline{V}_w$), that is,

$$\Psi = \frac{1}{V_w}\Delta\mu. \tag{6.4}$$

In general, Ψ is given by

$$\Psi = P + \pi + \rho g h, \tag{6.5}$$

where P is the pressure potential (the hydrostatic pressure), $\pi = RT/\overline{V}_w \ln X_w$ is called the osmotic potential (sometimes called osmotic pressure) and is approximated by $\pi = -RTC$ where C is the osmolal concentration of the solution, R is the gas constant, T is the Kelvin temperature, ρ is the density of water, g is the acceleration due to gravity, and h is the height aboveground level.

Water flows into a cell whenever the water potential outside the cell (Ψ_o) is greater than the water potential inside the cell (Ψ_c). Let us consider the water relations of a cell at ground level, that is, how water moves in and out of the cell in the course of a day. Water flows through plants in xylem conduits (vessels or tracheids), which are nonliving pipes, and the walls of the pipes are made of cellulose. Water can freely pass in and out of the conduits through the cellulose walls (Figure 6.1a). The water potential of the water in the xylem conduit is given by

$$\Psi_x = P_x + \pi_x. \tag{6.6}$$

During the course of a day, P_x might change from slightly negative values at sunrise (e.g., -0.05 MPa) to more negative values by early afternoon (e.g., -1.3 MPa) and then might return again by the next morning, as explained in Section "Cohesion–Tension Theory and Xylem Dysfunction". Because the concentration of solutes in xylem fluid is usually very low (i.e., plants transport nearly pure water in xylem), π_o is not very negative (e.g., -0.05 MPa). Ψ_x might change from -0.1 to -1.35 MPa in the example given in Figure 6.1b. This daily change in Ψ_x causes a daily change in Ψ_c as water flows out of the cell as Ψ_x falls and into the cell as Ψ_x rises.

The two primary factors that determine the water potential of a cell at ground level (Ψ_c) are turgor pressure P_t ($=P$ inside the cell) and π of the cell sap (π_c).

$$\Psi_c = P_t + \pi_c. \tag{6.7}$$

Plant cells generally have π_c values in the range of -1 to -3 MPa; consequently, P_t is often a large positive value whenever $\Psi_c = \Psi_x$. There are two reasons for cells that have P_t greater than 0: (1) The protoplasm of living cells is enclosed inside a semipermeable membrane (the plasmalemma membrane) that permits relatively rapid transmembrane movement of water and relatively slow transmembrane movement of solutes; therefore, the solutes inside the cell making π_c negative cannot move out to the xylem to make π_x more negative; and (2) The membrane-bound protoplasm is itself surrounded by a relatively rigid elastic cell wall. The cell wall must expand as water flows into the cell to accommodate the extra volume; the stretch of the elastic wall places the cell contents under a positive pressure (much like a tire pumped up by air puts the air in the tire under pressure). The rise in P_t raises Ψ_c until it reaches a value equal to Ψ_x, at which point water flow stops.

The effect of water movement into or out of a cell is described by a Höfler diagram (Figure 6.1a). Entry of water into the cell has two effects: it causes a dilution of cell contents,

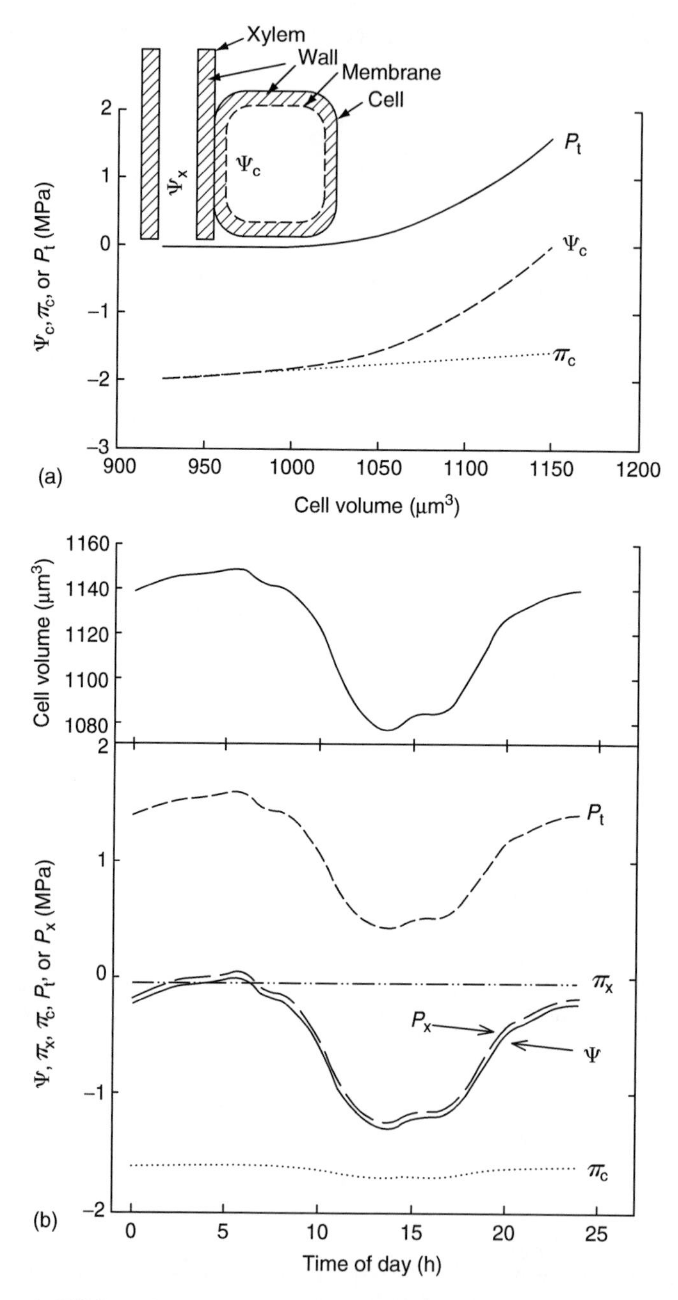

FIGURE 6.1 (a, inset) A living cell with plasmalemma membrane adjacent to a xylem conduit (shown in longitudinal section). The living cell is surrounded by a soft cell wall composed of cellulose that retards expansion of the cell and does not prevent the collapse of the cell as P_t falls. The xylem conduit is surrounded by a woody cell wall (cellulose plus lignin) that strongly retards both expansion and collapse when P_x becomes negative. (a) A Höfler diagram showing how the cell water potential (Ψ_c), the cell osmotic potential (π_c), and the turgor pressure (P_t) changes with cell volume. (b) A representative daily time course of how cell or xylem water potential (Ψ) changes during the course of a day. The component water potentials shown are xylem pressure potential (P_x), xylem osmotic potential (π_x), cell turgor pressure (P_t), and cell osmotic potential (π_c). In addition, how cell volume changes with time is also shown. See text for more details.

hence π_c becomes slightly less negative, and the turgor pressure, P_t, rises very rapidly with cell volume. The net effect of the increase in π_c and P_t is an increase in Ψ_c toward zero. Conversely, a loss of water makes Ψ_c, π_c, and P_t fall to increasingly negative values. If Ψ_c falls low enough, then the P_t falls to zero and remains at zero in cells in soft tissue. In woody tissues, that is, in cells with lignified cell walls, the lignification prevents cell collapse and P_t falls to negative values. The information in the Höfler diagram can be used to understand the water relations of cells in the course of a day.

A representative time course is shown in Figure 6.1b. Suppose the sun rises at 6 AM and sets at 6 PM. Radiant energy falling on leaves enhances the rate of evaporation above the rate at which roots can replace evaporated water, hence, both Ψ_c and Ψ_o falls to the most negative values in early afternoon (indicated by the solid line marked by Ψ in Figure 6.1b). As the afternoon progresses, the light intensity diminishes and the rate of water loss from the leaves falls below the rate of uptake of water from the roots, hence Ψ increases. Overnight, the value of Ψ returns to a value near 0 in wet soils or more negative values in drier soils; in either case, Ψ reaches a maximum value just before dawn. The value of Ψ before sunrise (called the predawn water potential) is often taken as a valid measure of soil dryness ($=\Psi_{soil}$) in the rooting zone of the plant. In the xylem, the osmotic potential (π_x) remains more or less constant and only slightly negative during the day; therefore, all change in Ψ_A is brought about by a large change in P_o, which closely parallels changes in Ψ. Similarly, in a living cell, changes in Ψ_c are brought about by large change in P_t, whereas the cell osmotic potential (π_c) changes only slightly and remains a large negative value.

WATER RELATIONS OF WHOLE PLANTS

Regulation of Water Loss by Leaves

The water relations of a whole plant can be understood in terms of the fundamental physiological role of the leaf. The leaf is an organ designed to permit CO_2 uptake at a rate needed for photosynthesis while keeping water evaporation from leaves at a reasonably low rate. The roots have the function of extracting water from the soil to replace water evaporated from leaves. The leaf structure of a typical plant is illustrated in Figure 6.2. The upper and lower epidermis of leaves are covered with a waxy cuticle that reduces water loss from the leaf to a negligible level. All gas exchange into and out of the leaf occurs via stomata. The guard cells of the stomata are capable of opening and closing air passages that provide pathways for diffusion of CO_2 into the leaf and for loss of water vapor form the leaf. Photosynthesis occurs primarily in the palisade and mesophyll cells of the leaf (Figure 6.2b). Leaves are thin enough (approximately 0.1 mm thick) to permit photon penetration to chloroplasts (shown as dots in Figure 6.2b). The chloroplasts absorb the energy of the photon and through a photochemical process use the energy to convert CO_2 and water into sugar. The consumption of CO_2 in the chloroplasts lowers the concentration of CO_2 in the cells and thus sets up a concentration gradient for the diffusion of more CO_2 into the cells via the stomatal pores and mesophyll air spaces. Because mesophyll and palisade cell surfaces are wet, water continuously evaporates from these surfaces and water vapor continuously diffuses out of the leaf by the same pathway taken by CO_2. There is a very high water cost for photosynthesis because C_3 and C_4 plants generally evaporate 600 H_2O for every CO_2 fixed. Evaporated water is continuously replaced by water flow through veins in the leaves. The veins contain xylem for water uptake and phloem for sugar export.

The physiology of guard cells has evolved to optimize photosynthesis when conditions are right for the photochemical reaction and to minimize water loss when conditions are wrong for photosynthesis. Stomata remain closed or only partly open when soils are dry; this is an advantage to the plant because roots are unable to extract water fast enough from dry soils to keep up with evaporation from leaves. Roots send a chemical signal to leaves, abscisic acid

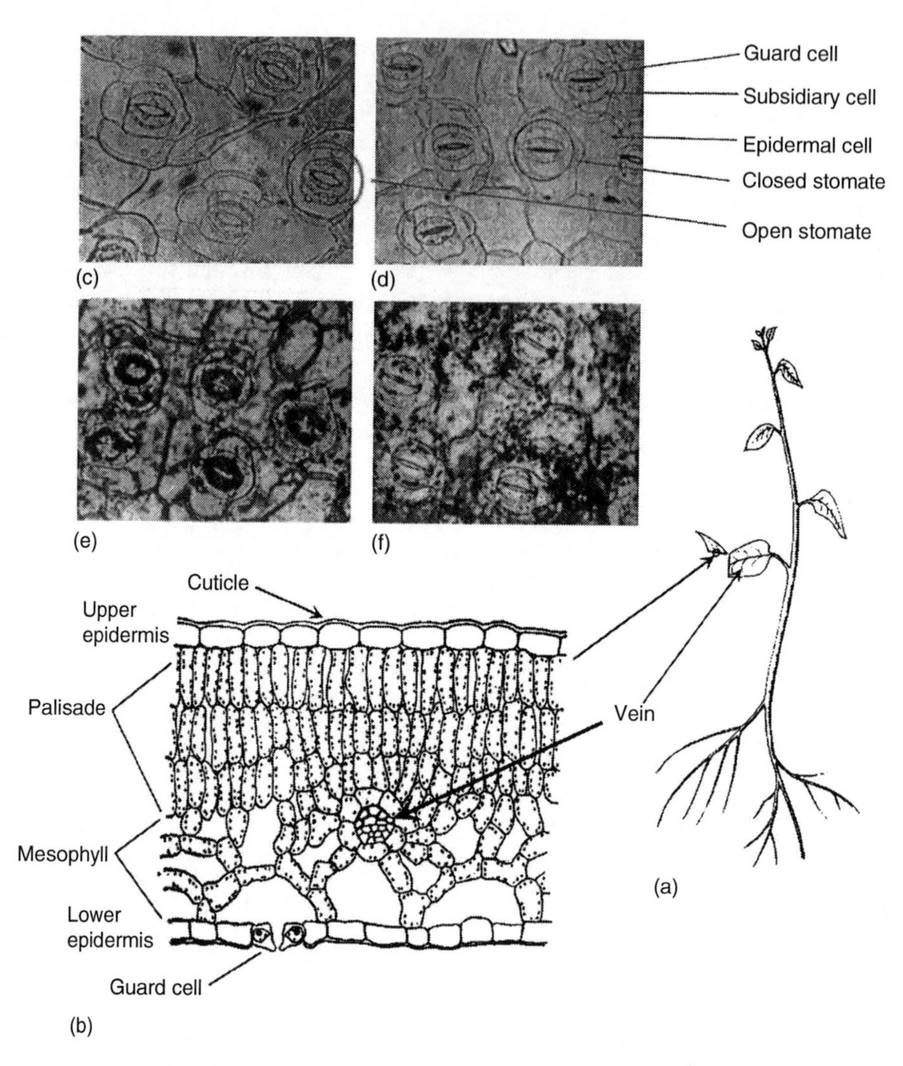

FIGURE 6.2 (a) A typical plant. (b) The lower leaf has been cut and the cross section is shown enlarged approximately 750×. (c–f) Photographs of the lower leaf surface enlarged approximately 1000×. (e) and (f) have been stained to show location of K^+. (c) and (e) illustrate the open state of stomata when the leaf is exposed to light. *Stoma* is from the Greek and means mouth (plural, stomata); the two guard cells for a structure looks much like a mouth. (d) and (f) are similar leaves in light but also exposed to abscisic acid (ABA), causing stomata to remain closed in light. Note that in open stomata, K^+ is concentrated in the guard cells (e) and the guard cells have moved apart to form an air passage for gas exchange (the stomatal pore). Note that the K^+ is located in the epidermal cells when the stomata are closed (f). (Adapted from Kramer, P.J., *Water Relations of Plants*, Academic Press, New York, 1983; Bidwell, R.G.S., *Plant Physiology*, Macmillan, New York, 1979.)

(ABA), which mediates stomatal closure (Davies and Zhang 1991). When soil water is not limited, and when the internal CO_2 concentrations fall below ambient levels in the atmosphere, stomata opens in sunlight. In approximately half of the known species, stomata have the additional capability of sensing the relative humidity (RH) of the ambient air and tend to progressively close as RH in the air adjacent to the leaves decreases. Rapid stomatal opening is mediated by movement of K^+ from the epidermal cells to the guard cells (Figure 6.2c–f). The movement of K^+ makes π_c less negative in the epidermal cells and more negative in the guard cells; consequently, water flows from the epidermal cells to the guard cell, P_t falls in the epidermal cells, and P_t rises in the guard cells. The mechanical effect of this water movement and change in P_t causes the guard cells to swell into the epidermal region and open a stomatal pore.

The evaporative flux density (E) of water vapor from leaves is ultimately governed by Fick's law of diffusion of gases in air. The control exercised by the plant is to change the area available for vapor diffusion through the opening and closing of stomatal pores. The value of E (mmol water s^{-1} m^{-2} of leaf surface) is given by

$$E = g_L(X_i - X_o), \tag{6.8}$$

where g_L is the diffusional conductance of the leaf (largely controlled by the stomatal conductance, g_s), X_i is the mole fraction of water vapor at the evaporative surface of the palisade and mesophyll cells, and X_o is the mole fraction of water vapor in the ambient air surrounding the leaf. The mole fraction is defined as $X = n_w/N$, where n_w is the number of moles of water vapor and N is the number of moles of all gas molecules, including water vapor, the most abundant gas molecules are N_2 and O_2. The dependence of g_L on some environmental and physiological variables is illustrated in Figure 6.3.

The maximum value of X occurs when RH is 100%, that is, when the air is saturated with water vapor; the maximum value of X increases exponentially with the Kelvin temperature of the air. The air at the evaporating surface of leaves is at the temperature of the leaf (T_L), and X_i is taken as the maximum value of X at saturation, which can be symbolized as $X_i = X[T_L]$.

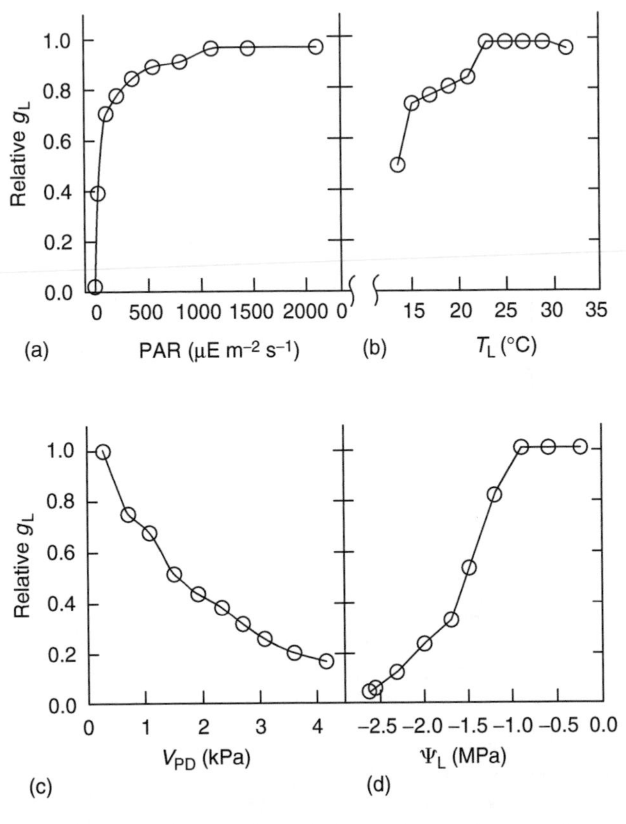

FIGURE 6.3 Relationship between change of g_L relative to the maximum value, $g_L/g_{L,max}$, and various environmental factors measured on *Acer saccharum* leaves. The environmental factors were: (a) photosynthetically active radiation (PAR), (b) leaf temperature (T_L), (c) vapor pressure deficit (V_{PD}), and (d) leaf water potential (Ψ_L). (Adapted from Yang, S., Liu, X., and Tyree, M.T., *J. Theor. Biol.*, 191, 197, 1998. With permission.)

The value of X_o depends on the microclimate near the leaf, that is, the air temperature and RH. However, the microclimate of the leaf is strongly influenced by the behavior of the plant community surrounding the leaf. Therefore, although the leaf has direct control over the value of g_s and g_L, it has less control over E than might appear from Equation 6.8.

The qualitative aspects of how leaves influence their own microclimate is easily explained. When the sun rises in the morning, the radiant energy load on the leaf increases. This has two effects: T_L rises as the sun warms the leaves, and hence, X_i rises and g_L increases as stomata open. However, the increased evaporation from the leaves causes X_o to rise as all the water vapor is added to the ambient air. Even changes of g_L under constant radiant energy load causes less change in E than might be expected from Equation 6.8. When g_L doubles, E also doubles, but only temporarily. The increased E lowers T_L because of increased evaporative cooling, and hence lowers X_i. The increased E from all of the leaves in a stand eventually increases X_o, hence $X_i - X_o$ declines, causing a decline in E. Consequently, Equation 6.8 is not very useful in predicting the value of E at the level of plant communities. Leaf-level behavior can be extrapolated to community-level equations if we take into account leaf-level solar energy budgets, that is, equations that describe light absorption by leaves at all wavelengths and the conversion of this energy to temperature and heat fluxes. Studies of solar energy budgets have been conducted at both the leaf and stand-level (Slatyer 1967, and Chang 1968; see also Section "Factors Controlling the Rate of Water Uptake and Movement").

Most of the changes in E at the leaf level can be explained in terms of net radiation absorbed at the stand level, which is relatively easy to measure. This relationship is illustrated in Figure 6.4, where daily values of E at the leaf level (kg m^{-2} day^{-1}) are correlated with daily

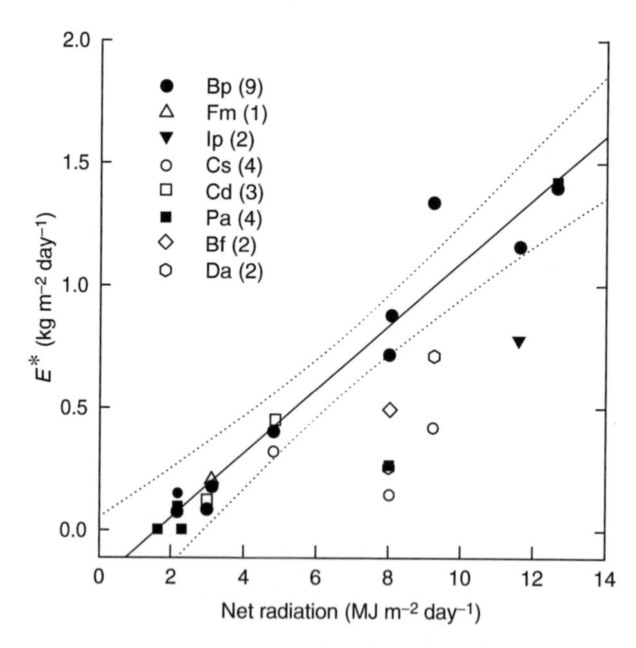

FIGURE 6.4 Correlation of daily water use of leaves (E^*) and net radiation (NR). Data are from potometer experiments with nine woody cloud forest species in Panama. The number of days per species is given in parenthesis. The regression (with 95% confidence intervals) is for Bp: *Baccharis pendunculata* ($E^* = -0.20 + 1.29$ NR, $r^2 = 0.92$). Even when the data for all species are pooled, NR proved to be a very good predictor of daily water use ($E^* = -0.2 + 1.04$ NR, $r^2 = 0.72$). Cm, *Citharexylum macradenium*; Cd, *Croton draco*; Fm, *Ficus macbridei*; Ip, *Inga punctata*; Pa, *Parathesis amplifolia*; Bf, *Blakea foliacea*; Cs, *Clusia stenophylla*; and Da, *Dendropanax arboreus*. (From Zotz, G., Tyree, M.T. Patiño, S., and Carlton, M.R., *Trees*, 12, 302, 1998. With permission.)

values of net radiation (MJ m^{-2} of ground per day) measured with an Eppley-type net radiometer. Equations presented elsewhere to describe stand-level E in terms of net radiation and other factors are validated by relationships such as shown in Figure 6.4. The other factors most commonly included account for plant control over E through leaf area index (leaf area per unit ground area), the effect of drought on g_L, and ambient temperature (see Section "Factors Controlling the Rate of Water Uptake and Movement").

Tissue–Water Relations (Pressure–Volume Curves)

From the preceding section, the reader might falsely conclude that plants can lose water from leaves without any negative impact on the growth and survival. However, there is more to water loss than is apparent from stomatal physiology and the energy interactions between leaves and their immediate environment. Whenever leaves lose water faster than the rate of water uptake by roots, the water potential in the xylem (Ψ_x) and of the leaf cells (Ψ_c) must also fall (see Section "Water Relations of Plant Cells"). Most of the decline in Ψ_x is due to a drop in P_x, and when P_x becomes too negative, cavitations occur that prevent water flow through vessels (see Section "Cohesion–Tension Theory and Xylem Dysfunction"). Most of the decline in Ψ_c is due to a drop in P_t, and when P_t becomes too small, cell growth stops. Growing cells are surrounded by relatively ridged "wooden boxes" consisting of cellulose cell walls. Cell walls must be stretched plastically to grow large, and the motive force on plastic stretch is the force of P_t against the cell walls. As leaves lose water, P_t and P_x must fall for the plant to extract water from the soil at a rate approximately equal to the rate of water loss. As soils become dry, values of P_t and P_x must fall even more for plants to extract water from soils. From Equation 6.7 it can be seen that $P_t = \Psi_c - \pi_c$, hence plants have some control over P_t through "adjustments" in π_c. Some species have evolved to have lower values of π_c than others, and some species can make π_c more negative in response to drought by increasing the osmolal concentration (C) of solutes in their cells (because $\pi_c = -RTC$).

The relationship between Ψ and its components (P and π) in leaves can be described by a leaf-level Höfler diagram (Figure 6.2a). Many studies have reported the comparative physiology of tissue–water relationship of leaves and have discussed how differences in these relationships might explain ecological adaptation of plants. Readers interested in learning more should consult the references contained in Tyree and Jarvis (1982). The following is a brief overview of how Höfler diagrams are measured in leaves and some ecological applications of this information.

The fastest way to derive a Höfler diagram is to measure the water potential of leaves (Ψ_{leaf}) versus water loss using a Scholander–Hammel pressure bomb (Scholander et al. 1965, Tyree and Hammel 1972). This is done by obtaining a series of balance pressure (P_B) versus weight loss of leaves or shoots; the P_B is an approximate measure of the Ψ_{leaf}. The pressure bomb is a metal chamber into which is placed an excised leaf or shot (Figure 6.5a) that is at an unknown water potential, Ψ_{leaf}. When gas pressure (P_{gas}) is applied to the leaf surface, the pressure of the fluid in the cells is increased by an equal amount so that $\Psi_{leaf} = \Psi_c = \pi_c + P_t + P_{gas}$. The xylem water potential and cell water potential are at equilibrium, $\Psi_c = \Psi_x$; from this it follows that $\pi_x + P_x = \pi_c + P_t + P_{gas}$ or $P_x = \pi_c + P_t + P_{gas} - \pi_x$. When the P_{gas} is at the balance pressure (P_B), P_x has risen to zero and xylem sap is squeezed out of the end of the branch or petiole protruding outside the pressure bomb (Figure 6.5). Therefore, we have $P_B = -(\pi_c + P_t) + \pi_x = -\Psi_{leaf}$, when the branch or leaf was outside the bomb) $+ \pi_x = P_x$. P_B is usually approximated by $-\Psi_{leaf}$ since π_x is usually much smaller than Ψ_{leaf}. However, P_B really measures P_x and the pressure bomb cannot measure pressure gradients within a leaf because all gradients dissipate during the P_B measurement. In some cases these distinctions are important to remember.

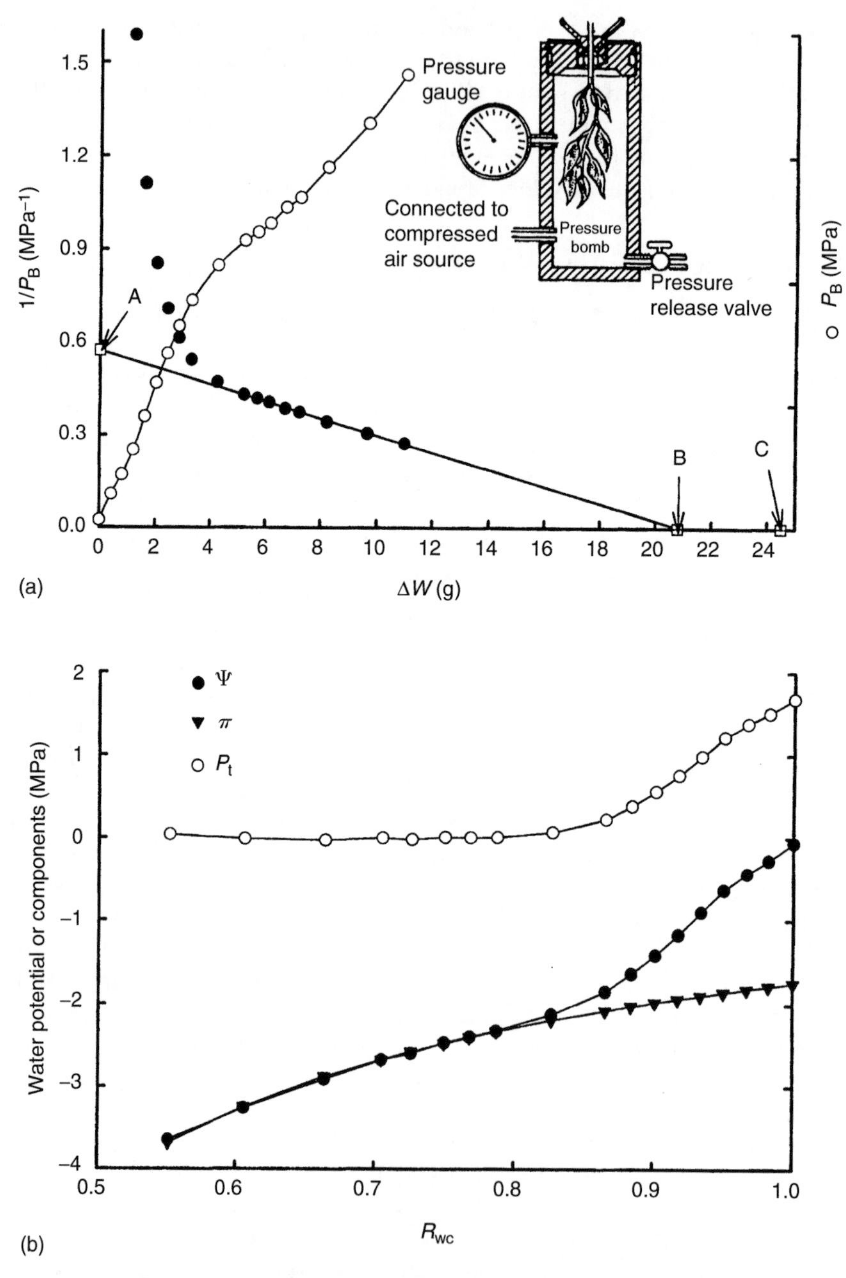

FIGURE 6.5 Scholander–Hammel pressure bomb and analysis of pressure bomb data. (a) A pressure bomb is shown in the inset. Open circles (*right* axis): Original data of balance pressure (P_B) versus weight loss of shoot in the bomb. Weight loss is induced by removing the shoot from the bomb and allowing water to evaporate. Closed circles (*left* axis): Transformed data of $1/P_B$ versus weight loss. The liner portion of the curve has been extrapolated to two points, A = y-intercept = $-1/\pi_o$, where π_o is the average osmotic pressure of the symplast at full hydration when $\Psi = 0$; B = x-intercept = weight of water in the symplast when $\Psi = 0$ [this value can also be computed from $(-1/p_0)$/slope]; C = total water content of the shoot = maximum weight loss when oven-dried. (b) The Höfler diagram for the whole shoot derived from the data in (a). The x-axis is relative water content of the shoot and the y-axis is π = osmotic potential of the symplasm (solid triangles), P_t = turgor pressure of symplasm (open circles), and Ψ = total water potential of the symplasm (solid circles).

A pressure–volume curve is obtained by slowly dehydrating a shoot and obtaining a series of weights, W, versus P_B values. If W_o was the original weight, then the cumulative weight loss is $\Delta W = W_o - W$. The pressure–volume curve is a plot of $1/P_B$ versus ΔW. Strictly speaking, the pressure–volume curve should be called a pressure–weight curve, but if ΔW is given in g, then that is the same as volume in mL since 1 mL of water weighs 1 g. The pressure–volume curve has a curved region for small ΔW values and a linear region for larger ΔW values (Figure 6.5b). When the linear portion of the plot is extrapolated back $\Delta W = 0$, the y-intercept (the point marked A in Figure 6.5a) is equal to $-1/\pi_o$, where π_o is the solute potential of the living cells at zero water potential. The point marked B in Figure 6.5a is the turgor loss point (Ψ_{tlp}), that is, the value of Ψ_{leaf} when P_t reaches zero. The x-intercept (point C) is the volume of water contained in the symplast (W_s = total water in the protoplasm and vacuoles of all living cells), and the difference in x-values (D–C) is the amount of water in the apoplast (W_a = total water in xylem and cell walls). Höfler diagrams for shoots or leaves are usually plots of Ψ, π, and P_t versus relative water content of the shoot or leaf. Relative water content (R_{WC}) is defined as (the current water content)/(the maximum water content at full hydration), $R_{WC} = (W_o - \Delta W)/(W_o - W_d)$, where W_d is the dry weight. Values of π at different R_{WC} are calculated from $\pi_o\ W_s/(W_o - \Delta W)$, values of Ψ are equated to $-P_B$, and values of P_t are calculated from $P_B - \pi_o/R_{WC}$. The justification for these relationships is given in a report by Tyree and Hammel (1972).

As plants dry, changes in π can be caused by changes in symplastic water content, W_s, or in the number of moles of solute in the symplasm, N_s, because $\pi = RTN_s/W_s$. Considerbale emphasis has been placed on demonstrating changes in π as a result to changes in N_s (Turner and Jones 1980). A change in π caused by a change in N_s is called an osmotic adjustment. Diurnal changes in π ranging from 0.4 to 1.6 MPa have been reported for some plants; the amount of change resulting from diurnal changes in N_s is in the range of 0.2–0.8 MPa. Medium-term changes in π induced by slow soil dehydration have also been attributed to osmotic adjustment in drought-stressed versus unstressed plants. Osmotic adjustments of 0.1–1 MPa have been reported over periods of 3 days to 3 weeks. Long-term or seasonal changes in π range from 0.2 to 1.8 MPa; some of the largest changes are recorded during the onset of winter in temperate plants and appear to be correlated to changes in frost tolerance. The degree of diurnal, medium-term, and long-term osmotic adjustment varies widely between species. There are some species that have shown little or no adjustment (Tyree and Jarvis 1982).

Low values of π in plants should enhance the ability of plants to take up soil water under dry or saline conditions (Tyree 1976). This advantage is probably marginal in sandy soils because the available water reserves at soil water potentials less than −0.4 MPa are very small; therefore, the plant's ability to grow deep roots (Section "Cohesion–Tension Theory and Xylem Dysfunction") is probably of greater advantage. In clay soils, however, there are considerable water reserves at water potentials less than −0.4 MPa, so that low leaf and root values of π may be as important as root growth in assisting water uptake. Low values of π in leaves also enable P_t to remain above zero at lower values of Ψ than otherwise would be possible as Ψ falls. This allows the maintenance of open stomata with larger apertures and high stomatal conductances and higher net rates of photosynthesis down to lower values of Ψ than would be the case if π were higher (less negative). Osmotic adjustments and/or lower π values also enable maintenance of turgor pressure for growth, since it has been shown that the rate of volume growth ($r = \mathrm{d}V/\mathrm{d}t$) of a cell is given by $m(P_t - Y)$, where m is the growth rate constant of a cell and Y is the yield point of the cell (Green et al. 1971).

Some attention in the past has been focused on the slope of the P_t line in the Höfler diagram in which the x-axis is R_{WC}. The bulk modulus of elasticity of a tissue is defined as

$$\varepsilon = \frac{\Delta P_t}{\Delta R_{WC}} R_{WC}. \tag{6.9}$$

The higher value of ε represents higher value of the slope of P_t versus R_{wc}. A large value of ε is seen as an advantage for water conservation. For a plant to extract water from the soil, it must first lose some water so that its Ψ_{leaf} falls below the soil water potential. Because most of the change in Ψ_{leaf} is due to the change in P_t, a large value of ε means that a plant would have to lose less water to lower Ψ_{leaf} than would a plant with a small value of ε.

Although values of ε have been reported to range from 0.5 to 20 MPa, the adaptive advantages of large versus small ε have never been clearly established (Tyree and Jarvis 1982) because the ecological advantage of a large ε may be rather marginal. All plants lose water during the day and regain most or all of the lost water during the night. Therefore, the advantage of less water loss means more constant concentrations of biochemical substrates during the day. Leaves can lose anywhere from 1% to 20% of their water content during the day, and this water loss would cause a corresponding 1%–20% diurnal variation in substrates. Because the concentration of reactants and products would normally change during the day, even without the influence of leaf dehydration, it is not clear how much is gained by keeping relatively constant cell volume because of large ε.

WATER ABSORPTION BY PLANT ROOTS

The primary factor affecting the pattern of water extraction by plants from soils is the rooting depth. Rooting depths can be extremely variable depending on soil conditions and species of plant producing the roots (Figure 6.6a). Many of the early studies of rooting depth and branching pattern of roots were performed in the 1920s and 1930s in deep, well-aerated prairie soils, where roots penetrate to great depths. At the extreme, roots have been traced to depths of 10–25 m, for example, alfalfa (10 m), longleaf pine, (17 m) (Kramer 1983), and drought-evading species in the California chaparral (25 m) (S. Davis, April 1991, personal communication). The situation is very different for plants growing in heavy soils, where 90% of the roots can be found in the upper 0.5–1.0 m.

In seasonally dry regions, for example, central Panama, the majority of the roots may be located in the upper 0.5 m, but it is far from clear if the majority of water absorption occurs in the upper 0.5 m during the dry season. Water use by many evergreen trees is higher in the dry season than in the wet season, although the upper 1 m of the soils is much drier than the leaves of the trees ($\Psi_{soil} < \Psi_{leaf}$) and excavations have shown that liana roots are less than 5 m in depth and hence extract water from quite wet soil during the dry season (M.T. Tyree, February 1989, personal observation). Hence, the role of shallow versus deep roots of woody species deserves more study.

In addition, deeply rooted species may contribute to the water supply of shallow-rooted species through a process called hydraulic lift (Richards and Caldwell 1987). In one study, it was found that shallow-rooted species growing within 1–5 m of the base of maple trees were in a better water balance than the same species growing more than 5 m away (Dawson 1993). Each night, the Ψ_{soil} at a depth of 0.5 m increased underneath maple trees, but Ψ_{soil} did not increase at distances greater than 5 m from the trees. This indicated that the deep maple roots were in contact with moist soil and were capable of transporting water from deep maple roots to shallow maple roots overnight. Since the water potential of the shallow maple roots exceeded the water potential of the adjacent soil, water flow from the shallow maple roots to the adjacent soil contributed to the overnight rehydration of shallow soils.

Roots absorb both water and mineral solutes found in the soil, and the flow of solutes and water interact with each other. The mechanism and pathway of water absorption by roots is more complex than in the case of a single cell (Section "Water Relations of Plant Cells"). Water must travel first radially from the epidermis to cortex, endodermis, and pericycle before it finally reaches the xylem vessels, from which point water flow is axial along

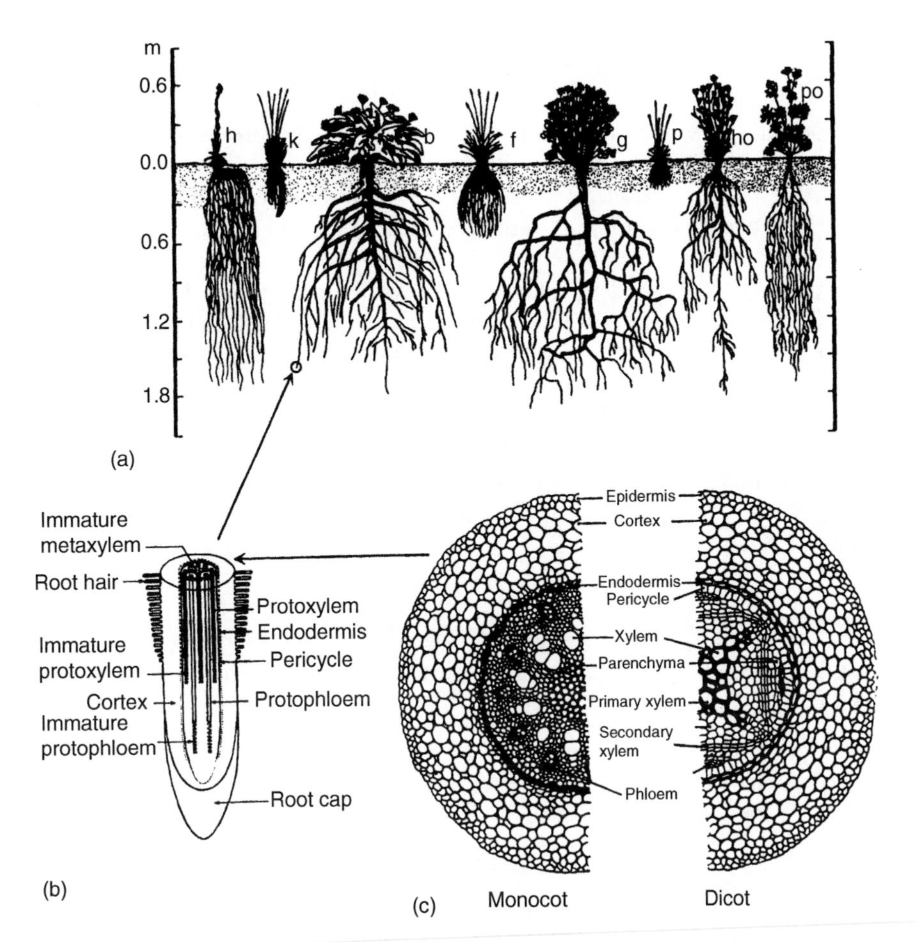

FIGURE 6.6 (a) Differences in root morphology and depth of root systems of various species of prairie plants growing in a deep, well-aerated soil. Species shown are: h, *Hieracium scouleri*; k, *Loeleria cristata*; b, *Balsamina sgittata*; f, *Festuca ovina ingrata*; g, *Geranium viscosissimum*; p, *Poa sandbergii*; ho, *Hoorebekia racemosa*; and po, *Potentilla blaschkeana*. (b) Enlargement of a dicot root tip enlarged approximately 50×. (c) Cross section of monocot and dicot roots enlarged approximately 400×. (Adapted from Kramer, P.J., *Water Relations of Plants*, Academic Press, New York, 1983; Steward, F.C., *Plants at Work*, Addison-Wesley, Reading, MA, 1964; Bidwell, R.G.S., *Plant Physiology*, Macmillan, New York, 1979.)

the root (Figures 6.6b and 6.6c). The radial pathway (typically 0.3 mm long in young roots) is usually much less conductive than the axial pathway (>1 m in many cases); therefore, whole-root conductance is generally proportional to the root surface area. The radial pathway can be viewed as a composite membrane separating the soil solution from the solution in the xylem fluid. The composite membrane consists of serial and parallel pathways made up of plasmalemma membranes, cell wall "membranes," and plasmodesmata (pores <0.5 μm diameter) that connect adjacent cells. The composite membrane is rather leaky to solutes; therefore, differences in osmotic potential between the soil (π_s) and the xylem (π_x) have less influence on the movement of water. At any given point along the axis, the water flux density across the root radius (J_r) is given by

$$J_r = L_r[(P_s - P_x) + \sigma(\pi_s - \pi_x)], \tag{6.10}$$

where L_r is the radial root conductance to water and σ is the solute reflection coefficient. For an ideal membrane in which water but not solutes may pass, $\sigma = 1$. For the composite membrane of roots, σ is usually between 0.1 and 0.8. The system of equations that describes water transport in roots is complex when all of the factors are taken into account, for example, axial and radial conductances, the fact that each solute has a different σ, and that the rate of water flow is influenced by the solute loading rate (J_s). Water and solute flow in roots can be described by a standing gradient osmotic flow model (readers interested in the details may consult Tyree et al. (1994b) and Steudle (1992).

Fortunately, the equations describing water flow become simple when the rate of water flow is high. The concentration of solutes in the xylem fluid is equal to the ratio of solute flux to water flux (J_s:J_w) during steady-state flux. Solute flux tend to be more or less constant with time, but water flux increases with increasing transpiration. When water flow is high, the concentration of solutes in the xylem fluid becomes small and approaches values comparable to that in the soil solution, and pressure differences become quite large, hence $(P_s - P_x) >> (\pi_s - \pi_x)$. Only at night or during rainy periods can values of $(P_s - P_x)$ approach those of $\sigma(\pi_s - \pi_x)$. Therefore, water flow (J_w, kg s^{-1}) through a whole-root system during the day can be approximated by

$$J_w \cong k_r(P_s - P_{x,b}) \tag{6.11}$$

where $P_{x,b}$ is the xylem pressure at the base of the plant and k_r is the total root conductance (combined radial and axial conductances).

HYDRAULIC ARCHITECTURE AND PATHWAY OF WATER MOVEMENT IN PLANTS

Van den Honert (1948) quantified water flow in plants in a classical paper in which he viewed the flow of water in the plant as a catenary process, where each catena element is viewed as a hydraulic conductance (analogous to an electrical conductance) across which water (analogous to electric current) flows. Thus, van den Honert proposed an Ohm's law analog for water flow in plants. The Ohm's analog leads to the following predictions; (1) the driving force of sap ascent is a continuous decrease in P_x in the direction of sap flow; and (2) evaporative flux density from leaves (E) is proportional to negative of the pressure gradient ($-\mathrm{d}P_x/\mathrm{d}x$) at any given point (cross section) along the transpiration gradient ($-\mathrm{d}P_x/\mathrm{d}x$) at any given point (cross section) along the transpiration stream. Thus, at any given point of a root, stem, or leaf vein, we have

$$-\mathrm{d}P_x/\mathrm{d}x = AE/K_h + \rho g\,\mathrm{d}h/\mathrm{d}x, \tag{6.12}$$

where A is the leaf-area supplied water by a stem segment with hydraulic conductivity K_h and $\rho g\,\mathrm{d}h/\mathrm{d}x$ is the gravitational potential gradient, where ρ is density of water, g is acceleration due to gravity, and $\mathrm{d}h/\mathrm{d}x$ is height gained, $\mathrm{d}h$, per unit distance, $\mathrm{d}x$, traveled by water in the stem segment.

In the context of stem segments of length (L) with finite pressure drops across ends of the segment, we have

$$\Delta P_x = \mathrm{LAE}/K_h + \rho g \Delta h. \tag{6.13}$$

Figure 6.7 illustrates water flow through a plant represented by a linear catena of conductance elements near the center and a branched catena of conductance elements on the left. The number and arrangement of catena elements are dictated primarily by the spatial precision

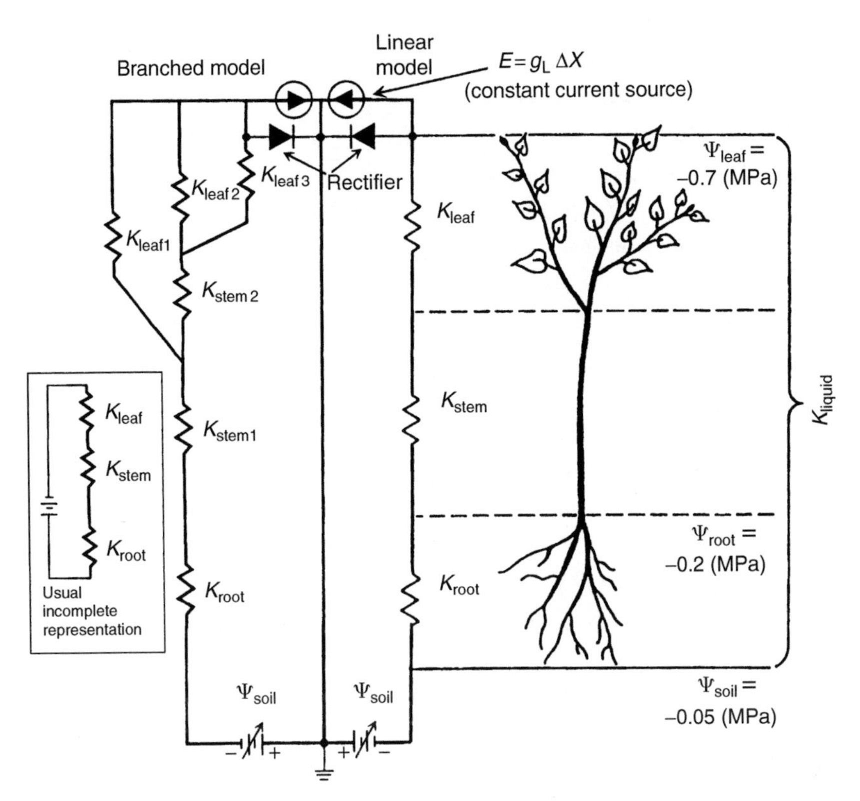

FIGURE 6.7 The Ohm's law analogy. The total conductance is seen as resultant conductance (K) of the root, stem, and leaf in series and parallel. Water flow is driven by evaporation of water from leaves, which creates a difference in water potential between the soil, Ψ_{soil}, and the water potential at the evaporating surface, Ψ_{evap}. On the *right* is the simplest Ohm's law analogy with conductances in series. On the *left* is a more complex conductance catena in which some conductance elements are in series and some in parallel.

desired in the representation of water flow through a plant; a plant can be represented by anywhere from one to thousands of conductance elements. Reviews of the hydraulic architecture of woody plants can be found in a report by Tyree and Ewers (1991 and 1996).

The usual way of representing the Ohm's law circuit is incomplete. The usual but incomplete representation of the Ohm's law circuit is shown in the boxed inset in the lower left side of Figure 6.7. The battery represents the water potential drop from root to leaf, but it is an incomplete representation of reality because no ground point is shown in the circuit. The ground point represents zero water potential. Water potential is always measured relative to pure water at the same temperature and pressure as the plant. The incomplete representation gives the correct drop in water potential across each conductance element but not the correct water potential relative to ground (the zero reference). The complete representation has a variable battery that gives the soil water potential (Ψ_{soil}) and has some analog components to represent the evaporation from the leaves. In Figure 6.8, the evaporation is shown by a constant current source and a rectifier in parallel. The amount of current, E, in the constant current source can be approximated by Equation 6.8. When $E = 0$ we expect leaf water potential to equal soil water potential and that there should be no water current flowing through the circuit; this condition is achieved in the circuit by the introduction of the rectifier that prevents backward current flow when $E = 0$ and $\Psi_{soil} < 0$. In some experimental conditions, Ψ_{soil} can be raised to positive values by placing a potted root system in a pressure chamber with

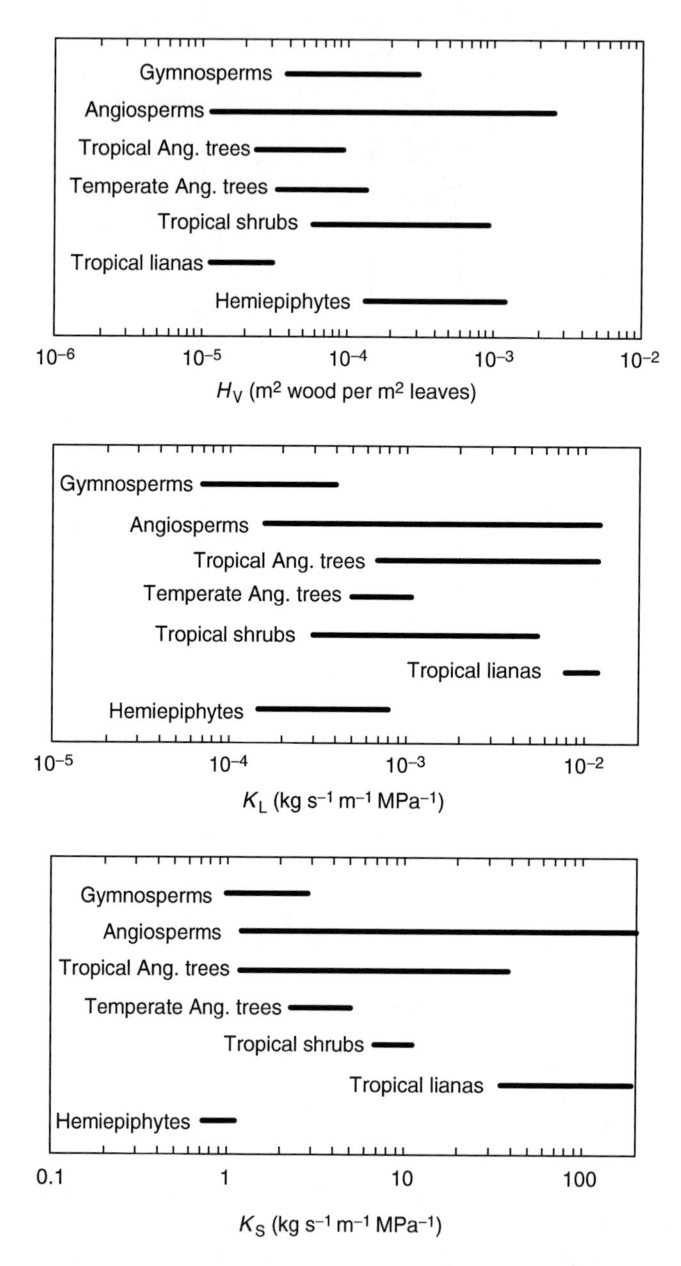

FIGURE 6.8 Ranges of hydraulic parameters by phylogeny or growth form. Horizontal bars demark the ranges read from the *bottom* axis of each parameter, where H_v is the Huber value, K_L is the leaf-specific conductivity, and K_s is the specific conductivity. Values represent ranges for 48 species. (Adapted from Patiño, S., Tyree, M.T., and Herre, E.A., *New Phytol.*, 129, 125, 1995. With permission.)

the shoot outside the chamber. When $E = 0$ and $\Psi_{soil} > 0$ we could expect gutation (exudation of water from leaves); this forward direction of current flow is permitted by the rectifier. When $E > 0$ and Ψ_{soil} is positive enough, gutation can occur simultaneously with evaporation when $\Psi_{leaf} > 0$; the rectifier permits this additional current flow (see also Wei et al. 1999).

An alternative representation of evaporation from leaves would be achieved by including an additional conductance element for the vapor phase conductance, K_{vapor} (analogous to $g_L \cong g_s$), with an additional water potential drop from the leaf to the ambient air outside the leaf (Ψ_{air}). K_{vapor} would be orders of magnitude smaller than the whole-plant conductance (K_{liquid}), and hence would dominate and control E and hence behave like a constant current source. There are two disadvantages to using this representation. First, g_L is never measured in the same units as K_{liquid}, making it difficult to compare values with those in the literature. Second, g_L measured in the units of K_{liquid} would no longer be a constant, that is, its value in theory would change with Ψ in the vapor phase. This can be seen immediately by looking at Equation 6.2 and Equation 6.4, from which we can get

$$J_{vapor} = \frac{C\overline{V}_w}{f}\frac{d\Psi}{dx}, \tag{6.14}$$

where C is the concentration of water. In the liquid phase, C is nearly a constant, but in the vapor phase, the concentration of water vapor can easily change by a factor of 10 from the evaporating surface in a leaf to the bulk air. The value of g_L in Equation 6.8 is superior in that it depends only on the geometry of the stomatal pores; this follows because in the vapor phase

$$\frac{d\Psi}{dx} = \frac{RT}{\overline{V}_c C}\frac{dC}{dx} \tag{6.15}$$

hence,

$$J_{vapor} = \frac{RT}{f}\frac{dC}{dx}, \tag{6.16}$$

which is Fick's law for water vapor transport in air. The coefficient RT/f is independent of dC/dx. Equation 6.8 can be viewed as derived from Fick's law with account taken of the specific geometry of water diffusion through stomates and replacing dC/dx with the gradient in mole fraction of water vapor.

Parameters and Concepts to Describe Hydraulic Architecture

The hydraulic architecture of a plant can be defined as a quantitative description of the plant in terms of the Ohm's law analog using a simple linear model of conductance elements or a complex branched catena of a few or even thousands of conductance elements (Figure 6.7). The conductance elements are quantified by measurements made on excised stem segments for branched catena models and by measurements on whole roots and shoots for simple linear models.

Measurements on stem segments are performed with a conductivity apparatus. Excised stem segments are fitted into water-filled lengths of plastic tubing. One end of the segment of length (L, m) is connected via tubing to an upper reservoir of water and the other end to a lower reservoir. The height difference between the reservoirs is usually set at 0.3–1 m to create a pressure drop ΔP of 3–10 kPa across the stem segment. The lower reservoir is usually placed on top of a digital balance to measure water flow rate, F, in kg s^{-1}. The fundamental parameter measured is the hydraulic conductivity (K_h), defined as

$$K_h = F/(\Delta P/L). \tag{6.17}$$

Values of K_h are usually measured for stems of different diameter, D, and regressions are used to obtain allometric relationships of K_h versus diameter of the form

$$K_h = A\ D^B, \tag{6.18}$$

where A and B are regression constants.

Stem segments can be viewed as bundles of conduits (vessels or tracheids) with a certain diameter and number of conduits per unit cross section. If all of the conduits were of the same diameter and number per unit cross section, then $B = 2$, because K_h would increase with number of conduits in parallel, which would increase with cross section, which is proportional to diameter squared. Usually, B is found to be more than 2 but less than 3 because the diameter of conduits tends to increase with stem diameter in most plants. According to the Hagen–Poiseuille law, the K_h of a single conduit of diameter, d, increases in proportion to d^4; therefore, the conductance of a stem segment with N conduits would be proportional to Nd^4. Although you cannot pack as many big-diameter conduits into a stem segment as small-diameter conduits, there is a net gain in K_h for a stem segment to have bigger diameter conduits, although fewer would fit into that available space. To prove this, let us imagine a stem segment of 1 mm diameter with 1000 conduits of $d_l = 0.01$ mm diameter each and another stem segment of 1 mm diameter with 250 conduits of $d_2 = 0.02$ mm diameter each. The cross-sectional area of each conduit is $\pi d^2/4$. Both segments would have the same cross-sectional area of conduits since $1000\ \pi d_l^{2/4} = 250\ \pi d_s{}^2/4$, but each conduit in the latter would be 16 times more conductive than the former because $d_1^4 = 16\, d_1^4$; therefore, although there are only one-fourth as many conduits of diameter d_2 and d_l, the stem is four times more conductive.

Since K_h of stem segments depends on stem cross section, one useful way of scaling K_h is to divide it by stem cross section to yield specific conductivity, K_s. Specific conductivity is a measure of the efficiency of stems to conduct water. The efficiency of stems increases with the number of conduits per unit cross section and with their diameter to the fourth power. In large woody stems, the central core is often nonconductive heartwood. It is often better to calculate K_s from K_h/A_{sw}, where A_{sw} is the cross-sectional area of conductive sapwood.

Leaf-specific conductivity (K_L), also known as LSC (Tyree and Ewers 1991), is equal to K_h divided by the leaf area distal to the segment (A_L, m^2). This is a measure of the hydraulic sufficiency of the segment to supply water to leaves distal to that segment. If we know the mean evaporative flux density (E, kg s^{-1} m^{-2}) from the leaves supplied by the stem segment and we ignore water storage capacitance, then the pressure gradient through the segment $(\mathrm{d}P/\mathrm{d}x) = E/K_L$. Therefore, the higher the K_L is, the lower the $\mathrm{d}P/\mathrm{d}x$ required to allow for a particular transpiration rate.

In simple linear Ohm's law models, whole-shoot and whole-root conductances (k_{sh} and k_r, respectively) are measured using a high-pressure flowmeter (HPFM; see later). These conductances are usually defined as the ratio of flow, F, across the whole root or shoot divided by the pressure drop, ΔP; hence, it differs from K_h, K_s, and K_L in that root or shoot length is not taken into account. The word *conductivity* is usually used when L is taken into account in the calculation, and *conductance* is used when L is not used in the calculation. Because a large plant or root becomes more conductive than small plants, some suitable means of normalization for plant size is needed. One way to do this is to calculate leaf-specific conductances, that is, $K_{sh} = K_{sh}/A_L$ and $K_r = k_r/A_L$. The advantage of this versus other kinds of scaling is discussed in Section "Root, Shoot, and Leaf Hydraulic Conductances".

The Huber value (H_v) is defined as the sapwood cross-section (or sometimes the stem cross-section) divided by the leaf area distal to the segment. Because, the H_v is in units of m^2 stem area per m^2 leaf area, if is often written without dimension. It is a measurement of the investment of stem tissue per unit leaf area fed. It follows from the definitions mentioned earlier that $K_L = H_v K_s$.

Figure 6.8 summarizes the known ranges of hydraulic architecture parameters in 48 taxa covering a range of growth forms and phylogenies. Comparisons of mean values between species are difficult because K_L, K_s, and H_v often change significantly with stem diameter, D. For example, K_L, and K_s can be 10–100 times greater when measured in the bole of trees ($D = 300$ mm) than when measured in young branches ($D = 3$ mm). Sometimes, differences between species in K_L or K_s measured at $D = 6$ mm may be reversed at $D = 60$ mm;

furthermore, stem morphologies may be such that the smallest segments bearing leaves were 20 mm in diameter in some species but just 3 mm in another. The mean values in Figure 6.8 were computed from regression values at $D = 15$ mm, except for three tropical species with large stems for which we used $D = 45$ mm because smaller branches did not exist.

Patterns of Hydraulic Architecture of Shoots

Values of K_L have been applied to complex "hydraulic maps" in which the aboveground portion of trees, were represented by hundreds to thousands of conductance elements (Figure 6.7). Using known values of K_L and E, it is possible to calculate P_x versus path length from the base of a tree to selected branch tips (Figure 6.9). Some species are so conductive (large K_L values) that the predicted drop in P_x is little more than needed to lift water against gravity (see Schefflera in Figure 6.9). In other species, gradients in P_x become very steep near branch tips (see Thuja in Figure 6.9). The predicted gradients of P_x become steeper in small branches of trees because K_L is often lower in small-diameter branches than in big-diameter branches. In 20 species of angiosperms, the value of K_L was found to be proportional to stem diameter, D, to the power of 0.5–2.0 depending on the species (Tyree and Ewers 1991, Patiño et al. 1995, Zotz et al. 1998). The increase in K_L with diameter could be due to an increase in H_v or K_s since $K_L = H_v K_s$, but in most cases, H_v was found to be approximately constant with D so the change was due to an increase in K_s with D. The exceptions seem to be gymnosperms with strong apical dominance, where H_v increases toward the apex, that is, decreases with increasing D (Ewers and Zimmermann 1984a,b). In one angiosperm, *Ficus dugandii*, H_v was found to increase with D to the power of 1.0 (Patiño et al. 1995).

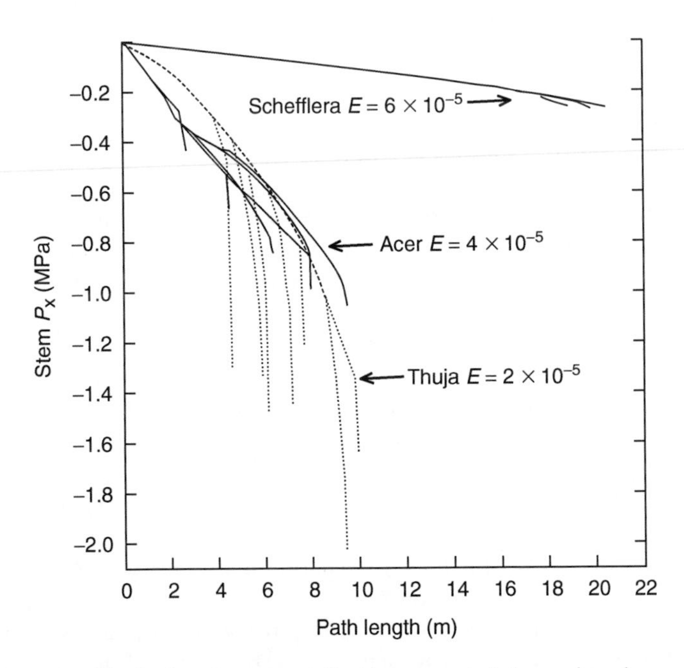

FIGURE 6.9 Pressure profiles in three large trees, that is, computed change in xylem pressure (P_x) versus path length. P_x values were computed from the base of each tree to a few randomly selected branch tips. The drooping nature of these plots near the apices of the branches is caused by the decline of leaf-specific conductivity (K_L) from base to apex of the trees. The pressure profiles do not include pressure drops across roots or leaves. In some species, pressure drop across leaves can be more than shown. The pressure drop across roots is generally equal to that across the shoots (including leaves). E, evaporative flux density in kg s^{-1} m^{-2}.

There appear to be no hydraulic constrictions at branch junctions in woody plants (Tyree and Alexander 1993), that is, the hydraulic conductance of water passing through branch junctions is approximately the same as in equal lengths of stem segments above- and below-branch junctions. This is contrary to earlier reports of major constrictions in branch junctions, but the number of measurements in these reports were too few to draw statistically provable conclusions (Ewers and Zimmermann 1984a,b, Zimmermann 1978). Unfortunately, a myth has arisen regarding the hydraulics of branch junctions because these earlier, preliminary studies have been frequently cited.

Root, Shoot, and Leaf Hydraulic Conductances

Most studies have focused on the hydraulic architecture of large woody tropical plants, for example, trees, shrubs, and vines (Ewers et al. 199, Patiño et al. 1995, Tyree and Ewers 1996). Parameters measured generally have been confined to hydraulic conductance of woody shoots. However, relatively little is known about total root conductance of plants relative to shoot conductances, and little is known about the conductivity of leaves relative to shoots. In large woody plants, the shoots of rapidly growing trees appear to be more conductive than slowly growing trees.

The recent development of an HPFM (Yang and Tyree 1994, Tyree et al. 1995) allows the rapid measurement of root and shoot hydraulic conductance of seedlings to saplings with stem diameters from 1 to 50 mm. Thus, it is now possible to study the growth dynamics of seedlings while monitoring dynamic changes in hydraulic conductance of roots and shoots. This ability raises the question of how best to scale conductance parameters to reveal ecological adaptation to light regimes. Because more thought has gone into scaling of shoot parameters (Tyree and Ewers 1996) than root parameters, a review of arguments for roots seems appropriate.

Root conductance (k_r) can be defined as water flow rate (kg s^{-1}) per unit pressure drop (MPa) driving flow through the entire root system. Values of k_r could be scaled by dividing by some measure of root size (root surface area, total root length, or mass) or by dividing k_r by leaf surface area. Division by root surface area (A_r) is justified by an analysis of axial versus radial resistances to water flow in roots. In the radial pathway, water flows from the root surface to the xylem vessels through nonvascular tissue. In the axial pathway, water flow is predominately through vessels. The resistance of the radial path is usually more than the axial path (French and Steudle 1989, North et al. 1992). Most water uptake is presumed to occur in fine roots (<2 mm diameter) and fine root surface area is usually greater than 90% of the total root surface area (M.T. Tyree, personal observation). Root uptake of water would appear to be limited by root surface area, and hence it is reasonable to divide k_r by A_r, yielding a measure of root efficiency. Some roots are more efficient than others. Division of k_r by total root length (L) is not as desirable, but is justified because A_r and L are correlated approximately and L can be estimated by a low-cost, line-intersection technique rather than a high-cost, image-analysis technique.

Scaling by root mass is justified by consideration of the cost of resource allocation. Plants must invest a lot of carbon into roots to grow and maintain them. The benefit derived from this carbon investment is enhanced scavenging for water and mineral nutrient resource. Total root dry weight (TRDW) is a measure of carbon investment into roots. Thus, the carbon efficiency of roots might be measured in terms of K_r/TRDW, A_r/TRDW, or L/TRDW. Scaling by TRDW provides information of ecological rather than physiological importance.

Scaling of k_r by leaf surface area (A_L) provides an estimate of the sufficiency of the roots to provide water to leaves. The physiological justification of scaling k_r to the leaf surface area (A_L) comes from an analysis of the Ohm's law analog for water flow from soil to leaf (van den

Honert 1948). The Ohm's law analog describes water flow rate (F, kg s^{-1}) in terms of the difference in water potential between the soil (Ψ_{soil}) and the leaf (Ψ_L)

$$\Psi_{soil} - \Psi_L = (1/k_{soil} + 1/k_r + 1/k_{sh})F, \quad (6.19a)$$

where k_{soil} is the hydraulic conductance of the soil. Because it is usually assumed that $k_{soil} \gg k_r$ and k_{sh} except in dry soils, $1/k_{soil}$ can be ignored. Leaf water potential is approximated by

$$\Psi_L \cong \Psi_{soil} - (1/k_r + 1/k_{sh})F. \quad (6.19b)$$

Or, if we wish to express Equation 6.19b in terms of leaf area and average evaporative flux density (E), we have

$$\Psi_L \cong \Psi_{soil} - (1/k_r + 1/k_{sh})A_L F. \quad (6.20)$$

This equation also can be rewritten so that root and shoot conductance are scaled to leaf surface areas, i.e., to give leaf-specific shoot and root conductances, $K_{sh} = k_{sh}/A_L$ and $K_r = k_r/A_L$, respectively:

$$\Psi_L \cong \Psi_{soil} - (1/k_r + 1/k_{sh})E. \quad (6.21)$$

Meristem growth and gas exchange are maximal when water stress is small, that is, when Ψ_L is near zero. From Equation 6.21 it can be seen that the advantage of high K_r and K_{sh} is that Ψ_L is closer to Ψ_{soil}. Because leaf-specific stem-segment conductivities K_L are high in adult pioneer trees, the water potential drop from soil to leaf is much smaller than in old-forest species (Machado and Tyree 1994). This may promote rapid extension growth of meristems in pioneers compared with old-forest species. In addition, stomatal conductance (g_s) and therefore net assimilation rate are reduced when Ψ_L is too low. During the first 60 days of growth of *Quercus rubra* L. seedlings, there was a strong correlation between midday g_s and leaf-specific plant conductance, $G = k_p/A_L$, where $k_p = k_r k_{sh}/(k_r + k_{sh})$ (Ren and Sucoff 1995). This suggests that whole-seedling hydraulic conductance is limiting g_s though its effect on Ψ_L. There is also reason to believe that whole-shoot conductance limits g_s in mature trees of *Acer saccharum* Marsh (Yang Tyree 1993). Thus, high values of K_r and K_{sh} may promote both rapid extension growth and high net assimilation rates in pioneers. See Tyree 2003 for a recent review of how whole-plant hydraulic conductance influences gas exchange and plant performance.

Scaling is always necessary to normalize for plant size. As seedlings grow exponentially in size, we would expect an approximately proportional increase in k_r and k_{sh}. Since roots and shoots both supply water to leaves and since an increase in leaf area means an increase in rate of water loss per plant, we would expect k_r and k_{sh} to be approximately proportional to A_L.

The major resistance (= inverse conductance) to liquid water flow in plants resides in the nonvascular pathways, that is, the radial pathway for water uptake in young roots and in the cells between leaf veins and the evaporating surfaces of leaves. Nonvascular resistance can be 60%–90% of the total plant resistance to water flow, although the total path length may be less than 1 mm. Although water flows through distances of 1–100 m in the vascular pathways from root vessel to stems to leaf vessels, the hydraulic resistance of this pathway rarely dominates. The biggest resistance to water flow (liquid and vapor) from the soil to the atmosphere resides in the vapor transport phase. The rate of flow from the leaf to the atmosphere is determined primarily by the stomata, as discussed in Section "Water Relations of Whole Plants."

Early successional (pioneer) species grow more rapidly and require higher light levels to survive than late successional (old-forest) species. A pattern is emerging that shows that the dry matter cost of root and shoot conductance is less in pioneer versus old-forest species, and the leaf-specific root and shoot conductances are higher. Hence, scaling hydraulic parameters by leaf area is ecologically meaningful. In Figure 6.10 (top), the root and shoot conductances are scaled by root and shoot dry weights, respectively. In Figure 6.10 (bottom), the root and shoot conductance are scaled by dividing both by leaf area. In both cases, the adaptive advantages of the pioneer species become evident. All pioneer species differed significantly from the nonpioneer species, although all species were growing in the same light regime. Figure 6.10 illustrates two advantages of pioneers versus other species in this study. The higher values of K_r and K_{sh} mean that the pioneer species can maintain less negative leaf water

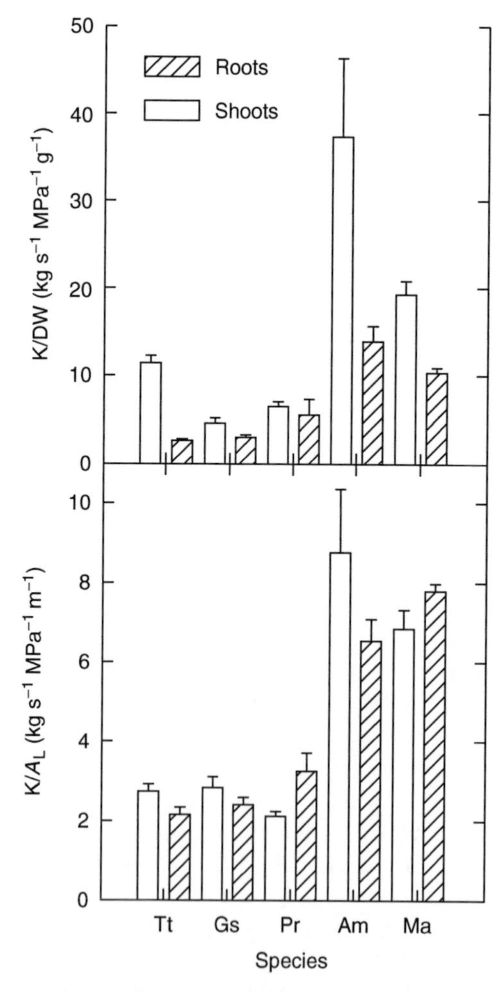

FIGURE 6.10 Hydraulic conductances of shoots and roots scaled to dry weight or leaf area (*top*) K_r per unit total root dry weight (TRDW) and K_{sh} per unit shoot dry weight. (*Bottom*) K_r and K_{sh}, both scaled to leaf are (A_L). Error bars are SEM, $n = 23$–36. Species are: Tt, *Trichilia tuberculata*; Gs, *Gustavia superba*; Pr, *Pouteria reticulata*; Am, *Apeiba membranacea*; and Ma, *Miconia argentea*. Ma, Am, and Pr are old-forest species, and Pr and Gs are pioneer species. Root and shoot means for Am and Ma were significantly different from corresponding root and shoot means for Tt, Pr, and Gs in both (*top*) and (*bottom*) (Tukey test, $P \leq 0.05$). (From Tyree, M.T., Velez, V., Dalling, J.W., *Oecologia*, 114, 293–298, 1998. With permission.)

potentials than the other species at any given transpiration rate. This might lead to higher rates of extension growth and net assimilation. The higher values of K_r/DW and K_{sh}/DW in pioneers means that pioneers spend less carbon to provide efficient hydraulic pathways than do the other species. Both of these advantages (Figure 6.10, top and bottom) mean that pioneers can be more competitive in gap environments than old forest species.

COHESION–TENSION THEORY AND XYLEM DYSFUNCTION

The cohesion–tension (C–T) theory was proposed more than 100 years ago by Dixon and Joly (1894), and some aspects of the C–T theory were put on a quantitative basis by van den Honert (1948) with the introduction of the Ohm's law analog of sap flow in the soil–plant–atmosphere continuum.

WATER TRANSPORT UNDER NEGATIVE PRESSURE

According to the C–T theory, water ascends plants in a metastable state under tension, that is, with xylem pressure (P_x) more negative than that of a perfect vacuum. The driving force is generated by surface tension at the evaporating surfaces of the leaf, and the tension is transmitted through a continuous water column from the leaves to the root apices and throughout all parts of the apoplast in every organ of the plant. Evaporation occurs predominately from the cell walls of the substomatal chambers due to the much lower water potential of the water vapor in air. The evaporation creates a curvature in the water menisci of apoplastic water within the cellulosic microfibril pores of cell wall. Surface tension forces consequently lower P_x in the liquid directly behind the menisci (the air–water interfaces). This creates a lower water potential, Ψ, in adjacent regions, including adjoining cell walls and cell protoplasts. The lowering of Ψ is a direct consequence of P_x, which is one of the two major components of water potential in plants, the other component is the solute potential π (see Equation 6.5 through Equation 6.7). The energy for the evaporation process ultimately comes from the sun, which provides the energy to overcome the latent heat of evaporation of the water molecules; that is, the energy to break hydrogen bonds at the menisci.

This tension (negative P_x) is ultimately transferred to the roots, where it lowers Ψ of the roots below the Ψ of the soil water. This causes water uptake from the soil to the roots and from the roots to the leaves to replace water evaporated at the surface of the leaves. The Scholander–Hammel pressure bomb (Scholander et al. 1965) is one of the most frequently used tools for estimating P_x. Typically, P_x can range down to -2 MPa (in crop plants) or to -4 MPa (in aridzone species), and in some cases -10 MPa (in California chaparral species). Readers interested in learning more about the C–T theory of sap ascent and recent experimental proofs should refer to the report by Tyree (1997), Wei et al. (1999, 2000), and Tyree and Zimmermann (2002).

CAVITATION, EMBOLISM, AND STABILITY OF BUBBLES

Water in xylem conduits is said to be in a metastable condition when P_x is below the pressure of a perfect vacuum, because the continuity of the water column, once broken, will not rejoin until P_x rise to values above that of vacuum. Metastable conditions are maintained by the cohesion of water to water and by adhesion of water to wall of xylem conduits. Both cohesion and adhesion of water are manifestations of hydrogen bonding. Although air–water interfaces can exist anywhere along the path of water movement, the small diameter of pores in cell walls and the capillary forces produced by surface tension within such pores prevent the passage of air into conduits under normal circumstances. When P_x becomes negative enough, the continuity of the water column in the conduit is rapidly broken, which is called a cavitation event. Water is drawn out of cavitated conduits by surrounding tissue, leaving a void filled with water vapor. Eventually, air diffuses into the void; when this happens, the conduit is said to be embolized.

A cavitation event in xylem conduits ultimately results in dysfunction. A cavitation occurs when a void of sufficient radius forms in water under tension. The void is filled with gas (water vapor and some air) and is inherently unstable, that is, surface tension forces makes it spontaneously collapse unless the water is under sufficient tension (negative pressure) to make it expand.

The chemical force driving the collapse is the energy stored in hydrogen bonds, the intermolecular force between adjacent water molecules. In ice, water is bound to adjacent water molecules by four hydrogen bonds. In the liquid state, each water molecule is bound by an average of 3.8 hydrogen bonds at room temperature. In the liquid state, hydrogen bonds are forming and breaking all the time, permitting more motion of molecules than in ice (Slatyer 1967). However, when an interface between water and air is formed, some of those hydrogen bounds are broken and the water molecules at the surface are at a higher energy state because of the broken bonds. The force (N = Newtons) exerted at the interface as hydrogen bonds break and reform can be expressed in pressure units (Pa) because pressure is dimensionally equal to energy (J = Joules) per unit volume of molecules, that is, $J\ m^{-3} = N\ m\ m^{-3} = N\ m^{-2} = Pa$. Stable voids in water tend to form spheres because spheres have the least surface area per unit volume, and thus a spherical void has the minimum number of broken hydrogen bonds per unit volume of void. The pressure tending to make a void collapse is given by $2\tau/r$, where r is the radius of the spherical void and τ is the surface tension of water (= 0.072 Pa m at 25°C).

For a void to be stable, its collapse pressure ($2\tau/r$) must be balanced by a pressure difference across its surface or meniscus = $P_v - P_w$, where P_w is the absolute pressure (= P_x + the atmospheric pressure) of the water and P_v is the absolute pressure of the void.

$$P_v - P_w = 2\tau/r. \tag{6.22}$$

P_v is always above absolute zero pressure (= perfect vacuum) since the void is usually filled with water vapor and some air. Relatively stable voids are commonplace in daily life, for example, the air bubbles that form in a cold glass of water freshly drawn from a tap. An entrapped air bubble is temporarily stable in a glass of water because P_w is a relatively constant 0.1 MPa, and P_v is determined by the ideal gas law, $P_v = nRT/V$, where n is the number of moles of air in the bubble, R is gas constant, T is absolute temperature, and V is the volume of the bubble. The tendency of the void to collapse ($2\tau/r$) makes V decrease, which causes P_v to increase according to the ideal gas law because P_v is inversely proportional to V. The rise in P_v provides the restoring force across the meniscus needed for stability. However, an air bubble in a glass of water is only temporarily stable, because according to Henry's law, the solubility of a gas in water increases with the pressure of the gas. Therefore, the increased pressure exerted by $2\tau/r$ makes the gas in the bubble more soluble in water, and it slowly collapses as the air dissolves, that is, as n decreases.

Air bubbles are rarely stable in xylem conduits because transpiration can draw P_w to values less than zero. As P_w falls toward zero, the bubble expands according to the ideal gas law, but because V can never grow larger than the volume of the conduit, P_v can never fall to or below zero to permit $P_v - P_w$ to balance $2\tau/r$ without a decline in P_w. Once the bubble has expanded to fill the lumen, the conduit is dysfunctional and no longer capable of transporting water. Fortunately for the plant, a dynamic balance at the meniscus in cell walls is ultimately achieved. This stability is discussed subsequently, first in the context of a vessel and its pit membranes.

As the air bubble is drawn up to the surface of the pit membrane in vessel cell walls, the pores in the pit-membrane break the meniscus into many small menisci at the opening of each pore. As the meniscus is drawn through the pores, the radius of curvature of the meniscus, r_m, falls toward the radius of the pores, r_p. As long as r_m exceeds r_p, the necessary conditions for stability are again achieved, that is,

$$P_v - P_w = 2\tau / r_m. \quad (6.23)$$

Usually, a dysfunctional conduit eventually fills with air at atmospheric pressure (as demanded by Henry's law); therefore, P_v eventually approaches 0.1 MPa as gas diffuses through water to the lumen and comes out of solution. When P_v equals 0.1 MPa, the conduit is said to be fully embolized. As P_w rises and falls as dictated by the demands of transpiration, r_m adjusts at the pit-membrane pores to achieve stability. When the conduit is fully embolized, both sides of Equation 6.23 can be expressed in terms of xylem pressure potential,

$$P_x = -(P_v - P_w) = -2\tau / r_m. \quad (6.24)$$

The minimum P_x that can be balanced by the meniscus is given when r_m equals the radius of the biggest pit-membrane pore bordering the embolized conduit. If the biggest pore is 0.1 or 0.05 μm, then the minimum stable P_x is −1.44 or −2.88 MPa, respectively. The porosity of the pit membrane is therefore critical to preventing dysfunction of vessels adjacent to embolized vessels (Sperry and Tyree 1988). When P_x falls below the critical value, the air bubble is sucked into an adjacent vessel, seeding a new cavitation.

Consequently, the genetics that determines pit morphology and pit-membrane porosity must be under strong selective pressure. A safe pit membrane is one with very narrow pores and thick enough and thus strong enough to sustain substantial pressure differences without rupturing. Recently substantial advances have been made in the above issues and readers should consult Sperry and Hacke (2004a,b), Hacke et al. (2004), and Wheeler et al. (2005).

The situation for tracheids of conifers is different because air movement from an embolized tracheid to an adjacent tracheid is prevented by the sealing (aspiration) of the torus against the overarching border of the pit. The porosity of the margo that supports the torus is too large to prevent meniscus passage at pressure difference exceeding 0.1 MPa in most cases (Sperry and Tyree 1990). However, because the margo is elastic, a pressure difference of just 0.03 MPa is sufficient to deflect the torus into the sealed position. Air bubbles pass between tracheids when the pressure difference becomes large enough to rip the torus out of its sealed position (Sperry and Tyree 1990).

Vulnerability Curves and the Air-Seeding Hypothesis

Water movement can occur at night or during rain when P_x is positive in some plants due to root pressure, that is, osmotically driven flow from roots. Water flow under positive P_x is often accompanied by gutation, that is, the formation of water droplet at leaf margins as frequently seen in many bamboo species. However, water is normally under negative pressure (tension) as it moves through the xylem toward the leaves during sunny days. The water is thus in a metastable condition and vulnerable to cavitation due to air entry into the water columns. Cavitation results in embolism (air blockage), thus disrupting the flow of water (Tyree and Sperry 1989). Cavitation in plants can result from water stress, and each species has a characteristic vulnerability curve, which is a plot of the percent loss k_h in stems versus the xylem pressure potential, P_x, required to induce the loss. Vulnerability curves are typically measured by dehydrating large excised branches to known P_x. Stem segments are then cut under water from the dehydrated branches; the air bubbles remain inside the conduits for the most part. An initial conductivity measurement is made and compared with the maximum K_h after air bubbles have been dissolved. Recently Cohcard et al. (2005) have invented a very

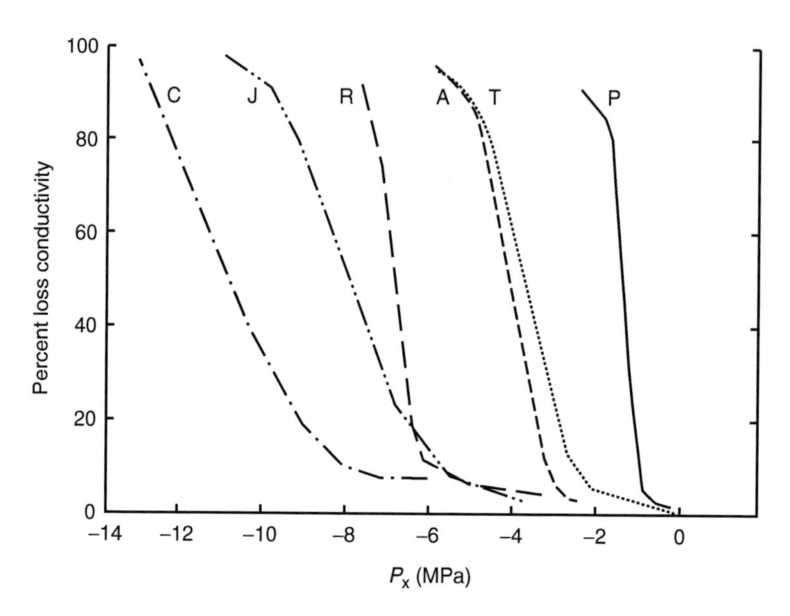

FIGURE 6.11 Vulnerability curves for various species. *Y*-axis is percent loss of hydraulic conductivity induced by the xylem pressure potential, P_x, shown on the *x*-axis. C, *Ceanothus megacarpus*; J, *Juniperus virginiana*; R, *Rhizophora mangle*; A, *Acer saccharm*; T, *Thuja occidentalis*; P, *Populus deltoids*.

clever way of measuring vulnerability curves using a centrifuge: the Cochard Cavitron. The Cochard Cavitron has the advantage of a very fast and accurate technique. The vulnerability curves of plants, in concert with their hydraulic architecture, can give considerable insight to drought tolerance and water relations "strategies."

The vulnerability curves for a number of species are illustrated in Figure 6.11 (Tyree et al. 1994a). These curves represent the range of vulnerabilities observed thus far in more than 120 species; 50% loss K_h occurs at P_x values ranging from −0.7 to −11 MPa. Many plants growing in areas with seasonal rainfall patterns (wet and dry periods) appear to be drought evaders, and others are drought tolerators. The drought evaders evade drought (low P_x and high percent loss K_h) by bearing deep roots and a highly conductive hydraulic system; alternatively, they evade drought by their deciduousness. The other species frequently reach very negative P_x for part or all of the year and are shallow-rooted or grow in saline environments. These species survive because they have a vascular system highly resistant to cavitations.

When air bubbles are sucked into xylem conduits from adjacent embolized conduits, the cavitation event is said to be air-seeded. Plants always have some embolized conduits to seed embolism into adjacent conduits. Embolisms are the natural consequence of foliar abscission, herbivory, wind damage, and other mechanical fates that might befall a plant. It is now appropriate to ask if all emboli are seeded from adjacent conduits or if some other mechanism occurs in some or most of the cases.

Four mechanisms for the nucleation of cavitations in plants have been proposed; these are illustrated in Figure 6.12, which shows for each mechanism the sequence of events that might occur as P_x declines in the lumen of a conduit. Readers are referred to Zimmermann (1983), Tyree et al. (1994a), and Pickard (1981) for a detailed discussion of the four mechanisms. Other air-seeding mechanisms have been proposed that apply to SCUBA divers (Yount 1989), and such mechanisms might occur in plants when gas solubility decreases in xylem water as it warms. However, little is known of the importance of this mechanism in plants. We need to be concerned only about which mechanism occurs most frequently in plants.

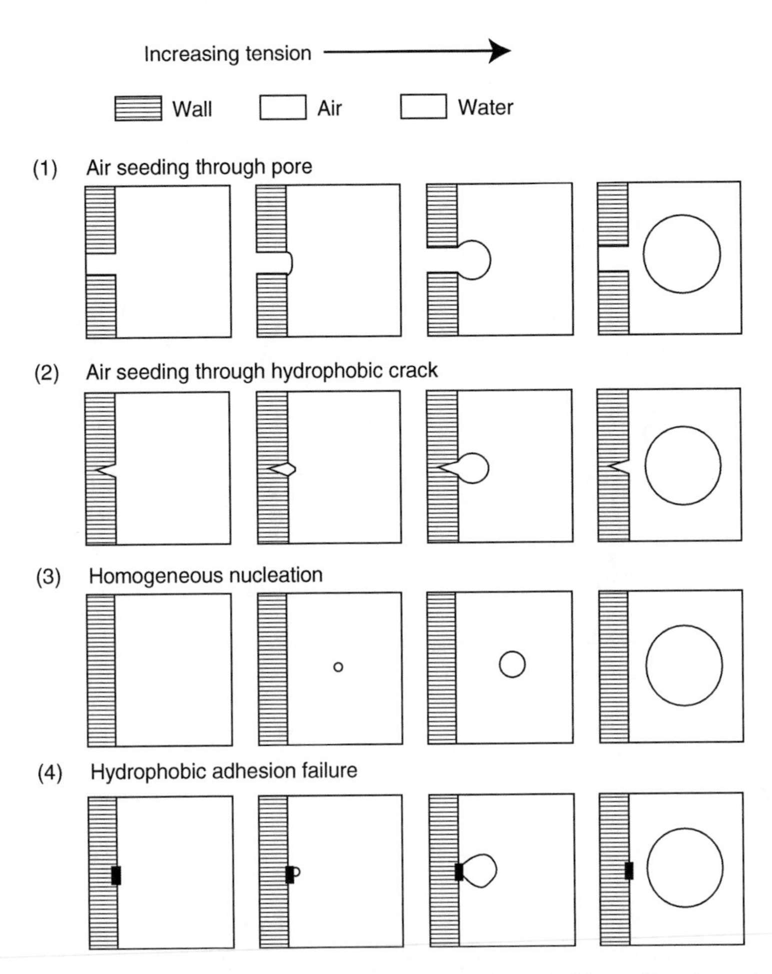

FIGURE 6.12 Four possible mechanisms of cavitation induction. (1) Air seeding through a pore occurs when the pressure differential across the meniscus is enough to allow the meniscus to overcome surface tension and pass through the pore. (2) Air seeding through a hydrophobic crack occurs when a stable air bubble resides at the base of a hydrophobic crack in the wall of a xylem conduit. When the P_x becomes negative enough, the bubble is sucked out of the crack. (3) Homogeneous nucleation involves the spontaneous generation of a void in a fluid. It is a random process requiring thermal motion of the water molecules. The hydrogen bonds at a specific locus are broken when all water molecules randomly move away from any locus at the same instant with sufficient energy to break all hydrogen bonds between water molecules. As the tension in the water increases, the hydrogen bonds are stretched and weakened so that the energy needed to break the bonds decreases, making a homogeneous nucleation more likely. (4) Hydrophobic adhesion failure is similar to homogeneous nucleation, except that hydrogen bonds are broken between water and a hydrophobic patch in the wall where the energy of binding between water and wall are reduced.

Experiments can discriminate between the air-seeding mechanism and the other three mechanisms (Figure 6.12; Yount 1989). All mechanisms predict cavitation when xylem fluid is under tension, but the air-seeding mechanism predicts that air can be blown into vessels while the fluid is under positive pressure. The air-seeding mechanism requires only a pressure differential ($P_a - P_w$), where P_a is the air pressure outside and P_w is the fluid pressure inside. It makes no difference if P_a is 0.1 and P_w is −3.0 or if P_a is 3.1 and P_w is 0.1. Experiments have shown that the same vulnerability curve results whether P_w is reduced by air dehydration or P_a is increased in a pressure bomb (Cochard et al. 1992). A vulnerability curve (VC) is a plot of percent loss

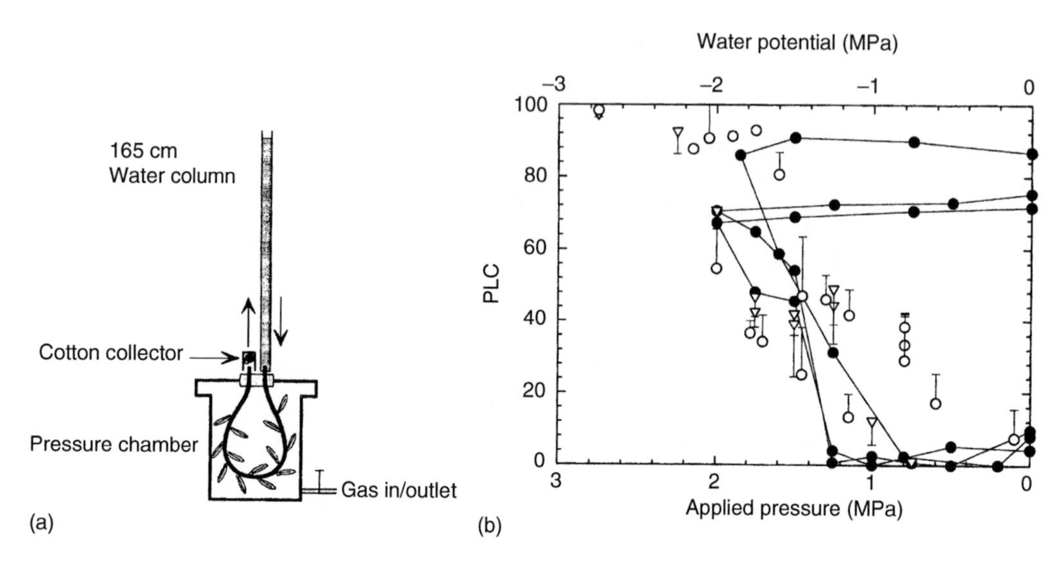

FIGURE 6.13 The first experimental test of the air-seeding hypothesis. (a) The experimental apparatus. A willow branch is bent around in a large pressure bomb so that both cut surfaces are outside the bomb. Water continually passes through the stem segment under positive pressure from a water column to a cotton collector. Flow rate is estimated by measuring the weight change of the cotton collector over known time intervals. (b) The results are shown as a vulnerability curve where the *y*-axis is the percent loss of hydraulic conductivity (PLC) and the *x*-axis is the negative P_x, or the air pressure in the bomb needed to cause the plotted PLC. Closed circles indicate PLC measured as in Figure 6.13a. Open circles indicate PLC induced by negative P_x in bench-dehydrated shoots. Open triangles indicate PLC induced by shoot dehydration in a pressure bomb but conductivity measured outside the pressure bomb on subsamples (stem segments excised from dehydrated shoot).

hydraulic conductivity (PLC) versus the P_x required to cause the PLC by cavitation events. The results of this experiment are illustrated in Figure 6.13b (see also Jarbeau et al. 1995).

Willow stem segments with leaves were enclosed in a pressure chamber with cut ends protruding into the open air. Water was passed continually through the xylem under positive pressure. While stem conductance was monitored, the gas pressure, P_a, in the pressure chamber was gradually increased. Initially, the hydraulic conductance of the stem segment did not decrease until a critical pressure of 1 MPa was applied. (Each solid circle in Figure 6.13b represents the application of pressure for 30–40 min.) When P_a was gradually increased beyond the critical value, the stem conductance began to fall (increased PLC). When P_a was gradually decreased, the PLC stopped decreasing. The VC from this experiment was identical to that found for similar branches dehydrated in the air.

How Plants Deal with Embolisms

Embolisms may be dissolved in plants if P_x in the xylem becomes positive or close to positive for adequate time periods (Tyree and Yang 1992, Lewis et al. 1994). Embolism disappears by dissolution of air into the water surrounding the air bubble. The solubility of air in water is proportional to the pressure of air adjacent to the water (Henry's law). Water in plants tends to be saturated with air at a concentration determined by the average atmospheric pressure of gas surrounding plants. Thus, for air to dissolve from a bubble into water, the air in the bubble has to be at a pressure in excess of atmospheric pressure. If the pressure of water (P_x) surrounding a bubble is equal to atmospheric pressure (P_a), bubbles naturally dissolve because surface tension (τ) of water raises the pressure of air in the bubble (P_b) above P_a. In general, $P_b = 2\,\tau/r + P_x$, where r is the radius of the bubble. According to the cohesion theory of sap ascent, P_x is drawn below P_a during transpiration. Since $2\,\tau/r$ of a dissolving bubble in a vessel is usually less than 0.03 MPa, and since P_x is in the range of −0.1 to −10 MPa during transpiration, P_b is usually $\leq P_a$ and hence bubbles, once formed in vessels, rarely dissolve. Repair (= dissolution) occurs only when P_x grows large via root pressure. One notable exception to this generality has recently been found in *Laurus nobilis* shrubs that appear to be able to refill embolized vessels even while P_x is at −1 MPa (Salleo et al. 1996, Tyree et al. 1999); the mechanism involved has evaded explanation (see also Holbrook and Zwieniecki 1999, Holbrook et al. 2001).

Not all plants deal with embolisms by dissolving them. Embolism dissolution is probably most important in monocots, because they lack the ability to grow new vessels by secondary growth. In dicots, however, stems can grow bigger in diameter by secondary growth of the cambium. Cambium divides the phloem (bark in woody plants) from the xylem (wood in woody plants). Cell division on the inner surface of the cambium ultimately results in the growth and differentiation of new, water-filled xylem conduits. So some species simply grow new conduits to replace conduits made dysfunctional by embolism.

FACTORS CONTROLLING THE RATE OF WATER UPTAKE AND MOVEMENT

Environmental conditions control the rate of water movement in plants. The dominant environmental factor is net solar radiation, as mentioned in Section "Water Relations of Whole Plants." Water movement can be explained by the solar energy budget of plants. We now take a more quantitative look at the solar energy budget of plants at both the leaf and stand level.

LEAF-LEVEL ENERGY BUDGETS

Objects deep in space far away from solar radiation tend to be rather cold (approximately 5 K), whereas objects on earth tend to be rather warm (270–310 K). The reason for the elevated temperature on earth is the warming effects of the sun's solar radiation. The net radiation absorbed by a leaf (R_{NL}, W m^{-2}) can be arbitrarily divided into shortwave and longwave radiation with a demarcation wavelength of 1 μm. Most shortwave radiation comes from the sun and most longwave radiation comes from the earth. As a leaf absorbs R_{NL}, the leaf temperature (T_L) rises, which increases the loss of energy from the leaf by three different mechanisms: black body radiation (B), sensible heat flux (H), and latent heat of vaporization of water (λE, where E is the evaporative flux density of water due to transpiration [mol of water s^{-1} m^{-2}] and λ is the heat required to evaporate a mole of water [J mol^{-1}]). R_{NL} can also be converted to chemical energy by the process of photosynthesis and energy storage, but these two factors are generally small for leaves and are ignored in the following equation:

$$R_{NL} = B + H + \lambda E. \qquad (6.25)$$

The B in Equation 6.25 is the black body radiation. All objects emit radiation. Very hot objects such as the sun emit mostly shortwave radiation, whereas cooler objects on earth emit mostly longwave radiation. The amount of radiation emitted increases with Kelvin temperature according to the Stefan–Boltzmann law: $B = e\,\sigma_{sb}\,T_L^4$, where e is the emisivity ($\geq$0.95 for leaves) and σ_{sb} is the Stefan–Boltzmann constant equal to 5.67×10^{-8} W m^{-2} K^{-4}.

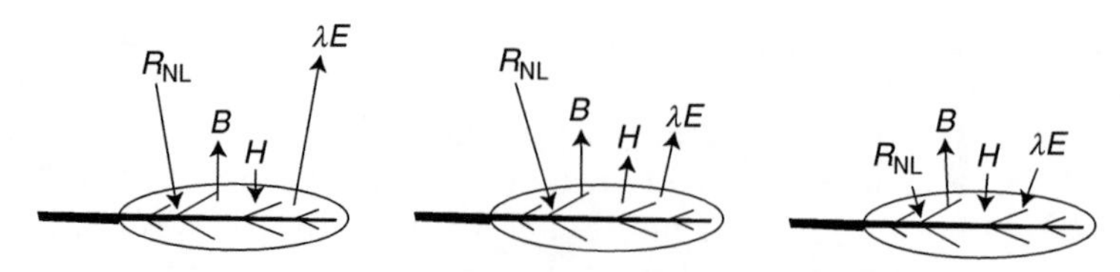

FIGURE 6.14 Solar energy budgets of leaves under different conditions. R_{NL}, net adsorbed shortwave and longwave radiation; B, black body radiation from leaves; H, sensible heat flux; λE, latent heat flux. Direction of arrow indicates direction of flux and length of arrow indicates relative magnitude. Specific conditions: (*left*) high transpiration rate so that leaf temperature is below air temperature because $B + \lambda E > R_{NL}$; (middle) intermediate transpiration rates so that leaf temperature is above air temperature; (*right*) during dew fall. Because $B > R_{NL}$ at night, leaf temperature is below the dew point temperature of the air.

H in Equation 6.25 is sensible heat flux density, that is, the heat transfer by heat diffusion between objects of different temperature. For leaves, the heat transfer is between the leaf and the surrounding air. The rate of heat transfer is proportional to the difference in temperature between the leaf and the air, so $H = k\ (T_L - T_a)$, where T_a is the air temperature and k is the heat transfer coefficient. Heat transfer is more rapid and hence k is large under windy conditions than in still air. However, the value of k is also determined by leaf size, shape, and orientation with respect to the wind direction. Readers interested in more detail should consult reports by Slatyer (1967) or Nobel (1991).

The factors determining E have already been given in Equation 6.8. If we combine Equation 6.8 with the other equations for B and H, we get

$$R_{NL} = e\ \sigma_{sb}\ T_L^4 + k(T_L - T_a) + \lambda g_L(X_i[T_L] - X_o). \tag{6.26}$$

Every term on the right side of Equation 6.25 is a function of leaf temperature. This gives the clue about how the balance is achieved in the solar energy balance equation. At any given R_{NL}, the value of T_L rises or falls until the sum of B, H, and λE equals R_{NL}. As R_{NL} increases or decreases, the value of T_L increases or decreases to reestablish equality. In practice, this equality is achieved with a leaf temperate near T_a; T_L is rarely less than 5°K below T_a or more than 15°K above T_a. Some examples of leaf energy budgets are shown in Figure 6.14. Once we know the value of T_L and solve the equation, we can calculate the value of E. If we sum the values of E for every leaf in a stand of plants, then we can calculate stand-level water flow.

Equation 6.25 provides a good qualitative understanding of the dynamics of energy balance for a community of plants (= the stand level) because the evaporation rate from a stand is the sum of the evaporation from all the leaves in the stand. However, Equation 6.25 is of little practical value because it is not possible to obtain values of R_{NL}, T_L k, and g_L for every leaf in a stand to compute the required sum.

Stand-Level Energy Budgets

Another approach to energy budgets is to measure energy and matter flux in a region of air above an entire stand of plants (Figure 6.15). A net radiometer can be used to measure net radiation (R_N) above the stand. R_N is the total radiation balance (incoming longwave and

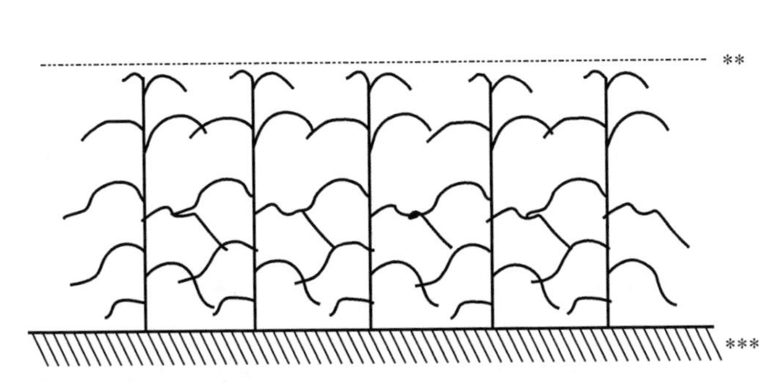

FIGURE 6.15 Points at which measurements are made for the solar energy budget of a uniform stand of plants. Net radiation (R_N) is measured at location marked*, air temperature and relative humidity are measured at locations marked* and **, and rate of heat storage in soil is measured at***.

shortwave radiation minus outgoing longwave and shortwave radiation, so at the leaf-level $R_N = R_{NL} - B$); the energy balance equation for this situation is

$$R_N = G + H + \lambda E. \quad (6.27)$$

The rate of heat storage in the soil G (W m^{-2}), can be significant because soil temperature can change a few degrees Kelvin on a daily basis in the upper few centimeters of soil. Heat storage rate is usually measured with soil-heat-flux sensor. The H and λE terms are similar to those in Equation 6.25, but the equations define the flux densities between the heights marked * and ** in Figure 6.15. The flux of heat and water vapor across the distance ΔZ is controlled by vertical air convection (eddy) and the defining equations are

$$H = -C_p \rho_a K_e \Delta T / \Delta Z \quad (6.28)$$

and

$$\lambda E = \lambda K_e \, \Delta[H_2O] / \Delta Z, \quad (6.29)$$

where K_e is the eddy transfer coefficient C_p is the heat capacity of air, ρ_a is the air density, ΔT is the difference in air temperature measured at the two levels in Figure 6.15 separated by a height difference of ΔZ, and $\Delta[H_2O]$ is the difference in water vapor concentration at the two levels in Figure 6.15 separated by a height difference of ΔZ.

It is easy to measure ΔT and $\Delta[H_2O]$, but difficult to assign a value for K_e, which changes dynamically with changes in wind velocity. The usual practice is to measure the ratio of $H/\lambda E = \beta$, which is called the Bowen ratio. From Equation 6.28 and Equation 6.29 it can be seen that

$$\beta = \frac{C_p \rho_a}{\lambda} \frac{\Delta T}{\Delta[H_2O]}. \quad (6.30)$$

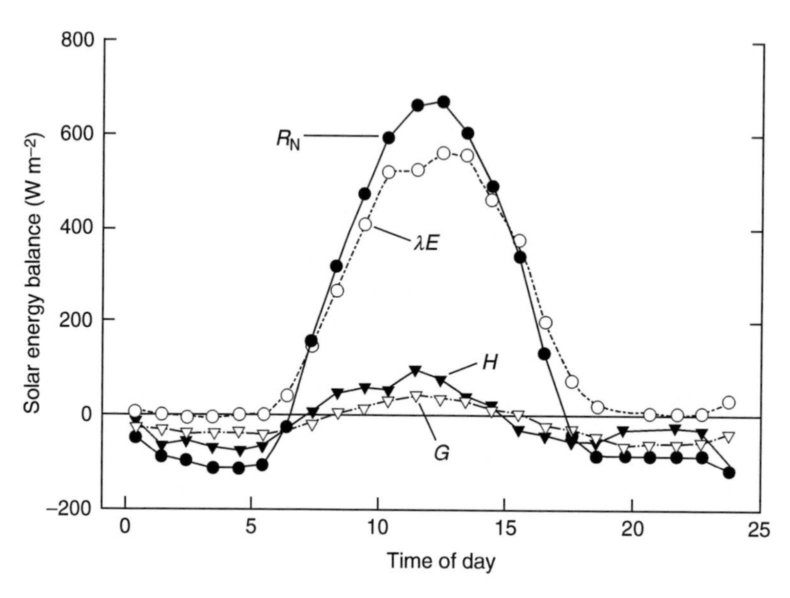

FIGURE 6.16 Solar energy budget values measured in a meadow. R_N, net solar radiation of the meadow; λE, latent heat flux; H, sensible heat flux; and G, rate of heat storage in soil. (Adapted from Slatyer, R.O., *Plant Water Relationships*, Academic Press, New York, 1967.)

Estimates of stand water use are obtained by solving Equation 6.29 and Equation 6.30 for E

$$E = \frac{R_N - G}{\lambda[\beta + 1]}. \tag{6.31}$$

An example of the total energy budget measured over a pasture is reproduced in Figure 6.16. The data in Figure 6.16 shows again that the main factor determining E is the amount of solar radiation, R_N, as in Figure 6.4. However, the leaves in a stand of plants do have some control over the value of E. As soils dry and leaf water potential falls, stomatal conductance falls. This causes a reduction in E, an increase in leaf temperature, and thus an increase in H.

Very intensive monitoring of climatic data is needed to obtain data for a solution to Equation 6.31. Temperature, R_N, and relative humidity has to be measured every second at several locations. Ecologists prefer to estimate E with a less complete data set. Fortunately, the Penman–Monteith formula (Monteith 1964) permits a relatively accurate estimate of E under some restricted circumstances. The Penman–Monteith formula is derived from energy budget equations together with a number of approximations to turn nonlinear functions into linear relations. After many obtuse steps in the derivations (Campbell 1981), a formula of the following form results:

$$E = \frac{e'(R_N - G) + \rho_a C_p V_{pd} g_a}{\lambda\left[e' + \gamma\left(1 + \frac{g_a}{g_c}\right)\right]}, \tag{6.32}$$

where e' is the rate of change of saturation vapor pressure with temperature at the current air temperature, V_{pd} is the difference between the vapor pressure of air at saturation and the current vapor pressure, γ is the psychometric constant, g_a is the aerodynamic conductance, and g_c is the canopy conductance of the stand. The problem with Equation 6.32 is that g_a and g_c are both difficult to estimate and are not constant. The value of g_a depends on wind speed and roughness parameters that describe the unevenness at the boundary between the canopy and the bulk air. Surface roughness affects air turbulence and hence the rate of energy transfer at any given wind speed. Surface roughness changes as stands grow and is difficult to estimate in terrain with hills or mountains. The value of g_c is under biological control and difficult to estimate from climatic data. In one study on an oak forest in France, E was estimated independently by Granier sap flow sensors in individual oak trees. This permitted the calculation of g_c versus season, morphological state of the forest, and climate (Granier and Bréda 1996). The value of g_c was found to be a function of global radiation, V_{pd}, leaf area per unit ground area, which changes with season, and relative extractable water, which is a measure of soil dryness. However, once all of these factors were taken into account, Equation 6.32 provided a reasonable estimate of half-hourly estimate of E over the entire summer.

PLANTS UNDER STRESS

WILTING

Wilting denotes the limp, flaccid, or drooping state of plants during drought. Wilting is most evident in leaves that depend on cell turgor pressure to maintain their shape; hence, wilting occurs when turgor pressure falls to zero. Many plants maintain leaf shape through rigid leaf fiber cells. Wilting in these species is considered to commence at the turgor loss point. Wilted plants generally have low E because stomata are closed and g_s is very small when leaf water potential Ψ_L falls during drought. Continued dehydration beyond the wilt point usually causes permanent loss of hydraulic conductance due to cavitation in the xylem. Complete loss of hydraulic conductance usually causes plant death, but there is a lot of species diversity in the water potentials causing loss of hydraulic conductance (Figure 6.11). Some plants in arid environments avoid drought either by short reproductive cycles confined to brief wet periods or by bearing deep roots that can access deep sources of soil water (Kramer 1983).

As plants approach the wilt point there is a gradual loss in stomatal conductance and, hence, a reduction in E and photosynthetic rate (Schulze and Hall 1982). Some species are much more sensitive than others (Figure 6.17b). The short-term effects of decreased Ψ_L on transpiration are less dramatic than long-term effects (Figure 6.17a). Long-term effects of drought are mediated by hormone signals from roots that cause a medium-term decline in g_s and by changes in root morphology, for example, loss of fine roots, suberization of root surfaces, and formation of corky layers (Ginter-Whitehouse et al. 1983). The morphological changes to roots cause a decrease in whole-plant hydraulic conductance (K_p); therefore, Ψ_L becomes $\Psi_L = \Psi_{soil} - E/K_p$. Very severe drought can further lower K_p due to cavitation of xylem vessels.

WATERLOGGING

Waterlogging denotes an environmental condition of soil water saturation or ponding of water that can last for just a few hours or for many months. Plants absent from flood-prone sites are damaged easily by waterlogging. On the other hand, plants that inhabit flood-prone sites include species that can grow actively in flooded soils or species that survive flooding in a quiescent or dormant state. Paradoxically, the most common sign of tobacco roots with an excess of water is the development of a water deficit in the leaves (Kramer 1983, Kramer and Jackson 1954), but flood-tolerant species are not so easily affected.

Flooding often affects root morphology and physiology. The wilting and defoliation that is found on flooding can be traced to an increased resistance to water flux in the roots (Mees and Weatherley 1957). In most flood-tolerant species, flooding induces morphological changes in the roots. The modifications usually involve root thickening with an increase in

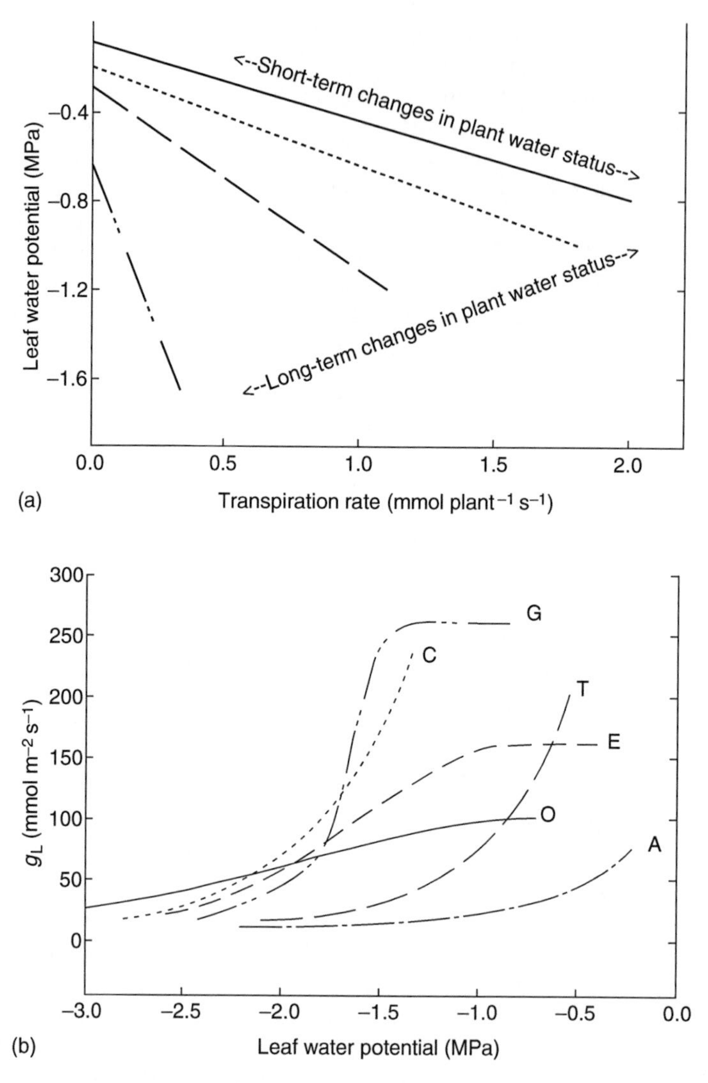

FIGURE 6.17 Effects of drought on transpiration, leaf water potential, and stomatal conductance (a) Short- and long-term effects of drought on leaf water potential and transpiration. The short-term effects are dynamic changes in leaf water potential that might occur in the course of 1 day. The long-term effects are associated with slow drying of soil over many days. The different lines show the short-term relation between leaf water potential versus transpiration rate measured in well-watered plants (—) or after several days without irrigation, for example, 5 days (· · ·), 10 days (– – –), or 20 days (– ·· – ·· –). (b) Short-term effects of leaf water potential on stomatal conductance. Species represented are: A, *Acer saccharum*; C, *Corylus avellana*; E, *Eucalyptus socialis*; G, *Glycine max*; T, *Triticum aestivum*; and O, *Olea europeae*. (Adapted from Schulze, E.D. and Hall, A.E., *Encyclopedia of Plant Physiology New Series*, O.L. Lange, P.S. Nobel, C.B. Osmond, H. Ziegler, eds, Springer-Verlag, New York, 1982.)

porosity. The increase in porosity increases the rate of oxygen diffusion to root tips and thus permits continued aerobic metabolism in the inundated roots. In flood-sensitive species, root and shoot growth are rapidly reduced on flooding, and root tips may be damaged. Growth of roots can be renewed only from regions proximal to the stem. The physiological responses and adaptations to waterlogging are numerous and beyond the scope of this chapter, but interested readers may consult a report by Crawford (1982).

REFERENCES

Bidwell, R.G.S., 1979. Plant Physiology. Macmillan, New York.

Campbell, G.S., 1981. Fundamentals of radiation and temperature relations. In: O.L. Lange, P.S. Nobel, C.B. Osmond, and H. Ziegler, eds. Encyclopedia of Plant Physiology New Series, Vol. 12A. Springer-Verlag, New York, pp. 11–40.

Chang, J.-H., 1968. Climate and Agriculture. Aldine Publishing, Chicago, IL.

Cochard, H., P. Cruiziat, and M.T. Tyree, 1992. Use of positive pressures to establish vulnerability curves: Further support for the air-seeding hypothesis and possible problems for pressure-volume analysis. Plant Physiology 100: 205–209.

Cochard H., G. Damour, C. Bodet, I. Tharwat, M. Poirier, and T. Ameglio, 2005. Evaluation of a new centrifuge technique for rapid generation of xylem vulnerability curves. Physiologia Plantarum 124: 410–418.

Crawford, R.M.M., 1982. Physiological response to flooding. In: O.L. Lange, P.S. Nobel, C.B. Osmond, and H. Ziegler, eds. Encyclopedia of Plant Physiology New Series, Vol. 12B. Springer-Verlag, New York, pp. 453–477.

Davies, W.J. and J. Zhang, 1991. Root signals and the regulation of growth and development of plants in drying soil. Annual Review of Plant Physiology and Plant Molecular Biology 42: 55–76.

Dawson, T.E., 1993. Hydraulic lift and water parasitism by plants: Implications for water balance, performance, and plant-plant interactions. Oecologia 95: 565–574.

Dixon, H.H. and J. Joly, 1894. On the ascent of sap. Philosophical Transactions of the Royal Society London, Series B, 186, 563–576.

Ewers, F.W. and M.H. Zimmermann, 1984a. The hydraulic architecture of balsam fir (*Abies balsamea*). Physiologia Plantarum 60: 453–458.

Ewers, F.W. and M.H. Zimmermann, 1984b. The hydraulic architecture of eastern hemlock (*Tsuga canadiensis*). Canadian Journal of Botany 62: 940–946.

Ewers, F.W., J.B. Fisher, and K. Fichtner, 1991. Water flux and xylem structure in vines. In: F.E. Putz and H.A. Mooney, eds. The Biology of Vines. Cambridge University, Cambridge, pp. 127–160.

Frensch, J. and E. Steudle, 1989. Axial and radial hydraulic resistance to roots of maize (*Zea mays* L.) Plant Physiology 91: 719–726.

Ginter-Whitehouse, D.L., T.M. Hinckley, and S.G. Pallardy, 1983. Spatial and temporal aspects of water relations of three tree species with different vascular anatomy. Forest Science 29: 317–329.

Granier, A. and N. Bréda, 1996. Modeling canopy conductance and stand transpiration of an oak forest from sap flow measurements. Annales des Sciences Forestières 53: 537–546.

Green, P.B., R.O. Erickson, and J. Buggy, 1971. Metabolic and physical control of cell elongation rate—in vivo studies of *Nitella*. Plant Physiology 47: 423–430.

Holbrook, N.M. and M.A. Zwieniecki, 1999. Embolism repair and xylem tension. Do we need a miracle? Plant Physiology 120: 7–10.

Holbrook, N.M., E.T. Ahrens, M.J. Burns, and M.A. Zwieniecki, 2001. In vivo observation of cavitation and embolism repair using magnetic resonance imaging. Plant Physiology 126: 27–31.

Jarbeau, J.A., F.W. Ewers, and S.D. Davis, 1995. The mechanism of water stress-induced embolism in two species of chaparral shrubs. Plant Cell and Environment 126: 695–705.

Kramer, P.J., 1983. Water Relations of Plants. Academic Press, New York.

Kramer, P.J. and W.T. Jackson, 1954. Causes of injury to flooded tobacco plants. Plant Physiology 29: 241–245.

Lewis, A.M., V.D. Harnden, and M.T. Tyree, 1994. Collapse of water-stress emboli in the tracheids of *Thuja occidentalis* L. Plant Physiology 106: 1639–1646.

Machado, J.-L. and M.T. Tyree, 1994. Patterns of hydraulic architecture and water relations of two tropical canopy trees with contrasting leaf phonologies: *Ochroma pyramidale* and *Pseudobombax septenatum*. Tree Physiology 14: 219–240.

Mees, G.C. and P.E. Weatherley, 1957. The mechanism of water absorption by roots. II. The role of hydrostatic pressure gradients. Proceedings of the Royal society London, Series B. 147: 381–391.

Monteith, J.L., 1964. Evaporation and environment. Symposium of the society for Experimental Biology 19: 205–234.

Nobel, P.S., 1991. Physiochemical and Environmental Plant Physiology. Academic Press, New York.

North, G.B., F.W. Ewers, and P.S. Nobel, 1992. Main root–lateral root junctions of two desert succulents: changes in axial and radial components of hydraulic conductivity during drying. American Journal of Botany 79: 1039–1050.

Patiño, S., M.T. Tyree, and E.A. Herre, 1995. Comparison of hydraulic architecture of woody plants of differing phylogeny and growth form with special reference to free-standing and hemi-epiphytic *Ficus* species from Panama. New Phytologist 129: 125–134.

Pickard, W.F., 1981. The ascent of sap in plants. Progress in Biophysics and Molecular Biology 37: 181–229.

Ren, Z. and E. Sucoff, 1995. Water movement through *Quercus rubra* L. Leaf water potential and conductance during polycyclic growth. Plant Cell and Environment 18: 447–453.

Richards, J.R. and M.M. Caldwell, 1987. Hydraulic lift: Substantial nocturnal water transport between soil layers by *Artemisia tridentate* roots. Oecologia 74: 486–489.

Salleo, S., M.A. LoGullo, D. De Paoli, and M. Zippo, 1996. Xylem recovery from cavitation-induced embolism in young plants of *Laurus nobilis*: a possible mechanism. New Phytologist 132: 47–56.

Scholander, P.F., H.T. Hammel, E.D. Bradstreet, and E.A. Hemmingsen, 1965. Sap pressures in vascular plants. Science 148: 339–346.

Schulze, E.D. and A.E. Hall, 1982. Stomatal responses, water loss and CO_2 assimilation rats of plants in contrasting environments. In: O.L. Lange, P.S. Nobel, C.B. Osmond, and H. Ziegler, eds. Encyclopedia of Plant Physiology New Series, Vol. 12B. Springer-Verlag, New York, pp. 181–230.

Slatyer, R.O., 1967. Plant Water Relationships. Academic Press, New York.

Sperry, J.S. and U.G. Hacke, 2004a. Analysis of circular border pit function. I. Angiosperm vessels with homogeneous pit membranes. American Journal of Botany 91: 369–385.

Sperry, J.S. and U.G. Hacke, 2004b. Analysis of circular border pit function. II. Gymnosperm tracheids with torus-margo pit membranes. American Journal of Botany 91: 386–400.

Sperry, J.S. and M.T. Tyree, 1988. Mechanism of water stress-induced xylem embolism. Plant Physiology 88: 581–587.

Sperry, J.S. and M.T. Tyree, 1990. Water-stress-induced xylem embolism in three species of conifers. Plant Cell and Environment 13: 427–436.

Steudle, E., 1992. The biophysics of plant water: Compartmentation, coupling with metabolic processes, and flow of water in plant roots. In: G.N. Somero, C.B. Osmond, and C.L. Bolis, eds. Water and Life: Comparative Analysis of Water Relationships at the Organismic, Cellular, and Molecular Levels. Springer-Verlag, Heidelberg, pp. 173–204.

Steward, F.C., 1964. Plants at Work. Addison-Wesley, Reading, MA.

Turner, N.C. and M.M. Jones, 1980. Turgor maintenance by osmotic adjustment: a review and evaluation. In: N.C. Turner and P.J. Kramer, eds. Adaptation of Plants to Water and High Temperature Stress. Wiley, New York, pp. 87–104.

Tyree, M.T., 1976. Physical parameters of the soil-plant-atmosphere system: Breeding for drought characteristics that might improve wood yield. In: M.G.R. Cannell and F.T. Last, eds. Tree Physiology and Yield Improvement. Academic Press, New York, pp. 329–348.

Tyree, M.T., 1997. The cohesion–tension theory of sap ascent: Current controversies. Journal of Experimental Botany 48: 1753–1765.

Tyree, M.T., 2003. Hydraulic limits on tree performance: Transpiration, carbon gain and growth of trees. Trees 17: 95–100.

Tyree, M.T. and J.D. Alexander, 1993. Hydraulic conductivity of branch junctions in three temperate tree species. Trees 7: 156–159.

Tyree, M.T. and F.W. Ewers, 1991. The hydraulic architecture of trees and other woody plants. New Phytologist 119: 345–360.

Tyree, M.T. and F.W. Ewers, 1996. Hydraulic architecture of woody tropical plants. In: A. Smith, K. Winter, and S. Mulkey, eds. Tropical Plant Ecophysiology. Chapman and Hall, New York, pp. 217–243.

Tyree, M.T. and H.T. Hammel, 1972. The measurement of the turgor pressure and the water relations of plants by the pressure-bomb technique. Journal of Experimental Botany 23: 267–282.

Tyree, M.T. and P.G. Jarvis, 1982. Water in tissues and cells. In: O.L. Lange, P.S. Nobel, C.B. Osmond, and H. Ziegler, eds. Physiological Plant Ecology II. Water Relations and Carbon Assimilation. Springer-Verlag, Berlin, pp. 35–77.

Tyree, M.T. and J.S. Sperry, 1989. The vulnerability of xylem to cavitation and embolism. Annual Review of Plant Physiology and Molecular Biology 40: 19–38.

Tyree, M.T. and S. Yang, 1992. Hydraulic conductivity recovery versus water pressure in xylem of *Acer saccharm*. Plant Physiology 100: 669–676.

Tyree, M.T. and M.H. Zimmermann, 2002. Xylem Structure and the Ascent of Sap. Springer, New York.

Tyree, M.T., S.D. Davis, and H. Cochard, 1994a. Biophysical perspective of xylem evolution: is there a tradeoff of hydraulic efficiency for vulnerability to dysfunction? Journal of the International Association of Wood Anatomists 15: 335–360.

Tyree, M.T., S. Yang, P. Cruiziat, and B. Sinclair, 1994b. Novel methods of measuring hydraulic conductivity of tree root systems and interpretation using AMAIZED: A maize-root dynamic model for water and solute transport. Plant Physiology 104: 189–199.

Tyree, M.T., S. Patiño, J. Bennink, and J. Alexander, 1995. Dynamic measurements of root hydraulic conductance using a high-pressure flowmeter in the laboratory and filed. Journal of Experimental Botany 46: 83–94.

Tyree, M.T., V. Velez, and J.W. Dalling, 1998. Growth dynamics of root and shoot hydraulic conductance in seedlings of five neotropical species: scaling to show possible adaptation to differing light regimes. Oecologia 114: 293–298.

Tyree, M.T., S. Salleo, A. Nardini, M.-A. LoGullo, R. Mosca, 1999. Refilling of embolized vessels in young stems of Laurel: Do we need a new paradigm? Plant Physiology 120: 11–21.

van den Honert, T.H., 1948. Water transport in plants as a caternary process. Discussions of the Faraday Society 3: 146–153.

Wei, C., M.T. Tyree, and E. Steudle, 1999. Direct measurement of xylem pressure in leaves of intact maize plants. A test of the cohesion-tension theory taking hydraulic architecture into consideration. Plant Physiology 121: 1191–1205.

Wei, C., M.T. Tyree, and J.P. Bennink, 2000. The transmission of gas pressure to xylem fluid pressure when plants are inside a pressure bomb. Journal of Experimental Botany 51: 309–316.

Wheeler J.K., J.S. Sperry, U.G. Hacke, and N. Hoang, 2005. Inter-vessel pitting and cavitation in Rosaceae and other plants: A basis for safety versus efficiency trade-off in xylem transport. Plant Cell and Environment 28: 800–812.

Yang, Y. and M.T. Tyree, 1993. Hydraulic resistance in the shoots of *Acer saccharum* and its influence on leaf water potential and transpiration. Tree Physiology 12: 231–242.

Yang, Y. and M.T. Tyree, 1994. Hydraulic architecture of *Acer saccharum* and *A. rubrum*: comparison of branches to whole trees and the contribution of leaves to hydraulic resistance. Journal of Experimental Botany 45: 179–186.

Yang, S., X. Liu, and M.T. Tyree, 1998. A model of stomatal conductance in sugar maple (*Acer saccharm* Marsh). Journal of Theoretical Biology 191: 197–211.

Yount, D.E., 1989. Growth of bubbles from nuclei. In: A.O. Brubakk, B.B. Hemmingsen, and G. Sundnes, eds. Supersaturation and Bubble Formation in Fluids and Organisms. Tapir Publishers, Trondheim, pp. 131–163.

Zimmermann, M.H., 1978. Hydraulic architecture of some diffuse-porous trees. Canadian Journal of Botany 56: 2286–2295.

Zimmermann, M.H., 1983. Xylem Structure and the Ascent of Sap. Springer-Verlag, Berlin.

Zotz, G., M.T. Tyree, S. Patiño, and M.R. Carlton, 1998. Hydraulic architecture and water use of selected species from a lower montane forest in Panama. Trees 12: 302–309.

7 Responses of Plants to Heterogeneous Light Environments

Robert W. Pearcy

CONTENTS

INTRODUCTION

Of all the environmental factors affecting plants, light is perhaps the most spatially and temporally heterogeneous. This heterogeneity takes on special importance in tropical forests in particular because here light is considered to be the single-most limiting resource for plant growth and reproduction. Accordingly, the life cycle and physiological responses of many tree and understory species have been shown to be closely keyed to changes in light availability

(Bazzaz and Pickett 1980, Denslow 1980, 1987, Strauss-Debenedetti and Bazzaz 1991, Canham et al. 1994, Sipe and Bazzaz 1994, Gratzer et al. 2004). Although less studied, herbaceous and shrub communities also exhibit heterogeneity in light environments (Tang et al. 1989, Washitani and Tang 1991, Ryel et al. 1994, 1996, Derner and Wu 2001). Much of the emphasis in ecology has been on the community and ecosystem consequences of environmental heterogeneity. For light in particular, the role of gaps in forest regeneration and diversity has received much attention. Several current models of forest dynamics have as their foundation differential species responses to heterogeneous light (Friend et al. 1993, Acevedo et al. 1996, Pacala et al. 1996, Shugart 1996, Bolker et al. 2003). Considerable attention has also been given to the responses of individuals in terms of their ability to capture and use light as a spatially and temporally heterogeneous resource (Caldwell et al. 1986, Chazdon 1992, Newell et al. 1993, Pearcy et al. 1994). At the level of the individual, the concepts of heterogeneity and plasticity are intertwined. Phenotypic plasticity has a central explanatory role in how individuals across populations differ in form and function in response to spatial heterogeneity. At the same time, physiological constraints on organism responses to temporal heterogeneity determine to a large extent the nature of the adaptive responses to it.

Although the concept of heterogeneity and its role in ecology has been widely discussed, its use has been somewhat imprecise (Stuefer 1996). A heterogeneous habitat is one where a resource is patchily distributed. This patchy distribution can be either spatial or temporal in nature and has a certain grain size. If the grain size is small or short relative to the scale of the functional response then the environment may be perceived as uniform. If, on the other hand, the grain size is large or long relative to the functional response then the environment may be perceived as nonuniform. Temporal gain size, for example, is important in determining whether there is a selective advantage to some plastic response to the heterogeneity. Since photosynthetic acclimation occurs over days, mechanisms that integrate and respond to daily totals rather than instantaneous photon fluxes would be favored (Chabot 1979). On the other hand, stomatal responses occur over minutes requiring sensing of the temporal heterogeneity on a much shorter timescale (Zeiger et al. 1985, Tinoco-Ojanguren and Pearcy 1993a). Patches also have contrast representing the sharpness of the transition from one patch to the other. Finally, predictability is an important but little-considered aspect of heterogeneity (Stuefer 1996). It is the accuracy with which a given resource availability can be predicted in space or time. The occurrence of one sunfleck is a reasonable predictor that others follow. Thus, a response to the first sunfleck, even if it is too delayed to significant carbon gain in that sunfleck, can improve the usage of subsequent sunflecks (Pearcy et al. 1994).

HETEROGENEITY AND PLASTICITY: SPATIAL AND TEMPORAL ASPECTS

Although it is often convenient to separate spatial and temporal heterogeneity as I have done in this chapter, the separation is rather artificial. Sunflecks cause the smallest scale of spatial heterogeneity and also the most rapid scale of temporal heterogeneity. Moving up the scales through gaps the spatial and temporal scales are highly correlated. Indeed in terms of analysis, spatial and temporal scales are frequently interchangeable. Wavelet analysis or autocorrelation, two commonly used techniques for analysis of heterogeneity (Baldocchi and Collineau 1994, Saunders et al. 2005), can be applied equally well to either spatial or temporal scales.

Differences do become apparent, however, when the nature of the responses is considered. For temporal heterogeneity, the dynamics of the system as determined by the time constants of the underlying processes must be considered. A dynamic system is one in which the current rate is dependent in some way not only on current conditions but also on the rate at some previous time. A system is at steady state when it depends only on the current input parameters. Studies of spatial heterogeneity almost invariably treat systems as steady state even though the spatial pattern observed may be a consequence of an underlying dynamics. Of course, essentially all physiological and ecological systems are dynamic if observed on the appropriate timescale. For example, characterizing photosynthetic responses to gap formation with a sequence of steady-state light curve measurements may be perfectly appropriate, whereas the growth response to a gap formation needs to consider the dynamics because of leaf turnover. Therefore, whether it is necessary to consider the dynamics depends on whether the system is adequately described by a series of steady states occurring in response to the changing temporal heterogeneity. If its dynamic properties are important, they cause a significant deviation from the steady-state predictions.

The concept in spatial heterogeneity most similar to dynamics in temporal heterogeneity is grain size (Levins 1968). Spatial heterogeneity is considered to be fine grained if the patch size is such that the individual plant may be influenced by more than one patch. It is coarse grained if the plant is wholly within one patch and not influenced significantly by adjacent patches. The gap–understory dichotomy is an example of a coarse-grained patchiness, whereas the understory itself with sunflecks and shade is an example of fine-grained patchiness. Plants within a coarse-grained patch generate a fine-grained heterogeneity within their canopy because of self-shading. Of course, grain size applies equally well to temporal heterogeneity, and indeed, as illustrated by the sunfleck example, spatial grain size often has a temporal origin. Therefore, the comparison of the effects of grain size to dynamics rests on the question of whether nearby patches influence the response to the environment in each other. Although this question has not been explored extensively it seems clear that hormonal and hydraulic signals provide a mechanism for such interactions.

It is also necessary to consider how heterogeneity is measured and how the plant perceives and responds to it. Measurements of the environment almost always involve linear scales, whereas the plant may respond in a distinctly nonlinear fashion. Nonlinear functions require sufficient temporal or spatial sampling to avoid sampling errors (Ruel and Ayres 1999). The response of photosynthesis to PFD is a particularly relevant example since it strongly increases as PFD increases at low values but then exhibits saturation at higher PFDs. Depending on the range of PFDs encountered, the apparent heterogeneity in photosynthetic rates may be greater or lesser than the apparent heterogeneity in the PFD. Since photoinhibition occurs in response to excess PFD, the apparent heterogeneity in it would be still different. Therefore, appropriate measures of heterogeneity, preferably based on the biological responses are important.

At the level of the individual, there can be several types of responses to environmental heterogeneity, depending on the scale. If the supply of a limiting resource is changed its acquisition is changed simply because of changes in diffusion rates and the kinetics of the enzymes involved in uptake. There can also be regulatory responses that may further increase or decrease the capacity for resource acquisition. An example would be a change in stomatal conductance (g_s) influencing the diffusion of CO_2 into the leaf and hence changing the CO_2 assimilation rate. Finally, there can be some change in the properties of the organism itself such as an increase in the concentration of enzymes in the leaf that influences the capacity for CO_2 assimilation. The latter falls clearly within the realm of a plastic response to the environment and if it improves the performance of the plant in heterogeneous environments then it would be considered an acclimation response. A change in g_s would not fall within a modern concept of plasticity. The modern concept includes those changes in an organism that enhance performance in a given environment rather than, as originally envisioned by Bradshaw (1965), as all the phenotypic changes observed when the environment is changed. Thus, a phenotypic change due to some stress injury would not be included in the modern concept of plasticity, but some adjustment that minimizes the stress injury would be included. Regulatory responses tend to function to fine-tune the capacity of component steps, such as g_s and the biochemical capacity for CO_2 fixation so that they are in balance with no one step greatly in excess or limitation. The regulated variations in a character such as g_s in response to changing environments are not an expression of plasticity, but differences in the maximum g_s are rightfully expressions of phenotypic plasticity.

The difference between regulatory responses and acclimation has its greatest consequence for the time constants of the response to environmental change. A change in concentrations of enzymes takes time for synthesis or degradation, whereas regulation of the activity of already present enzymes can occur on a much faster timescale. Moreover, much of the plasticity expressed in response to changing light environments involves development of new leaves that have sun and shade characteristics, requiring even more time. Acclimation is too slow to respond to or provide any significant advantage in sunflecks but is the dominant mechanism for phenotypic adaptation to gap formation and subsequent canopy closure.

The evolutionary and ecological role of plasticity has received increasing attention over the past decade or so (Schlichting 1986, Callahan et al. 1997, Sultan 2001, Schlichting and Smith 2002, Miner et al. 2005). In plants, phenotypic plasticity is most often expressed at the modular level where individual shoots, leaves, and so on respond to their own local environment (de Kroon et al. 2005). Historically, much of the focus in evolutionary theory has been on adaptive, habitat-based specialization with plasticity viewed as an alternative (the specialist or generalist argument). This view largely ignores the important role that modular-based expression of plasticity plays. Most of the focus in evolutionary theory has been on specialization. In fact, in heterogeneous environments, plasticity is not an alternative since it may be required for survival, or at least for maximizing performance. Moreover, plasticity itself may not be an alternative but a form of specialization. Although a characteristic arising as a result of plastic phenotypic variation cannot be the subject of selective forces, the capacity to express plastic variations in phenotype may be under genetic control and therefore subject to selection (Scheiner and Lyman 1991, Scheiner 1993, Via et al. 1995). This is the basis for the notion that species adapted to certain environments may have a greater capacity for acclimation than others from other environments. The role of plasticity and especially acclimation has been most commonly examined in terms of temporal heterogeneity. Paradoxically, they have been studied most commonly as a static process with plants grown for a long term in different light environments rather than as a dynamic process responding to temporal changes in the environment. Functionally, the mechanisms involved in temporal and spatial scales are similar, but a full understanding of the temporal scale requires understanding of the dynamics of the mechanisms and their interaction with development and not just their static end result.

Understanding the adaptive significance of plasticity of a character requires knowledge of how it scales up to whole-plant performance. If a plastic response functions to improve performance, most often the result at the next level up is a greater homeostasis. For example, sun shade acclimation at the leaf level does not compensate for the large difference in resource supply (light) between sun and shade environments, but it does result in relatively smaller differences in carbon gain and growth between the sun and the shade than would be expected in the absence of acclimation (Sims et al. 1994, Evans and Poorter 2001). The scaling needs to consider not only the benefits provided by the character itself but also the costs that impact the overall performance. Costs are often more difficult to assess and include both the direct (construction and maintenance) and opportunity costs.

SPATIAL HETEROGENEITY IN LIGHT ENVIRONMENTS

CAUSES AND SCALES OF SPATIAL HETEROGENEITY

Variation in overstory canopy structure is the primary factor causing the great spatial heterogeneity evident in forest understory light environments (Figure 7.1). Forest overstory canopies never close completely, in part because foliage tends to be clumped at the ends of

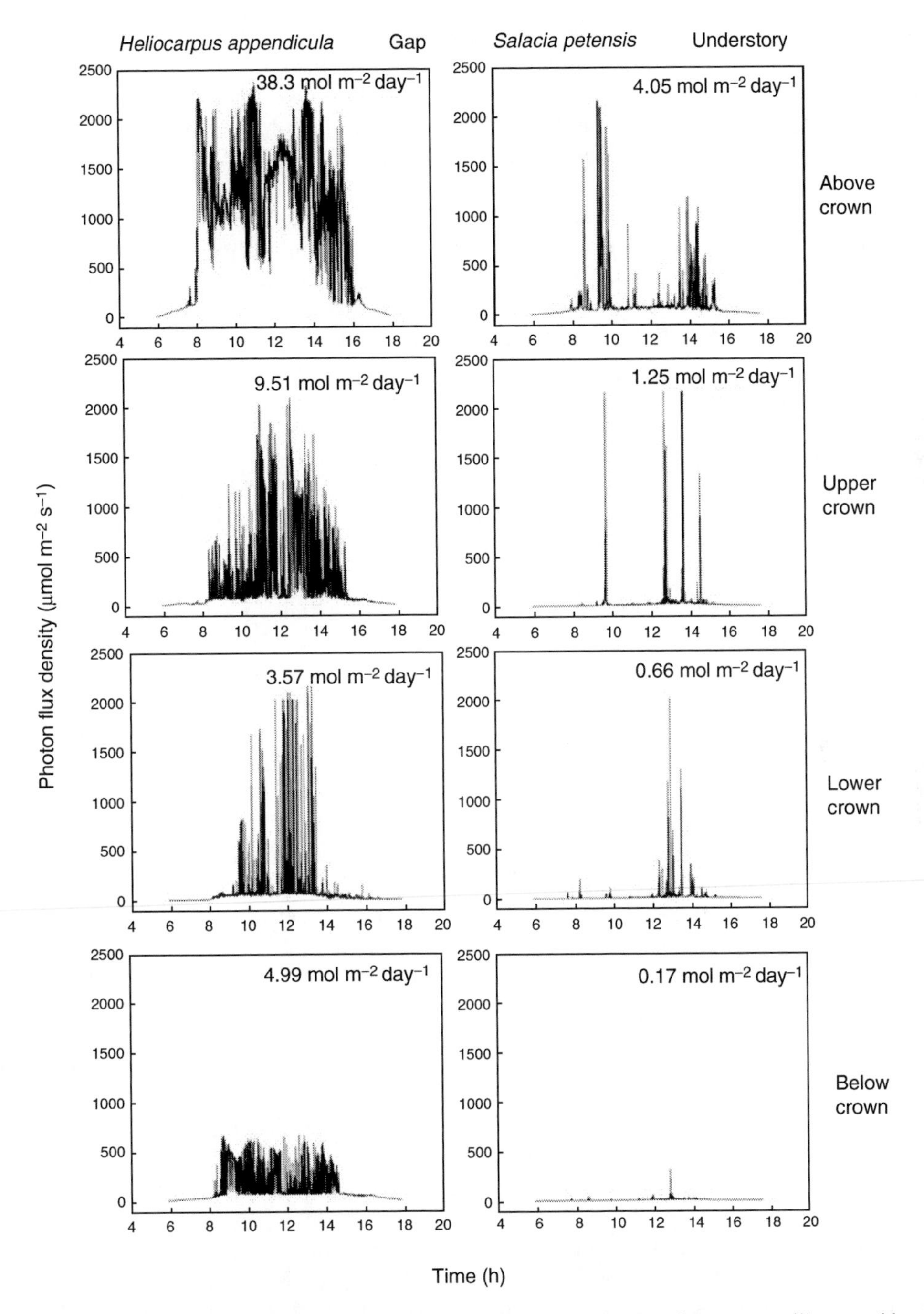

FIGURE 7.1 Patterns of spatial and temporal heterogeneity in the light environment as illustrated by the diurnal courses of PFD measured with sensors located at four positions in the crowns of a pioneer species, *Heliocarpus appendicula*, growing in a 900 m^2 gap and understory tree *Salicaia petensis* occuring under a closed canopy. The daily integrated PFD at each sensor is given in each box. (Adapted from Kuppers, M., Timm, H., Orth, F., Stegemann, J., Stober, R., Paliwal, K., Karunaichamy, K.S.T., and Ortez, R., *Tree Physiol.*, 16, 69, 1996. With permission.)

branches and because of the statistical probability that even at the maximum leaf area index, small gaps still exist. These small gaps create sunflecks, a major source of heterogeneity in understory light environments. Species differences in architecture influencing light transmission may be important in heterogeneity on a scale equivalent to a tree crown (Canham et al. 1994, Kabakoff and Chazdon 1996, Parker et al. 2001). Understory trees and shrubs then create further spatial heterogeneity. Crowns of adjacent trees rarely grow into one another so that between them gaps may be more prevalent. Wind-induced abrasion by adjacent crowns may be important in creating gaps between tree crowns especially in areas prone to high winds (Meng et al. 2006). At the next scale-up, branch falls create slightly larger canopy gaps. Individual tree falls create gaps from 50 to 600 m^{-2} depending on the size of the individual and the collateral damage. In most lowland tropical forests, 3%–15% of the land area is in gaps at various stages of recovery (Brokaw 1985). Estimated turnover rates (mean time between gaps at any one point) range from 60 to 450 years with a mean of about 100 years. Since the effect of a gap on the light environment extends into the adjacent understory for some distance because of lateral light transmission, it is likely that every plant in the forest is influenced by a gap at some time in its life cycle, and many may be influenced multiple times. Responses to even larger scales of landscape heterogeneity caused by fire, landslides, logging, and so on are beyond the scope of this chapter.

Within each of these scales of spatial heterogeneity, there exists a further fine-scale heterogeneity. Sunflecks exhibit strong gradients of PFD across them because of penumbral effects arising because of the approximately 0.27° arc diameter of the solar diskpoints from which part of the solar disk is occluded in partial shadow or penumbra (Cescatti and Niinemets 2004). Only when the full solar disk is visible (i.e., the gap is large enough and the solar disk is fully within it) is an area of full, direct-beam PFD present. Penumbra, while not changing the total PFD transmitted does even out the distribution and hence diminishes heterogeneity. Both peak values are diminished and the area receiving partial direct PFD is increased. The greater area means that an understory plant intercepts a sunfleck for a longer time, and since the penumbral light is at a lower PFD than the full direct-beam PFD, it is used more efficiently. Simulations reveal that this evening of the PFD can more than double the assimilation due to a sunfleck in an understory plant as compared with simulations where the sun is treated as a point source (Pearcy unpublished results). Under tall forest canopies, most sunflecks are entirely penumbral light with maximum PFDs of 0.1–0.5 of the full, direct-beam PFD. Deep crowns with many narrow leaves such as conifers cause most of the PFD within them to be penumbral. Under these circumstances, total crown photosynthesis can be enhanced by 10%–17% (Ryel et al. 2001), whereas photosynthesis of the lower branches can be enhanced by as much as 40% (Stenberg 1995). In contrast, penumbra are relatively unimportant within shoots of species with large leaves (Pearcy et al. 2004). The spatial scale of sunfleck size has been shown to vary from 0.1 to 1 m, still typically smaller than most plant crowns in the understory so that often only part of the crown is influenced by a given sunfleck (Baldocchi and Collineau 1994). Variability within the sunfleck and the self-shading within the crown creates an extremely heterogeneous light environment. Spatial autocorrelation analysis revealed that the correlation between light readings from arrays of photo sensors decreased to 0.4 within 0.2 m and to 0.03 by 0.5 m (Chazdon et al. 1988). Cloud cover greatly reduced the spatial heterogeneity both in the short and long terms confirming that most of the spatial heterogeneity was due to sunflecks. Since seasonal changes in solar angle change the location of sunflecks, the large spatial heterogeneity observed for a day could be considerably reduced if monthly or annual means are considered.

At the scale of a forest stand, geostatistical techniques such as spatial statistics, wavelet analysis (Bradshaw and Spies 1992), and lacunarity analysis (Frazer et al. 2005, Sinoquet et al.

2005) have been employed to estimate the scales of heterogeneity of forest understory light environments. These techniques are used to examine the inherent structure in spatial or temporal data sets and then to infer characteristics of the processes that imposed these patterns. For light, sampling is done along transects using hemispherical photographs or Li-Cor plant canopy analyzers (or similar instruments). Spatial statistical techniques have been the most widely used technique, and have revealed a significant spatial autocorrelation in light environments at scales of 1–5 or 10 m but spatial independence at scales of 20 m or more (Clark et al. 1996, Nicotra et al. 1999). The latter indicates a scale of sampling where more traditional statistical approaches can be used to analyze relationships between plant performance and light availability. In contrast to spectral analysis, which is sensitive to the periodicity present in a data set, wavelet analysis examines whether a particular data structural shape (the wavelet) that is chosen to meet the study objectives detects the occurrence of similar patterns at different scales in the data set (Bradshaw and Spies 1992). Lacunarity is sensitive to the texture or gapiness such that a pattern that shows substantial heterogeneity in the size distribution and spatial arrangement of gaps is considered to be highly lacunar (Saunders et al. 2005).

A major goal of investigations of heterogeneity of light environments is an understanding of the role of light availability in species diversity and forest regeneration. Numerous experimental studies have demonstrated a strong positive effect of light availability on seedling physiological and growth performance. From this lead, theoretical studies have proposed strong links between the spatial heterogeneity in light environments and tree regeneration (Pacala et al. 1996, Bolker et al. 2003). However, measurements reveal a much more complex pattern. For example, although spatial patterning at similar scales was found in both light availability and in seeding distribution in a Costa Rican tropical forest, the relationship between light availability and seedling abundance was weak and inconsistent, with as often as not a negative relationship (Montgomery and Chazdon 2001). Nevertheless, the similar scales of autocorrelation in seedling abundance and light availability suggest a more direct relationship. This could be explained if current patches of greater seedling abundance reflect former patches of greater light availability. However, patterns of seed rain, herbivory, and other factors may also act to obscure the relationship between light availability and seedling abundance.

Although the central portion of larger tree fall gaps may receive a relatively uniform light environment, pronounced gradients occur around the edges because of the shading by adjacent canopy trees. In the Northern Hemisphere, the southern edge of a gap is shaded even at midday, whereas the northern edge receives the full direct beam solar radiation. Similarly, the west side receives sunlight in the morning whereas the east side receives it in the afternoon. The west side of the gap should therefore be more favorable for photosynthesis because of the lower temperatures and higher humidities in the morning. Wayne and Bazzaz (1993a,b) found differences in the diurnal course of photosynthesis on the east and the west sides, and small differences in the daily carbon gain, but no differences in growth. Because of variations in solar elevation angle, the effect of gap geometry on the heterogeneity of light regimes in them depends on season and latitude. Gap size and geometry influence the total daily PFD available for photosynthesis because of the effect on hours of direct PFD and the transmission of diffuse PFD. Plants on the gap edges are usually in strongly directional light environments because of shading by adjacent plants on one side.

Within individual crowns, gradients of PFD occur within canopies because of self-shading, which depends on crown density and the spatial pattern of leaves and stems. Most of the foliage in tree crowns are concentrated in an outer 0.5–1 m thick shell, which may intercept 90% of the incident radiation. Therefore, gradients of available PFD in these crowns are quite steep (see Figure 7.1).

Responses to Spatial Heterogeneity

Photosynthesis and Growth

At the most basic level, spatial heterogeneity in light creates variation in the resources available and hence in photosynthesis and growth. Because light is highly variable but is often the most limiting resource in forest understories, correlations between light availability, photosynthetic parameters, and seedling growth have been demonstrated in many species (Montgomery and Chazdon 2002, Delagrange et al. 2004, Pearcy et al. 2004). Growth of understory saplings of two shade-tolerant Hawaiian tree species was highly correlated with spatial variation in the mean annual potential minutes of sunflecks received as estimated from fisheye photographs (Pearcy 1983). Similar results have been obtained with understory saplings in Costa Rica (Oberbauer et al. 1988) and oak seedlings in Japan (Washitani and Tang 1991). However, spatial heterogeneity in growth is not always correlated with estimates of sunfleck availability, and in some cases diffuse light availability has been a better predictor of growth (Tang et al. 1992). Pfitsch and Pearcy (1992) found no correlation between estimates of sunfleck light availability and growth of the redwood forest understory herb, *Adenocaulon bicolor*. Variation in daily carbon gain and in the proportion of the annual carbon gain contributed by photosynthetic utilization of sunflecks was, however, highly correlated with the variation in sunfleck availability (Pearcy and Pfitsch 1991). Moreover, removal of sunflecks with shadow bands designed to block sunfleck transmission along the solar track but pass most of the diffuse PFD, significantly reduced growth and reproduction (Pfitsch and Pearcy 1992). Thus, there is no doubt that sunfleck utilization was important to these plants but other limitations may have prevented translation of the spatial variation in sunfleck PFD into spatial variation in growth. There is evidence for only minor losses of carbon gain due to photoinhibition in sunflecks (Pearcy 1994) so other factors such as drought stress and increased competition in the brighter microsites need to be considered.

Plasticity of the Photosynthetic Apparatus in Response to Spatial Heterogeneity

One of the most universal responses to increased light availability is a plastic response of leaf photosynthetic properties that results in development of sun leaves that have higher photosynthetic capacities per unit area ($Amax_{area}$), greater leaf thickness, and a greater leaf mass per unit leaf area (LMA). The plasticity expressed by different species in different growth situations varies widely, ranging from 1.25- to a 3-fold or in a few cases a 5-fold difference in $Amax_{area}$ between sun and shade environments (Björkman 1981, Pearcy and Sims 1994, Chazdon et al. 1996, Strauss-Debenedetti and Bazzaz 1996, Huante and Rincon 1998). The plasticity of $Amax_{area}$ can be examined either in terms of the relative change in photosynthetic rates achieved under light-saturating conditions, or in terms of the range of light environments over which this plasticity is expressed. Most studies have focused on the former by comparing plants grown in low- and high-light environments designed to mimic shade and at least partial sun environments. These are useful for comparing the response to environments, either simulated or natural, such as a gap and understory but because of the nonlinear response of most phenotypic traits and the potential for combining acclamatory and inhibitory responses in a stressful environment, they are usually not satisfactory in identifying species differences in plasticity related to their regeneration niche. When compared in this way, fast-growing early successional species favoring high-light environments seem to exhibit greater plasticity of $Amax_{area}$, as compared with more slowly growing, shade-tolerant later successional shrub and tree species (Strauss-Debenedetti and Bazzaz 1996, Yamashita et al. 2000). Some care is required in interpreting plasticity since the response to the high-light environment may be a mixture of acclimation and photoinhibition with the latter acting to reduce $Amax_{area}$ (Mulkey and Pearcy 1992). In addition, extremely shaded light environments may interfere with the leaf development of shade-intolerant species, reducing their $Amax_{area}$. The way to avoid this problem is to examine photosynthetic responses over multiple light environments but this has been done in only a few studies (Sims and Pearcy 1989, Valladares et al. 2000, Muraoka et al. 2003). An interesting approach is to study transplanted seedlings along light gradient such as across a forest edge in a gap forest to pasture transition, giving a wide range of light environments (Chazdon 1992, Montgomery 2004). In general, results obtained from comparisons across multiple environments reveal a strong dependence of $Amax_{area}$ on the daily light availability up to moderate daily PFDs but little further increase. For example, Montgomery (2004) found that most growth and photosynthetic traits exhibited nonlinear responses to light availability with the greatest changes between 0.5% and 20% canopy transmittance of diffuse light with little additional change at higher canopy transmittances.

Given the interacting effects of leaf development, acclimation, and photoinhibition, the range of light environments over which plasticity is expressed can differ between species, and may be as important ecologically as the actual change in $Amax_{area}$ itself. Sims and Pearcy (1989) compared the shade-tolerant, rain-forest herb, *Alocasia macrorrhiza* to the crop species, *Colocasia esculenta* in five different light environments ranging from deep shade to 50% of full sun. Both exhibited a 2.2-fold difference in $Amax_{area}$ but the range of light environments over which this was expressed was higher in *Colocasia* than in *Alocasia*. Comparisons of 16 gap-dependent and shade-tolerant *Psychotria* species across three light environments from deep shade to 25% of full sun reveals the diversity of responses that can occur (Figure 7.2). Overall, the plasticity of $Amax_{area}$ differed little among the species but the responses to the changes in light environments contrasted markedly. Most species exhibited

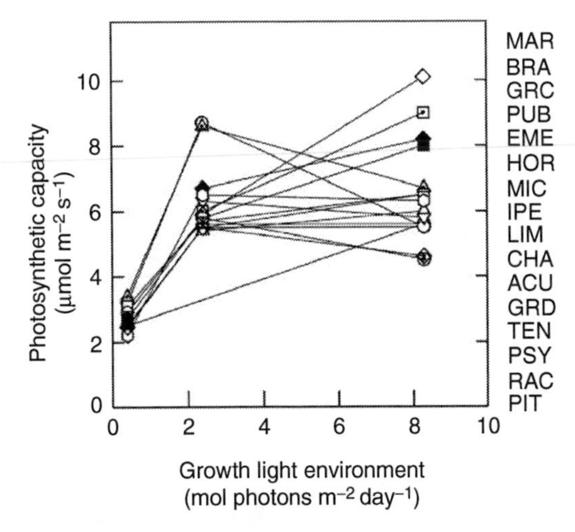

FIGURE 7.2 Response of photosynthetic capacity (Amax) of 16 *Psychotria* species to growth in three different light environments simulating understory shade to gap conditions. The abbreviations on the right side are for the different species and are ordered identically to the symbols shown for the highest light environment on the right side. The abbreviations are: MAR, *Psychotria marginata*; BRA, *Psychotria brachiata*; GRC, *Psychotria graciflora*; PUB, *Psychotria pubescens*; EME, *Psychotria emetica*; HOR, *Psychotria horizontalis*; MIC, *Psychotria micrantha*; IPE, *Psychotria ipecacuanha*; LIM, *Psychotria limonensis*; CHA, *Psychotria chagrensis*; ACU, *Psychotria acuminata*; GRD, *Psychotria granidensis*, TEN, *Psychotria tenuifolia*, PSY, *Psychotria psychotrifolia*; RAC, *Psychotria racemosa*; PIT. *Psychotria pittieri*. The abbreviations in boldface are for gap-dependent species and show that there is no relationship between gap versus understory species and plasticity of photosynthetic capacity. Each data point is the mean of at least three determinations on different plants. (Adapted from Valladares, F., Wright, S.J., Lasso, E., Kitajima, K., and Pearcy, R.W., *Ecology*, 81, 1925, 2000. With permission.)

the largest increase in $Amax_{area}$ between the low and the intermediate light environments. Some species exhibited a moderate increase between the intermediate and the high-light environments, others no increase and still others exhibited a decline in $Amax_{area}$ from the intermediate to the high-light environment. These responses of $Amax_{area}$ were not correlated with habitat requirement (gap-dependent vs. understory) of the species, though an overall measure of plasticity integrating both physiological and morphological characters showed that gap-dependent species had greater plasticity. Of course $Amax_{area}$ is only one of a suite of characters, and changes in leaf angle and so on may function to reduce the actual range of incident PFD experienced by the plant to avoid conditions that may cause photoinhibition. Indeed the PFD on the leaf surfaces is often close to the PFD that is just saturating for photosynthesis even though the incident PFD on a horizontal surface is much higher. It would be unrealistic to expect that increasing the PFD would result in a further increase in $Amax_{area}$. Indeed, the more common result may be photoinhibition.

Only a few studies have examined the plasticity of the photosynthetic apparatus to the spatial heterogeneity of light environments in the field. Chazdon (1992) found that $Amax_{area}$ of an early successionial shrub, *Piper sancti-felicis*, was closely related to light availability along transects across natural gaps. A later successional congener, *Piper arieianum*, exhibited a weaker association of $Amax_{area}$ with light availability, possibly because of a greater susceptibility to photoinhibition in the gap center (Figure 7.3). Comparisons of a habitat

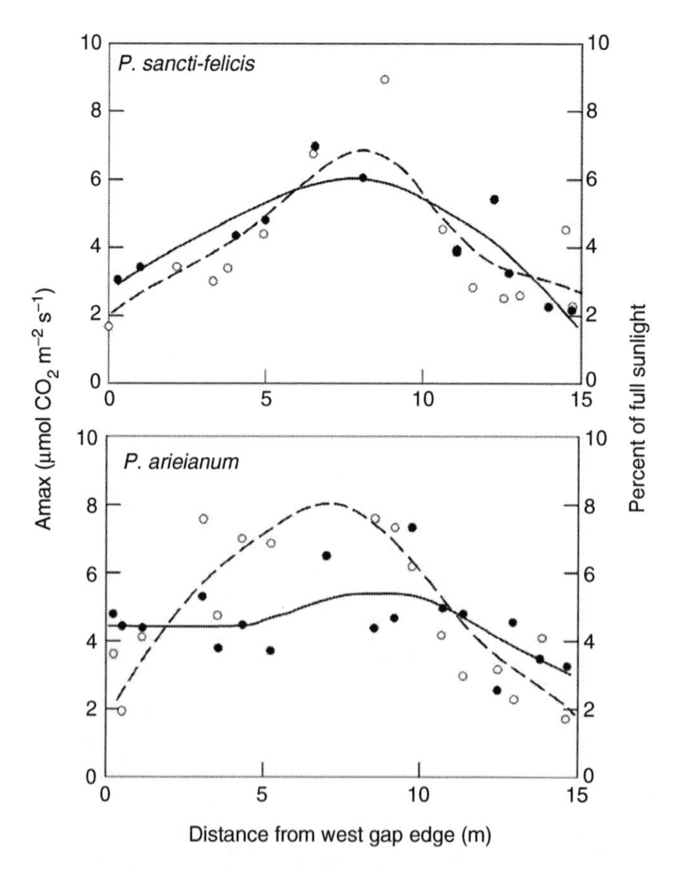

FIGURE 7.3 Gradients of light (open circles) and Amax (closed circles) of seedling of the gap species, *Piper sancti-felicis* (*top*), and the understory species, *P. arieanum* (*bottom*) across a small gap. Photosynthetic capacity tracks the light environment much better in the gap-dependent species than in the understory species. (Adapted from Chazdon, R.L., *Oecologia*, 92, 586, 1992. With Permission.)

specialist, the gap-dependent pioneer species, *Piper auritum*, with a habitat generalist species with respect to light, *Piper hispidum*, revealed similar responses of $Amax_{area}$ to light availability but less sensitivity of dark respiration and LMA to light in the specialist species. Although the net result was similar the two species may differ in the mode of acclimation. Within the narrower range of light environments in the understory, Chazdon and Field (1987) found no correlation between microsite light availability and $Amax_{area}$ in response to microsite variation in the understory but did find increasing $Amax_{area}$ with increasing light availability in gap environments. These results suggest that below a lower threshold PFD no further adjustments in photosynthetic capacity occur.

Sunflecks are a primary source of heterogeneity in daily PFD in the understory, but these are certainly too unpredictable for any advantage in daily carbon gain to accrue from adjustment of $Amax_{area}$. Sims and Pearcy (1993) found no differences in $Amax_{area}$ among *Alocasia macrorrhiza* plants grown under either short or long sunflecks or constant PFD environments where each environment had identical daily PFDs. Pearcy and Pfitsch (1991) found that $Amax_{area}$, LMA, and leaf N per unit area of the *A. bicolor* plants grown under shadow bands were all positively correlated with diffuse PFD levels. However, no correlation between these measures and either total PFD of sunfleck PFD was found for the adjacent control plants that received sunflecks. A filtering mechanism that ignored the unpredictable component of variation in daily PFD associated with sunflecks but allowed response to the much less variable but more predictable diffuse component would explain these results.

Maximizing Photosynthesis in High Light

Maximizing photosynthetic performance in high light requires investment in the carboxylation and electron transport capacity necessary to support high photosynthetic rates coupled with a sufficiently conductive pathway for CO_2 diffusion through the stomata and to the site of carboxylation in the chloroplasts. Growth in high versus low light has been shown to result in increases in activities (concentrations) per unit area of the primary carboxylating enzyme ribulose-1,5 bisphosphate carboxylase/oxygenase (Rubisco), Photosystem II (PSII) electron transport capacity, the electron carrier, cytochrome *f*, and chloroplast coupling factor that is involved in ATP synthesis (Björkman 1981, Anderson and Osmond 1987). It is likely that nearly all components in the photosynthetic carbon reduction cycle (PCRC) and electron transport must increase in a coordinated manner to maintain a balance between capacities of each component step. Indeed, acclimation seems to result in at most only a small increase in the ratio of electron transport to carboxylation capacity (Sims and Pearcy 1989, Thompson et al. 1992, Wullschleger 1993). Stomatal conductance must also increase to maintain a balance between the capacity for supply and for utilization of CO_2. Since internal diffusion limitations across cell walls and membranes to the Rubisco in the chloroplast are significant (Evans and Von Caemmerer 1996) these must also be adjusted to accommodate sufficient transport as photosynthetic capacity is increased.

Insight into the causes of the plasticity in photosynthetic capacity can be gained by comparing differences in photosynthetic capacities calculated on a per unit leaf area basis ($Amax_{area}$) with those calculated at per unit leaf mass basis ($Amax_{mass}$). Across ecologically and geographically broad species comparisons of high light-adapted species, thinner leaves with lower LMA have higher $Amax_{mass}$ than thicker, high LMA leaves (Reich et al. 1997). Across species and especially across life forms much of the difference in $Amax_{mass}$ is clearly due to genetically determined differences in concentrations of photosynthetic machinery per unit mass. From gas exchange analysis, Rubisco carboxylation capacity and electron transport capacity are higher in the thin-leaved species and the stomatal and liquid phase conductances to CO_2 also increase concurrently. The positive relationship between $Amax_{mass}$ and leaf nitrogen per unit mass is also consistent with a greater investment per unit mass in photosynthetic enzymes. Liquid phase diffusion limitations may also be important. Sclerophylous leaves that are characterized by low $Amax_{area}$ are typically hypostomatous, have been shown to have a lower CO_2 transfer conductance both through the intercellular air pathway and across the cell wall and liquid phase to the site of carboxylation than mesophytic leaves (Parkhurst and Mott 1990, Evans and Von Caemmerer 1996). Investment in thick cell walls characteristic of sclerophylous leaves dilutes the photosynthetic machinery resulting in low $Amax_{mass}$. Amphistomaty serves to overcome limitations to intercellular airspace diffusion in high photosynthetic capacity leaves by providing a pathway through both epidermi (Mott et al. 1982, Parkhurst and Mott 1990).

For light acclimation, the picture that emerges is somewhat different in that leaves that develop following a transfer of the plant from low to high light have higher LMA, are thicker, and have a higher $Amax_{area}$ and g_s than leaves that develop in the shade before the transfer. However, in many, perhaps most, species there is little or no change in $Amax_{mass}$ despite the change in leaf thickness. An increase in leaf thickness occurs because leaves developing in the sun typically have two layers of palisade, or longer palisade cells (Kamaluddin and Grace 1992), above a spongy mesophyll, whereas those developing in the shade have only a loosely organized and relatively undifferentiated and thin mesophyll (Chabot and Chabot 1977, Björkman 1981). This is consistent with a constant amount of photosynthetic machinery per unit mass but with the greater LMA in sun leaves resulting in a greater concentration of photosynthetic machinery per unit area. Similarly, significant relationships between $Amax_{area}$ per unit area and N per unit area have been commonly found in comparisons of plants grown in the sun and the shade (Seemann et al. 1987, Evans 1989b, Hikosaka and Terashima 1996), whereas there is no significant correlation between assimilation rates and N on a mass basis for most species. Indeed, for many species, increased leaf thickness resulting in increased cell and chloroplast volume per unit leaf area may be the only mechanism involved in the observed plasticity of $Amax_{area}$ since $Amax_{mass}$ remains nearly constant across light environments (Ellsworth and Reich 1992, Sims and Pearcy 1992). Since the leaf anatomy is largely specified early in leaf development (Sims and Pearcy 1992), mostly or fully developed leaves of many species show little or no capacity for adjusting $Amax_{area}$ (Mulkey and Pearcy 1992, Sims and Pearcy 1992, Newell et al. 1993, Oguchi et al. 2005). Some species, however, can increase their $Amax_{area}$ moderately even in fully developed leaves (Chazdon and Kaufmann 1993, Kursar and Coley 1999, Yamashita et al. 2000). Species that can respond may do so primarily by increases in chloroplast numbers or volume that fill cell wall locations unoccupied by chloroplasts when the leaf developed in the shade. On the other hand, species that are unable to respond may lack unoccupied locations for chloroplasts (Oguchi et al. 2003, 2005).

High-light environments, in addition to supporting higher photosynthetic carbon gains, also present circumstances where excess light may be potentially damaging. Moreover, this photoinhibition may be exacerbated by the co-occurrence of high temperatures and water stress (Mulkey and Pearcy 1992, Koniger et al. 1998). Upper canopy leaves in tropical forests, and also leaves in treefall gaps can reach temperatures 10°C above air temperature (Koniger et al. 1995, Niinemets and Valladares 2004). On the other hand, lower canopy leaves are usually near air temperatures. Thus, morphological and biochemical mechanisms that minimize photoinhibitory damage are important in acclimation to high light. Steep leaf angles and preferential orientation of leaves function to minimize the potential for photoinhibition yet maintain sufficient fluxes to also maximize photosynthetic carbon gain (Ehleringer and Forseth 1980, Ludlow and Björkman 1984, Gamon and Pearcy 1989). Within plant crowns, upper leaves often have steep leaf angles, decreasing the receipt of solar radiation at midday, whereas lower leaves have shallower leaf angles to maximize light capture (Valladares 1999). In addition to affording structural photoprotection in the upper canopy, changes in leaf angles decrease vertical heterogeneity in intercepted light.

Structural photoprotection is also accompanied by biochemical photoprotection involving down-regulation of PSII via the xanthophyll cycle where excess energy is dissipated by heat (Demmig-Adams and Adams 1992b, 1996, 2000). Sun leaves possess greater total pool sizes of xanthophyll-cycle pigments and a greater capacity for rapid conversion of violoxanthin to antheroxanthin and zeaxanthin that invokes this photoprotection (Thayer and Björkman 1990, Demmig-Adams and Adams 1992a, 1994). In addition, sun leaves have a greater capacity for scavenging of active oxygen free radicals that can cause photodamage (Grace and Logan 1996). When shade leaves are exposed to PFD in excess of the capacity of these mechanisms, photodamage to PSII may occur, requiring operation of a PSII repair cycle for recovery (Geiken et al. 1992, Oquist et al. 1992, Anderson et al. 1995).

Maximizing Photosynthesis in Shade

Maximizing leaf photosynthetic performance at the low PFDs inherent in shaded, diffuse-light conditions requires (1) maximizing light absorptance, (2) maximizing the quantum yield with which this light is used for CO_2 assimilation, and (3) minimizing respiratory losses. Careful comparisons under conditions where photorespiration is suppressed (low O_2 or high CO_2 partial pressures) have unequivocally established that there are no intrinsic differences in quantum yields between C_3 species adapted to sun and shade habitats or, provided that no stress effects interfere, among leaves on plants grown in high and low light (Björkman and Demmig 1987, Long et al. 1993, Singsaas 2001). The observed quantum yields are consistent to those derived from stoichiometric requirements for production of the ATP and NADPH necessary for carbon metabolism, small inefficiencies in energy transfer inherent within the chlorophyll antennae and reaction centers, and energy use for other processes such as nitrate reduction. Thus, there is essentially no scope for further improvements in shade as compared to sun leaves. Photorespiratory losses under normal atmospheric concentrations reduce the achieved quantum yields under natural conditions, but this is slightly different for sun and shade plants. Recently, it has been reported that quantum yields are reduced in extremely low light possibly because of insufficient buildup of the transthylakoid proton gradient required for ATP regeneration (Timm et al. 2002, Kirschbaum et al. 2004). Since quantum yields were greater than 85% restored by PFDs as low as 5 $\mu mol\ m^{-2}\ s^{-1}$, the loss of quantum yield would have an effect on carbon gain only early or late in the day, or only in the shadiest microsites.

Quantum yields are frequently reduced because of photoinhibition, especially when high light is coupled with high leaf temperatures, water stress, or low nitrogen nutrition (Björkman et al. 1981, Björkman and Powles 1984, Ferrar and Osmond 1986, Mulkey and Pearcy 1992, Koniger et al. 1998). Shade leaves are more susceptible to photoinhibition than sun leaves in this regard and observed differences in quantum yields may often be due to the occurrence of these other environmental factors and their interaction with high PFD. On the other hand, higher CO_2 concentrations in the understory and the high values of intercellular to ambient CO_2 pressures (c_i/c_a) characteristically observed under diffuse light conditions can enhance quantum yields for CO_2 uptake by 10%–20% due to reduced photorespiration (Pfitsch and Pearcy 1989a, DeLucia and Thomas 2000, Singsaas et al. 2000). This enhancement could be important under the conditions of rising atmospheric CO_2 in allowing plants to occur in shadier microsites than would be the case under current CO_2 concentrations.

Comparisons across many species also reveal no consistent differences in leaf absorptance or chlorophyll (Chl) concentration per unit leaf area between sun and shade leaves (Björkman 1981). Chl concentrations per unit area are typically sufficient to absorb 85%–92% of the incident photosynthetically active radiation with the remainder lost to either reflection or transmission. The lower LMA of shade leaves means that to maintain Chl per unit area constant there has to be an increased investment in Chl per unit mass. Although shade leaves in particular could benefit in terms of increased photosynthesis from increased Chl per unit area and hence leaf absorptance, the strongly diminishing returns of further investment in Chl–protein complexes required to affect a significant change in light capture may preclude it. A doubling of the Chl concentrations is required to increase leaf absorptance by just 5%–6% from the levels given earlier. Although there are only 4 mmol N in a mol of Chl, the associated proteins in the Chl–protein complexes contain 21–79 mmol N mol^{-1} Chl (Evans 1986, 1989a). Chlorophyll *a*/*b* ratios are much lower in shade than sun leaves and it was suggested that this may represent a mechanism for increasing absorption in the 640–660 nm wavelength range where PFD in the understory is somewhat greater than that at longer wavelengths where Chl in the overstory canopy has its maximum absorption. The advantage, however, may lie more in a significant savings of N since Chl *b* is part of the light-harvesting chlorophyll–protein complex II (LHCII), which contains only 25 mmol N mol^{-1} Chl, whereas Chl *a* in PSII represents an investment of 83 mmol N mol^{-1} Chl (Evans 1986). The lower Chl *a*/*b* ratio results from increased LHCII per unit leaf area coupled with decreased concentrations of the PSII core complex, which contains only Chl *a*. Since PSII complexes contain more N, the shift to more LHCII while maintaining total Chl per unit leaf area constant results in a significantly greater light capture per unit N invested. This savings is possible in shade, but not sun leaves because high capacities of PSII electron transport are needed in the high light.

Acclimation of Dark Respiration

To maintain a positive carbon return, gross assimilation rates must clearly exceed respiration rates. It is now widely documented that one of the most important differences between sun and shade leaves are the much lower respiration rates of the latter (Grime 1966, Björkman 1981). Leaf respiration itself and especially the mechanisms responsible for lower rates in shade leaves are not well understood. Respiration rates may be affected by the capacity of the respiratory pathways, substrate concentrations, rates of energy and reductant demand, and the extent of alternative pathway engagement. In addition, mitochondrial respiration is substantially inhibited by light at PFDs as low as 3 $\mu mol\ m^{-2}\ s^{-1}$ (Atkin et al. 1998, 2000). Thus, measurements done in the dark, the commonly used method for estimating respiration, rather than the more complex procedure of Laisk (1977), must be corrected in models exploring carbon balance in understory plants.

Mitochondrial respiration occurs in response to growth and maintenance processes. Since the low available light severely limits carbon gain and growth in the shade, respiration rates of the leaves of shade plants are correspondingly low, which is critical for maintaining a positive carbon balance in understory light regimes. The observed respiration rates of sun and shade plants reflect mostly the differing resource supplies and demands. Respiration rates appear to be differently regulated in sun and shade species (Noguchi et al. 1997, Noguchi and Terashima 1997). In the high-light species, *Spinacia oleracea*, respiration rates were limited by the carbohydrate supply in the leaves, whereas in the shade-adapted species, *Alocasia odorata*, they were limited by a low demand for ATP due to its slow growth rate. Transfer from high to low or from low to high light typically invokes a corresponding decrease or increase in dark respiration rates over the following 4–6 days (Sims and Pearcy 1992, Noguchi et al. 2001b). In shade plants such as *Alocasia odora*, feeding experiments involving either carbohydrates or respiratory inhibitors show that this decline in respiration on transfer from high to low light occurs because of the reduced demand for ATP. Conversely, in high light, more respiration may be needed for repair of PSII photodamage or for the greater phloem loading as growth accelerates and biosynthesizes. Overall, the maintenance costs of the photosynthetic apparatus are lower in shade than those in high light-acclimated leaves because of the lower soluble protein levels (Noguchi et al. 2001a) and likely lower costs for repair of photodamage. However, even in high light-acclimated leaves, maintenance respiration relative to photosynthetic capacity still appears to be a rather low cost measured relative to photosynthetic capacity and perhaps not much different between shade- and sun-acclimated leaves (Sims and Pearcy 1992).

Of more importance than maintenance respiration to the carbon balance of leaves and the payback time for a positive return on investment may be the construction costs, which occur principally early in the leaf's development, but which must be paid back over the leaf lifetime. On a per unit area basis, sun leaves are more expensive to construct because they are thicker and have higher per unit area concentrations of photosynthetic enzymes. On a per unit mass basis there is on average only a slightly greater (3%) construction cost of sun as compared with shade-acclimated leaves (Poorter et al. 2006). Construction costs depend on the composition of the leaf and vary depending on the amounts of more costly, reduced compounds such as soluble phenolics, lipids, proteins, and lignin as compared with carbohydrates. Sun leaves can recover construction costs within a few days, whereas shade leaves may require 60–150 or more days, depending on the light environment (Chabot and Hicks 1982, Jurik and Chabot 1986, Sims and Pearcy 1992).

Trade-Offs and Optimality in Sun Shade Acclimation

Modeling approaches have been used that explore the trade-offs involved in acclimation with respect to whether they maximize the return on investment in a given light environment. At the leaf level, one trade-off is how a given amount of biomass should be invested in either increasing the area of the leaf, thereby enhancing light capture, or in increasing leaf thickness thereby increasing the $Amax_{area}$ of the leaf. A second trade-off involves how N should be partitioned within the photosynthetic apparatus in a given light environment to maximize photosynthesis. This involves a trade-off between investment of N to maximize photosynthetic rate in high light versus investment in pigment–protein complexes that increase the slope of the light response curve in low light (Evans 1989b). Several studies have shown that nitrogen partitioning between light capture components (Chl-proteins) and components that increase photosynthetic capacity (Rubisco and electron transport carrier components) change with acclimation in a way that almost maximizes photosynthesis (Evans 1993, Hikosaka and Terashima 1995, 1996, Niinemets et al. 1998b). Moreover, daily carbon gain is quite sensitive to the changes in allocation observed in high versus low light-acclimated leaves, each with 20%–30% lower daily carbon gain in the light environment opposite to the one that they were grown in (Hikosaka and Terashima 1996). Central to these optimizations are first that light absorption is a hyperbolic function of chlorophyll content in agreement with observed changes (Gabrielsen 1948) and second that relatively small changes in the fraction of N allocated to pigment–protein complexes have a large effect on light absorption. Comparisons of the assimilation rates per unit leaf area at the simulated optimum partitioning to the actual measured assimilation rates revealed that the latter tracked the former quite closely with deviations only in the highest and lowest light environments. The changes for shade and sun species did not track in the highest and lowest light environments, respectively. Thus, nearly optimal partitioning could be maintained over a range of light environments, but this range differed between shade and sun species.

Changes in LMA mediate the other trade-off in that a given amount of biomass invested in a leaf can either be spread over a small area increasing its $Amax_{area}$ or over a large area increasing its light capture. Sims et al. (Pearcy and Sims 1994, Sims et al. 1994) modeled the sensitivity of whole-plant growth rate to changes in $Amax_{area}$ and LMA in *Alocasia macrorrhiza*. When just LMA was varied RGR was more sensitive to changes in LMA in the low-light environment than in the high-light environment. Conversely, when just $Amax_{area}$ was varied RGR was much less sensitive to changes in the low-light environment as compared with the high-light environment. When LMA and photosynthetic capacity were varied simultaneously, as occurs in acclimation, then RGR was much more sensitive in low as compared with high light. These simulations emphasize the central role of changes in LMA in acclimation to sun and shade and especially in improving performance in low light. This is further emphasized in simulations where changes in N partitioning and in LMA characteristic of acclimation were compared (Evans and Poorter 2001). This comparison revealed that daily photosynthesis per unit dry mass is far more sensitive to changes in LMA than to changes in allocation of N.

Plasticity of Whole-Plant Responses

Integration of the role of photosynthetic plasticity in enhancing plant function in heterogeneous light regimes requires an understanding of how leaf photosynthetic rates interact with and are determined by allocation patterns at the whole-plant level. Since the carbon content of biomass is nearly constant, growth and whole-plant photosynthesis are indeed closely linked by the conversion efficiency of fixed carbon into biomass (Dutton et al. 1988). Whole-plant photosynthesis is simply the photosynthetic rate of the leaves integrated over the leaf area. Of course, it is necessary to take into account that leaves are in different light microenvironments within the crown and, because of acclimation and aging phenomena, may have different capacities to use this light. Moreover, the carbon losses via respiration at the whole-plant level must be subtracted to arrive at a whole-plant carbon balance.

Enhanced whole-plant photosynthesis can therefore be obtained either by increasing the photosynthetic rate per unit area under the prevailing conditions, or by increasing the leaf area per plant. An important linkage in this process is the LMA since this determines how much leaf area is produced for a given investment of biomass in leaves. Most studies have found that the leaf weight ratio (LWR: mass of leaves per unit mass of plant) tends to be higher in sun as compared with shade-grown plants but the differences are usually rather small (Björkman 1981, Poorter and Nagel 2000). Instead, shade-grown plants invest more in stems and less in roots with little change in total leaf area (Reich et al. 1998). Greater investment in stems may be helpful in minimizing self-shading. Although the proportional biomass investment in leaves tends not to vary much with sun-shade acclimation, shade-grown plants typically have a much larger leaf area ratio (LAR: area of leaves per unit mass of plant) than sun plants because of the large differences in LMA. Since, as discussed earlier, LMA is also a determinant of $Amax_{area}$, it clearly mediates a trade-off between lower $Amax_{area}$ and greater leaf area for light capture in the shade, versus higher $Amax_{area}$ but less leaf area in the sun. The result in the shade is a large advantage in whole-plant carbon gain for the shade plant (Figure 7.4). Perhaps counterintuitively, experiments and simulations have shown no carbon gain advantage per unit biomass invested for the sun plant phenotype in high light since the reduced leaf area per plant more than offsets the advantage accrued due to the higher $Amax_{area}$ (Sims and Pearcy 1994, Sims et al. 1994). Indeed, shade-acclimated sunflower plants have been shown to have higher relative growth rates immediately after transfer to high light than do plants grown continuously in the high-light environment (Hiroi and Monsi 1963). Subsequent acclimation of the shade plant to the sun environment therefore actually causes the RGR to decrease. This does not mean that sun leaves provide no advantage in high light, especially over the long term, since they may well contribute to a greater stress resistance and greater water and nitrogen use efficiency.

Responses to within-Crown Spatial Heterogeneity

Within a plant crown, growth itself generates a heterogeneous light environment as younger branches and leaves overtop and shade lower branches and older leaves. This gradient has been studied in tree canopies (Niinemets et al. 1998b, Ellsworth and Reich 1993), herbaceous

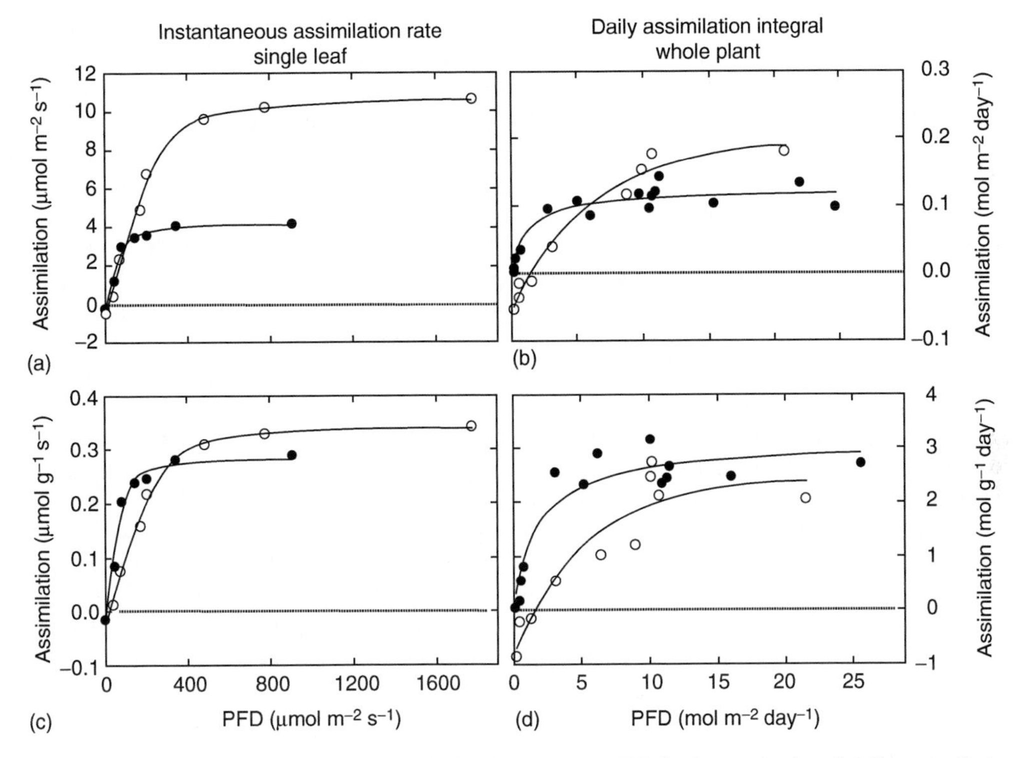

FIGURE 7.4 Response of instantaneous assimilation rates to PFD for leaves (a,c) and daily assimilation rates of whole plants (b,d) of *Alocasia macrorrhiza* acclimated to sun (open circles) and shade (closed circles) conditions. The curves for whole plants were obtained by growing individual plants in sun or shade conditions and then transferring to a whole-plant chamber for gas exchange conditions. Different daily PFDs were obtained with neutral-density screen and cloth filters under natural sunlight. (Adapted from Sims, D.A. and Pearcy, R.W., *Plant, Cell Environ.*, 17, 881, 1994. With permission.)

vegetation (Anten et al. 1995, Hirose et al. 1988), and crops (Gutschick and Wiegel 1988) and occurs both in closed canopies and in the crowns of isolated plants (c.f. Caldwell et al. 1986). In all, the gradient in light is primarily a function of the cumulative leaf area index and the architecture of the leaf and branch elements that controls the aggregation, inclination, and self-shading of leaves within and between shoots. Response to this highly dynamic gradient involves both senescence of older, more shaded leaves with reallocation of resources to the newly developing leaves in the higher-light environments as well as acclimation to the changed light environment. The modular nature of plants and the semiautonomy of these modules confers great flexibility to plants in responding to fine-grained differences in light availability (de Kroon et al. 2005).

Studies of the consequences of light attenuation within plant crowns have focused on the question of how resources, in particular nitrogen and also LMA, should be distributed between leaves at different positions in the crown to maximize photosynthesis. Optimal N use occurs when canopy photosynthesis is maximized for a given whole-canopy N content (Field 1983, Hirose and Werger 1987a). Optimal N use demands that more N be partitioned to the upper leaves located in high light, resulting in a higher $Amax_{area}$ of these leaves as compared with lower leaves in more shaded microenvironments. Differences in LMA have been shown to be central to this partitioning because of its pivotal role in determining photosynthetic capacity and N contents in sun and shade leaves. The optimal N distribution between leaves is the one that results in a gradient of photosynthetic capacity through the canopy such that light is equally limiting for leaves at all levels, giving a strict proportionality between the integrated PFD intercepted and $Amax_{area}$ (Field 1983, Farquhar 1989). Depending on crown architecture and the resulting light gradients an optimal versus a uniform N distribution among leaves can theoretically have a substantial impact on canopy photosynthesis. The increase ranges from 1% to 4% in open-canopied shrubs and trees with relatively low total canopy N concentrations (Field 1983, Leuning et al. 1991) to as much as 20%–40% in dense-canopied stands of with high-canopy N contents (Hirose and Werger 1987a, Hirose 1988, Evans 1993). Evans (1993) compared the consequences of nitrogen redistribution within a leaf due to acclimation (investment in chlorophyll–protein complexes versus carboxylation capacity) with that of nitrogen redistribution within a canopy. Leaf photosynthetic acclimation alone (optimal partitioning between photosynthetic components within a leaf at a constant N content) was shown to potentially increase canopy photosynthesis by about 4%. On the other hand, redistribution of N among leaves potentially increased canopy photosynthesis by 20% for the same canopy. Thus, in terms of maximizing photosynthesis in heterogeneous light environments within canopies, nitrogen redistribution among leaves is more important than nitrogen redistribution within leaves.

In the field, these theoretical returns from optimal resource distributions may never be attained. Measurements within a variety of canopies have found gradients in $Amax_{area}$ tending toward but not equal to an optimal gradient (Hirose and Werger 1987b, Hirose et al. 1988). Leaves have a minimum LMA and N content and a higher $Amax_{area}$ in the most shaded locations than predicted by a strict proportionality with light (Niinemets et al. 1998b, Meir et al. 2002). On the other hand, stresses, including high temperature, excessive light, and low water potentials may constrain full acclimation at the top of the canopy (Niinemets and Valladares 2004).

The patterns of leaf development within a canopy also have important consequences for the nature of the plastic responses to the resulting light heterogeneity. In species with a single, early-season leaf flush, light gradients are created simultaneously with and by leaf development. Gradients in $Amax_{area}$ in these species occur primarily in response to differences in partitioning of N between more rapidly growing leaves that develop high LMA in the sunnier crown positions and more slowly growing leaves with low LMA in the shadier crown positions (Ninnemets et al. 2004). In species that flush leaves continuously, lower leaves may either be those that developed in high light but were then overtopped, or leaves that developed in the shade (Pearcy and Seemann 1990). The former maintain the high LMAs from their earlier high-light existence and have limited capacity for acclimation as compared with those developing in the shade. Senescence therefore plays a greater role in establishing canopy gradients in these continually flushing canopies. Still another model is the evergreen canopies with leaf life spans of several years. In some Mediterranean species, leaves continue to accumulate mass for several years due to thickening and lignification of cell walls (Valladares and Pearcy 1999, Niinemets et al. 2006). Although N per unit mass decreases because of the mass dilution effect, N per unit area remains constant. Thus there is apparently little or no reallocation of N from lower to higher leaves as predicted by optimal N allocation. Wall thickening and lignification may be important in water stress resistance but a consequence is a decrease in $Amax_{area}$ due to greater wall diffusion limitations.

Responses to Directional Light Gradients

Plants near the edges of gaps or in competing plant populations often receive much more of their light on one side because of shading by nearby plants on the other. Phototropic reorientation of leaves and stems, which increases light capture from the prevailing direction, can often be observed in these plants (Ackerly and Bazzaz 1995). Additionally, preferential branch growth may occur to fill available space on the open side leading possibly to unbalanced crowns (Young and Hubbell 1991). Since branches are often thought to exhibit semiautonomy, the simplest explanation could be differential carbon gain on the two sides. Indeed a model of growth in crowded populations based on this principle adequately predicted the general patterns of branch growth and thinning (Takenaka 1994). However, hormonal signals may also be involved and branches are certainly limited in their extension perhaps by hydraulic constraints (Mencuccini 2002). Gilbert et al. (1995, 2001) found that reflected far-red light acting through phytochrome inhibited branch growth on the side nearest to a neighboring plant. Novoplansky (1990) showed that *Portulaca oleracea* preferentially branched in a direction away from a reflected or transmitted far-red light source. Additionally, for early successional herbaceous plants, reflected far-red radiation can be a signal of potential competitors, leading to internode elongation and accelerated height growth at the expense of tillering or branch growth (Ballare et al. 1987, Ballare 1994, Ballare and Casal 2000). This response occurs before any direct shading and therefore allows the plant to effectively anticipate a potential competitive interaction.

Differential Growth of Clonal Plants

Clonal plants can exhibit varying degrees of physiological integration, which may be important in their exploitation of patchy environments (Slade and Hutchings 1987, de Kroon and Hutchings 1995, de Kroon et al. 2005). A type of foraging behavior manifested as a greater concentration of ramets and leaves in favorable, high-light patches as compared with unfavorable, shaded patches can result from morphological plasticity. In particular, shorter petioles coupled with increased bud activation may be elicited when a clone reaches a favorable patch, whereas in unfavorable patches, longer internodes and petioles may help ramets and leaves escape (de Kroon and Hutchings 1995). Resource sharing via physiological integration of ramets in favorable versus unfavorable microenvironments may modify the responses of ramets. For example, Dong (1995) found that integration in the stoloniferous understory herb *Laminastrum galeobdolon* significantly evened out the morphological differences across high- and low-light patches. In contrast, there was no such dampening in the morphologically similar fenland species, *Hydrocotyle vulgaris*. Since a major source of heterogeneity in the understory is spatially unpredictable sunflecks, the evening out of responses across ramets may be beneficial in terms of whole-clone performance. Patches of light are spatially more predictable in the fenland, and therefore independence in morphological adaptation of ramets may be beneficial.

There is also evidence that clonal integration may enhance utilization of heterogeneous resources via spatial division of labor (Stuefer et al. 1994, 1996). In experiments where heterogeneous light and soil water environments were created with strictly negative covariance (high light always associated with low water and vice versa), interconnected ramets of *Trifolium repens* exhibited specialization for uptake of the locally most abundant resource, with resource sharing across ramets (Stuefer et al. 1996). Moreover, total clonal growth was significantly enhanced relative to that of clones subjected to homogeneous environments that were either well-watered but shaded or in high light but water-limited. Inherently, negative relationships between patchiness in resources such as light and water may be fairly common in natural environments. However, even in the absence of patchiness of one resource, spatial integration may prove beneficial. For example, in experiments with uniform soil moisture, integration enhanced the growth of ramets of the herbs *Potentilla replans* and *Potentilla anserina* in high light that were interconnected to ramets in low light, apparently because water uptake by the low-light ramets supported in part the high transpiration rates of the high-light ramets (Stuefer et al. 1994). Severing the connections reduced the growth of the low-light ramets in both species because of the restriction in carbon supply and also reduced the growth of the high-light ramets because of a reduction in water supply.

TEMPORAL HETEROGENEITY IN LIGHT ENVIRONMENTS

Timescales of Temporal Changes and Responses

Temporal changes in light of interest in ecophysiology involve timescales differing by several orders of magnitude. At one end of the spectrum, the formation and the closure of canopy gaps due to tree falls takes place over several to many years. Annual changes occur because of day length and solar elevation angle changes as well as seasonal variation in cloudiness. Diurnal changes occur because of the earth's rotation and the resulting path of the sun across the sky. Finally, at the other end of the spectrum, changes on the order of seconds to minutes occur in sunflecks under plant canopies and because of intermittent cloudiness.

The responses to these temporal changes can be conveniently divided into two categories, one involving the acclimation and developmental plasticity occurring in response to long-term changes in the light environment and the other involving regulatory responses to short-term changes that occur during the normal diurnal course of solar radiation or during more rapidly changing irradiance such as that due to sunflecks in an understory. On the order of a day appears to be the time dividing temporal changes for which acclimation and developmental plasticity are of paramount importance and those for which regulatory mechanisms are prominent. Acclimation to a change in light requires on the order of 4–5 days for leaves of fast-responding species such as peas (Chow and Anderson 1987a,b) and up to 45 days in some slow-responding species, such as the tropical tree *Bischofia javanica* (Kamaluddin and Grace 1992). These types of responses involve changes in enzyme concentrations within leaves that function to alter photosynthetic capacity. Both acclimation and developmental plasticity are sensitive to the level of other resources, especially nitrogen supply (Osmond 1983, Ferrar and Osmond 1986), and set in particular the maximum photosynthetic capacity that a leaf can achieve. Mature leaves have limited physiological plasticity so further change affecting whole-plant performance involves development of new leaves that have different structural properties and even changes in branching architecture. These types of changes obviously require even longer periods (weeks to months) with the maximal change in whole-plant physiological properties not occurring until all leaves have turned over. Indeed as discussed earlier, in the context of spatial heterogeneity, growth also generates a temporally heterogeneous light environment as newer branches and leaves overtop and shade lower branches and older leaves.

Although there are enzymes that show marked diurnal variation in concentration (Huffaker and Peterson 1976), photosynthetic enzymes involved in carbon metabolism and electron transport, and the light-harvesting chlorophyll–protein complexes in leaves, are typically already present in quite high concentrations. Affecting a significant relative change in concentrations due to either synthesis of degradation may require longer times than a day and would be energetically costly. Therefore, changes in photosynthetic enzyme concentrations may be evolutionarily precluded as too slow and too costly to be a mechanism for responding to diurnal changes in solar radiation or to sunflecks. Instead, responses to these faster changes involves modulation of enzyme activity by small regulatory molecules such as thioredoxin (Buchanan 1980), regulatory enzymes such as Rubisco activase (Portis 1992) binding of substrates to regulatory sites, chloroplast stromal pH, or energization of the thylakoid membranes (Foyer et al. 1990, Harbinson et al. 1990a,b). In addition, the regulatory mechanisms governing stomatal opening and closure are also important. These regulatory mechanisms appear to serve to match the capacity of the component steps to each other and to the supply of light energy and to the supply of CO_2. When the PFD is low, down-regulation of several enzymes in the PCRC occurs so that the capacity for carbon metabolism matches the supply of light. When light is in excess, regulatory mechanisms involving the xanthophyll cycle in the pigment beds cause the excess energy to be dissipated as heat, thereby affording protection from photoinhibition (Demmig-Adams and Adams 1992b).

Stomatal conductance also adjusts to maintain a balance between the supply of CO_2 and the capacity to use it, as well as to affect the compromise between water loss and carbon gain (Cowan and Farquhar 1977). The time constants for these regulatory changes typically range from 0.5 to 15 min to as long as an hour or so, depending on the particular process. Little is known about the quantitative energetic costs of these regulatory mechanisms but they are undoubtedly much less than those involved in acclimation and developmental plasticity.

Responses to Sunflecks

Temporal Nature of Sunfleck Light Regimes

Light environments in canopies and understories present a particularly complex case of temporal heterogeneity. Sunflecks in these environments occur because of the juxtaposition of the solar track for a particular day with a gap in the canopy. Since the sun moves along the track at 15° h^{-1}, sunflecks move across the forest floor at a rate determined by the height to the canopy gap. Penumbral effects in tall vegetation cause the PFD in a sunfleck to increase and decrease gradually as the solar disk moves across the gap. However, movement due to wind causes the edges of a canopy gap to continually change and even for gaps to open and close. Thus, what may be one gradually changing long sunfleck on a completely still day is more often broken up into many brief sunflecks in a cluster. Since the solar track shifts with time of the year, a sunfleck occurring at a particular time and place may do so for only a few days to weeks and then only when cloud cover does not intervene. Moreover, seasonal changes in solar elevation also influences the occurrence of sunflecks because most gaps are at high angles.

The complexity of sunfleck light regimes makes the task of defining a sunfleck difficult (Chazdon 1988, Smith 1989). Since the PFD transients are rapid, sampling rates must be high to adequately describe the record. Sampling theorem states that the sampling frequency must be double the rate of the most frequent event of interest if the objective is to reconstruct the record accurately. In understories, this dictates sampling of PFDs at intervals of 1–10 s, whereas in canopies, where leaf flutter creates highly dynamic light environments, sampling frequencies of 0.1 s may be in order (Pearcy et al. 1990, Roden and Pearcy 1993b). Once an adequate record is obtained, one approach is to define a sunfleck as a transient excursion of the PFD above a threshold level that is just above the background diffuse PFD. Alternatively, the threshold can be some value that is physiologically meaningful. This approach has been criticized since the threshold is arbitrary and different thresholds have to be used in different canopies and with different species (Baldocchi and Collineau 1994). Wavelet analysis provides an objective method for detecting sunfleck events in a data record and does not require a threshold (Baldocchi and Collineau 1994). Once sunflecks are detected, then they can be quantified as to their duration, PFD, and so on. However, wavelet analysis depends on selection of the appropriate wavelet transform function. Finding one that works well is difficult, given the highly variable and complex nature of sunflecks. So in the end it is not yet clear whether application of wavelet analysis provides a significant advantage over the threshold approach. Spectral analysis in which periodicity in the light environment represented by characteristic frequencies is detected in a record using fast Fourier transform techniques has been used in crop canopies (Desjardins 1967, Desjardins et al. 1973). Its usefulness is limited since usually sunflecks are so variable and irregular that no characteristic frequencies can be detected. The exceptions are canopies such as aspen or poplar where leaves flutter (Roden and Pearcy 1993a).

Measurements now completed in a wide variety of forest understories show that on clear days, from 10% to 80% of the daily PFD at any given location may be due to sunflecks (Pearcy 1983, Chazdon 1988, Pfitsch and Pearcy 1989b, Singsaas et al. 2000, Leakey et al.

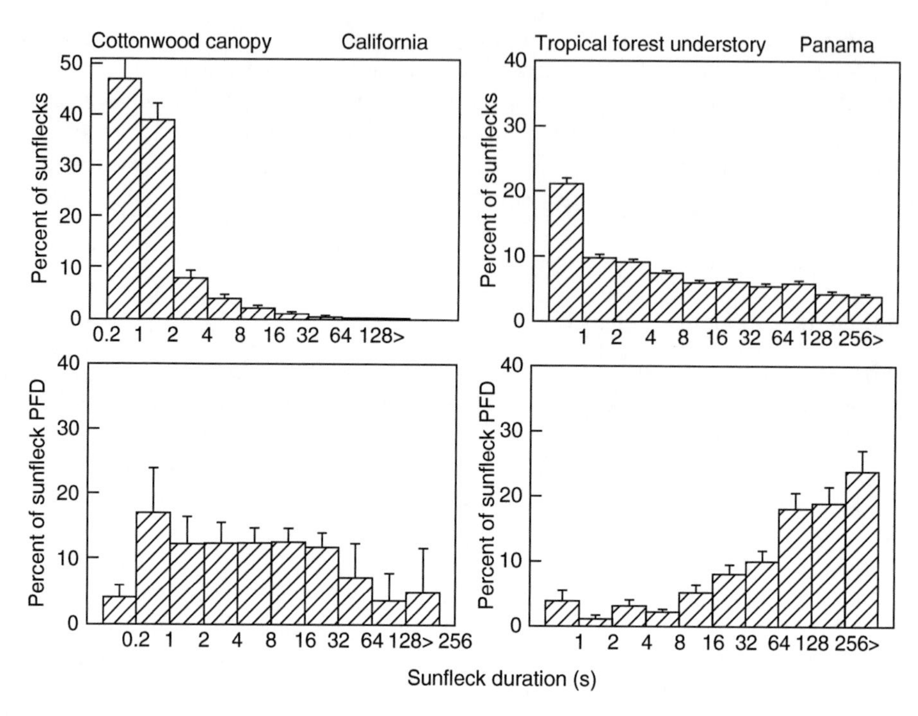

FIGURE 7.5 Histograms showing variation in sunfleck characteristics between a *Populus fremontii* canopy in California (*left side*) and a tropical forest understory on Barro Colorado Island, Panama (*right side*).

2003b). Thus sunflecks contribute a large fraction of the light available for photosynthesis and in turn could be expected to have a large effect on carbon balance and growth of understory plants. The characteristics of sunflecks depend on attributes such as canopy height, flexibility, and leaf size as well as weather conditions such as wind and cloudiness. Figure 7.5 shows histograms of sunfleck characteristics for a tropical forest on Barro Colorado Island (BCI), Panama, and for sensors mounted within a poplar canopy. Most sunflecks in forest understories are brief and have low-peak PFDs and contribute only a small proportion of daily PFD as compared with long-duration sunflecks. By contrast, short-duration sunflecks are a much more important contributor of PFD in aspen canopies. In the tropical forest understory on BCI, sunflecks tended to be shorter in the dry as compared with the wet season because of the more windy conditions in the former. In the wet season, most sunflecks occurred in the mornings because clouds typically built up and rain fell in the afternoons. Most days in the dry season were clear in both the mornings and the afternoons. Overall, the additional cloudiness in the wet season as compared with the dry season reduced the PFD contributed by sunflecks by about 50%. This is consistent with the reductions in PFD above the canopy in the wet versus the dry season.

Dynamic Responses of Photosynthesis to Sunflecks

Photosynthetic responses to sunfleck light regimes are complex because several components with markedly different time constants are involved. The photosynthetic response to an individual sunfleck, or its artificial counterpart—a lightfleck, is quite rapid but the carbon gain achieved by this response is determined by factors that change over a much longer timescale. If the leaf has been in the shade for a long time, photosynthetic induction (Figure 7.6) must occur in order for the full photosynthetic potential to be achieved (Walker 1981, Pearcy 1988).

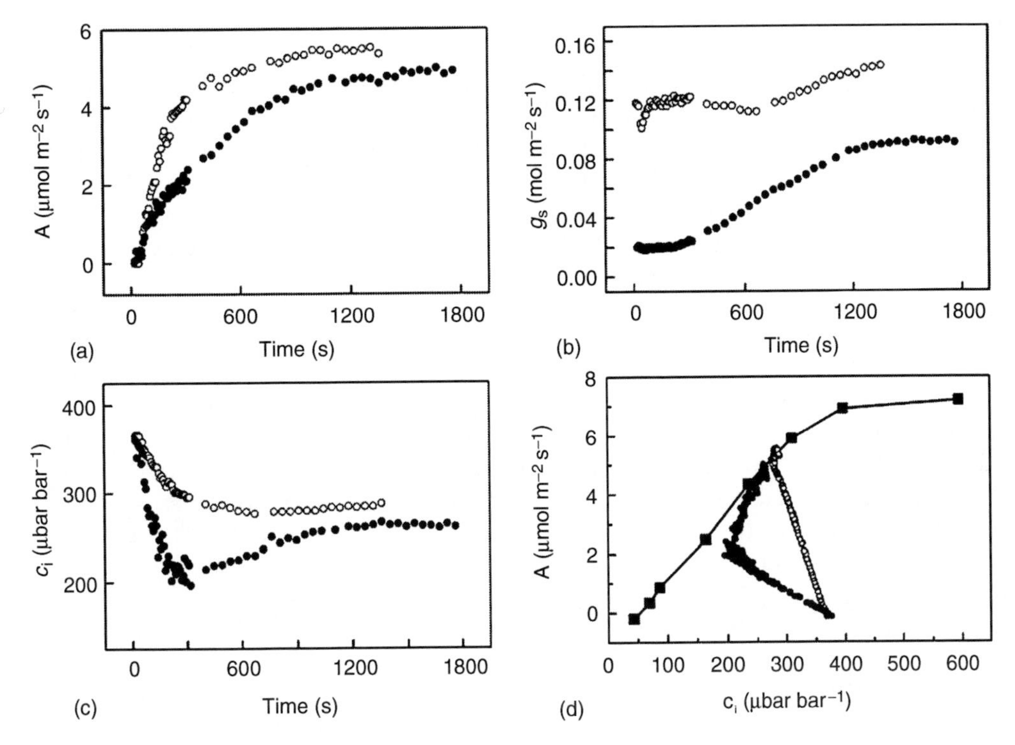

FIGURE 7.6 The time courses of (a) assimilation, (b) stomatal conductance, (c) c_i and (d) the relationship between assimilation and c_i during photosynthetic induction for a *P. marginata* leaf measured in the morning (open symbols) and the afternoon (closed symbols). PFD was increased from darkness to saturating levels at time zero. In addition, shown in (d) is the steady-state assimilation versus c_i curve (closed squares) measured after full induction. The open circles are from an induction response measured on a wet-season morning, whereas the closed circles are from one measured on the same leaf in the afternoon when the initial g_s was lower. (Adapted from Allen, M.T. and Pearcy, R.W., *Oecologia*, 122, 470, 2000a. With permission.)

Induction is important in determining the maximum assimilation rate that can be achieved during a sunfleck. A sunfleck received after a long period of shade results in only relatively low maximum photosynthetic rates during the sunfleck, whereas one received after a period of sunfleck activity results in much higher photosynthetic rates (Pearcy et al. 1985, Chazdon and Pearcy 1986). Thus, in effect, a sunfleck or a series of sunflecks primes the leaf so that it is better able to use subsequent sunflecks.

The actual transient response to a sunfleck measured with a fast-responding gas exchange system reveals further complex kinetics. Postillumination CO_2 fixation occurs as the primary CO_2 acceptor molecule, ribulose bisphosphate (RuBP), and the high-energy precursors to it that build up during the sunfleck are used for continued CO_2 fixation after the sunfleck. These pools of RuBP and its high-energy precursors essentially act as a capacitance that is charged as electron transport initially runs much faster than CO_2 fixation. Then the charge is used to support postlightfleck CO_2 fixation. In an uninduced leaf with a low capacity for utilization of RuBP, this capacitance is drained slowly and postillumination CO_2 fixation may continue at a decelerating rate for 20–60 s. In contrast, postillumination CO_2 fixation lasts for only a few seconds in an induced leaf. For brief lightflecks (less than 10 s), postillumination CO_2 fixation can enhance carbon gain by 50%–150% but in longer lightflecks it makes a diminishing contribution. A further kinetic complexity is introduced in relatively long lightflecks where a postillumination burst of CO_2 is released due to photorespiration. This burst occurs because some time is required to build up the photorespiratory metabolites and the shuttle of these metabolites from the chloroplasts to the glyoxysomes and mitochondria.

Mechanistically, the induction requirement, which strongly influences the capacity to use sunflecks, is primarily a consequence of three main factors: (1) the light activation of Rubisco, (2) the increase in g_s, and (3) the light activation of enzymes in the RuBP regeneration path. Up-regulation of the light-activated enzymes in RuBP regeneration (fructose-1,6-bisphosphatase, seduheptulose 1,7 bisphosphatase and ribulose-5-P kinase) occurs within 1–2 min following a light increase (Woodrow and Walker 1980, Sassenrath-Cole et al. 1994, Pearcy et al. 1996), thus allowing a sufficient capacity for RuBP regeneration so that RuBP concentrations are not rate-limiting for Rubisco. These enzymes are regulated by the redox state of thioredoxin, which in turn depends on photosynthetic electron flow (Buchanan 1980). Light activation of Rubisco itself is a slower process requiring 5–10 min for completion (Seemann et al. 1988, Woodrow and Mott 1988). Rubisco activity is regulated by covalent binding of Mg^{++} and CO_2 and by an enzyme, Rubisco activase, that removes bound sugar-phosphates from Rubisco, allowing catalytic activity (Campbell and Ogren 1992, Portis 2003). In addition, removal of a tight-binding inhibitor, carboxyarabinatol-1-phosphate, seems to be important in some species (Sage et al. 1993). Rubisco activase itself is light activated due to an ATP requirement (Campbell and Ogren 1992, 1995). Antisense plants with reduced levels of Rubisco activase have reduced rates of Rubisco activation and induction (Mott et al. 1997). Modeling studies suggest that there is an optimal balance between investment in Rubisco and Rubisco activase that may involve tradoffs between maintaining high steady-state photosynthetic rates and high rates of Rubisco activation, improving photosynthesis during sunflecks (Mott and Woodrow 2000).

In terms of assimilation rates, the key to understanding the limitations during induction is Rubisco, since the observed rates of CO_2 uptake are largely a mirror of its kinetics (Woodrow and Berry 1988). The relative limitations imposed by RuBP regneration, Rubisco activation, or g_s at any given time during induction determine the in vivo rate of Rubisco and hence the time course of CO_2 assimilation. This can be conveniently visualized by plotting assimilation versus c_i during induction (Figure 7.6d). Each point in the trajectory occurs on an imaginary assimilation versus c_i curve at a particular time during induction. As induction proceeds, the slope of this curve increases and gradually approaches the steady-state assimilation versus c_i curve, indicating an increasing carboxylation capacity of Rubisco. If the initial g_s is low then there is a significant decline in c_i, limiting assimilation, and a generally more sigmoidal increase in assimilation (Tinoco-Ojanguren and Pearcy 1993a, Allen and Pearcy 2000a). On the other hand, when the initial g_s is high, c_i remains high during induction, resulting in a faster increase in assimilation. Differences in the time required for induction are strongly dependent on the initial g_s (Figure 7.7). Lower initial g_s and slower induction in the afternoon as compared with the morning has been observed in tropical forest understory shrubs and redwood forest herbs (Pfitsch and Pearcy 1989a, Allen and Pearcy 2000a).

RuBP regeneration limitations are only important as long as the Rubisco activation and stomatal limitations are small. Thus, after a leaf has been in the shade for a long period so that g_s and the Rubisco activation state are low, the enzymes in RuBP regeneration impose almost no limitation early in induction. However, RuBP regeneration limitations can be quite significant in leaves that have been shaded for 5–10 min, since the RuBP regenerating enzymes are deactivated much more rapidly than Rubisco or the decline in g_s (Sassenrath-Cole and Pearcy 1994). This can lead to a rather prominent fast induction phase during which the photosynthetic rate increases over the first 1–2 min before a transition occurs to a slower increase that is due to stomatal opening and Rubisco activation (Kirschbaum and Pearcy 1988b, Tinoco-Ojanguren and Pearcy 1993b).

The relative roles of stomatal and biochemical limitations during induction have been the subject of some controversy. Early experiments were consistent with a greater role for

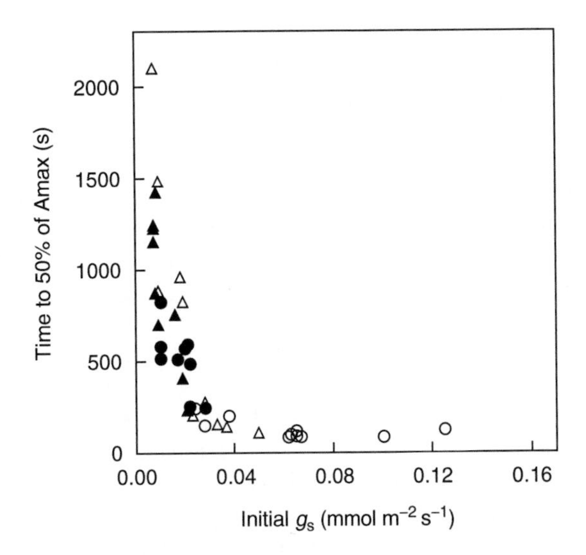

FIGURE 7.7 Relationship between initial stomatal conductance of *Psychotria marginata* leaves before the beginning of induction and time required to reach 50% of the final steady-state Amax. Circles represent measurements made in the morning, whereas triangles show afternoon measurements. Open symbols represent wet-season values, and closed symbols are values from the dry season. (Adapted from Allen, M.T. and Pearcy, R.W., *Oecologia*, 122, 470, 2000a. With permission.)

biochemical as compared with stomatal limitations (Usuda and Edwards 1984, Chazdon and Pearcy 1986a). This may well be true in some circumstances (Barradas and Jones 1996) but separation of the limitations using classical gas exchange approaches in which c_i is calculated is subject to a number of sources of error. When stomatal conductances are low, as they can be early in induction, it is necessary to consider the cuticular pathway for water loss, which when ignored can lead to a considerable overestimation in c_i (Kirschbaum and Pearcy 1988a). Moreover, considerable heterogeneity (patchiness) of stomatal conductance may be present during induction or even before it is in the shade, also leading to an overestimation of c_i (Cardon et al. 1994, Bro et al. 1996, Eckstein et al. 1996, Kuppers et al. 1999). However, computation of the errors due to patchiness in the estimate of c_i suggests that errors in partitioning the limitations between stomata and biochemistry are typically no more than 10%–15% (M.T. Allen and R.W. Pearcy unpublished results).

Since shade-tolerant plants in a forest understory receive 10%–70% of the available PFD in the form of sunflecks, these plants might be expected to have mechanisms that enhance sunfleck utilization by minimizing induction limitations. However, Naumberg and Ellsworth (2000) summarized the induction times from the literature and found that induction dynamics are not closely related to species shade tolerance, though individual studies with ecologically or phyllogenetically narrower groups of species often seem to be consistent with it (Poorter and Oberbauer 1993, Kuppers et al. 1996, Ogren and Sundlin 1996, Valladares et al. 1997, Rijkers et al. 2000, Hull 2002). The only clear trend from the literature survey of Naumberg and Ellsworth was that conifers tend to take longer to induce than angiosperms. Species differences in rates of Rubisco activation may play a minor role in differences in induction (Ogren and Sundlin 1996) but mostly they may be due to differences in stomatal behavior. The capacity for lightfleck utilization was much greater in a shade-tolerant species, *Piper aequale*, than in a pioneer species, *P. auritum*, when both were grown in the shade (Tinoco-Ojanguren and Pearcy 1992). These differences appeared to be related to stomatal behavior since stomatal conductance responded strongly to lightflecks in *P. aequale*, whereas almost no response occurred in *P. auritum*.

Differences in the utilization of transient light have also been found for species native to high-light habitats. Knapp and Smith (1987, 1990a,b) have identified species differences in the tracking response of stomatal conductance and hence photosynthesis to sun-shade transitions typical of those created by intermittent cloud cover. In some species, rapid stomatal opening and closure occurred in response to sun-shade transitions that limited the increase in photosynthesis during the light increase but overall enhanced water use efficiency. Other species exhibited reduced stomatal opening and closing responses and consequently were able to use the light increases more efficiently but at an overall lower water use efficiency. Although herbs, which tended to exhibit rather large water potential changes as compared with the woody species, were initially identified as having the more conservative response (Knapp and Smith 1989), this may not always be so (Knapp 1992). The high g_s of poplar species even in the shade may enhance the capability of these species to use the highly dynamic light environments found in their crowns due to leaf flutter (Roden and Pearcy 1993a,b). These species typically occur in habitats with abundant soil moisture so the high conductances and resulting low water use efficiency may cost relatively little as compared with the enhanced carbon gain. A similar lack of stomatal control of induction and consequently a fast induction response was reported for *Nothofagus cunninghamii* (Tausz et al. 2005). An alternative hypothesis to a sun shade-driven difference in induction times may be that species differences reflect different priorities in the trade-offs between carbon gain and water use (Naumburg and Ellsworth 2000).

The light environment during growth could also be expected to influence induction times and lightfleck utilization, but so far no clear picture has emerged. The higher photosynthetic capacity of high light-acclimated plants tends to increase carbon gain in sunflecks (Lei and Lechowicz 1997), but trade-offs between maximizing photosynthesis in low light versus maximizing it in sunflecks make this an unsuccessful strategy. Moreover, induction limitations would cause the realized carbon gain from the higher photosynthetic capacity to be lower. Rijkers et al. (2000) contrasted gap and understory plants of three species and found no difference in the rate of induction during a sequence of lightflecks. Despite higher photosynthetic capacities in the gap plants, the actual assimilation rate achieved during the lightflecks was either the same or was higher in the understory than in the gap plants. The understory plants did, however, maintain a higher induction state in each lightfleck in the sequence. Faster induction in shade as compared with sun leaves has been found (Kuppers and Schneider 1993, Kuppers et al. 1996) but other factors complicate the picture. For example sun-grown *P. auritum* leaves exhibit a large fast induction component, whereas shade-grown leaves do not (Tinoco-Ojanguren and Pearcy 1993b). Shade leaves have also been found to have higher lightfleck utilization efficiencies than sun-acclimated plants (Chazdon and Pearcy 1986b, Kuppers et al. 1996, Tang et al. 1994, Ogren Sundlin 1996). Pons and Pearcy (1992) found no difference in the capacity to use brief lightflecks (0.25 s) between sun- and shade-grown soybeans, but longer lightflecks were used less efficiently by the sun- as compared with the shade-grown plants. Induction should have little influence on utilization of brief lightflecks but would impact utilization of longer lightflecks.

Other environmental factors linked to shade and sun environments can also impact sunfleck utilization. High humidity characteristic of the understory caused slower decreases in stomatal conductance on shading and hence more carryover of induction from one lightfleck to the next (Tinoco-Ojanguren and Pearcy 1993a). Bright sunflecks can cause significant increases in leaf temperatures especially in large leaves with low stomatal conductance and these high temperatures can significantly inhibit utilization of sunflecks (Leakey et al. 2003a, 2005). Photoprotective systems in shade leaves appear to be fast responding (Logan et al. 1997), but still some photoinhibion cannot be avoided. This photoinhibition would be expected to primarily affect carbon gain in the low light following the lightfleck (Pearcy 1994, Zhu et al. 2004). Induction states have been found to be higher in the morning than the afternoon and in the wet versus the early dry season in a tropical forest (Allen and Pearcy 2000a,b). These effects are not readily explained by diurnal or seasonal environmental differences so their origin and significance remain elusive.

Models for Assessing the Consequences of Sunfleck Light Regimes

Since sunfleck light regimes are so temporally complex, a modeling strategy is needed to evaluate their overall consequences for carbon gain. Photosynthesis models such as either the simple Johnson–Thornley equation (Johnson and Thornley 1984) for the light dependence of photosynthesis or the more mechanistic Farquhar–von Cammerer (FvC) model are both steady-state models and thus do not provide an accurate simulation under dynamic light conditions. One approach is to measure assimilation during a sunfleck regime and then compare it to a steady-state model simulation as done by Pfitsch and Pearcy (1989a) and Schulte et al. (2003). In a steady-state model it is assumed that photosynthesis responds instantaneously to a change in PFD and therefore the limitations due to induction or the post-CO_2 assimilation are not present. If agreement between model and measurement is achieved it means that induction and postlightfleck CO_2 assimilation are of either little consequence for carbon gain or that the two cancel out. Using this type of approach Pfitsch and Pearcy (1989a) found that the steady-state Johnson–Thornley equation predicted about 20% more carbon gain than measured, indicating that induction limitations predominated and significantly lowered carbon gain in the understory. Essentially similar results were reported by Schulte et al. (2003).

To probe the consequences of sunflecks for carbon gain, more complex dynamic photosynthesis models are needed and several have now been developed. The models of Pearcy et al. (1997) and Kirschbaum et al. (1998) are both based on a combined dynamic stomatal model (Kirschbaum et al. 1988) and a dynamic assimilation model. The dynamic assimilation model is a modification of the FvC model (Gross et al. 1991), made dynamic by adding differential equations for activation or deactivation of Rubisco and RuBP regeneration and inclusion of metabolite pools. One difficulty with these models is that there are a large number of parameters, some of which require fairly elaborate gas exchange measurements for determination. Thus a more empirical model that simulates the essential dynamic features of the response to sunfleck light regimes (Stegemann et al. 1999) is an attractive approach.

Despite the considerable effort to parameterize them, mechanistic dynamic models do give good agreement between measured and modeled responses. Pearcy et al. (1997) found that assimilation of *Alocasia macrorrhiza* measured under a 3 h long sequence of lightflecks and simulated with their dynamic model agreed to within 3%. In contrast, a steady-state version of the model predicted 50% more carbon gain. The Pearcy et al. model can be run with input of PFD obtained with quantum sensors connected to data loggers that sample light at intervals as short as 0.1 s to make predictions of carbon gain and to visualize the time course of different limitations (Figure 7.8). The effects of induction on assimilation during sunflecks are clearly visible by comparing Figure 7.8a and Figure 8b. In addition, evident are the different time dependencies of g_s (Figure 7.8d), Rubisco (Figure 7.8e), and metabolite pool sizes (Figure 7.8f). Comparison of the dynamic and steady-state model predictions for carbon gain with light measurements in different microsites and days showed that the dynamic model gave a 3%–25% lower carbon gain (Table 7.1). The lowest values were for microsites and days with lower sunfleck activity but, because of the complex nature of the responses and the sunfleck regimes, no single characteristic of the sunfleck light regime such as numbers of sunflecks or their average duration was predictive of the limitations imposed. With defined lightfleck regimes, the dynamic limitation to carbon gain was much greater when there were just three 5 min lightflecks as compared with where there were one hundred and sixty nine 5 s lightflecks imposed on a low background PFD. This is consistent with the

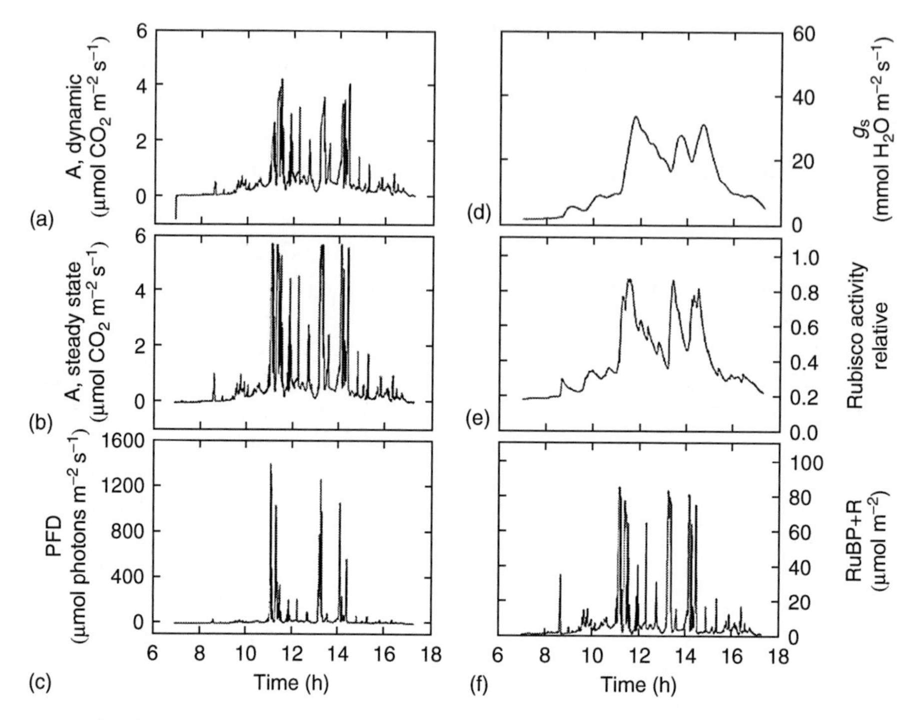

FIGURE 7.8 Simulated daily course of (a) assimilation of an *Alocasia macrorrhiza* leaf from the dynamic model of Pearcy et al. (1997) and (b) a steady-state version of the same model under (c) a natural sunfleck regime in the understory of a Queensland, Australia tropical forest. Additionally, shown are the simulated daily courses of (d) g_s, (e) Rubisco activity and (f) Pool sizes of RuBP and its high-energy precursors. Model inputs of PFD and the outputs were at 1 s intervals. (From Pearcy, R.W., Chazdon, R.L., Gross, L.J., and Mott, K.A., *Exploitation of Environmental Heterogeniety by Plants: Ecophysiological Processes above and below Ground*, Academic Press, San Diego, CA, 1994. With permission.)

fact that there are greater induction limitations for the long lightflecks as shown experimentally (Chazdon and Pearcy 1986b) and a greater relative contribution of postlightfleck CO_2 assimilation for short lightflecks.

Dynamic models are also useful for assessing the role of specific dynamic elements by selectively modifying the input parameters so that one or more elements become static while the others remain dynamic (Table 7.1). This is done by changing time constants such that g_s or Rubisco light activation or RuBP regeneration light activation, or any combination, respond instantly to a change in PFD, whereas the other limitations remain dynamic. Removing the dynamic stomatal limitations alone had a much larger effect than removing the Rubisco activation limitations. In the latter scenario, Rubisco responds instantaneously whereas stomata responded slowly, causing c_i to decrease and to more strongly limit carbon gain. Removing the dynamic stomatal limitations caused c_i to be higher increasing assimilation (see Figure 7.6). Removing both Rubisco activation and stomatal dynamic limitations caused assimilation to be nearly the same as predicted by the steady-state model. Finally, removing all dynamic limitations except for the metabolite pools responsible for postlightfleck CO_2 assimilation gave an identical carbon gain to the steady-state version of the model, except for the short 5 s lightflecks where carbon gain was increased. Postlightfleck CO_2 assimilation made no significant contribution to the simulated carbon gain in the understory because, as discussed earlier, short lightflecks themselves made only a very small contribution to the available PFD. However, the contributions may be significant or may at least offset

TABLE 7.1
Simulations of Lightfleck Use in Regimes with Either 5 min or 5 s Lightflecks or for Natural Light Regimes

Conditions		All Limitations	Stomatal Limitation	Rubisco Limitation	Stomatal and Rubisco Limitations	Stomatal, Rubisco, and FBPase Limitations
5 min lightflecks[a]		−22	−10	−21	−1	0
5 s lightflecks[a]		2	1	0	6	12
Understory light regimes[b]						
PFD mmol m^{-2} day^{-1}	Percent in sunflecks					
1.28	48	−3	−1	−2	0	0
0.52	23	−7	−2	−6	0	0
1.55	59	−17	−7	−15	−5	0
1.18	71	−25	−8	−22	−1	0

Note: Values in the table are percentage differences between a dynamic and a steady-state simulation using the model of Pearcy et al. (1997). A value of −10, for example, means that dynamic limitations inherent in leaf photosynthetic responses to lightflecks result in a carbon gain that is 10% less than that predicted by a steady-state model. The model was initially parameterized for an *Alocasia macrorrhiza* leaf and then various dynamic limitations were removed.

[a] Either three 5 min lightflecks or one hundred and sixty nine 5 s lightflecks uniformly distributed in an 8.6 h period. The lightfleck and shade PFDs were 500 and 15 μmol m^{-2} s^{-1}, respectively.

[b] Recorded at 1 s intervals in four sites in the understory of a tropical forest in Queensland Australia. PFD is the daily total and percent in sunflecks is the percentage of the daily total that was received as sunflecks.

some of the induction limitations in other circumstances. Simulations for crop, quaking aspen, and poplar canopies show that postlightfleck CO_2 assimilation can contribute to as much as a 6% enhancement of carbon gain under windy conditions when short sunflecks make a much more significant contribution (R.W. Pearcy and J. Roden unpublished results). In the understory, induction limitations predominate and significantly constrain carbon gain during sunflecks. Of these, dynamic stomatal limitations and the relationships between stomatal opening and Rubisco activation may be especially important.

Growth Responses to Sunflecks

At the ecological level, sunfleck responses, if they are of significance, should have consequences for the growth of plants. Such experiments are complicated since they require careful matching of light regimes in terms of the total PFD and differ only in their temporal variation. Comparisons of RGR for plants in microsites differing in sunfleck light regimes (Leakey et al. 2003b) are difficult because other factors may also differ. Thus experimental protocols in greenhouses using shutters to create lightflecks or in growth cabinets (Sims and Pearcy 1993, Watling et al. 1997) are better for unraveling the growth responses. Adding lightflecks onto a diffuse light background increases the total PFD, so comparisons back to assimilation under just the diffuse PFD include both the effects of the greater total PFD as well as any temporal effects due to the sunflecks. The PFD in the diffuse light regime can be increased to match that of the lightflecks plus diffuse light. However, in this case, assimilation is undoubtedly higher in the diffuse light because of the nonlinear nature of the photosynthetic response and because the diffuse light usually falls in the light-dependent part where PFD is used more efficiently. This approach was used by Watling et al. (1997) who found that in three of the four species tested, RGR was higher in the constant light than the lightfleck regime. Addition of lightflecks to a background of low PFD, resulting in additional total PFD, clearly increased RGR, except in *Alocasia macorrhiza* where RGR decreased for unknown reasons. Sims and Pearcy (1993) compared growth of *A. macorrhiza* under regimes of (1) long (8 min) but infrequent light flecks, (2) short (6 s) but frequent lightflecks, and (3) diffuse light only, where all were carefully matched with respect to daily PFD. They found that relative to the diffuse light only control, RGR was more inhibited under the long than the short lightfleck regime. This is consistent with the greater expected induction limitation in the long lightflecks.

Other growth experiments have yielded contrasting results, possibly because of species differences but also possibly different experimental protocols. Growth of dipterocarp seedlings was much greater in a long sunfleck regime created in a small gap than in a short sunfleck regime in the nearby understory (Leakey et al. 2003b). However, to match the total daily PFDs, shadecloth was used in the gap, also reducing the peak sunfleck PFD. The sunflecks therefore differed not only in their duration and maximum PFD, but also in their dispersion through the day. The experiment clearly shows that different sunfleck regimes have consequences for growth that can in part be attributed to differences in the dynamics of assimilation during sunflecks but the contrasting results as compared with those of Sims and Pearcy (1993) could be explained by any one of several factors.

Sunflecks can also be removed with shadow bands positioned over plants that block the view of the solar path. Growth of these plants can then be compared with the nearby control plants that receive sunflecks. This protocol revealed that sunfleck removal caused *A. bicolor* plants to decrease in size over 2 years as maintenance costs exceeded carbon supply, indicating that sunflecks were of great significance to the survival of this species (Pfitsch and Pearcy 1992). The individual pairs of plants were in different microsites receiving different sunfleck light regimes so that it was also possible to compare growth between the different microsites. An analysis of covariance, however, revealed no significant relationship between plant performance and the availability of sunflecks. Although additional sunflecks should increase carbon gain, competition for nutrients or greater water stress in the brighter microsites may have limited plant performance. Indeed, many of the characteristics such as the variation in LMA were better correlated with diffuse PFD than with sunfleck PFD.

Responses to Gap Dynamics

Temporal Nature of Gap Light Regimes

The most studied case of the responses to long-term temporal heterogeneity in light environments concerns tree-fall gaps in forests (Bazzaz and Carlson 1982, Canham 1989, Canham et al. 1990, Bazzaz 1991). When a tree falls there is an essentially instantaneous increase in PFD, followed by a gradual decrease as the gap is filled with new growth. The increase depends on gap geometry and cloudiness. Therefore, the dynamic response of an understory plant to the formation of a gap may depend on whether it occurs during a period of cloudy or clear weather.

Response to the increase in PFD following gap formation has received much more attention than the response to the much more gradual decrease that occurs as a gap fills. Gap filling occurs both from the bottom up as plants established in the gap gain height and from the sides as branches extend out into it. Significant shading of seedlings may develop within a year or two, whereas at higher levels the gap may not fill for several to many years. Thus, shading of saplings may be delayed as compared to seedlings. This gradual refilling stage may be quite important for understory shrubs that establish in gaps but persist and reproduce in the understory long after that gap has closed. Most tropical forest trees may require multiple gap events to make the transition from understory seedling to sapling to a canopy tree. Thus just how the photosynthetic apparatus and allocation patterns adjust to the gradual decrease in PFD as a gap closes could be quite important for persistence.

Dynamic Plasticity of Leaves and Plants to Long-Term Temporal Heterogeneity

Following a gap formation both acclimation of leaves and shifts in allocation to restore a functional balance at the whole plant level are important mechanisms for responding to the increased light availability. Kursar and Coley (1999) have identified two modes of acclimation, one where relatively short lived but inexpensive leaves are quickly dropped and replaced by a canopy of sun leaves. The other is where long lived, expensive leaves are retained and are only gradually replaced and exhibit moderate acclimation to the increased light. Long-lived leaves were initially more tolerant of high light stress and exhibited less photoinhibition measured as the reduced F_v/F_m values the following morning. These contrasting modes, which are probably representative of extremes along a spectrum of responses, are indicative of a trade-off between a capacity to rapidly accelerate growth with increased light availability and greater stress tolerance.

In terms of responses to gap formation the most revealing and ecologically relevant experiments are those that follow the dynamics of acclimation following a light increase (Ferrar and Osmond 1986, Mulkey and Pearcy 1992, Turnbull et al. 1993). These experiments typically reveal initial reductions in quantum yields and the fluorescence ratio F_v/F_m indicative of photoinhibition, and usually a reduction of light-saturated photosynthetic rates. This reduction in F_v/F_v can last for 1 day to a week or more but if conditions are not too severe, recovery from photoinhibition occurs. In peas, only a small, 1 day reduction in Amax occurred before Amax increases rapidly within 4 days to a new, threefold higher steady-state value (Anderson and Osmond 1987). In the evergreen tropical tree species *Bischofia javanensis*, Amax and the chlorophyll fluorescence ratio, F_v/F_m, remained depressed for nearly 8 days, and 30 days were required to reach the new steady-state Amax (Kamaluddin and Grace 1992). In the temperate vine, *Hedera helix*, 45 days were required to complete the acclimation process (Bauer and Thoni 1988). Low nitrogen status exacerbates the initial photoinhibitory reduction in Amax and slows or prevents full recovery when shade-grown plants are transferred to high light (Ferrar and Osmond 1986). High-leaf temperature also strongly interacts with the PFD level, causing a greater inhibition and slower recovery.

If the extant leaves are capable of acclimation, a further slow increase in Amax occurs. In mature leaves incapable of acclimation, an initial decline in photosynthetic rates occurs followed by at least a partial recovery if conditions are not too severe. New leaves then develop with a higher photosynthetic capacity. Ultimately acclimation at the whole-plant level is complete when there has been a complete turnover of the leaves and adjustments in LAR and root/shoot ratios have occurred. Depending on the species involved and its leaf longevity and production rates, full acclimation could take weeks to years.

A key to the initial response to gap formation is the dynamic interaction between acclimation and photoinhibition. As shown in Figure 7.9, exposure of shade-grown *Alocasia macrorrhiza* plants to 2 h of direct sunlight superimposed on a shade-light background resulted in a strong initial reduction in F_v/F_m and Amax followed by an almost complete recovery of F_v/F_m and only a partial recovery of Amax in 5–10 days (Mulkey and Pearcy 1992). The recovery during successive days with the same light treatment suggests development of an increased capacity for photoprotection. Leaves that developed under this gap simulation were completely resistant to photoinhibition and had a higher Amax. Thus complete acclimation required replacement of all the leaves, requiring in this species 60–80 days. Although the preexisting leaves had a reduced photosynthetic capacity as compared with later-developing leaves, they still provided substantial carbon gain because of the high

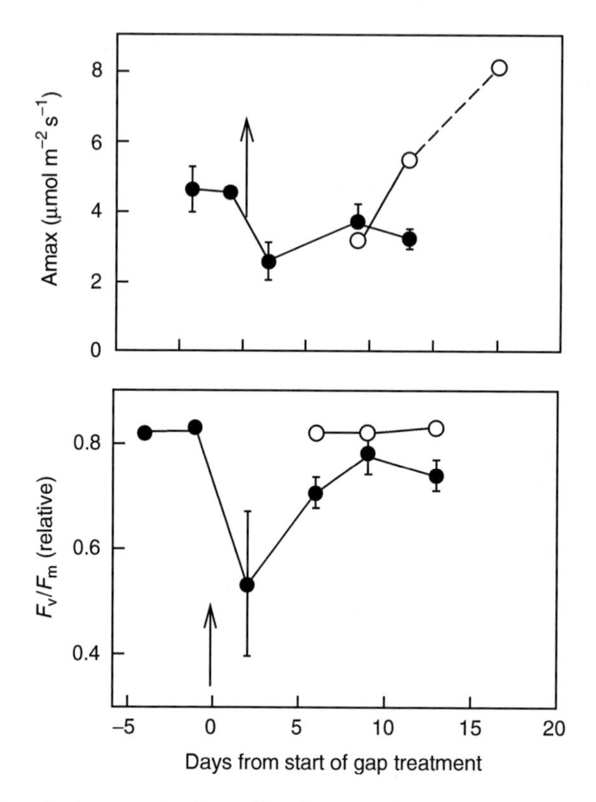

FIGURE 7.9 Dynamics of photosynthetic acclimation and photoinhibition as indicated by reduced F_v/F_m ratios following transfer of *Alocasia macorrhiza* plants from a shade environment simulating the understory to a gap environment in which 2 h of direct sunlight were received each day. The transfer occurred at the arrow. The filled circles show the response of leaves that had fully developed in the shade before the gap treatment. The open circles show the response of leaves that developed after the gap treatment commenced. The data point connected by the dashed line shows the maximum Amax achieved by leaves in the gap environment. Note the strong photoinhibitory decline and recovery (acclimation) in the shade developed leaves but complete absence of photoinhibition in the gap developed leaves. (Adapted from Mulkey, S.S. and Pearcy, R.W., *Funct. Ecol.*, 6, 719, 1992.)

light. Thus, protecting them through rapid acclimation of photoprotective mechanisms may be quite important for maintaining a carbon supply for growth of the new leaves. In this experiment, a strong interaction between high light and high leaf temperatures was found that greatly increased photoinhibition and delayed the recovery. This interaction is undoubtedly quite important in natural gap formation since leaf temperatures in gaps with bright sunlight can be 10°C–15°C above air temperatures and reach values that would be likely to strongly exacerbate photoinhibition. Different responses might therefore be expected when gaps form during cloudy versus clear periods.

CONCLUDING REMARKS

The response of individual plants to spatial and temporal heterogeneity depends on the scale of the heterogeneity. The temporal scale is paramount because it determines whether adjustments in physiological properties rely primarily on relatively rapid and low-cost regulatory mechanisms such as regulation of enzyme catalytic capacity and stomatal opening or whether

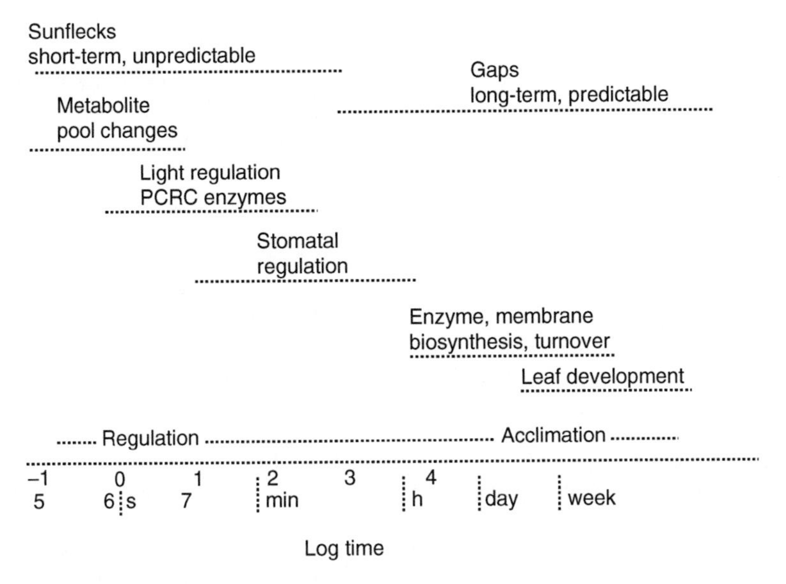

FIGURE 7.10 Ranges of time constants for various processes in relation to the timescales of sunflecks and gap dynamics.

plasticity involving regulation of the amounts of enzymes and plant morphological changes provide a benefit. Responses to temporal changes on a timescale of a day or less primarily involve regulation of activities and stomatal behavior, whereas responses to longer-term changes in the light environment involve acclimation and morphological plasticity (Figure 7.10). The available evidence suggests that acclimation of photosynthetic capacity depends on the mean daily integral of PFD rather than the peak PFDs. This provides a mechanism that minimizes potentially costly responses to unpredictable variations such as sunflecks. However, the mechanism by which plants sense the daily integral of PFD and then make adjustments leading to the proper acclimation state is not understood.

The response to the spatial scale of heterogeneity appears to be linked principally to the temporal scale. Sunflecks and light gradients within canopies have similar spatial scales but since the light gradient is more predictable, acclimation is the principal mechanism for adaptation to this heterogeneity, whereas for sunflecks regulatory mechanisms are overriding. There is evidence that interplant communication may play an important role in acclimation to spatial heterogeneity. The evidence is strongest for clonal plants, but patterns of branch growth and suppression in response to spatial light heterogeneity may also involve some branch to branch communication. The signals underlying these communications are poorly understood but could involve hormones or resource supplies such as carbohydrates. More research is needed to understand the nature and the role of these mechanisms in determining how plants respond to environmental heterogeneity.

Understanding physiological responses to heterogeneity requires understanding how its consequences scale-up. Acclimation of leaf gas exchange properties obviously influences carbon uptake per unit leaf area, but because it also involves changes in LMA it has consequences for carbon partitioning. Indeed, these latter effects may be more significant to whole-plant carbon gain than the direct effects of photosynthetic acclimation. A consequence of the scaling is that plasticity in photosynthetic properties such as Amax may lead to greater homeostasis in growth and resource use efficiency. It is never enough to offset the wide variation in light levels inherent in natural communities but nevertheless is important in survival spatially and temporally heterogeneous environments.

REFERENCES

Acevedo, M.F., D.L. Urban, and H.H. Shugart, 1996. Models of Forest Dynamics Based On Roles of Tree Species. Ecological Modelling 87: 267–284.

Ackerly, D.D. and F.A. Bazzaz, 1995. Seedling crown orientation and interception of diffuse radiation in tropical forest gaps. Ecology 76: 1134–1146.

Allen, M.T. and R.W. Pearcy, 2000a. Stomatal behavior and photosynthetic performance under dynamic light regimes in a seasonally dry tropical rain forest. Oecologia 122: 470–478.

Allen, M.T. and R.W. Pearcy, 2000b. Stomatal versus biochemical limitations to dynamic photosynthetic performance in four tropical rainforest shrub species. Oecologia 122: 479–486.

Anderson, J.M., W.S. Chow, and Y.I. Park, 1995. The grand design of photosynthesis: acclimation of the photosynthetic apparatus to environmental cues. Photosynthesis Research 46: 129–139.

Anderson, J.M. and C.B. Osmond, 1987. Shade-sun responses: compromises between acclimation and photoinhibition. In: D.J. Kyle, C.B. Osmond, and C.J. Arntzen, eds. Photoinhibition. Elsevier, Amsterdam, pp. 1–38.

Anten, N.P.R., F. Schieving, and M.J.A. Werger, 1995. Patterns of light and nitrogen distribution in relation to whole canopy carbon gain in C_3 and C_4 mono- and dicotyledonous species. Oecologia 101: 504–513.

Atkin, O.K., J.R. Evans, and K. Siebke, 1998. Relationship between the inhibition of leaf respiration by light and enhancement of leaf dark respiration following light treatment. Australian Journal of Plant Physiology 25: 437–443.

Atkin, O.K., J.R. Evans, M.C. Ball, H. Lambers, and T.L. Pons, 2000. Leaf respiration of snow gum in the light and dark. interactions between temperature and irradiance. Plant Physiology 122: 915–923.

Baldocchi, D. and S. Collineau, 1994. The physical nature of solar radiation in heterogeneous canopies: spatial and temporal attributes. In: M.M. Caldwell and R.W. Pearcy, eds. Exploitation of Environmental Heterogeneity by Plants: Ecophysiological Processes Above- and Belowgorund. Academic Press, San Diego, CA, pp. 21–72.

Baldocchi, D.D., B.A. Hutchinson, D.R. Matt, and R.T. McMillen, 1985. Canopy radiative transfer models for spherical and known leaf angle distributions: A test in an oak-hickory forest. Journal of Applied Ecology 22: 539–555.

Ballare, C.L., 1994. Light gaps: Sensing the light opportunities in highly dynamic canopy environments. In: M.M. Caldwell and R.W. Pearcy, eds. Exploitation of Environmental Heterogeneity by Plants: Ecophysiological Processes Above- and Below ground. Academic Press, San Diego, CA, pp. 73–110.

Ballare, C.L. and J.J. Casal, 2000. Light signals perceived by crop and weed plants. Field Crops Research 67: 149–160.

Ballare, C.L., R.A. Sanchez, A.L. Scopel, J.J. Casal, and C.M. Ghersa, 1987. Early detection of neighbor plants by phytochrome perception of spectral changes in reflected sunlight. Plant, Cell and Environment 10: 551–557.

Barradas, V.L. and H.G. Jones, 1996. Responses of CO_2 assimilation to changes in irradiance: Laboratory and field data and a model for beans (*Phaseolus vulgaris* L). Journal of Experimental Botany 47: 639–645.

Bauer, H. and W. Thoni, 1988. Photosynthetic light acclimation in fully developed leaves of the juvenile and adult life cycle phases of *Hedera helix*. Physiologia Plantarum 73: 31–37.

Bazzaz, F.A., 1991. Regeneration of tropical forests—physiological responses of pioneer and secondary species. Rain Forest Regeneration and Management 6: 91–118.

Bazzaz, F.A. and R.W. Carlson, 1982. Photosynthetic acclimation to variability in the light environment of early and late successional plants. Oecologia (Berlin) 54: 313–316.

Bazzaz, F.A. and S.T.A. Pickett, 1980. Physiological ecology of tropical succession: a comparative review. Annual Review of Ecology and Systematics 11: 287–310.

Björkman, O., 1981. Responses to different quantum flux densities. In: O.L. Lange, P.S. Nobel, C.B. Osmond, and H. Ziegler, eds. Physiological Plant Ecology I Physical Environment. Springer-Verlag, Berlin, pp. 57–107.

Björkman, O. and B. Demmig, 1987. Photon yield of O_2 evolution and chlorophyll flourescence characteristics at 77 K among vascular plants of diverse origins. Planta 170: 489–504.

Björkman, O. and S.B. Powles, 1984. Inhibition of photosynthetic reactions under water stress: Interaction with light level. Planta 161: 490–504.
Björkman, O., S.B. Powles, D.C. Fork, and G. Oquist, 1981. Interaction between high irradiance and water stress on photosynthetic reactions in *Nerium oleander*. Carnegie Institute of Washington Yearbook 80: 57–59.
Bolker, B.M., S.W. Pacala, and C. Neuhauser, 2003. Spatial dynamics in model plant communities: What do we really know? American Naturalist 162: 135–148.
Bradshaw, A.D., 1965. Evolutionary significance of phenotypic plasticity in plants. Advances in Genetics 13: 115–155.
Bradshaw, G.A. and T.A. Spies, 1992. Characterizing canopy gap structure in forests using wavelet analysis. Journal of Ecology 80: 205–215.
Bro, E., S. Meyer, and B. Genty, 1996. Heterogeneity of leaf CO_2 assimilation during photosynthetic induction. Plant, Cell and Environment 19: 1349–1358.
Brokaw, N.V.L., 1985. Treefalls, regrowth, and community structure in tropical forests. In: S.T.A. Pickett and P.S. White, eds. The Ecology of Natural Disturbance and Patch Dynamics. Academic Press, San Diego, CA, pp. 53–68.
Buchanan, B.B., 1980. Role of light in the regulation of chloroplast enzymes. Annual Review of Plant Physiology 31: 341–374.
Caldwell, M.M., H.-P. Meister, J.D. Tehunen, and O.L. Lange, 1986. Canopy structure, light microclimate and leaf gas exchange of *Quercus coccifera* L. in a Portugese macchia: Measurements in different canopy layers and simulations with a canopy model. Trees 1: 25–41.
Callahan, H.S., M. Pigliucci, and C.D. Schlichting, 1997. Developmental phenotypic plasticity: Where ecology and evolution meet molecular biology. Bioessays 19: 519–525.
Campbell, W.J. and W.L. Ogren, 1992. Light activation of Rubisco by Rubisco activase and thylakoid membranes. Plant and Cell Physiology 33: 751–756.
Campbell, W.J. and W.L. Ogren, 1995. Rubisco activase activity in spinach leaf extracts. Plant and Cell Physiology 36: 215–220.
Canham, C.D., 1989. Different responses to gaps among shade-tolerant tree species. Ecology 70: 548–550.
Canham, C.D., J.S. Denslow, W.J. Platt, J.R. Runkle, T.A. Spies, and P.S. White, 1990. Light regimes beneath closed canopies and tree-fall gaps in temperate and tropical forests. Canadian Journal of Forestry 20: 620–631.
Canham, C.D., A.C. Finzi, S.W. Pacala, and D.H. Burbank, 1994. Causes and consequences of resource heterogeneity in forests—interspecific variation in light transmission by canopy trees. Canadian Journal of Forest Research—Journal Canadien De La Recherche Forestiere 24: 337–349.
Cardon, Z.G., K.A. Mott, and J.A. Berry, 1994. Dynamics of patchy stomatal movements, and their contribution to steady-state and oscillating stomatal conductance calculated using gas-exchange techniques. Plant, Cell and Environment 17: 995–1007.
Cescatti, A. and U. Niinemets, 2004. Leaf to landscape. In: W.K. Smith, C. Critchley, and T.C. Vogelmann, eds. Photosynthetic Adaptation: Chloroplast to Landscape. Springer, New York, pp. 42–84.
Chabot, B.F., 1979. Influence of instantaneous and intergrated light-flux density on leaf anatomy and photosynthesis. American Journal of Botany 66: 940–945.
Chabot, B.F. and J.F. Chabot, 1977. Effects of light and temperature on leaf anatomy and photosynthesis in *Frageria vesca*. Oecologia 26: 363–377.
Chabot, B.F. and D.J. Hicks, 1982. The ecology of leaf life spans. Annual Review of Ecology Systems 13: 229–259.
Chazdon, R.L., 1988. Sunflecks and their importance to forest understory plants. Advances in Ecological Research 18: 1–63.
Chazdon, R.L., 1992. Photosynthetic plasticity of two rain forest shrubs across natural gap transects. Oecologia 92: 586–595.
Chazdon, R.L. and C.B. Field, 1987. Determinants of photosynthetic capacity in six rainforest *Piper* species. Oecologia 73: 222–230.
Chazdon, R.L. and S. Kaufmann, 1993. Plasticity of leaf anatomy of two rain forest shrubs in relation to photosynthetic light acclimation. Functional Ecology 7: 385–394.

Chazdon, R.L. and R.W. Pearcy, 1986a. Photosynthetic responses to light variation in rain forest species. I. Induction under constant and fluctuating light conditions. Oecologia 69: 517–523.
Chazdon, R.L. and R.W. Pearcy, 1986b. Photosynthetic responses to light variation in rain forest species. II. Carbon gain and light utilization during lightflecks. Oecologia 69: 524–531.
Chazdon, R.L., K. Williams, and C.B. Field, 1988. Interactions between crown structure and light environment in five rain forest Piper species. American Journal of Botany 75: 1459–1471.
Chazdon, R.L., R.W. Pearcy, D.W. Lee, and N. Fetcher, 1996. Photosynthetic responses of tropical forest plants to contrasting light environments. In: S.S. Mulkey, R.L. Chazdon, and A.P. Smith, eds. Tropical Forest Plant Ecophysiology. Chapman & Hall, New york, pp. 5–55.
Chow, W.S. and J.M. Anderson, 1987a. Photosynthetic responses of *Pisum sativum* to an increase in irradiance during growth II. Thylakoid membrane components. Australian Journal Plant Physiology 14: 9–19.
Chow, W.S. and J.M. Anderson, 1987b. Photosynthetic responses of *Pisum sativum* to an increase in irradiance during growth. I. Photosynthetic activities. Australian Journal of Plant Physiology 14: 1–8.
Clark, D.B., D.A. Clark, P.M. Rich, S. Weiss, and S.F. Oberbauer, 1996. Landscape scale evaluation of understory light and canopy structure: Methods and application in a neotropical lowland rain forest. Canadian Journal of Forest Research-Revue Canadienne De Recherche Forestiere 26: 747–757.
Cowan, I.R. and G.D. Farquhar, 1977. Stomatal function in relation to leaf metabolism and environment. Symposia of the Society for Experimental Biology 31: 471–505.
de Kroon, H., H. Huber, J.F. Stuefer, and J.M. van Groenendael, 2005. A modular concept of phenotypic plasticity in plants. New Phytologist 166: 73–82.
de Kroon, H. and M.J. Hutchings, 1995. Morphological plasticity in clonal plants: The foraging concept reconsidered. Journal of Ecology 83: 143–152.
Delagrange, S., C. Messier, M.J. Lechowicz, and P. Dizengremel, 2004. Physiological, morphological and allocational plasticity in understory deciduous trees: importance of plant size and light availability. Tree Physiology 24: 775–784.
DeLucia, E.H. and R.B. Thomas, 2000. Photosynthetic responses to CO_2 enrichment of four hardwood species in a forest understory. Oecologia 122: 11–19.
Demmig-Adams, B. and W.W. Adams, 1992a. Carotenoid composition in sun and shade leaves of plants with different life forms. Plant, Cell and Environment 15: 411–419.
Demmig-Adams, B. and W.W. Adams, 1992b. Photoprotection and other responses of plants to high light stress. Annual Review of Plant Physiology and Plant Molecular Biology 43: 599–626.
Demmig-Adams, B. and W.W. Adams, 1994. Capacity for energy dissipation in the pigment bed in leaves with different xanthophyll cycle pools. Australian Journal of Plant Physiology 21: 575–588.
Demmig-Adams, B. and W.W. Adams, 1996. The role of xanthophyll cycle carotenoids in the protection of photosynthesis. Trends in Plant Science 1: 21–26.
Demmig-Adams, B. and W.W. Adams, 2000. Photosynthesis—Harvesting sunlight safely. Nature 403: 371–374.
Denslow, J.S., 1980. Gap partitioning among tropical forest trees. Biotropica 12 (Suppl.): 47–55.
Denslow, J.S., 1987. Tropical rainforest gaps and tree species diversity. Annual Review of Ecology and Systematics 18: 431–451.
Derner, J.D. and X.B. Wu, 2001. Light distribution in mesic grasslands: Spatial patterns and temporal dynamics. Applied Vegetation Science 4: 189–196.
Desjardins, R.L., 1967. Time series analysis in agrometerological problems with emphasis on spectrum analysis. Canadian Journal of Plant Physiology 47: 477–491.
Desjardins, R.L., T.R. Sinclair, and E.R. Lemon, 1973. Light fluctuations in corn. Agronomy Journal 65: 904–907.
Dong, M., 1995. Morphological responses to local light conditions in clonal herbs from contrasting habitats, and their modification due to physiological integration. Oecologia 101: 282–288.
Dutton, R.G., J. Jiao, M.J. Tsujta, and B. Grodzinski, 1988. Whole plant CO_2 exchange measurements for nondestructive estimation of growth. Plant Physiology 86: 355–358.
Eckstein, J., W. Beyschlag, K.A. Mott, and R.J. Ryel, 1996. Changes in photon flux can induce stomatal patchiness. Plant, Cell and Environment 19: 1066–1074.

Ehleringer, J. and I. Forseth, 1980. Solar tracking by plants. Science 210: 1094–1098.
Ellsworth, D.S. and P.B. Reich, 1993. Canopy structure and vertical patterns of photosynthesis and related leaf traits in a deciduous forest. Oecologia 96: 169–178.
Ellsworth, D.S. and P.B. Reich, 1992. Leaf Mass Per Area, Nitrogen Content and Photosynthetic Carbon Gain in Acer-Saccharum Seedlings in Contrasting Forest Light Environments. Functional Ecology 6: 423–435.
Evans, J.R., 1986. A quantitative analysis of light distribution between the two photosystems, considering variation in both the relative amounts of the chlorophyll-protein complexes and the spectral quality of light. Photochemistry and Photobiophysics 10: 135–147.
Evans, J.R., 1989a. Partitioning of nitrogen between and within leaves grown under different irradiances. Australian Journal of Plant Physiology 16: 533–548.
Evans, J.R., 1989b. Photosynthesis and nitrogen relationships in leaves of C_3 plants. Oecologia 78: 9–19.
Evans, J.R., 1993. Photosynthetic acclimation and nitrogen partitioning within a lucerne canopy. II. Stability through time and comparison with a theoretical optimum. Australian Journal of Plant Physiology 20: 69–82.
Evans, J.R. and H. Poorter, 2001. Photosynthetic acclimation of plants to growth irradiance: the relative importance of specific leaf area and nitrogen partitioning in maximizing carbon gain. Plant, Cell and Environment 24: 755–767.
Evans, J.R. and S. Von Caemmerer, 1996. Carbon dioxide diffusion inside leaves. Plant Physiology 110: 339–346.
Farquhar, G.D., 1989. Models of integrated photosynthesis of cells and leaves. Philosophical transactions of the Royal Society of London. Series B, 323: 357–367.
Farquhar, G.D. and S. Von Caemmerer, 1982. Modelling of photosynthetic response to environmental conditions. In: A.P. Gottingen and M.H. Zimmerman, eds. Encyclopedia of Plant Physiology N.S. Springer-Verlag, Berlin, Germany, Vol. 12B, 549–588.
Ferrar, P.J. and C.B. Osmond, 1986. Nitrogen supply as a factor influencing photoinhibition and photosynthetic acclimation after transfer of shade-grown *Solanum dulcamara* to bright light. Planta 168: 563–570.
Field, C. 1983. Allocating leaf nitrogen for the maximization of carbon gain: Leaf age as a control on the allocation program. Oecologia 56: 341–347.
Foyer, C., R. Furbank, J. Harbinson, and P. Horton, 1990. The mechanisms contributing to photosynthetic control of electron transport by carbon assimilation in leaves. Photosynthesis Research 25: 83–100.
Frazer, G.W., M.A. Wulder, and K.O. Niemann, 2005. Simulation and quantification of the fine-scale spatial pattern and heterogeneity of forest canopy structure: A lacunarity-based method designed for analysis of continuous canopy heights. Forest Ecology and Management 214: 65–90.
Friend, A.D., H.H. Schugart, and S.W. Running, 1993. A physiology-based gap model of forest dynamics. Ecology 74: 792–797.
Gabrielsen, E.K., 1948. Effects of different chlorophyll concentrations on photosynthesis in foliage leaves. Physiologia Plantarum 1: 5–27.
Gamon, J.A. and R.W. Pearcy, 1989. Leaf movement, stress avoidance and photosynthesis in *Vitus californica*. Oecologia 79: 475–481.
Geiken, B., C. Critchley, and G. Renger, 1992. The turnover of photosystem II reaction centre proteins as a function of light. In: N. Murata, ed. Research in Photosynthesis. Kluwer Academic, Dordrecht, pp. 634–646.
Gilbert, I.R., P.G. Jarvis, and H. Smith, 2001. Proximity signal and shade avoidance differences between early and late successional trees. Nature 411: 792–795.
Gilbert, I.R., G.P. Seavers, P.G. Jarvis, and H. Smith 1995. Photomorphogenesis and canopy dynamics: Phytochromemediated proximity perception accounts for the growth dynamics of canopies of *Populus trichocarpa* × deltoides beaupre. Plant, Cell and Environment 18: 475–497.
Grace, S.C. and B.A. Logan, 1996. Acclimation of foliar antioxidant systems to growth irradiance in three broad-leaved evergreen species. Plant Physiology 112: 1631–1640.
Gratzer, G., C. Canham, U. Dieckmann, A. Fischer, Y. Iwasa, R. Law, M.J. Lexer, H. Sandmann, T.A. Spies, B.E. Splechtna, and J. Szwagrzyk, 2004. Spatio-temporal development of forests–current trends in field methods and models. Oikos 107: 3–15.

Grime, J.P., 1966. Shade avoidance and shade tolerance in flowering plants. In: G.C. Evans, R. Bainbridge, and O. Rackham, eds. Light as Ecological Factor. Blackwell, London, pp. 187–207.

Gross, L.J., M.U.F. Kirschbaum, and R.W. Pearcy, 1991. A dynamic model of photosynthesis in varying light taking account of stomatal conductance, C_3-cycle intermediates, photorespiration and RuBisCO activation. Plant, Cell and Environment 14: 881–893.

Gutschick, V.P. and F.W. Wiegel, 1988. Optimizing the canopy photosynthetic rate by patterns of investment in specific leaf mass. The American Naturalist 132: 67–86.

Harbinson, J., B. Genty, and N.R. Baker, 1990a. The relationship between CO_2 assimilation and electron transport in leaves. Photosynthesis Research 25: 213–224.

Harbinson, J., B. Genty, and C.H. Foyer, 1990b. Relationship between photosynthetic electron transport and stromal enzyme activity in pea leaves–toward an understanding of the nature of photosynthetic control. Plant Physiology 94: 545–553.

Hikosaka, K. and I. Terashima, 1995. A model of the acclimation of photosynthesis in the leaves Of C_3 plants to sun and shade with respect to nitrogen use. Plant, Cell and Environment 18: 605–618.

Hikosaka, K. and I. Terashima, 1996. Nitrogen partitioning among photosynthetic components and its consequences in sun and shade plants. Functional Ecology 10: 335–343.

Hiroi, T. and M. Monsi, 1963. Physiological and ecological analysis of shade tolerance of plants 3. Effect of shading on growth attributes of *Helianthus annuus*. The Botanical Magazine 77: 121–129.

Hirose, T., 1988. Modelling the relative growth rate as a function of plant nitrogen concentration. Physiolgia Plantarum 72: 185–189.

Hirose, T. and M.A. Werger, 1987a. Maximizing daily canopy photosynthesis with respect to the leaf nitrogen pattern in the canopy. Oecologia 72: 520–526.

Hirose, T. and M.A.J. Werger, 1987b. Nitrogen use efficiency in instantaneous and daily photosynthesis leaves in the canopy of a *Solidago altissima* stand. Physiologia Plantarum 70: 520–526.

Hirose, T., M.J.A. Werger, T.L. Pons, and J.W.A. van Rheenen, 1988. Canopy structure and leaf nitrogen distributon in a stand of *Lysimachia vulgaris* L. as influenced by stand destiny. Oecologia 77: 145–150.

Huante, P. and E. Rincon, 1998. Responses to light changes in tropical deciduous woody seedlings with contrasting growth rates. Oecologia 113: 53–66.

Huffaker, R.C. and L.W. Peterson, 1976. Protein turnover in plants and possible means of its regulation. Annual Review of Plant Physiology 25: 363–392.

Hull, J.C., 2002. Photosynthetic induction dynamics to sunflecks of four deciduous forest understory herbs with different phenologies. International Journal of Plant Sciences 163: 913–924.

Johnson, I.R. and J.H.M. Thornley, 1984. A model of instantaneous and daily canopy photosynthesis. Journal of Theoretical Biology 107: 531–545.

Jurik, T.W. and B.F. Chabot, 1986. Leaf dynamics and profitability in wild strawberries. Oecologia 69: 296–304.

Kabakoff, R.P. and R.L. Chazdon, 1996. Effects of canopy species dominance on understorey light availability in low-elevation secondary forest stands in Costa Rica. Journal of Tropical Ecology 12: 779–788.

Kamaluddin, M. and J. Grace, 1992. Photoinhibition and light acclimation in seedlings of *Bischofia javanica*, a tropical forest tree from Asia. Annals of Botany 69: 47–52.

Kirschbaum, M.U.F., L.J. Gross, and W.W. Pearcy, 1988. Observed and modelled stomatal responses to dynamic light environments in the shade plant *Alocasia macrorrhiza*. Plant, Cell and Environment 11: 111–121.

Kirschbaum, M.U.F., M. Kuppers, H. Schneider, C. Giersch, and S. Noe, 1998. Modelling photosynthesis in fluctuating light with inclusion of stomatal conductance, biochemical activation and pools of key photosynthetic intermediates. Planta 204: 16–26.

Kirschbaum, M.U.F., C. Ohlemacher, and M. Kuppers, 2004. Loss of quantum yield in extremely low light. Planta 218: 1046–1053.

Kirschbaum, M.U.F. and R.W. Pearcy, 1988a. Gas exchange analysis of the relative importance of stomatal and biochemical factors in photosynthetic induction in *Alocasia macrorrhiza*. Plant Physiology 86: 782–785.

Kirschbaum, M.U.F. and R.W. Pearcy, 1988b. Gas exchange analysis of the fast phase of photosynthetic induction in *Alocasia macrorrhiza*. Plant Physiology 87: 818–821.

Knapp, A.K., 1992. Leaf gas exchange in *Quercus macrocarpa* (Fagaceae): Rapid stomatal responses to variability in sunlight in a tree growth form. American Journal of Botany 79: 599–604.

Knapp, A.K. and W.K. Smith, 1987. Stomatal and photosynthetic responses during sun/shade transitions in subalpine plants: Influence on water use efficiency. Oecologia 74: 62–67.

Knapp, A.K. and W.K. Smith, 1989. Influence of growth form on ecophysiological responses to variable sunlight in subalpine plants. Ecology 70: 1069–1082.

Knapp, A.K. and W.K. Smith, 1990a. Contrasting stomatal responses to variable sunlight in 2 subalpine herbs. American Journal of Botany 77: 226–231.

Knapp, A.K. and W.K. Smith, 1990b. Stomatal and photosynthetic responses to variable sunlight. Physiologia Plantarum 78: 160–165.

Koniger, M., G.C. Harris, A. Virgo, and K. Winter, 1995. Xanthophyll-cycle pigments and photosynthetic capacity in tropical forest species: A comparative field-study on canopy, gap and understory plants. Oecologia 104: 280–290.

Koniger, M., G.C. Harris, and R.W. Pearcy, 1998. Interaction between photon flux density and elevated temperatures on photoinhibition in *Alocasia macrorrhiza*. Planta 205: 214–222.

Kuppers, M. and H. Schneider, 1993. Leaf gas exchange of beech (*Fagus sylvatica* L.) seedlings in lightflecks: Effects of fleck length and leaf temperature in leaves grown in deep and partial shade. Trees 7: 160–168.

Kuppers, M., H. Timm, F. Orth, J. Stegemann, R. Stober, K. Paliwal, K.S.T. Karunaichamy, and R. Ortez, 1996. Effects of light environment and successional status on lightfleck use by understory trees of temperate and tropical forests. Tree Physiology 16: 69–80.

Kuppers, M., I. Heiland, H. Schneider, and P.J. Neugebauer, 1999. Light-flecks cause non-uniform stomatal opening—studies with special emphasis on *Fagus sylvatica* L. Trees-Structure and Function 14: 130–144.

Kursar, T.A. and P.D. Coley, 1999. Contrasting modes of light acclimation in two species of the rainforest understory. Oecologia 121: 489–498.

Laisk, A., 1977. Kinetics of Photosyntheis and Photorespiration in C_3 plants. Nauka, Moscow, 198 p.

Leakey, A.D.B., M.C. Press, and J.D. Scholes, 2003a. High-temperature inhibition of photosynthesis is greater under sunflecks than uniform irradiance in a tropical rain forest tree seedling. Plant, Cell and Environment 26: 1681–1690.

Leakey, A.D.B., M.C. Press, and J.D. Scholes, 2003b. Patterns of dynamic irradiance affect the photosynthetic capacity and growth of dipterocarp tree seedlings. Oecologia 135: 184–193.

Leakey, A.D.B., J.D. Scholes, and M.C. Press, 2005. Physiological and ecological significance of sunflecks for dipterocarp seedlings. Journal of Experimental Botany 56: 469–482.

Lei, T.T. and M.J. Lechowicz, 1997. The photosynthetic response of 8 species of *Acer* to simulated light regimes from the centre and edges of gaps. Functional Ecology 11: 16–23.

Leuning, R., Y.P. Wang, and R.N. Cromer, 1991. Model simulations of spatial distributions and daily totals of photosynthesis in *Eucalyptus grandis* canopies. Oecologia 88: 494–503.

Levins, R., 1968. Evolution in Changing Environments. Princeton University Press, Princeton, NJ.

Logan, B.A., D.H. Barker, W.W. Adams, and B. Demmig-Adams, 1997. The response of xanthophyll cycle-dependent energy dissipation in *Alocasia brisbanensis* to sunflecks in a subtropical rainforest. Australian Journal of Plant Physiology 24: 27–33.

Long, S.P., W.F. Postl, and H.R. Bolhar-Nordenkampf, 1993. Quantum yields for uptake of carbon dioxide in C_3 vascular plants of contrasting habitats and taxonomic groupings. Planta 189: 226–234.

Ludlow, M.M. and O. Björkman, 1984. Paraheliotropic leaf movement in Siratro as a protective mechanism against drought-induced damage to primary photosynthetic reactions: damage by excessive light and heat. Planta 161: 505–518.

Meir, P., B. Kruijt, M. Broadmeadow, E. Barbosa, O. Kull, F. Carswell, A. Nobre, and P.G. Jarvis, 2002. Acclimation of photosynthetic capacity to irradiance in tree canopies in relation to leaf nitrogen concentration and leaf mass per unit area. Plant, Cell and Environment 25: 343–357.

Mencuccini, M., 2002. Hydraulic constraints in the functional scaling of trees. Tree Physiology 22: 553–565.

Meng, S.X., M. Rudnicki, V.J. Lieffers, D.E.B. Reid, and U. Silins, 2006. Preventing crown collisions increases the crown cover and leaf area of maturing lodgepole pine. Journal of Ecology 94: 681–686.

Miner, B.G., S.E. Sultan, S.G. Morgan, D.K. Padilla, and R.A. Relyea, 2005. Ecological consequences of phenotypic plasticity. Trends in Ecology and Evolution 20: 685–692.
Montgomery, R., 2004. Relative importance of photosynthetic physiology and biomass allocation for tree seedling growth across a broad light gradient. Tree Physiology 24: 155–167.
Montgomery, R.A. and R.L. Chazdon, 2001. Forest structure, canopy architecture, and light transmittance in tropical wet forests. Ecology 82: 2707–2718.
Montgomery, R.A. and R.L. Chazdon, 2002. Light gradient partitioning by tropical tree seedlings in the absence of canopy gaps. Oecologia 131: 165–174.
Mott, K.A., A.C. Gibson, and J.W. O'Leary, 1982. The adaptive significance of amphistomatic leaves. Plant, Cell and Environment 5: 455–460.
Mott, K.A., G.W. Snyder, and I.E. Woodrow, 1997. Kinetics of Rubisco activation as determined from gas-exchange measurements in antisense plants of *Arabidopsis thaliana* containing reduced levels of Rubisco activase. Australian Journal of Plant Physiology 24: 811–818.
Mott, K.A. and I.E. Woodrow, 2000. Modelling the role of Rubisco activase in limiting non-steady-state photosynthesis. Journal of Experimental Botany 51: 399–406.
Mulkey, S.S. and R.W. Pearcy, 1992. Interactions between acclimation and photoinhibition of photosynthesis of a tropical forest understorey herb, *Alocasia macrorrhiza*, during simulated canopy gap formation. Functional Ecology 6: 719–729.
Muraoka, H., H. Koizumi, and R.W. Pearcy, 2003. Leaf display and photosynthesis of tree seedlings in a cool-temperate deciduous broadleaf forest understorey. Oecologia 135: 500–509.
Naumburg, E. and D.S. Ellsworth, 2000. Photosynthesis sunfleck utilization potential of understory saplings growing under elevated CO_2 in FACE. Oecologia 122: 163–174.
Newell, E.A., E.P. McDonald, B.R. Strain, and J.S. Denslow, 1993. Photosynthetic responses of *Miconia* species to canopy openings in a lowland tropical rainforest. Oecologia 94: 49–56.
Nicotra, A.B., R.L. Chazdon, and S.V.B. Iriarte, 1999. Spatial heterogeneity of light and woody seedling regeneration in tropical wet forests. Ecology 80: 1908–1926.
Niinemets, U. and F. Valladares, 2004. Photosynthetic acclimation to simultaneous and interacting environmental stresses along natural light gradients: Optimality and constraints. Plant Biology 6: 254–268.
Niinemets, U., W. Bilger, O. Kull, and J.D. Tenhunen, 1998a. Acclimation to high irradiance in temperate deciduous trees in the field: Changes in xanthophyll cycle pool size and in photosynthetic capacity along a canopy light gradient. Plant, Cell and Environment 21: 1205–1218.
Niinemets, U., O. Kull, and J.D. Tenhunen, 1998b. An analysis of light effects on foliar morphology, physiology, and light interception in temperate deciduous woody species of contrasting shade tolerance. Tree Physiology 18: 681–696.
Niinemets, U., O. Kull, and J.D. Tenhunen, 2004. Within-canopy variation in the rate of development of photosynthetic capacity is proportional to integrated quantum flux density in temperate deciduous trees. Plant, Cell and Environment 27: 293–313.
Niinemets, U., A. Cescatti, M. Rodeghiero, and T. Tosens, 2006. Complex adjustments of photosynthetic potentials and internal diffusion conductance to current and previous light availabilities and leaf age in Mediterranean evergreen species *Quercus ilex*. Plant, Cell and Environment 29: 1159–1178.
Noguchi, K. and I. Terashima, 1997. Different regulation of leaf respiration between *Spinacia oleracea*, a sun species, and *Alocasia odora*, a shade species. Physiologia Plantarum 101: 1–7.
Noguchi, K., K. Sonoike, and I. Terashima, 1996. Acclimation of respiratory properties of leaves of *Spinacia oleracea* L, a sun species, and of *Alocasia macrorrhiza* (L) G. Don, a shade species, to changes in growth irradiance. Plant and Cell Physiology 37: 377–384.
Noguchi, K., C.S. Go, S.I. Miyazawa, I. Terashima, S. Ueda, and T. Yoshinari. 2001a. Costs of protein turnover and carbohydrate export in leaves of sun and shade species. Australian Journal of Plant Physiology 28: 37–47.
Noguchi, K., N. Nakajima, and I. Terashima, 2001b. Acclimation of leaf respiratory properties in *Alocasia odora* following reciprocal transfers of plants between high- and low-light environments. Plant, Cell and Environment 24: 831–839.
Novoplansky, A., D. Cohen, and T. Sachs, 1990. How portulaca seedlings avoid their neighbors. Oecologia 82: 490–493.

Oberbauer, S.F., D.B. Clark, D.A. Clark, and M.A. Quesada, 1988. Crown light environments of saplings of two species of rain forest emergent trees. Oecologia (Berlin) 75: 207–212.

Ogren, E. and U. Sundlin, 1996. Photosynthetic respnse to dynamic light: A comparison of species from contrasting habitats. Oecologia 106: 18–27.

Oguchi, R., K. Hikosaka, and T. Hirose, 2003. Does the photosynthetic light-acclimation need change in leaf anatomy? Plant, Cell and Environment 26: 505–512.

Oguchi, R., K. Hikosaka, and T. Hirose, 2005. Leaf anatomy as a constraint for photosynthetic acclimation: differential responses in leaf anatomy to increasing growth irradiance among three deciduous trees. Plant, Cell and Environment 28: 916–927.

Oquist, G., J.M. Anderson, S. Mccaffery, and W.S. Chow, 1992. Mechanistic differences in photoinhibition of sun and shade plants. Planta 188: 422–431.

Osmond, C.B., 1983. Interactions between irradiance, nitrogen nutrition, and water stress in the sun-shade responses of *Solanum dulcamara*. Oecologia 57: 316–321.

Pacala, S.W., C.D. Canham, J. Saponara, J.A. Silander, R.K. Kobe, and E. Ribbens, 1996. Forest models defined by field measurements—estimation, error analysis and dynamics. Ecological Monographs 66: 1–43.

Parker, G.G., M.A. Lefsky, and D.J. Harding, 2001. Light transmittance in forest canopies determined using airborne laser altimetry and in-canopy quantum measurements. Remote Sensing of Environment 76: 298–309.

Parkhurst, D.F. and K.A. Mott, 1990. Intercellular diffusion limits to CO_2 uptake in leaves. Plant Physiology 94: 1024–1032.

Pearcy, R.W., 1983. The light environment and growth of C_3 and C_4 tree species in the understory of a Hawaiian forest. Oecologia 58: 19–25.

Pearcy, R.W., 1988. Photosynthetic utilization of lightflecks by understory plants. Australian Journal of Plant Physiology 15: 223–238.

Pearcy, R.W., 1994. Photosynthetic responses to sunflecks and light gaps: mechanisms and constraints. In: N.R. Baker and J. Bowker, eds. Photoinhibition of Photosyntheis—Molecular Mechanisms to the Field. Bios Scientific Publishers, Oxford, pp. 255–271.

Pearcy, R.W., 1998. Acclimation to sun and shade. In: A.S. Raghavendra, ed. Photosynthesis a Comprehensive Treatise. Cambridge University Press, Cambridge, pp. 250–272.

Pearcy, R.W. and W.A. Pfitsch, 1991. Influence of sunflecks on the $\delta^{13}C$ of *Adenocaulon bicolor* plants occurring in contrasting forest understory microsites. Oecologia 86: 457–462.

Pearcy, R.W. and J.R. Seemann, 1990. Photosynthetic induction state of leaves in a soybean canopy in relation to light regulation of ribulose-1,5-bisphosphate carboxylase and stomatal conductance. Plant Physiology 94: 628–633.

Pearcy, R.W. and D.A. Sims, 1994. Photosynthetic acclimation to changing light environments: scaling from the leaf to the whole plant. In: M.M. Caldwell and R.W. Pearcy, eds. Exploitation of Environmental Heterogeneity by Plants: Ecophysiological Processes Above and Below Ground. Academic Press, San Diego, CA, pp. 223–234.

Pearcy, R.W., K. Osteryoung, and H.W. Calkin, 1985. Photosynthetic responses to dynamic light environments by Hawaiian trees. The time course of CO_2 uptake and carbon gain during sunflecks. Plant Physiology 79: 896–902.

Pearcy, R.W., J.S. Roden, and J.A. Gamon, 1990. Sunfleck dynamics in relation to canopy structure in a soybean (*Glycine max* (L.) Merr.) canopy. Agricultural and Forest Meteorology 52: 359–372.

Pearcy, R.W., R.L. Chazdon, L.J. Gross, and K.A. Mott, 1994. Photosynthetic utilization of sunflecks, a temporally patchy resource on a time scale of seconds to minutes. In: M.M. Caldwell and R.W. Pearcy, eds. Exploitation of Environmental Heterogeniety by Plants: Ecophysiological Processes Above and Below Ground. Academic Press, San Diego, CA, pp. 175–208.

Pearcy, R.W., G.F. Sassenrath-Cole, and J.P. Krall, 1996. Photosynthesis in fluctuating light environments. In: N.R. Baker, ed. Environmental Stress and Photosynthesis. Kleuwer Academic, The Hague.

Pearcy, R.W., L.J. Gross, and D. He, 1997. An improved dynamic model of photosynthesis for estimation of carbon gain in sunfleck light regimes. Plant, Cell and Environment 20: 411–424.

Pearcy, R.W., F. Valladares, S.J. Wright, and E. Lasso, 2004. A functional analysis of the crown architecture of tropical forest *Psychotria* species: Do species vary in light capture efficiency and consequently in carbon gain and growth? Oecologia 139: 163–167.

Pfitsch, W.A. and R.W. Pearcy, 1989a. Daily carbon gain by *Adenocaulon bicolor* (Asteraceae), a redwood forest understory herb, in relation to its light environment. Oecologia 80: 465–470.

Pfitsch, W.A. and R.W. Pearcy, 1989b. Steady-state and dynamic photosynthetic response of *Adenocaulon bicolor* (Asteraceae) in its redwood forest habitat. Oecologia 80: 471–476.

Pfitsch, W.A. and R.W. Pearcy, 1992. Growth and reproductive allocation of *Adenocaulon bicolor* following experimental removal of sunflecks. Ecology 73: 2109–2117.

Pons, T.L. and R.W. Pearcy, 1992. Photosynthesis in flashing light of soybean leaves grown in different conditions. II. Lightfleck utilization efficiency. Plant, Cell and Environment 15: 577–584.

Poorter, H. and O. Nagel, 2000. The role of biomass allocation in the growth response of plants to different levels of light, CO_2, nutrients and water: a quantitative review. Australian Journal of Plant Physiology 27: 595–607.

Poorter, L. and S.F. Oberbauer, 1993. Photosynthetic induction responses of two rainforest tree species in relation to light environment. Oecologia 96: 193–199.

Poorter, H., S. Pepin, T. Rijkers, Y. de Jong, J.R. Evans, and C. Korner, 2006. Construction costs, chemical composition and payback time of high- and low-irradiance leaves. Journal of Experimental Botany 57: 355–371.

Portis, A.R., 1992. Regulation of ribulose 1,5-bisphosphate carboxylase oxygenase activity. Annual Review of Plant Physiology and Plant Molecular Biology 43: 415–437.

Portis, A.R., 2003. Rubisco activase—Rubisco's catalytic chaperone. Photosynthesis Research 75: 11–27.

Reich, P.B., M.G. Tjoelker, M.B. Walters, D.W. Vanderklein, and C. Bushena. 1998. Close association of RGR, leaf and root morphology, seed mass and shade tolerance in seedlings of nine boreal tree species grown in high and low light. Functional Ecology 12: 327–338.

Reich, P.B., M.B. Walters, and D.S. Ellsworth, 1997. From tropics to tundra: Global convergence in plant functioning. Proceedings of the National Academy of Sciences 94: 13730–13734.

Rijkers, T., P.J. Jan de Vries, T.L. Pons, and F. Bongers, 2000. Photosynthetic induction in saplings of three shade-tolerant tree species: Comparing understorey and gap habitats in a French Guiana rain forest. Oecologia 125: 331–340.

Roden, J.S. and R.W. Pearcy, 1993a. Effect of leaf flutter on the light environment of poplars. Oecologia 93: 201–207.

Roden, J.S. and R.W. Pearcy, 1993b. Photosynthetic gas exchange response of poplars to steady-state and dynamic light environments. Oecologia 93: 208–214.

Ruel, J.J. and M.P. Ayres, 1999. Jensen's inequality predicts effects of environmental variation. Trends in Ecology and Evolution 14: 361–366.

Ryel, R.J., W. Beyschlag, and M.M. Caldwell, 1994. Light Field Heterogeneity Among Tussock Grasses—Theoretical Considerations of Light Harvesting and Seedling Establishment in Tussocks and Uniform Tiller Distributions. Oecologia 98: 241–246.

Ryel, R.J., W. Beyschlag, B. Heindl, and I. Ullmann, 1996. Experimental studies on the competitive balance between two central European roadside grasses with different growth forms. 1. Field experiments on the effects of mowing and maximum leaf temperatures on competitive ability. Botanica Acta 109: 441–448.

Ryel, R.J., E. Falge, U. Joss, R. Geyer, and J.D. Tenhunen, 2001. Penumbral and foliage distribution effects on *Pinus sylvestris* canopy gas exchange. Theoretical and Applied Climatology 68: 109–124.

Sage, R.F., C.D. Reid, B.D. Moore, and J.R. Seemann, 1993. Long-term kinetics of the light-dependent regulation of ribulose-1,5-bisphosphate carboxylase oxygenase activity in plants with and without 2-carboxyarabinitol 1-phosphate. Planta 191: 222–230.

Sassenrath-Cole, G.F. and R.W. Pearcy, 1992. The role of ribulose-1,5-bisphosphate regeneration in the induction requirement of photosynthetic CO_2 exchange under transient light conditions. Plant Physiology 99: 227–234.

Sassenrath-Cole, G.F. and R.W. Pearcy, 1994. Regulation of photosynthetic induction state by the magnitude and duration of low light exposure. Plant Physiology 105: 1115–1123.

Sassenrath-Cole, G.F., R.W. Pearcy, and S. Steinmaus, 1994. The role of enzyme activation state in limiting carbon assimilation under variable light conditions. Photosynthesis Research 41: 295–302.

Saunders, S.C., J.Q. Chen, T.D. Drummer, E.J. Gustafson, and K.D. Brosofske. 2005. Identifying scales of pattern in ecological data: a comparison of lacunarity, spectral and wavelet analyses. Ecological Complexity 2: 87–105.

Scheiner, S.M., 1993. Genetics and evolution of phenotypic plasticity. Annual Review of Ecology and Systematics 24: 35–68.

Scheiner, S.M. and R.F. Lyman, 1991. The genetics of phenotypic plasticity II. response to selection. Journal of Evolutionary Biology 4: 23–50.

Schlichting, C.D., 1986. The evolution of phenotypic plasticity in plants. Annual Review of Ecology and Systematics 17: 667–693.

Schlichting, C.D. and H. Smith, 2002. Phenotypic plasticity: Linking molecular mechanisms with evolutionary outcomes. Evolutionary Ecology 16: 189–211.

Schulte, M., C. Offer, and U. Hansen, 2003. Induction of CO_2-gas exchange and electron transport: comparison of dynamic and steady-state responses in *Fagus sylvatica* leaves. Trees—Structure and Function 17: 153–163.

Seemann, J.R., T.D. Sharkey, J.L. Wang, and C.B. Osmond, 1987. Environmental effects on photosynthesis, nitrogen-use efficiency, and metabolite pools in leaves of sun and shade plants. Plant Physiology 84: 796–802.

Seemann, J.R., M.U.F. Kirschbaum, T.D. Sharkey, and R.W. Pearcy, 1988. Regulation of ribulose 1,5-bisphosphate carboxylase activity in *Alocasia macrorrhiza* in response to step changes in irradiance. Plant Physiology 88: 148–152.

Sims, D.A., R. Gebauer, and R.W. Pearcy, 1994. Scaling sun and shade photosynthetic acclimation to whole plant performance. II Simulation of carbon balance and growth at different daily photon flux densities. Plant, Cell and Environment 17: 889–900.

Shugart, H.H. and T.M. Smith, 1996. A review of forest patch models and their application to global change research. Climatic Change 34: 131–153.

Sims, D.A. and R.W. Pearcy, 1989. Photosynthetic characteristics of a tropical forest understory herb, *Alocasia macrorrhiza*, and a related crop species, *Colocasia esculenta* grown in contrasting light environments. Oecologia 79: 53–59.

Sims, D.A. and R.W. Pearcy, 1992. Response of leaf anatomy and photosynthetic capacity in *Alocasia macrorrhiza* (Araceae) to a transfer from low to high light. American Journal of Botany 79: 449–455.

Sims, D.A. and R.W. Pearcy, 1994. Scaling sun and shade photosynthetic acclimation to whole plant performance. I Carbon balance and allocation at different daily photon flux densities. Plant, Cell and Environment 17: 881–887.

Sims, J.A. and R.W. Pearcy, 1993. Sunfleck frequency and duration affects growth rate of the understory plant, *Alocasia macrorrhiza* (L.) G. Don. Functional Ecology 7: 683–689.

Singsaas, E.L., D.R. Ort, and E.H. DeLucia, 2000. Diurnal regulation of photosynthesis in understory saplings. New Phytologist 145: 39–49.

Singsaas, E.L., D.R. Ort, and E.H. DeLucia, 2001. Variation in measured values of photosynthetic quantum yield in ecophysiological studies. Oecologia 128: 15–23.

Sinoquet, H., G. Sonohat, J. Phattaralerphong, and C. Godin, 2005. Foliage randomness and light interception in 3-D digitized trees: An analysis from multiscale discretization of the canopy. Plant, Cell and Environment 28: 1158–1170.

Sipe, T.W. and F.A. Bazzaz, 1994. Gap partitioning among maples (Acer) in central New England: Shoot architecture and photosynthesis. Ecology (Tempe) 75: 2318–2332.

Slade, A.J. and M.J. Hutchings, 1987. An analysis of the costs and benefits of physiological integration between ramets in the clonal perennial herb *Glechoma hederacea*. Oecologia 73: 425–432.

Smith, W.K., A.K. Knapp, and W.A. Reiners, 1989. Penumbral effects on sunlight penetration in plant communities. Ecology 70: 1603–1609.

Stegemann, J., H.C. Timm, and M. Kuppers, 1999. Simulation of photosynthetic plasticity in response to highly fluctuating light: An empirical model integrating dynamic photosynthetic induction and capacity. Trees—Structure and Function 14: 145–160.

Stenberg, P., 1995. Penumbra in within-shoot and between-shoot shading in conifers and its significance for photosynthesis. Ecological Modelling 77: 215–231.

Strauss-Debenedetti, S. and F.A. Bazzaz, 1991. Plasticity and acclimation to light in tropical Moraceae of different successional positions. Oecologia 87: 377–387.

Strauss-Debenedetti, S. and F. Bazzaz, 1996. Photosynthetic characteristics of tropical trees along successional gradients. In: S.S. Mulkey, R.L. Chazdon, and A.P. Smith, eds. Tropical Forest Plant Ecophysiology. Chapman & Hall, New York, pp. 162–186.

Stuefer, J.F., 1996. Potential and limitations of current concepts regarding the response of clonal plants to environmental heterogeneity. Vegetation 127: 55–70.

Stuefer, J.F., H.J. During, and H. Dekroon, 1994. High benefits of clonal integration in two stoloniferous species, in response to heterogeneous light environments. Journal of Ecology 82: 511–518.

Stuefer, J.F., H. De Kroon, and H.J. During, 1996. Exploitation of environmental heterogeneity by spatial division of labor in a clonal plant. Functional Ecology 10: 328–334.

Sultan, S.E., 2001. Phenotypic plasticity and ecological breadth in plants. American Zoologist 41: 1599–1599.

Takenaka, A., 1994. A simulation model of tree architecture development based on growth response to local light environment. Journal of Plant Research 107: 321–330.

Tang, Y., I. Washitani, and H. Iwaki, 1992. Effects of microsite light availability on the survival and growth of oak seedlings within a grassland. The Botanical Magazine, Tokyo 105: 281–288.

Tang, Y., K. Hiroshi, S. Mitsumasa, and W. Izumi, 1994. Characteristics of transient photosynthesis in *Quercus serrata* seedlings grown under lightfleck and constant light regimes. Oecologia 100: 463–469.

Tang, Y.-H., I. Washitani, T. Tsuchiya, and H. Iwaki, 1989. Spatial heterogenity of photosynthetic photon flux density in the canopy of *Miscanthus sinensis*. Ecological Research 4: 339–349.

Tausz, M., C.R. Warren, and M.A. Adams, 2005. Dynamic light use and protection from excess light in upper canopy and coppice leaves of *Nothofagus cunninghamii* in an old growth, cool temperate rainforest in Victoria, Australia. New Phytologist 165: 143–155.

Thayer, S.S. and O. Björkman, 1990. Leaf xanthophyll content and composition in sun and shade determined by HPLC. Photosynthesis Research 23: 331–343.

Thompson, W.A., P.E. Kriedemann, and I.E. Craig, 1992. Photosynthetic response to light and nutrients in sun-tolerant and shade-tolerant rainforest trees. I. Growth, leaf anatomy and nutrient content. Australian Journal of Plant Physiolology 19: 1–18.

Timm, H.C., J. Stegemann, and M. Kuppers, 2002. Photosynthetic induction strongly affects the light compensation point of net photosynthesis and coincidentally the apparent quantum yield. Trees—Structure and Function 16: 47–62.

Tinoco-Ojanguren, C. and R.W. Pearcy, 1992. Dynamic stomatal behavior and its role in carbon gain during lightflecks of a gap phase and an understory *Piper* species acclimated to high and low light. Oecologia 92: 222–228.

Tinoco-Ojanguren, C. and R.W. Pearcy, 1993a. Stomatal dynamics and its importance to carbon gain in two rainforest *Piper* species. I. VPD effects on the transient stomatal response to lightflecks. Oecologia 94: 388–394.

Tinoco-Ojanguren, C. and R.W. Pearcy, 1993b. Stomatal Dynamics and its importance to carbon gain in two rainforest Piper species. II. Stomatal versus biochemical limitations during photosynthetic induction. Oecologia 94: 395–402.

Turnbull, M.H., D. Doley, and D.J. Yates, 1993. The dynamics of photosynthetic acclimation to changes in light quantity and quality in three Australian rainforest tree species. Oecologia 94: 218–228.

Usuda, H. and G.E. Edwards, 1984. Is photosynthesis during the induction period in maize limited by the availability of intercellular carbon dioxide. Plant Science Letters 37: 41–45.

Valladares, F., 1999. Architecture, ecology and evolution of plant crowns. In: F.I. Pugnaire and F. Valladares, eds. Handbook of Functional Plant Ecology. Marcel Decker, New York, pp. 121–194.

Valladares, F., M.T. Allen, and R.W. Pearcy, 1997. Photosynthetic responses to dynamic light under field conditions in six tropical rainforest shrubs occurring along a light gradient. Oecologia 111: 505–514.

Valladares, F., S.J. Wright, E. Lasso, K. Kitajima, and R.W. Pearcy, 2000. Plastic phenotypic response to light of 16 congeneric shrubs from a Panamanian rainforest. Ecology 81: 1925–1936.

Via, S., R. Gomulkiewicz, G. Dejong, S.M. Scheiner, C.D. Schlichting, and P.H. Vantienderen, 1995. Adaptive phenotypic plasticity—consensus and controversy. Trends in Ecology and Evolution 10: 212–217.

Walker, D.A., 1981. Photosynthetic induction. In: G. Akoyonoglou, ed. Proceedings of the 5th International Congress on Photosynthesis Vol IV. Balaban International Sci. Series, Philadelphia, PA, pp. 189–202, 363–373.

Washitani, I. and Y. Tang, 1991. Microsite variation in light availability and seedling growth of *Quercus serrata* in a temperate pine forest. Ecological Research 6: 305–316.

Watling, J.R., M.C. Ball, and I.E. Woodrow, 1997. The utilization of lightflecks for growth in four Australian rain-forest species. Functional Ecology 11: 231–239.

Wayne, P.M. and F.A. Bazzaz, 1993a. Birch seedling responses to daily time courses of light in experimental forest gaps and shadehouses. Ecology 74: 1500–1515.

Wayne, P.M. and F.A. Bazzaz, 1993b. Morning vs afternoon sun patches in experimental forest gaps–consequences of temporal incongruency of resources to birch regeneration. Oecologia 94: 235–243.

Wirth, R., B. Weber, and R.J. Ryel, 2001. Spatial and temporal variability of canopy structure in a tropical moist forest. Acta Oecologica-International Journal of Ecology 22: 235–244.

Woodrow, I.E. and J.A. Berry, 1988. Enzymatic regulation of photosynthetic CO_2 fixation in C_3 plants. Annual Review of Plant Physiolology Plant Molecular Biology 39: 533–594.

Woodrow, I.E. and K.A. Mott, 1988. Quantitative assessment of the degree to which ribulose bisphosphate carboxylase/oxygenase determines the steady-state rate of photosynthesis during sun-shade acclimation in *Helianthus annus* L. Australian. Journal of Plant Physiology 15: 253–262.

Woodrow, I.E. and D.A. Walker, 1980. Light-mediated activation of stromal sedoheptulose bisphosphatase. Biochemical Journal 191: 845–849.

Wullschleger, S.D., 1993. Biochemical limitations to carbon assimilation in C_3 plants: a retrospective analysis of the a/ci curves from 109 species. Journal of Experimental Botany 44: 907–920.

Yamashita, N., A. Ishida, H. Kushima, and N. Tanaka, 2000. Acclimation to sudden increase in light favoring an invasive over native trees in subtropical islands, Japan. Oecologia 125: 412–419.

Yoda, K., 1978. Three-dimensional distribution of light intensity in a tropical rain forest of West Malaysia. Malaysia Nature Journal 30: 161–177.

Young, T.P. and S.P. Hubbell, 1991. Crown asymmetry, treefalls, and repeat disturbance of broad-leaved forest gaps. Ecology 72: 1464–1471.

Zeiger, E., M. Iino, and T. Ogawa, 1985. The blue light responses of stomata: pulse kinetics and some mechanistic implications. Journal of Photochemistry and Photobiology 42: 759–763.

Zhu, X.G., D.R. Ort, J. Whitmarsh, and S.P. Long, 2004. The slow reversibility of photosystem II thermal energy dissipation on transfer from high to low light may cause large losses in carbon gain by crop canopies: A theoretical analysis. Journal of Experimental Botany 55: 1167–1175.

8 Acquisition, Use, and Loss of Nutrients

Frank Berendse, Hans de Kroon, and Wim G. Braakhekke

CONTENTS

INTRODUCTION

In many natural environments, nutrient supply is one of the most important factors that affect the productivity and the species composition of plant communities (Kruijne et al. 1967, Elberse et al. 1983, Pastor et al. 1984, Tilman 1984). In many grassland, heathland, wetland and forest communities increased fertilizer gifts and increased nitrogen inputs through atmospheric deposition have caused not only dramatic changes in species composition, but also important losses of plant species diversity (Aerts and Berendse 1988, Berendse et al. 1992). To understand the changes in plant communities that occur after an increase in nutrient supply, it is essential to understand how plant species are adapted to environments with different nutrient availabilities. The relation between nutrient supply and long-term success of a plant individual in a natural ecosystem is determined by three important components of plant functioning:

1. the acquisition of nutrients in soils that are always more or less heterogeneous;
2. the use of absorbed nutrients for carbon assimilation and other plant functions;
3. the loss of nutrients determining the length of the time period that nutrients can be used.

In this chapter, we subsequently consider these three aspects and finally attempt to integrate them to conclude how plant species cope with nutrient-poor and nutrient-rich environments. We focus on plants growing in their natural habitat. Such plant individuals experience a heterogeneous substrate, they have to compete for soil resources and light with other plants, and they frequently lose large quantities of nutrients through abscission, disturbances, and herbivory.

NUTRIENT UPTAKE KINETICS: BASIC PRINCIPLES

Nutrient uptake is determined by both supply and demand at the root surface. Nutrients arrive at the root surface by the mass flow of water toward the root, which is driven by transpiration. Plants deplete the soil solution near the roots, when the nutrient uptake rate exceeds the rate at which nutrients arrive. By doing so they create concentration gradients around the roots that trigger diffusion of nutrients toward the root surface, which adds to the supply by mass flow. When the depletion at the root surface proceeds, uptake must come in pace with the supply rate. When the supply by mass flow exceeds the demand, nutrients (and other solutes) can either be excluded, accumulating near the root, or enter the root and accumulate in the plant to concentrations that may eventually become deleterious (Marschner 1995, Fitter and Hay 2002).

Depletion and accumulation at the root surface can occur simultaneously for different elements. Table 8.1 gives an indication of the relation between the demand of the major nutrients and their supply by mass flow. The listed concentrations in plant biomass are considered to be sufficient for adequate growth (Epstein 1965). There is a close relation between biomass production, nutrient demand, and water uptake. Based on the amount of water transpired during the production of a unit biomass (transpiration coefficient), we can calculate the nutrient concentration in the soil solution that would satisfy the nutrient demands as listed in the first column by means of mass flow alone. Actual concentrations in the soil solution of an average agricultural soil illustrate that mass flow rates of S, Mg, and Ca amply exceed the demand, whereas mass flow of P falls entirely short of the demanded rate of supply. In agricultural soils, mass flow rates of N and K are usually sufficient, but in most natural soils, concentrations of N, P, and K are much lower, so that the supply by mass flow alone is insufficient to satisfy the demand. Consequently, diffusion must play an important role in the supply of these nutrients to plants growing on natural soils. This calls for a root system that has the ability to take up nutrients selectively against a concentration gradient.

Selective uptake and transport through cell membranes is an energy-demanding process. Passive, nonselective uptake without energy expenses is only possible where nutrients do not have to pass a cell membrane on their way to the vascular cylinder of the root. Passive uptake can revert into efflux when the concentration in the soil solution falls below the concentration inside the root. Passive cation uptake through cell membranes can also proceed against a

TABLE 8.1
Average Nutrient Concentrations in Plant Biomass (Epstein 1965), Concentrations in Soil Solution Required to Satisfy the Demand by Mass Flow, Assuming the Transpiration Coefficient is 0.3 dm^{-3} g^{-1} d.wt., and Actual Concentrations in the Soil Solution in an Arable Field

Element	Concentration in Plant Biomass (mmol kg^{-1} d.wt.)	Sufficient Concentration in Mass Flow (mM)	Actual Concentration in Bulk Soil Solution (mM)
N	1000	3.33	3.1
K	250	0.83	0.5
Ca	125	0.42	1.7
Mg	80	0.27	0.5
P	60	0.20	0.002
S	30	0.10	0.6

Source: After Peters, M. in *Schriftenreihe des Institutes für Pflanzenernährung und Bodenkunde*, H.P. Blume, ed., Universität Kiel, Kiel, 1990.

concentration gradient, because cells can create an electrochemical gradient by actively pumping out protons across the membrane. Anions, on the other hand, have to be transported actively through cell membranes by means of carrier enzymes.

Passive uptake without energy expenses would be sufficient for nutrients that are required in low quantities and occur in relatively high concentrations in the soil solution, if it were not for the closed structure of the root. Since passive uptake is not selective and cannot be regulated, plants that rely too much on passive uptake can easily be overloaded with nutrients and toxic ions when concentrations in the soil solution are high. This makes it understandable why plant roots possess an endodermis that prevents passive nutrient transport (see Chapter 5, this volume). Solutes can enter the root via the apoplastic pathway between the cortex cells, but no further than the endodermis with its bands of Caspari. Solutes that are transported by mass flow to and into the root at a higher rate than the active uptake rate by rhizodermal, cortex, or endodermis cells accumulate between the cortex cells and at the root surface. This leads to diffusion in the direction opposite to the water flow, away from the root, back into the soil. Most nutrients enter the root across a cell membrane somewhere in the cortex by means of active transport and continue their way inside along the symplastic pathway, from one cell to another via intercellular cytoplasma connections (plasmodesmata), to pass through the endodermis and finally enter the vascular cylinder (Marschner 1995). A small fraction of the nutrients can circumvent the endodermis at the root tip where the bands of Caspari are not yet formed, and at places where the endodermis is pierced by lateral roots, torn, or damaged otherwise.

Active uptake against a concentration gradient, by means of an energy-demanding process, is the predominant uptake process for the major nutrients N, P, and K. The uptake rate depends on the nutrient concentration at the root surface. Usually an asymptotic relation is found between the concentration in a nutrient solution and the uptake rate, when measured in short-term experiments with excised roots from plants that have been deprived of nutrients for a few weeks. The uptake as a function of the concentration in the surrounding solution is generally described by:

$$V = V_{\text{max}} \frac{C_{\text{l}}}{C_{\text{l}} + K_{\text{m}}},$$

in which V is the gross uptake rate (μmol g^{-1} fw h^{-1}), V_{max} is the maximum uptake rate, C_l is the nutrient concentration in the soil solution at the root surface (mM), and K_m is the Michaelis constant, which is the value of C_l where V is half V_{max}. The Michaelis–Menten equation is typical for the kinetics of enzymatic processes and reflects the fact that the carrier enzymes in the cell membranes become saturated with increasing nutrient concentration. K_m^{-1} expresses the affinity of the carriers for the substrate ion (i.e., the slope of the curve in the origin, which is measured by V_{max}/K_m). Although different nutrient ions are transported by different carriers, the selectivity of the carriers is not perfect. Different nutrients may compete for the same carriers, so that K_m may be increased by the presence of other ions with the same electrical charge.

The maximum uptake rate (V_{max}) is realized when the concentration (C_l) is so high that all carrier enzymes are continuously occupied with a substrate ion. V_{max} depends on the density and the activity of carriers in the cell membranes. It depends also on the internal nutrient status of the plant, because the activity of the carriers can be suppressed by high nutrient concentrations in the root (compare V_{max} under deprived and well-fed conditions, Figure 8.1). This negative-feedback control operates when plants are growing under nutrient-rich conditions. It can reduce V_{max} to less than 20% of its value in a deprived plant (Loneragan and Asher 1967) and prevents accumulation of too high nutrient concentrations in the root. On the other hand, it has been found that plants growing under nutrient poor conditions can

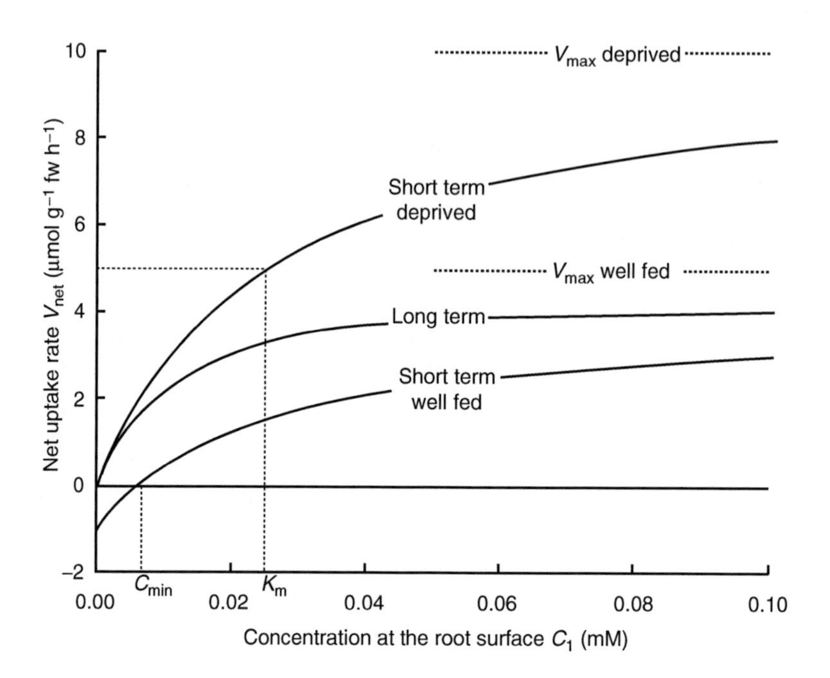

FIGURE 8.1 Relation between external nutrient concentration at the root surface (C_l) and net nutrient uptake rate ($V_{net} = V - E$). The upper curve represents the short-term uptake by excised roots from previously deprived plants ($V_{max} = 10$, $K_m = 0.025$, efflux = 0). The lower curve represents short-term uptake by excised roots of previously well-fed plants ($V_{max} = 5$, $K_m = 0.025$, efflux = 1). C_{min} is the concentration at which $V_{net} = 0$. The curve marked long term represents nutrient uptake by whole plants grown for several weeks on nutrient solutions with constant concentration, so that V_{max} and efflux are in steady state with the internal nutrient concentration and C_l.

temporarily increase V_{max} when the concentration in the soil is increased (Lefebvre and Glass 1982, Jackson et al. 1990).

Consequently, roots of a well-fed plant are operating far below their maximum uptake capacity. When the soil becomes depleted and the nutrient concentration in the plant starts to drop, the negative-feedback control on V_{max} is relaxed, so that V_{max} increases, which compensates for the decrease in supply rate. The responses of V_{max} to changes in internal and external nutrient concentrations allow a plant to regulate its nutrient uptake and rapidly use temporarily high nutrient concentrations that may occur locally in a predominantly poor soil. Under rich conditions, it allows a plant to maintain its overall uptake rate even when a large part of the root system is removed.

Besides uptake, efflux of nutrients may occur. When active uptake takes place, nutrient concentrations inside the root are usually higher than those outside the root. Since roots are not perfectly closed, leakage of nutrients can reduce the net uptake rate (V_{net}). When nutrient influx and efflux occur simultaneously, the plant can only decrease the nutrient concentration in the soil solution until it reaches a minimum concentration (C_{min}) at which influx and efflux are equal. At values of C_l lower than C_{min}, the efflux is larger than the gross influx, so that the net influx is negative and the roots lose nutrients to the solution until C_l equals C_{min} (Figure 8.1).

Like V_{max}, C_{min} is not a constant. Under nutrient poor conditions, when nutrient concentrations inside the root are low, the efflux is also low. Together with the release of the feedback control on V_{max}, this results in very low values of C_{min} under nutrient-poor conditions (Figure 8.1). It is unclear at present whether the K_m value is also able to respond to changes in internal or ambient nutrient concentrations (Marschner 1995). The reported changes in uptake kinetics of roots in response to localized nutrients (e.g., Drew and Saker 1975, Jackson et al. 1990) may be due to changes in V_{max} or efflux rate alone. In most plant species, the value of K_m for N, P, and K is so low that the concentration of these nutrients at the root surface can become virtually zero, for example, 0.35 μM for NO_3^- (Freijsen et al. 1989), 1 μM for K^+ (Drew et al. 1984), and less than 0.01 μM for $H_2PO_4^-$ (Breeze et al. 1984).

NUTRIENT ACQUISITION IN SOILS

The description of active nutrient uptake given earlier applies mainly to uptake by single roots in a well-mixed nutrient solution. However, the relation between uptake and concentration in the soil solution is of little consequence for the overall nutrient uptake by a whole plant growing in a poor soil. In soils, uptake of N, P, and K is almost always limited by the rate of transport toward the root surface and not by the capacity of the uptake mechanism. Concentrations of these nutrients in the soil solution are often so low, and the uptake is so efficient that all available nutrients near the root surface can be taken up within a few minutes. When nutrient uptake is not immediately compensated by nutrient transport from the bulk soil toward the root surface, the nutrient concentrations at the root surface fall and the uptake rate decreases. Even in nutrient solutions, where transport rates (TRs) are high, depletion at the root surface may reduce nutrient uptake rates, as appears from the stimulating effect of stirring (Freijsen et al. 1989). The importance of nutrient transport to the root in nutrient-poor soils is illustrated by the following analysis of the balance between nutrient supply and uptake at the root surface (Nye and Tinker 1977).

As explained earlier, nutrient uptake is determined by the concentration at the root surface (C_l), which, in its turn, is the resultant of nutrient transport to the root and the net nutrient uptake rate (V_{net}). The transport rate is the sum of the mass flow rate and the diffusion rate (DR). The mass flow rate is the product of water flow (V_w) and nutrient concentration in the bulk of the soil solution (C_b). The diffusion rate is the product of the concentration gradient toward the root surface (dC/dx) and the effective diffusion coefficient (D_e). The effective diffusion constant, in turn, depends on the moisture content of the soil, the tortuosity of the diffusion pathway, and the buffer power of the soil, which accounts for the degree to which nutrient transport is impeded by interaction with soil particles. Phosphate is strongly adsorbed to soil particles, which leads to low values of C_b resulting in low rates of mass flow and diffusion. At the other end of the spectrum, nitrate is much more mobile, because adsorption is negligible. Potassium takes an intermediate position.

The net uptake rate (V_{net}) is calculated as the gross uptake rate (V), minus the efflux rate (E). This leads to the following equations:

$$\mathrm{TR} = V_w\, C_b + D_e \frac{dC}{dx},$$

$$V_{net} = \left[V_{max} \frac{C_l}{C_l + K_m} \right] - E.$$

When the transport rate equals the net uptake rate, a dynamic equilibrium develops (with equilibrium concentration and uptake rate $C_l{}^*$ and $V_{net}{}^*$). When C_l is lower than $C_l{}^*$, the transport rate is higher than the net uptake rate, so that C_l increases until it reaches $C_l{}^*$ (cf. Figure 8.2). When C_l is higher than $C_l{}^*$, the transport rate is lower than the net uptake rate, so that C_l decreases until it reaches $C_l{}^*$. At a short term (hours), the equilibrium concentration ($C_l{}^*$) and the corresponding equilibrium uptake rate ($V_{net}{}^*$) are stable. When uptake proceeds for a few days, the equilibrium shifts gradually to lower C_l values, because the depletion zone grows, the diffusion gradient becomes less steep, and the transport rate decreases. Around thin roots, depletion proceeds slower than around thick roots, because of the radial geometry

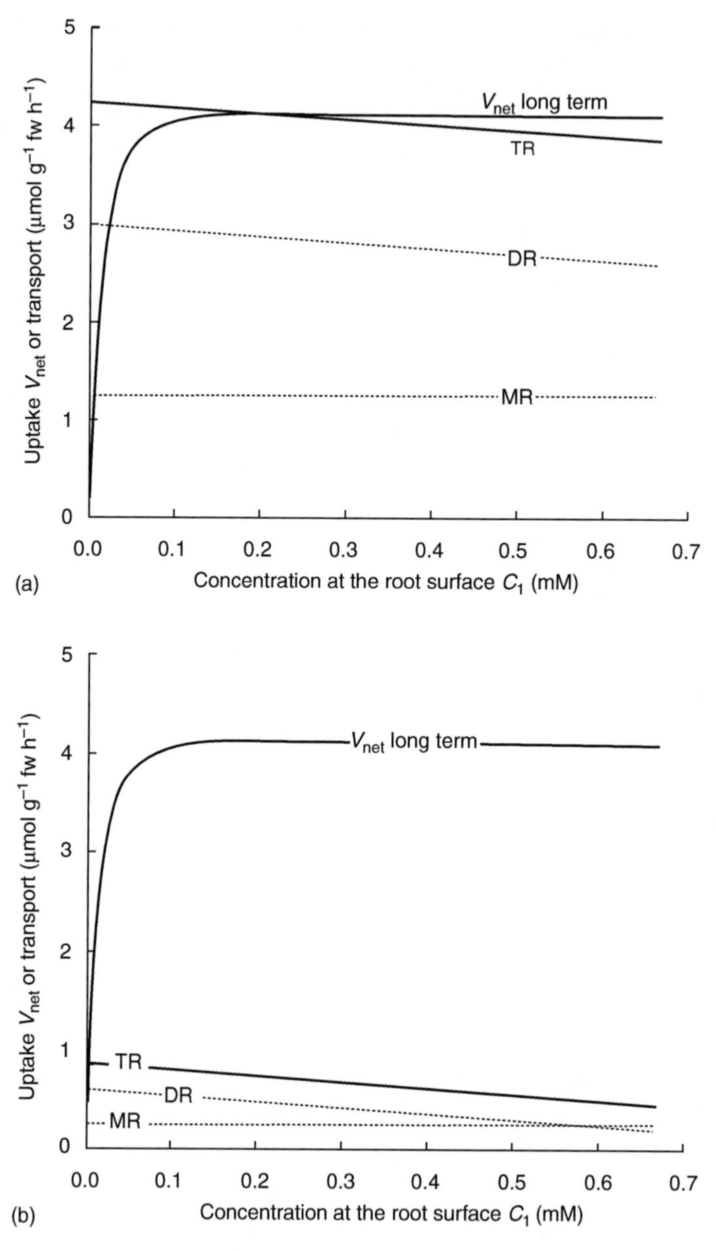

FIGURE 8.2 Net uptake rate (V_{net}) and nutrient transport rate to the root surface (TR) as a function of the nutrient concentration at the root surface (C_l). The diffusion rate (DR), the mass flow rate (MR), and their sum (TR) at (a) high ($C_b = 5$) and (b) low concentrations ($C_b = 0.3$) in the bulk soil. In either case, the equilibrium concentration (C_l*) and equilibrium uptake rate (V_{net}*) are found at the intersection of the lines V_{net} and TR. The parameter values used are hypothetical, keeping midway between NO_3^- and K^+. V_{net} is identical to the long-term uptake in Figure 8.1.

of roots. The amount of soil per unit root surface present within the same distance from the root surface is larger (and contains more nutrients) when the root diameter is smaller.

The depletion zone around the root continues to grow until the transport rate becomes so low that it equals the rate at which nutrients are released from the soil within the depletion zone by means of dissolution, desorption, or mineralization. Immobile nutrients, like phosphate, are released slowly and have narrow depletion zones, low uptake rates, and low C_l*. Eventually, when the nutrients adsorbed to soil particles are depleted up or when mineralization is interrupted due to low temperatures, the release rate falls and uptake stops.

To maximize uptake, plants can reduce the distance over which nutrients are transported through the soil, by increasing the density of their root system. Plants can realize higher root densities by increased allocation of carbon to their rooting system but also by reduced root diameters which leads to an increased root length and root surface per unit of root biomass. Root hairs are important in this respect, because they are thin and require little investment of biomass per unit of soil explored. However, their vulnerability and short life span make them less profitable when the bulk of the soil is already depleted, so that a plant has to "sit and wait" for nutrients that are released from the solid phase. In such nutrient-poor situations many plant species are living in symbiosis with fungi that form mycorrhizas. Such associations strongly increase the total surface area by which nutrients can be taken up. Fungal hyphae have much smaller diameters than roots (see Chapter 5, this volume).

Silberbush and Barber (1984) have studied the effect of changes in plant and soil characteristics on the equilibrium uptake rate of plants growing in soil. Figure 8.2 illustrates their results. The equilibrium uptake rate (V_{net}*) and nutrient concentration (C_l*) at the root surface are given by the intersection of the lines TR and V_{net}. When nutrient concentrations in the soil are high (Figure 8.2a, with $C_b = 5$), C_l* is situated at the horizontal part of the uptake curve. In this case, the equilibrium uptake rate V_{net}* is determined largely by the maximum uptake capacity of the roots (V_{max}) and not by the affinity of the uptake mechanism (K_m), nor by the mass flow or diffusion rate. Consequently, we may expect that natural selection on rich soils will favor plants with a high maximum uptake capacity (V_{max}).

When nutrient concentrations in the bulk soil are low (Figure 8.2b, with $C_b = 0.3$), C_l* is situated at the ascending part of the uptake curve. Here, V_{net}* is mainly determined by the effective diffusion coefficient D_e and by C_b, which determine the slope of the line that represents the diffusion rate and the intercept with the horizontal axis, respectively. In this case, the value of V_{net}* is relatively insensitive to changes in the kinetic parameters that rule the uptake process (K_m and V_{max}). Consequently, we may expect that natural selection of plants on poor soils will not lead to increased affinity or capacity of the nutrient uptake mechanism, but to properties that reduce the transport limitation, bringing the root surface closer to the nutrients (i.e., by fine and dense root systems and mycorrhizal associations). By increasing the root surface per unit plant biomass, a plant can sustain adequate growth rates with lower nutrient uptake rates per unit root surface and thus with lower nutrient concentrations at the root surface than competitors with a smaller root system.

The nutrient concentration in the bulk of the soil solution (C_b) can differ dramatically from the concentration at the root surface (C_l), implying that C_b is not a good indicator of nutrient availability. When the nutrient pool in the bulk of the soil solution is depleted, the nutrient supply to the root depends on the rate at which available forms of the nutrient are released from the organic and the mineral substrates. Plants can increase the release rate of nutrients by lowering the concentration in the soil solution or by affecting the chemical conditions or the microbial activity in the rhizosphere. Some species can use chemical forms or physical states (solid, dissolved, adsorbed, or occluded) of a nutrient that other species cannot use. Nitrogen, for example, can be taken up by most species only as NO_3^- and NH_4^+, but some species can also take up amino acids and other dissolved organic molecules that contain nitrogen (Kielland 1994, Northup et al. 1995, Schimel and Chapin 1996, Leadley et al. 1997). Some species can mobilize iron in calcareous soils by lowering the pH or by exudating chelating or reducing substances (Römheld and Marschner 1986). Phosphorus can be taken up by most species only as $H_2PO_4^-$, but some species are able to mobilize solid calcium phosphate by changing the chemical conditions in the rhizosphere, for example, by lowering pH, exudation of organic acids, or lowering the Ca concentration in the soil solution (Hoffland 1992). The ability to mobilize insoluble nutrients by altering the chemical conditions in the rhizosphere can be enhanced by forming dense clusters of lateral roots that intensify the rhizosphere effects. These morphological adaptations are called proteoid roots, after the *Proteaceae* family, but they occur also in species of other taxa (e.g., *Lupine*). An excellent review of the ability of plant species to use alternative nutrient sources is given by Marschner (1995).

Many plant species from nutrient-poor soils have special adaptations that enable them to use nutrients from other sources than the soil solution. The most widespread of these are associations with mycorrhizal fungi and symbiosis with N-fixing bacteria such as *Rhizobium* and *Frankia*. Mycorrhizal fungi are able to decompose dead organic material and to transport the mineralized nutrients to the plant root (see Chapter 5, this volume). Recently, Jongmans et al. (1997) suggested that mycorrhizal fungi are also able to penetrate rocky materials and to absorb P, Mg, Ca, and K from minerals by the excretion of organic acids and to transport these nutrients to connected roots. More peculiar adaptations that occur mainly in extremely nutrient-poor ecosystems are parasitism on other plants (e.g., *Rhinanthus*, *Pedicularis*) and carnivory (e.g., *Drosera, Pinguicula, Utricularia*).

UPTAKE OF ORGANIC NITROGEN COMPOUNDS

A few decades ago, it was assumed that most plant species absorbed nitrogen as nitrate and ammonium except for species with ecto- or ericoid mycorrhizal associations (Read 1991). It was long considered that the mineralization of organic nitrogen to ammonium and its subsequent oxidation were the major bottlenecks restricting the nitrogen supply to plants (Chapin 1995). However, these ideas were strongly disturbed by the observation that measured rates of net microbial production of inorganic nitrogen were often less than half the observed rates of nitrogen acquisition by plants (Fisk and Schmidt 1995, Kaye and Hart 1997). In the 1990s, it became clear that quite a few species can absorb amino acids from solution and can easily survive when no other nitrogen forms are supplied (Chapin et al. 1993, Kielland 1994). Schimel and Chapin (1996) showed that two tundra sedges, *Eriophorum vaginatum* and *Carex aquatilis*, which were unlikely to have any of these mycorrhizal associations, nevertheless absorbed amino acids (glycine and aspartate) under field conditions. The amino acids that they provided to the plants were labeled with ^{15}N and ^{13}C to test whether complete amino acids were taken up or that ammonium that was derived from decomposing amino acid molecules was absorbed. The authors did not detect any of the ^{13}C in the produced plant tissues and they attributed this failure to respiration of the labeled C atoms. Glycine and aspartate can be easily converted into glycolysis or TCA cycle intermediates and subsequently respired. Näsholm et al. (1998) solved this problem by ^{15}N and dual ^{13}C labeling of the glycine molecule. Here both C atoms were labeled instead of only the carbon in the carboxyl group, preventing that all ^{13}C were rapidly respired after decarboxylation. Their measurements in a boreal forest showed that at least 91%, 64%, and 42% of the nitrogen from the absorbed glycine was taken up as intact glycine molecules in the dwarfshrub *Vaccinium myrtillus*, the grass *Deschampsia flexuosa*, and the trees *Pinus sylvestris* and *Picea abies*, respectively. These results showed unambiguously that these different species, irrespective of their completely different mycorrhizal associations, can bypass nitrogen mineralization. The dwarfshrubs and trees have ericoid and ectomycorrhizal associations and were expected to absorb organic nitrogen. But it was surprising that the arbusco-mycorrhizal grass species also appeared to be able to absorb amino acids. A recent review (Aerts and Chapin 2000) mentioned that "the ability to take up organic N sources... hardly occurs in species with arbuscular mycorrhizas." But it seems that the ability to absorb dissolved organic nitrogen compounds is much more widespread among plant species than we earlier believed (Schmidt and Stewart 1997, Lipson et al. 1999, Persson et al. 2003). Nevertheless, the quantitative significance of the uptake of dissolved organic nitrogen as compared with the uptake of ammonium and nitrate is still to be assessed.

ROOT FORAGING IN HETEROGENEOUS ENVIRONMENTS

So far we have examined nutrient uptake and transport in a homogenous substrate. However, nutrient availability in soils may vary dramatically beyond the zone of influence of the roots themselves. Soil patches of different quality are created at various scales by abiotic factors (soil type differences, soil depth, microtopography) as well as by biotic factors such as treefalls and stemflow in forests (Gibson 1988a,b, Hook et al. 1991, Lechowicz and Bell 1991, Farley and Fitter 1999). In arid environments, organic matter accumulates in the vicinity of isolated trees, shrubs, and persistent turf grasses creating islands of fertility in a nutrient-deprived matrix (Jackson and Caldwell 1993, Alpert and Mooney 1996, Ryel et al. 1996, Schlesinger et al. 1996). Consequently, from the point of view of the plant individual, in many habitats the spatial distribution of water and nutrients is profoundly heterogeneous from scales as small as a few centimeters, to tens of meters and more. How effectively can plants capture the resources in such a heterogeneous world? What fraction of the growth achieved at a homogeneous supply of soil resources can be realized when similar amounts of resources are patchily distributed? Do species from habitats of different resource status have different foraging abilities?

Especially for the less-mobile ions such as phosphate, pockets of nutrients may only be captured by the plant if roots expand their surface area into the richer patch (Hutchings and de Kroon 1994, Robinson 1994, 1996). This foraging behavior may be very effective, as is perhaps best illustrated with the classical study by Drew and coworkers with barley (*Hordeum vulgare*). Single root axes of barley were grown into three compartments in which the concentration of nutrients could be controlled separately. High nitrate concentration in a given compartment promoted the formation of more first- and second-order laterals per unit of primary root length within that compartment and greater lateral root extension (Drew et al. 1973). When one-third of the entire root system received a nutrient-rich solution, total lateral root length per unit of length of the primary axis was 10 times higher, and the total root biomass 6 times higher, in the high-nutrient compartment than those in the low-nutrient compartments. Later in the experiment, when the lateral roots had grown out, whole-plant relative growth rate (RGR) under localized supply of nutrients approached the RGR of control plants growing under homogeneous nutrient supply (Drew and Saker 1975). When phosphate was supplied to 2 cm of the main root axis—a fraction amounting to only a few percent of its total length—whole-plant RGR was more than 80% of its value in control plants in which the whole root system received phosphate. When applied to 4 cm of the main root axis, the RGR was similar to that of controls. The higher local nutrient uptake from small pockets of nutrients to which part of the root system was exposed was not only due to an enlargement of the local root surface area. In addition, phosphate absorption rates per unit of root length increased in the enriched compartment, compared with both other parts of the root system in treated plants and the root system of control plants (Drew and Saker 1975).

The enhanced formation of roots in nutrient hotspots is now referred to as root proliferation, selective root placement, or root foraging precision (de Kroon and Mommer 2006). Local conditions determine where lateral root growth and uptake is promoted (Drew et al. 1973, Drew and Saker 1975), but the magnitude of the local response depends on the conditions experienced by the rest of the root system and the entire plant. An experiment by Drew (1975) illustrates this well. He subjected roots of barley plants to either a uniform or localized nutrient supply. Part of the root system given a high phosphate supply produced more and longer lateral roots when the rest of the root system was receiving low phosphate rather than high phosphate (Figure 8.3). This suggests that the local morphological response is stronger when phosphate is more limiting to the plant. Broadly similar effects were produced when the nitrate and ammonium supply to different sections of the root system was varied (Drew 1975). However, effects were less clear for nitrate (Drew et al. 1973,

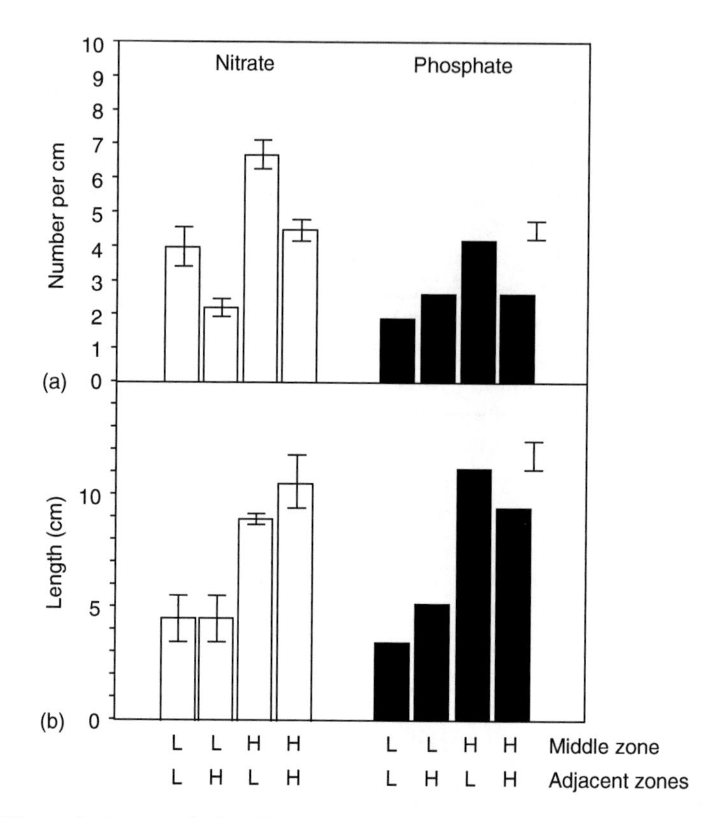

FIGURE 8.3 Effects of nitrate and phosphate supply on (a) the number of lateral roots per cm of main root axis and (b) the lengths of individual lateral roots in barley. Main root axes were divided into three zones and nutrients were supplied independently to each of these zones. Data given are those for the first-order laterals that developed in the middle zone. This zone experiences either a low (L) or a high (H) concentration of nitrate or phosphate. Adjacent rooting zones also grew in either a high- or low-nutrient solution. In the nitrate experiment, plants were grown hydroponically; in the phosphate experiment, they were grown in sand. Nitrate data are given as mean $\pm$SE, phosphate as means with separate bars showing the LSD at the 5% level. (After Drew, M.C., Saker, L.R., and Ashley, T.W., *J. Exp. Bot.*, 24, 1189, 1973; Drew, M.C., *New Phytol.*, 75, 479, 1975; adapted from Hutchings, M.J. and de Kroon, H., *Adv. Ecol. Res.*, 25, 159, 1994. Courtesy Academic Press. With permission.)

see Figure 8.3). In most studies in which nutrients were supplied heterogeneously, root growth was suppressed in the part of the root volume that experiences low nutrient supply (Robinson 1994).

The merits of the ability to forage for patchily distributed nutrients can perhaps best be illustrated by comparing the biomass production of plants grown on homogeneous and heterogeneous substrates each with the same overall nutrient availability. Fransen et al. (1998) created such treatments by mixing poor riverine sand with black humus-rich soil either homogeneously or by concentrating most of the black soil in a small column within the pot. Five grass species were grown individually in each of these treatments. Their roots readily reached and penetrated the enriched column but the responses were significant only for the three species characteristic of relatively nutrient-rich habitats. Combined for all species, whole-plant nutrient accumulation and biomass at the end of the experiment was significantly higher in the heterogeneous treatment than that in the homogeneous treatment. These results indicate that plant species may profit and grow faster at a heterogeneous distribution of soil nutrients, rather than slower. In this experiment with bunchgrasses the growth stimulus in the heterogeneous treatment compared with the homogeneous treatment was small. Clonal species that spread horizontally have the ability to take up nutrients locally and produce most of the biomass beyond the nutrient-rich patch. For such species, a several-fold increase in biomass production may occur if the distribution of nutrients is not homogeneous, but concentrated in small hotspots (Birch and Hutchings 1994, Hutchings and Wijesinghe 1997, Wijesinghe and Hutchings 1997).

To what extent these results on root foraging ability in heterogeneous soils are generally valid? In a recent meta-analysis covering the results of over 100 species, Kembel and Cahill (2005) showed that the responses are very variable, confirming earlier overviews (Robinson 1994, Hodge 2004). Species varied from little or no root proliferation at all, distributing their roots equally over the rich and poor parts of the soil, up to the very plastic responses as observed for barley in Drew's experiments. Species may also differ markedly in the time between nutrient application and response. For example, when exposed to nutrient enrichment, roots of the cold desert species *Agropyron desertorum* showed a fourfold increase in the RGR of root length within one day, whereas *Artemisia tridentata* and especially *Pseudoroegneria spicata* responded less vigorously (Jackson and Caldwell 1989). In the latter species, extension growth was not affected until several weeks after nutrient application.

For one of the three datasets analyzed, Kembel and Cahill (2005) found support for the notion of Grime et al. (1986) that plant species with a higher RGR place their roots more selectively in heterogeneous soils. When in a given experiment, plants of different growth rates are harvested after a fixed period of time, as is usually the case, the degree of selective root placement is indeed positively correlated to growth rate (Fransen et al. 1999, Aanderud et al. 2003). This correlation has its origin in the modular nature of the root system (*sensu* de Kroon et al. 2005), in which roots respond locally to the nutrient concentrations that they experience. Species with larger RGRs in terms of plant biomass are likely to also have larger root RGRs (more lateral root formation and higher root extension rates) in the richer microsites. However, when corrected for growth rate differences, the degree of selective root placement is the same for slow-growing and fast-growing species (Fransen et al. 1999, Aanderud et al. 2003).

However, it should be realized that species with higher root proliferation in enriched patches do not necessarily obtain more nutrients from heterogeneous soil than species with less root proliferation, unlike the results of Drew and Saker (1975) and Fransen et al. (1998) suggest. For their larger datasets, Kembel and Cahill (2005) found no significant correlation between the response to nutrient heterogeneity in terms of biomass production and the precision by which the roots were placed in the nutrient-richer patches. Some of these variable results may be explained by slow response of root proliferation that may come too late relative to nutrient release in the patches (Robinson 1996, Van Vuuren et al. 1996), suggesting a much more prominent role of enhanced maximum uptake capacity (i.e., physiological plasticity) for the acquisition of finite nutrient patches. The gain in biomass in heterogeneous versus homogeneous soils also becomes smaller when the experiments last longer because patches deplete and the precision of root placement reduces (Kembel and Cahill 2005). This makes sense because when all soil nutrients are taken up, no differences in biomass production are to be expected between homogeneous and heterogeneous soils if in both treatments the same total amount of nutrients is supplied. Despite these methodological caveats, our current understanding is that the root proliferation in enriched microsites is less important for nutrient acquisition in heterogeneous soils than previously thought (de Kroon and Mommer 2006), except when plants are in competition (Hodge et al. 1999, de Kroon et al. 2003; but see Fransen et al. 2001).

To evaluate the ecological significance of selective root placement, its benefits must be compared with its costs. The immediate benefits may be limited but if the costs of wrong placement are small, selective root placement may still be profitable. Jansen et al. (2006)

recently demonstrated for the herb *Rumex palustris* that the costs of placing roots at the wrong location may indeed be small. They created homogeneous and heterogeneous soils in pots with a dripping system and *R. palustris* roots developed rapidly and selectively in the nutrient hotspot supplied in one quadrant of the pot. Midway the experiment, in some of the pots, the nutrient supply pattern was changed from homogeneous to heterogeneous and vice versa, or the hotspot was replaced to another location at the opposite side of the pot. By analyzing the root RGRs in different quadrants, Jansen et al. (2006) were able to show that root growth responded immediately to the shifts in local nutrient supply. An increase in root biomass in response to an increase in nutrient supply was achieved faster than a decrease in root biomass when the nutrient supply was decreased. However, as significant root biomass was built up in the first part of the experiment the shifts in actual root placement were slow, and the plants in the switch treatments were confronted with most of their roots located in the quadrant with low nutrient supply in the second part of the experiment. Surprisingly, costs of this wrong placement were absent. Plants for which the nutrient patches were switched had similar total nutrient uptake and growth as those for which the homogeneous or heterogeneous supply of nutrients was unchanged. Jansen et al. (2006) explained this lack of costs by redistribution of stored nutrients to new biomass, reducing the demand on new nutrient uptake, and by high physiological plasticity, that is, elevated uptake kinetics especially of the young roots that rapidly developed in new nutrient patches after the switch. These results suggest that plants may have a remarkable flexibility to relocate their root placement pattern even if immediate returns are small.

The costs of selective root placement may be much higher on the long term when patches gradually deplete and if new patches do not appear. Fransen and de Kroon (2001) grew isolated plants of the fast-growing grass *Holcus lanatus* and the slow-growing grass *Nardus stricta* for two growing seasons in homogeneous poor and rich soil, and in a heterogeneous treatment consisting of a poor half and rich half. In the first few months after the start of the experiment *Holcus*, but not *Nardus*, proliferated its roots rapidly in the richer patch of the heterogeneous soil, as quantified by minirhizotron observations. This proliferation paralleled elevated growth of *Holcus* in the heterogeneous soils relative to the homogeneously poor and rich treatments. However, already in the course of the first growing season, the growth of *Holcus* started to decline and by the end of the second year its biomass in heterogeneous soil was almost as low as that in homogeneous poor soil. For *Nardus*, by contrast, biomass production in heterogeneous soils over the 2 years increased relative to the homogeneous controls. Fransen and de Kroon (2001) concluded that the fast-growing *Holcus* overproduced roots in the nutrient-rich microsite resulting in significant costs in the long term when nutrients deplete and roots die off. Under conditions of nutrient depletion, *Nardus* with hardly any selective root placement and much longer root life spans has larger long-term returns.

The data available to date suggest that slow-growing species from resource-poor versus fast-growing species from resource-rich habitats differ only little in root foraging abilities, although the higher growth rate itself give the species an advantage, especially in a competitive setting. Both morphological and physiological plasticities are important attributes. In extremely nutrient-poor habitats such as nutrient-poor tundra, where patches if they appear rapidly deplete, the ability of roots to survive periods of resource depletion seems to be of greater significance than high levels of morphological plasticity. The maintenance of a large viable root mass, despite long periods of low nutrient availability, and the ability to commence absorption of nutrients rapidly when conditions permit enable species to acquire nutrient pulses of short duration (Crick and Grime 1987, Campbell and Grime 1989, Kachi and Rorison 1990). The high carbon costs of maintaining viable roots (Eissenstat and Yanai 1997) may not be a great problem in these habitats because carbon is not the limiting resource. In very productive environments, however, carbon costs of root maintenance may be significant and roots generally have a shorter life span than species from less productive habitats (see Section "Allocation and Use of Absorbed Nutrients"). Rapid growth, high nutrient uptake rates, and a high turnover of roots may result in a more fugitive root behavior in these habitats. Enriched microsites are rapidly exploited after which the root system shifts its investments toward more profitable parts of the soil volume. This behavior is only profitable if such nutrient hotspots regularly reappear. The costs of switching foraging behavior continuously toward new patches may be limited, but the long-term costs of selective root placement is significant if patches deplete without getting replaced.

ALLOCATION AND USE OF ABSORBED NUTRIENTS

The acquisition of nutrients, their transport within the plant from the roots to the other organs, and their subsequent incorporation into organic compounds require a major carbon expense of the plant (Chapin et al. 1987, Farrar and Jones 2003). Vice versa, the assimilation of carbon requires nutrients, but especially N, in significant quantities. C_3 plants invest approximately 75% of their N in chloroplasts of which a major part is used in photosynthesis. About one-third of this chloroplast N is built into rubisco, the primary CO_2-fixing enzyme (Chapin et al. 1987). As a result, the photosynthetic capacity (the maximum rate of carbon assimilation) is highly positively correlated with leaf nitrogen concentration (Field and Mooney 1986, Evans 1989). The photosynthetic rate per unit of leaf nitrogen is referred to as the photosynthetic nitrogen use efficiency (PNUE) (Lambers and Poorter 1992, Fitter 1997). Beyond a critical level, photosynthesis does not increase further with increasing nitrogen concentration and may even decline. In such situations, other resources, such as light and water, may limit photosynthesis.

Although an important part of the assimilated N is allocated to the photosynthetic system, the plant requires N also for a whole variety of other plant functions (Lambers and Poorter 1992). The relationship between nitrogen concentration in the whole plant and RGR may be different for different plant species depending on—among other factors—the fraction of nitrogen that is allocated to the photosynthetic machinery. Such differences may be caused by variation in allocation to plant organs, such as leaves, roots, and stems, but also by differences in the allocation to the various organelles and compounds within the leaf. Some rapidly growing species such as *Lolium perenne* allocate an extremely large part of the leaf nitrogen to rubisco, whereas in other species part of the leaf nitrogen is used for the synthesis of defensive compounds or incorporated in supporting tissues. Ingestad (1979) characterized the relationship between RGR and whole-plant nitrogen concentration by the nitrogen productivity (A), defined as the rate of dry matter production per unit of nitrogen in the plant (g d.wt. g^{-1} N day^{-1}). Figure 8.4 gives the relationships between RGR and nitrogen concentration in the plant for three tree species, which appear to be linear with a virtually zero intercept over a broad range of nitrogen concentrations. The slopes of the regression lines represent the nitrogen productivities of each species. All three species increase their growth at higher internal nitrogen concentrations, but the faster-growing species make a more efficient use of the nitrogen that is present in the plant. Figure 8.4 also shows that the faster-growing species not only has a higher growth rate than the slower-growing species at higher nitrogen concentrations, but also at low concentrations. The difference in nitrogen productivity between the three species is probably caused by differences in allocation to the photosynthetic process, but may be explained as well by differences in costs of biosynthesis of plant tissues.

Whole-plant growth is optimized if all resources are equally limiting (Bloom et al. 1985). As a rule, new biomass is allocated to the plant organs that acquire the most strongly limiting resource. If nutrients are in short supply, there are several ways in which nutrient shortage in the plant may be avoided. More carbon may be invested in root biomass so that a larger soil volume can be explored and the competitive ability for soil nutrients is increased. Tilman

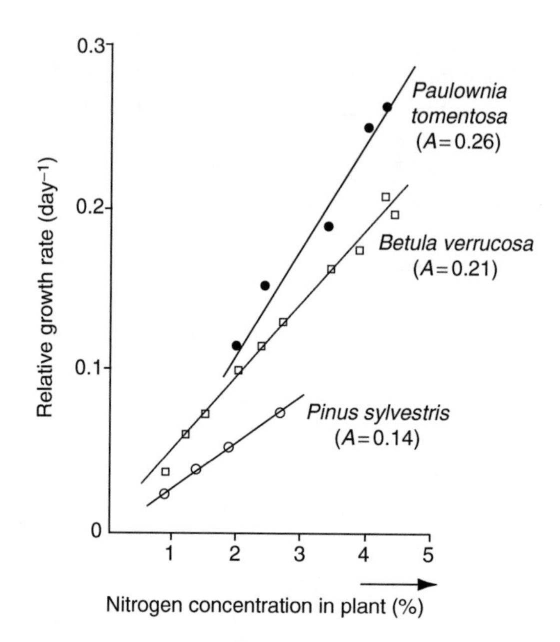

FIGURE 8.4 The relative growth rate of seedlings of three tree species versus nitrogen concentration in the total plant. The values of the nitrogen productivity A (g d.wt. g^{-1} N h^{-1}) are given by the regression coefficients of the presented lines. (After Ingestad, T., *Physiol. Plant.*, 45, 149, 1979; Hui-jun, J. and Ingestad, T., *Physiol. Plant.*, 45, 149, 1984; Ingestad, T. and Kähr, M., *Physiol. Plant.*, 65, 109, 1985; from Berendse, F. and Elberse, W.Th., *Perspectives on Plant Competition*, J.B. Grace and D. Tilman, eds, Academic Press, New York, 1990. With permission.)

(1988) suggested that increased allocation to root biomass would be one of the most important adaptations of plants to nutrient-poor soils. It has been known for a long time that the phenotypic response of all plant species to reduced nitrogen or water supply is an increased carbon and nitrogen allocation to roots (Brouwer 1962), but comparisons of species adapted to nutrient-rich and nutrient-poor sites do not confirm Tilman's hypothesis. Grass species adapted to nutrient-poor soils generally invest less or equal amounts of biomass in belowground parts than species characteristic of more fertile sites (Elberse and Berendse 1993). In a recent review of studies on plasticity in root weight ratio, Reynolds and D'Antonio (1996) showed that species from nutrient-poor and nutrient-rich habitats exhibit a similar increased root allocation in response to nitrogen shortage. The most important difference between species of nutrient-poor and nutrient-rich sites is that the roots of the former seem to have smaller diameters leading to an increased root length per unit root weight (Elberse and Berendse 1993, Fitter 1997).

The efficiency of nutrient utilization for growth also depends on other functions to which nutrients are allocated by the plant, such as support, defense, reproduction, and storage (Chapin et al. 1990). Allocation to supporting structures (such as woody tissue), chemical compounds for defense, or reproductive organs may curtail the growth rate of plants (Bazzaz et al. 1987). Plant species with a particularly high allocation to one or more of these functions, or plants in their reproductive phase, have low growth rates and low nitrogen productivity. Growth is also curtailed if a significant proportion of the resources is allocated to storage, that is, reserve formation that involves the metabolically regulated compartmentation or synthesis of storage compounds (Chapin et al. 1990, see Chapter 5, this volume). Although reserve formation directly competes for resources with growth, resources may accumulate because resource supply exceeds the demands for growth and other functions during a certain period. This accumulation is commonly referred to as luxury consumption (Chapin 1980) and should be distinguished from reserve formation (Chapin et al. 1990). Luxury consumption allows the slower-growing species to absorb nutrients in excess of immediate growth requirements during nutrient flushes. The reserves built up in this way may be used to support growth in periods of nutrient depletion.

Classical plant ecophysiology often depicts the growth of plants in natural environments simply as resulting from soil nutrient uptake and carbon assimilation. However, in many perennial plant species, growth strongly depends on amounts of nutrients and carbon that have been stored during preceding growing periods (see de Kroon and Bobbink 1997). In the alpine forb species *Bistorta bistortoides*, stored N reserves in the rhizomes accounted for 60% of the N allocation to the shoot during the growing season (Jaeger and Monson 1992). In this species N storage was largely accommodated by increased concentrations of amino acids (Lipson et al. 1996). Resources stored in perennial plant organs may support the aboveground biomass production of plants to a considerable degree, as illustrated by the study of Jonasson and Chapin (1985) with the sedge *E. vaginatum*. In extremely nutrient-poor tundra, they compared the growth of tillers (with attached belowground stems and roots) in bags without access to soil nutrients with the growth of unbagged tillers. They found that the bagged tillers accumulated as much leaf biomass during one growing season as the unbagged plants. Nutrients were transported from the belowground stems to the leaves during the first 2 months after snow melt. After senescence at the end of the growing season, nutrient contents in the belowground stems of the bagged tillers were only slightly lower than those in the unbagged ones.

LOSSES OF NUTRIENTS THROUGH ABSCISSION AND HERBIVORY

It is clear that the growth of a perennial plant individual is not only determined by the amount of nutrients that it acquires, but also by the amounts of stored nutrients that can be reused. In environments where nutrients limit plant growth, the long-term dynamics of perennial plant populations is largely determined by the balance between the uptake and the loss of nutrients. Losses of nutrients may occur in various ways: abscission of leaves and flowers, root death, mortality due to disturbance, nutrient capture by herbivores, leaching from leaves, seed or pollen production, and exudation from roots.

One of the most important pathways by which plants lose nutrients is the seasonal abscission of leaves, roots, and other organs. Several studies show that there is a huge variation in life spans of leaves among vascular plant species. Escudero et al. (1992) found that the life spans of leaves of tree and shrub species in the Pyrenees varied by a few orders of magnitude from a few months to more than 4 years. A similar large variation (from a few months to 10 years) was reported in a survey of several studies of leaf life spans (Reich et al. 1992). Nutrient losses due to leaf abscission are quantitatively significant, but are reduced by active retranslocation of nutrients in the period preceding abscission. Measurements of nutrient withdrawal should take into account that not only is the nutrient content reduced but that also the dry weight per leaf can decline because of respiration or retranslocation of carbohydrates, implying that nutrient withdrawal should be measured on a whole-leaf basis. In arctic ecosystems, 20%–80% of N and 20%–90% of P in leaves was withdrawn before abscission, whereas there did not seem to be important differences in percentage withdrawal between graminoids, forbs, and deciduous and evergreen dwarfshurbs (Chapin et al. 1975, Jonasson 1983, Chapin 1989, Chapin and Shaver 1989). Morton (1977) studied the decline in nutrient concentrations in leaves of the deciduous grass *Molinia caerulea* during abscission at the end of the growing season. He compared open plots with plots that during fall and winter were covered with a transparent roof, preventing leaching of nutrients by rain. The reduction in N and P concentrations in dying leaves was measured to be about 75% and occurred both in the open and the covered plots, but the decline in concentrations (ca. 90%) of K, Ca, and Mg took place only in the plots without cover. He concluded that the reduction in the N and P content of leaves occurred through active withdrawal from senescing leaves, but that the reduction in K, Ca, and Mg took place through leaching from the leaves to the soil.

Much less data are available about root life spans. Eissenstat and Yanai (1997) showed in their review that the time periods after which 50% mortality has occurred vary between 14 and 340 days, which correspond with life spans of 20–490 days (assuming a negative exponential decline in the number of living roots). Between-species comparisons are difficult because the life spans vary strongly among root cohorts produced in different seasons and most studies did not compare species at the same site. In a garden experiment we followed individual roots of 14 grassland species from birth to death using minirhizotrons. Average root life spans varied from 41 days in *Rumex obtusifolius* which occurs in very fertile habitats to 381 days in *Succisa pratensis* which is characteristic of nutrient-poor sites (Berendse, unpublished results). It is not yet clear whether nutrient losses due to root death can be significantly reduced through nutrient withdrawal preceding abscission. A few studies showed that nutrient resorption from dying roots is minimal (Nambiar 1986, Dubach and Russelle 1994), so that nutrient losses by root turnover might be very significant.

In addition to nutrient losses through seasonal abscission, an important pathway of nutrient loss occurs due to herbivory by a broad variety of organisms such as grazing mammals, phytophagous insects, parasitic fungi, and root nematodes. The quantities of nutrients that the plant loses because of the activities of herbivores aboveground and belowground have rarely been measured, but can probably be rather important. We simulated grazing through mammals by clipping plants with 8 week intervals at 5 cm above soil surface. We measured that at low levels of soil fertility the tall grass species *Arrhenatherum elatius* lost 57% of the total amount of nitrogen taken up, whereas the short grass *Festuca rubra* lost 24% (Berendse et al. 1992). These losses increased to more than 90% at higher soil nutrient levels.

The data presented strongly suggest that there is a wide variation in biomass and nutrient turnover among plant species depending on the life spans of plant organs, but they do not supply information about the quantitative significance of whole-plant nutrient loss rates as compared with nutrient supply rates. Especially, for plants growing in their natural environment such data are extremely difficult to collect. In the 1980s we carried out a comparative field study in which we attempted to quantify whole-plant nutrient losses and nutrient uptake in populations of the ericaceous dwarfshrub *Erica tetralix* and the perennial, deciduous grass *M. caerulea*. In recent decades, in many wet heathlands in Europe *E. tetralix* has been replaced by *M. caerulea*. In competition experiments in containers (Berendse and Aerts 1984) and in field fertilization experiments (Aerts and Berendse 1988, Aerts et al. 1990) we measured that at increased levels of nutrient supply *Molinia* is able to outcompete *Erica*, whereas *Erica* remains the dominant species under nutrient-poor conditions.

We measured nitrogen losses from populations of *Molinia* and *Erica* plants in adjacent sites for 2 years. Total losses of nitrogen from *Molinia* plants varied between 60% and 100% per year of the total amount of nitrogen present in the plants at the end of the growing season. We calculated the lower turnover rate assuming that 50% of the nitrogen in roots was withdrawn preceding abscission, whereas the higher figure was calculated assuming that no retranslocation took place. It is clear that losses of up to 100% have important consequences for the success of a population in an environment where nitrogen limits plant growth. Nitrogen losses from *Erica* were much smaller (ca. 30%). This seems to be an important adaptation to the nutrient-poor habitats that are dominated by this species. In Table 8.2, we compare the total N losses during 1982 with the N mineralization measured in the upper 10 cm of the soil during this year. The measured N mineralization is about equal to the total N supply rate. Almost all organic nitrogen is present in the upper 10 cm of the soil and the N input through atmospheric deposition is almost completely immobilized by the nutrient-poor

TABLE 8.2
The Relative Nitrogen Requirement, the Total (Aboveground and Belowground) Biomass at the End of the Growing Season and the Annual Nitrogen Loss from the Plant in Adjacent Populations of *Erica* and *Molinia* and the Annual N Mineralization on These Sites in 1982

	Erica	*Molinia*
Relative nitrogen requirement (mg N g^{-1} biomass $year^{-1}$)	2.3–3.4	7.4–11.7
Total biomass (g biomass m^{-2})	1270	919
Total N losses (g N m^{-2} $year^{-1}$)	2.9–4.3	6.8–10.8
N mineralization (g N m^{-2} $year^{-1}$)	11.5	10.1

litter layer, but is remineralized during the later phases of decomposition. In *Molinia*, the losses of N from the plant appear to be of the same order of magnitude as the rate of N supply to the plant, but in *Erica* N losses are less than 50% of the N that can be taken up.

If a plant loses a large part of the nutrients in its biomass annually, it must absorb more nutrients to maintain its biomass than a plant that is more economical with its acquired nutrients. To measure the nutrient uptake that plant species need in their natural environment, the concept of the relative nutrient requirement was introduced (Berendse et al. 1987). The relative nutrient requirement (L) is defined as the amount of nutrients that a plant population loses per unit of time and per unit of biomass. This amount of nutrients should be taken up again to maintain or replace each unit biomass during a given time period (mg N g^{-1} biomass $year^{-1}$). In the study referred to earlier, we measured that the relative nitrogen requirement was 2.3–3.4 mg N g^{-1} biomass $year^{-1}$ in *Erica* as compared with 7.4–11.7 mg N g^{-1} biomass $year^{-1}$ in *Molinia* depending on the assumption about N withdrawal from dying roots (Table 8.1). Apparently, *Erica* required much less nitrogen to be taken up to maintain its biomass than *Molinia* did. *Erica* plants appeared to require much less nitrogen because of the longer life spans of their leaves, stems and roots and not because of a higher retranslocation efficiency. The withdrawal of nitrogen from dying leaves in *Molinia* is even higher than that in *Erica*.

The differences between these two species seem to reflect a general pattern. Escudero et al. (1992) found that the life span of leaves of tree and shrub species in the Pyrenees was strongly correlated with the variation in soil fertility. Plant species dominant on infertile soils had leaves that lived longer than species that were abundant on more fertile soils. This positive correlation was not found between soil fertility and the fraction of N and P that was withdrawn from dying leaves. Recently, we carried out an experiment in which 14 plant species of Dutch grassland and heathland communities were grown in monocultures in experimental plots (Berendse, unpublished results). Here the direct effects of different soil characteristics were excluded. We found a significant inverse relationship between average leaf life span, as measured in these plots, and the nutrient index of each species which ranks the average soil fertility of the habitat in which the species involved is most frequently found (Figure 8.5).

ADAPTATION OF PLANTS TO NUTRIENT-POOR AND NUTRIENT-RICH ENVIRONMENTS

Plant species can increase their success in nutrient-poor habitats along three different lines. Firstly, they can maximize the acquisition of nutrients by increasing their competitive ability for soil nutrients or by exploring nutrient sources that are not available to competing plant

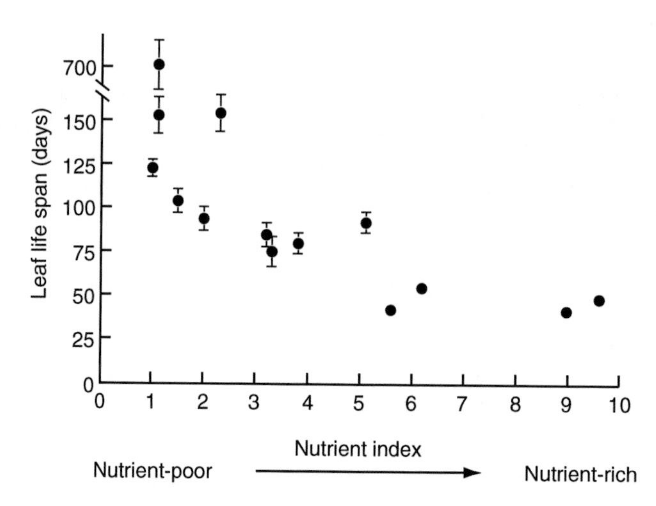

FIGURE 8.5 The average leaf life spans of 14 grassland and heathland species versus their nutrient index. Life spans were measured in plants growing in experimental plots under identical conditions. Measurements were carried out in 10 plants of each species by following marked leaves with 2 week intervals. The nutrient index is a descriptive parameter that ranks the average soil fertility of the habitats in which the species involved is most frequently found. Bars give standard errors of the mean. (From Berendse, unpublished results.)

populations. The affinity of the uptake system of most plants is sufficiently high to decrease the nutrient concentration at the root surface to practically zero. Further improvement of the uptake capacity is of little use. Uptake kinetics does not differ systematically between species from rich and poor soils. The competitive ability for soil nutrients can be increased by investing more carbon in fine root biomass or by changes in root morphology that increase root length or root surface area per unit biomass (by reduced root diameter or increased root hair density), so that a greater fraction of the available nutrients can be absorbed relative to competing plant species. In previous sections, we showed that, in general, plant species of nutrient-poor habitats do not invest more biomass in roots, but that some species of nutrient-poor habitats realize an increased absorbing root surface by producing thinner roots. In addition, species of nutrient-poor habitats do not seem to forage more effectively for patchily distributed nutrients, as compared with species from more productive habitats. However, many plant species of poor soils can explore additional organic nutrient sources by intimate associations with mycorrhizal fungi, and some genera (e.g., *Drosera*, *Pinguicula*, *Utricularia*) can even acquire and use living animal proteins. Deep-rooting species (such as forbs or trees) may absorb nutrients from deeper soil layers that are not available to competing species with a shallow rooting pattern.

The second line along which plant species may be adapted to nutrient-poor sites is by changes in the efficiency with which the nutrients that are present in the plant are used for carbon assimilation and subsequent growth. Different nutrients can be used for different plant functions affecting growth (e.g., N is mainly invested in rubisco, whereas K is required for stomata functioning). The parameter that measures this efficiency is the nutrient productivity A, as introduced in Section "Allocation and Use of Absorbed Nutrients". One would expect a strong selection in favor of an increased nutrient productivity in species adapted to nutrient-poor soils, but generally species of nutrient-poor habitats have lower nutrient productivities than species adapted to more fertile sites (e.g., Hui-jun and Ingestad 1984, Ingestad and Kähr 1985).

The third line of adaptation is increasing the length of the time period during which nutrients can be used. The length of this time period can be expanded by increased life spans of leaves, roots, and other organs. Life spans can be increased by investing in supporting tissues and in defensive compounds that reduce the risks of herbivory. The residence time of nutrients in the plant can also be increased by retranslocation of a large part of the nutrients in dying plant parts, but we showed earlier that the fraction of nutrients retranslocated from leaves before abscission is not clearly correlated with the soil fertility of the habitat in which the species occurs most frequently (Escudero et al. 1992). The earlier introduced relative nutrient requirement or relative nutrient loss rate measures nutrient losses per unit of biomass, but can also be expressed per unit of nutrient in the plant (Ln; g N g^{-1} N $year^{-1}$). Under steady-state conditions where nutrient losses are equal to nutrient uptake the inverse of this parameter (Ln^{-1}) measures the mean residence time of nutrients in the plant.

For a further analysis of the adaptation of plants to nutrient-poor substrates it is helpful to combine the instantaneous efficiency of nutrient utilization or nutrient productivity (A) with the mean residence time (Ln^{-1}) to the overall nutrient use efficiency (NUE) which measures the amount of biomass that can be produced per unit of nutrient taken up (g biomass produced/g nutrient absorbed):

$$\mathrm{NUE} = \frac{A}{\mathrm{Ln}}.$$

One would expect that the NUE would be higher in species of nutrient-poor habitats as compared with species from more fertile sites. Berendse and Aerts (1987) calculated this parameter for adjacent field populations of *Erica* and *Molinia* using the amount of biomass and the amount of nutrients in the plant as present at the end of the growing season (Table 8.3). It is striking that there is only a relatively small difference in NUE between the two species, but that the NUE values are composed of entirely different combinations of A and Ln^{-1}. *Erica* has a low N loss rate combined with a low N productivity, whereas *Molinia* has a much larger loss rate, but also a higher instantaneous utilization efficiency A. Apparently, the same overall NUE can be realized by various combinations of plant properties. These data suggest that especially the components A and Ln^{-1} are relevant in the adaptation of plant species to habitats with different nutrient supplies, rather than the NUE itself. Later, we consider the differences in growth rate between these two species in relation to their different nutrient loss rates.

The dwarfshrub *Erica* is able to maintain itself as the dominant species on nutrient-poor sites because it is much more economical with the nutrients that it has absorbed than the perennial grass species. But this difference does not explain why *Molinia* outcompetes *Erica*

TABLE 8.3
Nitrogen Productivity (A; g biomass g^{-1} N $year^{-1}$), Mean Residence Time (Ln^{-1}; year), and Nitrogen Use Efficiency (NUE; g biomass g^{-1} N) as Calculated for *Erica* and *Molinia*

	Erica	*Molinia*
A	24	94
Ln^{-1}	4.3	1.4
NUE	103	132

Source: Berendse, F. and Aerts, R., *Funct. Ecol.*, 1, 293, 1987.

after an increase in nutrient supply. In a field experiment with different nutrient supply rates, the potential growth rate of the grass *Molinia* was found to be much higher than that of the dwarfshrubs *Erica* and *Calluna* (Aerts et al. 1990). The higher potential growth rate of *Molinia* enabled this species to increase its biomass much more rapidly than *Erica* after an increase in nutrient supply. In communities where both species are present an increase or decrease in nutrient supply can result in complete dominance of *Molinia* and extinction of *Erica* or vice versa. The changes in such communities were calculated using our model for the competition between plant species (Berendse 1994). This model calculates nutrient and light competition and the losses of biomass and nutrients. The model predicts that species 1 outcompetes species 2 under nutrient-rich conditions, whereas species 2 replaces species 1 under nutrient-poor conditions if

$$\frac{\mathrm{Gmax}_1}{\mathrm{Gmax}_2} > \frac{\mathrm{Ln}_1}{\mathrm{Ln}_2} > 1,$$

in which Gmax_1 and Gmax_2 (g biomass m^{-2} $year^{-1}$) are the potential growth rates in a closed canopy of species 1 and 2, respectively, and Ln_1 and Ln_2 represent the relative nutrient loss rates of the two species, assuming that all other plant features are equal. This relationship leads us to conclude that interspecific competition is responsible for a strong selection pressure on the potential growth rate and the relative nutrient loss rate of species. Slight changes in these plant characteristics may lead to either complete disappearance or complete dominance. But we can conclude as well that species that combine a low nutrient loss rate with a high potential growth rate are superior at all nutrient supply rates. The question to be answered is whether plants can easily combine such characteristics.

The difference in potential growth rate (and nitrogen productivity) between the two species can be attributed to differences in allocation of nitrogen to the photosynthetic system and to differences in biosynthesis costs. At the end of the growing season *Erica* had allocated about 12% of total plant nitrogen to its leaves, whereas in *Molinia* plants 48% of the total plant N was present in leaves and green stems. This difference is not caused by differences in allocation to root biomass, but by the allocation of nitrogen to long-living, woody stems in *Erica*. Possibly, the two species differ as well in the allocation of nitrogen to the different compounds within the leaf. *Erica* leaves live about four times longer than *Molinia* leaves, thanks to their higher lignin content resulting in an increased toughness of the leaf. Lignin is much more expensive to biosynthesize than compounds such as cellulose. In a literature survey, Poorter (1994) did not find any systematic difference in biosynthesis costs of tissues produced by plant species from nutrient-poor and nutrient-rich soils. However, we found that the costs of biosynthesizing *Erica* tissue were higher than those of *Molinia* tissues (1.8 vs. 1.4 g glucose g^{-1} biomass). We conclude that the adaptation to nutrient-poor environments by minimizing the loss of nutrients has important negative side effects: the allocation of nitrogen to the photosynthetic system is reduced and the biosynthesis costs of tissues are increased, which results in reduced nitrogen productivity and reduced potential growth rate, which is an important disadvantage when soil fertility increases. Apparently, plant properties that determine nutrient losses and potential growth rates are strongly interconnected. The combinations low maximum growth rate–low loss rate and high maximum growth–high loss rate strongly correspond with, respectively, the stress-tolerant and competitive strategies that Grime (1979) distinguished much earlier. High maximum growth rates and low biomass losses cannot be easily combined for apparent physiological and morphological reasons.

This leads to the expectation that in plant species adapted to soils with different soil fertilities, biomass turnover rate and maximum growth rate are inversely correlated. In an experiment in growth chambers we measured maximum RGRs in the 14 grassland species, for which we had measured leaf life spans in a garden experiment (cf. Figure 8.5). We found a

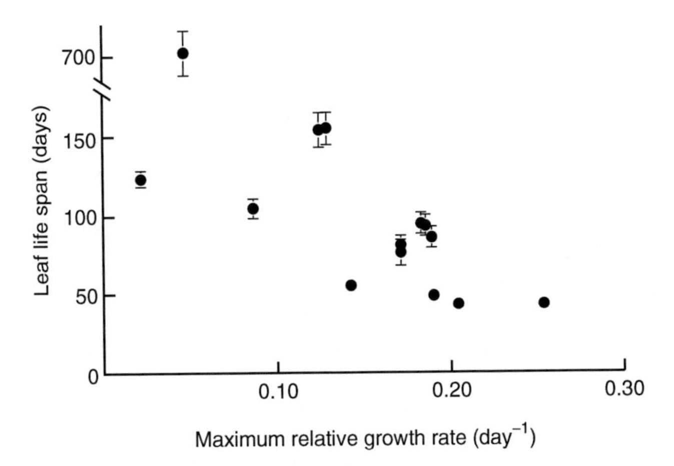

FIGURE 8.6 The average leaf life spans of 14 grassland and heathland species versus their maximum relative growth rate. Life spans were measured as given in the caption of Figure 8.5. Maximum relative growth rates were measured in a growth chamber experiment for seedlings at optimum nutrient supply rates. (From Berendse and Braakhekke, unpublished results.)

significant, negative relationship between leaf life span and maximum RGR which confirms our hypothesis (Figure 8.6). These results show that the differences between *Erica* and *Molinia* reflect a much more general pattern of adaptation of wild plant species to soils with low and high nutrient supplies.

REFERENCES

Aanderud, Z.T., C.S. Bledsoe, and J.H. Richards, 2003. Contribution of relative growth rate to root foraging by annual and perennial grasses from California oak woodlands. Oecologia 136: 424–430.

Aerts, R. and F. Berendse, 1988. The effects of increased nutrient availability on vegetation dynamics in wet heathlands. Vegetatio 76: 63–69.

Aerts, R. and F.S. Chapin III, 2000. The mineral nutrition of wild plants revisited: a re-evaluation of processes and patterns. Advances in Ecological Research 30: 1–62.

Aerts, R., F. Berendse, H. de Caluwe, and M. Schmitz, 1990. Competition in heathland along an experimental gradient of nutrient availability. Oikos 57: 310–318.

Alpert, P. and H.A. Mooney, 1996. Resource heterogeneity generated by shrubs and topography on coastal sand dunes. Vegetatio 122: 83–93.

Bazzaz, F.A., N.R. Chiariello, P.D. Coley, and L.F. Pitelka, 1987. Allocating resources to reproduction and defense. BioScience 37: 58–67.

Berendse, F., 1994. Competition between plant populations at low and high nutrient supply. Oikos 71: 253–260.

Berendse, F. and R. Aerts, 1984. Competition between *Erica tetralix* L. and *Molina caerulea* L. Moench as affected by the availability of nutrients. Acta Oecologica 5: 3–14.

Berendse, F. and R. Aerts, 1987. Nitrogen-use-efficiency: a biological meaningful definition? Functional Ecology 1: 293–296.

Berendse, F. and W.Th. Elberse, 1990. Competition and nutrient availability in heathland and grassland ecosystems. In: J.B. Grace and D. Tilman, eds. Perspectives on Plant Competition. Academic Press, NY, pp. 94–116.

Berendse, F., H. Oudhof, and J. Bol, 1987. A comparative study on nutrient cycling in wet heathland ecosystems. I. Litter production and nutrient losses from the plant. Oecologia 74: 174–184.

Berendse, F., W. Th. Elberse, and R.H.M.E. Geerts, 1992. Competition and nitrogen losses from plants in grassland ecosystems. Ecology 73: 46–53.

Birch, C.P.D. and M.J. Hutchings, 1994. Exploitation of patchily distributed soil resources by the clonal herb *Glechoma hederaceae*. Journal of Ecology 82: 653–664.

Bloom, A.J., F.S. Chapin, and H.A. Mooney, 1985. Resource limitation in plants—an economic analogy. Annual Review of Ecology and Systematics 16: 363–392.

Breeze, V.G., A. Wild, M.J. Hopper, and L.H.P. Jones, 1984. The uptake of phosphate by plants from flowing nutrient solution. II. Growth of *Lolium perenne* L. at constant phosphate concentrations. Journal of Experimental Botany 35: 1210–1221.

Brouwer, R., 1962. Nutritive influences on the distribution of dry matter in the plant. Netherlands Journal of Agricultural Science 10: 399–408.

Campbell, B.D. and J.P. Grime, 1989. A comparative study of plant responsiveness to the duration of episodes of mineral nutrient enrichment. New Phytologist 112: 261–267.

Chapin, F.S., 1980. The mineral nutrition of wild plants. Annual Review of Ecology and Systematics 11: 233–260.

Chapin, F.S., 1989. The costs of tundra plant structures: evaluation of concepts and currencies. American Naturalist 133: 1–19.

Chapin III, F.S., 1995. New cog in the nitrogen cycle. Nature 377: 199–200.

Chapin, F.S. and G.R. Shaver, 1989. Differences in growth and nutrient use among arctic plant growth forms. Functional Ecology 3: 73–80.

Chapin, F.S., K. Van Cleve, and L.L. Tieszen, 1975. Seasonal nutrient dynamics of tundra vegetation at Barrow, Alaska. Arctic and Alpine Research 7: 209–226.

Chapin, F.S., A.J. Bloom, C.B. Field, and R.H. Waring, 1987. Plant responses to multiple environmental factors. BioScience 37: 49–57.

Chapin, F.S., E.D. Schulze, and H.A. Mooney, 1990. The ecology and economics of storage in plants. Annual Review of Ecology and Systematics 21: 423–447.

Chapin III, F.S., L. Moilainen, and Kielland, K. 1993. Preferential use of organic nitrogen by a non-mycorrhizal arctic sedge. Nature 361: 150–153.

Crick, J.C. and J.P. Grime, 1987. Morphological plasticity and mineral nutrient capture in two herbaceous species of contrasted ecology. New Phytologist 107: 403–414.

de Kroon, H. and R. Bobbink, 1997. Clonal plant dominance under elevated nitrogen deposition, with special reference to *Brachypodium pinnatum* in chalk grassland. In: H. de Kroon and J. van Groenendael, eds. The Ecology and Evolution of Clonal Plants. Backhuys Publishers, Leiden, pp. 359–379.

de Kroon, H. and L. Mommer, 2006. Root foraging theory put to the test. Trends in Ecology & Evolution 21: 113–116.

de Kroon, H., L. Mommer, and A. Nishiwaki, 2003. Root competition: towards a mechanistic understanding. In: H. de Kroon and E.J.W. Visser, eds. Root Ecology. Springer, Berlin, pp. 215–234.

de Kroon, H., H. Huber, J.F. Stuefer, and J.M. van Groenendael, 2005. A modular concept of phenotypic plasticity in plants. New Phytologist 166: 73–82.

Drew, M.C., 1975. Comparison of the effects of a localized supply of phosphate, nitrate, ammonium and potassium on the growth of the seminal root system, and the shoot, in barley. New Phytologist 75: 479–490.

Drew, M.C. and L.R. Saker, 1975. Nutrient supply and the growth of the seminal root system in barley. II. Localized, compensatory increases in lateral root growth and rates of nitrate uptake when nitrate supply is restricted to only part of the root system. Journal of Experimental Botany 26: 79–90.

Drew, M.C., L.R. Saker, and T.W. Ashley, 1973. Nutrient supply and the growth of the seminal root system in barley. I. The effect of nitrate concentration on the growth of axies and laterals. Journal of Experimental Botany 24: 1189–1202.

Drew, M.C., L.R. Saker, S.A. Barber, and W. Jenkins, 1984. Changes in the kinetics of phosphate and potassium absorption in nutrient deficient barley roots measured by a solution-depletion technique. Planta 160: 490–499.

Dubach, M. and M.P. Russelle, 1994. Forage legume roots and nodules and their role in nitrogen transfer. Agronomy Journal 86: 259–266.

Eissenstat, D.M. and R.D. Yanai, 1997. The ecology of root lifespan. Advances in Ecological Research 27: 1–60.

Elberse, W.Th. and F. Berendse, 1993. A comparative study of the growth and morphology of eight grass species from habitats with different nutrient availabilities. Functional Ecology 7: 223–229.

Elberse, W. Th., J.P. van den Bergh, and J.G.P. Dirven, 1983. Effects of use and mineral supply on the botanical composition and yield of old grassland on heavy-clay soil. Netherlands Journal of Agricultural Science 31: 63–88.

Epstein, E., 1965. Mineral metabolism. In: J. Bonner and J.E. Varner, eds. Plant Biochemistry. Wiley, New York.

Escudero, A., J.M. Del Acro, I.C. Sanz, and J. Ayala, 1992. Effects of leaf longevity and retranslocation efficiency on the retention time of nutrients in the leaf biomass of different woody species. Oecologia 90: 80–87.

Evans, J.R., 1989. Photosynthesis and nitrogen relationships in leaves of C_3 plants. Oecologia 78: 9–19.

Farley, R.A. and A.H. Fitter, 1999. Temporal and spatial variation in soil resources in a deciduous woodland. Journal of Ecology 87: 688–696.

Farrar, J. and D.L. Jones, 2003. The control of carbon acquisition by and growth of roots. In: H. de Kroon and E.J.W. Visser, eds. Root Ecology. Berlin, Springer, pp. 91–124.

Field, C.B. and H.A. Mooney, 1986. The photosynthesis-nitrogen relationship in wild plants. In: T.J. Givnish, ed. On the Economy of Plant Form and Function. Cambridge University Press, Cambridge, pp. 25–55.

Fisk, M.C. and S.K. Schmidt, 1995. Nitrogen mineralization and microbial biomass nitrogen dynamics in three alpine tundra communities. Soil Scientific Society American 9: 1036–1043.

Fitter, A.H., 1997. Nutrient acquisition. In: M.J. Crawley, ed. Plant Ecology, second edition. Blackwell Scientific Publications, Oxford, pp. 51–72.

Fitter, A.H. and R.K.M. Hay, 2002. Environmental Physiology of Plants, third edition. Academic Press, London.

Fransen, B. and H. de Kroon, 2001. Long-term disadvantages of selective root placement: root proliferation and shoot biomass of two perennial grass species in a 2-year experiment. Journal of Ecology 89: 711–722.

Fransen, B., H. de Kroon, and F. Berendse, 1998. Root morphological plasticity and nutrient acquisition of perennial grass species from habitats of different nutrient availability. Oecologia 115: 351–358.

Fransen, B., H. de Kroon, C. de Kovel, and F. van den Bosch, 1999. Disentangling the effects of selective root placement and inherent growth rate on plant biomass accumulation in heterogeneous environments: a modelling study. Annals of Botany 84: 305–311.

Fransen, B., H. de Kroon, and F. Berendse, 2001. Soil nutrient heterogeneity alters competition between two perennial grass species. Ecology 82: 2534–2546.

Freijsen, A.H.J., S.R. Troelstra, H. Otten, and M.A. van der Meulen, 1989. The relationship between the specific absorption rate and extremely low ambient nitrate concentrations under steady-state conditions. Plant and Soil 117: 121–127.

Gibson, D.J., 1988a. The maintenance of plant and soil heterogeneity in dune grassland. Journal of Ecology 76: 497–508.

Gibson, D.J., 1988b. The relationship between sheep grazing and soil heterogeneity to plant spatial patterns in dune grassland. Journal of Ecology 76: 233–252.

Grime, J.P., 1979. Plant Strategies and Vegetation Processes. Wiley, Chichester.

Grime, J.P., J.C. Crick, and J.E. Rincon, 1986. The ecological significance of plasticity. In: D.H. Jennings and A.J. Trewavas, eds. Plasticity in Plants. Biologists Limited, Cambridge, pp. 5–29.

Hodge, A., 2004. The plastic plant: root responses to heterogeneous supplies of nutrients. New Phytologist 162: 9–24.

Hodge, A., D. Robinson, B.S. Griffiths, and A.H. Fitter, 1999. Why plants bother: root proliferation results in increased nitrogen capture from an organic patch when two grasses compete. Plant Cell and Environment 22: 811–820.

Hoffland, E., 1992. Quantitative evaluation of the role of organic acid exudation in the mobilisation of rock phosphate by rape. Plant and Soil 140: 279–289.

Hook, P.B., I.C. Burke, and W.K. Lauenroth, 1991. Heterogeneity of soil and plant N and C associated with individual plants and openings in North American shortgrass priarie. Plant and Soil 138: 247–256.

Hui-jun, J. and T. Ingestad, 1984. Nutrient requirements and stress response of *Populus simonii* and *Paulownia tomentosa*. Physiologia Plantarum 45: 149–157.
Hutchings, M.J. and H. de Kroon, 1994. Foraging in plants: the role of morphological plasticity in resource acquisition. Advances in Ecological Research 25: 159–238.
Hutchings, M.J. and D.K. Wijesinghe, 1997. Patchy habitats, division of labour and growth dividends in clonal plants. Trends in Ecology and Evolution 12: 390–394.
Ingestad, T., 1979. Nitrogen stress in Birch seedlings II. N, P, Ca, and Mg nutrition. Physiologia Plantarum 45: 149–157.
Ingestad, T. and M. Kähr, 1985. Nutrition and growth of coniferous seedlings at varied relative nitrogen addition rate. Physiologia Plantarum 65: 109–116.
Jackson, R.B. and M.M. Caldwell, 1989. The timing and degree of root proliferation in fertile-soil microsites for three cold-desert perennials. Oecologia 81: 149–153.
Jackson, R.B. and M.M. Caldwell, 1993. Geostatistical patterns of soil heterogeneity around individual perennial plants. Journal of Ecology 81: 683–692.
Jackson, R.B., J.H. Manwaring, and M.M. Caldwell, 1990. Rapid physiological adjustment of roots to localized soil enrichment. Nature 344: 58–60.
Jaeger, C.H. and R.K. Monson, 1992. Adaptive significance of nitrogen storage in *Bistorta bistortoides*, an alpine herb. Oecologia 92: 578–585.
Jansen, C., M.M.L. van Kempen, G.M. Bögemann, T.J. Bouma, and H. de Kroon, 2006. Limited costs of wrong root placement in *Rumex palustris* in heterogeneous soils. New Phytologist 171: 117–126.
Jonasson, S., 1983. Nutrient content and dynamics in north Swedish tundra areas. Holarctic Ecology 6: 295–304.
Jonasson, S. and F.S. Chapin, 1985. Significance of sequential leaf development for nutrient balance of the cotton sedge, *Eriophorum vaginatum* L. Oecologia 67: 511–518.
Jongmans, A.G., N. van Breemen, U. Lundstrom, P.W. van Hees, R.D. Finlay, M. Srinivasan, T. Unestam, R. Giesler, P.A. Melkerud, and M. Olsson, 1997. Rock-eating fungi. Nature 389: 682–683.
Kachi, N. and I.H. Rorison, 1990. Effects of nutrient depletion on growth of *Holcus lanatus* L. and *Festuca ovina* L. and on the ability of their roots to absorb nitrogen at warm and cool temperatures. New Phytologist 115: 531–537.
Kaye, J.P. and S.C. Hart, 1997. Competition for nitrogen between plants and soil microorganisms. Trends in Ecology and Evolution 12: 139–143.
Kembel, S.W. and J.F. Cahill, 2005. The evolution of the plant phenotypic plasticity belowground: a phylogenetic perspective on root foraging trade-offs. American Naturalist 166: 216–230.
Kielland, K., 1994. Amino acid absorption by arctic plants: Implications for plant nutrition and nitrogen cycling. Ecology 75: 2373–2383.
Kruijne, A.A., D.M. de Vries, and H. Mooi, 1967. Bijdrage tot de oecologie van de Nederlandse graslandplanten. Agricultural Research Reports 696: 1–65.
Lambers, H. and H. Poorter, 1992. Inherent variation in growth rate between higher plants: a search for physiological causes and ecological consequences. Advances in Ecological Research 23: 187–261.
Leadley, P.W., J.F. Reynolds, and F.S. Chapin, 1997. A model of nitrogen uptake by *Eriophorum vaginatum* roots in the field: ecological implications. Ecological Monographs 67: 1–22.
Lechowicz, M.J. and G. Bell, 1991. The ecology and genetics of fitness in forest plants. II. Microspatial heterogeneity of the edaphic environment. Journal of Ecology 79: 687–696.
Lefebvre, D.D. and A.D.M. Glass, 1982. Regulation of phosphate influx in barley roots; effects of phosphate deprivation and reduction of influx with provision of orthophosphate. Physiologia Plantarum 54: 199–209.
Lipson, D.A., W.D. Bowman, and R.K. Monson, 1996. Luxury uptake and storage of nitrogen in the rhizomatous alpine herb, *Bistorta bistortoides*. Ecology 77: 1277–1285.
Lipson, D.A., T.K. Raab, S.K. Schmidt, and R.K. Monson, 1999. Variation in competitive abilities of plants and microbes for specific amino acids. Biology and Fertility of Soils 29: 257–261.
Loneragan, J.F. and C.J. Asher, 1967. Response of plants to phosphate concentration in solution culture II. Role of phosphate absorption and its relation to growth. Soil Science 103: 311–318.

Marschner, H., 1995. Mineral Nutrition of Higher Plants, second edition. Academic Press, London.
Morton, A.J., 1977. Mineral nutrient pathways in a Molinietum in autumn and winter. Journal of Ecology 65: 993–999.
Nambiar, E.K.S., 1986. Do nutrients retranslocate from fine roots? Canadian Journal of Forest Research 17: 913–918.
Näsholm, T., Ekblad, A., Nordin, A., Giesler, R., Högberg, M., and Högberg, P., 1998. Boreal forest plants take up organic nitrogen. Nature 392: 914–916.
Northup, R.R., Z. Yu, R.A. Dahlgren, and K.A. Vogt, 1995. Polyphenol control of nitrogen release from pine litter. Nature 377: 227–229.
Nye, P.H. and P.B. Tinker, 1977. Solute Movements in the Root–Soil System. Blackwell, Oxford.
Pastor, J., J.D. Aber, and C.A. McClaugherty, 1984. Above ground production and N and P cycling along a nitrogen mineralization gradient on Blackhawk Island, Wisconsin. Ecology 65: 256–268.
Persson J., P. Högberg, A. Ekblad, M.N. Högberg, A. Nordgren, and T. Näsholm, 2003. Nitrogen acquisition from inorganic and organic sources by boreal forest plants in the field. Oecologia 137: 252–257.
Peters, M., 1990. Nutzungseinfluss auf die Stoffdynamik schleswig-holsteinischer Böden—Wasser-, Luft-, Nähr- und Schadstoffdynamik. In: H.P. Blume, ed. Schriftenreihe des Institutes für Pflanzenernährung und Bodenkunde. Universität Kiel, Kiel.
Poorter, H., 1994. Construction costs and payback time of biomass: a whole plant perspective. In: J. Roy and E. Garnier, eds. A Whole Plant Perspective on Carbon–Nitrogen Interactions. SPB Academic Publishing, the Hague, pp. 111–127.
Read, D.J., 1991. Mycorrhizas in ecosystems. Experientia 47: 376–391.
Reich, P.B., M.B. Walters, and D.S. Ellsworth, 1992. Leaf life-span in relation to leaf, plant, and stand characteristics among diverse ecosystems. Ecological Monographs 62: 365–392.
Reynolds, H.L. and C. D'Antonio, 1996. The ecological significance of plasticity in root weight ratio in response to nitrogen. Plant and Soil 185: 75–97.
Robinson, D., 1994. The responses of plants to non-uniform supplies of nutrients. New Phytologist 127: 635–674.
Robinson, D., 1996. Resource capture by localized root proliferation: Why do plants bother? Annals of Botany 77: 179–185.
Römheld, V. and H. Marschner, 1986. Mobilisation of iron in the rhizosphere of different plant species. In: P.B. Tinker and A. Läuchli, eds. Advances in Plant Nutrition. Vol. 2. Praeger Scientific, New York, pp. 155–204.
Ryel, R.J., M.M. Caldwell, and J.H. Manwaring, 1996. Temporal dynamics of soil spatial heterogeneity in sagebrush-wheatgrass steppe during a growing season. Plant and Soil 184: 299–309.
Schimel, J.P. and F.S. Chapin, 1996. Tundra plant uptake of amino acid and NH_4^+ nitrogen in situ: plants compete well for amino acid N. Ecology 77: 2142–2147.
Schlesinger, W.H., J.A. Raikes, A.E. Hartley, and A.E. Cross, 1996. On the spatial pattern of soil nutrients in desert ecosystems. Ecology 77: 364–374.
Schmidt, S. and G.R. Stewart, 1997. Waterlogging and fire impact on nitrogen availability and utilization in a subtropical wet heathland (wallum). Plant, Cell and Environment 20: 1231–1241.
Silberbush, M. and S.A. Barber, 1984. Phosphorus and potassium uptake of field grown soybean cultivars predicted by a simulation model. Soil Science Society of America Journal 48: 592–596.
Tilman, G.D., 1984. Plant dominance along an experimental nutrient gradient. Ecology 65: 1445–1453.
Tilman, D., 1988. Plant Strategies and the Dynamics and Structure of Plant Communities. Princeton University Press, Princeton.
van Vuuren, M.M.I., D. Robinson, and B.S. Griffiths, 1996. Nutrient inflow and root proliferation during the exploitation of a temporally and spatially discrete source of nitrogen in soil. Plant and Soil 178: 185–192.
Wijesinghe, D.K. and M.J. Hutchings, 1997. The effects of spatial scale of environmental heterogeneity on the growth of a clonal plant: an experimental study with *Glechoma hederacea*. Journal of Ecology 85: 17–28.

9 Functional Attributes in Mediterranean-Type Ecosystems

Richard Joffre, Serge Rambal, and Claire Damesin

CONTENTS

INTRODUCTION

Mediterranean-type environments share unique climatic pattern with a cool wet winter and a hot dry summer (Köppen 1900), constituting the core of what we call the Mediterranean climate. These climatic conditions can be coarsely considered as a transition between dry tropical and temperate climates, with an occurrence of a distinct summer drought. These conditions occur on the west coasts of all continents between latitudes 30° and 45°, because of very general reasons of air circulation as the dry subtropical high pressure cells move poleward in the warm season blocking the entry of midlatitude storm. One of the main corollary of this approach is to clearly define Mediterranean regions as zonal ones and to exclude limited geographical definitions considering only the Mediterranean Sea (Joffre and Rambal 2002). It is important to note that the Mediterranean climate is very recent in geological terms and first appeared approximately 3.2 million years ago during the Pliocene. Mediterranean climates have attained their greatest extent at present.

Since the beginning of the century, numerous studies have highlighted structural and functional affinities of vegetation communities in Mediterranean-type climate regions throughout the world, that is, the Mediterranean Basin itself, California, central Chile, the Cape region of south Africa, and parts of southwestern and southern Australia (see Specht 1969a for a summary of early works of Schimper 1903 and Warming 1909). In the 1970s, comparisons of physiological and structural properties of Mediterranean-type ecosystem (MTE) (Naveh 1967, Specht 1969a,b, Mooney and Dunn 1970) have shown that, in some cases, similarities of these plants extend to patterns of growth, morphology, and physiology. In contrast to similarities that explained invoking an evolutionary convergence driven by climate and perturbation (periodic fire), Herrera (1992) introduced a critical distinction in the analyses of Mediterranean flora between taxa present before the Pliocene versus those that have immigrated into the regions since the onset of the Mediterranean-type climate and showed that the two groups exhibit distinctive traits, the former dominated by sclerophylly, large vertebrate-dispersed fruits and seeds, and the latter by nonsclerophyllous leaves, anemochory, dry fruits, and small seeds. Analyzing convergent traits of Mediterranean woody plants belonging to pre-Mediterranean lineages, Verdú et al. (2003) supported the view that common morphological and life-history traits are due to phylogenetical inertia and cannot be merely interpreted as a pure consequence of adaptive processes.

Understanding the diversity of physiological and functional strategies among co-occurring species as evergreen and deciduous in MTE could be done through identifying major axes of ecological and life-history strategies (Ackerly 2004a,b) and the underlying adaptive trade-offs between the traits (Westoby et al. 2002). In MTE, perennial species have to cope with the high temporal variability of resources and perturbations. Studying the trade-offs between phenology, relative growth rate, life form, and seed mass among Mediterranean woody species, Castro-Diez et al. (2003) pointed out that Mediterranean woody plants exhibit a wide phenological diversity that cannot be explained just on the basis of climatic constraints. They suggested that duration of primary shoot growth allow to sort out the species between two extreme growth strategies: a conservative strategy characterized by a concentration of the primary shoot growth into a short period, free of frosts and droughts, and an opportunistic strategy defined by the allocation of resources to current growth whenever they are available. These strategies should have been selected for in environments of predictable and unpredictable resource availability, respectively.

The aim of this chapter is to discuss some aspects of the functioning of perennial woody species by (1) briefly reviewing the main features of MTE resources, (2) illustrating how individual and ecosystem cope with variability in water resource and control water loss, (3) presenting the assimilation characteristics in relation to constraints, and (4) evaluating the role of nutrients at plant and community levels.

MTEs CHARACTERISTICS

Human occupation as well as past and present land use differ strongly between the five MTEs regions of the world. This obvious fact has evident and important consequences on biodiversity, landscape structure, and ecosystem functioning and have been extensively reviewed recently (Blondel and Aronson 1995, Hobbs et al. 1995, Davis et al. 1996). For instance, the high resistance of the Mediterranean Basin ecosystems to invaders could be a result of the long history of close interactions between humans and ecosystems (Butzer 2005, Thompson 2005, Blondel 2006). Probably no other part of the world than the Mediterranean Basin has played a more fundamental role in the history of mankind with the development of numerous civilizations. Their legacy concerns the transformation and shaping of primitive ecosystems to the present man-made landscapes but also cultural attitudes such as our present position and perception vis-à-vis nature. Agriculture and animal husbandry began more than 10,000 BP in the Eastern Mediterranean, and around 8000 BP in Greece and western Mediterranean. It is important to highlight in agreement with Butzer (2003) and Blondel (2006) that the long-lasting management of Mediterranean ecosystems has not always resulted in a decrease in biodiversity but has been in fact beneficial for many components of biological diversity. In reshaping natural forest toward more diversified landscapes, Mediterranean civilizations have created in many regions land-use models whose aim was to achieve sustainable long-term ecosystem management (Joffre et al. 1999, Blondel 2006).

Climate Definition and Variability

The most distinctive feature of the Mediterranean climate involves the seasonality in air temperature and precipitation that lead to a hot drought period in summer and a cool wet period in winter (Köppen 1931, Bagnouls and Gaussen 1953, Aschmann 1973). This peculiarity of the Mediterranean climate has important implications for vegetation functioning and limits the most favorable season for growth to brief periods in spring and fall. As quoted by Rundel (1995), "dry summer conditions limit water availability and thus growth, while cool winter conditions limit growth during the season when water availability is generally highest." Based on the total amount of annual precipitation and a combination of averaged maximum temperature of the hottest month and the averaged minimum temperature for the coldest month, the Emberger pluviometric quotient allows a classification of climates from arid to super-humid and from cold to hot (Daget 1977, Le Houérou 1990).

From an ecological point of view, the variability or unpredictability of precipitation imposes strong constraints on plants that could be more important for the survival of individuals that the mean values. An extreme event is here defined as an event where a climate variable had a low relative frequency of occurrence or was lower than (or exceeded) a given critical threshold. This concern arises naturally, since the impacts of climate are realized largely through the incidence of variation about normal conditions or extreme events (Wigley 1985). Examining annual precipitation at 360 stations scattered over the earth, Conrad (1941) long ago found a relationship between interannual variability and mean precipitation. For 73 stations located in a 100 km × 100 km area around Montpellier, Rambal and Debussche (1995) found a linear relationship ($r = 0.87$, $P < 0.01$) between mean and standard deviation. Table 9.1 presents some estimations of the coefficient of variation for a 1000 mm mean. The values range between 0.16 and 0.29. The lowest values (0.16–0.18) are proposed by Le Houérou (1992) for a large set of locations in Africa and in the near East. Conrad (1941), identifying variability more than the rule or positive anomalies in some dry areas, showed in his paper that positive anomalies of 3%, 7%, and 10% were observed in Marseille (Southern France), San Diego (California, USA), and Alicante (Southern Spain), respectively. These findings agreed with those of Waggoner (1989) who observed

TABLE 9.1
Coefficient of Variations for Mean Annual Rainfall Amounts

References	Source	Number of Stations	Coefficient of Variation
Conrad (1941)	World	360	0.21
Waggoner (1989)	USA	55	0.20
Le Houérou (1992)	North Africa, Near East	407	0.18
	Sahel, Soudan	228	0.17
	East Africa	300	0.16
Rambal and Debussche (1995)	Languedoc (France)	73	0.29

that "frequency distributions with much larger variance than expected are in the Mediterranean climates" of California.

Identifying the responsiveness of several regions of the world to climate change, Giorgi and Bi (2005) and Giorgi (2006) showed that the Mediterranean Basin emerges among the primary climate change hot spots in Northern hemisphere. According to projections by 20 global climate models, a very pronounced increase in variability of summer temperature and precipitation as well as a strong decrease of summer precipitations (−22%) characterize the Mediterranean Basin climate at the end of twenty-first century.

Substrate

Different parent rocks and geological histories of the five Mediterranean climate regions have given rise to very distinct soils and fertility levels. Table 9.2 summarizes the more frequent lithological substratum in each region and the associated nutrient status dependent on leaching and water-driven erosion. We modify the original table first published by Groves et al. (1983) to take into account the presence of siliceous parent rock in large areas of the Mediterranean Basin (part of Iberian Peninsula, Corsica, part of Provence). South Africa and southern Australia comprise older landscapes, consisting of an inland mass of geologically old material, surrounded by discontinuous strip of more recent marine deposits (Hobbs et al. 1995). Soils of the upland of the Cape Mountains as well as southwestern Australia have been exposed to weathering since Paleozoic. In contrast, soils of the Mediterranean Basin, California, and Chile are much younger, and reflect Tertiary and Quaternary orogenic events. Analyzing mineral nutrient relations among MTE, Lamont (1994) concluded that "among Mediterranean regions, the soils in Chile are more fertile than those in California, which in turn are more fertile than soils in south-eastern Australia, which in turn are more fertile than those of southwestern Australia." It has been generally assessed that nutrient availability of soils in Mediterranean Basin and Chile is equivalent to that of South Africa and Australia (Groves et al. 1983).

Vegetation Types

In the five MT regions, evergreen species are more abundant than deciduous ones and the vegetation formations are characterized by the dominance of trees and woody shrubs with small, sclerophyllous leaves. Nevertheless, although the woody shrub sclerophyllous growth-form is the dominant element, they never represent the majority of the total floras. The main vegetation formations are usually called garrigue or maquis in France depending on the nature of the soil substrate calcareous or siliceous, chaparral in California, heath and mallee in Australia, matorral in Chile, and fynbos in South Africa. These formations are fairly similar in their physiognomy and have provided the ideal testing ground for the theory of ecological convergence from evolutionary, morphological, and physiological points of view.

TABLE 9.2
Main Lithological Substratum and Associated Nutrient Status from Soils of Mediterranean Climate Regions

Substratum	Nutrient Status	South Australia	Southwest Australia	South Africa	California	Chile	Mediterranean Basin
Siliceous rocks	Strongly leached nutrient-poor soils	•••	•••	•••	•	—	••
Argillaceous rocks	Strongly leached nutrient-poor soils and moderately leached nutrient-rich soils	••	Trace	••	•••	•••	Trace
Calcareous rocks	Moderately leached nutrient-rich soils and shallow, high pH soils	•••	Trace	•	Trace	—	•••
Ultramafic rocks (serpentines)	Mg-rich, Ca-poor soils	—	—	—	•	—	Trace

Source: Data from Groves, R.H., Beard, J.S., Deacon, H.J., Lambrechts, J.J.N., Rabinovitch-Vin, A., Specht, R.L., and Stock, W.D. in *Mineral Nutrients in Mediterranean Ecosystems*, J.A. Day, ed., Council for Scientific and Industrial Research, Pretoria, South Africa, 1983, 1–17. With permission.

•••, common.
••, frequent.
•, present but not widespread.
Trace, found in restricted localities.
—, not recorded.

Physiologically based models were presented to explain the distribution of broad-leaved evergreen species, and a number of ecological paradigms have been developed explaining the nature of plant adaptation to MT climates (for a comprehensive account see Keeley 1989). Nevertheless, strict climatic control of sclerophyll leaf morphologies is questioned by the presence of sclerophyllous vegetation types in non-Mediterranean environments, for example, chaparral in Arizona, heathland communities in eastern Australia (Specht 1979), fynbos-like vegetation in the Afromontane region of Africa (Killick 1979), and tropical zone of Mexico (Verdú et al. 2003). Moreover, the majority of ecophysiological works dealing with drought resistance of plants in Mediterranean ambient have been conducted on genera existing before the Pliocene, that is, the establishment of true Mediterranean climate conditions (e.g., *Ceratonia*, *Hedera*, *Olea*, *Phillyrea*, *Pinus*, *Quercus*, *Vitis*). It is therefore difficult to interpret the physiological and anatomical properties of these species as adaptive responses to Mediterranean ambient. The character syndromes in Mediterranean plants may largely be explained in relation to the age of the lineage (Tertiary pre-Mediterranean vs. Quaternary true Mediterranean) and thus do not represent convergent evolution. The role of nutrients in species and ecosystem convergence was questioned in the 1980s (Kruger et al. 1983) based on observations of the contrasting conditions between the oligotrophic soil conditions of South Africa and Western and South Australia, and the relatively richer soil conditions of the other three MTEs. The role of fire in convergence of traits was recently questioned by Lloret et al. (2005), Pausas and Verdú (2005), Keeley et al. (2006) and Pausas et al. (2006).

INDIVIDUAL AND ECOSYSTEM RESPONSES TO VARIABILITY IN WATER RESOURCE

Among the numerous mechanisms for drought resistance, Mediterranean plant species have three responses that act together to dampen the effects of variability in water resource (Figure 9.1). Change in the leaf area allows the plants to cope with low-frequency oscillations such as a decrease in the annual rainfall amount. The response of the root system dampens the medium-sized oscillations, for example, changes in seasonal distribution of a given annual rainfall distribution. Finally, the stomatal activity allows quasioptimization of water use at a daily timescale. Any change in the leaf area induces change in the root system. The new

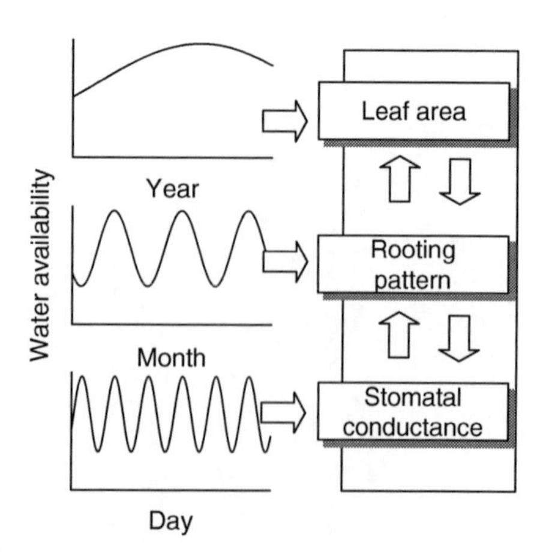

FIGURE 9.1 Conceptual diagram in which we proposed to associate the main mechanisms of drought resistance for Mediterranean plant species to the timescale of variations of the water availability from day to year. Vertical up and down arrows suggest that the mechanisms do not act independently.

functional equilibrium of the plant is controlled by carbohydrates and nitrogen or is hormonally mediated. Any change in the soil water deficit produces a change in the stomatal closure. The plant water status or a counterbalancing effect of phytohormones can act as indicators of the root stress. So, each mechanism is linked with the one immediately preceding it, providing an integrated strategy of the plant to improve its water balance under any given set of conditions (Rambal 1993, 1995).

WATER UPTAKE

Rooting Depth

Studies on the root distribution of several Californian chaparral shrub species established that deep root systems are characteristic for the dominant component of the Mediterranean-type vegetation (Hellmers et al. 1955). However, when examined in detail, these species have been shown to exhibit a great diversity in rooting depth (Kummerow et al. 1977) and consequently, in response to drought stress (Poole and Miller 1975). Based on these results, Mediterranean plant species can be roughly divided into two categories: deep-rooted species with root depth greater than 2 m, and shallow-rooted species with rooting depths less than 2 m. Mediterranean oaks are among the deepest-rooted plant species. For California oaks, Stone and Kalisz (1991) and Canadell et al. (1996) reported roots deeper than 8.5 m for the evergreen *Quercus agrifolia* (10.7 m), *Q. dumosa* (8.5 m), *Q. turbinella* (>9 m), *Q. wislizenii* (24.2 m), and the deciduous *Q. douglasii* (24.2 m). Similar or higher values have been estimated for species of the Chilean matorral (*Lithrea caustica* 5 m, *Quillaja saponaria* 8 m) and of the chaparral (*Adenostoma fasciculatum* 7.6 m). Lower values are found for dominant species of the Mediterranean Basin (*Pinus halepensis* 4.5 m, *Arbutus unedo* 3.5 m) or of the Australian mallee (*Casuarina* spp. >2.4 m, *Banksia* spp. 5 m). However, rooting depth is not the only functional trait helping the plant to access soil water. Root cover that is the maximum horizontal area colonized by the roots appears to be largely greater than the plant cover. This trait is particularly crucial in interpreting tree densities in savanna-like ecosystems, such as the ones we observed in dehesas of the Iberia peninsula (Moreno et al. 2005).

Soil Water Uptake Patterns

Deep root system as those presented earlier easily coped with low water availability. Rambal (1984) distinguished in *Q. coccifera* four patterns of water uptake throughout a drying cycle of 3 months (Figure 9.2). Late spring, water loss occurred exclusively from the top 0–50 soil layer, which lost about 4% of store water per day. The upper meter supplied three-quarter of the total. Early summer, root water uptake decreased in the upper layer, which then presents a daily loss of 1.2% of its reserve. Peak of water uptake was between 2 and 2.5 m depth. This layer supplied 0.62 mm day^{-1} of 2.64 mm day^{-1} of transpired water. Late summer was characterized by unevenness of water loss. All the upper layers were depleted and only the lower layers are able to supply water. During these two late periods, the deepest soil layers were contributing as much water as the top. Early fall, at the end of the dry period, all the layers were depleted. The flat profile did not allow the uptake of water at a rate greater than 0.62 mm day^{-1}. During the first three stages, the daily rates of transpiration were 2.84, 2.64, and 2.35 mm, respectively. Talsma and Gardner (1986) for various *Eucalyptus* observed the same patterning.

Rooting Patterns

In wet conditions, the major resistance to water uptake appears to be inside the root. Consequently, there is a good correlation between water uptake and rooting density.

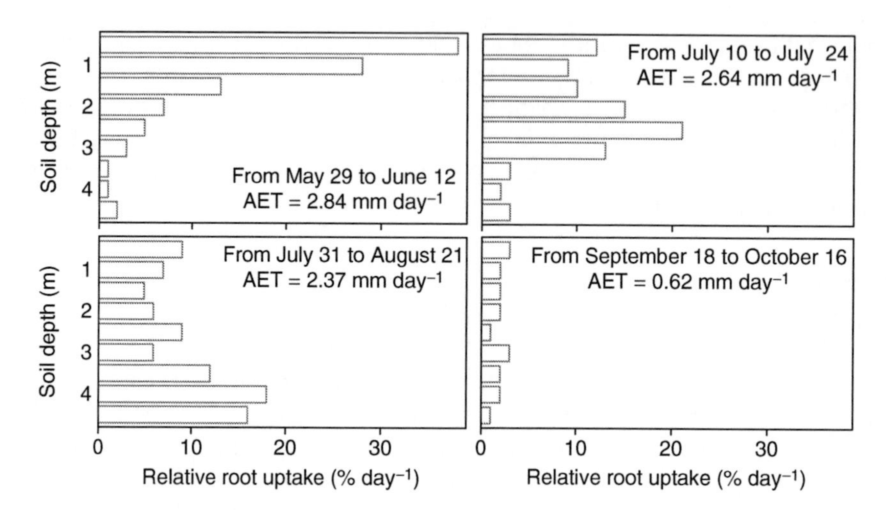

FIGURE 9.2 Patterns of water uptake with soil depth within four periods across a severe summer drought. For each 50 cm thick soil layer, the water uptake is expressed in percent of the extractable water used daily.

Hence, the late spring profile of water uptake we observed with *Q. coccifera* can be considered as a picture of its root density profile. The greatest accumulation of root mass was in the top meter. Below 1 m root mass decreased gradually with depth. The theoretical profile proposed by Jackson et al. (1996) for the sclerophyllous shrub group does not take into account the deep-rooted species we identified earlier. This sclerophyllous shrub group values included values from some chaparral, fynbos, heath, garrigue, and matorral stands. The root distribution with depth is described by the equation of Gale and Grigal (1987) $y = 1 - \beta^z$, where y is the cumulative root fraction from the soil surface to depth z (cm) and β is an extinction coefficient. For this largely Mediterranean group $\beta = 0.964$. This means that 67% of the root biomass are in the first 30 cm of soil and 90% in the 0–60 cm layer. These surprisingly high values are far to our estimates for *Q. coccifera* (48% for the 0–50 cm horizon, Rambal 1984) or values for *Q. turbinella* (53% for the 0–60 cm horizon). In the two species, the percent of roots deeper than 2 m were important (20% and 12%, respectively). Deep roots play a very important role during the summer period. As the soil dries, the layers below 2 m contribute an increasing fraction of the total water uptake, reaching between 12% and 23% of the total uptake depending of the severity of the drought (Rambal 1984).

Water Loss

Stomatal Regulation

Together with rooting pattern and leaf area, stomatal activity helps regulate water loss. According to their phenology, species may avoid or cope with water constraint. For instance, drought-deciduous species of the coastal sage community of California are drought evaders and chaparral species drought tolerators. Nevertheless, as stressed by Mooney (1989) "some coastal sage species that lose most of their leaves during the drought have been shown to tolerate very high water stress, whereas others somehow control stress while maintaining many of their leaves." Phenological variations among individuals could also be very high. Ne'eman (1993) found in *Q. ithaburensis*, a deciduous oak of Israel, that some trees were clearly deciduous, while others had only a short duration of leaflessness and, as a consequence, could be considered as evergreen. By analogy, there appears to be a continuum of stomatal behavior in response to water stress: no stomata closure, progressive closure of stomata (see Acherar et al. 1991 for Mediterranean oaks) and threshold effect leading to an early complete closure (see Aussenac and Valette 1982 for *Pinus* sp pl.). With a rapid rate of drying, for example, with a shallow-rooted system, an early response of stomata closure was observed in some cases whereas a delayed response was observed when a slower rate of drying, for example, with a deep-rooted system, was imposed. Studies such as those of Poole and Miller (1975, 1981) showed that some chaparral species are more sensitive to water stress than others, as indicated by their water potential after stomata closure. These authors assume that the degree of drought tolerance is linked to rooting depth. Thus, shallow-rooted species tend to close their stomata at high water potential and to bear tissues with the greatest drought tolerance.

Stomatal closing is considered to be under the control of leaf turgor pressure or leaf water potential. Recently, however, some authors showed that leaf conductance is not always closely coupled with leaf water potential or leaf turgor pressure. There is now some evidence that soil water deficit can also induce stomatal closure even when the leaf water status remains unchanged. These results suggest that leaf conductance is not only affected by leaf water potential, but also, more directly, by soil or root water potential. One much-discussed possibility is that the counterbalancing effects of two phytohormones, cytokinin and abscissic acid, might provide information on the water status of the roots and induce stomatal closing or opening, as the case may be (Zhang et al. 1987). But, for modeling purposes, the leaf water status still remains the main control variable of the leaf stomatal conductance (Tardieu and Davies 1993).

Daily curves of stomatal conductance in MTE species can be coarsely classified according to three patterns following Hinckley et al. (1983) (Figure 9.3). Type 1 curves are bell-shaped and represent situations in which soil water potential, leaf water potential, and vapor pressure deficit do not limit stomatal conductance. Type 2 curves have two maxima, one at the start of the day and the other in the afternoon, both separated by a depression at midday. They correspond to situations where one or more of the previously mentioned factors limit stomatal conductance. In type 3, the curves have a high point only at the start of the day as

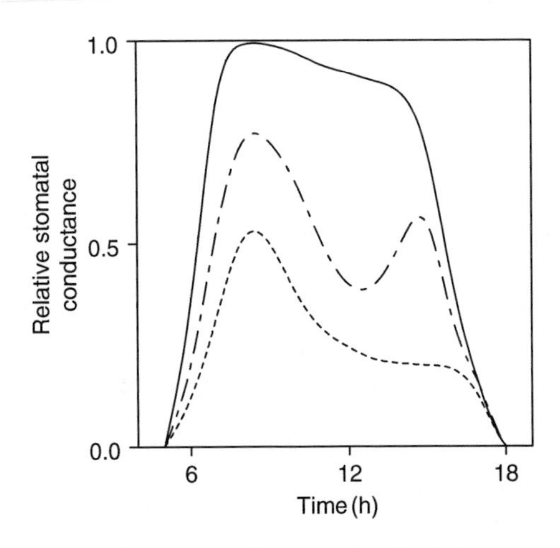

FIGURE 9.3 Typical patterns describing the daily courses of relative stomatal conductance during a drought period. Type 1 curve is bell-shaped and represents optimal situations in which soil water potential, leaf water potential, and vapor pressure deficit do not constraint transpiration. Curve of type 2 has two maximums, one early in the morning and the other in midafternoon, both separated by the so-called midday depression. They correspond to situations where water status and air dryness limit stomatal conductance. In type 3, the curve shows only a maximum early in the morning. Leaf water potential is at or below the turgor loss point for most part of the daylight period.

the leaf water potential of the plants, according to Hinckley, at or below the turgor loss point for part of the day. The midday stomatal closure was reported by Tenhunen et al. (1987) for many Mediterranean species. This drop in stomatal conductance has been interpreted as a feature, which allows Mediterranean species to limit water loss when the atmospheric evaporation is at its maximum. These authors suggested that midday stomatal closure is determined by the leaf-to-air vapor pressure deficit, whereas for Hinckley et al. (1983), this depends on the interaction among several factors, in particular the instantaneous water potential of the plant. Cowan and Farquhar (1977) showed theoretically how stomatal activity helps optimize water use on a daily timescale. This optimization pattern has been validated for few Mediterranean species (William 1983, see also Xu and Baldocchi 2003).

Leaf Area Index

The importance of adjustment in leaf area is emphasized by Passioura (1976): "It is the control of leaf area index and morphology which is often the most powerful means a mesophytic plant has for influencing its fate when subject to long term water stress in the field." Over a large range of climates, changes in leaf area indices have been studied at both individual and ecosystem scale along gradients from higher to lower rainfall amounts or from more moisture to drier habitats in broad- or needle-leaf tree or shrub communities (Ladiges and Ashton 1974, Specht and Specht 1989). In Southern Australia, *Eucalyptus viminalis* trees occur over a wide range of rainfall conditions and soil types. Ladiges and Ashton (1974) observed that at moist sites mature trees are tall and produce large leaves and, at drier sites, trees are shorter and tend to produce smaller leaves. Poole and Miller (1981) hold a similar view for Mediterranean shrub species of the California chaparral: "the main response of the shrubs to different precipitation regimes in the chaparral range is to change leaf-area index, not physiological parameters." The importance of this adjustment is largely species-dependent. The ranges of leaf-area index of some mature frequent MTE are summarized in Table 9.3. Understanding how the leaf-area index of a site comes into a predictable, dynamic equilibrium with the amount of water available is the target of researches in ecohydrology (Eagleson 2002, Eamus et al. 2006).

Water Transfer

Soil–Plant Resistance to Water Flow

A simple application of Ohm's Law analogy relating water potential difference from soil-to-leaf ($\Delta\Psi$) and transpiration rate (T) has been widely used to estimate total flow resistance (R)

TABLE 9.3
Leaf Area Index Ranges of Mature Stands of Some Frequent Vegetation Types

Vegetation Type	Dominant Species	LAI Range	References
Woodland	*Quercus ilex* (evergreen)	2.9–6.0	Damesin et al. (1998a)
Woodland	*Quercus pubescens* (deciduous)	2.0–4.2	Damesin et al. (1998a)
Shrubland	*Quercus coccifera* (evergreen)	1.5–4.0	Rambal and Leterme (1987)
Chaparral	*Adenostoma fasciculatum*	2.2–3.4	Rambal (2001)
	Ceanothus megacarpus	1.5–1.6	Rambal (2001)
Warm-temperate mallee	*Eucalyptus* spp.	1.5–6.0	Specht and Specht (1989)
Heathland south Australia	*Eucalyptus* spp.	2.5–4.0	Rambal (2001)

LAI, leaf area index.

expressed on a leaf-area basis: $\Delta\Psi = T\,R$ assuming firstly, little capacitance effect in the plant, and secondly, steady-state transpiration conditions. The pathway of water movement in soil and plant can be considered as comprising two main resistances in series, the soil-to-root resistance and the plant resistance. Partitioning soil and plant resistance is difficult. Generally, the plant resistance was assumed to be constant within a range of Ψ and so, the relative contribution of the soil resistance was estimated. For this, we applied Gardner (1964) who had showed that soil resistance is inversely proportional to the hydraulic conductance of the soil. As a consequence, the soil resistance is small at high water content and any observed difference in resistance should be largely attributable to differences in plant resistance. Under conditions of maximal transpiration (T_{max}) in well-watered conditions: $\Delta\Psi_{max} = T_{max}\,R_{min}$. Studying a large range of evergreen oak communities, we deduced a hierarchy of soil-to-leaf resistance from the highest R_{min} in xeric sites to the lowest in mesic sites (Rambal 1992). Thus, for *Q. ilex* alone the ratio of xeric/mesic R_{min} is 1.7. The presence and magnitude of differences in resistance suggest that this attribute could be an important component in drought tolerance. In the same way, Rambal and Leterme (1987) associated a decrease of leaf area index from 2.5 to 1.5 in the Mediterranean evergreen oak *Q. coccifera* growing across a rainfall gradient with changes in canopy structure and plant resistance. The role of the hydraulic resistance in the relative sensing of soil water deficit by roots has been emphasized. At a given rate of transpiration and soil water deficit, a plant with high hydraulic resistance lowers its leaf water potential to a greater degree than a plant with low resistance. This plant may further be more sensitive to maintain its rate of photosynthesis and growth. On the other hand, with a limited volume of water in the soil, an increase in hydraulic resistance saves water during the wetter periods for use during the drier ones.

Patterns of Changes in $\Delta\Psi$ with Increasing Water Stress

Richter (1976) observed that plant species "from sites with pronounced drought periods" did not undergo Ψ lower than that of desert plants. His analysis, for the first mentioned group, is largely based on works of Duhme (1974) conducted on 26 species of MTEs. In this study, Duhme measured Ψ of −4.4 MPa for *Q. coccifera*, a similar value of those we reported in a synthesis on Mediterranean evergreen oaks (Rambal and Debussche 1995). In this synthesis, whatever the study site and the amount of rain fallen during the measurement periods, minimum and predawn leaf water potentials were always higher than −4.4 and −3.8 MPa. Other examples well illustrated this assumption of lower bound of water stress even during a very dry year. Griffin (1973) observed at the end of the driest in 32 year period predawn potentials of the evergreen *Q. agrifolia*, in the more xeric location, between −2.5 and −3.1 MPa. These potentials remained also limited with the deciduous *Q. douglasii* and *Q. lobata*, −3.7 and −2.0 MPa, respectively (see also Damesin and Rambal 1995 for *Q. pubescens* values).

Nevertheless, the trajectories followed by minimum and predawn leaf water potentials to reach their limits were very different according to locations and species. This was particularly true for *Q. ilex* (Figure 9.4). $\Delta\Psi$ decreased to zero with decreasing soil water availability and predawn potential. As proposed by Ritchie and Hinckley (1975) "it is tempting to compare species based on these curves (....) as indicators of species differences." Waring and Cleary (1967) on Douglas fir first observed this pattern. But it was initially considered as marginal. Indeed, Hickman (1970), from measurements done on 44 species, concluded that the opposite pattern in which $\Delta\Psi$ increases with the water stress is the most common pattern. It corresponds to species characterized as conformers. Species with the same pattern as our observations was named regulators. Hickman (1970) suggested "this pattern is probably typical of most plant species in areas with modified (?) Mediterranean climates." It was also described by Aussenac and Valette (1982) for some trees (*Cedrus atlantica*, *Pinus* sp pl., *Q. pubescens*, and *Q. ilex*) and for the shrub *Buxus sempervirens*.

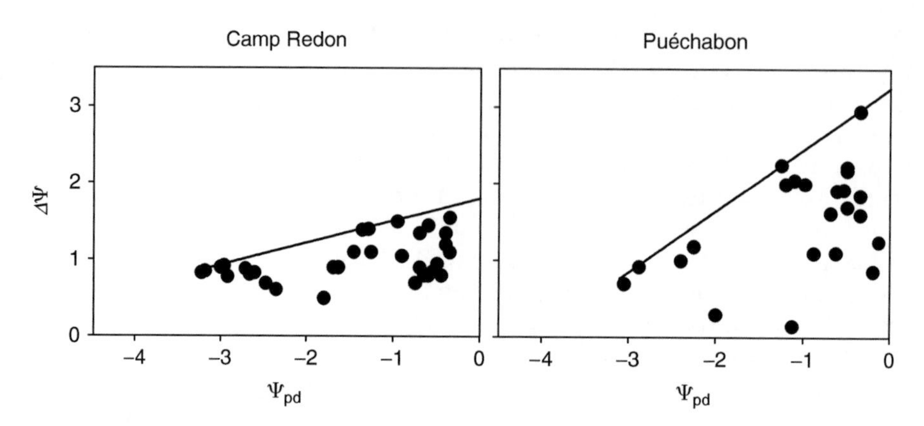

FIGURE 9.4 Scatter plots relating predawn leaf water potential Ψ_{pd}, MPa, and $\Delta\Psi$, that is predawn minus minimum potential from the same day, MPa. The boundary solid line gives information on how $\Delta\Psi$ declines with the drought. The data from Camp Redon (southern France) correspond to data from mature *Quercus ilex* trees growing in mesic conditions on a deep soil; those from Puéchabon (southern France) have been obtained on the same oak species growing in dry xeric conditions on karstic soil. In this location, *Q. ilex* functions as an isohydric plant (slope of $\Delta\Psi$ vs. $\Psi_{pd} \approx 1$) that can maintain nearly constant leaf water potential throughout the year despite changes in soil conditions.

An important question is asked by Reich and Hinckley (1989): "Does soil-to-leaf hydraulic conductance decrease with decreasing soil moisture due only to increased resistance in the soil, or is there a plant component as well?" (see also Bucci et al. 2005 for a substantial account). Changes in plant resistance under water limitation are attributable to effects on both roots and stems. As the soil dries, decreased permeability by root suberization and increased fine root mortality can reduce the balance between extraction capacity and transpiring leaf area. Xylem loss of vascular transport by cavitation might also cause an increase in plant resistance. There was evidence of embolism formation in Temperate and Mediterranean *Quercus*, one of the most common genera with ring-porous xylem anatomy (Cochard and Tyree 1990). For 1 year old twig segment of *Q. ilex*, the loss of conductivity began at −1.8 MPa and linearly increased to reach a total xylem cavitation at −4.35 MPa (Lo Gullo and Salleo 1993). When catastrophic xylem dysfunction occurs, Tyree and Sperry (1988) showed that minor branches begin to die, leading to a loss of leaf area and a reduction in the water flow, which improved water balance of the remaining living stems. Anatomical resistance to the formation and spread of air embolisms in the xylem may be of critical importance. There are a variety of features of xylem anatomy, which can increase the safety of water-conducting systems in Mediterranean species (Carlquist 1989). Vasicentric tracheids adjacent to many vessels act as subsidiary conducting system and occur in numerous Mediterranean genera such as *Quercus*, *Arctostaphylos*, *Phyllirea*, *Rhus*, and *Banksia*. Vascular tracheids also provide conductive tissues at high stress and are present in drought-deciduous and ericoid evergreen species such as *Cistus*, *Erica*.

Hydraulic Architecture

Zimmermann (1983) introduced the principle of plant segmentation stating that embolism should develop first in the terminal part of the trees (i.e., leaves and little branches), thus preserving the other parts of the crown from embolism damage. The risk of xylem dysfunction especially in the petioles may determine the ability to resist to drought. Are Mediterranean species less vulnerable than other species? Cochard et al. (1992) and Higgs and Wood (1995) compared drought susceptibility by examining hydraulic dysfunction of the xylem vessels in petioles of different oak species. The Mediterranean species (*Q. pubescens*, *Q. cerris*) did not show the lowest vulnerability (nor the highest) to embolism formation in comparison to the temperate ones (*Q. petraea*, *Q. robur*, *Q. rubra*). Early leaf senescence was also observed near −4 MPa on the Californian deciduous oak *Q. douglasii* (Griffin 1973); whereas all leaves of the deciduous *Q. pubescens* were yellowing at about −4.5 MPa in the Languedoc (Damesin and Rambal 1995). There is also some difference between Mediterranean species for their vulnerability to cavitation—partly explained by the distribution of xylem conduit diameter—which can be related to their different distribution within the Mediterranean Basin (Salleo et al. 1997, Cavender-Bares et al. 2005). As quoted by Preston et al. (2006), wood density and vessel characteristics are functionally interrelated but wood density was most strongly associated with soil water, and vessel traits showed contrasting relationships with plant height.

Maherali et al. (2004) compiled a database of 167 species and examined relationships among resistance to xylem cavitation and water transport capacity measured by the specific conductivity per unit of xylem area. For the Mediterranean woody species of this meta-analysis (Figure 9.5), we found mostly chaparral shrub species (*Ceanothus* and *Adenostoma* spp.) and some dominant trees growing around the Mediterranean Sea (*Quercus* spp.). Medians of Ψ_{50} for species occurring in the more arid environments were 6–7 times more negative than those growing in tropical rain forest. The Ψ_{50} medians are −5.3, −4.5, −2.8, −2.4, and −0.8 MPa for Mediterranean, desert, temperate forest, tropical dry forest, and tropical rain forest, respectively. Resistance increased significantly with decreasing precipitation and with the ratio of precipitation to potential evapotranspiration in evergreen angiosperms. The adaptive significance of increased resistance to cavitation with decreased precipitation as a mechanism of drought tolerance is of primary importance in such species because they need to preserve a water-conducting pathway for leaves year around. In contrast, Maherali et al. (2004) did not find association between water transport capacity and water availability suggesting that the evolutionary basis for a trade-off between cavitation resistance and water transport capacity is rather weak.

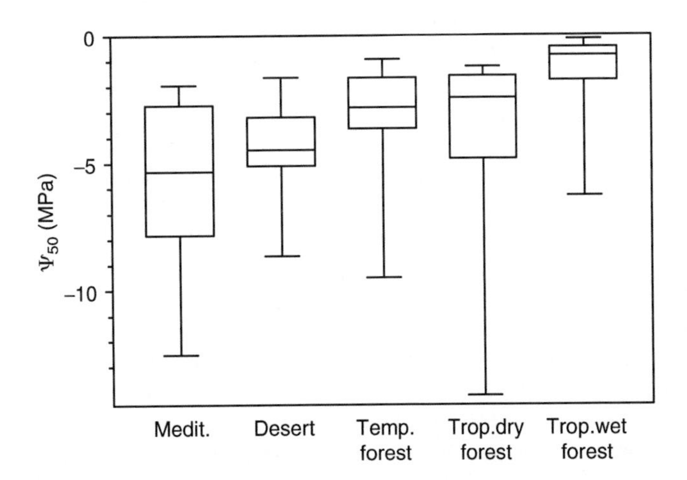

FIGURE 9.5 Box and whisker plots for the distribution of vulnerability to water stress-induced cavitation (as determined by plant water potential at which 50% cavitation occurred, Ψ_{50}) for 167 woody species grouped into five vegetation types (after Maherali et al. 2004). The number of species for the vegetation types are 16, 20, 71, 19, and 41 for Mediterranean, desert, temperate forest, tropical dry forest, and tropical rain forest, respectively. The corresponding medians are −5.3, −4.5, −2.8, −2.4, and −0.8 MPa.

CARBON ASSIMILATION: LEAF PHOTOSYNTHETIC PERFORMANCES

ASSIMILATION IN RELATION TO ENVIRONMENTAL CONDITIONS

Net CO_2 Assimilation in Optimal Conditions

The photosynthetic performance of Mediterranean species does not differ particularly from that of species from other biomes (Rambal 2001). For example, in the genus *Quercus*, Damesin et al. (1998a) analyzing the literature found that the Mediterranean species do not differ in their maximum assimilation—mean value of 16.3 $\mu mol\ m^{-2}\ s^{-1}$ calculated over five species—from non-Mediterranean species—mean value of 17.2 $\mu mol\ m^{-2}\ s^{-1}$ calculated over five species. Numerous studies have been conducted concerning species of the Californian chaparral and the Chilean matorral (Oechel et al. 1981) on the South African fynbos (Mooney et al. 1983, Van der Heyden and Lewis 1989) and on the shrublands and woodlands around the Mediterranean Sea (Tenhunen et al. 1987, Damesin et al. 1998a). Differences have been demonstrated between growth forms (Oechel et al. 1981) or guilds, and between restoid–ericoid and proteoid species (Van der Heyden and Lewis 1989). A robust relationship between nitrogen content of the leaves and their photosynthetic capacities has often been found in natural vegetation for a wide variety of plants (Wright et al. 2005). This correlation appears to be a consequence of the limitations on photosynthetic capacity imposed by the levels of the enzyme RuBP carboxylase and of the pigment–protein complexes. Mediterranean species do not deviate from this rule (Field 1991).

This relationship holds for species with low nitrogen content as observed with 36 evergreen sclerophyllous species growing at two sites in the coastal and mountain fynbos of South Africa. Herppich et al. (2002) averaged mass-based leaf nitrogen content of 0.52 ± 0.13, 0.76 ± 0.18, 0.55 ± 0.10, 1.00 ± 0.29 for proteoid, ericoid, restioid, and other sclerophylls in their coastal location and 0.74 ± 0.13, 1.29 ± 0.24, 0.60, 1.08 ± 0.31 for the same groups in their mountain location. They found a unique linear relationship between maximal light-saturated photosynthesis and nitrogen content for all the species except the coastal proteoid. Coastal proteoid too displayed a linear relationship with a steeper slope, that is, more photosynthesis with the same amount of nitrogen. Evergreen species might allocate less N to photosynthetic functions or they might allocate the same amount of N to photosynthetic functions but allocate that N inefficiently (Warren and Adams 2004).

Response to Water Constraint and Photoprotection

Mediterranean climate leaves have to cope with excess intercepted solar radiation when carbon assimilation is limited either by stomatal closure or a decrease of photosynthetic capacity due to water stress and high temperatures (summer) or low temperatures (winter). Indeed, absorption of light energy may be in excess of that required for carbon fixation and may result in damage to the photosystem. Different physiological regulatory mechanisms have been shown to occur during diurnal cycles to dissipate the excess of absorbed energy without any damage for cells: a downregulation of photosynthesis via a decrease of the photochemical efficiency of PSII (F_v/F_m) (Demmig-Adams et al. 1989, Damesin and Rambal 1995, Faria et al. 1996, Méthy et al. 1996) or via a decrease of chlorophyll content (Kyparissis et al. 1995), a change in the components of the xanthophyll cycle, a high antioxidative potential (Faria et al. 1996). It is difficult to assess if Mediterranean species have a more efficient photoprotection mechanism than temperate or tropical species. In particular, it would be interesting to examine if the Mediterranean species responses to high irradiance are acclimation or adaptation. On that account, a great proportion of the plant species growing in MTE produce and accumulate volatile organic compound (VOC) which may also serve as an excess energy dissipation system during a period of restricted growth

(Kesselmeier and Staudt 1999) although several studies pointed out that there is not necessarily a specific role for every VOC emitted (see Peñuelas and Llusia 2004 for a comprehensive account on this subject).

Modeling

Functioning at the biochemical level of the photosynthetic system, described by Farquhar and von Caemmerer (1982), can be summarized by both maximum carboxylation V_{cmax} and electron transport rates J_{max}. Mediterranean species fit in with the general scheme in terms of this functioning, if reference is made to the few Mediterranean species included in Wullschleger's (1993) review, although this author did not propose a separate grouping for these species. They do not deviate significantly from the empirical linear relation between V_{cmax} and J_{max} that he observed. Other data support this trend for Mediterranean oaks and *Arbutus unedo* (Hollinger 1992, Damesin et al. 1998a). This leaf photosynthesis model can be next integrated in a canopy level carbon balance model (Hollinger 1992, Sala and Tenhunen 1996). Relations between stomatal conductance (g_s) and assimilation (A) could be simulated following the empirical model proposed by Ball et al. (1987). That is the so-called coupled photosynthesis Ball–Berry stomatal model

$$g_s = g_0 + g_{fac}\frac{AR_h}{C_a},$$

in which A is net assimilation rate (μmol m^{-2} s^{-1}), g_0 is a value representing the stomatal conductance when $A = 0$ at the light compensation point (generally set to about 10 mmol m^{-2} s^{-1}), g_{fac} is a dimensionless empirical factor expressing the relation of g_s to A, to relative humidity at the leaf surface R_h (decimal fraction), and to CO_2 concentration at the leaf surface C_a (μmol mol^{-1}).

The functional dependency of g_{fac} on predawn leaf water potential (Ψ_{pd}) was measured in the field for *Q. ilex* (Sala and Tenhunen 1996) and *Q. pubescens* (Damesin and Rambal 1995). Both studies showed linear relationships for Ψ_{pd} equal or lower than −1 MPa, that is, during periods of reduced water availability (Figure 9.6). Xu and Baldocchi (2003) observed a strong correlation between g_s and AR_h/C_a during the growing season in *Quercus douglasii*. They suggested that leaf age and severe water stress did not alter g_{fac}; a constant g_{fac} of 8–10 seems typical for oaks (Goulden 1996, Xu and Baldocchi 2003). However, the effect of decreasing

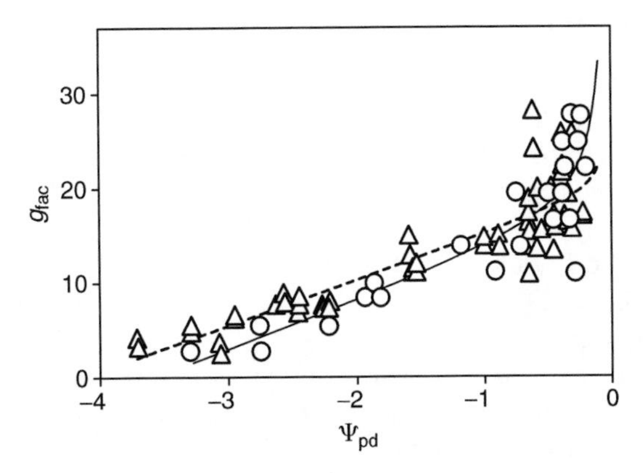

FIGURE 9.6 Relationships between g_{fac} and predawn water potential Ψ_{pd} for *Quercus ilex* and *Quercus pubescens*. The data for the former oak species have been obtained by Sala and Tenhunen (1996) (empty circle), and those for *Q. pubescens* by Damesin and Rambal (1995) (empty triangle up). For both data sets, we fitted the equation $g_{fac} = (a/\Psi_{pd}) + (b\Psi_{pd} + c)$. For *Q. ilex* (dotted), we obtained, $a = -0.31$, $b = 4.76$, $c = 19.6$, and $r^2 = 0.76$; for *Quercus pubescens* (solid line), $a = -1.64$, $b = 4.97$, $c = 17.35$, and $r^2 = 0.78$.

soil water content and consequently predawn leaf water potential Ψ_{pd} on g_{fac} is still in dispute. Some works assert that g_{fac} decreases as the soil dries. On the other hand, some others observed that g_{fac} remains constant and V_{cmax} decreases as the soil dries. A number of important gaps in our knowledge about photosynthetic downregulation and inhibition under water stress have been listed by Flexas et al. (2004).

Assimilation in Relation to Water Loss

One way to estimate water-use efficiency (ratio of photosynthesis and transpiration) in C_3 plants is to use leaf carbon isotope composition. Its measurement, easier than that of gaseous exchanges, allows the study of intraspecific and interspecific variability in field conditions (Damesin et al. 1998b).

Interspecific Variability

The creation of a superclass for Mediterranean species could be made on the basis of the discrimination (Δ) against $^{13}CO_2$ during the photosynthesis. For the xerophytic woods and scrubs superclass (which includes *Heteromeles arbutifolia*, *Nerium oleander*, and *Eucalyptus socialis*), Lloyd and Farquhar (1994) gave a surprising low Δ value of 12.9‰. By comparison, this discrimination reaches 18.3‰ for the superclass cool or cold mixed forest they proposed and averaged 17.8‰ for all the C_3 plants. This distinction in term of Δ values implies a segregation of the long-term estimates of the ratio c_i/c_a between the intercellular CO_2 concentration within leaves and the atmospheric CO_2 concentration and therefore of leaf performance and water use efficiency. This approach has been extended by Beerling and Quick (1995) who used it both for superclasses and at the scale of the individual plant.

They estimated V_{cmax} and J_{max} both from Δ and from maximum assimilation. What values of Δ should be adopted for Mediterranean species? From a study conducted at 25 stations in southern France (Damesin et al. unpublished results) with four species of co-occurring trees or shrubs the following values were obtained: 18.20 ± 0.65 for *Pinus halepensis*, 18.86 ± 1.53 for *Quercus pubescens*, 19.50 ± 0.52 for *Q. coccifera*, and 19.88 ± 0.68 for *Q. ilex*. Only *P. halepensis* deviated significantly from the other three species. A similar distinction was recorded by Williams and Ehleringer (1996) between *Q. gambelii* and *P. edulis*. Values obtained on Californian oaks (Goulden 1996) and on *Q. ilex* at other sites (Fleck et al. 1996) confirm these orders of magnitude. However, all these values are subject to averaging procedures to overcome the great individual variation in Δ (Damesin et al. 1997).

Intraspecific Variability

Differential responses of species to local variations of resources could be analyzed through isotopic measurements. For example, Williams and Ehleringer (1996) explained the between-site variability in Δ along a summer monsoon gradient in southwestern USA using a parameter that integrated both the water balance and the climatic demand throughout the growing season. Similarly, Damesin et al. (1998b) took into account both within- and between-site variability among two deciduous and evergreen Mediterranean oaks using the minimum seasonal leaf predawn potential. This type of response, which tends to optimize water resource utilization, can be extended to entire plant communities growing along a water availability gradient (Stewart et al. 1995).

Assimilation in Relation to Leaf Life Span

In Mediterranean communities, evergreen species are more abundant than deciduous ones (Mooney and Dunn 1970, Cody and Mooney 1978, di Castri 1981). As a consequence, ecological studies have largely focused on evergreen and sclerophyllous-leaved species and less attention was devoted to deciduous species. Nevertheless, Mooney and Dunn (1970) presented a conceptual model comparing the functioning of evergreen chaparral shrubs versus drought-deciduous species of the coastal sage community. The basic assumptions of this model were that the leaves of evergreen sclerophyllous have a relatively low photosynthetic capacity, compared with the malacophyllous, drought-deciduous leaves, but can amortize their cost of production over a longer period of time. Although the woody shrub growth-form is an important vegetation element in the MTE, the situation of co-occurrence of winter-deciduous and evergreen trees is not infrequent at least in California and Mediterranean Basin where many species of co-occurring *Quercus* exhibit the two habits (Hollinger 1992, Damesin et al. 1996, 1998a,b). How do these winter-deciduous species compete with their evergreen neighbors? Researches have been mainly oriented toward the mechanisms of seedling installation. However, trees generally have a long life span, and adult survival can thus also determine the abundance of a species in a given region. If the survival of a perennial species depends on its ability to maintain a positive carbon balance over the year (Givnish 1988), the winter deciduous habit could a priori present two disadvantages as compared with the evergreen one: (1) it implies a shorter photosynthetically active period and (2) the active period coincides with the most important constraint imposed by the Mediterranean climate, that is, the summer drought.

This distinction is in agreement with a validation of a cost-benefit model at the leaf level (Mooney and Dunn 1970). In this model, deciduous species compensate their lower leaf life span by a higher carbon assimilation by unit of time and a lower leaf production cost. This model was successfully tested for two coexisting Californian oaks (Hollinger 1992), but does not hold for two other oaks co-occurring in the North of the Mediterranean Basin where *Q. pubescens* has a lower area-based construction cost than *Q. ilex*, but does not have a higher photosynthetic capacity (Damesin et al. 1998a). Despite differences in biochemical composition, size, and mass per unit area, the leaves of the two species respond similarly to water-limited conditions and have similar intrinsic water use efficiencies (Damesin et al. 1997, 1998b). These results indicate that key factors distinguishing the functioning of the deciduous species from evergreen ones are more important at the higher level of organization (individual and ecosystem) than at the leaf level.

High level of secondary metabolites, essential oils, and resins, with their associated volatiles, characterize plants of semiarid regions (Kesselmeier and Staudt 1999). These secondary metabolites and VOCs represent an important part of the costs of construction of the leaves. Production of VOC has shown a great intraspecific variability (Staudt et al. 2001, 2004) and its regulation by environmental conditions have to be specified (Loreto et al. 1996, Lerdau et al. 1997). There has been rapid progress in understanding how the emission of volatiles is regulated, mostly focusing on enzymatic activity (Fischbach et al. 2002) and on biochemical control (Niinemets et al. 2004). The effect of stomatal conductance and drought still remains in debate (Staudt et al. 2002).

Potential Multieffect of an Increase of CO_2

Anticipating the responses of vegetation to increasing CO_2 concentration is critical attempting to predict the effect of global change on primary production, on vegetation spatial distribution, and on potential sinks for carbon, which may be provided by terrestrial ecosystems. Results obtained on greenhouse experiments shown that net CO_2 assimilation of *Q. suber* seedlings, in conditions of moderate drought, was at least twice as great in elevated atmospheric CO_2 conditions, but stomatal conductance was unchanged (Damesin et al. 1996). In addition, shoot and root biomass, stem height, and total leaf area were increased by elevated CO_2 as well as root and stem ramifications. Comparing tree ring chronologies of *Q. ilex* growing in natural CO_2 springs in Italy with those growing in ambient CO_2 control sites, Hättenschwiler et al. (1997) showed that trees grown under high CO_2 for 30 years showed a 12% greater final radial stem width. This stimulation was largely due to responses when trees were young in accordance with the previously cited study. Increasing CO_2 concentration could affect different leaf functional characteristics. The leaves of *Q. pubescens* grown under elevated CO_2 do not show changes in the number or size of stomata but on the size of the guard cells (Miglietta et al. 1995). In contrast, *Q. ilex* leaves (Paoletti and Gellini 1993) and *Pistacia lentiscus* (Peñuelas and Matamala 1990) collected in herbaria showed a reduction of stomatal density with increasing atmospheric CO_2 as shown for temperate species (Woodward 1987).

NUTRIENT REGULATION: INDIVIDUAL AND ECOSYSTEM LEVELS

NUTRIENT UPTAKE AND TROPHIC TYPES

The five regions supporting MTEs differ largely concerning their nutrient level (see Section "Substrate"). As a consequence, belowground structures do not show convergence, although majority of plants form vesicular–arbuscular (VA) mycorrhizae in all regions (Allen et al. 1995). In Mediterranean Basin forests, ectomycorrhizal (ECM) fungi are known to play a fundamental role in enhancing the acquisition of resources by the host plant. de Roman and de Miguel (2005) and Richard et al. (2005) found a striking diversity of ECM fungi in *Q. ilex* forests. Nevertheless, exhaustive inventory of the diversity is a real challenge due to the large number of rare types.

The benefits that ECM fungi confer on their host plants can lead to improved growth and greater drought tolerance. Nevertheless, very few field experiments have addressed how changes in climate, such as increased drought, might quantitatively or qualitatively influence these ECM communities. ECM fungal community composition also may shift in response to drought and soil warming (Rygiewicz et al. 2000, Shi et al. 2002, Swaty et al. 2004).

High proportion of the Proteaceae in South Africa are nonmycotrophic species and there appear to be no ECM plants. In Australia, nonmycorrhizal species are confined among woody species to root hemiparasites or species bearing proteoid cluster roots (Lamont 1984). The N_2-fixing plants carry strong root specialization of presumed significance in absorption of phosphorous (Pate 1994). The Australian species of Epacridaceae form very peculiar ericoid-type mycorrhizae in extremely nutrient-deficient soils (Read 1992). High incidence of ECM trees as well as VA mycorrhizal grasses and shrubs characterize the other MT regions. This pattern is probably related to contrast between the very old and phosphorous-deficient South African and Australian soils and the relatively high phosphorous Mediterranean, Chilean, and Californian soils (Allen et al. 1995). In Australian nutrient-impoverished heathlands, Pate (1994) emphasized that the mycorrhizal association is just one of many nutrient-acquiring specializations in natural ecosystems, including various forms of parasitism, epiparasitism, and autotrophy with and without mycorrhizal associations.

NUTRIENT USE EFFICIENCY, SCLEROPHYLLY, AND EVERGREENNESS

The dominance of sclerophyllous—leathery, rigid, and heavily cutinized leaves—evergreen plants in the five MTEs of the world has been interpreted since a long time as a convergent adaptation in responses to the unique environmental conditions associated with Mediterranean climate. Seddon (1974) presents a historical discussion of concepts of sclerophylly and xeromorphy. The possible functional role of sclerophylly has been interpreted in diverse directions: (1) adaptation to drought, thick cell walls can better resist negative turgor pressure under water stress (Lo Gullo and Salleo 1988, Salleo et al. 1997), (2) leaf hardness is a epiphenomenon of phosphorous deficiency in soils (Loveless 1961, Monk 1966, Rundel

1988), and (3) sclerophylly is an adaptation to herbivory. In fact, similarities of sclerophylly in MTE is presently interpreted as linked to the long-term persistence of tertiary lineages that evolved under a subtropical climate and is not interpreted as an adaptation to Mediterranean climate (Herrera 1992, Joffre and Rambal 2002, Verdú et al. 2003, Thompson 2005).

Rundel (1988) supported interpretations that leaf nutrition may be a critical factor in the evolution of leaf characteristics as evergreenness or leaf life span. Since the study of Monk (1966) postulating that the habit of evergreens an adaptation to low nutrient availability, several researches have emphasized that evergreens, due to their long leaf life-spans have a higher nutrient use efficiency (productivity per unit nutrient uptake) than deciduous species. Several researches (Pugnaire and Chapin 1993, Aerts 1995, 1996, Killingbeck 1996) show that nutrient resorption of nitrogen was significantly lower for evergreens than that for deciduous. In southern France, the proportion of nitrogen in fallen leaves was 78% of that in mature leaves in *Q. ilex* and 44% in *Q. pubescens* (Damesin et al. 1998a). This comparison suggests that the deciduous *Q. pubescens* have a more efficient mechanism for removing nitrogen from senescing leaves. This result is in agreement with that obtained by Del Arco et al. (1991) on five oak species. They found a positive relation between the percentage of nitrogen translocated and the nitrogen concentration of the mature leaves, which was itself negatively related to the life span of the leaves.

Leaf longevity appears more important than resorption efficiency as a nutrient conservation mechanism. Resorption efficiency has to be considered as one of the internal mechanisms of nutrient regulation involving the relative pool sizes of mobile and insoluble nutrients as well as the capacity to store and mobilize energy as carbohydrates and lipids (Escudero et al. 1992). Cherbuy et al. (2001) showed that remobilization of carbohydrates, lipids, nitrogen, and phosphorous from 1 year leaves of mature *Q. ilex* participates in supplying carbon and nutrients for the new growth in spring. Although substantial differences in seedling storage ability between seeder and resprouter species have been shown in five prevalent MTEs of southwestern Australia (Pate et al. 1990), there is actually no strong corpus of evidence showing clear differentiation between storage patterns of mature deciduous and evergreen plants in MTEs.

Nutrient Release and Decomposition

Nitrogen and phosphorous release patterns during leaf litter decomposition varied considerably between species. We illustrate the importance of growth form in the nutrient cycling at ecosystem level comparing the decomposition patterns of the deciduous and evergreen oaks (*Q. pubescens* and *Q. ilex*) in southern France. Gillon et al. (1994) showed that nitrogen was released quickly by *Q. ilex*. In this species, the first stage, in which there was a strong decrease in nitrogen amount, corresponded to the release of soluble nitrogen (Ibrahima et al. 1995). In contrast, nitrogen was strongly immobilized by *Q. pubescens* litter. Increases in the nitrogen amount immobilized in the litter during decomposition may be partly explained by microbial and fungal incorporation of nitrogen from soil organic matter. This net accumulation of nitrogen in litter during the early stages of decomposition would alter the rates and patterns of nitrogen uptake by trees and may limit tree production. The difference in litter composition is also associated with a difference in the timing of leaf fall. In *Q. ilex* canopies, there are typically two peaks, one in spring and the other in autumn, or sometimes a single peak in spring. In *Q. pubescens* canopies, leaf fall starts in autumn, but since the species is marcescent, there is also some leaf fall in winter. The difference in mineral input to the soil between both species is accentuated by a difference in the intensity of leaching by precipitation. These differences certainly imply difference in the regulation of nitrogen turnover in the forest floor. These results are in agreement with the hypothesis that functional differences between evergreen and deciduous species depend on the management of nutrients and particularly nitrogen (Monk 1966, Moore 1980, Aerts 1995).

SUMMARY

Five regions throughout the world, that is, the Mediterranean Basin itself, California, central Chile, the Cape region of south Africa, and parts of southwestern and southern Australia, are characterized by the same climatic regime marked by a strong seasonality in temperature and precipitation that leads to a hot drought period in summer and a cool wet period in winter. Their vegetation type presents numerous structural and functional affinities that has led to define MTEs. The aim of this chapter is to discuss some aspects of the functioning of the woody Mediterranean species.

The main climatic factor controlling the functioning of these ecosystems is water availability, which varies greatly in time and space, imposing strong constraints over the plant. Moreover, a general consensus of global circulation models indicates an increase of summer drought and of extreme events occurrences during the twenty-first century. Different mechanisms acting at several scales and levels illustrate how individuals and ecosystems cope with and control water uptake, water loss, and water transfer through the soil–plant–atmosphere continuum.

Carbon assimilation characteristics are presented in relation to environmental conditions. In optimal conditions, the photosynthetic performance of Mediterranean species does not differ particularly from that of species from other biomes. Nevertheless, their leaves have to tolerate high irradiance and they have to cope with excess intercepted solar radiation when carbon assimilation is limited either by stomatal closure or a decrease of photosynthetic capacity due to water stress and high temperatures (summer) or low temperatures (winter). A great proportion of the plant species growing in MTE produce and accumulate aromatic volatile oils that may also serve as an excess energy dissipation system during a period of restricted growth. The functioning of evergreen and deciduous species is presented and some assumptions based on a cost-benefit model explaining their coexistence are discussed.

The five regions supporting MTEs show very distinct nutrient level. As a consequence, some differences occur concerning the belowground structures and the role of mycorrhizae. In Australian and south African nutrient impoverished heathlands, some specific root adaptations are presented. The possible functional role of sclerophylly as a nutrient conservation mechanism is discussed. Relations between leaf life span, resorption efficiency, and nutrient-use efficiency differ between evergreen and deciduous species. Decomposition and release of mineral forms have contrasted patterns between these two groups of species. These differences certainly imply difference in the regulation of nitrogen turnover in the forest floor and seem to be in agreement with the hypothesis that functional differences between evergreen and deciduous species depend on the management of nutrients. In conclusion, the remarkable combination of gradient of natural resources and obviously different anthropic management in the five Mediterranean regions of the world leading to a great diversity of adaptive strategies and functional attributes in the MTEs is highlighted.

REFERENCES

Acherar, M., S. Rambal, and J. Lepart, 1991. Evolution du potentiel hydrique foliaire et de la conductance stomatique de quatre chênes méditerranéens lors d'une période de dessèchement. Annales des Sciences Forestières 48: 561–573.

Ackerly, D., 2004a. Functional strategies of chaparral shrubs in relation to seasonal water deficit and disturbance. Ecological Monographs 74: 25–44.

Ackerly, D.D., 2004b. Adaptation, niche conservatism, and convergence: Comparative studies of leaf evolution in the California chaparral. American Naturalist 163: 654–671.

Aerts, R., 1995. The advantages of being evergreen. Trends in Ecology and Evolution 10: 402–407.

Aerts, R., 1996. Nutrient resorption from senescing leaves of perennials: are there general patterns? Journal of Ecology 84: 597–608.

Allen, M.F., S.J. Morris, F. Edwards, and E.B. Allen, 1995. Microbe–plant interactions in Mediterranean-type habitats: shifts in fungal symbiotic and saprophytic functioning in response to global change. In: J.M. Moreno and W.C. Oechel, eds. Global Change and Mediterranean-Type Ecosystems. Ecological Studies 117. Springer-Verlag, Berlin, pp. 287–305.

Aschmann, H., 1973. Distribution and peculiarity of Mediterranean ecosystems. In: F. di Castri and H.A. Mooney, eds. Mediterranean–Type Ecosystems: Origin and Structure. Ecological Studies 7. Springer-Verlag, Berlin, pp. 11–19.

Aussenac, G. and J.C. Valette, 1982. Comportement hydrique estival de *Cedrus atlantica* Manetti, *Quercus ilex* L. et *Quercus pubescens* Willd. et de divers pins dans le Mont Ventoux. Annales des Sciences Forestières 39: 41–62.

Bagnouls, F. and H. Gaussen, 1953. Saison sèche et indice xérothermique. Bulletin Société d'Histoire Naturelle de Toulouse 88: 192–239.

Ball, J.T., I.E. Woodrow, and J.A. Berry, 1987. A model predicting stomatal conductance and its contribution to the control of photosynthesis under different environmental conditions. In: I.J. Biggins, ed. Progress in Photosynthesis Research. Vol IV.5. Dordrecht, The Netherlands: Martinus Nijhoff, pp. 221–244.

Beerling, D.J. and W.P. Quick, 1995. A new technique for estimating rates of carboxylation and electron transport in leaves of C_3 plants for use in dynamic global vegetation models. Global Change Biology 1: 289–294.

Blondel, J., 2006. The 'design' of Mediterranean landscapes: a millennial story of humans and ecological systems during the historic period. Human Ecology 34: 713–729.

Blondel, J. and J. Aronson. 1995. Biodiversity and ecosystem function in the Mediterranean basin: human and non-human determinants. In: G.W. Davis and D.M. Richardson, eds. Mediterranean-Type ecosystems: the function of biodiversity. Ecological Studies 109, Springer-Verlag, Berlin, pp. 43–119.

Bucci, S.J., G. Goldstein, F.C. Meinzer, A.C. Franco, P. Campanello, and F.G. Scholz, 2005. Mechanisms contributing to seasonal homeostasis of minimum leaf water potential and predawn disequilibrium between soil and plant water potential in neotropical savanna trees. Trees—Structure and Function 19: 296–304.

Butzer, K.W., 2003. Review of A. Grove and O. Rackham, the nature of Mediterranean Europe. Annals, Association of American Geographers 93: 494–498.

Butzer, K.W., 2005. Environmental history in the Mediterranean world: cross-disciplinary investigation of cause-and-effect for degradation and soil erosion. Journal of Archaeological Science 32: 1773–1800.

Canadell, J., R.B. Jackson, J.R. Ehleringer, H.A. Mooney, O.E. Sala, and E.-D. Schulze, 1996. Maximum rooting depth of vegetation types at the global scale. Oecologia (Berlin) 108: 583–595.

Carlquist, S., 1989. Adaptative wood anatomy of chaparral shrubs. In: S.C. Keeley, ed. The California Chaparral: Paradigms Reexamined. Science Series 34. Natural History Museum of Los Angeles County, Los Angeles, pp. 25–35.

Castro-Diez, P., G. Montserrat-Marti, and J.H.C. Cornelissen, 2003. Trade-offs between phenology, relative growth rate, life form and seed mass among 22 Mediterranean woody species. Plant Ecology 166: 117–129.

Cavender-Bares, J., P. Cortes, S. Rambal, R. Joffre, B. Miles, and A. Rocheteau, 2005. Summer and winter sensitivity of leaves and xylem to minimum freezing temperatures: a comparison of co-ccurring Mediterranean oaks that differ in leaf lifespan. New Phytologist 168: 597–611.

Cherbuy, B., R. Joffre, D. Gillon, and S. Rambal, 2001. Internal remobilization of carbohydrates, lipids, nitrogen and phosphorus in the Mediterranean evergreen oak Quercus ilex. Tree Physiology 21: 9–17.

Cochard, H. and M.T. Tyree, 1990. Xylem dysfunction in *Quercus*: vessel sizes, tyloses, cavitation and seasonal changes in embolism. Tree Physiology 6: 393–407.

Cochard, H., N. Bréda, A. Granier, and G. Aussenac, 1992. Vulnerability to air embolism of three European oak species (*Quercus petraea* (Matt) Liebl, *Q. pubescens* Willd, *Q. robur* L). Annales des Sciences Forestières 49: 225–233.

Cody, M.L. and H.A. Mooney, 1978. Convergence versus non convergence in Mediterranean-climate ecosystems. Annual Review of Ecology and Systematics 9: 265–321.

Conrad, V., 1941. The variability of precipitation. Monthly Weather Review 69: 5–11.

Cowan, I.R. and G.D. Farquhar, 1977. Stomatal function in relation to leaf metabolism and environment. Symposium Society of Experimental Biology 31: 471–505.

Daget, P., 1977. Le bioclimat méditerranéen, caractères généraux, modes de caractérisation. Vegetatio 34: 1–20.

Damesin, C. and S. Rambal, 1995. Field study of leaf photosynthetic performance by a Mediterranean deciduous oak tree (*Quercus pubescens*) during a severe summer drought. New Phytologist 131: 159–167.

Damesin, C., C. Galera, S. Rambal, and R. Joffre, 1996. Effects of elevated carbon dioxide on leaf gas exchange and growth of cork-oak (*Quercus suber* L.) seedlings. Annales des Sciences Forestières 53: 461–467.

Damesin, C., S. Rambal, and R. Joffre, 1997. Between-tree variations in leaf δ13C of *Quercus pubescens* and *Quercus ilex* among Mediterranean habitats with different water availability. Oecologia (Berlin) 111: 26–35.

Damesin, C., S. Rambal, and R. Joffre, 1998a. Cooccurrence of trees with different leaf habit: A functional approach on Mediterranean oaks. Acta Oecologica 19: 195–204.

Damesin, C., S. Rambal, and R. Joffre, 1998b. Seasonal and between-year changes in leaf δ13C in two co-occurring Mediterranean oaks: Relations to leaf growth and drought progression. Functional Ecology 12: 778–785.

Davis, G.W., D.M. Richardson, J.E. Keeley, and R.J. Hobbs, 1996. Mediterranean-type ecosystems: the influence of biodiversity on their functioning. In: H.A. Mooney, J.H. Cushman, E. Medina, O.E. Sala, and E.-D. Schulze, eds. Functional Roles of Biodiversity. A Global Perspective. Wiley, London, pp. 151–183.

de Roman, M. and A.M. de Miguel, 2005. Post-fire, seasonal and annual dynamics of the ectomycorrhizal community in a *Quercus ilex* L. forest over a 3-year period. Mycorrhiza 15: 471–482.

Del Arco, J.M., A. Escudero, and M.M. Vega Garrido, 1991. Effects of site characteristics on nitrogen retranslocation from senescing leaves. Ecology 72: 701–708.

Demmig-Adams, B., W.W. Adams III, K. Winter, A. Meyer, U. Schreiber, J.S. Pereira, A. Krüger, F.-C., Czygan, and O.L. Lange, 1989. Photochemical efficiency of photosystem II, photon yield of O_2 evolution, photosynthetic capacity, and carotenoid composition during the midday depression of net CO_2 uptake in *Arbutus unedo* growing in Portugal. Planta 177: 377–387.

di Castri, F., 1981. Mediterranean-type shrubland of the world. In: F. di Castri, D.W Goodall, and R.L Specht, eds. Mediterranean-Type Shrublands. Ecosystems of the World 11. Elsevier, Amsterdam, The Netherlands, pp. 1–52.

Duhme, F., 1974. Die Kennzeichneug des Ökologishen Konstitution von Gehölzen im Hinblick auf den Wasserhaushalt. Dissertationes Botanicae 28: 1–143.

Eagleson, P.S., 2002. Ecohydrology: Darwinian Expression of Vegetation Form and Function. Cambridge University Press, Cambridge.

Eamus, D., T. Hatton, P. Cook, and C. Colvin, eds. 2006. Ecohydrology. Vegetation Function, Water and Resource Management. CSIRO Publishing, Collingwood.

Escudero, A., J.M. del Arco, I.C. Sanz, and J. Ayala, 1992. Effects of leaf longevity and retranslocation efficiency on the retention time of nutrients in the leaf biomass of different woody species. Oecologia (Berlin) 90: 80–87.

Faria, T., J.I. Garcia-Plazaola, A. Abadia, S. Cerasoli, J.S. Pereira, and M.M. Chaves, 1996. Diurnal changes in photoprotective mechanisms in leaves of cork oak (*Quercus suber*) during summer. Tree Physiology 16: 115–123.

Farquhar, G.D. and S. von Caemmerer, 1982. Modeling of photosynthetic response to environmental conditions. In: O.L. Lange, P.S. Nobel, C.B. Osmond, and H. Ziegler, eds. Encyclopedia of Plant Physiology. Vol. 12B. Springer-Verlag, Berlin, pp. 549–587.

Field, C., 1991. Ecological scaling of carbon gain to stress and resource availability. In: H.A. Mooney, W.E. Winner, and E.J. Pell, eds. Response of Plant to Multiple Stresses. Academic Press, San Diego, pp. 35–65.

Fischbach, R.J., M. Staudt, I. Zimmer, S. Rambal, and J.-P. Schnitzler, 2002. Seasonal pattern of monoterperne synthase activities in leaves of the evergreen tree *Quercus ilex* L. Physiologia Plantarum 114: 354–360.

Fleck, I., D. Grau, M. Sanjosé, and D. Vidals, 1996. Carbon isotope discrimination in *Quercus ilex* resprouts after fire and tree-fell. Oecologia (Berlin) 105: 286–292.

Flexas, J., J. Bota, J. Cifre, J.M. Escalona, J. Galmés, J. Gulías, L. El-Kadri, S.F. Martínez-Cañellas, M.T. Moreno, M. Ribas-Carbó, D. Riera, B. Sampol, and H. Medrano, 2004. Understanding down regulation of photosynthesis under water stress: future prospects and searching for physiological tools for irrigation management. Annals of Applied Biology 144: 273–283.

Gale, M.R. and D.F. Grigal, 1987. Vertical root distributions of northern tree species in relation to successional plants. Canadian Journal of Forest Research 17: 829–834.

Gardner, W.R., 1964. Relation of root distribution to water uptake and availability. Agronomy Journal 56: 41–45.

Gillon, D., R. Joffre, and A. Ibrahima, 1994. Initial litter properties and decay rate: a microcosm experiment on Mediterranean species. Canadian Journal of Botany 72: 946–954.

Giorgi, F., 2006. Climate change hot-spots. Geophysical Research Letters 33: L08707, doi:10.1029/2006GL025734.

Giorgi, F. and X. Bi, 2005. Regional changes in surface climate interannual variability for the 21st century from ensembles of global model simulations. Geophysical Research Letters 32: L21715, doi:10.129/2005GL024288.

Givnish, T.J., 1988. Adaptation to sun and shade: a whole plant perspective. Australian Journal of Plant Physiology 15: 63–92.

Goulden, M.L., 1996. Carbon assimilation and water-use efficiency by neighboring Mediterranean-climate oaks that differ in water access. Tree Physiology 16: 417–424.

Griffin, J.R., 1973. Xylem sap tension in three woodland oaks of central California. Ecology 54: 152–159.

Groves, R.H., J.S. Beard, H.J. Deacon, J.J.N. Lambrechts, A. Rabinovitch-Vin, R.L. Specht, and W.D. Stock, 1983. Introduction: the origins and characteristics of mediterranean ecosystems. In: J.A. Day, ed. Mineral Nutrients in Mediterranean Ecosystems. South African National Scientific Programmes Report 71. Council for Scientific and Industrial Research, Pretoria, South Africa, pp. 1–17.

Hättenschwiler S., F. Miglietta, A. Raschi, and C. Körner, 1997. Thirty years of *in situ* growth under elevated CO_2: a model for future responses? Global Change Biology 3: 463–471.

Hellmers, H., J.S. Horton, G. Juhren, and J. O'Keefe, 1955. Root systems of some chaparral plants in southern California. Ecology 36: 667–678.

Herppich, M., W.B. Herppich, and D.J. von Willert, 2002. Leaf nitrogen content and photosynthetic activity in relation to soil nutrient availability in coastal and mountain fynbos plants (South Africa). Basic Applied Ecology 3: 329–337.

Herrera, C.M., 1992. Historical effects and sorting processes as explanations for contemporary ecological patterns—character syndromes in Mediterranean woody-plants. American Naturalist 140: 421–446.

Hickman, J.C., 1970. Seasonal course of xylem sap tension. Ecology 51: 1052–1056.

Higgs, K.H. and V. Wood, 1995. Drought susceptibility and xylem dysfunction in seedlings of 4 European oak species. Annales des Sciences Forestières 52: 507–513.

Hinckley, T.M., F. Duhme, A.R. Hinckley, and H. Richter, 1983. Drought relations of shrub species: assessment of the mechanism of drought resistance. Oecologia (Berlin) 59: 344–350.

Hobbs, R.J., D.M. Richardson, and G.W. Davis, 1995. Mediterranean-type ecosystems: opportunities and constraints for studying the function of biodiversity. In: G.W. Davis and D.M. Richardson, eds. Mediterranean-Type ecosystems: the function of biodiversity. Ecological Studies 109. Springer-Verlag, Berlin, pp. 1–32.

Hollinger, D.Y., 1992. Leaf and simulated whole-canopy photosynthesis in two co-occurring tree species. Ecology 73: 1–14.

Ibrahima, A., R. Joffre, and D. Gillon, 1995. Changes in litter during the initial leaching phase: an experiment on the leaf litter of Mediterranean species. Soil Biology Biochemistry 27: 931–939.

Jackson, R.B., J. Canadell, J.R. Ehleringer, H.A. Mooney, O.E. Sala, and E.-D. Schulze, 1996. A global analysis of root distributions for terrestrial biomes. Oecologia (Berlin) 108: 389–411.

Joffre, R. and S. Rambal, 2002. Mediterranean ecosystems. Encyclopedia of Life Sciences. Macmillan, Nature Publishing Group, London, www.els.net, pp. 1–7.

Joffre, R., S. Rambal, and J.P. Ratte, 1999. The dehesa system of southern Spain and Portugal as a natural ecosystem mimic. Agroforestry Systems 45: 57–79.

Keeley, S.C., ed., 1989. The California Chaparral: Paradigms Reexamined. Science Series 34. Natural History Museum of Los Angeles County, Los Angeles.

Keeley, J.E., C.J. Fotheringham, and M. Baer-Keeley, 2006. Demographic patterns of postfire regeneration in Mediterranean-climate shrublands of California. Ecological Monographs 76: 235–255.

Kesselmeier, J. and M. Staudt, 1999. Biogenic volatile organic compounds (VOC): An overview on emission, physiology and ecology. Journal of Atmospheric Chemistry 33: 23–88.

Killick, D.J.B., 1979. African mountain heathlands. In: R.L. Specht, ed. Heathlands and Related Shrubland of the World. Elsevier, Amsterdam, pp. 97–116.

Killingbeck, K.T., 1996. Nutrients in senesced leaves: keys to the search for potential resorption and resorption proficiency. Ecology 77: 1716–1727.

Köppen, W., 1900. Versuch einer Klassification der Klimate vorsugsweise nach ihren Bezichungen zur Pflanzenwelt. Geograph. Zeirsehr 6: 593–611, 657–679.

Köppen, W., 1931. Die Klimate der Erde, Grundriss der Klimakunde, second edition. De Gruyter, Berlin, Leipzig.

Kruger, F.J., D.T. Mitchell, and J.U.M. Jarvis, eds., 1983. Mediterranean–type ecosystems: the role of nutrients. Ecological Studies 43, Springer-Verlag, Berlin.

Kummerow, J.D., D. Krause, and W. Jow, 1977. Root systems of chaparral shrubs. Oecologia (Berlin) 29: 163–177.

Kyparissis, A., Y. Petropoulou, and Y. Manetas, 1995. Summer survival of leaves in a soft-leaved shrub (*Phlomis fruticosa* L., Labiatae) under Mediterranean field conditions: avoidance of photoinhibitory damage through decreased chlorophyll contents. Journal of Experimental Botany 46: 1825–1831.

Ladiges, P.Y. and D.H. Ashton, 1974. Variation in some central Victorian populations of *Eucalyptus viminalis* Labill. Australian Journal of Botany 22: 81–102.

Lamont, B.B., 1984. Specialized modes of nutrition. In: J.S. Pate and J.S. Beard, eds. Kwongan—Plant Life of the Sandplain. University of Western Australia, Nedlands, pp. 236–245.

Lamont, B.B., 1994. Mineral nutrient relations in Mediterranean regions of California, Chile, and Australia. In: M.T.K. Arroyo, P.H. Zedler, and M.D. Fox, eds. Ecology and biogeography of Mediterranean ecosystems in Chile, California, and Australia. Ecological Studies 108. Springer-Verlag, Berlin, pp. 211–238.

Le Houérou, H.N., 1990. Global change: vegetation, ecosystems and land use in the southern Mediterranean basin by the mid twenty-first century. Israel Journal of Botany 39: 481–508.

Le Houérou, H.N., 1992. Relations entre la variabilité des précipitations et celle des productions primaire et secondaire en zone aride. In: E. Le Floc'h, M. Grouzis, and A. Cornet, eds. L'aridité: une contrainte au développement. ORSTOM, Paris, pp. 197–220.

Lerdau M., A. Guenther, and R. Monson, 1997. Plant production and emission of volatile organic compounds. Bioscience 47: 373–383.

Lloret, F., H. Estevan, J. Vayreda, and J. Terradas, 2005. Fire regenerative syndromes of forest woody species across fire and climatic gradients. Oecologia 146: 461–468.

Lloyd, J. and G.D. Farquhar, 1994. ^{13}C discrimination during CO_2 assimilation by the terrestrial biosphere. Oecologia (Berlin) 99: 201–215.

Lo Gullo, M.A. and S. Salleo, 1988. Different strategies of drought resistance in three Mediterranean sclerophyllous trees growing in the same environmental conditions. New Phytologist 108: 267–276.

Lo Gullo, M.A. and S. Salleo, 1993. Different vulnerabilities of *Quercus ilex* L. to freeze and summer drought-induced xylem embolism: an ecological interpretation. Plant, Cell and Environment 16: 511–519.

Loreto, F., P. Ciccioli, A. Cecinato, E. Brancaleoni, M. Frattoni, and D. Tricoli, 1996. Influence of environmental factors and air composition on the emission of α-pinene from *Quercus ilex* L. leaves. Plant Physiology 110: 267–275.

Loveless, A.R., 1961. A nutritional interpretation of sclerophylly based on difference in the chemical composition of sclerophyllous and mesophytic leaves. Annals of Botany 25: 168–184.

Maherali, H., W.T. Pockman, and R.B. Jackson, 2004. Adaptative variation in the vulnerability of woody plants to xylem cavitation. Ecology 85: 2184–2199.

Méthy, M., C. Damesin, and S. Rambal, 1996. Drought and photosystem II activity in two Mediterranean oaks. Annales des Sciences Forestières 53: 255–263.

Miglietta, F., M. Badiani, I. Bettarini, P. van Gardingen, F. Selvi, and A. Raschi, 1995. Preliminary studies of the long-term CO_2 response of Mediterranean vegetation around natural CO_2 events. In: J.M. Moreno and W.C. Oechel, eds. Global change and Mediterranean-type ecosystems. Ecological Studies 117. Springer-Verlag, Berlin, pp. 102–120.

Monk, C.D., 1966. An ecological significance of evergreenness. Ecology 47: 504–505.

Mooney, H.A., 1989. Chaparral physiological ecology – paradigms revisited. In: S.C. Keeley, ed. The California Chaparral: Paradigms Reexamined. Science Series 34. Natural History Museum of Los Angeles County, Los Angeles, pp. 85–90.

Mooney, H.A. and E.L. Dunn, 1970. Photosynthetic systems of Mediterranean-climate shrubs and trees of California and Chile. American Naturalist 104: 447–453.

Mooney, H.A., C. Field, S.L. Gulmon, P. Rundel, and F.J. Kruger, 1983. Photosynthetic characteristics of South African sclerophylls. Oecologia (Berlin) 58: 398–401.

Moore, P., 1980. The advantages of being evergreen. Nature 285: 535.

Moreno, G., J.J. Obrador, E. Cubera, and C. Dupraz, 2005. Fine root distribution in dehesas of Central-Western Spain. Plant Soil 277: 153–162.

Naveh, Z., 1967. Mediterranean ecosystems and vegetation types in California and Israel. Ecology 48: 345–359.

Ne'eman, G., 1993. Variation in leaf phenology and habit in *Quercus ithaburensis*, a Mediterranean deciduous tree. Journal of Ecology 81: 627–634.

Niinemets, Ü., F. Loreto, and M. Reichstein, 2004. Physiological and physicochemical controls on foliar volatile organic emission. Trends in Plant Science 9: 180–186.

Oechel, W.C., W. Lawrence, J. Mustafa, and J. Martinez, 1981. Energy and carbon acquisition. In: P.C. Miller, ed. Resource use by chaparral and matorral. A comparison of vegetation function in two Mediterranean type ecosystems. Ecological Studies 39. Springer-Verlag, Berlin, pp. 151–182.

Paoletti, E. and R. Gellini, 1993. Stomatal density variation in beech and holm oak leaves collected over the past 200 years. Acta Oecologica 14: 173–178.

Passioura, J.B., 1976. Physiology of grain yield in wheat growing on stored water. Australian Journal of Plant Physiology 3: 559–565.

Pate, J.S., 1994. The mycorrhizal association: just one many nutrient acquiring specializations in natural ecosystems. Plant and Soil 159: 1–10.

Pate, J.S., R.H. Froend, B.J. Bowen, A. Hansen, and J. Kuo, 1990. Seedling growth and storage characteristics of seeder and resprouter species of Mediterranean-type ecosystems of S.W. Australia. Annals of Botany 65: 585–601.

Pausas, J.G. and M. Verdú, 2005. Plant persistence traits in fire-prone ecosystems of the Mediterranean basin: a phylogenetic approach. Oikos 109: 196–202.

Pausas, J.G., J.E. Keeley, and M. Verdú, 2006. Inferring differential evolutionary processes of plant persistence traits in Northern Hemisphere Mediterranean fire-prone ecosystems. Journal of Ecology 94: 31–39.

Peñuelas, J. and R. Matamala, 1990. Changes in N and S leaf content, stomatal density and specific leaf area of 14 plant species during the last three centuries of CO_2 increase. Journal of Experimental Botany 41: 1119–1124.

Peñuelas, J. and J. Llusia, 2004. Plant VOC emissions: making use of the unavoidable. Trends in Ecology & Evolution 19: 402–404.

Poole, D.K. and P.C. Miller, 1975. Water relations of selected species of chaparral and coastal sage communities. Ecology 56: 1118–1128.

Poole, D.K. and P.C. Miller, 1981. The distribution of plant water stress and vegetation characteristics in southern California chaparral. American Midland Naturalist 105: 32–43.

Preston, K.A., W.K. Cornwell, and J.L. DeNoyer, 2006. Wood density and vessel traits as distinct correlates of ecological strategy in 51 California coast range angiosperms. New Phytologist 170: 807–818.

Pugnaire, F.I. and F.S. Chapin III, 1993. Controls over nutrient resorption from leaves of evergreen Mediterranean species. Ecology 74: 124–129.

Rambal, S., 1984. Water balance and pattern of root water uptake by a *Quercus coccifera* L. evergreen scrub. Oecologia (Berlin) 62: 18–25.

Rambal, S., 1992. *Quercus ilex* facing water stress: a functional equilibrium hypothesis. Vegetatio 99–100: 147–153.

Rambal, S., 1993. The differential role of mechanisms for drought resistance in a Mediterranean evergreen shrub: a simulation approach. Plant, Cell and Environment 16: 35–44.

Rambal, S., 1995. From daily transpiration to seasonal water balance: an optimal use of water? In: J. Roy, J. Aronson, and F. di Castri, eds. Times Scales of Biological Responses to Water Constraints. SPB Academic Publishing, Amsterdam, pp. 37–51.

Rambal, S., 2001. Hierarchy and productivity of Mediterranean-type ecosystems. In: H.A. Mooney, B. Saugier, and J. Roy, eds. Terrestrial Global Productivity: Past, Present, Future. Academic Press, San Diego, pp. 315–344.

Rambal, S. and J. Leterme, 1987. Changes in aboveground structure and resistances to water uptake in *Quercus coccifera* along a rainfall gradient. In: J.D. Tehnunen, F.M. Catarino, O.L. Lange, and W.D. Oechel, eds. Plant Response to Stress. Functionnal Analysis in Mediterranean Ecosystems. NATO-ASI series, Vol. G15. Springer-Verlag, Berlin, pp. 191–200.

Rambal, S. and G. Debussche, 1995. Water balance of Mediterranean ecosystems under a changing climate. In: J.M. Moreno and W.C. Oechel, eds. Global change and Mediterranean-type ecosystems. Ecological Studies 117. Springer-Verlag, Berlin, pp. 386–407.

Read, D.J., 1992. The mycorrhizal mycelium. In: M.F. Allen, ed. Mycorrhizal Functioning. Chapman and Hall, London, pp. 102–133.

Reich, P.B. and T.M. Hinckley, 1989. Influence of pre-dawn water potential and soil-to-leaf hydraulic conductance on maximum daily leaf diffusive conductance in two oak species. Functional Ecology 3: 719–726.

Richard, F., S. Millot, M. Gardes, and M.-A. Selosse, 2005. Diversity and specificity of ectomycorrhizal fungi retrieved from an old-growth Mediterranean forest dominated by Quercus ilex. New Phytologist 166: 1011–1023.

Richter, H., 1976. The water status in the plant. Experimental evidence. In: O.L. Lange, L. Kappen, and E.-D. Schulze, eds. Water in Plant Life. Ecological Studies 19. Springer-Verlag, Berlin, pp. 42–58.

Ritchie, G.A. and T.M. Hinckley, 1975. The pressure chamber as an instrument for ecological research. Advances in Ecological Research 9: 165–254.

Rundel, P.W., 1988. Leaf structure and nutrition in mediterranean-climate sclerophylls. In: R.L. Specht, ed. Mediterranean-Type Ecosystems: A Data Source Book. Kluwer, Dordrecht, pp. 157–167.

Rundel, P.W., 1995. Adaptative significance of some morphological and physiological characteristics in mediterranean plants: facts and fallacies. In: J. Roy, J. Aronson, and F. di Castri, eds. Times Scales of Biological Responses to Water Constraints. SPB Academic Publishing, Amsterdam, pp. 119–139.

Rygiewicz, P.T., K.J. Martin, and A.R. Tuininga, 2000. Morphotype community structure of ectomycorrhiza on Douglas-fir (*Pseudotsuga menziesii* (Mirb.) Franco) seedlings grown under elevated atmospheric CO_2 and temperature. Oecologia 124: 299–308.

Sala, A. and J.D. Tenhunen, 1996. Simulations of canopy net photosynthesis and transpiration in *Quercus ilex* L. under the influence of seasonal drought. Agricultural and Forest Meteorology 78: 203–22.

Salleo, S., A. Nardini, and M.A. Lo Gullo, 1997. Is sclerophylly of Mediterranean evergreens an adaptation to drought? New Phytologist 135: 603–612.

Schimper, A.F.W., 1903. Plant-Geography upon a Physiological Basis. Clarendon Press, Oxford.

Seddon, G., 1974. Xerophytes, xeromorphs and sclerophylls: the history of some concepts in ecology. Biological Journal of the Linnean Society 6: 65–87.

Shi, L.B., M. Guttenberger, I. Kottke, and R. Hampp, 2002. The effect of drought on mycorrhizas of beech (*Fagus sylvatica* L.): Changes in community structure, and the content of carbohydrates and nitrogen storage bodies of the fungi. Mycorrhiza 12: 303–311.

Specht, R.L., 1969a. A comparison of the sclerophyllous vegetation characteristic of Mediterranean type climates in France, California, and Southern Australia I. Structure, morphology, and succession. Australian Journal of Botany 17: 277–292.

Specht, R.L., 1969b. A comparison of the sclerophyllous vegetation characteristic of Mediterranean type climates in France, California, and southern Australia. II. Dry matter, energy, and nutrient accumulation. Australian Journal of Botany 17: 293–308.

Specht, R.L., 1979. Heathlands and related heathlands of the world. In: R.L. Specht, ed. Heathlands and Related Shrubland of the World. Elsevier, Amsterdam, pp. 1–18.

Specht, R.L. and Specht, A., 1989. Canopy structure in Eucalyptus-dominated communities in Australia along climatic gradients. Acta Oecologica, Oecologia Plantarum 10: 191–213.

Staudt, M., N. Mandl, R. Joffre, and S. Rambal, 2001. Intraspecific variability of monoterpene composition emitted by Quercus ilex leaves. Canadian Journal of Forestry 31: 174–180.

Staudt, M., S. Rambal, R. Joffre, and J. Kesselmeier, 2002. Impact of drought on seasonal monoterpene emissions from Quercus ilex in Southern France. Journal of Geophysical Research—Atmosphere 107 (D21): 4602.

Staudt, M., C. Mir, R. Joffre, S. Rambal, A. Bonin, D. Landais, and R. Lumaret, 2004. Isoprenoid emissions of *Quercus spp.* (*Q. suber* and *Q. ilex*) in mixed stands contrasting in interspecific genetic introgression. New Phytologist 163: 573–584.

Stewart, G.R., M.H. Turnbull, S. Schmidt, and P.D. Erskine, 1995. ^{13}C natural abundance in plant communities along a rainfall gradient: a biological integrator of water availability. Australian Journal of Plant Physiology 22: 51–55.

Stone, E.L. and P.J. Kalisz, 1991. On the maximum extent of tree roots. Forest Ecology and Management 46: 59–102.

Swaty, R.L., R.J. Deckert, T.G. Whitham, and C.A. Gehring, 2004. Ectomycorrhizal abundance and community composition shifts with drought: Predictions from tree rings. Ecology 85: 1072–1084.

Talsma, T. and E.A. Gardner, 1986. Soil water extraction by mixed *Eucalyptus* forest during a drought period. Australian Journal of Soil Research 24: 25–32.

Tardieu, F. and W.J. Davies, 1993. Integration of hydraulic and chemical signalling in the control of stomatal regulation and water status of droughted plants. Plant, Cell and Environment 16: 341–349.

Tenhunen, J.D., W. Beyschlag, O.L. Lange, and P.C. Harley, 1987. Changes during summer drought in leaf CO_2 uptake rates of macchia shrubs growing in Portugal: Limitation due to photosynthetic capacity, carboxylation efficiency, and stomatal conductance. In: J.D. Tehnunen, F.M. Catarino, O.L. Lange, and W.D. Oechel, eds. Plant Response to Stress. Functionnal Analysis in Mediterranean Ecosystems. NATO-ASI series, Vol. G15. Springer-Verlag, Berlin, pp. 305–327.

Thompson, J.D., 2005. Plant Evolution in the Mediterranean. Oxford University Press, Oxford.

Tyree, M.T. and J.S. Sperry, 1988. Do woody plants operate near the point of catastrophic xylem dysfunction caused by dynamic water stress? Answers from a model. Plant Physiology 88: 574–580.

Van der Heyden, F. and O.A.M. Lewis, 1989. Seasonal variation in photosynthetic capacity with respect to plant water status of five species of the Mediterranean climate region of South Africa. South African Journal of Botany 55: 509–515.

Verdú, M., P. Davila, P. Garcia-Fayos, N. Flores-Hernandez, and A. Valiente-Banuet, 2003. 'Convergent' traits of mediterranean woody plants belong to pre-mediterranean lineages. Biological Journal of the Linnean Society 78: 415–427.

Waggoner, P.E., 1989. Anticipating the frequency distribution of precipitation if the climate change alters its mean. Agricultural and Forest Meteorology 47: 321–337.

Waring, R.H. and B.D. Cleary, 1967. Plant moisture stress: evaluation by pressure bomb. Science 155: 1248, 1953–1954.

Warming, E., 1909. Oecology of Plants. Clarendon Press, Oxford.

Warren, C.R. and M.A. Adams, 2004. Evergreen trees do not maximize instantaneous photosynthesis. Trends in Plant Science 9: 270–274.

Westoby, M., D.S. Falster, A.T. Moles, P.A. Vesk, and I.J. Wright, 2002. Plant ecological strategies: Some leading dimensions of variation between species. Annual Review of Ecology and Systematics 33: 125–159.

Wigley, T.M.L., 1985. Impact of extreme events. Nature 316: 106–107.
William, W.E., 1983. Optimal water-use efficiency in a California shrub. Plant, Cell and Environment 6: 145–151.
Williams, D.G. and J.R. Ehleringer, 1996. Carbon isotope discrimination in three semi-arid woodland species along a monsoon gradient. Oecologia (Berlin) 106: 455–460.
Woodward, F.I., 1987. Stomatal numbers are sensitive to increases in CO_2 from pre-industrial levels. Nature 127: 617–618.
Wright, I.J., P.B. Reich, J.H.C. Cornelissen, D.S. Falster, P.K. Groom, K. Hikosaka, W. Lee, C.H. Lusk, U. Niinemets, J. Oleksyn, N. Osada, H. Poorter, D.I. Warton, and M. Westoby, 2005. Modulation of leaf economic traits and trait relationships by climate. Global Ecology and Biogeography 14: 411–421.
Wullschleger, S.D., 1993. Biochemical limitations to carbon assimilation in C3 plants-a retrospective analysis of the A/ci curves from 109 species. Journal of Experimental Botany 44: 907–920.
Xu, L. and D.D. Baldocchi, 2003. Seasonal trends in photosynthetic parameters and stomatal conductance of blue oak (*Quercus douglasii*) under prolonged summer drought and high temperature. Tree Physiology 23: 865–877.
Zhang, J., U. Schurr, and W.J., Davies, 1987. Control of stomatal behaviour by abscissic acid which apparently originates in the roots. Journal of Experimental Botany 38: 1174–1181.
Zimmerman, M.H., 1983. Xylem Structure and the Ascent of Sap. Springer-Verlag, Berlin.

10 Tropical Forests: Diversity and Function of Dominant Life-Forms

Ernesto Medina

CONTENTS

INTRODUCTION

Tropical forests are geographically limited to areas located between the tropics of Cancer and Capricorn. They harbor the largest diversity of species and ecosystem types on Earth. Their importance for the maintenance of stable trace of gas concentrations in the atmosphere is increasingly recognized. As a result, large international projects are currently underway throughout the tropics to understand the global interactions derived from biogeochemical cycling in tropical forests and the consequences of large-scale deforestation for loss of biological diversity and stability of the world climate. Numerous fundamental contributions on the ecology of tropical forests were produced during the last century, describing composition, structure, and functioning of those communities (Richards 1952, 1996, Odum and Pigeon 1970, Walter 1973, Unesco 1978, Vareschi 1980, Leigh et al. 1982, Sutton et al. 1983, Whitmore 1984, 1990, Jordan 1985, Jordan et al. 1989, Proctor 1989, Gómez-Pompa et al. 1991, Mulkey et al. 1996, Kellman and Tackaberry 1997). These contributions served as a conceptual basis for this chapter, aimed at highlighting the functional characteristics of tropical forest plant components. The chapter gives an outline of the distribution and edaphoclimatic characteristics of tropical forest as a basis for the more detailed discussion of diversity and function of main life-forms. The relationships between water and nutrient availability as regulators of the photosynthetic performance of tropical trees are also discussed to assess the role of tropical forests in the carbon balance of the atmosphere.

Climate

Climatic delimitation of tropical regions has been discussed by numerous investigators (Holdridge 1967, Walter 1973, Whitmore 1984, Richards 1996, Schultz 1995). Climatic characterization emphasizes the lack of seasonal variations in temperature and daylight duration, but these parameters increase linearly from the equator toward the tropics. Using temperature criteria, Walter (1973) defined the tropics as areas of the world, where diurnal variations in temperature are more pronounced than the differences in average temperatures between the hottest and the coldest month. This characterization applies to the whole range of tropical forests, from the lowlands to the high-altitude formations well represented in all tropical regions of the world. Other treatments separate tropical from nontropical systems using the occurrence of freezing temperatures at altitudes below 100 m above sea-level (Holdridge 1967).

Tropical ecosystems are affected by seasonality of rainfall distribution, a climatic trait associated with the displacement of the intertropical convergence zone. Most ecological descriptions of tropical ecosystems distinguish humid from dry tropics. The distinction is based on the annual ratio between precipitation and potential evapotranspiration (*P*/*E* ratio). The actual value of this ratio varies from less than 0.5 in very dry regions to more than five in wet tropical regions (Holdridge 1967). Practically all tropical climates present varying degree of seasonality in rainfall, measured as absolute rainfall or as frequency of dry days (daily rainfall less than evapotranspiration), dry season lasting from few days to several months.

In the last 20 years more attention has been given to climate variability as a driving factor in tropical latitudes. This approach is emphasized because the emerging understanding of the impact of global climate change. Generally aseasonal tropical regions may be subjected to occasional, at times severe, drought that increase tree growth and mortality. These processes were well documented during the 1997–1998 El Niño (Walsh and Newbery 1999, Slik 2004). In addition to those drought events long-term changes in climate have been observed in tropical rainforest regions. Malhi and Wright (2004) conducted a worldwide analysis of temperature and rainfall records in these areas. They showed that since 1970 average temperature increased 0.26°C per decade, in probable association with global temperature rise, whereas rainfall decreased approximately 1% per decade, decline being more pronounced in northern tropical Africa than in south-east Asia and the Amazon basin. These tendencies of increasing temperature and declining rainfall may offset the CO_2 effect on photosynthesis (Clark 2004).

Dendrochronology is producing much needed information on growth rhythms associated with climatic events.These studies have been made possible by the finding that many tropical trees form annual rings in areas where growth is limited during part of the year by drought or flooding. The advent of new approaches such as the analysis of ^{14}C and stable carbon and oxygen isotope ratios has generated hard data on growth dynamics in several tropical forests systems (Worbes and Junk 1989, Verheyden et al. 2006).

Brienen and Zuidema (2006) conducted tree ring analyses on six tree species from the Bolivian lowland moist forest forming annual rings. They found that age of trees with the same diameter may differ widely as a consequence of the variation of suppression and release events during the life time of the tree. They concluded that the species studied reach the canopy passing through periods of rapid and slow growth, in a pattern probably determined by light availability.

Schöngart et al. (2002) and Fichtler et al. (2003) used a combination of the ^{14}C method of Worbes and Junk (1989) with conventional wood anatomy to study seasonality of growth species in wet environments with seasonal flooding in the upper Amazon or in rainforests in Costa Rica. These studies showed that many trees form annual rings, so that tree growth may be correlated with environmental conditions. Indeed, Schöngart et al. (2004) showed that the growth of *Piranhea trifoliata* (Euphorbiaceae) is a reliable indicator of El Niño events during the last 200 years.

Ecoclimatic Classifications

Rainfall and temperature account for the large majority of changes in forest structure and species composition in the tropics. Within a certain temperature–rainfall regime, soil structure and fertility regulate the differentiation of forest communities (Medina and Cuevas 1994, Schultz 1995, Swaine 1996).

Temperature and rainfall interact to determine water availability at a given site. Temperature is directly correlated with evaporative demand of the atmosphere, whereas rainfall provides water supply through the roots and modify evaporative demand through changes in atmospheric water content. Humidity indices are frequently calculated using temperature and rainfall data alone, mainly because of the lack of direct measurement of evaporation (see discussions on climate classification including tropical areas in Lal 1987, Cramer and Leemans 1993, Schulz 1995, Olson et al. 2001).

A relatively simple approach to separate large groups of climates in the tropics uses average values of rainfall (as a measure of water availability) and temperature (as a measure of evaporative demand of the atmosphere, and also a regulator of plant activity). With mean annual data series, climate types can be separated with increasing water availability (moisture provinces in Figure 10.1) and with altitudinal belts related to reductions in temperatures (altitudinal belts in Figure 10.1). One widely used system to classify climatic units in connection with forest ecosystems is the Life Zone System of Holdridge (1967). Such a comprehensive summary of climatic data is useful to evaluate patterns of forest distribution on a large scale. For specific studies, the analysis of seasonal variations of environmental conditions is essential to understand the ecological process, because in the tropics seasonal variations in rainfall are far more important than variations in temperature. The seasonality of a certain climate is better observed in synthetic displays of the climatic variables (climate diagrams; see Walter 1973 for details). Figure 10.2 shows the climatic characterization of a number of meteorological stations in northern South America using the criteria of Bailey (1979) for separating humid and dry months. Temperature decreases with altitude, and so does the amount of monthly rainfall required to compensate the atmospheric evaporative demand.

Soil Fertility

The separation of humid and dry tropics is frequently paraleled by changes in edaphic characteristics. Soils of the humid tropics tend to be severely leached by heavy rainfall and

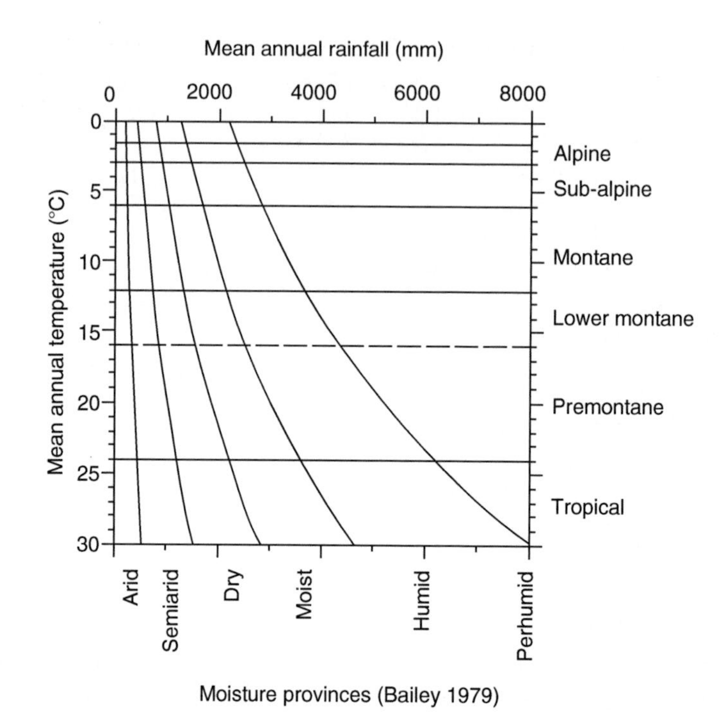

FIGURE 10.1 Basic categorization of ecoclimates in the tropics based on humidity provinces after Bailey (1979) and altitudinal belts after Holdridge (1967), Humidity provinces are separated according to rainfall and evaporation ratios, whereas the altitudinal belts are separated according to annual temperature averages. The dotted horizontal line indicates the altitudinal belt at which frosts are expected to occur.

show slight to strong acidity, and a tendency to high Al-mobility. These characteristics represent a condition in which one or more nutrients may limit plant growth and productivity. In these areas, lower geomorphological settings lead to water logging, compromising nutrient availability by low oxygen supply to the roots during periods of variable duration. In humid tropical forests, water and temperature are not limiting, and nutrients acquire the leading role in differentiating forests types.

In the dry tropics, the main limiting factor for forest development is low water availability during periods ranging from 2 to 7 months, with soils generally with neutral pH, and cation saturations of the soils exchange complex generally above 50%.

Large-scale gradients of rainfall and soil fertility put forward the covariance of these two factors in the tropics. Fertile soils, measured with a combination of total sum of bases, percentage of base saturation, phosphorus availability, and pH, are more frequently found in areas with rainfall below 1500 mm. The large-scale studies conducted in Ghana by Swaine and others (Hall and Swaine 1976, Swaine 1996) showed a good separation of the influence of rainfall (water availability) and soil fertility (nutrient availability). In humid areas (less than 1500 mm rainfall), soil fertility appeared as a strong determinant of vegetation, whereas in drier areas, other factors gained more relevance. Investigators were able to separate the group of 48 tree species surveyed throughout the forest zone of Ghana in four distributions groups established by a χ^2 technique: (1) biased to dry-fertile sites; (2) biased to wet-infertile sites; (3) biased to wet-fertile sites; and (4) nonbiased to any specific site. The majority of the species studied were within the wet-infertile and the nonbiased groups.

Laurance et al. (1999) analyzed the correlation of forest biomass and soil texture and nutrient availability using 65 one hectare plots within an area covering 1000 km^2 in central

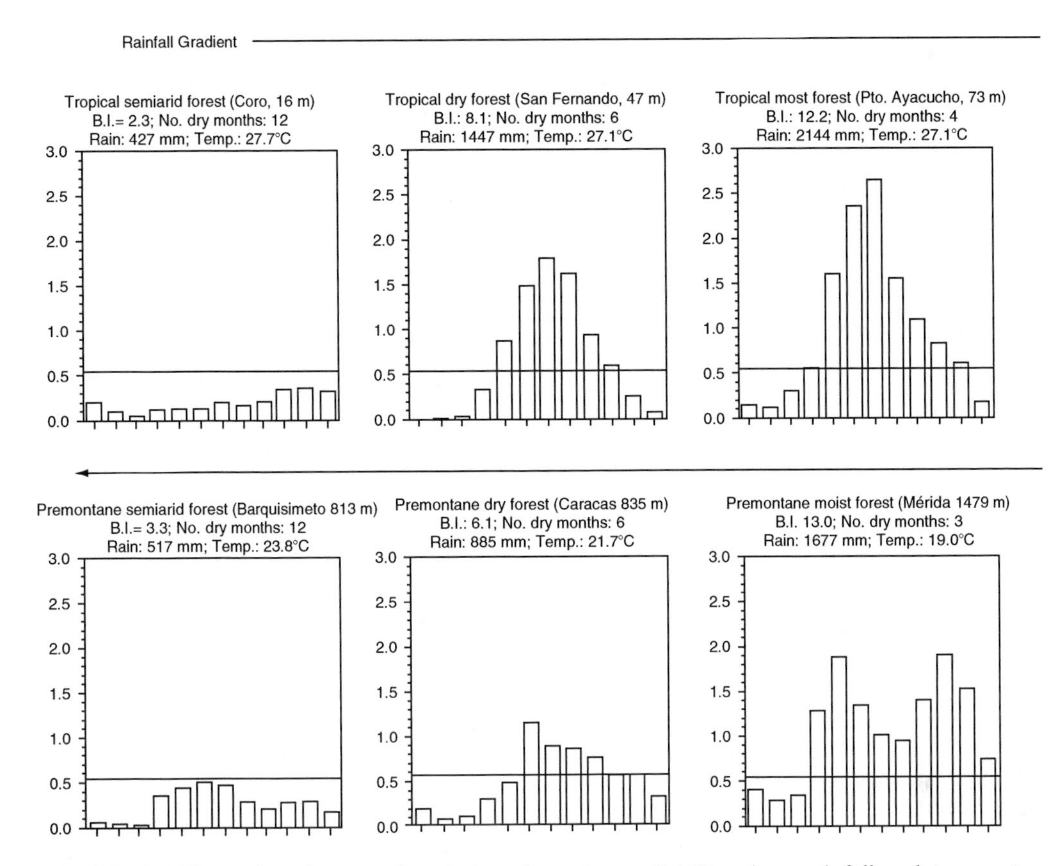

FIGURE 10.2 Examples of seasonal variations in water availability along rainfall and temperature gradients in tropical forest sites in northern South America. The *Y*-axes give the monthly *S* index of Bailey (1979). The horizontal thick line gives the value of the *S* index separating the humid from the dry realms.

Amazonia. The aboveground biomass ranged from 231 to 492 metric tons ha^{-1}, and it was significantly correlated with total N, total exchangeable bases, K^+, Mg^{2+}, clay, and organic C in soils, and negatively correlated with sand and Al saturation. The authors concluded that parameters of soil fertility account for more than a third of the variation in forest biomass. On a smaller scale in Costa Rica, Clark et al. (1999), showed that soil factors are responsible for nonrandom distribution of trees in upland soils.

Similar analyses conducted in drier areas in the tropics may render much insight into the ecological factors determining ecophysiological demands for seedling establishment in these areas (see reviews in Bullock et al. 1995). This differentiation may prove to be useful in designing experiments to establish nutritional and hydric dependence of seedling establishment and development under natural conditions.

Distribution and Extension

The accurate estimation of distribution and extension of tropical forests is essential to: (1) determine the potential carbon fixation capacity of tropical areas of the earth; and (2) assess the potential effects of increasing concentration of atmospheric CO_2 on global climate change. The literature before 1975 has been analyzed and used by Lieth and Whittaker (1975)

TABLE 10.1
Area and Net Primary Production of Organic Matter Estimated by Direct Measurements and Using a Process-Based Ecosystem Simulation Model

Vegetation Units	Area ($\times 10^6$ km^2)	%	Net Primary Production (g C m^{-2} year^{-1})	Total Net Primary Production (10^{15} g C year^{-1})	%
World total	127.3			53.2	
Tropical evergreen forest	17.4	13.7	1098	19.1	35.9
Tropical deciduous forest	4.6	3.6	871	4.0	7.5
Tropical savanna	13.7	10.8	393	5.4	10.2
Xeromorphic forests	6.8	5.3	461	3.1	5.8
Total tropical	42.5	33.0		31.6	59.4

Source: Adapted from Melillo, J.M., McGuire, A.D., Kicklighter, D.W., Moore III B., Vörösmarty, C.J., and Schloss, A.L., *Nature*, 363, 234, 1993.

to calculate the productivity of organic matter by terrestrial ecosystems. For the estimation of productivity of terrestrial ecosystems as a whole, the point values of productivity at specific sites were expanded to an area basis using the extension of each ecosystem unit (Lieth 1975). More recently, estimations using both climatic and geographic criteria based on satellite estimation of Normalized Vegetation Indices have been published (see summarizing reports including distribution and areal extension: Cramer and Leemans 1993, Melillo et al. 1993, Schultz 1995).

Tropical evergreen forests occupy approximately 17 million km^2, or nearly 14% of all land surface of the planet, but account for more than one-third of total net primary production of terrestrial ecosystems (Mellilo et al. 1993, 1998, Field et al. 1998; Table 10.1). As a whole, tropical regions of the earth occupy 33% of the total surface, but their total net primary production accounts for up to 59% of the total primary production of terrestrial ecosystems. This bigger contribution to total organic matter production is largely the result of longer growing periods.

SPECIES DIVERSITY

PATTERNS OF SPECIES DIVERSITY

Tropical forests contain the largest set of higher plants species known. Gentry (1988) reported values as high as 250 tree species 0.1 ha^{-1} in several wet tropical forests in South America. Phillips et al. (1994) analyzed tree species surveys in tropical rain forests throughout the world, and reported species richness of trees greater than or equal to 10 cm diameter at breast height (DBH) surpassing 200 species ha^{-1} in Malaysia, Sarawak, Ecuador, and Perú, while plots in Uganda, Perú, Venezuela, and Ghana were relatively poor, with approximately 60–90 species ha^{-1}. Most sites showed values between 100 and 200 species ha^{-1}. The highest species richness values reached 235 (Lambir, Sarawak) and 267 (Yanamomo, Perú) trees greater than or equal to10 cm DBH per 500 stems, while the lowest values measured in Kibale, Uganda, and Tambopata, Perú, were approximately 50 species per 500 stems.

Gentry (1988) found that the number of tree species 0.1 ha^{-1} increased with rainfall, from approximately 50 species at rainfall levels near 1000 mm to 250 at rainfall levels of greater than or equal to 4000 mm. In addition, vascular species diversity decreased strongly with altitude (from 1500 to 3000 m altitude). No clear relationship was found between soil fertility and diversity, although species composition generally changed strongly in association with soil fertility. Forests without relatively dry months (rainfall $\leq$100 mm) and good soil conditions, such as La Selva and Mersing, had tree-species richness of approximately 100 species ha^{-1}, while sites in Lambir, Sarawak, and Mishana, Perú, with no rainfall limitation but low fertility, had numbers well above 200 species ha^{-1} (Phillips et al. 1994; Figure 10.3). Using data from Holdridge et al. (1971) for Costa Rica, Huston (1980) showed that tree-species richness appeared to be inversely correlated with soil fertility. There were significant negative correlations with available P, exchangeable K and Ca, total bases, base saturation, and cation exchange capacity. These results were interpreted as a support for the hypothesis that higher tree diversity in tropical forests occurs under nutritional conditions that prevent dominance of single, highly competitive species (Huston 1994). The assessment of soil fertility using conventional soil chemical analysis, such as nutrient concentration without considering rooting depth, bulk density, and nutrient availability, is frequently misleading. Therefore,

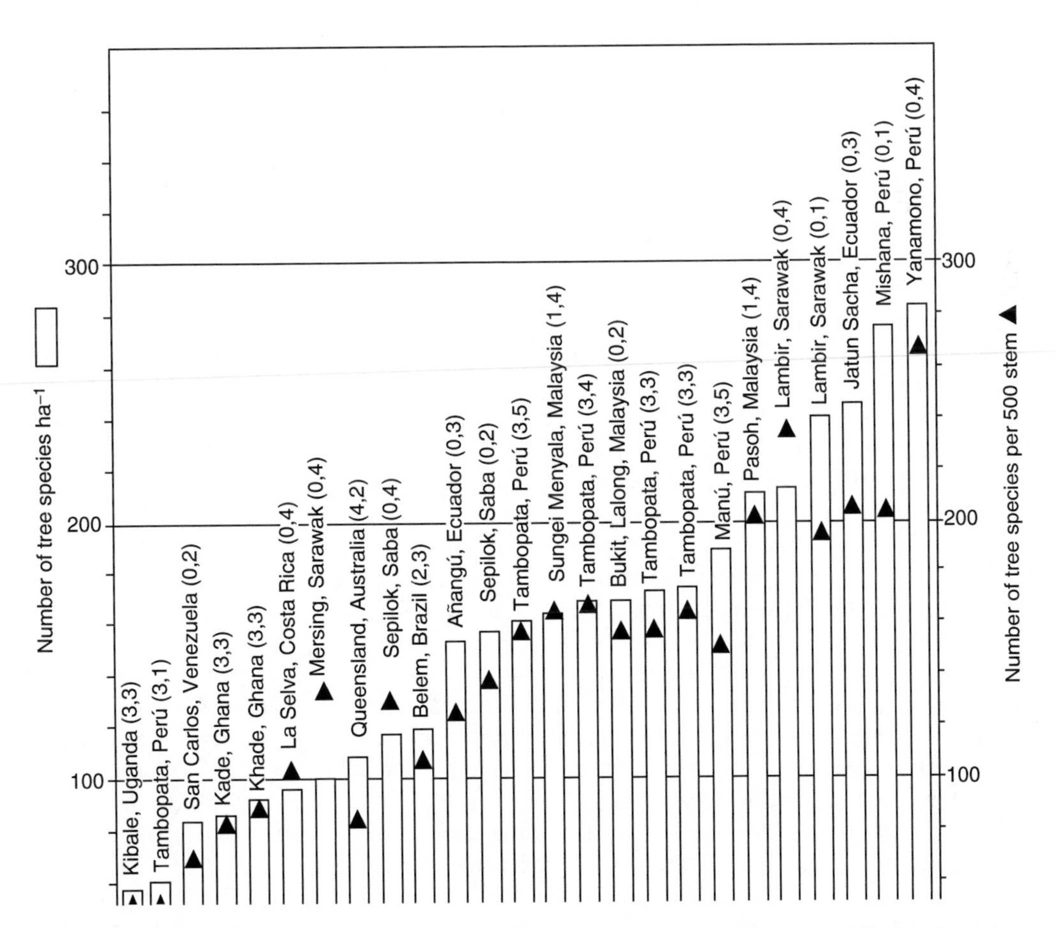

FIGURE 10.3 Diversity of tree species in tropical forest sites throughout the world. Sites have been ordered a1ong-increasing diversity as number of species per hectare or number of species per 500 stems. The first number after the site name indicates the number of dry months (0–4), and the second indicates site fertility increasing from 1 to 5. (From Phillips, O.L., Hall, P., Gentry, A.H., Sawyer, S.A., and Vasquez, R., *Proc. Natl. Acad. Sci. USA*, 91, 2805, 1994. With permission.)

further experimental work is necessary to reach a definitive picture of the role of nutrients in the regulation of diversity patterns in the tropics.

The contribution of nontrees (herbs and epiphytes) to total diversity of vascular plants is substantial, particularly in the neotropics, and increases with rainfall and possibly with soil fertility (Gentry and Dodson 1987a,b, Gentry and Emmons 1987).

There is voluminous literature on the mechanisms explaining the development and maintenance of diversity on local and temporal scales (see review by Barbault and Sastrapadja 1995). Models of particular significance to understand patterns of variations of plant in the tropics are those of Tilman (1988) and Huston (1994). These models provide a framework for experimental testing of predictions based on physiological and behavioral characteristics of plant species.

Low-Diversity Tropical Forests

Of considerable theoretical importance is the occurrence of low-diversity tropical forests, because it may provide explanations for the patterns of diversity in the tropics (Newbery et al. 1988, Connell and Lowman 1989, Torti et al. 1997). Connell and Lowman (1989) point out that the majority of those low-diversity forests are constituted by Caesalpinioid legumes, forming ectomycorrhizal symbiosis. Examples of these forests include the following: (1) in Africa: *Gilbertiodendron dewevrei* and *Brachystegia laurentii* (Zaire), *Cynometra alexandri* (Uganda), and *Tetraberlinia tubmaniana* (Liberia); and (2) in tropical South America: *Mora excelsa, Mora gonggrijpi*, and *Eperua falcata* (Trinidad and Guyana); and *Pentaclethra macroloba* (Costa Rica).

The Connell–Lowman hypothesis of monodominance is based on the fact that mycorrhizae improve the nutrient and water relations of the host plant (this subject is discussed further in Section "Dicotyledonous Woody Plants"). Ectomycorrhizae (EM) tend to be more specific; therefore, their benefits are limited to fewer host species. On the other hand, vesicular–arbuscular mycorrhizae (VAM) are more promiscuous and widely distributed in tropical forests; therefore, their predominance promotes diversity. A relevant ecological situation regarding the argument of mycorrhizal specificity is that of the Dipterocarpaceae, a predominantly ectomycorrhizal family that dominates large tracts of tropical forests in south-east Asia. In this case, ectomycorrhizal fungi are able to form associations with several species. This promiscuity promotes diversity, but only among species of the family that are susceptible to the same fungus. Hart et al. (1989) discussed other possible mechanisms such as reduced fertility, dominance restricted to a certain successional state, gradients of mortality, and herbivory based on high recruitment and herbivory pressure on the dominant species, and finally the low disturbance regime that may lead to the dominance of a few more competitive species.

It has been shown that in several forests with the tendency to monodominance of caesalpinioid species, the species involved were either exclusively VAM or had a mixture of VAM and EM. Moyersoen (1993) found that *Eperua leucantha*, forming nearly monodominant forests in the upper Rio Negro basin, forms only VAM, whereas *Aldina kundhartiana* (Caesalpiniopid) and species of *Guapira* and *Neea* (Nyctagynaceae) formed both VAM and EM. Torti et al. (1997) showed that two notorious caesalpinioid species forming monodominant forests in Trinidad (*Mora excelsa*) and Panama (*Prioria copaifera*) are exclusively VAM in the areas sampled. These findings reopen the issue of the fundamental causes for the existence of monodominant forests in the lowland tropics. Gross et al. (2000) tested the hypothesis of resistance to herbivory in mixed forests and *Gilbertiodendron dewevrei* dominated forests in Africa, and their results showed that the dominant species suffered the highest damage level of the species surveyed. In this context, it is worthwhile to mention the contrasting patterns of diversity of mycorrhizal fungi and higher plants (Allen et al. 1995).

The coniferous forests in the northern hemisphere have more than 1000 species of ecto-mycorrhizal fungi, but are dominated by a limited number of higher plants forming EM mycorrhizae. In the tropics, the predominant form of mycorrhiza is endophytic (VAM), but the number of fungi species is very limited despite the high diversity of higher plants. Allen et al. (1995) propose that while EM are taxonomically diverse and tend to be species-specific, VAM are physiologically diverse and tend to be more promiscuous, capable of forming mycorrhizae with a large variety of higher plant species.

DIVERSITY AND FUNCTIONAL ECOLOGY OF LIFE-FORMS

Species richness of tropical humid forests is paraleled by the diversity of co-existing life-forms, presumably indicating the variety of ecological niches that can be occupied by individuals with contrasting morphology and physiology. Life-forms may be described by a number of morphological features associated with physiological properties of significance for competitive ability and reproductive capacity (Solbrig 1993, Ewel and Bigelow 1996). The association of morpho-physiological properties allows a working organization of the species-richness characteristic of tropical forests in a small number of categories that can be associated with ecosystem function (Ewel and Bigelow 1996) and response to environmental change (Denslow 1996).

A number of life-form systems have been used as a classification scheme to differentiate forest types throughout the globe (Box 1981, Schultz 1995). Some of those systems focus mainly on tropical forest communities (Halle et al. 1978, Vareschi 1980, Ewel and Bigelow 1996). The most relevant functional aspects of the larger groups of vascular plants are discussed here, including: (1) trees and shrubs (including palms); (2) vines, lianas, and hemiepiphytes; and (3) epiphytes.

DICOTYLEDONOUS WOODY PLANTS

Dicot trees constitute the dominant life-form of the majority of tropical forest communities. Predominance of gymnosperms is only observed in some montane forests near the equator (with several species of *Podocarpus*), or toward the limits between the tropical, subtropical, and temperate realms (with species of *Pinus* and *Araucaria*) (Hueck 1966). Palm-dominated forests are restricted to seasonal or permanently flooded sites and are discussed later.

Forest dynamic is characterized by events of disturbance and regeneration that may be categorized as follows: (1) gap phase: the opening of gaps within the forest matrix caused by natural disturbances; (2) recovery phase: seedling establishment, tree resprouting, sapling, and pole development; and (3) mature phase: canopy closure and slow disappearance of pioneer species established during the gap and recovery phases (Whitmore 1991). This successional process is complex in nature due to the space-temporal variability of the interactions between physicochemical factors, the availability of seed sources, and biotic constraints such as seed predation, herbivory, and disease (Bazzaz 1984, Whitmore 1991).

Trees follow several stages during forest regeneration: (1) seedling, (2) sapling, (3) pole, and (4) mature (Oldeman and van Dijk 1991). The first and last phase are better known ecophysiologically and both extremes are discussed to highlight the present knowledge of environmental constrains and ecophysiological adaptations of tropical forest trees.

Studies of population dynamics under natural conditions have established that tropical rain forest trees present a large range of shade tolerance. The extremes of this continuum are early successional or pioneer trees that grow rapidly in gaps under high-light intensity, and late successional (mature forest trees) that are found in the forest understory and persist for prolonged periods, showing slow or nil growth rates (Swaine and Whitmore 1988). Those differences are also associated to population properties such as reproductive effort and frequency, intrinsic growth rate, and sensitivity to nutrient and water availability.

The universe of morphological characteristics, growth patterns, and light requirements may be used for a broad characterization of tree types in tropical forests. They have been described as temperaments that encompass the range of behavior observed under natural conditions and may serve as a guide for the study of forest conservation and regeneration. The main types as identified by Oldeman and van Dijk (1991) include species that reproduce frequently, producing large number of seedlings, with little tolerance to low light intensity, and rapid growth rates. Those are the light demander pioneers (Swaine and Whitmore 1988), early successional species, or gamblers in the nomenclature of Bazzaz and Picket (1980). At the other extreme are the species that reproduce less frequently, producing a small number of shade-tolerant seedlings, capable of enduring prolonged periods under the forest canopy. Those are the climax species of Swaine and Whitmore (1988), late successional species, or the strugglers of Bazzaz and Picket (1980). There is a range of intermediate types, and it is not simple to characterize fully the behavior of a certain species, because frequently their characteristics may also change with age and successional status (Oldeman and van Dijk 1991) (Table 10.2).

Establishment and Development

Forest regeneration requires the establishment of seedlings and saplings within the same (or similar) environment where the parent trees grow. However, in rain forests, environmental conditions determining performance of adult trees contrast with those under which their seeds germinate and develop. Adult trees occupy a volume of the forest canopy, with levels of light availability at least one order of magnitude higher than those prevalent in the forest understory (Chazdon and Fetcher 1984). Photosynthetically active radiation is the driving force in the production of organic matter; therefore, it has been assumed that openings of the canopy (gaps) allowing enough light to reach the understory are required for the regeneration of forest trees (Strauss-Debenedetti and Bazzaz 1996).

Light intensity has been identified as the main limitation for seedling establishment in tropical rain forests. Nutrient availability may also be critical as a regulator of seedling growth rate. Root competition and mutualistic symbiosis also affect seedling establishment, at least in forests growing on acidic, highly leached soils. Their influence is mainly related to the regulation of nutrient availability. Genetic factors regulating growth rate determine the demand for environmental resources, and therefore may influence the carbon balance in the shady understory of tropical forests. Herbivory is frequently critical for seedling survival, but it is not within the scope of the present analysis.

Plasticity and Acclimation of Photosynthesis

The bulk of literature on the effect of light intensity on growth of rainforest tree seedlings has been reviewed elsewhere (Kitajima 1994, Strauss-Debenedetti and Bazzaz 1996). A summary from those reviews indicates the following: (1) independently of successional status, survival, and growth rates of all species are lower under light regimes similar to those prevalent in the understory of tropical rain forests; (2) photosynthetic traits of species classified as late successional are generally less plastic than those of the species classified as early successional or pioneers; and (3) there is a continuum in the tolerance to shade among tropical forest trees.

Growth under the low-light regimes of the forest understory represents a significant stress jeopardizing seedling establishment and regeneration of the tree species that dominate those forests. It is frequently assumed that some kind of forest disturbance that results in gap formation of varying sizes might be indispensable for forest regeneration. However, seedling of trees classified as shade tolerant may differ markedly in their relative growth rates under low (less than 1% full sun light) that may translate into differential competitivity when gaps

TABLE 10.2
Main Morphological and Ecophysiological Characteristics Separating Types of Tree "Temperaments"

Temperament Classification	Leaf Arrangement	Successional Status	Light Requirement	Examples
Hard gambler	Monolayer, spherical to hemispherical, phyllomorphic	Early successional	Light demanding, shade intolerant	*Cecropia, Jacaranda, Didymopanax*, fan palms, *Annona, Apeiba*
Gambler	Multilayer	Later successional	Light demanding, some tolerance	*Trema, Macaranga, Goupia, Ceiba*
Gambling strugglers, struggling gamblers	Umbrella, pagoda, paucilayer	Later to late successional	Shade tolerant to intolerant, favored by light	Feather palms, *Terminalia, Manilkara Pouruma, Iryanthera Aspidosperma*
Strugglers	Elongate, diffuse	Late successional	Shade tolerant, favored by moderate light levels	*Anaxagorea, Duguetia, Hirtella, Perebea*
Hard strugglers	Monolayer	Late successional	Very shade tolerant, favored by moderate light levels	*Pavonia, Guarea, Casearia*

Source: Adapted from Oldeman, R.A.A. and van Dijk, J., *Rain Forest Regeneration and Management*, A. Gómez-Pompa, T.C. Whitmore, and M. Hadley, eds, *Man and the Biosphere Series, Vol. 6.*, UNESCO, Paris, 1991.

occur (Bloor and Grubb 2003). These differences were documented in two shade-tolerant shrub species in Costa Rica, *Ouratea lucens*, with long-lived leaves, and *Hybanthus prunifolius* with short-lived leaves (Kursar and Coley 1999). When exposed to light intensity in a gap after cultivation in the forest understory for 2 years, *O. lucens* maintain their leaves that increase their photosynthetic capacity by 50%, whereas *H. prunifolius* shed all leaves within a few weeks after transfer and produced new leaves with a 2.5 times higher photosynthetic capacity.

Among the multiple responses of the photosynthetic machinery of different species to contrasting light regimes, the interactions between regulation of maximum photosynthetic rate (A_{max}) and the allocation of biomass to photosynthetic structures or roots still have to be analyzed and explained in detail. Several studies showed that the responses of seedlings from trees of different successional status may be related to both increased photosynthesis and increased allocation into leaf biomass when the plants are grown under relatively high intensity (Ramos and Grace 1990, Tinoco-Oranjuren and Pearcy 1995).

Plants developed under a certain light regime adapt their morphology and biochemical characteristics so that carbon gain is maximized. The degree of change in those properties for a species cultivated under contrasting light regimes represents its photosynthetic plasticity (Fetcher et al. 1983, Oberbauer and Strain 1985, Ramos and Grace 1990, Riddoch et al. 1991a,b, Strauss-Debenedetti and Bazzaz 1991, Turnbull 1991, Ashton and Berlyn 1992, Kammaludin and Grace 1992a,b, Turnbull et al. 1993, Tinoco-Ojanguren and Pearcy 1995). Changes in light conditions during growth, from low to high or in the reverse direction, generate strong physiological changes that can be measured as responses in growth and leaf photosynthetic rates. The speed and efficiency with which those changes take place constitute the acclimation capacity of the species considered (Bazzaz and Carlson 1982). Photosynthesis acclimation can take place within the same leaves developed in the previous light regime, or in new modules that develop after the shift of light conditions (Kamaluddin and Grace 1992a,b, Strauss-Debenedetti and Bazzaz 1996). In general, acclimation of leaves formed under a certain light regime is not as complete as the one observed in the newly formed leaves. These results support the view of Bazzaz et al., that light-demanding species are more plastic and acclimate faster to contrasting light regimes than late successional species.

Both A_{max} and dark respiration (R_{dark}) are reduced when plants are transferred from high- to low-light intensity, regardless of the successional status of the species considered (Turnbull et al. 1993). Transfers from low to high light result in larger increases in A_{max} and R_{dark} in pioneer and early successional species. Pioneer species can modify their allocation of nitrogen to photosynthetic proteins according to levels of light energy available during growth; therefore, they are more plastic, as reported by Bazzaz and Pickett (1980). Late successional or shade types cannot take full advantage of the higher light availability because the content of RuBP-carboxylase does not increase as much as in the sun types (Gauhl 1976, Björkman 1981). Regulation of leaf respiration in seedlings developed under low light, may have a more important role in shade tolerance than has been appreciated to date.

Most studies on the photosynthetic plasticity and acclimation capacity of tree seedlings have been conducted at the leaf level only. Conclusions regarding light requirements for survival based on this type of analysis may be erroneous because leaf area development has a number of costs associated with its own construction and maintenance and with support and transport of water and nutrients (Givnish 1988). Ecological significant light compensation points must be calculated on a daily basis, taking into account the photosynthetic surplus of the leaves and the consumption of assimilates by leaf night respiration, and amortization of construction costs. When whole-plant gas exchange is considered, the light requirements for a positive carbon gain increase rapidly. This type of analysis is required for a more fundamental understanding of seedling survival in the understory of tropical forests.

For many tropical rainforests, the light environment in the understory is very dynamic because of the occurrence of sunflecks. Pearcy and others showed that seedlings and saplings growing in the understory of natural forests respond to the frequency and intensity of sunflecks. These short-term pulses of light contribute to improve the carbon balance of plants in the shade (Chazdon and Pearcy 1986, Pearcy 1988). Photosynthetic capacity of shade plants is induced by sunflecks of several minutes duration (light activation of ribulose biphosphate [RuBP]-carboxylase). Afterward, photosynthetic responses to subsequent sunflecks are rapid. Fully induced leaves show higher carbon gain than expected with steady-state assimilation rates because of the postillumination CO_2 fixation (Pearcy 1988). Models showed that frequency of light flecks is critical to attain higher carbon balances than those expected under steady state of photosynthesis.

In the understory of rain forests, CO_2 concentrations are frequently above the average concentrations in the atmospheric air above the forest (Medina et al. 1986, Wofsy et al. 1988). The higher CO_2 concentration in the forest understory is a result of the CO_2 production by soil respiration (root respiration + respiration of decomposer organisms in the soil). Higher CO_2 concentration may improve the carbon gain of light-limited leaves by increasing the CO_2 gradient between air and carboxylating sites in the chloroplasts. Plants in the understory of tropical forests have a lower abundance of ^{13}C in their tissues as measured by the δ ^{13}C value (Medina et al. 1986). This results from the contribution of carbon 13-depleted CO_2 from organic matter decomposition and tree respiration to the photosynthesis of the shade flora (Medina et al. 1986, van der Merwe and Medina 1989, Sternberg et al. 1989, Buchmann et al. 1997), and to a reduction in the ratio of internal to external CO_2 concentration (C_i/C_a) in the leaves (Farquhar et al. 1989).

Radiation Load and Photoprotection

Light energy utilization by the photosynthetic apparatus involves light absorption, electron transport against electrochemical gradients leading to energy accumulation in phosphorylated compounds, and synthesis of reduced compounds (photochemical component). This "reducing" power (adenosine triphosphate [ATP] + reduced nicotinamide adenine dinucleotide phosphate [NADPH]) is used to incorporate carbon into energy-rich organic compounds (enzymatic component) and in photorespiration. The use efficiency of absorbed light energy depends on the electron transport capacity of the photochemical component, and the consumption of the ATP and NADPH molecules generated in the process by the enzymatic component. Under natural conditions, when light absorption surpasses the capacity of the photosynthetic machinery to process it, the phenomenon of photoinhibition is observed. It is defined as a reduction in the quantum yield of photosystem II that may lead to a decrease in the CO_2 uptake rate (Long et al. 1994, Osmond and Grace 1995). Higher plants differ in these capacities according to genetic constraints and ecological conditions such as light climate during growth and drought stress. The study of photoinhibition in nature has expanded quickly, resulting from theory development and the availability of portable instrumentation for assessment of leaf fluorescence under natural illumination fields (Schreiber et al. 1994).

Plants occupying different strata in an undisturbed forest are submitted to contrasting light conditions during the day. There is a vertical, logarithmic reduction in integrated light intensity that may vary at least by one order of magnitude (Fetcher et al. 1983, Chazdon 1986). The photochemical efficiency of photosystem II is affected by changes in the light climate in a dynamic fashion. In leaves of fully sun-exposed plants, intrinsic quantum yield of photosystem II (measured as reduction in the quotient of variable fluorescence to maximum fluorescence in dark adapted leaves, F_v/F_m) decreases as the light energy absorbed by the leaf increases. The reduction in quantum yield of photosystem II works as a protective mechanism of the photosynthetic machinery, and it is accompanied by a highly effective mechanism that dissipate excess light energy as heat, the xantophyll cycle (Björkman and Demming-Adams 1995). Photoinhibition processes under natural conditions are, in general, readily reversed under low light or darkness in the order of minutes (dynamic photoinhibition) or hours (chronic photoinhibition) (Osmond and Grace 1995).

Sensitivity toward photoinhibiton in tropical forests may be assessed by measuring the amount of pigments of the xantophyll cycle (Königer et al. 1995). Canopy species showed higher maximum photosynthetic rates and had a xanthophyll cycle pool (violaxanthin + anteraxanthin + zeaxanthin) averaging 87 mmol mol^{-1} chlorophyll, leaves of the gap plants had intermediate photosynthetic rates and a xanthophyll cycle pool of 35 mmol mol^{-1} chlorophyll, whereas the understory plants had both the lowest photosynthetic rates and the lowest concentration of the xanthophyll cycle pool, 22 mmol mol^{-1} chlorophyll.

In fully exposed leaves, either in the canopy or in large gaps, the efficiency of photosystem II measured as the ratio F_v/F_m is inversely related to the light intensity before the time of measurement; however, this reduction is generally recovered overnight (Königer et al. 1995). The reversion of fluorescence quenching produced by photoinhibition during the day recovers in the shade within 1–2 h in several gap species in Panama (Krause and Winter 1996). Young leaves appear to be more sensitive to photoinhibition, but they have a similar recovery capacity to that of adult leaves (Krause et al. 1995).

Leaf longevity of shade-tolerant species in the understory of rain forests differs from few months to several years and short-lived leaves are more sensitive to photoinhibition than long-lived leaves (Lovelock et al. 1998). The same species grown in forest gaps are less sensitive to photoinhibition, and leaves with different life spans have similar properties in their photosynthetic machinery. These results contrast with those reported by Kursar and Coley (1999) emphasizing the variety of growth responses to changes in light conditions within the group of shade-tolerant species.

Ishida et al. (2005) showed that leaves of pioneer and climax species when developed under full sun light maintain their physiological differences. Similarly, leaves of drought-deciduous species and evergreen trees from tropical dry forests perform quite differently throughout their leaf life spans (Ishida et al. 2006). Leaves of deciduous trees were characterized by shorter leaf life spans, higher N concentration, and higher mass-based photosynthesis compared with leaves of evergreen trees. At the beginning of the dry season photosynthetic rates decreased in both leaf types, but deciduous trees maintained their photochemical capacity and nonphotochemical quenching (NPQ) relatively constant, whereas the evergreen leaves maintained a higher water use efficiency and became relatively photoinhibition-tolerant through the increase in NPQ.

When plants belonging to different successional groups are experimentally grown under low-light intensity and then transferred to full sunlight, the result is generally an abrupt reduction in F_v/F_m. This reduction is more pronounced and takes longer to recover in species usually found in understory environments (Lovelock et al. 1994). Other studies did not find differences among species of different successional status regarding chronic photoinhibition when cultivated without water or nutrient stress (Castro et al. 1995).

The intensity of light impinging on the leaf surface is a function of the cosine of the incidence angle. Therefore, the simplest mechanism to regulate excess of light energy is the change in the angle of leaf inclination. Pronounced leaf angles are frequent in canopy trees in the humid tropics, and had been shown to be critical for leaf energy balance under eventual water stress (Medina et al. 1978). Lovelock et al. (1994) showed that for a set of species growing under varying light climates, degree of leaf inclination is frequently large enough to compensate for differences in leaf exposure to sunlight, resulting in similar F_v/F_m ratios in sun and shade plants.

Nutrient Availability and the Role of Mycorrhizal Symbiosis

Soon after exhaustion of seed reserves, seedling growth is strongly dependent on the availability of nutrients from the forest soils (Kitajima 1996). In most tropical forests, P is the most common limiting nutrient (Grubb 1977, Vitousek 1984, Vitousek and Sanford 1986, Medina and Cuevas 1994), but seedling growth and survival may be affected by simultaneous limitations of several nutrients.

In *Melastoma malabathricum* mixtures of P plus Ca, micronutrients, Mg, K, and N stimulated growth far beyond that provoked by P alone (Burslem et al. 1994, 1995). Particularly important are the light-intensity–nutrient-availability interactions. Increasing light intensity elevates nutrient demands, most notably in the case of N (Medina 1971). Increased nutrient supply leads to higher maximum photosynthetic rates in seedlings of plants of different successional status, irrespective of the light intensity of cultivation (Thompson et al. 1988, 1992, Riddoch et al. 1991b). High-light intensity may even negatively affect photosynthetic performance under conditions of low-nutrient availability (Thompson et al. 1988).

Mycorrhizal symbiosis is widespread in tropical humid forests (Janos 1983, St John and Uhl 1983, Hopkins et al. 1996). Most species develop VAM, but an important group constituted by the Dipterocarpaceae and legumes of the subtribes Amherstieae and Detarieae (Caesalpinioid), and such important genera as *Aldina* and *Swartzia* in the Papilionoid, are ectomycorrhizal and can reach dominance over large areas in the humid tropics (Janos 1983, Alexander and Högberg 1986, Alexander 1989). The occurrence of mycorrhiza is essential to understand the nutrient balance of humid tropical forests. The widespread limitation of P availability over vast tropical areas emphasizes the importance of this biological interaction, which is considered to increase the capability of water and nutrient uptake, particularly P, by higher plants. Went and Stark (1968) proposed the occurrence in tropical forests of a "closed" nutrient cycle mediated by mycorrhiza preventing or reducing nutrient leaching. It is now generally accepted that predominance of mycorrhizal symbiosis (both ecto- and endomycorrhiza) in the majority of humid tropical forests is certainly associated to low P availability in the soil (Newbery et al. 1988). The frequency and percentage of mycorrhizal infection is inversely related to soil pH and available P (van Noordwijk and Hairiah 1986). In tropical forests with a thick root mat, phosphate solutions sprayed on the soil surface are effectively taken up by VAM, thereby preventing nutrient leaching in these forests (Jordan et al. 1979).

The improvement in nutrient supply brought about by mycorrhizal symbioses is related to the increase of surface for nutrient absorption, by penetrating the soil beyond the zone of nutrient depletion around the fine roots. In addition, some mycorrhizal fungi are capable of using organic P sources directly or through previous digestion by extracellular phosphatases (Alexander 1989).

Diversity of VAM is generally low to very low compared with the diversity of host vascular plants in tropical forests, and little information exists about the differential efficiency of VAM species in the process of nutrient uptake and the specific relationships between VAM and host species. Lovelock et al. (2003) showed that VAM communities in a tropical forest at La Selva, Costa Rica, are affected by host tree species and soil fertility. These observations were confirmed by the analysis of VAM community diversity in plantations on relatively fertile soils (Lovelock and Ewel 2005). In plots planted alone or in combination of *Cedrela odorata*, *Cordia alliodora*, and *Hyeronima alcorneoides* they showed that host tree species and host plant diversity had strong effects on the VAM fungal community. VAM fungal diversity (Shannon-index) was positively correlated with ecosystem net primary productivity, whereas evenness was correlated with P-use effciency. The authors concluded that in tropical forests diversity of VAM fungi and ecosystem NPP are correlated.

Nutrient availability in a given soil may be markedly affected by root competition. Particularly in forests growing on nutrient poor soils, fine roots tend to accumulate near and above the soil surface, developing a root mat that can exert a strong competitive pressure for water and nutrients (Stark and Jordan 1977, Jordan et al. 1979, Cuevas and Medina 1988). Root trenching experiments in a tropical rain forest of the upper Orinoco basin resulted in increased concentration of N and P in the trenched saplings, and also a marked increase in leaf area development and branching. These experiments show that root competition for nutrients with established trees reduces sapling growth and survival in the understory (Coomes and Grubb 1996, 1998, Coomes 1997).

Efficiency of Nutrient Use in Carbon Uptake and Organic Matter Production

Productivity of tropical humid forests is frequently limited by the nutrient supply from the soil substrate. Main limitations have been described for N, P, K, and Ca (Vitousek 1984, Medina and Cuevas 1994, Laurance et al. 1999). Nutrient limitation is reflected in the photosynthetic capacity of trees, N being the most common limiting nutrient, followed by P, K, and less frequently Ca. The impact of nutrient limitation on the development and performance of the photosynthetic machinery of trees is complex. Nutrient supply affects leaf expansion, leaf thickness, leaf weight/area ratios, and intrinsic capacity for photosynthesis (Medina 1984, Reich et al. 1991, 1995). Efficiency of nutrient use in photosynthesis may be conveniently expressed measuring the relationship between the specific nutrient concentration and the maximum rate of photosynthesis under natural or laboratory conditions. Studies in tropical humid forests on sandy soils with strong nutrient limitations confirmed previous findings, showing that photosynthetic rate (μmol CO_2 per unit area or weight) and leaf N concentration are often linearly related (Reich et al. 1991, Raaimakers et al. 1995). These studies indicate that within a certain habitat, the use efficiencies of N and P in photosynthesis (μmol CO_2 mol^{-1} nutrient s^{-1}) are highly correlated (Figure 10.4). In this data set, photosynthetic N/P efficiency ratios vary from approximately 50 to 70. However, in communities apparently not limited by P, with leaf P contents much higher than those of

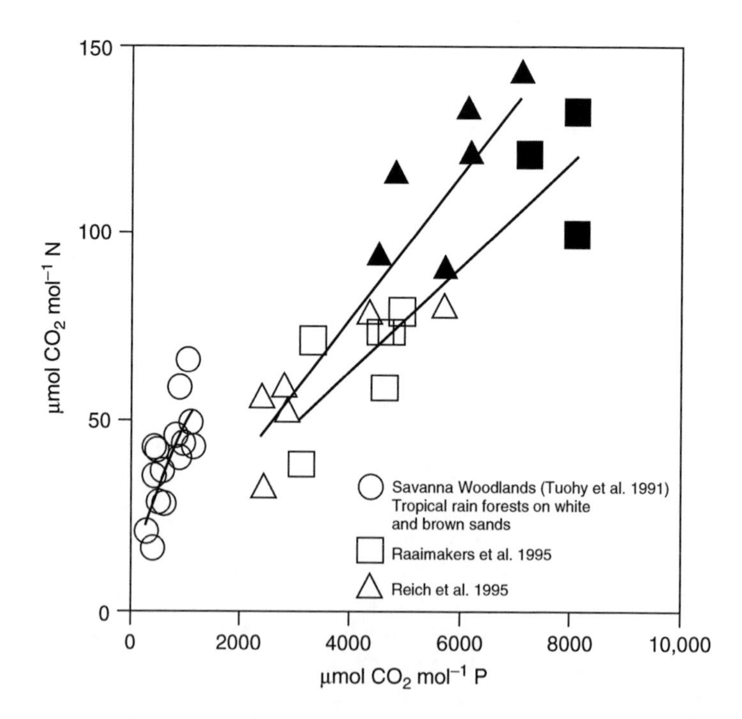

FIGURE 10.4 Instantaneous N- and P- use efficiencies in photosynthesis measured in leaves of native trees under natural conditions. Shaded symbols indicate pioneer and early succession species. (Data from Raaimakers, D., Boot, R.G.A., Dijkstra, P., Pot, S., and Pons, T., *Oecologia*, 102, 120, 1995; Reich, P.B., Elsworth, D.S., and Uhl, C., *Funct. Ecol.*, 9, 65, 1995; Tuohy, J.M., Prior, J.A.B., and Stewart, G.R., *Oecologia*, 88, 378, 1991.)

the Amazonian trees previously discussed, the P use efficiencies are much lower, and the N/P ratios vary around 20 (Tuohy et al. 1995) (Figure 10.4).

Both P- and N- use efficiencies in photosynthesis (and overall plant growth) are significantly higher in species belonging to the pioneer and early succession categories. These species are characterized by larger growth rates, shorter life spans, and higher leaf nutrient concentrations. One significant generalization on the relationships between photosynthetic capacity and nutrient concentration has been achieved combining measurements of photosynthetic performance with demographic studies aimed to measure leaf life spans (Reich et al. 1991). Leaf life span increased along the successional status of the species considered from pioneer, early, mid, and late successional. Along this gradient, photosynthetic capacity decreased, together with leaf structural properties such as leaf area/weight ratios and N and P concentrations (Reich et al. 1995). The relationships between structural, nutritional, and functional properties of leaves described for tropical humid forests apply for a wide variety of forest communities across large climatic gradients (Reich et al. 1997).

For the forest as a whole, the efficiency of nutrient use for organic matter production increases in nutrient-limited environments. This has been shown using data on fine litter fall (mainly leaves) and litter nutrient concentration. The inverse of the nutrient concentration (mass per unit nutrient content) is higher for N, P, and Ca in forests with a limited supply of these nutrients (Vitousek 1984, Medina and Cuevas 1994). As in the case of the nutrient use efficiency for photosynthesis, P shows much higher values than N, as a consequence of the differences in physiologically active concentrations of these nutrients. The P/N efficiency ratios measured in large litter fall data sets vary from 59 (Vitousek 1984) to 72 (Proctor 1984), numbers remarkably similar to the ratios reported for instantaneous use efficiency in photosynthesis (Figure 10.4).

PALMS

Palms are a ubiquitous component of tropical humid forests that can reach dominance in areas with a tendency to water logging. They are botanically well understood and more than 200 well-separated genera have been recognized (Tomlinson 1979). In spite of a relative restricted growth habit, palms have species capable of occupying every habitat in periodically or permanently flooded soils of swamp equatorial forests (from canopy to understory), and constitute nearly monospecific stands in permanently water-logged soils (Kahn and de Granville 1992).

Architecture and Growth Patterns

Palms have a precise growth programming characterized by an essentially continuous vegetative growth, lacking dormancy mechanisms that restrict them to climatically predictable environments such as those predominant in tropical and subtropical environments (Tomlinson 1979).

Their architecture is restricted to two essential types (Halle et al. 1978, Tomlinson 1979):

1. Monoaxial trees, characterized by a single vegetative shoot meristem, without reiteration. This type includes hapaxanthic (monocarpic) species with terminal inflorescences (Holtum's model), such as *Caryota urens, Corypha umbraculifera, Metroxylon salomonense, Raphia regalis*; and species that grow from a single aerial meristem producing one unbranched axis with lateral inflorescences, therefore policarpic (Comer's model). This second group includes the majority of single-stemmed palms; examples include *Areca catechu, Borassus aethiopium, Cocos nucifera, Mauritia fiexuosa, Oenocarpus distichus, Phytelephas macrocarpa, Roystonea oleracea*, and *Socratea exorrhiza*.

2. Polyaxial trees, characterized by repeated development of equivalent orthotropic modules in the form of basal branches initially restricted to the epicotiledonary region of the seedling axis and the basal nodes in the subsequent axes (Tomlinson's model), or those that grow from meristems producing orthotropic or plagiotropic trunks forking at regular but distant intervals by equal dichotomy, with lateral inflorescences but without vegetative lateral branches (Shoute's model). The former group includes almost all multistemmed palms, with species such as *Bactris gasipaes, Euterpe oleracea, Hyphaene guineensis, Oncosperma tigilaria, Phoenix dactilifera*, and *Raphia gigantea*. The second group of this type is less frequent and includes species such as *Allagoptera arenaria, Chamaedorea cataractarum, Hyphaene thebaica, Nannorrhops ritchiana, Nypa fruticans*, and *Vonitra utilis*.

Light and Water Relations

Palms are a taxonomically uniform group constituting the only monocots that contribute to the canopy, subcanopy, and forest floor layers of most tropical humid forests. Their ecophysiological differentiation is considerable and excluded only from the driest tropical forest communities. In the Amazon region they occupy the whole range of available habitats to such an extent that a key for the forest formations in this large expanse of tropical forest communities has been developed on the basis of the palm species (Kahn and de Granville 1992; Table 10.3). The range of habitats of ecophysiological interest vary from permanently water-logged forests (*Mauritia flexuosa* or *minor*), and seasonally flooded palm savannas with sandy or clay soils (*Sabal mauritiaeformis, Orbignya martiana*), to hyperseasonal sites in savannas with 3–5 month dry periods and water logging during the rainy season (*Copernicia tectorum, Copernicia cerifera*, and *Cocos schizophylla*) (Hueck 1966, Walter 1973, Kahn and de Granville 1992, Kalliola et al. 1993).

The diffuse distribution of the vascular tissue within the stem renders these plants resistant to burning, critical for survival of palm species occurring in savannas of Africa (*Borassus*) and South America (*Copernicia, Acrocomia, Sabal*). Their tolerance to flooding and comparatively long periods of drought is also related to their anatomy. Holbrook and Sinclair (1992) showed in *Sabal palmetto* that the amount of water accumulated in the stem per unit of living leaf area increases linearly with plant height. In addition, leaf epidermal conductance was in the range reported for xerophytes. Tolerance to prolonged drought is reduced by the fact that palms do not have a resting phase during their life cycle.

Tropical palms frequently constitute a dense understory layer in tropical moist forests (Gentry and Emmons 1987, Kahn and de Granville 1992, Kalliola et al. 1993). In these environments, light is the main factor influencing plant growth rate and time of reproduction. Photosynthesis studies on understory palms showed positive carbon balances at the leaf level when exposed to light intensities as low as 25 μmol m^{-2} s^{-1}. However, their growth under natural conditions was stimulated significantly when growing in gap edges compared with plants under full shade in the forest understory (Chazdon 1986). Computer simulations of carbon gain of full-shade-grown seedlings indicate that 24 h positive carbon balance is reached whenever daily irradiation is greater than 0.2 mol m^{-2}.

The long-lasting leaf scars on the palm stem have been used as a marker to age palm populations and to detect changes in growth conditions (Piñero et al. 1984). A gap formed by the fall of one or a few trees not only opens the forest canopy, allowing more light to reach the forest floor, but turns down small undercanopy palms without uprooting them. The tilted stems of *Astrocaryum mexicanum* in this study continued to grow vertically, forming a sharp bend that indicated the occurrence of a disturbance. The number of leaf scars after the bending of the stems multiplied by the average leaf life span was found to be an accurate measure of the time elapsed since the disturbance. However, although growth is continuous,

TABLE 10.3
Forest Types and Palm Genus Diversity in the Amazon Basin

Forest Type	Arborescent: Single or Multistemmed Large (Leaf > 4 m) or Slender (Leaf < 4 m)	Medium Sized: Large or Slender, Single or Multistemmed, Erect to Creeping	Small: Erect, Prostrate, or Climbing; Single, Multistemmed or Acaulescent	Subterranean Stemmed: Large Leaved
Terra firme forests	*Astrocaryum, Dyctiocaryum, Iriartea, Jessenia, Maximiliana, Oenocarpus, Orbygnia, Socratea*	*Astrocaryum, Oenocarpus, Socratea, Syagrus, Wettinia*	*Aiphanes, Astrocaryum, Bactris, Chamaedorea, Chelyocarpus, Desmoncus, Geonoma, Hyospathe, Iriartella, Pholidostachys*	*Astrocaryum, Orbignya Scheelea*
Forests on periodically flooded alluvial soils	*Astrocaryum, Attalea, Iriartea, Euterpe, Socratea, Oenocarpus, Orbignya, Scheelea*	*Astrocaryum, Chelyocarpus Itaya, Phytelephas*	*Bactris, Geonoma, Desmoncus*	
Forests periodically flooded by Blackwater	*Astrocaryum*	*Leopoldinia*	*Bactris, Leopoldinia, Desmoncus*	
Swamp forests on organic, permanently flooded soils	*Mauritia, Euterpe Socratea*	*Oenocarpus*	*Bactris, Desmoncus*	
Seasonal swamp forest on water-logged, irregularly flooded soils	*Euterpe, Jessenia, Mauritia, Mauritiella Socratea*	*Astrocaryum, Eleaeis, Manicaria, Oenocarpus, Wettinia*	*Asterogyne, Bactris Catoblastus, Desmoncus, Geonoma, Hyospathe*	
Forests on dry, white-sandy soils		*Mauritiella*	*Bactris, Desmoncus*	
Forests on water-logged, white-sandy soils	*Jessenia, Mauritia, Mauritiella*	*Elaeis, Euterpe*	*Bactris, Desmoncus Lepidocaryum, Pholidostachys*	*Astrocaryum, Orbignya Scheelea*
Submontane and montane forests	*Dictyocaryum, Euterpe Iriartea, Prestoea*	*Geonoma, Oenocarpus Wettinia*	*Aiphanes, Chamaedorea Geonoma*	
Savannahs	*Acrocomia, Astrocaryum Copernicia, Mauritia, Syagrus*	*Mauritiella, Syagrus*	*Bactris, Desmoncus*	

Source: Kahn, F. and de Granville, J.-J. *Palms in Forest Ecosystems of Amazonia. Ecological Studies Vol. 95*, Springer-Verlag, Berlin, 1992.

the rates of height growth and leaf production may change significantly with ecological conditions. In populations of *Prestoea montana* in a lower montane forest in Puerto Rico, average leaf production rate after a 36 year observation period was 4 leaves year^{-1}, but the production rate changed with the age of the population sampled and was also lower in the suppressed individuals compared with those that reached the canopy (Lugo and Rivera-Battle 1987).

Climbing Plants and Hemiepiphytes

A characteristic structural feature of tropical forests is the occurrence of woody plants that use other trees as a support for expanding their photosynthetic area under full sun. These species either climb from the forest floor to the upper canopy (climbers) or descend with their roots from the upper canopy to the forest floor (hemiepiphytes) (Richards 1952, 1996, Walter 1973, Whitmore 1990). Climbing plants may be herbaceous (mostly called vines) or woody (frequently called lianas), although, there may be changes in the structural characteristics of the stem during the lifetime of certain species. They are conveniently distinguished by the method of climbing: (1) scandent climbers (with thorns or hooks); (2) tendrillar climbers (with tendrils); (3) root or bole climbers (with adhesive organs); and (4) twiners (twisting apical organs around the host) (Hegarty 1989).

Biomass allocation in these types of plants is characterized by a larger investment in photosynthetic surface at the expense of the production of self-supporting tissues, at least during a significant portion of their life cycle. The consequences of this pattern of biomass allocation for the ecophysiology of these plants vary in lianas and hemiepiphytes. The former are connected to the forest floor throughout their life cycle, no more limited in their water and nutrient supply than the host trees, whereas the hemiepiphytes have to cope with water and nutrient stress during the establishment phase, until they are able to establish contact with the forest floor through adventitious roots.

Lianas

Woody vines (lianas) contribute significantly to the total leaf area of lowland forests (Gentry 1983, Putz 1983), and their number and diversity in certain forest types may be comparable to those of tree species (Rollet 1969, Pérez-Salicrup et al. 2001, Burnham 2002). Lianas species are capable of rapid growth and thereby affect the development of leaf area and branching of the host trees and may increase mortality (Putz 1984). In fact, the presence of lianas does not greatly affect the total leaf area of the forest, the development of liana leaf area is compensated by a reduction in leaf area production by the host tree (Hegarty 1989). The larger ratio of leaf litter production to wood production observed in tropical forests compared with temperate forests may simply be the result of the frequency of lianas in the tropics (Gentry 1983).

The life cycle of a liana begins by germinating in the forest floor, either in a process of natural regeneration or in explosive growth events after natural or man-made disturbances in forest gaps. Light requirements of lianas are similar in variety to those reported for rain forest trees. Rollet (1969) documented the regeneration behavior of 95 liana species in an evergreen forest in the venezuelan Guayana. The most abundant species (8) were absent from the regeneration plots and were categorized as shade intolerant. Among the rest there were shade-tolerant species with weak regeneration but long survival, and intermediate-illumination species with abundant regeneration but small survivorship. It is therefore erroneous, to consider all liana species as shade intolerant.

During the initial phases lianas are self-supporting, but after a few years they begin to climb on neighboring trees in a process that is tightly coupled to the gap dynamics of the forest. Gaps promote development of lianas, which take advantage of their rapid growth in length to gain height in the forest using neighboring trees as a support (Putz 1984).

Lianas are characterized by a larger leaf biomass/stem biomass ratio than self-supporting trees (Putz 1983), and their xylem vessels are large, offering little resistance to water transport (Ewers 1985). In fact, lianas and vines from evergreen and deciduous tropical forests are characterized by higher specific conductivity of the stem than those of the trees of the same communities, at least when stems of similar diameter are compared (Ewers 1985, Gartner et al. 1990). Putz and Windsor (1987) hypothesized that the large lumen of the xylem vessels was more prone to cavitation; therefore, lianas should be quite sensitive to drought, shedding their leaves at the beginning of the dry season. However, lianas were found to be evergreen, with a longer leaf-producing period than that of the tree species used for comparison. In the study site (Barro Colorado, Panamá), water availability during the dry season was not as severely curtailed as expected, or there were other mechanisms preventing cavitation in these plants. Despite the anatomical differences, Andrade et al. (2004) found that both lianas and co-occurring trees follow the same linear relationship between sap flow and DBH. Besides, measurements of D/H ratios in xylem water indicated that lianas compete with self-supporting trees tapping water at similar soil depths. Lianas inhibit saplings and seedling growth of self-supporting tree species, the main cause is the belowground competition (Pérez-Salicrup 2001, Schnitzer et al. 2005). Therefore, the water status of trees should be affected by the presence of lianas. Experimental removal of lianas during the dry season in seasonally dry forests in Bolivia produced contradictory results. *Senna multijuga* trees increased their water potential by liana removal as soon as 1 day after cutting, and grew twice as fast compared with controls (Pérez-Salicrup and Baker 2000). Similar experiments with *Swietenia macrophylla* trees showed that lianas had higher water potential and leaf conductance than the host tree, but host trees with and without lianas had similar water status, indicating that lianas do not reduce the water availability.

Hemiepiphytes

Hemiepiphytes are plants that germinate on branches or stems of trees and that grow and develop leaves and roots using nutrient and water resources that accumulate within the crevices of the stem bark (Richards 1952, 1996, Walter 1973). Their roots eventually reach the soil and develop there in a similar fashion as roots of their host trees. In this process, the plants pass from a truly epiphytic stage, characterized mainly by frequent drought stress, to a self-supporting stage in which the plant functions as a normal tree. A large group of the hemiepiphytic species behave as "stranglers" because their roots grow around the stem of the host tree. As root diameter increases, they compress the stem underneath. This process eventually continues until the host tree stem cannot transport water and nutrient upward and therefore dies, leaving the strangler as a self-supporting independent tree. The group of strangler species is rather limited and is frequently found among the genus *Ficus*, while permanent hemiepiphytes remain host-dependent for support despite extensive root systems in contact with the soil (Todzia 1986).

Epiphytic and tree forms of hemiepiphytes differ in their growth habit and water relations. Holbrook and Putz (1996) found that the epiphytic form of hemiepiphytic *Ficus* strangler species produced thinner leaves with larger specific leaf areas and water contents (%), but lower stomatal densities than the leaves produced at the tree stage. In correspondence with their higher water content, osmotic potential at full turgor was lower in the epiphytic forms, but the rate of cuticular water loss was smaller than that measured in excised tree leaves. In this group of hemiepiphytes, the functional properties of the epiphytic stage is directed to conservative water use according to the uncertainty of water supply in the epiphytic habitat, particularly during the dry season. Although nutrient supply is linked to water availability, it did not appear to be a limiting factor for the development of plants at this stage.

One outstanding group among the hemiepiphytes in the neotropics is constituted by the species of the genus *Clusia* that perform a combination of C3 and crassulacean acid metabolism (CAM) photosynthesis. This process was originally described in *Clusia lundelli* (Clusiaceae) (Tinoco-Oranjuren and Vásquez-Yanes 1983), a true strangler, beginning its life cycle as an epiphyte and then developing roots that eventually anchor in the soil. The same photosynthetic metabolism has been thoroughly documented in the large tree strangler *Clusia rosea* in the United States. Virgin Islands (Ting et al. 1985, Ball et al. 1991) and confirmed in several species of this genus (Popp et al. 1987, Sternberg et al. 1987, Ting et al. 1987, Borland et al. 1992, Winter et al. 1992).

Gas exchange of *Clusia* CAM species has been shown to be highly flexible, going from typical CAM to typical C3, including continuous net fixation of CO_2 during a 24 h cycle, depending on light intensity during growth, day–night temperature cycles, drought, and even nutrition (Schmitt et al. 1988, Lee et al. 1989, Zotz and Winter 1993). Drought induces contrasting effects on *Clusia* species that fix CO_2 during the night. In *Clusia uvitana*, daytime uptake decreased rapidly without irrigation, whereas nocturnal fixation increased throughout the treatment (Winter et al. 1992).

VASCULAR EPIPHYTES

Epiphytes are plants that can complete their life cycle living on the stem, branches, or even leaves of other plants. They acquire water and nutrients from rain water and dust, and from accumulation of organic debris accumulated on the body of the plant or on the surface of the host tree. This life-form has been successful from the evolutionary point of view. Within the higher vascular plants (Magnoliophyta), approximately 6% of the genera and 9% of the species have been recorded as epiphytes (Madison 1977, Kress 1986, Benzing 1990). Epiphytic genera and species are more frequent among monocots (Liliopsida) than among dicots (Magnoliopsida). Benzing (1990) registered 520 genera and 16,608 species of epiphytic monocots compared with 262 genera and 4521 species of epiphytic dicots. The Orchidaceae account for the predominance of epiphytes within the monocots with 67% of the vascular epiphytes.

Epiphytic habitats restrict plant growth mainly through drought stress. Nutrient supply is also restricted because accumulation of organic matter and flow of nutrients in rain water and dust is scarce and erratic. Many epiphytes therefore depend on their capacity to accumulate organic residues produced by the supporting trees, or by animals associated with them (ants, termites), and by their capacity to accumulate water internally via a specialized morphology.

Water and nutrient availability vary with the location of the epiphyte within the forest. Those located in the topmost branches depend on the interception of dust particles and the absorption of dissolved nutrients in rain water, and are exposed to nearly full sunlight. Exposed epiphytic habitats are characterized by a high evapotranspiration demand. Drought spells may be of short duration, but frequent. The duration of the dry spells increases from wet rain forests to dry seasonal forests.

In the middle to low canopy, evapotranspiration demands are lower, and nutrient availability increases because of tree leaf litter accumulation and enrichment of rain water through nutrient leaching. On the other hand, light availability may be limiting for photosynthesis.

Patterns of vascular epiphytes diversity have been documented in many tropical mountains since the original observations of Gentry and Dodson (1987a). Epiphytes reach their peak in biomass and species diversity at middle elevations in a fashion closely correlated with rainfall and air humidity (Küper et al. 2004). Cardelús et al. (2006) showed that larger diversity at middle elevations results at the overlap of large-ranged species at these elevations.

Nutrient N is a limiting factor for most epiphytes. Stewart et al. (1995) used natural abundance of ^{15}N ($\delta\ ^{15}N$) to investigate the source of N for epiphytes compared with their host trees. Samples from six rain forests in several continents showed that epiphytes were consistently depleted in ^{15}N compared with trees at each site. Authors hypothesized that epiphytes were using ^{15}N depleted N from atmospheric deposition and perhaps also from N_2 fixation. Hietz et al. (1999) showed in an altitudinal transect in Mexico that variations in $\delta\ ^{15}N$ were caused by the N acquisiton from different sources. Atmospheric bromeliads presented lower $\delta\ ^{15}N$ values than tank-forming bromeliads. Ground-rooted hemiepiphytes showed the highest N concentrations and $\delta\ ^{15}N$ values. Analyses of distribution of $\delta\ ^{15}N$ values in different positions within the forests led to the conclusion that differences are not only due to the sources, but also due to different patterns of ^{15}N discrimination during N acquisition (Wania et al. 2002). Before a definitive statement is made in this direction it is necessary to analyze the form of available N in each position (proportions of NO_3 and NH_4).

Main adaptations of vascular plants to the epiphytic habitat are related to their water and nutrient economy. They may be summarized as follows:

1. Development of foliose structures capable of retaining rain water and accumulating organic detritus, the so-called tanks and nests. Many bromeliads are tank formers and develop short roots that penetrate into accumulated organic matter and absorb water and nutrients (Pittendrigh 1948). Nests are less efficient than tanks at retaining rain water, but litter accumulation provides a substrate rich in nutrients that are absorbed by the roots of the nest-forming plants. This structure is found among ferns, orchids, and aroids.
2. Development of water-accumulating tissues (succulence). Many epiphytes possess succulent roots, stems, and leaves. These organs accumulate amounts of water and are protected against evaporation by epidermises and highly impermeable cuticles. Water accumulated in these tissues maintains turgor of photosynthetic tissues during prolonged dry spells. Succulent leaves are rather frequent in epiphytic species belonging to the families Piperaceae, Gesneriaceae, Orchidaceae (developing also succulents pseudobulbs), and Bromeliaceae.
3. Sclerophyllous photosynthetic organs resistant to desiccation and with high-nutrient use efficiency. These characteristics are found within the epiphytic Ericaceae and Rubiaceae that form fiber-rich leaves with highly impermeable cuticles. These leaves are long-lived, have low-nutrient concentration per unit dry weight, and are desiccation resistant.
4. Specialized mechanisms for water and nutrient uptake. The bromeliad leaves, particularly within the subfamily Tillandsioideae, possess specialized trichomes for the uptake of water and nutrients. Orchid roots, on the other hand, are covered by a hygroscopic tissue operative in the absorption of water and nutrients, the velamen. According to Benzing et al. (1983) this structure may be important in explaining the predominance of orchids in epiphytic tropical habitats.
5. Mutualistic symbiosis. Mycorrhizae are a common feature of vascular plants in the tropics (Janos 1983). Among the richest group of epiphytic plants, the orchids, a highly specialized mycorrhiza, dominate. Orchid seeds are not nutritionally independent and associate with fungi to complete their germination process until they reach photosynthetic independence. Later, they develop permanent mycorrhizal associations.
6. High water use efficiency and nocturnal CO_2 fixation. A large number of epiphytes among the Orchidaceae, Bromeliaceae, and Cactaceae, particularly those of strongly seasonal forests, are capable of nocturnal CO_2 fixation via CAM. CO_2 taken up during the night is accumulated in the vacuoles as malic acid. This acid is remobilized during the following day, decarboxylated in the cytoplasm, and the evolving CO_2 is taken up via the photosynthetic reduction cycle in the chloroplasts. In this process, stomata open during the night and remain closed during part of the day, depending on the water status of the photosynthetic tissue and the actual amount of malic acid accumulated. The consequence of this inverted stomatal rhythm is a high water use efficiency (amount of water transpired per unit carbon taken up).

Ecophysiology of CAM Epiphytes

CAM plants may be constitutive or facultative (Winter 1985, Medina 1990). The former show CAM characteristics under a wide range of environmental conditions, whereas the latter only show CAM under conditions of water or salt stress. Most CAM plants in the tropics are constitutive, as the number of facultative species described is limited. There are two variations in the nocturnal organic acid accumulation process that are considered to be related to CAM and are frequently found in epiphytes of humid forests: CAM-idling and CAM-cycling (Kluge and Ting 1978, Benzing 1990). In both processes, nocturnal increases of acid concentration are observed without net CO_2 uptake from the atmosphere. CAM-idling occurs in water-stressed constitutive CAM plants and results from the refixation of respiratory CO_2 (Szarek et al. 1973, Ting 1985), whereas CAM-cycling occurs in tissues without water stress (Ting 1985). CAM-cycling has been hypothesized to be a precursor of full-CAM plants (Monson 1989, Martin 1994), but it is still not clear why the reduction in internal CO_2 partial pressure resulting from fixation of respiratory CO_2 during the night does not influence stomatal behavior.

The following ecophysiological features are essential to understand the abundance and ecological significance of CAM in humid tropical forests:

1. High water use efficiency. Constitutive CAM plants take up CO_2 mainly during the night, when the leaf-air water vapor saturation deficit is lower. High water use efficiency associated with the succulent morphology of photosynthetic organs allows the maintenance of a positive carbon balance in drought-prone habitats, and is therefore crucial for the occupation of exposed epiphytic habitats in rainforests.
2. Respiratory CO_2 recycling. In many constitutive CAM plants, simultaneous measurements of CO_2 exchange and organic acid accumulation during the night frequently reveal that the increase in acid concentration is larger than that expected from the amount of CO_2 taken up by PEP-carboxylase. Dark CO_2 fixation should yield two acid equivalents (malic acid) per CO_2 molecule taken up. However, this is not the case with several species of the Bromeliaceae and Orchidaceae, of the genus *Clusia*, and in CAM ferns. This difference is apparently due to refixation of respiratory CO_2 (Griffiths 1988). Recycling of respiratory CO_2 may be a significant carbon conservation mechanism under stressful environmental conditions.
3. Adaptation to shady environments. The numerous taxa with CAM in humid tropical forests that can grow and compete with C3 plants in shady environments indicate that CAM by itself does not appear to be a liability in low-light environments. In humid forest environments, shade-adapted CAM plants have growth rates and net carbon gains similar to those of C3 plants that are limited by low light (Martin et al. 1986, Medina 1990). It seems that CO_2 fixation during the night by CAM plants might be favored by the high CO_2 concentrations of the air under the canopy of humid tropical forests (Martin 1994).

Occurrence of CAM in Specific Groups of Epiphytes

The number of epiphytic genera with reported CAM species is substantial (Table 10.4) and is clearly dominated by orchids.

Pteridophytes. Only species of the genera *Pyrrosia* (Hew and Wong 1974, Ong et al. 1986, Winter et al. 1986, Kluge et al. 1989b) and *Microsorium* (Earnshaw et al. 1987) have been reported as CAM plants among ferns. *Pyrrosia longifolia* can be found growing under fully exposed or shaded conditions in humid tropical forests, performing CAM in both conditions. In both populations, the proportion of total carbon gain derived from night CO_2 fixation is similar as indicated by their $\delta\ ^{13}C$ values (Winter et al. 1986).

TABLE 10.4
Families and Genera in Which Epiphytic CAM Species of Humid Tropical Forests Have Been Reported

Family	Genera with CAM Species		
Polypodiaceae	*Pyrrosia*		
	Microsorium		
Bromeliaceae	*Acantostachys*	*Canistrum*	*Tillandsia*
	Aechmea	*Hohenbergia*	*Portea*
	Araeococcus	*Neoregelia*	*Quesnelia*
	Billbergia	*Nidularium*	*Wittrockia*
	Bromelia	*Streptocalyx*	
Orchidaceae (~500 genera and more than 20,000 epiphytic species)	*Bulbophyllum*	*Luisia*	*Robiquetia*
	Cadetia	*Micropera*	*Saccolabiopsis*
	Campylocentrum	*Mobilabium*	*Saccolabium*
	Cattleya	*Oberonia*	*Sarchochilus*
	Chilochista	*Oncidium*	*Schoenorchis*
	Cymbidium	*Phalenopsis*	*Taeniophyllum*
	Dendrobium	*Pholidota*	*Thrixspermum*
	Epidendrum	*Plectorrhiza*	*Trachoma*
	Eria	*Pomatocalpa*	*Trichoglottis*
	Flickingeria	*Rhinerrhiza*	*Vanda*
Asclepiadaceae	*Hoya*		
	Dischidia		
Cactaceae (25 epiphytic genera, most probably all CAM)	*Epiphyllum*		
	Hylocereus		
	Rhipsalis		
	Strophocactus		
	Zygocactus		
Clusiaceae	*Clusia*		
	Oedematopus		
Crassulaceae	*Echeveria*		
	Sedum		
Gesneriaceae	*Codonanthe*		
Piperaceae	*Peperomia*		
Rubiaceae	*Myrmecodia*		
	Hydnophytum		

Sources: Data from Coutinho, L.M., *Bol. Fac. Filosofia*, 288, 81, 1963, Coutinho, L.M., Novas observaçoes sobre a ocorrencia do "Efeito de DeSaussure" e suas relaçoes com a suculencia, a temperatura folhear e os movimentos estomáticos. Boletin Faculdáde de Filosofia, Ciencias e Letras, Universidade de Sao Paulo, 331, 1969; McWilliams, E.L., *Bot. Gazette*, 131, 285, 1970; Medina, E., *Evolution*, 28, 677, 1974; Medina, E., Delgado, M., Troughton, J.H., and Medina, J.D., *Flora*, 166, 137, 1977, Medina, E. and Cuevas, E., *Mineral Nutrients in Tropical Forest and Savanna Ecosystems*, J. Proctor, ed., Blackwell Scientific Publications, Oxford, 1989; Griffiths, H. and Smith, J.A.E., *Oecologia*, 60, 176, 1983; Ting, L.P., Bates, L., Sternberg, L.S.L., and DeNiro, M.J., *Plant Physiol.*, 78, 246, 1985; Winter, K., Wallace, B.J., Stocker, G.C., and Roksandic, Z., *Oecologia*, 57, 129, 1983, Winter, K., Medina, E., García, V., Mayoral, M.L., and Muñiz, R., *J. Plant Physiol.*, 111, 73, 1985; Earnshaw, M.J., Winter, K., Ziegler, H., Stichler, W., Cruttwell, N.E.G., Kerenga, K., Cribb, P.J., Wood, J., Croft, J.R., Carver, K.A., and Gunn, T.C., *Oecologia*, 73, 566, 1987; Griffiths, H., *Ecol. Stud.*, 76, 42, 1989; Benzing, D.H., *Vascular Epiphytes*, Cambridge University Press, Cambridge, 1990; Lüttge, U., ed., *Vascular Plants as Epiphytes. Ecological Studies, Vol. 76*, Springer-Verlag, Berlin, 1991; Martin, C.E., *Bot. Rev.*, 60, 1, 1994; Franco, A.C., Olivares, E., Ball, E., Lüttge, U., and Haag-Kerwer, A., *New Phytol.*, 126, 203, 1994.

The question of how CAM may have evolved within the Polypodiaceae has been discussed by Kluge et al. (1989a) and Benzing (1990). All relatives of the genus *Pyrrosia* grow in humid, shady environments; therefore, it seems improbable that the CAM species are derived from xerophytic ancestors. In this particular case, it appears that high light and frequent low humidity, characteristic of epiphytic environments, may have been the selective factors leading to the development of CAM in this genus. In addition, ferns are a group of plants highly dependent on continuous water availability during their gametophytic stage. It seems more feasible that CAM ferns are derived from shade forms in humid forests, and that CAM evolved secondarily in the epiphytic environment (Kluge et al. 1989a).

Gesneriaceae and Piperaceae. The Gesneriaceae and Piperaceae have a large number of epiphytic species, many of which are succulent. In a few species (*Codonanthe crassifolia* in the Gesneriaceae and approximately 15 species within the genus *Peperomia*), CAM activity has been detected only as acid accumulation during the night period (Ting et al. 1985, Guralnick et al. 1986). Their isotopic composition indicates that all carbon used for growth derives from daytime CO_2 uptake, but their δD values are more similar to those of CAM plants than to those of C3 or C4 plants (Ting et al. 1985).

Orchidaceae and Bromeliaceae. Orchids and bromeliads contain the largest number of epiphytic species with constitutive CAM metabolism, and many grow under partial- or heavy-shade conditions. The orchids occupy pantropical epiphytic habitats, whereas the bromeliads are restricted to the neotropics. Both families contain epiphytic and terrestrial species with typical CAM, expressed in gas exchange patterns, nocturnal malic acid accumulation, and reduction of nonstructural carbohydrate content during the night (Coutinho 1969, McWilliams 1970, Medina 1974, Neales and Hew 1975, Goh et al. 1977, Griffiths and Smith 1983, Lüttge et al. 1985, Goh and Kluge 1989). There are several accounts on the morphology, physiology, and ecology of epiphytes in these two families (Benzing 1990, Martin 1994, Medina 1990, Lüttge 1991).

Distribution of epiphytic species of orchids and bromeliads is correlated with atmospheric humidity. In montane forests, the proportion of CAM epiphytes decreases with altitude in direct correlation with air humidity (Earnshaw et al. 1987, Smith 1989).

The larger genera of bromeliads and orchids contain both C3 and CAM species. The genus *Tillandsia* in the Bromeliaceae, for instance, has species ranging from dry to humid forests. CAM species predominate in the former, while C_3 species predominate in the latter. Another bromeliad genus, *Aechmea*, contains only CAM species, some of which are restricted to the floor of humid forests. In the Orchidaceae, similar photosynthetic diversity can be observed in the genera *Bulbophyllum* (1000 epiphytic species) and *Dendrobium* (900 epiphytic species). In these genera, there is an almost continuous variation in $\delta\ ^{13}C$ values, from less than −30‰ to −12‰. CAM activity in these orchid genera is correlated with the development of succulence. This morphological property is associated both with the presence of cells with big vacuoles, where malic acid is accumulated during the night, and with the development of water-storage parenchyma, often photosynthetically inactive. For example, the leaf thickness of high-altitude orchids in upper montane forests of Papua New Guinea, is due to the development of a chlorophyll-free hypodermis (Earnshaw et al. 1987), very similar to that described for *Peperomia* spp. (Ting et al. 1985) and *Codonanthe* spp. (Guralnick et al. 1986, Medina et al. 1989).

Roots of many orchid species contain chloroplasts that are photosynthetically active (Benzing et al. 1983). Chloroplast-bearing cells, specialized in photosynthesis, and mycorrhizal cells, specialized in nutrient uptake, coexist in the roots of some leafless species. These roots are capable of performing CAM (Winter et al. 1985).

The process of shade adaptation of CAM plants in view of the relatively higher energetic requirement merits more research at the biochemical level. A number of epiphytic CAM bromeliads grow in a wide range of light conditions, although their abundance tends to increase under full sun exposure (Pittendrigh 1948, Smith et al. 1986).

Cactaceae. Differentiation between terrestrial and epiphytic species in the Cactaceae is more pronounced. There is no genus with both terrestrial and epiphytic species; however, several climbing and decumbent species are found within the genera *Acanthocereus* and *Heliocereus*. The epiphytic cacti are confined within the tribes Rhipsalidanae (8 genera), Epiphyllanae (9 genera), and Hylocereanae (9 genera) within the subfamily Cereoideae (Britton and Rose 1963). The photosynthetic metabolism of epiphytic cacti has been studied in only a few genera (Table 10.4), but there is little doubt that all the species are most likely constitutive CAM plants. Species growing in deep shade, such as *Strophocactus witti* (Medina et al. 1989), are damaged when exposed to high-light intensity, yet they show typical CAM gas exchange and acid accumulation.

FOREST DYNAMICS AND CLIMATE CHANGE

What are the roles tropical forests are playing in the carbon balance of the atmosphere? Answers to this questions are in the center of a debate to identify where is the sink for approximately 2 Pg of carbon that are produced by the combustion of fossil fuels and deforestation in the tropics (Tian et al. 1998, Grace 2004). The answer requires (1) an accurate assessment of the impact of increasing atmospheric CO_2 concentrations and temperature on the forest productive capacity and the accumulation and turnover of organic matter and (2) a precise estimation of the magnitude of deforestation in the tropics. The first part may be accomplished using experimental approaches such as measuring the increase in photosynthetic carbon fixation in isolated plants or whole sections of natural ecosystems, or using forest inventories to estimate biomass. The second aspect is dealt with successfully using remote sensing techniques (http://www-gvm.jrc.it/glc2000/; http://lpdaac.usgs.gov/modis/).

Grace et al. (1995) published a report showing that a forest site in the Amazon basin was acting as a carbon sink during the period of measurements (55 days including wet and dry seasons). The report generated a strong response in the scientific community involved in global change and biogeochemical research. The study was based on the use of micrometeorological techniques to measure CO_2 fluxes into the forest, and has been subjected to severe scrutiny regarding the accuracy and the capability of this methodological approach to be expanded to large forests tracts.

In the assessment of forest C balance long-term measurements and climate variations has to be taken into account before concluding if a certain forest tract is a source or a sink. The impact of seasonal and episodic events on the forest net primary productivity was analyzed by Tian et al. (1998) using a transient process-based biogeochemical model of terrestrial ecosystems. During El Niño years the hot, dry weather caused most of the Amazon basin to act as a source (0.2 Pg in 1987 and 1992). In wet years the same systems acts as sink (0.7 Pg C in 1981 and 1983). During the same period deforestation accounted for 0.3 Pg C per year.

Mahli and Grace (2000) argue that the lack of any significant tropical source detected in the distribution of atmospheric CO_2, in spite of the large CO_2 efflux brought about by deforestation indicates the presence of a large tropical sink amounting to 1–3 Pg C per year. The accumulation of micrometeorological data in several undisturbed forests sites within the Amazon basin shows that carbon balance in generally negative, that is, forest are acting as carbon sinks (Andreae et al. 2002). A long-term study using eddy covariance methods in Santarem, Brazil gave unexpected results (Saleska et al. 2003). The forest studied lost C during the wet season and gained it in the dry season. Calculated losses over the 3 year period were 1.3×10^6 Mg ha^{-1} $year^{-1}$. This net loss was attributed to recent disturbances superimposed on long-term C balances, but the authors indicate that this kind of disturbance is characteristic of old-growth forests and probably reduce the carbon sequestration values reported by other studies using similar methodology.

The assessment of biomass changes in permanent plots throughout the Amazon basin led to the conclusion that forests, in the basin as a whole, were acting as a carbon sink (Phillips et al. 1998). This paper was severely criticized based on the unreliability of data (Clark 2002). The same group presented a new analysis of biomass data using updated forest inventories across 59 sites in the Amazon basin (Baker et al. 2004). In this analysis corrections were made for unreliable data obtained by different inventory methods, and the authors arrived to the same conclusion, that over the last two decades Amazonian old-growth forests have been acting as a carbon sink on a regional scale. Another biometric study conducted in the Amazon basin near Santarem, Brasil, showed an increase in live wood biomass after 2 years of 1.4 Mg ha^{-1} $year^{-1}$. This gain in woody biomass was exceeded by respiration losses from coarse woody debris, the net loss of carbon at the end of the observation period amounting to 1.9 Mg ha^{-1} $year^{-1}$. This study confirmed the results obtained using micrometeorological techniques in the same site (Saleska et al. 2003). The conclusion is that for accurate assessments of carbon balances, biometrics studies have to include the coarse woody debris.

Biomass changes over 22 years were assessed in permanent plots of undisturbed Atlantic tropical moist forest (Rolim et al. 2005). The authors concluded that during the observation period the climatic events associated with El Niño reduced forest biomass followed by gradual biomass accumulation. Present results indicate that the plots have been experimenting a small reduction averaging 1.2 Mg ha^{-1} $year^{-1}$.

In the analysis of effects of climatic change (CO_2 and temperature) on tropical forests little has been done on the changes in forest composition and structure (Slik 2004). Phillips et al. (2002) reports that in nonfragmented Amazon forests there is an overall increase in density, basal area, and mean size of lianas (woody climbers) at a rate reaching 1.7%–4.6% $year^{-1}$. Wright et al. (2004) measured liana litter production and proportion of liana litter forest-wide during 14 years in Barro Colorado, Panamá, and concluded that the proportion of lianas in this forest increased steadily during the observation period. This phenomenon needs to be further documented as lianas compete for aerial space with host trees and may increase mortality and suppress tree growth.

CONCLUSIONS

The analysis of the functional properties of tropical forests components, and of the forest as a whole, continues to develop at a fast pace as a result of the international efforts to study the forest–atmosphere interactions within the context of global climate and land-use change. These efforts are directed to develop operative models capable of quantifying the role of tropical forests in the dynamics of trace greenhouse gases in the atmosphere. The discussion on the impact of increasing atmospheric CO_2 concentrations on the productive performance of tropical forests goes on strongly. A number of groups maintain the notion that elevated CO_2 concentrations in the atmosphere may be stimulating the photosynthetic capacity of tropical forests resulting from a "fertilization" effect. However, others maintain that these effects may be counteracted by water limitations (in water-limited systems) or by nutrient limitations as most humid tropical forests grow on substrates of limited fertility. Some recent developments show that tropical humid forests in the Amazon basin may be acting as a sink for CO_2. In addition, there are already some promising modeling efforts integrating the physiological and ecological knowledge on the effects of CO_2 on photosynthesis and the capability for nutrient uptake under natural conditions (Lloyd and Farquhar 1996). We can expect dramatic developments in the understanding of the physiological ecology of the production processes in tropical forests during the near future. This will be of utmost importance for future restoration efforts considering the rate of land-use change in the tropics.

REFERENCES

Alexander, I., 1989. Mycorrhizas in tropical forests. In: J. Proctor, ed. Mineral Nutrients in Tropical Forest and Savanna Ecosystems. Blackwell Scientific, Oxford, pp. 169–188.

Alexander, U. and Högberg, P., 1986. Ectomycorrhizas of tropical angiospermous trees. New Phytologist 102: 541–549.

Allen, E.B., M.F. Allen, D.J. Helm, J.M. Trappe, R. Molina, and E. Rincón, 1995. Patterns and regulation of mycorrhizal plant and fungal diversity. Plant and Soil 170: 47–62.

Andrade, J.L., F.C. Meinzer, G. Goldstein, and S.A. Schnitzer, 2004. Water uptake and transport in lianas and co-occurring trees of a seasonally dry tropical forest. Trees 19: 282–289.

Andreae, M.O., P. Artaxo, C. Brandão, F.E. Carswell, P. Ciccioli, A.L. da Costa, A.D. Culf, J.L. Esteves, J.H.C. Gash, J. Grace, P. Kabat, J. Lelieveld, Y. Malhi, A.O. Manzi, F.X. Meixner, A.D. Nobre, C. Nobre, M.D.L.P. Ruivo, M.A. Silva-Dias, P. Stefani, R. Valentini, J. von Jouanne, and M.J. Waterloo, 2002. Biogeochemical cycling of carbon, water, energy, trace gases, and aerosols in Amazonia: The LBA-EUSTACH experiments. Journal of Geophysical Research 107 (D20): 8066.

Ashton, P.M.S. and G.P. Berlyn, 1992. Leaf adaptations of some *Shorea* species to sun and shade. New Phytologist 121: 587–596.

Bailey, H.P., 1979. Semi-arid climates: their definition and distribution. In: A.E. Hall, G.H. Cannell, and H.W. Lawton, eds. Agriculture in Semi-arid Environments. Springer-Verlag, Berlin, pp. 73–97.

Baker, T.R., O.L. Phillips, Y. Malhi, S. Almeida, L. Arroyo, A. Di Fiore, T. Erwin, N. Higuchi, T.J. Killeen, S.G. Laurance, W.F. Laurance, S.L. Lewis, A. Monteagudo, D.A. Neill, P. Núñez-Vargas, N.C.A. Pitman, J.N.M. Silva, and R. Vázquez-Martínez, 2004. Increasing biomass in Amazonian forest plots. Philosophical Transactions of the Royal Society of London B 359: 353–365.

Ball, E., J. Hann, M. Kluge, H.S.J. Lee, U. Lüttge, B. Orthen, M. Popp, A. Schmitt, and I. Ting, 1991. Ecophysiological comportment of the tropical CAM-tree clusia in the field. II. Modes of photosynthesis in trees and seedlings. New Phytologist 117: 483–491.

Barbault, R. and S. Sastrapradja, 1995. Generation, maintenance and loss of biodiversity. In: V.H. Heywood, ed. Global Biodiversity Assessment. Cambridge University Press, Cambridge, pp. 192–274.

Barker, M.G. and D. Pérez-Salicrup, 2000. Comparative water relations of mature mahogany (*Swietenia macrophylla*) trees with and without lianas in a subhumid, seasonally dry forest in Bolivia. Tree Physiology 20: 1167–1174.

Bazzaz, P.A., 1984. Dynamics of wet tropical forests and their species strategies. In: E. Medina, H.A. Mooney, and C. Vázquez-Yanes, eds. Physiological Ecology of Plants of the Wet Tropics. Junk, The Hague, pp. 233–243.

Bazzaz, P.A. and R.W. Carlson, 1982. Photosynthetic acclimation to variability in the light environment of early and late successional plants. Oecologia 54: 313–316.

Bazzaz, F.A. and S.T.A. Pickett, 1980. Physiological ecology of tropical succession: a comparative review. Annual Review of Ecology and Systematics 11: 287–310.

Benzing, D.H., 1990. Vascular Epiphytes. Cambridge University Press, Cambridge.

Benzing, D., W.E. Friedman, G. Peterson, and A. Renfrow, 1983. Shootlessness, velamentous roots and the pre-eminence of Orchidaceae in the epiphytic biotope. American Journal of Botany 70: 121–133.

Björkman, O., 1981. Responses of different quantum flux densities. In: O.L. Lange, P.S. Nobel, C.B. Osmond, and H. Ziegler, eds. Encyclopedia of Plant Physiology. New Series Vol 12 A. Physiological Plant Ecology I. Springer-Verlag, Berlin, pp. 57–107.

Björkman, O. and B. Demming-Adams, 1995. Regulation of photosynthetic light energy capture, conversion and dissipation in leaves of higher plants. In: E.D. Schulze and M.M. Caldwell, eds. Ecophysiology of Photosynthesis. Springer-Verlag, Berlin, pp. 17–47.

Bloor, J.M.G. and P.J. Grubb, 2003. Growth and mortality in high and low light: Trends among 15 shade-tolerant tropical rain forest tree species. Journal of Ecology 91: 77–85.

Borland, A.M., H. Griffiths, C. Maxwell, M.S.J. Broadmeadow, N.M. Griffiths, and J.D. Barnes, 1992. On the ecophysiology of the Clusiaceae in Trinidad: expression of CAM in Clusia minor during the transition from wet to dry season and characterisation of three endemic species. New Phytologist 122: 349–357.

Box, E.O., 1981. Predicting physiognomic vegetation types with climate variables. Plant Ecology 45: 127–139.
Brienen, R.J.W. and P.A. Zuidema, 2006. Lifetime growth patterns and ages of Bolivian rain forest trees obtained by tree ring analysis. Journal of Ecology 94: 481–493.
Britton, N.L. and J.N. Rose, 1963. The Cactaceae. Republication of the 2nd original edition of 1937, Carnegie Institution of Washington. Dover Publications Inc., New York.
Buchmann, N., J.-M. Guehl, T.S. Barigah, and J.R. Ehleringer, 1997. Interseasonal comparison of CO_2 concentrations, isotopic composition, and carbon dynamics in an Amazonian rainforest (French Guiana). Oecologia 110: 120–131.
Bullock, S.H., H.A. Mooney, and E. Medina, eds., 1995. Seasonally Dry Tropical Forests. Cambridge University Press, Cambridge.
Burnham, R.J., 2002. Dominance, diversity and distribution of lianas in Yasuní, Ecuador: Who is on top? Journal of Tropical Ecology 18: 845–864.
Burslem, D.F.R.P., I.M. Tumer, and P.J. Grubb, 1994. Mineral nutrient status of coastal hill dipterocarp forest and adinandra belukar in Singapore: Bioassays of nutrient limitation. Journal of Tropical Ecology 10: 579–599.
Burslem, D.F.R.P., P.J. Grubb, and M. Turner, 1995. Responses to nutrient addition among shade-tolerant tree seedlings of lowland rain forest in Singapore. Journal of Ecology 83: 113–122.
Cardelús, C.L., R.K. Colwell, and J.E. Watkins Jr., 2006. Vascular epiphyte distribution patterns: explaining the mid-elevation richness peak. Journal of Ecology 94: 144–156.
Castro, Y., N. Fetcher, and D.S. Fernández, 1995. Chronic photoinhibition in seedlings of tropical trees. Physiologia Plantarum 94: 560–565.
Chazdon, R.L., 1986. Light variation and carbon gain in rainforest understory palms. Journal of Ecology 74: 995–1012.
Chazdon, R.L. and N. Fetcher, 1984. Photosynthetic light environments in a lowland tropical rainforest in Costa Rica. Journal of Ecology 72: 553–564.
Chazdon, R.L. and R.W. Pearcy, 1986. Photosynthetic responses to light variation in rainforest species. II. Carboxylation gain and photosynthetic efficiency during lightflecks. Oecologia 69: 524–531.
Clark, C.B., 2002. Are tropical forests an important carbon sink? reanalysis of the long-term plot data. Ecological Applications 12: 3–7.
Clark, D.A., 2004. Tropical forests and global warming: slowing it down or speeding it up? Frontiers in Ecology and the Environment 2: 73–80.
Clark, D.B., M.W. Palmer, and D.A. Clark, 1999. Edaphic factors and the landscape-scale distributions of tropical rain forest trees. Ecology 80: 2662–2675.
Connell, J.H. and M.D. Lowman, 1989. Low-diversity tropical rain forests: Some possible mechanisms for their existence. American Naturalist 134: 88–119.
Coomes, D.A., 1997. Nutrient status of Amazonian caatinga forests in a seasonally dry area: nutrient fluxes, litterfall and analyses of soils. Canadian Journal of Forest Research 27: 831–839.
Coomes, D.A. and P.J. Grubb, 1996. Amazonian caatinga and related communities at La Esmeralda, Venezuela: Forest structure, physiognomy and floristics, and control by soil factors. Vegetatio 122: 167–191.
Coomes, D.A. and P.J. Grubb. 1998. Responses of juveniles to above- and belowground competition in nutrient-starved Amazonian rainforest. Ecology 79: 768–782.
Coutinho, L.M., 1963. Algumas informações sôbre a ocorrência do "Efeito de DeSaussure" em epifitas e herbáceas terrestres da mata pluvial. Bol Fac Filosofia, Ciencias e Letras Universidade de São Paulo No. 288: 81–98.
Coutinho, L.M., 1969. Novas observaçoes sobre a ocorrencia do "Efeito de DeSaussure" e suas relaçoes com a suculencia, a temperatura folhear e os movimentos estomáticos. Boletin Faculdáde de Filosofia, Ciencias e Letras, Universidade de Sao Paulo, No. 331: 79–102.
Cramer, W.P. and R. Leemans, 1993. Assessing impacts of climate change on vegetation using climate classification systems. In: A.M. Solomon and H.H. Shugart, eds. Vegetation Dynamics and Global Change. Chapman and Hall, New York, pp. 190–217.
Cuevas, E. and E. Medina, 1988. Nutrient dynamics within Amazonian forest ecosystems. II. Root growth, organic matter decomposition, and nutrient release. Oecologia (Berlin) 76: 222–235.

Denslow, J., 1996. Functional group diversity and responses to disturbance. In: G.H. Orians, R. Dirzo, and J.H. Cushman, eds. Biodiversity and Ecosystem Processes in Tropical Forests. Ecological Studies, Vol. 122. Springer-Verlag, Berlin, pp. 127–151.

Earnshaw, M.J., K. Winter, H. Ziegler, W. Stichler, N.E.G. Cruttwell, K. Kerenga, P.J. Cribb, J. Wood, J.R. Croft, K.A. Carver, and T.C. Gunn, 1987. Altitudinal changes in the incidence of crassulacean acid metabolism in vascular epiphytes and related life forms in Papua New Guinea. Oecologia 73: 566–572.

Ewel, J.J. and S.W. Bigelow, 1996. Plant life-forms and tropical ecosystem functioning. In: G.H. Orians, R. Dirzo, and J.H. Cushman, eds. Biodiversity and Ecosystem Processes in Tropical Forests. Ecological Studies, Vol. 122. Springer-Verlag, Berlin, pp. 101–126.

Ewers, F.W., 1985. Xylem structure and water conduction in conifer trees, dicot trees, and lianas. IAWA Bulletin (New Series) 6: 309–317.

Farquhar, G.D., J.R. Ehleringer, and K.T. Hubick, 1989. Carbon isotope discriination and photosynthesis. Annual Review of Plant Physiology 40: 503–537.

Fetcher, N., B.R. Strain, and S.F. Oberbauer, 1983. Effects of light regime on the physiology and growth of seedlings of tropical trees: A comparison of gap and pioneer species. Oecologia 58: 314–319.

Fichtler, E., D.A. Clark, and M. Worbes, 2003. Age and Long-term Growth of Trees in an Old-growth Tropical Rain Forest, Based on Analyses of Tree Rings and ^{14}C. Biotropica 35: 306–317.

Field, C.B., M.J. Behrenfeld, J.T. Randerson, and P. Falkowski, 1998. Primary production of the biosphere: integrating terrestrial and oceanic components. Science 281: 237–240.

Franco, A.C., E. Olivares, E. Ball, U. Lüttge, and A. Haag-Kerwer, 1994. *In situ* studies of Crassulacean acid metabolism in several sympatric species of tropical trees of the genus Clusia. New Phytologist 126: 203–213.

Gartner, B.L., S.R. Bullock, H.A. Mooney, V.B. Brown, and J.L. Whitbeck, 1990. Water transport properties of vine and tree stems in a tropical deciduous forest. American Journal of Botany 77: 742–749.

Gauhl, E., 1976. Photosynthetic responses to varying light intensity in ecotypes of *Solanum dulcamara* L from shaded and exposed environments. Oecologia 22: 275–286.

Gentry, A.H., 1983. Lianas and the "paradox" of contrasting latitudinal gradients in wood and litter production. Tropical Ecology 24: 63–67.

Gentry, A.H., 1988. Changes in plant community diversity and floristic composition on environmental and geographic gradients. Annals of Missouri Botanical Garden 75: 1–34.

Gentry, A.H. and C. Dodson, 1987. Contribution of nontrees to species richness of a tropical rain forest. Biotropica 19:149–156.

Gentry, A.H. and C. Dodson, 1987a. Contribution of nontrees to species richness of a tropical rain forest. Biotropica 19: 149–156.

Gentry, A.H. and C. Dodson, 1987b. Diversity and biogeography of neotropical vascular epiphytes. Annals of Missouri Botanical Garden 74: 205–233.

Gentry, A.H. and L.H. Emmons, 1987. Geographical variation in fertility, phenology and composition of the understory of neotropical forests. Biotropica 19: 216–227.

Givnish, T.J., 1988. Adaptation to sun and shade: A whole plant perspective. Australian Journal of Plant Physiology 15: 63–92.

Goh, C.J. and M. Kluge, 1989. Gas exchange and water relations in epiphytic orchids. Vascular plants as epiphytes (Lüttge, U., ed.) Ecological Studies 76: 139–166. Springer-Verlag, Berlin.

Goh, C.J., P.N. Avadhani, C.S. Loh, C. Hanegraf, and J. Arditti, 1977. Diurnal stomata and acidity rhythms in orchid leaves. New Phytologist 78: 365–372.

Gómez-Pompa, A., T.C. Whitmore, and M. Hadley, eds., 1991. Rain Forest Regeneration and Management. Man and the Biosphere Series, Vol. 6. Unesco and The Parthenon, Paris.

Grace, J., 2004. Understanding and managing the global carbon cycle. Journal of Ecology 92: 189–202.

Grace, J., J. Lloyd, J.A. Macintyre, A.C. Miranda, P. Meir, H.S. Miranda, C.A. Nobre, J. Moncrief, J. Massheder, I.R. Wright, and J.H.C. Gash, 1995. Carbon dioxide uptake by undisturbed tropical forest. Science 270: 778–780.

Griffiths, H., 1988. Crassulacean acid metabolism: a re-appraisal of physiological plasticity in form and function. Advances in Botany Research 15: 43–92.

Griffiths, H., 1989. Carbon dioxide concentrating mechanisms and the evolution of CAM in vascular epiphytes. Vascular plants as epiphytes (Lüttge, U., ed.) Ecological Studies 76: 42–86. Springer-Verlag, Berlin.
Griffiths, H. and J.A.E. Smith, 1983. Photosynthetic pathways in the Bromeliaceae of Trinidad: relations between life-forms, habitat preference and the occurrence of CAM. Oecologia 60: 176–184.
Gross, N.D., S.D. Torti, D.H. Feener Jr., and P.D. Coley, 2000. Monodominance in an African Rain Forest: Is Reduced Herbivory Important? Biotropica 32: 430–439.
Grubb, P.J., 1977. Control of forest growth and distribution on wet tropical mountains. Annual Review of Ecological Systems 8: 83–107.
Guralnick, L.J., I.P. Ting, and E.M. Lord, 1986. Crassulacean acid metabolism in the *Gesneriaceae*. American Journal of Botany 73: 336–345.
Hall, J.B. and M.D. Swaine, 1976. Classification and ecology of closed-canopy forest in Ghana. Journal of Ecology 64: 913–951.
Halle, F., R.A.A. Oldeman, and P.B. Tornlinson, 1978. Tropical Trees and Forests: An Architectural Analysis. Springer-Verlag, Berlin.
Hart, T.B., J.A. Hart, and P.G. Murphy, 1989. Monodominant and species-rich forests of the humid tropics: causes for their co-occurrence. American Naturalist 133: 613–633.
Hegarty, E.E., 1989. The climbers-Lianes and vines. In: H. Lieth, and M.J.A. Werger, eds. Ecosystems of the World, Vol. 14B. Elsevier, New York, pp. 339–353.
Hew, C.S. and Y.S. Wong, 1974. Photosynthesis and respiration of ferns in relation to their habitat. American Fern Journal 64: 40–48.
Hietz, P., W. Wanek, and M. Popp, 1999. Stable isotopic composition of carbon and nitrogen and nitrogen content in vascular epiphytes along an altitudinal gradient. Plant, Cell and Environment 22: 1495–1449.
Holbrook, N.M. and F.E. Putz, 1996. From epiphyte to tree: differences in leaf structure and leaf water relations associated with the transition in growth form in eight species of hemiepiphytes. Plant, Cell and Environment 19: 631–642.
Holbrook, N.M. and T.R. Sinclair, 1992. Water balance in the arborescent palm *Sabal palmetto*. 1. Stem structure, tissue water release properties and leaf epidermal conductance. Plant, Cell and Environment 15: 393–399.
Holdridge, L.R., 1967. Life Zone Ecology. Tropical Science Center, San José, Costa Rica.
Holdridge, L.R., W.C. Grenke, W.H. Hatheway, T. Liang, and J.A. Tosi, 1971. Forest Environments in Tropical Life Zones: A Pilot Study. Pergamon Press, New York.
Hopkins, M.S., P. Reddell, R.K. Hewett, and A.W. Graham, 1996. Comparison of root and mycorrhizal characteristics of primary and secondary rainforest on metamorphic soil in North Queensland, Australia. Journal of Tropical Ecology 12: 871–885.
Hueck, K., 1966. Die Wälder Südamerikas. Gustav Fischer-Verlag, Stuttgart.
Huston, M., 1980. Soil nutrients and tree species richness in Costa Rican forests. Journal of Biogeography 7: 147–157.
Huston, M., 1994. Biological Diversity. Cambridge University Press, Cambridge.
Ishida, A., T. Toma, and Marjenah, 2005. A comparison of *in situ* leaf photosynthesis and chlorophyll fluorescence at the top canopies in rainforest mature trees. Japan Agricultural Research Quarterly 39: 57–67.
Ishida, A., S. Diloksumpun, P. Ladpala, D. Staporn, S. Panuthai, M. Gamo, K. Yazaki, M. Ishizuka, and L. Puangchit, 2006. Contrasting seasonal leaf habits of canopy trees between tropical dry-deciduous and evergreen forests in Thailand. Tree Physiology 26: 643–656.
Janos, D.P., 1980. Vesicular-arbuscular mycorrhizae affect lowland tropical rain forest plant growth. Ecology 61: 151–162.
Janos, D.P., 1983. Tropical mycorrhizas, nutrient cycles and plant growth. In: S.L. Sutton, T.C. Whitmore, and A.E. Chadwick, eds. Tropical Rain Forest: Ecology and Management. Blackwell Scientific, Oxford, pp. 327–345.
Jordan, C.F., 1985. Nutrient Cycling in Tropical Forest Ecosystems. Wiley, Chichester.

Jordan, C.F., ed., 1989. An Amazonian Rain Forest: The Structure and Function of a Nutrient Stressed Ecosystem and the Impact of Slash-and-Bum Agriculture. Man and the Biosphere Series, Vol. 2. UNESCO and The Partenon, London.

Jordan, C.F., R.L. Todd, and G. Escalante, 1979. Nitrogen conservation in a tropical rain forest. Oecologia 39: 123–128.

Kahn, F. and J.-J. de Granville, 1992. Palms in Forest Ecosystems of Amazonia. Ecological Studies Vol. 95. Springer-Verlag, Berlin.

Kalliola, R., M. Puhakka, and W. Danjoy, eds., 1993. Amazonia Peruana: Vegetación húmeda tropical en el llano subandino. PAUT (Finland) and ONERN (Peru). Jyväskylä, Finnland.

Kamaluddin, M. and J. Grace, 1992a. Photoinhibition and light acclimation in seedlings of *Bischofia javanica*, a tropical forest tree from Asia. Annals of Botany 69: 47–52.

Kamaluddin, M. and J. Grace, 1992b. Acclimation in seedlings of a tropical tree, *Bischofia javanica*, following a stepwise reduction in light. Annals of Botany 69: 557–562.

Kellman, M. and R. Tackaberry, 1997. Tropical Environments. Routledge, London.

Kitajima, K., 1994. Relative importance of photosynthetic traits and allocation pattems as correlates of seedling shade tolerance of 13 tropical trees. Oecologia 98: 419–428.

Kitajima, K., 1996. Ecophysiology of tropical tree seedlings, In: S.S. Mulkey, R.L. Chazdon, and A.P. Smith, eds. Tropical Forest Plant Ecophysiology. Chapman and Hall, New York, pp. 559–596.

Kluge, M., P.N. Avadhani, and C.J. Goh, 1989a. Gas exchange and water relations in epiphytic tropical ferns. Vascular plants as epiphytes (Lüttge, U., ed.) Ecological Studies 76: 87–108. Springer Verlag, Berlin.

Kluge, M., V. Friemert, B.C. Ong, J. Brulfert, and C.J. Goh, 1989b. *In situ* studies of the crassulacean acid metabolism in *Drymoglossum piloselloides*, an epiphytic fern of the humid tropics. Journal of Experimental Botany 40: 441–452.

Königer, M., G.C. Harris, A. Virgo, and K. Winter, 1995. Xanthophyll-cycle pigments and photosynthetic capacity in tropical forest species: A comparative field study in canopy, gap and understory plants. Oecologia 104: 280–290.

Kluge, M. and I.P. Ting, 1978. Crassulacean Acid Metabolism: Analysis of an ecological adaptation. Springer-Verlag, Berlin.

Krause, G.H. and K. Winter, 1996. Photoinhibition of photosynthesis in plants growing in natural tropical forests gaps. A chlorophyll fluorescence study. Botanica Acta 109: 456–462.

Krause, G.H., A. Virgo, and K. Winter, 1995. High susceptibility to photoinhibition of young leaves of tropical forest trees. Planta 197: 583–591.

Kress, W.J., 1986. The systematic distribution of vascular epiphytes: an update. Selbyana 9: 2–22.

Küper, W., H. Kreft, J. Nieder, N. Köster, and W. Barthlott, 2004. Large-scale diversity patterns of vascular epiphytes in Neotropical montane rain forests. Journal of Biogeography 31: 1477–1487.

Kursar, T.A. and P.D. Coley, 1999. Contrasting modes of light acclimation in two species of the rainforest understory. Oecologia 121: 489–498.

Lal, R., 1987. Tropical Ecology and Physical Edaphology. Wiley, Chichester.

Laurance, W.F., P.M. Fearnside, S.G. Laurance, P. Delamonica, T.E. Lovejoy, J.M. Rankin-de Merona, J.Q. Chambers, and C. Gascon, 1999. Relationship between soils and Amazon forest biomass: A landscape-scale study. Forest Ecology and Management 118: 127–138.

Lee, H.S.J., A.K. Schmitt, and U. Lüttge, 1989. The response of the C_3-CAM tree, *Clusia rosea*, to light and water stress II. Internal CO_2 concentration and water use efficiency. Journal of Experimental Botany 40: 171–179.

Leigh, E.G. Jr., A.S. Rand, and D.M. Windsor, eds., 1982. The Ecology of a Tropical Rain Forest: Seasonal Rhythms and Long-Term Changes. Smithonian Institution Press, Washington, D.C.

Lieth, H., 1975. Historic survey of primary productivity research. In: H. Lieth and R. Whittaker, eds. Primary Productivity of the Biosphere. Ecological Studies 14. Springer-Verlag, Berlin, pp. 7–16.

Lieth, H. and R. Whittaker, eds., 1975. Primary Productivity of the Biosphere. Ecological Studies 14. Springer-Verlag, Berlin.

Lloyd, J. and G.D. Farquhar, 1996. The CO_2 dependence of photosynthesis, plant growth responses to elevated atmospheric CO_2 concentrations and their interaction with soil nutrient status. I. General principles and forest ecosystems. Functional Ecology 10: 4–32.
Long, S.P., S. Humphries, and P.G. Falkowski, 1994. Photoinhibition of photosynthesis in nature. Annual Review of Plant Physiology and Molecular Biology 45: 633–662.
Lovelock, C.E. and J.J. Ewel, 2005. Links between tree species, symbiotic fungal diversity and ecosystem functioning in simplified tropical ecosystems. New Phytologist 167: 219–228.
Lovelock, C.E., M. Jebb, and C.B. Osmond, 1994. Photoinhibition and recovery in tropical plant species: response to disturbance. Oecologia 97: 297–307.
Lovelock, C.E., T.A. Kursar, J.B. Skillman, and K. Winter, 1998. Photoinhibition in tropical forest understory species with short- and long-lived leaves. Functional Ecology 12: 553–560.
Lovelock, C.E., K. Andersen, and J.M. Morton, 2003. Host tree and environmental control on arbuscular mycorrhizal spore communities in tropical forests. Oecologia 135: 268–279.
Lugo, A.E. and C. Rivera Batelle, 1987. Leaf production, growth rate, and age of the palm *Prestoea montana* in the Luquillo Experimental Forest. Journal of Tropical Ecology 3: 151–161.
Lüttge, U., K.-H. Stimmel, J.A.C. Smith, and H. Griffiths, 1985. Comparative ecophysiology of CAM and C3 bromeliads II. Field measurements of gas exchange of CAM bromeliads in the humid tropics. Plant, Cell and Environment 9: 377–384.
Lüttge, U., ed., 1991. Vascular Plants as Epiphytes. Ecological Studies, Vol. 76. Springer-Verlag, Berlin.
Madison, M., 1977. Vascular epiphytes: their systematic occurrence and salient features. Selbyana 2: 1–13.
Malhi, Y. and J. Grace, 2000. Tropical forests and atmospheric carbon dioxide. TREE 15: 332–337.
Malhi, Y. and J. Wright, 2004. Spatial patterns and recent trends in the climate of tropical rainforest regions. Philosophical Transactions Royal Society London B 359: 311–329.
Martin, C.E., 1994. Physiological ecology of Bromeliaceae. Botanical Review 60: 1–82.
Martin, C.E., C.A. Eades, and R.A. Pitner, 1986. Effects of Irradiance on Crassulacean Acid Metabolism in the Epiphyte *Tillandsia usneoides* L. (Bromeliaceae). Plant Physiology 80: 23–26.
McWilliams, E.L., 1970. Comparative rates of dark CO_2 uptake and acidification in Bromeliaceae, Orchidaceae and Euphorbiaceae. Botanical Gazette 131: 285–290.
Medina, E., 1971. Effects of nitrogen supply and light intensity during growth on the photosynthetic capacity and carboxidismutase activity of *Atriplex patula* leaves. Carnegie Year Book 70: 551–559.
Medina, E., 1974. Dark CO_2 fixation, habitat preference and evolution within the Bromeliaceae. Evolution 28: 677–686.
Medina, E., 1984. Nutrient balance and physiological processes at the leaf level. In: E. Medina, H.A. Mooney, and C. Vazquez-Yánes, eds. Physiological Ecology of Plants of the Wet Tropics. Junk, The Hague, pp. 139–154.
Medina, E., 1990. Eco-fisiología y evolución de las Bromeliaceae. Boletin de la Academia Nacional de Ciencias, Córdoba 59: 71–100.
Medina, E., 1996. CAM and C_4 plants in the humid tropics. In: S.S. Mulkey, R.L. Chazdon, and A.P. Smith, eds. Tropical Forest Plant Eco-physiology. Chapman and Hall, London, pp. 56–88.
Medina, E. and E. Cuevas, 1989. Pattems of nutrient accumulation and release in Amazonian forests of the upper Río Negro basin. In: J. Proctor, ed. Mineral Nutrients in Tropical Forest and Savanna Ecosystems. Blackwell Scientific Publications, Oxford, pp. 217–240.
Medina, E. and E. Cuevas, 1994. Mineral nutrition: tropical forests. Progress in Botany 55: 115–129.
Medina, E., M. Delgado, J.H. Troughton, and J.D. Medina, 1977. Physiological ecology of CO_2 fixation in Bromeliaceae. Flora 166: 137–152.
Medina, E., M. Sobrado, and R. Herrera, 1978. Significance of leaf orientation for leaf temperature in an Amazonian sclerophyll vegetation. Radiation and Environmental Biophysics 15: 131–140.
Medina, E., G. Montes, E. Cuevas, and Z. Rokzandic, 1986. Profiles of CO_2 concentration and $\delta\ ^{13}C$ values in tropical rain forests of the Upper Rio Negro basin, Venezuela. Journal of Tropical Ecology 2: 207–217.
Medina, E., M. Delgado, and V. Garda, 1989. Cation accumulation and leaf succulence in *Codonanthe macradenia* J.D. Smith (Gesneriaceae) under field conditions. Amazoniana 11: 13–22.
Medina, E., V. García, and E. Cuevas, 1990. Sclerophylly and oligotrophic environments: Relationships between leaf structure, mineral nutrient content and drought resistance in tropical rain forests of the upper Río Negro region. Biotropica 22: 51–64.

Melillo, J.M., A.D. McGuire, D.W. Kicklighter, B. Moore III, C.J. Vörösmarty, and A.L. Schloss, 1993. Global climate change and terrestrial net primary production. Nature 363: 234–240.

Monson, R.K., 1989. On the evolutionary pathways resulting in C[4] photosynthesis and Crassulacean acid metabolism (CAM). Advances in Ecological Research 19: 57–110.

Moyersoen, B., 1993. Ectomicorrizas y micorrizas vesículo-arbusculares en Caatinga Amazónica al sur de Venezuela. Scientia Guaianae 3: 82.

Mulkey, S.S., R.L. Chazdon, and A.P. Smith, eds., 1996. Tropical forest plant eco-physiology. Chapman and Hall, New York.

Neales, T.F. and C.S. Hew, 1975. Two types of carbon fixation in tropical orchids. Planta 123: 303–306.

Newbery, D.M., L.J. Alexander, D.W. Thomas, and J.S. Gartlan, 1988. Ectomycorrhizal rain-forest legumes and soil phosphorus in Korup National Park, Cameroon. New Phytologist 109: 433–450.

Oberbauer, S.F. and B.R. Strain, 1985. Effects of light regime on the growth and physiology of *Pentaclethra macroloba* (Mimosaceae) in Costa Rica. Journal of Tropical Ecology 1: 303–320.

Odum, H.T. and R.F. Pigeon, eds., 1970. A Tropical Rain Forest: A Study of Irradiation and Ecology at El Verde, Puerto Rico. US Atomic Energy Commission, Oak Ridge, Tennessee.

Oldeman, R.A.A. and J. van Dijk, 1991. Diagnosis of the temperament of tropical rain forest trees. In: A. Gómez-Pompa, T.C. Whitmore, and M. Hadley, eds. Rain Forest Regeneration and Management. Man and the Biosphere Series, Vol. 6. UNESCO, Paris, pp. 21–65.

Olson, D.M., E. Dinerstein, E.D. Wikramanayake, N.D. Burgess, G.V.N. Powell, E.C. Underwood, J.A. D'Amico, I. Itoua, H.E. Strand, J.C. Morrison, C.J. Loucks, T.F. Allnutt, T.H. Ricketts, Y. Kura, J.F. Lamoreux, W.W. Wettengel, P. Hedao, and K.R. Kassem, 2001. Terrestrial ecoregions of the world: A new map of life on Earth. BioScience 51: 933–938.

Osmond, C.B. and S.C. Grace, 1995. Perspectives on photoinhibition and photorespiration in the field: quintessential inefficiencies of the light and dark reactions of photosynthesis? Journal of Experimental Botany 46: 1351–1362.

Pearcy, R.W., 1988 Photosynthetic utilization of sunflecks. Australian Journal of Plant Physiology 15: 223–238.

Pérez-Salicrup, D.R., 2001. Effect of liana cutting on tree regeneration in a liana forest in Amazonian Bolivia. Ecology 82: 389–396.

Pérez-Salicrup, D.R. and M.G. Barker, 2000. Effect of liana cutting on water potential and growth of adult *Senna multijuga* (Caesalpinioideae) trees in a Bolivian tropical forest. Oecologia 124: 469–475.

Pérez-Salicrup, D.R., L. Victoria, V.L. Sork, and F.E. Putz, 2001. Lianas and trees in a liana forest of Amazonian Bolivia. Biotropica 33: 34–47.

Phillips, O.L., P. Hall, A.H. Gentry, S.A. Sawyer, and R. Vasquez, 1994. Dynamics and species richness of tropical rain forests. Proceedings of the National Academy of Science USA 91: 2805–2809.

Phillips, O.L., Y. Malhi, N. Higuchi, W.F. Laurance, P.V. Núñez, R.M. Vásquez, S.G. Laurance, L.V. Ferreira, M. Stern, S. Brown, and J. Grace, 1998. Changes in the Carbon Balance of Tropical Forests: Evidence from Long-Term Plots. Science 282: 439–442.

Phillips, O.L., R. Vásquez Martínez, L. Arroyo, T.R. Baker, T. Killeen, S.L. Lewis, Y. Malhi, A. Monteagudo-Mendoza, D. Neillq, P. Núñez-Vargas, M. Alexiades, C. Cerón, A. Di Fiore, T. Erwinkk, A. Jardim, W. Palacios, M. Saldias, and B. Vinceti, 2002. Increasing dominance of large lianas in Amazonian forests. Nature 418: 770–774.

Piñero, D., M. Martinez-Ramos, and J. Sarukhan, 1984. A population model of *Astrocaryum mexicanum* and a sensitivity analysis of its finite rate of increase. Joumal of Ecology 72: 977–991.

Pittendrigh, C.S., 1948. The *bromeliad-Anopheles-malaria* complex in Trinidad. 1. The bromeliad flora. Evolution 2: 58–89.

Popp, M., D. Kramer, H. Lee, M. Diaz, H. Ziegler, and U. Lüttge, 1987. Crassulacean acid metabolism in tropical dicotyledonous trees of the genus *Clusia*. Trees 1: 238–247.

Proctor, J., 1984. Tropical forests litterfall. II. The data set. In: S.L. Sutton and A.C. Chadwick, eds. Tropical Rain Forest: The Leeds Symposium. Leeds Phylosophical and Literary Society, Leeds, UK, pp. 83–113.

Proctor, J., ed., 1989. Mineral Nutrients in Tropical Forest and Savanna Ecosystems. Blackwell Scientific, Oxford.

Putz, P.E., 1983. Liana biomass and leaf area of a "tierra firme" forest in the Río Negro basin, Venezuela. Biotropica 15: 185–189.
Putz, F.E., 1984. The natural history of lianas on Barro Colorado Island, Panamá. Ecology 65: 1713–1724.
Putz, F.E. and D.M. Windsor, 1987. Liana phenology on Barro Colorado Island, Panamá. Biotropica 9: 334–341.
Raaimakers, D., R.G.A. Boot, P. Dijkstra, S. Pot, and T. Pons, 1995. Photosynthetic rates in relation to leaf phosphorus content in pioneer versus climax tropical species. Oecologia 102: 120–125.
Ramos, J. and J. Grace, 1990. The effects of shade on the gas exchange of seedlings of four tropical trees from Mexico. Functional Ecology 4: 667–677.
Reich, P.B., C. Uhl, M.B. Walters, and D.S. Ellsworth, 1991. Leaf lifespan as a determinant of leaf structure and function among 23 Amazonian tree species. Oecologia 86: 16–24.
Reich, P.B., D.S. Elsworth, and C. Uhl, 1995. Leaf carbon and nutrient assimilation and conservation in species differing status in an oligotrophic Amazonian forest. Functional Ecology 9: 65–76.
Reich, P.B., M.B. Walters, and D.S. Ellsworth, 1997. From tropics to tundra: Global convergence in plant functioning. Proceedings of the National Academy of Science, USA 94: 13730–13734.
Rice, A.H., E. Hammond Pyle, S.R. Saleska, L. Hutyra, M. Palace, M. Keller, P.B. de Camargo, K. Portilho, D.F. Marques, and S.C. Wofsy, 2004. Carbon balance and vegetation dynamics in an old-growth Amazonian forest. Ecological Applications 14 (Supplement): S55–S71.
Richards, P.W., 1952, 1996. The Tropical Rainforest. Cambridge University Press, Cambridge.
Riddoch, I., J. Grace, F.E. Fasehun, B. Riddoch, and O. Ladipo, 1991a. Photosynthesis and successional status of seedlings in a tropical semi-deciduous rain forest in Nigeria. Journal of Ecology 79: 491–503.
Riddoch, I., H. Lehto, and J. Grace, 1991b. Photosynthesis of tropical tree seedlings in relation to light and nutrient supply. New Phytologist 119: 137–147.
Rolim, S.G., R.M. Jesus, H.E.M. Nascimento, H.T.Z. do Couto, and J.Q. Chambers, 2005. Biomass change in an Atlantic tropical moist forest: the ENSO effect in permanent sample plots over a 22-year period. Oecologia 142: 238–246.
Rollet, B., 1969. La regénération naturelle en foret dense humide sempervirente de plaine de la Guyane Vénézuélienne. Bois et Forets des Tropiques 124: 19–38.
Saleska, S.R., S.D. Miller, D.M. Matross, M.L. Goulden, S.C. Wofsy, H.R. da Rocha, P. de Camargo, P. Crill, B.C. Daube, H.C. de Freitas, L. Hutyra, M. Keller, V. Kirchhoff, M. Menton, J.W. Munger, E. Hammond Pyle, A.H. Rice, and H. Silva, 2003. Carbon in Amazon Forests: Unexpected Seasonal Fluxes and Disturbance-Induced Losses. Science 302: 1554–1557.
Schmitt, A.K., H.S.J. Lee, and U. Lüttge, 1988. The response of the C_3-CAM tree, *Clusia rosea*, to light and water stress. I. Gas exchange characteristics. Journal of Experimental Botany 39: 1581–1590.
Schnitzer, S.A., M.E. Kuzee, and F. Bongers, 2005. Disentangling above- and below-ground competition between lianas and trees in a tropical forest. Journal of Ecology 93: 115–1125.
Schöngart, J., M.T.F. Piedade, S. Ludwigshausen, V. Horna, and M. Worbes, 2002. Phenology and stem-growth periodicity of tree species in Amazonian floodplain forests. Journal of Tropical Ecology 18: 581–597.
Schöngart J., W.J. Junk, M.T.F. Piedade, J.M. Ayres, A. Shüttermann, and M. Worbes, 2004. Teleconnection between tree growth in the Amazonian floodplains and the El Niño–Southern Oscillation effect. Global Change Biology 10: 1–10.
Schreiber, U., W. Bilger, and C. Neubauer, 1994. Chlorophyll fluorescence as a non intrusive indicator for rapid assessment of *in vivo* photosynthesis. In: E.-D. Schulze and M.M. Caldwell, eds. Ecophysiology of Photosynthesis. Ecological Studies 100. Springer-Verlag, Berlin, pp. 49–70.
Schultz, J., 1995. The Ecozones of the World: The Ecological Divisions of the Geosphere. Springer-Verlag, Berlin.
Slik, J.W., 2004. El Niño droughts and their effects on tree species composition and diversity in tropical rain forests. Oecologia 141: 114–20.
Smith, J.A.C., 1989. Epiphytic bromeliads. In: U. Lüttge, ed. Vascular Plants as Epiphytes. Ecological Studies 76. Springer-Verlag, Berlin, pp. 109–138.
Smith, J.A.C., H. Griffiths, and U. Lüttge, 1986. Comparative ecophysiology of CAM and C3 bromeliads 1. The ecology of Bromeliaceae in Trinidad. Plant, Cell and Environment 9: 359–376.

Solbrig, O.T., 1993. Plant traits and adaptive strategies: Their role in ecosystem function. In: E.D. Schulze and H.A. Mooney, eds. Biodiversity and ecosystem function. Ecological Studies, Vol. 99. Springer-Verlag, Berlin.

St. John, T.V. and C. Uhl, 1983. Mycorrhizae in the rain forest of San Carlos de Río Negro, Venezuela. Acta Científica Venezolana 34: 233–237.

Stark, N. and C.F. Jordan, 1977. Nutrient retention in the root mat of an Amazonian rain forest. Ecology 59: 434–437.

Sternberg, L.S.L., L.P. Ting, D. Price, and J. Hann, 1987. Photosynthesis in epiphytic and rooted *Clusia rosea* Jacq. Oecologia 72: 457–460.

Sternberg, L.S.L., S.S. Mulkey, and S.J. Wright, 1989. Ecological interpretation of leaf carbon isotope ratios: Influence of recycled carbon dioxide. Ecology 70: 1317–1324.

Stewart, G.R., S. Schmidt, L.L. Handley, M.H. Turnbull, P.D. Erskine, and C.A. Joly, 1995. ^{15}N natural abundance of vascular rainforest epiphytes: implications for nitrogen source and acquisition. Plant, Cell and Environment 18: 85–90.

Strauss-Debenedetti, S. and F.A. Bazzaz, 1991. Plasticity and acclimation to light in tropical Moraceae of different successional positions. Oecologia 87: 377–387.

Strauss-Debenedetti, S. and F.A. Bazzaz, 1996. Photosynthetic characteristics of tropical trees along successional gradients. In: S.S. Mulkey, R.L. Chazdon, and A.P. Smith, eds. Tropical Forest Plant Ecophysiology. Chapman and Hall, New York, pp. 162–186.

Sutton, S.L., T.C. Whitmore, and A.C. Chadwick, eds., 1983. Tropical Rain Forest: Ecology and Management. Blackwell Scientific, Oxford.

Swaine, M.D., 1996. Rainfall and soil fertility factors limiting forest species distribution in Ghana. Journal of Ecology 84: 419–428.

Swaine, M.D. and T.C. Whitmore, 1988. On the definition of ecological species groups in tropical rain forests. Vegetatio 75: 81–86.

Szarek, S.R., H.B. Johnson, and I.P. Ting, 1973. Drought Adaptation in *Opuntia basilaris*. Significance of recycling carbon through crassulacean acid metabolism. Plant Physiology 52: 539–541.

Thompson, W.A., G.C. Stocker, and P.E. Kriedemann, 1988. Growth and photosynthesis response to light and nutrients of *Flindersia brayleyana* F. Muell., a rainforest tree with broad tolerance to sun and shade. Australian Journal of Plant Physiology 15: 299–315.

Thompson, W.A., P.E. Kriedemann, and L.E. Craig, 1992. Photosynthetic response to light and nutrients in sun-tolerant and shade-tolerant rainforest trees. II. Leaf gas exchange and component processes of photosynthesis. Australian Journal of Plant Physiology 19: 19–42.

Tian, H., J.M. Melillo, D.W. Kicklighter, A.D. McGuire, J.V.K. Helfrioch III, B. Morre III, and Vörösmarty, 1998. Effect of interannual climate variability on carbon storage in Amazonian ecosystems. Nature 396: 664–667.

Tilman, D., 1988. Plant Strategies and the Dynamics and Structure of Plant Communities. Princeton University Press, Princeton, NJ.

Ting, L.P., L. Bates, L.S.L. Sternberg, and M.J. DeNiro, 1985. Physiological and isotopical aspects of photosynthesis in *Peperomia*. Plant Physiology 78: 246–249.

Ting, L.P., J. Hann, N.M. Holbrook, P.E. Putz, L.S.L. Stemberg, D. Price, and G. Goldstein, 1987. Photosynthesis in hemiepiphytic species of *Clusia* and *Ficus*. Oecologia 74: 339–346.

Tinoco-Ojanguren, C. and R.W. Pearcy, 1995. A comparison of light quality and quantity effects on the growth and steady-state and dynamic photosynthetic characteristics of three tropical tree species. Functional Ecology 9: 222–230.

Tinoco-Ojanguren, C. and C. Vázquez-Yanes, 1983. Especies CAM en la selva tropical húmeda de los Tuxtlas, Veracruz. Boletin de la Sociedad Botanica de Mexico 45: 150–153.

Todzia, C., 1986. Growth habits, host tree species, and density of hemiepiphytes on Barro Colorado island. Panamá. Biotropica 18: 22–27.

Tomlinson, P.B., 1979. Systematics and ecology of the Palmae. Annual Review of Ecology and Systematics 10: 85–107.

Torti, S.D., P.D. Coley, and D.P. Janos, 1997. Vesicular-arbuscular mycorrhizae in two tropical monodominant trees. Journal of Tropical Ecology 13: 623–629.

Turnbull, M.H., 1991. The effect of light quantity and quality during development on the photosynthetic characteristics of six Australian rainforest species. Oecologia 87: 110–117.

Turnbull, M.H., D. Doley, and D.J. Yates, 1993. The dynamics of photosynthetic acclimation to changes in light quantity and quality in three Australian rainforest tree species. Oecologia 94: 218–228.

Tuohy, J.M., J.A.B. Prior, and G.R. Stewart, 1991. Photosynthesis in relation to leaf nitrogen and phosphorus content in Zimbabwean trees. Oecologia 88: 378–382.

UNESCO, 1978. Tropical Forest Ecosystems. A state-of-knowledge report prepared by Unesco, UNEP, FAO. Natural Resources Research Series 14. UNESCO, Paris.

van der Merwe, N.J. and E. Medina, 1989. Photosynthesis and $^{13}C/^{12}C$ ratio in Amazonian rain forests. Geochimica et Cosmochimica Acta 53: 1091–1094.

van Noordwijk, M. and K. Hairiah, 1986. Mycorrhizal infection in relation to soil pH and soil phosphorus content in a rain forest of northem Sumatra. Plant and Soil 96: 299–302.

Vareschi, V., 1980. Vegetationsökologie der Tropen.Verlag Eugen Ulmer, Stuttgart.

Verheyden, A., G. Helle, G.H. Schleser, F. Dehairs, H. Beeckman, and N. Koedam, 2004. Annual cyclicity in high-resolution stable carbon and oxygen isotope ratios in the wood of the mangrove tree *Rhizophora mucronata*. Plant, Cell and Environment 27: 1525–1536.

Vitousek, P.M., 1984. Litterfall, nutrient cycling, and nutrient limitation in tropical forests. Ecology 65: 285–298.

Vitousek, P.M. and R.L. Sanford, 1986. Nutrient cycling in moist tropical forest. Annual Review of Ecology and Systematics 17: 137–167.

Walsh, R.P. and D.M. Newbery, 1999. The ecoclimatology of Danum, Sabah, in the context of the world's rainforest regions, with particular reference to dry periods and their impact. Philosophical Transactions Royal Society London B 354(1391): 1869–83.

Walter, H., 1973. Die Vegetation der Erde. l. Die tropischen und subtropishen Zonen. VEB Gustav Fischer Verlag, Jena.

Wania R., P. Hietz, and W. Wanek, 2002. Natural ^{15}N abundance of epiphytes depends on the position within the forest canopy: Source signals and isotope fractionation. Plant, Cell and Environment 25: 581–589.

Went, F.W. and N. Stark, 1968. The biological and mechanical ro1e of soil fungi. Proceedings of the National Academy of Science 60: 497–504.

Whitmore, T.C., 1984. Tropical Rain Forests of the Far East. Clarendon Press, Oxford.

Whitmore, T.C., 1990. An lntroduction to Tropical Rain Forests. Oxford University Press, Oxford.

Whitmore, T.C., 1991. Tropical rain forest dynarnics and its implications for management. In: A. Gómez-Pompa, T.C. Whitmore, and M. Hadley, eds. Rain Forest Regeneration and Management. Man and the Biosphere Series, Vol. 6. UNESCO, Paris, pp. 67–89.

Winter, K., 1985. Crassulacean acid metabolism. In: J. Barber and N.R. Baker, eds. Photosynthetic Mechanisms and the Environment. Elsevier Science Publishers, B.V. Amsterdam. pp. 328–387.

Winter, K., B.J. Wallace, G.C. Stocker, and Z. Roksandic, 1983. Crassulacean acid metabolism in Australian vascular epiphytes and some related species. Oecologia 57: 129–141.

Winter, K., E. Medina, V. García, M.L. Mayoral, and R. Muñiz, 1985. Crassulacean acid metabolism in roots of a leafless orchid, *Campylocentrum tyrridion* Garay and Dunsterv. Journal of Plant Physiology 111: 73–78.

Winter, K., C.B. Osmond, and K.T. Hubick, 1986. Crassulacean acid metabolism in the shade. Studies on an epiphytic fern, *Pyrrosia longifolia*, and other rainforest species from Australia. Oecologia 68: 224–230.

Winter, K., G. Zotz, B. Baur, and K.J. Dietz, 1992. Light and dark CO_2 fixation in *Clusia uvitana* and the effects of plant water status and CO_2 availability. Oecologia 91: 47–51.

Wofsy, S.C., R.C. Harriss, and W.A. Kaplan, 1988. Carbon dioxide in the atmosphere over the Amazon basin. Journal of Geophysical Research 93: 1377–1387.

Worbes, M. and W.J. Junk, 1989. Dating tropical trees by means of ^{14}C from bomb tests. Ecology 70: 503–507.

Wright, S.J., O. Calderón, A. Hernández, and S. Paton, 2004. Are lianas increasing in importance in tropical forests? A 17-Year Record from Panama. Ecology 85: 484–489.

Zotz, G. and K. Winter, 1993. Short-term regulation of crassulacean acid metabolism in a tropical hemiepiphyte, *Clusia uvitana*. Plant Physiology 102: 835–841.

11 Plant Diversity in Tropical Forests

S. Joseph Wright

CONTENTS

INTRODUCTION

Plant species diversity is greatest in tropical forest. The highest plant species density yet recorded, 365 species in just 1000 m^2, is for the only complete enumeration of tropical forest plants (Gentry and Dodson 1987). The diversity of lowland tropical forest at a local scale may exceed the diversity of extratropical forests at a continental scale. The combined temperate forests of Europe, North America, and Asia support 1166 tree species in 4.2×10^6 km^2 (Latham and Ricklefs 1993). In contrast, 1175 tree species occur in just 0.5 km^2 of lowland dipterocarp forest in Borneo (LaFrankie 1996). Trees comprise just 25% of the plant species in tropical rain forests (Gentry and Dodson 1987). There are 65% of all flowering plant species, 92% of fern species, and 75% of moss species that are tropical (Prance 1977).

These levels of diversity challenge modern ecology (Rosenzweig 1995). Theory and experiment confirm that two species cannot coexist if they use resources in an identical manner (Gause 1934). This competitive exclusion principle has become the touchstone against which theory addresses the coexistence of species. Plants compete for light, water, and perhaps a dozen of mineral nutrients. Plants must also disperse their progeny and survive the depredations of their pests. Are interspecific differences in these few parameters sufficient to explain the levels of plant diversity observed in tropical forests? Do other mechanisms contribute to the coexistence of tropical forest plants? Or is the competitive exclusion principle an inappropriate touchstone?

This chapter addresses these questions in three steps. First, five mechanisms are evaluated that are frequently invoked to explain the coexistence of extraordinary numbers of plant species in tropical forests. Second, four appendices present significant attributes of plant diversity within and among tropical forests. A successful theory must explain these observations. Finally, the origination of tropical forest plant diversity is considered. The extraordinary species densities of tropical forests may have more to do with the evolutionary history of plants, especially angiosperms, than with any unique ecological attribute of tropical forests.

COEXISTENCE OF TROPICAL FOREST PLANTS

This chapter focuses on five mechanisms that may contribute to the coexistence of very large numbers of species of plants and is not intended to be exhaustive. Many potential mechanisms are omitted (see Palmer 1994). The five mechanisms were selected for two reasons: (1) they dominate the literature for tropical forests and (2) each mechanism is amenable to experimental and comparative tests. The relationship to plant species density is briefly described for each mechanism. Principal tests in tropical forests are then evaluated. Finally, the mechanisms are related to one or more of the patterns of plant diversity observed among tropical forests (Appendix 11.1 through Appendix 11.4).

NICHE DIFFERENTIATION

Species may exploit limiting resources in ways that are sufficiently distinctive to permit stable coexistence in equilibrial communities (Ashton 1969, MacArthur 1969). Coexistence occurs when resources vary spatially, and each species occurs where it is a superior competitor.

Treefalls create spatial heterogeneity. Light intensities are highest near the center of a treefall and lowest nearby, beneath the intact forest canopy. Treefalls also create soil heterogeneity. Clays from depth are exposed when roots tip up, when rich humic material accumulates where fallen canopies decompose. Tropical forest plants may segregate along both light and soil gradients created by treefalls (Ricklefs 1977, Connell 1978, Denslow 1980, 1987, Orians 1982, Platt and Strong 1989).

This treefall gap hypothesis has been evaluated thoroughly. A few species do segregate treefall gaps by size and light intensity (Brokaw 1987). However, in large mapped plots, the spatial distributions of saplings of most tree species were indifferent to treefall gaps (Hubbell and Foster 1986, Lieberman et al. 1995). Saplings of just 5% of the tree species evaluated were specialized to treefall environments in repeated surveys of the performance of 250,000 trees in Panama (Welden et al. 1991). The great majority of species survived well and grew slowly, both in treefall gaps and in the deeply shaded understory. Interspecific differences in performance in response to gaps are necessary to promote coexistence. Recent experimental studies have failed to identify such differences (Uhl et al. 1988, Denslow et al. 1990, Brown and Whitmore 1992, Osunkjoya et al. 1992). The emerging conclusion is that spatial heterogeneity associated with treefalls maintains a relatively small number of light-demanding, shade-intolerant species in the forest landscape and makes a limited contribution to tropical forest plant diversity.

An intriguing extension of the treefall gap hypothesis relates the frequency of treefalls to forest productivity. Tree mortality increases with rainfall in tropical forests and with soil fertility in temperate-zone forests (Grubb 1986, Phillips et al. 1994). More productive forests with higher tree mortality rates have higher rates of gap formation and are predicted to have higher spatial heterogeneity of forest light environments. To the extent that plants differentiate light environments, this enhanced spatial heterogeneity contributes to the increase in plant species diversity with rainfall in tropical forests and with soil fertility in temperate forests (Grubb 1986, Phillips et al. 1994). However, we have seen that tropical forest plants show limited niche differentiation with respect to light environments. Still, the hypothesis of Phillips et al. (1994) is one of the few that addresses the ubiquitous increase in tropical-forest plant diversity with rainfall (Appendix 11.1) and merits further examination.

Variation in soil resources related to topography and underlying geological formations may also introduce spatial variation and promote species coexistence. Many tropical forest plant species have restricted distributions along soil moisture and soil fertility gradients (Hubbell and Foster 1983, Lieberman et al. 1985, Baillie et al. 1987, Tuomisto and Ruokolainen 1994, Swaine 1996, Sollins 1998). However, in most of these studies, many more species were indifferent to the edaphic gradient and occurred everywhere. Most tropical plant species are generalists with respect to both edaphic gradients and treefall microhabitats.

A modal relationship has been predicted between plant species density and soil fertility when three conditions are met (Tilman 1982): (1) few species survive the least fertile soils; (2) soil resources vary spatially; and (3) the most fertile soils provide ample resources everywhere, effectively negating spatial heterogeneity. Under these conditions, species diversity increases with the pool of tolerant species from infertile to intermediate soils and then decline on the most fertile soils as effective spatial heterogeneity declines and competitive dominance develops. The predicted modal relationships between tropical tree species' densities and soil fertility have been reported (Appendix 11.2). The postulated mechanisms remain untested, however, and the declining portion of the diversity–fertility relationship may be an artifact (Appendix 11.2).

In conclusion, spatially heterogeneous resources and ecologically segregated plant species are obvious to even casual observers of tropical forests. Niche differentiation undoubtedly contributes to tropical forest plant diversity. However, there are many ecologically similar plant species (Appendix 11.4). Moreover, there is no indication that levels of ecological differentiation among species distinguish dry from wet tropical forest or tropical forest from other forest biomes. Additional mechanisms must contribute to the diversity of tropical forest plants.

Pest Pressure

Microbes, fungi, and animals consume living plant tissue. These pests may contribute to the coexistence of their host plant species if the more abundant plant species or the superior competitors suffer disproportionately high levels of damage (Gillett 1962).

Pest pressure is severe in tropical forests. Insect herbivores alone consume an average of 11% of the leaf area produced in the understories of wet tropical forest (Coley and Barone 1996). Granivorous and browsing mammals reduce seedling recruitment, survivorship, and growth in a wide range of tropical forests (Dirzo and Miranda 1991, Osunkjoya et al. 1992, Terborgh and Wright 1994, Asquith et al. 1997). Pathogens kill entire seedling crops as well as significant numbers of adult trees (Augspurger 1984, Gilbert et al. 1994). Pathogens that impair but do not kill the host plant have received less attention, but are likely to be very important. Herbivores consume a larger proportion of leaf production in tropical forests than in temperate forests, despite greater plant investment in antiherbivore defenses in tropical forests (Coley and Aide 1991, Coley and Barone 1996). High levels of activity of plant pests may set tropical forests apart from other forested biomes.

Pest pressure may vary on several spatial scales. At the smallest scale, pests may cause disproportionately high levels of damage on the concentrations of conspecific seeds or seedlings found near fruiting trees. Conspecific may be prevented from recruiting successfully near one another, freeing space for other plant species, and potentially raising plant diversity

(Janzen 1970, Connell 1971). Seeds, seedlings, and saplings experience high mortality rates near conspecific adults (Clark and Clark 1984, Hubbell et al. 1990, Condit et al. 1992). However, the mechanism is not necessarily pest pressure. The alternative is intraspecific competition, occurring either among juvenile plants or between juveniles and the nearby adult (Clark and Clark 1984).

The relationship between pest pressure and the local population density of host plants becomes critical at larger spatial scales (Appendix 11.3). Pest pressure is positively density dependent when pest pressure increases more rapidly than host density. Positively density-dependent pests cause the highest levels of damage on the most abundant hosts and potentially contribute to the coexistence of host species. Alternatively, pest pressure is negatively density dependent when pests are unable to keep in pace with host densities. High host densities satiate pests (Schupp 1992). Negatively density-dependent pests cause the greatest damage on rarer hosts and potentially reduce host diversity. Negative density dependence occurs when factors other than host density control pest populations. Possibilities include control of pest populations by their own predators and parasites or by social behavior, especially territorial behavior. Thus, at spatial scales over which the densities of reproductive plants vary, pest pressure may be positively or negatively density dependent with the potential to increase or decrease plant diversity, respectively.

Keystone predators control community organization through their impact on prey species (Paine 1966). Terborgh (1992) hypothesized that felids and raptors are keystone predators in tropical forests. The hypothesis has two components. First, felids and raptors collectively limit midsized terrestrial mammals. Second, when their numbers are not checked by predation, these prey species alter forest regeneration. This hypothesis has profound implications. Felids and raptors may indirectly control plant diversity in tropical forests just as starfish control the diversity of sessile organisms in some marine environments.

The second component of this hypothesis has recently been falsified. Fences were used to exclude mammals from seeds and seedlings at Cocha Cashu, Peru, where the biota is intact, and at Barro Colorado Island, Panama, where several large predators are missing (Terborgh and Wright 1994). The exclosures had large and virtually identical effects at both sites, enhancing all indices of seedling performance. Browsing and granivorous mammals clearly alter plant regeneration, but the presence or absence of large felids and raptors had no additional effect.

Pest pressure may contribute to the increase in plant species diversity with rainfall (Appendix 11.1). Wet forests have higher and more constant pest pressure if seasonal temperature and rainfall reduce pest populations in drier forests. Plant diversity is, in turn, enhanced in wet forests if the more abundant plant species or the superior competitors suffer disproportionately high levels of damage. The premise of this hypothesis, higher pest pressure in wetter forests, is yet to be established. Levels of herbivory are actually higher in dry forests than in wet forests in the tropics (Coley and Barone 1996). Realized levels of herbivory result from the interaction between herbivores and plant defenses. In dry deciduous forests, leaves live less than 1 year and are poorly defended. In wet evergreen forests, long-lived leaves are better defended. The full cost of pests includes the diversion of resources to pest defenses as well as direct pest damage. Evaluation of pest pressure along tropical rainfall gradients awaits this information, as well as information on levels of activity of fungal and microbial pests.

In conclusion, pest pressure is greater in tropical forests than in other forest biomes. Positively density-dependent pest pressure increases plant diversity. At very small spatial scales, pest pressure almost certainly reduces recruitment near conspecific adults (Clark and Clark 1984). For larger spatial scales, there have been almost no tests for density-dependent pest pressure for tropical forests (see Schupp 1992). Pest pressure remains a promising but unproven mechanism of plant species coexistence in tropical forests.

Intermediate Disturbance

Windstorms, lightning, fire, landslides, and other disturbances kill trees. Connell (1978) predicted that species diversity varies during succession after such disturbances. Diversity is low immediately after a disturbance as just those species with the greatest ability to disperse their seeds arrive. Diversity reaches a maximum at intermediate times after a disturbance as many more species arrive. Diversity then declines with time as the best competitor or the species most resistant to pests or physical stress comes to dominate the forest. Species diversity over entire landscapes is low at (1) high disturbance frequencies, because recently disturbed patches and the few species with the greatest dispersal abilities dominate the landscape; and (2) low disturbance frequencies, because old patches and the few species best able to persist dominate the landscape. Diversity is greatest at intermediate disturbance frequencies because the landscape includes patches of a great variety of ages supporting a wide mix of species.

Connell (1978) applied the intermediate disturbance hypothesis to treefall gaps. He predicted that species best able to persist would perform poorly when transplanted into large gaps. These experiments have since been conducted with generally negative results. Performance improves for all species when seedlings are transplanted into gaps (Denslow et al. 1990, Brown and Whitmore 1992, Osunkjoya et al. 1992). Connell (1978) also failed to recognize that saplings of species capable of persisting in deep shade survive the formation and closure of treefall gaps (Uhl et al. 1988). Treefalls do not wipe the ground clean as envisioned by Connell (1978).

Intermediate disturbance may be important at larger spatial scales. Blowdowns caused by downburst of wind associated with severe storms are an important disturbance. Blowdowns larger than 30 ha can be identified from satellite images. The frequency of large blowdowns over Amazaonian Brazil increases with storm activity and annual rainfall (Nelson et al. 1994). Such large disturbances may contribute to the ubiquitous increases in plant species density with rainfall in tropical forests (Appendix 11.1).

Huston (1979, 1994) extended the intermediate disturbance hypothesis by incorporating environmental factors postulated to affect rates of competitive exclusion. Huston reasoned that productivity controls the rate of competitive exclusion. Given a constant frequency of disturbances, diversity is low in the least productive environments where few species can tolerate physical stress, highest in intermediate environments where many species can tolerate the environment but recurrent disturbance prevent competitive exclusion, and low again in the most productive environments where competitive exclusion occurs more rapidly than disturbance. Huston (1980, 1993, 1994) used soil fertility as a proxy for productivity in tropical forests. This proxy is evaluated in Appendix 11.2. Chesson and Huntly (1997) recently overturned the postulated connection between disturbance and rates of competitive exclusion. They demonstrated theoretically that disturbance either has no effect on or enhances rates of competitive exclusion. Huston's extension of the intermediate disturbance hypothesis must be reconsidered.

In conclusion, the intermediate disturbance hypothesis remains largely untested for tropical forests. One reason is overlap among hypotheses (see Palmer 1994). For example, the mechanisms postulated by Connell (1978) for treefall gaps incorporate interspecific differences in response to light intensity considered here under niche differentiation (see Section "Niche Differentiation"). The intermediate disturbance hypothesis also implicitly assumes a trade-off between the abilities to disperse and persist so that the species with the greatest dispersal abilities are poor competitors and vice versa (Petraitis et al. 1989; see Section "Life History Trade-Offs"). Finally, the intermediate disturbance hypothesis in most likely to be important for large disturbances that kill all advanced regeneration. These types of disturbances await further study.

LIFE HISTORY TRADE-OFFS

Sessile organisms live in communities with rigid spatial structure. Each individual occupies a space. Direct interactions are limited to colonists of the same space and near neighbors. The ability to colonize empty space, compete within a space, and persist once established jointly determines success. Trade-offs in these abilities, such that a superior colonist was an inferior competitor, might facilitate species coexistence.

Life history trade-offs occur among tropical forest plants. The number of seeds produced and the mass of individual seeds are inversely related. Presumably, species producing many small seeds disperse to rare high-resource environments, whereas species producing a few large seeds tolerate frequent low-resource environments (Hammond and Brown 1995). At the leaf level, maximum photosynthetic potentials and leaf longevity are inversely related (Reich et al. 1992). Presumably, species with long-lived leaves tolerate low-light environments, and species with short-lived leaves are superior competitors in high-light environments. At the level of individuals, maximum potential growth rates and survivorship in deep shade are inversely related (Kitajima 1994). Presumably, allocation to defense against pests limits growth rates but enhances shade survivorship. These trade-offs may facilitate species coexistence.

An infinite number of species can coexist when there is a strictly ordered trade-off between longevity, dispersal ability, and competitive ability (Tilman 1994). However, a critical assumption of this spatial competition hypothesis is violated in tropical forests. Tilman assumed a strict trade-off between longevity and competitive ability, such that superior competitors had short longevity and vice versa. Superior competitors in tropical forests are likely to tolerate deep shade as seedlings and cast deep shade as adults. Such species invest heavily in defenses against pests, have long-lived tissues including leaves, and are characterized by low growth and mortality rates (Kitajima 1994). Tilman acknowledged the possibility that long lifetimes and superior competitive ability may co-occur, falsifying the spatial competition hypothesis.

A very large number of species can also coexist if dispersal limits recruitment (Hurtt and Pacala 1995). Hurtt and Pacala modeled spatially heterogeneous environments and environmentally dependent competitive abilities. Sites vacated by the death of an adult were won by the best competitor among the species that reached the site. When the absolutely best competitor for the environment at the site did not arrive, the site was won by forfeit, and dispersal limitation occurred. Hurtt and Pacala proposed a positive feedback between species diversity and dispersal limitation. Dispersal limitation becomes more important as diversity increases because all species become rarer, produce fewer propagules, and reach fewer sites. However, a negative feedback may disrupt dispersal limitation. If a superior competitor became abundant, its seeds would approach ubiquity and only the rarer, less-competitive species would be dispersal limited. The two most common tree species on Barro Colorado Island, Panama, have dispersed seeds to each and every one of 200 tiny (0.5 m^2) litter traps randomly located throughout a 50 ha forest plot. Both species are superior competitors whose abundant, widespread seedlings grow and survive well both in the deeply shaded understory and in treefall gaps. Such abundant species have escaped dispersal limitation, even in a species-rich tropical moist forest.

Empirically driven simulation models may elucidate the relationship between species diversity and life history traits. Pacala et al. (1996) have successfully applied this approach to North American forests with nine tree species. Model inputs determined empirically for each species included seed dispersal, light-dependent growth rates, growth-dependent death rates, and light interception. The model tracked the performance of individual trees, where each tree competed for light with its immediate neighbors. Important trade-offs were evident between maximum potential growth rates and (1) dispersal, (2) light interception, and (3) survivorship under low light. A similar approach for a species-rich tropical forest has the potential to identify life history trade-offs that facilitate the coexistence of tropical forest plants.

Chance

A recurrent hypothesis postulates that many tropical forest plants are ecologically equivalent (Appendix 11.4; Aubréville 1938, Hubbell 1979, Hubbell and Foster 1986, Gentry 1989, Wright 1992, Federov 1966). Ecologically equivalent species might coexist for a very long time, given the large population sizes and random birth and death processes (Hubbell 1979, Hubbell and Foster 1986). Ecological equivalence might arise through common descent (Federov 1966), convergent evolution for a generalized ability to tolerate diffuse competition (Hubbell and Foster 1986), or the chance dynamics of life in the deeply shaded understory (Gentry and Emmons 1987, Wright 1992). The latter possibility is developed here.

Competitive exclusion requires that species compete and that the better competitor consistently wins. Consider competition among forest understory plants. There is an asymmetry between strata in tall forests. Large trees and lianas form the canopy and dominate understory environments. Trees and lianas intercept up to 99.5% of the photosynthetically active radiation reaching tropical forest canopies (Chazdon and Pearcy 1991), and tree and liana roots dominate the underground environment. Trees and lianas suppress understory plants, and, as a result, direct interactions among understory plants are unlikely. Understory environments may violate the two requisites for competitive exclusion.

First, competition among understory plants is unimportant. Pests maintain low understory stem densities that minimize direct competition. For example, the densities of understory plants increase dramatically after the experimental removal of mammalian herbivores (Dirzo and Miranda 1991, Osunkjoya et al. 1992, Terborgh and Wright 1994). Manipulations of understory stem densities provide direct evidence for limited competition among understory plants. The removal of all understory plants (less than 5 cm diameter at breast height [dbh]) had no effect on seedling recruitment and survival in a wet tropical forest (Marquis et al. 1986). Suppression by canopy plants limits direct competitive interactions in the deeply shaded understory.

Second, chance, not competitive ability, determines which individuals succeed in the understory. Chance is introduced by severe light limitation. Understory irradiance has two components. Dim, diffuse irradiance occurs throughout the understory, whereas direct solar irradiance occurs in sunflecks. Sunflecks contribute 32%–65% of the daily carbon gain of understory plants in closed canopy forests (Chazdon and Pearcy 1991). The occurrence of a sunfleck depends on the juxtaposition of the solar path and a canopy opening, cloud cover that diffuses solar irradiance, and wind that moves the vegetation, altering canopy openings. Most sunflecks are small in tall tropical forests (hundreds to thousands of square centimeters). Sunfleck location and intensity vary on a variety of timescales as wind, cloud cover, solar declination, and canopy openings change (Chazdon and Pearcy 1991). Sunflecks introduce chance as a primary determinant of performance in the understory.

Treefalls reinforce the importance of chance. Understory plants are temporarily released from canopy suppression when canopy trees die and open large gaps (Denslow 1987). Canopy gaps occur at random with respect to understory plants. Most importantly, canopy gaps close before competition causes mortality among previously suppressed understory plants. For example, more than 80% of the stems present 4 years after gap formation were present before gaps were formed in an Amazonian rain forest (Uhl et al. 1988). Likewise, more than 80% of tree saplings survived the opening and the closure of experimental treefall gaps in central Panama (N. Brokaw and A.P. Smith, unpublished data, 1993). Treefall gaps close before competition reduces species richness. Both sunflecks and treefalls introduce chance as an important determinant of the performance of understory plants.

To summarize, pests maintain low stem densities that prevent competition among understory plants, and chance sunflecks and treefall gaps largely determine which individuals succeed. Differences in competitive abilities are never realized among herbs and shrubs that spend their entire lives in the understory or among the seedling and saplings that form the advanced regeneration for the canopy. The absence of competition and the important role of chance in the understory contributes to species coexistence in all forest strata.

An extreme form of the chance hypothesis has been advanced by Hubbell (Hubbell 1979, Hubbell and Foster 1986). Species composition is predicted to fluctuate randomly as identical species experience random births and deaths. Species composition is, in fact, very similar for forests in similar abiotic environments, falsifying this prediction (Leigh et al. 1993, Terborgh et al. 1996). A more sophisticated version of the chance hypothesis must incorporate the ecological differences that so obviously exist among species. Chance enhances the potential for coexistence, but is not the sole explanation of population dynamics. It is clear that chance plays a large role in the deeply shaded understory. A renewed theoretical effort is required to explore the consequences for species coexistence.

ORIGINS OF TROPICAL PLANT DIVERSITY

Perhaps the most robust rule in ecology is that species richness and area increase together (Rosenzweig 1995). Terborgh (1973) ascribed high tropical species richness to the large areas with tropical climates. Terborgh noted that the area within each degree of latitude is greatest at the equator and decreases poleward, and that mean temperatures are uniformly high within 25° of the equator and then decrease poleward. As a consequence, tropical biomes cover four times more area than do subtropical, temperate, or boreal biomes (Rosenzweig 1992). Considerations of area alone suggest that species richness should be exceptionally high in the tropics.

Tropical biomes have been even more important in the past. In particular, the earth was largely tropical while the angiosperms radiated. Angiosperms appeared approximately 140 million years (Myr) ago in the early Cretaceous. The first angiosperms were tropical and required 20–30 Myr to spread beyond the tropics (Friis et al. 1987).The mean duration of angiosperm species in the fossil record exceeds 2 Myr (Niklas et al. 1983), making the last few tens of millions of years most relevant to modern angiosperm diversity. Tropical forests extended to 65° latitude in the Eocene (50 Myr ago), contracted to 15° latitude in the Oligocene (30 Myr), expanded to 35° latitude in the Miocene (20 Myr), and finally contracted to 25° latitude today (Behrensmeyer et al. 1992). The angiosperm radiation had a 20–30 Myr head start in the tropics and occurred when tropical climates covered up to 80% of all land (Eocene). The great species richness of tropical angiosperms may have ancient origins.

CONCLUSIONS

Tree and epiphyte diversity is greatest in everwet tropical forests (Appendix 11.1), and when terrestrial herbs and shrubs are included, it is probable that plant diversity is greatest in everwet tropical forests on fertile soils (Appendix 11.2). Several factors enhance plant diversity in these forests. First, temperature, moisture, and nutrients do not limit the pool of species able to survive in the most diverse tropical forests. Second, negative dependence characterizes common species, limiting competitive exclusion (Appendix 11.3). Both intraspecific competition and pest pressure may contribute to negative density dependence. In particular warm, moist conditions favor pathogens and small insects, and year-round pest pressure may reinforce negative density dependence in moist tropical forests.

High productivity may also indirectly enhance plant diversity in everwet tropical forests. Wet tropical forests are the most productive tierra firme biome and fertile soils may further enhance production (Vitousek 1984, Grubb 1986, Silver 1994). High tree turnover rates are associated with high production and year-round growth, and high frequencies of large blowdowns are associated with high rainfall (Nelson et al. 1994, Phillips et al. 1994). The most productive forests may be characterized by a mosaic of successional microhabitats on spatial scales ranging from single treefall gaps to blowdowns covering tens to thousands of hectares. This may permit coexistence through microhabitat specialization and life history trade-offs wherein inferior competitors are superior colonists and vice versa.

These possibilities are all mutually compatible, and there is ample evidence for each of the postulated mechanisms. Performance is habitat dependent (Section "Niche Differentiation"), pests do attack concentrations of conspecific plants (Section "Pest Pressure"), disturbances do kill tress (Section "Intermediate Disturbance"), and life history traits do covary (Section "Life History Trade-Offs"). What remains unclear is how these events affect species coexistence.

Future research in three areas is particularly valuable. First, are plant pests specific to particular hosts, are pest depredations density dependent, and does pest pressure vary among forest biomes? Second, do life history traits covary among the 905 tropical forest species that persist in deep shade as seedlings and saplings? And, third, does that 90% include species that require the spatial heterogeneity introduced by treefalls and other disturbances to recruit successfully? Until these questions are answered, the chance events that determine the success of understory plants and a tacit denial of the competitive exclusion principle continue to fascinate tropical ecologists.

ACKNOWLEDGMENT

I thank John Barone, Richard Condit, Kyle Harms, Egbert Leigh, and many others for stimulating discussion.

APPENDIX 11.1: PLANT SPECIES DENSITY INCREASES WITH RAINFALL

Plant species densities increase with rainfall in tropical forests. The increase is greatest for epiphytes, intermediate for terrestrial herbs and shrubs, and least for lianas and trees (Gentry and Dodson 1987, Gentry and Emmons 1987, Gentry 1988). Even for trees, however, species densities increase several-fold along rainfall gradients in Ghana, the Neotropics, and Southeast Asia (Whitmore 1975, Hall and Swaine 1976, 1981, Gentry 1988, Clinebell et al. 1995). The gross primary production of tropical forests also increases with rainfall (Brown and Lugo 1982, Jordan 1983, Medina and Klinge 1983). Plant species densities are greatest in the most productive tropical forests.

At least these mechanisms may contribute to the increase in plant species densities with rainfall. First, rates of tree turnover and treefall formation are higher in more productive forests (Grubb 1986, Phillips et al. 1994). Treefalls create a mosaic of forest patches recovering from past treefalls. This may increase diversity by increasing the spatial heterogeneity of forest light environments (see Section "Niche Differentiation"), by preventing competitive dominance (see Section "Intermediate Disturbance"), or by introducing spatiotemporal variation (see Section "Life History Trade-Offs").

Pest pressure provides a second mechanism that may contribute to the increase in plant species densities with rainfall. Many microbes, fungi, and small insects are vulnerable to desiccation. Density-dependent attacks by these pests may be more severe in high-rainfall forests that lack a dry, desiccating season (see Section "Pest Pressure").

Finally, the most productive forests occur where rainfall and soils are most favorable for plant growth. Here, the physiological requirements for moisture and mineral nutrients are fulfilled for the greatest number of species. Allocation can be shifted from roots to photosynthetic functions permitting the greatest number of species to maintain a positive carbon balance and regenerate from the shaded understory (see Section "Chance"). Regardless of the mechanism, it is clear that the most productive tropical forests also support the greatest plant diversity.

APPENDIX 11.2: PLANT SPECIES DENSITY AND SOIL FERTILITY

The relationship between plant species density and soil fertility has variously been reported to be positive, modal, negative, or absent for tropical forests. This diversity of results has at least two causes.

First, soils and climate covary. Plant species densities increase (Appendix 11.1) and soil nutrient concentrations decrease with rainfall (Clinebell et al. 1995). Apparent negative relationships between plant species density and soil fertility may be caused by parallel variation in rainfall. The negative relationship between tree species density and soil fertility reported by Huston (1980, 1993, 1994) are suspected for this reason. Huston (1980) omitted rainfall from a multiple regression analysis even though rainfall was the single best predictor of tree species density. Huston (1993, 1994) reproduced a relationship between plant species density and the first axis of an ordination performed by Hall and Swaine (1976, 1981). Hall and Swaine (1981, p. 31) state that this axis was "closely correlated with the moisture gradient." Huston (1993, 1994) refers to the same axis as a "composite soil fertility index." In a recent multiple regression analysis, annual rainfall and rainfall seasonality were the most important variables explaining tree species richness; soil nutrient concentrations were negatively correlated with rainfall; and, after rainfall was included, soil attributes explained little additional variation in plant species richness. Clinebell et al. (1995) concluded that "tropical forest species richness is surprisingly independent of soil quality."

The second problem suggests that this conclusion is premature. Only surface soils (0–10 cm depths) and shallow subsurface soils (most often 30–40 cm depths) are considered. Forest trees have roots to 12 m depths in Amazonia, and lianas have roots to at least 5 m depths in Panama (Nepstad et al. 1994; M. Tyree and S.J. Wright, personal observation, 1994). Deep roots reach decomposing rock, a rich and overlooked source of mineral nutrients. Many analyses of the relationship between tropical plant species densities and soils have been limited to the most deeply rooted life-forms: either trees more than 10 cm dbh (Holdridge et al. 1971, Hall and Swaine 1976, 1981, Huston 1980) or trees and lianas more than 2.5 cm dbh (Clinebell et al. 1995). These analyses omit a potentially large source of nutrients from deeper soils.

Shrubs and herbs have relatively shallow roots (Wright 1992), and surface and shallow subsurface soils are more relevant for shrubs and herbs than for trees and lianas. The species densities of ferns and melastomes (mostly shrubs and small treelets) increase with soil fertility in Amazonian Peru (Tuomisto and Ruokolainen 1994, Tuomista and Poulsen 1995). More generally, the species densities of fertile understory plants increase with soil fertility throughout the Neotropics (Gentry and Emmons 1987). The number of rain forest genera also increases with soil phosphorus in Australia (Beadle 1966). Species densities increase with soil fertility when fertility is measured for the soil volume reached by roots (i.e., for ferns, herbs, and shrubs). We await comparable analyses that include nutrients available from decomposing rock in the very deep soil horizons reached by tree and liana roots.

APPENDIX 11.3: DENSITY DEPENDENCE

Negative density dependence occurs when high local densities of conspecifics impair performance. One possible mechanism, intraspecific allelopathy, is unknown for tropical forests.

Pest pressure and the competition for resource implicit to niche differentiation also mediate interspecific coexistence through negative density dependence. Negative density dependence prevents any species from becoming dominant and excluding others. The search for negative density dependence has been intense.

Only weak tests for density dependence are possible for long-lived organisms. Strong tests evaluated population fluctuations over many generations, searching for density-dependent temporal variation. Tests for long-lived organisms are limited to spatial variations, specifically to comparisons of performance among sites that differ in conspecific density. An implicit assumption is that population fluctuations are asynchronous among sites. Spatial tests fail to detect density dependence if populations vary synchronously among sites. Spatial tests are also compromised if resources vary spatially. Spatially variable resources introduce apparent positive density dependence. High resource sites support high performance and high population densities. Spatially variable resources are important in tropical forests (see Section "Niche Differentiation"). Negative density dependence must overcome spatial variation in resources and performance. Therefore, we expect evidence for negative density dependence from spatial tests to be rare.

However, this is not the case. Strong evidence for negative density dependence comes from the 50 ha forest dynamics plot on Barro Colorado Island, Panama. The performance of more than 300,000 stems greater than or equal to 1 cm dbh of 314 species was monitored in 1982, 1985, 1990, and 1995. Negative density dependence is evident at scales ranging up to 100 m for recruitment, growth, and survivorship of several of the more abundant species (Hubbell et al. 1990, Condit et al. 1994). Tests for density dependence in rarer species were usually not significant; however, trends were consistent in the direction of negative density dependence. A very local negative density dependence, reduced performance near a conspecific adult, has also been documented in many tropical forests (see Section "Pest Pressure"). Negative density dependence is a fact for the more abundant trees in tropical forests. It only remains to determine the contribution of negative density dependence to interspecific coexistence (Hubbell et al. 1990, Condit et al. 1994).

Negative density dependence has also been reported for much rarer trees (Connell et al. 1984, Wills et al. 1997). The evidence consists of negative correlations between the number of recruits per conspecific adult (R/A) and the number of conspecific adults (A) or the basal area of conspecifics (BA). These correlations are suspected because the independent variable (A) is also the denominator of the dependent variable (R/A). Note that A and BA are closely related.

Wills et al. (1997) performed Monte Carlo simulations to evaluate the significance of R/A versus BA correlations. The size (dbh) and intercensus performance of randomly chosen pairs of conspecific trees were switched while maintaining the mapped location of each tree. Performance included death versus survival for trees present in the first census and recruitment for trees that appeared only in the second census. Adults were larger than a species-specific threshold dbh. After a large number of random switches, simulated correlations between R/A and BA were calculated. This procedure was repeated 99 times for each of a wide range of quadrat sizes, and the observed correlation coefficient for each quadrat size was compared with the distribution of simulated correlations. A subtle bias compromises these simulations for analyses of recruitment.

Two facts unrelated to density dependence affect the relative values of observed and simulated correlations. First, recruits are always small. Second, recruits are never adults. Consider a quadrat for which R/A is large and BA is small. This quadrat includes a relatively large number of recruits. The net effect of the randomization is likely to switch a recruit for an adult. This decrease is small. This quadrat includes a relatively large number of recruits. The net effect of the randomization is likely to switch a recruit for an adult. This decreases R/A and increases BA because the adult gained is larger than the recruit lost. The opposite tends to be true for quadrats for which *R*/*A* is small and BA is large. The randomization tends to move quadrats in *R*/*A*–BA space so that simulated correlations between *R*/*A* and BA are less negative than the observed correlations. This bias is unrelated to density.

This bias explains the spatial pattern of significant result reported by Wills et al. (1997). Recruitment was negatively density dependent for small quadrats but not for large quadrats. This bias identified only arises when there is a net loss or gain of recruits. Net changes are likely for small quadrats with small numbers of individuals. The randomization is less likely to cause net changes for large quadrats where larger numbers of individuals are switched back and forth. Artifact explains the spatial scale of significant results for recruitment.

A size-stratified randomization would avoid this artifact. Small recruits would only be switched with small survivors. This would eliminate the correlated change in *R*/*A* and BA that occurs whenever a small recruit and large adult are switched. Until a size-stratified randomization is performed, we must conclude that negative density dependence only affects recruitment for the most common trees in tropical forests (Hubbell et al. 1990, Condit et al. 1994).

APPENDIX 11.4: SYNTOPIC CONGENERS

As species of the same genus have usually, though by no means invariably, some similarity in habits and constitution, and always in structure, the struggle is generally more severe between species of the same genus, when they come into competition with each other, than between species of distinct genera (Darwin 1859).

Interactions between syntopic congeners often select for character divergence (Grant 1986). Coexistence is then possible through niche differentiation. Hubbell and Foster (1986) proposed instead that syntopic tropical forest plants are selected to converge on a generalized ability to tolerate diffuse competition. Which outcome is more likely? A full answer requires a phylogenetic analysis because convergence and evolutionary stasis are indistinguishable unless the character state of the common ancestor is known (Harvey and Pagel 1991). Still, we can inquire whether syntopic congeners are now ecologically similar or dissimilar.

Large numbers of congeneric plants coexist in tropical forests. Gentry (1988) and Gentry et al. (1987) drew attention to this phenomenon, describing species swarms in *Piper* (Piperaceae), *Miconia* (Melastomataceae), *Psychotria* (Rubiaceae), and several herbaceous genera in southern Central America and northwestern South America. Other examples include *Eugenia* (Myrtaceae) with 45 tree species in a 50 ha plot at Pasoh, Malaysia; *Pouteria* (Sapotaceae) with 22 tree species in a single hectare in Amazonian Ecuador; and *Inga* (Mimosoideae) with 22 tree species in a 0.16 ha plot also in Amazonian Ecuador (Manokaran et al. 1992, Valencia et al. 1994, R. Foster, personal communication, 1996). Are these syntopic congeners ecologically distinct or ecologically similar?

Reproductive phenologies illustrate both possibilities. Cross-pollination and competition for pollinators and fruit dispersal agents select for divergent phenologies that minimize temporal overlap. The fruiting phenologies of Tinidadian *Miconia* (Melastomataceae) and the flowering phenologies of Malaysian *Shorea* (Dipterocarpaceae) and Costa *Heliconia* (Musaceae) are segregated in time (Snow 1965, Stiles 1977, Ashton et al. 1988). These congeners have diverged despite recent common descent. The attraction of larger numbers of pollinators or fruit dispersal agents and the satiation of pests select for coincident reproductive displays that maximize temporal overlap. Flowering phenologies overlap completely for different species of Panamanian *Costus* (Zingiberaceae) with morphologically similar flowers that attract the same pollinator species to the same microhabitats (Schemske 1981). Both divergence and convergence or stasis are evident among the reproductive phenologies of syntopic congeners. Which is more likely?

When all genera in a local flora are evaluated, a clear answer emerges. Phenologies are remarkably similar in an overwhelming proportion of genera (Wright and Calderon 1995). Strong demonstrable similarities among species motivate the hypothesis that chance population dynamics contributes to the maintenance of tropical forest plant diversity (see Section "Chance").

REFERENCES

Ashton, P.S., 1969. Speciation among tropical forest trees: some deductions in the light of recent evidence. In: R.H. Lowe-McConnell, ed. Speciation in Tropical Environments. Academic Press, London, pp. 155–196.

Ashton, P.S., T.J. Givnish, and S. Appanah, 1988. Staggered flowering in the Dipterocarpaceae: new insights into floral induction and the evolution of mast fruiting in the aseasonal tropics. American Naturalist 132: 44–66.

Asquith, N.M., S.J. Wright, and M.J. Claus, 1997. Does mammal community composition control seedling recruitment in neotropical forests? Evidence from islands in central Panama. Ecology 78: 941–946.

Aubréville, A., 1938. La foret de la Cote d'Ivoire. Bulletin du Comité d'Études Historiques et Scientifiques de 1' Afrique Occidentale Francaise 15: 205–261.

Augspurger, C.K., 1984. Seedling survival of tropical tree species: interactions of dispersal distance, light-gaps and pathogens. Ecology 65: 1705–1712.

Baillie, I.C., P.S. Ashton, M.N. Court, J.A.R. Anderson, E.A. Fitzpatrick, and J. Tinsley, 1987. Site characteristics and the distribution of tree species in mixed dipterocarp forest on Tertiary sediments in central Sarawak, Malaysia. Journal of Tropical Ecology 3: 201–220.

Beadle, N.C.W., 1966. Soil phosphate and its role in molding segments of the Australian flora and vegetation, with special reference to xeromorphy and sclerophylly. Ecology 47: 992–1007.

Behrensmeyer, A.K., J.D. Damuth, W.A. DiMichele, R. Potts, H. Sues, and S.L. Wing, 1992. Terrestrial Ecosystems through Time. University of Chicago Press, Chicago.

Brokaw, N., 1987. Gap-phase regeneration of three pioneer tree species in a tropical forest. Journal of Ecology 75: 9–19.

Brown, S. and A.E. Lugo, 1982. The storage and production of organic matter in tropical forests and their role in the global carbon cycle. Biotropica 14: 161–187.

Brown, N.D. and T.C. Whitmore, 1992. Do dipterocarp seedling really partition tropical rain forest gaps? Philosophical Transaction of the Royal Society, London B 335: 369–378.

Chazdon, R.L. and R.W. Pearcy, 1991. The importance of sunflecks for forest understory plants. Bioscience 41: 760–766.

Chesson, P. and N. Huntly, 1997. The roles of harsh and fluctuating conditions in the dynamics of ecological communities. American Naturalist 150: 519–553.

Clark, D.A. and D.B. Clark, 1984. Spacing dynamics of a tropical rain-forest tree: evaluation of the Janzen-Connell model. American Naturalist 124: 769–788.

Clinebell, R.R., O.L. Phillips, A.H. Gentry, N. Stark, and H. Zuuring, 1995. Prediction of neotropical tree and liana species richness from soil and climatic data. Biodiversity and Conservation 4: 56–60.

Coley, P.D. and T.M. Aide, 1991. A comparison of herbivory and plant defenses in temperate and tropical broad-leaved forests. In: P.W. Price, T.M. Lewinsohn, G.W. Fernades, and W.W. Benson, eds. Plant–Animal Interactions: Evolutionary Ecology in Tropical and Temperate regions. Wiley, New York, pp. 25–49.

Coley, P.D. and J.A. Barone, 1996. Herbivory and plant defenses in tropical forests. Annual Review of Ecology and Systematic 27: 305–335.

Condit, R., S.P. Hubbell, and R.B. Foster, 1992. Recruitment near conspecific adults and the maintenance of tree and shrub diversity in a neotropical forest. American Naturalist 140: 261–286.

Condit, R., S.P. Hubbell, and R.B. Foster, 1994. Density dependence in two understory tree species in a neotropical forest. Ecology 75: 671–680.

Connell, J.H., 1971. On the role of natural enemies in preventing competitive exclusion in some marine animals and in rain forest trees. In: P.J. den Boer and G.R. Grad-well, eds. Dynamics of Populations. Centre for Agricultural Publishing and Documentation, Wageningen, The Netherlands, pp. 298–313.
Connell, J.H., 1978. Diversity in tropical rain forests and coral reefs. Science 199: 1302–1309.
Connell, J.H., J.G. Tracey, and L.J. Webb, 1984. Compensatory recruitment, growth and mortality as factors maintaining rain forest tree diversity. Ecological Monographs 54: 141–164.
Darwin, C.R., 1859. On the Origin of Species by Means of Natural Selection. John Murray, London.
Denslow, J.S., 1980. Gap partitioning among tropical rainforest trees. Biotropica 12(suppl): 47–55.
Denslow, J.S., 1987. Tropical rainforest gaps and tree species diversity. Annual Review of Ecology and Systematics 18: 431–451.
Denslow, J.S., J.C. Schultz, P.M. Vitousek, and B.R. Strain, 1990. Growth responses of tropical shrubs to treefall gap environments. Ecology 71: 165–179.
Dirzo, R., and A. Miranda, 1991. Altered patterns of herbivory and diversity in the forest understory: a case study of the possible consequences of contemporary defaunation. In: P.W. Price, P.W. Lewinsohn, G.W. Fernades, and W.W. Benson, eds. Plant–Animal Interactions: Evolutionary Ecology in Tropical and Temperate Regions. Wiley, New York, pp. 273–287.
Federov, A.A., 1966. The structure of the tropical rain forest and speciation in the humid tropics. Journal of Ecology 54: 1–11.
Friis, E.M., W.G. Chaloner, and P.R. Crane, eds., 1987. The Origins of Angiosperms and Their Biological Consequences. Cambridge University Press, Cambridge.
Gause, G.F., 1934. The Struggle for Existence. Hafner, New York.
Gentry, A.H., 1988. Changes in plant community diversity and floristic composition on environmental and geographical gradients. Annals of the Missouri Botanical Garden 75: 1–34.
Gentry, A.H. and C. Dodson, 1987. Contribution of nontrees to species richness of a tropical rain forest. Biotropica 19: 149–156.
Gentry, A.H. and L.H. Emmons, 1987. Geographical variation in fertility, phenology, and composition of the understory of neotropical forests. Biotropica 19: 216–227.
Gilbert, G.S., S.P. Hubbell, and R.B. Foster, 1994. Density and distance-to-adult effects of a canker disease of trees in a moist tropical forest. Oecologia (Berlin) 98: 100–108.
Gillett, J.B., 1962. Pest pressure, an underestimated factor in evolution. Systematics Association Publication 4: 37–46.
Grant, P.R., 1986. Ecology and Evolution of Darwin's Finches. Princeton University Press, Princeton, NJ.
Grubb, P.J., 1986. Global trends in species-richness in terrestrial vegetation: a view from the northern hemisphere. In: J.H.R. Gee and P.S. Giller, eds. Organization of Communities Past and Preset. Blackwell Scientific, London, pp. 99–118.
Hall, J.B. and M.D. Swaine, 1976. Classification and ecology of closed-canopy forest in Ghana. Journal of Ecology 64: 913–951.
Hall, J.B. and M.D. Swaine, 1981. Distribution and Ecology of Vascular Plants in a Tropical Rain Forest. Dr. W. Junk Publishers, The Hague.
Hammond, D.S. and V.K. Brown, 1995. Seed size of woody plants in relations to disturbance, dispersal, soil type in wet neotropical forests. Ecology 76: 2544–2561.
Harvey, P.H. and M.D. Pagel, 1991. The Comparative Method in Evolutionary Biology. Oxford University Press, Oxford.
Holdridge, L.R., W.C. Grenke, W.H. Hatheway, T. Liang, and J.A. Tosi Jr., 1971. Forest Environments in Tropical Life Zones: A Pilot Study. Pergamon Press, New York.
Hubbell, S.P., 1979. Tree dispersion, abundance, and diversity in a tropical dry forest. Science 203: 1299–1309.
Hubbell, S.P. and R.B. Foster, 1983. Diversity of canopy trees in a neotropical forest and implications for conservation. In: S.L. Sutton, T.C. Whitmore, and A.C. Chadwick, eds. Tropical Rain Forest: Ecology and Management. Blackwell Scientific, Oxford, pp. 25–41.

Hubbell, S.P. and R.B. Foster, 1986. Biology, chance, and history and the structure of tropical rain forest tree communities. In: J. Diamond and T.J. Case, eds. Community Ecology. Harper and Row, New York, pp. 314–329.
Hubbell, S.P., R. Condit, and R.B. Foster, 1990. Presence and absence of density dependence in a neotropical tree community. Philosophical Transactions of the Royal Society, London (B) 330: 269–281.
Hurtt, G.C. and S.W. Pacala, 1995. The consequences of recruitment limitation: reconciling chance, history and competitive differences between plants. Journal of Theoretical Biology 176: 1–12.
Huston, M., 1979. A general hypothesis of species diversity. American Naturalist 113: 81–101.
Huston, M., 1980. Soil nutrients and tree species richness in Costa Rican forests. Journal of Biogeography 7: 147–157.
Huston, M., 1993. Biological diversity, soils, and economics. Science 262: 1676–1680.
Huston, M.A., 1994. Biological Diversity. Cambridge University Press, Cambridge.
Janzen, D.H., 1970. Herbivores and the number of tree species in tropical forests. American Naturalist 104: 501–528.
Jordan, C.F., 1983. Productivity of tropical rain forest ecosystems and the implications for their use as future wood and energy sources. In: F.B. Golley, ed. Ecosystems of the World. 14A. Tropical Rain forest Ecosystems. Elsevier, Amsterdam, pp. 117–136.
Kitajima, K., 1994. Relative importance of photosynthetic traits and allocation patterns as correlates of seedling shade tolerance of 13 tropical trees. Oecologia 98: 419–428.
LaFrankie, J., 1996. Initial findings from Lambir: trees, soils and community dynamics. Center for Tropical Forest Science 1995: 5.
Latham, R.E. and R.E. Ricklefs, 1993. Continental comparisons of temperate-zone tree species diversity. In: R.E. Ricklefs and D. Schluter, eds. Species Diversity in Ecological Communities. University of Chicago Press, Chicago, pp. 294–314.
Leigh, E.G. Jr., S.J. Wright, and F.E. Putz, 1993. The decline of tree diversity on newly isolated tropical islands: a test of a null hypothesis and some implications. Evolutionary Ecology 7: 76–102.
Lieberman, M., D. Lieberman, G.S. Hartshorn, and R. Peralta, 1985. Small-scale altitudinal variation in lowland wet tropical forest vegetation. Journal of Ecology 73: 505–516.
Lieberman, M., D. Lieberman, R. Peralta, and G.S. Hartshorn, 1995. Canopy closure and the distribution of tropical forest trees of La Selva, Costa Rica. Journal of Tropical Ecology 11: 161–177.
MacArthur, R.H., 1969. Patterns of communities in the tropics. Biological Journal of the Linnean Society 1: 19–30.
Manokaran, N., J.V. La Frankie, K.M. Kochummen, E.S. Quah, J.E. Klahn, P.S. Ashton, and S.P. Hubbell, 1992. Stand table and distribution of species in the 50-ha research plot at Pasoh Forest Reserve. Forestry Research Institute of Malaysia, Research Data 1: 1–454.
Marquis, R.J., H.J. Young, and H.E. Braker, 1986. The influence of understory vegetation cover on germination and seeding establishment in a tropical lowland wet forest. Biotropica 18: 273–278.
Medina, E. and H. Klinge, 1983. Productivity of tropical forests and tropical woodlands. In: O.L. Lange, P.S. Nobel, C.B. Osmond, and H. Ziegler, eds. Physiological Plant Ecology IV. Springer-Verlag, Berlin, pp. 281–303.
Nelson, B.W., V. Kapos, J.B. Adams, W.J. Oliverira, O.P.G. Braun, and I.L. do Amaral, 1994. Forest disturbance by large blowdowns in the Brazilian Amazon. Ecology 75: 853–858.
Nepstad, D.C., C.R. de Carvalho, E.A. Davidson, P.H. Jipp, P.A. Lefebvre, G.H. Negreiros, E.D. da Silva, T.A. Stone, S.E. Trumbore, and S. Vieria, 1994. The role of deep roots in the hydrological and carbon cycles of Amazonian forests and pastures. Nature 372: 666–669.
Niklas, K.J., B.H. Tiffney, and A.H. Knoll, 1983. Patterns in vascular land versification. Nature 303: 614–616.
Orians, G.H., 1982. The influence of tree-falls in tropical forests in tree species richness. Tropical Ecology 23: 255–279.
Osunkjoya, O.O., J.E. Ash, M.S. Hopkins, and A.W. Graham, 1992. Factors affecting survival of tree seedlings in North Queensland rainforests. Oecologia (Berlin) 91: 569–578.

Pacala, S.W., C.D. Canham, J. Saponara, J.A. Silander, R.K. Kobe, and E. Ribbens, 1996. Forest models defined by field measurements: estimation, error analysis and dynamics. Ecological Monographs 66: 1–43.
Paine, R.T., 1966. Food web complexity and species diversity. American Naturalist 100: 65–75.
Palmer, M.W., 1994. Variation in species richness; towards a unification of hypotheses. Folia Geobotanica & Phytotaxonomica 29: 511–530.
Petraitis, P.S., R.E. Latham, and R.A. Niesenbaum, 1989. The maintenance of species diversity by disturbance. Quarterly Review of Biology 64: 393–418.
Phillips, O.L., P. Hall, A.H. Gentry, S.A. Sawyer, and R. Váquez, 1994. Dynamics and species richness of tropical rain forests. Proceedings of the National Academy of Science (USA) 91: 2805–2809.
Platt, W.J. and D.R. Strong, eds., 1989. Special feature: gaps in forest ecology. Ecology 70: 535–576.
Prance, G.T., 1977. Floristic inventory of the tropics: where do we stand? Annals of the Missouri Botanical Garden 64: 659–684.
Reich, P.B., M.B. Walters, and D.S. Ellsworth, 1992. Leaf life-span in relation to leaf, plant and stand characteristics among diverse ecosystems. Ecological Monographs 62: 365–392.
Ricklefs, R.E., 1977. Environmental heterogeneity and plant species diversity: a hypothesis. American Naturalist 376–381.
Rosenzweig, M.L., 1992. Species diversity gradients: we know more and less than we thought. Journal of Mammalogy 73: 715–730.
Rosenzweig, M.L., 1995. Species Diversity in Space and Time. Cambridge University Press, Cambridge.
Schemske, D.W., 1981. Floral convergence and pollinator sharing in two bee-pollinated tropical herbs. Ecology 62: 946–954.
Schupp, E.W., 1992. The Janzen-Connell model for tropical tree diversity: population implications and the importance of spatial scale. American Naturalist 140: 526–530.
Silver, W.L., 1994. Is nutrient availability related to plant nutrient use in humid tropical forests? Oecologia (Berlin) 98: 336–343.
Snow, D.W., 1965. A possible selective factor in the evolution of fruiting seasons in tropical forest. Oikos 15: 274–281.
Sollins, P., 1998. Factors influencing species composition in tropical lowland rain forest: does soil matter? Ecology 79: 23–30.
Stiles, F.G., 1977. Coadapted pollinators: the flowering seasons of hummingbird-pollinated plants in a tropical forest. Science 198: 1177–1178.
Swaine, M.D., 1996. Rainfall and soil fertility as factors limiting forest species distributions in Ghana. Journal of Ecology 84: 419–428.
Terborgh, J., 1973. On the notion of favorableness in plant ecology. American Naturalist 107: 481–501.
Terborgh, J., 1992. Maintenance of diversity in tropical forests. Biotropica 24: 283–292.
Terborgh, J. and S.J. Wright, 1994. Effects of mammalian herbivores on plant recruitment in two neotropical forests. Ecology 75: 1829–1833.
Terborgh, J., R.B. Foster, and V.P. Nunez, 1996. Tropical tree communities: a test of the nonequilibrium hypothesis. Ecology 77: 561–567.
Tilman, D., 1982. Resource Competition and Community Structure. Princeton University Press, Princeton, NJ.
Tilman, D., 1994. Competition and biodiversity in spatially structured habitats. Ecology 75: 2–16.
Tuomisto, H. and A.D. Poulsen, 1996. Influence of edaphic specialization on pteridophyte distribution in neotropical rain forests. Journal of Biogeography 23: 283–293.
Tuomisto, H. and K. Ruokolainen, 1994. Distribution of Pteridophyta and Melastomataceae along an edaphic gradient in an Amazonian rain forest. Journal of Vegetation Science 5: 25–34.
Uhl, C., K. Clark, N. Dezzeo, and P. Maquirino, 1988. Vegetation dynamics in Amazonian treefall gaps. Ecology 69: 751–763.
Valencia, R., H. Balslev, and G. Paz y Miño C., 1994. High tree alpha-diversity in Amazonian Ecuador. Biodiversity and Conservation 3: 21–28.
Vitousek, P.M., 1984. Litterfall, nutrient cycling and nutrient limitation in tropical forests. Ecology 65: 285–298.

Welden, C.W., S.W. Hewett, S.P. Hubbell, and R.B. Foster, 1997. Sapling survival, growth, and recruitment: relationship to canopy height in a neotropical forest. Ecology 72: 35–50.

Whitmore, T.C., 1975. Tropical Rain Forest of the Far East. Clarendon Press, Oxford.

Wills, C., R. Condit, R.B. Foster, and S.P. Hubbell, 1997. Strong density- and diversity-related effects help to maintain tree species diversity in a neotropical forest. Proceedings of the National Academy of science (USA) 94: 1252–1257.

Wright, S.J., 1992. Seasonal drought, soil fertility, and the species diversity of tropical forest plant communities. Trends in Ecology and Evolution 7: 260–263.

Wright, S.J. and O. Calderon, 1995. Phylogenetic constraints on tropical flowering phenologies. Journal of Ecology 83: 937–948.

12 Arctic Ecology

Sarah E. Hobbie

CONTENTS

INTRODUCTION

The Arctic encompasses the region of the globe that lies north of the latitudinal tree line. Its environment is extreme in myriad ways, with low mean annual temperatures and precipitation, short growing seasons, and cold soils that are often underlain by permafrost, thawing only incompletely during the growing season. Consistent with the extreme environment, productivity is lower in the Arctic than in most other regions of the world (Figure 12.1).

The Arctic comprises a variety of vegetation types at both small and large scales (Bliss and Matveyeva 1992). The High Arctic, consisting primarily of the islands of the Canadian Arctic Archipelago, is characterized by polar desert and semidesert—large patches of bare, unvegetated ground are interspersed with vegetation dominated by cushion plants, forbs, and dwarf shrubs. In contrast, the Low Arctic has near-continuous cover of tundra vegetation with varying proportions of dwarf shrubs, sedges, mosses, lichens, and forbs. Within both the High Arctic and the Low Arctic, large variation in both species composition and plant productivity occurs at relatively small spatial scales (meters to kilometers) because of variation in topography and parent material, and consequently in drainage, nutrient availability, and snow cover (Walker et al. 1994).

Because of its severe environment, the Arctic has long been of interest to plant physiological ecologists (Billings and Mooney 1968). The Arctic has received much less attention from other areas of plant ecology, particularly population and community ecology, for both practical and theoretical reasons. The long-lived nature of most tundra species and their low rates of sexual reproduction (see later) make study of plant populations and manipulation of plant densities in situ difficult. Furthermore, some plant ecologists (Savile 1960, Grime 1977, Callaway et al. 2002) have suggested that competitive interactions are unimportant relative to

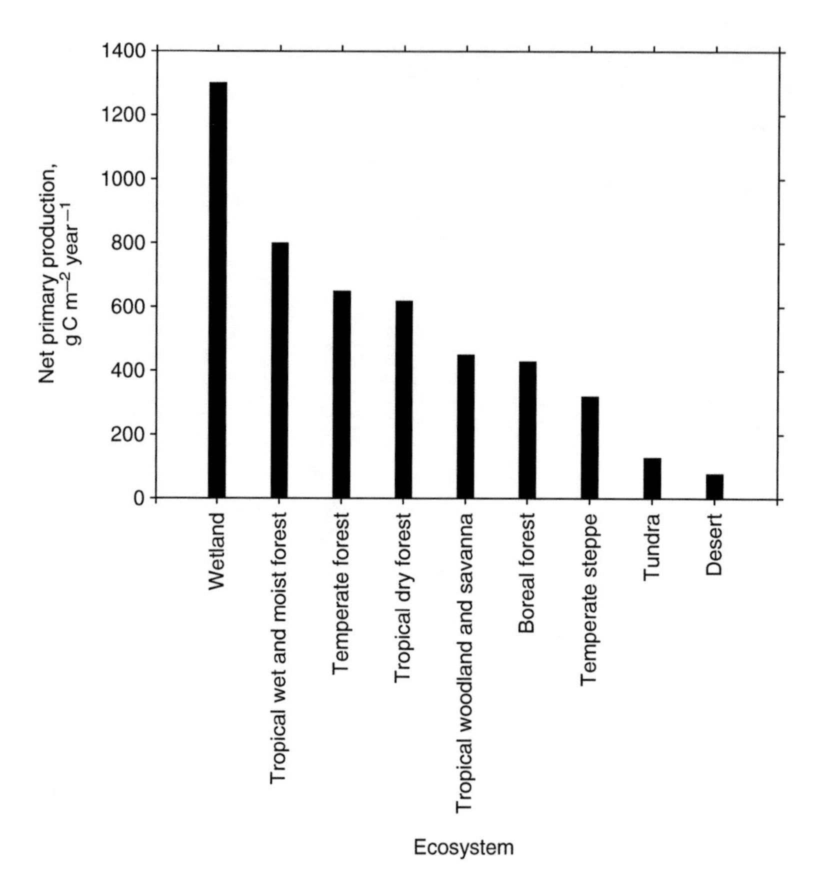

FIGURE 12.1 Net primary production (NPP) in various ecosystems of the world. (Data from Houghton, R.A. and Skole, D.L., *The Earth as Transformed by Human Action*, B.L. Turner, W.C. Clark, R.W. Kates, J.F. Richards, J.T. Matthews, and W.B. Meyer, eds, Cambridge University Press, Cambridge, UK, 1990. With permission.)

abiotic factors in structuring communities in high-stress environments, such as the Arctic, perhaps steering population and community ecologists away from working at high latitudes.

In this chapter, I take a broad view of arctic plant ecology. I touch on the unique characteristics of arctic plant populations and discuss the accumulating evidence regarding the importance of interspecific interactions in structuring arctic ecosystems. I briefly review the physiological ecology of arctic plants, but I refer the reader to numerous previous reviews and books on this topic (Billings and Mooney 1968, Bliss et al. 1981, Chapin and Shaver 1985a, Körner and Larcher 1988, Chapin et al. 1992). The major emphasis of this chapter is on the synthesis of a large amount of recent experimental work in tundra to evaluate the hypotheses that were formulated during these earlier ecophysiological studies.

CHARACTERISTICS OF ARCTIC PLANT POPULATIONS

Relatively little is known about arctic plant populations, perhaps because of the long-lived nature and the infrequent sexual reproduction of arctic plant species. In the closed communities of the Low Arctic, reproduction from seed is relatively rare (Callaghan and Emanuelsson 1985, McGraw and Fetcher 1992, Gough 2006), and few annual plant species occur (Hultén 1968). Low rates of sexual reproduction may result from low allocation to total reproductive effort, although allocation to viable seed is not necessarily lower in arctic species than in temperate ones (Chester and Shaver 1982). Low rates of recruitment from seed may also result from the paucity of suitable germination sites in these closed communities and from high mortality of young seedlings (McGraw and Shaver 1982, Callaghan and Emanuelsson 1985, Gough 2006). Recruitment from seed increases after natural and human-caused disturbance, particularly for graminoid species, often from seed stored in the seed bank (Chapin and Chapin 1980, McGraw 1980, Freedman et al. 1982, Gartner et al. 1983, 1986, Grulke and Bliss 1988, Ebersole 1989, McGraw and Vavrek 1989). In the more open vegetation of the High Arctic, recruitment from seed is more common (Callaghan and Emanuelsson 1985) and increases with fertilization (Robinson et al. 1998).

In contrast with sexual reproduction, asexual reproduction is nearly ubiquitous in the Arctic. Many arctic plant species possess rhizomes (belowground stems) or stolons, or produce adventitious roots from aboveground stems that are overgrown by mosses and subsequently become belowground stems. Ramets can ultimately become independent of one another, making the recognition of individual genets in intact tundra difficult. Although demographic analysis at the level of individual genets is therefore problematic, understanding the demography of plant parts within ramets can enlighten studies of the response of arctic plants to environmental perturbations (McGraw and Fetcher 1992). For example, demographic analysis of shoots or tillers can indicate whether changes in production or biomass at the ecosystem level result from changes in branching, shoot growth, or shoot mortality (McGraw 1985a, Chapin and Shaver 1985c, Bret-Harte et al. 2001); alternatively, demographic analyses may indicate when turnover of plant parts has changed with no change in total biomass (Fetcher and Shaver 1983).

PLANT SPECIES INTERACTIONS IN ARCTIC ECOSYSTEMS

Species interactions have been studied much less in arctic ecosystems than in temperate ecosystems, presumably because of the difficulty of manipulating densities of long-lived, slow-growing, often woody, perennial species that occur only rarely as seedlings (Gough 2006). Despite the lack of experimental evidence, many have hypothesized that direct limitation by stressful abiotic factors (e.g., low temperature) is much more important than are species interactions in structuring plant communities in the Arctic (Savile 1960, Warren Wilson 1966, Billings and Mooney 1968, Grime 1977). Others have suggested that competitive interactions are important because of low nutrient availability and help explain community responses to environmental manipulations (Chapin and Shaver 1985b, Chapin et al. 1995).

Most studies of competitive interactions in the Arctic have found little evidence for strong negative interactions in tundra with a few exceptions (Fetcher 1985, McGraw 1985a, Jonasson 1992, Shevtsova et al. 1995, Hobbie et al. 1999, Bret-Harte et al. 2004, Gough 2006). For example, in subarctic Scandinavia, only one of the three dwarf shrubs (*Vaccinium vitis-idaea*) responded positively to removal of other dwarf shrubs (Shevtsova et al. 1995), and a separate study found no positive effects of dwarf shrub removal on any of a number of species in several different tundra types (Jonasson 1992). In Alaskan tussock tundra, where seven species were removed individually in separate treatments and numerous species responses measured, only two pairs of species exhibited negative interactions (Hobbie et al. 1999). In a separate study, various combinations of shrub and moss removals had little effect on abundance of remaining species or growth forms (Bret-Harte et al. 2004). Similarly, *Dryas octopetala* showed no response to removal of all neighbors after 3 years in Alaskan tundra (McGraw 1985a).

A few studies have demonstrated negative interactions among species in the Arctic. *Eriophorum vaginatum* responded to dwarf shrub removal with increased tillering (Fetcher 1985).

Another study that manipulated ramets by outplanting, rather than removal, demonstrated competitive interactions between two species of *Eriophorum* in Alaska (McGraw and Chapin 1989). In the High Arctic, moss removal increased growth of forbs (Sohlberg and Bliss 1986), and shoot biomass of two species, *Luzula confusa* and *Salix polaris*, was reduced in the presence of heterospecific neighbors (Dormann et al. 2004). None of the studies demonstrating negative interactions between tundra species distinguished competitive interactions from other kinds of negative interactions such as allelopathy or effects of species on abiotic conditions such as soil temperature.

Species removal studies have found as much evidence for positive interactions among plant species in arctic ecosystems as for negative interactions. Several dwarf shrub species responded negatively to removal of other dwarf shrubs in Sweden and Finland (Jonasson 1992, Shevtsova et al. 1995). Mosses in tussock tundra and in boreal Canada also responded negatively to shrub removal, which was attributed to photoinhibition of photosynthesis under high light intensities (Murray et al. 1993) and increased evaporative stress (Busby et al. 1978) when the shrub canopies were removed. Thus, facilitation may be important in determining performance of some species in the Arctic, particularly at high elevations where temperatures are coldest and abiotic stresses like wind scouring most severe (Callaway et al. 2002), and could explain the clustering of plant species in some tundra environments (Callaghan and Emanuelsson 1985, Carlsson and Callaghan 1991).

The lack of evidence for strong competitive or other kinds of negative interactions in tundra supports Grime's (1977) contention that in high-stress environments plants are directly limited by abiotic factors, and that competitive interactions are relatively unimportant. However, several other explanations exist for the general failure to demonstrate strong competitive interactions in the Arctic. Arctic species respond individualistically to manipulations of various environmental factors (Chapin and Shaver 1985b), suggesting that growth of co-occuring species may be limited by different factors (see later). In addition, arctic plants may minimize competition for nitrogen (N) by partitioning their use of the N pool (e.g., into inorganic and organic N, see later) (McKane et al. 2002). Arctic plant species may also partition their uptake of nutrients in time and space. The timing of root initiation in the spring differs among species and among growth forms (Shaver and Billings 1977, Kummerow et al. 1983). In particular, evergreens begin root growth earlier in the spring than do deciduous species that initiate roots only after leaf expansion has begun. Such interspecific differences in the timing of root growth could have important consequences for nutrient acquisition in arctic ecosystems that are often characterized by a pulse of nutrient mineralization in the spring (Kielland 1990, Nadelhoffer et al. 1992). Different species also show characteristic rooting depths (Shaver and Billings 1975), perhaps partitioning their use of nutrients in space. Consistent with these observations, a study in Alaskan tussock tundra that added labeled forms of N to the soil at different depths and in different seasons found that most species showed greater N uptake earlier than later in the growing season, and some species showed greater uptake from shallow than from deep soil depths (McKane et al. 2002).

Even if it does occur, competition may be difficult to demonstrate in arctic tundra where annual uptake of nutrients is a relatively small proportion of the annual nutrient requirement of a species and plants grow relatively slowly. Removing a single species may have only a minor effect on nutrient availability for the remaining species over the timescale of most experiments. Indeed, in one study that measured inorganic nutrient availability in removal treatments, removal demonstrably increased soil N availability only in treatments in which several codominant species were removed simultaneously (Bret-Harte et al. 2004). Studying competitive interactions by increasing plant densities (e.g., by outplanting or seed-sowing) may be a more effective way of demonstrating the importance of competitive interactions than removals; however, such manipulations are problematic with arctic species, since they are generally slow growing, perennial, and ramets may suffer high mortality from transplant shock (Gough 2006). The true reason for the lack of strong competitive interactions in tundra deserves more exploration.

ECOPHYSIOLOGY OF ARCTIC PLANTS

Ecophysiological research on arctic plants generally indicates that resource acquisition is not directly limited by cold temperatures. This observation has led to the hypothesis that growth and productivity in arctic ecosystems are limited more by the indirect effects of cold temperatures (i.e., short growing season length and low nutrient availability) than by their direct effects (Chapin 1983). After reviewing the observations that led to these generalizations, I evaluate how well recent experimental evidence supports these generalizations.

CARBON

In terms of carbon gain, arctic plants may be limited more by the length of the growing season than by low irradiance or low photosynthetic rates due to cold temperatures per se. Although average irradiance during daytime hours is less in the Arctic than those at lower latitudes, the average daily irradiance is comparable, because of the 24 h photoperiod of the northern summer (Chapin and Shaver 1985a). However, many of the long days occur early in the growing season when the ground is still snow-covered or the soil is still frozen. Thus growing season length, rather than light intensity per se, may limit plant growth. Evergreen vascular and nonvascular plants effectively extend the growing season by photosynthesizing under the spring snow pack when days are long, taking advantage of elevated CO_2 concentrations, sufficient light, and near-freezing air temperatures under snow (Kappen 1993, Starr and Oberbauer 2003).

The photosynthetic temperature optima and carbohydrate status of tundra plants provide additional indirect evidence that arctic species are not carbon-limited during the growing season. Although photosynthetic temperature optima are often above ambient air temperatures (Tieszen 1973, Oechel 1976), many tundra species have relatively broad photosynthetic temperature optima, allowing them to achieve near-maximum rates at fairly low temperatures (Tieszen 1973, Johnson and Tieszen 1976, Limbach et al. 1982, Semikhatova et al. 1992). Relatively large pools of total nonstructural carbohydrate in arctic plants also suggest that they are able to acquire carbon in excess of their growth requirements (Chapin and Shaver 1985a). Interestingly, photosynthesis in arctic plants may be more sensitive to soil than to air temperature. For two arctic sedges, photosynthesis and stomatal conductance increased greatly with soil warming between 0°C and 10°C (Starr et al. 2004), suggesting that carbon gain could be sensitive to future soil warming, particularly warming associated with loss of permafrost.

NUTRIENTS

Nutrient uptake in tundra plants is also relatively insensitive to the direct effects of low temperature. Rather, cold temperatures limit nutrient uptake indirectly, by causing low rates of nutrient supply. For example, cold temperatures reduce nutrient inputs from N_2 fixation (Chapin and Bledsoe 1992) and weathering and recycling of nutrients during decomposition (Figure 12.2) (Nadelhoffer et al. 1992). Although temperatures in arctic soils are colder than optimum temperatures for nutrient uptake (Chapin and Bloom 1976), tundra species have a higher potential for nutrient uptake at low temperatures than do warm-adapted species (Chapin 1974).

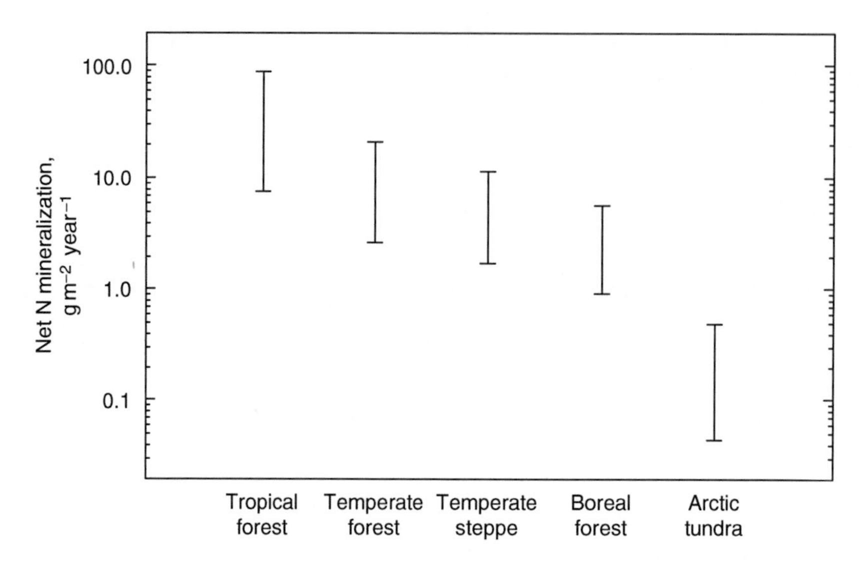

FIGURE 12.2 Net nitrogen mineralization rates measured in various ecosystems of the world. (Reprinted from Nadelhoffer, K.J., Giblin, A.E., Shaver, G.R., and Linkins, A.E., *Arctic Ecosystems in a Changing Climate: An Ecophysiological Perspective*, F.S. Chapin III, R.L. Jefferies, J.F. Reynolds, G.R. Shaver, and J. Svoboda, eds, Academic Press, San Diego, CA, 1992. With permission.)

Tundra species use acquired nutrients efficiently, relying heavily on stored nutrients to support growth and retranslocating much of the nutrients from senescing tissues. Stored nutrients supply the majority of the nutrients to current growth (Berendse and Jonasson 1992) and may allow arctic plants to grow during times when soil nutrients are frozen and unavailable (Shaver and Kummerow 1992). Many arctic plant species retranslocate a relatively high percentage of nutrients from senescing tissues, although their retranslocation efficiencies are not consistently higher than those of temperate species (Chapin and Shaver 1989, Jonasson 1989, Berendse and Jonasson 1992).

One potential way that arctic plants effectively increase the pool of nutrients available for uptake is by using organic N directly, rather than by relying on soil microbes to mineralize N from organic compounds. As in other ecosystems, there is growing evidence that arctic plants may short-circuit the N cycle in this way (Lipson and Nasholm 2001). Such use of organic N would help explain the large discrepancy between annual N uptake and annual net N mineralization rates in the Arctic (Giblin et al. 1991, Kielland 1994). Concentrations and turnover of free amino acids are relatively high in tundra soils (Kielland 1995, Weintraub and Schimel 2005). Both mycorrhizal and nonmycorrhizal species from tussock tundra take up and grow on amino acids in solution culture (Chapin et al. 1993, Kielland 1994) and in situ (Schimel and Chapin 1996, McKane et al. 2002, Nordin et al. 2004). Furthermore, both ericoid and ectomycorrhizal species have proteolytic capabilities and can transfer organic N from soil to the host plant (Read 1991) and many of the common arctic species have these types of mycorrhizal associations. Estimates of N acquisition via ericoid and ectomycorrhizal species range between 61% and 86% of plant N, much of which is likely accessed in organic form (Hobbie and Hobbie 2006).

Growth

Whether cold temperatures directly limit growth of arctic plants is unclear. The relative growth rates of arctic plants are similar to those of temperate species, suggesting that length of the growing season, rather than cold temperatures during the growing season itself, limits biomass accumulation in the Arctic (Chapin and Shaver 1985b, Semikhatova et al. 1992). On the other hand, ambient temperatures are often suboptimal for growth (Kummerow and Ellis 1984, Körner and Larcher 1988) and manipulations of temperature in situ sometimes increase plant growth (see later). Thus, growth of some species may be directly limited by cold temperature.

INDIVIDUAL PLANT RESPONSES TO ENVIRONMENTAL FACTORS

During the past decade, numerous studies in various arctic habitats have examined the response of plant growth, biomass, reproduction, phenology, and production to manipulations of environmental factors. These experiments allow us to evaluate the generalizations and hypotheses proposed earlier. Three general patterns emerge from studies that examined individual plant responses (primarily growth) to environmental manipulations. First, of the various factors manipulated (including nutrients, temperature, water, light, and CO_2), plant growth most often responds to increased nutrient availability. Second, in almost all studies demonstrating a positive response of species growth to nutrient addition, there are exceptions—species that either do not respond or respond negatively to nutrient addition. Third, manipulation of environmental factors besides nutrients has less predictable results. Later, I explore each of these points in detail. For simplicity, I have subsumed a number of different kinds of measurements (e.g., current year's shoot length, branching, tillering) under the word *growth*. I have also included changes in total biomass in growth although many studies did not determine whether change in biomass resulted from changes in individual shoot growth or from changes in shoot demography (branching or mortality).

Fertilization studies in both the Low Arctic and the High Arctic in North America and in Europe have demonstrated generally positive biomass responses of deciduous shrubs and graminoids (grasses and sedges) and negative responses of lichens and mosses (Dormann and Woodin 2002, van Wijk et al. 2003a). In subarctic Sweden, addition of N, phosphorus (P), and potassium (K) increased the growth or biomass of numerous species in heath tundra (Havström et al. 1993, Parsons et al. 1994, 1995, Michelsen et al. 1996, Press et al. 1998), in a fellfield (a relatively open, rocky community, Michelsen et al. 1996), and in graminoid and shrub tundra (Jonasson 1992). Fertilization increased growth and biomass of many species in upland tussock tundra, wet meadow (sedge-dominated) tundra, and heath tundra in Alaska (McKendrick et al. 1978, 1980, Shaver and Chapin 1980, 1986, 1995, Chapin and Shaver 1985b, 1996, Shaver et al. 1986, 1996, 1998, Gough et al. 2002, Hobbie et al. 2005). Although most studies added N, P, and K simultaneously, when nutrients were added separately, this response was usually to N rather than P in upland tundra (Shaver and Chapin 1980, Gebauer et al. 1995, but see McKendrick et al. 1980). Some studies showed little response to N or P alone, but large responses to their combined addition (Gough et al. 2002, Gough and Hobbie 2003). Wet meadow tundra, on the other hand, is more often P-limited (Shaver and Chapin 1995, Shaver et al. 1998). In the High Arctic, fertilization also increased growth, biomass, reproduction, and seedling establishment of some species (Henry et al. 1986, Wookey et al. 1994, 1995, Robinson et al. 1998).

Despite numerous examples of positive growth or biomass responses to nutrient addition in a variety of tundra ecosystems, many studies demonstrated exceptions—species that did not respond or responded negatively to nutrient addition (Shaver and Chapin 1980, 1986, Chapin and Shaver 1985b, 1996, Havström et al. 1993, Shaver et al. 1996, 1998, Press et al. 1998, Robinson et al. 1998, Cornelissen et al. 2001, Graglia et al. 2001). The unresponsive species often differed among studies, and intercomparison is further complicated because fertilization experiments varied in duration, in amount or types of fertilizer applied, in the timing of fertilization relative to plant phenology of nutrient uptake, and in the species whose responses were measured. Several possible reasons for these lack of responses to fertilization exist. Some species in tundra may be limited by factors other than low nutrient availability (see later) (Chapin and Shaver 1985b). Species may not be able to respond to additional nutrients in the presence of superior competitors, and understory species may be vulnerable to increased shading by overstory species whose biomass increases with nutrient addition (Chapin and Shaver 1985b, 1996, McGraw 1985b). Indeed, in a study that combined species removals and fertilization, some species responded more to nutrient addition when neighbors were removed (Bret-Harte et al. 2004). Thus, while many species may respond positively to fertilization in the short term, long-term fertilization often reduces species richness and leads to dominance by one or a few species, such as *Betula nana* in Alaskan tussock tundra (Shaver et al. 2001).

Nonvascular species (bryophytes and lichens) often showed negative responses to fertilization, particularly in Low Arctic studies (Cornelissen et al. 2001, van Wijk et al. 2003a, see Robinson et al. 1998 for contrasting results in the High Arctic). These species may be moisture-rather than nutrient-limited (Murray et al. 1989a,b, Tenhunen et al. 1992), perhaps explaining their lack of (or negative) response to nutrient addition (Jonasson 1992, Chapin et al. 1995). They may also be negatively affected by greater shading and litter accumulation in fertilized plots (Cornelissen et al. 2001, van Wijk et al. 2003a).

After nutrients, increased temperature is the environmental factor that most often increases growth or biomass in the Arctic. Warming increased growth of about half of the dominant species in tussock tundra (Chapin and Shaver 1985b, 1996, Shaver et al. 1986). Warming also increased growth of some species in Swedish subarctic tundra (Havström et al. 1993, Parsons et al. 1994, 1995, Michelsen et al. 1996) and resulted in greater growth and reproduction and earlier phenology in the High Arctic (Havström et al. 1993, Welker et al. 1993, Wookey et al. 1995). Meta-analyses of warming experiments throughout the Arctic suggest that shrubs and graminoids in particular grow more and exhibit earlier phenology with experimental warming (Arft et al. 1999, Dormann and Woodin 2002, Walker et al. 2006). In Alaska, increased shrub biomass in response to experimental warming is consistent with greater shrub biomass during past warming events in the Holocene (Brubaker et al. 1995, Hu et al. 2002) and in the past century as inferred from repeat aerial photography (Sturm et al. 2001, Hinzman et al. 2005). Changes in shrub biomass with warming have regional-scale implications for climate, as decreased albedo associated with greater shrub biomass could significantly amplify future high-latitude warming (Chapin et al. 2005).

Besides their sensitivity to growing season temperature, arctic plant species also respond to increased growing season length. For example, species exhibit earlier growth and earlier senescence, and leaf area index is higher throughout the growing season when the snow-free period begins earlier and ends later in the season (Oberbauer et al. 1998, Starr et al. 2000).

Far fewer manipulations of light, water, and CO_2 have been done in the Arctic than those of nutrients or temperature. Shade decreased growth and biomass particularly of overstory species in both the Low Arctic (Chapin and Shaver 1985b, 1996) and the High Arctic (Havström et al. 1993). Water is rarely limiting in tundra (Oberbauer and Dawson 1992, Gold and Bliss 1995) and water addition had few effects on growth or biomass in subarctic Sweden (Parsons et al. 1994) or in the High Arctic (Henry et al. 1986, Wookey et al. 1994, 1995, Dormann and Woodin 2002). Addition of flowing water to a tussock tundra slope increased the growth of some species (Murray et al. 1989b, Oberbauer et al. 1989), but this might have been a result of greater flow of nutrients, rather than a direct water effect (Chapin et al. 1988). In studies in arctic and subarctic tundra, respectively, elevated CO_2 did not alter photosynthetic rate or growth of individual tillers of *E. vaginatum* in Alaskan tussock tundra (Tissue and Oechel 1987) and had little or negative effects on dwarf shrub growth in subarctic Sweden (Gwynn-Jones et al. 1997).

The response of biomass or production at the shoot or whole-plant level to environmental manipulations is not readily predicted from the physiological response of leaves or roots. For example, photosynthetic responses to treatments generally do not translate directly into growth responses (Bigger and Oechel 1982, Wookey et al. 1994, Gebauer et al. 1995, Chapin and Shaver 1996). Similarly, the response of phenology and nutrient uptake provides little indication of how species respond at the community level (Chapin and Shaver 1996).

In summary, at the individual level, arctic plants are most often nutrient-limited; however, some species in some regions may be limited by other environmental factors such as low temperature. Some studies have suggested that the indirect effects of temperature (i.e., low nutrient availability) are more likely to limit growth at the southern extent of species' ranges in the Arctic, whereas cold temperatures directly limit growth at the northern extent of species' ranges (Havström et al. 1993). Although this idea is compelling, too few species have been studied experimentally over a range of latitudes to determine whether this pattern is general.

ECOSYSTEM RESPONSES TO ENVIRONMENTAL FACTORS

Although individual plants respond to both nutrients and temperature, net primary production (NPP) and total plant biomass respond more consistently to nutrient addition alone than to manipulation of other environmental factors. However, fewer studies have measured total production and biomass than individual plant growth or biomass, and even measurements of total production have mostly excluded root production. Vascular aboveground net primary production (ANPP) and total biomass increased with nutrient addition in Low Arctic moist tundra (Chapin and Shaver 1985b, Shaver and Chapin 1986, Chapin et al. 1995, Hobbie et al. 2005), wet tundra (Shaver et al. 1998), and heath tundra (ANPP only) (Gough et al. 2002), subarctic heath, graminoid, and shrub tundra (Jonasson 1992, Press et al. 1998), and High Arctic sites on Ellesmere Island and Svalbard (Henry et al. 1986, Robinson et al. 1998). Fine root production also increased, although not significantly, in wet and moist arctic tundra (Nadelhoffer et al. 2002) and root biomass increased in a moist nonacidic tundra (van Wijk et al. 2003b) with fertilization. However, where mosses are a significant component of the community, decreased moss biomass largely offset increased vascular plant biomass, resulting in little change in total plant biomass after long-term fertilization (Chapin et al. 1995, Press et al. 1998, Hobbie et al. 2005).

Although vascular plant biomass generally increases with long-term fertilization, one study demonstrated a decline in total ecosystem carbon storage because nutrient addition increased losses from soil carbon pools (Mack et al. 2004). The exact mechanism responsible for depleted soil carbon pools is unknown. However, greater soil organic matter or litter decomposition with added nutrients and a shifts in species composition to more shallow-rooted species—whose roots decompose more quickly than deeply rooted species because of warmer soil temperatures near the soil surface—could have contributed to this. In contrast with nutrients, elevated CO_2 has little effect on net ecosystem production in tundra because of downregulation (Grulke et al. 1990, Oechel et al. 1994).

Experimental warming led to a significant but much smaller increase in ANPP than did nutrients in Alaskan tussock tundra, although this increase was attributed to the indirect effects of temperature on ANPP acting through increased soil nutrient availability (Figure 12.3) (Chapin et al. 1995). By contrast, increased temperature had little effect on total plant biomass and production in Alaskan wet sedge (Shaver et al. 1998) and moist nonacidic (Gough and Hobbie 2003) tundras, and no effect on total biomass in a subarctic heath tundra in Sweden (Press et al. 1998). Most studies that increase air temperature also increase soil temperature as well, making it difficult to attribute changes in ANPP to the direct or indirect effects of warmer temperatures. However, one study in Alaskan tussock

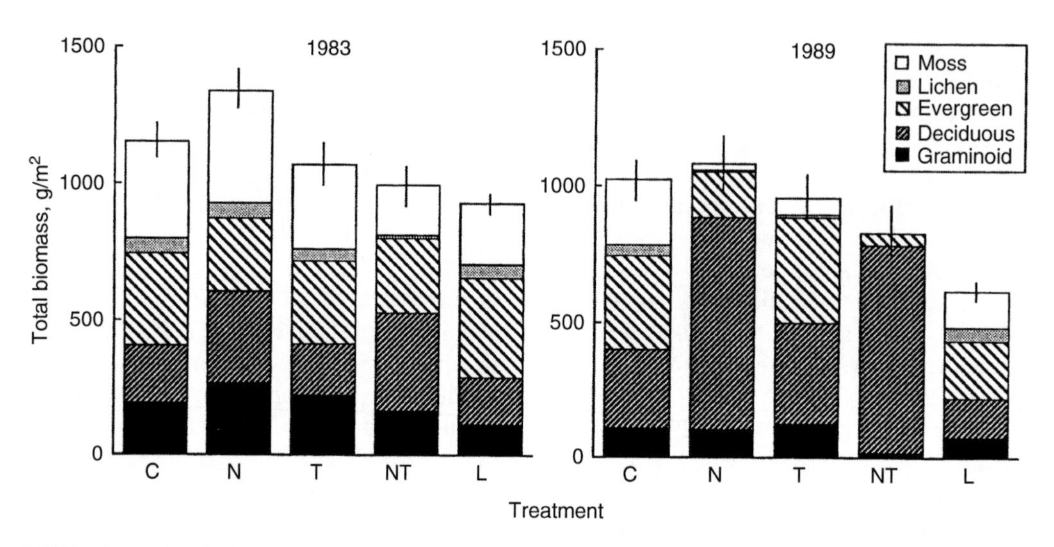

FIGURE 12.3 Total aboveground net primary production (NPP) by growth form of tussock tundra in response to environmental manipulations near Toolik Lake, Alaska, measured 3 (1983) and 9 (1989) years after initiation of treatments. Treatments are control (C), nutrient addition (N), warming (T), nutrient addition-warming (NT), and light attenuation (L). (From Chapin III, F.S., Shaver, G.R., Giblin, A.E., Nadelhoffer, K.J., and Laundre, J.A., *Ecology*, 76, 694, 1995. With permission.)

tundra that manipulated air temperature alone found no response of ANPP (Hobbie and Chapin 1998).

One potential indirect consequence of warming, extended growing season, also seems to increase NPP. Experimental extension of the growing season via snow removal and exclusion increased leaf area index throughout the growing season (Oberbauer et al. 1998). Consistent with this, earlier snowmelt and longer growing season length at high latitudes correspond with greater plant growth at high latitudes, as inferred from satellite data (Myneni et al. 1997, Zhou et al. 2001).

Some evidence suggests that responses to warming are greatest where soil resources are not limiting. Meta-analyses of warming responses across a number of sites suggest that tundra responds more to warming in the Low Arctic than the High Arctic where nutrient availability is presumably higher. Tundra also responds more to warming where soil moisture is optimum (i.e., in mesic sites more than in dry or very wet sites) (Walker et al. 2006). Interestingly, however, experimental studies have not shown significant positive interactions between warming and nutrient addition for NPP (Chapin et al. 1995, Gough and Hobbie 2003), although artifacts associated with plastic greenhouses (shading, greater herbivory) may explain this lack of positive interaction (McKane et al. 1997, Gough and Hobbie 2003).

A significant issue that has motivated much of the earlier research in recent decades is whether high-latitude ecosystems respond to warming in ways that exacerbate (positive feedback) or ameliorate (negative feedback) rising atmospheric CO_2 and warming. Because of the large stocks of carbon stored in tundra and boreal ecosystems, these ecosystems have received a great deal of attention as potential sources of CO_2 to the atmosphere as climate warms and decomposition rates increase (Post 1990). However, empirical and modeling studies suggest that climate warming could ultimately increase carbon sequestration. Recent climate warming in Alaska stimulated greater carbon efflux from tundra initially, but the stimulation diminished over time (Oechel et al. 1993, 2000). Biogeochemical models of tundra suggest that long-term warming combined with elevated CO_2 increases NPP and soil organic matter decomposition, but that increased carbon uptake more than offsets greater carbon losses from decomposition, resulting in greater net ecosystem production and carbon sequestration, primarily because warming stimulates decomposition of soil organic matter and increases N availability, leading to a net transfer of N from soils to vegetation, with their higher C:N ratio (McKane et al. 1997, Rastetter et al. 1997, Le Dizes et al. 2003). In addition, in some tundra ecosystems, greater nutrient availability leads to increased woody production as shrubs become more abundant (Shaver et al. 2001) and wood has a relatively high C:N ratio. That, combined with elevated CO_2, results in higher vegetation C:N ratios overall, also contributing to greater ecosystem carbon sequestration.

In summary, experimental evidence supports the contention that NPP in the Arctic is limited primarily by the indirect effects of low temperature, namely low soil nutrient availability resulting from slow decomposition in cold soils and short growing season length. In contrast, tundra shows little capacity to respond to changes in aboveground conditions or resources (i.e., warmer air temperature or elevated CO_2) without an accompanying increase in nutrient availability. Although NPP is decreased by light attenuation, it is unknown, but unlikely, that increased light availability would stimulate NPP.

These conclusions based on experimental evidence are generally consistent with patterns of NPP across Arctic landscapes. Primary production can vary up to 10-fold among different vegetation types within relatively small distances in the Arctic (Figure 12.4) (Shaver and Chapin 1991). Gently sloping mesic areas of tussock tundra have intermediate levels of production (Shaver and Chapin 1991, Shaver et al. 1996), although even within mesic tundra, NPP can vary significantly (Hobbie et al. 2005). In contrast, the lowest productivity is found on well-drained ridge-tops supporting heath tundra and poorly drained low areas

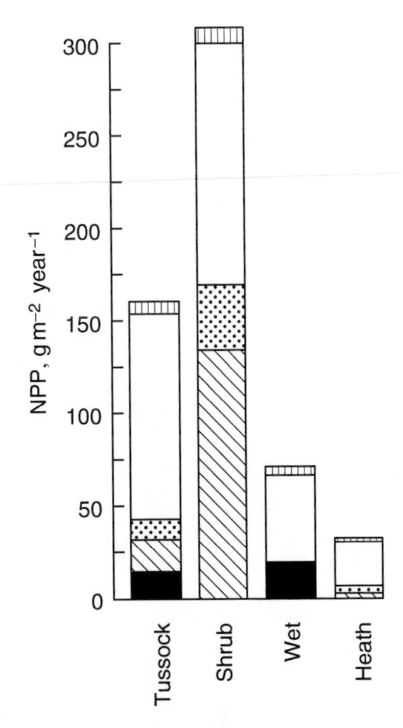

FIGURE 12.4 Total net primary production (NPP, excluding roots) of the vascular plants in each of four tundra vegetation types near Toolik Lake, Alaska. Total NPP is indicated by the height of the bar. Within each bar, inflorescence production is indicated by vertical stripes, leaf production by open, apical stem growth (current year's twigs) by dots, secondary stem growth by diagonal stripes, and belowground rhizome growth by solid. (From Shaver, G.R. and Chapin III, F.S., *Ecol. Monogr.*, 61, 1, 1991. With permission.)

supporting wet meadow tundra (Shaver and Chapin 1991). The highest productivity is found in areas of flowing water (riparian areas and water tracks) that support more productive graminoid or shrub tundra (Chapin et al. 1988, Hastings et al. 1989, Shaver and Chapin 1991).

The variation in productivity across arctic landscapes is proximately related to variation in nutrient availability. For example, the highest production is found on deeply thawed soils that offer protection from wind, in sites dominated by N fixers (Shaver et al. 1996), in sites influenced by animals that disturb the soil and import nutrients, increasing nutrient availability (McKendrick et al. 1980), and in sites influenced by flowing water, which increases bulk flow of nutrients and stimulates net N mineralization (Chapin et al. 1988). Additionally, edaphic factors, such as parent material, can influence nutrient availability and NPP. For example, Alaskan moist acidic tundra has 10-fold greater net N mineralization rates and is twice as productive as moist nonacidic tundra (Hobbie and Gough 2002, Hobbie et al. 2005). In general, vegetation types with the greatest productivity are associated with soils that have relatively high rates of net N mineralization (Chapin et al. 1988, Kielland 1990, Giblin et al. 1991, Hobbie and Gough 2002).

ACKNOWLEDGMENTS

Syndonia Bret-Harte, Laura Gough, and Dave Hooper provided helpful comments on an earlier version of this chapter.

REFERENCES

Arft, A.M., M.D. Walker, J. Gurevitch, J.M. Alatalo, M.S. Bret-Harte, M. Dale, M. Diemer, F. Gugerli, G.H.R. Henry, M.H. Jones, R.D. Hollister, I.S. Jonsdottir, K. Laine, E. Levesque, G.M. Marion, U. Molau, P. Molgaard, U. Nordenhall, V. Raszhivin, C.H. Robinson, G. Starr, A. Stenstrom, M. Stenstrom, O. Totland, P.L. Turner, L.J. Walker, P.J. Webber, J.M. Welker, and P.A. Wookey, 1999. Responses of tundra plants to experimental warming: Meta-analysis of the International Tundra Experiment. Ecological Monographs 69: 491–511.

Berendse, F. and S. Jonasson, 1992. Nutrient use and nutrient cycling in northern ecosystems. In: F.S. Chapin III, R.L. Jefferies, J.F. Reynolds, G.R. Shaver, and J. Svoboda, eds. Arctic Ecosystems in a Changing Climate. Academic Press, San Diego, CA, pp. 337–356.

Bigger, C.M. and W.C. Oechel, 1982. Nutrient effect on maximum photosynthesis in arctic plants. Holarctic Ecology 5: 158–163.

Billings, W.D. and H.A. Mooney, 1968. The ecology of arctic and alpine plants. Biological Review 43: 481–529.

Bliss, L.C. and N.V. Matveyeva, 1992. Circumpolar arctic vegetation. In: F.S. Chapin III, R.L. Jefferies, J.F. Reynolds, G.R. Shaver, and J. Svoboda, eds. Arctic Ecosystems in a Changing Climate: An Ecophysiological Perspective. Academic Press, San Diego, CA, pp. 59–89.

Bliss, L.C., O.W. Heal, and J.J. Moore, ed, 1981. Tundra Ecosystems: A Comparative Analysis. Cambridge University Press, Cambridge, UK.

Bret-Harte, M.S., G.R. Shaver, J.P. Zoerner, J.F. Johnstone, J.L. Wagner, A.S. Chavez, R.F. Gunkelman IV, S.C. Lippert, and J.A. Laundre, 2001. Developmental plasticity allows *Betula nana* to dominate tundra subjected to an altered environment. Ecology 82: 18–32.

Bret-Harte, M.S., E.A. Garcia, V.M. Sacre, J.R. Whorley, J.L. Wagner, S.C. Lippert, and F.S. Chapin III, 2004. Plant and soil responses to neighbor removal and fertilisation in Alaskan tussock tundra. Journal of Ecology 92: 635–647.

Brubaker, L.B., P.M. Anderson, and F.S. Hu, 1995. Arctic tundra biodiversity: A temporal perspective from late Quaternary pollen records. In: F.S. Chapin III and C. Körner, eds. Arctic and Alpine Biodiversity: Patterns, Causes and Ecosystem Consequences. Springer-Verlag, Berlin, pp. 111–125.

Busby, J.R., L.C. Bliss, and C.D. Hamilton, 1978. Microclimate control of growth rates and habitats of the boreal forest mosses, *Tomenthypnum nitens* and *Hylocomium splendens*. Ecological Monographs 48: 95–110.

Callaghan, T.V. and U. Emanuelsson, 1985. Population structure and processes of tundra plants and vegetation. In: J. White, ed. The Population Structure of Vegetation. Junk, Dordrecht, pp. 399–439.

Callaway, R.M., R.W. Brooker, P. Choler, Z. Kikvidze, C.J. Lortie, R. Michalet, L. Paolini, F.I. Pugnaire, B.J. Cook, E.T. Aschehong, C.Y. Armas, and B. Newingham, 2002. Positive interactions among alpine plants increase with stress: a global experiment. Nature 417: 844–848.

Carlsson, B.A. and T.V. Callaghan, 1991. Positive plant interactions in tundra vegetation and the importance of shelter. Journal of Ecology 79: 973–983.

Chapin, F.S., III, 1974. Morphological and physiological mechanisms of temperature compensation in phosphate absorption along a latitudinal gradient. Ecology 55: 1180–1198.

Chapin, F.S., III, 1983. Direct and indirect effects of temperature on arctic plants. Polar Biology 2: 47–52.

Chapin, F.S., III and A. Bloom, 1976. Phosphate absorption: adaptation of tundra graminoids to low temperature, low-phosphorus environment. Oikos 26: 111–121.

Chapin, D.M. and C.S. Bledsoe, 1992. Nitrogen fixation in arctic plant communities. In F.S. Chapin, III, R.L. Jefferies, J.F. Reynolds, G.R. Shaver, and J. Svoboda, eds. Arctic Ecosystems in a Changing Climate: An Ecophysiological Perspective. Academic Press, San Diego, CA, pp. 301–320.

Chapin, F.S., III and M.C. Chapin, 1980. Revegetation of an arctic disturbed site by native tundra species. Journal of Applied Ecology 17: 449–456.

Chapin, F.S., III and G.R. Shaver, 1985a. Arctic. In: B.F. Chabot and H.A. Mooney, eds. Physiological Ecology of North America. Chapman and Hall, New York, pp. 16–40.

Chapin, F.S., III and G.R. Shaver, 1985b. Individualistic growth response of tundra plant species to environmental manipulations in the field. Ecology 66: 564–576.

Chapin, F.S., III and G.R. Shaver, 1985c. Individualistic growth response of tundra plants to environmental manipulations in the field. Ecology 66: 564–576.

Chapin, F.S., III and G.R. Shaver, 1989. Differences in growth and nutrient use among arctic plant growth forms. Functional Ecology 3: 73–80.

Chapin, F.S., III and G.R. Shaver, 1996. Physiological and growth responses of arctic plants to a field experiment simulating climatic change. Ecology 77: 822–840.

Chapin, F.S., III, N. Fetcher, K. Kielland, K.R. Everett, and A.E. Linkins, 1988. Productivity and nutrient cycling of Alaskan tundra: Enhancement by flowing soil water. Ecology 69: 693–702.

Chapin, F.S., III, R.L. Jefferies, J.F. Reynolds, G.R. Shaver, and J. Svoboda, eds, 1992. Arctic Ecosystems in a Changing Climate: An Ecophysiological Perspective. Academic Press, San Diego, CA.

Chapin, F.S., III, L. Moilanen, and K. Kielland, 1993. Preferential use of organic nitrogen for growth by a non-mycorrhizal arctic sedge. Nature 361: 150–153.

Chapin, F.S., III, G.R. Shaver, A.E. Giblin, K.J. Nadelhoffer, and J.A. Laundre, 1995. Responses of arctic tundra to experimental and observed changes in climate. Ecology 76: 694–711.

Chapin, F.S., III, M. Sturm, M.C. Serreze, J.P. McFadden, J.R. Key, A.H. Lloyd, A.D. McGuire, T.S. Rupp, A.H. Lynch, D.S. Schimel, J. Beringer, W.L. Chapman, H.E. Epstein, E.S. Euskirchen, L.D. Hinzman, G.J. Jia, C.-L. Ping, K.D. Tape, C.D.C. Thompson, D.A. Walker, and J.M. Welker, 2005. Role of land-surface changes in arctic summer warming. Science 310: 657–660.

Chester, A.L. and G.R. Shaver, 1982. Reproductive effort in cottongrass tussock tundra. Holarctic Ecology 5: 200–206.

Cornelissen, J.H.C., T.V. Callaghan, J.M. Alatalo, A. Michelsen, E. Graglia, A.E. Hartley, D.S. Hik, S.E. Hobbie, M.C. Press, C.H. Robinson, G.H.R. Henry, G.R. Shaver, G.K. Phoenix, D. Gwynn Jones, S. Jonasson, F.S. Chapin III, U. Molau, C. Neill, J.A. Lee, J.M. Melillo, B. Sveinbjornsson, and R. Aerts, 2001. Global change and arctic ecosystems: is lichen decline a function of increases in vascular plant biomass? Journal of Ecology 89: 984–994.

Dormann, C.F. and S.J. Woodin, 2002. Climate change in the arctic: using plant functional types in a meta-analysis of field experiments. Functional Ecology 16: 4–17.

Dormann, C.F., R. Van Der Wal, and S.J. Woodin, 2004. Neighbor identity modifies effects of elevated temperature on plant performance in the High Arctic. Global Change Biology 10: 1587–1598.

Ebersole, J.J., 1989. Role of seed bank in providing colonizers on a tundra disturbance in Alaska. Canadian Journal of Botany 67: 466–471.

Fetcher, N., 1985. Effects of removal of neighboring species on growth, nutrients, and microclimate of *Eriophorum vaginatum*. Arctic and Alpine Research 17: 7–17.

Fetcher, N. and G.R. Shaver, 1983. Life histories of tillers of *Eriophorum vaginatum* in relation to tundra disturbance. Journal of Ecology 71: 131–147.

Freedman, B., N. Hill, J. Svoboda, and G. Henry, 1982. Seed banks and seedling occurrence in a high arctic oasis at Alexandra Fjord, Ellesmere Island, Canada. Canadian Journal of Botany 60: 2112–2118.

Gartner, B.L., F.S. Chapin III, and G.R. Shaver, 1983. Demographic patterns of seedling establishment and growth of native graminoids in an Alaskan tundra disturbance. Journal of Applied Ecology 20: 965–980.

Gartner, B.L., F.S. Chapin III, and G.R. Shaver, 1986. Reproduction of *Eriophorum vaginatum* by seed in Alaskan tussock tundra. Journal of Ecology 74: 1–18.

Gebauer, R.L.E., J.F. Reynolds, and J.D. Tenhunen, 1995. Growth and allocation of the arctic sedges *Eriophorum angustifolium* and *E. vaginatum*: effects of variable soil oxygen and nutrient availability. Oecologia 104: 330–339.

Giblin, A.E., K.J. Nadelhoffer, G.R. Shaver, J.A. Laundre, and A.J. McKerrow, 1991. Biogeochemical diversity along a riverside toposequence in arctic Alaska. Ecological Monographs 61: 415–435.

Gold, W.G. and L.C. Bliss, 1995. Water limitations and plant community development in a polar desert. Ecology 76: 1558–1568.

Gough, L., 2006. Neighbor effects on germination, survival, and growth in two arctic tundra plant communities. Ecography 29: 44–56.

Gough, L. and S.E. Hobbie. 2003. Responses of moist non-acidic arctic tundra to altered environment: Productivity, biomass and species richness. Oikos 103: 204–216.

Gough, L., P.A. Wookey, and G.R. Shaver, 2002. Dry heath arctic tundra responses to long-term nutrient and light manipulations. Arctic, Antarctic and Alpine Research 34: 211–218.

Graglia, E., S. Jonasson, A. Michelsen, I.K. Schmidt, M. Havstrøm, and L. Gustavsson, 2001. Effects of environmental perturbations on abundance of subarctic plants after three, seven and ten years of treatments. Ecography 24: 5–12.

Grime, J.P., 1977. Evidence for the existence of three primary strategies in plants and its relevance to ecological and evolutionary theory. American Naturalist 111: 1169–1194.

Grulke, N.E. and L.C. Bliss, 1988. Comparative life-history characteristics of two high arctic grasses, Northwest Territories. Ecology 69: 484–496.

Grulke, N.E., G.H. Reichers, W.C. Oechel, U. Hjelm, and C. Jaeger, 1990. Carbon balance in tussock tundra under ambient and elevated CO_2. Oecologia 83: 485–494.

Gwynn-Jones, D., J.A. Lee, and T.V. Callaghan, 1997. Effects of enhanced UV-B radiation and elevated carbon dioxide concentrations on a sub-arctic forest heath ecosystem. Plant Ecology 128: 243–249.

Hastings, S.J., S.A. Luchessa, W.C. Oechel, and J.D. Tenhunen, 1989. Standing biomass and production in water drainages of the foothills of the Philip Smith Mountains, Alaska, USA. Holarctic Ecology 12: 304–311.

Havström, M., T.V. Callaghan, and S. Jonasson, 1993. Differential growth responses of *Cassiope tetragona*, an arctic dwarf-shrub, to environmental perturbations among three contrasting high- and sub-arctic sites. Oikos 66: 389–402.

Henry, G.H.R., R. Freedman, and J. Svoboda, 1986. Effects of fertilization on three tundra plant communities of a polar desert oasis. Canadian Journal of Botany 64: 2502–2507.

Hinzman, L.D., N.D. Bettez, W.R. Bolton, F.S. Chapin III, M.B. Dyurgerov, C.L. Fastie, B. Griffith, R.D. Hollister, A. Hope, H.P. Huntington, A.M. Jensen, G.J. Jia, T. Jorgenson, D.L. Kane, D.R. Klein, G. Kofinas, A.H. Lynch, A.H. Lloyd, A.D. McGuire, F.E. Nelson, W.C. Oechel, T.E. Osterkamp, C.H. Racine, V.E. Romanovsky, R.S. Stone, D.A. Stow, M. Sturm, C.E. Tweedie, G.L. Vourlitis,

M.D. Walker, D.A. Walker, P.J. Webber, J.M. Welker, K.S. Winker, and K. Yoshikawa, 2005. Evidence and implications of recent climate change in northern Alaska and other arctic regions. Climatic Change 72: 251–298.

Hobbie, S.E. and F.S. Chapin III, 1998. The response of tundra plant biomass, aboveground production, nitrogen, and CO_2 flux to temperature manipulation. Ecology 79: 1526–1544.

Hobbie, S.E. and L. Gough, 2002. Foliar and soil nutrients in tundra on glacial landscapes of contrasting ages in northern Alaska. Oecologia 131: 453–463.

Hobbie, J.E. and E.A. Hobbie, 2006. ^{15}N in symbiotic fungi and plants estimates nitrogen and carbon flux rates in arctic tundra. Ecology 87: 816–822.

Hobbie, S.E., A. Shevtsova, and F.S. Chapin III, 1999. Plant responses to species removal and experimental warming in Alaskan tussock tundra. Oikos 84: 417–434.

Hobbie, S.E., L. Gough, and G.R. Shaver, 2005. Species compositional differences on different-aged glacial landscapes drive contrasting responses of tundra to nutrient addition. Journal of Ecology 93: 770–782.

Houghton, R.A. and D.L. Skole, 1990. Carbon. In: B.L. Turner, W.C. Clark, R.W. Kates, J.F. Richards, J.T. Matthews, and W.B. Meyer, eds. The Earth as Transformed by Human Action. Cambridge University Press, Cambridge, UK, pp. 393–408.

Hu, F.S., B.Y. Lee, D.S. Kaufman, S. Yoneji, D.M. Nelson, and P.D. Henne, 2002. Response of tundra ecosystem in southwestern Alaksa to Younger-Dryas climatic oscillation. Global Change Biology 8: 1156–1163.

Hultén, E., 1968. Flora of Alaska and Neighboring Territories. Stanford University Press, Stanford, CA.

Johnson, D.A. and L.L. Tieszen, 1976. Aboveground biomass allocation, leaf growth, and photosynthesis patterns in tundra plant forms in arctic Alaska. Oecologia 24: 159–173.

Jonasson, S., 1989. Implications of leaf longevity, leaf nutrient reabsorption, and translocation for the resource economy of five evergreen species. Oikos 56: 121–131.

Jonasson, S., 1992. Plant responses to fertilization and species removal in tundra related to community structure and clonality. Oikos 63: 420–429.

Kappen, L., 1993. Plant activity under snow and ice, with particular reference to lichens. Arctic 46: 297–302.

Kielland, K., 1990. *Processes Controlling Nitrogen Release and Turnover in Arctic Tundra*. Ph.D. thesis. University of Alaska, Fairbanks.

Kielland, K., 1994. Amino acid absorption by arctic plants: implications for plant nutrition and nitrogen cycling. Ecology 75: 2373–2383.

Kielland, K., 1995. Landscape patterns of free amino acids in arctic tundra soils. Biogeochemistry 31: 85–98.

Körner, C. and W. Larcher, 1988. Plant life in cold climates. Symposium of the Society of Experimental Biology 42: 25–57.

Kummerow, J. and B. Ellis, 1984. Temperature effect on biomass production and root/shoot biomass ratios in two arctic sedges under controlled environmental conditions. Canadian Journal of Botany 62: 2150–2153.

Kummerow, J., B.A. Ellis, S. Kummerow, and F.S. Chapin III, 1983. Spring growth of shoots and roots in shrubs of an Alaskan muskeg. American Journal of Botany 70: 1509–1515.

Le Dizes, S., B.L. Kwiatkowski, E.B. Rastetter, A. Hope, J.E. Hobbie, D.A. Stow, and S. Daeschner, 2003. Modeling biogeochemical responses of tundra ecosystems to temporal and spatial variations in climate in the Kuparuk River Basin (Alaska). Journal of Geophysical Research D—Atmospheres 108: 8165. doi: 8110.1029/2001JD000960.

Limbach, W.E., W.C. Oechel, and W. Lowell, 1982. Photosynthetic and respiratory responses to temperature and light of three Alaskan tundra growth forms. Holarctic Ecology 5: 150–157.

Lipson, D. and T. Nasholm, 2001. The unexpected versatility of plants: organic nitrogen use and availability in terrestrial ecosystems. Oecologia 128: 305–316.

Mack, M.C., E.A.G. Schuur, M.S. Bret-Harte, G.R. Shaver, and F.S. Chapin III, 2004. Ecosystem carbon storage in arctic tundra reduced by long-term nutrient fertilization. Nature 431: 440–443.

McGraw, J.B., 1980. Seed bank size and distribution of seeds in cottongrass tussock tundra. Canadian Journal of Botany 58: 1607–1611.

McGraw, J.B., 1985a. Experimental ecology of *Dryas octopetala* ecotypes. III. Environmental factors and plant growth. Arctic and Alpine Research 17: 229–239.

McGraw, J.B., 1985b. Experimental ecology of *Dryas octopetala* ecotypes: Relative response to competitors. New Phytologist 100: 233–241.
McGraw, J.B. and F.S. Chapin III, 1989. Competitive ability and adaptation to fertile and infertile soils in two *Eriophorum* species. Ecology 70: 736–749.
McGraw, J.B. and N. Fetcher, 1992. Response of tundra plant populations to climatic change. In F.S. Chapin III, R.L. Jefferies, J.F. Reynolds, G.R. Shaver, and J. Svoboda, eds. Arctic Ecosystems in a Changing Climate: An Ecophysiological Perspective. Academic Press, San Diego, CA, pp. 359–376.
McGraw, J.B. and G.R. Shaver, 1982. Seedling density and seedling survival in Alaskan cotton grass tussock tundra. Holarctic Ecology 5: 212–217.
McGraw, J.B. and M.C. Vavrek, 1989. The role of buried viable seeds in arctic and alpine plant communities. In: M.A. Leck, V.T. Parker, and R.L. Simpson, eds. Ecology of Soil Seed Banks. Academic Press, San Diego, CA, pp. 91–106.
McKane, R.B., E.B. Rastetter, G.R. Shaver, K.J. Nadelhoffer, A.E. Giblin, J.A. Laundre, and F.S. Chapin III, 1997. Climatic effects on tundra carbon storage inferred from experimental data and a model. Ecology 78: 1170–1187.
McKane, R.B., L.C. Johnson, G.R. Shaver, K.J. Nadelhoffer, E.B. Rastetter, B. Fry, A.E. Giblin, K.J. Kielland, B.L. Kwiatkowski, J.A. Laundre, and G. Murray, 2002. Resource-based niches provide a basis for plant species diversity and dominance in arctic tundra. Nature 415: 68–71.
McKendrick, J.D., V.J. Ott, and G.A. Mitchell, 1978. Effects of nitrogen and phosphorus fertilization on carbohydrate and nutrient levels in *Dupontia fisheri* and *Arctagrostis latifolia.* In: L.L. Tieszen, ed. Vegetation and Production Ecology of an Alaskan Arctic Tundra. Springer-Verlag, New York, pp. 565–578.
McKendrick, J.D., G.O. Batzli, K.R. Everett, and J.C. Swanson, 1980. Some effects of mammalian herbivores and fertilization on tundra soils and vegetation. Arctic and Alpine Research 12: 565–578.
Michelsen, A., S. Jonasson, D. Sleep, M. Havström, and T.V. Callaghan, 1996. Shoot biomass, $\partial^{13}C$, nitrogen, and chlorophyll responses of two arctic dwarf shrubs to in situ shading, nutrient application and warming simulating climatic change. Oecologia 105: 1–12.
Murray, K.J., P.C. Harley, J. Beyers, H. Walz, and J.D. Tenhunen, 1989a. Water content effects on photosynthetic response of *Sphagnum* mosses from the foothills of the Philip Smith Mountains, Alaska. Oecologia 79: 244–250.
Murray, K.J., J.D. Tenhunen, and J. Kummerow, 1989b. Limitations on moss growth and net primary production in tussock tundra areas of the foothills of the Philip Smith Mountains, Alaska. Oecologia 80: 256–262.
Murray, K.J., J.D. Tenhunen, and R.S. Nowak, 1993. Photoinhibition as a control on photosynthesis and production of *Sphagnum* mosses. Oecologia 96: 200–207.
Myneni, R., C.D. Keeling, C.J. Tucker, G. Asrar, and R.R. Nemani, 1997. Increased plant growth in the northern high latitudes from 1981 to 1991. Nature 386: 698–702.
Nadelhoffer, K.J., A.E. Giblin, G.R. Shaver, and A.E. Linkins, 1992. Microbial processes and plant nutrient availability in arctic soils. In: F.S. Chapin III, R.L. Jefferies, J.F. Reynolds, G.R. Shaver, and J. Svoboda, eds. Arctic Ecosystems in a Changing Climate: An Ecophysiological Perspective. Academic Press, San Diego, CA, pp. 281–300.
Nadelhoffer, K.J., L. Johnson, J. Laundre, A.E. Giblin, and G.R. Shaver, 2002. Fine root production and nutrient use in wet and moist arctic tundras as influenced by chronic fertilization. Plant and Soil 242: 107–113.
Nordin, A., I.K. Schmidt, and G.R. Shaver, 2004. Nitrogen uptake by arctic soil microbes and plants in relation to soil nitrogen supply. Ecology 85: 955–962.
Oberbauer, S.F. and T.E. Dawson, 1992. Water relations of arctic vascular plants. In: F.S. Chapin III, R.L. Jefferies, J.F. Reynolds, G.R. Shaver, and J. Svoboda, eds. Arctic Ecosystems in a Changing Climate: An Ecophysiological Perspective. Academic Press, San Diego, CA, pp. 259–279.
Oberbauer, S.F., S.J. Hastings, J.L. Beyers, and W.C. Oechel, 1989. Comparative effects of downslope water and nutrient movement on plant nutrition, photosynthesis, and growth in Alaskan tundra. Holarctic Ecology 12: 324–334.

Oberbauer, S.F., G. Starr, and E.W. Pop, 1998. Effects of extended growing season and soil warming on carbon dioxide and methane exchange of tussock tundra. Journal of Geophysical Research 103: 29075–29082.

Oechel, W.C., 1976. Seasonal patterns of temperature response of CO_2 flux and acclimation in arctic mosses growing in situ. Photosynthetica 10: 447–456.

Oechel, W.C., S.J. Hastings, G. Vourlitis, M. Jenkins, G. Riechers, and N. Grulke, 1993. Recent change of Arctic tundra ecosystems from a net carbon dioxide sink to a source. Nature 361: 520–523.

Oechel, W.C., S. Cowles, N. Grulke, S.J. Hastings, W. Lawrence, T. Prudhomme, G. Riechers, B. Strain, D. Tissue, and G. Vourlitis, 1994. Transient nature of CO_2 fertilization in arctic tundra. Nature 371: 500–503.

Oechel, W.C., G.L. Vourlitis, S.J. Hastings, R.C. Zulueta, L. Hinzman, and D.L. Kane, 2000. Acclimation of ecosystem CO_2 exchange in the Alaskan Arctic in response to decadal climate warming. Nature 406: 978–981.

Parsons, A.N., J.M. Welker, P.A. Wookey, M.C. Press, T.V. Callaghan, and J.A. Lee, 1994. Growth responses of four sub-arctic dwarf shrubs to simulated environmental change. Journal of Ecology 82: 307–318.

Parsons, A.N., P.A. Wookey, J.M. Welker, M.C. Press, T.V. Callaghan, and J.A. Lee, 1995. Growth and reproductive output of *Calamagrostis lapponica* in response to simulated environmental change in the subarctic. Oikos 72: 61–66.

Post, W.M., 1990. Report of a workshop on climate feedbacks and the role of peatlands, tundra, and boreal ecosystems in the global carbon cycle. 3289, Oak Ridge National Laboratory, Oak Ridge, TN.

Press, M.C., J.A. Potter, M.J. Burke, T.V. Callaghan, and J.A. Lee, 1998. Responses of a subarctic dwarf shrub heath community to simulated environmental change. Journal of Ecology 86: 315–327.

Rastetter, E.B., R.B. McKane, G.R. Shaver, K.J. Nadelhoffer, and A.E. Giblin, 1997. Analysis of CO_2, temperature, and moisture effects on carbon storage in Alaskan arctic tundra using a general ecosystem model. In: W.C. Oechel, T. Callaghan, T. Gilmanov, J.I. Holten, B. Maxwell, U. Molau, and B. Sveinbjörnsson, eds. Global Change and Arctic Terrestrial Ecosytems. Springer, New York, pp. 437–451.

Read, D.J., 1991. Mycorrhizas in ecosystems. Experientia 47: 376–391.

Robinson, C.H., P.A. Wookey, J.A. Lee, T.V. Callaghan, and M.C. Press, 1998. Plant community responses to simulated environmental change at a High Arctic polar semi-desert. Ecology 79: 856–866.

Savile, D.B.O., 1960. Limitations of the competitive exclusion principle. Science 132: 1761.

Schimel, J.P. and F.S. Chapin III, 1996. Tundra plant uptake of amino acid and NH_4^+ nitrogen in situ: plants compete well for amino acid N. Ecology 77: 2142–2147.

Semikhatova, O.A., T.V. Gerasimenko, and T.I. Ivanova, 1992. Photosynthesis, respiration, and growth of plants in the Soviet Arctic. In: F.S. Chapin III, R.L. Jefferies, J.F. Reynolds, G.R. Shaver, and J. Svoboda, eds. Arctic Ecosystems in a Changing Climate: An Ecophysiological Perspective. Academic Press, San Diego, CA, pp. 169–190.

Shaver, G.R. and W.D. Billings, 1975. Root production and root turnover in a wet tundra ecosystem, Barrow, Alaska. Ecology 56: 401–410.

Shaver, G.R. and D.W. Billings, 1977. Effects of daylength and temperature on root elongation in tundra graminoids. Oecologia 28: 57–65.

Shaver, G.R. and F.S. Chapin III, 1980. Response to fertilization by various plant growth forms in an Alaskan tundra: nutrient accumulation and growth. Ecology 61: 662–675.

Shaver, G.R. and F.S. Chapin III, 1986. Effect of fertilizer on production and biomass of tussock tundra, Alaska, U.S.A. Arctic and Alpine Research 18: 261–268.

Shaver, G.R. and F.S. Chapin III, 1991. Production: biomass relationships and element cycling in contrasting arctic vegetation types. Ecological Monographs 61: 1–31.

Shaver, G.R. and F.S. Chapin III, 1995. Long-term responses to factorial NPK fertilizer treatment by Alaskan wet and moist tundra sedge species. Ecography 18: 259–275.

Shaver, G.R. and J. Kummerow, 1992. Phenology, resource allocation, and growth of arctic vascular plants. In: F.S. Chapin III, R.L. Jefferies, J.F. Reynolds, G.R. Shaver, and J. Svoboda, eds. Arctic Ecosystems in a Changing Climate: An Ecophysiological Perspective. Academic Press, San Diego, CA, pp. 193–211.

Shaver, G.R., F.S. Chapin III, and B.L. Gartner, 1986. Factors limiting seasonal growth and peak biomass accumulation in *Eriophorum vaginatum* in Alaskan tussock tundra. Journal of Ecology 74: 257–278.

Shaver, G.R., J.A. Laundre, A.E. Giblin, and K.J. Nadelhoffer, 1996. Changes in live plant biomass, primary production, and species composition along a riverside toposequence in Arctic Alaska, U.S.A. Arctic and Alpine Research 28: 363–379.

Shaver, G.R., L.C. Johnson, D.H. Cades, G. Murray, J.A. Laundre, E.B. Rastetter, K.J. Nadelhoffer, and A.E. Giblin, 1998. Biomass accumulation and CO_2 flux in wet sedge tundras: responses to nutrients, temperature, and light. Ecological Monographs 68: 75–97.

Shaver, G.R., M.S. Bret-Harte, M.H. Jones, J. Johnstone, L. Gough, J. Laundre, and F.S. Chapin III, 2001. Species composition interacts with fertilizer to control long-term change in tundra productivity. Ecology 82: 3162–3181.

Shevtsova, A., A. Ojala, S. Neuvonen, M. Vieno, and E. Haukioja, 1995. Growth and reproduction of dwarf shrubs in subarctic plant community: annual variation and above-ground interactions with neighbors. Journal of Ecology 83: 263–275.

Sohlberg, E.H. and L.C. Bliss, 1986. Responses of *Ranunculus sabinei* and *Papaver radicatum* to removal of the moss layer in a high-arctic meadow. Canadian Journal of Botany 65: 1224–1228.

Starr, G. and S.F. Oberbauer, 2003. Photosynthesis of arctic evergreens under snow: implications for tundra ecosystem carbon balance. Ecology 84: 1415–1420.

Starr, G., S.F. Oberbauer, and E.W. Pop, 2000. Effects of lengthened growing season and soil warming on the phenology and physiology of *Polygonum bistorta*. Global Change Biology 6: 357–369.

Starr, G., D.S. Neuman, and S.F. Oberbauer, 2004. Ecophysiological analysis of two arctic sedges under reduced root temperatures. Physiologia Plantarum 20: 458–464.

Sturm, M., C. Racine, and K. Tape, 2001. Climate change: Increasing shrub abundance in the Arctic. Nature 411: 546–547.

Tenhunen, J.D., O.L. Lange, S. Hahn, R. Siegwolf, and S.F. Oberbauer, 1992. The ecosystem role of poikilohydric tundra plants. In: F.S. Chapin III, R.L. Jefferies, J.F. Reynolds, G.R. Shaver, and J. Svoboda, eds. Arctic Ecosystems in a Changing Climate: An Ecophysiological Perspective. Academic Press, San Diego, CA, pp. 213–237.

Tieszen, L.L., 1973. Photosynthesis and respiration in arctic tundra grasses: field light intensity and temperature responses. Arctic and Alpine Research 5: 239–251.

Tissue, D.T. and W.C. Oechel, 1987. Response of *Eriophorum vaginatum* to elevated CO_2 and temperature in the Alaskan tussock tundra. Ecology 68: 401–410.

van Wijk, M.T., K.E. Clemmensen, G.R. Shaver, M. Williams, T.V. Callaghan, F.S. Chapin III, J.H.C. Cornelisen, L. Gough, S.E. Hobbie, S. Jonasson, J.A. Lee, A. Michelsen, M.C. Press, S.J. Richardson, and H. Rueth, 2003a. Long-term ecosystem level experiments at Toolik Lake, Alaska, and at Abisko, Northern Sweden: generalizations and differences in ecosystem and plant type responses to global change. global Change Biology 10: 105–123.

van Wijk, M.T., M. Williams, L. Gough, S.E. Hobbie, and G.R. Shaver, 2003b. Luxury consumption of soil nutrients: a possible competitive strategy in aboveground and belowground biomass allocation for slow-growing arctic vegetation? Journal of Ecology 91: 664–676.

Walker, M.D., D.A. Walker, and N.A. Auerbach, 1994. Plant communities of a tussock tundra landscape in the Brooks Range Foothills, Alaska. Journal of Vegetation Science 5: 843–866.

Walker, M.D., C.H. Wahren, R.D. Hollister, G.H.R. Henry, L.E. Ahlquist, J.M. Alatalo, M.S. Bret-Harte, M.P. Calef, T.V. Callaghan, A.B. Carroll, H.E. Epstein, I.S. Jønsdøttir, J.A. Klein, B. Magnusson, U. Molau, S.F. Oberbauer, S.P. Rewa, C.H. Robinson, G.R. Shaver, K.N. Suding, C. Thompson, A. Tolvanen, Ø. Totland, P.L. Turner, C.E. Tweedie, P.J. Webber, and P.A. Wookey, 2006. Plant community response to experimental warming across the tundra biome. Proceedings of the National Academy of Sciences 103: 1342–1346.

Warren Wilson, J., 1966. An analysis of plant growth and its control in arctic environments. Annals of Botany 30: 383–482.

Weintraub, M.N. and J.P. Schimel, 2005. The seasonal dynamics of amino acids and other nutrients in Alaskan Arctic tundra soils. Biogeochemistry 73: 359–380.

Welker, J.M., P.A. Wookey, A.P. Parsons, T.V. Callaghan, M.C. Press, and J.A. Lee, 1993. Comparative responses of subarctic and high arctic ecosystems to simulated climate change. Oikos 67: 490–502.

Wookey, P.A., J.M. Welker, A.N. Parsons, M.C. Press, T.V. Callaghan, and J.A. Lee, 1994. Differential growth, allocation and photosynthetic response of *Polygonum viviparum* to simulated environmental change at a high arctic polar semi-desert. Oikos 70: 131–139.

Wookey, P.A., C.H. Robinson, A.N. Parsons, J.M. Welker, T.V. Callaghan, and J.A. Lee, 1995. Environmental constraints on the growth, photosynthesis and reproductive development of *Dryas octopetala* at a high Arctic polar semi-desert, Svalbard. Oecologia 102: 478–489.

Zhou, L., C.J. Tucker, R.K. Kaufman, D. Slayback, N.V. Shabanov, and R.B. Myneni, 2001. Variations in northern vegetation activity inferred from satellite data of vegetation index during 1981 to 1999. Journal of Geophysical Research 106: 20069–20084.

13 Plant Life in Antarctica

T.G. Allan Green, Burkhard Schroeter, and Leopoldo G. Sancho

CONTENTS

INTRODUCTION

Antarctica is the coldest, driest, highest and windiest continent, its plants grow where it is warm, wet, low and calm. Ecophysiologists should be grateful.

Research on terrestrial plants in Antarctica has been most intense for just over four decades. The International Geophysical Year (IGY 1958, p. 59) led to the establishment of many national stations in Antarctica and, although research concentrated on the physical sciences, there was always a proportion of natural science. Progress has, however, certainly been spasmodic as can be seen in the summary of the history of terrestrial biota research in the western part of the peninsula (Smith 1996). Initial studies, in the 1950s and 1960s, coincided with a growth in interest in stress survival mechanisms in organisms and, as a result, there has been a considerably higher proportion of plant ecophysiological research in Antarctica than in work on the same groups elsewhere. Taxonomy, to some extent, languished but the situation has been rectified by the appearance of substantial good reviews on some major groups and areas, for example, *Usnea* (Walker 1985), *Bryum* (Seppelt and Kanda 1986), *Umbilicaria* (Filson 1987), *Stereocaulon* (Smith and Övstedal 1991), Cladoniaceae (Stenroos 1993), *Caloplaca* (Söchting and Olech 1995), lichens of the Terra Nova area (Castello 2003), bryophytes of Southern Victoria Land (Seppelt and Green 1998), together with important floras for liverworts (Bednarek-Ochyra et al. 2000), lichens (Övstedal and Smith 2001), and mosses of King George Island (Ochyra 1998). This improvement of knowledge has resulted in a better understanding of the geographical relationships of the flora (Castello and Nimis 1995, 1997, Seppelt 1995; Smith 2000, Peat et al. 2007) and the appearance of searchable online databases such as VICTORIA for the lichens of Victoria Land (Castello et al. 2006), the online searchable herbarium database of the British Antarctic Survey, and Australian Antarctic Data Centre (Australian Antarctic Programme).

Despite the mixture of research approaches we are still far from understanding well both the distribution and functioning of the terrestrial plants and animals. This is certainly a reflection of the difficulties of working in the region and also of the patchy nature of research in some national programs. This has now started to be rectified by the development of multidisciplinary, long-term research using standardized methodologies, for example: Long Term Ecological Research site in the Taylor Valley (McMurdo LTER, National Science Foundation, USA), Evolutionary Biology of Antarctica (EBA, a SCAR initiative), and the Latitudinal Gradient Project (LGP, Antarctica New Zealand). The first State of the Environment Report has also appeared for the Ross Sea region (Waterhouse 2001). However, as is true for all research, new data reveal new puzzles and new questions and, as a result, ideas about the vegetation in Antarctica now appear to be in a greater state of flux than a decade ago.

In this chapter we try to bring out the major features of the ecophysiology of terrestrial plants in Antarctica with some emphasis on what appears to be controlling the distribution and performance. We do this by linking information about distribution and abundance with knowledge of the ecophysiological performance including growth rates, and by considering whether the plants show special adaptations to the Antarctic environment. This knowledge is of growing importance because of the needs to both conserve and manage the communities as well as the suggested potential to use the vegetation to detect global change processes such as climate warming (Kennedy 1995). Several excellent review articles exist that can provide more detail about specific plant groups or locations (Holdgate 1964, 1977, Ahmadjian 1970, Smith 1984, 1996, 2000, Longton 1988a,b, Kappen 1988, 1993a, Vincent 1988). Some recent reviews and literature compilations have addressed the possible effects of global climate change in Antarctica (Robinson et al. 2003, Frenot et al. 2004, Barnes et al. 2006). A history of botanical exploration has also been published (Senchina 2005) and a comprehensive study of contaminants of the vegetation has been prepared (Bargagli 2005).

Definitions

In this chapter, plants are defined to include higher plants (only two angiosperm species occur), all bryophytes (predominantly mosses since liverworts are rare south of about 70° S latitude), and lichens. The latter, associations between fungi as host and algae or cyanobacteria as symbionts, are nowadays correctly referred to as the fungi; however, their prominent role in Antarctica and their phototrophic lifestyle supports the tradition of including them as members of the terrestrial vegetation. The algae and cyanobacteria, which are also common in wet, terrestrial sites (Vincent 1988), are not covered except when in symbiotic association in a lichen.

Antarctica includes the main Antarctic continent (continental Antarctica), the Antarctic Peninsula, and its closely associated islands to the west and north (South Shetland Islands, South Orkney Islands). Subantarctic islands are not included.

Climate Zones

Antarctica spans a large latitudinal range of around 27° latitude from the north of the peninsula (63° S) to the pole and an associated large climate range. It is certainly true to say that Antarctica is the coldest, windiest, driest, and highest continent, the latter feature contributing to the very low inland temperatures (Table 13.1). Several authors have produced subdivisions based on the climate and these are summarized in Smith (1996). There is general agreement that two zones can be separated reflecting both climate and vegetation. The first is the maritime Antarctic (Holdgate 1964) or cold–polar zone of Longton (1988a,b). This comprises the Antarctic Peninsula: west side north of *c.*70° S (from southern Marguerite Bay) and offshore islands (including northern Alexander I.), north-eastern side north of *c.*64° S; South Shetland Is, South Orkney Is., South Sandwich I., Bouvetøya. The zone is characterized by warmest months with mean temperatures above freezing point (0–2°C) and mean winter temperatures rarely less than −10°C (see data for Faraday Station and Bellingshausen Station in Table 13.1). Precipitation, which occurs as rain in summer, is between 350 and 750 mm

TABLE 13.1
Climatic Information for Various Locations in the Maritime Zone, Continental Zone, and Polar Plateau in Antarctica

Location	Mean Temperature Warmest Month	Mean Temperature Coldest Month	Precipitation (mm Rain Equivalent)
Maritime			
South Shetland Islands (Bellingshausen) (62° 12′ S)	+1.6	−6.4	729
North-west Peninsula (Faraday) (63–69° S)	+0.7	−9.4	380–500
Continent			
Casey (66° 17′ S)	+0.1	−14.8	330
Syowa (69° 00′ S)	−0.7	−19.4	250
Mawson (67° 26′ S)	0.1	−18.6	100
Davis (68° 35′ S)	+0.9	−17.5	350
Cape Hallett (72° 19′ S)	−1.4	−26.6	225
Taylor Valley (77° 35′ S)	+1.0	−39.3	10 (floor) 50 (mountains)
Scott Base (77° 51′ S)	−4.8	−26.3	130
Polar Plateau			
Dome Fuji (77° 31′ S) (altitude 3810 m)	−26.1	−67.8	100

(rain equivalent) and the area is, essentially, an oceanic cold tundra. The climate is controlled by the strong westerly, maritime influences, and is markedly different from the eastern side of the peninsula. The vegetation includes the only two Antarctic angiosperms (*Colobanthus quitensis* (Kunth.)) Bartl. and *Deschampsia antarctica* Desv., and substantial numbers of lichens and bryophytes (around 350 and 100 species, respectively) including around 27 species of liverworts (Peat et al. 2007). Analysis of published collections and herbarium records suggests that the Antarctic Peninsula could be validly subdivided into two parts that better reflect the species present (Peat et al. 2007). The suggested zones are the northwestern Peninsula including adjacent islands, and the southern and eastern Peninsula. This new subdivision appears to be a better representation of the vegetation and is used in the remainder of this chapter.

The remaining zone is continental Antarctic (Holdgate 1964) or frigid Antarctic of Longton (1988a,b). Geographically this includes the whole of the main Antarctic continent, the east of the Antarctic peninsula and the west of the peninsula as far north as, and including, southern Alexander Island; it is approximately delineated as south of the Antarctic polar circle (Figure 13.1). The zone is characterized by the mean temperature of the warmest month, which is below freezing point, often substantially below it (Table 13.1). The mean winter temperatures are much lower than in the maritime Antarctic as they are typically around −20°C or below (Table 13.1). Temperatures, wind, and precipitation are all governed by the circumpolar vortex. The circular shape of the continent encourages cyclones to circle rather than moving onto it. Entry of maritime weather systems is also discouraged by the height of the polar plateau, which is predominantly above 2000 m. Precipitation, 300 mm (rain equivalent) as snow, occurs mainly in the coastal belt and the net result is growing aridity with increase in latitude so that annual precipitation can be very low, around 120 mm (rain equivalent) at McMurdo (77° 51′ S) and 50 mm near the pole (Table 13.1). Precipitation shadows can lead to even lower, local values such as the 225 mm at Cape Hallett (72° 19′ S) and the extreme 10–50 mm in the Dry Valley region (around 77° 00′–77° 50′ S). Cold air descending from the polar plateau causes strong katabatic winds at the continent margins

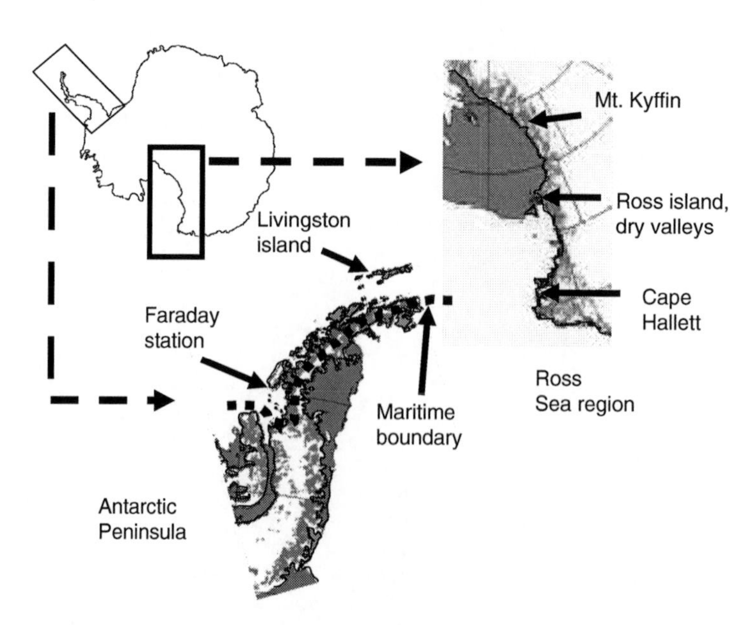

FIGURE 13.1 Map of Antarctica, with insets of the Antarctic Peninsula and the Ross Sea areas, showing the division into maritime and Continental zones, with the boundary most clearly seen in the Peninsula insert map. The named locations are referred to in the text or in the tables.

making some areas exceptionally windy (e.g., Cape Denison, mean wind speed of 24.9 m s^{-1} for July 1913, in Mawson, The Home of the Blizzard). The vegetation in this zone is composed entirely of lichens and mosses with the rare occurrence of one species of liverwort [*Cephaloziella exiliflora* (Tayl.) Steph.]. The vegetation is confined to ice-free areas of varying size. There are around 20, so-called oases (Pickard 1986), and even in those the vegetation is very scattered and only locally abundant, as it is dependent on moisture supply, warmth, and protection from the wind. Lichens and mosses have been reported from as far south as 86° 30″ S and 84° S, respectively (Wise and Gressit 1965, Claridge et al. 1971, Broady and Weinstein 1998), and, although some represent isolated occurrences there is evidence of unexpected biodiversity at high latitudes (Cawley and Tyndale-Biscoe 1960, Tuerk et al. 2004). Desiccation over winter becomes very important in this zone and plants are not only frozen but can become completely freeze-dried as temperatures reach −50°C at low ambient humidity.

VEGETATION TRENDS

Biodiversity

There is a marked decline in biodiversity with increase in latitude; however, it is not linear but shows disjunctions. The number of species of all groups is at their highest in the northern maritime zone with about 350 species of lichen, 100–115 mosses, 27 liverworts, and 2 higher plants (Table 13.2). There is a sharp decline in lichens and mosses to about a third of these levels and in liverworts to about one-tenth. The continental Antarctic has no phanerogams; it has only one liverwort and about 20% of the number of lichen and moss species. However, it should be noted that no more than around 30–40 species of lichens are found at any particular location (Table 13.2). The two higher plants are confined to the Antarctic Peninsula and reach their absolute southern limit at 68° 43′ S. Over the past 40 years there has been evidence of local population increase in line with a warming climate trend in the region (Smith 1994, Day et al. 1999). Neither of the two species occurs in the continental Antarctic. All the remaining components of the vegetation are poikilohydric but show difference in their distribution that appears to reflect water availability rather than temperature (although desiccation and cold resistance are tightly linked). The liverworts have less resistance to desiccation than mosses and are only of significance in the north-western Peninsula (the maritime zone). Their presence elsewhere must indicate exceptional local conditions such as the occurrence on heated soil on Mt Melbourne (Broady et al. 1987). The mosses show a distinct decline along the Peninsula and only about 20 species are found in the continent. Their southernmost records are close to those of lichens at around 84° S. Outside the northern Peninsula, the vegetation is dominated by lichens with around 100 species still present in continental Antarctica. Some less obvious trends exist, such as lichens with cyanobacterial photobionts confined to the maritime Antarctic (Kappen 1993a, Schroeter et al. 1994) and a general increase in the proportion of crustose lichen species as conditions become more severe (Kappen 1988). Kappen (1988) refers to a group of 13 circumpolar lichen species that tend to occur around the continent with other species occurring only locally.

Although about 100 species of lichens are found on the main continent, it is interesting that only about 30–40 species are found at any particular location (Table 13.2). Although a cline in lichen biodiversity occurs in the Peninsula, it appears that this is not true for the main continent. Distributions in the Ross Sea region confirm this with the richest site being Edmonson Point, Central Victoria Land but with about 27 species still present at Mt Kyffin, 84° S. A more detailed analysis revealed that there was only between 20% and 35% similarity in species between sites in the Ross Sea. Essentially each site has a different sampling of about 100 species that occur in the region and only around 4 species are common to all sites

TABLE 13.2
Numbers of Species of Flowering Plants, Lichens, Mosses, and Liverworts, Community Types, Maximal Biomass, and Presence of Peat at Various Locations in the Maritime and Continental Antarctica

Location	Flowering Plants	Lichens	Mosses	Liverworts	Community Types	Maximal Biomass (g m^{-2})	Peat
Maritime							
North-west peninsula 65–68° S	2	*c.*350	100–115	27	1.1; 1.2; 1.3; 1.4; 1.5; 1.6; 2.1	1,000–46,000	1.5 m
Southern peninsula 68–72° S	2	*c.*120	40–50	2	1.1; 1.2; 1.3; 2.1	?	0
Continental							
Mawson 67° 26′ S	0	9	2	0	1.1; 1.2; 1.3	<1100	0
Davis 68° 35′ S	0	34	6	1	1.1; 1.2; 1.3	?	0
Casey 66° 17′ S	0	35	5	1	1.1; 1.2; 1.3	900	0
Birthday Ridge 70° 48′ S	0	33	5	0	1.1; 1.2; 1.3	50–950	0
Botany Bay[a] 77° 00′ S	0	34	8	1	1.1; 1.2; 1.3	>1000	0
Dry Valleys[a] 77° 45′ S	0	30	7	0	1.1; 1.3	?	0
Mt Kyffin[a] 84° 45′ S	0	27	2	0	1.1; 1.2; 1.3	?	0

Sources: Communities (Crustaceous and foliose lichen sub-formation (1.1), Fruticose and foliose lichen sub-formation (1.2), Short moss cushion and turf sub-formation (1.3), Tall moss turf sub-formation (1.4), Tall moss cushion sub-formation (1.5), Bryophyte carpet and mat sub-formation (1.6), Grass and cushion chaemophyte sub-formation (2.1), are from Smith, R.I.L., *Foundations for Ecological Research West of the Antarctic Peninsula*, Koss, R.M., Hofmann, E., and Quetin, L.B., eds, Antarctic Research Series, Volume 70, American Geophysical Union, Washington, 1996). Definition of maritime zones and numbers of species are from Peat, H.J., Clarke, A., and Convey, P., *J. Biogeo.*, 34, 132, 2007.

[a] Species numbers from data of authors collected during LGP research.

(*Acarospora gwynii*, *Buellia frigida*, *Umbilicaria aprina*, and a *Lecidea* species. This is a complete contrast to the Peninsula where there is a gradual loss of species with increase in latitude. One contributor to this situation is the loss of species south of Ross Island, which are usually taken as indicating enrichment by birds (*Caloplaca* species, *Candelariella* species, *Physcia caesia*, *Physcia dubia*, *Xanthoria elegans*, and *Xanthoria mawsonii*). This indicates a blocking of marine influence in the southern Ross Sea where the Ross Ice Shelf is present. A second possible explanation is that conditions have become so dry in the continent and the Ross Sea region that lichens are confined to sporadic microsites with suitable climates and that it is a lottery as to which species can effect the initial colonization.

Biomass

Biomass data from a range of Antarctic locations are given in Table 13.2. Within continental Antarctica there is a remarkable agreement between the maximal biomasses from most sites when expressed as 100% cover. A typical maximal value of around 1000–1500 g m^{-2} is found both for lichens and mosses. Kappen (1993a) gives a data summary showing that such maximal biomasses occur for lichens from Victoria Land (70° S) to King George Island (South Shetland islands) and Signy Island in the northern maritime Antarctic (61–62° S). This seems to be the maximal biomass for this thin, almost two-dimensional vegetation. In the maritime there can be extensive peat build-up to depths of 50 cm or more leading to greatly increased values, to 46,000 g m^{-2} in some cases under *Chorisodontium aciphyllum* (Hook f. et Wils.) Broth. and *Polytrichum alpestre* Hoppe. The age of such peat banks is still uncertain but some have been carbon dated to 5350 years in the South Orkney Islands (Bjork et al. 1991). Biomass is about 10,000 g m^{-2}, in the 20 cm deep layer above the permafrost. A value of 300–1000 g m^{-2} is given by Longton (1988a) for the green, photosynthetic shoots, a number similar to that for maximal biomass without peat production. It is difficult to explain the lack of peat production in the continental Antarctic where, due to the colder temperatures, decomposition processes would be expected to be even slower (Collins 1976, Davis 1980, 1986); however, it could simply reflect much lower production rates.

The data must be treated with considerable caution since they are often adjusted to 100% cover equivalence and, even if this is not done, they only normally apply to where the plants are present and not to the overall ice-free area. The data are, therefore, unrepresentative since vegetation occurrence is sporadic. Kennedy (1993b) has calculated that, in Ablation Valley on Alexander Island, terrestrial vegetation is confined almost entirely to seven, discrete patches totaling 2300 m^2 in a total ice-free area of 400,000 m^2 (Light and Heywood 1975). Similarly, the rich Canada Glacier flush totaling some 10,000 m^2 is the only extensive patch of vegetation in an ice-free area of 25,000,000 m^2 (Schwarz et al. 1992). In many ways the situation is similar to describing the vegetation of the Sahara Desert by extrapolation from surveys at oases. It is unfortunate that the focusing of scientific attention on the plants disguises their relative rarity, if a typical area had to be preserved in ice-free areas it would be almost entirely bare ground. This would not be inappropriate because recent studies have shown this apparently barren ground to have substantial microbial biomasses (Cowan et al. 2001, Cowan and Tow 2004).

Community Structure

There have been several detailed accounts of vegetation and community structure in the Antarctic (see Longton 1988a,b, Smith 1996) and only the major trends are considered here. The plant association confined to the maritime is that dominated by higher plants, the grass, and cushion chaemophyte sub-formation. Although the two phanerogams can be found occasionally in other formations, their best development occurs on moist to dry soil especially in sheltered, north-facing coastal habitats where closed swards can sometimes develop. Subdivisions of cryptogamic communities, here from Smith (1996), depend on whether lichens or bryophytes (mosses) are dominant. In the bryophyte carpet and mat sub-formation (pleurocarpous mosses and liverworts on wet ground), lichens are sparse or absent, in the tall moss cushion sub-formation (tall moss cushions and some deep carpet forms along melt-stream courses) and in the tall moss turf sub-formation (predominantly *Polytrichum–Chorisodontium* species), fruticose lichens of the genera *Stereocaulon*, *Cladonia*, and *Sphaerophorus* are also very common. These formations are only found in the maritime zone. They are much more luxuriant and rich than communities in continental Antarctica where the predominant association is the short moss cushion and turf sub-formation (mainly a *Bryum–Ceratodon–Pottia* association) occurs. Overall the bryophytes become shorter, less able to develop continuous stands and form less extensive patches (Table 13.2). Location becomes ever more important and mosses are confined to highly protected sites where regular melt-water can occur such as cracks in rocks, adjacent to permanent snow banks and in areas with running water usually close to glaciers or in depressions. In some areas the shoots are restricted to just below or at the surface of the substrate.

FIGURE 13.2 Photograph of a rich Fruticose and Foliose Lichen sub-formation (1.2 in Smith 1996) at Livingston Island in the maritime Antarctic. The most visible lichens are specimens of the fruticose genus *Usnea*.

Lichen communities show similar trends. In the maritime zone there are extensive areas where fruticose lichens are dominant, the fruticose and foliose lichen formation, typically on the sides of rocks or boulders forming marine benches or similar areas (Figure 13.2). This distribution seems to be determined by improved water relations and it is easy to see how snow gathers in such areas, how wind is reduced and how light is high but rarely direct sunlight. Such sites are well illustrated in Kappen (1993a). The genera *Usnea* (Figure 13.3) and *Umbilicaria* are typically dominant and plants can reach large sizes, 20 + cm across for *Umbilicaria* specimens. These formations are rarer in the continental zone but do occur, for example, Birthday Ridge (Kappen 1985a) and Botany Bay, Granite Harbour (unpublished, in preparation). The second lichen formation, *crustaceous* and *foliose lichens*, is extensive in both the maritime and continental Antarctic and is the only association on bare rock faces. It is the predominant formation in the continental zone and can form anywhere where there is protection and supply of water either as melt or blown snow. However, there are also considerable limitations to its presence and the lichens are often confined to a particular rock face or to crevices depending on wind, light, and snow occurrence (Figure 13.3). It is a patchy formation and rarely is 100% cover approached, more typically the cover is very low. Crustaceous species become almost the only lichens under the more extreme conditions and, under dry conditions but with a water supply and light, the endolithic association appears (Friedmann 1982). In the mountains of the Dry Valley region this is the most common community but it is present over almost the entire continent where other growth forms cannot occur. In both the maritime and continental zones the occurrence of some species is strongly dependent on a rich nutrient supply from perching birds (Olech 1990).

Overall there is a clear trend for smaller plants, lower biodiversity, and greater confinement to protected sites as latitude increases (Table 13.2). Where there is a coincidence of shelter, warmth, light, and reliable water supply, rich communities can develop at high latitudes. A good example is the exceptional community at Botany Bay, Granite Harbour which, even though at 77° S, is richer than almost all other continental sites and even has the liverwort *C. exiliflora* present (Seppelt and Green 1998). The occurrence of this one species

FIGURE 13.3 Upper photograph: Lichen community at Botany Bay, Granite Harbour showing the well developed *Umbilicaria aprina* thalli in the water fall area and the surrounding crustose species. Lower phtograph: *U. aprina* thalli in a crack in granite in Lower Taylor Valley showing the formation of a suitable microclimate by water gathering in the rock crack.

would suggest that the site has summer months with mean temperatures close to or above freezing although no data are available to test this hypothesis at present.

PLANT PERFORMANCE: CO_2-EXCHANGE

Higher plants are considered only briefly because they are confined to the maritime Antarctic. The majority of the studies have been on lichens and bryophytes, which dominate most communities. In the case of the bryophytes, the mosses are clearly the major group; liverworts are not significant in continental Antarctica and are only locally common in the maritime zone; unfortunately, they have been little studied. Plant performance has mostly been studied as photosynthetic activity measured mainly as CO_2 exchange in the field or on samples returned to laboratories. CO_2 exchange is reported as net photosynthesis (NP), dark respiration (DR), photorespiration (PR), and gross photosynthesis (GP calculated as NP + DR). There is a growing use of chlorophyll *a* fluorescence techniques but these have some important limitations; in particular they are not a reliable indicator of CO_2 exchange.

CO_2-Exchange Response to Major Environmental and Plant Factors

Higher Plants

The ecophysiology of *C. quitensis* and *D. antarctica* has been reviewed by Alberdi et al. (2002) although the plants have not been extensively studied. No morphological adaptations specifically for Antarctica were found. Both species tend to be xeromorphic, which is almost certainly a response to the generally cold environment, especially the soils. *D. antarctica* has a high water use efficiency, about 60–120 mol H_2O mol^{-1} CO_2 fixed compared with values of 300–500 for normal C_3 plants (Montiel et al. 1999). Apart from conserving water this might also be expected to raise leaf temperatures due to lowered transpirational cooling. Plants studied in the laboratory were found to have optimal temperatures for NP of 13°C for *C. quitensis* and 19°C for *D. antarctica*. Both species retained NP of about 30% of maximal rate at 0°C and it is suggested that this ability is the reason for their success in the maritime Antarctic (Edwards and Smith 1988). Light levels for saturation were low, 30 μmol m^{-2} s^{-1} PPFD (Photosynthetic Photon Flux density) at 0°C and 150 μmol m^{-2} s^{-1} PPFD at 10°C, indicating shade-adapted plants. This could also be considered as acclimation for the generally cloudy conditions in this area. Both plants show little photoinhibition and seem to be well adapted to natural radiation and UV-B levels. Some studies have shown improved growth when incident UV-B was reduced but this is most likely a secondary effect rather than a direct impact of the UV radiation (Day et al. 1999). Overall, Edwards and Smith (1988) suggested that the plants showed no particular special adaptation for growth in Antarctica and Convey (1996) suggests that their presence could be a matter of chance. However, when it is considered how long there has been contact between South America and the maritime Antarctic without strong quarantine controls until recently, with only minor new introductions, also how extensive the range is of these two plants, then perhaps they do have some special property or properties that are yet to be identified. Fowbert and Smith (1994) demonstrated that the populations are increasing and have been doing so since the 1940s, an increase that is suggested to reflect higher summer temperatures in the area (Smith 1994).

Liverworts

The few studies on liverworts have provided some interesting insights into their poor performance in Antarctica. *Marchantia berteroana* Lehm. et Lindenb., studied in Signy Island, was extremely sensitive to both desiccation and low temperatures (Davey 1997a). Photosynthesis was only 16% of original rate at 10°C after 24 h at −5°C and there was no activity at all after 24 h at −25°C. Five freeze/thaw cycles to −5°C caused NP to fall to around 3% of original value. The plants required a high water content of 10 g $g(d.wt.)^{-1}$ (1000%) for maximal GP and desiccation for 1–6 months led to an almost complete loss of photosynthetic ability. However, in the field the plants were always found to be fully hydrated and never suffered desiccation. Maximal GP and NP were slightly higher than the published values for a related species *Marchantia foliacea* Mitt., in New Zealand (Green and Snelgar 1982). *C. exiliflora*, the only liverwort to grow in continental Antarctica, loses its typical purple color in shaded areas and the pigment probably functioned as a form of light protection (Post and Vesk 1992). This agrees with studies showing rapid adjustment of UV-B filtering compounds to incident UV-B (Newsham 2002). Green plants, without the pigment, had lower light compensation and saturation values and a greater apparent quantum efficiency (Post and Vesk 1992). These features, together with the similar Chl *a*/Chl *b* ratios in the two forms are typical for bryophytes utilizing some form of light filter (Green and Lange 1994). Maximal GP, 100 μmol O_2(mg $Chl)^{-1}$ h^{-1}, was similar to those of other bryophytes. Unfortunately there were no studies of low temperature tolerance but the distribution of the plant suggests that this will not be high.

Mosses

Mosses show a normal saturation response to PPFD but there is contrasting evidence of adaptation to incident light. Some early studies suggested low compensation and saturation PPFD (Rastorfer 1970, 1972) with saturation at around 15% of full sunlight (about 300 $\mu mol\ m^{-2}\ s^{-1}$ Longton 1988a). However, recent studies with measurements made in the field suggest that saturation may not occur even at full sunlight under some conditions (Figure 13.4). *Bryum subrotundifolium* can show a range of forms depending on the light climate of its environment and also change rapidly between forms (Green et al. 2000). Post (1990) found a similar situation with ginger and green forms of *Ceratodon purpureus* (Hedw.) Brid. Although these seem to be similar to the sun/shade adaptations of higher plant leaves, they represent a different form of response. In contrast to higher plants the sun forms can have both lower maximal NP and lower apparent quantum efficiency (Green et al. 2000). It appears that adaptation to high light is by putting filters in place, so that they are protected shade forms, rather than by acclimation of the photosynthetic apparatus. There is little evidence that photoinhibition occurs when the plants are in their natural habitat (Schlensog et al. 2003). The ability to rapidly change form suggests that measurements made in the laboratory may not be representative of the field situation.

Net photosynthetic response to temperature shows an optimum (Topt) between 5°C and 25°C (Longton 1988a). Topt appears to be variable between species and even within the same species at different times (Pannewitz et al. 2005). Topt for *C. purpureus* was around 6.5°C at Granite Harbour (77° S) in 2000 and 2001, whereas that of *Bryum pseudotriquetrum* changed from 9.1°C to 15.9°C. Smith (1999) found Topt for three species, *C. purpureus*, *B. pseudotriquetrum*, and *Bryum argenteum*, at Edmonson Point (74° S) to all be around 15°C. At present the high variability makes it difficult to find any trends with latitude. Topt for NP depends strongly on PPFD and is lower at PPFD below saturation (Rastorfer 1970, Pannewitz et al. 2005). At temperatures below Topt, NP declines due to limitation by low temperature and excess light energy is handled by increased nonphotosynthetic quench mechanisms (Green et al. 1998). Above Topt, the depression has always been explained as that due to increased DR since that rises almost exponentially with increase in temperature, however,

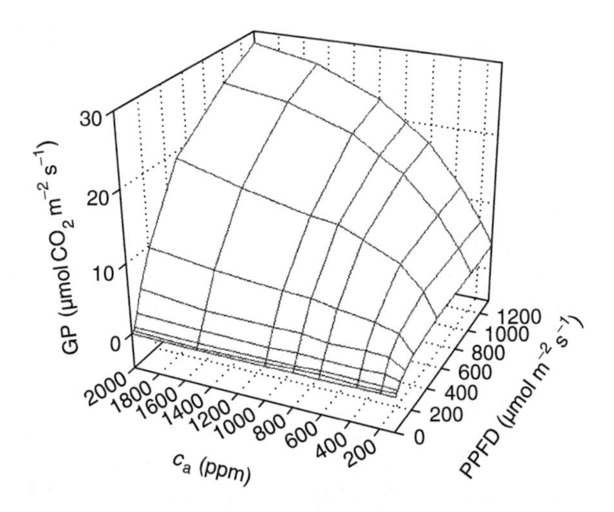

FIGURE 13.4 Response of gross photosynthesis (net photosynthesis plus dark respiration, vertical axis) of the moss *Bryum pseudotriquetrum* to light (*right* axis, PPFD in μmol Photons $m^{-2}\ s^{-1}$) and carbon dioxide concentration (front axis in ppm—parts per million) at 20°C. There is no saturation by light up to 1200 $\mu mol\ m^{-2}\ s^{-1}$ at all CO_2 concentrations of ambient and above.

there is little evidence for this and the depression could easily represent increased PR as in higher plants. Many mosses show negative NP at higher temperatures, over about 20°C (Gannutz 1970; Longton 1988; Kappen 1989; Noekes and Longton 1989a; Wilson 1990; Convey 1994, Davey and Rothery 1997). At 0°C NP can still be substantial but is always very low by −5°C and ceases at around −7°C (Kappen et al. 1989, Pannewitz et al. 2005).

Data on the photosynthetic response to WC for Antarctic mosses are not extensive but it seems unlikely that they will differ greatly from temperate mosses. Most data are for species in the northern, maritime Antarctic (Davey 1997b) but little difference was found between these plants and others studied in temperate or northern tundra areas (Fowbert 1996). At Cape Hallett (72° S) maximal WC for *B. subrotundifolium* and *B. pseudotriquetrum* were around 750% and 600% d.wt., respectively; there was a depression in NP at WC above that optimal for NP (about 400% and 300%, respectively. Thallus water contents for half maximal NP for the two species were 150% and 120%. These values for *B. pseudotriquetrum* fall in the mid-range of those found for the same species in wet and dry sites at Windmill Islands (66° 17′ S, Robinson et al. 2000). In their study on *Grimmia. antarctici*, *B. pseudotriquetrum*, and *C. purpureus*, Robinson et al. (2000) found correlations between water loss rate under standard conditions, maximal water content, WC for half maximal photosynthetic efficiency, desiccation tolerance, and water status of the habitat. Samples and species from the drier habitat lost water less rapidly, had a lower WCmax, and remained photosynthetically active to lower WC than those from wetter sites.

A strong relationship is reported between moss species distribution and water flow for several species (Schwarz et al. 1992, Okitsu et al. 2004, Robinson et al. 2003), and Kappen et al. (1989) found photosynthetic differences between mesic and xeric ecodemes of *Schistidium (Grimmia) antarctici* Card.in the Windmill Islands. However, more detailed investigations are needed because the moss species appear to vary in this characteristic as well. At Granite Harbour, *C. purpureus* occupies the wetter sites, the exact opposite as that found for Windmill Islands. Additionally, drying rates, water contents, and NP are commonly expressed on a dry weight basis and it is an unfortunate fact that a large proportion of the samples can be inorganic, for example, *Pottia heimii* samples that came from an apparently clean moss hummock in the Dry Valleys had an inorganic content of a surprising 66% (unpublished results). Response to CO_2 concentration has been little investigated for Antarctic mosses but appears to be similar to that found for other C_3 species with saturation occurring at levels well above ambient (Pannewitz et al. 2005). In studies at Cape Hallett at 20°C, *B. pseudotriquetrum* was not CO_2 saturated at 2000 ppm (Figure 13.4) compared with *B. subrotundifolium* that was saturated at about 1000 ppm. The actual CO_2 concentration around the mosses is a matter of debate but very high values, several times normal ambient levels, have been found in *Grimmia antarctici* in continental Antarctica, (Tarnawski et al. 1992) and in *B. subrotundifolium* at Cape Hallett (Green et al. 2000). The source of the high CO_2, its seasonal change and its effect on overall productivity are not yet known. *B. subrotundifolium* measured at normal (360 ppm) and 2000 ppm CO_2 under identical conditions showed a 60%–80% increase in daily carbon gain so carbon budgets modeled on normal ambient levels may be in error (Pannewitz et al. 2005).

Lichens

Considerably more work has been done on Antarctic lichens than the mosses (Kappen 1993a, 2000). However, as they are poikilohydric the lichen and moss photosynthetic responses are similar in form although differing in detail (see Longton 1988a). PPFD required for saturation can be very high, often around 1000 μmol m^{-2} s^{-1}, and light compensation values are dependent on the thallus temperature, for example: saturation at 1300 μmol m^{-2} s^{-1} above 1°C and compensation from 5 μmol m^{-2} s^{-1} at −2°C to 128 μmol m^{-2} s^{-1} at 20°C for *Leptogium puberulum* Hue (Figure 13.5 and see Schlensog et al. 1997a). Sun and shade

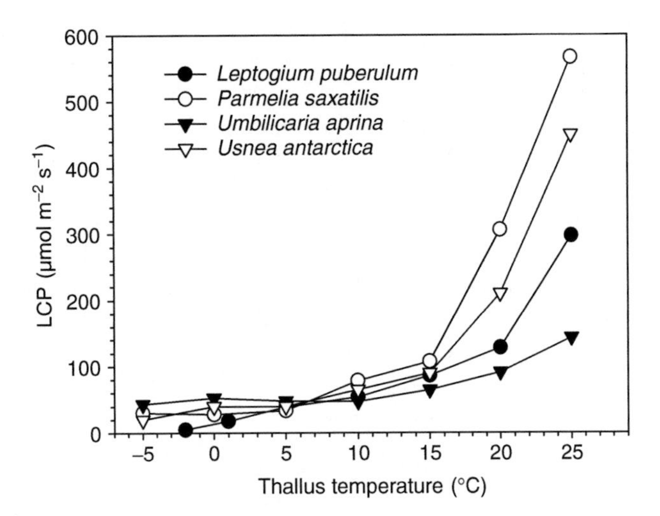

FIGURE 13.5 The typical rise in light compensation point with temperature for four lichens measured in the field in Antarctica. The most continental species, *Umbilicaria aprina*, shows the least change. (Data from Schlensog, M., Schroeter, B., Sancho, L.G., Pintado, A., and Kappen, L., *Secc. Biol.* 93, 107, 1997a [*Leptogium puberulum*]; Sancho, L.G., Parmelia, A., Valladares, F., Schroeter, B., and Schlensog, M., *Bibl. Lichenol.* 67, 197, 1997 [*Parmelia saxatilis*], Schroeter, B., Kappen, L., and Moldaenke, C., *Lichenologist*, 23, 253, 1991 [*Usnea antarctica*], and Schroeter, B., Grundlagen der Stoffproduktion von Kryptogamen unter besonderer Berücksichtigung der Flechten—eine Synopse—Habilitationsschrift der Mathematisch—Naturwissenschaftliche Fakultät der Christian—Albrechts—Universität zu Kiel, 1997 [*U. aprina*].)

forms of *Usnea sphacelata* R. Br. differed little in the form of their photosynthetic responses to PPFD at identical temperatures although the sun form had considerably larger NP at higher temperatures (Kappen 1983). Topt tends to be lower and more consistent than for mosses and values around 5°C–15°C are common (Longton 1988a, Kappen 1993a). In mosses, Topt declines with fall in PPFD below saturation (Figure 13.6) (Schroeter et al. 1995,

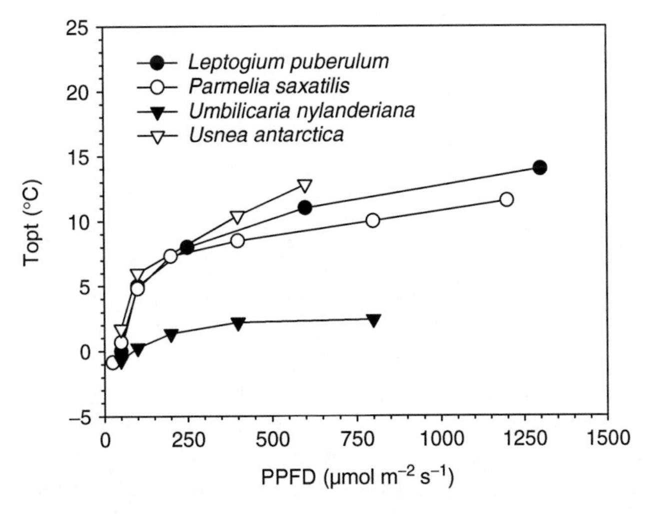

FIGURE 13.6 The typical change in Topt (optimal temperature for net photosynthesis) with increase in irradiance (PPFD in µmol Photons m^{-2} s^{-1}) for four lichens measured in the field in Antarctica. Sources as for Figure 13.5 except *Umbilicaria nylanderiana*. (From Sancho, L.G., Pintado, A., Valladares, F., Schroeter, B., and Schlensog, M., *Bibl. Lichenol.*, 67, 197, 1997.)

Sancho et al. 1997). In complete contrast to mosses, positive NP has been found to temperatures as low as −17°C for *U. aprina* Nyl. measured in the field (Schroeter et al. 1994) although this ability is not confined to Antarctic lichens, for example: −24°C for *Cladonia alcornis* (Lightf.) Rabh. measured in the laboratory (Lange 1965).

Fewer studies of the effect of WC on photosynthesis exist but those that do show the characteristic low WC range (200%–400% maximal WC) compared with mosses and often with depressed NP at high WC due to increased CO_2 diffusion resistances (Kappen 1985b, Harrisson et al. 1989, Kappen and Breuer 1991, Schroeter 1991). Lichens in the Antarctic normally occupy much drier habitats than mosses and are not found in wet areas unless on the drier tops of mosses. Response to CO_2 concentration seems to have not been measured but would be expected to be similar to that of temperate lichens that is, not saturated at ambient CO_2 levels. The increased ambient CO_2 reported around mosses would not be expected around lichens because of their elevation and lack of substantial organic substrates below them (Tarnawski et al. 1992).

Maximal Rates of Photosynthesis

Longton (1988a) gives an extensive list of maximal NP and Topt for a wide range of polar bryophytes and lichens although only a few are from the areas covered in this article. Kappen (1988, 1993a) gives additional, similar data for lichens. The brevity of both lists indicates the lack of knowledge that exists at present. Considerable variation in rates certainly exists for lichens, from 0.08 mg CO_2 g^{-1} h^{-1} for *Rhizoplaca melanophthalma* (Ram.) Leuck and Poelt, to 0.8 mg CO_2 g^{-1} h^{-1} for *U. aprina*. Although it is occasionally suggested that the abundance of particular lichen species may be related to photosynthetic performance we do not really yet have the data to be certain. Lechowicz (1982) has analyzed the relationship between several photosynthetic parameters and latitude for the Northern Hemisphere. From his analysis maximal NP are certainly substantially lower at latitudes greater than 65° N and these far-northern values are similar to those found in the Antarctic. Kappen (1988) states that maximal rates of NP are lower for lichens in continental Antarctic than in the maritime zone. No similar analysis exists for mosses but, in general, the rates found seem to be close to the higher rates reported elsewhere (1–2 mg CO_2 g.d.wt.$^{-1}$ h^{-1}, Rastorfer 1972, Longton 1988a, Kappen et al. 1989) and no depression is obvious.

Diel and Long-Term Photosynthetic Performance

Continuous measurement of photosynthetic performance over a day, or several days, has only recently become common in Antarctica and a few extensive studies have been made in the northern maritime (e.g., Kappen et al. 1991, Schroeter 1991, Schroeter et al. 1991, Sancho et al. 1997 for lichens, Collins 1977 for mosses). A more common approach to obtain an estimate of seasonal production is to log microclimate parameters and to interpolate plant performance from photosynthetic response to PPFD and temperature (e.g., Kappen et al. 1991, Schroeter 1991, Schroeter et al. 1995). The present situation is probably a reflection of the difficulties of working in Antarctica with large and expensive equipment. Kappen et al. (1991) obtained eight diurnal records for *U. sphacelata* near Casey Station and recorded positive photosynthesis through the entire light period of the day even though temperatures were mostly below 0°C. They were able to demonstrate relatively good agreement between their model, based on PPFD and temperature, and in situ rates of photosynthesis. Sancho et al. (1997) obtained diurnal courses for three cosmopolitan lichen species, *Parmelia saxatilis* (L.) Ach., *Pseudophebe pubescens* (L.) M. Choisy, and *Umbilicaria nylanderiana* (Zahlbr.) H. Magn., over three months of the summer, 1995, in the maritime Antarctic (Livingston Island) and found activity only for 182 h. Typical examples of diel patterns for two lichens

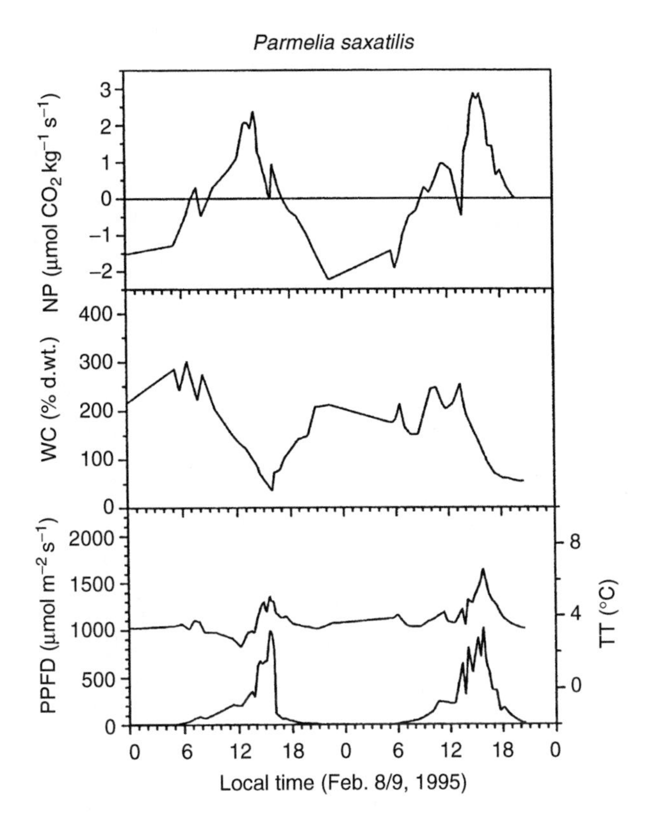

FIGURE 13.7 Diel pattern of net photosynthesis (*upper* panel, μmol CO_2 kg^{-1} s^{-1}), water content (*middle* panel in % d.wt.) and thallus temperature and irradiance (*lower* panel, PPFD in μmol Photons m^{-2} s^{-1}) for *Parmelia saxatilis* on February 8 and 9, 1995 measured at Livingston Island. Because this is a maritime site there is a brief period of darkness each day leading to a period of negative NP. Measurements were made in the field at a natural growth site using a porometer system and the lichen was rewetted on the 8th by rainfall. (From Sancho, L.G., Pintado, A., Valladares, F., Schroeter, B., and Schlensog, M., *Bib. Lichenol*, 67, 197, 1997.)

(*P. saxatilis* and *U. aprina*) and one moss (*B. subrotundifolium*) are given in Figure 13.7, Figure 13.8, and Figure 13.9. The two examples from continental Antarctica (Figure 13.8 and Figure 13.9) show positive NP 24 h a day, whereas the maritime example (*P. saxatilis*, Figure 13.7) had negative NP overnight and the daily carbon balance was often negative as, also, was the cumulative balance calculated over 3 months (Sancho et al. 1997).

Schroeter et al. (1991) demonstrated that basal chlorophyll *a* fluorescence could be used to monitor photosynthetic activity of *Usnea antarctica* Du Rietz. Photosynthetic carbon fixation could be estimated using microclimate records of temperature and PPFD (Schroeter 1991, Schroeter et al., 1997a,b). Schroeter et al. (1992) demonstrated that the photosynthetic activity of the crustose lichen, *B. frigida* (Darb.) Dodge at Granite Harbour could easily be monitored with a chlorophyll *a* fluorescence system and this has now become a common investigative method. Concurrent measurements of CO_2 exchange and chlorophyll *a* fluorescence revealed complex relationships between ETR (relative electron transport rate calculated from the fluorescence signal) and photosynthetic rate for lichens, and simpler but still not linear relationships for mosses (Green et al. 1998). Schroeter et al. (1997c) measured fluorescence activity of *B. frigida* over several days at Granite Harbour. The studies revealed the extremely erratic nature of thallus moistening, which depended on small-scale topography, proximity to snow, degree of snow melt, and snow fall (Figure 13.10). Even though air

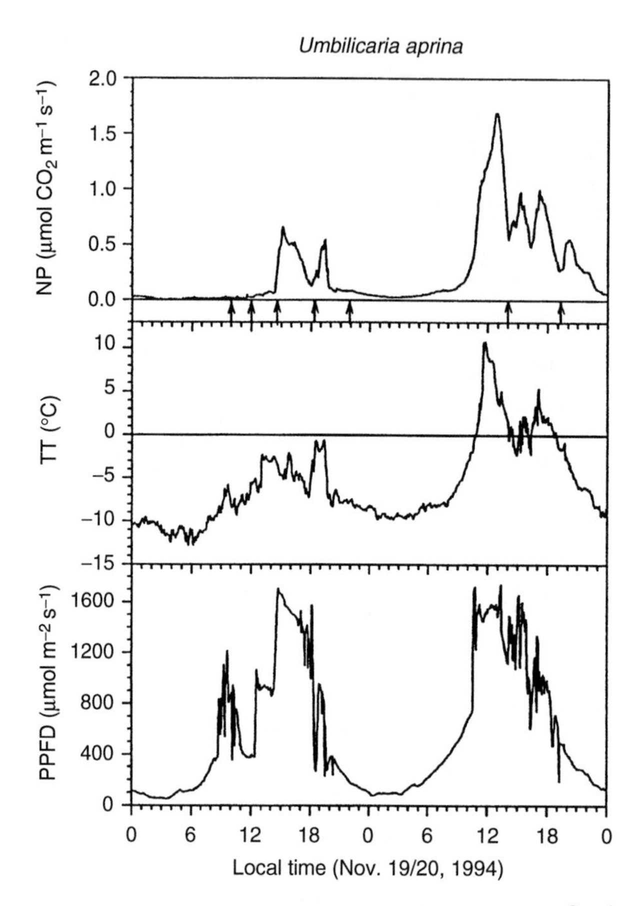

FIGURE 13.8 Diel pattern of net photosynthesis (*upper* panel, μmol CO_2 m^{-2} s^{-1}), thallus temperature (*middle* panel, °C), and irradiance (*lower* panel, PPFD in μmol Photons m^{-2} s^{-1}) for *Umbilicaria aprina* on November 19 and 20, 1994 at Botany Bay, Granite Harbour. The lichen was rewetted at the points marked with arrows in the upper panel. Although this is early in the season at a continental site (77° S) the irradiance reaches high values each day and there is no period of complete darkness. Hence positive NP can occur through the full 24 h and also, on the 19th at subzero thallus temperatures. (Data from Schroeter, B., Grundlagen der Stoffproduktion von Kryptogamen unter besonderer Berücksichtigung der Flechten—eine Synopse—Habilitationsschrift der Mathematisch—Naturwissenschaftliche Fakultät der Christian—Albrechts—Universität zu Kiel, 1997.)

temperatures remained between −7.5°C and −12.0°C lichen thallus temperatures were much higher, often around 15°C higher, even when wet. Thus, most of the photosynthetic activity actually occurred at positive temperatures, but was detectable down to −8°C, and no saturation was evident at 1500 μmol m^{-2} s^{-1}. These results, which were made in November at 77° S, indicate that the growing season could be much longer than previously thought, even at these high latitudes. The introduction of imaging systems for chlorophyll *a* fluorescence has allowed even finer analysis of the responses to water and drying (Barták et al. 2005).

A feature of all the diurnal studies is the elevation of thallus temperatures above ambient air temperatures because of the absorbed irradiation. Kappen (1993a) noted that thallus temperatures of wet lichens were often about 8°C–10°C at many maritime and continental sites as a result of shelter and sun exposure, although *B. frigida* at Granite Harbour could reach 15°C (Schroeter et al. 1997c, Kappen et al. 1998b). Monitoring of the photosynthetic activity of *P. heimii* on the Canada Glacier flush (Taylor Valley) showed that, in late

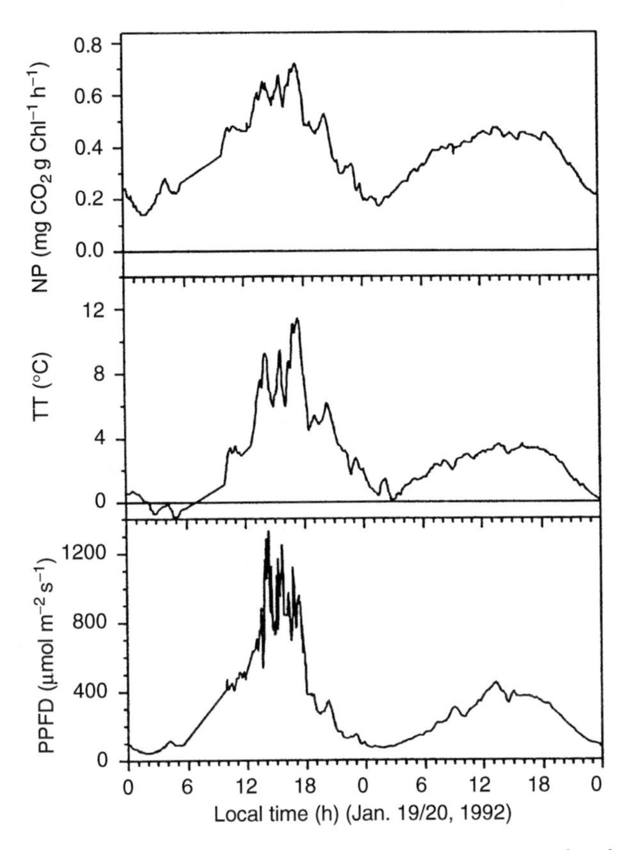

FIGURE 13.9 Diel pattern of net photosynthesis (*upper* panel mg CO_2 Chl^{-1} h^{-1}), thallus temperature (*middle* panel, °C), and irradiance (*lower* panel, PPFD in μmol Photons m^{-2} s^{-1}) for moist *Bryum subrotundifolium* on January 19 and 20, 1992 at Botany Bay, Granite Harbour. This is just after midsummer and NP is positive through the entire period. Thallus temperatures are almost always positive and are linked to the irradiance. (From Schroeter, B., Grundlagen der Stoffproduktion von Kryptogamen unter besonderer Berücksichtigung der Flechten—eine Synopse—Habilitationsschrift der Mathematisch—Naturwissenschaftliche Fakultät der Christian—Albrechts—Universität zu Kiel, 1997.)

December, the plant was continuously active and did not face subzero temperatures. The wet plants were buffered from the colder air by ice formation at zero degrees (Figure 13.12).

Nutrition Effects

The role of nutrients in the abundance and diversity of Antarctic plants is not well known and requires more detailed studies. Elemental analyses for mosses (Smith 1996) show considerable variation depending on proximity to birds and species. Total nitrogen ranged from a low 0.55%–0.77% for *P. alpestre* to over 3% for *Drepanocladus (Sanionia) uncinatus* (Hedw.) Loesk. Rastorfer (1972) found 2.95%, 2.75%, 1.50%, and 2.35% for *Calliergidium austrostramineum*, *Drepanocladus uncinatus*, *Polytrichum strictum* Brid. (syn. *P. alpestre*), and *Pohlia nutans* (Hedw.) Lindb., respectively. These values would not seem to be limiting and are not likely to be the major control on production. Lichens, in contrast, can have very low nitrogen contents, for example: 0.36% and 0.83% for *U. sphacelata* and *Usnea aurantiaca-atra* (Jacq.) Bory, respectively (Kappen 1985b). Lichens with green algal photobionts seem to almost always have below 1% total nitrogen (Green et al. 1980) and fruticose species are the lowest (Green et al. 1997). However, chlorophyll content, which can be related to

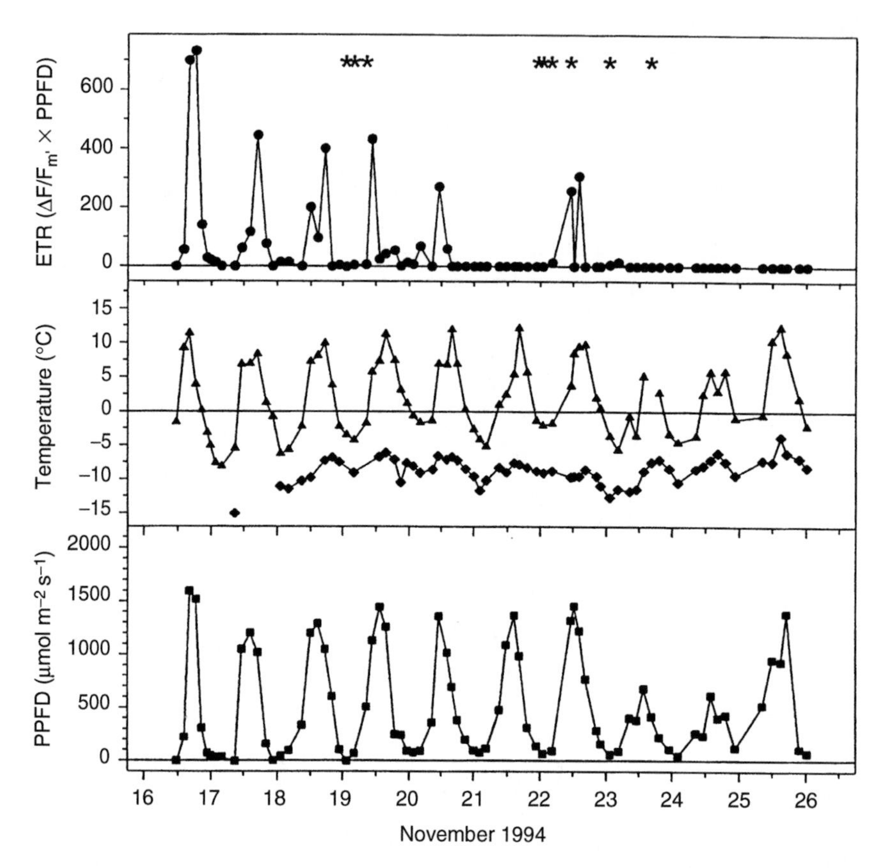

FIGURE 13.10 Diel pattern of relative electron transport rate through Photosystem II, ETR, (*upper* panel, μmol e^- m^{-2} s^{-1}), thallus and air temperatures (*middle* panel, °C), and irradiance (*lower* panel, PPFD in μmol Photons m^{-2} s^{-1}) for a *Buellia frigida* thallus over a 10 day period, November 16–25, 1994, at Botany Bay, Granite Harbour. Although air temperatures remain around −10°C through the whole period the thallus temperature is positive at times of high irradiance and several degrees above air temperature during the night because of heat storage in the rock surface. ETR depends on both irradiance and moistening of the thalli. Initially the lichen received water from melting snow on the rock but, as the snow patch retreated away from the lichen, it was active only after snow falls (indicated by the stars in the upper panel). ETR activity is least or zero in the late morning when the lichen has dried out, highest in the afternoon, after wetting by melt, and continues overnight at subzero temperatures—the so-called reverse diel pattern (see text). (From Schroeter, B., Grundlagen der Stoffproduktion von Kryptogamen unter besonderer Berücksichtigung der Flechten—eine Synopse—Habilitationsschrift der Mathematisch—Naturwissenschaftliche Fakultät der Christian—Albrechts—Universität zu Kiel, 1997.)

nitrogen content (Green et al. 1997), is much lower in Antarctic than in temperate species (Sancho et al. 1997). In essence, the values are little different to those found for temperate lichens and mosses.

Nutrient supply, especially nitrogen and phosphorous, is strongly influenced by the presence of birds and other animals. Leishman and Wild (2001) showed that soil nutrient were enriched near to sea bird nests and declined rapidly with increasing distance. The number of lichen species and mean lichen species abundance increased with soil nutrient levels, with total P having a stronger influence than total N. In contrast, moss diversity and abundance showed little correlation with soil nutrient levels but was positively correlated with soil water content. Although nutrient deficiency has rarely been shown (Kappen and Schroeter

2002), addition of nutrients causes increased photosynthetic rates, ETR (relative electron transport rate from chlorophyll fluorescence), and chlorophyll content (Smith 1993, Wasley et al. 2006). Although it is normally claimed that this indicates nutrient limitation, it may be that the plants are not carbon limited and the enhancement may represent utilization of storage materials over several years. Longer-term increases in nutrient supply lead to changes in community structure and biodiversity rather than just increased growth. This can be clearly seen for mosses in the Ross Sea region. At low nutrient sites such as Taylor Valley, the moss patches contain only two main species. At a more nutrient rich site, Granite Harbour or Edmonson Point, four or five species can be common. Finally, where there is an overwhelming presence of birds, Cape Hallett or Beaufort Island, then only *B. subrotundifolium* is found. Overall, the major determinant for the occurrence of mosses, in particular, and lichens is the availability of liquid water. Extra water may not enhance photosynthesis as much as when nutrients are also present (Wasley et al. 2006). Finally, there are also suggestions that salt can control distributions close to the coast (Broady 1989).

SURVIVING ANTARCTICA'S EXTREMES

There are few articles, and possibly even fewer grant applications, written that do not mention some extreme aspects of Antarctica, usually the climate, and the accepted fact that the organisms show adaptations to the stresses that they face. Of course the Antarctic climate is extreme to us, but is it extreme to the organisms that survive there, and do they show special adaptations; these are important questions in our understanding of potential responses to change.

DESICCATION

All Antarctic terrestrial biota face a long period of cold, darkness, and desiccation over the winter. Conditions are particularly extreme in continental Antarctica (Table 13.1) and there can be little doubt that this limits the vegetation to mainly species of lichens and bryophytes. Both groups are poikilohydric, meaning that their thallus hydration tends to equilibration with the water status of the environment (Green and Lange 1994). Water uptake tends, with the exception of the endohydric mosses like *Polytrichum*, to occur over the entire surface. Typically, these groups also show exceptional resistance to desiccation and can survive very low water contents for long periods. Such resistance is, however, not inevitable and some lichens, inhabitants of consistently moist rainforest, are very sensitive to desiccation (Green et al. 1991). The Antarctic liverwort *M. berteroana* (see earlier Section "Liverworts") is also very sensitive (Davey 1997a).

Schlensog et al. (2004) studied the ability of lichens and mosses to recover from desiccation after an Antarctic winter at Granite Harbour (77° S). They found that, although all species showed recovery starting within minutes as measured by chlorophyll fluorescence, the lichen *U. aprina*, from open rock surfaces, reached full photosynthetic activity in just over an hour, whereas the mosses, *B. subrotundifolium* and *Hennediella heimii* from running water sites, took almost 24 h (Figure 13.11). This difference cannot be generalized to all bryophytes as various mosses from xeric environments are known to reactivate within minutes (Proctor 2000). Moreover, the lichen *P. caesia*, characteristic of submerged water channel sides, took considerably longer than *U. aprina* to recover (Figure 13.11). The results fit with the concept from non-Antarctic bryophytes that the tolerance and activation of poikilohydric organisms can be correlated to their normal active environment. That is, plants from xeric environments are strongly desiccation-tolerant and activate rapidly, and the opposite occurs for plants from consistently wet environments.

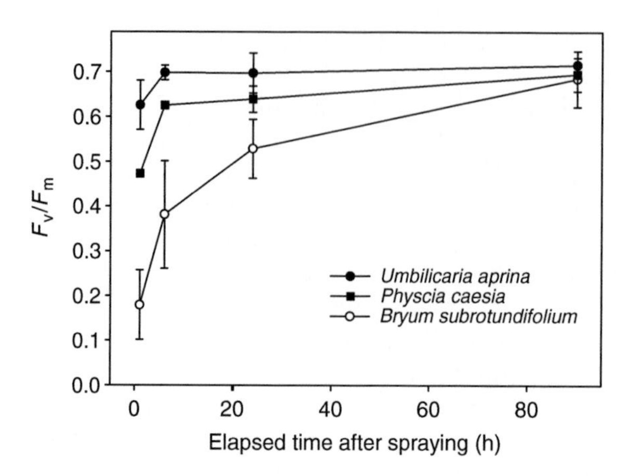

FIGURE 13.11 Recovery of F_v/F_m (optimal quantum efficiency of Photosystem II measured after darkening) for specimens that had been rehydrated after collection in a dehydrated state under snow in early summer. Note the rapid recovery of *Umbilicaria aprina* (rock surface lichen), slow recovery of *Bryum subrotundifolium* (moss from continually wet areas), and intermediate rate for *Physcia caesia* (a lichen species that borders intermittent water flows and can be submerged for some time each day).

Extreme Cold

In the dry state, poikilohydric organisms tend to be resistant to environmental extremes. Dry lichens all survived liquid nitrogen temperatures (−196°C, Kappen 1973) and moist thalli of Antarctic species (*Xanthoria candelaria* (L.) Th. Fr., *R. melanophthalma*) fully tolerated slow or rapid freezing to −196°C (Kappen and Lange 1972). Extended periods of cold and dryness also had little effect, *Alectoria ochroleuca* (Hoffm.) Massal recovered totally after 3.5 years at −60°C (Larson 1978) and the moss *Schistidium antarctici* withstood 18 months at −18°C (Kappen et al. 1989). Despite the impressive figures given above it is equally clear that considerable differences do exist between species, both lichens and mosses, in their abilities to withstand cold. Liverworts, based on the evidence from *M. berteroana*, seem to be excluded from continental Antarctica, because of their lack of cold tolerance, as also are cyanobacterial lichens (Lange 1965, Schroeter et al. 1994). Lichens also show a range of abilities to photosynthesize below 0°C (Lange 1965).

There are, however, several reports of damage through subzero temperatures to young shoot apices in mosses (Longton and Holdgate 1967, Collins 1976) and it does appear that these highly hydrated shoots are at risk. Generally, however, it must be accepted that the plants in Antarctica can survive the winter cold of their habitat.

Freeze–Thaw Cycles

Freeze–thaw cycles can be extremely common in Antarctica with up to 110 in 1 year recorded in the northern maritime (Longton 1988a). The cycles, although rarely falling to temperatures more than a few degrees below freezing point, are thought to provide a severe stress to the plants through intracellular freezing damaging tissue, extracellular freezing disrupting structure, dehydration by withdrawal of water to external ice and phase changes in membranes leading to loss of cell contents. Lovelock et al. (1995a,b) studied the effects of freeze/thaw cycles on the photosynthesis of *G. antarctici* as monitored using chlorophyll *a* fluorescence. Every subzero cycle caused decreased photosynthetic efficiency but recovery at low light was also rapid.

The most thorough analysis has been by Kennedy (1993a) for a northern maritime species, *P. alpestre*, an endohydric moss in contrast to the entirely exohydric species of the continental Antarctic. The species proved to be sensitive to freeze/thaw cycles with reduction in GP almost directly proportional to the temperature reached and with no recovery at −5°C or below. Almost all damage occurred on the first freeze cycle and plants with lower thallus water contents were less sensitive. The mechanism of damage was not clarified but could well have been membrane disruption since increased nitrogen loss from the plants followed the freezing (Greenfield 1988). Kennedy (1993b) was of the opinion that freeze/thaw cycles could easily limit species distribution, however, it must be questioned as to how severe this stress actually is, especially to exohydric mosses. Continual recordings of photosynthetic activity of *P. heimii* in Taylor Valley (Pannewitz et al. 2003a) showed that freezing of surrounding water buffered the moss so that it faced temperatures only a few tenths of a degree below freezing (Figure 13.12). Actual freezing temperature of the moss initiates about −1.8°C (personal research observations). It appears that, for much of the summer season, mosses in wet areas are rarely exposed to subzero temperatures and that freeze–thaw events may be confined to drier environments.

Wetting/drying cycles, also common in the Antarctic, are also thought to be disruptive and release of carbohydrates has been detected (Davey 1997b, Greenfield 1993, Melick and Seppelt 1992). Friedmann et al. (1993) estimated that examples of the cryptoendolithic community in the Ross Desert would pass through an active/inactive cycle (either wet/dry or freeze/thaw) a minimum of 120–150 days each year with substantial loss of metabolites (Greenfield 1988). The stresses associated with these cycles contributed to the difference between a modeled net production of 106 mg C m^{-2} $year^{-1}$ and actual growth, estimated using a variety of techniques, to be 3 mg C m^{-2} $year^{-1}$.

Studies on *U. aprina*, going through a regular daily wetting/drying event near a retreating snow patch showed no deleterious effects on the lichen (Schroeter et al. 1997c). A fuller understanding of these transients is still needed before we can better gauge their effects on

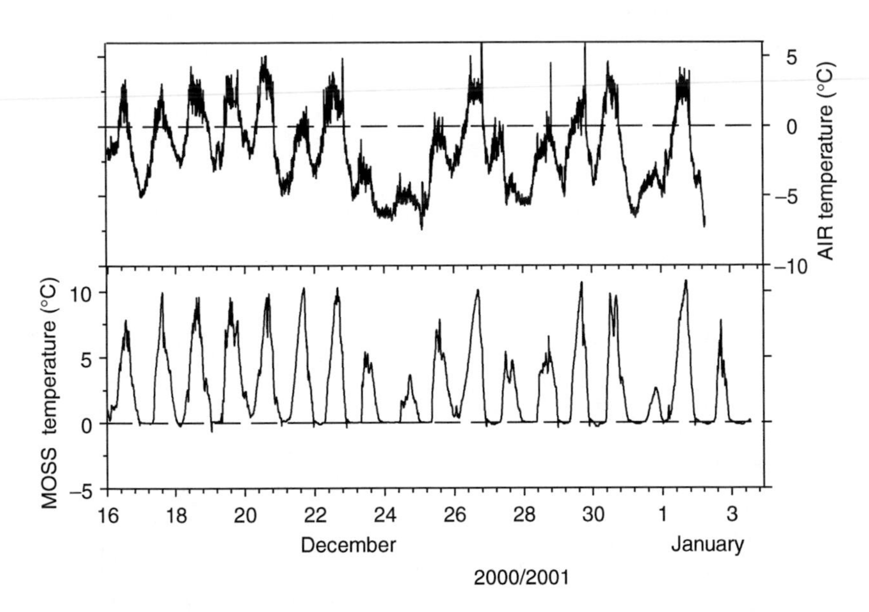

FIGURE 13.12 Uncoupling of thallus temperature (*lower* panel, °C) from air temperature (*upper* panel, °C) for *Pottia heimii* measured over an 18 day period in mid-summer at the Canada Glacier flush, Taylor Valley, 77° S. Note how the air temperature is substantially negative each day, whereas the wet and active moss never goes subzero. The temperature maxima and minima for air/moss were 4.0/10.0 and −7.0/−0.1°C, respectively. (Data modified from Pannewitz, S., Green, T.G.A., Scheidegger, C., Schlensog, M., and Schroeter, B., *Pol. Biol.*, 26, 545, 2003a.)

productivity. However, it appears certain that the severity of any effect depends on the normal habitat of the plant, in particular how xeric it is.

Photosynthesis at Subzero Temperatures

Lichens have long been known to carry out photosynthesis at subzero temperatures. Lange (1965) surveyed a wide range of species including some from Antarctica and found photosynthesis to −24°C by *C. alcornis*, a temperate lichen. Other species, *Umbilicaria decussata* (Vill.) Zahlbr., and *P. caesia* (Hoffm.) Fuernr. (*Parmelia coreyi* in article) from Cape Hallett, had limits of −11°C and −14°C, respectively, values that are similar to species from hot deserts (Lange and Kappen 1972). Schroeter et al. (1994) have since measured photosynthesis of *U. aprina* to −17°C in the field. In their analysis of the response of NP to subzero temperatures they proposed that tolerance of subzero temperatures had the same physiological basis as tolerance to low water potentials. This impression was gained from the very similar response shown by photosynthesis to low water potentials whether generated by subzero temperatures or by equilibration with low atmospheric humidity (Kappen 1993b). In both cases water is removed from the cells either to intercellular ice, lichens have external ice-nucleating agents with freezing initiating at around −5°C (Ashworth and Kieft 1992, Schroeter and Scheidegger 1995), or to the atmosphere. Schroeter et al. (1994) suggested that this explained the absence of cyanobacterial lichens in continental Antarctica because they cannot photosynthesize in equilibrium with humid air but need liquid water (Lange et al. 1988, Schroeter 1994). In addition, Schroeter and Scheidegger (1995) demonstrated convincingly that lichen thalli could rehydrate at subzero temperatures only in the presence of ice. Thallus water content depended on the temperature being only 7% of maximal WC at −21°C, 18% at −4.5°C, and nearly 100% at 8°C (Figure 13.13). Concurrent use of a low temperature scanning electron microscope showed the progressive refilling of the algal and fungal cells as the temperature was increased with the water coming, via the atmosphere, from extracellular ice (Schroeter and Scheidegger 1995). At the same time photosynthesis commenced and increased in line with thallus WC.

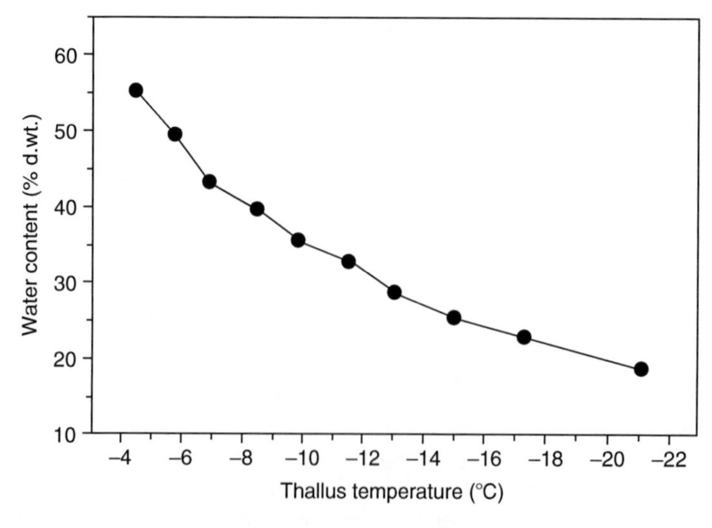

FIGURE 13.13 The relationship between thallus water content (% d.wt.) and thallus temperature for *Umbilicaria aprina* (Schroeter and Scheidegger 1995). Dry lichen thalli where equilibrated in the dark for 24 h at the selected temperature and in the presence of ice.

TABLE 13.3
A Summary of Physiological Abilities of Cyanobacterial Lichens, Green Algal Lichens, Liverworts, and Mosses with Respect to Photosynthetic Performance at Subzero Temperatures

	Plant Type			
	Cyanobacterial Lichens	Liverworts	Mosses	Green Algal Lichens
Distribution	Maritime only	Mainly maritime	Throughout Antarctica	
Subzero photosynthesis	to −2°C	Little	to −8°C	to −24°C
Positive NP In humid air	None	?	Little ability	Above 80% rh
Desiccation/ Cold tolerance: WET	High	Low	Low/medium	High
DRY	Very high	Low	High	Very high

Note: Positive NP in humid air refers to the ability of the group to attain positive net photosynthesis from a dry condition only in the presence of humid (90%–95% relative humidity) air. Desiccation/cold tolerances are given in two rows, the upper when wet and the lower when dry.

A common mechanism would also explain the correlation between cold resistance and dry habitats, such as deserts, found for lichens. The concept has even more far-reaching interest when it is realized that their distribution pattern of Antarctic plants better reflects their rank order for subzero photosynthesis than their desiccation or cold tolerance (Table 13.3). For instance, whereas cyanobacterial lichens have very good resistance to desiccation and to low temperatures, they have nearly no ability to photosynthesize at subzero temperatures, and are excluded from continental Antarctica.

Photosynthesis under Snow

Kappen (1989) demonstrated that rehydration and reactivation of *U. sphacelata* occurred in the field under snow and at subzero temperatures in the maritime Antarctic. This was an important observation because, taken with the later studies of Schroeter and Scheidegger (1995) it appeared that green algal lichens can reactivate photosynthesis under snow without liquid water present that is, without the need of a thaw cycle. The influence of snow cover has been reviewed by Kappen (1993b, 2000). Snow can provide an efficient insulation against wind and extreme temperatures. Considerable light, equivalent to around 10%–30% of incident values, can penetrate to the base of a 15 cm snow pack (Kappen and Breuer 1991), and this is more than sufficient to saturate NP at the low temperatures. The performances of northern maritime Antarctic mosses under snow (Collins and Callaghan 1980) and *Cetraria nivalis* in Sweden (Kappen et al. 1996) were modeled and considerable photosynthetic production was found. It seems almost certain that substantial photosynthesis can occur long before the temperatures reach above freezing and while the lichens are still covered with snow.

The situation, however, appears to be different in continental Antarctica. In an innovative experiment Pannewitz et al. (2003b) constructed fiber optics to measure lichen and moss activity in late summer at Granite Harbour (77° S) and left them to be covered by winter snow. In the following early summer it was then possible to follow the chlorophyll *a* fluorescence activity of the samples under the snow without disturbance. In a complete contrast to the maritime Antarctic there was no evidence of reactivation until liquid water formed when temperatures reached freezing point (Figure 13.14). Temperatures under the snow had, over the long winter, equilibrated with those above the snow and the insulating

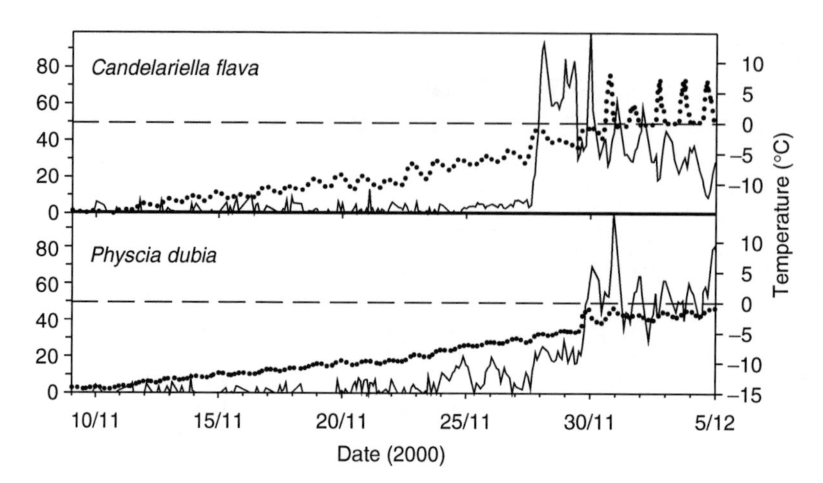

FIGURE 13.14 Insulating effect of a snow bank at Cape Geology, Granite Harbour (77° S): chlorophyll *a* fluorescence (solid line) and thallus temperature (dotted line) were measured for the two lichens *Candelariella flava* (*upper* panel) and *Physcia dubia* (*lower* panel) using probes put in place the previous year and allowed to become naturally covered with snow. Note how the thallus temperature slowly rises (ambient temperatures were around zero degrees) showing a slight daily fluctuation due to incident radiation. Photosynthetic activity does not start until the temperature is close to zero and free water appears.

properties of the snow then prevented rapid warming in the early summer of the buried plants. Far from providing a suitable environment for production the snow delayed any activity until it was completely melted. Increases in snow fall could, therefore, have a very negative effect on lichen and moss biodiversity by shortening the productive season. The effect of large snow banks is clearly visible at Cape Hallett, where there is high lichen abundance at the outer edges of the snow banks but a zone with no lichens appears when the snow bank edge retreats later in the season.

Reverse Diel Cycle of Photosynthesis and High Light Stress

Being poikilohydric, lichens and mosses depend on water to rehydrate and become active. Typically, even in Antarctica, this means that liquid water is required (Hovenden and Seppelt 1995); the exception would be the rehydration by lichens at subzero temperatures. For mosses, in particular, liquid water is certainly required and the larger biomasses are found where consistent water flows occur. In temperate zones water is normally provided by rain, which typically occurs at cloudy times. Mosses and lichens rapidly dry out as soon as brighter light conditions and full sunshine reoccur. Desiccation has often been given as one of the means used by poikilohydric plants to avoid high light and high temperature stress (Kappen 1988).

The situation in the Antarctic tends to be the opposite, melt water only occurs when insolation and temperatures are high so that the lichens and mosses are active when the light is brightest. This can be seen in Figure 13.10 where the photosynthetic activity of *B. frigida* was monitored using chlorophyll *a* fluorescence (Schroeter et al. 1997). Initially when sunlight reaches the thalli in the early morning they rapidly dry and become inactive. Then, about noon, the snowmelt rehydrates the thalli and maximal activity occurs. The thalli then remain photosynthetically active, even when frozen overnight, until drying again occurs at sunrise the next morning. Mosses also display this reverse pattern since water flow from snow or glacier melt is always at its maximum during the day and when irradiance is greatest.

TABLE 13.4
Mean Values for Incident Radiation and Thallus Temperature for Three Species of Lichens in Spain, the Maritime Antarctic and the Continental Antarctic

		Spain, Guadarrama Summit, Madrid, (2000 m) 41° N *Lasallia hispanica*	Antarctica Livingston I. (10 m) 62° S *Usnea aurantiaco-atra*	Antarctica Granite Harbour (10 m) 77° S *Umbilicaria aprina*
Radiation mean	All values	17.5	9.8	38
PAR (mol m^{-2} day^{-1})	Active only	2.7	3.8	77
Mean temperature (°C)	All values	9.7	−2.2	−9.74
	Active only	4.5	1.1	2.9

Note: The mean values have been calculated in each case either for the entire measuring period (1, 14, and 3 years, respectively) and for times when the lichens were active (Active only).

High light stress is, therefore, of unexpected importance in Antarctica. The severity of the conditions can be seen in Table 13.4 that shows the mean conditions when the lichens studied were active. There is a major difference between the continental site at Granite Harbour, mean PPFD about 55% of full sunlight, and the temperate and maritime sites, mean PPFD about 5% of full sunlight.

A combination of cold temperatures and high light has been found to be particularly likely to cause photoinhibition in some higher plants and a similar response is sometimes suggested for Antarctic mosses and lichens. In mosses, photoinhibition has, indeed, been reported several times in studies in both the maritime and continental Antarctica. A midday depression in NP was consistently found under high light by Collins (1977) and was built into models of photosynthetic production by Collins and Callaghan (1980) and Davis (1983). Adamson et al. (1988) demonstrated severe photoinhibition in *Grimmia* (*Schistidium*) *antarctici*. Depressed photosynthesis occurred after 100 min at 500 μmol m^{-2} s^{-1} and CO_2 exchange was negative after 150 min at 1000–1800 μmol m^{-2} s^{-1} and there was a corresponding fall in F_v, the variable fluorescence.

Later studies have tended not to confirm these earlier results and especially not so in continental Antarctica. Lovelock et al. (1995a,b) demonstrated the occurrence of photoinhibition, measured with chlorophyll *a* fluorescence, during freeze/thaw cycles. Plants that were originally under snow showed severe photoinhibition if the snow was removed and they became exposed to full irradiance. Lovelock et al. (1995a,b) showed full recovery to occur under warmer temperatures and under low light, and they suggested that photoinhibition was more of a protective process than one of damage. Post et al. (1990) showed similar photoinhibition and recovery for *C. purpureus*. Post (1990) also showed that the ginger pigment found in exposed plants of *C. purpureus* appeared to serve a protective function against high irradiance. In studies on sun- and shade-adapted *B. subrotundifolium*, Green et al. (2000) showed that only the shade-adapted form, which was bright green, was susceptible to high light. It appeared that the silvery shoot points and nonphotochemical quenching within the photosystems could protect the sun form against several hours of full sunlight.

Although Kappen et al. (1991) reported photoinhibition, indicated by depressed NP, for *U. sphacelata* under high light, no signs of photoinhibition were found when *U. aprina* was monitored for 2 days under full natural sunlight even though the thalli had been dug from under 70 cm of snow before receiving any sunlight in that summer (Kappen et al. 1998a). Similarly, Schlensog et al. (1997b) found no signs of photoinhibition for *L. puberulum* after

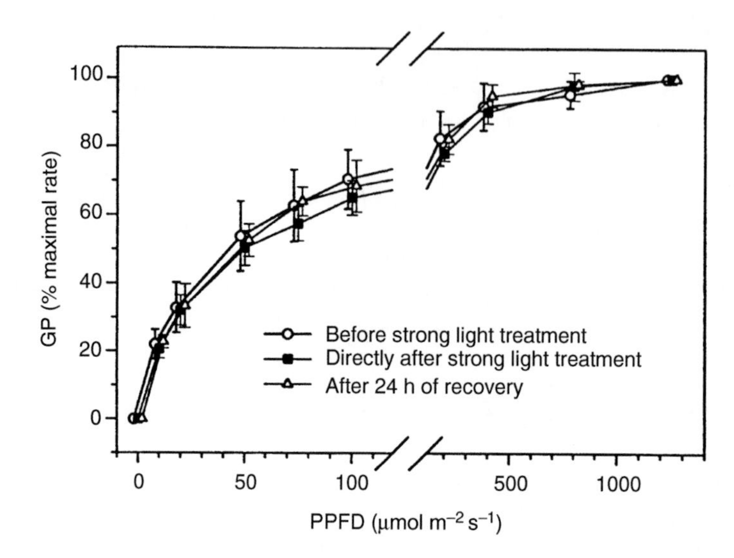

FIGURE 13.15 The almost complete lack of photoinhibition in the cyanobacterial lichen *Leptogium puberulum* despite treatment with strong light (3 h at 1600 μmol m^{-2} s^{-1} PPFD). Immediately after the treatment there was a slight but insignificant decline in NP at intermediate irradiances but no change in apparent quantum efficiency (initial slope of the response to irradiance). (Modified from Schlensog, M., Schroeter, B., Sancho, L.G., Pintado, A., and Kappen, L., *Bibliotheca Lichenologica,* 67, 235, 1997b.)

extensive treatment with high light (Figure 13.15). It seems that pigments in lichens act to reduce the light level within the thallus and thus protect the photobionts (Schroeter et al. 1992, Buedel and Lange 1994, Rikkinen 1995, Schlensog et al. 1997b).

UV Radiation

There has been considerable interest in the possible effects of UV (especially UV-B) radiation on Antarctic plants since the presence of the ozone hole in spring over Antarctica since the 1980s (Robinson et al. 2003, Convey and Smith 2006). Many experiments have been carried out usually involving the removal of UV-B by filters (Robinson et al. 2003). The typical response is subtle and by changed growth suggesting reallocation of resources (Ruhland and Day 2000, Lud et al. 2002, Robinson et al. 2005). Despite earlier worries that Antarctic plants would suffer extensively because they had evolved under very low UV-B levels (not actually correct because levels are high around midday) and because they were simple in structure (Gehrke 1998, Gwynn-Jones et al. 1999) recent studies indicate otherwise. Newsham et al. (2002) have shown that several bryophytes on the Antarctic Peninsula can rapidly (within a day) adjust their UV-B protection to match UV-B incidence. Green et al. (2005) have shown that the sun form of *B. subrotundifolium* is well protected against incident UV-A and that, when this protection is lowered by shading, it is reinstated within a few days following reexposure with little obvious effect on plant performance (Figure 13.16). Studies on DNA damage by the formation of cyclobutyl pyrimidine dimer formation showed that the moss *Sanionia uncinata* (Hedw.) at Leonie Island in the maritime Antarctic (67° 35′ S) showed that any negative effects were transitory and that the species appeared to be well adapted to ambient levels of UV-B radiation (Lud et al. 2002, 2003). When these results from mosses are coupled with the extreme protection to incident radiation possessed by lichens, no effect of full sunlight for 2 days was found on *U. aprina* samples that were immediately exposed after uncovering from snow following winter (Kappen et al. 1998a), it must now be concluded that UV radiation is unlikely to be a stress for these plants except for manipulated situations where UV levels have been rapidly changed.

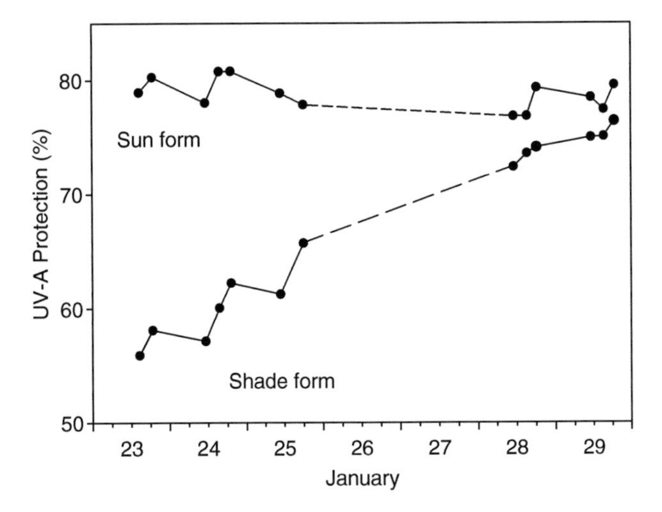

FIGURE 13.16 Changes in UV-A protection (% depression of chlorophyll *a* fluorescence) of the moss *Bryum subrotundifolium* measured in situ with a UV-A PAM chlorophyll *a* fluorometer (Gademann. Instruments, Würzburg, Germany). The UV-A protection of the sun form of the moss remained constant, whereas the protection of a shade form (created by shading the moss for 10 days) rapidly increased back to normal over 6 days. (Modified from Green, T.G.A., Kulle, D., Pannewitz, S., Sancho, L.G., and Schroeter, B., *Pol. Biol.*, 28, 822, 2005.)

Endolithic Lichen Community

Under extremely dry conditions in Antarctica, and in other deserts, lichens can adopt an endolithic growth form where they live within the pores of rocks composed of materials like sandstone and limestone (Friedmann and Galun 1974). The most studied endoliths are the "cryptoendoliths" that penetrate to 2 cm in the Beacon Sandstone of the Dry Valleys, Southern Victoria Land, 77° S, (Friedmann 1982, Kappen 1988). Other species such as *Lecidea phillipsiana* also grow widely in East Antarctica in granites where they produce a prominent brown color on the surface and cause extensive rock flaking (Hale 1987). The lichens form a layered structure within the rock with the upper layer, containing fungal hyphae and the lichen green algal symbiont *Trebouxia*, that had a dark color, possibly to reduce light intensities. The endoliths grow only on the faces of rocks where higher insolation is received, north-facing or horizontal. On a sunny day the temperature of the rock can reach + 8°C, whereas the air temperature is still lower than 5°C below freezing (Kappen et al. 1981). Humidity within the rock is around 80%–90%, which is considerably higher than air, which is normally around 30%–40% rh. Because the temperature is so strongly dependent on insolation it can fluctuate markedly during a day and freeze/thaw transitions are common. A typical daily pattern of the internal rock environment is shown in Figure 13.17 and a freeze/thaw or wet/dry transition is expected to occur at least once on all days when metabolic activity occurs, around 120–150 per year. Water is provided to the rock by blowing snow which then melts into the rock when it is warmed by sunlight. The endoliths are wetted either by equilibration with the high humidity within the rock or by direct moisture uptake after snow melt. The latter method is thought to be the most important in the Dry Valley region (Friedmann 1978).

Carbon metabolism (incorporation of $H^{14}CO_3$) was saturated at 150 μmol m^{-2} s^{-1} at 0°C, and had an optimum at 15°C (Vestal 1988). Measurements of CO_2 exchange showed it to be maximal at 3°C–6°C and to be still positive at −10°C. In a classic long-term study of the nanoclimate using automated recording systems reporting over satellites, 3 years

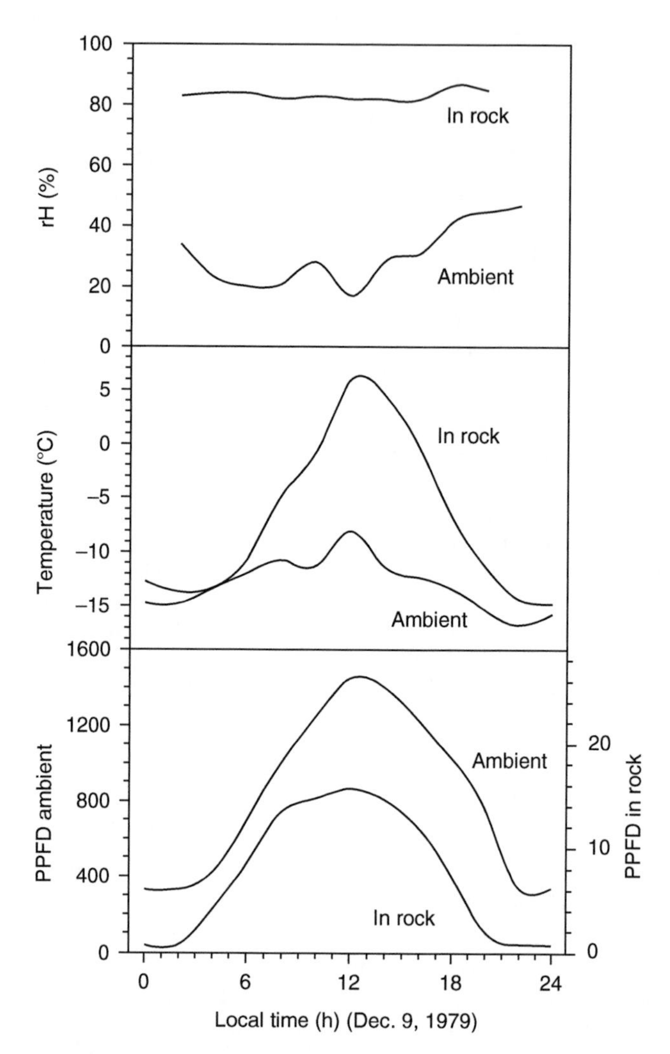

FIGURE 13.17 An example of the diel pattern of humidity (*upper* panel, % RH), temperature (*middle* panel, °C), and irradiance (*lower* panel), PPFD in μmol Photons m^{-2} s^{-1}) both within the rock and in the surroundings of a north-facing endolithic lichen community on December 9, 1979 at Linnaeus Terrace, Aasgard Range, McMurdo Dry Valleys. (Modified from Kappen, L., Friedmann, E.I., and Garty, J., *Flora* 171, 216, 1981.) In the lower panel note the very different scales for the incident PPFD (*left*-hand axis) and PPFD within the rock (*right*-hand axis).

of continuous data were obtained by Friedmann et al. (1987) allowing modeling of aspects like the thermal environment (Nienow et al. 1988). This was then connected to the CO_2 exchange studies to produce estimates by modeling of the community productivity throughout the year (Friedmann et al. 1993). The cryptoendoliths showed positive photosynthesis for around 13 h day^{-1} with annual totals reaching around 1000 h depending on aspect and slope (Kappen and Friedmann 1983). Net productivity was greatest around −2 to +2°C with significant gains down to −8 to −10°C. Horizontal surfaces proved more productive than north-facing sloping surfaces, although colder, because of better water relations. One unusual feature of the community is the relative unimportance of respiratory carbon loss during the dark winter, the system is simply too cold. Despite the long productive periods and the low respiration the estimated carbon gain of 106 mg C m^{-2} $year^{-1}$ only translates into an actual gain of around 3 mg C m^{-2} $year^{-1}$ (estimated from carbon dating techniques and metabolic turnover rates) probably because of the high stresses including possible loss of metabolites in the freeze/thaw or wet/dry cycles (Greenfield 1988). Because the endoliths can grow on almost all north-facing rock faces (porous rock types), where epilithic species are excluded because of the low temperatures and, in particular the abrasive wind, they are the most common terrestrial vegetation community in the Dry Valleys (Friedmann 1982). Endolithic communities are one of the better demonstrations of the importance of shelter, aspect, and water supply in continental Antarctica. As a contrast, it appears that endolithic communities may also contribute extensively to weathering of their host rock by dissolving the silica matrix by the production of locally high pH during photosynthesis (Buedel et al. 2004).

More recently, studying the lithobiontic microorganisms has concentrated in some ultrastructural aspects. It has been demonstrated that these lithobiontic communities can be defined as complex biofilms (Figure 13.18) exhibiting a high diversity and different ecological requirements (De los Ríos et al. 2005b). Accurate identification of lithobiontic microorganisms was possible by means of molecular techniques (De los Ríos et al. 2005a) and unexpectedly high genetic diversities have been found (De la Torre et al. 2003). It was also demonstrated that some minerals in Antarctic rocks are biogenically transformed to generate inorganic biomarkers—traces left by living microorganisms due to their biological activity (Wierzchos and Ascaso, 2001, Wierzchos et al., 2003, 2006). Further, in situ microscopy studies have for the first time demonstrated the presence of microbial fossils within Antarctic sandstone rocks from the Dry Valleys (Ascaso and Wierzchos, 2003, Wierzchos et al. 2005, Ascaso et al. 2005), which has important implications for the detection of endolithic microfossils everywhere in the world or even in extraterrestrial probes (Friedmann et al. 2001, Ascaso and Wierzchos 2002, Wierzchos and Ascaso 2002). These studies have been concentrated in the McMurdo Sound to date. However, the lithobiontic habitats of many important Antarctic areas, such as the Antarctic Peninsula and Transantarctic Mountains and some ice-free oasis on land remain almost completely unexplored.

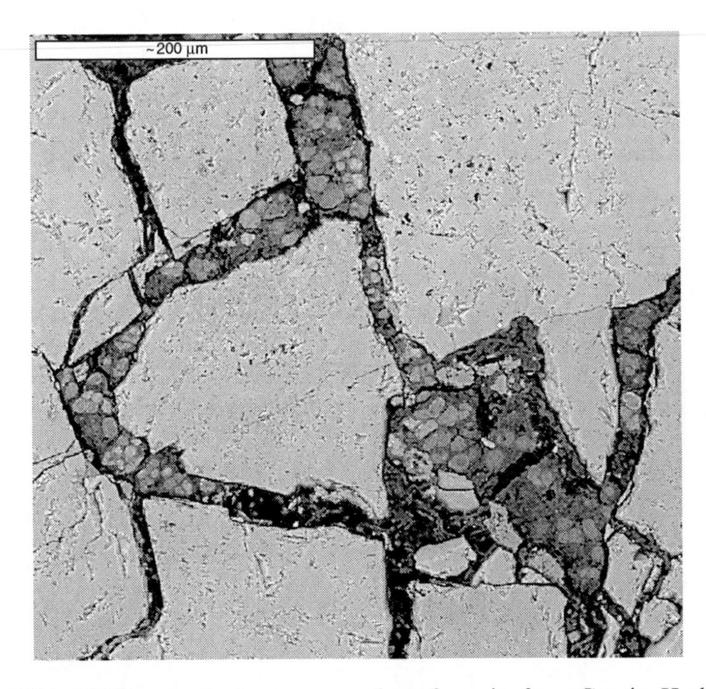

FIGURE 13.18 SEM–BSE image of a transverse section of granite from Granite Harbour showing an algae-rich biofilm colonizing the rock fissures. Scale = 200 µm.

Active versus Inactive

Typically, when looking at performance by plants across Antarctica, it is necessary to use normal meteorological data that do not reflect local microclimates. In addition, even when microclimate data are available, all lichens and mosses in Antarctica are poikilohydric and are inactive for variable, but often large, periods and, at those times, are almost totally protected from the extremes of climate. Data sets are now available that have recorded activity of lichens using chlorophyll *a* fluorescence techniques as well as thallus parameters such as temperature and incident PPFD. It is now possible to make preliminary analyses of the conditions when these organisms are active and to see how substantially they differ from the annual means that include inactive periods as well. Mean annual temperature ranges from +9.7°C, at the summit of Guadarrama Mountains near Madrid, to −9.7°C Granite Harbour, at a range of 19.4 K. In contrast, the mean temperatures during the active periods show only a 3.4 K range with the coldest being the maritime Antarctic site, Livingston Island, and the difference between the Guadarrama summit and Granite Harbour being only 1.6 K. For total daily radiation during the active period the Guadarrama and maritime sites are practically identical, whereas the Granite Harbour site stands out with an exceptionally high value equivalent to a mean instantaneous rate of about 55% full sunlight. This confirms the high light stress that comes from the reverse diel pattern (Section "The Reverse Diel Cycle of Photosynthesis and High Light Stress") but shows that it is a feature of continental Antarctica and not of the maritime Antarctic.

The net result is that the habitats of the active plants are by no means as extreme as the ambient conditions would suggest and, in fact, may be remarkably constant over large latitudinal ranges, the relative constancy of habitat conditions, first demonstrated by Poelt (1987) with lichens growing from the Mediterranean to Greenland. If this concept proves correct for Antarctica then it means that restriction to suitable habitats controls distributions and adaptation may not play as major role a originally expected. This is certainly an area requiring future investigation.

Metabolic Agility

The majority of the vegetation in Antarctica is composed of lichens and bryophytes, which have considerably simpler morphology than higher plants. However, it is a mistake to equate simpler morphology with simpler metabolism as has been done in the past, for example: for UV-B protection (Gehrke 1998, Gwynn-Jones et al. 1999). There is growing evidence that lichens and mosses can be agile in their metabolism. The rapid change in protection against UV-B found by Newsham et al. (2002) and Green et al. (2005) are two examples. The moss *B. subrotundifolium*, and no doubt other mosses like C. *purpureus*, are capable of changing from sun to shade forms and back, again within days to weeks. Lichens are known to be able to alter their dark respiration rate at such a rate that they are almost fully acclimated (Lange and Green 2005). This agility poses obvious problems if samples are taken from the field and kept in some form of storage before use. The extreme shade response to PPFD of photosynthesis by the mosses studied by Rastorfer (1970) is a probable result of prestorage of the material in the laboratory before use.

INTEGRATING PERFORMANCE

Annual Productivity

Considerable efforts have been made to obtain values for the productivity, the seasonal net carbon gain, for Antarctic plants. Unfortunately most estimates have been made in the northern maritime area (Signy Island). However, since no species have had their photosynthesis monitored for complete years the estimates have been produced by the application of models constructed by linking CO_2 exchange and microclimate data sets. The two largest data sets are those for the cryptoendolithic community in the Dry Valleys (see Section "The Endolithic Lichen Community") and for *Usnea aurantiaco-atra* on Livingston Island in the South Shetlands where microclimate, CO_2 exchange and activity (from chlorophyll fluorescence) data sets have been linked (Schroeter et al. 1991, 1997a). In the latter case productivity gains are predicted throughout the year with spring and autumn with the higher rates and winter and summer limited by cold and drought, respectively. Considerable variation in annual total production was also found from year to year (Schroeter et al. 1997a). When modeled CO_2 gain is plotted against actual PPFD and temperature data it is clear that the lichen is rarely active under conditions suitable for optimal photosynthesis but is usually limited by drying out at high PPFD and high temperatures (Figure 13.19). Most other productivity estimates are from small data sets and with many and varied assumptions. One common assertion is that increased winter temperatures will reduce productivity because of greater respiration rates. However, despite substantial winter warming in the northern Antarctic Peninsula there appears to have been no effect on lichen growth rates, in fact some seem to have slightly increased (Sancho and Pintado. 2004). It is known from a temperate study that some lichens can fully acclimate their respiration to seasonal temperature and it appears that this may also be happening in Antarctica (Lange and Green 2005).

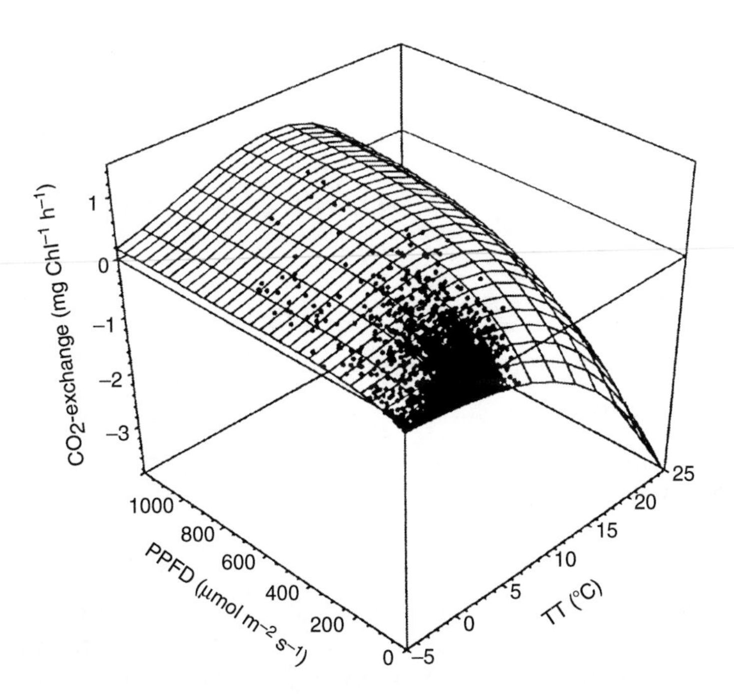

FIGURE 13.19 The relationship between thallus temperature, PPFD and CO_2 exchange for *Usnea aurantiaco-atra* at Livingston Island, maritime Antarctica. Thallus temperature (°C), PPFD (μmol Photons m^{-2} s^{-1}), and photosynthetic activity (from chlorophyll *a* fluorescence measurements) were recorded at 10 min intervals for 1 year. Combinations of temperature and PPFD when the lichen was active were then plotted on a NP response surface to PPFD and temperature generated in the laboratory. The lichen was rarely active under optimal conditions for PPFD and temperature. (Modified from Schroeter, B., Grundlagen der Stoffproduktion von Kryptogamen unter besonderer Berücksichtigung der Flechten—eine Synopse—Habilitationsschrift der Mathematisch—Naturwissenschaftliche Fakultät der Christian—Albrechts—Universität zu Kiel, 1997.)

TABLE 13.5
Estimates of Annual Production for Higher Plants, Lichens and Mosses at Several Locations in Antarctica

Species	Location	Annual Production mg gd.wt.$^{-1}$(1) g m^{-2} y^{-1}(2) mg CO_2 gd.wt.$^{-1}$(3)	Comments
Higher plants			
Deschampsia antarctica	Signy Island (60° 43″ S)	1700 (2) 390 (2)	Estimated from model based on microclimate and NP data. Gravimetric (Edwards and Smith 1988; Edwards 1972)
Lichens			
Cladonia rangifera Usnea antarctica, Usnea aurantiaco-atra	Signy Island	80–200 (1)	Gravimetric (Hooker 1980)
U. antarctica	King George Island (62° 09′ S)	<300 (3)	Estimated from model and year-round microclimate studies (Schroeter et al. 1995)
U. aurantiaco-atra	Signy Island Livingston Island (62° 40′ S)	250 (2) 85 (3)	Smith (1984), Schroeter (1997)
Usnea sphacelata	Casey (66° 17′ S)	14 (1)	Estimated from model and assumed season length (Kappen et al. 1991)
Cryptoendolithic lichens	Dry Valleys (77° 35′ S)	3 (2)	Estimated from NP data and microclimate (Kappen 1993a)
Mosses			
Chorisodontium Polytrichum spp.	Signy Island	315–660 (2)	Harvest plus prediction (Collins 1973, 1977, Longton 1970)
Calliergidium + *Drepanocladus* spp.	Signy Island	223–893 (2)	Davis (1983)
Bryum pseudotriquetrum	Syowa (69° 00′ S)	−16 to +13 (mean + 3.7)	Modelling from microclimate (Ino 1983)
Bryum argenteum	Ross Island (77° 50′ S)	100 (2)	Modelling from microclimate + NP reponses (Longton 1974)

Note: The sources of the data are given in the comments column. The numbers in brackets in the annual production column refer to the units for the production.

Some estimates of annual production are given in Table 13.5. There is also little difference between the higher rates for both lichens and mosses and both show a strong decline from the maritime to the higher latitudes. The extreme climate conditions of the continent depress production, which suggests that there is little adaptation by the plants. It is unfortunate that so many estimates come from the maritime although, in defense, it is one of the major cryptogam-dominated areas of the world. Overall, however, the data are very thin and more and better measurements are needed, especially in the continental Antarctic.

Growth Rates

It is difficult to be sure just how reliable the productivity data are because of the assumptions involved. One check is to compare what is known about growth rates in the same areas.

TABLE 13.6
Estimates of Growth Rates for Lichens and Mosses from Various Locations in Antarctica

Species	Location	Growth Rate (mm y^{-1} unless Otherwise Stated)	Comments
Lichens			
Crustose spp.	King George Island (on whale bones)	40 mm in 78 year	Kappen 1993a
Caloplaca sublobulata	Livingston Island	0.86	Sancho and Pintado 2004
Acarospora macrocyclos	Signy Island/ Livingston Island	1–3/0.72	Lindsay 1973/Sancho and Pintado 2004
Xanthoria elegans	Signy Island	0.2–0.5	Lindsay 1973
Rhizocarpon geographicum	Signy Island/ Livingston Island	0.34/0.5	Lindsay 1973/Sancho and Pintado 2004
Buellia latemarginata	Livingston Island	0.87	Sancho and Pintado 2004
Buellia frigida	Cape Hallett	0.06	Brabyn et al. 2005
B. frigida	Asgaard Range (Dry Valleys)	<0.01	Green unpublished
Mosses			
Polytrichum alpestre	Signy Island	2–5	Longton 1970
Calliergidium austrostramineum	Signy Island	10–32	Collins 1973
Drepanocladus uncinatus	Signy Island	11–16	Collins 1973
Bryum inconnexum	Syowa Coast	<1	Matsuda 1968
Bryum algens	Mawson	1.2	Seppelt and Ashton 1978
Pottia heimii	Taylor Valley	<0.2	Green unpublished

Note: Sources of the data, which are representative, are given in the Comments column.

Examples are given in Table 13.6 from a variety of sources. Once again the vast majority of estimates are from the northern maritime zone and there are very few data in total. However, sufficient data exist to show the same sharp difference between the maritime and continental Antarctica. For the mosses, growth rates decline from 10 mm $year^{-1}$ in the northern maritime to around 1 or <1 mm $year^{-1}$ at the continental margin to an estimated <0.2 mm $year^{-1}$ in the Dry Valleys. A similar trend exists for lichens. Growth rates can be measured relatively easily in the maritime (Sancho and Pintado 2004) and reach rates that are similar to the largest known from anywhere in the world (Table 13.5, Figure 13.20). However, growth is scarcely detectable in continental Antarctica. Figure 13.20 shows two photographs of *B. frigida* taken 23 years apart in the Asgaard Range, Dry Valleys; growth is detectable but is less than 0.01 mm $year^{-1}$. An intermediate rate is found for *B. frigida* of 0.06 mm $year^{-1}$ over 42 years at Cape Hallett (Brabyn et al. 2005) and it appears that a large cline in lichen growth rate exists from the northern Peninsula to central Victoria Land. While there are photographic records showing colonization of man-made substrates by lichens in the maritime, no such photographs exist for the continent (Kappen 1993a). The longest available photographic record for analysis of change seems to be the map produced by Rudolph in 1962 at Cape Hallett. This site has been remapped in 2003 and substantial changes found in the cover of the three main vegetation types, algae, lichens, and mosses (Brabyn et al. 2006). The largest change is an increase in the abundance of algae indicating an improvement in water availability at the site. A series of long-term monitoring sites have been established in the Ross Sea region to allow future monitoring of change (Cannone 2006).

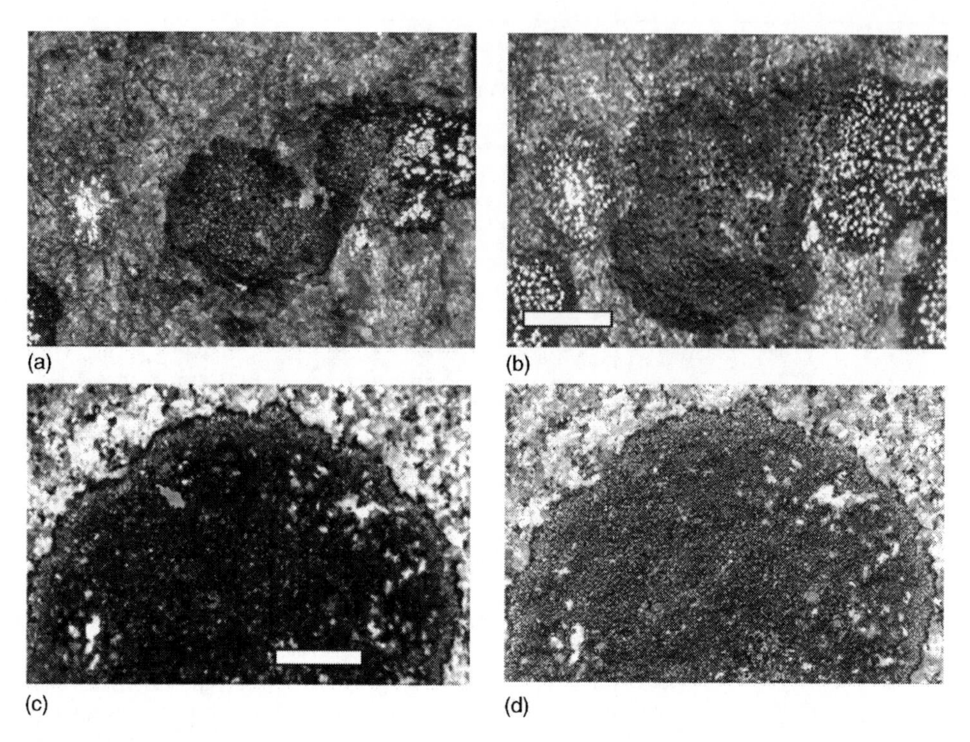

FIGURE 13.20 Photographs of lichens taken several years apart in the maritime and continental Antarctic. *Upper* panels, *Buellia latemarginata* at Livingston Island (62° 39′ S) photographed in January 1991 (a), and January 2002 (b) (from Sancho and Pintado 2004); *Lower* panels, *Buellia frigida* at Mt Falconer, Taylor Valley (77° 35′ S) photographed in December 1980 (c), and January 2003 (d). The scale is about 1 cm.

ANTARCTIC PLANTS AS INDICATORS OF GLOBAL CLIMATE CHANGE

SITUATION

There has been increasing interest recently into the possibility of plants in Antarctica providing early indications of global climate change, especially for temperature where any increase would act to relieve both the geographical isolation and climatic constraints (Kennedy 1995, Robinson et al. 2003, Hennion et al. 2006). This interest has been driven by the accepted increases in temperature in Antarctica.

Global temperature rise is suggested to be greater and more rapid in polar regions. A global increase of 0.03°C year^{-1} (0.2–0.5°C year^{-1} range) is predicted to be 0.5–0.7 times larger in Antarctica. Temperature data from the past 45 years also show that a rise has occurred; 0.067°C year^{-1} at Marguarite Bay (67° 30′ S), 0.056°C year^{-1} at Faraday Station (65° S), 0.038°C year^{-1} at South Shetland Islands (62° 30′ S), 0.022°C year^{-1} at Signy Island (63° 43′ S), and 0.032°C year^{-1} at Halley Station (76° S). As predicted by global change models, the rise has been particularly large in winter temperatures. Although there is large interannual variability the trend seems to be confirmed by glacial and ice shelf retreat. In the remainder of Antarctica, particularly in the Ross Sea region, there is little evidence of any increase in temperature (Doran et al. 2002).

Changes in vegetation have also been reported from the western Antarctic Peninsula. Improved colonization by bryophytes and lichens has been seen, particularly in areas of newly exposed ground however, the most impressive evidence comes from the expansion in abundance of the two phanerogams (Fowbert and Smith 1994, Smith 1994). A study over 27 years has found large increases in populations of *C. quitensis* and *D. antarctica* both in Signy Island (63° 43′ S) and the Argentine Islands (65° 15′ S). At Skua Island, Argentine Islands, as an example, the population of *D. antarctica* was 190 plants in 1964 and 4868 plants in 1990 (Smith 1994). It is also clear that the rise in summer temperatures has led to increased reproduction, germination, and survival.

Problems

At present the approach to the expected effects of global climate change appears to be fairly simplistic and many papers simply demonstrate the effect of some factor on a lichen or moss and then suggest how the species or community might change with altered climate. Change might not be so simple to either predict or detect because several complexities are now known to exist that compromise the simple approaches.

First, it is important that we know that change can actually be detected. Growth of mosses and lichens is rapid in the maritime Antarctic but is much slower in the continent. Recent studies show that change can be detected in both individual plants and communities (Brabyn et al. 2005, Clarke et al. 2006, Brabyn et al. 2006) but that decades are required, not single years. This is a major constraint on modeling change.

Biodiversity patterns in Antarctica are also more complex than originally thought. In the Peninsula it is now confirmed (Peat et al. 2007) that a strong cline exists and that there is a steady loss of species with increase in latitude. Such a situation is amenable in predicting the effects of climate change, such as temperature rise, and the proof is in the already detected changes in the phanerogams. The situation in the main continent is very different. The analysis of lichen populations in the Ross Sea region suggests that the low precipitation has resulted in a series of local microsites available for colonization and that this has been achieved by a sample from the potential species. This is a more complex situation because the controlling factor appears to be water availability and this is much more difficult to model than temperature.

An underlying assumption in any predictions of vegetation change with climate is that the vegetation is in equilibrium with the local climate. At present there is little evidence to support this assumption and some worrying indicators that it might not apply at least in some areas. Work on collembolids has clearly demonstrated that population structure is determined by past events rather than present conditions; for instance the formation of Ross Island (Stevens and Hogg 2003), the presence of Lake Washburn in Taylor Valley 6000 years ago (Nolan et al. 2006), and the presence of relic populations that precede present glaciations and have Gondwana affiliations (Stevens et al. 2006). The high endemic rate of lichens suggest an ancient rather than recent origin (Peat et al. 2006), whereas the presence of species known only from the Peninsula at Mt Kyffin, 84° S, Ross Sea, suggest relic populations in the same areas as the collembolids (Tuerk et al. 2004). The somewhat enigmatic moss, *B. subrotundifolium* (Seppelt and Green 1998) also has a Gondwana distribution known from the Ross Sea, New Zealand, and Australia, all land masses that were earlier in contact. Soil communities in the Taylor Valley are driven by legacy carbon also deposited at the time of Lake Washburn (Lyons et al. 2000).

To the above uncertainties must also be added our lack of knowledge about dispersal and colonization. Assuming that dispersal is possible (Marshall 1996) then the most likely place to show the effects of warming is the Antarctic Peninsula, where a biodiversity cline and warming are present. The second possibility is the ice-free areas that fringe the coast of the continent at about the Antarctic polar circle. These are not warming but are so close to having months with positive mean temperatures that community changes do not seem unlikely in the future.

At present our best monitor for climate change seems to be the growth of lichens. This shows two orders of magnitude change from 77° S to 62° S, a range matched by no other parameter (Table 13.6). This wide range almost certainly reflects the fact that the lichens and mosses are not performing at their physiological optima under present ambient conditions so that any increase in temperature would produce a concomitant rise in productivity.

There is growing evidence that a considerable propagule bank exists in Antarctic soils, especially in the maritime zone. Use of cloches that can both increase the temperature and water relations has resulted in substantial plant growth where none could previously occur (Smith 1993, Convey and Smith 2006). Thus any area influenced by an improvement in temperature or related water supply, especially in the Peninsula, will almost inevitably become rapidly vegetated.

SUMMARY

It might appear, after an initial look at the literature, that there is very little evidence of special adaptations by the lichens and mosses to Antarctic conditions. Lack of special adaptation means that Antarctic species show much the same abilities as can be found in temperate species. In addition, it appears that lichens in Antarctica are forced into the same, but much rarer, microhabitat as those in temperate areas (Section "Active versus Inactive"), again suggesting little adaptation. This point has been made by previous authors (e.g., Longton 1988a, Kappen 1993a). However, the rapid decline in species numbers from the maritime to the continental zone and even more at higher latitudes indicates that, even if no new adaptations are occurring, there is at least strong selection for a group of species that can tolerate Antarctic conditions.

Ability to carry out photosynthesis at low and subzero temperatures and the related tolerance to low water potentials seems to limit the major groups of plants (Schroeter et al. 1994). Liverworts and cyanobacterial lichens seem to be excluded because they lack this ability. It is particularly startling that cyanobacterial lichens are excluded because cyanobacteria are extremely common in Antarctica producing substantial communities as far south as the Scott Glacier, 84° S (Vincent 1988). However, the cyanobacteria are confined to habitats where liquid water occurs, particularly the sides of ponds and streams. These habitats cannot be occupied by lichens so the Antarctic is an unusual example of a situation where the formation of a symbiosis (the lichen) does not extend the distribution of the photobiont.

The large fall in productivity and growth rates between the maritime and the continent in line with the decrease in precipitation, however, does indicate that the lichens and mosses are predominantly controlled by the availability of water. The Antarctic is a cold desert with water availability controlled not just by location but also by remaining frozen for long periods. Plant growth is confined to those locations where precipitation is concentrated and made available by warm conditions. This distribution control by water rather than by temperature has been well explained by Kennedy (1993b). In essence, plants grow wherever liquid water is reliably present as a result of glacier melt stream, by snow melt from stable snow banks and by trapping of snow (e.g., by warm rock surfaces or fruticose lichens). It is important that water, warmth, and light all coincide for plant growth. Kennedy (1993b) makes the point that the Dry Valleys are a relatively warm area but no mosses or lichens (even cryptoendoliths) can grow in them except where water is consistently present (Green et al. 1992). Antarctica is the coldest, driest, highest and windiest continent; the lichens and mosses grow where it is warm, wet, low and protected.

Overall productivity is strongly influenced by the length of period when water is available and the plants become, therefore, increasingly confined to areas of exceptionally good microclimate. It is this strong link between microclimate, water availability and productivity/growth that makes the system so potentially useful for monitoring global climate change, especially temperature increase. The large change in biodiversity along the Antarctic Peninsula covers the occurrence of months with mean temperatures above freezing. Even a small increase in temperature will markedly alter the areas over which such months occur and bring with it a marked community shift (Kennedy 1995).

It is becoming clear that there are two Antarcticas, the Peninsula and the main continent. These differ in the controls on biodiversity distribution, there is a probably water unlimited but temperature-determined biodiversity cline in the Peninsula compared to a, water-controlled, temperature-independent, fragmented vegetation in the continent. The reverse diel pattern of activity with the presence of very high light stress also seems to be confined to the continent. Superimposed on this, and of particular worry to predicting climate change, is the probable occurrence of population distributions on the continent that reflect past events rather than present climate.

Antarctica will, therefore, remain in the future an important natural laboratory for studying ecosystems under stress and the ecophysiology of cryptogams.

REFERENCES

Adamson, H., M. Wilson, P. Selkirk, and R.D. Seppelt, 1988. Photoinhibition in Antarctic mosses. Polarforschung 58: 103–112.

Ahmadjian, V., 1970. Adaptations of Antarctic terrestrial plants. In: M.W. Holdgate, ed. Antarctic ecology, Academic Press, London, pp. 801–811.

Alberdi, M., L.A. Bravo, A. Gutiérrez, M. Gidekel, M., and L.J. Corcuera, 2002. Ecophysiology of antarctic vascular plants. Physiologia Plantarum 115: 479–486.

Ascaso, C. and J. Wierzchos, 2002. New approaches to the study of Antarctic lithobiontic microorganisms and their inorganic traces, and their application in the detection of life in Martian rocks. International Microbiology 5: 215–222.

Ascaso, C. and J. Wierzchos, 2003. The search for biomarkers and microbial fossils in Antarctic rock microhabitats. Geomicrobiology Journal 20: 439–450.

Ascaso, C., J. Wierzchos, M. Speranza, J.C. Gutierrez, A. Martin Gonzalez, and J. Alonso, 2005. Fossil protists and fungi in amber and rock substrates. Micropaleontology 51: 436–443.

Ashworth, E.N. and T.L. Kieft, 1992. Measurement of ice nucleation in lichen using thermal analysis. Cryobiology 29: 400–406.

Bargagli, R., 2005. Antarctic ecosystems: environmental contamination, climate change and human impact. Ecological Studies 175. Springer, Berlin, Heidelberg, pp. 395.

Barnes, D.K.A., D.A. Hodgson, P. Convey, C.S. Allen, and A. Clarke, 2006. Incursion and excursion of Antarctic biota: past, present and future. Global Ecology and Biogeography, 15: 121–142.

Barták, M., J. Gloser, and J. Hájek, 2005. Visualised photosynthetic characteristics of the lichen *Xanthoria elegans* related to daily courses of light, temperature and hydration: A field study from Galindez Island, maritime Antarctica. Lichenologist 37: 433–443.

Bednarek-Ochyra, H., J. Vána, R. Ochyra, and R.I.L. Smith, 2000. The Liverwort Flora of Antarctica. Polish Academy of Sciences, Institute of Botany, Cracow.

Bjork, S., C. Hjort, O. Ingolfsson, and G. Skog, 1991. Radiocarbon dates from the Antarctic Peninsula region—problems and potential. Quaternary Proceedings 1: 55–65.

Brabyn, L., T.G.A. Green, C. Beard, R. Seppelt, 2005. GIS goes nano: Vegetation studies in Victoria Land, Antarctica. N Z Geographer 61: 139–147.

Brabyn, L., C. Beard, R.D. Seppelt, R. Tuerk, and T.G.A. Green, 2006. Quantified vegetation change over 42 years (1962–2004) at Cape Hallett, continental Antarctica. Antarctic Science 18: 561–572.

Broady, P.A., 1989. Broadscale patterns in the distribution of aquatic and terrestrial vegetation ar three ice-free regions on Ross Island, Antarctica. Hydrobiologia 172: 77–95.

Broady, P.A. and R.N. Weinstein, 1998. Algae, lichens and fungi in the La Gorce Mountains, Antarctica. Antarctic Science 10: 376–385.

Broady, P., D. Given, L. Greenfield, and K. Thompson, 1987. The biota and environment of fumaroles on Mt. Melbourne, northern Victoria Land. Polar Biology 7: 97–113.

Buedel, B. and O.L. Lange, 1994. The role of cortical and eprinacral layers in the lichen genus *Peltula*. Cryptogamic Botany 4: 262–269.

Buedel, B., B. Weber, M. Kühl, H. Pfanz, D. Sültemeyer, and D. Wessels, 2004. Reshaping of sandstone surfaces by cryptoendolithic cyanobacteria: bioalkalization causes chemical weathering in arid landscapes. Geobiology 2: 261–268.

Cannone, N., 2006. A monitoring network of terrestrial ecosystems across a latitudinal gradient in continental Antarctica. Antarctic Science 18: 549–560.

Castello, M., 2003. Lichens of the Terra Nova Bay area, northern Victoria Land (Continental Antarctica). Studia Geobotanica 22: 3–54.

Castello, M. and P.L. Nimis, 1995. A critical revision of antarctic lichens described by C.W. Dodge. Bibliotheca Lichenologica 57: 71–92.

Castello, M. and P.L. Nimis, 1997. Diversity of lichens in Antarctica. In: B. Battaglia, J. Valencia, and D.W.H. Walton, eds. Antarctic Communities, Species, Structure and Survival. Cambridge University Press, pp. 15–21.

Castello, M., S. Martillos, and P.L. Nimis, 2006. VICTORIA: An on-line information system on the lichens of Victoria Land (Continental Antarctica). Polar Biology 29: 604–608.

Cawley, R.W. and C.H. Tyndale-Biscoe, 1960. N.Z. Alpine Club Antarctic expedition 1959/60. The New Zealand Alpine Journal 18: 253–268.

Claridge, G.G.C., I.B. Campbell, J.D. Stout, M.E. Dutch, and E.A. Flint, 1971. The occurrence of soil organisms in the Scott Glacier region, Queen Maud Range. New Zealand Journal of Science 14: 306–312.

Clarke, L.J., Robinson, S.A., Hua, Q., and Fink, D. 2006. Watching moss grow: radiocarbon reveals growth rate of Antarctic mosses. SCAR XXIX, Hobart, July 2006, Extracts, p. 422.

Collins, N.J., 1973. Productivity of selected bryophytes in the maritime Antarctic. In: L.C. Bliss and F.E. Wiegolaski, eds. Proceedings of the Conference on Primary Production and Production Processes, Tundra Biome, Edmonton, IBP Tundra Biome Steering Committee, pp. 177–183.

Collins, N.J., 1976. Growth and population dynamics of the moss *Polytrichum alpestre* in the maritime Antarctic. Oikos 27: 389–401.

Collins, N.J., 1977. The growth of mosses in two contrasting communities in the maritime Antarctic: Measurement and prediction of net annual production. In G.A. Llano, ed. Adaptations within Antarctic Ecosystems. Washington, Smithsonian Institution, pp. 921–933.

Collins, N.J. and T.V. Callaghan, 1980. Predicted patterns of photosynthetic production in maritime antarctic mosses. Annals of Botany 45: 601–620.

Convey, P., 1994. Photosynthesis and dark respiration in Antarctic mosses—an initial comparative study. Polar Biology 14: 65–69.

Convey, P., 1996. Reproduction of antarctic flowering plants. Antarctic Science 8: 127–134.

Convey, P. and R.I.L. Smith, 2006. Responses of terrestrial Antarctic ecosystems to climate change. Plant Ecology 18: 1–10.

Cowan, D.A. and L.A. Tow, 2000. Endangered Antarctic environments. Annual Review of Microbiology 58: 649–690.

Cowan, D.A., N.J. Russell, A. Mamais, and D.M. Sheppard, 2002. Antarctic Dry Valley mineral soils contain unexpectedly high levels of microbial biomass. Extremophiles 6: 431–436.

Davey, M.C. 1997a. Effects of physical factors on photosynthesis by the Antarctic liverwort *Marchantia berteroana*. Polar Biology 17: 219–227.

Davey, M.C., 1997b. Effects of short-term dehydration and rehydration on photosynthesis and respiration by antarctic bryophytes. Environment and Experimental Botany 37: 187–198.

Davey, M.C. and P. Rothery, 1997. Seasonal variation in respiratory and photosynthetic parameters in three mosses from the maritime Antarctic. Annals of Botany 78: 719–728.

Davis, R.C., 1980. Peat respiration and decomposition in antarctic terrestrial moss communities. Botanical journal of the Linnean Society. Linnean Society of London 14: 39–49.

Davis, R.C., 1983. Prediction of net primary production in two antarctic mosses by two models of net CO_2 fixation. Bulletin of British Antarctic Survey 59: 47–61.

Davis, R.C., 1986. Environmental factors influencing decomposition rates in two antarctic moss communities. Polar Biology 5: 95–103.

Day, T.A., C.T. Ruhland, C.W. Grobe, and F. Xiong, 1999. Growth and reproduction of Antarctic plants in response to warming and UV radiation reductions in the field. Oecologia 119: 24–35.

De la Torre, J.R., B.M. Goebel, E.I. Friedmann, and N.R. Pace, 2003. Microbial diversity of cryptoendolithic communities from the Mcmurdo Dry Valleys, Antarctica. Applied and Environmental Microbiology, 69: 3858–3867.

De los Ríos, A., L.G. Sancho, M. Grube, J. Wierzchos, and C. Ascaso, 2005a. Endolithic growth of two *Lecidea* lichens in granite from continental Antarctica detected by molecular and microscopy techniques. New Phytologist, 165: 181–190.

De los Ríos, A., J. Wierzchos, L.G. Sancho, T.G.A. Green, and C. Ascaso, 2005b. Ecology of endolithic lichens colonizing granite in continental Antarctica. The Lichenologist 37: 383–395.

Doran, P.T., J.C. Priscu, W.B. Lyons, A.G. Fountain, D.M. McKnight, M. Walsh, D.L., Moorhead, R.A. Virginia, D.H, Wall, G.D. Clow, C.H. Fritsen, C.P. McKay, and A.N. Parsons, 2002. Antarctic climate cooling and terrestrial ecosystem response. Nature 415: 517–520.

Edwards, J.A., 1972. Studies in *Colobanthus quitensis* (Kunth) Bartl. and *Deschampsia antarctica* Desv.: V. Distribution, ecology and vegetative performance on Signy Island. Bulletin of British Antarctic Survey 28: 11–28.

Edwards, J.A. and R.I.L. Smith, 1988. Photosynthesis and respiration of *Colobanthus quitensis* and *Deschampsia antarctica* from the maritime Antarctic. Bulletin of British Antarctic Survey 81: 43–63.

Filson, R.B., 1987. Studies in Antarctic lichens 6: Further notes on *Umbilicaria*. Muelleria 6: 335–347.

Fowbert, J.A., 1996. An experimental study of growth in relation to morphology and shoot water content in maritime antarctic mosses. New Phytologist 133: 363–373.

Fowbert, J.A. and R.I.L. Smith, 1994. Rapid population increases in native vascular plants in the Argentine Islands, Antarctic peninsula. Arctic and Alpine Research 26: 290–296.

Frenot, Y., S.L. Chown, J. Whinam, P.M. Selkirk, P. Convey, M. Skotnicki, and D.M. Bergstrom, 2004. Biological invasions in the Antarctic: Extent, impacts and implications. Biological Review 79: 1–28.

Friedmann, E.I., 1978. Melting snow in the dry valleys is a source of water for endolithic microoganisms. Antarctic Journal of the United States 13: 162–163.

Friedmann, E.I., 1982. Endolithic microorganisms in the Antarctic cold desert. Science 215: 1045–1053.

Friedmann, E.I. and M. Galun, 1974. Desert algae, lichens and fungi. In: G.W. Brown, ed. Desert Biology. Academic Press, New York.

Friedmann, E.I., C.P. McKay, and J.A. Nienow, 1987. The cryptoendolithic microbial environment in the Ross Desert of Antarctica: Satellite-transmitted continuous nanoclimate data, 1984–1986. Polar Biology 7: 273–287.

Friedmann, E.I., L. Kappen, M.A. Meyer, and J.A. Nienow, 1993. Long-term productivity in the cryptoendolithic community of the Ross Desert, Antarctica. Microbial Ecology 25: 25–51.

Friedmann, E.I., J. Wierzchos, C. Ascaso, and M. Winklhofer, 2001. Chains of magnetite crystals in the meteorite ALH84001: Evidence of biogenous origin. Proceedings of the National Academy of Science of US, 98: 2176–2181.

Gannutz, T.P., 1970. Photosynthesis and respiration of plants in the Antarctic Peninsula Area. Antarctic Journal of the United States 5: 49–52.

Gehrke, C., 1998. Affects of enhanced UV-B radiation on production-related properties of a *Sphagnum fuscum* dominated subarctic bog. Functional Ecology 12: 940–947.

Green, T.G.A. and O.L. Lange, 1994. Photosynthesis in poikilohydric plants: A comparison of lichens and bryophytes. In E.-D. Schulze and M.M. Caldwell, ed. Ecological Studies, Volume 100, Ecophysiology of Photosynthesis. Springer Verlag, Berlin, Heidelberg, New York, London, Paris, Tokyo, Hong Kong, Barcelona, Budapest, pp. 319–341.

Green, T.G.A. and W.P. Snelgar, 1982. A comparison of photosynthesis in two thalloid liverworts. Oecologia 54: 275–280.

Green, T.G.A., J. Horstmann, H. Bonnett, A. Wilkins, and W.B. Silvester, 1980. Nitrogen fixation by members of the Stictaceae (Lichenes) of New Zealand. New Phytologist 84: 339–348.

Green, T.G.A., E. Kilian, and O.L. Lange, 1991. *Pseudocyphellaria dissimilis*: A desiccation-sensitive, highly shade-adapted lichen from New Zealand. Oecologia 85: 498–503.

Green, T.G.A., R.D. Seppelt, and A.M.J. Schwarz, 1992. Epilithic lichens on the floor of the Taylor Valley, Ross Dependency, Antarctica. Lichenologist 24: 57–61.

Green, T.G.A., B. Buedel, A. Meyer, H. Zellner, and O.L. Lange, 1997. Temperate rainforest lichens in New Zealand: Light response of photosynthesis. New Zealand Journal of Botany 35: 493–504.

Green, T.G.A., B. Schroeter, K.S. Maseyk, R.D. Seppelt, L. Kappen, 1998. An assessment of the relationship between carbon dioxide exchange and chlorophyll a fluorescence in an antarctic moss and lichen. Planta 206: 611–618.

Green, T.G.A., B. Schroeter, and R.D. Seppelt, 2000. Effect of temperature, light and ambient UV on the photosynthesis of the Moss *Bryum argenteum* Hedw. in Continental Antarctica. In W. Davison C. Howard-Williams, P. Broady, eds. Antarctic Ecosystems: Models for a wider ecological understanding, Canterbury University, pp. 165–170.

Green, T.G.A., D. Kulle, S. Pannewitz, L.G. Sancho, and B. Schroeter, 2005. UV-A protection in mosses growing in continental Antarctica. Polar Biology 28: 822–827.

Greenfield, L.G., 1988. Forms of nitrogen in Beacon Sandstone rocks containing endolithic microbial communities in southern Victoria Land, Antarctica. Polarforschung 58: 211–218.

Greenfield, L.G., 1993. Decomposition studies on New Zealand and Antarctic lichens. Lichenologist 25: 73–82.

Gwynn-Jones, D., U. Johnson, G. Phoenix, et al., 1999. UV-B impacts and interactions with other co-occurring variables of environmental change: An arctic perspective. In: J. Rozema, ed. Stratospheric ozone depletion: The effects of enhanced UV-B radiation. Backhuys Publishers, Leiden, The Netherlands, pp. 187–201.

Hale, M.E. 1987. Epilithic lichens in the beacon sandstone formation, Victoria Land, Antarctica. Lichenologist 19: 269–287.

Harrisson, P.M., D.W.H. Walton, and P. Rothery, 1989. The effects of temperature and moisture on CO_2 uptake and total resistance to water loss in the antarctic foliose lichen *Umbilicaria antarctica*. New Phytologist 111: 673–682.

Hennion, F., A.H. Huiskes, S.A. Robinson, and P. Convey, 2006. Physiological traits of organisms in a changing environment. In: D.M. Bergstrom, P. Convey, and A. Huiskes, eds. Trends in Antarctic Terrestrial and Limnetic Ecosystems: Antarctica as a Global Indicator. Springer, Berlin, Heidelberg, pp. 127–157.

Holdgate, M.W., 1964. Terrestrial ecology in the maritime Antarctic. In: R. Carrick, M.W. Holdgate, and J. Prévost, eds. Biologie Antarctique. Hermann, Paris, pp. 181–194.

Holdgate, M.W., 1977. Terrestrial ecosystems in the Antarctic. In: V. Fuchs and R.M. Laws, eds. Scientific Research in Antarctica , Philosophical Transactions of Royal Society London, Series B, Biological Sciences 279: 5–25.

Hooker, T.N., 1980. Growth and production of *Usnea antarctica* and *U. fasciata* on Signy Island, South Orkney Islands. Bulletin of British Antarctic Survey 50: 35–49.

Hovenden, M.J. and R.D. Seppelt, 1995. Uptake of water from the atmosphere by lichens in continental Antarctica. Symbiosis 18: 111–118.

Ino, Y., 1983. Estimation of primary production in moss community on East Ongul Island; Antarctica. Antarctic Recorder 80: 30–38.

Kappen, L., 1973. Response to extreme environments. In: V. Ahmadjian and M.E. Hale, eds. The Lichens. Academic Press, New York London, pp. 310–380.

Kappen, L., 1983. Ecology and physiology of the antarctic fruticose lichen *Usnea suphurea* (Koenig) Th. Fries. Polar Biology 1: 249–255.

Kappen, L., 1985a. Vegetation and ecology of ice–free areas of Northern Victoria Land, Antarctica. I. The lichen vegetation of Birthday Ridge and an inland mountain. Polar Biology 4: 213–225.

Kappen, L. 1985b. Water relations and net photosynthesis of *Usnea*. A comparison between *Usnea fasciata* (maritime Antarctic) and *Usnea sulphurea* (continental Antarctic). In: D.H. Brown, ed. Lichen Physiology and Cell Biology. Plenum Press, New York London, pp. 41–56.

Kappen, L., 1988. Ecophysiological relationships in different climatic regions. In: M. Galun, ed. CRC Handbook of Lichenology. CRC Press, Boca Raton, Florida, pp. 37–100.

Kappen, L., 1989. Field measurements of carbon dioxide exchange of the Antarctic lichen *Usnea sphacelata* in the frozen state. Antarctic Science 1: 31–34.

Kappen, L., 1993a. Lichens in the Antarctic region. In: E.I. Friedmann, ed. Antarctic Microbiology, Wiley–Liss, pp. 433–490.

Kappen, L., 1993b. Plant activity under snow and ice, with particular reference to lichens. Arctic 46: 297–302.
Kappen, L., 2000. Some aspects of the great success of lichens in Antarctica. Antarctic Science 12: 314–324.
Kappen, L. and M. Breuer, 1991. Ecological and physiological investigations in continental Antarctic cryptogams. II. Moisture relations and photosynthesis of lichens near Casey Station, Wilkes Land. Antarctic Science 3: 273–278.
Kappen, L. and E.I. Friedmann, 1983. Ecophysiology of lichens of the Dry Valleys of Southern Victoria Land, Antarctica. II. CO_2 gas exchange in cryptoendolithic lichens. Polar Biology 1: 227–232.
Kappen, L. and O.L. Lange, 1972. Die Kälteresistenz einiger Macrolichenen. Flora 161: 1–29.
Kappen, L. and B. Schroeter, 2002. Plants and lichens in the Antarctic, their way of life and their relevance to soil formation. In: L. Beyer, and M. Bölter eds. Geoecology of Antarctic Ice-Free Coastal Landscapes. Ecological Studies 154. p. 327. Springer-Verlag, Berlin, Heidelberg, pp. 427.
Kappen, L., E.I. Friedmann, and J. Garty, 1981. Ecophysiology of lichens of the Dry Valleys of Southern Victoria Land, Antarctica. I. Microclimate of the cryptoendolithic lichen habitat. Flora 171: 216–235.
Kappen, L., R.I.L. Smith, and M. Meyer, 1989. Carbon dioxide exchange of two ecodemes of *Schistidium antarctici* in continental Antarctica. Polar Biology 9: 415–422.
Kappen, L., M. Breuer, and M. Bölter, 1991. Ecological and physiological investigations in continental Antarctic cryptogams. 3. Photosynthetic production of *Usnea sphacelata*: diurnal courses, models, and the effect of photoinhibition. Polar Biology 11: 393–401.
Kappen, L., B. Schroeter, C. Scheidegger, M. Sommerkorn, and G. Hestmark, 1996. Cold resistance and metabolic activity of lichens below 0°C. Advances in Space Research 18: 119–129.
Kappen, L., B. Schroeter, T.G.A. Green, and R.D. Seppelt, 1998a. Chlorophyll a fluorescence and CO_2 exchange of *Umbilicaria aprina* under extreme light stress in the cold. Oecologia 113: 325–331.
Kappen, L., B. Schroeter, T.G.A. Green, and R.D. Seppelt, 1998b. Microclimatic conditions, meltwater moistening, and the distributional pattern of *Buellia frigida* on rock in a southern continental Antarctic habitat. Polar Biology 19: 101–106.
Kennedy, A.D., 1993a. Photosynthetic response of the antarctic moss *Polytrichum alpestre* Hoppe to low temperatures and freeze–thaw stress. Polar Biology 13: 271–279.
Kennedy, A.D., 1993b. Water as a limiting factor in the Antarctic terrestrial environment: A biogeographical synthesis. Arctic and Alpine Research 25: 308–315.
Kennedy, A.D., 1995. Antarctic terrestrial ecosystem response to global environmental change. Annual Reviews of Ecology and Systematics, 26: 683–704.
King, J.C., 1994. Recent climate variability in the vicinity of the Antarctic Peninsula. International Journal Climatology 14: 357–369.
Lange, O.L., 1965. Der CO_2–Gaswechsel von Flechten bei tiefen Temperaturen. Planta 64: 1–19.
Lange, O.L. and L. Kappen, 1972. Photosynthesis of lichens from Antarctica. In: G.A. Llano, ed. Antarctic Research Series 20. American Geophysical Union, Washington D.C., pp. 83–95.
Lange, O.L. and T.G.A. Green, 2005. Lichens show that fungi can acclimate their respiration to seasonal change in temperature. Oecologia 142: 11–19.
Lange, O.L., T.G.A. Green, and H. Ziegler, 1988. Water status related photosynthesis and carbon isotope discrimination in species of the lichen genus *Pseudocyphellaria* with green or blue—green photobionts and in photosymbiodemes. Oecologia 75: 494–501.
Larson, D.W., 1978. Patterns of lichen photosynthesis and respiration following prolonged frozen storage. Canadian Journal of Botany 56: 2119–2123.
Lechowicz, M.J., 1982. Ecological trends in lichen photosynthesis. Oecologia 53: 330–336.
Leishman, M.R. and C. Ad Wild, 2001. Vegetation abundance and diversity in relation to soil nutrients and soil water content in Vestfold Hills, East Antarctica. Antarctic Science 13: 126–134.
Light, J.J. and R.B. Heywood, 1975. Is the vegetation of continental Antarctica predominantly aquatic? Nature 156: 199–200.
Lindsay, D.C., 1973. Estimates of lichen growth rates in the maritime Antarctic. Arctic and Alpine Research 5: 341–346.
Longton, R.E., 1970. Growth and productivity of the moss *Polytrichum alpestre* Hoppe in Antarctic regions. In: M.W. Holdgate, ed. Antarctic Ecology. Academic Press, London, pp. 818–837.

Longton, R.E., 1974. Microclimate and biomass in communities of the *Bryum* association on Ross Island, continental Antarctica. Bryologist 77: 109–127.

Longton, R.E., 1988a. The biology of polar bryophytes and lichens. Cambridge University Press, Cambridge.

Longton, R.E., 1988b. Adaptations and strategies of polar bryophytes. Botanical journal of the Linnean Society. Linnean Society of London 98: 253–268.

Longton, R.E. and M.W. Holdgate, 1967. Temperature relations of antarctic vegetation. Philosophical Transactions of the Royal Society, London B. 252: 237–250.

Lovelock, C.E., C.B. Osmond, and R.D. Seppelt, 1995a. Photoinhibition in the antarctic moss *Grimmia antarctici* Card. when exposed to cycles of freezing and thawing. Plant Cell and Environment 18: 1395–1402.

Lovelock, C.E., A.G. Jackson, D.R. Melick, and R.D. Seppelt, 1995b. Reversible photoinhibition in antarctic moss during freezing and thawing. Plant Physiology 109: 955–961.

Lud, D., T. Moerdijk, W. Van der Poll, A.G.J. Buma, and A.H.L. Huiskes, 2002. DNA damage and photosynthesis in Antarctic and Arctic *Sanionia uncinata* (Hedw.) Loeske. under ambient and enhanced levels of UV-B radiation. Plant Cell and Environment 25: 1579–1589.

Lud, D., M. Schlensog, B. Schroeter, and A.H.L. Huiskes, 2003. The influence of UV-B radiation on light-dependent photosynthetic performance in *Sanionia uncinata* (Hedw.) Loeske in Antarctica. Polar Biology, 26: 225–232.

Lyons, W.B., A. Foutain, P. Doran, J.C. Priscu, K. Neumann, and K.A. Welch, 2000. Importance of landscape position and legacy: the evolution of the lakes in Taylor Valley, Antarcica. Freshwater Biology 43: 355–367.

Marshall, W.A., 1996. Aerial dispersal of lichen soredia in the maritime Antarctic. New Phytologist 134: 523–530.

Matsuda, T., 1968. Ecological study of the moss community and microorganisms in the vicinity of Syowa Station, Antarctica. Japanese Antarctic Research Expedition Science Reports. E29: 1–58.

Melick, D.R. and R.D. Seppelt, 1992. Loss of soluble carbohydrates and changes in freezing point of antarctic bryophytes after leaching and repeated freeze–thaw cycles. Antarctic Science 4: 399–404.

Montiel, P., A. Smith, and D. Keiler, 1999. Photosynthetic responses of selected Antarctic plants to solar radiation in the southern maritime Antarctic. Polar Research 18: 229–235.

Newsham, K., D. Hodgson, and A. Murray et al., 2002. Response of two Antarctic bryophytes to stratospheric ozone depletion. Global Change Biology 8: 1–12.

Nienow, J.A., C.P. McKay, and E.I. Friedmann, 1988. The cryptoendolithic microbial environment in the Ross Desert of Antarctica. Mathematical models of the thermal regime. Microbial Ecology 16: 253–270.

Noakes, T.D. and R.E. Longton, 1989. Studies on water relations in mosses from the cold—Antarctic. In: R.B. Heywood, ed. University Research in Antarctica. Antarctic Special Topic. British Antarctic Survey, Cambridge, pp. 103–116.

Nolan, L., I.D. Hogg M.I. Stevens, and M. Haase, 2006. Fine scale distribution of mtDNA haplotypes for the springtail *Gomphocephalus hodgsoni* (Collembola) corresponds to an ancient shoreline in Taylor Valley, continental Antarctica. Polar Biology 29: 813–819.

Ochyra, R., 1998. The moss flora of King George Island, Antarctica. Polish Academy of Sciences, Cracow, Poland pp. 279.

Okitsu, S., S. Imura, and E. Ayukawa, 2004. Micro-relief distribution of major mosses in ice-free areas along the Sôya Coast, The Syowa Station area, East Antarctica. Polar Bioscience 17: 69–82.

Olech, M., 1990. Preliminary studies on ornithocoprophilous lichens of the arctic and Antarctic regions. NIPR Symposium on Polar Biology 3: 218–223.

Øvstedal, D.O. and R.I.L. Smith, 2001. *Lichens of Antarctica and South Georgia.* A guide to their identification and ecology. Cambridge University Press, Cambridge.

Pannewitz, S., T.G.A. Green, C. Scheidegger, M. Schlensog, and B. Schroeter, 2003a. Activity pattern of the moss *Hennediella heimii* (Hedw.) Zand. in the Dry Valleys, Southern Victoria Land, Antarctica during the mid-austral summer. Polar Biology 26: 545–551.

Pannewitz, S., M. Schlensog, T.G.A. Green, S. Sancho, and B. Schroeter, 2003b. Are lichens active under snow in continental Antarctica? Oecologia 135: 30–38.

Pannewitz, S., T.G.A. Green, K. Masek, M. Schlensog, R.D. Seppelt, L.G. Sancho, R. Tuerk, and B. Schroeter, 2005. Photosynthetic responses of three common mosses from continental Antarctica. Antarctic Science 17: 341–352.

Peat, H.J., A. Clarke, and P. Convey, 2007. Diversity and biogeography of the Antarctic flora. Journal of Biogeography 34: 132–146.

Pickard, J., 1986. Antarctic oases, Davis station and the Vestfold Hills. In J. Pickard, ed. Antarctic Oasis; Terrestrial Environments and History of the Vestfold Hills. Academic Press Australia, Sydney, pp. 1–19.

Poelt, J., 1987. Das Gesetz der relativen Standortkonstnz bei den Flechten. Botanische Jahrbuecher der Systematik 108: 363–371.

Post, A., 1990. Photoprotective pigment as an adaptive strategy in the Antarctic moss *Ceratodon purpureus*. Polar Biology 10: 241–245.

Post, A. and M. Vesk, 1992. Photosynthesis, pigments, and chloroplast ultrastructure of an antarctic liverwort from sun-exposed and shaded sites. Canadian Journal of Botany 70: 2259–2264.

Post, A., E. Adamson, and H. Adamson, 1990. Photoinhibition and recovery of photosynthesis in antarctic bryophytes under field conditions. Current Research in Photosynthesis 4: 635–638.

Proctor, M.C.F., 2000. Physiological ecology. In: A.J. Shaw and B. Goffninet, eds. Bryophyte Biology. Cambridge University Press, Cambridge, pp. 225–247.

Rastorfer, J.R., 1970. Effects of light intensity and temperatre on photosynthesis and respiration of two east Antarctic mosses, *Bryum argenteum* and *Bryum antarcticum*. Bryologist 73: 544–556.

Rastorfer, J.R., 1972. Comparative physiology of four west antarctic mosses. In: G.A. Llano, ed. Antarctic Terrestrial Biology. Antarctic Research Series Vol. 20. American Geophysical Union, Washington, pp. 143–161.

Rikkinen, J., 1995. What's behind the pretty colours? A study on the photobiology of lichens. Bryobrothera 4: 1–239.

Robinson, S.A., J. Wasley, M. Popp, and C.E. Lovelock, 2000. Desiccation tolerance of three moss species from continental Antarctica. Australian Journal of Plant Physiology 27: 379–388.

Robinson, S.A., J. Wasley, and A.K. Tobin, 2003. Living on the edge—plants and global change in continental and maritime Antarctica. Global Change Biology 9: 1681–1717.

Robinson, S.A., J.D. Turnbull, and C.E. Lovelock, 2005. Impact of changes in natural ultraviolet radiation on pigment composition, physiological and morphological characteristics of the Antarctic moss, *Grimmia antarctici*. Global Change Biology 11: 476–489.

Ruhland, C. and T. Day, 2001. Size and longevity of seed banks in Antarctica and the influence of ultraviolet-B radiation on survivorship, growth and pigment concentration of *Colobanthus quitensis* seedlings. Environmental and Experimental Biology 45: 143–154.

Sancho, L.G. and A. Pintado, 2004. Evidence of high annual growth rate for lichens in the maritime Antarctic. Polar Biology 27: 312–319.

Sancho, L.G., A. Pintado, F. Valladares, B. Schroeter, and Schlensog, M., 1997. Photosynthetic performance of cosmopolitan lichens in the maritime Antarctic. Bibliotheca Lichenologica 67: 197–210.

Schlensog, M., B. Schroeter, L.G. Sancho, A. Pintado, and L. Kappen, 1997a. Photosynthetic performance of the cyanobacterial lichen *Leptogium puberulum* in the maritime Antarctic Biology of the Royal Society Española de Historia Natural. Sección biológica 93: 107–113.

Schlensog, M., B. Schroeter, L.G. Sancho, A. Pintado, and L. Kappen, 1997b. Effect of strong irradiance on photosynthetic performance of the melt-water dependent cyanobacterial lichen *Leptogium puberulum* (Collemataceae) Hue from the maritime Antarctic. Bibliotheca Lichenenologica 67:235–246.

Schlensog, M., B. Schroeter, S. Pannewitz, and T.G.A. Green, 2003. Adaptations of mosses and lichens to irradiance stress in maritime and continental Antarctic habitats. In: Huiskes, Gieskes, Rozema, Schorno, van der Vries and W. Wolff, eds. Antarctic Biology in a Global Context. Backhuys, Leiden, 352 pp.

Schlensog, M., S. Pannewitz, T.G.A. Green, and B. Schroeter, 2004. Metabolic recovery of continental antarctic cryptogams after winter. Polar Biology 27: 399–408.

Schroeter, B., 1991. Untersuchungen zu Primärproduction und Wasshaushalt von Flechten der maritimen Antarktis unter besonderer Berücksichtigung von *Usnea antarctica* Du Rietz. Dissertation Universitaet Kiel, pp. 1–148.
Schroeter, B., 1994. In situ photosynthetic differentiation of the green algal and the cyanobacterial photobiont in the crustose lichen *Placopsis contortuplicata*. Oecologia 98: 212–220.
Schroeter, B., 1997. Grundlagen der Stoffproduktion von Kryptogamen unter besonderer Berücksichtigung der Flechten—eine Synopse—Habilitationsschrift der mathematisch—Naturwissenschaftlichen Fakultät der Christian—Albrechts—Universität zu Kiel, pp. 130.
Schroeter, B. and C. Scheidegger, 1995. Water relations in lichens at subzero temperatures: Structural changes and carbon dioxide exchange in the lichen *Umbilicaria aprina* from continental Antarctica. New Phytologist 131: 273–285.
Schroeter, B., L. Kappen, and C. Moldaenke, 1991. Continuous in situ recording of the photosynthetic activity of Antarctic lichens–established methods and a new approach. Lichenologist 23: 253–265.
Schroeter, B., T.G.A. Green, R.D. Seppelt, and L. Kappen, 1992. Monitoring photosynthetic activity of crustose lichens using a PAM–2000 fluorescence system. Oecologia 92: 457–462.
Schroeter, B., T.G.A. Green, L. Kappen, and R.D. Seppelt, 1994. Carbon dioxide exchange at subzero temperatures. Field measurements on *Umbilicaria aprina* in Antarctica. Cryptogamic Botany 4: 233–241.
Schroeter, B., A. Olech, L. Kappen, and W. Heitland, 1995. Ecophysiological investigations of *Usnea antarctica* in the maritime Antarctic. I. Annual microclimatic conditions and potential primary production. Antarctic Science 7: 251–260.
Schroeter, B., L. Kappen, and F. Schulz, 1997a. Long–term measurements of microclimate conditions in the fruticose lichen *Usnea aurantiaco-atra* in the maritime Antarctic. In: J. Cacho, and D. Serrat, eds. Actas del V Simposio de Estudios Antarcticos (1993). CICYT, Madrid, pp. 63–69.
Schroeter, B., F. Schulz, and L. Kappen, 1997b. Hydration-related spatial and temporal variation of photosynthetic activity in antarctic lichens. In: B. Battaglia, J. Valencia, and D.W.H. Walton, eds. Antarctic Communities. Species, Structure and Survival. Cambridge University Press, Cambridge, pp. 221–225.
Schroeter, B., L. Kappen, T.G.A. Green, and R.D. Seppelt, 1997c. Lichens and the antarctic environment: effects of temperature and water availability on photosynthesis. In: W.B. Lyons, C. Howard-Williams, and I. Hawes, eds. Ecosystem Processes in Antarctic Ice-free Landscapes. Balkema, Rotterdam, Brookfield, pp. 103–118.
Schwarz, A.M.J., T.G.A. Green, and R.D. Seppelt, 1992. Terrestrial vegetation at Canada Glacier, Southern Victoria Land, Antarctica. Polar Biology 12: 397–404.
Senchina, D.S., 2005. A historical survey of botanical exploration in Antarctica. Huntia 12: 31–69.
Seppelt, R.D., 1995. Phytogeography of continental antarctic lichens. Lichenologist 27: 417–431.
Seppelt, R.D. and D.H. Ashton, 1978. Studies on the ecology of the vegetation of Mawson Station Antarctica. Australian Journal of Ecology 3: 373–388.
Seppelt, R.D. and H. Kanda, 1986. Morphological variation and taxonomic interpretation in the moss genus Bryum in Antarctica. Member of the National Institute of Polar Research 37: 27–42.
Seppelt, R.D. and T.G.A. Green, 1998. A bryophyte flora for Southern Victoria Land, Antarctica. New Zealand Journal of Botany 36: 617–635.
Smith, R.I.L., 1984. Terrestrial plant biology of the sub-Antarctic and Antarctic. In: R.M. Laws, ed. Antarctic Ecology. Academic Press, London, pp. 61–162.
Smith, R.I.L., 1993. The role of bryophyte propagule banks in primary succession: case-study of an antarctic fellfield soil. In: J. Miles and D.W.H. Walton, eds. Primary Succession on Land. Oxford, Blackwells, pp. 55–78.
Smith, R.I.L., 1994. Vascular plants as bioindicators of regional warming in Antarctica. Oecologia 99: 322–328.
Smith, R.I.L. 1996. Terrestrial and freshwater biotic components of the western Antarctic Peninsula. In: R.M. Koss, E. Hofmann, and L.B. Quetin, eds. Foundations for Ecological Research West of the Antarctic Peninsula. Antarctic Research Series, Volume 70, pp. 15–59. American Geophysical Union, Washington.

Smith, R.I.L., 1999. Biological and environmental characteristics of three cosmopolitan mosses dominant in continental Antarctica. Journal of Vegetation Science 10: 231–242.

Smith, R.I.L., 2000. Plants of extreme habitats in Antactica. Bibliotheca Lichenologica 75: 405–419.

Smith, R.I.L. and D.O. Övstedal, 1991. The lichen genus *Stereocaulon* in Antarctica and South Georgia. Polar Biology 11: 91–112.

Smith, V.R., 1993. Effect of nutrients on CO_2 assimilaton by mosses on a sub-antarctic island. New Phytologist 123: 693–697.

Söchting, U. and M. Olech, 1995. The lichens genus *Caloplaca* in polar regions. Lichenologist 27: 463–471.

Stenroos, S., 1993. Taxonomy and distribution of the lichen family Cladoniaceae in the Antarctic and peri–Antarctic regions. Cryptogamic Botany 3: 310–344.

Stevens, M.I. and I.D. Hogg, 2003. Long-term isolation and recent expansion from glacial refugia revealed for the endemic springtail *Gomphiocephalus hodgsoni* from Victoria Land, Antarctica. Molecular Ecology 12: 2357–2369.

Stevens, M.I., P. Greenslade, I.D. Hogg, and P. Sunnucks, 2006. Southern hemisphere springtails: Could they have survived glaciation of Antarctica. Molecular Biology and Evolution 23: 874–882.

Tarnawski, M., D. Melick, D. Roser, E. Adamson, H. Adamson, and R. Seppelt, 1992. In situ carbon dioxide levels in cushion and turf forms of *Grimmia antarctici* at Casey Station, east Antarctica. Journal of Bryology 17: 241–249.

Tuerk, R., T.G.A. Green, L.G. Sancho, and R. Seppelt, 2004. Unexpectedly high number of lichen species at 84° S, Southern Victoria Land: Flattening the biodiversity cline. SCAR Open Science Conference, Bremen, Germany, 25–31 July.

Vestal, J.R., 1988. Primary production of cryptoendolithic microbiota from the Antarctic desert. Polarforschung 58: 193–198.

Vincent, W.F., 1988. Microbial systems of Antarctica. Cambridge University Press, Cambridge.

Walker, F.J., 1985. The lichen genus *Usnea* subgenus *Neuropogon*. Bulletin of the British Museum Botany Series 13: 1–130.

Wasley, J., S.A. Robinson, C.E. Lovelock, and M. Popp, 2006. Climate change manipulations show Antarctic flora is more strongly affected by elevated nutrients than water. Global Change Biology, 12: 1800–1812.

Waterhouse, E.J., ed., 2001. Ross Sea Region: A state of the environment report for the Ross Sea Region of Antarctica. New Zealand Antarctic Institute, Christchurch.

Wierzchos, J. and C. Ascaso, 2001. Life, decay and fossilisation of endolithic microorganisms from the Ross Desert, Antarctica. Polar Biology 24: 863–868.

Wierzchos, J. and C. Ascaso, 2002. Microbial fossil record of rocks from the Ross Desert, Antarctica: implications in the search for past life on Mars. International Journal of Astrobiology 1: 51–59.

Wierzchos, J., C. Ascaso, L. Garcia Sancho, and A. Green, 2003. Iron-rich diagenetic minerals are biomarkers of microbial activity in Antarctic rocks. Geomicrobiology Journal 20: 15–24.

Wierzchos, J., L. Garcia-Sancho, and C. Ascaso, 2005. Biomineralization of endolithic microbes in rocks from the McMurdo Dry Valleys of Antarctica: Implications for microbial fossil formation and their detection. Environmental Microbiology 7: 566–575.

Wierzchos, J., C. Ascaso, F.J. Ager, I. García-Orellana, A. Carmona-Luque, and M.Á. Respaldiza, 2006. Identifying elements in rocks from the Dry Valleys desert (Antarctica) by ion beam proton induced X-ray emission. Nuclear Instruments and Methods in Physics Research B 249: 571–574.

Wilson, M.E., 1990. Morphology and photosynthetic physiology of *Grimmia antarctici* from wet and dry habitats. Polar Biology 10: 337–341.

Wise, K.A.J. and J.L. Gressit, 1965. Far southern animals and plants. Nature 207: 101–102.

14 Facilitation in Plant Communities

Ragan M. Callaway and Francisco I. Pugnaire

CONTENTS

INTRODUCTION

Until the last decade, the importance of positive relationships among plant species have been underestimated in community ecology, perhaps based on broader perspectives on the individualistic nature of communities (Callaway 1997). However, rapidly accumulating evidence for common and strong facilitative effects in communities (Callaway 1995, Bruno et al. 2003) have led to reconsideration of formal community theory and the rejection of strict individualistic theory by some ecologists (Lortie et al. 2004).

Traditional models for plant interactions have emphasized competition, assuming that plants must compete for always-limiting resources, such as water, nutrients, light, space, or pollinators, and try to overthrow each other when in proximity. However, a large amount of experimental evidence for interspecific positive interactions has accumulated, suggesting that old conceptual models for the forces that structure communities are inadequate. The fact that different species of plants may benefit each other when growing together has attracted the interest of ecologists, and the number of papers reporting, or somehow involved with, positive interactions has grown considerably in the recent years.

Neighboring plant species may compete with one another for resources, but they may also provide benefits for neighbors such as shade, higher nutrient levels, more available moisture, soil oxygenation, protection from herbivores, a more favorable soil microflora, shared resources via mycorrhizae, and increased pollinator visits (Callaway 1995), resulting in interactions that can be cumulatively positive for at least one of the species involved. Positive interactions, or facilitation, occur when one plant species enhances the survival, growth, or fitness of another, and have been demonstrated in virtually all biome types (DeAngelis et al. 1986, Hunter and Aarssen 1988, Bertness and Callaway 1994, Callaway 1995, Callaway and Walker 1997, Tirado and Pugnaire 2005) though they have been regarded in the past as interesting anecdotal curiosities rather than principles fundamentally important to communities (Bertness and Callaway 1994).

Facilitation affects plant community structure and diversity in very different ways than competition. Competitive interactions limit coexistence among species, and competition-based theory focuses on how species avoid competitive exclusion (Lotka 1932, Gause 1934, Hardin 1990, Hutchinson 1961). Coexistence in a world dominated by competition has been attributed to niche partitioning (Parrish and Bazzaz 1976, Cody 1986), variation in the physical environment and subtle differences in competitive advantages, disturbance that continuously provides competition-free microhabitat and alters competitive hierarchies (McNaughton 1985), heterogeneity in the ratios of limiting resources that alter competitive hierarchies (Tilman 1976, 1985, 1988), the development of local and species-specific resource depletion zones that, under certain conditions, do not strongly affect the resources available to neighbors (Huston and DeAngelis 1994, Grace 1995), and spatial structures that suggest niche facilitation (Van der Maarel et al. 1995). In contrast to the suite of theories that attempt to explain species coexistence despite competition for the few resources that are shared by all plants, positive interactions may directly promote coexistence and increase community diversity (Kikvidze et al. 2005, Michalet et al. 2006).

MECHANISMS

SHADE

The benefits of shade include maintenance of plant tissues below lethal or near-lethal temperatures, decreasing respiration costs, lowering transpirational demands by decreasing the vapor pressure difference between leaves and air, reduction of ultraviolet irradiation, and increased soil moisture due to lower evaporative demand. These benefits may, but do not have to, come at the cost of reduced energy for photosynthesis. For the cost of shade to exceed its benefit, shade must be below the photosynthetic saturation point of the beneficiary species. If not, the beneficiary can get a free lunch because it cannot use any photosynthetically active radiation (PAR) above the saturation point anyway. Many species reach their maximum photosynthetic rates at PAR levels far below the general natural maximum ($\approx$2000 μmol m^{-2} s^{-1}). These species may benefit from the effects of taller neighbors on leaf temperature and transpirational loss without any cost of decreased carbon gain. For example, *Arnica cordifolia* is a perennial herb that is commonly found in the understory of conifer forests in the northern Rocky Mountains. Young and Smith (1983) found that a 30% decrease in light on the forest floor during cloudy days in the Medicine Bow Mountains of Wyoming resulted in a 37% increase in carbon gain for *Arnica* and an 84% reduction in transpiration. They found that the photosynthetic rates of *Arnica* remained near saturation even on cloudy days because of its low PAR saturation point, thus *Arnica* gained from the lower transpiration rates associated with decreased light levels without a carbon cost.

Milena Holmgren and colleagues (1997) took these potential cost-benefit trade-offs into account by modeling the relative importance of shade from a nurse plant as a facilitative effect or a competitive effect depending on the synergy between shade and moisture in a plant's environment. Assuming that plant growth is a relatively simple product of light and moisture, and that soil moisture is affected by canopies to a greater degree in xeric environments than in mesic environments, they argued that plant growth increases as light increases. However, in xeric environments increasing light is correlated with rapidly decreasing soil moisture, and plant growth decreases at higher light intensities. This sets the stage for shade-derived facilitation in water-limited environments and decreased likelihood for such facilitation in mesic environments.

Indeed, the importance of shade has been emphasized in deserts where many species of cacti and other succulents may depend on nurse plants when they are young because seedlings have low surface-to-volume ratios and dissipate heat inefficiently, and exposure of these seedlings to exceptionally hot temperatures on the desert floor may be fatal (Figure 14.1). For example, saguaro seedlings are sheltered and facilitated by many different species of perennial plants, but predominantly by *Cercidium microphyllum* (Niering et al. 1963, Turner et al. 1966, 1969, Steenberg and Lowe 1969, 1977). Young *Ferrocactus acanthodes* cacti may experience 11°C decrease in maximum stem surface temperatures in the shade of nurse plants (Nobel 1984). Valiente-Banuet and Ezcurra (1991) compared the relative importance of protection from predation and shade in the nurse plant relationship between *Neobuxbaumia tetetzo*, a columnar cactus, and *Mimosa luisana* in the Viscaino Desert and the Gran Desierto de Altar in Mexico. They found that cages improved survival, but that long-term survival was restricted to shade treatments. In a perennial community in the Chihuahua Desert, Silvertown and Wilson (1994) found more positive associations between species than expected for a random assemblage, and concluded that community structure relied on species whose establishment was dependent on conditions provided by other species. This is also evident in a wide variety of communities, from deserts to tropical savannas to high mountains (Tirado and Pugnaire 2005).

Facilitative shade effects are not restricted to deserts. For example, herbaceous species in California oak savannas also show strong preferences to either understory or open microhabitats. Parker and Muller (1982) demonstrated higher shade tolerance of *Bromus diandrus* and *Pholistoma auritum*, two species apparently facilitated by *Quercus agrifolia* (an evergreen oak), than that of *Avena fatua*, a species common in the open. In laboratory experiments, *B. diandrus* had a higher relative growth rate at low light than *A. fatua* (Mahall et al. 1981). Marañón and Bartolome (1993) conducted experiments in which whole soil blocks were

FIGURE 14.1 Young saguaro cactus under the canopy of a paloverde tree in the Sonoran Desert, USA. Much older saguaros can be seen in the open, beyond the tree canopy. (Photo by Alfonso Valiente-Banuet.)

transplanted between the understory of coast live oak, adjacent open grassland, and artificial shade treatments. They found that herbaceous species typically found in open habitats in California oak savannas were limited by canopy shade. In some cases, the overstory oaks themselves require shade from nurse plants when they are young. Callaway (1992) tested the facilitative effects of shade and seedling predation on the survival of *Quercus douglasii* (a winter-deciduous oak) seedlings using unshaded and shaded cages. Unshaded cages reduced mortality rates, but eventually all unshaded seedlings died, whereas 35% of shaded and caged seedlings survived for 1 year.

Gradients of both radiation reaching the soil and soil temperature provided by canopies may allow different species to coexist in the more heterogeneous understory environment. Moro et al. (1997) demonstrated how several herbaceous species positioned themselves differently under the canopy of *Retama sphaerocarpa*, a leguminous shrub, in response to several gradients including irradiance and temperature. Radiation at soil level in a central position under the canopy was 60% of that outside, and temperature difference reached 7°C between both the positions.

Shade may also have indirect facilitative effects. Shade provided by salt-tolerant species in salt marshes appears to reduce evaporation from subcanopy soils and consequently maintain lower soil salinities than in those soils exposed to direct insolation (Bertness and Hacker 1994, Callaway 1994). Bertness (1991) and Bertness and Shumway (1993) experimentally manipulated soil salinity in the upper zones of a New England salt marsh and found that *Distichlis spicata*, a salt-tolerant colonizer of saline bare patches, facilitated the growth of *Juncus gerardi* when soil salinities were high. When soil salinities were artificially reduced by watering, growth of *Juncus* was not improved by *Distichlis*. Similarly, Bertness and Hacker (1994), Bertness and Yeh (1994), and Hacker and Bertness (1996) found that shade provided by marsh elder shrubs (*Iva frutescens*) decreased soil salinity and facilitated the growth and survival of *J. gerardi*.

Soil Moisture

The water relations of one species may be enhanced by the presence of another species. Joffre and Rambal (1993) measured a significant delay in soil water loss under *Quercus rotundifolia* and *Quercus suber* in Spanish savannas relative to soils in open areas between the trees. They argued that increased soil moisture was accountable for large differences in species composition under trees and in the open. Generally, moisture conditions in the understory are enhanced by overstory plants (Petranka and McPherson 1979, Maranga 1984, Vetaas 1992), but sometimes the reverse is true. Pugnaire et al. (1996a) found that *R. sphaerocarpa* shrubs in southern Spain had higher water potentials when *Marrubium vulgare* occupied the understory than that when the understory was bare ground. This effect was observed in spring but not in summer, suggesting that the presence of a dense understory helped to retain rainfall water in the soil mound that accumulated under the shrub.

In North American shrub steppe, *Artemisia tridentata* (Great Basin sage) transports water from deep, moist soils to dry surface soils during the night via hydraulic lift (Richards and Caldwell 1987). Additional experiments using stable isotope analysis indicated that water from hydraulic lift was distributed to neighboring plants, although the magnitude of water transferred was small Caldwell (1990) and Dawson (1993), however, used similar techniques to investigate the magnitude of hydraulic lift conducted by *Acer saccharum* and the effects of the hydraulically lifted water on understory plants. He found hydraulically lifted water in all understory plants he examined, with the proportional use of hydraulically lifted water ranging from 3% to 60%. Plants that used large proportions of hydraulically lifted water had more favorable water relations and growth than those that did not.

Soil Nutrients

Soils directly beneath the canopies of perennials are often richer in nutrients than those in surrounding open spaces without perennial cover. Subcanopy soil enrichment may occur as a result of nutrient pumping (sensu Richards and Caldwell 1987), as deep-rooted perennials take up nutrients unavailable to shallowly rooted plants and deposit them on the soil surface via litterfall and throughfall. In addition, perennial canopies may trap airborne particles that are eventually deposited at the base of the plant (Pugnaire et al. 1996b, Whitford et al. 1997). Nutrient enrichment may also occur indirectly via nitrogen fixation.

Enrichment of soil nutrients (and often a corresponding change in species composition and increased productivity) has been reported under perennials in many systems throughout the world, but these effects have been described most often in savannas and other semiarid regions with clearly demarked understory and open microhabitat (Patten 1978, Schmida and Whittaker 1981, Yavitt and Smith 1983, Weltzin and Coughenhour 1990, Callaway et al. 1991, Vetaas 1992, Belsky 1994). Nutrient enrichment has also been reported in more mesic systems including alder shrublands (Goldman 1961), Patagonian shrublands (Rostagno et al. 1991), and forest-pasture ecotonehs of Central America (Kellman 1985).

Pugnaire et al. (1996a) reported a facilitative mutualism between the shrub *R. sphaerocarpa* and the herb *M. vulgare* based on improvements in soil nutrients and water. *Marrubium* plants under *Retama* had greater leaf-specific area, leaf mass, shoot mass, leaf area, flowers, higher leaf nitrogen concentration, and more nitrogen per plant than those that were not near a *Retama* shrub. Conversely, *Retama* shrubs with *Marrubium* underneath them had larger cladodes, greater total biomass, nitrogen content, and higher shoot water potentials than those without *Marrubium* in the understory, suggesting that facilitation could be bidirectional rather than unidirectional.

Experimental studies of the relationship between soil enrichment via litterfall and enhanced growth of plants or species shifts are less common. Turner et al. (1966) found that saguaro seedlings survived better on soil collected from under paloverde trees (*C. microphyllum*) than on those from under either mesquite (*Prosopis juliflora*) or ironwood (*Olneya tesota*). Monk and Gabrielson (1985) found that addition of litter from the floor of deciduous forests in South Carolina generally improved the overall productivity in understory plots, but the effects of litter on individual species ranged from facilitation to interference. Soils under *Q. douglasii* canopies are much more nutrient rich than those in the surrounding open grassland (Callaway et al. 1991), and improve the growth of *B. diandrus*, an annual grass in the understory, in greenhouse tests when compared with open grassland soils. Walker and Chapin (1986) demonstrated facilitative effects of nitrogen-rich soil from under *Alnus tenuifolia* on *Salix alaxensis* and *Populus balsamifera* seedlings. In a second experiment in Glacier Bay, Alaska, they found that *Alnus sinuata* and *Dryas drummondii*, both nitrogen fixers, had strong positive effects on the growth of *Picea sitchensis* (Chapin et al. 1994).

Moro et al. (1997) showed that the interaction between *Retama* and its understory vegetation was strongly affected by gradients of litter accumulation and decomposition, both of which influenced species composition. The abundance of annual herbs produced a more favorable habitat for soil micro-organisms that increased mineralization rate, enhanced litter decomposition, and increased nutrient dynamics under *Retama* shrubs. Litter has both positive and negative effects on plant growth, as has been shown also elsewhere (Facelli 1994, Hoffmann 1996).

Soil Oxygenation

Wetland emergent plants often passively transport oxygen from leaves to roots through aerenchymous tissues to alleviate belowground oxygen limitation (Armstrong 1979). In some cases, oxygen appears to leak out of submerged roots and oxidize toxic substances and nutrients in the rhizosphere and oxygenate marsh sediments (Howes et al. 1981, Armstrong et al. 1992). Castellanos et al. (1994) reported that *Spartina maritima* aerates surface sediments in wetlands of southern Spain, creating conditions favorable for the invasion of *Arthrocnemum perenne*. In eastern salt marshes in North America, Hacker and Bertness (1996) found that the aerenchymous *Juncus maritimus* increased the redox potential in its rhizosphere, which corresponded with increased growth of *I. frutescens*, a woody-stemmed perennial, and the extension of *Iva*'s distribution to lower elevations in the marsh.

Plantago coronopus and *Samolus valerandi* are clumped with tussocks of the aerenchymous *J. maritimus* in dune slacks on the coast of Holland, where survival rates appeared to be enhanced by soil oxygenation and oxidation of iron, manganese, and sulfide (Schat and Van Beckhoven 1991). When *P. coronopus* and *Centaurium littorale* were grown with *J. maritimus* in the greenhouse, growth and nutrient uptake were improved (Schat 1984).

Substrate oxygenation and facilitation may also occur in freshwater marshes. In greenhouse experiments, undrained pots containing *Typha latifolia* (cattail) had dissolved oxygen contents over four times greater than those without *Typha*, and other marsh plants grown with *Typha* survived longer and grew larger than those in pots without *Typha* when pot substrates were kept between 11°C and 12°C (Callaway and King 1996). In the field, *Myosotis laxa* (herbaceous perennials) growing next to transplanted *Typha* were larger and produced more fruits that those isolated from *Typha*.

Protection from Herbivores

Positive interactions may be indirectly mediated through herbivores. *Themeda triandra*, an East African savanna grass, is preferred by many herbivores, and suffers ≈80% mortality from grazers when not associated with less-palatable grass species (McNaughton 1978). As codominance with unpalatable species increases, mortality of *Themeda* rapidly decreases. Similar examples of associational defenses have been reported by others (Attsat and O'Dowd 1976).

In the Sonoran Desert, paloverde seedlings are protected by various shrub species (McAuliffe 1986). Ninety-two percent of naturally occurring *C. microphyllum* in the open was eaten by herbivores; but only 14% of the seedlings that were touching shrubs were consumed. McAuliffe (1984a) also found that young *Mammillaria microcarpa* and *Echinocereus englemannii* cacti were much more common under live and dead *Opuntia fulgida* where spine-covered stem joints from the nurse plant protected them from herbivores.

In south-east Spain, unpalatable *Artemisia barrelieri* shrubs facilitate seed germination and seedling establishment of more palatable *Anthyllis cytisoides* shrubs in addition to providing shelter from herbivory during early stages of growth, and before competitive displacement by *Artemisia* (Haase et al. 1997).

In northern Sweden, *Betula pubescens* experiences high herbivory when associated with plants of higher palatability, *Sorbus aucuparia* or *Populus tremuloides*, but low herbivory when associated with plants of lower palatability, *Alnus incana* (Hjalten et al. 1993). Hay (1986) found that several species of highly palatable seaweeds were protected from grazers in safe microsites provided by unpalatable seaweed species.

As noted earlier, the positive effect of coastal scrub shrubs on *Q. douglasii* seedling recruitment (Callaway 1992) was caused in part by shade from the shrub canopies; however, analysis of the mortality of individual seedlings showed that the causes and the timing of mortality differed significantly under shrubs and in the open grassland. Predation on acorns appeared to be due to rodents and was much higher under shrubs than in the grassland only 1 m away. Emergent shoots, however, were eaten by deer and experienced much higher predation in the grassland than under shrubs. Consequently, blue oak acorns that survived early rodent predation under shrubs became seedlings that occupied sites relatively free from deer predation on shoots. Similarly, seed and seedling predation appears to be a major factor limiting *Pinus ponderosa* to shrub-free, hydrothermally altered soils in the Great Basin (Callaway et al. 1996).

In most cases referred to earlier, benefactor species appear to physically shelter or hide beneficiary species from herbivores (see Chapter 15). A similar phenomenon, called associational resistance by Root (1972) and associational plant refuges by Pfister and Hay (1988), may occur when some species experience less herbivory as a function of the visual or olfactory complexity of the surrounding vegetation. A number of ecologists have found that community complexity serves as an impediment to search efficiency, in contrast to physical protection from herbivores or close association with unpalatable species (Root 1972, Risch 1981).

Pollination

Plants that are highly attractive to pollinators may facilitate their less-attractive neighbors by enticing insects into the vicinity. Thomson (1978) found that *Hieracium florentinum* received more visits from pollinators when it was mixed with *Hieracium auranticum* than when alone. Moeller (2004) found that populations occurring with multiple congeners had higher pollinator availability and lower pollen limitation than those occurring alone. Similarly, Ghazoul (2006) found that pollinator visits to *Raphanus raphanistrum*, a self-incompatible herbaceous plant, increased when it occurred with one or a combination of *Cirsium arvense, Hypericum perforatum*, and *Solidago canadensis* than when it occurred alone. In the understory of deciduous forests in Ontario, Laverty and Plowright (1988) recorded higher fruit and seed set in *Podophyllum peltatum* (mayapples) that were associated with *Pedicularis canadensis* (lousewort) than those that were not near *Pedicularis* plants. In later studies, Laverty (1992) found that higher fruit and seed set in mayapple, which produces no nectar, depends on infrequent visits from queen bumble bees that accidentally encounter mayapple while collecting nectar from *Pedicularis*, the magnet species.

Johnson et al. (2003) experimentally evaluated the potential for species that do not attract large numbers of pollinators to benefit from neighbors using the nonrewarding bumblebee-pollinated orchid, *Anacamptis morio*, and associated nectar-producing plants in Sweden. Pollen receipt and pollen removal for *A. morio* was significantly greater for individuals translocated to patches of nectar-producing plants (*Geum rivale* and *Allium choenoprasum*) than for those placed outside (>20 m away) patches. Their results provide strong support for the existence of facilitative magnet species effects. Geer et al. (1995) have shown that three co-occurring species of *Astragalus* facilitate each other's visitation rate of pollinators rather than competing for them.

There may be even facilitation to attract pollinators between a rust fungus (*Puccinia monoica*) and *Anemone patens* (Ranunculaceae). The presence of rust's pseudoflowers may increase visitation rate to *Anemone*, though the positive effect could be counterbalanced because sticky pseudoflowers remove pollen from visiting insects and fungal spermatia deposited on flower stigmas reduce seed set (Roy 1996).

Mycorrhizae and Root Grafts

Woody plants may form intraspecific and interspecific root grafts in which resources and photosynthate move among individuals (Bormann and Graham 1959, Graham 1959, Bormann 1962). Bark girdling of one *Pinus strobus* sapling grafted to a conspecific neighbor resulted in a significant drop in the diameter growth of the ungirdled member (Bormann 1966). He reported that the intact tree supported root growth in the girdled tree for a period of 3 years. Bark-girdling of nongrafted white pines resulted in death within a year, but grafted trees remained alive for 2 or more years after girdling. Bormann argued that the development of naturally occurring white pine stands is shaped by both competition and "a noncompetitive force governed by inter-tree food translocation." However, because both independently acquired and shared resources would require intact transport tissues, it is not clear how grafting overcame the effect of girdling.

Mycorrhizal fungi also facilitate the exchange of carbon and nutrients (Chiarello et al. 1982, Francis and Read 1984, Walter et al. 1996). Grime et al. (1987) found that labeled $^{14}CO_2$ was transferred from *Festuca ovina* to many other plant species in artificial microcosms that shared a common mycorrhizal network, but not to others that did not share the network. The presence of mycorrhizae led to decreased biomass of the dominant species, *F. ovina*, and increased biomass of otherwise competitively inferior species, including *Centaurea nigra*, and ultimately experimental microcosms that were infected with mycorrhizae were more diverse than those that were not infected. Marler et al. (1999) studied the role of mycorrhizae on interactions between *Festuca idahoensis* and *Centaurea maculosa*, a major invasive weed in North America. They found that mycorrhizae mediated strong positive effects of *Festuca* on *Centaurea*. There were no direct effects of the mycorrhizae on the growth of either species, but when *Centaurea* was grown with large *Festuca* in the presence of mycorrhizae, they were 66% larger than in the absence of mycorrhizae. The fact that mycorrhizae mediated much stronger growth responses from large *Festuca* than small *Festuca* suggests that mycorrhizae mediated parasite-like interactions, such as those that occur between myco-heterotrophic parasites (Leake 1993).

Mycorrhizal networks connecting roots of neighboring plants transfer nutrients from one plant to another, in such a way that may counterbalance the negative effects of competition, as in *P. strobus* on seedlings recruiting underneath them (Booth 2004). However, beneficial influence of trees acting as a source of ectomycorrhizal infections to seedlings depend on the distance to the tree, because growth is maximized at intermediate distances from the trunk (Dickie et al. 2005).

INTERACTIONS BETWEEN FACILITATION AND INTERFERENCE

In the central Rocky Mountains of the USA, Ellison and Houston (1958) showed that herbaceous species in the understory of *P. tremuloides* (quaking aspen) were stunted unless the tree roots were excluded. However, after trenching, the growth of understory species exceeded that of the surrounding open areas. Facilitative mechanisms of shade or nutrient inputs appeared to have the potential to facilitate understory species, but their effects were outweighed by root interference. More recently, a large number of studies have demonstrated that positive and negative interactions operate at the same time. Walker and Chapin (1986) demonstrated that *A. tenuifolia* litter had the potential to facilitate *S. alaxensis* and *P. balsamifera* in greenhouse experiments and in the field (see earlier). However, under natural conditions, *S. alaxensis*, *P. balsamifera*, and *Picea glauca* grew poorer in *A. tenuifolia* stands than in other vegetation. In other experiments, they found that root interference and shading in alder stands was more influential on the other species than nutrient addition via litter, and overrode the effects of facilitation. In a similar study in Glacier Bay, Chapin et al. (1994) found the reverse: that early to mid-successional species, such as *A. sinuata*, affected the late-successional *P. sitchensis* in positive ways (nutrient uptake and growth) and in negative ways (germination and survivorship). *P. sitchensis* seedlings that were planted in *A. sinuata* stands accumulated over twice the biomass and acquired significantly higher leaf concentrations of nitrogen and phosphorus than seedlings planted in *P. sitchensis* forests. However, as found at the other site, root trenching in *A. sinuata* stands further increased growth and nutrient acquisition, demonstrating that competitive and facilitative mechanisms were operating simultaneously. In contrast to the interior floodplain, however, the facilitative effects of *Alnus* on its neighbors overrode root interference in natural conditions.

In the Patagonian steppe, Aguiar et al. (1992) found that various shrub species protected grasses from wind and desiccation, but strong facilitative effects were only expressed when root competition was experimentally reduced. In the same system, Aguiar and Sala (1994) found that young shrubs had stronger facilitative effects on grasses, but the positive effects declined as grass densities increased near the shrubs.

During primary succession on volcanic substrates, Morris and Wood (1989) measured both negative and positive effects of *Lupinus lepidus*, the earliest colonist, on species that arrived later the successional process, but concluded that the net effect of *Lupinus* was facilitative. On the island of Hawaii, the exotic tree *Myrica faya* is a successful invader on volcanic soils where it is replacing the native tree, *Metrosideros polymorpha* (Whiteaker and Gardener 1985). Walker and Vitousek (1991) found that direct effects of the invading *Myrica* on the native *Metrosideros* were both facilitative and interfering. *Myrica* enriched the nitrogen content of soils and improved *Metrosideros* growth in greenhouse experiments, and shade from *Myrica* improved *Metrosideros* seedling germination and survival. However, *Metrosideros* germination was sharply decreased by litter from *Myrica*, and the growth of young *Metrosideros* did not improve in the shade of *Myrica* in the field. They concluded that the lack of regeneration of *Metrosideros* under the canopies of *Myrica* in the field was the result of these negative mechanisms dominating the positive mechanisms.

Q. douglasii deposits large amounts of nutrients to the soil beneath their canopies and soil, and litter bioassays demonstrate strong facilitative effects of these components on the growth of a dominant understory grass, *B. diandrus* (Callaway et al. 1991). In the field, however, the expression of this facilitative mechanism is determined by the root architecture of individual trees. *Q. douglasii* with low fine root biomass in the upper soil horizons and those that appeared to have roots that reached the water table (much higher predawn water potentials) elicited strong positive effects on understory biomass. In contrast, trees with high fine root biomass in the upper soil horizons and those that did not appear to root at the water table (much lower water potentials) elicited strong negative effects on understory productivity. In field experiments, they demonstrated that soil from beneath *Q. douglasii*, regardless of root architecture, had strong positive effects; but that root exclosures only improved understory growth under trees with dense shallow roots. Thus in this ecosystem, as in the Alaskan floodplain studied by Walker and Chapin (1986), root interference, when present, outweighed the positive effects of nutrient addition.

Palatable intertidal seaweed species that depend on less-palatable species for protection (Hay 1986, see earlier) are poor competitors with their benefactors. Hay found that, in the absence of herbivores, several highly palatable seaweed species grew 14%–19% less when in mixtures with their benefactors than when alone. In the presence of herbivores, however, palatable species survived only when mixed with competitively superior, but unpalatable species. In this system, the effects of competition with neighbors were outweighed by the protection they provided.

The balance of facilitation and interference may change with the lifestages of the interacting plants. Patterns of nurse plant mortality observed in numerous systems indicate that species may begin their lives as the beneficiaries of nurse plants and later become significant competitors with their former benefactors as they mature. McAuliffe (1988) found that young *Larrea tridentata* plants were positively associated with dead *Ambrosia dumosa*, a species critical to the initial establishment of *Larrea*. Similarly, mature saguaros were associated disproportionally with dead paloverde trees that commonly function as nurse plants to seedling saguaros (McAuliffe 1984b). However, in both of these cases, young *Larrea* may do better in the shade of *Ambrosia* that are already dead because the positive effects come without any competitive cost. In the Tehuacan Valley of Mexico, *N. tetetzo* that is nursed by *M. luisana* (Valiente-Banuet et al. 1991) eventually suppresses the growth and reproduction of its benefactor (Flores-Martinez et al. 1994). The same occurs with *Opuntia rastrera* (Silvertown and

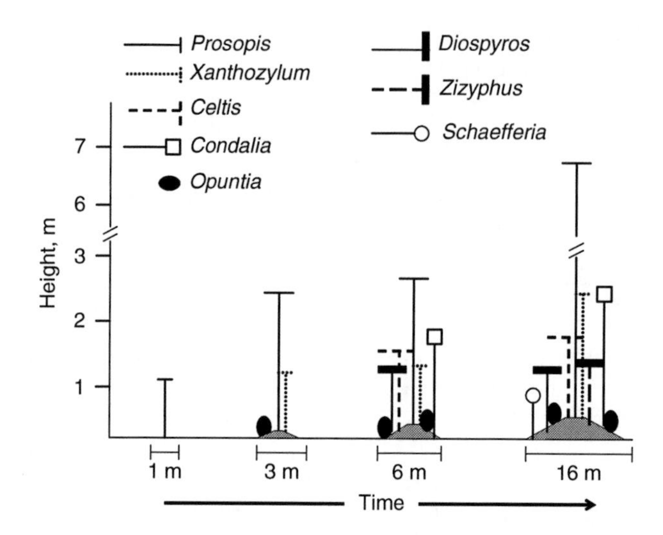

FIGURE 14.2 Illustration of the development of shrub clusters facilitated by *Prosopis glandulosa* in Texas grasslands. (Redrawn from Archer, S., Scifres, C., Bassham, C.R., and Maggio, R., *Ecol. Monogr.*, 58, 111, 1988. With permission.)

Wilson 1994). Archer et al. (1988) found that *Prosopis glandulosa* trees provide a focal point for the regeneration of many other species in southcentral Texas (Figure 14.2), but *Prosopis* regenerated very poorly in these clusters once they were established. In dry systems in south east Spain, Armas and Pugnaire (2005) found that the short-term balance of the interaction between *Cistus clusii* shrubs and *Stipa tenacissima* grasses shifted easily in response to environmental variability and had important consequences for the structure of the plant community.

Other studies have shown that a particular benefactor species may have facilitative effects on some species, but competitive effects on other similar species. *A. tridentata* dominates large regions of the Great Basin of Nevada, USA, but is completely absent from outcrops of phosphorus-poor hydrothermally altered andesite where *P. ponderosa* is abundant (DeLucia et al. 1988, 1989). Field experiments demonstrated that *A. tridentata* outcompetes *P. ponderosa* for water off of the altered andesite refuges, and prevents *P. ponderosa* from expanding onto typical soils of the Great Basin. Another pine species, *Pinus monophylla*, is often common in mixtures with *A. tridentata,* but found only as seedlings on altered andesite (Callaway et al. 1996). Experiments indicated that *P. monophylla* is not prevented from maturing on altered andesite by soil conditions, but by the absence of *A. tridentata* that acts as a nurse plant for *P. monophylla* (DeLucia et al. 1989, Callaway et al. 1996). Both species of pines have similar leaf-level physiological characteristics and nutrient requirements.

In the upper zones of southern California salt marshes, the perennial subshrub *Arthrocnemum subterminale* exists in a matrix of winter ephemeral species that emerge when soil salinity drops at the beginning of the rainy season. *Arthrocnemum* has strong facilitative effects on some codominant ephemerals and they tend to occur in its understory (Callaway 1994). Another, othewise similar annual is competitively reduced by *Arthrocnemum* and is more common in the open between the shrubs.

Bertness and Callaway (1994) hypothesized that the balance between facilitation and interference is affected by the harshness of physical conditions, and that the importance of facilitation in plant communities increases with increasing abiotic stress or increasing consumer pressure. Alternatively, they hypothesized that the importance of competition in communities would increase when physical stress and consumer pressure were relatively low because neighbors buffer one another from extremes of the abiotic environment (e.g., temperature or salinity) and herbivory.

In support of the abiotic stress hypothesis, Bertness and Yeh (1994) found that the effects of *I. frutescens* shrubs on conspecific seedlings were positive because soil salinity was moderated by the shade of the benefactors in a New England salt marsh. When patches were watered, however, strong competitive interactions developed between adults and seedlings, and among seedlings. Interactions between *I. frutescens* plants were dependent on environmental conditions, competing more intensely under mild abiotic conditions and facilitating each other when abiotic conditions were harsh. Bertness and Shumway (1993) also eliminated the facilitative effects of *D. spicata* and *Spartina patens* on *J. gerardi* by watering experimental plots and reducing soil salinity. In the same marsh, the fitness of *Iva* shrubs associated with *J. gerardi* was enhanced by neighbors at lower elevations but strongly suppressed by the same species at higher elevations where soil salinity was lower (Bertness and Hacker 1994).

The relationship between abiotic stress and shifts in competition and facilitation has been shown in other systems. Pugnaire and Luque (2001) found that facilitation effects decreased quickly from strongly arid to more mesic sites, whereas competition remained constant but changed from underground in the dry site to aboveground in the more mesic site. In intermountain grasslands of the northern Rockies, *Lesquerella carinata* is positively associated with bunchgrass species in relatively xeric areas, but negatively associated with the same species in relatively mesic locations (Greenlee and Callaway 1996). They studied these spatial patterns experimentally and documented shifts between interference and facilitation between years at a xeric site. Bunchgrasses interfered with *L. carinata*, in a wet and cool (low-stress) year, but facilitated *Lesquerella* in a dry and hot year (Figure 14.3). Other, nonexperimental, studies indicate that competitive effects are stronger in wet, cool years and facilitative in dry, hot years (Fuentes et al. 1984, De Jong and Klinkhamer 1988, Frost and McDougald 1989, McClaran and Bartolme 1989, Belsky 1994). In contrast to these studies, Casper (1996) examined survival, growth, and flowering of *Cryptantha flava* in experiments explicitly designed to test for shifts in positive and negative interactions with neighboring shrubs and found no evidence that competition or facilitation changed as soil water varied between years.

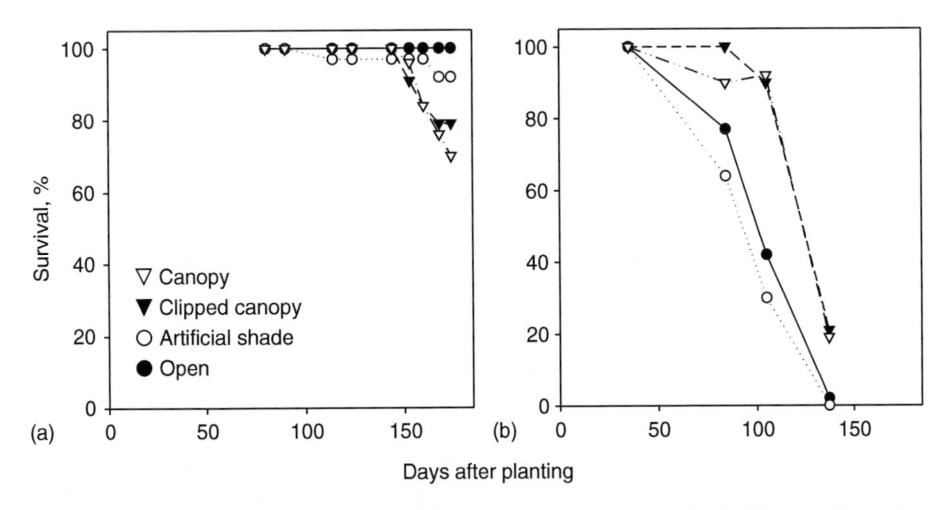

FIGURE 14.3 Effects of bunchgrasses and artificial canopies on the survival of *Lesquerlla carinata* in a wet year (a) and in a dry year (b) in western Montana. (From Greenlee, J. and Callaway, R.M., *Am. Nat.*, 148, 386, 1996. With permission.)

Strong shifts between competition and facilitation with stress have been shown on alpine elevational gradients (Callaway et al. 2002). They found that competition generally, but not exclusively, dominated interactions at lower elevations where conditions are less physically stressful. In contrast, at high elevations where abiotic stress is high the interactions among plants were predominantly positive (Figure 14.4).

Shifts in interspecific interactions have also been found to occur at different temperatures in anaerobic substrates. *M. laxa* appeared to benefit from soil oxygenation when grown with *T. latifolia* at low soil temperatures (Callaway and King 1996). But at higher soil temperatures, *Typha* had no effect on soil oxygen (presumably due to increased microbial and root respiration), and the interaction between *Typha* and *Myosotis* became competitive.

Changes in interspecific interactions may occur in a different sequence. Haase et al. (1996) have shown that coexisting *A. barrelieri* and *R. sphaerocarpa* shrubs compete during early stages of succession in abandoned semiarid fields in south-east Spain, but after a time period during which *Artemisia* is competitively displaced, this subshrub is found preferentially under the canopy of larger *Retama* shrubs. The pattern of the interaction suggested that facilitation prevailed over competition because of niche separation that developed over time.

Complex combinations of negative and positive interactions operating simultaneously between species appear to be widespread in nature. Such concomitant interactions in which current conceptual models of interplant interactions are based on resource competition alone may not accurately depict processes in natural plant communities.

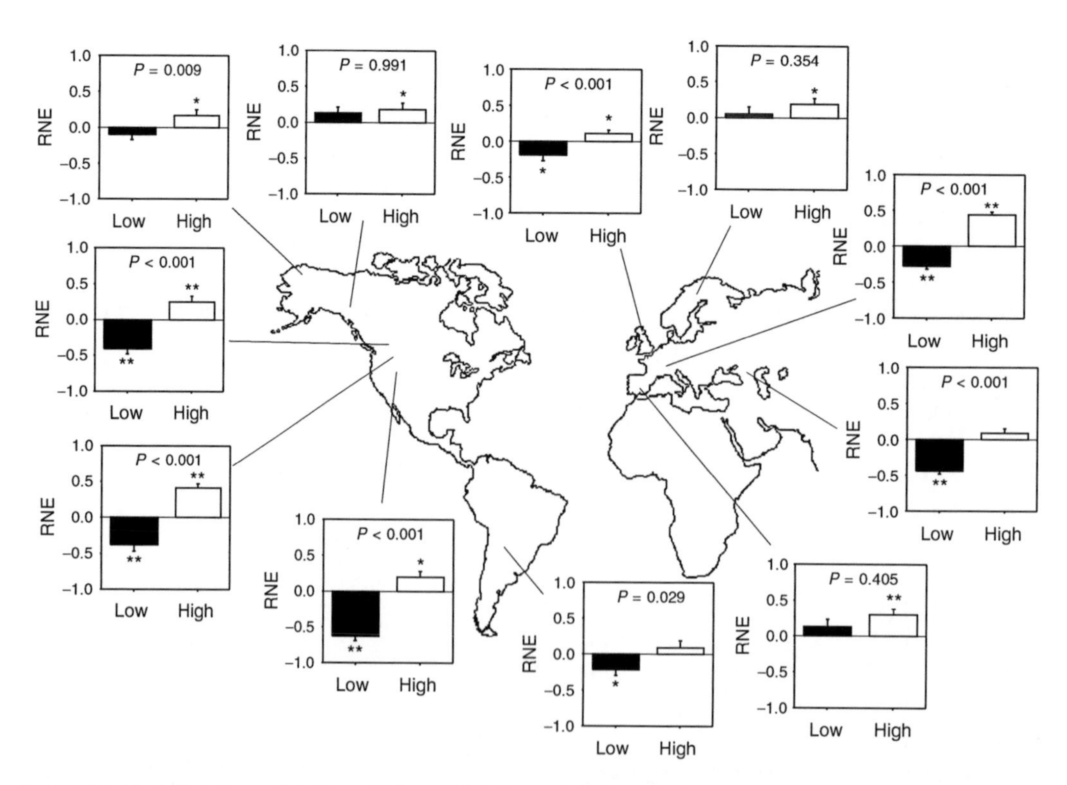

FIGURE 14.4 Relative Neighbor Effect (RNE) at the 11 experimental sites. Error bars represent one standard error, P values denote significance of differences between the two sites (ANOVA with site and species as main effects), and asterisks denote a site effect that was significantly different from zero ($P > 0.01$). (Reprinted from Callaway, R.W., Brooker, C.P.Z., Kikvidze, C.J., Lortie, R., Michalet, L., Paolini, F.I., Pugnaire, B.J., Cook, E.T., Aschehong, C. Armas, and Newingham, B., *Nature*, 417, 844, 2002. With permission.)

ARE BENEFACTOR SPECIES INTERCHANGEABLE?

The species-specificity of positive interactions among plants, or whether or not benefactor species are interchangeable, is crucial to understanding of positive interactions and interdependence in plant communities (Callaway 1998a). In other words, are the positive effects of plants simply due to the alteration of the biophysical environment that can be imitated by inanimate objects? Or can facilitation depend on the species, with some species eliciting strong positive effects and other, morphologically similar, species producing no effect?

Spatial associations between beneficiaries and benefactors are often disproportional to the abundance of potential beneficiaries. For example, Hutto et al. (1986) reported that saguaros were distributed nonrandomly among potential nurse plants at two locations in Organ Pipe National Monument in the Sonoran Desert, USA. They found that saguaros were proportionally more abundant under many species of shrubs and trees, but significantly more saguaros were associated with *P. juliflora* (mesquite) and paloverde trees and less saguaros were associated with *L. tridentata* (creosote bush) than expected based on the proportional cover of these species.

Suzán et al. (1996) identified a large number of aborescent, shrub, and cacti species as nurse plants of *O. tesota* (ironwood) in the Sonoran Desert of Mexico and the USA, and argued that, as a habitat modifier, *Olneya* is a keystone species for biodiversity (also see Burqúez and Quintana 1994). They described 30 species as shade-dependent, with 5 preferring *Cercidium* species, 4 preferring *Prosopis* species, and 22 preferring *O. tesota*. Franco and Nobel (1989) found that most saguaro seedlings in their Sonoran Desert study sites were associated with *C. microphyllum* and *Ambroisa deltoidea*. In contrast, a second cactus species, the exceptionally heat-tolerant *F. acanthodes* was preferentially associated with a bunchgrass, *Hilaria rigida*.

In Californian shrubland and woodland *Q. agrifolia* and *Q. douglasii* seedlings are disproportionally associated with shrub species, and experimental manipulations have demonstrated the importance of nurse shrubs for the survival of both species (Callaway and D'Antonio 1991, Callaway 1992). However, not all shrubs had positive effects on *Q. agrifolia*; 43% of germinating seedlings survived under *Ericameria ericoides*, 34% under *Artemisia californica*, 5% under *Mimulus auranticus*, and 0% under *Lupinus chamissonis* (Callaway and D'Antonio 1991).

In other studies, the species of the benefactor plant does not seem to matter. For example, Steenberg and Lowe (1969) reported that 15 different species can apparently act as nurse plants for saguaro and are found associated with saguaro seedlings in proportion to their frequencies, indicating that no specific biotic factor is involved, and that the nurse plant association is only the by-product of microclimatic changes under canopies. Greenlee and Callaway (1996) found that *L. carinata*, a small perennial herb, was commonly found under the canopies of bunchgrasses on xeric sites in western Montana. However, *Lesquerella* was distributed among bunchgrass species in proportion to their abundances.

Potential benefactor species (those with similar morphologies) may have specific effects on beneficiaries simply because they alter the physical environment differently. For example, Suzán et al. (1996) attributed the disproportional positive affect of *O. tesota* to the fact that it is the only tall evergreen aborescent in the region and ameliorates climatic conditions throughout the year.

Because facilitation occurs in complex combinations with competition, potential benefactor species may have the same positive effect, but vary in their negative effects. For example, Suzán et al. (1996) suggested that *Olneya*'s superior facilitative ability may be due to its phreatophytic life history and deeply distributed root architecture, thus reducing niche overlap (e.g., Cody 1986) and accentuating positive mechanisms.

The consistently poor performance of creosote bush as a nurse plant in the Sonoran Desert (Hutto et al. 1986, McAuliffe 1988) may be due to the strong negative effects this species has on perennial neighbors (Fonteyn and Mahall 1981). Mahall and Callaway (1991, 1992) found that creosote bush substantially inhibited the root elongation rates of *A. dumosa*, and that these negative effects were reduced by the addition of small amounts of activated carbon, a strong adsorbent to organic molecules (Cheremisinoff and Ellerbusch 1978). Thus creosote bush canopies may have the potential to facilitate shade-requiring plants, but prospective beneficiaries may be eliminated by root allelopathy. In the Chihauhan Desert, however, *L. tridentata* has been reported to faciliate *Opuntia leptocaulis* (Yeaton 1978). Muller (1953) documented strong positive associations between *A. dumosa* and many species of desert annuals; however, *Encelia farinosa* shrubs did not harbor any annual species. He attributed this difference to the allelopathic effect of *Encelia* leaves.

Pugnaire et al. (1996b) showed that edaphic conditions and productivity improved in the understory of *R. sphaerocarpa* shrubs with age and that there was a replacement of species under *Retama* canopies with a clear successional trend. The species found under the youngest shrubs were also found in gaps and, as soil fertility increased and aspects of microclimate (particularly irradiance and temperature) were ameliorated, different species colonized the understory, including agricultural weeds and perennials. Growth conditions improved for *Retama* itself, and there was an increase in cladode mass, photosynthetic area, and fruit production with age, supporting the hypothesis that overstory shrubs benefit from the understory herbs, and that indirect interactions between *R. sphaerocarpa* and its understory herbs could be considered as a two-way facilitation in which both partners benefit from their association (Pugnaire et al. 1996a).

Species that have similar affects on understory microclimate may vary in their affects on soil nutrients creating species-specific effects. Turner et al. (1966) found that saguaro seedlings survived better on soil collected from under paloverde trees than on those from under either mesquite or *O. tesota*; however, these differences were confounded by soil albedo and temperature.

Some facilitative mechanisms are produced by unusual plant traits, and thus infer a high degree of species-specificity. For example, some of the strongest positive effects are produced by protection from herbivores (Atsatt and O'Dowd 1976, McAuliffe 1984a, 1986, Hay 1986). Repelling consumers requires specific morphological traits such as spines or tough tissues, or the possession of chemical defenses. Not all potential benefactors in a community have these traits.

Many of the traits described earlier such as oxygen transport (Callaway and King 1996), shared pollinator attraction (Thomson 1978, Laverty 1992), hydraulic lift (Richards and Caldwell 1987, Dawson 1993), and positive effects mediated through fungal intermediaries (Grime et al. 1987, Marler et al. 1999) are not shared by all large number of species in a community, suggesting that benefactors are not highly interchangeable.

POSITIVE INTERACTIONS AND COMMUNITY THEORY

The consistent observation of independent distributions of plant species along environmental gradients (the individualistic-continuum) is a cornerstone of plant community theory (Goodall 1963, McIntosh 1967, Austin 1990). Plant species are almost never completely associated with another species at all points on environmental gradients, and so community ecologists have long assumed that the distributions of plant species are determined by their idiosyncratic responses to the abiotic environment and interspecific competition. However, some evidence suggests that plant species may have some degree of interdependence at some points on gradients, yet interact individualistically at others (Callaway 1998b).

Pinus albicaulis (whitebark pine) and *Abies lasiocarpa* (subalpine fir) dominate the upper end of many xeric elevation gradients in the Northern Rockies (Daubenmire 1952, Pfister et al. 1977). In many places, *P. albicaulis* is the dominant species at or near timberline, but it also overlaps with *A. lasiocarpa* at lower elevations where the latter is more abundant. Thus, these two species exhibit the classic continuum that is the cornerstone of the individualistic-continuum theory. However, a more detailed examination of interactions between these species creates a more complex picture. *P. albicaulis* appears to have a cumulative competitive effect on *A. lasiocarpa* at lower elevations, where there are no significant spatial associations and the death of *P. albicaulis* corresponds with higher *A. lasiocarpa* growth rates (Callaway 1998a). But at timberlines in xeric areas, *A. lasiocarpa* is highly clumped around *P. albicaulis*, and the death of the latter is associated with decreased *A. lasoicarpa* growth rates. Similar processes are also apparent in alpine communities of the central Caucasus Mountains. Kikvidze (1993, 1996) measured numerous significant positive spatial associations and found evidence for facilitation via amelioration of abiotic stress. At high elevations, significant positive spatial associations are four times more common than negative associations. However, at low elevations gradient-positive spatial associations were four times less common than negative associations.

Variations for virtually all important types of interspecific interactions have been shown to vary with changes in abiotic conditions. These include predation (Martin, in press), herbivory (Maschinski and Whitham 1989), parasitism (Gibson and Watkinson 1992, Pennings and Callaway 1996), mutualism (Bronstein 1994), competition (Connell 1983, Kadmon 1995), and allelopathy (Tang et al. 1995).

Shifting positive and negative interactions on environmental gradients indicates that nodes, or fully overlapping discreet groups of species, are not required to demonstrate some level of interdependence among plants in a community. Because plants can have neutral or negative effects on neighbors at one point on an environmental gradient and positive effects at another, and the positive effect of benefactors are often not interchangeable among species, a continuum does not necessarily infer fully individualistic relationships among plant species. The ubiquity of facilitative interactions in plant communities indicates the necessity of explicit reconsideration of formal community theory. Lortie et al. (2004) proposed the concept of the integrated community (IC) allowing natural plant communities to range from highly individualistic to highly interdependent depending on synergism among: (i) stochastic processes, (ii) the abiotic tolerances of species, (iii) positive and negative interactions among plants, and (iv) indirect interactions within and between trophic levels. Rejecting strict individualistic theory may allow ecologists to better explain variation occurring at different spatial scales, synthesize more general predictive theories of community dynamics, and develop models for community-level responses to global change.

CONCLUSION

Positive interactions are common in nature, and are caused by many different mechanisms that are substantially different from those involved in competition. Positive interactions co-occur with competition and the overall effect of one plant on another is often determined by the balance of several mechanisms in a particular abiotic environment. In many cases, positive interactions are highly species-specific and benefactors are not interchangeable, suggesting that plant communities may be more interdependent than has been thought since the widespread acceptance of Gleasonian individualistic communities. Furthermore, plants may have strong competitive effects on a neighbor at one end of its distribution along a physical gradient, but strong positive effects on the same neighbor at the other end. This confounds interpretation of continuous distributions of plants in gradient analyses as evidence for fully individualistic plant communities and supports the concept that plant communities are real entities (Van der Maarel 1996). The growing body of evidence for positive interactions in plant communities, and its theoretical framework, suggests that facilitation plays important roles in determining the structure, diversity, and dynamics of many plant communities.

REFERENCES

Aguiar, M.R. and O.E. Sala, 1994. Competition, facilitation, seed distribution and the origin of patches in a Patagonian steppe. Oikos 70: 26–34.

Aguiar, M.R., A. Soriano, and O.E. Sala, 1992. Competition and facilitation in the recruitment of seedlings in Patagonian steppe. Functional Ecology 6: 66–70.

Archer, S., C. Scifres, C.R. Bassham, and R. Maggio, 1988. Autogenic succession in a subtropical savanna: conversion of grassland to woodland. Ecological Monographs 58: 111–127.

Armas, C. and F.I. Pugnaire, 2005. Plant interactions govern population dynamics in a semiarid plant community. Journal of Ecology 93: 978–989.

Armstrong, W., 1979. Aeration in higher plants. Advances in Botanical Research 7: 226–332.

Armstrong, J., W. Armstrong, and P.M. Becket, 1992. *Phragmites australis*: Venturi- and humidity-induced pressure flows enhance rhizome aeration and rhizosphere oxidation. New Phytologist 120: 197–207.

Atsatt, P.R. and D.J. O'Dowd, 1976. Plant defense guilds. Science 193: 24–29.

Austin, M.P., 1990. Community theory and competition in vegetation. In: J.B. Grace and D. Tilman, eds, Perspectives on Plant Competition. Academic Press, New York, NY, USA, pp. 215–238.

Belsky, A.J., 1994. Influences of trees on savanna productivity: tests of shade, nutrients, and tree-grass competition. Ecology 75: 922–932.

Bertness, M.D., 1991. Interspecific interactions among high marsh perennials in a New England salt marsh. Ecology 72: 125–137.

Bertness, M.D. and R.M. Callaway, 1994. Positive interactions in communities. Trends in Ecology and Evolution 9: 191–193.

Bertness, M.D. and S.D. Hacker, 1994. Physical stress and positive associations among marsh plants. The American Naturalist 144: 363–372.

Bertness, M.D. and S.W. Shumway, 1993. Competition and facilitation in marsh plants. The American Naturalist 142: 718–724.

Bertness, M.D. and S.M. Yeh, 1994. Cooperative and competitive interactions in the recruitment of marsh elders. Ecology 75: 2416–2429.

Booth, M.G., 2004. Mycorrhizal networks mediate overstorey-understorey competition in a temperate forest. Ecology Letters 7: 538–546.

Bormann, F.H., 1962. Root grafting and noncompetitive relationships between trees. In: T.T. Kozlowski, ed. Tree Growth. Ronald Press, New York, pp. 237–245.

Bormann, F.H., 1966. The structure, function, and ecological significance of root grafts in *Pinus strobus* L. Ecological Monographs 36: 1–26.

Bormann, F.H. and B.F. Graham Jr., 1959. The occurrence of natural root grafting in eastern white pine, *Pinus strobus* L., and its ecological implications. Ecology 40: 677–691.

Bronstein, J., 1994. Conditional outcomes in mutualistic interactions. Trends in Ecology and Evolution 9: 214–217.

Bruno, J.F., J.J. Stachowicz, and M.D. Bertness, 2003. Inclusion of facilitation into ecological theory. Trends in Ecology and Evolution 18: 119–125.

Búrquez, A. and M.A. Quintana, 1994. Islands of diversity: Ironwood ecology and the richness of perennials in a Sonoran Desert biological reserve. In: G.P. Nabhan and J.L. Carr, eds. Ironwood: An Ecological and Cultural keystone of the Sonoran Desert. Conservation International, Washington D.C.

Caldwell, M.M., 1990. Water parasitism stemming from hydraulic lift: a quantitative test in the field. Israel Journal Botany 39: 395–402.

Callaway, R.M., 1992. Effect of shrubs on recruitment of *Quercus douglasii* and *Quercus lobata* in California. Ecology 73: 2118–2128.

Callaway, R.M. 1994. Facilitative and interfering effects of *Arthoenemum subterminate* on winter annuals. Ecology 75: 681–686.

Callaway, R.M., 1995. Positive interactions among plants. Botanical Review 61: 306–349.

Callaway, R.M., 1997. Positive interactions and the individualistic-continuum concept. Oecologia 112: 143–149.

Callaway, R.M., 1998a. Are positive interactions among plants species-specific? Oikos 82: 202–207.

Callaway, R.M., 1998b. Competition and facilitation on elevational gradients in subalpine forests of the northern Rocky Mts., USA. Oikos 82: 561–573.

Callaway, R.M. and C.M. D'Antonio, 1991. Shrub facilitation of coast live oak establishment in central California Madrono 38: 158–169.

Callaway, R.M. and L. King, 1996. Oxygenation of the soil rhizosphere by *Typha latifolia* and its facilitative effects on other species. Ecology 77: 1189–1195.

Callaway, R.M. and L.R. Walker, 1997. Competition and facilitation: a synthetic approach to interactions in plant communities. Ecology 78: 1958–1965.

Callaway, R.M., N.M. Nadkarni, and B.E. Mahall, 1991. Facilitation and interference of Quercus douglasii on understory productivity in central California. Ecology 72: 1484–1499.

Callaway, R.M., E.H. DeLucia, D. Moore, R. Nowak, and W.H. Schlesinger, 1996. Competition and facilitation: contrasting effects of *Artemisia tridentata* on desert vs. montane pines. Ecology 77: 2130–2141.

Callaway, R.M., R.W. Brooker, C.P.Z. Kikvidze, C.J. Lortie, R. Michalet, L. Paolini, F.I. Pugnaire, B.J. Cook, E.T. Aschehong, C. Armas, and B. Newingham, 2002. Positive interactions among alpine plants increase with stress: a global experiment. Nature 417: 844–848.

Casper, B., 1996. Demographic consequences of drought in the herbaceous perennial *Cryptantha flava*: effects of density, associations with shrubs, and plant size. Oecologia 106: 144–152.

Castellanos, E.M., M.E. Figueroa, and A.J. Davy, 1994. Nucleation and facilitation in saltmarsh succession: interactions between *Spartina maritima* and *Arthrocnemum perenne*. Journal of Ecology 82: 239–248.

Chapin, F.S., III., L.R. Walker, C.L. Fastie, and L.C. Sharman, 1994. Mechanisms of primary succession following deglaciation at Glacier Bay, Alaska. Ecological Monographs 64: 149–175.

Cheremisinoff, P.N. and F. Ellerbusch, 1978. Carbon Adsorption Handbook. Ann Arbor Science Publishers, Ann Arbor, MI.

Chiariello, N.R., J.C. Hickman, and H. Mooney, 1982. Endomycorrhizal role for interspecific transfer of phosphorus in a community of annual plants. Science 217: 941–943.

Cody, M.L., 1986. Structural niches in plant communities. In: J. Diamond and T.J. Case, eds. Community Ecology. Harper and Row, NY, pp. 381–405.

Connell, J.H., 1983. On the prevalence and relative importance of interspecific competition: evidence from field experiments. American Naturalist 122: 661–696.

Daubenmire, R.F., 1952. Forest vegetation of northern Idaho and adjacent Washington and its bearing on concepts of vegetation classification. Ecological Monographs 22: 301–330.

Dawson, T.E., 1993. Hydraulic lift and water use by plants: implications for water balance, performance and plant-plant interactions. Oecologia 95: 565–574.

De Jong, T.J. and P.G.L. Klinkhamer, 1988. Population ecology of the biennials *Cirsium vulgare* and *Cynoglossum officinale* in a coastal sand-dune area. Journal of Ecology 73: 147–167.

DeAngelis, D.L., W.M. Post, and C.C. Travis, 1986. Positive Feedback in Natural Systems. Springer-Verlag, New York, NY.

DeLucia, E.H., W.H. Schlesinger, and W.D. Billings, 1988. Water relations and the maintenance of Sierran conifers on hydrothermally altered rock. Ecology 69: 303–311.

DeLucia, E.H., W.H. Schlesinger, and W.D. Billings, 1989. Edaphic limitations to growth and photosynthesis in Sierran conifers and Great Basin Vegetation. Oecologia 78: 184–190.

Dickie, I.A., S.A. Schnitzer, P.B. Reich, and S.E. Hobbie, 2005. Spatially disjunct effects of co-occurring competition and facilitation. Ecology Letters 8: 1191–1200.

Ellison, L. and W.R. Houston, 1958. Production of herbaceous vegetation in openings and under canopies of western aspen. Ecology 39: 338–345.

Facelli, J.M., 1994. Multiple indirect effects of plant litter affect the establishment of woody seedlings in old fields. Ecology 75: 1727–1735.

Flores-Martinez, A., E. Ezcurra, and S. Sanchez-Colon, 1994. Effect of *Neobuxbaumia tetetzo* on growth and fecundity of its nurse plant *Mimosa luisana*. Journal of Ecology 82: 325–330.

Fonteyn, P.J. and B.E. Mahall, 1981. An experimental analysis of structure in a desert plant Community. Journal of Ecology 69: 883–896.

Francis, R. and D.J. Read, 1984. Direct transfer of carbon between plants connected by vesicular-arbuscular mycelium. Nature 307: 53–56.

Franco, A.C. and P.S. Nobel, 1989. Effect of nurse plants on the microhabitat and growth of cacti. Journal of Ecology 77: 870–886.
Frost, W.E. and N.K. McDougald, 1989. Tree canopy effects on herbaceous production of annual rangeland during drought. Journal of Range Management 42: 281–283.
Fuentes, E.R., R.D. Otaiza, M.C. Alliende, A. Hoffman, and A. Poiani, 1984. Shrub clumps of the Chilean matorral vegetation: structure and possible maintenance mechanisms. Oecologia 62: 405–411.
Gause, G.F., 1934. The Struggle for Existence. Hafner, NY.
Geer, S.M., V.J. Tepedino, T.L. Griswold, and V.R. Bowlin, 1995. Pollinator sharing by 3 sympatric milkvetches, including the endangered species Astragalus montii. Great Basin Naturalist 55: 19–28.
Ghazoul, J., 2006. Floral diversity and the facilitation of pollination. Journal of Ecology 94: 295–304.
Gibson, C.C. and A.R. Watkinson, 1992. The role of the hemiparasitic annual *Rhinanthus minor* in determining grassland community structure. Oecologia 89: 62–68.
Goldman, C.R., 1961. The contribution of alder trees (*Alnus tenuifolia*) to the primary productivity of Castle Lake, California. Ecology 42: 282–288.
Goodall, D.W., 1963. The continuum and the individualistic association. Vegetatio 11: 297–316.
Grace, J.B., 1995. In search of the Holy Grail: explanations for the coexistence of plant species. Trends in Ecology and Evolution 10: 263–264.
Graham, B.F. Jr. 1959. Transfer of dye through natural root grafts of *Pinus strobus* L. Ecology 41: 56–64.
Greenlee, J. and R.M. Callaway, 1996. Effects of abiotic stress on the relative importance of interference and facilitation. The American Naturalist 148: 386–396.
Grime, J.P., J.M.L. Mackey, S.H. Hillier, and D.J. Read, 1987. Floristic diversity in a model system using experimental microcosms. Nature 328: 420–422.
Haase, P., F.I. Pugnaire, S.C. Clark, and L.D. Incoll, 1996. Spatial patterns in a two-tiered semi-arid shrubland in southeastern Spain. Journal of Vegetation Science 7: 527–534.
Haase, P., F.I. Pugnaire, S.C. Clark, and L.D. Incoll, 1997. Spatial patterns in a two-tiered semi-arid shrubland in southeastern Spain. Journal of Vegetation Science 8: 627–634.
Hacker, S.D. and M.D. Bertness, 1996. Trophic consequences of a positive plant interaction. American Naturalist, 148: 559–575.
Hardin, G., 1990. The competitive exclusion principle. Science 131: 1292–1297.
Hay, M.E., 1986. Associational plant defenses and the maintenance of species diversity: turning competitors into accomplices. American Naturalist 128: 617–641.
Hjalten, J., K. Danell, and P. Lundberg, 1993. Herbivore avoidance by association: vole and hare utilization of woody plants. Oikos 68: 125–131.
Hoffmann, W.A., 1996. The effects of fire and cover on seedling establishment in a neotropical savanna. Journal of Ecology 84: 383–393.
Holmgren, M., M. Scheffer, and M.A. Huston, 1997. The interplay of facilitation and competition in plant communities. Ecology 78: 1966–1975.
Howes, B.L., R.W. Howarth, J.M. Teal, and I. Valiela. 1981. Oxidation–reduction potentials in a salt marsh: spatial patterns and interactions with primary production. Limnology and Oceanography 26: 350–360.
Hunter, A.F. and L.W. Aarssen, 1988. Plants helping plants. Bioscience 38: 34–40.
Huston, M.A. and D.L. DeAngelis, 1994. Competition and coexistence: the effects of resource transport and supply rates. American Naturalist 144: 954–977.
Hutchinson, G.E., 1961. The paradox of the plankton. American Naturalist 95: 137–145.
Hutto, R.L., J.R. McAuliffe, and L. Hogan, 1986. Distributional associates of the saguaro (*Carnegia gigantea*). Southwestern Naturalist 31: 469–476.
Joffre, R. and S. Rambal, 1993. How tree cover influences the water balance of Mediterranean rangelands? Ecology 74: 570–582.
Johnson, S.D., C.I. Peter, L.A. Nilsson, and J. Agren, 2003. Pollination success in a deceptive orchid is enhanced by co-occurring rewarding magnet plants. Ecology 84: 2919–2927.
Kadmon, R., 1995. Plant competition along soil moisture gradients: a field experiment with the desert annual *Stipa capensis*. Journal of Ecology 83: 253–262.
Kellman, M., 1985. Forest seedling establishment in Neotropical savannas: transplant experiments with *Xylopia frutescens* and *Calophyllum brasiliense*. Journal of Biogeography 12: 373–379.

Kikvidze, Z., 1993. Plant species associations in alpine-subnival vegetation patches in the Central Caucasus. Journal Vegetation Science 4: 297–302.

Kikvidze, Z., 1996. Neighbour interactions and stability in subalpine meadow communities. Journal Vegetation Science 7: 41–44.

Kikvidze, Z., F.I. Pugnaire, R. Brooker, P. Choler, C.J. Lortie, R. Michalet, and R.M. Callaway, 2005. Linking patterns and processes in alpine plant communities: a global study. Ecology 86: 1395–1400.

Laverty, T.M., 1992. Plant interactions for pollinator visits: A test of the magnet species effect. Oecologia 89: 502–508.

Laverty, T.M. and R.C. Plowright, 1988. Fruit and seed set in Mayapple (*Podophyllum peltatum*): influence of intraspecific factors and local enhancement near *Pedicularis canadensis*. Canadian Journal of Botany 66: 173–178.

Leake, J.R., 1993. Tansley Review Number 69: The biology of myco–heterotrophic (saprophytic) plants. New Phytologist 127: 171–216.

Lortie, C.J., R.W. Brooker, P. Choler, Z. Kikvidze, R. Michalet, F.I. Pugnaire, and R.M. Callaway, 2004. Rethinking plant community theory. *Oikos* 107: 433–438.

Lotka, A.J. 1932. The growth of mixed populations: two species competing for a common food supply. Journal of the Washington Academy of Science 22: 461–469.

Mahall, B.E. and R.M. Callaway, 1991. Root communication among desert shrubs. Ecology 88: 874–876.

Mahall, B.E. and R.M. Callaway, 1992. Root communication mechanisms and intracommunity distributions of two Mojave Desert shrubs. Ecology 73: 2145–2151.

Mahall, B.E., V.T. Parker, and P.J. Fonteyn, 1981. Growth and photosynthetic responses of *Avena fatua* L. and *Bromus diandrus* Roth. and their ecological significance in California savannas. Photosynthetica 15: 5–15.

Maranga, E.K., 1984. Influence of *Acacia tortilis* Trees on the Distribution of *Panicum maximum* and *Digitaria macroblephara* in South Central Kenya. MS Thesis, Texas A & M University, College Station, TX.

Marañón, T. and J.W. Bartolome, 1993. Reciprocal transplants of herbaceous communities between *Quercus agrifolia* woodland and adjacent grassland. Journal of Ecology 81: 673–682.

Marler, M., Zabinski, C., and R.M. Callaway, 1999. Mycorrhizal mediation of competition between an exotic forb and a native bunchgrass. Ecology 80: 1180–1186.

Martin, T.E. Are microhabitat preferences of coexisting species under selection and adaptive? Ecology (in press).

Maschinski, J. and T.G. Whitham, 1989. The continuum of plant responses to herbivory: the influence of plant association, nutrient availability, and timing. American Naturalist 134: 1–19.

McAuliffe, J.R., 1984a. Prey refugia and the distributions of two Sonoran Desert cacti. Oecologia 65: 82–85.

McAuliffe, J.R., 1984b. Saguaro-nurse tree associations in the Sonoran Desert: competitive effects of sahuaros. Oecologia 64: 319–321.

McAuliffe, J.R., 1986. Herbivore-limited establishment of a Sonoran Desert tree: *Cercidium microphyllum*. Ecology 67: 276–280.

McAuliffe, J.R., 1988. Markovian dynamics of simple and complex desert plant communities. American Naturalist 131: 459–490.

McClaran, M.P. and J.P. Bartolome, 1989. Effect of *Quercus douglasii* (Fagaceae) on herbaceous understory along a rainfall gradient. Madrono 36: 141–153.

McIntosh, R.P., 1967. The continuum concept of vegetation. Botanical Review 33: 130–187.

McNaughton, S.J., 1978. Serengeti ungulates: feeding selectivity influences the effectiveness of plant defense guilds. Science 199: 806–807.

McNaughton, S.J., 1985. Ecology of a grazing ecosystem: the Serengeti. Ecological Monographs 55: 259–294.

Michalet, R., R.W. Brooker, L.A. Cavieres, Z. Kikvidze, C.J. Lortie, F.I. Pugnaire, A. Valiente-Banuet, and R.M. Callaway, 2006. Do biotic interactions shape both sides of the humped-back model of species richness in plant communities? Ecology Letters 9: 767–773.

Moeller, D.A., 2004. Facilitative interactions among plants via shared pollinators. Ecology 85: 3289–3301.

Monk, C.D. and F.C. Gabrielson, 1985. Effects of shade, litter and root competition on old-field vegetation in South Carolina. Bulletin of the Torrey Botanical Club 112: 383–392.
Moro, M.J., F.I. Pugnaire, P. Haase, and J. Puigdefábregas, 1997. Mechanisms of interaction between *Retama sphaerocarpa* and its understory layer in a semi-arid environment. Ecography 20: 175–184.
Morris, W.F. and D.M. Wood, 1989. The role of lupine in succession on Mount St. Helens: facilitation or inhibition? Ecology 70: 697–703.
Muller, C.H. 1953. The association of desert annuals with shrubs. American Journal of Botany 40: 53–60.
Niering, W.A., R.H. Whittaker, and C.H. Lowe, 1963. The saguaro: a population in relation to environment. Science 142: 15–23.
Nobel, P.S., 1984. Extreme temperatures and thermal tolerances for seedlings of desert succulents. Oecologia 62: 310–317.
Parker, V.T. and C.H. Muller, 1982. Vegetational and environmental changes beneath isolated live oak trees (*Quercus agrifolia*) in a California annual grassland. American Midlands Naturalist 107: 69–81.
Parrish, J.A.D. and F.A. Bazzaz, 1976. Underground niche separation in successional communities. Ecology 57: 1281–1288.
Patten, D.T., 1978. Productivity and production efficiency of an upper Sonoran Desert community. American Journal of Botany 65: 891–895.
Pennings, S.C. and R.M. Callaway, 1996. Impact of a parasitic plant on the structure and function of salt marsh vegetation. Ecology 77: 1410–1419.
Petranka, J.W. and J.K. McPherson, 1979. The role of *Rhus copallina* in the dynamics of the forest-prairie ecotone in north-central Oklahoma. Ecology 60: 956–965.
Pfister, C.A. and M.E. Hay, 1988. Associational plant refuges: convergent patterns in marine and terrestrial communities result from differing mechanisms. Oecologia 77: 118–129.
Pfister, R.D, B.L. Kovalchik, S.E. Arno, and R. Presby, 1977. Forest habitat types of western Montana. USDA, Forest Service Technical Report, INT-34. Ogden, UT.
Pugnaire, F.I. and M.T. Luque, 2001. Changes in plant interactions along a gradient of environmental stress. Oikos 93: 42–49.
Pugnaire, F.I., P. Hasse, and J. Puigdefabregas, 1996a. Facilitation between higher plant species in a semiarid environment. Ecology 77: 1420–1426.
Pugnaire, F.I., P. Haase, J. Puigdefábregas, M. Cueto, L.D. Incoll, and S.C. Clark, 1996b. Facilitation and succession under the canopy of the leguminous shrub, *Retama sphaerocarpa,* in a semi-arid environment in south-east Spain. Oikos 76: 455–464.
Richards, J.H. and M.M. Caldwell, 1987. Hydraulic lift: substantial nocturnal water transport between soil layers by *Artemisia tridentata* roots. Oecologia 73: 486–489.
Risch, S.J, 1981. Insect herbivore abundance in tropical monocultures and polycultures: an experimental test of two hypotheses. Ecology 62: 1325–1340.
Root, R.B., 1972. The influence of vegetational diversity on the population ecology of a specialized herbivore, *Phyllotreta cruciferae* (Coleoptera: Chrysomelidae). Oecologia 10: 321–346.
Rostagno, C.M., H.F. del Valle, and L. Videla, 1991. The influence of shrubs on some chemical and physical properties of an aridic soil in north-eastern Patagonia, Argentina. Journal of Arid Environments 20: 179–188.
Roy, B.A., 1996. A plant pathogen influences pollinator behavior and may influence reproduction of nonhosts. Ecology 77: 2445–2457.
Schat, H., 1984. A comparative ecophysiological study on the effects of waterlogging and submergence on dune slack plants: growth, survival and mineral nutrition in sand culture experiments. Oecologia 62: 279–286.
Schat, H. and K. Van Beckhoven, 1991. Water as a stress factor in the coastal dune system. In: J. Rozema and J.A.C. Verkleij, eds. Ecological Responses to Environmental Stresses. Kluwer Academic Publishers, Amsterdam, The Netherlands, pp. 76–89.
Schmida, A. and R.H. Whittaker, 1981. Pattern and biological microsite effects in two shrub communities in southern California. Ecology 62: 234–251.
Silvertown, J. and Wilson J.B., 1994. Community structure in a desert perennial community. Ecology 75: 409–417.

Steenberg, W.F. and C.H. Lowe, 1969. Critical factors during the first year of life of the saguaro (*Cereus giganteus*) at Saguaro National Monument. Ecology 50: 825–834.
Steenberg, W.F. and C.H. Lowe, 1977. Ecology of the Saguaro. II. Reproduction, Germination, Establishment, Growth, and Survival of the Young Plant. National Park Service Science Monograph Series No. 8. NPS, Washington, D.C., USA.
Suzán, H., G.P. Nablan, and D.T. Patten, 1996. The importance of *Olneya tesota* as a nurse plants in the Sonoran Desert. Journal of Vegetation Science 7: 635–644.
Tang, C., W. Cai, K. Kohl, and R.K. Nishimoto, 1995. Plant stress and allelopathy. In: Inderjit, K.M.M. Dakshini, and F.A. Einhellig, eds. Allelopathy. American Chemical Society, Washington, D.C., USA, pp. 142–157.
Thomson, J.D., 1978. Effects of stand composition on insect visitation in two-species mixtures of *Hieracium*. American Midlands Naturalist 100: 431–440.
Tilman, D., 1976. Ecological competition between algae: experimental confirmation of resource-based competition and predation. Science 192: 463–465.
Tilman, D., 1985. The resource ratio hypothesis of succession. American Naturalist 125: 827–852.
Tilman, D., 1988. Plant Strategies and the Dynamics and Structure of Plant Communities. Princeton Monograph Series. Princeton University Press, Princeton, NJ.
Tirado, R. and F.I. Pugnaire, 2005. Community structure and positive interactions in constraining environments. Oikos 111: 437–444.
Turner, R.M., S.M. Alcorn, G. Olin, and J.A. Booth, 1966. The influence of shade, soil, and water on saguaro seedling establishment. Botanical Gazette 127: 95–102.
Turner, R.M., S.M. Alcorn, and G. Olin, 1969. Mortality of transplanted saguaro seedlings. Ecology 50: 835–844.
Valiente-Banuet, A. and E. Ezcurra, 1991. Shade as a cause of the association between the cactus *Neobuxbaumia tetetzo* and the nurse plant *Mimosa luisana* in the Tehuacan Valley, Mexico. Journal of Ecology 79: 961–971.
Valiente-Banuet, A., F. Vite, and J.A. Zavala-Hurtado, 1991. Interaction between the cactus *Neobuxbaumia tetetzo* and the nurse shrub *Mimosa luisana*. Journal of Vegetation Science 2: 11–14.
Van der Maarel, E., 1996. Pattern and process in the plant community: Fifty years after A.S. Watt. Journal of Vegetation Science 7: 19–28.
Van der Maarel, E., V. Noest, and M.W. Palmer, 1995. Variation in species richness on small grassland quadrats—Niche structure or small-scale plant mobility. Journal of Vegetation Science 6: 741–752.
Vetaas, O.R., 1992. Micro-site effects of trees and shrubs in dry savannas. Journal of Vegetation Science 3: 337–344.
Walker, L.R. and F.S. Chapin III, 1986. Physiological controls over seedling growth in primary succession on an Alaskan floodplain. Ecology 67: 1508–1523.
Walker, L.R. and P.M. Vitousek, 1991. An invader alters germination and growth of a native dominant tree in Hawaii. Ecology 72: 1449–1455.
Walter, L.E., D.C. Fisher, D. Hartnett, B.A.D. Hetrick, and A.P. Schwab, 1996. Interspecific nutrient transfer in a tallgrass prairie plant community. American Journal of Botany 83: 180–184.
Weltzin, J.F. and M.B. Coughenhour, 1990. Savanna tree influence on understorey vegetation and soil nutrients in northwestern Kenya. Journal of Vegetation Science 1: 325–332.
Whiteaker, L.D. and D.E. Gardner, 1985. The distribution of *Myrica faya* Ait. in the state of Hawaii. Tech. Rep. No. 55. Cooperative National Park Resources Studies Unit, University of Hawaii at Manoa, Honolulu, HI.
Whitford, W.G., J. Anderson, and P.M. Rice, 1997. Stemflow contribution to the "fertile island" effect in creosote bush, *Larrea tridentata*. Journal of Arid Environments 35: 451–457.
Yavitt, J.B. and L. Smith Jr., 1983. Spatial patterns of mesquite and associated herbaceous species in an Arizona desert upland. American Midlands Naturalist 109: 89–93.
Yeaton, R.I., 1978. A cyclical relationship between *Larrea tridentata* and *Opuntia leptocaulis* in the northern Chihuahuan Desert. Journal of Ecology 66: 651–656.
Young, D.R. and W.K. Smith, 1983. Comparison of intraspecific variations in the reproduction and photosynthesis of an understory herb, *Arnica cordifolia*. American Journal of Botany 70: 728–734.

15 Plant Interactions: Competition

Heather L. Reynolds and Tara K. Rajaniemi

CONTENTS

Not until we reach the extreme confines of life, in the arctic regions or on the borders of an udder desert, will competition cease. The land may be extremely cold or dry, yet there will be competition between some few species, or between the individuals of the same species, for the warmest or dampest spots.

—Darwin (1859)

INTRODUCTION

Organisms are said to compete when shared resource needs have mutual negative effects (– –) on survival, growth, or reproduction. This happens when resources are limiting, that is, when organismal demand for resources exceeds the supply of resources from the environment. Resources are quantities of matter or energy whose consumption promotes positive per capita growth or reproduction of an organism over at least some range of resource availability (Tilman 1982). As autotrophs, plants require the inorganic resources carbon dioxide (CO_2), water, light, and mineral nutrients. The major mineral nutrients are nitrogen, phosphorus, potassium, iron, calcium, magnesium, and sulfur; the minor ones are molybdenum, copper, zinc, manganese, boron, chlorine, sodium (some Chenopodiaceae), aluminum (ferns), cobalt (legumes that symbiotically fix nitrogen), and silicon (diatoms) (Banister 1976). Because plants are sessile and do not exhibit behavior, the main mechanism of plant competition is exploitation; competitors interact solely by consuming (i.e., depleting or preempting) resources (Figure 15.1; see Weiner and Thomas 1986 and Thomas and Weiner 1989 for discussion of resource depletion vs. resource preemption). Conversely, interference competition involves direct interactions between competitors that prevent access to resources, such as fighting between animals. In plants, interference competition can occur by one individual overgrowing another. Allelopathy, or production of

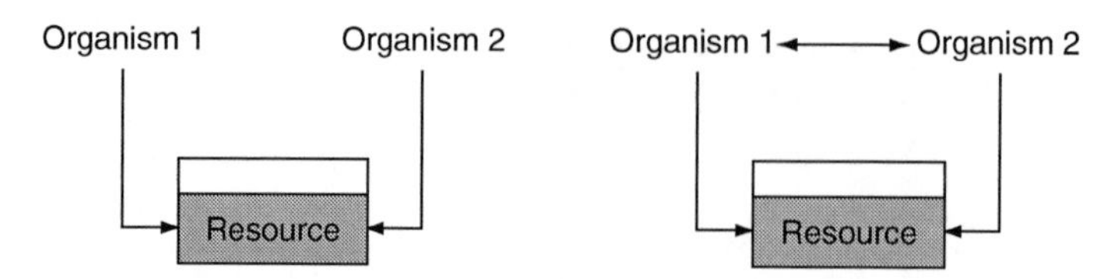

FIGURE 15.1 Mechanisms of competition. (*Left*) In exploitation competition, organisms interact only indirectly, via shared resource use. (*Right*) In interference competition, organisms interact directly for the shared resource. Interference competition is common in animals (e.g., fighting over mates or food), although plants can sometimes interfere by growing over one another, or via chemical production (allelopathy).

chemicals that inhibit plant function, may also be (but is not necessarily) an evolved mechanism of interference competition (Williamson 1990). Allelopathy has received renewed attention of late, particularly for its relevance to understanding exotic plant invasions (e.g., Wardle et al. 1998, Mallik and Pellissier 2000, Bais et al. 2003, Callaway and Ridenour 2004).

The relative importance of competition versus other biotic interactions (e.g., mutualism or facilitation, predation, herbivory) and abiotic factors (e.g., fire, pH, soil fertility) to the ecology and evolution of species and the structure of communities is an old debate in ecology (Roughgarden and Diamond 1986). Related issues concern whether the intensity of competition varies with environmental conditions and what traits are important to competitive ability. During about a 20 year period from the late 1970s to the mid-1990s, debate about these issues among plant ecologists crystallized in the contrasting perspectives of the competition–stress–ruderal (C–S–R) model of Grime (1977, 1979, 1988) and the resource competition model of Tilman (1982, 1988, 1990). After a number of stimulating exchanges and a substantial associated literature, the Grime–Tilman debate largely faded away, as attention turned to a number of new pressing issues, including the relationship between biodiversity and ecosystem functioning (Naeem et al. 2002, and see Chapter 10) and the causes and consequences of exotic species invasions (Levine et al. 2003, Hierro et al. 2005). Yet C–S–R and resource competition theory continue to be used as guiding frameworks in these and other areas of plant ecology (e.g., Wardle et al. 1997, Shea and Chesson 2002, MacDougall and Turkington 2005). This chapter reviews C–S–R and resource competition theory and the associated literature on competition intensity (CI) and competitive traits. We then turn to a relatively new area of focus for plant ecologists who study competition—soil heterogeneity, asking how current work in this area is informed by, and informs, C–S–R and resource competition theory. We conclude with a brief summary and suggestions for future research. The appendix to this chapter provides an overview of the various approaches to designing competition experiments and calculating competition indices.

C–S–R AND RESOURCE COMPETITION MODELS

It has sometimes been asserted that different definitions of competition are at the root of any disagreement between the C–S–R and resource competition models (Thompson 1987, Grace 1991a). However, both Grime and Tilman define competition in terms of exploitation (Tilman 1987): "Here, competition is defined as the tendency of neighboring plants to utilize the same quantum of light, ion of a mineral nutrient, molecule of water, or volume of space" (Grime 1973) and "The mechanism of competitive displacement is resource consumption" (Tilman 1988).

Competition can only occur for resources when they are limiting. Some resources required by plants (e.g., CO_2) are generally superabundant, and use by one organism has no effect on

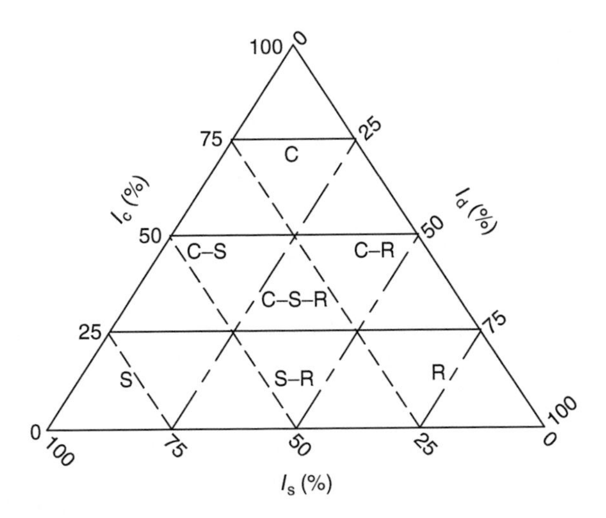

FIGURE 15.2 Model showing the various equilibria between competition (C), stress (S), and disturbance (d) in vegetation and the location of primary and secondary strategies. I_c, relative importance of competition (——); I_s, relative importance of stress (– – – – –); I_d, relative importance of disturbance (—— – – ——). (*Source*: Grime, J.P., *Am. Nat.*, 111, 1169, 1977. With permission.)

other organisms requiring the same resource. Resources become limiting when potential demands on resources by organisms exceed resource supply. The major source of disagreement between the C–S–R and resource competition models is when resource demand has the potential to exceed resource supply.

C–S–R

The C–S–R model is a conceptual model that categorizes vegetation into general strategies based on adaptation to combinations of two types of selection pressures: stress and disturbance (Figure 15.2). Stress is defined as any environmental condition that limits plant production (e.g., extremes of pH or temperature), and disturbance as any environmental condition that removes plant biomass (e.g., herbivory, pathogens, wind, fire, human activities) (Grime 1977). Low availability of mineral nutrients (low soil fertility) is most often associated with stress in this model. Stress and disturbance are thought to reduce competition by preventing development of a dense vegetation and thereby reducing demand for resources. Competition is hypothesized to be of highest intensity and greatest importance in highly productive vegetation, such as that which develops on fertile soil under relatively undisturbed conditions. Proponents of the C–S–R model (Grime 1977, Campbell et al. 1991) reason that dense, highly productive vegetation result in increased demand for both light and belowground resources (water, mineral nutrients). Therefore, traits important to the competitive strategy are considered to be those that maximize acquisition of all resources. These traits include high growth rates, large size and height, and high plasticity in roots and shoots to allow active foraging for undepleted resource pockets. Investment in resource acquisition comes at the expense of investment in traits important for adapting to stress and disturbance. Such trade-offs in allocation to competition, stress, and disturbance mean that each species cannot be adapted to the extremes of more than one strategy, although adaptation to intermediate intensities of competition, stress, and disturbance may occur (Grime 1977, 1988).

Resource Competition

The resource competition model is a mechanistic, mathematical model of plant population dynamics that considers essentially similar selection pressures (or environmental constraints):

Rate of biomass change = Growth − Loss

(a) $$\frac{dB_i}{B_i dt} = f_i(R) - m_i$$

Rate of resource change = Supply rate − Sum of consumption rates

(b) $$\frac{dR}{dt} = y(R) - \sum_{i=1}^{n} [Q_i B_i f_i(R)]$$

FIGURE 15.3 Simple consumer-resource model of exploitative competition for a limiting soil nutrient. (a) Per-unit biomass rate of change of a population, where B_i is biomass of species i; $f_i(R)$ is a function that describes the dependence of net growth for species i on the resource (R); and m_i is loss rate of species i. (b) Resource dynamics, where $y(R)$ is a function that describes resource supply; Q_i is the nutrient content per unit biomass of species i, and n is the total number of consumer species. This simple model can be extended to include more than one limiting resource. (*Source*: Tilman, D., *Perspectives on Plant Competition*, J.B. Grace and D. Tilman, eds, Academic Press, San Diego, CA, 1990.)

soil resource supply rates and loss (disturbance, mortality) rates. The model consists of two types of coupled equations, one describing plant dynamics as a function of resource-dependent growth and loss and the other describing resource dynamics as a function of resource supply and plant consumption (Figure 15.3). The model is explicit about the fact that plants grow by consuming resources, and exploitative competition is an integral component of the model because the resources consumed (light, soil resources) are depletable (i.e., plant consumption affects availability) and can therefore become limiting.

Competition is not restricted mainly to dense vegetation of high productivity habitats, as in the C–S–R model, but occurs under every combination of resource supply and loss rate. The reason for this is that the model considers plant demand for resources relative to resource supply. It is the ratio of supply to demand that determines whether resources are limiting (and consequently whether competition occurs) (Taylor et al. 1990, see also a related discussion by Chesson and Huntly 1997). The absolute demand for soil resources may be high in high productivity sites, but soil resource supply rates are also high, and thus the ratio of supply to demand is not necessarily different from that in low productivity sites (in which absolute demand is lower, but so is supply).

In fact, the resource competition model predicts that competition is equally important at both low and high soil fertility (low and high plant productivity). Only the nature of the resources for which there is competition is predicted to change along fertility gradients, as the limiting resources change from soil nutrients at low soil resource supply rates (causing strong belowground competition) to light at high soil resource supply rates (causing strong aboveground competition). Loss rates can similarly influence the nature of competition (below- vs. aboveground) through their influence on availabilities of soil nutrients and light (Tilman 1988, Smith and Huston 1989).

Traits important to competitive ability are also predicted to vary with resource supply and loss rates. For example, at low soil resource supply rates, allocation to acquisition and retention of soil resources is predicted (e.g., traits such as high root to shoot ratio, long-lived tissues). At high soil resource supply rates, allocation to acquisition of light may become important (e.g., traits such as high stem allocation) (Tilman 1988). Investment in below-ground resource acquisition and use necessarily comes at the expense of investment in traits important for acquisition and use of aboveground resources. Such trade-offs impose limits on the ability of any one organism to dominate in all habitat types (Tilman 1988, 1990, Smith and Huston 1989).

By acquiring depletable resources for itself, an individual reduces availability of those resources to other species. The level to which the individuals of a population deplete a given resource at equilibrium is known as R* (Tilman 1982, 1988, 1990). Tilman (1990) and Reynolds and Pacala (1993) have shown analytically how traits can influence R*s. Resource reduction is the mechanism of exploitation competition, and thus, the lower the R* of a species, the more resource it has gained for itself and the better is its competitive ability for that resource. At equilibrium, the species with the lowest R* (i.e., greatest ability to consume the resource) is predicted to displace all competitors (Tilman 1977, 1982, 1988, 1990).

It is important to note that the resource competition model can also be solved numerically to predict the competitive dominants under nonequilibrium conditions (Tilman 1987, 1988). However, these so-called transient dominants are not considered to be the true superior competitors for a given set of soil resource supply and loss rates. According to resource competition theory, the true superior competitor for a given set of soil resource supply and loss rates is defined as the species that dominates at equilibrium (Tilman 1987, 1988). For practical reasons, many competition experiments are of very short term, and thus the competitive outcomes may sometimes reflect transient dynamics rather than equilibrium conditions. Failure to distinguish transient from equilibrium outcomes has likely contributed to debate between the C–S–R and resource competition perspectives (Tilman 1987).

COMPETITION INTENSITY

In the 1980s and 1990s, experimental work on competition moved from demonstrating the existence of plant competition in the field (Connell 1983, Schoener 1983) toward a focus on testing theory, including the contrasting predictions of the C–S–R and resource competition models (Goldberg and Barton 1992, Gurevitch et al. 1992). Two questions are typically asked: (1) Does competition intensity (competitive ability) change along gradients of soil fertility and disturbance? (2) What physiological and morphological traits are important to competitive ability? No consensus has been reached on either of these questions. Especially for the first question, several issues have been raised that might lead to conflicting results and obscure a clear answer. These include:

1. *Choice of an appropriate competition index*. Competition intensity may be measured as absolute reduction in performance (physiological state, growth rate, fecundity, size, fitness) due to the presence of competition or as relative reduction in performance (Appendix 15.1). Absolute CI will almost inevitably be greater as plant biomass increases (e.g., with fertilization), simply because the maximum possible difference between monoculture and mixture performance increases as biomass increases (Campbell and Grime 1992, Grace 1993). Relative CI avoids this effect because it is standardized to maximum possible performance. Most recent competition studies employ relative measures of CI.
2. *Failure to clearly distinguish between the intensity of competition and its importance*. Intensity (CI, as earlier) measures the reduction below optimum in some measure of performance due to competition, whereas importance measures the relative degree that competition affects performance compared with other processes (e.g., herbivory, disturbance, direct abiotic effects, facilitation) (Weldon and Slauson 1986). Weldon and Slauson argue that high intensity of competition does not necessarily translate into high importance (or vice versa), as is typically assumed, and that the two concepts are often confused. Although it is sometimes argued that the predictions of the C–S–R model are for competition importance, and those of the resource competition model are for competition intensity, the vast majority of experiments measure only intensity

(Brooker et al. 2005, but see Welden et al. 1988, Mclellan et al. 1997, Sammul et al. 2000).

3. *Inattention to the selection regime commonly experienced by the study species.* A common expectation in competition experiments is that species act as phytometers of competition intensity that are dependent only on the experimental conditions (e.g., soil fertility) and independent of the environmental conditions to which the study species are adapted. This expectation is reasonable only if the C–S–R model holds, in which case species should experience greater competition intensity at high versus low soil fertility regardless of whether they are adapted to low- or high-fertility soil (Taylor et al. 1990). However, if the resource competition model holds, a species adapted to a lower-fertility soil (i.e., a good competitor for soil resources) should experience greater competition intensity when competing on a higher-fertility soil against better light competitors than when competing on its own soil type (Aarssen 1984, Tilman 1988, Smith and Huston 1989). The converse is true for a species adapted to high-fertility soil and similar reasoning can be applied to the relationship between competition intensity and other important environmental factors (e.g., disturbance).

This section discusses the experimental evidence from field studies on competition intensity across soil fertility (or productivity) and disturbance gradients in light of the selection regime-dependent predictions of the resource competition model. The focus is on field studies conducted in natural or partially natural soil because artificial soil mixes are likely to have depleted microbe communities compared with natural soils or natural soil mixes (Diaz et al. 1993). Only results based on relative CI are discussed here.

Soil Fertility or Productivity Gradients

A number of studies measuring the interactions between target species and surrounding vegetation have found that competition was more intense at high compared with low soil fertility or productivity, as predicted by the C–S–R model. Some of these studies examined target species most common at (and presumably most adapted to) low-fertility environments (Wilson and Keddy 1986, Reader and Best 1989, Reader 1990, Bonser and Reader 1995). As previously argued, the resource competition model predicts higher competition intensity at high versus low soil fertility for species adapted to low-fertility soil when competing against species adapted to high-fertility soil. These results could therefore be consistent with either model. Other studies have examined target species most abundant at high fertility or productivity (Kadmon and Schmida 1990, Kadmon 1995, Keddy et al. 2000) and calculated competition intensity by averaging across a range of target species (Wilson and Keddy 1986a, Twolan-Strutt and Keddy 1996). These studies tend to provide clearer support for the C–S–R model. A study by Sammul et al. (2000) found that both competition intensity and importance were greater at higher versus lower plant community productivity for a short-statured grass species (*Anthoxanthum odoratum*, whose pattern of abundance along the productivity gradient was unclear).

In contrast, other studies have found that competition intensity did not vary with soil fertility for at least some target species (Wilson and Keddy 1986a, Wilson and Shay 1990, Wilson and Tilman 1991, 1993, Reader et al. 1994, Belcher et al. 1995, Twolan-Strutt and Keddy 1996, Peltzer et al. 1998, Cahill 1999). Several studies that also measured below- versus aboveground competition found that belowground competition was stronger at low soil fertility, whereas aboveground competition was stronger at high soil fertility for these target species (Wilson and Tilman 1991, 1993, Twolan-Strutt and Keddy 1996, Peltzer et al. 1998, Cahill 1999, Belcher et al. 1995 found that aboveground competition was consistently insignificant). These results can be interpreted as support for the resource competition model, with the caveat that some species can be plastic in their competitive abilities, such that they are able to compete well at both low and high soil fertility. Differences in plasticity among species could explain why some studies find that the relationship between competition intensity and soil fertility varies with target species (DiTommaso and Aarssen 1991, Reader 1992, Wilson and Tilman 1995).

A meta-analysis of target competition experiments (Goldberg et al. 1999) revealed unexpected patterns. Competition intensity was constant along productivity gradients, as predicted by the resource competition model, when competition was measured as the effect of surrounding vegetation on growth. When competition was measured as effects on final biomass or survival, on the other hand, competition intensity decreased with increasing productivity, which is consistent with neither model.

Results from studies examining competition within experimental mixtures of low and high fertility-adapted species tend to support the resource competition model. A number of these studies found the predicted reversals of competitive ability among low and high fertility-adapted species with increases in soil fertility (low fertility-adapted species favored on low-fertility soil, high fertility-adapted species favored on high-fertility soil) (Helgadóttir and Snaydon 1985, McGraw and Chapin 1989, Aerts et al. 1990; see also Aerts et al. 1991). Similarly, species common to the high-fertility end of a shoreline gradient were found to be stronger competitors at high fertility than those more common to the low-fertility end of the shoreline gradient (Wilson and Keddy 1986b). Two other studies again suggested the existence of plasticity in competitive ability, finding that for most study species, relative (although not absolute) competition intensity was constant over a range of soil fertilities (Campbell and Grime 1992, Turkington et al. 1993).

One of the most straightforward ways of addressing whether competition intensity varies with soil fertility is to compare the intensity of aboveground, belowground, and total competition among low fertility-adapted species growing on low-fertility soil with that among high fertility-adapted species growing on high-fertility soil (Taylor et al. 1990). The same approach could be used to examine competition intensity along disturbance or other environmental gradients. This type of experiment has hardly ever been conducted. A partial example is provided by Wilson (1993), who examined belowground competition intensity in forest (higher fertility) versus prairie (lower fertility). In support of resource competition theory, Wilson found that belowground competition was stronger where soil resources were the lowest. More recently, Callaway et al. (2002) examined competition intensity for a total of 115 target species at low versus high alpine elevation (higher vs. lower primary productivity, interpreted as a stress gradient) at 11 locations around the world. Although it is not clear whether aboveground or both above- and belowground competition were examined, Callaway et al. found that competition was strongest where productivity was highest and that plant interactions were in fact facilitative on balance at high elevations, providing support for the C–S–R model.

In conducting this type of experiment, it is important to consider the soil heterogeneity of the study sites. For example, a site may have a matrix of low soil fertility but contain patches of higher-nutrient soil (and the converse may be true for a site of mostly high soil fertility). We consider the role of soil heterogeneity later in this chapter.

Disturbance Gradients

Effects of disturbance on competition intensity have been examined using a variety of experimentally imposed treatments (fire, tilling, clipping, trampling) and natural gradients (herbivory). The study species used in such studies vary in the extent to which they are adapted to each disturbance. Annual species are presumably adapted to tilling; perennial prairie grasses to fire, grazing (clipping), and possibly trampling; and species typically found where herbivory is intense are presumably adapted to herbivory. The C–S–R model (see also Taylor et al. 1990) predicts that species experience lower competition intensity in the presence of disturbance regardless of how well they or their competitors are adapted to the disturbance. In support of this, Reader (1992) found that exposure to herbivory appeared to decrease the intensity of competition (as measured by flower production) experienced by pasture species. In addition, annual tilling reduced the relative intensity of belowground (Wilson and Tilman 1995) and total competition (Wilson and Tilman 1993) for old-field species, regardless of whether they were annuals or perennials.

However, accounting for differences in species' adaptations to disturbance could be important in interpreting experimental results. Disturbance is thought to reduce competition intensity by reducing population density or biomass and so reducing demand for resources. But adaptation to disturbance (e.g., defense against herbivores) may require increased expenditures of energy and resources by organisms, that is, disturbance may result in fewer organisms that require more resources, so that the effective demand for resources may not really be changed at all by disturbance (see Holt 1985 and Chesson and Huntly 1997 for related discussion). Thus, whether disturbance reduces competition intensity may depend on whether the competing species are adapted to the disturbance (equivalently, on how novel the disturbance is). Competition intensity might be lower in the presence versus absence of a novel disturbance if the disturbance knocks species abundances far enough below the resource-supplying power of the environment. Alternatively, competition intensity among species similarly adapted to a disturbance may show no change in the presence versus absence of that disturbance. Competition among species similarly adapted to grazing (Turkington et al. 1993) and fire (Wilson and Shay 1990) may explain why no change in relative competition intensity with disturbance was typically observed in two studies of grassland species.

Of course, in the field, disturbance gradients may often be correlated with soil fertility or productivity gradients, with important consequences for understanding patterns in the intensity of plant competition. For example, degree of wave exposure along shorelines creates a gradient from nutrient-poor, disturbed (via wave action) conditions to nutrient-rich, undisturbed conditions (Wilson and Keddy 1986a). Patterns in competition intensity may reflect one or the other, or both of the underlying fertility and disturbance gradients. As another example, the exploitation ecosystem hypothesis (Fretwell 1977, Oksanen et al. 1981, Oksanen and Oksanen 2000) predicts that the degree of plant competition versus herbivory alternate along gradients of fertility or primary productivity, depending on whether conditions are productive enough to support carnivores that limit the impact of herbivores. Work in arctic-alpine plant communities has supported the exploitation ecosystem hypothesis, finding that plant competition is more intense and herbivory less intense at higher habitat fertilities and productivities (Olofsson et al. 2002, see also Sammul et al. 2006).

TRAITS

In most cases, it is not feasible to experimentally manipulate traits within any one species. The influence of traits on competitive ability must therefore be examined by correlating traits (e.g., R*s, root to shoot ratio, height, seed size) across species with some measure of species performance in competition. Traits are usually measured on species grown in the absence of interspecific competition. Relatively few studies have examined the relationship between traits and competitive ability; therefore, this section includes greenhouse studies and studies conducted in artificial soils as well as field studies.

An approach that allows comparison of a large number of species is the neighborhood design of Gaudet and Keddy (1988). The competitive ability of different neighbor species is measured as relative ability to suppress the growth of a common target species, or phytometer. Multiple regression is used to determine which morphological and physiological traits of each neighbor species correlate with competitive ability. The generality of results can be assessed by testing the neighbor species (or some subset of them) against different phytometer species. Thus far, this approach has been used to compare species of highly productive wetland communities growing in high-fertility soil (Gaudet and Keddy 1988). Results showed that competitive ability was most strongly correlated with aboveground biomass, followed by total and belowground biomass. Plant height, canopy diameter and area, and leaf shape were of secondary importance. These results could be interpreted as supporting either the C–S–R or resource competition model. The C–S–R model predicts that large size and height are among traits that allow acquisition of both above- and belowground resources at high soil fertility, whereas the resource competition model emphasizes large aboveground size as important for light acquisition at high soil fertility.

To date, relatively few traits and species have been evaluated in most other studies of plant competitive ability. As a group, however, these studies support resource competition theory in suggesting that traits related to acquisition of aboveground resources are important at high soil fertility, and traits related to acquisition of belowground resources are important at low soil fertility.

For example, size (biomass, seed size, height) is often a good predictor of competitive ability when light competition becomes important (i.e., conditions of high soil fertility and consequently high productivity) (Black 1958, Schoener 1983, Dolan 1984, Gross 1984, Stanton 1984, Weiner and Thomas 1986, Goldberg 1987, Miller and Werner 1987, Bazzaz et al. 1989, Reekie and Bazzaz 1989, Houssard and Escarré 1991, Wilson and Tilman 1991, Grace et al. 1992). A reason suggested for this is that light is a directional resource, and thus competition for light is asymmetric; large plants are often able to obtain proportionately more light than smaller plants (Weiner and Thomas 1986, Thomas and Weiner 1989). Asymmetric competition may explain the existence of competitive hierarchies and transitivity (i.e., consistent rankings of competitive ability; A > B, B > C, therefore A > C) in some competition studies (Keddy and Shipley 1989; but see Silvertown and Dale 1991 for a critique of these studies).

However, size is not always a good indicator of ability to preempt light. The degree to which biomass or productivity negatively correlates with light interception may, instead, depend on canopy architecture. For example, Tremmel and Bazzaz (1993) found that neighbor biomass was a poor predictor of the ability of neighbors to suppress targets, but that an index of neighbor light interception was a good predictor of neighbor competitive ability. In addition, Wilson (1994) found that size made no difference in seedling competitive ability when seedlings were competing against mature (larger) vegetation.

Conversely, when competition is mostly for soil resources (i.e., conditions of low soil fertility and consequently low productivity), root allocation, specific root length, and ability to retain or to efficiently use nutrients become good predictors of competitive success (Eissenstat and Caldwell 1988, McGraw and Chapin 1989, Tilman and Wedin 1991, Berendse et al. 1992, Wilson 1993a,b). R*s for soil resources also correspond with competitive success at low soil fertility (Tilman and Wedin 1991), and Tilman (1990) has shown analytically how R* summarizes the effects of these other traits on ability to acquire resources.

SOIL HETEROGENEITY

Resource heterogeneity is a feature of all natural environments, and its potential importance for plant competition and coexistence has been a recent focus of competition research. Soil resource availability differs at large scales of several meters or more (Lechowicz and Bell 1991, Jackson and Caldwell 1993a,b, Robertson et al. 1993, Gross et al. 1995, Ryel et al. 1996, Cain et al. 1999, Lister et al. 2000, Guo et al. 2002), and at small scales of less than a meter

(Ryel et al. 1996, Farley and Fitter 1999b). Soil resources also change temporally, within a growing season (Ryel et al. 1996, Cain et al. 1999, Farley and Fitter 1999b) or across successional time (Grace 1991b, Gross et al. 1995). Although large-scale, long-term resource variation contributes to patterns such as succession and gradients in species composition and diversity, small-scale heterogeneity detectable by individual plants has the potential to influence competitive interactions between individuals. Thus, this discussion focuses primarily on small-scale variation. It also focuses on heterogeneity of soil resources, rather than light, since most studies of heterogeneity at this scale address only soil resources.

The effects of temporal heterogeneity of resources on competitive interactions have largely been neglected. Goldberg and Novoplansky (1997) proposed a model of two-phase resource dynamics, in which the intensity of competition differs between periods of resource pulses (e.g., periods of rainfall in a water-limited system) and interpulse periods. This model attempts to reconcile the C–S–R and resource competition models. During pulses, competition is intense as plants attempt to preempt available resources (despite the influx of resources, pulse periods can thus be interpreted as times when resource demand by plants outweighs resource supply). During interpulse periods, tolerance of low resource levels determines survival. In productive environments, pulses are frequent and competition during pulses dominates. In unproductive environments, interpulse periods are longer. Competition may be the main influence on the community (as predicted by resource competition) if interpulse resource levels are determined primarily by plant uptake and if plant survival in interpulses depends on growth during pulses (i.e., resource demand on average outweighs resource supply). On the other hand, the community may be influenced mostly by stress tolerance (as predicted by C–S–R) if interpulse resource levels are determined more by physical processes such as drainage or leaching and if plant survival in interpulses is uncorrelated with competitive success during pulses (i.e., resource supply on average outweighs resource demand, which might occur when a species is better adapted to pulse than to interpulse periods, such that the onset of an interpulse period greatly reduces survival or growth, and thus demand for resources). To date, little data exist with which to test the two-phase resource dynamics hypothesis, and Goldberg and Novoplansky's (1997) call for more studies that focus on competition under pulsed resource regimes is well justified.

Most research on heterogeneity and plant competition has focused on effects of spatial heterogeneity. Predictions regarding the effects of spatial heterogeneity within a community depend on the size of resource patches, which may be smaller or larger than an individual plant's rooting zone.

For patches larger than (or equal to) an individual rooting zone, heterogeneity may generate diversity. In the resource competition model, heterogeneity promotes coexistence of species at any point along a soil fertility or productivity gradient because different patches represent areas where different species are the best competitors (Tilman 1982, Tilman and Pacala 1993). Although the C–S–R model does not explicitly address heterogeneity at this scale, it would presumably make a similar prediction, at least for productive environments where competition is predicted to be most intense. Several observational studies have reported positive correlations between environmental heterogeneity and species diversity (Harner and Harper 1976, Lundholm and Larson 2003, Davies et al. 2005). In addition, field studies have reported shifts in competitive abilities with naturally occurring local-scale soil heterogeneity (e.g., Reynolds et al. 1997). Experiments that have manipulated resource heterogeneity, on the other hand, find that increasing average soil resource levels (or decreasing light) reduces diversity, but heterogeneity has no effect (Collins and Wein 1998, Stevens and Carson 2002, Baer et al. 2004, Reynolds et al. in press, but see Vivian-Smith 1997). For instance, Stevens and Carson (2002) manipulated both average availability and heterogeneity of light in an old field community, and found that average light had a strong effect on diversity whereas degree of heterogeneity had no effect. Similarly, Reynolds et al. (submitted) found that spatial heterogeneity of soil nutrients did not affect species diversity, and that species did not sort into areas of preferred nutrient availability. Instead, dominance of clonal species increased with both homogeneous and heterogeneous fertilizer applications. The authors suggest that, by foraging over areas larger than the patches created, clonal species may constrain the community response to heterogeneity.

Patches within a single plants' rooting zone may also affect the outcome of competition, although such patches are treated differently under C–S–R versus resource competition models. According to the resource competition model, these patches are not relevant. An individual plant uniformly depletes resources within its rooting zone, and the effects of varying resource levels on the plant are averaged out across the entire rooting zone (Tilman and Pacala 1993). In contrast, the C–S–R model predicts that morphological plasticity and active foraging for resource-rich patches are important traits contributing to competitive ability (Grime 1977, Grime et al. 1991). Therefore, ability to respond to fine-scale heterogeneity should be favored in productive environments, where both the resources available for growing new roots and the reserves they can tap into are high (but see Reynolds and D'Antonio 1996 for a different prediction). Furthermore, plasticity is predicted to be coarse-grained (altering root:shoot ratio, for example) for dominant species and fine-grained (altering within root allocation) for subordinate species (Campbell et al. 1991, Grime et al. 1991). There has been little work testing the prediction that foraging for resource patches is most adaptive in productive environments. However, many studies have addressed the C–S–R predictions that foraging influences competition and that species may employ different strategies for foraging.

Clearly, many species do exhibit plastic responses to resource-rich patches in the soil. Generally, both root growth and nutrient uptake rates increase in resource-rich patches, although the degree of increase varies greatly among species (Robinson 1994). In isolated plants, proliferation of roots does not appear to be adaptive at first: because nitrate is highly mobile in the soil, intense proliferation is not necessary to access nitrogen and does not in fact increase total nitrogen uptake (Robinson 1996, Robinson et al. 1999). However, proliferation does offer an advantage in a competitive situation: when plants of different species compete, the species with the greater proliferation response captures more nitrogen from a patch (Hodge et al. 1999, Robinson et al. 1999, Hodge 2003, Kembel and Cahill 2005).

Campbell et al. (1991) offered a framework for classifying species' morphological responses to heterogeneity by defining three aspects of foraging: scale (ability to explore a large soil volume), precision (ability to proliferate roots extensively in resource-rich patches), and rate (ability to reach patches quickly). They also proposed that there should be a trade-off between foraging scale and rate, with dominant species using high-scale foraging and subordinate species using high-precision foraging (Campbell et al. 1991). Several studies have measured foraging scale and precision for sets of species. These studies confirm that foraging traits are highly variable between species, although trade-offs have not always been found (Campbell et al. 1991, Einsmann et al. 1999, Farley and Fitter 1999a, Wijesinghe et al. 2001, Rajaniemi and Reynolds 2004, Kembel and Cahill 2005).

The effects of resource heterogeneity and species' foraging strategies on competition are inconsistent. The outcome of two-species competition sometimes differs between homogeneous and heterogeneous soils (Fransen et al. 2001, Bliss et al. 2002, Day et al. 2003b), and sometimes does not (Cahill and Casper 1999). In those pairwise experiments in which heterogeneity matters, the more precise forager tends to be the better competitor in patchy soils (Fransen et al. 2001, Bliss et al. 2002). However, in a study testing for correlations between competitive effect and foraging traits, competitive hierarchies were not affected by resource heterogeneity, and a species' foraging scale was a better predictor of its competitive ability than its foraging precision (Rajaniemi, in press). Any changes in competition in heterogeneous soils are also predicted to affect population and community structure

(Hutchings et al. 2003). For example, if large individuals can preempt resources, soil heterogeneity might lead to size-asymmetric competition (Schwinning and Weiner 1998), which would increase size inequality in populations. Again, some studies support this prediction, finding that size inequality is greater on patchy soils (Fransen et al. 2001, Facelli and Facelli 2002, Day et al. 2003a) whereas others do not (Casper and Cahill 1996, 1998, Blair 2001). Finally, effects of heterogeneity on competition might ultimately influence community structure. In the single study addressing this question, Wijeshinghe et al. (2005) showed that composition in an experimental community of herbaceous plants differed between homogeneous and heterogeneous soils, but species diversity did not.

Some of the inconsistencies described earlier might be resolved by considering the effects of patch characteristics. Patches may differ in size, contrast with background soil, spatial distribution, duration, and temporal predictability (Fitter 1994, Hutchings et al. 2003). The effects of heterogeneity on competitive interactions and population and community structure may depend on the nature of the patches (Hutchings et al. 2003), but researchers have only begun to address these factors. A few studies have failed to find effects of patch characteristics. Fransen et al. (2001) showed that the presence of patches influenced competition between two grass species, but that the size of patches had no effect. In this case, total nutrient levels were equal in all treatments, meaning that smaller patches also had greater nutrient concentration (and therefore greater contrast). Number of patches (also confounded with contrast) did not alter species' competitive effects in a target-neighbor experiment (Rajaniemi, submitted). On the other hand, a study of an experimental community provides intriguing evidence that patch characteristics may affect competitive outcomes. In this study, patch number did not affect diversity, but did affect species composition (Wijesinghe et al. 2005). In addition, temporal predictability of patches had one subtle effect on community structure: the predictable treatment had a greater biomass of species considered colonists. The predictably low-nutrient patches may have served as refuges from competition for these species (Wijesinghe et al. 2005). Future research on the importance of soil resource heterogeneity should further pursue the effects of patch characteristics, to unravel when and where we might expect strong effects of heterogeneity. Other issues still to be resolved include the importance of temporal resource pulses, the role of heterogeneity in promoting diversity in environments of differing productivity, and the adaptive value of foraging in environments of differing productivity.

CONCLUSIONS AND FUTURE DIRECTIONS

Plants compete when demands on resources (water, light, mineral nutrients, CO_2) exceed resource supply. Plant competition has been measured in a variety of habitats, from desert to temperate grassland, yet controversy remains about its significance for the ecology and evolution of species and the structure of communities, about whether the intensity of competition varies with environmental conditions, and about what traits are important to competitive ability. The need to understand these issues has only increased in the face of global changes, such as nitrogen enrichment, exotic invasions, and altered climate, that have the potential to alter competitive interactions and change community structure. By definition, exotic invasive species competitively exclude existing species in the communities to which they are introduced. Invasions are also generally associated with predator release, disturbance, and increased resource availability (Tilman 1999, Shea and Chesson 2002, Hierro et al. 2005). Invasions thus provide a particularly good opportunity to test competition theory as well as to assess the relative role of competition versus other factors, such as herbivory, in structuring plant communities.

Further progress in resolving controversy over the importance of plant competition, its intensity in different habitats, and the traits most important in predicting competitive outcomes requires studies that (1) distinguish between importance and intensity, (2) examine a wide range of species (including native, noninvasive exotic, and invasive exotic species) and traits (including R*, a potentially useful summary trait), (3) include as many environmental constraints (e.g., soil fertility, disturbance) as possible, including spatially heterogeneous and temporally pulsed resources, and (4) pay close attention to the extent to which each study species is adapted to the particular environmental constraints used. It may be useful to distinguish two aspects of the question of whether competition intensity changes along fertility (or disturbance) gradients: (1) Is competition equally intense among low fertility-adapted species growing on low-fertility soil and among high fertility-adapted species growing on high-fertility soil (low disturbance-adapted species growing in undisturbed environments vs. high disturbance-adapted species growing in disturbed environments)? and (2) Does competition play a role in maintaining community boundaries, for example, in preventing low fertility-adapted species from invading high-fertility soil and vice versa?

ACKNOWLEDGMENTS

The National Science Foundation supported this research through grants DEB-0235767 to H.L.R. and DEB-0129493 to T.K.R. and H.L.R. Thanks to Kay Gross for comments that improved an earlier version of this chapter and to Karen Haubensak for fruitful discussions.

APPENDIX 15.1: EXPERIMENTAL DESIGN AND COMPETITION INDICES

There are four main types of experimental designs for measuring competition between two species: (1) the replacement series (or substitutive design); (2) the additive design; (3) the bivariate factorial design; and (4) the neighborhood design. These designs have produced a variety of competition indices (Table 15.A1). In a replacement series (de Wit 1960), species are grown in pure stands and in mixtures, with one species at proportion p of its pure stand density and the other species at proportion $(1-p)$ of its pure stand density (Connolly 1986). The pure stand densities of each species need not be equal; however, the standard replacement design consists of one pure stand of each species grown at the same density, and a single 1:1 mixture of the two species, with each at half the pure stand density (Figure 15.A4a). The replacement approach thus attempts to measure interspecific competition relative to intraspecific competition.

The additive design also uses pure stands and mixtures of two species, but the density of each species is the same in pure and mixed stands. The standard additive design consists of pure stands of each species at the same density and a single 1:1 mixture of the two species, with each at its pure stand density (Figure 15.A4b). The additive approach thus attempts to measure the magnitudes of interspecific competition while holding intraspecific competition constant. Knowledge of the magnitudes of both types of competition is necessary for predicting the effects of competition on species distribution and abundance (Roughgarden 1979, Underwood 1986). Additive designs can easily be extended to measure absolute magnitudes of intraspecific competition (Underwood 1986).

For both designs, many indices of competition are based on performance of species in mixture compared with their performances in pure stand. Replacement designs have been criticized because this comparison confounds change in intraspecific density with change in interspecific density (Firbank and Watkinson 1985, Connolly 1986, Underwood 1986, Silvertown and Dale 1991, Snaydon 1991). Such confounding causes dependence of the competition indices on species densities in pure stands versus mixtures (Connolly 1986, Snaydon 1991), as well as on species' proportions and patterns of response to density (Snaydon 1991). Inherent differences in species' sizes can also cause dependence of competition indices

TABLE 15.A1
Common Indices of Competition for Replacement, Additive, Simplified Neighbor, Neighbor, and Bivariate Factorial Designs

Design[a]	Index	Abbreviation	Formula[b]		Reference
			Interspecific[c]	Intraspecific	
r, a	Relative yield total	RYT	$(Y_{ij}/Y_{ii}) + (Y_{ji}/Y_{jj})$	—	de Wit (1960)
r, a	Relative crowding coefficient[d]	RCC	$(Y_{ij}/Y_{ji})/(Y_{ii}/Y_{jj})$	—	Harper (1977)
r, a	Aggressivity	A	$1/2[(W_{ij}/W_{ii}) - (W_{ji}/W_{jj})](r)$	—	McGilchrist and Trenbath (1971)
			$(W_{ij}/W_{ii}) - (W_{ji}/W_{jj})(a)$		
			or		
			$(Y_{ij}/Y_{ii}) - (Y_{ji}/Y_{jj})(r, a)$		Snaydon (1991)
r, a	Absolute severity of competition	ASC	$\log_{10}(W_{i0}/W_{ij})$	$\log_{10}(W_{i0}/W_{ii})$	
r, a	Relative severity of competition	RSC	$\log_{10}(W_{ii}/W_{ij})$	—	Snaydon (1991)
a, sn	Absolute competition intensity	Absolute CI	$Y_{ii} - Y_{ij}$ (a)		Campbell and Grime (1992)
			$Y_{i0} - Y_{ij}$ (a, sn)	$Y_{i0} - Y_{ii}$ (a, sn)	
a, sn	Relative competition intensity	Relative CI	$(Y_{ii} - Y_{ij})/Y_{ii}$ (a)		Campbell and Grime (1992)
			$(Y_{i0} - Y_{ij})/Y_{i0}$ (a, sn)	$(Y_{i0} - Y_{ii})/Y_{i0}$ (a, sn)	Wilson and Tilman (1991)
a, sn	Relative neighbor effect	RNE	$Y_{i0} - Y_{ij}/(\mathrm{Max}[Y_{i0}, Y_{ij}])$		Markham and Chanway (1996)
a, sn	Log response ratio	Ln(RR)	$\mathrm{Ln}\,(Y_{i0}/Y_{ij})$		Hedges et al. (1999)
a, sn	Relative interaction index	RII	$Y_{ij} - Y_{i0}/(Y_{ij} + Y_{i0})$		Armas et al. (2004)
n	Per-amount competitive effect	X	$W_i = W_{i0}/(1 + XB_j)$	$W_i = W_{i0}/(1 + XB_i)$	Goldberg (1987)
a	Competition coefficient	α	$W_i = W_{i0}[1 + a_i(N_i + \alpha_{ij}N_j)]$	—	Firbank and Watkinson (1990)
b	Competition coefficient	c	$W_i = W_{i0}/(1 + N_i^{c_{ii}} + N_j^{c_{ij}})$	Same	Law and Watkinson (1987)
n	Interference coefficient[e]	c	$W_i = W_{i0}/(1 + c_{ii}N_i + c_{ij}N_j)$	Same	Silander and Pacala (1990)

[a] r, replacement design; a, additive design; sn, simplified neighbor design; n, neighbor design; b, bivariate factorial design.
[b] Variables: Y, yield per unit area; W, yield or weight per plant; B, total neighbor biomass; N, total number of neighbors; a and b, fitted regression parameters. Subscripts: i, species i; j, species j; i0, species i grown with no competition; ii, species i grown in pure stand; jj, species j grown in pure stand; ij, for Y and W, refers to species i when grown in mixture with species j, whereas for α and c, refers to effect of species j on species i; ji, for Y and W, refers to species j when grown in mixture with species i, whereas for α and c, refers to effect of species i on species j.
[c] Note that interspecific formulas are given in terms of species i when grown with species j.
[d] For 1:1 ratios only (see Willey and Rao 1980, for application to other component ratios). Snaydon (1991) recommends $\log_{10}$ or loge transformations to facilitate statistical and biological analysis.
[e] See Pacala and Silander (1985, 1990), and Silander and Pacala (1985, 1990), for methods of determining the best neighborhood size.

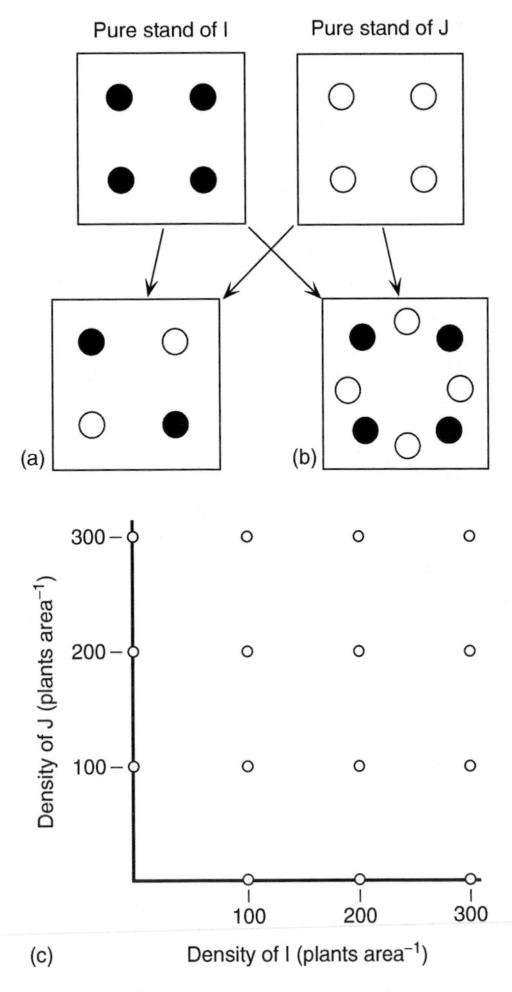

FIGURE 15.A4 Planting arrangements for pure stands of I (filled circle) and J (open circles) and for (a) replacement mixtures and (b) additive mixtures of I with J. (c) Diagrammatic representation of a bivariate factorial design, in which densities of the two species are varied independently. (*Source*: Snaydon, R.W., *J. Appl. Ecol.*, 28, 930, 1991. With permission.)

on species' densities (Connolly 1986, Grace et al. 1992). Sackville Hamilton (1994) defends replacement designs, but only for a restricted set of conditions (Snaydon 1994).

Because species' densities are the same in pure stands and mixtures in additive designs, competition indices are unaffected by pure stand densities or by species' patterns of response to density (Snaydon 1991). However, additive (and replacement) designs are, of course, sensitive to total density. If total densities are low enough, resources will not be limiting, and species will not compete even if they would experience strong competition at higher densities (Taylor and Aarssen 1989). Additive designs may also be sensitive to species proportions, that is, it is possible that relative competitive abilities change as species proportions in mixture change. To deal with these issues, additive designs may be repeated over a range of densities or proportions, yielding additive series (Snaydon 1991). When species' densities are varied factorially, producing a complete range of total densities and species' proportions, the resulting design has been termed both an addition series (Firbank and Watkinson 1985, 1990) and a bivariate factorial design (Snaydon 1991).

Additive series and bivariate factorial designs can be analyzed by regressing weight per plant in mixture on the density and proportion of each component species (Firbank and Watkinson 1985, 1990) or by calculating separate competition indices for each mixture (Snaydon 1991).

A modification of the additive design that has become popular is the neighborhood design, in which the density of a neighbor species around a single individual of a target species is varied (Goldberg and Werner 1983, Goldberg 1987) (Figure 15.A5). Performance (e.g., growth rate, survival) of the target species can be regressed on the amount (e.g., biomass, density) of neighbor species. The slope of this regression measures the per-amount competitive effect of the neighbor species on the target species (Goldberg and Werner 1983), and the coefficient of determination measures the importance of competition (the proportion of variation in performance that is explained by variation in neighbor amount) (Welden and Slauson 1986). The total competitive effect of a neighbor species is given by the product of its per-amount effect and its abundance. Per-biomass effects are indicative of physiological and morphological effects, whereas per-individual effects are often related to inherent size (Goldberg 1987). This approach has also been developed for populations of two species whose densities vary randomly, and can also be used to measure intraspecific competition, to compare the effects of different neighbor species on one target species or the responses of different target species to one neighbor species, and to account for the effects of neighbor distance or angular dispersion (Mack and Harper 1977, Weiner 1982, Pacala and Silander 1985, 1987, Silander and Pacala 1985). Gaudet and Keddy (1988) have used a simplified neighborhood approach to identify traits important to competitive ability (see "Conclusions and Future Directions").

All of the designs mentioned earlier measure competition between only two species at a time, yet natural communities are typically composed of many more than two species. The number of experiments necessary to study all possible pairs of species in a multispecies assemblage quickly becomes unreasonably large—even without considering intraspecific interactions, different species' densities or proportions, or the effect of environmental

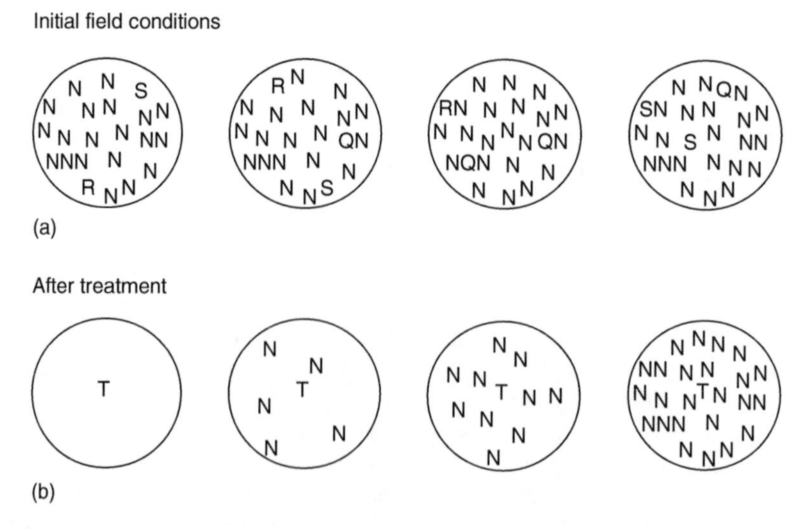

FIGURE 15.A5 Example of the experimental design for evaluating competitive effects of one neighbor species (N) on a target species (T). R, Q, and S represent individuals not belonging to the neighbor species selected for study. Only four steps of the neighbor density gradient (after treatment) are shown: the experiments must include a much wider range of densities to estimate accurately the slope of the regression of target performance on amount of neighbors. (*Source*: Goldberg, D.E. and Werner, P.A., *Am. J. Bot.*, 70, 1098, 1983. With permission.)

heterogeneity on competitive interactions (Keddy 1990, Tilman 1990). Ability to predict the outcome of competition in a community from pairwise competition experiments may also be complicated by indirect effects. Indirect interactions between two species arise via direct interactions with a third species (Connell 1990).

Many studies have used a simplified neighborhood approach to multispecies competition: comparing the performance of target individuals in the presence versus the absence of neighbor vegetation (Reader and Best 1989, Reader 1990, Reader et al. 1994, Wilson 1994), or in the presence versus absence of neighbor roots (Wilson 1993a), or roots and shoots (Aerts et al. 1991, Wilson and Tilman 1991, 1993, 1995, Wilson 1993b, Belcher et al. 1995, Gerry and Wilson 1995). This approach has been useful in examining changes in total, above- and belowground competition along environmental gradients (such as soil fertility gradients; see the section on "Soil Heterogeneity").

Another approach to multispecies competition is to compare the performance of each species when grown in a single, multispecies community to that when grown in pure culture (additive design: Campbell and Grime 1992, replacement series design: Turkington et al. 1993). Alternatively, species can be removed from existing communities, and the performance of the remaining species can be compared with that in no-removal controls (Pinder 1975, Abul-Fatih and Bazzaz 1979, Fowler 1981, Silander and Antonovics 1982, Gurevitch and Unnasch 1989, Keddy 1989). Goldberg et al. (1995) suggested a design called the community density series that additionally varies neighbor abundance. In this design, community density is varied while initial relative abundances of component species are held constant. The slope of a regression of eventual relative abundance on initial community density for each species yields a measure of its community-context competitive ability. Although these multispecies approaches do not separate the effects of different neighbor species, or the effects of total neighbor biomass or density from per-amount effects of neighbors, they provide information about species' competitive performance under conditions more similar to that in the field.

Tilman (1990) suggested that a focus on the mechanism of competition—resource reduction (see the section on "Competition Intensity")—leads to the simplest and the most predictive approach to understanding multispecies competition. Theoretically, the species that reduces the limiting resource (e.g., available nitrogen, water, light) to the lowest level in monoculture is eventually able to displace all competitors (Tilman 1982, 1988). Many fewer experiments are required to determine the R* (or R*s, if more than one resource is limiting) of each species in a community than to run pairwise competition experiments (Tilman 1990). R*s for soil nitrogen were found to predict the outcome of pairwise competition among four perennial grass species (Tilman and Wedin 1991). Similarly, greater rates of water extraction and greater depletion of available water were correlated with the superior competitive ability of one aridland tussock grass compared with another (Eissenstat and Caldwell 1988). This approach is yet to be applied to other terrestrial plant systems or to competition among multispecies assemblages.

REFERENCES

Aarssen, L.W., 1984. On the distinction between niche and competitive ability: implications for coexistence theory. Acta Biotheoretica 33: 67–83.

Aarssen, L.W., 1989. Competitive ability and species coexistence: a 'plant's-eye' view. Oikos 56: 386–401.

Abul-Fatih, H.A. and F.A. Bazzaz, 1979. The biology of *Ambrosia trifida.* L.I. Influence of species removal on the organization of the plant community. New Phytologist 83: 813–816.

Aerts, R., F. Berendse, H. de Caluwe, and M. Schmitz, 1990. Competition in heathland along an experimental gradient of nutrient availability. Oikos 57: 310–318.

Aerts, R., R.G.A. Boot, and P.J.M. van der Aart, 1991. The relation between above- and belowground biomass allocation patterns and competitive ability. Oecologia 87: 551–559.

Armas, C., R. Ordiales, and F.I. Pugnaire, 2004. Measuring plant interactions: a new comparative index. Ecology 85: 2682–2686.

Baer, S.G., J.M. Blair, S.L. Collins, and A.K. Knapp, 2004. Plant community responses to resource availability and heterogeneity during restoration. Oecologia 139: 617–629.

Bais, H.P., R. Vepachedu, S. Gilroy, R.M. Callaway, and J.M. Vivanco, 2003. Allelopathy and exotic plant invasion: from molecules and genes to species interactions. Science 301: 1377–1380.

Bannister, P., 1976. Introduction to Physiological Plant Ecology. Blackwell Scientific, Oxford.

Bazzaz, F.A., K. Garbutt, E.G. Reekie, and W.E. Williams, 1989. Using growth analysis to interpret competition between a C_3 and a C_4 annual under ambient and elevated CO_2. Oecologia 79: 223–235.

Belcher, J.W., P.A. Keddy, and L. Twolan-Strutt, 1995. Root and shoot competition intensity along a soil depth gradient. Journal of Ecology 83: 673–682.

Berendse, F., W.Th. Elberse, and R.H.M.E. Geerts, 1992. Competition and nitrogen loss form plants in grassland ecosystems. Ecology 73: 46–53.

Black, J.N., 1958. Competition between plants of different initial seed sizes in swards of subterranean clover (*Trifolium subterraneum* L.) with particular reference to leaf area and the light microclimate. Australian Journal of Agricultural Research 9: 299–318.

Blair, B., 2001. Effect of soil nutrient heterogeneity on the symmetry of belowground competition. Plant Ecology 156: 199–203.

Bliss, K.M., R.H. Jones, R.J. Mitchell, and P.P. Mou, 2002. Are competitive interactions influenced by spatial nutrient heterogeneity and root foraging behavior? New Phytologist 154: 409–417.

Bonser, S.P. and R.J. Reader, 1995. Plant competition and herbivory in relation to vegetation biomass. Ecology 76: 2176–2183.

Brooker, R., Z. Kikvidze, F.I. Pugnaire, R.M. Callaway, P. Choler, C.J. Lortie, and R. Michalet, 2005. The importance of importance. Oikos 109: 63–70.

Cahill, J.F. and B.B. Casper, 1999. Growth consequences of soil nutrient heterogeneity for two old-field herbs, *Ambrosia artemisiifolia* and *Phytolacca americana*, grown individually and in combination. Annals of Botany 83: 471–478.

Cain, M.L., S. Subler, J.P. Evans, and M.J. Fortin, 1999. Sampling spatial and temporal variation in soil nitrogen availability. Oecologia 118: 397–404.

Callaway, R.M. and W.M. Ridenour, 2004. Novel weapons: invasive success and the evolution of increased competitive ability. Frontiers in Ecology and the Environment 8: 436–443.

Callaway, R.M., R.W. Brooker, P. Choler, Z. Kikvidze, C.J. Lortie, R. Michale, L. Paolini, F.I. Pugnaire, B. Newingham, E.T. Aschehoug, C. Armas, D. Kikodze, and B.J. Cook, 2002. Positive interactions among alpine plants increase with stress. Nature 417: 844–848.

Campbell, B.D. and J.P. Grime, 1992. An experimental test of plant strategy theory. Ecology 73: 15–29.

Campbell, B.D., J.P. Grime, and J.M.L. Mackey, 1991. A trade-off between scale and precision in resource foraging. Oecologia 87: 532–538.

Casper, B.B. and J.F. Cahill Jr., 1996. Limited effects of soil nutrient heterogeneity on populations of *Abutilon theophrasti* (Malvaceae). American Journal of Botany 83: 333–341.

Casper, B.B. and J.F. Cahill, 1998. Population-level responses to nutrient heterogeneity and density by *Abutilon theophrasti* (Malvaceae): an experimental neighborhood approach. American Journal of Botany 85: 1680–1687.

Chesson, P. and N. Huntly, 1997. The roles of harsh and fluctuating conditions in the dynamics of ecological communities. The American Naturalist 150: 519–553.

Collins, B. and G. Wein, 1998. Soil heterogeneity effects on canopy structure and composition during early succession. Plant Ecology 138: 217–230.

Connell, J.H., 1983. On the prevalence and relative importance of interspecific competition: evidence form field experiments. American Naturalist 122: 661–696.

Connell, J.H., 1990. Apparent versus "real" competition in plants. In: J.B. Grace and D. Tilman, eds. Perspectives on Plant Competition. Academic Press, San Diego, CA, pp. 9–26.

Connolly, J., 1986. On difficulties with replacement-series methodology in mixture experiments. Journal of Applied Ecology 23: 125–137.

Darwin, C., 1859. On the Origin of Species. Harvard University Press, Cambridge, MA.

Davies, K.F., P. Chesson, S. Harrison, B.D. Inouye, B.A. Melbourne, and K.J. Rice, 2005. Spatial heterogeneity explains the scale dependence of the native-exotic diversity relationship. Ecology 86: 1602–1610.

Day, K.J., M.J. Hutchings, and E.A. John, 2003a. The effects of spatial pattern of nutrient supply on yield, structure and mortality in plant populations. Journal of Ecology 91: 541–553.

Day, K.J., E.A. John, and M.J. Hutchings, 2003b. The effects of spatially heterogeneous nutrient supply on yield, intensity of competition and root placement patterns in *Briza media* and *Festuca ovina.* Functional Ecology 17: 454–463.

de Wit, C.T., 1960. On competition. Verslagen van Landouwkundige Onderzoekingen 66: 1–82.

Diaz, S., J.P. Grime, J. Harris, and E. McPherson, 1993. Evidence of a feedback mechanism limiting plant response to elevated carbon dioxide. Nature 364: 616–617.

DiTommaso, A. and L.W. Aarssen, 1991. Effect of nutrient level on competition intensity in the field for three coexisting grass species. Journal of Vegetation Science 2: 513–522.

Dolan, R.W., 1984. The effect of seed size and maternal source on individual seed size in a population of *Ludwigia leptocarpa* (Onagraceae). American Journal of Botany 71: 1302–1307.

Einsmann, J.C., R.H. Jones, M. Pu, and R.J. Mitchell, 1999. Nutrient foraging traits in 10 co-occurring plant species of contrasting life forms. Journal of Ecology 87: 609–619.

Eissenstat, D.M. and M.M. Caldwell, 1988. Competitive ability is linked to rates of water extraction. A field study of two aridland tussock grasses. Oecologia 75: 1–7.

Facelli, E. and J.M. Facelli, 2002. Soil phosphorus heterogeneity and mycorrhizal symbiosis regulate plant intra-specific competition and size distribution. Oecologia 133: 54–61.

Farley, R.A. and A.H. Fitter, 1999a. The responses of seven co-occurring woodland herbaceous perennials to localized nutrient-rich patches. Journal of Ecology 87: 849–859.

Farley, R.A. and A.H. Fitter, 1999b. Temporal and spatial variation in soil resources in a deciduous woodland. Journal of Ecology 87: 688–696.

Firbank, L.G. and A.R. Watkinson, 1985. On the analysis of competition within two species mixtures of plants. Journal of Applied Ecology 22: 503–517.

Firbank, L.G. and A.R. Watkinson, 1990. On the effects of competition: from monocultures to mixtures. In: J.B. Grace and D. Tilman, eds. Perspectives on Plant Competition. Academic Press, San Diego, CA, pp. 165–192.

Fitter, A.H., 1994. Architecture and biomass allocation as components of the plastic response of root systems to soil heterogeneity. In: M.M. Caldwell and R.W. Pearcy, eds. Exploitation of Environmental Heterogeneity by Plants: Ecophysiological Processes Above- and Belowground. Academic Press, San Diego CA, pp. 305–323.

Fowler, N., 1981. Competition and coexistence in a North Carolina grassland. II. The effects of the experimental removal of species. Journal of Ecology 69: 843–854.

Fransen, B., H. de Kroon, and F. Berendse, 2001. Soil nutrient heterogeneity alters competition between two perennial grass species. Ecology 82: 2534–2546.

Fretwell, S.D., 1977. The regulation of plant communities by food chains exploiting them. Perspectives in Biology and Medicine 20: 169–185.

Gaudet, C.L. and P.A. Keddy, 1988. A comparative approach to predicting competitive ability from plant traits. Nature 334: 242–243.

Gerry, A.K. and S.D. Wilson, 1995. The influence of initial size on the competitive responses of six plant species. Ecology 76: 272–279.

Goldberg, D.E., 1987. Neighborhood competition in an old-field plant community. Ecology 68: 1211–1223.

Goldberg, D.E. and P.A. Werner, 1983. Equivalence of competitors in plant communities: a null hypothesis and field experimental approach. American Journal of Botany 70: 1098–1104.

Goldberg, D.E. and A.M. Barton, 1992. Patterns and consequences of interspecific competition in natural communities: a review of filed experiments with plants. American Naturalist 139: 771–801.

Goldberg, D.E. and A. Novoplansky, 1997. On the relative importance of competition in unproductive environments. Journal of Ecology 85: 409–418.

Goldberg, D.E., R. Turkington, and L. Olsvig-Whittaker, 1995. Quantifying the community-level consequences of competition. Folia Geobotanica Phytotaxonomica 30: 231–242.

Goldberg, D.E., T. Rajaniemi, J. Gurevitch, and A. Stewart-Oaten, 1999. Empirical approaches to quantifying interaction intensity: competition and facilitation along productivity gradients. Ecology 80: 1118–1131.

Grace, J.B., 1991a. A clarification of the debate between Grime and Tilman. Functional Ecology 5: 583–587.

Grace, J.B., 1991b. Physical and ecological evaluation of heterogeneity. Functional Ecology 5: 192–201.

Grace, J.B., 1993. The effects of habitat productivity on competition intensity. Tree 8: 229–230.

Grace, J.B., J. Keough, and G.R. Guntenspergen, 1992. Size bias in traditional analyses of substitutive experiments. Oecologia 90: 429–434.

Grime, J.P., 1973. Reply to Newman. Nature 244: 311.

Grime, J.P., 1977. Evidence for the existence of three primary strategies in plants and its relevance to ecological and evolutionary theory. American Naturalist 111: 1169–1194.

Grime, J.P., 1979. Plant Strategies and Vegetation Processes. Wiley, Chichester.

Grime, J.P., 1988. The C-S-R model of primary plant strategies—origins, implications and tests. In: L.D. Gottlieb and S.K. Jain, eds. Plant Evolutionary Biology. Chapman and Hall, London, pp. 371–393.

Grime, J.P., B.D. Campbell, J.M.L. Mackey, and J.C. Crick, 1991. Root plasticity, nitrogen capture and competitive ability. In: D. Atkinson, ed. Plant Root Growth: An Ecological Perspective. Blackwell Scientific Publications, Oxford, pp. 381–397.

Gross, K.L., 1984. Effects of seed size and growth form on seedling establishment of six monocarpic perennial plants. Journal of Ecology 72: 369–387.

Gross, K.L., K.S. Pregitzer, and A.J. Burton, 1995. Spatial variation in nitrogen availability in three successional plant communities. Journal of Ecology 83: 357–367.

Guo, D., P. Mou, R.H. Jones, and R.J. Mitchell, 2002. Temporal changes in spatial patterns of soil moisture following disturbance: an experimental approach. Journal of Ecology 90: 338–347.

Gurevitch, J. and R.S. Unnasch, 1989. Experimental removal of a dominant species at two levels of soil fertility. Canadian Journal of Botany 67: 3470–3477.

Gurevitch, J., L.L. Morrow, A. Wallace, and J.S. Walsh, 1992. A meta-analysis of competition in field experiments. American Naturalist 140: 539–572.

Harner, R.F. and K.T. Harper, 1976. The role of area heterogeneity and favorability in plant species diversity of pinyon juniper ecosystems. Ecology 57: 1254–1263.

Harper, J.L., 1977. Population Biology of Plants. Academic Press, London.

Hedges, L.V., J. Gurevitch and P.S. Curtis, 1999. The meta-analysis of response ratios in experimental ecology. Ecology 80: 1150–1156.

Helgadóttir, A. and R.W. Snaydon, 1985. Competitive interactions between populations of *Poa pratensis* and *Agrostis tenuis* from ecologically-contrasting environments. Journal of Applied Ecology 22: 525–537.

Hierro, J.L., J.L. Maron, and R.M. Callaway, 2005. A biogeographical approach to plant invasions: the importance of studying exotics in their introduced and native range. Journal of Ecology 93: 5–15.

Hodge, A., 2003. Plant nitrogen capture from organic matter as affected by spatial dispersion, interspecific competition and mycorrhizal colonization. New Phytologist 157: 303–314.

Hodge, A., D. Robinson, B.S. Griffiths, and A.H. Fitter, 1999. Why plants bother: root proliferation results in increased nitrogen capture from an organic patch when two grasses compete. Plant Cell and Environment 22: 811–820.

Holt, R.D., 1985. Density-independent mortality, non-linear competitive interactions, and species coexistence. Journal of Theoretical Biology 116: 479–493.

Houssard, C. and J. Escarré, 1991. The effects of seed weight on growth and competitive ability of *Rumex acetosella* from two successional old-fields. Oecologia 86: 236–242.

Hutchings, M.J., E.A. John, and D.K. Wijesinghe, 2003. Toward understanding the consequences of soil heterogeneity for plant populations and communities. Ecology 84: 2322–2334.

Jackson, R.B. and M.M. Caldwell, 1993a. Geostatistical patterns of soil heterogeneity around individual perennial plants. Journal of Ecology 81: 683–692.

Jackson, R.B. and M.M. Caldwell, 1993b. The scale of nutrient heterogeneity around individual plants and its quantification with geostatistics. Ecology 74: 612–614.

Kadmon, R., 1995. Plant competition along soil moisture gradients: a field experiment with the desert annual *Stipa capensis*. Journal of Ecology 83: 253–262.
Kadmon, R. and A. Schmida, 1990. Competition in a variable environment: an experimental study in a desert annual plant population. Israel Journal of Botany 39: 403–412.
Keddy, P.A., 1989. Effects of competition form shrubs on herbaceous wetland plants: a 4-year field experiment. Canadian Journal of Botany 67: 708–716.
Keddy, P.A., 1990. Competitive hierarchies and centrifugal organization in plant communities. In: J.B. Grace and D. Tilman, eds. Perspectives on Plant Competition. Academic Press, San Diego, CA, pp. 265–290.
Keddy, P.A. and B. Shipley, 1989. Competitive hierarchies in herbaceous plant communities. Oikos 54: 234–241.
Keddy, P., C. Gaudet, and L.H. Fraser, 2000. Effects of low and high nutrients on the competitive hierarchy of 26 shoreline plants. Journal of Ecology 88: 413–423.
Kembel, S.W. and J.F. Cahill Jr., 2005. Plant phenotypic plasticity belowground: a phylogenetic perspective on root foraging trade-offs. American Naturalist 166: 216–230.
Law, R. and A.R. Watkinson, 1987. Response-surface analysis of two-species competition: An experiment with *Phleum arenarium* and *Vulpia fasciculata*. Journal of Ecology 75: 871–886.
Lechowicz, M.J. and G. Bell, 1991. The ecology and genetics of fitness in forest plants II. Microspatial heterogeneity of the edaphic environment. Journal of Ecology 79: 687–696.
Levine, J.M., M. Vilà, C.M. D'Antonio, J.S. Dukes, K. Grigulis, and S. Lavorel, 2003. Mechanisms underlying the impacts of exotic plant invasions. Proceedings of the Royal Society of London B 270: 775–781.
Lister, A.J., P.P. Mou, R.H. Jones, and R.J. Mitchell, 2000. Spatial patterns of soil and vegetation in a 40-year-old slash pine (*Pinus elliottii*) forest in the Coastal Plain of South Carolina, USA. Canadian Journal of Forest Research 30: 145–155.
Lundholm, J.T. and D.W. Larson, 2003. Relationships between spatial environmental heterogeneity and plant species diversity on a limestone pavement. Ecography 26: 715–722.
MacDougall, A.S. and R. Turkington, 2005. Are invasive species the drivers or passengers of change in degraded ecosystems? Ecology 86: 42–55.
Mack, R.N. and J.L. Harper, 1977. Interference in dune annuals: spatial pattern and neighborhood effects. Journal of Ecology 65: 345–363.
Mallik, A.U. and F. Pellissier, 2000. Effects of *Vaccinium myrtillus* on spruce regeneration: testing the notion of coevolutionary significance of allelopathy. Journal of Chemical Ecology 26: 2197–2209.
Markham, J.H. and C.P. Chanway, 1996. Measuring plant neighbour effects. Functional Ecology 10: 548–549.
McGilchrist, C.A. and B.R. Trenbath, 1971. A revised analysis of plant competition experiments. Biometrics 27: 659–671.
McGraw, J.B. and F.S. Chapin III, 1989. Competitive ability and adaptation to fertile and infertile soils in two *Eriophorum* species. Ecology 70: 736–749.
Mclellan, A.J., R. Law, and A.H. Fitter, 1997. Response of calcareous grassland plant species to diffuse competition: results from a removal experiment. Journal of Ecology 85: 479–490.
Miller, T.E. and P.A. Werner, 1987. Competitive effects and responses between plant species in a first-year old-field community. Ecology 68: 1201–1210.
Naeem, S., M. Loreau, and P. Inchausti, 2002. Biodiversity and ecosystem functioning: the emergence of a synthetic ecological framework. In: M. Loreau, S. Naeem, and P. Inchausti, eds. Biodiversity and Ecosystem Functioning. Synthesis and Perspectives. Oxford University Press Inc., New York, pp. 3–11.
Newman, E.I., 1973. Competition and diversity in herbaceous vegetation. Nature 244: 310.
Oksanen, L., S.D. Fretwell, J. Arruda, and P. Niemela, 1981. Exploitation ecosystems in gradients of primary productivity. American Naturalist 118: 240–261.
Oksanen, L. and T. Oksanen, 2000. The logic and realism of the hypothesis of exploitation ecosystems (EEH). American Naturalist 155: 703–723.
Olofsson, J., J. Moen, and L. Oksanen, 2002. Effects of herbivory on competition intensity in two arctic-alpine tundra communities with different productivity. Oikos 96: 265–272.

Pacala, S.W. and J.A. Silander Jr., 1985. Neighborhood models of plant population dynamics. I. Single-species models of annuals. American Naturalist 125: 385–411.
Pacala, S.W. and J.A. Silander Jr., 1987. Neighborhood interference among velvet leaf, *Abutilon theophrasti*, and pigweed, *Amaranthus retroflexus*. Oikos 48: 217–224.
Peltzer, D.A., S.D. Wilson, and A.K. Gerry, 1998. Competition intensity along a productivity gradient in a low-diversity grassland. American Naturalist 151: 465–476.
Pinder, J.E., III, 1975. Effects of species removal on an old-field plant community. Ecology 56: 747–751.
Rajaniemi, T.K. Root foraging traits and competitive ability in heterogeneous soils. Oecologia (in press).
Rajaniemi, T.K. and H.L. Reynolds, 2004. Root foraging for patchy resources in eight herbaceous species. Oecologia 141: 519–525.
Reader, R.J., 1990. Competition constrained by low nutrient supply: an example involving *Hieracium floribundum* Wimm Y Grab. (Compositae). Functional Ecology 4: 573–577.
Reader, R.J., 1992. Herbivory, competition, plant mortality and reproduction on a topographic gradient in an abandoned pasture. Oikos 65: 414–418.
Reader, R.J. and B.J. Best, 1989. Variation in competition along an environmental gradient: *Hieracium floribundum* in an abandoned pasture. Journal of Ecology 77: 673–684.
Reader, R.J., S.D. Wilson, J.W. Belcher, I. Wisheu, P.A. Keddy, D. Tilman, E.C. Morris, J.B. Grace, J.B. McGraw, H. Olff, R. Turkington, E. Klein, Y. Leung, B. Shipley, R. van Hulst, M.E. Johansson, C. Nilsson, J. Gurevitch, K. Grigulis, and B.E. Beisner, 1994. Plant competition in relation to neighbor biomass: an intercontinental study with *Poa pratensis*. Ecology 75: 1753–1760.
Reekie, E.G. and F.A. Bazzaz, 1989. Competition and patterns of resource use among seedlings of five tropical trees grown at ambient and elevated CO_2. Oecologia 79: 212–222.
Reynolds, H.L. and C.M. D'Antonio, 1996. The ecological significance of plasticity in root weight ratio in response to nitrogen: Opinion. Plant and Soil 185: 75–97.
Reynolds, H.L. and S.W. Pacala, 1993. An analytical treatment of plant competition for soil nutrient and light. American Naturalist 141: 51–70.
Reynolds, H.L., B.A., Hungate, F.S. Chapin III, and C.M. D'Antonio, 1997. Soil heterogeneity and plant competition in an annual grassland. Ecology 78: 2076–2090.
Reynolds, H.L., G. Mittelbach, T.L. Darcy-Hall, G.R. Houseman, and K.L. Gross, No effect of varying soil resource heterogeneity on plant species richness in a low fertility grassland. Journal of Ecology (in press).
Robertson, G.P., J.R. Crum, and B.G. Ellis, 1993. The spatial variability of soil resources following long-term disturbance. Oecologia 96: 451–456.
Robinson, D., 1994. The responses of plants to non-uniform supplies of nutrients. New Phytologist 127: 635–674.
Robinson, D., 1996. Resource capture by localized root proliferation: why do plants bother? Annals of Botany 77: 179–185.
Robinson, D., A. Hodge, B.S. Griffiths, and A.H. Fitter, 1999. Plant root proliferation in nitrogen-rich patches confers competitive advantage. Proceedings of the Royal Society of London Series B. Biological Sciences 266: 431–435.
Roughgarden, J., 1979. Theory of Population Genetics and Evolutionary Ecology: An Introduction. Macmillan, New York; Collier Macmillan, London.
Roughgarden, J. and J. Diamond, 1986. Overview: the role of species interactions in community ecology. In: J. Diamond and T.J. Case, eds. Community Ecology, Harper & Row Publishers, Inc., New York, pp. 333–343.
Ryel, R.J., M.M. Caldwell, and J.H. Manwaring, 1996. Temporal dynamics of soil spatial heterogeneity in sagebrush-wheatgrass steppe during a growing season. Plant and Soil 184: 299–309.
Sackville Hamilton, N.R., 1994. Replacement and additive designs for plant competition studies. Journal of Applied Ecology 31: 599–603.
Sammul, M., K. Kull, L. Oksanen, and P. Veromann, 2000. Competition intensity and its importance: results of field experiments with *Anthoxanthum odoratum*. Oecologia 125: 18–25.
Sammul, M., L. Oksanen, and M. Magi, 2006. Regional effects on competition–productivity relationship: a set of field experiments in two distant regions. Oikos 112: 138–148.

Schoener, T.W., 1983. Field experiments on interspecific competition. American Naturalist 122: 240–285.
Schwinning, S. and J. Weiner, 1998. Mechanisms determining the degree of size asymmetry in competition among plants. Oecologia 113: 447–455.
Shea, K. and P. Chesson, 2002. Community ecology theory as a framework for biological invasions. Tree 17: 170–176.
Silander, J.A. and J. Antonovics, 1982. Analysis of interspecific interactions in a coastal plant community—a perturbation approach. Nature 298: 557–560.
Silander, J.A., Jr. and S.W. Pacala, 1985. Neighborhood predictors of plant performance. Oecologia 66: 266–263.
Silander, J.A., Jr. and S.W. Pacala, 1990. The application of plant population dynamic models to understanding plant competition. In: J.B. Grace and D. Tilman, eds. Perspectives on Plant Competition. Academic Press, San Diego, CA, USA, pp. 67–91.
Silvertown, J. and P. Dale, 1991. Competitive hierarchies and the structure of herbaceous plant communities. Oikos 61: 441–444.
Smith, T. and M. Huston, 1989. A theory of the spatial and temporal dynamics of plant communities. Vegetatio 83: 49–69.
Snaydon, R.W., 1991. Replacement or additive designs for competition studies? Journal of Applied Ecology 28: 930–946.
Snaydon, R.W., 1994 Replacement and additive designs revisited: comments on the review paper by N.R. Sackville Hamilton. Journal of Applied Ecology 31: 784–786.
Stanton, M.L., 1984. Seed variation in wild radish: effect of seed size on components of seedling and adult fitness. Ecology 65: 1105–1112.
Stevens, M.H.H. and W.P. Carson, 2002. Resource quantity, not resource heterogeneity, maintains plant diversity. Ecology Letters 5: 420–426.
Taylor, D.R. and L.W. Aarssen, 1989. On the density dependence of replacement-series competition experiments. Journal of Ecology 77: 975–988.
Taylor, D.R., L.W. Aarssen, and C. Loehle, 1990. On the relationship between r/K selection and environmental carrying capacity: a new habitat templet for plant life history strategies. Oikos 58: 239–250.
Thomas, S.C. and J. Weiner, 1989. Including competitive asymmetry in measures of local interference in plant populations. Oecologia 80: 349–355.
Thompson, K., 1987. The resource ratio hypothesis and the meaning of competition. Functional Ecology 1: 297–315.
Tilman, D., 1977. Resource competition between planktonic algae: an experimental and theoretical approach. Ecology 58: 338–348.
Tilman, D., 1982. Resource Competition and Community Structure. Monographs in Population Biology 17. Princeton University Press, Princeton, NJ.
Tilman, D., 1987. On the meaning of competition and the mechanisms of competitive superiority. Functional Ecology 1: 304–315.
Tilman, D., 1988. Plant Strategies and the Dynamics and Structure of Plant Communities. Monographs in Population Biology 26. Princeton University Press, Princeton, NJ.
Tilman, D., 1990. Mechanisms of plant competition for nutrients: the elements of a predictive theory of competition. In: J.B. Grace and D. Tilman, eds. Perspectives on Plant Competition. Academic Press, San Diego, CA, pp. 117–141.
Tilman, D., 1999. The ecological consequences of changes in biodiversity: a search for general principles. Ecology 80: 1455–1474.
Tilman, D. and S.W. Pacala, 1993. The maintenance of species richness in plant communities. In: R.E. Ricklefs and D. Schluter, eds. Species Diversity in Ecological Communities: Historical and Geographic Perspectives. University of Chicago Press, Chicago, pp. 13–25.
Tilman, D. and D. Wedin, 1991. Plant traits and resource reduction for five grasses growing on a nitrogen gradient. Ecology 72: 685–700.
Tremmel, D.C. and F.A. Bazzaz, 1993. How neighbor canopy architecture affects target plant performance. Ecology 74: 2114–2124.

Turkington, R., E. Klein, and C.P. Chanway, 1993. Interactive effects of nutrients and disturbance: an experimental test of plant strategy theory. Ecology 74: 863–878.

Twolan-Strutt, L. and P.A. Keddy, 1996. Above- and belowground competition intensity in two contrasting wetland plant communities. Ecology 77: 259–270.

Underwood, T., 1986. The analysis of competition by field experiments. In: J. Kikkawa and D.J. Anderson, eds. Community Ecology: Pattern and Process. Blackwell Scientific, Boston, pp. 240–268.

Vivian-Smith, G., 1997. Microtopographic heterogeneity and floristic diversity in experimental wetland communities. Journal of Ecology 85: 71–82.

Wardle, D.A., O. Zackrisson, G. Hörnberg, and C. Gallet, 1997. The influence of island area on ecosystem properties. Science 277: 1296–1299.

Wardle, D.A., M.-C. Nilssoon, C. Gallet, and O. Zackrisson, 1998. An ecosystem-level perspective of allelopathy. Biological Reviews 73: 305–319.

Weiner, J., 1982. A neighborhood model of annual-plant interference. Ecology 63: 1237–1241.

Weiner, J. and S.C. Thomas, 1986. Size variability and competition in plant monocultures. Oikos 47: 211–222.

Welden, C.W. and W.L. Slauson, 1986. The intensity of competition versus its importance: an overlooked distinction and some implications. Quarterly Review of Biology 61: 23–44.

Welden, C.W., W.L. Slauson, and R.T. Ward, 1988. Competition and abiotic stress among trees and shrubs in Northwest Colorado. Ecology 69: 1566–1577.

Wijesinghe, D.K., E.A. John, S. Beurskens, and M.J. Hutchings, 2001. Root system size and precision in nutrient foraging: responses to spatial pattern of nutrient supply in six herbaceous species. Journal of Ecology 89: 972–983.

Wijesinghe, D.K., E.A. John, and M.J. Hutchings, 2005. Does pattern of soil resource heterogeneity determine plant community structure? An experimental investigation. Journal of Ecology 93: 99–112.

Willey, R.W. and M.R. Rao, 1980. A competitive ratio for quantifying competition between intercrops. Experimental Agriculture 16: 117–125.

Williamson, G.B., 1990. Allelopathy, Koch's postulates, and the neck riddle, In: J.B. Grace and D. Tilman, eds. Perspectives on Plant Competition. Academic Press, San Diego, CA, pp. 143–162.

Wilson, S.D., 1993a. Belowground competition in forest and prairie. Oikos 68: 146–150.

Wilson, S.D., 1993b. Competition and resource availability in heath and grassland in the Snowy Mountains of Australia. Journal of Ecology 81: 445–451.

Wilson, S.D., 1994. Initial size and the competitive responses of two grasses at two levels of soil nitrogen: a field experiment. Canadian Journal of Botany 72: 1349–1354.

Wilson, S.D. and P.A. Keddy, 1986a. Measuring diffuse competition along an environmental gradient: results from a shoreline plant community. American Naturalist 127: 862–869.

Wilson, S.D. and P.A. Keddy, 1986b. Species competitive ability and position along a natural stress/disturbance gradient. Ecology 67: 1236–1242.

Wilson, S.D. and J.M. Shay, 1990. Competition fire, and nutrients in a mixed-grass prairie. Ecology 71: 1959–1967.

Wilson, S.D. and D. Tilman, 1991. Components of pant competition along an experimental gradient of nitrogen availability. Ecology 72: 1050–1065.

Wilson, S.D. and D. Tilman, 1993. Plant competition and resource availability in response to disturbance and fertilization. Ecology 74: 599–611.

Wilson, S.D. and D. Tilman, 1995. Competitive responses of eight old-field plant species in four environments. Ecology 76: 1169–1180.

16 Plant–Herbivore Interaction: Beyond a Binary Vision

Elena Baraza, Regino Zamora, José A. Hódar, and José M. Gómez

CONTENTS

INTRODUCTION

Herbivory is currently defined as the interaction that results when an animal consumes the live tissues of a plant (i.e., a heterotroph preying on an autotroph), usually without causing the plant's death (Crawley 1983). It is an antagonistic interaction, in which the animal gets food whereas the plant loses live tissues. This makes herbivory the most basic trophic interaction in the food chain. Herbivory has given rise to the appearance of marvelous phenotypic traits both in plants and animals, has molded the vegetation as well as entire landscapes of virtually all the ecosystems known on earth, and, finally, has determined the success of many species, including our own. Herbivory, thus, deserves firm attention from ecologists.

In the previous version of this chapter, we conducted a review to identify the features that drive current research in plant–herbivore interactions in terrestrial ecosystems (Zamora et al. 1999). In general, the systems traditionally studied were simple pairs (one plant vs. one herbivore of interacting elements), and research on herbivory typically concerns adult plants (woody or herbaceous) affected by defoliation either by insects or vertebrates and analyzing mainly plant chemistry or growth in a simple pair of protagonists. Consequently, herbivory has been viewed mainly as a binary interaction, and the aim of this chapter is to transcend this limited view, examining from a phytocentric perspective the effect of herbivory within a broader ecological framework.

The literature on plant–herbivore interactions in terrestrial ecosystems is vast. The readers can consult many excellent reviews on more specific aspects of plant–herbivore interactions published in such journals as *Annual Review of Ecology, Evolution and Systematics*, *Trends in Ecology and Evolution,* and the minireview section in *Oikos*.

PLANT TRAITS THAT DETERMINE HERBIVORY

In most habitats, plants are abundant, and therefore food quantity is, in principle, not a problem for herbivores. However, plants are not a highly profitable food because they bear compounds of low nutritive value (high content in cellulose and low in proteins) and substances that reduce digestibility, thus hampering the herbivore's nutrient acquisition (Hartley and Jones 1997). Moreover, there are many ways that a plant may avoid herbivores. Some avoidance mechanisms have been extensively studied in early and current plant–animal interaction research, whereas other mechanisms are relatively unexplored (Milchunas and Noy-Meir 2002).

PROBABILITY OF BEING FOUND

As the first step in herbivory, the plant must be found by the herbivore. There are several traits that reduce the plant's probability of being discovered, as for example, remaining inconspicuous within a given habitat, in terms of both morphology and abundance, occupying enemy-free sites, mimicking another well-defended plant (Lev-Yadun and Ne'eman 2004 and references therein) or damaged tissue, completely lacking chemical attraction (odorless), having a short life cycle, and asynchronous timing—that is, the plant produces edible tissues in the period least likely to coincide with herbivore presence or feeding (Chew and Courtney 1991, Aide 1992, Tikkanen and Julkunen-Tiitto 2003).

Another way to escape detection at the individual level includes synchronizing as closely as possible the production of tissues subject to attack (flowers, leaves, or fruit), both within the same plant and among plants of a population. This strategy, termed mass flowering or fruiting, attempts to overwhelm the herbivore's capacity to consume all the tissues available. Plants with mass flower or fruit production benefit from dense populations. In this way, the effect of time is multiplied by the effect of space (Kelly and Sork 2002, Crone and Lesica 2004, Russell and Louda 2004).

Herbivore-free sites could be considered as refuges, providing a degree of physical barrier against the herbivore. Milchunas and Noy-Meir (2002) considered two types of refuges: biotic, when a plant protects a target plant, and geologic. A rock outcrop, mesas, buttes, or islands can act as geological refuges that permitted the subsistence of more palatable species. An intensive literature survey was conducted on studies examining plant-community composition of geologic refuges compared to similar grazed communities reporting increases in diversity inside the refuge (Milchunas and Noy-Meir 2002).

Physical Barriers

After finding a host plant, herbivores must overcome other barriers. Plants have a myriad of structures to repel prospective herbivores (Lucas et al. 2000). These include scales, barbs, thorns, and spines, which are particularly effective against mammals (Grubb 1992, Young et al. 2003), and trichomes, leaf hairs, and similar structures meant to discourage invertebrate approach (Bernays and Chapman 1994, Van Dam and Hare 1998). In the same way, some plants (e.g., Cariophyllaceae) have glands that secrete an adhesive to hamper small pests from crawling freely over the plant; these glands, which trap primarily small arthropods, are thought to have given rise to carnivory in plants. Recent studies in conifer stems showed calcium oxalate crystals as a constitutive defense, which in combination with fiber rows provides an effective barrier against small bark-boring insects (Hudgins et al. 2003, Franceschi et al. 2005). In addition, leaf toughness is considered as mechanical defense, since tougher leaves are avoided by herbivores (Lucas et al. 2000, Teaford et al. 2006).

Spines appear more in zones subject to heavy herbivore pressure, and even within an individual plant, parts less accessible to mammal herbivores tend to bear fewer spines than do vulnerable parts (Grubb 1992). Several recent studies had found a direct relation between herbivory and spines density and size (Takada et al. 2001, Young et al. 2003). Furthermore, it has been demonstrated that spines have a negative effect on herbivore performance, reflected by consumption rate (Cooper and Owen-Smith 1986, Milewsky et al. 1991, Gowda 1997).

Quality of Plants as Food

Herbivores are faced with a poor-quality food resource not only because plants are low in nutrients but also because they produce plant secondary defense compounds that have wide-ranging physiological effects from direct toxicity to digestion impairment (Hartley and Jones 1997). Some of the chemical characteristics that reduce the quality of plants as food are a direct result of its functioning (Coleman and Jones 1991). For instance, the main reason for which the cell wall of higher plants consists mainly of cellulose, hemicellulose, and lignin is the necessity of support for their tissues, to overgrow the neighbors, and transport resources. In fact, these three substances comprise the primary component of the biomass of trees (*c*.90%) and grasses (*c*.65%). However, most herbivore animals cannot directly use this superabundant food resource because they cannot produce the enzymes necessary for decomposing them. Not only do some compounds of primary metabolisms act as digestibility reducers, but also some secondary compounds (e.g., cutins, tannins, and silicate particles—Howe and Westley 1988) increase the large fraction of vegetal tissues consisting of nondigestible vegetal material.

In general, they are quantitative (or chronic) metabolites, which exert their effects according to their concentration within the plant, and do not act instantaneously, but rather gradually depress the growth or fecundity of the herbivore. In addition to reducing digestibility, silicate particles also accelerate tooth wear by the abrasion of mouthparts, contributing to the development of esophageal canker, and may cause fatal urolithiasis due to the formation of calculi in the urinary tracts (McNaughton et al. 1985). Consequently, herbivores generally reject plant parts containing high concentrations of this compound.

Thus, herbivores are limited not by the energy available but by the nutritional quality of plant tissues (White 1993). This is because there are strong stoichiometric differences between the composition of plant and animal tissues; that is, plant matter, compared with animal tissues, is higher in carbon and lower in other essential elements, such as N, P, and S (Sterner and Hessen 1994). As a whole, animal herbivores contain nearly 10 times more nitrogen than do the plants that they eat. Thus, the plant–animal interface is characterized by a marked disparity in the biochemical makeup of consumer and resource. Furthermore, not all mineral nutrients present in vegetable tissues are equally available to herbivores. For instance, such nitrogen sources as proteins can be easily assimilated, whereas several nitrogenous compounds (alkaloids, cyanogenic glycosides) may even be poisonous. In this case, the total nitrogen concentration may not reflect the nutritional values of the plant tissue (Bentley and Johnson 1992).

Consequently, herbivores live in a "green desert" and are critically dependent (especially for female reproduction, Moen et al. 1993) on relatively rare, high-quality plants, or plant organs. Herbivores must, therefore, selectively ingest and assimilate essential limiting minerals. For example, McNaughton (1988) found that the heterogeneous distributions of African ungulates correlated significantly between animal density and levels of minerals such as magnesium, sodium, and phosphorous in vegetation. Similar cases have been found for many different herbivores (insects: Mattson 1980; sea urchins: Renaud et al. 1990; small mammals: Batzli 1983). Despite this dietary selectivity, the chemical composition of herbivore diets is generally unbalanced. The result is that herbivores must consume large quantities of carbon to obtain enough nitrogen and other essential nutrients. The resulting nutritional imbalance can decrease growth efficiency for herbivores, and could filter down from the top of the food chain, causing reduced production at all trophic levels. For example, assimilation efficiency of herbivores is lower (20%–50%) than that of carnivores (nearly 80%, Begon et al. 1990). This nutritional limitation has forced some herbivore species to develop opportunistic feeding strategies to obtain alternative and complementary food nutritionally richer than vegetal tissues (see White 1993 for a thorough revision).

Other types of compounds that reduce the quality of vegetal food are secondary defense compounds. These compounds, accumulated from enzyme catalysis in biosynthesis (Harborne 1997), wield their influence by their sheer presence (qualitative secondary metabolites), striking with immediacy and often inflicting death on the herbivore, many of them acting as inhibitors of tissue digestibility or as poison for herbivores. Normally, these compounds accumulate in tissues unprotected by quantitative substances—new leaves, immature fruits, or flower buds. There is a great variety of secondary defense compounds that, together with the structures of mechanical defense (spines, thorns, barbs, hairs), constitute the main traits that reduce the preference or performance of herbivores, this translating as the resistance of plants (Strauss and Agrawal 1999). The principal chemical compounds of this type are alkaloids, glucosinolates, toxic amino acids, terpenoids, and cyanogenic compounds (e.g., Rosenthal and Berenbaum 1991, Harborne 1997).

Variability of Plants as Food: Theory of Plant Defense

The herbivores are exposed to temporal and spatial variations in the abundance and quality of their food (Hunter and Price 1992, Danell and Bergström 2002). Variations in abundance are intimately linked to annual plant phenology, and vegetation response to climatic variations, whereas internal processes are related to the growth, development, and reproduction of the plant, as well as response to environmental variations (Hartley and Jones 1997). For example, it is well established that plant chemical characteristics change in the course of a season in the same plant (Dement and Money 1974, Schultz et al. 1982, Riipi et al. 2002). Moreover, plants may differ in their quality as food for herbivores whether between different species, between individuals of the same species, or between parts of the same plant (Orians and Jones 2001). These variations can be measured as differences in nutritional value, concentration of chemical and physical defenses, as well as morphological characteristics of twigs and tissues (Hartley and Jones 1997, Danell and Bergström 2002). Dissimilarities in environmental growing conditions (Bryant et al. 1983, Coley et al. 1985, Larsson et al. 1985, Walls et al. 2005, MacDonald and Bach 2005), differing histories of relationships with herbivores (Provenza and Malechek 1984), or genetic differences (Graglia et al. 2001, Oiser and Lindroth 2001) provoke this high variability in plant characteristics, creating a nutritional mosaic for herbivores. For example, plant secondary defense compounds exhibit some of the highest diversities seen in nature.

During the last 30 years several theories have tried to explain the pattern of distribution observed in the concentration and distribution of secondary defense compounds in the different plant species. Berenbaum (1995), for instance, lists as many as 12 different theories to account for the allocation of chemical defenses in plants, and a few more hypotheses have been added to this list recently. Many studies have tried to test the most plausible of these theories, uncovering evidences in favor of and against most of them, without a clear prevalence of one above the others (Stamp 2003). Some authors emphasize the risk of herbivory as the main factor determining the quantity and type of defensive compounds in plants. Thus, according to the apparency theory (Feeny 1976, Rhoades and Cates 1976), "apparent" species (long-lived plant species, with a dominant presence in landscapes) have a greater damage probability and thus require a greater investment in defense, favoring digestibility-reducing chemicals (e.g., tannins), whereas less apparent species are defended by toxic chemicals (e.g., alkaloids). Grubb (1992) developed this theory, taking into account the nutritive value of tissues exposed to herbivore damage.

Other authors consider resource availability as the key factor determining the amount and kind of defenses that plants produce. The carbon/nutrient balance theory (Bryant et al. 1983) is based on the idea that the relationship between the availability of carbon and nitrogen in the environment determines the kind and the amount of resources that a plant invests in defense or growth. The growth/differentiation balance hypothesis assumes that the synthesis of defensive compounds is constrained not only by the external availability of resources but also by internal trade-offs in resource allocation between growth and defense (Herms and Mattson 1992). The resource-availability hypothesis (Coley et al. 1985) predicts that differences in toxicity might depend on the costs versus benefits of defending plant parts in a way that low resource availability favors plants with inherently slow growth rates, which in turn favors large investments in anti-herbivore defense. More recent hypotheses attempt to explain allocation patterns in terms of biosynthetic differences between types of defensive compounds sink/source hypothesis (Honkanen and Haukioja 1998), particularly among terpenoids and different classes of phenolics (Muzika and Pregitzer 1992, Haukioja et al. 1998, Koricheva et al. 1998a), and competition between protein and phenolic synthesis for the common precursor, L-phenylalanine (Jones and Hartley 1999). Some authors defend the evolutionary origin of the resistance characteristics on plant. The optimal-defense theory (Rhoades 1979, Hamilton et al. 2001) argues that the evolution of the defenses in a plant is governed mainly by the balance between the metabolic cost of the defense production and the benefit obtained from the reduction of the loss of tissues by herbivory. Assuming a genetic control of plant defense, herbivores represent a selective pressure that favors the production of secondary defense compounds depending on the cost or benefit for the plant fitness (Hamilton et al. 2001).

None of the above theories of chemical defense have ever been definitively rejected; they all coexist by virtue of supportive evidence in one system or another (Stamp 2003). Only in the case of carbon/nutrient balance hypothesis, several authors consider that the fundamental assumptions have proved to be incorrect (Berenbaum 1995, Koricheva et al. 1998a), considering the use of carbon/nutrient balance hypothesis no longer logically or philosophically justifiable (Hamilton et al. 2001, Koricheva 2002a, Nitao et al. 2002).

Variability of Plants as Food: Effects of Plant Stress

There are two contrary theories that seek to explain how the effect of environmental stresses on plant quality as food can affect the preference and performance of herbivores, especially insects. The plant-stress hypothesis (White 1993) predicts that environmental stresses on plants decrease plant resistance to insect herbivory by altering biochemical source–sink relationships and foliar chemistry, to increase the availability of nutrients for herbivores (e.g., soluble amino acids in phloem). Such changes in the nutritional landscape for insects may facilitate insect population outbreaks during periods of moderate stress on host plants. However, not every kind of stress provokes the same response in plants, and there are cases in which stressed plants represent suboptimal food with respect to control plants (e.g., see Mopper and Whitham 1992). According to Mopper and Whitham (1992), sustained environmental stress (poor soils or persistent drought), by causing numerous metabolic changes (e.g., increasing soluble nitrogen availability, reducing secondary compounds), can be beneficial to insects, whereas a brief drought period during insect oviposition may harm herbivore performance. By contrast, Price (1991) stated that some specialized herbivores feed preferentially on vigorous plants or plant modules (plant-vigor hypothesis). It appears that the latter applies especially to insect herbivores most closely associated with plant growth processes, as endophytic gallers and shoot borers (Price 1991, but see Rehill and Schultz 2001). Recent studies also found a better performance of other types of insects in more vigorous plants (Inbar et al. 2001). For example, in the leaf miner *Agromyza nigripes* (Agromyzidae), larva performance (density, survival, and development) was highest on vigorous plants (De Bruyn et al. 2002). In addition, the grass miner *Chromatomyia milii* (Agromyzidae) prefers and performs better on vigorously growing plants not exposed to excess nutrients (Scheirs and De Bruyn 2004) or water stress (Joern and Mole 2005). Many insect species feed specifically on certain plant tissues, which are likely to be differentially affected by stressful conditions. Consequently, the effect of stress on insect performance would vary depending on local plant response to this source of stress (Koricheva et al. 1998b). The revision of diverse studies by Koricheva et al. (1998b) showed that in general, boring and sucking insects performed better on stressed plants, whereas plant stress adversely affected gall-makers and chewing insects. Reduction in performance of chewers was greater on stressed slow-growing plants than on stressed fast growers. Reproductive potential of sucking insects was increased by pollution but reduced by water stress. Furthermore, Huberty and Denno (2004) described consistent positive effects of water stress for borers, negative responses for gall-formers, and inconsistent responses for free-living species and leaf miners.

EFFECT ON PLANT PERFORMANCE AND POPULATIONS

Herbivory and Plant Performance

In natural systems, herbivory is often detrimental to plants as it removes resources through loss of nutrients and photosynthetic area. However, the effects of the herbivory plants are not the simple result of the foraging behavior of the herbivores on vegetation, but also the result of the different capacities of the plants to react against herbivory (Augustine and McNaughton 1998). A plant may respond to herbivore attack via induced defense, which is important in understanding plant life-history traits (Iwasa 2000). Induced defense takes two basic forms: resistance and tolerance. Tolerance is a plant's ability to reduce the negative effects of consumers (e.g., herbivores, pathogens) on plant fitness (Rosenthal and Kotanen 1994, Stowe et al. 2000), in some cases not only compensating for tissue loss but also overcompensating by increase in plant growth and fitness after herbivory (Strauss and Agrawal 1999). Such responses to herbivory are called induced resistance when they are known to decrease rates of herbivory (Karban and Baldwin 1997). Constitutive resistance is always expressed, whereas induced resistance appears only after an individual has been damaged, serving to reduce additional damage.

Factors Affecting Tolerance to Damage

The tolerance capacity of plants is species-specific and highly dependent on the type of tissue damage, on plant age, and on the amount, pattern, timing, and frequency of herbivory, and the environmental conditions in which the plant is growing.

The impact that herbivory exerts on plant performance is more dependent on the role of the tissue damaged than on the total biomass lost. Although defoliation signifies the loss of photosynthetic tissues, the damage of other tissues such as meristem can have more detrimental effects (Spotswood et al. 2002). Flower and fruit damage implies a greater loss of female reproductive success than does injury to other tissues (Dirzo 1984, Hendrix 1988). This loss may be reflected directly by the diminished number of ovules after herbivore feeding on flowers (Dirzo 1984, Wise and Cummins 2006). Root herbivory is very difficult to investigate and is often a neglected facet of plant–herbivore interaction, despite the fact that belowground herbivory can have dramatic consequences for plant performance (Prins et al. 1992, Müller-Schärer and Brown 1995, Bebber et al. 2002, Blossey and Hunt-Joshi 2003).

The impact of herbivory also depends on the age of the plant. Plant tolerance to herbivore damage appears to be high in the cotyledon stage, declines in later seedling stages, but then increases as plants reach the sapling and the reproductive stage (Boege and Marquis 2005). This pattern can change when root damage is analyzed, since seedlings have less root development than do saplings (Stout et al. 2002). Herbivores influence plant mortality primarily at the seed and seedling stages (Harper 1977, Hulme 1996). In this case, herbivory is analogous to predation.

Foliar and stem loss can have a detrimental or stimulatory effect, or no effect at all, on plant growth or reproduction, depending on the intensity and timing of the loss (Maschinski and Whitham 1989, Hester et al. 2004). Even within a modular unit (such as a leaf), the timing of herbivory can have differential effects. For the plant, herbivory losses are more harmful when leaves are young than when leaves are older, especially toward senescence (Harper 1977). Browsing damage to woody plants is also less detrimental at the beginning of the growing season than later (Guillet and Bergström 2006). Moreover, while seedlings may overcompensate for tissue browsed during the dormant season, they may not compensate for tissue lost while a plant is actively growing (Canham et al. 1994, Bergström and Danell 1995).

The frequency of damage is also important for plant tolerance capacity. In saplings of some tree species, the repeated loss of the apical meristem forcibly modifies the architecture of the tree, from tall to dwarf individuals, and retards the timing of the first reproductive season. For example, in the Sierra Nevada, some wild and domestic ungulates feeding on saplings of Scots pine *Pinus sylvestris nevadensis* (Pinaceae) and other tree species develop stunted morphologies due to browsing by Spanish ibex and domestic goats (Zamora et al. 2001).

Capacity of Compensation

Compensation enables damaged plants to maintain their fitness through extra growth and reproduction and may involve a variety of physiological and morphological mechanisms, including increased photosynthesis and altered patterns of resource allocation, differential balance in vegetative and reproductive tissues, changes in nutrient uptake, and altered hormonal balance (see Rosenthal and Kotanen 1994). Several studies have found that compensation follows herbivore attack (Hendrix and Trapp 1989) or experimental clipping (Ehrlén 1992, Obeso and Grubb 1994), and at low levels of herbivory, plants may even overcompensate for damage (Belsky 1986).

Compensation capacity after herbivory is very different between species (Rosenthal and Kotanen 1994, Hawkes and Sullivan 2001). For example, deciduous woody plants tend to have more flexible growth patterns and thus a greater ability for compensatory growth than certain evergreen species such as pines (Vanderklein and Reich 2000, Millard et al. 2001, Hester et al. 2004). Different plant characteristics can be crucial to offset herbivore damage depending on the tissue removed. For example, since browsing damage is more likely to lead to loss of the nutrients stored in foliage or aboveground woody tissues than in trunk or roots, species that store nutrients in the trunk and roots may have more tolerance capacity to browsing damage (Baraza 2004).

Furthermore, many extrinsic factors influence a plant's physiological state and, as a consequence, its ability to compensate for damage (Herms and Mattson 1992). It has long been assumed that a plant's tolerance to herbivory should be greater in low-stress, resource-rich environments, and this assumption has been formalized in what has become known as the compensatory-continuum hypothesis (Maschinski and Whitham 1989). For example, light availability affects the plant's capacity to tolerate herbivory, by determining to a great extent the carbon resources for the synthesis of chemicals and growth, so that, when light is a limiting resource, tolerance diminishes (Lentz and Cipollini 1998, Harmer 1999, Saunders and Puettmann 1999, Baraza et al. 2004). However, resource availability does not affect all species in the same way. Hawkes and Sullivan (2001) in a review by meta-analysis showed that basal meristem monocots in general grew significantly more after herbivory in high resources, while both dicot herbs and woody plants grew significantly more after herbivory in low resources. Wise and Abrahamson (2005) proposed an alternative model, called the limiting-resource model, which specifically identifies the factors that limit plant fitness and the resources that are affected by particular herbivores.

Induced Resistance

Induced plant resistance refers to any active or passive change in the plant after herbivory or infection, which reduces preference, performance, or pathogenicity of the attacker on attack compared with controls (Karban and Baldwin 1997). As in the case of tolerance, resistance induction differs between species and is subject to environmental conditions as well as intensity, timing and pattern of damage (Karban and Balwin 1997, Nykänen and Koricheva 2003). The common induced-resistance trait analyzed is biochemical change (e.g., increase of toxin or decrease in nutrient value) in the tissues remaining after herbivory (Nykänen and Koricheva 2003). However, there are other types of induced resistance, including heavier mechanical defense (Gómez and Zamora 2002, Young et al. 2003) or changes in morphology that reduce herbivore intake rates (Massei et al. 2000, Martínez and López-Portillo 2003).

Induced plant responses to herbivory have been shown to be a cost-saving strategy (Cipollini 1998, Agrawal et al. 2002). That is, given the costly nature of plant allocation to defense, plant fitness could be enhanced by only allocating to resistance in the presence of herbivory. However, Koricheva (2002b) analyzing diverse studies by meta-analysis detected similar costs of both types of resistance, when constitutive resistance is associated mainly with metabolically cheap defenses, whereas inductibility evolved primarily in expensive defenses. In addition, cost saving represents only one of many possible reasons why induced resistance may be favored or maintained by selection (reviewed in Agrawal and Karban 1999). For instance, another important but less explored potential benefit of induced resistance may be increased spatial and temporal variability in food quality for herbivores (Shelton 2000).

HERBIVORY AND PLANT POPULATION DYNAMICS

It has been demonstrated repeatedly that the performance (growth, reproductive output, and survival) of many plant species is negatively affected by the severe impact inflicted by vertebrate and invertebrate herbivores (Marquis 1992a,b, Guretzky and Louda 1997). Contrasting with this copious literature concerning the effect of herbivores on individual plants, much less empirical information exists on the real effect that these organisms have on the abundance and density of plants (Maron and Crone 2006). Consequently, the importance of herbivory for plant populations remains a controversial issue (Hendrix 1988, Andersen 1989, Crawley 1989, Eriksson and Ehrlén 1992, Marquis 1992a,b, Osem et al. 2004). Although herbivores can theoretically affect plant abundance by arresting recruitment as a consequence of their detrimental effect on plant performance, only a few studies have been able to experimentally demonstrate this effect. Carson and Root (2000) demonstrated that phytophagous insects control the abundance of goldenrod *Solidago altissima* (Asteraceae), whereas Maron and Simms (2001) showed that granivory by rodents affects *Lupinus arboreus* (Fabaceae) recruitment and thereby adult abundance. Unfortunately, the effect of herbivores on plant-population dynamics is usually inferred from their mere effect on some demographic components, such as seed germination, seedling emergence, or juvenile recruitment (Crawley and Long 1995, Louda and Potvin 1995, Hulme 1996, 1997, Curran and Webb 2000, Maron and Gardner 2000, Wenny 2000, Ehrlén 2003).

However, two main nonexclusive reasons suggest that it is not accurate to infer any effect on populations merely from the effects on demographic components. First, incomplete components of the performance of plants, such as seed production or seedling survival, cannot be universally used as a substitute for total performance, since total performance in many plant species represents the integration of the effects occurring during different phases of the life cycle (Ehrlén 2003). Focusing on a single component overlooks important trade-offs between different demographic components (Ehrlén 2002). Indeed, a strong effect of herbivores on a particular component of the life cycle of plants can be counteracted by compensatory effects on other components (Ehrlén 2002, 2003). Second, a significant herbivore effect on host-population dynamics takes place only when the number of propagules entering the adult stage is smaller than the number of deaths by herbivory (i.e., the recruitment rate is lower than the death rate, Harper 1977). Under field conditions, plant recruitment depends on the availability of seeds (seed limitation) as well as on the availability of suitable microsites for seed germination and seedling survival (establishment limitation, Eriksson and Ehrlén 1992, Clark et al. 1998, Edwards and Crawley 1999, Nathan and Müller-Landau 2000). When plant populations are limited mainly by the availability of microsites to germinate and become established, rather than by the production of seeds, an increase in propagule production (performance) due to a release from herbivores does not automatically translate into an increase in plant abundance (Hulme 1998, Turnbull et al. 2000). In this scenario, the outcome of the herbivory interactions can be neutral at the population level irrespective of their outcome at the plant–individual level. Consequently, when no information exists on the extent to which subsequent density-dependent compensation counteracts the effect of seed and seedling reduction caused by herbivores, a common mistake is to assume that a strong herbivore effect on plant performance implies an equally intense effect on plant population dynamics (Louda and Potvin 1995, Edwards and Crawley 1999, Hickman and Hartnett 2002).

Herbivory and Plant Distribution

Several studies have suggested that some herbivores are able to shape the habitat distribution of their host plant (Bruelheide and Scheidel 1999, Kleijn and Steinger 2002, DeWalt et al. 2004, and references therein). Two pieces of information have been used to support this proposal: the mere existence of habitat-dependence in the activity of herbivores (Boyd 1988, Herrera 1991, 1993, Gómez 1996, Louda and Rodman 1996, Cabin and Marshall 2000, Sipura and Tahvanainen 2000) and the effect of herbivore release in the habitat expansion of invasive plants (enemy-release hypothesis, Keane and Crawley 2002, DeWalt et al. 2004). Under these circumstances, the habitat distribution of many plant species inhabiting heterogeneous landscapes can be a direct consequence of the activity of their major herbivores (Jordano and Herrera 1995, Schupp 1995, Schupp and Fuentes 1995, Louda and Rodman 1996, Cabin and Marshall 2000, Rey and Alcántara 2000, Sipura and Tahvanainen 2000). Despite its crucial importance, the role that herbivores play in shaping the spatial distribution pattern of the plant populations remains unclear, most studies simply reporting the advantage for individual plants of growing close to neighbors or in specific microhabitats (Danell et al. 1991, Hjältén et al. 1993, Hjältén and Price 1997, WallisDeVries et al. 1999, Rebollo et al. 2002). More recent works concerning herbivores in affecting local plant distribution have been undertaken. For example, Gómez (2005a) demonstrated that herbivores influence the spatial distribution of two species of *Erysimum* (Brassicaceae), the distribution of which is limited under shrubs when ungulate herbivores are present (see *Case study* 3). Fine et al. (2004) demonstrated that heavy insect herbivory on tropical tree seedlings might be responsible for limiting the local distribution of particular tree species to sites with specific soil conditions.

EVOLUTIONARY PLAY

Plant–Herbivore Coevolution?

A traditional assumption among evolutionary ecologists is that herbivory tends to lead to coevolution (Erhlich and Raven 1964), implying that there is simultaneous evolution of ecologically interacting populations, which means synchronous reciprocal adaptation. Contrary to this coevolutionary thinking, the theory of sequential evolution (Jermy 1993) states that plants evolve by selective pressures far more imposing than those exerted by herbivores. Thus, according to this idea, plants shape herbivore evolution, not vice versa. Although paired plant–animal coevolution is theoretically possible, such pairing remains unlikely in nature because most plants interact with an array of herbivorous species and vice versa. A plant species must often respond to the selective pressures exerted by a multispecific system (Simms and Rausher 1989, Meyer and Root 1993). The result can be a dilution of all selective pressures, because the pressure of one herbivore species on plant traits is often opposed, constrained, or modified by pressures of other herbivore species. In this context, diffuse coevolution (Janzen 1980) was suggested as alternative to pairwise convolution when selection imposed reciprocally by one species on another is dependent on the presence or absence of other species. In a recent review, Strauss et al. (2005) outlined a quantitative genetic approach for understanding and quantifying diffuse evolution, taking it as the more plausible evolutive relationship between plants and herbivores (Rausher 1996, Iwao and Rausher 1997, Agrawal 2000a, Rausher 2001).

For a herbivore to be said to select plant traits, a heritability base of the variability of those proposed defensive traits in plants and correlation between the trait (and the damage, of course) and plant fitness must be demonstrated. A genetic basis of the individual variation in damage has been recently shown as stronger than previously thought, since the relative contribution of plant genotype with respect to environmental variability to determine resistance traits seems to be important in many of the studied systems (Agrawal and Van Zandt 2003). Increased fitness in defended plants when the herbivore is present has been demonstrated for plant-induced resistance to herbivory (Agrawal 1998). However, there are still doubts about the herbivore's role in the evolution of some plant-resistance traits, which could represent a secondary phenomenon that fortuitously benefits the plant (Tuomi et al. 1990). In fact, most of the traits currently related to herbivory would have evolved in a world without animals as a result of abiotic selection for vegetative growth and survival. For example, sclerophyllous leaves may be an adaptive mechanism related to water and nutrient conservation (Turner 1994). Even secondary compounds may now act exclusively as a defense against herbivores, since it has been demonstrated that they play a role in pollination and seed dispersal, pathogen interactions, alellopathic processes, and protection against ultraviolet rays (Bennett and Wallsgrove 1994, Waterman and Mole 1994, Close et al. 2003).

Cost of Defense

The idea that adaptation is costly is a deeply entrenched principle in evolutionary biology. In an evolutionary context, the incremental fitness benefit associated with genotypes conferring greater defense on plants is accompanied by a forfeit in fitness associated with reallocation of resources away from other fitness-enhancing functions (Fritz and Simms 1992). This means that defense has a cost. Costs of defense production may arise by many mechanisms, including allocation trade-offs, ecological interactions, and genetic effects such as pleiotropy (Rausher 2001, Heil and Baldwin 2002). The empirical evidence for the existence of such cost is conflicting, suggesting that significant fitness costs of defense arise in some circumstances but not in others, depending on the environment in which they are measured, the resources available to the plant, and the ecological interactions of the community (Rausher 2001, Heil and Baldwin 2002, Strauss et al. 2002, Koricheva 2002b). For example, high resource availability may diminish allocation costs, thereby allowing for both growth and defense (Siemens et al. 2003, Walls et al. 2005, Donaldson et al. 2006).

Evolution of Plant Tolerance versus Plant Resistance

The joint evolution of plant tolerance and resistance to herbivores has attracted substantial theoretical attention over the last decade (Rosenthal and Kotanen 1994, Strauss and Agrawal 1999, Mauricio 2000, Stowe et al. 2000, Tiffin 2000a, Fornoni et al. 2003, 2004a,b). Resistance and tolerance have been considered for years as incompatible strategies of resistance against herbivores. However, Leimu and Koricheva (2006), reviewing the empirical evidence for tolerance-resistance trade-offs by means of meta-analysis, found that conditions under which a negative association between resistance and tolerance occurs and, thus, the evolution of multiple resistance strategies in plants is constrained, are much more restrictive than previously assumed. Mixed defense strategies, with resistance and tolerance, both maintained at intermediate levels, are possible when the cost of each defense rises disproportionately with its effectiveness (Fornoni et al. 2004a,b), and strongly negative genetic correlations between resistance and tolerance can promote polymorphism in each (Tiffin 2000b).

MULTISPECIFIC CONTEXT OF HERBIVORY

Herbivory has traditionally been viewed as a binary interaction focusing on a simple pair of interacting elements (one plant vs. one herbivore; see previous edition of this chapter). This species-to-species view of plant–herbivore interactions has been progressively challenged by an increasing body of studies showing that plant–herbivore interactions are strongly affected in a predictable way by the community context (Björkman and Hambäck 2003, Strauss and Irwin 2004). Plants compete against other plants for substrate and nutrients at the same time as they may be simultaneously eaten by many herbivores and pollinated by many species of floral visitors. The importance of the community context becomes apparent with the observation that the strength and even the sign of the interaction between two species may change in the presence of others by the action of the so-called *indirect effects* (Strauss 1991, Strauss and Irwin 2004). Under these circumstances, scenarios in which only two or three species interact generally offer an overly simplistic and even inappropriate view of what in fact occurs in nature. In this section, we provide examples of multispecies systems in plant–herbivore interaction. A useful approach to the study of the enormous complexity of ecological communities is the "community modules" proposed by Holt (1997), involving a small number of species (3–6) linked in a specified structure of interactions. Most ecological studies on herbivory venture beyond the paired species traditionally analyzed in these types of subsystems (see later).

EFFECT OF HERBIVORES ON PLANT–PLANT INTERACTION

Affecting Competition between Plants

Competition from plant neighbors and herbivory are two factors that determine the growth, survival, and reproduction of plant individuals, and subsequently the abundance of plant populations (Harper 1977, Crawley 1983, Gurevitch et al. 2000). Herbivory influences the effect of competitive interactions between plants and vice versa.

Herbivores can impair the competitive abilities of their host plants (Harper 1977, Edwards 1989, Figure 16.1, *left*). The selective consumption of individual plants can result in a hierarchy of sizes within a given plant population and thereby increase the likelihood of asymmetrical competition between the plants (Weiner 1993). McEvoy et al. (1993) showed that herbivory on ragwort *Senecio jacobaea* (Asteraceae) by the beetle *Longitarsus jacobaeae* (Chrysomelidae) intensifies the competition of the plant with other species, speeding the ragwort's elimination, which would otherwise come about slowly. Moreover, herbivory by livestock can alter competition between plants. *Cirsium obalatum* (Asteraceae) and *Veratrum lobelianum* (Liliaceae), two large unpalatable native perennial herbs, had strong positive effects on the growth of two more palatable species, *Anthoxanthum odoratum* and *Phleum alpinum* (Poaceae) and no effects on unpalatable species *Luzula pseudosudetica* (Juncaceae) when livestock were present. Contrarily, inside exclosures they had no effect on palatable species and had competitive effects on *L. pseudosudetica* (Callaway et al. 2005).

Herbivory can also produce apparent competition among plants that share herbivores (Figure 16.1, *center*). An increase in density of one plant species results in a decrease in density

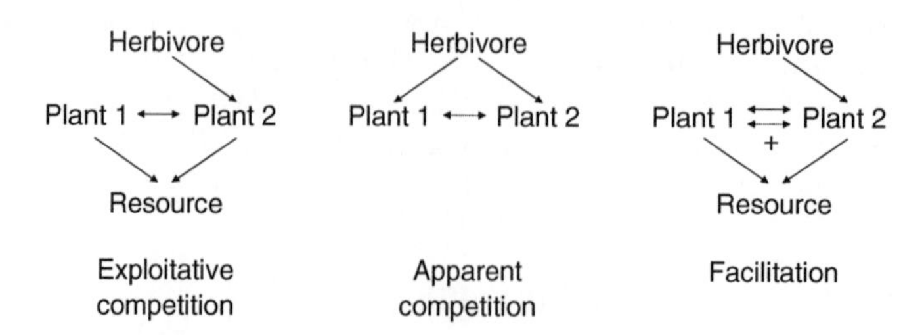

FIGURE 16.1 Schematic representation of the modules referring to more than one plant interacting with one herbivore. Solid arrows mean direct effects, whereas dotted arrows refer to indirect effects. All effects are negative except those marked with "+."

of another, not because they compete for the same resources, but because they are consumed by the same herbivore (Huntly 1991). For example, Rand (2003) found that the presence of *Salicornia* when insect herbivores are excluded has no effect on *Atriplex* (both Chenopodiaceae). Meanwhile, when herbivores appeared, the presence of *Salicornia* resulted in a pronounced decrease in plant survivorship and fruit production of *Atriplex*. Thus, shared herbivory resulted in a strong apparent competitive effect of *Salicornia* on *Atriplex*.

Competition between plants can affect their relationship with their herbivores. Competition may limit resource availability for plants and, in turn, this may influence the resistance to herbivores of plants (van Dam and Baldwin 1998, Agrawal 2000b). In addition, competition and herbivory produce additive effects for plant growth (e.g., Fowler and Rausher 1985, Mutikainen and Walls 1995, Reader and Bonser 1998, Erneberg 1999, but see Fowler 2002, Agrawal 2004, Haag et al. 2004), decreasing the herbivore tolerance when plants compete for limited resources. Moreover, intraspecific competitive interactions with herbivory can affect components of fitness and mating system (Steets et al. 2006).

Associations among Plants Sharing Herbivores

The probability that a plant will be attacked by a herbivore depends not only on the characteristics of the individual plant, but also on the quality and abundance of the neighbors. A plant species may have a positive net effect on another species by deterring the amount of herbivory that would otherwise be inflicted on the other species (Figure 16.1c, *right*). For example, palatable plants in a matrix of unpalatable vegetation may remain undetected by the herbivore and thereby escape consumption. Moreover, neighboring plants may affect the local resource abundance to polyphagous herbivores in ways that reduce the attack rate or the time herbivores remain on their host plant. These processes are called associational resistance, associational defense, associational refuge, or plant-defense guilds (Tahvanainen and Root 1972, Pfister and Hay 1988, Holmes and Jepson-Innes 1989, Hjältén et al. 1993, Hambäck et al. 2000). For example, Russell and Louda (2005) found a marked decline in head weevil (*Rhinocyllus conicus*, Curculionidae) attack of wavyleaf thistle (*Cirsium undulatum*, Asteraceae) flower heads in the presence of successful flowering by an alternate, newly adopted native host plant, the platte thistle (*Cirsium canescens*, Asteraceae).

Conversely, when the herbivore selects within the patch, the result of the association of a palatable plant with unpalatable ones can shift to greater consumption or damage of the edible species, which was preferred by the herbivore. Moreover, an unpalatable plant surrounded by palatable plants can be damaged by the herbivore attracted by its neighbors. These processes are called associational susceptibility, associational damage, or shared doom (Atsatt and O'Dowd 1976, McNaughton 1978, Karban 1997). For instance, White and Whitham (2000) found strong indications for associational susceptibility of cottonwoods (*Populus angustifolia–Populus fremontii*, Salicaceae) to cankerworms (*Alsophila pometaria,* Geometridae) when growing under the most preferred species (*Acer negundo*, Aceraceae), since it was colonized by two- to threefold more cankerworms, and suffered two- to threefold greater defoliation than cottonwoods growing in the open or under mature cottonwoods.

Facilitation can also result from physical protection provided by nurse plants. There is a considerable number of studies that demonstrate a grazing protection component of woody and perennial plants harboring other species growing under them (Milchunas and Noy-Meir 2002, and references therein), enhancing community diversity (Olff et al. 1999, Callaway et al. 2000, Rebollo et al. 2002). Shrubs can protect saplings against herbivores (Callaway 1995, García and Obeso 2003) facilitating the regeneration of palatable tree species that would be untenable without shrub presence (Rousset and Lépart 1999, Meiners and Martinkovic 2002, Smit et al. 2006) The advantage of facilitation increases parallel to herbivore pressure (Bertness and Callaway 1994, Baraza et al. 2006), to the point that in some situations the only seedlings and juvenile trees that survive remain within the islands formed by spiny shrubbery. Beyond these protectorates, natural regeneration can be completely arrested by strong herbivore pressure.

More than One Herbivore

Interspecific relationships between two herbivorous species can range from mutually competitive to mutually beneficial (Crawley 1983, Strauss 1991). When one plant becomes the host of several different herbivore species, it is difficult to understand the result of an interacting pair of species without taking into account the effect of the other herbivores.

Above and Belowground Multitrophic Interactions

Plants are frequently attacked by both above- and belowground arthropod herbivores. Aboveground and belowground herbivores influence each other indirectly via changes in biomass and the nutritional quality of host plants (Blossey and Hunt-Joshi 2003, and references therein). For example, root-feeding herbivores can induce changes in plant secondary chemistry (increase induced defenses), which reduce the performance of the foliage-feeding insect (Bezemer and van Dam 2005). Moreover, belowground herbivores can affect not only secondary metabolisms but also primary plant compounds. In fact, Bezemer et al. (2005) found a significant reduction in offspring production of aphids (*Rhopalosiphum padi*, Aphididae) in the presence of a nematode, probably by a decrease in foliar nitrogen and amino acid concentrations in the preferred host plant *A. odoratum* (Poaceae).

Although belowground decomposers are not directly associated with plant roots, they can influence aboveground plant-defense levels as a result of differences in the availability of nutrients to the plant (Wurst and Jones 2003, Wurst, et al. 2004a). Through decomposition, earthworms can increase nitrogen availability in the soil, resulting in the plant investing more in growth and less in direct defense compounds (Wurst et al. 2006). For example, Wurst et al. (2004b) found a decline in foliar catalpol concentration of *Plantago lanceolata* (Plantaginaceae) in the presence of earthworms (*Aporrectodea caliginos*, Lumbricidae), documenting the potential of decomposers to influence concentrations of plant secondary metabolites.

In the same way, aboveground herbivores can alter subterranean organisms and processes through plant responses. Two principal mechanisms have been proposed by which this occurs—through herbivore effects on patterns of root exudation and carbon allocation and through altering the quality of input of plant litter (Bardgett et al. 1998). Positive effects arise when herbivores promote compensatory plant growth, returning organic matter to the soil as labile fecal material (rather than as recalcitrant plant litter), inducing greater concentrations of nutrients in remaining plant tissues and impairing plant succession, thereby inhibiting the ingress of plant species with poorer litter quality. Negative effects arise through the impairment of plant productivity by tissue removal, induced production of secondary defenses, and promotion of succession by favoring the dominance of unpalatable plant species with poor litter quality. Whether net effects are positive or negative depends on the context (Wardle et al. 2004). In general, positive effects of herbivory on soil biota and soil processes are most common in ecosystems of high soil fertility and high consumption rates, whereas negative effects are most common in unproductive ecosystems with low consumption rates (Bardgett and Wardle 2003).

Interactions between Herbivores and Pathogens

The interaction between herbivores, such as insects, slugs, snails, birds, or mammals, and pathogens (e.g., bacteria, fungi, and viruses) has been studied only recently (Faeth and Wilson 1997). Bowers and Sachi (1991) recorded an increase in disease levels of the rust *Uromyces trifolii* (Pucciniaceae) on clover (*Trifolium pratense*, Fabaceae) in fenced exclosure plots compared with control plots. This increase results from an increase in host plant density in the exclosures. On the other hand, some studies have reported that macro-herbivores prefer plants bearing micro-herbivores. Molluscs graze more heavily on rust-infested plants than on healthy ones (Ramsell and Paul 1990). Similarly, Ericson and Wennström (1997) have analyzed the interaction between the fungus *Urocystis tridentalis* (Ustilaginales), its host plant *Trientalis europaea* (Primulaceae), and two herbivores (scale insects and voles) in a 2 year experiment. The results indicate that both the scale insects and the voles preferred smut-infected shoots to healthy shoots. Fencing out the voles resulted in a significant boost in host density and a significantly higher disease level. On the contrary, the willow leaf beetle *Plagiodera versicolora* (Chrysomelidae) significantly avoided feeding and oviposition on leaves of the willow hybrid *Salix* × *cuspidata* (Salicaceae) when they are infected by the rust fungus *Melampsora allii-fragilis* (Uredinales et al. 2005).

The interaction between herbivores and pathogens is not always antagonistic. For example, herbivores can transmit diseases to the host plant, increasing its harmful effect without substantial direct consumption of plant tissue. European elm (*Ulmus* spp., Ulmaceae) forests have declined in the last 20 years because of a parasitic fungus that produces graphiosis. This fungus is transmitted by some species of herbivorous beetles belonging to the family Scolytidae (Gil 1990). The detrimental effect of the beetle increases elm mortality not by direct consumption of cambium, but by acting as vector of the parasite.

Effect of Herbivores on Mutualism Involving the Host Plant

Plants are involved in a diverse array of mutualistic interactions, including pollination, seed dispersal, or mycorrhiza symbiosis. Herbivores can influence any of these mutualistic interactions displayed by the host plant.

Effect on Pollen-Dispersal System

By reducing plant resources, herbivory may have direct consequences on the mating system. Resource limitation caused by herbivory can affect flower production (e.g., Lehtilä and Strauss 1997, Mothershead and Marquis 2000), flowering phenology (Juenger and Bergelson 1997), and seed mass and number (e.g., Stephenson 1981, Koptur et al. 1996, Agrawal 2001, Hódar et al. 2003). For example, leaf damage decreases pollen production and performance in *Cucurbita texana* (Cucurbitaceae, Quesada et al. 1995) and produces selective fruit abortion in *Lindera benzoin* (Lauraceae, Niesenbaum 1996).

Leaf damage can reduce the number of simultaneously open flowers on a plant (Strauss et al. 1996, Elle and Hare 2002) and, thus, decrease the potential for pollinators to affect geitonogamy (selfing among flowers on a plant) (Harder and Barrett 1995). Herbivory can also modify flower morphology and reward, which in turn may reduce pollinator visitation (Strauss et al. 1996, Mothershead and Marquis 2000), Moreover, flower consumption by herbivores also affects the pollen-dispersal system indirectly, by altering the visitation rate of pollinators in entomophilous plants (Marquis 1992a).

Effect on Plant–Mycorrhiza Interaction

The relationships among plants, their mycorrhizal fungi, and their herbivores are likely to be complex and can be observed from different standpoints:

1. Herbivore effects on mycorrhiza. Foliage removal by insect herbivores can reduce arbuscular mycorrhizal (AM) colonization levels of herbaceous plants (Gange et al. 2002a), and ectomycorrhizal (ECM) colonization levels in trees (Gehring and

Whitham 2002). Conversely, moderate grazing in tallgrass prairie microcosms seems to improve AM colonization levels (Kula et al. 2005), perhaps because aboveground herbivory may increase the nutrient demand of host plants (Eom et al. 2001). In an experiment with three AM species, Klironomos et al. (2004) found that clipping of *Bromus inermis* (Poaceae) affects certain mycorrhizal characteristics, depending on the fungal species involved. As a result, any mycorrhizal feedback that may occur in response to herbivory is not simple to predict, either (Wamberg et al. 2003).

2. Mycorrhiza effect on herbivores. Both AM and ECM fungi are known to alter plant physiology and chemistry, and, as a result, can affect herbivores that feed on them. Several works have reported resistance to insect herbivory in plants inoculated with mycorrhiza (Gange and West 1994, Borowicz 1997, Gange et al. 2005). However, the interaction between mycorrhizal infection and herbivory is complex, depending on the species not only of herbivore, but also of the fungi and plant (Gehring and Whitham 2002, Gange et al. 2005). In general, changes provoked by AM on plants boosted the growth of specialist chewing as well as specialist and generalist sucking insects, but decreased the growth of generalist chewers (Gange and West 1994, Borowicz 1997, Gange et al. 1999, Goverde et al. 2000, Gange et al. 2002b). The underlying mechanism by which AM fungi affect an insect community has been linked to mycorrhizal-induced changes in plant chemistry, either through changes in secondary metabolites (Gange and West 1994) or alterations in plant-nitrogen content (Gange and Nice 1997, Rieske 2001).
3. Mycorrhiza effect on plant tolerance to herbivores. Mycorrhiza improvement in plant nutrients supplied could also confer a greater capacity for recovering from herbivory (Hokka et al. 2004, Kula et al. 2005). Nevertheless, there is controversy concerning the effect of mycorrhiza on damaged plants, since, on the one hand, symbiosis imposes a cost of carbon that cannot be used for plant growth, and, on the other hand, mycorrhiza contributes necessary nutrients for plant growth (Borowicz 1997). Moreover, plant response to herbivory depends on environmental conditions, mycorrhizal symbiosis becoming more important in the case of high-intensity light and low water and nutrients availability (Gehring and Whitham 1994). In addition, in this case, the plant species is determinant. For example, Allsopp (1998) found that *Lolium* and *Digitaria* (Poaceae), which are pasture species, are better able to maintain an external AMF hyphal network following fairly frequent defoliation, whereas *Themeda* (also Poaceae), a rangeland grass, which is more intolerant of grazing, has a lower capacity for sustaining its hyphal network when defoliated.

MULTISPECIFIC INTERACTIONS

The basic food chain is composed of a plant, its herbivore, and the predator of the herbivore (Figure 16.2). The most widely studied tritrophic systems consist of a plant or a seed, a parasitic herbivore (seed predator, gall-maker, or the like), and parasitoids, although

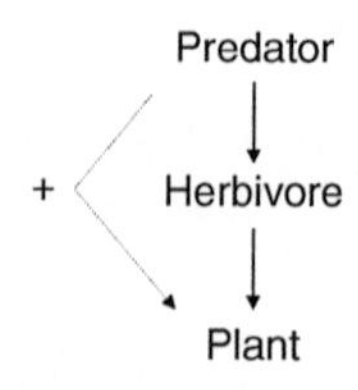

FIGURE 16.2 Schematic representation of a basic food chain.

interest is growing with respect to the types of tritrophic systems, such as those in which insectivorous vertebrates (birds, reptiles, or mammals) intervene in relationship between herbivores and plants (see Tscharntke 1997). For instance, Van Bael et al. (2003) observed that birds decreased local arthropod densities on canopy branches and reduced consequent damage to leaves for three Neotropical tree species. However, this effect of birds on plant damage does not always exert an effect on plant-biomass production (Strong et al. 2000 and references therein).

Parasitism represents a crucial mortality factor for many species of herbivorous insects. For this reason, parasitoids can improve plant performance. Gómez and Zamora (1994) tested the totality of direct and indirect forces in a tritrophic system composed of a guild of three parasitoid species, a single weevil seed predator, and the host plant. When parasitoids were experimentally excluded, the percentage of attacked fruits rose from 20% to 43%, the parasitoids thus enhancing plant reproductive performance. The effect of parasitoids in herbivore population is influenced by characteristics of the host plant. For example, von Zeipel et al. (2006) found an important effect of plant population size on the results of the tritrophic system formed by a perennial plant, *Actaea spicata*, the associated specialist moth seed predator, *Eupithecia immundata,* and a guild of parasitoids. In large plant populations, parasitoids reduced the level of seed predation, thereby enhancing plant fitness. In small populations, usually either a high proportion of seeds was preyed on because of seed predator presence and parasitoid absence or there was no seed predation when the seed predator was absent. Finally, when plant population was of intermediate size, there was intense seed predation, since the seed predator was present but parasitoids were often absent.

Plants are not passive elements in these tritrophic interactions. In response to herbivore damage, several plant species emit volatile chemicals that attract natural enemies (predators and parasitoids), which attack herbivores (Dicke and van Loon 2000 and references therein). Moreover, plants can adaptively react to the chemical information emitted by their neighbors by two types of responses: the induction of a direct defense that makes them resistant to subsequent herbivore attack and an indirect defense that involves the recruitment of carnivorous arthropods as "bodyguards" (Arimura et al. 2000, Dicke et al. 2003).

Multispecific interactions can occur throughout guilds as interactive units, when there are functionally equivalent animals or plants (i.e., from the plant's or the herbivore's perspective). Plants may interact with a guild of ecomorphologically similar herbivore species rather than with a particular species. The degree of generalization determines the breadth of the filter of the interaction and the real possibility that the system might be facultative (different species with the same role). For example, Maddox and Root (1990), studying the trophic organization of the herbivorous insect community (more than 100 species distributed among 5 orders) of *Solidago altissima* (Asteraceae), suggest that the functionally similar herbivore groups may constitute selective units more powerful than individual species. This opens the possibility of synergetic responses as opposed to the same blocks of selective pressures (broad-spectrum responses). In this way, Krischik et al. (1991) indicated that nicotine was inhibitory to the growth both of herbivores and of pathogens, suggesting that certain secondary plant chemicals with high toxicity are of a generalized nature and affect multiple species. Adler and Kittelson (2004) determine how different environmental effects influence alkaloid profiles and resistance to multiple herbivores in *Lupinus arboreus* (Leguminosaceae), showing a highly complex response by the different herbivores analyzed. For instance, the density of the leaf galler *Dasineura lupinorum* (Cecidomyiidae) and the fungus *Colletotrichium* spp. (Nectriodaceae) was affected by total alkaloid concentration and alkaloid profiles, whereas the density of apical flies and bud gallers was not affected by any alkaloid measure.

PLANT–HERBIVORE INTERACTION: A MULTISPECIFIC VISION

CASE STUDY 1. CLIMATE EFFECTS ON INSECT OUTBREAKS: THE PINE PROCESSIONARY

The pine processionary moth (*Thaumetopoea pityocampa*, hereafter PPC) is a good example of how climate and plant characteristics interact to provide a given kind of life cycle in a herbivorous insect. PPC is a serious defoliator in the Mediterranean area, which attacks different species of the genus *Pinus* (see e.g., Dajoz 1998). Traditionally, it has been assumed that the incidence of PPC defoliation depended on winter temperatures (Demolin 1969, Hódar et al. 2003). This is due to its particular life cycle; that is, while most arthropods develop as larvae or nymphs during spring and summer, with abundant food and warm temperatures, PPC develops as larvae during winter. For this, larvae of the same egg batch develop together in a communal silk nest that allows them to save heat and continue development (Breuer et al. 1989, Breuer and Devkota 1990, Halperin 1990). However, very low temperatures (−10 to −15°C) can be lethal, and above +30°C larvae cannot stay together in the communal nest. Despite the importance of temperature, it has long been recognized that food quality for larval development is also an important issue in the population dynamics of the PPC. This suggestion is based on the different incidence of defoliation by PPC in the different pine species: while White pine *Pinus pinea* is particularly resistant and defoliation is usually low, others such as Black pine *Pinus nigra* or exotic species are heavily defoliated. In Spain, the more resistant species, such as White pine, inhabit low altitudes, whereas Aleppo pine *Pinus halepensis* and Cluster pine *Pinus pinaster* do so to a lesser degree. On the contrary, Black pine and Scots pine *P. sylvestris*, inhabiting middle or high altitudes in mountains, or introduced species as Canary Island pine *Pinus canariensis* or Monterey pine *Pinus radiata*, are particularly susceptible to PPC attack. Many works have tried to identify the features in pine needles that affect PPC larval development (Schopf and Avtzis 1987, Battisti 1988, Devkota and Schmidt 1990, Tiberi et al. 1999, Petrakis et al. 2001, Hódar et al. 2002, 2004) but none have found conclusive evidence.

The distribution of the pine species in altitude, depending on its palatability, suggests that the most resistant pines, living at lower altitudes with mild winters that enhance PPC development, have acquired constitutive chemical defenses against defoliation. By contrast, pines living at high altitudes, with cold winters that rarely allow the development of PPC larvae (or exotic pines never defoliated by PPC), did not develop these defenses and have a very limited capacity for chemical response (Hódar et al. 2004). When the winter is warm and pines are planted in zones not adequate for their defense, outbreaks of PPC can be frequent. This situation is worsening for two main reasons. The first is the massive forestation with exotic pine species in zones with high PPC incidence, such as *P. radiata* in coastal northern Spain. The second is the increase in temperatures due to climatic change, which is giving PPC the opportunity of thriving in pine woodlands belonging to palatable species, which, until now, were free of PPC attack for climatic reasons. In particular, rising winter temperatures are favoring the progression of PPC in altitude (Hódar et al. 2003, Hódar and Zamora 2004, Battisti et al. 2006) and in latitude (Battisti et al. 2005).

Abundant scientific literature provides analyses of specialized cases of an insect herbivore feeding on a plant depending on nutritional characteristics (see Section "Introduction"). The case of PPC is more complex, because the same insect species feeds on different (but related) pine species with different abilities to tolerate defoliation and because the development of PPC is strongly modulated by the climatic conditions at the pine woodland where PPC lives. The best hosts live where temperatures are inhospitable for PPC. The best temperature for PPC occurs where pines are not a good food, and this interaction between PPC and their food determines the alternation between years of low infestation and years of severe outbreaks.

Case Study 2. Conditional Outcomes in Plant–Herbivore Interactions: Neighbors Matter

Although herbivores try to select more nutritive plants and avoid excessive toxin consumption, other numerous factors influence their foraging behavior, this necessary to be considered when analyzing plant–animal interactions (Provenza et al. 2002). As shown in Section "Associations among Plants Sharing Herbivores" differences in the palatability of coexisting plant species can affect the interaction of a particular herbivore species with a particular plant. Moreover, other conditions such as climate or herbivore density can alter herbivore foraging behavior. For example, in Sierra Nevada in wet years, only 20% of Scots pine saplings undergo some herbivore attack, while in dry years, with low pasture production, up to 80% of saplings suffer browsing damage (Hódar et al. 1998). In this scenario, herbivore foraging behavior, plant characteristics, the surrounding vegetation palatability, and the environmental conditions could interact to determine the probability of damage to a given plant (Provenza et al. 2002).

Baraza et al. (2006) in an experimental reforestation planted two tree species (a palatable tree and unpalatable one), under four experimental microhabitats: highly palatable shrub, palatable but spiny shrub, unpalatable spiny shrub, and control. The finding was that three factors determine the damage probability of saplings. Palatable species were usually attacked, whereas unpalatable species were only rarely attacked. As surrounding vegetation, highly palatable shrubs can promote high herbivory in the sapling beneath it, whereas an unpalatable shrub reduces the probability of attack (Callaway 1992, Rousset and Lepart 2002, Smit et al. 2006). These two factors can interact in a way that the degree of protection offered by the shrub is greater as its palatability decreases with respect to sapling palatability (Baraza et al. 2006). In addition, herbivore pressure acts as one of the most important and potentially variable factors affecting the degree of sapling protection by shrubs (Baraza et al. 2006). With high herbivore pressure, only unpalatable shrubs can protect palatable saplings, whereas for unpalatable saplings the probability of attack tends to increase when growing near shrubs (Figure 16.3). On the contrary, with low herbivore pressure, shrubs of intermediate palatability

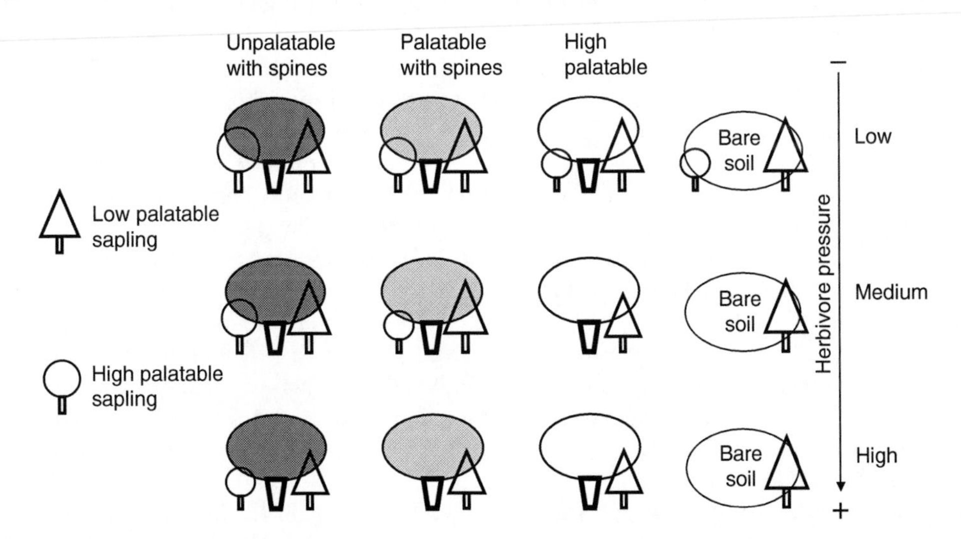

FIGURE 16.3 Sapling damage probability depends on sapling palatability, microhabitat of growth, and herbivore pressure. Smaller sapling figures represent more probability of being eaten and, as a result, less probability of establishment. Probability of damage is higher for palatable saplings than for unpalatable ones in all conditions, while the protective role of shrubs depends on herbivore pressure and the palatability of the shrub. (Reproduced from Baraza, E., Zamora, R., and Hódar, J.A., *Oikos*, 113, 148, 2006. With permission.)

may protect palatable saplings from herbivores, whereas the protective role of unpalatable shrubs increases, and unpalatable saplings are safe in any microhabitat. With intermediate herbivore pressure, palatable saplings decrease their probability of attack by growing under shrubs of intermediate or low palatability, whereas unpalatable saplings start to undergo damage when growing under shrubs (Figure 16.3). These findings show that the quality of a microhabitat for recruitment depends strongly on the degree of herbivore pressure, together with relative palatability of saplings and shrub. Consequently, when herbivory pressure increase, the landscape can change from a high-quality matrix for sapling recruitment to becoming a low-quality matrix where only unpalatable plants constitute available patches for recruitment for palatable species (Callaway et al. 2005, Baraza et al. 2006). These examples show the complexity of plant–herbivore interactions that is affected not only by neighbors (see Section "Associations among Plants Sharing Herbivores") but also by the environmental context (Baraza et al. 2006). Moreover, this effect of neighbors can determine the consequence of herbivory on vegetation structure (Olff et al. 1999, Calaway et al. 2000, Rebollo et al. 2002).

Case Study 3. Ungulates Affect Populations of Both Plants and Other Co-Occurring Herbivores

Erysimum mediohispanicum (Cruciferae) is a monocarpic herb found in many montane regions of SE Spain from 1000 to 2000 m a.s.l. Plants usually grow for 2–3 years as vegetative rosettes, then die after producing one (up to eight) reproductive stalk, which can display between a few and several hundred hermaphroditic, slightly protandrous bright yellow. At the SE Spain, reproductive individuals are fed by many different species of herbivores. Several species of sap-suckers (outstanding the bugs *Corimeris denticulatus*, *Eurydema fieberi,* and *Eurydema ornata*, Pentatomidae) feed on the reproductive stalks, both during flowering and fruiting. In addition, stalks are bored into by a weevil species (*Melanobaris erysimi erysimoides)*, which consume the inner tissues, whereas another weevil species (*Ceutorhynchus chlorophanus*, both Curculionidae) develops inside the fruits, living on developing seeds. Some floral buds do not open because they are galled by flies (*Dasineura* sp., Cecidomyiidae). However, the main herbivores in the study zones are domestic (sheep) and wild ungulates (Spanish ibex *Capra pyrenaica*, Bovidae). Postdispersed seeds of both species are consumed by woodmice (*Apodemus sylvaticus*, Muridae), several species of birds (*Fringilla coelebs*, *Serinus serinus*, *Carduelis cannabina* [Frigillidae], among others), several species of medium-sized granivorous beetles (*Iberozabrus* sp. [Carabidae], among others) and ants (*Lasius niger*, *Tetramorium caespitum* and probably *Cataglyphis velox* and *Leptothorax tristis*, Formicidae).

Ungulates damage exclusively the inflorescences and infructescences of the plants, cutting them and consuming the flowers/fruits plus the stalks. Damage to plants occurred before they had dispersed the seeds, increasing the potential detrimental effect of herbivory. Indeed, ungulates heavily affect many components of *Erysimum* perfomance, such as fecundity, seed survival to postdispersal seed predation, seedling emergence, and survival. Consequently, ungulates had a significant detrimental effect on the population dynamics of *Erysimum mediohispanicum* (Gómez 2005a).

The occurrence of ungulates has additional indirect effects in the interaction that plants maintain with invertebrate pollinators and herbivores between. Thus, ungulates deplete the abundance and diversity of pollinators (Gómez 2005b). Plants frequently grazed by ungulates are pollinated by a less diverse assemblage of pollinators than nondamaged plants, presumably because grazing decreases flower size and number. This fact suggests an indirect negative effect of ungulates on plants through a decrease in mutualistic interactions. In addition, ungulates also outcompete all the phytophagous insects co-occurring in the plants, since the experimental exclusion of ungulates provokes a significant increase in the abundance of predispersal seed predators, sapsuckers, gall-makers, and stem-borers. Most important, different mechanisms account for the increase in the abundance of different herbivores. Thus, predispersal seed predators are affected mainly by density-mediated indirect interactions, the negative effect of ungulates provoked mostly by a decrease in the shared resources, the seeds. However, ungulates have a significant trait-mediated indirect effect on gall-makers, sapsuckers, and stem-borers, as suggested by the change in the per-capita interaction strength between these insects and the plant when ungulates were present. In fact, in addition to decrease abundance, ungulates modify the phenotype of the plants. Damaged plants produced less and smaller flowers, shorter and narrower flowering stalks, and more trichomes. Sapsuckers, stem-borers, and gall-makers are deterred by these phenotypic traits, and thereby, they show a negative preference for damaged plants (Gómez and Gónzalez-Megías 2007). Ungulates also affect seed-predators and gall-makers directly by incidentally ingesting them while browsing (Gómez and González-Megías 2007). These two species are endophagous, as they are unable to avoid the interaction with ungulates by directly leaving the plant and usually located in the upper part of the plant, the one more likely ingested by ungulates.

Finally, ungulates also affect postdispersal seed predators. Thus, the rate of seed removal by predators increases after experimentally excluding ungulates. Several nonexclusive reasons can account for this result (Gómez 2005a). First, it could be a consequence of exploitative competition occurring between seed predators and ungulates, which means that after removing ungulates seed-predator abundance could increase. Competition between ungulates and granivorous species has been widely reported (Davidson et al. 1984, 1985, Baines et al. 1994, and references therein) and may occur as a consequence of resource removal (flowers, fruits, and seeds) as well as by the negative effect of vegetative tissue removal on flower and fruit production (Meyer 1993, Meyer and Root 1993, Müller-Scharer and Brown 1995). In the studied systems, by consuming flower and fruits, sheep and ibex surely deplete the resources used by granivorous mammals and invertebrates.

In brief, this example illustrates that the assemblage of herbivores living on the same host plant can maintain complex indirect interactions among them. A full and deep understanding of these interactions can help to accurately find out the net effect that herbivores have on the functioning of plant individuals and populations.

ACKNOWLEDGMENTS

We thank the Agencia de Medio Ambiente, Junta de Andalucía, for permitting to conduct field work in the Parque Natural de Sierra Nevada and in the Parque Natural de la Sierra de Baza. Thanks to Blackwell publishing for permission to use Figure 16.3. The research that supported most part of the work shown here was financed by projects AGF99-0618, 1FD97-0743-CO3-02, and REN2002-04041-CO2-01/GLO (to R.Z.) and UGR2002-30P3176 and BOS2003-09045 (to J.M.G.). E.B. is currently supported by a postdoctoral grant from the Fundación Ramón Areces. David Nesbitt read through the English version of the text.

REFERENCES

Adler, L.S. and P.M. Kittelson, 2004. Variation in *Lupinus arboreus* alkaloid profiles and relationships with multiple herbivores. Biochemical Systematics and Ecology 32: 371–390.

Agrawal, A.A., 1998. Induced responses to herbivory and increased plant performance. Science 279: 1201–1202.

Agrawal, A.A., 2000a. Specificity of induced resistance in wild radish: causes and consequences for two specialist and two generalist caterpillars. Oikos 89: 493–500.

Agrawal, A.A., 2000b. Benefits and costs of induced plant defence for *Lepidium virginicum* (Brassicaceae). Ecology 81: 1804–1813.

Agrawal, A.A., 2001. Transgenerational consequences of plant responses to herbivory: An adaptive maternal effect? American Naturalist 157: 555–569.

Agrawal, A.A., 2004. Resistance and susceptibility of milkweed: competition, root herbivory, and plant genetic variation. Ecology 85: 2118–2133.

Agrawal, A.A. and R. Karban, 1999. Why induced defences may be favored over constitutive strategies in plants. In: R. Tollrian and C.D. Harvell, eds. The Ecology and Evolution of Inducible Defences. Princeton University Press, Princeton, NJ, pp. 45–61.

Agrawal, A.A. and P.A. Van Zandt, 2003. Ecological play in the coevolutionary theatre: genetic and environmental determinants of attack by a specialist weevil on milkweed. Journal of Ecology 91: 1049–1059.

Agrawal, A.A., J.K. Conner, M.T.J. Johnson, and R. Wallsgrove, 2002. Ecological genetics of an induced plant defence against herbivores: additive genetic variance and costs of phenotypic plasticity. Evolution 56: 2206–2213.

Aide, T.M., 1992. Dry season leaf production—an escape from herbivory. Biotropica 24: 532–537.

Arimura, G.-I., K. Tashiro, S. Kuhara, T. Nishioka, R. Ozawa, and J. Takabayashi, 2000. Gene responses in bean leaves induced by herbivory and by herbivore-induced volatiles. Biochemical and Biophysical Research Communications 277: 305–310.

Allsopp, N., 1998. Effect of defoliation on the arbuscular mycorrhizas of three perennial pasture and rangeland grasses. Plant and Soil 202: 117–124.

Andersen, A.N., 1989. How important is seed predation to recruitment in stable populations of long-lived perennials? Oecologia 81: 310–315.

Atsatt, P.R. and D.J. O'Dowd, 1976. Plants defence guilds. Science 193: 24–29.

Augustine, D.J. and S.J. McNaughton, 1998. Ungulate effects on the functional species composition of plant communities. Herbivore selectivity and plant tolerance. Journal of Wildlife Management 62: 1165–1183.

Baines, D., R.B. Sage, and M.M. Baines, 1994. The implications of red deer grazing to ground vegetation and invertebrate communities of Scottish native pinewoods. Journal of Applied Ecology 31: 776–783.

Baraza, E., 2004. Efecto de los pequeños ungulados en la regeneracion del bosque mediterráneo de montaña: desde la química hasta el paisaje. PhD-thesis, Universidad de Granada, Spain (in Spanish).

Baraza, E., J.M. Gómez, J.A. Hódar, and R. Zamora, 2004. Herbivory has a greater impact in shade than in sun: response of *Quercus pyrenaica* seedlings to multifactorial environmental variation. Canadian Journal of Botany 82: 357–364.

Baraza, E., R. Zamora, and J.A. Hódar, 2006. Conditional outcomes in plant–herbivore interactions: neighbours matter. Oikos 113: 148–156.

Bardgett, R.D. and D.A. Wardle, 2003. Herbivore-mediated linkages between aboveground and below-ground communities. Ecology 84: 2258–2268.

Bardgett, R.D., D.A. Wardle, and G.W. Yeates, 1998. Linking above-ground and below-ground interactions: how plant responses to foliar herbivory influence soil organisms. Soil Biology and Biochemistry 30: 1867–1878.

Battisti, A., 1988. Host-plant relationships and population dynamics of the pine processionary moth *Thaumetopoea pityocampa* (Denis and Schiffermüller). Journal of Applied Entomology 105: 393–402.

Battisti, A., M. Stastny, S. Netherer, C. Robinet, A. Schopf, A. Roques, and S. Larsson, 2005. Expansion of geographic range in the pine processionary moth caused by increased winter temperatures. Ecological Applications, 15: 2084–2096.

Battisti, A., M. Stastny, E. Buffo, and S. Larsson, 2006. A rapid altitudinal range expansion in the pine processionary moth produced by the 2003 climatic anomaly. Global Change Biology 12: 662–671.

Batzli, G.O., 1983. Responses of arctic rodent populations to nutritional factors. Oikos 40: 396–406.

Bebber, D., N. Brown, and M. Speight, 2002. Drought and root herbivory in understorey *Parashorea* Kurz (Dipterocarpaceae) seedlings in Borneo. Journal of Tropical Ecology 18: 795–804.

Begon, M., J.L. Harper, and C.R. Townsend, 1990. Ecology: individuals, populations and communities, 2nd edition. Blackwell Scientific Publications, Oxford, England.
Belsky, A.J., 1986. Does herbivory benefit plants? A review of the evidence. American Naturalist 127: 870–892.
Bennett, R.N. and R.M. Wallsgrove, 1994. Secondary metabolites in plant defence mechanisms New Phytologist 127: 617–633.
Bentley, B.L. and N.D. Johnson, 1992. Plants as food for herbivores: The roles of nitrogen fixation and carbon dioxide enrichment. In: P.W. Price, T.M. Lewinsohn, G.W. Fernandes, and W.W. Benson, eds. Plant–Animal Interactions: Evolutionary Ecology in Tropical and Temperate Regions. John Wiely and Sons, New York, pp. 257–272.
Berenbaum, M.R., 1995. The chemistry of defence: Theory and practice. Proceedings of the National Academy of Sciences of the United States of America 92: 2–8.
Bergström, R. and K. Danell, 1995. Effects of simulated summer browsing by moose on leaf and shoot biomass of birch, *Betula pendula*. Oikos 72: 132–138.
Bernays, E.A. and R.F. Chapman, 1994. Host plant selection by phytophagous insects. Chapman and Hall, New York.
Bertness, M.D. and R. Callaway, 1994. Positive interactions in communities. Trends in Ecology and Evolution 9: 191–193.
Bezemer, T.M. and N.M. Van Dam, 2005. Linking aboveground and belowground interactions via induced plant defences. Trends in Ecology and Evolution 20: 617–624.
Bezemer, T.M., G.B. De Deyn, T.M. Bossinga, N.M. van Dam, J.A. Harvey, and W.H. Van der Putten, 2005. Soil community composition drives aboveground plant–herbivore–parasitoid interactions. Ecology Letters 8: 652–661.
Björkman, C. and P. Hambäck, 2003. Context-dependence in plant–herbivore interactions. Oikos 101: 3–5.
Blossey, B. and T.R. Hunt-Joshi, 2003. Belowground herbivory by insects: Influence on plants and aboveground herbivores. Annual Review of Entomology 48: 521–547.
Boege, K. and R.J. Marquis, 2005. Facing herbivory as you grow up: the ontogeny of resistance in plants. Trends in Ecology and Evolution 20: 441–448.
Borowicz, V.A., 1997. A fungal root symbiont modifies plant resistance to an insect herbivore. Oecologia 112: 534–542.
Bowers, M.A. and C.F. Sacchi, 1991. Fungal mediation of a plant–herbivore interaction in an early successional plant community. Ecology 72: 1032–1037.
Boyd, R.S., 1988. Herbivory and species replacement in the west-coast searockets (*Cakile*, Brassicaceae). American Midland Naturalist 119: 304–317.
Breuer, M. and B. Devkota, 1990. Studies on the importance of nest temperature of *Thaumetopoea pityocampa* (Den. and Schiff.) (Lep. Thaumetopoeidae). Journal of Applied Entomology 109: 331–335.
Breuer, M., B. Devkota, E. Douma-Petridou, A. Koutsaftikis, and G.H. Schmidt, 1989. Studies on the exposition and temperature of nests of *Thaumetopoea pityocampa* (Den. and Schiff.) (Lep. Thaumetopoeidae) in Greece. Journal of Applied Entomology 107: 370–375.
Bruelheide, H. and U. Scheidel, 1999. Slug herbivory as a limiting factor for the geographical range of *Arnica montana*. Journal of Ecology 87: 839–848.
Bryant, J.P., F.S. Chapin III, and D.R. Klein, 1983. Carbon/nutrient balance of boreal plants in relation to vertebrate herbivory. Oikos 40: 357–368.
Cabin, R.J. and D.L. Marshall, 2000. The demographic role of soil seed banks. I. Spatial and temporal comparisons of below- and above-ground populations of the desert mustard *Lesquerella fendleri*. Journal of Ecology 88: 283–292.
Callaway, R.M., 1992. Effect of shrubs on recruitment of *Quercus douglasii* and *Quercus lobata* in California. Ecology 73: 2118–2128.
Callaway, R.M., 1995. Positive interactions among plants. The Botanical Review 61: 306–349.
Callaway, R.M., Z. Kikvidze, and D. Kikodze, 2000. Facilitation by unpalatable weed may conserve plant diversity in overgrazed meadows in the Caucasus Mountains. Oikos 89: 275–282.
Callaway, R.M., D. Kikodze, M. Chiboshvili, and L. Khetsuriani, 2005. Unpalatable plants protect neighbors from grazing and increase plant community diversity. Ecology 86: 1856–1862.

Canham, C.D., J.B. McAninch, and D.M. Wood, 1994. Effect of the frequency, timing, and intensity of simulated browsing on growth and mortality of tree seedlings. Canadian Journal of Forest Research 24: 817–825.

Carson, W.P and R.B. Root, 2000. Herbivory and plant species coexistence: community regulation by an outbreaking phytophagous insect. Ecological Monographs 70: 73–99.

Chew, F.S. and S.P. Courtney, 1991. Plant apparency and evolutionary escape from insect herbivory. American Naturalist 138: 729–750.

Cipollini, D., 1998. Induced defences and phenotypic plasticity. Trends in Ecology and Evolution 13: 200–200.

Clark, J.S., E. Macklin, and L. Wood, 1998. Stages and spatial scales of recruitment limitation in southern Appalachian forests. Ecological Monographs 68: 213–235.

Close, D., C. McArthur, S. Paterson, H. Fitzgerald, and A. Walsh, 2003. Photoinhibition: A link between effects of the environment on eucalypt leaf chemistry and herbivory. Ecology 84: 2952–2966.

Coleman, J.S. and C.G. Jones, 1991. Phytocentric perspective of phytochemical induction by herbivores. In: D.W. Tallamy and M.J. Raupp, eds. Phytochemical Induction by Herbivores. John Wiley and Sons, New York, pp. 3–45.

Coley, P.D., J.P. Bryant, and F.S. Chapin III, 1985. Resource availability and plant antiherbivore defence. Science 230: 895–899.

Cooper, S.M. and N. Owen-Smith, 1986. Effects of plant spinescence on large mammalian herbivores. Oecologia 68: 446–455.

Crawley, M.J., 1983. Herbivory: The Dynamics of Animal–Plant Interactions. Blackwell Scientific Publications, Oxford, UK.

Crawley, M.J., 1989. Insect herbivory and plant population dynamics. Annual Review of Entomology 34: 531–564.

Crawley, M.J. and C.R. Long, 1995. Alternate bearing, predator satiation and seedling recruitment in *Quercus robur* L. Journal of Ecology 83: 683–696.

Crone, E.E. and P. Lesica, 2004. Causes of synchronous flowering in *Astragalus scaphoides*, an iteroparous perennial plant. Ecology 87: 1944–1954.

Curran, L.M. and C.O. Webb, 2000. Experimental tests of the spatiotemporal scale of seed predation in mast-fruiting Dipterocarpaceae. Ecological Monographs 70: 129–148.

Dajoz, R., 1998. Les insectes et la fôret. Technique et documentation, S.A. Paris:

Danell, K. and R. Bergström, 2002. Mammalian herbivory in terrestrial environments. In: C.M. Herrera, and O. Pellmyr, eds. Plant–Animal Interactions: An Evolutionary Approach. Blackwell, Oxford, UK, pp. 107–131.

Dannell, K., P. Niemelä, T. Varvikko, and T. Vuorisalo, 1991. Moose browsing on Scots pine along a gradient of plant productivity. Ecology 72: 1624–1633.

Davidson, D.W., R.S. Inouye, and J.H. Brown, 1984. Granivory in a desert ecosystem—experimental evidence for indirect facilitation of ants by rodents. Ecology 65: 1780–1786.

Davidson, D.W., D.A. Samson, and R.S. Inouye, 1985. Granivory in the Chihuahuan desert—interactions within and between trophic levels. Ecology 66: 486–502.

De Bruyn, L., J. Scheirs, and R. Verhagen, 2002. Nutrient stress, host plant quality and herbivore performance of a leaf-mining fly on grass. Oecologia 130: 594–599.

Dement, W.A. and H.A. Mooney, 1974. Seasonal variation in the production of tannins and cyanogenic glucosides in the chaparral shrub, *Heteromeles arbutiflora*. Oecologia 15: 65–76.

Demolin, G., 1969. Bioecología de la procesionaria del pino *Thaumetopoea pityocampa* Schiff. Incidencia de los factores climáticos. Boletín del Servicio de Plagas Forestales 12: 9–24.

Devkota, B. and G.H. Schmidt, 1990. Larval development of *Thaumetopoea pityocampa* (Den. and Schiff.) (Lepidoptera: Thaumetopoeidae) from Greece as influenced by different host plants under laboratory conditions. Journal of Applied Entomology 109: 321–330.

DeWalt, S.J., J.S. Denslow, and K. Ickes, 2004. Natural-enemy release facilitates habitat expansion of the invasive tropical shrub *Clidemia hirta*. Ecology 85: 471–483.

Dirzo, R., 1984. Herbivory: a phytocentric overview. In: R. Dirzo and J. Sarukhán, eds. Perspectives in Plant Population Ecology. Sinauer Associates, Sunderland, MA, pp. 141–165.

Dicke, M. and J.J.A. van Loon, 2000. Multitrophic effects of herbivore-induced plant volatiles in an evolutionary context. Entomologia Experimentalis et Applicata 97: 237–249.

Dicke, M., A.A. Agrawal, and J. Bruin, 2003. Plants talk, but are they deaf? Trends in Plant Science 8: 403–405.

Donaldson, J.R., E.L. Kruger, and R.L. Lindroth, 2006. Competition-and resource-mediated tradeoffs between growth and defensive chemistry in trembling aspen *Populus tremuloides*. New Phytologist 169: 561–570.

Edwards, P.J., 1989. Insect herbivory and plant defence theory. In: P.J. Grubb and J.B. Whittaker, eds. Toward a More Exact Ecology. Blackwell Scientific Publications, Oxford, UK, pp. 275–297.

Edwards, G.R. and M.J. Crawley, 1999. Rodent seed predation and seedling recruitment in mesic grassland. Oecologia 118: 288–296.

Ehrlén, J., 1992. Proximate limits to seed production in a herbaceous perennial legume, *Lathyrus vernus*. Ecology 73: 1820–1831.

Ehrlén, J., 2002. Assessing the lifetime consequences of plant–animal interactions for the perennial herb *Lathyrus vernus* (Fabaceae). Perspectives in Plant Ecology Evolution and Systematics 5: 145–163.

Ehrlén, J., 2003. Fitness components versus total demographic effects: Evaluating herbivore impacts on a perennial herb. American Naturalist 162: 796–810.

Elle, E. and J.D. Hare, 2002. Environmentally induced variation in floral traits affects the mating system in *Datura wrightii*. Functional Ecology 16: 79–88.

Eom, A., G.W.T. Wilson, and D.C. Hartnett, 2001. Effects of ungulate grazers on arbuscular mycorrhizal symbiosis and fungal community structure in tallgrass prairie. Mycologia 93: 233–242.

Erhlich, P.R. and P.H. Raven, 1964. Butterflies and plants: a study of coevolution. Evolution 18: 586–608.

Ericson, L. and A. Wennström, 1997. The effect of herbivory on the interaction between the clonal plant *Trientalis europaea* and its smut fungus *Urocystis trientalis*. Oikos 80: 107–111.

Eriksson, O. and J. Ehrlén, 1992. Seed and microsite limitation of recruitment in plant populations. Oecologia 91: 360–364.

Erneberg, M., 1999. Effects of herbivory and competition on an introduced plant in decline. Oecologia 118: 203–209.

Faeth, S.H. and D. Wilson, 1997. Induced responses in trees: mediators of interactions among macro- and micro-herbivores? In: A.C. Gange and V.K. Brown, eds. Multitrophic Interactions in Terrestrial Ecosystems. Blackwell Scientific Publications, Oxford, UK, pp. 201–215.

Feeny, P., 1976. Plant apparency and chemical defence. Recent Advances in Phytochemistry 10: 1–40.

Fine, P.V.A., I. Mesones, and P.D. Coley, 2004. Herbivores promote habitat specialization by trees in Amazonian forest. Science 305: 663–665.

Fornoni, J., P.L. Valverde, and J. Núñez-Farfán, 2003. Evolutionary ecology of tolerance to herbivory: advances and perspectives. Comments on Theoretical Biology 8: 1–21.

Fornoni, J., J. Núñez-Farfán, P.L. Valverde, and M.D. Rausher, 2004a. Evolution of mixed strategies of plant defence allocation against natural enemies. Evolution 58: 1685–1695.

Fornoni, J., P.L. Valverde, and J. Núñez-Farfán, 2004b. Population variation in the cost and benefit of tolerance and resistance against herbivory in *Datura stramonium*. Evolution 58: 1696–1704.

Fowler, N.L., 2002. The joint effects of grazing, competition, and topographic position on six savanna grasses. Ecology 83: 2477–2488.

Fowler, N.L. and M.D. Rausher, 1985. Joint effects of competitors and herbivores on growth and reproduction in *Aristolochia reticulata*. Ecology 66: 1580–1587.

Franceschi, V.R., P. Krokene, and E. Christiansen, 2005. Anatomical and chemical defences of conifer bark against bark beetles and other pests. New Phytologist 167: 353–375.

Fritz, R.S. and E.L. Simms, 1992. Plant resistance to herbivores and pathogens. Ecology, evolution, and genetics. Chicago University Press, Chicago, IL.

Gange, A.C. and H.E. Nice, 1997. Performance of the thistle gall fly, *Urophora cardui*, in relation to host plant nitrogen and mycorrhizal colonization. New Phytologist 137: 335–343.

Gange, A.C. and H.M. West, 1994. Interactions between arbuscular-mycorrhizal fungi and foliar-feeding insects in *Plantago lanceolata* L. New Phytologist 128: 79–87.

Gange, A.C., E. Bower, and V.K. Brown, 1999. Positive effects of mycorrhizal fungi on aphid life history traits. Oecologia 120: 123–131.
Gange, A.C., E. Bower, and V.K. Brown, 2002a. Differential effects of insect herbivory on arbuscular mycorrhizal colonization. Oecologia 131: 103–112.
Gange, A.C., P.G. Stagg, and L.K. Ward, 2002b. Arbuscular mycorrhizal fungi affect phytophagous insect specialism. Ecology Letters 5: 11–15.
Gange, A.C., D.R.J. Gane, Y.L. Chen, and M.Q. Gong, 2005. Dual colonization of *Eucalyptus urophylla* ST Blake by arbuscular and ectomycorrhizal fungi affects levels of insect herbivore attack. Agricultural and Forest Entomology 7: 253–263.
García, D. and J.R. Obeso, 2003. Facilitation by herbivore-mediated nurse plants in a threatened tree, *Taxus baccata*: Local effects and landscape level consistency. Ecography 26: 739–750.
Gehring, C.A., and T.G. Whitham, 1994. Comparisons of ectomycorrhizae on pinyon pines (*Pinus edulis*, Pinaceae) across extremes of soil type and herbivory. American Journal of Botany 81: 1509–1516.
Gehring, C.A. and T.G. Whitham, 2002. Mycorrhizae–herbivore interactions: Population and community consequences. In: M.G.A. van der Heijden and I.R. Sanders, eds. Mycorrhizal Ecology. Springer-Verlag, Germany, pp. 295–320.
Gil, L., 1990. Los olmos y la grafiosis en España. Colección Técnica Ministerio de Agricultura, Pesca y Alimentación, Madrid, Spain.
Gómez, J.M., 1996. Predispersal reproductive ecology of an aridland crucifer, *Moricandia moricandioides*: effect of herbivory by mammals on seed production. Journal of Arid Environments 33: 425–437.
Gómez, J.M., 2005a. Ungulate effect on the performance, abundance and spatial structure of two montane herbs: A 7-yr experimental study. Ecological Monographs 75: 231–258.
Gómez, J.M., 2005b. Non-additive effects of pollinators and herbivores on *Erysimum mediohispanicum* (Cruciferae) fitness. Oecologia 143: 412–418.
Gómez, J.M. and A. González-Megías, 2007. Trait-mediated indirect interactions, density-mediated indirect interactions, and direct interactions between mammalian and insect herbivores. In: T. Ohgushi, T.P. Craig, and P.W. Price, eds. Ecological Communities: Plant Mediation in Indirect Interaction Webs. Cambridge University Press, Cambridge, UK.
Gómez, J.M. and R. Zamora, 1994. Top-down effects in a tritrophic system: Parasitoids enhance plant fitness. Ecology 75: 1023–1030.
Gómez J.M. and R. Zamora, 2002. Thorns as induced mechanical defence in a long-lived shrub (*Hormathophylla spinosa*, Cruciferae). Ecology 83: 885–890.
Goverde, M., M.G.A. van der Heijden, A. Wiemken, I.R. Sanders, and A. Erhardt, 2000. Arbuscular mycorrhizal fungi influence life history traits of a lepidopteran herbivore. Oecologia 125: 362–369.
Gowda, J.H. 1997. Physical and chemical response of juvenile *Acacia tortilis* trees to browsing. Experimental evidence. Funtional Ecology 11: 106–111.
Graglia E., R. Julkunen-Tiitto, G.R. Shaver, I.K. Schmidt, S. Jonasson, and A. Michelsen, 2001. Environmental control and intesite variations of phenolic in Betula nana in tundra ecosystems. New Phytologist 151: 227–236.
Grubb, P.J., 1992. A positive distrust in simplicity—lessons from plant defences and from competition among plants and among animals. Journal of Ecology 80: 585–610.
Guillet, C. and R. Bergström, 2006. Compensatory growth of fast-growing willow (*Salix*) coppice in response to simulated large herbivore browsing. Oikos 113: 33–42.
Guretzky, J.A. and S.M. Louda, 1997. Evidence for natural biological control: Insects decrease survival and growth of a native thistle. Ecological Applications 7: 1330–1340.
Gurevitch, J., J.A. Morrison, and L.V. Hedges, 2000. The interaction between competition and predation: a meta-analysis of field experiments. American Naturalist 155: 435–453.
Haag, J.J., M.D. Coupe, and J.F.J. Cahill, 2004. Antagonistic interactions between competition and insect herbivory on plant growth. Journal of Ecology 92: 156–167.
Halperin, J., 1990. Life history of *Thaumetopoea* spp. (Lep. Thaumetopoeidae) in Israel. Journal of Applied Entomology 110: 1–6.

Hambäck P.A., J. Ågren, and L. Ericson, 2000. Associational resistance: Insect damage to purple loosestrife reduced in thickets of sweet gale. Ecology 81: 1784–1794.
Hamilton, J.G., A.R. Zangerl, E.H. Delucia, and M.R. Berenbaum, 2001. The carbon–nutrient balance hypothesis: Its rise and fall. Ecology Letters 4: 86–95.
Harborne, J.B., 1997. Plant secondary metabolism. In: M.J. Crawley, ed. Plant Ecology, 2nd edn. Blackwell Scientific Publications, Oxford, UK, pp. 132–155.
Harder L.D. and S.C.H. Barrett, 1995. Mating cost of large floral displays in hermaphrodite plants. Nature 373: 512–515.
Harmer, R., 1999. Survival and new shoot production by artificially browsed seedlings of ash, beech, oak and sycamore grown under different levels of shade. Forest Ecology and Management 116: 39–50.
Harper, J.L., 1977. Population Biology of Plants.Academic Press, New York.
Hartley, S.E. and C.G. Jones, 1997. Plant chemistry and herbivory, or why the world is green. In: M.J. Crawley, ed. Plant Ecology, 2nd edition. Blackwell Scientific Publications, Oxford, UK, pp. 284–324.
Haukioja, E., V. Ossipov, J. Koricheva, T. Honkanen, S. Larsson, and K. Lempa, 1998. Biosynthetic origin of carbon-based secondary compounds: cause of variable responses of woody plants to fertilization? Chemoecology 8: 133–139.
Hawkes, V.H. and J.J. Sullivan, 2001.The impact of herbivory on plant in different resources conditions: A meta analysis. Ecology 82: 2045–2058.
Heil, M. and I.T. Baldwin, 2002. Fitness costs of induced resistance: Emerging experimental support for a slippery concept. Trends in Plant Science 7: 61–67.
Hendrix, S.D., 1988. Herbivory and its impact on plant reproduction. In: J. Lovett-Doust and L. Lovett-Doust, eds. Plant Reproductive Ecology: Patterns and Strategies. Oxford University Press, Oxford, UK, pp. 246–263.
Hendrix, S.D. and E.J. Trapp, 1989. Floral herbivory in *Pastinaca sativa*: do compensatory responses offset reductions in fitness? Evolution 43: 891–895.
Herms, D.A. and W.J. Mattson, 1992. The dilemma of plants: To grow or defend. The Quarterly Review of Biology 67: 283–335.
Herrera, J., 1991. Herbivory, seed dispersal, and the distribution of a ruderal plant living in a natural habitat. Oikos 62: 209–215.
Herrera, C.M., 1993. Selection on floral morphology and environmental determinants of fecundity in a hawk-moth-pollinated violet. Ecological Monographs 63: 251–275.
Hester, A., P. Millard, G.J. Baillie, and R. Wendler, 2004. How does timing of browsing affect above- and below-ground growth of *Betula pendula*, *Pinus sylvestris* and *Sorbus aucuparia*? Oikos 105: 536–550.
Hickman, K.R. and D.C. Hartnett, 2002. Effects of grazing intensity on growth, reproduction, and abundance of three palatable forbs in Kansas tallgrass prairie. Plant Ecology 159: 23–33.
Hjältén, J. and P.W. Price, 1997. Can plantas gain protection from herbivory by association with unpalatable neighbours? A field experiement in a willow-sawfly system. Oikos 78: 317–322.
Hjältén, J., K. Danell, and P. Lundberg, 1993. Herbivore avoidance by association: Vole and hare utilization of woody plants. Oikos 68: 125–131.
Hódar, J.A. and R. Zamora, 2004. Herbivory and climatic warming: A Mediterranean outbreaking caterpillar attacks a relict, boreal pine species. Biodiversity and Conservation 13: 493–500.
Hódar, J.A., J. Castro, J.M. Gómez, D. García, and R. Zamora, 1998. Effects of herbivory on growth and survival of seedlings and saplings of *Pinus sylvestris nevadensis* in SE Spain. In: V.P. Papanastasis and D. Peter, eds. Ecological Basis of Livestock Grazing in Mediterranean Ecosystems. Official Publications of the European Community, Luxembourg, pp. 264–267.
Hódar, J.A., R. Zamora, and J. Castro, 2002. Host utilisation by moth and larval survival of pine processionary caterpillar *Thaumetopoea pityocampa* in relation to food quality in three *Pinus* species. Ecological Entomology 27: 292–301.
Hódar, J.A., J. Castro, and R. Zamora, 2003. Pine processionary caterpillar *Thaumetopoea pityocampa* as a new threat for relict Mediterranean Scots pine forests under climatic warming. Biological Conservation 110: 123–129.

Hódar, J.A., R. Zamora, J. Castro, and E. Baraza, 2004. Feast and famine: Previous defoliation limiting survival of pine processionary caterpillar *Thaumetopoea pityocampa* in Scots pine *Pinus sylvestris*. Acta Oecologica 26: 203–210.

Hokka, V., J. Mikola, M. Vestberg, and H. Setala, 2004. Interactive effects of defoliation and an AM fungus on plants and soil organisms in experimental legume-grass communities. Oikos 106: 73–84.

Holmes, R.D. and K. Jepson-Innes, 1989. A nieghborhood analysis of herbivory in *Bouteloua gracilis*. Ecology 70: 971–976.

Holt, R.D., 1997. Community modules. In: A.C. Gange and V.K. Brown, eds. Multitrophic Interactions in Terrestrial Ecosystems. Blackwell Scientific Publications, Oxford, UK, pp. 333–350.

Honkanen, T. and E. Haukioja, 1998. Intra-plant regulation of growth and plant–herbivore interactions. Écoscience 5: 470–479.

Howe, H.F. and L.C. Westley, 1988. Ecological Relationships of Plants and Animals. Oxford University Press, New York.

Huberty, A.F. and R.F. Denno, 2004. Plant water stress and its consequences for herbivorous insects: A new synthesis. Ecology 85: 1383–1398.

Hudgins, J.W., T. Krekling, and V.R. Franceschi, 2003. Distribution of calcium oxalate crystals in the secondary phloem of conifers: a constitutive defence mechanism? New Phytologist 159: 677–690.

Hulme, P.E., 1996. Herbivory, plant regeneration, and species coexistence. Journal of Ecology 84: 609–615.

Hulme, P.E., 1997. Post-dispersal seed predation and the establishment of vertebrate dispersed plants in Mediterranean scrublands. Oecologia 111: 91–98.

Hulme, P.E., 1998. Post-dispersal seed predation and seed bank persistence. Seed Science Research 8: 513–519.

Hunter, M.D. and P.W. Price, 1992. Natural variability in plants and animals. In: M.D. Hunter, T. Ohgushi, and P.W. Price, eds. Effects of resource distribution on animal–plant interactions. Academic Press, San Diego, CA, pp. 1–12.

Huntly, N., 1991. Herbivores and the dynamics of communities and ecosystems. Annual Review of Ecology and Systematics 22: 477–503.

Inbar, M., H. Doostdar, and R.T. Mayer, 2001. Suitability of stressed and vigorous plants to various insect herbivores. Oikos 94: 228–235.

Iwao, K. and M.D. Rausher, 1997. Evolution of plant resistance to multiple herbivores: Quantifying diffuse coevolution. American Naturalist 149: 316–335.

Iwasa, Y., 2000. Dynamic optimization of plant growth. Evolutionary Ecology Research 2: 437–455.

Janzen, D.H., 1980. When is it coevolution? Evolution 34: 611–612.

Jermy, T., 1993. Evolution of insect-plant relationships—a devil's advocate approach. Entomologia Experimentalis et Applicata 66: 3–12.

Joern, A. and S. Mole, 2005. The plant stress hypothesis and variable responses by blue grama grass (*Bouteloua gracilis*) to water, mineral nitrogen, and insect herbivory. Journal of Chemical Ecology 31: 2069–2090.

Jones, C.G. and S.E. Hartley, 1999. A protein competition model of phenolic allocation. Oikos 86: 27–44.

Jordano, P. and C.M. Herrera, 1995. Shuffling the offspring—uncoupling and spatial discordance of multiple stages in vertebrate seed dispersal. Écoscience 2: 230–237.

Juenger, T. and J. Bergelson, 1997. Pollen and resource limitation of compensation to herbivory in scarlet gilia, *Ipomopsis aggregata*. Ecology 78: 1684–1695.

Karban, R., 1997. Neighbourhood affects a plant's risk of herbivory and subsequent success. Ecological Entomology 22: 433–443.

Karban, R. and I.T. Baldwin, 1997. Induced Responses to Herbivory. University of Chicago Press, Chicago, IL.

Keane, R.M. and M.J. Crawley, 2002. Exotic plant invasions and the enemy release hypothesis. Trends in Ecology and Evolution 17: 164–170.

Kelly, D. and V.L. Sork, 2002. Mast seeding in perennial plants: Why, how, where? Annual Review of Ecology and Systematics 33: 427–447.

Kleijn, D. and T. Steinger, 2002. Contrasting effects of grazing and hay cutting on the spatial and genetic population structure of *Veratrum album*, an unpalatable, long-lived, clonal plant species. Journal of Ecology 90: 360–370.

Klironomos, J.N., J. McCune, and P. Moutoglis, 2004. Species of arbuscular mycorrhizal fungi affect mycorrhizal responses to simulated herbivory. Applied Soil Ecology 26: 133–141.

Koptur, S., C.L. Smith, and J.H. Lawton, 1996. Effects of artificial defoliation on reproductive allocation in the common vetch, *Vicia sativa* (Fabaceae: Papilionoideae). American Journal of Botany 83: 886–889.

Koricheva, J., 2002a. The Carbon-Nutrient Balance Hypothesis is dead; long live the carbon–nutrient balance hypothesis? Oikos 98: 537–539.

Koricheva, J., 2002b. Meta-analysis of sources of variation in fitness costs of plant antiherbivore defences. Ecology 83: 176–190.

Koricheva, J., S. Larsson, E. Haukioja, and M. Keinanen, 1998a. Regulation of woody plant secondary metabolism by resource availability: hypothesis testing by means of meta-analysis. Oikos 83: 212–226.

Koricheva, J., S. Larsson, and E. Haukioja, 1998b. Insect performance on experimentally stressed woody plants: a meta-analysis. Annual Review of Entomology 43: 195–216.

Krischik, V.A., R.W. Goth, and P. Barbosa, 1991. Generalized plant defence: Effects on multiple species. Oecologia 85: 562–571.

Kula, A.A.R., D.C. Hartnett, and W.T. Wilson, 2005. Effects of mycorrhizal symbiosis on tallgrass prairie plant–herbivore interactions. Ecology Letters 8: 61–69.

Larsson, S., A. Wirén, L. Lundgren, and T. Ericsson, 1985. Effects of light and nutrient stress on leaf phenolic chemestry in *Salix dasyclados* and susceptibility to *Galerucella lineola* (Coleoptera). Oikos 47: 205–210.

Lehtilä, K. and S.Y. Strauss, 1997. Leaf damage by herbivores affects attractiveness to pollinators in wild radish, *Raphanus raphanistrum*. Oecologia 111: 396–403.

Leimu, R. and J. Koricheva, 2006. A meta-analysis of tradeoffs between plant tolerance and resistance to herbivores: Combining the evidence from ecological and agricultural studies. Oikos 112: 1–9.

Lentz, K.A. and D.F. Cipollini Jr., 1998. Effect of light and simulated herbivory on growth of endangered northeastern bulrush, *Scirpus ancistrochaetus* Schuyler. Plant Ecology 139: 125–131.

Lev-Yadun, S. and G. Ne'eman, 2004. When may green plants be aposematic? Biological Journal of the Linnean Society 81: 413–416.

Louda, S.M. and M.A. Potvin, 1995. Effects of inflorescence-feeding insects on the demography and lifetime fitness of a native plant. Ecology 76: 229–245.

Louda, S.M. and J.E. Rodman, 1996. Insect herbivory as a major factor in the shade distribution of a native crucifer (*Cardamine cordifolia* A. Gray, bittercress). Journal of Ecology 84: 229–237.

Lucas, P.W., I.M. Turner, N.J. Dominy, and N. Yamashita, 2000. Mechanical defences to herbivory. Annals of Botany 86: 913–920.

MacDonald, K.P. and C.E. Bach, 2005. Resistance and tolerance to herbivory in Salix cordata are affected by different environmental factors. Ecological Entomology 30: 581–589.

Maddox, G.D. and R.B. Root, 1990. Structure of the selective encounter between goldenrod (*Solidago altissima*) and its diverse insect fauna. Ecology 71: 2115–2124.

Maron, J.L. and E. Crone, 2006. Herbivory: Effects on plant abundance, distribution and population growth. Proceedings of the Royal Society B: Biological Sciences First Cite Early Online Publishing Paper 1471–2954.

Maron, J.L. and S.L. Gardner, 2000. Consumer pressure, seed versus safe-site limitation, and plant population dynamic. Oecologia 124: 260–269.

Maron, J.L. and E.L. Simms, 2001. Rodent-limited establishment of bush lupine: Field experiments on the cumulative effect of granivory. Journal of Ecology 89: 578–588.

Marquis, R.J., 1992a. The selective impact of herbivores. In: R.S. Fritz and E.L. Simms, eds. Plant Resistance to Herbivores and Pathogens. Ecology, Evolution, and Genetics. Chicago University Press, Chicago, IL, pp. 301–325.

Marquis, R.J., 1992b. A bite is a bite is a bite? Constraints on response to folivory in *Piper aireianum* (Piperaceae). Ecology 73: 143–152.

Martinez, A.J. and J. López-Portillo, 2003. Growth and architecture of small honey mesquites under jackrabbit browsing: Overcoming the disadvantage of being eaten. Annals of Botany 92: 365–375.

Maschinski, J. and T.G. Whitham, 1989. The continuum of plant responses to herbivory: the influence of plant association, nutrient availability, and timing. American Naturalist 134: 1–19.

Massei, G., S.E. Hartley, and P.J. Bacon, 2000. Chemical and morphological variation of Mediterranean woody evergreen specie: Do plants respond to ungulate browsing? Journal of Vegetation Science 11: 1–8.
Mattson, W.J. Jr, 1980. Herbivory in relation to plant-nitrogen content. Annual Review of Ecology and Systematics 11: 119–161.
Mauricio, R., 2000. Natural selection and joint evolution of tolerance and resistance as plant defences. Evolutionary Ecology 14: 491–507.
McEvoy, P.B., N.T. Rudd, C.S. Cox, and M. Huso, 1993. Disturbance, competition, and herbivory effects on ragworth *Senecio jacobaea* populations. Ecological Monographs 63: 55–75.
McNaughton, S.J., 1978. Serengeti ungulates—feeding selectivity influences effectiveness of plant defence guilds. Science 199: 806–807.
McNaughton, S.J., 1988. Mineral nutrition and spatial concentrations of African ungulates. Nature 334: 343–345.
McNaughton, S.J., J.L. Tarrants, M.M. McNaughton, and R.H. Davis, 1985. Silica as a defence against herbivory and a growth promotor in African grasses. Ecology 66: 528–535.
Meiners, S.J. and M.J. Martinkovic, 2002. Survival of and herbivore damage to a cohort of *Quercus rubra* planted across a forest—old-field edge. American Midland Naturalist 147: 247–255.
Meyer, G.A., 1993. A comparison of the impacts of leaf- and sap-feeding insects on growth and allocation of goldenrod. Ecology 74: 1101–1116.
Meyer, G.A. and R.B. Root, 1993. Effects of herbivorous insects and soil fertility on reproduction of goldenrod. Ecology 74: 1117–1128.
Milchunas, D.G. and I. Noy-Meir, 2002. Grazing refuges, external avoidance of herbivory and plant diversity. Oikos 99: 113–130.
Milewsky, A.V., T.P. Young, and D. Madden, 1991. Thorns as induced defences: Experimental evidence. Oecologia 86: 70–75.
Millard, P., A. Hester, R. Wendler, and G. Baillie, 2001. Interspecific defoliation responses of trees depend on sites of winter nitrogen storage. Funtional Ecology 15: 535–543.
Moen, J., H. Gardfjell, L. Oksanen, L. Ericson, and P. Ekerholm, 1993. Grazing by food-limited microtine rodents on a productive experimental plant community: Does the "green desert" exist? Oikos 68: 401–413.
Mopper, S. and T.G. Whitham, 1992. The plant stress paradox: effects on pinyon sawfly sex ratios and fecundity. Ecology 73: 515–525.
Mothershead, K. and R.J. Marquis, 2000. Fitness impacts of herbivory through indirect effects on plant–pollinator interactions in *Oenothera macrocarpa*. Ecology 81: 30–40.
Müller-Schärer, H. and V.K. Brown, 1995. Direct and indirect effects of above- and below-ground insect herbivory on plant density and performance of *Tripleurospermum perforatum* during early plant succession. Oikos 72: 36–41.
Mutikainen, P. and M. Walls, 1995. Growth, reproduction and defence in nettles: Responses to herbivory modified by competition and fertilization. Oecologia 104: 487–495.
Muzika, R.M. and K.S. Pregitzer, 1992. Effect of nitrogen-fertilization on leaf phenolic production of grand fir seedlings. Trees: Structure and Function 6: 241–244.
Nathan, R. and H.C. Müller-Landau, 2000. Spatial patterns of seed dispersal, their determinants and consequences for recruitment. Trends in Ecology and Evolution 15: 278–285.
Niesenbaum, R.A., 1996. Linking herbivory and pollination: defoliation and selective fruit abortion in *Lindera benzoin*. Ecology 77: 2324–2331.
Nitao, J.K., A.R. Zangerl, and M.R. Berenbaum, 2002. CNB: requiescat in pace? Oikos 98: 540–546.
Nykänen, H. and J. Koricheva, 2003. Damage-induced changes in woody plants and their effects on insect herbivore performance: a meta-analysis. Oikos 104: 247–268.
Obeso, J.R. and Grubb, P.J., 1994. Interactive effects of extent and timing of defoliation, and nutrient supply on reproduction in a chemically protected annual *Senecio vulgaris*. Oikos 71: 506–514.
Oiser, T.D. and R.L. Lindroth, 2001. Effects of genotype, nutrient availabilllity, and defoliation on aspen phytochemistry and insect performance. Journal of Chemical Ecology 27: 1289–1313.
Olff, H., F.W.M. Vera, J. Bokdam, E.S. Bakker, J.M. Gleichman, K. de Maeyer, and R. Smit, 1999. Shifting mosaics in grazed woodlands driven by the alternation of plant facilitation and competition. Plant Biology 1: 127–137.

Orians, C.M. and C.G. Jones, 2001. Plants as resource mosaics: A functional model for predicting patterns of within-plant resource heterogeneity to consumers based on vascular architecture and local environmental variability. Oikos 94: 493–504.

Osem, Y., A. Perevolotsky, and J. Kigel, 2004. Site productivity and plant size explain the response of annual species to grazing exclusion in a Mediterranean semi-arid rangeland Journal of Ecology 92: 297–309.

Petrakis, P.V., V. Roussis, and A.H. Ortiz, 2001. Host selection by *Thaumetopoea pityocampa* (Den. and Schif.): the relative importance of needle terpenoid and morpho-anatomical profiles. In: K. Radoglou, ed. Forest Research: A Challenge for an Integrated European Approach, Vol. I. NAGREF, Forest Research Institute, Thessaloniki, pp. 343–348.

Pfister, C.A. and M.E. Hay, 1988. Associational plant refuges—convergent patterns in marine and terrestrial communities result from differing mechanisms. Oecologia 77: 118–129.

Price, P.W. 1991. The plant vigor hypothesis and herbivore attack. Oikos 62: 244–251.

Prins, A.H., H.W. Nell, and G.L. Klinkhammer, 1992. Size-dependent root herbivory on *Cynoglossum officinale*. Oikos 65: 409–413.

Provenza, F.D. and J.C. Malechek, 1984. Diet selection by domestic goat in relation to blackbrush twig chemistry. Journal of Applied Ecology 21: 831–841.

Provenza, F.D., J.J.Y. Villalba, and J.P. Bryant, 2002. Foraging by herbivores: linking the biochemical diversity of plants to herbivore culture and landscape diversity. In: J.A. Bissonette and I. Storch, eds. Landscape Ecology and Resource Management: Linking Theory with Practice. Island Press, New York, pp. 387–421.

Quesada, M., K. Bollman, and A.G. Stephenson, 1995. Leaf damage decreases pollen production and hinders pollen performance in *Cucurbita texana*. Ecology 76: 437–443.

Ramsell, J. and N.D. Paul, 1990. Preferential grazing by molluscs of plants infected by rust fungi. Oikos 58: 145–150.

Rand, T.A., 2003. Herbivore-mediated apparent competition between two salt marsh forbs. Ecology 84: 1517–1526.

Rausher, M.D., 1996 Genetic analysis of coevolution between plants and their natural enemies. Trends in Genetics 12: 212–217.

Rausher, M.D., 2001. Co-evolution and plant resistance to natural enemies. Nature 411: 857–864.

Reader, R.J. and S.P. Bonser, 1998. Predicting the combined effect of herbivory and competition on a plant's shoot mass. Canadian Journal of Botany 76: 316–320.

Rebollo, S., D.G. Milchunas, I. Noy-Meir, and P.L. Chapman, 2002. The role of a spiny plant refuge in structuring grazed shortgrass steppe plant communities. Oikos 98: 53–64.

Rehill, B.J., J.C. Schultz, 2001. Hormaphis hamamelidis and gall size: a test of the plant vigor hypothesis. Oikos 95: 94–104.

Renaud, P.E., M.E. Hay, and T.M. Schmitt, 1990. Interactions of plant stress and herbivory: intraspecific variation in the susceptibility of a palatable versus an unpalatable seaweed to sea urchin grazing. Oecologia 82: 217–226.

Rey, P.J. and J.M. Alcántara, 2000. Recruitment dynamics of a fleshy-fruited plant (*Olea europaea*): connecting patterns of seed dispersal to seedling establishment. Journal of Ecology 88: 622–633.

Rhoades, D.F., 1979. Evolution of chemical defence against herbivores, In: G.A. Rosenthal and D.H. Janzen, eds. Herbivores: Their Interaction with Secondary Plant Metabolites. Academic Press, New York, pp. 3–54.

Rhoades, D.F. and R.G. Cates, 1976. Toward a general theory of plant antiherbivore chemistry. Recent Advances in Phytochemistry 10: 168–213.

Rieske, L.K., 2001. Influence of symbiotic fungal colonization on oak seedling growth and suitability for insect herbivory. Environmental Entomology 30: 849–854.

Riipi, M., V. Ossipov, K. Lempa, E. Haukioja, J. Koricheva, S. Ossipova, and K. Pihlaja, 2002. Seasonal changes in birch leaf chemistry:are there trade-offs between leaf growth and accumulation of phenolics? Oecologia 130: 380–390.

Rosenthal, G.A. and M.R. Berenbaum, 1991. Herbivores: Their Interactions with Secondary Plant Metabolites, Vol. I: The Chemical Participants. Academic Press, New York.

Rosenthal, J.P. and P.M. Kotanen, 1994. Terrestrial plant tolerance to herbivory. Trends in Ecology and Evolution 9: 145–148.

Rousset, O. and J. Lépart, 1999. Shrub facilitation of *Quercus humilis* regeneration in succession on calcareous grasslands. Journal of Vegetation Science 10: 493–502.

Rousset, O. and J. Lépart, 2002. Neighbourhood effects on the risk of an unpalatable plant being grazed. Plant Ecology 165: 197–206.

Russell, F.L. and S.M. Louda, 2004. Phenological synchrony affects interaction strength of an exotic weevil with Platte thistle, a native host plant. Oecologia 139: 525–534.

Russell, F.L. and S.M. Louda, 2005. Indirect interaction between two native thistles mediated by an invasive exotic floral herbivore. Oecologia 146: 373–384.

Saunders, M.R. and K.J. Puettmann, 1999. Effects of overstory and understory competition and simulated herbivory on growth and survival of white pine seedlings. Canadian Journal of Forest Research 29: 536–546.

Scheirs, J. and L. de Bruyn, 2004. Excess of nutrients results in plant stress and decreased grass miner performance. Entomologia Experimentalis et Applicata 113: 109–116.

Schopf, R. and N. Avtzis, 1987. Die bedeutung von Nadelinhalsstoffen für die disposition von fünf kiefernarten gegenüber *Thaumetopoea pityocampa* (Schiff.). Journal of Applied Entomology 103: 340–350.

Schultz, J.C., P.J. Nothnagle, and I.T. Baldwin, 1982. Seasonal and individual variation in leaf quality of two northern hardwoods tree species. American Journal of Botany 69: 753–759.

Schupp, E.W., 1995. Seed-seedling conflicts, habitat choice, and patterns of plant recruitment. American Journal of Botany 82: 399–409.

Schupp, E.W. and M. Fuentes, 1995. Spatial patterns of seed dispersal and the unification of plant-population ecology. Écoscience 2: 267–275.

Shelton, A.L., 2000. Variable chemical defences in plants and their effects on herbivore behaviour. Evolutionary Ecology Research 2: 231–249.

Siemens, D.H., H. Lischke, N. Maggiulli, S. Schurch, and B.A. Roy, 2003. Cost of resistance and tolerance under competition: The defence-stress benefit hypothesis. Evolutionary Ecology 17: 247–263.

Simms, E.L. and M.D. Rausher, 1989. The evolution of resistance to herbivory in *Ipomoea purpurea*. II: Natural selection by insects and cost of resistance. Evolution 43: 573–585.

Simon, M. and M. Hilker, 2005. Does rust infection of willow affect feeding and oviposition behavior of willow leaf beetles? Journal of Insect Behavior 18: 115–129.

Sipura, M. and J. Tahvanainen, 2000. Shading enhances the quality of willow leaves to leaf beetles—but does it matter? Oikos 91: 550–558.

Smit, C., J.A.N. Den Ouden, and H. Müller-Scharer, 2006. Unpalatable plants facilitate tree sapling survival in wooded pastures. Journal of Applied Ecology 43: 305–312.

Spotswood, E., K.L. Bradley, and J.M.H. Knops, 2002. Effects of herbivory on the reproductive effort of 4 prairie perennials. BMC Ecololy 2: 2.

Stamp, N., 2003. Out of the quagmire of plant defence hypotheses. The Quarterly Review of Biology 78: 23–55.

Steets, J.A., R. Salla, and T.-L. Ashman, 2006. Herbivory and competition interact to affect reproductive traits and mating system expression in *Impatiens capensis*. American Naturalist 167: 591–600.

Stephenson, A.G., 1981. Flower and fruit abortion: proximate causes and ultimate functions. Annual Review of Ecology and Systematics 12: 253–279.

Sterner, R.W. and D.O. Hessen, 1994. Algal nutrient limitation and the nutrition of aquatic herbivores. Annual Review of Ecology and Systematics 25: 1–29.

Stout, M.J., W.C. Rice, and D.R. Ring, 2002. The influence of plant age on tolerance of rice to injury by the rice water weevil, *Lissorhoptrus oryzophilus* (Coleoptera: Curculionidae). Bulletin of Entomological Research 92: 177–184.

Stowe, K.A., R.J. Marquis, C.G. Hochwender, and E.L. Simms, 2000. The evolutionary ecology of tolerance to consumer damage. Annual Review of Ecology and Systematics 31: 565–595.

Strauss, S.Y., 1991. Indirect effects in community ecology: their definition, study and importance. Trends in Ecology and Evolution 6: 206–210.

Strauss, S.Y. and A.A. Agrawal, 1999. The ecology and evolution of plant tolerance to herbivory. Trends in Ecology and Evolution 14: 179–185.

Strauss, S.Y. and R.E. Irwin, 2004. Ecological and evolutionary consequences of multispecies plant–animal interactions. Annual Review of Ecology, Evolution and Systematics 35: 435–466.

Strauss S.Y., J.K. Conner, and S.L. Rush, 1996. Foliar herbivory affects floral characters and plant attractiveness to pollinators: Implications for male and female plant fitness. American Naturalist 147: 1098–1107.

Strauss, S.Y., H. Sahli, and J.K. Conner, 2005. Toward a more trait-centered approach to diffuse (co)evolution. New Phytologist 165: 81–89.

Strauss, S.Y., J.A. Rudgers, J.A. Lau, and R.E. Irwin, 2002. Direct and ecological costs of resistance to herbivory. Trends in Ecology and Evolution 17: 278–285.

Strong, A.M., T.W. Sherry, and R.T. Holmes, 2000. Bird predation on herbivorous insects: Indirect effects on sugar maple saplings. Oecologia 125: 370–379.

Tahvanainen, J.O. and R.B. Root, 1972. The influence of vegetation diversity on the population ecology of a specialized herbivore, *Phyllotreta cruciferae* (Coleoptera: Chrysomelidae). Oecologia 10: 321–346.

Takada, M., M. Asada, and T. Miyashita, 2001, Regional differences in the morphology of a shrub Damnacanthus indicus: An induced resistance to deer herbivory? Ecological Research 16: 809–813.

Teaford, M.F., P.W. Lucas, P.S. Ungar, and K.E. Glander, 2006. Mechanical defences in leaves eaten by Costa Rican howling monkeys (*Alouatta palliata*). American Journal of Physical Anthropology 129: 99–104.

Tiberi, R., A. Niccoli, M. Curini, F. Epifanio, M.C. Marcotullio, and O. Rosati, 1999. The role of monoterpene composition in *Pinus* spp. needles, in host selection by the pine processionary caterpillar *Thaumetopoea pityocampa*. Phytoparasitica 27: 263–272.

Tiffin, P., 2000a. Mechanisms of tolerence to herbivore demage: what do we know? Evolutionary Ecology 14: 523–536.

Tiffin, P., 2000b. Are tolerance, avoidance, and antibiosis evolutionarily and ecologically equivalent responses of plants to herbivores? American Naturalist 155: 128–138.

Tikkanen, O.P. and R. Julkunen-Tiitto, 2003. Phenological variation as protection against defoliating insects: the case of *Quercus robur* and *Operophtera brumata*. Oecologia 136: 244–251.

Tscharntke, T., 1997. Vertebrate effects on plant-invertebrate food webs. In: A.C. Gange and V.K. Brown, eds. Multitrophic Interactions in Terrestrial Ecosystems. Blackwell Scientific Publications, Oxford, UK, pp. 277–297.

Tuomi, J., P. Niëmëla, and S. Sirén, 1990. The Panglossian paradigm and delayed inducible accumulation of foliar phenolics in mountain birch. Oikos 59: 399–410.

Turnbull, L.A., M.J. Crawley, and M. Rees, 2000. Are plant populations seed-limited? A review of seed sowing experiments. Oikos 88: 225–238.

Turner, J.M., 1994. Sclerophylly: primarily protective? Functional Ecology 8: 669–675.

Van Bael, S.A., J.D. Brawn, and S.K. Robinson, 2003. Birds defend trees from herbivores in a Neotropical forest canopy. Proceedings of the National Academy of Sciences of the United States of America 100: 8304–8307.

Van Dam, N.M. and Baldwin, I.T., 1998. Costs of jasmonate-induced responses in plants competing for limited resources. Ecology Letters 1: 30–33.

Van Dam, N.M. and D.J. Hare, 1998. Differences in distribution and performance of two sap-sucking herbivores on glandular and non-glandular *Datura wrightii*. Ecological Entomology 23: 22–32.

Vanderklein, D.W. and P.B. Reich, 2000. European larch and eastern white pine respond similarly during three years of partial desfoliation. Tree Physiology 20: 283–287.

von Zeipel, H., O. Eriksson, and J. Ehrlén, 2006. Host plant population size determines cascading effects in a plant–herbivore–parasitoid system. Basic and Applied Ecology 7: 191–200.

WallisDeVries, M.F., E.A. Laca, and M.W. Demment, 1999. The importance of scale of patchiness for selectivity in grazing herbivores Oecologia 121: 355–363.

Walls, R., H. Appel, M. Cipollini, and J. Schultz, 2005. Fertility, root reserves and the cost of inducible defences in the perennial plant *Solanum carolinense*. Journal of Chemical Ecology 31: 2263–2288.

Wamberg, C., S. Christensen, and I. Jakobsen, 2003. Interaction between foliar-feeding insects, mycorrhizal fungi, and rhizosphere protozoa on pea plants. Pedobiologia 47: 281–287.

Wardle, D.A., R.D. Bardgett, J.N. Klironomos, H. Setälä, W.H. van der Putten, and D.H. Wall, 2004. Ecological linkages between aboveground and belowground biota. Science 304: 1629–1633.

Waterman, P.J.Y. and S. Mole, 1994. Analysis of Phenolic Plant Metabolites. Blackwell, Oxford, UK.

Weiner, J., 1993. Competition, herbivory and plant size variability: *Hypochaeris radicata* grazed by snails (*Helix aspersa*). Functional Ecology 7: 47–53.

Wenny, D.G., 2000. Seed dispersal, seed predation, and seedling recruitment of a neotropical montane tree. Ecological Monographs 70: 331–351.

White, T.C.R., 1993. The inadequate environment: Nitrogen and the abundance of animals. Springer-Verlag, Berlin, Germany.

White, J.A. and T.G. Whitham, 2000. Associational susceptibility of cottonwood to a box elder herbivore. Ecology 81: 1795–1803.

Wise, M.J. and W.G. Abrahamson, 2005. Beyond the compensatory continuum: Environmental resource levels and plant tolerance of herbivory. Oikos 109: 417–428.

Wise, M.J. and J.J. Cummins, 2006. Strategies of *Solanum carolinense* for regulating maternal investment in response to foliar and floral herbivory. Journal of Ecology 94: 629–636.

Wurst, S. and T.H. Jones, 2003. Indirect effects of earthworms (*Aporrectodea caliginosa*) on an above–ground tritrophic interaction. Pedobiologia 47: 91–97.

Wurst, S., D. Dugassa-Gobena, and S. Scheu, 2004a. Earthworms and litter distribution affect plant-defensive chemistry. Journal of Chemical Ecology 30: 691–701.

Wurst, S., D. Dugassa-Gobena, R. Langel, M. Bonkowski, and S. Scheu, 2004b. Combined effects of earthworms and vesicular arbuscular mycorrhizas on plant and aphid performance. New Phytologist 163: 169–176.

Wurst, S., R. Langel, S. Rodger, and S. Scheu, 2006. Effects of belowground biota on primary and secondary metabolites in *Brassica oleracea*. Chemoecology 16: 69–73.

Young, T.P., M.L. Stanton, and C.E. Christian, 2003. Effects of natural and simulated herbivory on spine lengths of *Acacia drepanolobium* in Kenya. Oikos 101: 171–179.

Zamora, R., J.A. Hódar, and J.M. Gómez, 1999. Plant–herbivore interactions: Beyond a binary vision. In: F.A. Pugnaire and F. Valladares, eds. Handbook of functional plant ecology, 1st edition. Marcel Dekker, New York, pp. 677–718.

Zamora, R., J.M. Gómez, J.A. Hódar, J. Castro, and D. García, 2001. Effect of browsing by ungulates on sampling growth of Scots pine in a Mediterranean enviroment: Consequences for forest regeneration. Forest Ecology and Management 144: 33–42.

17 Ecology of Plant Reproduction: Mating Systems and Pollination

Anna Traveset and Anna Jakobsson

CONTENTS

... That these and other insects, while pursuing their food in the flowers, at the same time fertilize them without intending and knowing it and thereby lay the foundation for their own and their offspring's future preservation, appears to me to be one of the most admirable arrangements of nature.

—Sprengel (1793)

INTRODUCTION

One of the main differences between most plants and animals is that the former cannot move in search of a partner to mate and thus needs a vector, which can be inanimate, such as wind or water, or an animal, vertebrate or invertebrate, to transport the male gametes (pollen) among flowers. This passivity has caused plants to evolve a great variety of adaptations, either to disperse the pollen, for instance by attracting animal pollinators with a reward, or to become independent of pollen vectors, that is, by reproducing asexually or by self-pollinating.

This chapter focuses on the mechanisms by which plants are able to accomplish reproduction. We first describe how plants reproduce asexually and the advantages of sexual reproduction. Then we briefly review the different kinds of plant mating systems and what is known about their evolution, maintenance, and lability. The study of plant breeding systems addresses questions on the genetics of mating patterns, mainly associated with inbreeding depression and, until the last three decades, it was considered as a separate research line from that of pollination biology. The two areas have begun to merge into what has been called a new synthesis (Lloyd and Barrett 1996) or a new plant reproductive biology (Morgan and Schoen 1997) as floral biologists have enlarged their backgrounds with natural history, ecology, genetics, and theoretical approaches. The different systems of self-incompatibility, widespread among flowering plants, are treated briefly and the reader is referred to Nettancourt (1977), Barrett (1992), and Charlesworth et al. (2005) to explore this topic further. The paternal side of plant reproduction is increasingly receiving more attention in studies of reproductive success, and here we synthesize existing information on this subject, giving some directions for future research. For further readings about plant reproductive strategies and breeding systems we recommend Richards (1997) and de Jong and Klinkhammer (2005). Finally, we briefly review studies on the influence of pollinators on the evolution of floral traits and diversification of angiosperms.

ASEXUAL REPRODUCTION

Asexual reproduction is fairly common in plants and allows them to persist in their habitats with complete independence of pollinating vectors. Two types are distinguished, both quite similar from the genetic viewpoint, although their mechanisms are different: (a) vegetative reproduction, that is, asexual multiplication of an individual (genet)—which has originally arisen from a zygote—into physiologically independent units (ramets) (Harper 1977, Abrahamson 1980) and (b) agamospermy, the production of fertile seeds without sexual fusion of gametes. Advantages of asexual reproduction include the possibility to exploit larger areas and new locations, provided that vegetative propagules are widely dispersed (Janzen 1977, Lovett Doust 1981), and the preservation of successful genotypes since they are not lost during sexual recombination, which would be the case for example during heterozygote advantage (Peck and Waxman 1999). In many perennial plants, both asexual and sexual reproduction take place, the latter usually occurring once a growth threshold has been attained (e.g., Weiner 1988, Schmid and Weiner 1993, Worley and Harder 1996). A trade-off between asexual and sexual reproduction has been reported in a number of studies (e.g., Sohn and Policansky 1977, Law et al. 1983, Westley 1993) and can be influenced by plant size (Worley and Harder 1996), ramet density (Humprey and Pyke 1998), resource state of the growing site (Gardner and Mangel 1999), and population age (Sun et al. 2001).

VEGETATIVE REPRODUCTION

Vegetative reproduction is widespread among the angiosperms, especially in herbaceous perennials, but rare among the gymnosperms (possibly due to the predominantly woody habit of this group). Among woody plants, it is much more common in dwarf or creeping shrubs, climbers, and vines than in trees, although there are exceptions as, for example, the English elm (*Ulmus procera*) in Britain where all individuals are derived from one single cone (Gil et al. 2004). Vegetative reproduction is also quite conspicuous in anemophilous monocotyledons, and some species such as *Phragmites* and *Ammophila* occur in a specialized habitat throughout the world and are among the most widespread plant species known

(Heywood 1993). Vegetative reproduction is particularly successful in hydrophytes, probably because water is an adequate environment for the dispersal of relatively unprotected propagules, and invasive hydrophytes often cause severe environmental and economic problems. An example is *Caulerpa taxifolia*, a tropical green alga accidentally introduced into the western Mediterranean Sea in 1984, which has rapidly spread over a large area because of its efficient reproduction through stolons (Ceccherelli and Cinelli 1999).

The usual organs developed by plants to reproduce asexually are modifications of stems or axillary buds, which are stem initials. However, underground bulbs and corms are also common and have a protective function, especially during dormancy (hibernation or aestivation). Vegetative reproduction may be disadvantageous when a single clone occupies a large area, as the distance between individuals can be large and genetic variation is much reduced. The whole population may fail to set seed if the species is self-incompatible as in the case of bamboos or if it is diclinous as in the case of *Elodea canadensis* in Britain where all individuals are females. Clonal reproduction may also lose vigor with age, either due to an increased viral load through viral multiplication and reinfection or due to the accumulation of disadvantageous somatic mutations (Richards 1997). Furthermore, clonal reproduction is often more common in the margins of a species geographical range where environmental conditions limit seed set (Eckert 2002b).

Agamospermy

Agamospermy, asexual production of seeds, is a phenomenon absent in gymnosperms and limited to a small group (34 families) of angiosperms, occurring mainly in the Compositae, Gramineae, and Rosaceae (Asker and Jerling 1992, Richards 1997). It is highly polyphyletic, arisen on many occasions from sexual taxa, and examples of genera including both sexual and agamospermous species are *Taraxacum*, *Crepis*, *Hieracium*, *Sorbus*, and *Crataegus* (Nygren 1967). There are a few documented cases of evolution of agamospermy from different types of breeding systems such as autogamy (e.g., *Aphanes*), dioecy (e.g., *Antennaria*, *Lindera*), or heteromorphy (e.g., *Limonium, Erythroxylum*) (Berry et al. 1991, Richards 1997, Dupont 2002). Agamospermy can be sporophytic as in the case with *Citrus*, and the sporophyte embryo is then budded directly from the old sporophyte ovular tissue, usually the nucellus (adventitious embryony). However, more commonly is gametophytic agamospermy, where a female gametophyte is produced with the sporophytic chromosome number. Then the nonreduction of chromosome number results either from a complete avoidance of female meiosis (apospory and mitotic diplospory) or by a failure in it (meiotic diplospory) (Richards 2003).

It might seem as if production of seeds is assured in agamospermous species in the absence of pollination, but actually most species with adventitious embryony and apospory require the stimulus of pollination to fertilize the endosperm nucleus (pseudogamy). The seed habit, however, gives them the advantage of dispersal and the potential for extended dormancy, added to the possibility of fixing a successful genotype through asexual reproduction. Most agamosperms with apospory or adventitious embryony retain good pollen function, which can also be used in sexual reproduction. Within agamospermic species, both diploid individuals that reproduce sexually and polyploidy individuals that reproduce by agamospermy are usually found, but the capacity of both sexual and asexual seed production is very seldom found in the same individual (Bengtsson and Ceplitis 2000, Van Baarlen et al. 2000). The main disadvantage of agamospermy is that the cell line forms a gigantic linkage in which the advantageous genes cannot escape from the accumulated harmful ones. Moreover, such a cell line is unable to recombine novel advantageous mutants and thus cannot adapt to the new conditions after an environmental change, although some genetic variation can exist through somatic recombination (chromosome breakage and fusion), meiotic recombination, chromosome lose and gain, and accumulation of mutants (Richards 1997). That is probably why truly obligate agamospermy, in which all possibility of sexuality has been lost, is rare (Asker 1980) and appears to be limited to a few diplosporous genera in which pollen is absent (unusual, as male-sterile mutants cannot be recombined). Even though a great deal of information has been accumulated on the origin, distribution, and mechanisms of agamospermy (e.g., Darlington 1939, Gustafsson 1946, Asker 1980, Berry et al. 1991, Richards 2003), much needs to be done yet to understand the evolution of this phenomenon and for the adequate interpretation of the observed patterns. Currently there is a great interest on the mechanisms underlying agamospermy because the possibility to select highly productive individuals and reproduce them asexually by seeds would imply an enormous potential for crop improvement (Ramulu et al. 1999, Bhat et al. 2005).

ADVANTAGES OF SEXUAL REPRODUCTION

The two most important characteristics of sexuality are (1) it creates genetic variability, through sexual fusion of gametes, chromosome segregation, and allele recombination and (2) it allows gene migration, so successful mutations can spread between generations and move within and between populations. Moreover, sexuality, and thus meiotic mechanisms, dissipates Müller's Ratchet (accumulation of harmful mutations), breaks up linkage disequilibrium, and also engenders zygotes that are free of virus (Richards 1997). Sexual reproduction is a primitive trait of nearly all eukaryotic organisms and has probably contributed to their success and long-term survival. The genetic variability gives sexual lines evolutionary potential to adapt to new conditions after an environmental change, a feature absent in asexual organisms as mentioned earlier. Sexuality is absent only in a few groups of animals that reproduce parthenogenetically, in agamospermous plants and in sterile (usually hybrid) plant clones. Here, we refer only to seed plants. The reproductive ecology of algae, bryophytes, and pteridophytes has been reviewed in Lovett Doust and Lovett Doust (1988).

The whole process of embryology in angiosperms (flowering plants) was already described in detail nearly half century ago by Maheshwari (1950). A good introductory chapter to the anatomy and physiology of sexual reproduction in both gymnosperms and angiosperms can be found in Richards (1997), and recent reviews on the origin and evolution of flowers are found in Doyle (1994) and Friis et al. (2005). The transition from a free-sporing heterosporous pteridophyte to a plant with gymnospermous reproduction, assessing adaptive explanations for the origin of seeds, is dealt with in Haig and Westoby (1989). According to these two authors, the first seeds would have originated from heterosporous species, the megaspores of which would have been selected for a larger size; the decisive character in their success would have been related to pollination, by evolving traits to capture microspores before dispersal of the megaspore. In pteridophytes, fertilization always takes place after gametes have been dispersed.

The gymnosperms, composed of five polyphyletic groups, are characterized by the ovule or seed borne externally (gymnosperm means naked seed), although they are greatly diverse in most reproductive structures. Two general features of their reproduction, relevant to the genetic structure of plant populations, are as follows:

1. There are no hermaphrodite cones. Thus, plants are either monoecious (separate sexes on the same individual plant; e.g., Pinaceae, Taxodiaceae) or dioecious (an individual plant has either all female cones or all male cones; e.g., Cycadaceae, Ginkgoaceae, Taxaceae), although some species have populations with both monoecious and dioecious members (Givnish 1980) and some previously reported monoecious Cupressaceae, such as *Juniperus phoenicea*, have shown to depart significantly from cosexuality (Jordano 1991 and references therein). If monoecious, there is usually dichogamy (separation in time of anther dehiscence from stigma receptivity), so outcrossing is always promoted.

FIGURE 17.1 The lizard *Podarcis lilfordi* pollinating the gymnosperm *Ephedra fragilis* on Dragonera Island (Balearic Archipelago). (Photo by Javier Rodríguez-Pérez. With permission.)

2. Pollination is almost always by wind (anemophily). Pollen grains in the Pinaceae even have two lateral air-filled sacs that act as wings, which allow them to fly very long distances. The genus *Ephedra* is an exception as it can be pollinated by insects (entomophily) and even by lizards in insular systems (Bino and Meeuse 1981, A. Traveset, personal observation of lizards and syrphid flies feeding on *Ephedra* flowers on Cabrera Island SE off Majorca) (Figure 17.1).

The reproductive ecology of gymnosperms has in general received less attention than that of angiosperms and we still need much more information on the former to infer about the genetic control of mating patterns within or between species. Ellstrand et al. (1990) reviewed the available data for genetic structure of gymnosperms, and concluded that they are generally highly diverse but have a low spatial differentiation. So far, there seems to be no evidence that nonangiosperms have low genetic diversity or that they are characterized by low gene flow (Midgley and Bond 1991, Brown et al. 2004).

Both gymnosperms and angiosperms have two major advantages over pteridophytes: (1) they do not depend on external water for sexual reproduction and (2) the zygote is protected within a seed, which in turn can be dispersed far from the parent plant. However, only about 750 species of gymnosperms exist today, in contrast to 10,000 pteridophytes. The major speciation has indeed occurred in the angiosperms, a group represented by over 220,000 species (Cronquist 1981). Such species richness is due to different factors (see Section "Diversification of Angiosperms"), of which perhaps the most important is the wider range of growth forms that allow them to inhabit a wider range of habitats than either pteridophytes or gymnosperms. The latter are mostly trees, which restrict the number of habitats where they can live, have limited breeding systems, limited pollination systems, and unspecialized seed dispersal (Givnish 1980). The development of a gynoecium (pistil) in the angiosperms (term meaning enclosed seeds) has probably been one of the most important steps in the evolution of plants.

Seed plants have a wide array of reproductive options that have evolved under particular environmental conditions (e.g., scarcity or absence of pollinators) and that are maintained or changed through the process of natural selection. Next, we focus on such reproductive options, on the factors that select for them and, briefly, on the genetic consequences for the plant population.

SELF-POLLINATION

Most angiosperms bear perfect flowers (containing both anthers and stigmas) and a large fraction of them are self-compatible and thus potentially selfing species (Bertin and Newman 1993, Vogler and Kalisz 2001). Estimates suggest that 62%–84% of temperate plants (mostly herbs) and 35%–78% of tropical plants (including shrubs, trees, vines, and herbs) are at least partially selfers (e.g., Arroyo and Uslar 1993). Comprehensive reviews of self-fertilization can be found in Jarne and Charlesworth (1993), Holsinger (1996), and Goodwillie et al. (2005) and several recent models have been developed to explain its evolution (e.g., Morgan et al. 2005, Porcher and Lande 2005, Scofield and Schultz 2006). Self-pollination can take place within a flower (autogamy) or between flowers of the same genet (geitonogamy). It is termed autonomous if the pollen is transferred to the stigma by various mechanisms not involving pollinators and facilitated if the pollen is transferred by a pollinator. The level of autogamy depends on the degree of separation between anthers and stigma (Table 17.1). Neither herkogamy nor dichogamy prevents geitonogamy, although they may reduce it considerably.

The level of geitonogamous crosses varies greatly both among and within species and depends mainly on pollinator foraging behavior and number of flowers of the same genet simultaneously open, since the time a pollinator spends in a patch usually increases with flower numbers (reviewed by Ohasi and Yahara 2001). It can also be affected by the architectural structure of the inflorescence (Jordan and Harder 2006). Geitonogamy reduces male and female fitness by reducing pollen export to other individuals, so-called pollen discounting, and by reducing the number of ovules available for outcrossing, so-called seed discounting. Geitonogamous pollination was for a long time called the neglected side of selfing (de Jong et al. 1993), but during the last decade it has received more attention by plant ecologists. Recent examples are studies of the effects of geitonogamy on evolution of dioecy (e.g., de Jong and Geritz 2002), sex allocation in hermaphrodites (de Jong et al. 1999), degree of selfing (Karron et al. 2004), and fruit set (Finer and Morgan 2003).

Plants have a variety of mechanisms to promote or prevent selfing. For example, the rapid wilting of pollinated flowers in some species has been suggested to be an adaptive regulation of the size of floral display to prevent geitonogamy (Harder and Johnson 2005). Herkogamy has evolved in plants that depend on animals for their pollination and dichogamy is found both in animal- and wind-pollinated species (for more details see Tammy et al. 2006). Dichogamy is found as often in species with self-incompatibility systems as in species without such systems (Bertin 1993). The reason for this could be that the different forms of dichogamy, protandry and protogyny, serve different functions. Protogyny is expected to be more efficient in preventing self-pollination and thus reducing inbreeding depression, whereas both protogyny and protandry are expected to decrease the interference between the male and the female functions (i.e., pollen and seed discounting). This theory was confirmed by Routely et al. (2004) who found protandry to be correlated to self-incompatible species and protogyny to be correlated to self-compatible species. Self-incompatibility is a mechanism preventing selfing that is known in about 30% of the angiosperm families; it is apparently controlled by a few loci, and it seems to have evolved independently several times (Jarne and Charlesworth

TABLE 17.1
Mechanisms of Plants to Avoid Overlapping of Male and Female Functions

Herkogamy: separation in space between anthers and stigma position
Dichogamy: separation in time between stamen dehiscence and stigma receptivity
 I. Protandry: the male phase is first
 II. Protogyny: the female phase is first

1993 and references therein). Male sterility (gynodioecy) and female sterility (androdioecy) are also mechanisms that reduce selfing and both might represent early steps in the evolution of dioecy (see Section "Sex Expression"). Cleistogamy is a mechanism that promotes selfing, as flowers do not open and can only self-fertilize. All species with cleistogamous flowers also produce hermaphrodite open-pollinated (chasmogamous) flowers (Lord 1981), and this appears to be evolutionarily stable under certain restrictive conditions (Schoen and Lloyd 1984, Masuda et al. 2001).

The immediate genetic consequences of selfing, and especially of obligate selfing, are a decrease in genetic variability commonly associated with high levels of homozygosity and, in the long term, the elimination of unfavorable recessive and partially recessive alleles, so-called purging (e.g., Barrett and Charlesworth 1991, Byers and Waller 1999, Crnokrak and Barrett 2002). In contrast to outcrossing organisms in which recombination generates variance among progeny, selfing species respond to changes in environments by interline selection (Jarne and Charlesworth 1993). The high levels of homozygosity usually cause a decrease in offspring quality, compared with the progeny of outcrossers. Such decrease is termed inbreeding depression, δ, and is considered as the major factor preventing self-fertilization. It was systematically studied by Darwin (1876), and much information has been gathered on its evolutionary consequences (e.g., Charlesworth and Charlesworth 1987, Holsinger 1991, Husband and Schemske 1996, Charlesworth and Charlesworth 1999). An increase in selfing is selectively favored if the progeny of selfing has a fitness greater than half that of the progeny produced by outcrossing.

There are three main factors that promote selfing and that are considered as explanations for the evolution of this breeding system:

1. *Reproductive assurance*. This is the factor that Darwin (1876) thought was the most important one for the evolution of selfing. Selfing has been classified as (1) prior, (2) competing, or (3) delayed, depending on the timing relative to a possible outcrossing event (Wyatt 1983, Lloyd and Schoen 1992). Delayed selfing takes place when the possibilities of cross-pollination have past and is therefore selected for despite the variation in outcross pollen availability, whereas the invasion ability of a prior selfing gene in a population depends on the variation in outcross pollination success (Morgan and Wilson 2005). Delayed selfing has been shown to increase seed set when pollinator visits are infrequent (Kalisz et al. 2004).
2. *Mating costs*. There are two kinds of outcrossing costs: (1) those referring to the transmission of genes and (2) those referring to the resources needed for copulation and pollination. Due to the higher parent–offspring relatedness in selfing compared with random mating, selfing alleles have a 50% transmission advantage (Jain 1976). This advantage, however, can be reduced by factors such as pollen discounting (Lloyd 1979). The energetic costs of producing large quantities of pollen, mainly in wind-pollinated plants, plus rewards such as nectar or oil for animal-pollinated ones, are relatively high in most species, and much greater in outcrossing than in highly autogamous plants, in which attractive structures (e.g., petals) and male reproductive functions are reduced (Jarne and Charlesworth 1993, but see Damgaard and Abbott 1995).
3. *Preservation of successful genotypes*. When environmental conditions are stable, selfing preserves the genotype adapted to those conditions. Evidence for the local adaptive hypothesis, which postulates that individuals perform better at their native site, whereas fitness of transplanted individuals declines with increasing distance, has been found for several plant species (e.g., Schmitt and Gamble 1990, Galloway and Fenster 2000, Joshi et al. 2001) though results are inconsistent with other plant species (references in Jarne and Charlesworth 1993, Jakobsson and Dinnetz 2005).

Data available so far on the breeding system of different species suggest that most outbreeders can also self-pollinate and most selfers can outcross as well, that is, that mixed mating systems are rather common in nature (Barrett and Eckert 1990). Furthermore, as might be expected, mixed mating systems appear to be more common in biotic than in abiotic pollinated species (Vogler and Kalilsz 2001). The range of outcrossing rates can vary greatly among populations of the same species, because of both genetic and environmental causes. For example, outcrossing rates in populations of *Aquilegia coerulea* vary with the abundance of pollinator groups (Brunet and Sweet 2006). Several models predict evolutionary stability of intermediate levels of selfing (reviewed in Goodwillie et al. 2005), although it is not clear yet how often evolutionary stability of mixed systems occurs in nature (Barrett and Eckert 1990, Plaistow et al. 2004). Intermediate selfing rates are expected to evolve in plants where selfing reduces either male or female fitness, for example, when there is pollen discounting (Cheptou 2004), or when competing selfing reduces the number of fertilized ovules (seed discounting) (Lloyd 1979). Mixed mating systems can also be maintained when there is an optimum pollen dispersal distance due to local adaptation (Campbell and Waser 1987) or when inbreeding depression affects dispersed progeny more than nondispersed progeny (Holsinger 1986). Further studies of correlations between flower traits, environmental variables, and mating systems are needed, and experimental approaches are crucial to discern if a character is a cause or an evolutionary consequence of the breeding system (e.g., Herlihy and Eckert 2004). In addition, further molecular studies (DNA sequence data, in particular) help to assess the consequences of selfing and outcrossing on genetic variability within and between populations.

SEXUAL EXPRESSION

There are a number of possibilities of how male and female organs can be distributed within a plant species, and this determines the levels of selfing and outcrossing (Table 17.2). For the last three decades, plant biologists have tried to understand the different evolutionary pathways that have led to the large variation in sexual systems found within flowering plants (reviewed by Barret 2002). Models of sex allocation (gain curves) have been used to investigate how male and female fitness change with increases in the allocation of limited resources to each sexual function (reviewed by Charlesworth and Morgan 1991). Sex allocation

TABLE 17.2
Classification of the Different Possibilities by Which Male and Female Organs Are Distributed in a Plant Species

Hermaphroditism: all individuals (genets) have perfect flowers, all bearing functional stamens and pistils
Monoecy: the two sexes are found on all individuals, but in separate flowers
Andromonoecy: the same individual bears both perfect and male flowers
Gynomonoecy: the same individual bears both perfect and female flowers
Dioecy[a]: male and female flowers are on separate genets
Androdioecy[a]: male and hermaphrodite flowers are on separate genets
Gynodioecy[a]: female and hermaphrodite flowers are on separate genets
Subdioecy[a]: intermediate stage between monoecy and dioecy in which sex expression of males and females is not constant
Polygamy[a]: different combinations of males, females, and hermaphrodites are possible

[a] In these cases, when both male and female functions are not regularly found on the same genet, dicliny is said to occur.

theory alone, however, cannot explain all aspects of sex expression, such as, the spatio-temporal variation in sex expression found in nondiclinous plants (e.g., Solomon 1985, Emms 1993). The evolution of the various sexual systems found in plants may also be affected by gamete packaging (Lloyd and Yates 1982, Burd 1995) and selection for certain pollination modes (Golonka et al. 2005). Furthermore, the combination of biogeographical data on diversity of sexual systems with phylogenies can help understand patterns of sexual diversification (Gross 2005).

Monoecy

Monoecy is widespread, especially in large wind-pollinated plants such as trees, sedges, and aquatic plants, and rarer in insect-pollinated plants (Richards 1997). At least in some floras this breeding system is associated with trees and shrubs that produce dry many-seeded fruits (Flores and Schemske 1984). One of the benefits of separate sexes on the same individual is that plants have the capacity to invest more on one sex or the other, depending on environmental conditions, to maximize the efficiency of both pollen dispersal and pollen capture. Moreover, monoecious plants benefit from a reduction of inbreeding depression, due to the spatial—and often temporal—segregation of sexes (Freeman et al. 1981). Evolutionary theories based on relative costs and benefits of male and female reproductive structures predict that plants growing under favorable conditions (larger in size, a greater resource supply, or a greater total reproductive effort) should invest relatively more in female than in male function (e.g., Freeman et al. 1981, Klinkhamer et al. 1997, Méndez and Traveset 2003). The opposite is often found for wind-pollinated plants, which have been found to increase relative maleness as patch quality improves (e.g., Burd and Allen 1988, Traveset 1992, Fox 1993). An explanation for this could be that large wind-pollinated plants may benefit from a relatively greater male investment if pollen is carried for longer distances (e.g., Smith 1981, Solomon 1989, Traveset 1992). However, Sakai and Sakai (2003) showed in a model that size and height in wind-pollinated cosexual plants may increase allocation to either male or female sex depending on several conditions, as for example plant density and number of small and large plants in the pollen dispersal area.

Models of sex allocation predict that the evolution of self-fertilization should result in a reduced allocation to (1) male function and (2) pollinator attraction (Charlesworth and Morgan 1991). In selfing monoecious plants, however, the investment to male function cannot be much reduced compared with hermaphroditic plants, as separate structures (petals, sepals, and pedicels) for male flowers and a higher production of pollen (to be transferred between flowers) are needed. Moreover, evolutionary changes in allocation patterns may be constrained by lack of genetic variation or by genetic correlations among characters (e.g., Ross 1990, Mazer 1992, Agren and Schemske 1995). More data on the importance of these genetic constraints, on the genetic and phenotypic correlations between allocations to both sex functions, and on the relationships between sex allocation, mating system, and reproductive success of the two sex functions are needed to understand the evolutionary dynamics of sex allocation. Long-term data on gender variation in natural populations are also necessary in studies of the evolution of sex expression (e.g., Primack and McCall 1986, Jordano 1991). There is much individual variation in patterns of sex allocation, and a variety of factors (reviewed by Goldman and Willson 1986) can cause a lack of consistent results. Variations at spatial and temporal scales in environmental conditions need to be considered, as gender expression of a species may vary, for instance, across a climatic gradient (Costich 1995). Documenting such variation at the individual, within-, and between-population level in the field is crucial to understand the selective pressures involved in the evolution of gender expression.

Andromonoecy

Andromonoecy is a breeding system that has been of particular interest in the study of sex expression patterns. It is uncommon, occurring in less than 2% of plant species (Yampolsky and Yampolsky 1922) and has probably evolved from hermaphrodite ancestors, by means of a mutation removing pistils from some perfect flowers and a subsequent regulation of male flower number (Spalik 1991) or by the production of staminate (male) flowers (Anderson and Symon 1989). According to resource allocation models, andromonoecy occurs in species in which the cost of maturing a fruit is great and the optimal number of male flowers is greater than the number of flowers that can set fruit (Bertin 1982, Anderson and Symon 1989, Spalik 1991). Pollen from male flowers can be more fertile than pollen from hermaphrodite flowers, as found in *Cneorum tricoccon* (Traveset 1995), representing an advantage to andromonoecy as this may increase both male and female fitness (Bertin 1982). By producing less-expensive staminate flowers, andromonoecious species may also increase floral display and hence attractiveness to pollinators (Anderson and Symon 1989). However, staminate flowers must not necessarily be less expensive, as in *Sagittaria guyanensis* ssp. *lappula* in which staminate flowers had more and larger anthers and also longer petals than hermaphrodite flowers (Huang 2003). Temporal differences in the functioning of male and female organs are a common feature of monoecious and andromonoecious taxa (e.g., Anderson and Symon 1989, Emms 1993). In the andromonoecious *Zigadenus paniculatus*, for instance, male flowers are produced at the end of the blooming period, when the returns on female allocation are small or nonexistent (Emms 1996). We need more data to test if these temporal patterns are adaptive and to answer questions such as: (1) how frequent are the mutations causing pistil loss? (2) is the production of surplus pistils advantageous (thus selecting against andromonoecy)? (see review in Ehrlén 1991), (3) does the rechanneling of resources from pistils to other structures (e.g., male flowers) increase fitness in hermaphroditic species? As claimed by Emms (1996), rather than asking why andromonoecy has evolved, it may be more interesting to ask why it is so rare. Moreover, to fully understand this breeding system, we also need to identify the factors that control male fitness. We need more data, for instance, on variation in pollen production per flower. We do not yet know if total pollen output in andromonoecious species is regulated through an increase in flower number or through the amount of pollen per flower. Data on a few species reveal that pollen grain number does not differ between male and hermaphroditic flowers (Solomon 1985, Traveset 1995, Cuevas and Polito 2004) or is even lower in males (Spalik 1991). Some authors have suggested that andromonoecy restricts outcrossing (Primack and Lloyd 1980, Bertin 1982, Narbona et al. 2002) whereas others argue that, depending on the pollinator, it may serve to reduce selfing (Anderson and Symon 1989).

Sex expression in andromonoecious species can be quite variable, among individuals, within and among populations, and through time (Diggle 1993 and references therein; Traveset 1995). Such variation can either be genetic or phenotypically plastic, varying with resource availability (e.g., light, water, nutrients available) (e.g., Solomon 1985, Diggle 1993). In andromonoecious species, staminate flowers are hermaphroditic in their early development and become mainly male by slower growth of the gynoecium compared with the androecium (Diggle 1992), and studies suggest that when resource levels are low the production of staminate flowers is favored (Calvino and Garcia 2005). Further supporting the connection between low resource status and staminate flowers is the finding that *Olea europea* produces staminate flowers in positions of the inflorescence that are less nurtured (Cuevas and Polito 2004).

Gynomonoecy

Gynomonoecy is much rarer than andromonoecy, occurring only in about a dozen families, and there seems to be no satisfactory explanation yet for the difference in frequencies of these two breeding systems. One possible reason may be the more expensive production of fruits compared with flowers (Charlesworth and Morgan 1991). Recent studies suggest that the benefits of this breeding system lies in the promotion of outcrossing and increase in pollinator attractiveness, rather than in the flexibility in allocation of resources to either female or male function (Bertin and Gwisc 2002, Davis and Delph 2005).

Dioecy

Dioecy is found in a large proportion of gymnosperms (*c*.52%; Givnish 1980) compared with angiosperms (*c*.6%; Renner and Ricklefs 1995), and appears to be strongly associated with woodiness in certain tropical floras (e.g., Givnish 1980, Sakai et al. 1995). The incidence of dioecy varies notably among regional floras, ranging from values as low as 2.6% (Balearic Islands) or 2.8% (in California; Fox 1985) to *c*.15% in the Hawaiian flora (Sakai et al. 1995). This mating system has evolved independently many times, as suggested by its scattered systematic distribution (Lloyd 1982). Dioecy has been found to be associated with monoecy, wind and water pollination, and climbing growth (Renner and Ricklefs 1995) as well as with tropical distribution, woody growth form, plain flowers, and fleshy fruits (Vamosi and Vamosi 2004). The two major evolutionary pathways for the origin of dioecy are via monoecy (Yampolsky and Yampolsky 1922, Dorken and Barrett 2004) and gynodioecy (Freeman et al. 1997, Weiblen et al. 2000), although it has also evolved from androdioecy, as in the genus *Acer* (Gleiser and Verdú 2005) and from distyly in several angiosperm genera (Beach and Bawa 1980). The monoecy pathway has been described as a gradual divergence in the relative proportions of male and female flowers in the two incipient sexes (Charlesworth and Charlesworth 1978, Ross 1978, 1982, Lloyd 1982) and is presumably easier than the evolution of dioecy via other mating systems, as mutations affecting pollen or ovule production have already occurred in the unisexual flowers (Renner and Ricklefs 1995).

The classic hypothesis on the mechanism underlying the evolution of dioecy states that it has evolved to overcome the negative effects of inbreeding depression (Thomson and Barrett 1981, Charlesworth 2001, de Jong and Klinkhammer 2005). The frequently documented association between dioecy and abiotic pollination with its imprecise pollen movement supports this view (references in Renner and Ricklefs 1995). Another school of thought believes that dioecy is the outcome of sexual selection (Willson 1979, Armstrong and Irvine 1989), suggesting that separation of sexes may result in a more efficient use of resources for both male and female functions. Freeman et al. (1997) argue that both schools are correct but that the mechanisms act on taxa with different life histories and different historical contexts. These authors hypothesize that in self-incompatible species dioecy has resulted from selection for sexual specialization, whereas in self-compatible species dioecy would have evolved via gynodioecy, a route that involves a genetic control of gender. According to them, the pathway toward dioecy via monoecy (especially common in wind-pollinated species) might be more controlled by ecological factors, since sex-changing and sexual lability occur mostly among species that arose by this pathway and not via gynodioecy.

The possibility that differential predation (specifically, seed predation or flower herbivory) could be another force selecting for dioecy was hypothesized a long time ago (Janzen 1971), and sex-related differences in herbivore damage have been reported for some species (see reviews in Watson 1995, Ashman 2002, Cornelissen and Stiling 2005). Seed dispersal has also been suggested to influence evolution of dioecy (Thomson and Brunet 1990), and a recent model showed that dioecy has negative effects on seed dispersal, resulting in a more clumped distribution of seeds since they are only produced by females (Heilbuth et al. 2001). In addition, pollen dispersal can be negatively affected in those dioecious species that have been found to have segregated spatial distribution for the different sexes (Eppley 2005).

Some plants are cryptically dioecious, that is, they are morphologically hermaphrodite but functionally dioecious, since either males or females, or both, have sterile or disfunctional opposite-sex structures (reviewed by Mayer and Charlesworth 1991, Verdú et al. 2004). Other species are classified as subdioecious, with populations possessing strictly male or female functions and a variable proportion of hermaphrodites; such proportion may vary depending on how favorable growing conditions are, as found in *Schiedea globosa* (Sakai and Weller 1991).

Gynodioecy

The overall frequency of gynodioecy is generally considered to be low (Yampolsky and Yampolsky 1922, Lloyd 1975), although recent studies have shown that it has been overlooked. For females to be maintained in the population, their lack of male function has to be compensated with a higher female fitness (a higher quantity and quality of seed production). Sex allocation models predict that female fitness has to be at least twofold of that of hermaphrodites, if inheritance of male sterility is governed by nuclear genes, though it can be less than double when both nuclear and cytoplasmic genes control gender (Lloyd 1975). This compensation has been found in several gynodioecious species such as *Cucurbita foetidissima* (Kohn 1988), *Geranium maculatum* (Agren and Willson 1991), *Chionographis japonica* (Maki 1993), *Prunus mahaleb* (Jordano 1993), and *Opuntia quimilo* (Díaz and Cocucci 2003), but was not found in *Kallstroemnia grandifolia* (García et al. 2005). Cytoplasmic genes that produce females by causing male sterility is the most common cause of gynodioecy, and the balance between such genes and nuclear restorer genes that restore pollen production is crucial for the maintenance of gynodioecy in populations (reviewed by Jacobs and Wade 2003 and modeled by Bailey et al. 2003). The evolution of this breeding system has usually been interpreted as an escape from inbreeding depression, and this may actually be the main selective factor in species such as *C. japonica* (Maki 1993). However, factors other than inbreeding avoidance select for gynodioecy in other species (e.g., *Ocotea tenera*; Gibson and Wheelwright 1996). Variation in resource allocation to floral organs as corolla size (Eckhart 1992), anther size, and nectar production (Delph and Lively 1992) between females and hermaphrodites might also be important in the evolution of gynodioecy.

Androdioecy

One of the first observations of functional androdioecy was made in *Datisca glomerata* (Liston et al. 1990) and it is still being documented in fewer occasions than gynodioecy (Pannell 2002). Androdioecy is most likely to evolve from dioecy (Pannell 2002), and pollen limitation has been suggested as the mechanism underlying the transfer between these breeding systems (Wolf and Takebayashi 2004). In a recent model, Pannell and Verdú (2006) point out that androdioecy also could evolve from heterodichogamic hermaphrodite populations. It is probably the requirements for its maintenance, predicted by sex allocation theory, which makes it a rare and evolutionarily unstable breeding system. Models predict that the frequency of male plants should be much lower than that of hermaphrodite plants and the siring success of the former should be at least twice as high (Lloyd 1975, Charlesworth 1984). *Phillyrea angustifolia* is reported to have both functional androdioecious and functional dioecious populations in southern France, and contrary to predictions no difference in number of seeds sired has been found between hermaphrodite males and pure males (Traveset 1994, Vassiliadis et al. 2002). In *Mercurialis annua*, androdioecy was found to be stable within a metapopulation context, females within populations were always selected for, but during founding of new populations the hermaphrodite individuals had advantages over females as these were not able to breed properly (Pannell 2001).

Many of the previously cited androdioecious species have shown to be dioecious after closer inspection of the functionality of the breeding system. Thus, it is necessary to go deeper into the functionality of this breeding system and to document that hermaphrodites produce viable pollen that sire a significant number of progeny.

SELF-INCOMPATIBILITY SYSTEMS

Systems of self-incompatibility are widely distributed among flowering taxa and have been recorded from approximately 20 orders and over 70 families of dicots and monocots, with different life-forms, and from tropical as well as temperate zones (Barrett 1988). Here we briefly review the major classes of self-incompatibility, their general properties, and the hypotheses to explain the evolution of some of these systems.

Self-incompatibility systems can be heteromorphic where morphological differences can be seen on the sporophyte as two (distyly) or three (tristyly) mating types that differ in style length, anther height, pollen size, and pollen production. They can also be monomorphic where the preventing of selfing relies on a chemical–physiological response. Monomorphic systems (see reviews in Franklin-Tong and Franklin 2003, Hiscock and McInnis 2003) can be (1) gametophytic and expressed during pollen tube growth coded by the haploid genotype of the pollen tube, as found for instance in Solanaceae and Papaveraceae or (2) sporophytic and governed by the genotype of the pollen-producing plant and transferred as proteins to the pollen grain coat, as found in the Brassicaceae (Castric and Vekemans 2004 and references therein). Monomorphic and heteromorphic systems do not seem to co-occur in the same plant families, except for the Rubiaceae (Wyatt 1983).

In monomorphic systems, pollen and pistil incompatibility is controlled by different but tightly linked genes, the S-locus (self-incompatibility locus) that should rather be called the S-genes complex (Schopfer et al. 1999, Takasaki et al. 2000, Castric and Vekemans 2004). Pollen tube growth is inhibited in the style (in most gametophytic systems) or in the stigmatic surface (in most sporophytic systems). Inhibition in the ovary, so-called late-acting incompatibility is also common, although when the rejection is postzygotic it is difficult to discern its effect from inbreeding effects (Seavey and Bawa 1986). It is also difficult to separate between inbreeding effects and self-incompatibility in species where self-incompatibility is cryptic, that is, where tube growth rate is greater for cross- than for self-pollen, such as *Cheiranthus cheiri* (Bateman 1956 in Barrett 1988), and *Dianthus chinensis* (Aizen et al. 1990). Gametophytic incompatibility systems have evolved independently several times in the angiosperms (Steinbachs and Holsinger 2002, Charlesworth et al. 2005) and work with very different mechanisms (Franklin-Tong and Franklin 2003).

Heterostyly is governed by a single locus with two alleles, in distylous species, or by two loci each with two alleles and epistasis operating between them in tristylous plants. Heterostyly also has a polyphyletic origin and has been reported from about 25 families of flowering plants (Barrett 1990). The most visible trait in heterostylous plants is the significant difference between morphs in the height at which stigma and anthers are positioned within the flowers. This polymorphism is usually associated with a sporophytically controlled diallelic self-incompatibility system that prevents self- and intra-morph fertilizations, but not all heteromorphic species are self-incompatible (e.g., Casper 1985, Barrett et al. 1996). Herbaceous heterostylous taxa such as *Primula*, *Oxalis*, *Linum*, and *Lythrum* have received much attention in molecular studies since long ago (references in Barrett 1990), although mostly in controlled experimental conditions. In the last decades, a number of studies on population biology, and on structural, developmental, and physiological aspects of heterostylous species have been carried out, and much information has been accumulated on the function and evolution of heterostyly (comprehensively reviewed in Barrett 1992 and in de Jong and Klinkhammer 2005). Different hypotheses have been formulated on the sequence of evolutionary events in heterostylous plants. The classic model (Charlesworth and Charlesworth 1979) assumes that inbreeding avoidance has selected for a diallelic self-incompatibility, followed by evolution of reciprocal herkogamy and appearance of the different floral polymorphisms to increase the efficiency of pollen transfer between incompatible morphs. This was challenged by Lloyd and Webb (1992), who believe that reciprocal herkogamy evolved first as a result of selection to increase the efficiency of pollen transfer, and that self-incompatibility appears later as a gradual adjustment of pollen tube growth in the different morphs. The hypothesis of Lloyd and Webb, in fact, supports Darwin's idea that the style–stamen polymorphism acts as promoter of disassortative pollination, and evidence for it is accumulated in studies of pollen deposition patterns on the stigmas of the different morphs (Kohn and Barrett 1992, Lloyd and Webb 1992). Some authors (e.g., Olmstead 1986), however, see self-incompatibility independent of the level of inbreeding in the population as a whole, and argue that inbreeding is more influenced by small effective population sizes than by selfing avoidance.

PATERNAL SUCCESS

For many years, studies of plant reproductive success were strongly biased by examining only the female function (see review in Willson 1994, Schlichting and Delesalle 1997). However, in the last three decades, different aspects of male reproductive success, such as pollen production, pollen removal, and paternity of offspring, have been examined in a number of studies (see reviews in Snow and Lewis 1993, Ashman and Morgan 2004). Male fitness is usually expressed in terms of the number of sired offspring surviving to reproductive age. As this is very difficult to measure, and has to be indirectly estimated from genetic markers, correlates of fitness such as pollen germination ability, pollen tube growth rate, ability of pollen to affect fertilization, weight and number of seeds sired, seed germination, and performance of sired seedlings are usually evaluated. This far, most pollen competition studies with heritable markers have been hand-pollination studies with known pollen donors, and allozymes have been used for diagnosing parental identity (reviewed in Bernasconi 2003). The development of more variable molecular markers in combination with statistical models to assess male reproductive success will hopefully help to understand fitness returns from investment in male function (e.g., Barrett and Harder 1996, Smouse et al. 1999, Burczyk et al. 2002).

Resources allocated to male function are in turn divided among number and size of pollen grains, male accessory structures (e.g., petals, sepals, bracts), and substances (e.g., nectar). Such resource allocation may be linked to male fitness, although we still have little experimental evidence supporting this (see examples in Bertin 1988, Young and Stanton 1990). When measuring male fitness, it is important to quantify pollen removal and also to monitor the success of removed pollen as these two variables may not be positively correlated (e.g., Wilson and Thomson 1991, but see Conner et al. 1995). For instance, a bee removing much pollen from a nectar-rich plant may fly short distances or promote much geitonogamy, which may limit potential gains in male fitness. The success of the removed pollen is influenced by the percentage of pollen grains germinating, by the rapidity of germination on the stigma, and by pollen tube growth rate in the style, all of which in turn are affected by abiotic factors, especially temperature (Bertin 1988, Murcia 1990). The success of a particular pollen grain also depends on the composition and size of the whole pollen load on the stigma. Several studies on pollen tube growth rate have found that the presence of self- or incompatible pollen has a negative effect on tube growth of cross-compatible pollen (e.g., Shore and Barrett 1984). The competitive ability of a pollen grain is influenced by its own genotype, which differs among individuals and among pollen grains from the same individual. Thus, the genotype of all pollen grains on the stigma influences the success of a particular one (e.g., Bookman 1984). Large pollen loads can be advantageous over small loads because the former are more likely to enhance pollen germination as well as tube growth rate (e.g., Ter-Avanesian 1978). However, large pollen loads may yield fewer pollen tubes per pollen grain than small loads (Snow 1986) and, thus, the probability that a certain pollen grain is represented in the seed crop can also be lower for large pollen loads. The sequence of pollen deposition on the stigma has also shown to be important determining the proportion of seeds sired by the different pollen grains (e.g., Mulcahy et al. 1983) and it affects the potential for interaction (competition) among grains, which may have been brought by different pollinators (e.g., Murcia 1990). In a study on *Hibiscus moscheutos*, Snow and Spira (1996) gave strong evidence that pollen tube competitive ability varies among coexisting plants, arguing that it may be a relevant component of male fitness in plants. Pollen grains from different donors on the stigma not only race for access to ovules (exploitation competition) but can also interfere with the germination and growth of each other (interference competition), as it has been found in wild radish (Marshall et al. 1996) and in *Palicourea* (Murcia and Feinsinger 1996).

For hermaphroditic plants, the male function has been predicted to be limited by mating opportunities and not by resources, whereas the opposite is expected for the female function. This hypothesis has been termed the fleurs-du-mâle hypothesis (Queller 1983), also known as the male function or pollen donation hypothesis (PDH) (e.g., Fishbein and Venable 1996, Broyles and Wyatt 1997). According to the PDH, large floral displays would especially benefit the male function as they would have a greater fraction of their pollen exported. Some authors even believe that flower number in the angiosperms has been selected by such male function (e.g., Sutherland and Delph 1984). Several variants of the PHD have been formulated and are reviewed by Burd and Callahan (2000). These authors propose that the PHD should explain the evolution of excessive (nonfruiting) flowers, not total flower number, and that studies should consider the whole plant fitness, not only the fitness of single flowers or inflorescences. It is also possible that excessive flowers have a positive effect on the female function, by enhancing the reception of larger amounts of outcross pollen (Burd 2004). More studies with adequate experimental designs and controlling for variables such as level of resources are needed to determine whether male function does select for large floral displays. We must also know the consequences of self- versus cross-pollination, as large floral displays may be less efficient at exporting pollen if pollinators promote geitonogamy (de Jong et al. 1993). Theoretically, if female fitness (achieved via fruit production) is less affected by geitonogamy than male fitness (achieved via siring of fruits on other plants), we would predict that small plants invest more in male reproduction whereas large plants emphasize more on the female function. Some data seem to support this prediction (de Jong et al. 1993, 1999).

By evaluating both female and male reproductive success, it is possible to examine (1) whether they are correlated or not, (2) the genetic variation for female and male components of fitness, and (3) whether the components of male and female reproductive success are equally affected by environmental factors. Some studies have documented genetic variation in both male and female functions, and evidence for a male–female trade-off was found in *Collinsia parviflora* when flower size was controlled for (Parachnowitsch and Elle 2004) although other studies have found no consistent pattern of such trade-offs (e.g., Schlichting and Devlin 1992, Mutikainen and Delph 1996, Strauss et al. 1996). A survey on the consequences of herbivory on male and female functions shows that these are neither equal nor proportional (Mutikainen and Delph 1996, Thomson et al. 2004), although we still need more data that evaluate the plastic responses of male and female components to different environmental factors. Studies on functional architecture are also necessary to estimate the genetic and phenotypic correlations between both quantitative and qualitative aspects of male and female functions.

ROLE OF POLLINATORS ON THE EVOLUTION OF FLORAL TRAITS AND DISPLAY

The evolution of plant mating systems has undoubtedly been linked to the evolution of traits that influence the type of pollination (animal vs. wind pollination) and pollinator attraction to flowers (e.g., quantity and quality of floral rewards, petal coloration, flower size, flowering time). By producing large floral displays, or great amounts of nectar, for instance, plants can affect the behavior of pollinators, which in turn influences gene flow among plants and, ultimately, plant fitness (see reviews in Zimmerman 1988 and Pellmyr 2002). The goal of numerous studies on pollination biology has been to identify pollination syndromes, that is, suites of structural and functional floral traits that presumably reflect adaptations to different types of pollinating agents (Proctor and Yeo 1973, Faegri and van der Pijl 1979, Hingston and McQuillan 2000, Wilson et al. 2004). The variation in floral characters within a species, and its association with the variation in reproductive success, has so far received less attention. Several studies have demonstrated phenotypic selection on floral traits (Galen 1989, Schemske and Horvitz 1989, Herrera 1993, Johnson and Steiner 1997, Hansen et al. 2000, Medel et al. 2003), whereas others have found no evidence of the fact that floral differences are the outcome of adaptation to pollinators (Herrera 1996, Wilson and Thomson 1996, Armbruster 2002). Some floral characters may not represent adaptations to current pollinators, but exaptations (Gould and Vrba 1982, Lamborn and Ollerton 2000), evolved as a consequence of selection by pollinators that are now extinct or not present in the current scenario. Studies of correlated trait shifts represent another way to reveal how frequently pollinators exert selection pressures on floral characters in nature, but have to be combined with experiments on the adaptive basis of the traits (e.g., Lamborn and Ollerton 2000, Tadey and Aizen 2001, Castellanos et al. 2003). In a review on pollination syndromes, Fenster et al. (2004) summarized studies of correlated phylogenetic and ecotypic shifts in flower traits and functional groups of pollinators (phylogenetic shifts implying that closely related plant species show different traits and rely on different functional groups of pollinators, and ecotypic shifts implying a correlation between variation in floral trait and pollinators within a species). More than 50% of the studies of the traits reward, morphology, and color had detected a correlated trait change, whereas less than 50% of the studies of the trait fragrance had detected a correlated change.

To determine the contribution of a pollinator to plant fitness (i.e., the pollinator effectiveness), it is essential (1) to quantify the number of flowers it pollinates (quantitative component) and (2) to evaluate its efficiency as a pollinator (qualitative component). The former depends on the frequency of pollinator visits to a plant and on the flower visitation rate whereas the latter is a function of the pollen delivered to stigmas, the foraging patterns, and the selection of floral sexual stage by the pollinator (Herrera 1988). Pollination effectiveness is determined by the product of (1) frequency of visitation and (2) efficiency, and somewhat counterintuitive there is not always a positive correlation between these two factors (e.g., Herrera 1988, Schemske and Horvitz 1989, Pellmyr and Thompson 1996, Gómez and Zamora 1999, Mayfield et al. 2001). However, such a positive correlation has been found (Olsen 1997, Fenster and Dudash 2001), and Vázquez et al. (2005) have developed a model that shows that the most frequent mutualists often contribute most to reproduction regardless of their efficiency on a per-interaction basis.

The strength of selection of floral traits by pollinators and the plant's response to such selection may be limited by factors that are either intrinsic (genetic or life history) or extrinsic (environmental) to the plant (Herrera 1996, 2005). Among the latter, the spatio-temporal variation in the composition of pollinator assemblages and in their relative abundance is probably the most important factor precluding or strongly reducing selection on floral traits by pollinators. Differences at a spatial and at a temporal scale in the assemblage of pollinators have been frequently documented (Herrera 1995, Traveset and Sáez 1997, Gómez and Zamora 1999, Maad 2000, Thompson 2001, Eckert 2002a, Minckley et al. 1999), but see Cane et al. (2005) for a high similarity in pollinator assemblage over time. Such spatial and temporal differences can create a mosaic of selective regimes (Thompson 2005), and if the mosaic is at a small scale, for example, within the same geographical area where there is gene flow among plants, selection on floral traits is probably much weakened. Large variation has been found in the pollinator assemblage visiting *Lavandula latifolia* in southeastern Iberian Peninsula, both between individuals within a single population and among populations (Herrera 2005). Abiotic conditions such as shade and vicinity to streams accounted for much of the observed variation and were positively related to pollinator diversity, as expected in a dry Mediterranean habitat. Selection on floral traits may also be weakened if the effect of a specific pollinator is context dependent, as in the case of *Penstemons* where bees are less-effective pollinators than hummingbirds, so in the presence of the latter, traits attracting bees are selected against (Wilson et al. 2006). In *Ipomopsis agregata*, the presence of another plant species nearby increases competition for pollinators and affects selection on floral traits (Caruso 2000).

To date, most studies of evolution of floral traits have, explicitly or implicitly, been founded on an assumption of the most effective pollinator principle (Stebbins 1970), that is, that the most effective pollinators are driving the evolution of floral traits. An alternative view is that floral traits must not represent adaptations to the most common or the most effective pollinators as long as the trait provides a marginal increase in fitness (Aigner 2006). We should then expect to find adaptations to rare or inefficient pollinators as long as those adaptations do not reduce the effectiveness of common pollinators. This view is supported by the observation that many species tend to be pollinated by several types of pollinators despite very specialized floral traits (Ollerton 1996), and that the unexpected pollinators—given the pollination syndrome—may even be the most effective, as in the case of bumblebees that are five times more effective than hummingbirds in pollinating the hummingbird flower *Ipomopsis aggregata* (Mayfield et al. 2001).

Even if phenotypic selection on a floral trait occurs, it may have a small effect on individual variation in maternal fitness relative to that of other factors, such as plant size, herbivory, and seed dispersal success. For instance, individual variation in floral morphology of different species (*Calathea ovandensis* (Schemske and Horvitz 1989), *Viola cazorlensis* (Herrera 1993), and *Hormathophylla spinosa* (Gómez and Zamora 2000)) accounted for less than 10% of the variance in fruit production, and in a combined pollination and herbivore experiment the presence of pollinators had a positive effect on recruitment only when herbivores of flowers and fruits were absent (Herrera et al. 2002). As mentioned in the previous section, however, selection may occur via the male function, and thus it is necessary to examine both female and male fitness to determine if phenotypic selection is important (e.g., Primack and Kang 1989, Conner et al. 1996, Maad and Alexandersson 2004, Caruso et al. 2005). In species that produce more than one flower, the operational unit of either male or female function—when determining the effect of phenotypic selection of a floral trait on maternal fitness—has to be the whole floral display because, as mentioned earlier, the level of geitonogamy determines the incidence of self-pollination and pollen discounting, and ultimately the plant's mating success (Harder and Barrett 1996).

Reproductive assurance, that is, an increase in autonomous self-pollination when pollinators are rare or absent (Harder and Barrett 1996), has been shown both for populations (Fausto et al. 2001, Kalisz et al. 2004) and single flowers (Kalisz and Vogler 2003). Even if common, pollinators may also affect the levels of selfing, and thus offspring quality, by their inefficiency (Harder and Barrett 1996). For instance, they may move frequently among different plant species transporting pollen between them so that fertilization by conspecific pollen is interfered with (e.g., Thomson et al. 1981, Harder et al. 1993, Caruso and Alfaro 2000,

Brown et al. 2002) and pollen is lost on foreign stigmas (Campbell 1985, Feinsinger and Tiebout 1991). Pollinators may also be inefficient by making several visits to the same plant individual and thus promoting geitonogamy (e.g., Brunet and Sweet 2006). Although pollen limitation has been frequently demonstrated (reviewed in Ashman et al. 2004, Knight et al. 2005), only a few studies have investigated its consequences for progeny fitness in the field (Brown and Kephart 1999, Colling et al. 2004) and the exploration of effects on population persistence has only begun (Ashman et al. 2004).

During the last decade, there has been a debate over the degree of generalizations versus specialization in pollination systems (Waser et al. 1996, Johnson and Steiner 2000, Ollerton and Cranmer 2002, Vázquez and Aizen 2003, Fenster et al. 2004, Herrera 2005, Waser and Ollerton 2006). The classical view that pollination systems tend toward specialization and that pollinator specialization is critical to plant speciation has been implicit in many pollination studies (Grant 1949, Baker 1963, Grant and Grant 1965, Stebbins 1970, Crepet 1983), but was questioned by Waser et al. (1996) who argued that pollination systems are more generalized and dynamic than previously believed. Supporting this view is, for example, the rareness of a complete match of geographical ranges of plants and pollinators indicating nonobligate interactions (Thompson 2005) and the invasion of new areas by pollinators (references in Traveset and Richarson 2006). The conclusions about the prevailing generalization level in a system may to some extent be a question of definition, and can change depending on how pollinator generalization is measured. One can, for example, use raw counts of the number of pollinators, or consider the phyletic or functional diversity of them and estimate the fraction of pollinators used of the total available species pool (Gómez and Zamora 2006). A striking example of the importance of the method is a plant–pollinator system in Illinois (Robertson 1928), which has been defined both as generalized (Waser et al. 1996) and specialized (Fenster et al. 2004), depending on the classification of pollinators. The interest in degree of generalization of pollination systems has emphasized the importance of investigating whole pollination networks, that is, all interactions between plants and pollinating animals in a system (Memmot 1999, Olesen and Jordano 2002, Bascompte et al. 2003). One conclusion from such studies is that reciprocal specializations, that is, when a pollinator species and a plant species are exclusively interacting with each other, are rare in plant–pollinator systems (Minckley and Roulston 2006). Rather it seems that plant–pollinator interactions often are asymmetric so that specialized species often interact with generalist species (Bascompte et al. 2003, Vázquez and Aizen 2004). The concept of pollination syndrome, in fact, has sometimes proven to be of little use when predicting the pollinators of a certain plant species and when explaining interspecific variation in pollinator composition (Herrera 1996 and references therein; Ollerton and Watts 2000, Mayfield et al. 2001). In habitats where pollinators are uncommon, it is not rare to find plants with both abiotic and biotic pollinating agents (e.g., Gómez and Zamora 1996, Lázaro and Traveset 2005). Future studies that examine the spatio-temporal variation in pollinator assemblages and in pollen limitation (Dudash and Fenster 1997, Kay and Schemske 2004, Knight et al. 2005 and references therein) and which assess not only one but several of the interactions an organism experiences (Irwin 2006) are crucial to determine how often and in what conditions plants specialize to particular pollinator agents.

INFLUENCE OF BIOTIC POLLINATION IN ANGIOSPERM DIVERSIFICATION

A long-standing question in the study of plant evolution is how and to what extent the emergence of animal pollination has driven the great and rapid early speciation of flowering plants. Different authors (e.g., Raven 1977, Regal 1977, Burger 1981, Crepet et al. 1991, Eriksson and Bremer 1992) have argued that animal pollinators, referring mostly to insects, may have influenced the rate of angiosperm diversification by (1) promoting genetic isolation of plant populations, through mechanical or ethological mechanisms (see review in Grant 1994), (2) promoting outcrossing, so genetically diverse populations may undergo rapid phyletic evolution, and (3) reducing extinction rates, as they move pollen across long distances among sparse populations. Similarly, it has been suggested that the biotic dispersal of seeds, mainly referring to dispersal by vertebrates, also has contributed to some extent to angiosperm diversification (e.g., Tiffney and Mazer 1995 and references therein, Smith 2001). However, there is still controversy whether animal pollination increases speciation, and using phylogenetic data recent studies have found both a decrease in speciation in wind-dispersed species (Dodd et al. 1999) as well as no evidence for species richness to be higher in animal-pollinated than in wind-pollinated groups (Bolmgren et al. 2003). It is also important to note that lineages other than the angiosperms, such as Gnetales, Benettitales, Cheirolepidiaceae, and Medullosales, were insect pollinated but never underwent species radiations (Gorelick 2001), and that numerous shifts in diversification rates have taken place within the angiosperms, and some quite recently, as shown by the construction of a supertree of all angiosperm families (Davies et al. 2004). In contrast to the view that biotic dispersal of pollen and seeds has caused or favored the speciation of angiosperms, other authors (e.g., Midgley and Bond 1991, Stebbins 1981, Doyle and Donoghue 1993, Ricklefs and Renner 1994) believe that morphological and physiological characters in flowering plants have played a more important role in their diversification, although both biotic and abiotic mechanisms may be acting simultaneously (Verdú 2002). The reason why angiosperms have been more successful than gymnosperms may lie more on factors such as the greater plasticity in (1) growth forms, (2) type of habitats they can inhabit, (3) ways to exploit the environmental resources, (4) types of reproduction (vegetative reproduction is very rare in gymnosperms), and (5) possibly even in types of breeding system, which is usually less complex in gymnosperms. As Ricklefs and Renner (1994) point out, however, it is important to consider that factors affecting the displacement of gymnosperms by angiosperms may not be the same as those affecting their diversification rate. Angiosperms may be competitively superior for different causes: efficiency of water use in particular dry environments, efficiency of insect pollination in habitats where wind is nearly or totally absent, rapid growth, double fertilization, capacity of vegetative reproduction, and so on. It is plausible, though, that angiosperm diversification has promoted their proliferation in some habitats and under some circumstances (by their diversity, angiosperms may be more likely to survive and propagate after an environmental stress such as a period of drought, which may be devastating for a species of gymnosperm). Ricklefs and Renner concluded that the major factor contributing to speciation is probably the capacity of taxa to exploit a wide range of ecological opportunities by adopting different growth forms and life histories and by differentiating morphologically to be pollinated and dispersed by different vectors (biotic and abiotic). The study by Tiffney and Mazer (1995), in contrast, does show an important effect of biotic dispersal of seeds in angiosperm diversification (they do not include pollination systems in their analysis). The reason for such conflicting results is attributed, by these two authors, to the pooling of angiosperms with different growth forms or other traits, which masks differences among various groups. They perform separate analyses for woody and herbaceous monocots and dicots, finding that dispersal by vertebrates contributes to species richness in woody dicots, and that abiotically dispersed families exhibit higher levels of diversification in herbaceous monocots and dicots than vertebrate-dispersed families. The possibility exists, therefore, that the effect of biotic dispersal of both pollen and seeds was underestimated in the analyses of Ricklefs and Renner. Other potential problems with this kind of analyses, pointed out by Bawa (1995), are (1) the use of families, rather than genera or species, as independent units and (2) the broad classification of pollination and seed dispersal into biotic and abiotic categories, as both categories are very heterogeneous. Further analyses that include more variables that might influence diversification (capacity of asexual reproduction, size of flower, fruit and seed, specificity of pollinators, etc.) will certainly reveal new patterns and probably contribute to explain a larger fraction of the variation in species richness among taxa.

With the information gathered to date, most ecologists believe that insect pollination has relevantly contributed to the diversification of some of the most speciose families (e.g., orchids), but we need much more data to determine its role on the massive mid-Cretaceous angiosperm diversification (Crane et al. 1995). We also know that insect pollination was already present when angiosperms originated (early Cretaceous, about 130 and 90 million years ago), as shown by Jurassic fossils of Bennetitales, the closest fossil group to angiosperms, which suggest the presence of a plant–pollinator interaction (Crepet et al. 1991). The androecium in early angiosperms probably served as the only reward for insects, as it occurred in the Bennetitales, and flowers were presumably small, apetalous, with few structures, either asymmetric or cyclically arranged (Crepet et al. 1991). Such early flowers co-occurred with a greater variety of insects than previously thought. According to these authors, the idea that Coleoptera were the main early pollinators needs to be reviewed, as other insect groups (e.g., pollen-chewing flies and micropterygid moths) were also present at that time. Nectaries appeared later, and were present in many of the late Cretaceous rosids, when a rapid radiation of bees took place (Crepet et al. 1991). The Cretacean radiation of major pollinator groups such as bees, pollen wasps, brachyceran flies, and butterflies coincided with the appearance of entomophilous syndromes in Cretacean flowers (Grimaldi 1999). Similarly, the radiation of Lepidoptera coincides with patterns of accelerating radiation in angiosperms (Pellmyr 1992). However, there is still very little knowledge about the causes and effects of these events, and even though there exists a reliable phylogeny and information on pollinator function for Lepidoptera, Pellmyr (1992) found no evidence that the evolution of this group of insects caused radiation in flowering plants.

CONCLUDING REMARKS

To understand the evolutionary dynamics of plant reproduction, a unified approach between the study of (1) mating systems and (2) pollination biology is crucial. The study of factors that influence pollen transfer (floral morphology, timing of self- vs. outcross pollination, pollinator's effectiveness, etc.) gives valuable information to faithfully model the pollen movement within and among flowers, which reflect the outcome of various plant–pollinator interactions. Such modeling certainly helps to understand and compare the evolutionary dynamics in different pollination systems. The growing DNA sequence data help to assess the consequences of selfing and outcrossing on genetic variability within and between populations, and we need more information on how often evolutionary stability of mixed mating systems occurs in nature. More studies designed to detect natural and sexual selection on floral traits and display will allow determining the frequency of occurrence of floral adaptations to pollinators. In addition, further experimental studies of the relationships between flower traits, environmental variables, and mating systems are needed if we are to discern if a character is a cause or an evolutionary consequence of the breeding system. More data on whole pollinator assemblages visiting a plant species, and their spatio-temporal variation in composition and effectiveness, also permit evaluation of the degree of plant specialization to pollinators and knowledge of the extent to which biotic pollination may influence angiosperm diversification. The latter will be assessed as more reliable phylogenetic trees of plants and pollinators are built with both morphological and genetic data, and as more information on the other factors affecting diversification are available.

ACKNOWLEDGMENTS

We especially thank Mary F. Willson for her comments on the first edition of the chapter, Marta Macíes for efficiently supplying some of the needed references, and Miguel Verdú for his comments on the new version of the chapter. We also thank Rodolfo and Patrik, for their patience during the period we have been working on this.

REFERENCES

Abrahamson, W.G., 1980. Demography and vegetative reproduction. In: O.T. Solbrig, ed. Demography and Evolution in Plant Populations. Blackwell Scientific Publications, London, England, pp. 89–106.

Agren, J. and D.W. Schemske, 1995. Sex allocation in the monoecious herb *Begonia semiovata*. Evolution 49: 121–130.

Agren, J. and M.F. Willson, 1991. Gender variation and sexual differences in reproductive characters and seed production in gynodioecious *Geranium maculatum*. American Journal of Botany 78: 470–480.

Aigner, P.A., 2006. The evolution of specialized floral phenotypes in a fine-grained pollination environment. In N.M. Waser and J. Ollerton, eds. Plant–Pollinator Interactions: From Specialization to Generalization. The University of Chicago Press, Chicago, pp. 23–46.

Aizen, M., K.B. Searcy, and D.L. Mulcahy, 1990. Among- and within-flower comparisons of pollen tube growth following self-and cross-pollinations in *Dianthus chinensis* (Caryophyllaceae). American Journal of Botany 77: 671–676.

Anderson, G.J. and D.E. Symon, 1989. Functional dioecy and andromonoecy in *Solanum*. Evolution 43: 204–219.

Armbruster, W.S., 2002. Can indirect selection and genetic context contribute to trait diversification? A transition-probability study of blossom-colour evolution in two genera. Journal of Evolutionary Ecology 15: 468–486.

Armstrong, J.E. and A.K. Irvine, 1989. Flowering, sex ratios, pollen-ovule ratios, fruit set, and reproductive effort of a dioecious tree, *Myristica insipida* (Myristicaceae), in two different rain forest communities. American Journal of Botany 76: 74–85.

Arroyo, M.T.K. and P. Uslar, 1993. Breeding systems in a temperate Mediterranean-type climate montane sclerophyllous forest in central Chile. Botanical Journal of the Linnean Society 111: 83–102.

Ashman, T.-L., 2002. The role of herbivores in the evolution of separate sexes from hermaphroditism. Ecology 83: 1175–1184.

Ashman, T.L. and M.T. Morgan, 2004. Explaining phenotypic selection on plant attractive characters: male function, gender balance or ecological context? Proceedings of the Royal Society B-Biological Sciences 271: 553–559.

Ashman, T.-L., T.M. Knight, J.A. Steets, P. Amarasekare, M. Burd, D.R. Campbell, M.R. Dudash, M.O. Johnston, S.J. Mazer, R.J. Mitchell, M.T. Morgan, and W. Wilson, 2004. Pollen limitation of plant reproduction: ecological and evolutionary causes and consequences. Ecology 85: 2408–2421.

Asker, S., 1980. Gametophytic apomixis: elements and genetic regulation. Hereditas 93: 277–293.

Asker, S. and L. Jerling, 1992. Apomixis in Plants. CRC Press, Boca Raton.

Bailey, M.F., L.F. Delph, and C.A. Lively, 2003. Modeling gynodioecy: Novel scenarios for maintaining polymorphism. American Naturalist 161: 762–776.

Baker, H.G., 1963. Evolutionary mechanisms in pollination biology. Science 139: 877–883.

Barrett, S.C.H., 1988. The evolution, maintenance, and loss of self-incompatibility systems. In: J. Lovett Doust and L. Lovett Doust, eds. Plant Reproductive Ecology. Patterns and Strategies. Oxford University Press, Oxford, England.

Barrett, S.C.H., 1990. The evolution and adaptive significance of heterostyly. Trends in Ecology and Evolution 5: 144–148.

Barrett, S.C.H., 1992. Evolution and Function of Heterostyly. Springer-Verlag, Berlin, Germany.

Barrett, S.C.H., 2002. The evolution of plant sexual diversity. Nature Reviews Genetics 3: 274–284.
Barrett, S.C.H. and D. Charlesworth, 1991. Effects of a change in the level of inbreeding on the genetic load. Nature 352: 522–524.
Barrett, S.C.H. and C.G. Eckert, 1990. Variation and evolution of plant mating systems. In: S. Kawano, ed. Biological Approaches and Evolutionary Trends in Plants. Academic Press, NY, pp. 229–254.
Barrett, S.C.H. and L.D. Harder, 1996. Ecology and evolution of plant mating. Trends in Ecology and Evolution 11: 73–82.
Barrett, S.C.H., L.D. Harder, and A.C. Worley, 1996. The comparative biology of pollination and mating in flowering plants. Philosophical Transactions of the Royal Society of London (Series B) 351: 1271–1280.
Bascompte, J., P. Jordano, C.J. Melián, and J.M. Olesen, 2003. The nested assembly of plant-animal mutualistic networks. Proceedings of the National Academy of Sciences (USA) 100: 9383–9387.
Bateman, A.J., 1956. Cryptic self-incompatibility in the wallflower. I. Theory. Heredity 10: 257–261.
Bawa, K.S., 1980. Evolution of dioecy in flowering plants. Annual Review of Ecology and Systematics 11: 15–39.
Bawa, K.S., 1995. Pollination, seed dispersal and diversification of angiosperms. Trends in Ecology and Evolution 10: 311–312.
Beach, J.H. and K.S. Bawa, 1980. Role of pollinators in the evolution of dioecy from distyly. Evolution 34: 1138–1142.
Bengtsson, B.O. and A. Ceplitis, 2000. The balance between sexual and asexual reproduction in plants living in variable environments. Journal of Evolutionary Biology 13: 415–422.
Bernasconi, G., 2003. Seed paternity in flowering plants: an evolutionary perspective. Perspectives in Plant Ecology, Evolution and Systematics 6: 149–158.
Berry, P.E., H. Tobe, and J.A. Gómez, 1991. Agamospermy and the loss of distyly in *Erythroxylum undulatum* (Erythroxilaceae) from northern Venezuela. American Journal of Botany 78: 595–600.
Bertin, R.I., 1982. The evolution and maintenance of andromonoecy. Evolutionary Theory 6: 25–32.
Bertin, R.I., 1988. Paternity in plants. In: J. Lovett Doust and L. Lovett Doust, eds. Plant Reproductive Ecology. Patterns and Strategies. Oxford University Press, Oxford, England, pp. 30–59.
Bertin, R.I. and G.M. Gwisc, 2002. Floral sex ratios and gynomonoecy in Solidago (Asteraceae). Biological Journal of the Linnean Society 77: 413–422.
Bertin, R.I. and C.M. Newman, 1993. Dichogamy in angiosperms. Botanical Review 59: 112–152.
Bhat, V., K.K. Dwivedi, J.P. Khurana, and S.K. Sopory, 2005. Apomixis: An enigma with potential applications. Current Science 89: 1879–1893.
Bino, R.J. and A.D.J. Meeuse, 1981. Entomophily in dioecious species of *Ephedra*: a preliminary report. Acta Botanica Neerlandica 30: 151–153.
Bolmgren, K., O. Eriksson, and P. Linder, 2003. Contrasting flowering phenology and species richness in abiotically and biotically pollinated angiosperms. Evolution 57: 2001–2011.
Bookman, S.S., 1984. Evidence for selective fruit production in *Asclepias*. Evolution 38: 72–86.
Brown, E. and S. Kephart, 1999. Variability in pollen load: implications for reproduction and seedling vigor in a rare plant, *Silene douglasii* var. *oraria*. International Journal of Plant Sciences 160: 1145–1152.
Brown, B.J., J.M. Randall, and S.A. Graham, 2002. Competition for pollination between an invasive species (*Purple loosestrife*) and a native congener. Ecology 83: 2328–2336.
Brown, G.R., G.P. Gill, R.J. Kuntz, C.H. Langley, and D.B. Neale, 2004. Nucleotide diversity and linkage disequilibrium in Loblolly pine. Proceedings of the National Academy of Sciences (USA) 42: 15255–15260.
Broyles, S.B. and R. Wyatt, 1997. The pollen donation hypothesis revisited: a response to Queller. American Naturalist 149: 595–599.
Brunet, J. and H.R. Sweet, 2006. Impact of insect pollinator group and floral display size on outcrossing rate. Evolution 60: 234–246.
Burczyk, J., T. Adams, G.F. Moran, and R. Griffin, 2002. Complex patterns of mating revealed in a *Eucalyptus regnans* seed orchard using allozyme markers and the neighbourhood model. Molecular Ecology 11: 2379–2391.

Burd, M., 1995. Ovule packaging in stochastic pollination and fertilization environments. Evolution 49: 100–109.

Burd, M., 2004. Offspring quality in relation to excess flowers in *Pultenaea gunni* (Fabaceae). Evolution 58: 2371–2376.

Burd, M. and T.F.H. Allen, 1988. Sexual allocation strategy in wind-pollinated plants. Evolution 42: 403–407.

Burd, M. and H.S. Callahan, 2000. What does the male function hypothesis claim? Journal of Evolutionary Biology 13: 735–742.

Burger, W.C., 1981. Why are there so many kinds of flowering plants? Bioscience 31: 572–581.

Byers, D.L. and D.M. Waller, 1999. Do plant populations purge their genetic load? Effects of population size and mating history on inbreeding depression. Annual Reviews of Ecology and Systematics 30: 479–513.

Calvino, A. and C.C. Garcia, 2005. Sexual dimorphism and gynoecium size variation in the andromonoecious shrub *Caesalpinia gilliesii*. Plant Biology 7: 195–202.

Campbell, D.R., 1985. Pollen and gene dispersal: the influences of competition for pollination. Evolution 39: 419–431.

Campbell, D.R. and N.M. Waser, 1987. The evolution of plant mating systems: multilocus simulations of pollen dispersal. American Naturalist 129: 593–609.

Cane, J.H., R.L. Minckley, L. Kervin, and T.H. Roulston, 2005. Temporally persistent patterns of incidence and abundance in a pollinator guild at annual and decadal scales: the bees of *Larrea tridentata*. Biological Journal of the Linnean Society 85: 319–329.

Caruso, C.M., 2000. Competition for pollination influences selection on floral traits of *Ipomopsis aggregata*. Evolution 54: 1546–1557.

Caruso, C.M. and M. Alfaro, 2000. Interspecific pollen transfer as a mechanism of competition: effect of *Castilleja linariaefolia* pollen on seed set of *Ipomopsis aggregata*. Canadian Journal of Botany 78: 600–606.

Caruso, C.M., D.L.D. Remington, and K.E. Ostergren, 2005. Variation in resource limitation of plant reproduction influences natural selection on floral traits of *Asclepias syriaca*. Oecologia 146: 68–76.

Casper, B.B., 1985. Self-compatibility in distylous *Cryptantha flava* (Boraginaceae). New Phytologist 99: 149–154.

Castellanos, M.C., P. Wilson, and J.D. Thomson, 2003. Pollen transfer by hummingbirds and bumblebees, and the divergence of pollination modes in *Penstemon*. Evolution 57: 2742–2752.

Castric, V. and X. Vekemans, 2004. Plant self-incompatibility in natural populations: a critical assessment of recent theoretical and empirical advances. Molecular Ecology 13: 2873–2889.

Ceccherelli, G. and F. Cinelli, 1999. The role of vegetative fragmentation in dispersal of the invasive alga *Caulerpa taxifolia* in the Mediterranean. Marine Ecology Progress 182: 299–303.

Charlesworth, D., 1984. Androdioecy and the evolution of dioecy. Biological Journal of the Linnean Society 23: 333–348.

Charlesworth, D., 2001. Evolution: An exception that proves the rule. Current Biology 11: 13–15.

Charlesworth, B. and D. Charlesworth, 1978. A model for the evolution of dioecy and gynodioecy. American Naturalist 112: 975–997.

Charlesworth, B. and D. Charlesworth, 1979. The maintenance and breakdown of heterostyly. American Naturalist 114: 499–513.

Charlesworth, D. and B. Charlesworth, 1987. Inbreeding depression and its evolutionary consequences. Annual Review of Ecology and Systematics 18: 237–268.

Charlesworth, B. and D. Charlesworth, 1999. The genetic basis of inbreeding depression. Genetic Research 74: 329–340.

Charlesworth, D. and M.T. Morgan, 1991. Allocation of resources to sex functions in flowering plants. Philosophical Transactions of the Royal Society of London (Series B) 332: 91–102.

Charlesworth, D., X. Vekemans, V. Castric, and S. Glemin, 2005. Plant self-incompatibility systems: a molecular evolutionary perspective. New Phytologist 168: 61–69.

Cheptou, P.O., 2004. Allele effect and self-fertilization in hermaphrodites: reproductive assurance in demographically stable populations. Evolution 58: 2613–2621.

Colling, G., C. Reckinger, and D. Matthies, 2004. Effects of pollen quantity and quality on reproduction and offspring vigor in the rare plant *Scorzonera humilis* (Asteraceae). American Journal of Botany 91: 1774–1782.

Conner, J.K., R. Davis, and S. Rush, 1995. The effect of wild radish floral morphology on pollination efficiency by four taxa of pollinators. Oecologia 104: 234–245.
Conner, J.K., S. Rush, S. Kercher, and P. Jennetten, 1996. Measurements of natural selection on floral traits in wild radish (*Raphanus raphanistrum*). II. Selection through lifetime male and total fitness. Evolution 50: 1137–1146.
Cornelissen, T. and P. Stiling, 2005. Sex-biased herbivory: a meta-analysis of the effects of gender on plant-herbivore interactions. Oikos 111: 488–500.
Costich, D.E., 1995. Gender specialization across a climatic gradient: experimental comparison of monoecious and dioecious *Ecballium*. Ecology 76: 1036–1050.
Crane, P.R., E.M. Friis, and K.R. Pedersen, 1995. The origin and early diversification of angiosperms. Nature 374: 27–33.
Crepet, W.L., 1983. The role of insect pollination in the evolution of the angiosperms. In: L. Real, ed. Pollination and Biology. Academic Press, OL, pp. 29–50.
Crepet, W.L., E.M. Friis, and K.C. Nixon, 1991. Fossil evidence for the evolution of biotic pollination. Philosophical Transactions of the Royal Society of London (Series B) 333: 187–195.
Crnokrak, P. and S.C.H. Barrett, 2002. Purging the genetic load: a review of the experimental evidence. Evolution 56: 2347–2358.
Cronquist, A., 1981. An Integrated System of Classification of Flowering Plants. Columbia University Press, NY.
Cuevas, J. and V.S. Polito, 2004. The role of staminate flowers in the breeding system *of Olea europaea* (Oleaceae): an andromonoecious, wind-pollinated taxon. Annals of Botany 93: 547–553.
Damgaard, C. and R.J. Abbott, 1995. Positive correlations between selfing rate and pollen-ovule ratio within plant populations. Evolution 49: 214–217.
Darlington, D.D., 1939. The evolution of genetic systems. Cambridge University Press, Cambridge, England.
Darwin, C.R., 1876. The Effects of Cross and Self Fertilization in the Vegetable Kingdom. John Murray, London, England.
Davies, T.J., T.G. Barraclough, M.W. Chase, P.S. Soltis, D.E. Soltis, and V. Savolainen, 2004. Darwin's abominable mystery: Insights from a supertree of the angiosperms. Proceedings of the National Academy of Sciences (USA) 101: 1904–1909.
Davis, S.L. and I.F. Delph, 2005. Prior selfing and gynomonoecy in *Silene noctiflora* L. (Caryophyllaceae): opportunities for enhanced outcrossing and reproductive assurance. International Journal of Plant Sciences 166: 475–480.
de Jong, J.H. and S.A.H. Geritz, 2002. The role of geitonogamy in the gradual evolution towards dioecy in cosexual plants. Selection 2: 133–146.
de Jong, J.H. and P. Klinkhammer, 2005. Evolutionary Ecology of Plant Reproductive Strategies. Cambridge University Press, Cambridge.
de Jong, T.J., N.M. Waser, and P.G.L. Klinkhamer, 1993. Geitonogamy: the neglected side of selfing. Trends in Ecology and Evolution 8: 321–325.
de Jong, T.J., P.G.L. Klinkhamer, and M.C.J. Radermarker, 1999. How geitonogamous selfing affects sex allocation in hermaphrodite plants? Journal of Evolutionary Ecology 14: 213–231.
Delph, L.F. and C.M. Lively, 1992. Pollinator visitation, floral display, and nectar production of the sexual morphs of a gynodioecious shrub. Oikos 63: 161–170.
Diaz, L. and A.A. Cocucci, 2003. Functional gynodioecy in *Opuntia quimilo* (Cactaceae), a tree cactus pollinated by bees and hummingbirds. Plant Biology 5: 531–539.
Diggle, P.K., 1992. Development and the evolution of plant reproductive characters. In: R. Wyatt, ed. Ecology and Evolution of Plant Reproduction: New Approaches. Chapman and Hall, NY.
Diggle, P.K., 1993. Developmental plasticity, genetic variation, and the evolution of andromonoecy in *Solanum hirtum* (Solanaceae). American Journal of Botany 80: 967–973.
Dodd, M.E., J. Silvertown, and M.W. Chase, 1999. Phylogenetic analysis of trait evolution and species diversity variation among angiosperm families. Evolution 53: 732–744.
Dorken, M.E. and S.C.H. Barrett, 2004. Sex determination and the evolution of dioecy from monoecy in *Sagittaria latifolia* (Alismataceae). Proceedings of the Royal Society B-Biological Sciences 271: 213–219.

Doyle, J.A., 1994. Origin of the angiosperm flower: a phylogenetic perspective. Plant Systematics and Evolution Suppl. 8: 7–29.
Doyle, J.A. and M.J. Donoghue, 1993. Phylogenies and angiosperm diversification. Paleobiology 19: 141–167.
Dudash, M.R. and C.B. Fenster, 1997. Multiyear study of pollen limitation and cost of reproduction in the iteroparous *Silene virginica*. Ecology 78: 484–493.
Dupont, Y.L., 2002. Evolution of apomixis as a strategy of colonization in the dioecious species *Lindera glauca* (Lauraceae). Population Ecology 44: 293–297.
Eckert, C.G., 2002a. Effect of geographical variation in pollinator fauna on the mating system of *Decodon verticillatus* (Lythraceae). International Journal of Plant Sciences 163: 123–132.
Eckert, C.G., 2002b. The loss of sex in clonal plants. Evolutionary Ecology 15: 501–520.
Eckhart, V.M., 1992. The genetics of gender and the effects of gender on floral characters in gynodioecious *Phacelia linearis* (Hydrophyllaceae). American Journal of Botany 79: 792–800.
Ehrlén, J., 1991. Why do plants produce surplus flowers? A reserve-ovary model. American Naturalist 138: 918–933.
Ellstrand, N.C., R. Ornduff, and J.M. Clegg, 1990. Genetic structure of the Australian Cycad, *Macrozamia communis* (Zamiaceae). American Journal of Botany 77: 677–681.
Emms, S.K., 1993. Andromonoecy in *Zigadenus paniculatus* (Liliaceae): spatial and temporal patterns of sex allocation. American Journal of Botany 80: 914–923.
Emms, S.K., 1996. Temporal patterns of seed set and decelerating fitness returns on female allocation in *Zigadenus paniculatus* (Liliaceae), an andromonoecious lily. American Journal of Botany 83: 304–315.
Eppley, S.M., 2005. Spatial segregation of the sexes and nutrients affect reproductive success in a dioecious wind-pollinated grass. Plant Ecology 181: 179–190.
Eriksson, O. and B. Bremer, 1992. Pollination systems, dispersal modes, life forms, and diversification rates in angiosperm families. Evolution 46: 258–266.
Faegri, K. and L. van der Pijl, 1979. The Principles of Pollination Ecology, 3rd rev. edn. Oxford University Press, Oxford, England.
Fausto, J.A., V.M. Eckhart, and M.A. Geber, 2001. Reproductive assurance and the evolutionary ecology of self-pollination in *Clarkia xantiana* (Onagraceae). American Journal of Botany 88: 1794–1800.
Feinsinger, P. and H.M. III. Tiebout, 1991. Competition among plants sharing hummingbird pollinators: laboratory experiments on a mechanism. Ecology 72: 1946–1952.
Fenster, C.B. and M.R. Dudash, 2001. Spatiotemporal variation in the role of hummingbirds as pollinators of *Silene virginica*. Ecology 82: 844–851.
Fenster, C.B., W.S. Armbruster, P. Wilson, M.R. Dudash, and J.D. Thomson, 2004. Pollination syndromes and floral specialization. Annual Reviews of Ecology, Evolution and Systematics 35: 375–403.
Finer, M.S. and M.T. Morgan, 2003. Effects of natural rates of geitonogamy fruit set in *Asclepias speciosa* (Apocynaceae): evidence favoring the plant's dilemma. American Journal of Botany 90: 1746–1750.
Fishbein, M. and D.L. Venable, 1996. Evolution of inflorescence design: theory and data. Evolution 50: 2165–2177.
Flores, S. and D.W. Schemske, 1984. Dioecy and monoecy in the flora of Puerto Rico and the Virgin Islands: ecological correlates. Biotropica 16: 132–139.
Fox, J.F., 1985. Incidence of dioecy in relation to growth form, pollination and dispersal. Oecologia 67: 244–249.
Fox, J.F., 1993. Size and sex allocation in monoecious woody plants. Oecologia 94: 110–113.
Franklin-Tong, V.E. and F.C.H. Franklin, 2003. Gametophytic self-incompatibilityt inhibits pollen tube growth using different mechanisms. Trends in Plant Science 8: 598–605.
Freeman, D.C., E.D. McArthur, K. Harper, and A.C. Blauer, 1981. Influence of environment on the floral sex ratio of monoecious plants. Evolution 35: 194–197.
Freeman, D.C., J.L. Doust, A. El-Keblawy, K.J. Miglia, and E.D. McArthur, 1997. Sexual specialization and inbreeding avoidance in the evolution of dioecy. The Botanical Review 63: 65–94.

Friis, E.M., K. Raunsgaard Pedersen, and P.R. Crane, 2005. When earth started blooming: insights from the fossil record. Current Opinion in Plant Biology 8: 5–12.

Galen, C., 1989. Measuring pollinator-mediated selection on morphometric floral traits: bumblebees and the alpine sky pilot, *Polemonium viscosum*. Evolution 43: 882–890.

Galloway, L.F. and C.B. Fenster, 2000. Population differentiation in an annual legume: Local adaptation. Evolution 54: 1173–1181.

García, E.C., J. Marquez, C.A. Dominguez, and F. Molina-Freaner, 2005. Evidence of gynodioecy in Kallstroemia grandiflora (Zygophyllaceae): Microsporogenesis in hermaphrodite and female plants and lack of reproductive compensation. International Journal of Plant Sciences 166: 481–491.

Gardner, S.N. and M. Mangel, 1999. Modeling investment in seeds, clonal offspring, and translocation in a clonal plant. Ecology 80: 1202–1220.

Gibson, J.P. and N.T. Wheelwright, 1996. Mating system dynamics of *Ocotea tenera* (Lauraceae), a gynodioecious tropical tree. American Journal of Botany 83: 890–894.

Gil, L., P. Fuentes-Utrilla, M. Álvaro Soto, T. Cervera, and C. Collada, 2004. English elm is a 2,000-year-old Roman clone. Nature 431: 1053.

Givnish, T.J., 1980. Ecological constrains on the evolution of breeding systems in seed plants: dioecy and dispersal in gymnosperms. Evolution 34: 959–972.

Gleiser, G. and M. Verdú, 2005. Repeated evolution of dioecy from androdioecy in Acer. New Phytologist 165: 633–640.

Goldman, D.A. and M.F. Willson, 1986. Sex allocation in functionally hermaphroditic plants: a review and critique. The Botanical Review 52: 157–194.

Golonka, A.M., A.K. Sakai, and S.G. Weller, 2005. Wind pollination, sexual dimorphism, and changes in floral traits of *Schiendea* (Carophyllaceae). American Journal of Botany 1492–1502.

Gómez, J.M. and R. Zamora, 1996. Wind pollination in high-mountain populations of *Hormathophylla spinosa* (Cruciferae). American Journal of Botany 83: 580–585.

Gómez, J.M. and R. Zamora, 1999. Generalization vs. specialization in the pollination of *Hormathophylla spinosa* (Cruciferae). Ecology 80: 796–805.

Gómez, J.M. and R. Zamora, 2000. Spatial variation in the selective scenarios of *Hormathophylla spinosa* (Cruciferae). American Naturalist 155: 657–668.

Gómez, J.M. and R. Zamora, 2006. Ecological factors that promote the evolution of generalization in pollination systems. In: N. Waser and J. Ollerton, eds. Plant–Pollinator Interactions: From Specialization to Generalization. The University of Chicago Press, Chicago, pp. 145–166.

Goodwillie, C., S. Kalisz, and C.G. Eckert, 2005. The evolutionary engima of mixed mating systems in plants: Occurrence, theoretical explanations, and empirical evidence. Annual Reviews of Ecology, Evolution and Systematics 36: 47–79.

Gorelick, R., 2001. Did insect pollination cause increased seed plant diversity? Biological Journal of the Linnean Society 74: 407–427.

Gould, S.J. and E.S. Vrba, 1982. Exaptation—a missing term in the science of form. Paleobiology 8: 4–15.

Grant, V., 1949. Pollination systems as isolating mechanisms in angiosperms. Evolution 3: 82–97.

Grant, V., 1994. Modes and origins of mechanical and ethological isolation in angiosperms. Proceedings of the National Academy of Sciences 91: 3–10.

Grant, V. and K.A. Grant, 1965. Flower Pollination in the Phlox Family. Columbia University Press, NY.

Grimaldi, D., 1999. The co-radiation of pollinating insects and angiosperms in the Cretaceous. Annals of the Missouri Botanical Garden 86: 373–406.

Gross, C.L., 2005. A comparison of the sexual systems in the trees from the Australian tropics with other tropical biomes—more monoecy but why? American Journal of Botany 92: 907–919.

Gustafsson, A., 1946. Apomixis in higher plants. I–III. Lunds Univ. Arsskr. 42: 1–67.

Haig, D. and M. Westoby, 1989. Selective forces in the emergence of the seed habit. Biological Journal of the Linnean Society 38: 215–238.

Hansen, T.F., W.S. Armbruster, and L. Antonsen, 2000. Comparative analysis of character displacement and spatial adaptations as illustrated by the evolution of Dalechampia blossoms. American Naturalist 157: 245–261.
Harder, L.D. and S.C.H. Barrett, 1996. Pollen dispersal and mating patterns in animal-pollinated plants. In D.G. Lloyd and S.C.H. Barrett, eds. Floral Biology. Studies on Floral Evolution in Animal-Pollinated Plants. Chapman and Hall, NY, pp. 140–190.
Harder, L.D. and S.D. Johnson, 2005. Adaptive plasticity of floral display size in animal-pollinated plants. Proceedings of the Royal Society B-Biological Sciences 272: 2651–2657.
Harder, L.D., M.B. Cruzan, and J.D. Thomson, 1993. Unilateral incompatibility and the effects of interspecific pollination for *Erythronium americanum* and *Erythronium albidum* (Liliaceae). Canadian Journal of Botany 71: 353–358.
Harper, J.L., 1977. Population Biology of Plants. Academic Press, London, England.
Heilbuth, J.C., K.L. Ilves, and S.P. Otto, 2001. The consequences of dioecy for seed dispersal: Modeling the seed-shadow handicap. Evolution 55: 880–888.
Herlihy, C.H. and C.G. Eckert, 2004. Experimental dissection of inbreeding and its adaptive significance in a flowering plant, *Aquilegia canadensis* (Ranunculaceae). Evolution 58: 2693–2703.
Herrera, C.M., 1988. Variation in mutualisms: the spatio-temporal mosaic of a pollinator assemblage. Biological Journal of the Linnean Society 35: 95–125.
Herrera, C.M., 1993. Selection on floral morphology and environmental determinants of fecundity in a hawk moth-pollinated violet. Ecological Monographs 63: 251–275.
Herrera, C.M., 1995. Microclimate and individual variation in pollinators: flowering plants are more than their flowers. Ecology 76: 1516–1524.
Herrera, C.M., 1996. Floral traits and plant adaptation to insect pollinators: a devil's advocate approach. In D.G. Lloyd and S.C.H. Barrett, eds. Floral Biology. Studies on Floral Evolution in Animal-Pollinated Plants. Chapman and Hall, NY, pp. 65–87.
Herrera, C.M., 2005. Plant generalization on pollinators: species property or local phenomenon? American Journal of Botany 92: 13–20.
Herrera, C.M., M. Medrano, P.J. Rey, A.M. Sánchez-Lafuente, M.B. Garcia, J. Guitián, and A.J. Manzaneda, 2002. Interaction of pollinators and herbivores on plant fitness suggests a pathway for correlated evolution of mutualism- and antagonism-related traits. Proceedings of the National Academy of Sciences (USA) 99: 16823–16828.
Heywood, V.H., 1993. Flowering Plants of the World. B T Batsford, Oxford.
Hingston, A.B. and P.B. McQuillan, 2000. Are pollination syndromes useful predictors of floral visitors in Tasmania? Austral Ecology 25: 600–609.
Hiscock, S.J. and S.M. McInnis, 2003. Pollen recognition and rejection during the sporophytic self-incompatibility response: *Brassica* and beyond. Trends in Plant Science 8: 606–613.
Holsinger, K.E., 1986. Dispersal and plant mating systems: the evolution of self-fertilization in subdivided populations. Evolution 40: 405–413.
Holsinger, K.E., 1991. Inbreeding depression and the evolution of plant mating systems. Trends in Ecology and Evolution 6: 307–308.
Holsinger, K.E., 1996. Pollination biology and the evolution of mating systems in flowering plants. Evolutionary Biology 29: 107–149.
Huang, S.Q., 2003. Flower dimorphism and the maintenance of andromonoecy in *Sagittaria guyanensis* ssp. *lappula* (Alismataceae). New Phytologist 157: 357–364.
Humphrey, L.D. and D.A. Pyke, 1998. Demographic and growth responses of a guerrilla and a phalanx perennial grass in competitive mixtures. Journal of Ecology 86: 854–865.
Husband, B.C. and D.W. Schemske, 1996. Evolution of the magnitude and timing of inbreeding depression in plants. Evolution 50: 54–70.
Irwin, R.E., 2006. The consequences of direct versus indirect species interactions to selection on traits: pollination and nectar robbing in *Ipomopsis aggregata*. American Naturalist 167: 315–328.
Jacobs, M.S. and M.J. Wade, 2003. A synthetic review of the theory of gynodioecy. American Naturalist 161: 837–851.
Jain, S.K., 1976. The evolution of inbreeding in plants. Annual Review of Ecology and Systematics 7: 469–495.

Jakobsson, A. and P. Dinnetz, 2005. Local adaptation and the effects of isolation and population size-the semelparous perennial *Carlina vulgaris* as a study case. Evolutionary Ecology 19: 449–466.
Janzen, D.H., 1971. Seed predation by animals. Annual Review of Ecology and Systematics 2: 465–492.
Janzen, D.H., 1977. What are dandelions and aphids? American Naturalist 111: 586–589.
Jarne, P. and D. Charlesworth, 1993. The evolution of the selfing rate in functionally hemaphrodite plants and animals. Annual Review of Ecology and Systematics 24: 441–466.
Johnson, S.D. and K.E. Steiner, 1997. Long-tongued fly pollination and evolution of floral spur length in the *Disa draconis* complex (Orchidaceae). Evolution 51: 45–53.
Johnson, S.D. and K.E. Steiner, 2000. Generalization versus specialization in plant pollination systems. Trends in Ecology and Evolution 15: 140–143.
Jordan, C.Y. and L.D. Harder, 2006. Manipulation of bee behavior by inflorescence architecture and its consequences for plant mating. American Naturalist 167: 496–509.
Jordano, P., 1991. Gender variation and expression of monoecy in *Juniperus phoenicea* (L.) (Cupressaceae). Botanical Gazette 152: 476–485.
Jordano, P., 1993. Pollination biology of *Prunus mahaleb* L.: deferred consequences of gender variation for fecundity and seed size. Biological Journal of the Linnean Society 50: 65–84.
Joshi, J., B. Schmid, M.C. Caldeira, P.G. Dimitrakopoulos, J. Good, R. Harris, A. Hector, K. Huss-Danell, A. Jumpponen, A. Minns, C.P.H. Mulder, J.S. Pereira, A. Prinz, M. Scherer-Lorenzen, A.C. Terry, A.Y. Troumbis, and J.H. Lawton, 2001. Local adaptation enhances performance of common plant species. Ecology Letters 4: 536–544.
Kalisz, S. and D.W. Vogler, 2003. Benefits of autonomous selfing under unpredictable pollinator environments. Ecology 84: 2928–2942.
Kalisz, S., D.W. Vogler, and K.M. Hanley, 2004. Context-dependent autonomous self-fertilization yields reproductive assurance and mixed mating. Nature 430: 884–887.
Karron, J.D., R.J. Mitchell, K.G. Holmquist, J.M. Bell, and B. Funk, 2004. The influence of floral display size on selfing rates in *Mimulus ringens*. Heredity 92: 242–248.
Kay, K.M. and D.W. Schemske, 2004. Geographic patterns in plant–pollinator mutualistic networks: comment. Ecology 85: 875–878.
Klinkhamer, P.G.L., T. de Jong, and H. Metz, 1997. Sex and size in cosexual plants. Trends in Ecology and Evolution 12: 260–265.
Knight, T.M., J.A. Steets, J.C. Vamosi, S.J. Mazer, M. Burd, D.R. Campbell, M.R. Dudash, M.O. Johnston, R.J. Mitchell, and T.-L. Ashman, 2005. Pollen limitation of plant reproduction: pattern and process. Annual Reviews of Ecology, Evolution and Systematics 36: 467–497.
Kohn, J.R., 1988. Why be female? Nature 335: 431–433.
Kohn, J.R. and S.C.H. Barrett, 1992. Experimental studies on the functional significance of heterostyly. Evolution 46: 43–55.
Lamborn, E. and J. Ollerton, 2000. Experimental assessment of the functional morphology of inflorescences of *Daucus carota* (Apiaceae): testing the 'fly catcher effect'. Functional Ecology 14: 445–454.
Law, R., R.E.D. Cook, and R.J. Manlove, 1983. The ecology of flower and bulbil production in *Polygonum viviparum*. Nordic Journal of Botany 3: 559–565.
Lázaro, A. and A. Traveset, 2005. Spatio-temporal variation in the pollination mode of *Buxus balearica* (Buxaceae), an ambophilous and selfing species: mainland-island comparison. Ecography 28: 640–652.
Liston, A., L.H. Rieseberg, and T.S. Elias, 1990. Functional androdioecy in the flowering plant *Datisca glomerata*. Nature 343: 641–642.
Lloyd, D.G., 1975. The maintenance of gynodioecy and androdioecy in Angiosperms. Genetica 45: 325–339.
Lloyd, D.G., 1979. Some reproductive factors affecting the selection of self-fertilization in plants. American Naturalist 113: 67–79.
Lloyd, D.G., 1982. Selection of combined versus separate sexes in seed plants. American Naturalist 120: 571–585.
Lloyd, D.G. and S.C.H. Barrett, 1996. Floral Biology. Studies on Floral Evolution in Animal-Pollinated Plants. Chapman and Hall, NY.

Lloyd, D.G. and D.J. Schoen, 1992. Self and cross fertilization in plants. I. Functional dimensions. International Journal of Plant Science 153: 358–369.

Lloyd, D.G. and C.J. Webb, 1992. The evolution of heterostyly. In: S.C.H. Barrett, ed. Evolution and Function of Heterostyly. Springer-Verlag, Berlin, Germany, pp. 151–178.

Lloyd, D.G. and J.M.A. Yates, 1982. Intersexual selection and the segregation of pollen and stigmas in hermaphrodite plants, exemplified by *Wahlenbergia albomarginata* (Campanulaceae). Evolution 36: 903–913.

Lord, E.M., 1981. Cleistogamy: a tool for the study of floral morphogenesis, function and evolution. The Botanical Review 47: 421–449.

Lovett Doust, J., 1981. Population dynamics and local specialization in a clonal perennial (*Ranunculus repens*). I. The dynamics of ramets in contrasting habitats. Journal of Ecology 69: 743–755.

Lovett Doust, J. and L. Lovett Doust, 1988. Plant Reproductive Ecology. Patterns and Strategies. Oxford University Press, NY.

Maad, J., 2000. Phenotypic selection in hawkmoth-pollinated *Platanthera bifolia*: targets and fitness surfaces. Evolution 54: 112–123.

Maad, J. and R. Alexandersson, 2004. Variable selection in *Platanthera bifolia* (Orchidaceae): phenotypic selection differed between sex functions in a drought year. Journal of Evolutionary Biology 17: 642–650.

Maheshwari, P., 1950. An Introduction to the Embryology of Angiosperms. McGraw-Hill, NY.

Maki, M., 1993. Outcrossing and fecundity advantage of females in gynodioecious *Chionographis japonica var. kurohimensis* (Liliaceae). American Journal of Botany 80: 629–634.

Marshall, D.L., M.W. Folson, C. Hatfield, and T. Bennett, 1996. Does interference competition among pollen grains occur in wild radish? Evolution 50: 1842–1848.

Masuda, M., T. Yahara, and M. Maki, 2001. An ESS model for the mixed production of cleistogamous and chasmogamous flowers in a facultative cleistogamous plant. Evolutionary Ecology Research 3: 429–439.

Mayer, S.S. and D. Charlesworth, 1991. Cryptic dioecy in flowering plants. Trends in Ecology and Evolution 6: 320–324.

Mayfield, M.M., N.M. Waser, and M.V. Price, 2001. Exploring the "most effective pollinator principle" with complex flowers: bumblebees and *Ipomopsis aggregata*. Annals of Botany 88: 591–596.

Mazer, S.J., 1992. Environmental and genetic sources of variation in floral traits and phenotypic gender in wild radish: consequences for natural selection. In: R. Wyatt, ed. Ecology and Evolution of Plant Reproduction. Chapman and Hall, NY, pp. 281–325.

Medel, R., C. Botto-Mahan, and M. Kalin-Arroyo, 2003. Pollinator-mediated selection on the nectar guide phenotype in the Andean Monkey Flower, *Mimulus luteus*. Ecology 84: 1721–1732.

Memmot, J., 1999. The structure of a plant–pollinator food web. Ecology Letters 2: 276–280.

Méndez, M. and A. Traveset, 2003. Sexual allocation in single-flowered hermaphroditic individuals in relation to plant and flower size. Oecologia 137: 69–75.

Midgley, J.J. and W.J. Bond, 1991. How important is biotic pollination and dispersal to the success of the angiosperms? Philosophical Transactions of the Royal Society of London 333: 209–216.

Minckley, R.L. and T.H. Roulston, 2006. Incidental mutualisms and pollen specialization among bees. In: N.M. Waser and J. Ollerton, eds. Plant–Pollinator Interactions. From Specialization to Generalization. The University of Chicago Press, Chicago, pp. 69–98.

Minckley, R.L., J.H. Cane, L. Kervin, and T.H. Roulston, 1999. Spatial predictability and resource specialization of bees (Hymenoptera: Apoidea) at a superabundant, widespread resource. Biological Journal of the Linnean Society 67: 119–147.

Morgan, M.T. and D.J. Schoen, 1997. The role of theory in an emerging new plant reproductive biology. Trends in Ecology and Evolution 12: 231–234.

Morgan, M.T. and G.W. Wilson, 2005. Self-fertilization and the escape from pollen limitation in variable pollination environments. Evolution 59: 1143–1148.

Morgan, M.T., W.G. Wilson, and T.M. Knight, 2005. Plant population dynamics, pollinator foraging, and the selection of self-fertilization. American Naturalist 166: 169–183.

Mulcahy, D.L., P.S. Curtis, and A.A. Snow, 1983. Pollen competition in a natural population. In: C.E. Jones and R.J. Little, eds. Handbook of Experimental Pollination and Biology. Van Nostrand, NY, pp. 330–337.

Murcia, C., 1990. Effect of floral morphology and temperature on pollen receipt and removal in *Ipomoea trichocarpa*. Ecology 71: 1098–1109.
Murcia, C. and P. Feinsinger, 1996. Interspecific pollen loss by hummingbirds visiting flower mixtures: effects of floral architecture. Ecology 77: 550–560.
Mutikainen, P. and L.F. Delph, 1996. Effects of herbivory on male reproductive success in plants. Oikos 75: 353–358.
Narbona, E., Ortiz, P.L. and Artista, M., 2002. Functional Andromonoecy in *Euphorbia* (Euphorbiaceae). Annals of Botany 89: 571–577.
Nettancourt, D., 1977. Self-Incompatibility in Angiosperms. Springer-Verlag, Berlin, Germany.
Nygren, A., 1967. Apomixis in the angiosperms. Handb. der Pflanzenphys. 18: 551–596.
Ohasi, K. and T. Yahara, 2001. Behavioral responses of pollinators to variation in floral display size and their influence on the evolution of floral display. In: L. Chittka and J.D. Thompson, eds. Cognitive Ecology of Pollination: Animal Behavior and Floral Evolution. Cambridge University Press, Cambridge, pp. 274–296.
Olesen, J.M. and P. Jordano, 2002. Geographic patterns in plant–pollinator mutualistic networks. Ecology 83: 2416–2424.
Ollerton, J., 1996. Reconciling ecological processes with phylogenetic patterns: The apparent paradox of plant–pollinator systems. Journal of Ecology 84: 767–769.
Ollerton, J. and L. Cranmer, 2002. Latitudinal trends in plant–pollinator interactions: are tropical plants more specialised? Oikos 98: 340–350.
Ollerton, J. and S. Watts, 2000. Phenotype space and floral typology: Towards an objective assessment of pollination syndromes. Det Norske Videnskaps-Akademi 1. Matematisk Naturvidenskaplige Klasse; Skrifter, Ny Serie 39: 149–159.
Olmstead, R.G., 1986. Self-incompatibility in light of population structure and inbreeding. In D.L. Mulcahy et al., eds. Springer-Verlag, NY, p. 239.
Olsen, K.M., 1997. Pollination effectiveness and pollinator importance in a population of *Heterotheca subaxillaris* (Asteraceae). Oecologia 109: 114–121.
Pannell, J.R., 2001. A hypothesis for the evolution of androdioecy: the joint influence of reproductive assurance and local mate competition in a metapopulation. Evolutionary Ecology 14: 195–211.
Pannell, J.R., 2002. The evolution and maintenance of androdioecy. Annual Reviews of Ecology, Evolution and Systematics 33: 397–425.
Pannell, J.R. and M. Verdú, 2006. The evolution of gender specialization from dimorphic hermaphroditism: paths from heterodichogamy to gynodioecy and androdioecy. Evolution 60: 660–673.
Parachnowitsch, A.L. and E. Elle, 2004. Variation in sex allocation and male-female trade-offs in six populations of *Collinsia parviflora* (Scrophulariaceae s.l.). American Journal of Botany 91: 1200–1207.
Peck, J.R. and D. Waxman, 1999. What's wrong with a little sex? Journal of Evolutionary Biology 13: 63–69.
Pellmyr, O., 1992. Evolution of insect pollination and angiosperm diversification. Trends in Ecology and Evolution 7: 46–49.
Pellmyr, O., 2002. Pollination by animals. In: C.M. Herrera and O. Pellmyr, eds. Plant–Animal Interactions. Blackwell Science Ltd, Oxford, pp. 157–185.
Pellmyr, O. and J.N. Thompson, 1996. Sources of variation in pollinator contribution within a guild. The effects of plant and pollinator factors. Oecologia 107: 595–604.
Plaistow, S.J., R.A. Johnstone, N. Colegrave, and M. Spencer, 2004. Evolution of alternative mating tactics: conditional versus mixed strategies. Behavioral Ecology 15: 534–542.
Porcher, E., and R. Lande, 2005. The evolution of self-fertilization and inbreeding depression under pollen discounting and pollen limitation. Journal of Evolutionary Biology 18: 497–508.
Primack, R.B. and H. Kang, 1989. Measuring fitness and natural selection in wild plant populations. Annual Review of Ecology and Systematics 20: 367–396.
Primack, R.B. and D.G. Lloyd, 1980. Andromonoecy in the New Zealand montane shrub Manuke, *Leptospermum scoparium* (Myrtaceae). American Journal of Botany 67: 361–368.
Primack, R.B. and C. McCall, 1986. Gender variation in a red maple population (*Acer rubrum*; Aceraceae): a seven-year study of a "polygamodioecious" species. American Journal of Botany 73: 1239–1248.

Proctor, M. and P. Yeo, 1973. The Pollination of Flowers. Collins, London.
Queller, D.C., 1983. Sexual selection in a hermaphroditic plant. Nature 305: 706–707.
Ramulu, K.S., V.K. Sharma, T.N. Naumova, P. Dijkhuis, and M.M. van Lookeren Campagne, 1999. Apomixis for crop improvement. Protoplasma 208: 196–205.
Raven, P.H., 1977. A suggestion concerning the cretaceous rise to dominance of the angiosperms. Evolution 31: 451–452.
Regal, P.J., 1977. Ecology and evolution of flowering plant dominance. Science 196: 622–629.
Renner, S.S. and R.E. Ricklefs, 1995. Dioecy and its correlates in the flowering plants. American Journal of Botany 82: 596–606.
Richards, R.J., 1997. Plant Breeding Systems. Chapman and Hall, NY.
Richards, A.J., 2003. Apomixis in flowering plants: an overview. Philosophical Transactions of the Royal Society 358: 1085–1093.
Ricklefs, R.E. and S.S. Renner, 1994. Species richness within families of flowering plants. Evolution 48: 1619–1636.
Robertson, C., 1928. Flowers and Insects. Lists of Visitors of Four Hundred and Fifty-Three Flowers. Carlinville, IL.
Ross, M.D., 1978. The evolution of gynodioecy and subdioecy. Evolution 32: 174–188.
Ross, M.D., 1982. Five evolutionary pathways to dioecy. American Naturalist 119: 297–318.
Ross, M.D., 1990. Sexual asymmetry in hermaphroditic plants. Trends in Ecology and Evolution 5: 43–47.
Routley, M.B., R.I. Bertin, and B.C. Husband, 2004. Correlated evolution of dichogamy and self-incompatibility: A phylogenetic perspective. International Journal of Plant Sciences 165: 983–993.
Sakai, A. and S. Sakai, 2003. Size-dependent ESS sex allocation in wind-pollinated cosexual plants: Fecundity vs. stature effects. Journal of Theoretical Biology 222: 283–295.
Sakai, A.K. and S.G. Weller, 1991. Ecological aspects of sex expression in subdioecious *Schiedea globosa* (Caryophyllaceae). American Journal of Botany 78: 1280–1288.
Sakai, A.K., W.L. Wagner, D.M. Ferguson, and D.R. Herbst, 1995. Origins of dioecy in the Hawaiian flora. Ecology 76: 2517–2529.
Schemske, D.W. and C.C. Horvitz, 1989. Temporal variation in selection on a floral character. Evolution 43: 461–465.
Schlichting, C.D. and V.A. Delesalle, 1997. Stressing the differences between male and female functions in hermaphroditic plants. Trends in Ecology and Evolution 12: 51–52.
Schlichting, C.D. and B. Devlin, 1992. Pollen and ovule sources affect seed production of *Lobelia cardinalis* (Lobeliaceae). American Journal of Botany 79: 891–898.
Schmid, B. and J. Weiner, 1993. Plastic relationships between reproductive and vegetative mass in *Solidago altissima*. Evolution 47: 61–74.
Schmitt, J. and S.E. Gamble, 1990. The effect of distance from the parental site on offspring performance in *Impatiens capensis*: a test of the local adaptation hypothesis. Evolution 44: 2022–2030.
Schoen, D.J. and D.G. Lloyd, 1984. The selection of cleistogamy and heteromorphic diaspores. Botanical Journal of the Linnean Society 23: 303–322.
Schopfer, C.R., M.E. Nasrallah, and J.B. Nasrallah, 1999. The male determinant of self-incompatibility in *Brassica*. Science 286: 1697–1700.
Scofield, D.G. and S.T. Schultz, 2006. Mitosis, stature and evolution of plant mating systems: low-Phi and high-Phi plants. Proceedings of the Royal Society B-Biological Sciences 273: 275–282.
Seavey, S.R. and K.S. Bawa, 1986. Late-acting self-incompatibility in angiosperms. Botanical Review 52: 195–219.
Shore, J.S. and S.C.H. Barrett, 1984. The effect of pollination intensity and incompatible pollen on seed set in *Turnera ulmifolia* (Turneraceae). Canadian Journal of Botany 62: 1298–1303.
Smith, C.C., 1981. The facultative adjustment of sex ratio in lodgepole pine. American Naturalist 118: 291–305.
Smith, J.F., 2001. High species diversity in fleshy-fruited tropical understory plants. American Naturalist 157: 646–653.

Smouse, P.E., T.R. Meagher, and C.J. Kobak, 1999. Parentage analysis in *Chamaelirium luteum* (L.) Gray (Liliaceae): why do some males have higher reproductive contributions? Journal of Evolutionary Biology 12: 1069–1077.

Snow, A.A., 1986. Pollination dynamics of *Epilobium canum* (Onagraceae): consequences for gametophytic selection. American Journal of Botany 73: 139–157.

Snow, A.A. and P.O. Lewis, 1993. Reproductive traits and male fertility in plants: empirical approaches. Annual Review of Ecology and Systematics 24: 331–351.

Snow, A.A. and T.P. Spira, 1996. Pollen-tube competition and male fitness in *Hibiscus moscheutos*. Evolution 50: 1866–1870.

Sohn, J.J. and D. Policansky, 1977. The costs of reproduction in the mayapple, *Podophyllum peltatum* (Berberidaceae). Ecology 58: 1366–1374.

Solomon, B.P., 1985. Environmentally influenced changes in sex expression in an andromonoecious plant. Ecology 66: 1321–1332.

Solomon, B.P., 1989. Size-dependent sex ratios in the monoecious, wind-pollinated annual, *Xanthium strumarium*. American Midland Naturalist 121: 209–218.

Spalik, K., 1991. On evolution of andromonoecy and "overproduction" of flowers: a resource allocation model. Biological Journal of the Linnean Society 42: 325–336.

Sprengel, C.K., 1793. Discovery of the secret nature in the structure and fertilization of flowers. [Translated from German in 1996 by Peter Haase.] In: D.G. Lloyd and S.C.H. Barrett, eds. Floral Biology. Studies on Floral Evolution in Animal-Pollinated Plants. Chapman and Hall, NY, pp. 3–43.

Stebbins, G.L., 1970. Adaptive radiation of reproductive characteristics in angiosperms, I: pollination mechanisms. Annual Review of Ecology and Systematics 1: 307–326.

Stebbins, G.L., 1981. Why are there so many species of flowering plants? Bioscience 31: 573–577.

Steinbachs, J.E. and K.E. Holsinger, 2002. S-RNase-mediated gametophytic self-incompatibility is ancestral in Eudicots. Molecular Biology and Evolution 19: 825–829.

Strauss, S.Y., J.K. Conner, and S.L. Rush, 1996. Foliar herbivory affects floral characters and plant attractiveness to pollinators: implications for male and female plant fitness. American Naturalist 147: 1098–1107.

Sun, S., X. Gao, and Y. Cai, 2001. Variations in sexual and asexual reproduction of *Scirpus mariqueter* along an elevational gradient. Ecological Research 16: 263–274.

Sutherland, S. and L.F. Delph, 1984. On the importance of male fitness in plants: patterns of fruit-set. Ecology 65: 1093.

Tadey, M. and M.A. Aizen, 2001. Why do flowers of a hummingbird-pollinated mistletoe face down? Functional Ecology 15: 782–790.

Takasaki, T., K. Hatakeyama, G. Suzuki, M. Watanabe, A. Isogai, and K. Hinata, 2000. The S-receptor kinase determines self-incompatibility in *Brassica* stigma. Nature 403: 913–916.

Tammy, L.S., B.C. Husband, and M.B. Routley, 2006. Plant breeding systems and pollen dispersal. In: A. Dafni, P.G. Kevan, and B.C. Husband, eds. Practical Pollination Biology. Enviroquest, Onatario.

Ter-Avanesian, D.V., 1978. The effect of varying the number of pollen grains used in fertilization. Theoretical and Applied Genetics 52: 77–79.

Thompson, J.D., 2001. How do visitation patterns vary among pollinators in relation to display and floral design in a generalist pollination system? Oecologia 126: 386–394.

Thompson, J.N., 2005. The Geographic Mosaic of Coevolution. University of Chicago Press, Chicago.

Thomson, J.D. and S.C.H. Barrett, 1981. Selection for outcrossing, sexual selection, and the evolution of dioecy in plants. American Naturalist 118: 443–449.

Thomson, J.D. and J. Brunet, 1990. Hypotheses for the evolution of dioecy in seed plants. Trends in Ecology and Evolution 5: 11–16.

Thomson, J.D., B.J. Andrews, and R.C. Plowright, 1981. The effect of a foreign pollen on ovule development in *Diervilla lonicera* (Caprifoliaceae). New Phytologist 90: 777–783.

Thomson, V.P., A.B. Nicotra, and S.A. Cunningham, 2004. Herbivory differentially affects male and female reproductive traits of *Cucumis sativus*. Plant Biology 6: 621–628.

Tiffney, B.H. and S.J. Mazer, 1995. Angiosperm growth habit, dispersal and diversification reconsidered. Evolutionary Ecology 9: 93–117.

Traveset, A., 1992. Sex expression in a natural population of the monoecious annual, *Ambrosia artemisiifolia* (Asteraceae). American Midland Naturalist 127: 309–315.
Traveset, A., 1994. Reproductive biology of *Phillyrea angustifolia* L. (Oleaceae) and effect of galling-insects on its reproductive output. Botanical Journal of the Linnean Society 114: 153–166.
Traveset, A., 1995. Reproductive ecology of *Cneorum tricoccon* L. (Cneoraceae) in the Balearic Islands. Botanical Journal of the Linnean Society 117: 221–232.
Traveset, A. and D.M. Richardson, 2006. Biological invasions as disruptors of plant reproductive relationships. Trends in Ecology and Evolution 21: 208–216.
Traveset, A. and E. Sáez, 1997. Pollination of *Euphorbia dendroides* by lizards and insects. Spatio-temporal variation in flower visitation patterns. Oecologia 111: 241–248.
Vamosi, J.C. and S.M. Vamosi, 2004. The role of diversification in causing the correlates of dioecy. Evolution 58: 723–731.
Van Baarlen, P., P.J. Van Dijk, R.F. Hoekstra, and J.H. de Jong, 2000. Meiotic recombination in sexual diploid and apomictic triploid dandelions (*Taraxacum officinale* L.). Genome 43: 827–835.
Vassiliadis, C., P. Saumitou-Laprade, J. Lepart, and F. Viard, 2002. High male reproductive success of hermaphrodites in the androdioecious *Phillyrea angustifolia*. Evolution 56: 1362–1373.
Vázquez, D.P. and M.A. Aizen, 2003. Null model analyses of specialization plant–pollinator interactions. Ecology 84: 2493–2501.
Vázquez, D.P. and M.A. Aizen, 2004. Asymmetric specialization: a pervasive feature of plant–pollinator interactions. Ecology 85: 1251–1257.
Vazquez, D.P., W.F. Morris, and P. Jordano, 2005. Interaction frequency as a surrogate for the total effect of animal mutualists on plants. Ecology Letters 8: 1088–1094.
Verdú, M., 2002. Age at maturity and diversification in woody angiosperms. Evolution 56: 1352–1361.
Verdú, M., A.I. Montilla, and J.R. Pannell, 2004. Paternal effects on functional gender account for cryptic dioecy in a perennial plant. Proceedings of the Royal Society B-Biological Sciences 271: 2017–2023.
Vogler, D.W. and S. Kalisz, 2001. Sex among the flowers: The distribution of plant mating systems. Evolution 55: 202–204.
Waser, N.M. and J. Ollerton, eds, 2006. Plant Pollinator Interactions: From Specialization to Generalization. The University of Chicago Press, Chicago.
Waser, N.M., L. Chittka, M.V. Price, N.M. Willias, and J. Ollerton, 1996. Generalization in pollination systems, and why it matters. Ecology 77: 1043–1060.
Watson, M.A., 1995. Sexual differences in plant developmental phenology affect plant-herbivore interactions. Trends in Ecology and Evolution 10: 180–182.
Weiblen, G.D., R.K. Oyama, and M.J. Donoghue, 2000. Phylogenetic analysis of dioecy in monocotyledons. American Naturalist 155: 46–58.
Weiner, J., 1988. The influence of competition on plant reproduction. In: J. Lovett Doust and L. Lovett Doust, eds. Plant Reproductive Ecology: Patterns and Strategies. Oxford University Press, NY, pp. 228–245.
Westley, L.C., 1993. The effect of inflorescence bud removal on tuber production in *Helianthus tuberosus* L. (Asteraceae). Ecology 74: 2136–2144.
Willson, M.F., 1979. Sexual selection in plants. American Naturalist 113: 777–790.
Willson, M.F., 1994. An overview of sexual selection in plants and animals. American Naturalist 144 (Suppl.): 13–39.
Wilson, P. and J.D. Thomson, 1991. Heterogeneity among floral visitors leads to discordance between removal and deposition of pollen. Ecology 72: 1503–1507.
Wilson, P. and J.D. Thomson, 1996. How do flowers diverge? In: D.G. Lloyd and S.C.H. Barrett, eds. Floral Biology. Studies on Floral Evolution in Animal-Pollinated Plants. Chapman and Hall, NY, pp. 88–111.
Wilson, P., M.C. Castellanos, J.N. Hogue, J.D. Thomson, and W.S. Armbruster, 2004. A multivariate search for pollination syndromes among penstemons. Oikos 104: 345–361.
Wilson, P., M.C. Castellanos, A.D. Wolfe, and J.D. Thompson, 2006. Shifts between bee and bird pollination in penestemons. In: N. Waser and J. Ollerton, eds. Plant Pollinator Interactions: From Specialization to Generalization. University of Chicago Press, Chicago.

Wolf, D.E. and N. Takebayashi, 2004. Pollen limitation and the evolution of androdioecy from dioecy. American Naturalist 163: 122–137.
Worley, A.C. and L.D. Harder, 1996. Size-dependent resource allocation and costs of reproduction in *Pinguicula vulgaris* (Lentibulariaceae). Journal of Ecology 84: 195–206.
Wyatt, R., 1983. Pollinator–plant interactions and the evolution of breeding systems. In: L. Real, ed. Pollination Biology. Academic Press, NY, pp. 51–95.
Yampolsky, C. and H. Yampolsky, 1922. Distribution of sex forms in the phanerogamic flora. Bibliotheca Genetica 3: 1–62.
Young, H.J. and M.L. Stanton, 1990. Influences of floral variation on pollen removal and seed production in wild radish. Ecology 71: 536–547.
Zimmerman, M., 1988. Nectar production, flowering phenology, and strategies for pollination. In: J. Lovett Doust and L. Lovett Doust, eds. Plant Reproductive Ecology. Patterns and Strategies. Oxford University Press, NY, pp. 157–178.

18 Seed and Seedling Ecology

Kaoru Kitajima

CONTENTS

INTRODUCTION

Many features of plant communities are strongly influenced by events surrounding reproduction by seeds. Characteristics such as the relative abundance of the species, their annual fluctuation in number, and their spatial pattern are all influenced by the ability of each species to reproduce. The diversity and dynamics of plant communities hinge on the ability of species to regenerate successfully, resist local extinction, as well as disperse from neighboring communities. Grubb (1977) suggests that the coexistence of so many plant species that appear to have indistinguishable niches as adults can be explained by their distinct requirements in early stages of their life histories.

The goal of this chapter is to demonstrate how species-specific traits of seeds and seedlings are related to life history traits and regeneration strategies of species. The importance of seed and seedling ecology has been increasingly recognized both in basic and applied research during the last few decades as is evident in a literature search (Figure 18.1). Within the limited space here, only a limited number of important new perspectives can be discussed in addition to the key processes summarized in the previous version prepared with Michael Fenner in 1997. Those interested in further details should consult books that provide a more comprehensive coverage (Baskin and Baskin 1998, Fenner 2002, Fenner and Thompson 2005). The following sections follow largely chronological order of events surrounding regeneration from seeds. The take-home message, however, is that each step of seedling regeneration is intimately linked to other events surrounding reproduction, as well as to the overall life history of the species.

SEED SIZE AND ITS CORRELATES

Seed size is a key trait that has evolved in association with a multitude of other species-specific traits (Moles and Westoby 2006). Among present-day seed plants, seed size varies over

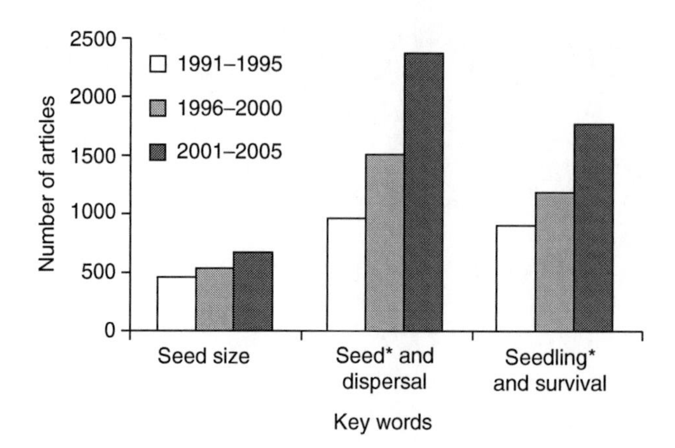

FIGURE 18.1 Number of journal articles indexed by Web of Science search engine for three key words, "seed size," "seed* and dispersal," and "seedling* and survival" for three 5 year periods, 1991–1995, 1996–2000, and 2001–2005. The asterisk allows searching of both plural and singular forms. The number during the 8 year period since the original preparation of this review (1998–2005) was 989, 3329, and 2498 in each of these topic areas, respectively.

10^{11}-fold, but within a given local community, the range is typically 10^6 excluding extremes such as the dust-sized seeds of orchids and the double-coconuts that weigh 20 kg (Moles et al. 2005). The size adopted by a particular species is partly determined by phylogenetic influences; big (or small) seeds run in families. Flowering plant species of the early Cretaceous had small seeds and fruits, but their eventual dominance in closed forest vegetation was apparently conducive for the evolution of larger seeds and fruits (Eriksson et al. 2000). A recent analysis of seed size evolution in seed plants reveals that the largest divergence occurred as an overall reduction of seed size from gymnosperms to angiosperms (Moles et al. 2005). The same analysis also confirms a widely observed pattern found in a range of floras that seed size is associated with growth form and height of parents (Leishman et al. 1995, Poorter and Rose 2005). Daisies cannot produce seeds the size of coconuts; adult height and terminal twig diameter set an upper limit on seed size (Grubb et al. 2005) (Figure 18.2). This may be the main reason for the latitudinal gradient of seed

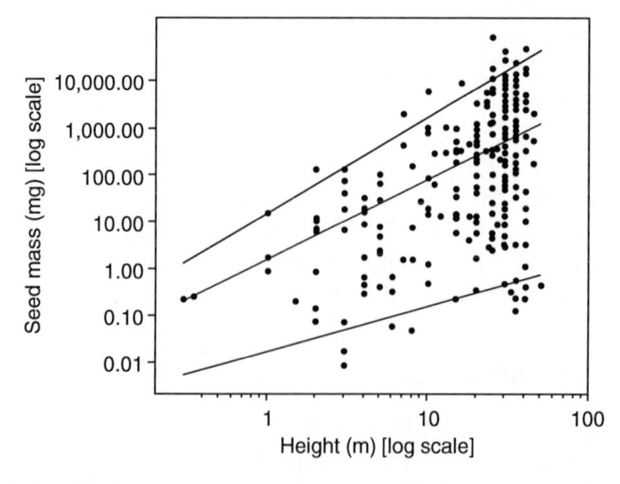

FIGURE 18.2 The relationship between mean seed dry mass and mature plant height for 226 species in tropical lowland rainforest in Australia. (Adapted from Figure 1 of Grubb, P.J., Metcalfe, D.J., and Coomes, D., *Science* 310, 783A, 2005. With permission.)

size, as tropical communities dominated by trees should have larger seed size on average than temperate communities represented proportionally more by herbaceous species. Across latitudes, mean seed mass decreases by 10-fold for every *c*. 23° moved toward the poles (Moles and Westoby 2003). An obvious question to functional ecologists, then, is how seed size variation among species is associated with their differences in life history and habitat preferences within and across plant communities.

Within the constraints of the genetic makeup and size of the plants, the characteristic seed size of each species is presumed to be the result of natural selection. There is usually some variation in mass among and within populations, and within the progeny of individual plants. However, the variation within a species is generally much less than that of the vegetative parts (Harper 1977). The selection pressures influencing seed size are likely to have been numerous, often operating in opposite directions, resulting in a size that may represent the best compromise.

Seed size should be viewed in relation to the overall reproductive strategy and life history of the species. For a given allocation of resources to reproduction, the plant can either invest in a small number of large seeds or a large number of small ones, or at some intermediate combination of number and size. The compromise size adopted is conventionally thought to represent the conflicting requirements of dispersal (favoring small seeds) and establishment (favoring large seeds). Seed size variation is often interpreted in terms of a trade-off between dispersability and establishment (Ganeshaiah and Uma 1991, Geritz 1995, Ezoe 1998). Short-lived early colonizers of disturbed sites and open ground typically produce numerous light, easily dispersed seeds; long-lived species of less-disturbed sites tend to have larger, less widely dispersed seeds. Seed augmentation experiments demonstrate that dispersal poses a greater constraint for colonization in large-seeded species than in small-seeded species (Leishman 2001, Makana and Thomas 2004, McEuen and Curran 2004). Large-seeded species tend to be more competitive, and depend less on disturbance for seedling establishment than small-seeded species in grasslands (Reader 1993, Burke and Grime 1996, Lindsay et al. 2004). Improved seedling establishment associated with large seed size helps compensate for dispersal limitation, but it does not appear sufficient to overcome the average reduction of fecundity per unit crown area associated with increase in seed size (Moles and Westoby 2004).

Numerous additional life history correlates, such as long-term survival of established seedlings, time to reach reproductive maturity, and life-time seed production, must be taken into account to understand the seed number–size trade-offs. Furthermore, residual variation among species arises from the methods of dispersal adopted, influence of particular species of animals as seed dispersers and predators, establishment conditions (degree of shade, drought, nutrient level), formation of persistent seed banks, cotyledon functional morphologies, and adoption of a parasitic or hemi-parasitic way of life (Leishman et al. 1995).

Ecologists have long identified the advantage of large seeds for successful seedling establishment in shade (Salisbury 1942, Grime and Jeffrey 1965). Recent meta-analyses confirm this pattern for temperate and tropical species (Hewitt 1998, Hodkinson et al. 1998, Moles and Westoby 2004, Poorter and Rose 2005). Across and within many taxa, seed size is correlated with the ability to survive and establish in shade (Osunkoya et al. 1994, Saverimuttu and Westoby 1996, Paz and Martinez-Ramos 2003), although there are exceptions of small-seeded shade-tolerant species and large-seeded light-demanders (Augspurger 1984b, Metcalfe and Grubb 1995). There are three possible ways in which large seed size may contribute to seedling establishment in shade (Westoby et al. 1992). First, a large seed can create a large seedling that can successfully display leaves above litter (Molofsky and Augspurger 1992). Second, a significant fraction of resources in a large seed may remain in storage instead of being used for immediate seedling development (Garwood 1996, Green and Juniper 2004). Third, the advantage of a large seed may be indirect via association of seed size with seedling morphology, development types, and growth rates

(Hladik and Miquel 1990, Kitajima 1996). Across many floras, large-seeded species tend to have storage cotyledons, whereas small-seeded species tend to have thin photosynthetic cotyledons (Ibarra-Manriquez et al. 2001, Zanne et al. 2005). Phylogeny exerts a strong influence on both seed size and cotyledon functional morphology, which have diverged in concert (Wright et al. 2000, Zanne et al. 2005). The three pathways for large-seed advantage are not mutually exclusive, and the relative importance of these mechanisms must be evaluated experimentally.

All else being equal, the greater the seed reserve mass, the greater the initial seedling mass. As a rule of thumb, seed reserves alone appear sufficient to construct the seedling up to development of the first photosynthetic organs (cotyledons or leaves, depending on the species). Certainly, within a species, bigger seeds produce larger seedlings initially (Stanton 1984, Wulff 1986, Gonzalez 1993). When different species are compared, lipid content in seeds is a small modifier to this rule; a unit mass of oil-rich seed is converted to a greater mass of seedling than a unit mass of starchy seed (Penning de Vries and Van Laar 1977, Kitajima 1992a). Yet, interspecific variation in seedling size due to variation in seed lipid content (up to twofold) is completely dwarfed by the variation due to seed mass (typically up to 10^6-fold). Other factors, such as whether the seedling sets aside a part of its seed reserves in storage, or what type of seedling tissue is created, appear to be greater modifiers of the relationship between seed size and seedling size.

Do larger-seeded species depend exclusively on seed reserves for a longer duration than smaller-seeded species? Initially, seedlings depend completely on seed reserves for both supply of energy and mineral nutrients, but gradually following the development of photosynthetic organs and roots, seedlings start utilizing externally supplied resources. Reserves remaining in large-storage cotyledons may be utilized for a rapid recovery from shoot loss to herbivory in very early stages (Dalling et al. 1997a). Logically, this advantage lasts only as long as the reserves last. For example, Saverimuttu and Westoby (1996) found that the large-seed advantage of seedling longevity in shade exists only during the cotyledon stage, but not for seedlings transferred to deep shade after full expansion of leaves. In California tan oak, transfer of energy reserves from storage cotyledons occurs before leaf expansion (Kennedy et al. 2004). After the first leaf expansion, seedlings of five tropical trees experiencing negative carbon balance due to defoliation or shading do not rely on energy reserves in cotyledons, but instead on starch and sugar stored in stems and roots (Myers and Kitajima 2007). However, storage cotyledons that remain attached to the seedling axis may continue to support seedling demands for mineral nutrients (Oladokun 1989, Milberg et al. 1998), even after energy reserves cease to be exported. Functional growth analysis of seedlings raised with and without deprivation of light or nitrogen demonstrates that complete seed reserve dependency lasts longer for nitrogen than for light in all three species tested (Kitajima 2002). Seedling size achievable without external supply of an individual mineral element can also be indicative of the relative duration of seed reserve dependency for that element (Fenner and Lee 1989, Hanley and Fenner 1997).

If prolonged support enabled by large seed size is more important for mineral nutrients than for energy, large seeds should enhance seedling establishment in infertile soils. Interestingly, experimental support for this idea comes largely from fire-prone communities on infertile soils (Jurado and Westoby 1992, Hanley and Fenner 1997, Milberg et al. 1998, Vaughton and Ramsey 1998). Lee et al. (1993) found that among species in the grass genus *Chionochloa*, there is a negative correlation between seed size and soil fertility of the habitats of the species. In contrast, Maranon and Grubb (1993) found that in a selection of 27 Mediterranean annuals, the species with the largest seeds tend to occupy the soils with a richer nutrient supply. A higher seed concentration of a particular mineral element also extends the dependency on seed reserves for that element, as shown for a prolonged nitrogen dependency in a Bignoniaceae species (Kitajima 2002). Concentrating particular mineral elements in seeds should be preferred to increasing seed size when there is a selective pressure to enhance dispersal. Indeed, there is a negative correlation between seed mass and nitrogen concentration across species (Fenner 1983, Pate et al. 1985, Grubb 1998). However, plants from more infertile habitats do not necessarily have greater mineral nutrient concentrations (Lee et al. 1993, Grubb and Coomes 1998). Yet, it is possible that high concentrations of particular mineral elements in seeds may complement the deficiencies of these elements in the environments (Stock et al. 1990). Perhaps, imbalance of nitrogen and phosphorus supplies in postfire soils may select for fire-dependent species to concentrate nitrogen reserves in seeds.

There is also some evidence that plants from dry habitats tend to have larger seeds. Baker (1972) carried out a survey of 2490 species in California and showed a fairly consistent relationship between seed size and dry conditions. It is thought that greater seed reserves might enable the seedling to establish roots quickly and so exploit a greater volume of soil for moisture than would otherwise be possible. However, in a survey of dunes in Indiana, Mazer (1989) was not able to show any significant relationship between seed size and water availability. Jurado and Westoby (1992) in a test involving Australian species found that seedlings from heavier-seeded species do not (as they hypothesized) allocate a greater proportion of their resources to roots than lighter-seeded species. Glasshouse experiments on seeds of semiarid species by Leishman and Westoby (1994b) indicate an advantage to larger seeds in dry soil, but their field experiments failed to confirm this. Further surveys of the type carried out by Baker (1972), on a range of floras, would help to clarify the relationship between seed size and dry habitats.

There is no doubt that a myriad of complex natural selective pressures have acted on plants resulting in the seed sizes observed in contemporary floras. Many ecological traits at seed and seedling stages discussed in the subsequent sections could not have evolved independent of seed size.

NATURAL ENEMIES

Seeds and young seedlings represent attractive resources to a broad array of consumers. In general, seed tissue has a much higher concentration of nitrogen, phosphorus, sulfur, and magnesium than other plant tissues, in addition to being a rich source of carbohydrates and, in some cases, oils (Vaughan 1970, Barclay and Earl 1974). It is not surprising therefore to find that in many plant species a large proportion of seed production is lost to predation. Crawley (1992) provides a useful list of examples from the literature of percentage loss of seeds to predators in different plants. The proportion averages at about 45%–50%, but often approaches 100%. Two distinct groups of seed eaters exist. Predispersal seed predators are typically highly specialized sedentary larvae of beetles, flies, moths, or wasps that mature within the seed or seedhead. In contrast, postdispersal seed predators are usually vertebrates, more mobile, less-specialized feeders, although some tropical insect seed predators attack seeds postdispersally. Whole taxa of granivorous birds and mammals have evolved (e.g., finches, rodents) to exploit this rich food source. In the seasonally inundated forests of Amazonia nearly all the seeds that fall into the water are eaten by fish (Kubitzki and Ziburski 1994). In addition, there are many invertebrates that act as predators of dispersed seeds: various species of ants (Gross et al. 1991), earwigs (Lott et al. 1995), slugs (Godnan 1983), and even crabs (O'Dowd and Lake 1991). Soil-borne fungi are also important consumers of seeds after dispersal (Dalling et al. 1998, O'Hanlon-Manners and Kotanen 2004, Schaffer and Kotanen 2004). Many of these organisms also act as predators of young seedlings attracted to their soft and less-defended tissues.

Rather few experimental studies have been carried out to determine the long-term demographic effect of seed predation. In some cases, there is no doubt that seed eaters reduce recruitment. Louda (1982) excluded seed-eating insects from the Californian shrub

Haplopappus squarrosus by the use of insecticide, and found that the mean number of seedlings established per adult after 1 year was greater in the treated plots by a factor of 23. Further proof that seed predators can reduce subsequent recruitment (and hence lifetime fitness) is provided by a demographic study in which insecticide was applied to the thistle *Cirsium canescens* by Louda and Potvin (1995). Generally, there are significant increases in recruitment when seeds were protected from predators (Molofsky and Fisher 1993, Terborgh and Wright 1994, Asquith et al. 1997). However, the consequences of seed predation for a plant population depend on whether regeneration is limited by seed numbers or by some other factor such as dispersal and availability of safe sites, which may change from season to season (Edwards and Crawley 1999). Even when seed number is not limiting, predators may still influence the genetic makeup of the plant population by differential selection of the seeds. They can also affect the evolution of the structural defenses of the seeds. Benkman (1995) compared the allocation of putative seed defenses in limber pine (*Pinus flexilis*) in sites where tree squirrels are present (in the Rocky Mountains) with sites where they are absent (in the Great Basin). He found that allocation of energy to cone, resin, and seed coat relative to the kernel is greater by a factor of 2 where the predators are present. This difference in allocation may be a relatively recent evolutionary development since tree squirrels became extinct in the Great Basin only within the last 12,000 years.

Seed predation by animals may have had an evolutionary influence on seed size. One means by which a plant could reduce loss to predation would be to reduce seed size (with corresponding increase in seed number), thereby increasing the foraging cost/benefit ratio of potential predators. Janzen (1969) cites the case of two groups of Central American legumes, which adopt contrasting means of coping with predation by beetle larvae. The small-seeded group escapes predation by subdivision of their reproductive allocation, whereas the large-seeded group is defended by toxic compounds. A study of predispersal predation of seeds in a number of *Piper* species in Costa Rica found that the large-seeded species lost a much greater proportion of their seeds to insects (Greig 1993). Within a species, the larger seeds may be more vulnerable to attack by predispersal predators. For example, bruchid beetles preferentially oviposit on larger seeds in the sabal palm (Moegenburg 1996). At the same time, extremely large seeds of some tropical trees (seed reserve mass >50 g) appear to have ample reserve to germinate even after consumed by up to eight bruchid larvae (Dalling et al. 1997a). Differential loss is also seen in vertebrate grazers that consume seeds as part of their forage. Among legume seeds likely to be eaten by grazing livestock, small seeds may be at an advantage. Tests with sheep found that small seeds have the highest survival rate after passage through the gut (Russi et al. 1992). Large seeds may thus need to devote more of their resources to structural defense. Fenner (1983) showed a consistent trend among 24 herbaceous Compositae for relatively greater seed coats in larger seeds. The proportion of seed weight allocated to seed coat varies from 15% in *Erigeron canadense* (seed weight 0.072 mg) to 61% in *Tragopogon pratense* (seed weight 10.3 mg). Thus, defense against seed predation may be another factor in determining the balance between seed size and number.

Seed predation (mainly by insects, rodents, or birds) is widely thought to select for masting, that is, bumper crops at irregular intervals with a light seed crop (or total crop failure) in the intervening years (Kelly and Sork 2002). Recently published examples of long-term studies on seed production for individual tree species include rimu (Norton and Kelly 1988), southern beech (Allen and Platt 1990), oak (Crawley and Long 1995), and ash (Tapper 1996). Multiple species may participate in community-level masting by synchronizing to climate cues or simply tracking favorable climate. Because climatic variation is greater in temperate latitudes than in the tropics, Kelly and Sork (2002) hypothesized that masting is more likely in temperate than in tropical forests. In support of this view, interannual variability of seed production is lower in a tropical forest in Panama (Wright et al. 2005) than in a temperate forest in Japan (Shibata et al. 2002). Yet, community-level masting occurs in the tropics, most famously in the SE Asian forests dominated by Dipterocarpaceae (Janzen 1974, Curran and Leighton 2000). It is hypothesized that masting results in the alternate starvation and satiation of the seed predators; in lean years the predators eat most of the seeds produced, but are overwhelmed by the bounty in bumper years, leaving a surfeit available for regeneration.

Swamping predators may not be easy. Species-specificity, mobility, and generation time of seed predators affect whether they can be successfully satiated or not. Even if species-specific predators are satiated, natural enemies that can attack multiple species, such as damping-off pathogens, may cause greater seed and seedling mortality when seed crop is high. Seed predator populations can respond markedly, at least in some cases, to the level of mast and remove the entire crop in most years (Wolff 1996). Seeds unconsumed by resident predators may be eventually eaten by nomadic animals that become attracted to masting localities (Curran and Webb 2000). There are alternative explanations for the benefits of masting, such as greater pollination efficiency and the need for large-seeded species to accumulate sufficient reserves for reproduction (Fenner 1991, Kelly and Sork 2002). Masting may also be a way to track favorable climate for seed production (Wright et al. 1999) and seedling establishment (Williamson and Ickes 2002). Although it is difficult to exclude these alternative explanations, there is a large body of evidence in support of the general applicability of the predator satiation hypothesis. For example, species most prone to seed predation show masting behavior most strongly (Silvertown 1980a). Seedling establishment can be virtually confined to those following mast years (Jensen 1985, Forget 1997). Rogue individuals that produce seed in a nonmasting year are targeted by seed predators, thus selecting for synchronicity, as found for pinyon pine populations (Ligon 1978), the cycad *Macrozamia* (Ballardie and Whelan 1986), and *Acacia* spp. (Auld 1986). These observations are at least consistent with the predator satiation hypothesis.

In addition to obvious population effects, predators and other natural enemies affect spatial patterns within a community. Janzen (1970) and Connell (1971) put forward the idea that seed predation near trees in tropical rainforests may prevent regeneration of the same species in the immediate vicinity of the parent plant, reducing intraspecific clumping and so promoting diversity. The high mortality near parents may occur either as a direct result of distance to parents (because the parent and its offspring share the same natural enemies) or an indirect result of high density of offspring near parents (because the natural enemies are either attracted to or can spread easily in a dense populations). Because seed density is almost always confounded with distance from the parent plant, an experimental approach is necessary to tease apart whether it is density or distance that is responsible for the observed patterns (Augspurger and Kitajima 1992). This is an important distinction, as modeling studies found that dispersal patterns of not only seeds, but also of natural enemies, affect whether such interactions would yield greater plant species richness (Nathan and Casagrandi 2004, Adler and Muller-Landau 2005). The natural enemies that operate in a density-dependent manner include not only seed predators, but also pathogenic microbes (Augspurger 1983b, Dalling et al 1998, Bell et al. 2006) and leaf-eating herbivores (Sanchez-Hidalgo et al. 1999). Such negative density dependency is observed not only in tropical rain forests but also in less-species rich temperate forests (Packer and Clay 2000). Interestingly, a recent meta-analysis of distance effects on seed and seedling survival using 152 published data sets found a significant effect of distance on seedling survival but not for seed survival (Hyatt et al. 2003).

Testing the Janzen–Connell hypothesis requires two steps: (1) demonstration of distance or density dependency of juvenile survival, and (2) demonstration that such effects promote species diversity. From comparison of community-wide analysis of seeds collected in traps and seedlings in plots adjacent to these traps, Harms et al. (2000) conclude that density-dependent natural enemies increase species diversity between seed and seedling stage. It is important to remember, however, that the overall level of seed mortality is determined by interactions of multiple predators that often exhibit contrasting functional responses to seed density. A given seed density may be high enough to satiate one predator species, but may promote consumption by another. Furthermore, the availability of alternative food sources, phenologies, clumping of adult trees, and other environmental factors affect the spatial patterns of seed predation (Forget et al. 1997, Hammond and Brown 1998, Kwit et al. 2004). Nevertheless, the net result is negative density dependency for many coexisting tree species in terms of seedling recruitment from seeds (Harms et al. 2000, Wright et al. 2005) and seedling survival (Webb and Peart 1999).

DISPERSAL

Seed dispersal is important for avoiding competition from the parent, escape from localized natural enemies, arrival in safe sites, successful colonization of other communities to avoid extinction, and so determining plant diversity and distribution at both local and regional scales (Wang and Smith 2002, Vormisto et al. 2004, Muller-Landau and Hardesty 2005). Some of these processes clearly hinge on rare long-distance dispersal events that are important but hard to quantify (Cain et al. 2000). Understanding seed dispersal is also important for conservation of endangered species and management of invasive exotic species. A species may be absent at a given locality simply because seeds do not arrive there (dispersal limitation) or because it is not a safe site for seedling establishment (establishment limitation). The relative importance of these processes can be evaluated experimentally by planting seeds to overcome dispersal limitation. Dispersal limitation appears ubiquitous across biomes (e.g., Tilman 1997, Maron and Gardner 2000, Dalling et al. 2002, Makana and Thomas 2004, McEuen and Curran 2004, Svenning and Wright 2005) and is considered important for species coexistence (Tilman 1994, Hubbell et al. 1999).

The means by which seeds are transported varies from species to species. Many appear to have no particular adaptation for dispersal. They may be carried in mud on the feet of animals and birds, as was shown in experiments by Darwin (1859), or eaten as part of the forage of grazers and survive passage through the gut and deposition some distance from their source (Janzen 1984, Sevilla et al. 1996). The seed itself may be the reward in many scatter-hoarded species, such as oaks and many tropical tree species with fruits and seeds that lack any apparent dispersal appendages. Other plant species provide an attractive reward for their dispersers in the form of a fleshy fruit (or aril) in which the seeds are imbedded. Another large group exploits the wind as a means of transport, with wings or feathers that decrease the rate of descent, thereby increasing the horizontal distance traveled in a given time (Augspurger 1986). The distance traveled is also a function of the height of release. Techniques for quantifying the rate of descent of seeds under standardized conditions allow comparisons of dispersal potentials among species (Askew et al. 1997).

The interaction of dispersers with a species results in a characteristic spatial pattern of distribution of its seeds, called its seed shadow or dispersal kernel. Much progress has been made in statistical techniques to describe seed shadows in recent years (Okubo and Levin 1989, Clark et al. 1999, Nathan and Muller-Landau 2000, Levin et al. 2003, Greene et al. 2004). Yet, how to model the tail of dispersal shadows, that is dispersal beyond 100 m from the parent, continues to pose an important challenge to ecologists (Cain et al. 2003). Genetic methods are increasingly recognized to be useful for quantification of long-distance dispersal events (Cain et al. 2000, Wang and Smith 2002, Jones et al. 2005, Hardesty et al. 2006). Regardless of the methods employed, spatial patterns of seed dispersal are easier to model for wind-dispersed species than for animal-dispersed species. Wind-dispersal is usually skewed toward the down-wind direction, often peaking at a short distance from the source (Augspurger 1983a). Steep slopes can also influence the skewness (Lee et al. 1993). Animal-dispersed seeds tend to be more clumped because they are deposited beneath roosting sites (by birds and bats,

Russo and Augspurger 2004), in caches (by rodents, Howe 1989, Forget 1990, Willson 1993), or in latrines (by tapirs, Fragoso et al. 2003). Some dispersal agents not only help seeds escape negative density dependency in the vicinity of the parent, but also help deliver seeds to specific safe sites, such as treefall gaps (directed dispersal, Wenny 2001). Examples include bellbirds in tropical cloud forests (Wenny and Levey 1998), ants in lowland tropical forests (Horvitz and Schemske 1994), and mice in temperate forests (Seiwa et al. 2002). Even wind may preferentially deliver seeds into treefall gaps by their interaction with canopy roughness (Augspurger and Franson 1988, but see Jones et al. 2005). Effectiveness of dispersal not only depends on the identity of the dispersers, but also their interaction with fruit and seed size (Seiwa et al. 2002, Alcantara and Rey 2003, Jansen et al. 2004). Loss of effective dispersal animals due to hunting and habitat fragmentations are likely to result in a large proportion of seeds undispersed near parents, all of which may be killed by density-dependent natural enemies (Wright and Duber 2001, Chapman et al. 2003).

The range of animals involved in seed dispersal is very wide. The most important groups are birds and mammals, but cases of seed dispersal by other vertebrates are known, for example, fish (Goulding 1980, Horn 1997), amphibians (Silva et al. 1989), and reptiles (Hnatiuk 1978). Seed dispersal by earthworms has also been recorded (McRill and Sagar 1973, Piearce et al. 1994). Some seeds may be dispersed more than once: first deposited by birds, monkeys, and bats, and then removed by secondary dispersers such as ants (Hughes and Westoby 1992, Levey and Byrne 1993), dung beetles (Chapman et al. 2003), and scatter-hoarding rodents (Forget and Milleron 1991). Survival of seeds may be negligible if they remain in clumps under bat or bird roosts. Ants are the only invertebrate group that disperses seeds in any appreciable number (Stiles 2000). Dispersal by ants (myrmecochory) is especially prevalent in warm dry climates and on infertile soils (Beattie and Culver 1982, Westoby et al. 1991). Ant-dispersed seeds are typically provided with an oil body (elaiosome), which the ants eat. They retrieve the seed from the ground, carry them off to their nests, remove the elaiosome, and deposit the seed in a refuse heap. Not all seeds survive ant transport, and in some cases a proportion of the seeds are eaten as well (Hughes and Westoby 1992, Levey and Byrne 1993). The advantages to the plant are thought to be (a) dispersal, though usually only within a few meters of the source; (b) protection from rodents by being burried out of sight; (c) protection from fire; and (d) deposition in a favorable microsite for germination and establishment (Bennet and Krebs 1987). Not all of these features may be equally important in all cases. The importance of the mutualism for the plant can be seen in cases where native ants have been replaced by less well adapted invaders, as in the case of fynbos species in South Africa pushed out by the Argentine ant (Bond and Slingby 1984) and native ants in North America pushed out by fire ants (Zettler et al. 2001).

Long-distance dispersal is clearly important for movement of plants after major climate changes, migration to oceanic islands and fragmented habitats, and invasion by exotic species (Cain et al. 2000, 2003). Yet, there are selective pressures against long-distance dispersal, because a seed transported to very long distances is likely to face a risk of removal from its natural habitat, which may be patchily distributed. Comparisons between related plants on mainlands and islands show that dispersabilty of wind-dispersed species is often reduced on islands, presumably because of selective survival of the less-mobile seeds (Cody and Overton 1996). Remote islands are more likely to be colonized by seeds carried by birds than by wind or sea drift, as in the case of the Pacific Islands (Carlquist 1965). In contrast to the random action of wind and sea, bird movement is from island to island, often on migration routes, and so targeting the islands effectively with seeds deposited in feces and preened from feathers. Birds are also important in dispersing seeds to other types of islands including forest fragments (Johnson and Adkisson 1985) and isolated trees in the middle of pastures (Holl 1999, Zahawi and Augspurger 1999, Slocum and Horvitz 2000).

Variety of traits that influence dispersal patterns must have evolved in relation to life history and regeneration strategies of the species. For example, wind-dispersed species tend to be smaller and more common among pioneer species. Animal-dispersed species that are dispersed in clumps may be selected to have greater resistance against fungal pathogens, which can cause density-dependent mortalities. Thus, dispersal affects distribution and abundance of seedlings not only in terms of the initial spatial pattern, but also through its relationship with functional traits that modify seed and seedling survival after dispersal.

DORMANCY AND GERMINATION

Another strategy for escaping from the parental plant is the formation of long-lived reservoirs of seeds in the soil, thus undergoing dispersal in time rather than space. This is an effective strategy especially in environments in which likelihood of seedling establishment varies greatly from year to year (Chesson 1985) or season to season (Baskin and Baskin 1998). Persistent seed banks consist of buried seeds that have the ability to remain viable for at least several years. They will only germinate if they are brought to the surface by some chance disturbance such as a tree-fall, an animal digging, or a farmer plowing. Although viable seed populations have patchy distribution and show large seasonal fluctuations (Thompson and Grime 1979, Thompson 1986, Dessaint et al. 1991, Dalling et al. 1997b), rough generalizations can be made for typical seed bank sizes across biomes: 20,000–40,000 m^{-2} in arable fields, typically below 1000 m^{-2} in mature tropical forests, and only 10–100 m^{-2} in subarctic forests (Leck et al. 1989, Fenner 1995). Persistent seed banks are the most characteristic of habitats that are prone to frequent but unpredictable disturbance such as cultivation, fire, and floods. Examples of plant communities with large soil seed banks are agricultural fields, heathlands, chaparral, and disturbed wetlands (Thompson and Mason 1977, Leck et al. 1989). However, in many less-disturbed communities, those species that are characteristic of the early stages of succession and habitually the first colonizers of gaps, also form persistent seed banks. Although these species often dominate the seed bank, they usually form only a very small part of the current aboveground vegetation (e.g., Kitajima and Tilman 1996, Dalling and Denslow 1998). They represent both the past and the potential future species composition of the community (Fenner 1995). Within each species, genetic makeup of a soil seed population must be the result of selection in different years over a period of time, and the appearance of old gene combinations may put a damper on genetic change in the population (Templeton and Levin 1979, Brown and Venable 1986).

Survival of seeds in soil differs greatly among species and biotic and abiotic environments. Some temperate weeds are known to survive in soil for decades (Roberts and Feast 1973, Kivilaan and Bandurski 1981). Among tropical pioneer tree species, persistence of buried seeds range widely from species dying within a few months to species that do not exhibit any detectable mortality over a few years (Dalling et al. 1997b). Buried dormant seeds may suffer high mortality from fungal pathogens (Crist and Fruesem 1993, Dalling et al. 1998). All else being equal, the greater the depth of burial, the better the survival (Toole 1946, Roberts and Feast 1972, Dalling et al. 1997b), as attack from pathogens may be more active in shallow well-oxygenated soil. Small, round, smooth seeds can infiltrate more easily to greater depths in the soil by percolating into crevices. In contrast, large, elongated seeds with appendages such as awns or hairs would need an external agent to be buried. For a range of British grasses, species that form persistent soil seed banks mostly possess smooth and round seeds less than 0.3 mg, whereas those that do not form soil seed banks tend to have elongated bigger seeds with appendages (Thompson et al. 1993). Bekker et al. (1998) extend these generalizations, indicating that seed size and shape can be used in a predictive way as a guide to probable persistence. However, the same trend does not exist in Australia, possibly because of differences in burial regimes and disturbance (Leishman and Westoby 1998).

Dormancy prevents seeds from germinating at times, which would be unfavorable for growth and establishment. Some seeds possess absolute dormancy and do not germinate until certain developmental processes (such as after-ripening) have occurred. However, dormancy can often be a matter of degree. A dormant seed may be induced to germinate, but only under a very restricted set of conditions. The narrower the required conditions, the greater the level of dormancy. This is well illustrated by the cyclical changes in the level of dormancy, which occur in the seeds of many annual species (Baskin and Baskin 1985). During summer and autumn months, the seeds of summer annuals in the soil are fully dormant. However, the seeds are gradually released from dormancy by the chilling temperatures experienced during winter (Washitani and Masuda 1990). This is shown by the fact that if the seeds are taken from the field and tested for germinability in the laboratory, they germinate over an increasingly wider range of temperatures as spring approaches. As spring then advances into summer, the range of permitted germination temperatures narrows, eventually resulting in complete dormancy again. This mechanism of cyclical dormancy thus ensures that the seeds germinate only in spring, the most favorable germination for plants to complete their life cycle in a temperate environment. A similar mechanism ensures that winter annuals germinate only in autumn, in this case with seeds that require high temperatures to release them from dormancy (Vegis 1964, Baskin and Baskin 1980, Bouwmeester 1990). It is important to note that in these examples there is a clear distinction between the conditions required to overcome dormancy and the conditions needed for germination.

Another type of dormancy uses physiological mechanisms to ensure germination only in a gap in vegetation and near soil surface. If a seed germinates when buried below a given critical depth, it will not be able to emerge. Some seeds are indeed lost in this way (Fenner and Thompson 2005), but most seeds remain dormant at depth. Exposure of freshly dispersed and imbibed seeds to low red/far red ratio under leaf-canopy is important in inducing secondary dormancy to prevent fatal germination after burial (Washitani 1985). Once they are brought to (or near) the surface, usually by some unpredictable disturbance, it is advantageous for them to ensure that their dormancy is not broken unless they are in a suitable gap in the vegetation.

Some of the responses of seeds to various environmental stimuli may act as gap detection mechanisms. The requirement for light with a high red/far red ratio means that many seeds will not germinate if shaded by other plants (Gorski et al. 1977, Fenner 1980, Silvertown 1980b). The frequent requirement for fluctuating temperatures (Thompson and Mason 1977) or high temperature (Daws et al. 2006) could act as both a gap-detecting and a depth-sensing mechanism. Which of these gap-detection mechanisms is employed must reflect species specializations to different sizes and positions of gaps, as well as seed size (Pearson et al. 2002). Four shrub species within the genus *Piper* in a neotropical forest differ in their sensitivity to red/far red ratios, temperature, and nitrate (Daws et al. 2002a). The positive response to nitrate seen in many species (Hilhorst and Karssen 2000) could also be related to germinating in gaps, where the disturbed soil releases a flush of nitrate (Pons 1989). Some species in fire-prone communities respond to favorable conditions by requirement of high temperature or smoke for breaking dormancy (Keeley 1991, Hanley and Fenner 1998, Keeley and Fotheringham 1998, Brown et al. 2003). These various specific responses likely help seeds to identify favorable sites in which to germinate. Certainly, the seeds of many parasitic species such as *Orobanche* and *Striga* can detect the presence of their host plant by a root secretion in the soil (Joel et al. 1995). The concept of gap-detection is in principle no different from host-detection, though the latter is considerably more specific.

The opposite of the seed-banking strategy is exhibited by recalcitrant seeds. Recalcitrant seeds completely lack dormancy, and must germinate immediately after they are shed. They also have to be dispersed in rainy months because they do not survive desiccation (Pritchard et al. 2004). The majority of nonpioneer tree species in the tropics, as well as some large-seeded temperate species, fall into this category (Ng 1978, Hopkins and Graham 1987, Garwood and Lighton 1990, Pammenter and Berjak 2000, Rodriguez et al. 2000). In general, larger seeds tend to be more sensitive to desiccation (Daws et al. 2005). Complex selective pressures may be responsible for evolution of large seeds that lack dormancy. The advantage of large seed size for seedling establishment must be balanced against the risks of seed predation and desiccation. Without burial by scatter-hoarding rodents, the seeds would remain near the soil surface, exposed to a host of pests. Lack of dormancy and fast germination are advantageous to escape seed predators. The smaller surface-to-volume ratios of larger seeds also make them less likely to reabsorb water (Kikuzawa and Koyama 1999). Thus, quick radicle emergence is also advantageous for avoiding the risk of losing water. Yet, some large tropical seeds, including palms and legumes, tolerate dry conditions before radicle emergence. Many such seeds exhibit delayed germination, remaining viable in soil for months and years (Garwood 1983). It remains completely unknown what controls germination timing of such seeds.

SEEDLING RECRUITMENT

Transformation of a germinating seed to a seedling represents the most vulnerable phase in the life of a plant. Plant species differ greatly in probability to recruit seedlings per capita of seeds shed. In a moist tropical forest in Panama, Wright et al. (2005) counted seeds in 200 traps placed over a 50 ha area weekly, and enumerated seedlings recruited in adjacent plots yearly (in the dry season to count all seedlings that germinated and survived throughout the previous rainy season). The number of seedlings recruited per seed ranged from 0.0003 to 0.15 among 32 species of trees, lianas, and shrubs based on the total sums across the 8 years. These differences among species may reflect their differences in susceptibility to postdispersal seed predation and disease, as well as probability of seed survival in soils, seedling emergence, and early seedling survival. The importance of natural enemies was strongly suggested in this data set, because recruitment probability was negatively dependent on local seed density in all 32 species. Moles and Westoby (2006) demonstrate the significant advantage of large seed size in postdispersal seed survival, survival in soil, and early seedling survival by compiling a large data set mainly from Australia. However, there is a large unexplained variation at a given seed size. This is perhaps not surprising because species show considerable variation within a seed-size category in allocation patterns of seed reserves to construct seedlings, and so suffer differently from different potential causes of mortality.

Young seedlings are susceptible to many hazards including predators attracted to seeds that remain attached to seedlings (Smyth 1978), desiccation (Miles 1972, Maruta 1976, Engelbrecht et al. 2006), pathogens (Augspuger 1983a,b, 1984a,b, Packer and Clay 2000), winter death and grazing (Mack and Pyke 1984), competition from existing vegetation (Fenner 1978, Aguilera and Lauenroth 1993, Tyler and Dantonio 1995, Kolb and Robberecht 1996), and damage caused by litterfall (Clark and Clark 1989, Scariot 2000, Gillman et al. 2004). These hazards may operate sequentially. For example, during the first 2–4 weeks following germination, seedlings of a neotropical tree, *Tachigalia versicolor*, suffer a high mortality rate from mammalian grazers; however, after the fourth week, as hypocotyls become woody and mammalian attack ceases, and pathogens become the main source of mortality (Kitajima and Augspurger 1989). There are many trade-offs that affect evolution of species traits at early seedling stages. Large seeds attract seed-eating animals, but they create large seedlings that emerge above litter and herbaceous vegetation. Small seeds can have better contact with soil to absorb water, but the limited root length of small seedlings makes them more vulnerable to drought. Fast growth and development help reduce the duration of this vulnerable phase, but fast growth also requires soft tissue vulnerable to various natural enemies.

Physical and chemical defenses are important for avoiding predation by various grazing animals. Comparing eight tropical tree species, Alvarez-Clare (2005) found that tissue toughness of seedling stems and leaves was positively correlated with their first year survival. High tissue densities (= dry mass per unit volume of stem or leaves) are strongly and positively correlated with tissue toughness, and thus important for defense against not only grazing animals but also pathogenic microbes that cause damping-off disease (Augspurger 1984b). Mollusc grazing kills many seedlings in temperate grasslands (Barker 1989, Hulme 1994, Fenner and Thompson 2005), and may affect number of recruits (Hanley et al. 1995) and eventual species composition (Hanley et al. 1996). Across species, palatability to molluscs is not correlated between adult and seedling leaves, but seedling leaves are always more palatable than adult leaves of the same species (Fenner et al. 1999).

An important determinant of a seedling's likelihood of survival and establishment is whether the seed is deposited in a safe site, defined as a place where the seed is provided with (a) the stimuli for breaking dormancy, (b) the conditions and resources required for germination, and (c) the absence of predators, competitors, pathogens, and toxins (Harper 1977, Fenner and Thompson 2005). Since different species have different requirements and tolerances as seedlings, a safe site for one species may not be safe for another. For example, in a temperate forest in Chile, seedling distribution of small-seeded species is more biased toward elevated microsites, such as logs, than large-seeded species (Lusk and Kelly 2003). Grain size of the substrate affects seedling emergence and establishment in an Alpine environment (Chambers 1995). Flood plain forests cannot be colonized by terra firme species whose seeds are not buoyant (Lopez 2001). Slopes provide microsites free of litter as well as greater moisture availability during the dry season, which are favored by small-seeded species (Daws et al. 2002b). A more common requirement for many seedlings, however, is the absence of competition from larger plants within the immediate vicinity. Closed vegetation provides an inhospitable arena for seedling establishment. Breaks in continuous vegetation cover mean not only higher availability of light, water, and nutrients, but also difference in activity of natural enemies. Mortality due to fungal pathogens is typically lower in gaps than that in the shaded forest understory, possibly because fungi prefer moist and shaded environments in general (Augspurger 1984a, Hood et al. 2004). In contrast, insect herbivores tend to be more abundant and active in gaps (e.g., Chacom and Armesto 2006), but higher photosynthetic income makes it easier for seedlings to tolerate herbivory.

Gaps of different sizes, as well as different positions within a gap, variably affect seedling emergence, growth, and survival. When fates of seedlings are followed in experimental gaps of different sizes, species often differ in their responses in relation to seed size (Gross 1984, McConnaughay and Bazzaz 1987, Bullock et al. 1995, Gray and Spies 1996, Dalling et al. 1999, Pearson et al. 2003). Differential responses of seed emergence in relation to gap size (Daws et al. 2002a) as well as seedling survival (Brokaw 1987) seem to explain species differences in minimum gap size requirements for seedling recruitment. On the other hand, growth rates of 12 tropical pioneer species in large- and small-gap environments are positively correlated, and thus, cannot explain their differences in gap-size preference observed in the field (Dalling et al. 2004). Due to the large environmental gradient from the center to the edge of a given gap (Brown 1993), a seedling's position within a gap may be more important to its survival than gap size per se (Brown and Whitmore 1992). One species may be favored in the center, whereas others survive better near the margins. Gap shape (which determines the ratio of margin to area) may be important for this reason.

There is clearly also a large stochastic element governing regeneration, influenced by unpredictable factors such as the presence of the parent in the vicinity, the absence of grazers, the occurrence of suitable weather conditions, all coinciding at the right place and time. Hence, it is difficult to predict species composition of a given gap, and gaps do not increase species diversity on a per-stem basis (Hubbell et al. 1999, Brokaw and Busing 2000).

Yet, within this haphazard framework, it is still possible to show that certain types of gaps favor the establishment of certain species from seed. A practical example of this is found in forestry management practice in which natural regeneration is encouraged for timber species (Fredericksen and Mostacedo 2000, van Rheenen et al. 2004, Makana and Thomas 2005). Effective treatments to enhance regeneration of timber species, many of which are light demanding as seedlings, include enlargement of logging gaps, soil surface scarification, and maintenance of seed parents near the gaps.

In small gaps or shaded environments, the attainment of a minimum size may be necessary for a seedling to secure an independent existence. An increased height should be most useful to seedlings in conditions where there is a steep gradient of light (due to shade from surrounding vegetation; Grime and Jeffrey 1965, Leishman and Westoby 1994a), or for seeds germinating below litter (Molofsky and Augspurger 1992). The hypocotyl or epicotyl may elongate in response to a low red/far red ratio of light coming through surrounding vegetation (Ballaré et al. 1988, Leishman and Westoby 1994a). In relation to their habitats and phylogeny, species and ecotypes differ in sensitivity of hypocotyl extension to red/far red ratios (Morgan and Smith 1979, Corré 1983). In a study of 15 tropical tree species, hypocotyls elongate in response to red/far red ratio only in Bombacaceae, a family largely represented by pioneer species (Kitajima 1994). Similarly, ecotypes from open habitats but not from forests exhibit stem elongation response to red/far red ratios (Dudley and Schmitt 1995). Stem elongation, however, is perhaps a maladaptive response in forest understories because it does not allow escape from shade; it only makes stems weak and more likely to topple. Initial seedling morphology of each species, including the degree of stem elongation and leaf display patterns, must have evolved in relation to its regeneration niche.

Clearly, how seed reserves are allocated and how long they support the energy and nutrient demands of seedlings are important for seedling recruitment (Kitajima 2002). However, that is not the end of the regeneration phase. Eventually, seedlings must become independent of seed reserves, initiating a completely autotrophic way of life to keep growing toward reproductive maturity.

SEEDLING GROWTH AND SURVIVAL

After initial construction of leaves and roots with seed reserves, seedlings of some species continue to grow, whereas others appear to wait with little visible growth. In general, seed size is negatively correlated with seedling relative growth rate (RGR) between species (e.g., Figure 18.3 for tropical tree species; see Shipley and Peters 1990 for review). This has been found in a wide range of plant families and habitats, for example, pasture grasses and legumes (Fenner and Lee 1989); species from variety of climatic conditions in Australia (Jurado and Westoby 1992, Swanborough and Westoby 1996); and woody plants in temperate (Cornelissen et al. 1996, Reich et al. 1998a) and tropical climates (Kitajima 1994, Huante et al. 1995, Poorter and Rose 2005); Mediterranean annuals (Maranon and Grubb 1993). This negative correlation is stronger under greater light availability in which seed reserve mass represents a smaller proportion of the seedling mass. Thus, it is not a mere reflection of allometry or autocorrelation (which is indeed expected because seed mass or initial seedling mass is the denominator in RGR calculation).

Most likely, the negative interspecific correlation between seed mass and seedling RGR stems from among-species differences in allocation patterns that evolved in relation to their life-history strategies. First, small-seeded species tend to have thin, epigeal, leaf-like cotyledons (Kitajima 1996, Wright et al. 2000, Zanne et al. 2005). Photosynthetic cotyledons allows them to start using light as the main source of energy earlier than they could if they had storage-type cotyledons (Kitajima 1992b, 2002). Second, small-seeded species tend to have a higher specific leaf area (SLA, leaf area divided by leaf mass) and leaf area ratio (LAR, leaf

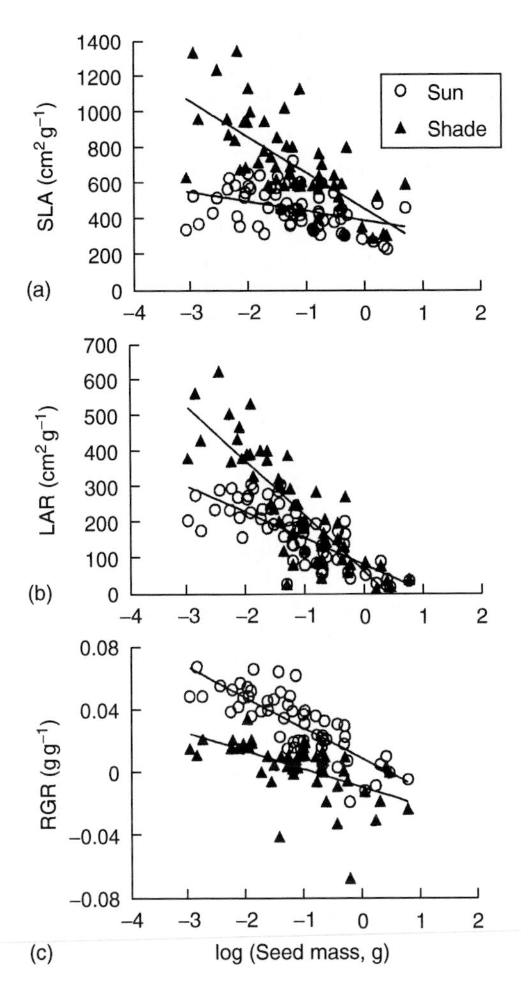

FIGURE 18.3 Associations of seed mass with morphological traits and relative growth rates of seedlings for neotropical woody species in a seasonal, moist forest in Panama. Seedlings of each species were raised from seeds in controlled sun and shade conditions (23% and 1% of open sky). (a) SLA and (b) LAR were determined at the full expansion of the first true leaves, and (c) subsequent RGR was determined from harvest 10 weeks later. Large-seeded species had significantly smaller SLA, LAR, and RGR in both sun and shade (linear regressions shown; $P < 0.002$ for SLA in sun, $P < 0.0001$ for all others; $n = 57$–61 and 51–58 for sun and shade, respectively). (Data from Kitajima, K., The Importance of Cotyledon Functional Morphology and Patterns of Seed Reserve Utilization for the Physiological Ecology of Neotropcial Tree Seedlings. Ph.D. thesis, University of Illinois, Urbana, Illinois, 1992a.)

area divided by total plant dry mass), the two traits known to be important for explaining difference in RGR (see Chapter 3). Although seedlings of different herbaceous and woody plants differ in RGR, the relationship of RGR with SLA is very similar across growth forms (Wright and Westoby 2001). Additional correlations of early seedling morphology and seed size, such as the relationship between root architecture and seed size (Huante et al. 1992, Kohyama and Grubb 1994, Reich et al. 1998a) also contribute to the relationship between seed mass and relative growth rates.

Another general rule that has gained increasing support during the last 10 years is the negative interspecific correlation between growth rates and survival of seedlings. In general, plant species from resource-poor habitats have slower maximum growth rates than those from resource-rich habitats (Chapter 3). In high-light environments conducive to fast growth (e.g., gaps), shade-tolerant seedlings more abundant in shaded understory exhibit lower RGR than gap specialists. What is more interesting is that shade-tolerant species also exhibit lower RGR than gap specialists when both are grown in shade. In other words, species that grow fast in one light environment also tend to grow fast in another. Such significant concordance of RGR between sun and shade is demonstrated in a number of other studies (Ellison et al. 1993, Kitajima 1994, Osunkoya et al. 1994, Kobe et al. 1995, Poorter 1999, Valladares et al. 2000, Walters and Reich 2000, Bloor and Grubb 2003, Dalling et al. 2004, Baralotos et al. 2005), with few studies showing a lack of a relationship (Popma and Bongers 1988) or the opposite pattern (Agyeman et al. 1999). However, in terms of survival, shade-tolerant tree species tend to outperform shade-intolerant tree species in both shade and gaps in Neotropical forests (Augspurger 1984a, Kobe 1999). Thus, the general functional basis for high-shade survival is not the ability to grow fast in shade, but the ability to survive through avoidance or tolerance of inevitable tissue loss to various hazards. In sum, natural enemies appear to mediate niche specialization of seedlings along light gradients.

Natural enemies are also important in mediating growth-survival trade-off and specializations to rich and poor soils. This is elegantly demonstrated by Fine et al. (2004) in the Peruvian Amazon, where fertile alluvial soil and nutrient-poor white sand support contrasting tree communities. In a factorial experiment, they demonstrate that seedlings of alluvial-soil specialists grew faster than those of white-sand specialists in both soil types when they are protected from herbivores. However, alluvial-soil specialists suffer greater herbivory and lower leaf area growth rates than white-sand specialists when grown in white sand. Other studies also found that fast-growing species in rich soil also grow fast in poor soil in environments protected from herbivores (Huante et al. 1995, Lusk et al. 1997, Schreeg et al. 2005). Species from infertile soil tend to have lower SLA and greater leaf life span than those from rich soils (Wright et al. 2002), reflecting characteristics of leaves better protected against herbivores (Wright and Cannon 2001). These results suggest that avoidance of tissue loss is an important selective pressure in resource-limited environments, in which replacement of lost tissue requires a long time (Coley et al. 1985).

Allocation to support mutualistic microbes may be equally costly but as important as allocation to defense against natural enemies. Mycorrhizal fungi are perhaps the most important group of mutualists, which not only help seedlings acquire limiting soil nutrients especially phosphorus (e.g., Allsopp and Stock 1995), but also help them defend against soil-borne pathogens (Hood et al. 2004). Since mycorrhizal fungi differ in their effectiveness, seedling performance may be significantly altered by availability of beneficial mycorrhizal fungi (Kiers et al. 2000, Bray et al. 2003). Likewise, fungi that live inside leaves (endophytes) include beneficial species that help seedlings defend against pathogenic fungi (Arnold et al. 2003). The potential benefits, as well as energy costs of supporting such microbes for seedlings (Lovelock et al. 1997), are yet to be quantified in most systems.

Maintenance of positive carbon balance is a prerequisite for long-term survival, as well as continuous growth of seedlings. However, this does not necessarily mean maximization of the rate of net carbon gain and growth. The net carbon gain rate of a seedling is a function of (a) total photosynthetic rate minus (b) total respiration rate minus (c) tissue loss rate to natural enemies, disturbance, and tissue senescence. Thus, the maintenance of positive carbon balance can be achieved by maximization of (a), minimization of (b), and minimization of (c). When multiple tree species are compared, seedlings of all species exhibit acclimation responses to shade by decreasing respiration rates at the whole-plant level, but differences in respiration rates alone do not often explain difference in shade survival among species (Kitajima 1994, Reich et al. 1998b, Kaelke et al. 2001). The negative correlation between growth rates and survival observed among species differing in seedling shade tolerance, instead, points to the importance of difference in allocation to defense and storage. Fast-growing species tend to have high SLA and LAR, which tend to make them more susceptible to herbivory (Kitajima 1994, Cornelissen et al. 1996). Shorter leaf life span inherent to species with high SLA leads to a greater tissue loss rate in the long term, making it more difficult for them to maintain carbon balance as well (Sack and Grubb 2003). Cornelissen et al. (1996) also have shown that high tissue density is negatively correlated with RGR, and presumably positively correlated with high survival. The most direct demonstration of the importance of physical defense for seedling survival is provided by Alvarez-Clare (2005), who showed a positive correlation between tissue toughness and first-year survival among eight neotropical tree species.

Carbohydrate storage is also important for maintenance of positive carbon balance. No matter how well seedlings are defended, seedlings in canopy shade are likely to experience negative carbon balance from time to time because of variation in weather, physical disturbance, and attacks by natural enemies. For survival through such episodes of negative carbon balance, as well as to recover from it, seedlings must rely on stored carbohydrate reserves in the form of starch and sugar. In a comparison of seven neotropical tree species, Myers and Kitajima (2007) experimentally demonstrated that species that survive well in shade have greater total amount of sugar and starch in stems and roots, especially after receiving the additional stress of defoliation and heavy shading (0.08% of open sky condition). Interestingly, seedlings in stress treatments did not use carbohydrate reserves remaining in cotyledons, and cotyledon carbohydrate reserve size was uncorrelated with seedling survival. Carbohydrate reserves are also important for over-winter survival of temperate deciduous tree species (Canham et al. 1999), as well as survival of savanna tree species to fire (Hoffmann et al. 2004). Thus, carbohydrate storage is important for survival when seedlings experience negative carbon balance because of stress.

The small size of seedlings ultimately constrains strategies for the maintenance of positive carbon balance. In the studies cited in the previous paragraph, seedling survival was correlated with the total carbohydrate pool size (gram glucose equivalent), but not with tissue concentration (milligram glucose equivalent per gram dry mass) of starch and sugar. Since the total pool size is the product of concentration and biomass, there is an upper limit to the carbohydrate pool size for small seedlings. As a result, seedlings in resource-limited environments must avoid tissue loss, which would be difficult to replace (Coley et al. 1985). Thus, at the seedling stage, shade-tolerant species have leaves with low SLA and high tissue density, even though such leaves are not efficient in capture and conversion of light for photosynthetic production. However, larger plants can take a more opportunist strategy for survival in shade, by leaves with high SLA that allow high net carbon gain rates, even though faster leaf turnover poses frequent carbon demands. Thus, it is expected that a species with high SLA has a high light requirement for survival as seedlings, but once they achieve large size, they may be able to tolerate shading. For example, as seedlings, *Alseis blackiana* can grow and survive only in large gaps; however, as saplings, they can persist in shaded understories (Dalling et al. 2001). Growth, however, also brings about an increase in support biomass to leaf area, which causes a decrease in the ratio of photosynthesis to respiration at the whole plant level (Veneklaas and Poorter 1998, Delagrange et al. 2004). Thus, if they manage to survive and grow, seedlings have to deal with not only temporal changes in external environmental factors such as opening and closure of gaps, but also changes in physiological and morphological constraints associated with size.

When do plants graduate from the seedling stage? In other words, how long does the influence of seed and early seedling traits last? From a proximate perspective, seedling phase ends when the relative contribution of seed reserves becomes negligible relative to the cumulative autotrophic resource gain from their independent way of life. However, from the life-history perspective, seed and seedling traits are intimately associated with the overall life-history strategies and habitat preference of species. This point is well illustrated by the negative correlation of seed size with not only early seedling RGR, but also with sapling and adult traits, such as RGR and maximum height. Some correlations exist because of concordance in traits at different stages; thus, a reasonable null expectation is that the relative position of species along the growth-survival trade-off line is positively correlated between seedling and sapling stages (Gilbert et al. 2005). Likewise, preferences for light environments appear to be largely concordant between small and large juvenile stages (Poorter et al. 2005), even though they are decoupled from preferred light environment as adults. Yet, ontogenetic changes may lead to diversity of overall regeneration strategies among species, significantly contributing to the maintenance of species diversity (Baralotos et al. 2005). Transition from seedlings to larger juveniles is an understudied topic particularly important for long-lived woody species.

CONCLUSION

Regeneration from seeds is influenced by a wide range of environmental factors, plant characteristics, and stochastic events. The species composition of a plant community is a consequence of the successful regeneration of a selection of the potential species available. The long-term maintenance of each species requires the recurrent creation of suitable regeneration opportunities at appropriate intervals. At any one site, these opportunities are unlikely to remain constant with time, due to natural disturbance, human influences, and even climate change. Knowledge of regeneration requirements of key species is of great practical importance in vegetation management, either for commercial or conservation purpose. Comparative studies of the relationships between seed and seedling traits and regeneration requirements of the species are particularly useful in this context. Seed and seedling traits are the products of natural selection operating through life history trade-offs. How selective pressures on seed and seedling traits vary in contrasting environments provides insight into niche specialization, density-dependency, and colonization limitation, which are considered critical for community assembly processes at local and regional scales.

ACKNOWLEDGMENTS

I warmly acknowledge the contribution by Michael Fenner, who was the lead author for this chapter in its earlier edition. His perspective on seed and seedling ecology had a profound influence on my research interests over the years. This revision was prepared during sabbatical in Panama under sponsorship of the Smithsonian Tropical Research Institute and financial support from NSF Grant 0093303 to KK. I would like to thank Helene Muller-Landau, Kelly Anderson, Joseph Phillips, Jim Dalling, and Amy Zanne for their constructive comments.

REFERENCES

Adler, F.R. and H.C. Muller-Landau, 2005. When do localized natural enemies increase species richness? Ecology Letters 8: 438–447.

Aguilera, M.O. and W.K. Lauenroth, 1993. Seedling establishment in adult neighbourhoods intraspecific constraints in the regeneration of the bunchgrass *Bouteloua gracilis*. Journal of Ecology 81: 253–261.

Agyeman, V.K., M.D. Swaine, and J. Thompson, 1999. Responses of tropical forest tree seedlings to irradiance and the derivation of a light response index. Journal of Ecology 87: 815–827.

Alcantara, J.M. and P.J. Rey, 2003. Conflicting selection pressures on seed size: evolutionary ecology of fruit size in a bird-dispersed tree, *Olea europaea*. Journal of Evolutionary Biology 16: 1168–1176.

Alvarez-Clare, S., 2005. Biomechanical properties of tropical tree seedlings as a functional correlate of shade tolerance. MS thesis, University of Florida, Gainesville, USA.
Allen, R.B. and K.H. Platt, 1990. Annual seedfall variation in Nothofagus solandri (Fagaceae), Canterbury, New Zealand. Oikos 57: 199–206.
Allsopp, N. and W.D. Stock, 1995. Relationships between seed reserves, seedling growth and mycorrhizal responses in 14 related shrubs (Rosidae) from a low-nutrient environment. Functional Ecology 9: 248–254.
Arnold, A.E., L.C. Mejia, D. Kyllo, E.I. Rojas, Z. Maynard, N. Robbins, and E.A. Herre, 2003. Fungal endophytes limit pathogen damage in a tropical tree. Proceedings of the National Academy of Sciences of the United States of America 100: 15649–15654.
Askew, A.P., D. Corner, D.J. Hodkinson, and K. Thompson, 1997. A new apparatus to measure the rate of fall of seeds. Functional Ecology 11: 121–125.
Asquith, N.M., S.J. Wright, and M.J. Clauss, 1997. Does mammal community composition control recruitment in neotropical forests? Evidence from Panama. Ecology 78: 941–946.
Augspurger, C.K., 1983a. Offspring-recruitment around tropical trees: changes in cohort distance with time. Oikos 40: 189–196.
Augspurger, C.K., 1983b. Seed dispersal of the tropical tree, *Platypodium elegans*, and the escape of its seedlings from fungal pathogens. Journal of Ecology 71: 759–771.
Augspurger, C.A., 1984a. Seedling survival of tropical tree species; interactions of dispersal distance, light gaps, and pathogens. Ecology 65: 1705–1712.
Augspurger, C.K., 1984b. Light requirements of neotropical tree seedlings: a comparative study of growth and survival. Journal of Ecology 72: 777–795.
Augspurger, C.K., 1986. Morphological and dispersal potential of wind-dispersed diaspores of neotropical trees. American Journal of Botany 73: 353–363.
Augspurger, C.K. and S.E. Franson, 1988. Input of wind-dispersed seeds into light-gaps and forest sites in a neotropical forest. Journal of Tropical Ecology 4: 239–252.
Augspurger, C.K. and K. Kitajima, 1992. Experimental studies of seedling recruitment from contrasting seed distributions. Ecology 73: 1270–1284.
Auld, T.D., 1986. Variation in predispersal seed predation in several Australian *Acacia* spp. Oikos 47: 319–326.
Baker, H.G., 1972. Seed weight in relation to environmental condition in California. Ecology 53: 997–1010.
Ballardie, R.T. and R.J. Whelan. 1986. Masting, seed dispersal and seed predation in the cycad *Macrozamia communis*. Oecologia 70: 100–105.
Ballaré, C.L., R.A. Sánchez, A.L. Scopel, and C.M. Chersa, 1988. Morphological responses of *Datura ferox* L. seedlings to the presence of neighbours. Oecologia 76: 288–293.
Baralotos, C., D.E. Goldberg, and D. Bonal, 2005. Performance trade-offs among tropical tree seedlings in contrasting microhabitats. Ecology 86: 2461–2472.
Barclay, A.S. and F.R. Earl, 1974. Chemical analyses of seeds. III. Oil and protein content of 1253 species. Economic Botany 28: 178–236.
Barker, G.M., 1989. Slug problems in New Zealand pastoral agriculture. In: I.F. Henderson, ed. Slugs and Snails in World Agriculture. British Crop Protection Council, BCPC Monograph, Thornton Heath, UK, pp. 59–68.
Baskin, J.M. and C.C. Baskin, 1980. Ecophysiology of secondary dormancy in seeds of *Ambrosia artemisiifolia*. Ecology 61: 475–480.
Baskin, J.M. and C.C. Baskin, 1985. The annual dormancy cycle in buried weed seeds: a continuum. BioScience 35: 492–498.
Baskin, C.C. and J.M. Baskin, 1998. Seeds: Ecology, Biogeography and Evolution of Dormancy and Germination. Academic Press, San Diego, CA.
Beattie, A.J. and D.C. Culver, 1982. Inhumation: How ants and other invertebrates help seeds. Nature 297: 627.
Bekker, R.M., J.P. Bakker, U. Grandin, R. Kalamees, P. Milberg, P. Poschlod, K. Thompson, and J.H. Willems, 1998. Seed size, shape and vertical distribution in the soil: Indicators of seed longevity. Functional Ecology 12: 834–842.

Bell, T., R.P. Freckleton, and O.T. Lewis, 2006. Plant pathogens drive density-dependent seedling mortality in a tropical tree. Ecology Letters 9: 569–574.

Benkman, C.W., 1995. The impact of tree squirrels (*Tamiasciurus*) on limber pine seed dispersal adaptations. Evolution 49: 585–592.

Bennet, A. and J. Krebs, 1987. Seed dispersal by ants. Trends in Ecology & Evolution 2: 291–292.

Bloor, J.M.G. and P.J. Grubb, 2003. Growth and mortality in high and low light: Trends among 15 shade-tolerant tropical rain forest tree species. Journal of Ecology 91: 77–85.

Bond, W. and P. Slingby, 1984. Collapse of an ant-plant mutualism: The Argentine ant (*Iridomyrmex humilis*) and myrmecochorous Proteaceae. Ecology 65: 1031–1037

Bouwmeester, H.J., 1990. The Effect of Environmental Conditions on the Seasonal Dormancy Pattern and Germination of Weed Seeds, PhD thesis, Agricultural University of Wageningen, The Netherlands.

Bray, S.R., K. Kitajima, and D.M. Sylvia, 2003. Mycorrhizae differentially alter growth, physiology, and competitive ability of an invasive shrub. Ecological Applications 13: 565–574.

Brokaw, N.V.L., 1987. Gap-phase regeneration of three pioneer tree species in a tropical forest. Journal of Ecology 75: 9–19.

Brokaw, N. and R.T. Busing, 2000. Niche versus chance and tree diversity in forest gaps. Trends in Ecology & Evolution 15: 183–188.

Brown, N., 1993. The implications of climate and gap microclimate for seedling growth conditions in a Gornean lowland rain forest. Journal of Tropical Ecology 9: 153–168.

Brown, N.A.C., J. van Staden, M.I. Daws, and T. Johnson, 2003. Patterns in the seed germination response to smoke in plants from the Cape Floristic Region, South Africa. South African Journal of Botany 69: 514–525.

Brown, J.S. and D.L. Venable, 1986. Evolutionary ecology of seed bank annuals in temporally varying environments. American Naturalist 127: 31–47.

Brown, N.D. and T.C. Whitmore, 1992. Do dipterocarp seedlings really partition tropcial rainforest gaps? Philosophical Transactions of the Royal Society of London Series B 335: 369–378.

Bullock, J.M., B.C. Hill, J. Silvertown, and M. Sutton, 1995. Gap colonization as a source of grassland community change: effects of gap size and grazing on the rate and mode of colonization by different species. Oikos 72: 273–282.

Burke, M.J.W. and J.P. Grime, 1996. An experimental study of plant community invasibility. Ecology 77: 776–790.

Cain, M.L., B.G. Milligan, and A.E. Strand, 2000. Long-distance seed dispersal in plant populations. American Journal of Botany 87: 1217–1227.

Cain, M.L., R. Nathan, and S.A. Levin, 2003. Long-distance dispersal. Ecology 84: 1943–1944.

Canham, C.D., R.K. Kobe, E.F. Latty, and R.L. Chazdon, 1999. Interspecific and intraspecific variation in tree seedling survival: Effects of allocation to roots versus carbohydrate reserves. Oecologia 121: 1–11.

Carlquist, S., 1965. Island Life. New York, Natural History Press.

Chacon, P. and J.J. Armesto, 2006. Do carbon-based defences reduce foliar damage? Habitat-related effects on tree seedling performance in a temperate rainforest of Chiloe Island, Chile. Oecologia 146: 555–565.

Chambers, J.C., 1995. Relationships between seed fates and seedling establishment in an alpine ecosystem. Ecology 76: 2124–2133.

Chapman, C.A., L.J. Chapman, K. Vulinec, A. Zanne, and M.J. Lawes, 2003. Fragmentation and alteration of seed dispersal processes: An initial evaluation of dung beetles, seed fate, and seedling diversity. Biotropica 35: 382–393.

Chesson, P.L., 1985. Coexistence of competitors in spatially and temporally varying environments: A look at the combined effects of different sorts of variability. Theoretical Population Biology 28: 263–287.

Clark, D.B. and D.A. Clark, 1989. The role of physical damage in the seedling mortality regime of a neotropical rain forest. Oikos 55: 225–230.

Clark, J.S., M. Silman, R. Kern, E. Macklin, and J. HilleRisLambers, 1999. Seed dispersal near and far: Patterns across temperate and tropical forests. Ecology 80: 1475–1494.

Cody, M.L. and J.M. Overton, 1996. Short-term evolution of reduced dispersal in island plant populations. Journal of Ecology 84: 53–61.
Coley, P.D., J.P. Bryant, and F.S. Chapin III, 1985. Resource availability and plant anti-herbivore defense. Science 230: 895–899.
Connell, J.H., 1971. On the role of natural enemies in preventing competitive exclusion in some marine animals and in rain forest trees. In: P.J.D. Boer and G.R. Gradwell, eds. Dynamics of Populations. Pudoc, Wageningen, pp. 290–310.
Cornelissen, J.H.C., 1999. A triangular relationship between leaf size and seed size among woody species: allometry, ontogeny, ecology and taxonomy. Oecologia 118: 248–255.
Cornelissen, J.H.C., P.C. Diez, and R. Hunt, 1996. Seedling growth, allocation and leaf attributes in a wide range of woody plant species and types. Journal of Ecology 84: 755–765.
Corré, W.J., 1983. Growth and morphogenesis of sun and shade plants. II. The influence of light quality. Acta Botanica Neerlandica 32: 185–202.
Crawley, M.J., 2000. Seed predators and plant population dynamics. In: M. Fenner, ed. Seeds: The Ecology of Regeneration in Plant Communities, 2nd edition. CAB International, Wallingford, pp. 167–182.
Crawley, M.J. and C.R. Long, 1995. Alternate bearing, predator satiation and seedling recruitment in *Quercus robur*. Journal of Ecology 83: 683–696.
Crist, T.O. and Friese, C.F., 1993. The impact of fungi on soil seeds: implications for plants and granivores in a semiarid shrub-steppe. Ecology 74: 2231–2239.
Curran, L.M. and M. Leighton, 2000. Vertebrate responses to spatiotemporal variation in seed production of mast-fruiting dipterocarpaceae. Ecological Monographs 70: 101–128.
Curran, L.M. and C.O. Webb, 2000. Experimental tests of the spatiotemporal scale of seed predation in mast-fruiting Dipterocarpaceae. Ecological Monographs 70: 129–148.
Dalling, J.W. and J.S. Denslow, 1998. Changes in soil seed bank composition along a chronosequence of lowland secondary tropical forest, Panama. Journal of Vegetation Science 9: 669–678.
Dalling, J.W., K.E. Harms, and R. Aizprua, 1997a. Seed damage tolerance and seedling resprouting ability of *Prioria copaifera* in Panama. Journal of Tropical Ecology 13: 481–490.
Dalling, J.W., M.D. Swaine, and N.C. Garwood, 1997b. Soil seed bank community dynamics in seasonally moist lowland tropical forest, Panama. Journal of Tropical Ecology 13: 659–680.
Dalling, J.W., M.D. Swaine, and N.C. Garwood, 1998. Dispersal patterns and seed bank dynamics of two pioneer tree species in moist tropical forest, Panama. Ecology 79: 564–578.
Dalling, J.W., K. Winter, J.D. Nason, S.P. Hubbell, D.A. Murawski, and J.L. Hamrick, 2001. The unusual life history of *Alseis blackiana*: A shade- persistent pioneer tree? Ecology 82: 933–945.
Dalling, J.W., C.E. Lovelock, and S.P. Hubbell, 1999. Growth responses of seedlings of two neotropical pioneer species to simulated forest gap environments. Journal of Tropical Ecology 15: 827–839.
Dalling, J.W., H.C. Muller-Landau, S.J. Wright, and S.P. Hubbell, 2002. Role of dispersal in the recruitment limitation of neotropical pioneer species. Journal of Ecology 90: 714–727.
Dalling, J.W., K. Winter, and S.P. Hubbell, 2004. Variation in growth responses of neotropical pioneers to simulated forest gaps. Functional Ecology 18: 725–736.
Darwin, C., 1859. The Origin of Species. Murray, London.
Daws, M.I., D. Burslem, L.M. Crabtree, P. Kirkman, C.E. Mullins, and J.W. Dalling, 2002a. Differences in seed germination responses may promote coexistence of four sympatric *Piper* species. Functional Ecology 16: 258–267.
Daws, M.I., C.E. Mullins, D. Burslem, S.R. Paton, and J.W. Dalling, 2002b. Topographic position affects the water regime in a semideciduous tropical forest in Panama. Plant and Soil 238: 79–90.
Daws, M.I., N.C. Garwood, and H.W. Pritchard, 2005. Traits of recalcitrant seeds in a semi-deciduous tropical forest in Panama: some ecological implications. Functional Ecology 19: 874–885.
Daws, M.I., D. Orr, D. Burslem, and C.E. Mullins, 2006. Effect of high temperature on chalazal plug removal and germination in *Apeiba tibourbou* Aubl. Seed Science and Technology 34: 221–225.
Delagrange, S., C. Messier, M.J. Lechowicz, and P. Dizengremel, 2004. Physiological, morphological and allocational plasticity in understory deciduous trees: Importance of plant size and light availability. Tree Physiology 24: 775–784.

Dessaint, F., R. Chadoeuf, and G. Barralis, 1991. Spatial pattern analysis of weed seeds in the cultivated soil seed bank. Journal of Applied Ecology 28: 721–730.

Dudley, S.A. and J. Schmitt, 1995. Genetic differentiation in morphological responses to simulated foliage shade between populations of *Impatiens capensis* from open and woodland sites. Functional Ecology 9: 655–666.

Edwards, G.R. and M.J. Crawley, 1999. Rodent seed predation and seedling recruitment in mesic grassland. Oecologia 118: 288–296.

Ellison, A.M., J.S. Denslow, B.A. Loiselle, and M.D. Brenes, 1993. Seed and seedling ecology of neotropical. Melastomataceae. Ecology 74: 1733–1749.

Engelbrecht, B.M.J., J.W. Dalling, T.R.H. Pearson, R.L. Wolf, D.A. Galvez, T. Koehler, M.T. Tyree, and T.A. Kursar, 2006. Short dry spells in the wet season increase mortality of tropical pioneer seedlings. Oecologia, 148: 258–269.

Eriksson, O., E.M. Friis, and P. Lofgren, 2000. Seed size, fruit size, and dispersal systems in angiosperms from the early cretaceous to the late tertiary. American Naturalist 156: 47–58.

Ezoe, H., 1998. Optimal dispersal range and seed size in a stable environment. Journal of Theoretical Biology 190: 287–293.

Fenner, M., 1978. Susceptability to shade in seedlings of colonizing and closed turf species. New Phytologist 81: 739–744.

Fenner, M., 1980. Germination tests on thirty-two East African weed species. Weed Research 20: 135–138.

Fenner, M., 1983. Relationships between seed weight, ash content and seedling growth in twenty-four species of compositae. New Phytologist 95: 697–706.

Fenner, M., 1991. Irregular seed crops in forest trees. Quarternary Journal of Forestry 85: 166–172.

Fenner, M., 2000. Seeds: The ecology of regeneration in plant communities, 2nd edition. CABI, Wallingford.

Fenner, M. and W.G. Lee, 1989. Growth of seedlings of pasture grasses and legumes deprived of single mineral nutrients. Journal of Applied Ecology 26: 223–232.

Fenner, M. and K. Thompson, 2005. Ecology of Seeds. Cambridge University Press, Cambridge.

Fenner, M., M.E. Hanley, and R. Lawrence, 1999. Comparison of seedling and adult palatability in annual and perennial plants. Functional Ecology 13: 546–551.

Fine, P.V.A., I. Mesones, and P.D. Coley, 2004. Herbivores promote habitat specialization by trees in Amazonian forests. Science 305: 663–665.

Forget, P.-M., 1990. Seed-dispersal of *Vacapoua americana* (Caesalpiniaceae) by caviomorph rodents. Journal of Tropical Ecology 6: 459–468.

Forget, P.M., 1997. Ten-year seedling dynamics in *Vouacapoua americana* in French Guiana: A hypothesis. Biotropica 29: 124–126.

Forget, P.M. and T. Milleron, 1991. Evidence for secondary seed dispersal by rodents in Panama. Oecologia 87: 596–599.

Forget, P.-M., K. Kitajima and R.B. Foster, 1999. Pre- and post-dispersal seed predation in a tropical tree, *Tachigalia versicolor* (Caesalpinaceae): effects of fruiting timing and among-tree variation. Journal of Tropical Ecology 15: 61–81.

Fragoso, J.M.V., K.M. Silvius, and J.A. Correa, 2003. Long-distance seed dispersal by tapirs increases seed survival and aggregates tropical trees. Ecology 84: 1998–2006.

Fredericksen, T.S. and B. Mostacedo, 2000. Regeneration of timber species following selection logging in a Bolivian tropical dry forest. Forest Ecology and Management 131: 47–55.

Ganeshaiah, K.N. and S.R. Uma, 1991. Seed size optimization in a wind dispersed tree *Butea monosperma* a trade-off between seedling establishment and pod dispersal efficiency. Oikos 60: 3–6.

Garwood, N.C., 1983. Seed germination in a seasonal tropical forest Panama: A community study. Ecological Monographs 53: 159–181.

Garwood, N.C., 1996. Functional morphology of tropical tree seedlings. In: M.D. Swaine, ed. The Ecology of Tropical Forest Tree Seedlings. UNESCO, Paris, pp. 59–129

Garwood, N.C. and J.R.B. Lighton, 1990. Physiological ecology of seed respiration in some tropical species. New Phytologist 115: 549–558.

Geritz, S.A., 1995. Evolutionarily stable seed polymorphism and small-scale spatial variation in seedling density. American Naturalist 146: 685–707.

Gilbert, B., S.J. Wright, H. Muller-Landau, K. Kitajima, and A. Hernandes, 2005. Life history trade-offs in tropical trees and lianas. Ecology 87: 1271–1288.

Gillman, L.N., J. Ogden, S.D. Wright, K.L. Stewart, and D.P. Walsh, 2004. The influence of macro-litterfall and forest structure on litterfall damage to seedlings. Austral Ecology 29: 305–312.

Godnan, D., 1983. Pest Slugs and Snails. Springer-Verlag, Berlin.

Gonzalez, E., 1993. Effect of seed size on germination and seedling vigor of *Virola koschny* Warb. Forest Ecology Management 57: 275–281.

Gorski, T., K. Gorska, and J. Nowicki, 1977. Germination of seeds of various species under leaf canopy. Flora 166: 249–259.

Goulding, M., 1980. The Fishes and the Forest. University of California Press, Berkeley, CA.

Gray, A.N. and T.A. Spies, 1996. Gap size, within-gap position and canopy structure effects on conifer seedling establishment. Journal of Ecology 84: 635–645.

Green, P.T. and P.A. Juniper, 2004. Seed-seedling allometry in tropical rain forest trees: Seed mass-related patterns of resource allocation and the 'reserve effect'. Journal of Ecology 92: 397–408.

Greene, D.F., C.D. Canham, K.D. Coates, and P.T. Lepage, 2004. An evaluation of alternative dispersal functions for trees. Journal of Ecology 92: 758–766.

Greig, N., 1993. Predispersal seed predation on five *Piper* species in tropical rainforest. Oecologia 93: 412–420.

Grime, J.P. and D.W. Jeffrey, 1965. Seedling establishment in vertical gradients of sunlight. Journal of Ecology 53: 621–642.

Gross, K., 1984. Effects of seed size and growth form on seedling establishment of six monocarpic perennial plants. Journal of Ecology: 369–387.

Gross, C.L., M.A. Whalen, and M.H. Andrew, 1991. Seed selection and removal by ants in a tropical savanna woodland in northern Australia. Journal of Tropical Ecology 7: 99–112.

Grubb, P.J., 1977. The maintenance of species-richness in plant communities: the importance of the regeneration niche. Biological Review 52: 107–145.

Grubb, P.J., D.J. Metcalfe, and D. Coomes, 2005. Comment on "A brief history of seed size." Science 310: 783A.

Hammond, D.S. and V.K. Brown, 1998. Disturbance, phenology and life-history characteristics: Factors influencing distance/density-dependent attack on tropical seeds and seedlings. In: D.M. Newberry, H.H.T. Prins, and N.D. Brown, eds. Dynamcis of Tropical Communities. Blackwell Scientific, Oxford, pp. 51–78.

Hanley, M.E. and M. Fenner, 1997. Seedling growth of four fire-following Mediterranean plant species deprived of single mineral nutrients. Functional Ecology 11: 398–405.

Hanley, M.E. and M. Fenner, 1998. Pre-germination temperature and the survivorship and onward growth of Mediterranean fire-following plant species. Acta Oecologica 19: 181–187.

Hanley, M.E., M. Fenner, and P.J. Edwards, 1995. The effect of seedling age on the likelihood of herbivory by the slug *Deroceras reticulatum*. Functional Ecology 9: 754–759.

Hanley, M.E., M. Fenner, and P.J. Edwards, 1996. Mollusc grazing and seedling survivorship of four common grassland plant species: the role of gap size, species and season. Acta Oecologica 17: 331–341.

Hardesty, B.D., C.W. Dick, A. Kremer, S. Hubbell, and E. Bermingham, 2005. Spatial genetic structure of Simarouba amara Aubl. (Simaroubaceae), a dioecious, animal-dispersed Neotropical tree, on Barro Colorado Island, Panama. Heredity 95: 290–297.

Harms, K.E., S.J. Wright, O. Calderon, A. Hernandez, and E.A. Herre, 2000. Pervasive density-dependent recruitment enhances seedling diversity in a tropical forest. Nature 404: 493–495.

Harper, J.L., 1977. Population Biology of Plants. Academic Press, London.

Hewitt, N., 1998. Seed size and shade-tolerance—a comparative analysis of north American temperate trees. Oecologia 114: 432–440.

Hilhorst, H.W.M. and Karssen, C.M., 2000. Effect of chemical environment on seed germination. In: M. Fenner, ed. Seeds. The Ecology of Regeneration in Plant Communities. CAB International, Wallingford, UK, pp. 293–310.

Hladik, A. and S. Miquel, 1990. Seedling types and plant establishment in an African rain forest. In: K.S. Bawa and M. Hadley, eds. Reproductive Ecology of Tropical Forest Plants, Parthenon, Carnforth, UK, pp. 261–282.

Hnatiuk, S.H., 1978. Plant dispersal by the Aldabran giant tortoise *Geochelone gigantea* (Scheigger). Oecologia 36: 345–350.
Hodkinson, D.J., A.P. Askew, K. Thompson, J.G. Hodgson, J.P. Bakker, and R.M. Bekker, 1998. Ecological correlates of seed size in the British flora. Functional Ecology 12: 762–766.
Hoffmann, W.A., B. Orthen, and A.C. Franco, 2004. Constraints to seedling success of savanna and forest trees across the savanna-forest boundary. Oecologia 140: 252–260.
Holl, K.D., 1999. Factors limiting tropical rain forest regeneration in abandoned pasture: Seed rain, seed germination, microclimate, and soil. Biotropica 31: 229–242.
Hood, L.A., M.D. Swaine, and P.A. Mason, 2004. The influence of spatial patterns of damping-off disease and arbuscular mycorrhizal colonization on tree seedling establishment in Ghanaian tropical forest soil. Journal of Ecology 92: 816–823.
Hopkins, M.S. and A.W. Graham, 1987. The viability of seeds of rainforest species after experimental soil burials under wet lowland forest in northeastern Australia. Australian Journal of Ecology 12: 97–108.
Horn, M.H., 1997. Evidence for dispersal of fig seeds by the fruit-eating characid fish *Brycon guatemalensis regan* in a Costa Rican tropical rain forest. Oecologia 109: 259–264.
Horvitz, C.C. and D.W. Schemske, 1994. Effects of dispersers, gaps, and predators on dormancy and seedling emergence in a tropical herb. Ecology 75: 1949–1958.
Howe, H.F., 1989. Scatter and clump-dispersal and seedling demography: Hypothesis and implications. Oecologia 79: 417–426.
Huante, P., E. Rincon, and M. Gavito, 1992. Root system analysis of seedlings of seven tree species from a tropical dry forest in Mexico. Trees: Structure and Function 6: 77–82.
Huante, P., E. Rincón, and I. Acosta, 1995. Nutrient availability and growth rate of 34 woody species from a tropical deciduous forest in Mexico. Functional Ecology 9: 849–858.
Hubbell, S.P., R.B. Foster, S.T. O'Brien, K.E. Harms, R. Condit, B. Wechsler, S.J. Wright, and S.L. de Lao, 1999. Light-gap disturbances, recruitment limitation, and tree diversity in a neotropical forest. Science 283: 554–557.
Hughes, L. and M. Westoby, 1992. Fate of seeds adapted for dispersal by ants in Australian sclerophyll vegetation. Ecology 73: 1285–1299.
Hulme, P.E., 1994. Seedling herbivory in grassland: relative impact of vertebrate and invertebrate herbivores. Journal of Ecology 82: 873–880.
Hyatt, L.A., M.S. Rosenberg, T.G. Howard, G. Bole, W. Fang, J. Anastasia, K. Brown, R. Grella, K. Hinman, J.P. Kurdziel, and J. Gurevitch, 2003. The distance dependence prediction of the Janzen-Connell hypothesis: A meta-analysis. Oikos 103: 590–602.
Ibarra-Manriquez, G., M.M. Ramos, and K. Oyama, 2001. Seedling functional types in a lowland rain forest in Mexico. American Journal of Botany 88: 1801–1812.
Jansen, P.A., F. Bongers, and L. Hemerik, 2004. Seed mass and mast seeding enhance dispersal by a neotropical scatter-hoarding rodent. Ecological Monographs 74: 569–589.
Janzen, D.H., 1969. Seed-eaters vs seed size, number, toxicity and dispersal. Evolution 23: 1–27.
Janzen, D.H., 1970. Herbivores and the number of tree species in tropical forests. American Naturalist 104: 501–528.
Janzen, D.H., 1974. Tropical Blackwater rivers, animals, and mast fruiting by the Dipterocarpaceae. Biotropica, 6: 69–103.
Janzen, D.H., 1984. Dispersal of small seeds by big herbivores: foliage is the fruit. American Naturalist 123: 338–353.
Jensen, T.S., 1985. Seed–seed predator interactions of European beech, *Fagus sylvatica*, and forest rodent, *Clethrionomys glareolus* and *Apodemus flavicollis*. Oikos 44: 149–156.
Joel, D.M., J.C. Steffens, and D.E. Matthews, 1995. Germination of weedy root parasites. In: J. Kigel and G. Galili, eds. Seed Development and Germination. Marcel Dekker, New York, pp. 567–597.
Johnson, W.C. and C.S. Adkisson, 1985. Dispersal of beech nuts by blue jays in fragmented landscapes. American Midland Naturalists 113: 319–324.
Jones, F.A., J. Chen, G.J. Weng, and S.P. Hubbell, 2005. A genetic evaluation of seed dispersal in the neotropical tree Jacaranda copaia (Bignoniaceae). American Naturalist 166: 543–555.
Jurado, E. and M. Westoby, 1992. Seedling growth in relation to seed size among species of arid Australia. Journal of Ecology 80: 407–416.

Kaelke, C.M., E.L. Kruger, and P.B. Reich, 2001. Trade-offs in seedling survival, growth, and physiology among hardwood species of contrasting successional status along a light-availability gradient. Canadian Journal of Forest Research 31: 1602–1616.

Keeley, J.E., 1991. Seed germination and life history syndromes in the California Chaparral. Botanican Review 57: 81–116.

Keeley, J.E. and C.J. Fotheringham, 1998. Mechanism of smoke-induced seed germination in a post-fire chaparral annual. Journal of Ecology 86: 27–36.

Kelly, D. and V.L. Sork, 2002. Mast seeding in perennial plants: Why, how, where? Annual Review of Ecology and Systematics 33: 427–447.

Kennedy, P.G., N.J. Hausmann, E.H. Wenk, and T.E. Dawson, 2004. The importance of seed reserves for seedling performance: An integrated approach using morphological, physiological, and stable isotope techniques. Oecologia 141: 547–554.

Kiers, E.T., C.E. Lovelock, E.L. Krueger, and E.A. Herre, 2000. Differential effects of tropical arbuscular mycorrhizal fungal inocula on root colonization and tree seedling growth: implications for tropical forest diversity. Ecology Letters 3: 106–113.

Kikuzawa, K. and H. Koyama, 1999. Scaling of soil water absorption by seeds: An experiment using seed analogues. Seed Science Research 9: 171–178.

Kitajima, K., 1992a. The Importance of Cotyledon Functional Morphology and Patterns of Seed Reserve Utilization for the Physiological Ecology of Neotropcial Tree Seedlings. Ph.D. thesis, University of Illinois, Urbana, Illinois.

Kitajima, K., 1992b. Relationship between photosynthesis and thickness of cotyledons for tropical tree species. Functional Ecology 6: 582–589.

Kitajima, K., 1994. Relative importance of photosynthetic traits and allocation patterns as correlates of seedling shade tolerance of 13 tropical trees. Oecologia 98: 419–428.

Kitajima, K., 1996. Cotyledon functional morphology, seed reserve utilization, and regeneration niches of tropical tree seedlings. In: M.D. Swaine, ed. The Ecology of Tropical Forest Tree Seedlings. UNESCO, Paris, pp. 193–208.

Kitajima, K., 2002. Do shade-tolerant tropical tree seedlings depend longer on seed reserves? Functional growth analysis of three Bignoniaceae species. Functional Ecology 16: 433–444.

Kitajima, K. and C.K. Augspurger, 1989. Seed and seedling ecology of a monocarpic tropical tree, *Tachigalia versicolor*. Ecology 70: 1102–1114.

Kitajima, K. and D. Tilman, 1996. Seed banks and seedling establishment on an experimental productivity gradient. Oikos, 76:381–391.

Kivilaan, A. and R.S. Bandurski, 1981. The one hundred-year period for Dr. Beal's seed viability experiment. American Journal of Botany 68: 1290–1292.

Kobe, R.K., 1999. Light gradient partitioning among tropical tree species through differential seedling mortality and growth. Ecology 80: 187–201.

Kobe, R.K., S.W. Pacala, J.A.J. Silander, and C.D. Canham, 1995. Juvenile tree survivorship as a component of shade tolerance. Ecological Applications 5: 517–532.

Kohyama, T. and P.J. Grubb, 1994. Below- and above-ground allometries of shade-tolerant seedlings in a Japanese warm-temperate rain forest. Functional Ecology 8: 229–236.

Kolb, P.F. and R. Robberecht, 1996. *Pinus ponderosa* seedling establishment and the influence of competition with the bunchgrass *Agropyron spicatum*. International Journal of Plant Science 157: 509–515.

Kubitzki, K. and A. Ziburski, 1994. Seed dispersal in flood plain forests of Amazonia. Biotropica, 26: 30–43.

Kwit, C., D.J. Levey, and C.H. Greenberg, 2004. Contagious seed dispersal beneath heterospecific fruiting trees and its consequences. Oikos 107: 303–308.

Leck, M.A., V.T. Parker, and R.L. Simpson, 1989. Ecology of soil seed banks. Academic Press, San Diego, CA.

Lee, W.G., M. Fenner, and R.P. Duncan, 1993. Pattern of natural regeneration of narrow leaved snow tussock *Chionochloa rigida* ssp. rigida in Central Otago, New Zealand. New Zealand Journal of Botany 31: 117–125.

Leishman, M.R., 2001. Does the seed size/number trade-off model determine plant community structure? An assessment of the model mechanisms and their generality. Oikos 93: 294–302.

Leishman, M.R. and M. Westoby, 1994a. The role of large seed size in shaded conditions: Experimental evidence. Functional Ecology 8: 205–214.
Leishman, M.R. and M. Westoby, 1994b. The role of seed size in seedling establishment in dry soil conditions: experimental evidence from semi-arid species. Journal of Ecology 82: 249–258.
Leishman, M.R. and M. Westoby, 1998. Seed size and shape are not related to persistence in soil in Australia in the same way as in Britain. Functional Ecology 12: 480–485.
Leishman, M.R., M. Westoby, and E. Jurado, 1995. Correlates of seed size variation—a comparison among 5 temperate floras. Journal of Ecology 83: 517–529.
Levey, D.J. and M.M. Byrne, 1993. Complex ant plant interactions—rain-forest ants as secondary dispersers and postdispersal seed predators. Ecology 74: 1802–1812.
Levin, S.A., H.C. Muller-Landau, R. Nathan, and J. Chave, 2003. The ecology and evolution of seed dispersal: A theoretical perspective. Annual Review of Ecology Evolution and Systematics 34: 575–604.
Ligon, D.J., 1978. Reproductive interdependence of pinyon jays and pinyon pines. Ecological Monographs 48: 111–126.
Lopez, O.R., 2001. Seed flotation and postflooding germination in tropical terra firme and seasonally flooded forest species. Functional Ecology 15: 763–771.
Lott, R.H., G.N. Harrington, A.K. Irvine, and S. McIntyre, 1995. Density-dependent seed predation and plant dispersion of the tropical palm *Normanbya normanbyi.* Biotropica 27: 87–95.
Louda, S.M., 1982. Limitations of the recruitment of the shrub *Haplopappus squarrosus* (Asteraceae) by flower- and seed-feeding insects. Journal of Ecology 70: 43–53.
Louda, S.M. and M.A. Potvin, 1995. Effect of inflorescence-feeding insects on the demography and lifetime fitness of a native plant. Ecology 76: 229–245.
Lovelock, C.E., D. Kyllo, M. Popp, H. Isopp, A. Virgo, and K. Winter, 1997. Symbiotic vesicular-arbuscular mycorrhizae influence maximum rates of photosynthesis in tropical tree seedlings grown under elevated CO2. Australian Journal of Plant Physiology 24: 185–194.
Lusk, C.H., O. Contreras, and J. Figueroa, 1997. Growth, biomass allocation and plant nitrogen concentration in Chilean temperate rainforest tree seedlings: effects of nutrient availability. Oecologia 109: 49–58.
Lusk, C.H. and C.K. Kelly, 2003. Interspecific variation in seed size and safe sites in a temperate rain forest. New Phytologist 158: 535–541.
Mack, R.N. and D.A. Pyke, 1984. The demography of Bromus tectorum: The role of microclimate, grazing and disease. Journal of Ecology 72: 731–748.
Makana, J.R. and S.C. Thomas, 2004. Dispersal limits natural recruitment of African mahoganies. Oikos 106: 67–72.
Makana, J.R. and S.C. Thomas, 2005. Effects of light gaps and litter removal on the seedling performance of six African timber species. Biotropica 37: 227–237.
Maranon, T. and P.J. Grubb, 1993. Physiological basis and ecological significance of the seed size and relative growth rate relationship in Mediterranean annuals. Functional Ecology 7: 591–599.
Maron, J.L. and S.N. Gardner, 2000. Consumer pressure, seed versus safe-site limitation, and plant population dynamics. Oecologia 124: 260–269.
Maruta, E., 1976. Seedling establishment of *Polygonum cuspidatum* on Mt. Fuji. Japanese Journal of Ecology 26: 101–105.
Mazer, S.J., 1989. Ecological, taxonomic, and life history correlates of seed mass among Indiana dune angiosperms. Ecological Monographs 59: 153–175.
McConnaughay, K.D.M. and F.A. Bazzaz, 1987. The relationships between gap size and performance of several colonizing annuals. Ecology 68: 411–416.
McEuen, A.B. and L.M. Curran, 2004. Seed dispersal and recruitment limitation across spatial scales in temperate forest fragments. Ecology 85: 507–518.
McRill, M. and G.R. Sagar, 1973. Earthworms and seeds. Nature 243: 482.
Metcalfe, D.J. and P.J. Grubb, 1995. Seed mass and light requirements for regeneration of Southeast Asian rain forest. Canadian Journal of Botany 73: 817–826.
Milberg, P., M.A. Perez-Fernandez, and B.B. Lamont, 1998. Seedling growth response to added nutrients depends on seed size in three woody genera. Journal of Ecology 86: 624–632.

Miles, J., 1972. Early mortality and survival of self-sown seedlings in Glenfeshie, Inverness-shire. Journal of Ecology 61: 93–98.

Moegenburg, S.M., 1996. Sabal palmetto seed size: Causes of variation, choices of predators, and consequences for seedlings. Oecologia 106: 539–543.

Moles, A.T. and M. Westoby, 2003. Latitude, seed predation and seed mass. Journal of Biogeography 30: 105–128.

Moles, A.T. and M. Westoby, 2004. Seedling survival and seed size: a synthesis of the literature. Journal of Ecology 92: 372–383.

Moles, A.T. and M. Westoby, 2006. Seed size and plant strategy across the whole life cycle. Oikos 113: 91–105.

Moles, A.T., D.D. Ackerly, C.O. Webb, J.C. Tweddle, J.B. Dickie, and M. Westoby, 2005. A brief history of seed size. Science 307: 576–580.

Molofsky, J. and C.K. Augspurger, 1992. The effect of leaf litter on early seedling establishment in a tropical forest. Ecology 73: 68–77.

Molofsky, J. and B.L. Fisher, 1993. Habitat and predation effects on seedling survival and growth in shade-tolerant tropical trees. Ecology 74: 261–264.

Morgan, D.C. and H. Smith, 1979. A systematic relationship between phytochrome-controlled development and species habitat, for plants grown in simulated natural radiation. Planta 145: 253–258.

Muller-Landau, H.C. and B.D. Hardesty, 2005. Seed dispersal of woody plants in tropical forests: concepts, examples, and future directions. In: D. Burslem, M. Pinar, and S. Hartley, eds. Biotic Interactions in the Tropics: Their Role in the Maintenance of Species Diversity. Cambridge University Press, Cambridge, pp. 267–309.

Myers, J. and K. Kitajima, 2007. Carbohydrate storage enhances seedling shade and stress tolerance in a neotropical forest. Journal of Ecology 95: 383–395.

Nathan, R. and R. Casagrandi, 2004. A simple mechanistic model of seed dispersal, predation and plant establishment: Janzen-Connell and beyond. Journal of Ecology 92: 733–746.

Nathan, R. and H.C. Muller-Landau, 2000. Spatial patterns of seed dispersal, their determinants and consequences for recruitment. Trends in Ecology & Evolution 15: 278–285.

Ng, F.S.P., 1978. Strategies of establishment in Malayan forest trees. In: T.B. Tomlinson and H.M. Zimmerman, eds. Tropical Trees as Living Systems. Cambridge University Press, Oxford, pp. 129–162.

Norton, D.A. and D. Kelly, 1988. Mast seedling over 33 years by *Dacrydium cupressinum* Lamb. (rimu) (Podocarpaceae) in New Zealand: the importance of economies of scale. Functional Ecology 2: 399–408.

O'Dowd, D.J. and P.S. Lake, 1991. Red crabs in rain forest, Christmas Island: Removal and fate of fruits and seeds. Journal of Tropical Ecology 7: 113–122.

Okubo, A. and S.A. Levin, 1989. A theoretical framework for data analysis of wind dispersal of seeds and pollen. Ecology 70: 329–338.

Oladokun, M.A.O., 1989. Nut weight and nutrient contents of *Cola acuminata* and *C. nitida* (Sterculiaceae). Economic Botany 43: 17–22.

Osunkoya, O.O., J.E. Ash, M.S. Hopkins, and A.W. Graham, 1994. Influence of seed size and seedling ecological attributes on shade-tolerance of rain forest tree species in Northern Queensland. Journal of Ecology 82: 149–163.

Packer, A. and K. Clay, 2000. Soil pathogens and spatial patterns of seedling mortality in a temperate tree. Nature 404: 278–281.

Pammenter, N.W. and P. Berjak, 2000. Some thoughts on the evolution and ecology of recalcitrant seeds. Plant Species Biology 15: 153–156.

Pate, J.S., E. Rasins, J. Rullo, and J. Kuo, 1985. Seed nutrient reserves of Proteaceae with special reference to protein bodies and their inclusions. Annals of Botany 57: 747–770.

Paz, H. and M. Martinez-Ramos, 2003. Seed mass and seedling performance within eight species of *Psychotria* (Rubiaceae). Ecology 84: 439–450.

Pearson, T.R.H., D. Burslem, C.E. Mullins, and J.W. Dalling, 2002. Germination ecology of neotropical pioneers: Interacting effects of environmental conditions and seed size. Ecology 83: 2798–2807.

Pearson, T.R.H., D. Burslem, R.E. Goeriz, and J.W. Dalling, 2003. Interactions of gap size and herbivory on establishment, growth and survival of three species of neotropical pioneer trees. Journal of Ecology 91: 785–796.

Penning de Vries, F.W.T. and H.H. Van Laar, 1977. Substrate utilization in germinating seeds. In: J.J. Landsberg and C.V. Cutting, eds. Environmental Effects on Crop Physiology. Academic Press, London, pp. 217–228.

Piearce, T.G., N. Roggero, and R. Tipping, 1994. Earthworms and seeds. Journal of Biological Education 28: 195–202.

Pons, T.L., 1989. Breaking of seed dormancy by nitrate as a gap detection mechanism. Annals of Botany 63: 139–143.

Poorter, L., 1999. Growth responses of 15 rain-forest tree species to a light gradient: The relative importance of morphological and physiological traits. Functional Ecology 13: 396–410.

Poorter, L. and S. Rose, 2005. Light-dependent changes in the relationship between seed mass and seedling traits: a meta-analysis for rain forest tree species. Oecologia 142: 378–387.

Poorter, L., F. Bongers, F.J. Sterck, and H. Woll, 2005. Beyond the regeneration phase: differentiation of height-light trajectories among tropical tree species. Journal of Ecology 93: 256–267.

Popma, J. and F. Bongers, 1988. The effect of canopy gaps on growth and morphology of seedlings of rain forest species. Oecologia 75: 625–632.

Pritchard, H.W., M.I. Daws, B.J. Fletcher, C.S. Gamene, H.P. Msanga, and W. Omondi, 2004. Ecological correlates of seed desiccation tolerance in tropical African dryland trees. American Journal of Botany 91: 863–870.

Reader, R.J., 1993. Control of seedling emergence by ground cover and seed predation in relation to seed size for some old-field species. Journal of Ecology 81: 169–175.

Reich, P.B., M.G. Tjoelker, M.B. Walters, D.W. Vanderklein, and C. Bushena, 1998a. Close association of RGR, leaf and root morphology, seed mass and shade tolerance in seedlings of nine boreal tree species grown in high and low light. Functional Ecology 12: 327–338.

Reich, P.B., M.B. Walters, M.G. Tjoelker, D. Vanderklein, and C. Buschena, 1998b. Photosynthesis and respiration rates depend on leaf and root morphology and nitrogen concentration in nine boreal tree species differing in relative growth rate. Functional Ecology 12: 395–405.

Roberts, H.A. and P.M. Feast, 1972. Fate of seed of some annual weeds in different depths of cultivated and undisturbed soil. Weed Research 4: 296–307.

Roberts, H.A. and P.M. Feast, 1973. Emergence and longevity of seeds of annual weeds in cultivated and undisturbed soil. Journal of Applied Ecology 10: 133–143.

Rodriguez, M.D., A. Orozco-Segovia, M.E. Sanchez-Coronado, and C. Vazquez-Yanes, 2000. Seed germination of six mature neotropical rain forest species in response to dehydration. Tree Physiology 20: 693–699.

Russi, L., P.S. Cocks, and E.H. Roberts, 1992. The fate of legume seeds eaten by sheep from a mediterranean grassland. Journal of Applied Ecology 29: 772–778.

Russo, S.E. and C.K. Augspurger, 2004. Aggregated seed dispersal by spider monkeys limits recruitment to clumped patterns in *Virola calophylla*. Ecology Letters 7: 1058–1067.

Sack, L. and P.J. Grubb, 2003. Crossovers in seedling relative growth rates between low and high irradiance: analyses and ecological potential (reply to Kitajima and Bolker 2003). Functional Ecology 17: 281–287.

Salisbury, E.J., 1942. The Reproductive Capacity in Plants: Sutdies in Quantitative Biology. Bell, London.

Sanchez-Hidalgo, M.E., M. Martinez-Ramos, and J. Espinosa-Garcia, 1999. Chemical differntiation between leaves of seedlings and spatially close adult trees from the tropical rain-forest species *Nectandra ambigens* (Lauraceae): An alternative test of the Janzen-Connel model. Functional Ecology 13: 725–732.

Saverimuttu, T. and M. Westoby, 1996. Seedling longevity under deep shade in relation to seed size. Journal of Ecology 84: 681–689.

Scariot, A., 2000. Seedling mortality by litterfall in Amazonian forest fragments. Biotropica 32: 662–669.

Schreeg, L.A., R.K. Kobe, and M.B. Walters, 2005. Tree seedling growth, survival and morphology in response to landscape-level variation in soil resource availability in northern Michigan. Canadian Journal of Forest Research 35: 263–273.

Seiwa, K., A. Watanabe, K. Irie, H. Kanno, T. Saitoh, and S. Akasaka, 2002. Impact of site-induced mouse caching and transport behaviour on regeneration in Castanea crenata. Journal of Vegetation Science 13: 517–526.

Sevilla, G.H., O.N. Fernandez, D.P. Minon, and L. Montes, 1996. Emergence and seedling survival of *Lotus tenuis* in *Festuca arundinacea* pastures. Journal of Range Management 49: 509–511.

Shibata, M., H. Tanaka, S. Iida, S. Abe, T. Masaki, K. Niiyama, and T. Nakashizuka, 2002. Synchronized annual seed production by 16 principal tree species in a temperate deciduous forest, Japan. Ecology 83: 1727–1742.

Shipley, B. and R.H. Peters, 1990. The allometry of seed weight and seedling relative growth rate. Functional Ecology 4: 523–529.

Silva, H.R.D., M.C.D. Britto-Pereira, and U. Caramaschi, 1989. Frugivory and seed dispersal by *Hyla truncata*; a neotropical tree frog. Copeia 3: 781–783.

Silvera, K., J.B. Skillman, and J.W. Dalling, 2003. Seed germination, seedling growth and habitat partitioning in two morphotypes of the tropical pioneer tree *Trema micrantha* in a seasonal forest in Panama. Journal of Tropical Ecology 19: 27–34.

Silvertown, J.W., 1980a. The evolutionary ecology of mast seeding in trees. Biological Journal of the Linnean Society 14: 235–250.

Silvertown, J.W., 1980b. Leaf-canopy-induced seed dormancy in a grassland flora. New Phytologist 85: 109–118.

Slocum, M.G. and C.C. Horvitz, 2000. Seed arrival under different genera of trees in a neotropical pasture. Plant Ecology 149: 51–62.

Smyth, N., 1978. The natural history of the Central American agouti (*Dasypracta punctata*). Smithsonian Contribution to Zoology 257: 20.

Stanton, M., 1984. Seed variation in wild radish: Effect of seed size on components of seedling and adult fitness. Ecology 65: 1105–1112.

Stiles, E.W., 2000. Animals as seed dispersers. In: M. Fenner, ed. Seeds: The Ecology of Regeneration in Plant Communities, 2nd edition. CAB International, Wallingford, Oxford, pp. 111–124.

Stock, W.D., J.S. Pate, and J. Delfs, 1990. Influence of seed size and quality on seedling development under low nutrient conditions in five Australian and South African members of the Proteaceae. Journal of Ecology 78: 1005–1020.

Svenning, J.C. and S.J. Wright, 2005. Seed limitation in a Panamanian forest. Journal of Ecology, 93: 853–862.

Swanborough, P. and M. Westoby, 1996. Seedling relative growth rate and its components in relation to seed size: Phylogenetically independent contrasts. Functional Ecology 10: 176–184.

Tapper, P.G., 1996. Irregular fruiting in *Fraxinus excelsior*. Journal of Vegetation Science 3: 41–46.

Templeton, A.R. and D.R. Levin, 1979. Evolutionary consequences of seed pools. American Naturalist 114: 232–249.

Terborgh, J. and S.J. Wright, 1994. Effects of mammalian herbivores on plant recruitment in two neotropical forests. Ecology 75: 1829–1833.

Thompson, K., 1986. Small-scale heterogeneity in the seed bank of an acidic grassland. Journal of Ecology 74: 733–738.

Thompson, K. and J.P. Grime, 1979. Seasonal variation in the seed banks of herbaceous species in ten contrasting habitats. Journal of Ecology 67: 893–921.

Thompson, K., J.P. Grime, and A.G. Mason, 1977. Seed germination in response to diurnal fluctuations of temperature. Nature 267: 147–149.

Thompson, K., S.R. Band, and J.G. Hodgson, 1993. Seed size and shape predict persistence in soil. Functional Ecology 7: 236–241.

Tilman, D., 1994. Competition and biodiversity in spatially structured habitats. Ecology 75: 2–16.

Tilman, D., 1997. Community invasibility, recruitment limitation, and grassland biodiversity. Ecology 78: 81–92.

Toole, E.H., 1946. Final results of the Duval buried seed experiment. Journal of Agricultural Research. 72: 201–210.

Tyler, C.M. and C.M. Dantonio. 1995. The effects of neighbors on the growth and survival of shrub seedlings following fire. Oecologia 102: 255–264.

Valladares, F., S.J. Wright, E. Lasso, K. Kitajima, and R.W. Pearcy, 2000. Plastic phenotypic response to light of 16 congeneric shrubs from a Panamanian rainforest. Ecology 81: 1925–1936.
van Rheenen, H., R.G.A. Boot, M.J.A. Werger, and M.U. Ulloa, 2004. Regeneration of timber trees in a logged tropical forest in North Bolivia. Forest Ecology and Management 200: 39–48.
Vaughan, J.G., 1970. The Structure and Utilization of Oil Seeds.Chapman and Hall, London.
Vaughton, G. and M. Ramsey, 1998. Sources and consequences of seed mass variation in *Banksia marginata* (Proteaceae). Journal of Ecology 86: 563–573.
Vegis, A., 1964. Dormancy in higher plants. Annual Review of Plant Physiology 15: 185–224.
Veneklaas, E.J. and L. Poorter, 1998. Growth and carbon partitioning of tropical tree seedlings in contrasting light environments. In: H. Lambers, H. Poorter, and M.M.I. Van Vuuren, eds. Inherent variation in plant growth. Physiological mechanisms and ecological consequences. Backhuys Publishers, Leiden, pp. 337–361.
Vormisto, J., J.C. Svenning, P. Hall, and H. Balslev, 2004. Diversity and dominance in palm (Arecaceae) communities in terra firme forests in the western Amazon basin. Journal of Ecology 92: 577–588.
Walters, M.B. and P.B. Reich, 2000. Seed size, nitrogen supply, and growth rate affect tree seedling survival in deep shade. Ecology 81: 1887–1901.
Wang, B.C. and T.B. Smith, 2002. Closing the seed dispersal loop. Trends in Ecology & Evolution 17: 379–385.
Washitani, I., 1985. Field fate of *Amaranthus patulus* seeds subjected to leaf-canopy inhibition of germination. Oecologia 66: 338–342.
Washitani, I. and M. Masuda, 1990. A comparative study of the germination characteristics of seeds from a moist tall grassland community. Functional Ecology 4: 543–557.
Webb, C.O. and D.R. Peart, 1999. Seedling density dependence promotes coexistence of Bornean rain forest trees. Ecology 80: 2006–2017.
Wenny, D.G., 2001. Advantages of seed dispersal: A re-evaluation of directed dispersal. Evolutionary Ecology Research 3: 51–74.
Wenny, D.G. and D.J. Levey, 1998. Directed seed dispersal by bellbirds in a tropical cloud forest. Proceedings of the National Academy of Sciences of the United States of America 95: 6204–6207.
Westoby, M., 1998. A leaf-height-seed (LHS) plant ecology strategy scheme. Plant and Soil 199: 213–227.
Westoby, M., K. French, L. Hughes, B. Rice, and L. Rogerson, 1991. Why do more plant species use ants for dispersal on infertile compared with fertile soils? Australian Journal of Ecology 16: 445–455.
Westoby, M., E. Jurado, and M. Leishman. 1992. Comparative evolutionary ecology of seed size. Trends in Ecology & Evolutioni 7: 368–372.
Williamson, G.B. and K. Ickes. 2002. Mast fruiting and ENSO cycles—does the cue betray a cause? Oikos 97: 459–461.
Willson, M.F., 1993. Dispersal mode, seed shadows and colonization patterns. Vegetatio 107/108: 261–281.
Wolff, J.O., 1996. Population fluctuations of mast-eating rodents are correlated with production of acorns. Journal of Mammalogy 77: 850–856.
Wright, I.J. and K. Cannon, 2001. Relationships between leaf lifespan and structural defences in a low-nutrient, sclerophyll flora. Functional Ecology 15: 351–359.
Wright, S.J. and H.C. Duber. 2001. Poachers and forest fragmentation alter seed dispersal, seed survival, and seedling recruitment in the palm Attalea butyraceae, with implications for tropical tree diversity. Biotropica 33: 583–595.
Wright, I.J. and M. Westoby, 1999. Differences in seedling growth behaviour among species: Trait correlations across species, and trait shifts along nutrient compared to rainfall gradients. Journal of Ecology 87: 85–97.
Wright, I.J. and M. Westoby, 2001. Understanding seedling growth relationships through specific leaf area and leaf nitrogen concentration: Generalisations across growth forms and growth irradiance. Oecologia 127: 21–29.
Wright, S.J., C. Carrasco, O. Calderon, and S. Paton. 1999. The El Nino Southern Oscillation variable fruit production, and famine in a tropical forest. Ecology 80: 1632–1647.

Wright, I.J., H.T. Clifford, R. Kidson, M.L. Reed, B.L. Rice, and M. Westoby, 2000. A survey of seed and seedling characters in 1744 Australian dicotyledon species: cross-species trait correlations and correlated trait-shifts within evolutionary lineages. Biological Journal of the Linnean Society 69: 521–547.

Wright, I.J., M. Westoby, and P.B. Reich, 2002. Convergence towards higher leaf mass per area in dry and nutrient-poor habitats has different consequences for leaf life span. Journal of Ecology 90: 534–543.

Wright, S.J., H.C. Muller-Landau, O. Calderon, and A. Hernandez. 2005. Annual and spatial variation in seedfall and seedling recruitment in a neotropical forest. Ecology 86: 848–860.

Wulff, R.D., 1986. Seed Size variation in *Desmodium paniculatum*. II. Effects on seedling growth and physiological performance. Journal of Ecology 74: 99–114.

Zahawi, R.A. and C.K. Augspurger, 1999. Early plant succession in abandoned pastures in Ecuador. Biotropica 31: 540–552.

Zanne, A.E., C.A. Chapman, and K. Kitajima, 2005. Evolutionary and ecological correlates of early seedling morphology in East African trees and shrubs. American Journal of Botany 92: 972–978.

Zettler, J.A., T.P. Spira, and C.R. Allen, 2001. Ant-seed mutualisms: can red imported fire ants sour the relationship? Biological Conservation 101: 249–253.

19 Biodiversity and Interactions in the Rhizosphere: Effects on Ecosystem Functioning

Susana Rodríguez-Echeverría, Sofia R. Costa, and Helena Freitas

CONTENTS

INTRODUCTION

Understanding the implications for ecosystem function of soil biodiversity and processes is the last frontier in terrestrial ecology. Research on this field is lagging behind aboveground studies mainly because soil is such a complex matrix. Some soil processes, such as decomposition and mineralization of organic matter and biogeochemical cycles, have long been recognized as key components of ecosystems. In addition, recent studies in natural ecosystems have revealed that organisms from the rhizosphere—plant pathogens, parasites, herbivores, and mutualists—have a significant impact on natural plant communities (Van der Putten and Peters 1997, Klironomos 2002, De Deyn et al. 2004). The rhizosphere is a hot spot of soil biodiversity driven primarily by plant roots. The exudations of these roots provide nutrients for microbes, and may also attract or repel some organisms (van Tol et al. 2001, Rasmann et al. 2005). The interactions between plants and rhizosphere organisms can range from mutualistic to pathogenic, including direct competition for resources. In general, nitrogen-fixers and mycorrhizal fungi enhance plant growth and survival, and pathogenic fungi and root-feeders decrease plant fitness. These and other interactions with nonmycorrhizal fungi, rhizosphere bacteria, protozoa, and viruses can also modify the effect of soil-borne pathogens, herbivores, and mutualists in plant populations.

In this chapter, we briefly describe the main groups of organisms that are closely associated with plant roots and their effect on plant growth and survival. We also review the biological and chemical interactions that occur in the rhizosphere and how this changes the outcome for the associated plant. The last part of the chapter is devoted to the implications of these interactions for ecosystem functioning.

MAJOR GROUPS OF ORGANISMS AND DIRECT INTERACTIONS WITH PLANTS

This section focuses on four groups of organisms that live in very close association with plant roots and are thought to have the greatest impact on plant performance and ecosystem processes (Figure 19.1). In focusing on particular groups of organisms, others are necessarily left out, even though they may play an important role. We, however, refer to these when appropriate throughout this chapter. Certainly, a plant is exposed to more than one of these groups at any time and the interactions between them can change the outcome for the plant. This is also discussed in the following section.

SYMBIOTIC NITROGEN-FIXERS

Nitrogen is the most limiting nutrient for plant growth in terrestrial ecosystems. Although molecular nitrogen is very abundant in the atmosphere, eukaryotes have not evolved the ability to fix atmospheric nitrogen into ammonia (Eady 1991). In fact, this capacity is limited to a number of bacteria and archaea species with very different life strategies. Some of them are free-living in soil and water (i.e., *Azotobacter*, *Clostridium*), others occupy the rhizosphere, phyllosphere, or intercellular spaces of plants (i.e., *Azospirillum*, *Azoarcus*, *Gluconacetobacter*), and still others are highly specialized symbionts (like *Frankia,* associated with species of *Alnus*, *Myrica*, *Ceanothus*, *Eleagnus*, and *Casuarina*; and legume symbionts collectively known as rhizobia).

The symbiotic diazotrophs are the main contributors to biological nitrogen fixation in terrestrial ecosystems. Research has focused mainly in the legume symbionts because of the importance of this plant family in agriculture. However, they also play a key role in natural

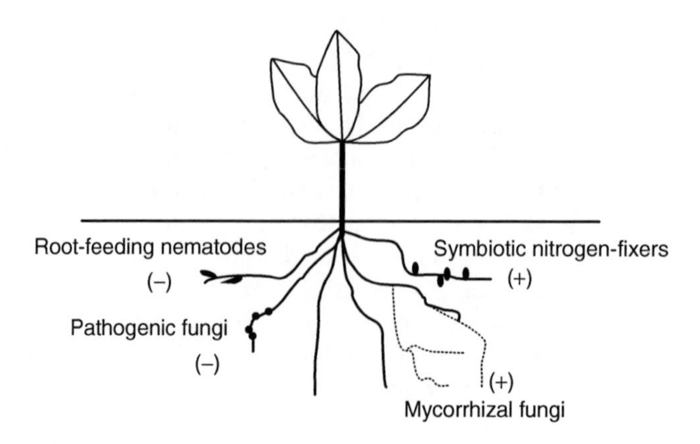

FIGURE 19.1 Rhizosphere organisms that have the greatest impact on plant performance. Positive interactions are indicated by (+); negative interactions are indicated with (–).

ecosystems because of the wide distribution of legumes in temperate, tropical, and arid regions (Lafay and Burdon 1998, Ulrich and Zaspel 2000, Rodríguez-Echeverría et al. 2003). Most of the known legume symbionts belong to the order Rhizobiales, but there are also some species that nodulate legumes in the order Burkholderiales. Currently all the species within the genera *Rhizobium*, *Sinorhizobium*, *Mesorhizobium*, *Bradyrhizobium*, *Azorhizobium*, and *Allorhizobium* have the ability to nodulate legumes and fix nitrogen. Some other genera, like *Ensifer*, *Blastobacter*, *Burkholderia*, and *Ralstonia*, contain both legume symbionts and nonsymbiotic species (Sawada et al. 2003). Nevertheless, the taxonomy of the bacterial symbionts of legumes is under revision and could result in the amalgamation of some genera. The ability to nodulate legumes and to fix nitrogen is encoded in mobile genetic elements such as transmissible plasmids or conjugative transposon-like sequences. The taxonomical diversity of legume symbionts might therefore be explained by the horizontal transfer of those elements between soil bacteria.

The specificity of the association between legumes and their bacterial symbionts depends on a very fine molecular communication between the plants and the rhizobia. Nodulating bacteria respond to the flavonoids produced by legume roots by producing *N*-acylated oligomers of *N*-acetyl-D-glucosamine, known as Nod-factors, which initiate the physiological changes in the host roots leading to nodulation. The basic structure of Nod-factors has variations that are dependent on each strain or species and determine the host-specificity (Perret et al. 2000). Although the symbiotic association between legumes and their symbionts was considered to be highly specific, it is now believed that this only applies to the tribes Trifolieae, Viceae, and Cicereae (Perret et al. 2000). Symbiotic promiscuity is common in nature and could be an advantage for colonizing new soils. In fact, some studies suggest that highly promiscuous legumes are successful invasive species (Richardson et al. 2000, Ulrich and Zaspel 2000). The association is crucial for the establishment and growth of many pioneer leguminous species. Soil enrichment in nitrogen due to these associations subsequently facilitates the growth of other plant species, thus promoting plant succession. In turn, increasing levels of nitrogen can also lead to the displacement of other species promoting spatial heterogeneity. This is, therefore, a key symbiosis for the functioning of terrestrial ecosystems.

Mycorrhizal Fungi

A mycorrhiza is a symbiotic, nonpathogenic, permanent association between a plant root and a specialized fungus, both in the natural environment and in cultivation. This is the most common and ancient symbiotic association to be found in plants and evolved with the colonization of land by primitive plants (Brundrett 2002). In this symbiosis, plants exchange carbohydrates for mineral nutrients—mainly phosphorus, nitrogen, potassium, calcium, and zinc—retrieved by the fungal mycelium from large soil volumes. Mycorrhizal fungi are also involved in many other processes such as plant protection against abiotic stresses (Allen and Allen 1986) or root pathogens and herbivores (Newsham et al. 1995, de la Peña et al. 2006); the degradation of complex and organic molecules, making essential nutrients available to the plant (Cairney and Meharg 2003); and the synthesis or stimulation of plant-growth hormones like auxins, citokinins, and gibberellins. However, not all mycorrhizal associations are positive. When one of the partners does not receive a quantitative benefit, they can become exploitative. In fact, the mycorrhizal symbiosis is in the mutualism–parasitism continuum, depending on the identity of plant and fungus species and abiotic factors (Johnson et al. 1997).

The classification of mycorrhizas is primarily based on the morphology and physiology of the association. There are three main morphological groups of mycorrhizas: (i) the ectomycorrhizas, with fungal mycelia surrounding the root and penetrating the intercellular spaces; (ii) the endomycorrhizas (which can be either arbuscular or ericoid mycorrhizas), in which the mycelium does not coat the root, yet there is an intimate contact between the fungi and the root through structures inside the root cells that are specialized for nutrient exchange and storage; and (iii) the intermediate types that share characteristics with both ecto- and endomycorrhizas, and include the ectendo-, arbutoid, monotropoid, and orchid mycorrhizas. The most widespread mycorrhizal associations by far are the ectomycorrhizas and the arbuscular mycorrhizas.

The ectomycorrhizal association occurs in 140 genera of seed plants belonging to the families Betulaceae, Fagaceae, Pinaceae, Rosaceae, Myrtaceae, Mimosaceae, and Salicaceae. Although there are much fewer species of ectomycorrhizal plants than of endomycorrhizal plants, the association is ecologically significant, as it involves the dominant species of boreal, temperate, and many subtropical forests. The fungi involved in this symbiosis are almost exclusively basidiomycetes and ascomycetes. Common genera of Basidiomycetous fungi include both hypogeous and epigeous genera such as *Amanita*, *Boletus*, *Leccinium*, *Suillus*, *Hebeloma*, *Gomphidius*, *Paxillus*, *Clitopilus*, *Lactarius*, *Russula*, *Laccaria*, *Thelephora*, *Rhizopogon*, *Pisolithus*, and *Scleroderma* (Smith and Read 1997).

The arbuscular mycorrhizas are ubiquitous, occurring over a broad ecological range with almost all natural and cultivated plant species. With few exceptions, species from all angiosperm families can form endomycorrhizal associations. A few gymnosperms such as species of *Taxus* and *Sequoia* also show infection. Phylogenetically, these fungi are the oldest symbionts infecting also bryophytes and pteridophytes. The fungi that form these associations (arbuscular mycorrhizal fungi or AMF) belong to the Glomeromycota phylum (Schüßler et al. 2001) and are obligate symbionts. Little specificity has traditionally been recognized in this association, but more recent studies have shown a higher genetic and functional diversity than previously estimated (Sanders et al. 1996, Helgason et al. 2002, Munkvold et al. 2004). The presence of AMF can increase plant diversity and ecosystem productivity (Grime et al. 1987, van der Heijden et al. 1998). This could be explained by the high functional diversity of AMF and the specificity of the outcome of the interaction with different plant species. A rich AMF community is more competent at exploiting soil resources and it is more likely to benefit a wider range of plant species (van der Heijden et al. 1998). There is, however, an alternative explanation for the positive correlation between AMF and plant diversity, and that comes from the observation that AMF can also have a detrimental effect on plant growth. According to this hypothesis, a richer fungal community increases plant diversity because no plant has a greater advantage with all AMF at the site (Klironomos 2003).

In some circumstances, the absence of mycorrhizal fungi can lead to an increase in plant diversity. This is the case with plant communities that are dominated by highly mycotrophic species, or by one mycorrhizal type, that is, ectomycorrhizal species. The removal of mycorrhizal fungi leads to a decrease of the dominant species and the consequent competitive release of the subordinate species (Connell and Lowman 1989, Hartnett and Wilson 1999).

The external mycelium of mycorrhizal fungi establishes an underground network that links different plants. This fungal network also reduces nutrient losses by sequestering nitrogen, phosphorus, and carbon within their biomass (Simard et al. 2002). Nutrients move within the external mycelium according to fungal needs, but there is also a nutrient transfer between plants through the hyphal network (Simard et al. 2002). Carbon transfers between plants are better known in ectomycorrhizas (Smith and Read 1997), but they also occur through arbuscular mycorrhizas (Carey et al. 2004). The transfer of N and P between live, intact plants has been documented mainly for arbuscular mycorrhizal plants (Simard et al. 2002). The net transfer of nutrients between plants varies with mycorrhizal colonization, soil nutrient content, and the plant physiological status. Therefore, the results obtained in greenhouse studies have been very variable. A high rate of nutrient transfer between plants through the external hyphae of mycorrhizal fungi would have important ecological consequences. For example, nutrient transfer can enhance the establishment and growth of new seedlings of mycorrhizal plants, allowing a quick recovery after disturbance and also affecting plant competition. Little is known about the specificity of this mechanism in natural systems, whether some plants species are mainly donors or sinks of nutrients, or whether the transfer is species-specific.

Pathogenic Fungi

The research about soil fungi that have deleterious effect on plant growth has historically focused on agricultural systems for obvious economic reasons. Only in the last two decades have ecologists started to explore the diversity and the role of pathogenic fungi in natural ecosystems.

The majority of soil fungal pathogens that attack plants are ascomycetes. There are many genera of pathogenic ascomycetes that have been identified from plants in agricultural systems and later isolated from natural systems. In coastal sand dune studies that focused on the degeneration of pioneer plant species, *Verticillium* and *Fusarium* species were isolated from declining stands of the dune grass *Ammophila arenaria* in The Netherlands (Van der Putten et al. 1990); and species of *Fusarium*, *Cladosporium*, *Phoma*, and *Sporothrix* were involved in the degeneration of *Leymus arenarius* in Iceland (Greipsson and El-Mayas 2002). Another example is the dieback of the endemic Hawaiian tree koa (*Acacia koa*), a keystone species in upper-elevation forests, caused by the systemic wilt pathogen *Fusarium oxysporum* f. sp. *koae* (Anderson et al. 2002). Other root rot fungi play a significant role in the dynamics of temperate forests by killing big trees and opening gaps in the forest. A well-studied example is the basidiomycete *Phellinus weirii* that attacks specifically *Pseudotsuga menziensii* in temperate forests of North America (Hansen 2000).

There is also a fungal-related group of organisms that attacks plant species in both natural and agricultural systems: the oomycotan genera *Pythium* and *Phytophthora*. Species of *Pythium* are responsible for the mortality of seedlings in tropical and temperate forests (Augspurger 1983, Packer and Clay 2000, Reinhart et al. 2005, Bell et al. 2006). The proximity to parent trees causes a high mortality of new seedlings, which is correlated with the build-up of pathogenic *Pythium* spp. on the rhizosphere of the parent trees.

Among the *Phytophthora* species isolated in natural systems, we would highlight *Phytophthora cinnamomi*, identified as the cause of die-backs of native tree species in North America (Zentmyer 1980), Australia (Wills and Kinnear 1993), and Southern Europe (Brasier et al. 1993). In North America, the most affected species were *Pinus echinata*, *Abies fraseri*, and *Castanea dentata*. In Australia, it has caused the sudden death of plants belonging to more than 20 genera including *Acacia*, *Banksia*, *Eucalyptus*, and *Grevillea* species. In Southern Europe, this species, in combination with other *Phytophthora* spp., has been suggested to contribute to oak decline since the beginning of the twentieth century.

The impact of pathogenic fungi and oomycetes depends not only on the life-stage of the plants, but also on the specificity, virulence, and overall life history of the pathogen. Gilbert (2002) classifies the fungal pathogens of noncrop plants as (a) seed decay, (b) seedling diseases, (c) foliage diseases, (d) systemic infections, (e) cankers, wilts, and diebacks, (f) root and butt rots, and (g) floral diseases, and these are good descriptors of the many ways these organisms can interfere (and interact) with plants. The impact that fungal pathogens can have on plant populations is thought to contribute to plant genetic diversity, species diversity, and succession in natural systems (Gilbert 2002, Van der Putten 2003).

Nematodes

Nematodes are the most abundant metazoans. Some 20,000 species of nematodes have been described, a small proportion of the estimated 10^5 or 10^6 likely to exist, and they can be found in any environment where decomposition occurs. In ecological studies, they are usually classified by their feeding habit. Nematodes can be bacterial feeders, fungal feeders, omnivores, or plant feeders (Bongers and Bongers 1998). This is a relatively simplified classification, as other authors consider nematodes to be functionally divided into eight groups (Yeates et al. 1993). In this section, we focus on the plant feeders, a group of nematodes that have specialized mouth structures (stylet) to feed on plant roots. Plant-feeding nematodes are highly specialized obligate parasites that have evolved through close interactions with plants, and this explains the high impacts on the plant populations they attack. According to Stirling (1991), the belowground plant–parasitic nematodes can be subdivided into four different groups: sedentary endoparasites, sedentary semiendoparasites, migratory endoparasites, and ectoparasites.

- Sedentary endoparasites (e.g., *Meloidogyne* spp., *Heterodera* spp.) are completely surrounded and protected by their host's root tissue for most of their life cycle (Stirling 1991). They interact with the plant root to develop permanent and highly specialized feeding sites within the root tissues that act as nutrient sinks (Zacheo 1993).
- Sedentary semiendoparasites (e.g., *Rotylenchulus* spp.) are partially exposed in the root tissue for part of their life cycles, and juveniles and young females feed ectoparasitically, spending much time in the rhizosphere.
- Migratory endoparasites (e.g., *Pratylenchus* spp.) can hatch and develop to maturity inside the root tissue of their hosts, and are rarely found in soil unless their host plant is under stress (Stirling 1991). They do not establish a permanent feeding site, but migrate within roots, causing extensive damage. *Pratylenchus* nematodes have been reported to feed ectoparasitically on some grasses (Timper, personal communication).
- Ectoparasites (e.g., *Xiphinema* spp.) only penetrate root tissue with the stylet; their body is outside the root tissue at all times. Ectoparasites are not protected by roots and feed on epidermal and cortical root tissues (Zacheo 1993).

Sedentary endoparasites and migratory endoparasites are the main nematode groups implicated in disease complexes, or additive effects on disease incidence or severity on the host plant by association with bacteria or fungi (Hillocks 2001). Plant-feeding nematodes can also develop additive and synergistic interactions with pathogenic fungi and bacteria and some (e.g., *Xiphinema*, *Longidorus*) are vectors of plant viruses.

Plant-feeding nematodes and their host plants are involved in a coevolutionary arms race. One of the classical examples of nematode resistance in plants is that of *Tagetes erecta* and *Tagetes patula* (Goff 1936). The research to discover the causes of resistance led to the isolation of various nematicidal polythienyl compounds from *Tagetes* plants, the first of which is thiphene R—terthienyl (Ulenbroek and Brijloo 1958). It was later discovered that endoroot bacteria in both *T. patula* and *T. erecta* roots produced nematotoxic compounds that reduced nematode populations in soil. These bacteria were successfully transferred to potato, *Solanum tuberosum*, and effectively reduced the numbers of nematode parasites of this plant (Sturz and Kimpinski 2004).

Nematodes can detect (and react to) chemical gradients in soil, and plant metabolites in roots, through their chemoreceptors. A well-illustrated example is that of *Globodera rostochiensis*, an obligate parasite of potato. Nematode eggs exposed to the potato root exudates are stimulated to hatch (Jones et al. 1997). These juveniles increase their activity in response to the exudates, and orientate themselves, following the exudation gradient, to the roots (Perry 1997). Then they invade the roots and alter the physiology of the root cells to form a syncytium on which the nematode feeds.

Entomologists and nematologists have tried to identify semiochemicals (signalling compounds) in the rhizosphere that would help insects and nematodes to locate roots (Perry 1996, Johnson and Gregory 2006). According to the physical soil structure, both volatiles and water-soluble compounds could be involved, but volatile compounds can potentially travel faster (Young and Ritz 2005). A common molecule to which both nematodes and insects are attracted is CO_2. The main problem with considering CO_2 a semiochemical is that it is ubiquitous in soil, and therefore could only be of potential importance to generalist root herbivores (Johnson and Gregory 2006).

Plant-feeding nematodes are very responsive to changes in vegetation (Korthals et al. 2001), and plant identity greatly influences their population densities (Yeates 1987). Plant-parasitic nematodes can have dramatic effects in agricultural systems, where they cause estimated losses of US$ 100 billion every year due to yield reductions or overall damage to crops (Oka et al. 2000). There is not much information about this interaction from natural systems (Van der Putten and Van der Stoel 1998). Nevertheless, a disease complex of plant-feeding nematodes and fungal pathogens has been implicated in the degeneration of *A. arenaria* (marram grass) in coastal sand dunes (Van der Putten et al. 1990). If nematode herbivory is low, however, plant growth might be enhanced through changes in the exudation pattern and release of nutrients from damaged roots. These changes promote soil nutrient influx, which increases soil microbial biomass and root growth of the attacked and neighboring plants (Bardgett et al. 1999a,b).

INTERACTIONS IN THE RHIZOSPHERE

In this section, we describe some of the interactions that occur between organisms in the rhizosphere. At this point, it seems important to take a holistic approach, and although we have divided the section into subheadings, all these interactions are likely to occur at the same time, arguably with different ecological importance for different systems. We include here trophic interactions but also other ecological and chemical interactions.

All organisms produce chemicals and respond to chemical release by others, in a vast network of communications (Eisner and Meinwald 1995). Plants themselves are involved in this communication system, although this has only recently been recognized (D'Alessandro and Turlings 2006, Schnee et al. 2006). They constantly release not just primary compounds (CO_2, sugars), but also secondary metabolites through root exudations and leaf volatiles, which are indicative of their physiological state. These can act as cues for their herbivores and for the natural enemies of these herbivores.

The term allelopathy has classically been used to describe strictly plant–plant direct interactions. But these often cannot be dissected out and are very difficult to prove, as the effect of allelochemicals can be modified or influenced by both abiotic and biotic factors (Inderjit and Weiner 2001, Inderjit 2005). Therefore, the allelopathy concept has been extended to encompass microorganism-mediated processes of plant interference (Inderjit and Weiner 2001). An even broader concept is that of the International Allelopathy Society, which includes the effects and activities of not only plants and algae, but also of fungi and bacteria. In this chapter, we refer to the biological, ecological, and behavioral effects of such chemical interactions.

Unfortunately, chemical and biological interactions have mostly been studied separately, despite their intrinsic links. We have tried to reunite them in describing the rhizosphere interactions included in this section, namely belowground plant–plant interactions, the effect of soil organisms on resource availability and uptake in plants, and interactions between the soil organisms previously described.

BELOWGROUND PLANT–PLANT INTERACTIONS

Belowground, plant roots explore the soil heterogeneity and patchiness and compete for nutrient resources (Hodge 2006). This competition effect apparently occurs only between different plants, as recent root physiology studies suggest that plant roots can distinguish between self and nonself, changing their growth patterns accordingly (Gruntman and Novoplansky 2004). This mechanism should act to avoid competition between roots of the same plant and maximize root exploratory potential.

Plants can interact negatively through the production of phytotoxic compounds. For a recent review on aspects of plant interference see Weston and Duke (2003). As an example, we mention the case of mugwort (*Artemisia vulgaris*), which has been extensively studied. This plant has a range of reported biological activities and its chemical composition has been studied in detail. Mugwort is a ruderal nitrophylic plant, a noxious and highly successful weed that interferes with the growth and development of neighboring plants. Its root leachates act by inducing chemical changes in soil, a process mediated by microorganisms (Inderjit and Foy 1999). Phytotoxicity in mugwort has also been attributed to compounds of the rhizome of this plant which significantly inhibit germination and seedling development of other plants (Onen and Ozer 2002). Incidentally, rhizome compounds also have nematotoxic, including nematicidal, effects (Costa et al. 2003).

Resource Uptake and Partitioning

Soil microbes might enhance plant coexistence by resource partitioning (Reynolds et al. 2003). In addition, mycorrhizal fungi increase plant availability of phosphorus and nitrogen from organic and inorganic pools mainly through enzymatic activities (Marschner 1995, Turnbull et al. 1996). The high multifunctional diversity observed for AMF and the specificity of ectomycorrhizal associations might also be related to nutrient partitioning among different plant species. This partitioning has been confirmed for different nitrogen sources in some Australian ectomycorrhizal isolates from *Eucalyptus maculate* (Turnbull et al. 1995) and in AMF isolates studied in vitro (Hawkins et al. 2000). Resource partitioning could also be related to the preferential association with rhizosphere bacteria. For instance, the ability of using molecular nitrogen as a source depends on the association with nitrogen-fixing bacteria. In the same way, the ability of some plants to selectively use ammonium, nitrate, or amino acids as source of nitrogen (McKane et al. 2002) could be related to the differential recruitment of microbial communities in their rhizosphere (Reynolds et al. 2003).

There is evidence that ectomycorrhizal fungi can mobilize complex and organic forms of nitrogen and phosphorus making them available to their plant partners. Ectomycorrhizal fungi can mobilize organic forms of nitrogen from litter and pollen grains transferring them to associated plants (Bending and Read 1995, Northup et al. 1995, Pérez-Moreno and Read 2000, 2001b). The species *Paxillus involutus* can transfer nitrogen and phosphorus from dead nematodes to symbiotic seedlings of *Betus pendula* (Pérez-Moreno and Read 2001a). Furthermore, the hyphae from *Laccaria bicolor* can even act as a predator of springtails, immobilizing the animals, colonizing their bodies, and subsequently transferring nitrogen to the symbiotic seedlings of *Pinus strobus* (Klironomos and Hart 2001).

Interactions between Mycorrhizal Fungi and Soil Fauna

In spite of the great diversity of soil animals, we focused in the previous section on plant-feeding nematodes because they can have a strong impact on plant performance. The description of other soil invertebrates is not within the scope of this chapter but we refer here to some groups that are known to interact, directly or indirectly, with mycorrhizal fungi.

Some of these interactions are positive for the plant, for example, earthworms, isopods, diplopods, and insects can act as vectors of AMF by ingesting hyphal fragments or spores and transporting them in their movements (Gange and Brown 2002). Mycophagous mites, collembolan, and nematodes feed preferentially on nonmycorrhizal fungi, thereby releasing mycorrhizal fungi from competition with other fungi (Gange and Brown 2002). Soil invertebrates can in theory have a negative impact on mycorrhizal fungi, by disrupting or feeding on the external mycelia, but these interactions have not been shown to have a major impact in natural systems.

It has been proposed that the main benefit that plants obtain from mycorrhizal fungi in natural systems is protection against pathogens and herbivores (Fitter and Garbaye 1994). Root-feeding insects and nematodes can have a serious negative impact on plant growth and performance, and, in general, AMF reduce plant damage by root herbivores, although this effect can be highly variable (Table 19.1). The outcome of the interaction seems to depend on several factors such as soil characteristics and the genotypes of plants, herbivores, and AMF.

TABLE 19.1
Summary of Available Data on the Effect of Mycorrhizal Colonization for Plant Feeding Nematodes and Host Plants in Natural Systems

Nematode Species	Plant Species	Mycorrhizal Fungi Species	Effect on Plant	Effect on Nematodes	Reference
Pratylenchus spp.	*A. brevigulata*	*Glomus etunicatum*, *Glomus aggregatum*, *Glomus geosporum*, *Gigaspora albida*, *Acaulospora scrobiculata*, *Acaulospora spinosa*, *Scutellospora calospora*	Positive	Not described	Little and Maun 1996
Heterodera spp. *M. incognita*	*Trifolium repens*	*Glomus mosseae* *Glomus intraradices* *G. aggregatum*	Positive	*G. intraradices* reduced number of nematodes and galls	Habte et al. 1999
Pratylenchoides magnicauda *Paratylenchus microdorus* *Rotylenchus goodeyi* *Merlinius joctus*	*L. arenarius*	*Glomus fasciculatum* *Glomus caledonium* *G. mosseae*	Positive	Not described	Greipsson and El-Mayas 2002
Tylenchorhynchus gladiolatus *Pratylenchus pseudopratensis*	*Afzelia africana*	Six strains of *Scleroderma* and other native EM	None (Nematodes did not affect plant growth)	None	Villenave and Cadet 1998
Pratylenchus penetrans	*A. arenaria*	Mixed native inocula of AMF	None (Nematodes did not affect plant growth)	Suppression	de la Peña et al. 2006

Most of the research in this field, especially for natural systems, has been done on root-feeding nematodes.

The presence of AMF can offer plant protection through increased host resistance (Azcón-Aguilar and Barea 1996). Many of the chemicals present in mycorrhizal roots—phenolics, isoflavonoids, terpenoids—can negatively affect root-feeding insects and suppress sedentary nematode parasite reproduction and feeding (Gange and Brown 2002).

Root colonization by AMF also changes the quality of root exudates, an effect that depends on the identity of the fungi. Since root exudates are the main source for microbial activity, these changes would also affect the rhizosphere communities. In some cases, the presence of AMF is correlated with a reduction in the number of pathogens in the rhizosphere and an increase in the number of beneficial organisms (Azcón-Aguilar and Barea 1996). In addition, shifts in root exudates can affect chemotatic attraction of nematodes and egg hatching (Smith and Kaplan 1988).

Mycorrhizal protection against nematodes might occur also through increased host tolerance, because of the improvement in plant health due to an increased uptake of P, Ca, Cu, Mn, S, and Zn (Smith and Kaplan 1988). This protection is only effective if roots are mycorrhizal before they are attacked by the nematodes. In this case, AMF might be considered an extension of the plant that can compensate for plant damage. In agriculture, plants can be preinoculated with mycorrhizal fungi before planting, but in natural communities the outcome of the interaction depends on which organism colonizes the root first (Gange and Brown 2002). AMF might compete with endoparasitic nematodes for space and photosynthates inside plant roots. Competition for photosynthates, especially affecting sedentary nematodes, has received little support from experimental data. Competition for space is well documented, and several studies have shown that both groups of organisms can negatively affect each other, that is, the presence of one reduces infection by the other (Roncadori 1997). Finally, some studies have suggested that *Glomus* can be a weak parasite of the sedentary endoparasite *Heterodera glycines* (Francl and Dropkin 1985).

Most of our understanding about the interaction between mycorrhiza and root herbivores comes from studies with agronomic species. Research on natural systems has mainly focused in coastal dune systems. Greipsson and El-Mayas (2002) found that a commercial AMF inoculum protected the dune grass *L. arenarius* against migratory endoparasitic nematodes. In addition, Little and Maun (1996) showed that mycorrhizal protection of *Ammophila brevigulata* against *Pratylenchus* and *Heterodera* spp. was effective if sand burial occurred simultaneously. De la Peña et al. (2006) demonstrated that AMF can also protect *A. arenaria* through the suppression of *Pratylenchus penetrans* colonization and reproduction. The data suggest that AMF can indeed directly outcompete migratory endoparasitic nematodes in the roots of the plant host. Root colonization by *P. penetrans* and nematode multiplication were drastically reduced by AMF through local mechanisms that were more efficient in premycorrhizal plants. The authors could not detect mutual inhibition between AMF and nematodes, and further conclude that root colonization by AMF was not inhibited by the nematodes.

Interactions between Plant-Feeding Nematodes, Legumes, and Bacterial Symbionts

Plant-parasitic nematodes and rhizobia can interact in the rhizosphere and inside the roots of host legumes, although the outcome of these interactions for legumes is not clear. Some studies suggest that plant-feeding nematodes may reduce nodule formation (Villenave and Cadet 1998, Duponnois et al. 1999). But rhizobial strains have also been shown to elicit plant-induced resistance against plant-feeding nematodes (Reitz et al. 2000, Mitra et al. 2004). In addition, ectomycorrhizal fungi can increase plant tolerance to sedentary endoparasitic nematodes like *Meloidogyne javanica* (Duponnois et al. 2000b). There is also evidence of horizontal gene transfer from rhizobia to plant-feeding nematodes, which might have conferred parasitic nematodes the ability to successfully invade the roots of some plant species (Scholl et al. 2003).

AMF and rhizobial symbionts also interact in the roots of legumes and this can affect plant growth and nutrient content and nodulation. Flavonoids exuded by the legume roots play a key role in the establishment of both rhizobial and mycorrhizal associations. In fact, the establishment of one of the symbiotic partners in the root can change root flavonoid concentrations and stimulate root colonization by the other (Antunes et al. 2006). In general, colonization by AMF enhances nodulation and nitrogen fixation. But the outcome of the interaction varies with the identity of the host plant and the symbiotic partners (Xavier and Germida 2003). Compatible species or strains of AMF and rhizobia interact synergistically to improve N and P content and plant growth, but incompatible strains can also lead to a reduction of the efficacy of nitrogen fixation (Xavier and Germida 2003).

Interactions between Nematodes and Their Microbial Enemies

Research on nematode interactions with other soil microorganisms has increased dramatically in an attempt to develop biologically based control systems in agriculture (Whipps and Davies 2000). Nematode natural enemies include fungi and bacteria, and with a smaller effect, predatory nematodes, protozoans, and soil microarthropods (Rodriguez-Kabana 1991). There are complex communities of fungal and bacterial natural enemies with high intraspecific variability, and the role of this biodiversity is poorly understood (Kerry and Hominick 2002).

The nematode-destroying fungi are notoriously diverse, occurring throughout all fungal groups with the possible exception of ascomycetes (Barron 1977) and can be divided in three main groups (Siddiqui and Mahmood 1996): endoparasites, predatory (or trapping) fungi, and opportunists (or facultative parasites). The majority of endoparasitic fungi are obligate parasites (e.g., *Nematophtora gynophila, Hirsutella rhosiliensis*). They spend most of their life cycles inside the body of their host and do not produce extensive hyphae in the rhizosphere. Frequently, these fungi are only detected in soil in the form of spores (Barron 1977, Siddiqui and Mahmood 1996).

Although the primary ecological role of predatory fungi appears to be that of wood decay (Barron 2003), their ability to capture nematodes using trapping devices has led to further specification and diversification (Ahren and Tunlid 2003). The traps formed by these fungi can consist of adhesive branches, hyphal networks or knobs, and constricting and nonconstricting rings. Although they can colonize the rhizosphere, their sensitivity to environmental changes makes them poor competitors in soil (Siddiqui and Mahmood 1996).

Opportunistic fungi can use both living and nonliving matter as sources of nutrients (Jaffee 1992). These fungi can infect sedentary stages of endoparasitic nematodes within the roots or when exposed on the root surface or in soil. Their saprophytic ability makes them good colonizers of the rhizosphere, even in the absence of their nematode hosts. The opportunistic fungi *Pochonia chlamydosporia* (formerly known as *Verticillium chlamydosporium*) and *Paecilomyces lilacinus* have been studied extensively and are being developed as biological control agents of sedentary nematode parasites (Jatala 1985, Stirling 1991).

Nematophagous fungi can also produce nematode-antagonistic compounds. A crude extract of the nematode parasitic fungus *Myrothecium* sp. has nematicidal activity against a wide range of plant parasitic nematodes, and is now under development as a bionematicide (DiTera TM) (Warrior et al. 1999, Twomey et al. 2000).

Perhaps one of the best studied bacterial enemies of nematodes is the endospore-forming *Pasteuria penetrans*, a potential biocontrol agent of sedentary endoparasitic nematodes (Stirling 1991). As an example, infection of root-knot nematodes by an isolate of *P. penetrans* allowed a better development of *Acacia holosericea* seedlings (Duponnois et al. 2000a).

The genus *Pasteuria* includes four species that have been found parasitizing several different genera of nematodes (Sturhan 1988). However, most nematode antagonistic bacteria are not natural enemies, but act by producing metabolic by-products with nematode toxicity (Siddiqui and Mahmood 1996). A surprising example comes from an apparently obscure interaction between two nematodes that are obligate parasites: the root-feeding nematode *Meloidogyne incognita* and the insect parasite *Steinernema glaseri*. Bird and Bird (1986) demonstrated that *M. incognita* feeding on the roots of tomato plants was suppressed by the addition of the insect parasite. Further research demonstrated that *S. glaseri* has a symbiotic relation with the bacterium *Xenorhabdus* sp., which produces a nematotoxic allelochemical that induces mortality in the root-knot nematode and also inhibits egg hatching (Grewal et al. 1999).

Although several fungal and bacterial natural enemies of nematodes have been described, knowledge is lacking on their population biology and dynamics. Community diversity and population dynamics can be influenced not only by nematode identity, but also indirectly by the nematode host plant, in a tritrophic interaction (Kerry and Hominick 2002).

Field studies on the population dynamics of *Hirsutella rhossiliensis* have concluded that high levels of parasitism can occur in soil, but build up very slowly, resulting in a time-lagged density-dependent effect (Jaffee 1992). A similar conclusion was reached for *P. chlamydosporia* and *N. gynophila* after a 10-year monitoring study of their populations and that of their sedentary endoparasitic nematode host, *Heterodera avenae*, in four field sites. After 3–4 years of the establishment of fungal enemies in the soil, the nematode populations dropped to almost nondetectable levels, even under monoculture of their cereal host (Kerry and Crump 1998). On the other hand, the population dynamics of the interaction between *P. penetrans* and *Xiphinema diversicaudatum* has been described as a typical predator–prey system, in which the number of *Pasteuria* endospores in soil is dependent on the number and activity of available hosts (Ciancio 1995).

In summary, by reducing plant-feeding nematode populations, these microbial enemies contribute to improved plant growth, and could further influence their distribution.

ECOLOGICAL IMPLICATIONS

The importance of soil biodiversity for ecosystem processes is an ongoing debate, but unraveling the impact of soil diversity on ecosystems is extremely difficult. Soils are highly heterogeneous and complex, which makes it complicated to reproduce in laboratory or greenhouse experiments. Experimental studies have failed to show a uniform pattern of the effect of removing groups or species of soil organisms. For example, the removal of soil fauna affects plant community composition through its effects on the rhizosphere microbial biomass that alters decomposition rates and nutrient cycling, but major ecosystem functions such as net ecosystem productivity were reportedly unchanged (Bardgett et al. 1999a,b). As a result, different theories based on the concept of key species, functional traits, or functional dissimilarity have been proposed. These theories are not discussed further here (for more information on the topic see Bradford et al. (2002), Bardgett (2004), and Heemsbergen et al. (2004)). Instead, we move a step further to link the described interactions between rhizosphere organisms and single plant species with the dynamics of plant communities and terrestrial ecosystems.

The first point to be mentioned is that the impact of rhizosphere organisms on plant community processes is a function of environmental factors (Reynolds et al. 2003). For example, the availability of phosphorus and nitrogen determines plant dependence on mycorrhizal fungi and symbiotic nitrogen-fixers. In addition, the deleterious effect of pathogens can be reduced under optimal nutrient and light conditions; conversely, such conditions can

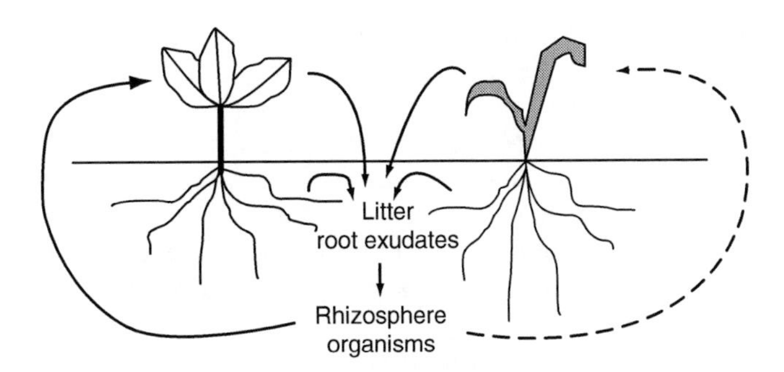

FIGURE 19.2 Representation of feedbacks between plants and soil organisms. Plants change the composition of rhizosphere communities through the quantity and the quality of litter and root exudates. In turn, the rhizosphere communities affect different plant species in different ways.

switch mycorrhizal outcomes for the plant from beneficial to detrimental. In addition, most studies have been done with a fairly limited set of AMF species from culture banks. This raises the question of the relevance of the results obtained in greenhouse or laboratory experiments for ecosystems functioning. Undoubtedly, valid data have been obtained from those experiments that help the understanding of the different mechanisms involved in the plant–rhizosphere interactions. The challenge now is to obtain a broader knowledge from field experiments or mathematical models that integrate that information with the naturally changing environment.

The existence of feedbacks between plants and soil communities that affect above- and belowground processes is now widely accepted (Ehrenfeld et al. 2005) (Figure 19.2). Plants modify the composition of the rhizosphere mainly through the quality and the quantity of exudates and litter that they produce (Bardgett 2004) (and references therein). Therefore, the existence of different plant functional types is more important for the rhizosphere biota than plant diversity per se. Rhizosphere organisms can in turn affect plant community composition and dynamics because they change the outcome of competition between plants. In this way, specific feedbacks between plants and soil organisms are established and can determine the functioning of many terrestrial ecosystems (Van der Putten 2003). Soil feedbacks are considered positive if beneficial organisms for the plant accumulate in the rhizosphere over time. Soil feedbacks are negative when plant growth is hampered by its own soil community, either due to the increase of detrimental organisms or because other plant species have a higher benefit from that soil community.

Positive feedbacks lead to the dominance of the associated plant, and therefore, to the reduction of plant diversity (Bever 2003). This mechanism is evident for plant species with a high dependency on mycorrhizal or rhizobial symbiosis. Plant establishment depends on the initial abundance of the symbiotic partners and fails without these organisms in the rhizosphere. As an example, the introduction of pine trees in the tropics was only successful when the trees were inoculated with the compatible ectomycorrhizal fungi (Reynolds et al. 2003). There is also evidence that plants select for AMF that benefit them most, creating patches of positive feedbacks between them and those AMF species (Hart and Klironomos 2002, Klironomos 2002). Thus, positive feedbacks lead to a reduction of plant diversity in local patches but to an increase of landscape heterogeneity.

Negative feedbacks can increase plant coexistence and, therefore, diversity. Negative feedbacks that happen due to the accumulation of deleterious organisms are very obvious in agricultural soils, and can be alleviated by crop rotation. There are more examples of negative feedbacks shaping natural plant communities than positive feedbacks. Sometimes these feedbacks are caused by a single group of organisms like those described for pathogenic fungi or plant-feeding nematodes. In addition, complexes of pathogenic fungi and plant-feeding nematodes can drive the decline of the pioneer dune species *A. arenaria* and plant succession in general (Van der Putten et al. 1993, Van der Putten and Peters 1997, Greipsson and El-Mayas 2002). Negative feedbacks can also occur if the mutualistic fungi that accumulate in the rhizosphere of one species translocate nutrients to a competing neighboring species. This mechanism occurs in the invasion of North-American grasslands by invasive *Centaurea* species, where the growth of some native species decreases in the presence of the invader only when growing with the native soil community (Callaway et al. 2004).

Soil feedbacks contribute to both plant rarity and invasiveness (Klironomos 2002). In an experiment conducted to compare soil feedbacks between coexisting rare and abundant plants, Klironomos (2002) demonstrated how rare plants have strong negative soil feedbacks as a consequence of the accumulation of fungal pathogens in the rhizosphere. In addition, positive feedbacks between plants and mutualistic rhizosphere organisms can increase the ability of the plant species to colonize and invade new areas (Richardson et al. 2000). The importance of positive and negative feedbacks in large-scale vegetation patterns might also shift over temporal and spatial gradients (Reynolds et al. 2003). In the early stages of plant succession, mutualists play an important role in the establishment of their host plants and facilitate the establishment of new plant species (Figure 19.3). Since nitrogen is more limiting in the early stages of succession, the positive rhizobia–legume interaction plays a key role at that stage. With the progression in succession and the accumulation of organic matter, ecto- and ericoid mycorrhizas are favored due to their ability to break down complex organic compounds. Negative feedbacks should be more important at those later stages of succession, when there is also an accumulation of soil pathogens. Some authors postulate that early successional species are quick-growers and, therefore, poorly defended against pathogens. Over time, the accumulation of soil pathogens leads to a replacement of those by other slow-growing, better-defended species. Progression in succession also leads to more specific interactions since better-defended species are usually less susceptible to the attack of generalist pathogens and herbivores.

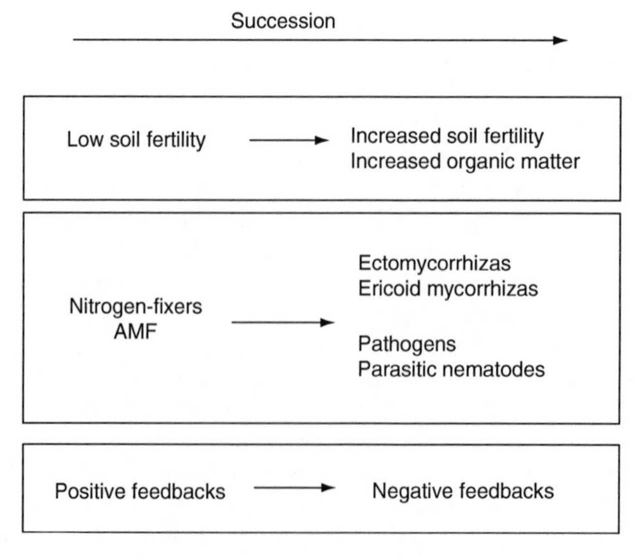

FIGURE 19.3 Changes in soil fertility, relative abundance of rhizosphere organisms, and feedbacks between plants and the rhizosphere communities that occur with succession.

CONCLUSION

The organisms associated with the rhizosphere can greatly influence plant performance. They establish close associations that have a positive or negative effect on plant establishment, growth, and fecundity, but the net effect for a plant species also depends on the interactions that occur between those organisms in the rhizosphere. The implications of belowground interactions for ecosystem functioning are widely documented. They affect plant richness and abundance, landscape heterogeneity, and plant succession. In turn, the rhizosphere is also a dynamic part of the ecosystem that changes with plant diversity and identity and with successional processes. Therefore, the interactions between plants and their rhizosphere should not be ignored in ecological studies.

REFERENCES

Ahren, D. and A. Tunlid, 2003. Evolution of parasitism in nematode-trapping fungi. Journal of Nematology 35: 194–197.

Allen, E.B. and M.F. Allen, 1986. Water relations of xeric grasses in the field: interactions of mycorrhizae and competition. New Phytologist 104: 559–571.

Anderson, R.C., D.E. Gardner, C.C. Daehler, and F.C. Meinzer, 2002. Dieback of *Acacia koa* in Hawaii: ecological and pathological characteristics of affected stands. Forest Ecology and Management 5576: 1–14.

Antunes, P.M., I. Rajcan, and M.J. Goss, 2006. Specific flavonoids as interconnecting signals in the tripartite symbiosis formed by arbuscular mycorrhizal fungi, *Bradyrhizobium japonicum* (Kirchner) Jordan and soybean (*Glycine max* (L.) Merr.). Soil Biology and Biochemistry 38: 533–543.

Augspurger, C.K., 1983. Seed dispersal of the tropical tree, *Platypodium elegans*, and the escape of its seedlings from fungal pathogens. Journal of Ecology 71: 759–771.

Azcón-Aguilar, C. and J.M. Barea, 1996. Arbuscular mycorrhizas and biological control of soil-borne plant pathogens—An overview of the mechanisms involved. Mycorrhiza 6: 457–464.

Bardgett, R., 2004. The Biology of Soil. A Community and Ecosystem Approach. Oxford University Press, Oxford.

Bardgett, R.D., C.S. Denton, and R. Cook, 1999a. Below-ground herbivory promotes soil nutrient transfer and root growth in grassland. Ecology Letters 2: 357–360.

Bardgett, R.D., R. Cook, G.W. Yeates, and C.S. Denton, 1999b. The influence of nematodes on below-ground processes in grassland ecosystems. Plant and Soil 212: 23–33.

Barron, G.L., 1977. The Nematode Destroying Fungi. Canadian Publications Ltd, Ontario.

Barron, G.L., 2003. Predatory fungi, wood decay, and the carbon cycle. Biodiversity 4: 3–9.

Bell, T., R.P. Freckleton, and O.T. Lewis, 2006. Plant pathogens drive density-dependent seedling mortality in a tropical tree. Ecology Letters 9: 569–574.

Bending, G.D. and D.J. Read, 1995. The structure and function of the vegetative mycelium of ectomycorrhizal plants. VI. Activities of nutrient mobilizing enzymes in birch litter colonized by *Paxillus involutus* (Fr.) Fr. New Phytologist 130: 411–417.

Bever, J.D., 2003. Soil community feedback and the coexistence of competitors: conceptual framework and empirical tests. New Phytologist 157: 465–473.

Bird, A.F. and J. Bird, 1986. Observations on the use of insect parasitic nematodes as a means of biological-control of root-knot nematodes. International Journal for Parasitology 16: 511–516.

Bongers, T. and M. Bongers, 1998. Functional diversity of nematodes. Applied Soil Ecology 10: 239–251.

Bradford, M.A., T.H. Jones, R.D. Bardgett, H.I.J. Black, B. Boag, M. Bonkowski, R. Cook, T. Eggers, A.C. Gange, S.J. Grayston, E. Kandeler, A.E. McCaig, J.E. Newington, J.I. Prosser, H. Setala, P.L. Staddon, G.M. Tordoff, D. Tscherko, and J.H. Lawton, 2002. Impacts of soil faunal community composition on model grassland ecosystems. Science 298: 615–618.

Brasier, C.M., F. Robredo, and J.F.P. Ferraz, 1993. Evidence for *Phytophtora cinnamomi* involvement in Iberian oak decline. Plant Pathology 42: 140–145.

Brundrett, M.C., 2002. Coevolution of roots and mycorrhizas of land plants. New Phytologist 154: 275–304.
Cairney, J.W.G. and A.A. Meharg, 2003. Ericoid mycorrhiza: a partnership that exploits harsh edaphic conditions. European Journal of Soil Science 54: 735–740.
Callaway, R.M., G.C. Thelen, S. Barth, P.W. Ramsey, and J.E. Gannon, 2004. Soil fungi alter interactions between the invader *Centaurea maculosa* and North American natives. Ecology 85: 1062–1071.
Carey, E.V., M.J. Marler, and R.M. Callaway, 2004. Mycorrhizae transfer carbon from a native grass to an invasive weed: evidence from stable isotopes and physiology. Plant Ecology 172: 133–141.
Ciancio, A., 1995. Density-dependent parasitism of *Xiphinema diversicaudatum* by *Pasteuria penetrans* in a naturally infested field. Phytopathology 85: 144–149.
Connell, J.H. and M.D. Lowman, 1989. Low-diversity tropical rain forests: some possible mechanisms for their existence. American Naturalist 134: 88–119.
Costa, S.R., M.S.N.A. Santos, and M.F. Ryan, 2003. Effect of *Artemisia vulgaris* rhizome extracts on hatching, mortality, and plant infectivity of *Meloidogyne megadora*. Journal of Nematology 35: 437–442.
D'Alessandro, M. and T.C.J. Turlings, 2006. Advances and challenges in the identification of volatiles that mediate interactions among plants and arthropods. Analyst 131: 24–32.
De Deyn, G.B., C.E. Raaijmakers, and W.H. van der Putten, 2004. Plant community development is affected by nutrients and soil biota. Journal of Ecology 92: 824–834.
de la Peña, E., S. Rodriguez-Echeverria, W.H. van der Putten, H. Freitas, and M. Moens, 2006. Mycorrhizal fungi control migratory endoparasitic nematodes in *Ammophila arenaria*. New Phytologist 169: 829–840.
Duponnois, R., M. Neyra, K. Senghor, and A.M. Ba, 1999. Effects of the root-knot nematode *Meloidogyne javanica* on the symbiotic relationships between different strains of *Rhizobia* and *Acacia holosericea* (A Cunn. ex G. Don). European Journal of Soil Biology 35: 99–105.
Duponnois, R., M. Fargette, S. Fould, J. Thioulouse, and K.G. Davies, 2000a. Diversity of the bacterial hyperparasite *Pasteuria penetrans* in relation to root-knot nematodes (*Meloidogyne* spp.) control on *Acacia holosericea*. Nematology 2: 435–442.
Duponnois, R., H. Founoune, A. Ba, C. Plenchette, S. El Jaafari, M. Neyra, and M. Ducousso, 2000b. Ectomycorrhization of *Acacia holosericea* A. Cunn. ex G. Don by *Pisolithus* spp. in Senegal: Effect on plant growth and on the root-knot nematode *Meloidogyne javanica*. Annals of Forest Science 57: 345–350.
Eady, R.R., 1991. The dinitrogen-fixing bacteria. In: A. Balows, H.G. Trüper, M. Dworkin, W. Harder, and K.H. Schleifer, eds. The Prokaryotes. Springer-Verlag, New York, pp. 534–553.
Ehrenfeld, J.G., B. Ravit, and K. Elgersma, 2005. Feedback in the plant–soil system. Annual Review of Environment and Resources 30: 75–115.
Eisner, T. and J. Meinwald, 1995. Chemical Ecology. Proceedings of the National Academy of Sciences of the United States of America 92: 1–1.
Fitter, A.H. and J. Garbaye, 1994. Interactions between mycorrhizal fungi and other soil organisms. Plant and Soil 159: 123–132.
Francl, L.J. and V.H. Dropkin, 1985. *Glomus fasciculatum*, a weak pathogen of *Heterodera glycines*. Journal of Nematology 17: 470–475.
Gange, A.C. and V.K. Brown, 2002. Actions and interactions of soil invertebrates and arbuscular mycorrhizal fungi in affecting the structure of plant communities. In: M.G.A. van der Heijden and I.R. Sanders, eds. Mycorrhizal Ecology. Springer-Verlag, Berlin, pp. 321–344.
Gilbert, G., 2002. Evolutionary ecology of plant diseases in natural ecosystems. Annual Review of Phytopathology 40: 13–43.
Goff, C., 1936. Relative susceptibility of some annual ornamentals to root-knot. University of Florida Agricultural Experimental Station Bulletin 291: 15.
Greipsson, S. and H. El-Mayas, 2002. Synergistic effect of soil pathogenic fungi and nematodes reducing bioprotection of arbuscular mycorrhizal fungi on the grass *Leymus arenarius*. Biocontrol 47: 715–727.
Grewal, P.S., E.E. Lewis, and S. Venkatachari, 1999. Allelopathy: a possible mechanism of suppression of plant-parasitic nematodes by entomopathogenic nematodes, pp. 735–743.

Grime, J.P., J.M.L. Mackey, S.H. Hillier, and D.J. Read, 1987. Floristic diversity in a model system using experimental microcosms. Nature 328: 420–422.

Gruntman, M. and A. Novoplansky, 2004. Physiologically mediated self/non-self discrimination in roots. Proceedings of the National Academy of Sciences of the United States of America 101: 3863–3867.

Habte, M., Y.C. Zhang, and D.P. Schmitt, 1999. Effectiveness of *Glomus* species in protecting white clover against nematode damage. Canadian Journal of Botany 77: 135–139.

Hansen, E.M., 2000. *Phellinus weirii* and other native root pathogens as determinants of forest structure and process in western North America. Annual Review of Phytopathology 38: 515–539.

Hart, M.M. and J.N. Klironomos, 2002. Diversity of arbuscular mycorrhizal fungi and ecosystem functioning. In: M.G.A. van der Heijden and I.R. Sanders, eds. Mycorrhizal Ecology. Berlin, Springer-Verlag, pp. 225–242.

Hartnett, D.C. and G.W.T. Wilson, 1999. Mycorrhizae influence plant community structure and diversity in tallgrass prairie. Ecology 80: 1187–1195.

Hawkins, H.-J., A. Johansen, and E. George, 2000. Uptake and transport of organic and inorganic nitrogen by arbuscular mycorrhizal fungi. Plant and Soil 226: 275–285.

Heemsbergen, D.A., M.P. Berg, M. Loreau, J.R. van Haj, J.H. Faber, and H.A. Verhoef, 2004. Biodiversity effects on soil processes explained by interspecific functional dissimilarity. Science 306: 1019–1020.

Helgason, T., J.W. Merryweather, J. Denison, P. Wilson, J.P.W. Young, and A.H. Fitter, 2002. Selectivity and functional diversity in arbuscular mycorrhizas of co-occurring fungi and plants from a temperate deciduous woodland. Journal of Ecology 90: 371–384.

Hillocks, R.J., 2001. The implications for plant health of nematode-fungal interactions in the root zone. In: M.J. Jeger and N.J. Spence, eds. Biotic Interactions in Plant–Pathogen Associations. CAB International, Oxford, pp. 269–283.

Hodge, A., 2006. Plastic plants and patchy soils. Journal of Experimental Botany 57: 401–411.

Inderjit, 2005. Soil microorganisms: An important determinant of allelopathic activity. Plant and Soil 274: 227–236.

Inderjit and C.L. Foy, 1999. Nature of the interference mechanism of mugwort (*Artemisia vulgaris*). Weed Technology 13: 176–182.

Inderjit and J. Weiner, 2001. Plant allelochemical interference or soil chemical ecology? Perspectives in Plant Ecology Evolution and Systematics 4: 3–12.

Jaffee, B.A., 1992. Population biology and biological control of nematodes. Canadian Journal of Microbiology 38: 359–364.

Jatala, P., 1985. Biological control of nematodes. In: J. Sasser and C. Carter, eds. An Advanced Treatise on *Meloidogyne*. North Carolina State University Graphics, Raleigh, pp. 303–308.

Johnson, S.N. and P.J. Gregory, 2006. Chemically-mediated host-plant location and selection by root-feeding insects. Physiological Entomology 31: 1–13.

Johnson, N.C., J.H. Graham, and F.A. Smith, 1997. Functioning of mycorrhizal associations along the mutualism-parasitism continuum. New Phytologist 135: 575–585.

Jones, J.T., L. Robertson, R.N. Perry, and W.M. Robertson, 1997. Changes in gene expression during stimulation and hatching of the potato cyst nematode *Globodera rostochiensis*. Parasitology 114: 309–315.

Kerry, B.R. and D.H. Crump, 1998. The dynamics of the decline of the cereal cyst nematode, *Heterodera avenae*, in four soils under intensive cereal production. Fundamental and Applied Nematology 21: 617–625.

Kerry, B.R. and W.M. Hominick, 2002. Biological Control. Taylor & Francis, London.

Klironomos, J.N., 2002. Feedback with soil biota contributes to plant rarity and invasiveness in communities. Nature 417: 67–70.

Klironomos, J.N., 2003. Variation in plant response to native and exotic arbuscular mycorrhizal fungi. Ecology 84: 2292–2301.

Klironomos, J.N. and M.M. Hart, 2001. Food-web dynamics—Animal nitrogen swap for plant carbon. Nature 410: 651–652.

Korthals, G.W., P. Smilauer, C. Van Dijk, and W.H. Van der Putten, 2001. Linking above- and below-ground biodiversity: abundance and trophic complexity in soil as a response to experimental plant communities on abandoned arable land. Functional Ecology 15: 506–514.

Lafay, B. and J.J. Burdon, 1998. Molecular diversity of *Rhizobia* occurring on native shrubby legumes in Southeastern Australia. Applied and Environmental Microbiology 64: 3989–3997.

Little, L.R. and M.A. Maun, 1996. The "*Ammophila* problem" revisited: a role for mycorrhizal fungi. Journal of Ecology 84: 1–7.

Marschner, H., 1995. Mineral Nutritions of Higher Plants. Academic Press, San Diego, CA.

McKane, R.B., L.C. Jonhson, G.R. Shaver, K.J. Nadelhoffer, E.B. Rastetter, B. Fry, A.E. Giblin, K. Kielland, B.L. Kwiatkowski, J.A. Laundre, and G. Murray, 2002. Resource-based niches provide a basis for plant species diversity and dominance in the arctic tundra. Nature 415: 68–71.

Mitra, R.M., S.L. Shaw, and S.R. Long, 2004. Six non nodulating plant mutants defective for Nod factor-induced transcriptional changes associated with the legume-rhizobia symbiosis. Proceedings of the National Academy of Sciences of the United States of America 101: 10217–10222.

Munkvold, L., R. Kjøller, M. Vestberg, S. Rosendahl, and I. Jakobsen, 2004. High functional diversity within species of arbuscular mycorrhizal fungi. New Phytologist 164: 357–364.

Newsham, K.K., A.H. Fitter, and A.R. Watkinson, 1995. Arbuscular mycorrhiza protect an annual grass from root pathogenic fungi in the field. Journal of Ecology 83: 991–1000.

Northup, R.R., Z. Yu, R. Dahlgren, and K. Vogt, 1995. Polyphenol control of nitrogen release from pine litter. Nature 377: 227–229.

Oka, Y., H. Koltai, M. Bar-Eyal, M. Mor, E. Sharon, I. Chet, and Y. Spiegel, 2000. New strategies for the control of plant-parasitic nematodes. Pest Management Science 56: 983–988.

Onen, H. and Z. Ozer, 2002. Study of allelopathic influence of mugwort (*Artemisia vulgaris* L.) on several crops. Zeitschrift Fur Pflanzenkrankheiten Und Pflanzenschutz-Journal of Plant Diseases and Protection 18(special issue): 339–347.

Packer, A. and K. Clay, 2000. Soil pathogens and spatial patterns of seedling mortality in a temperate tree. Nature 404: 278–281.

Pérez-Moreno, J. and D.J. Read, 2000. Mobilization and transfer of nutrients from litter to tree seedlings via the vegetative mycelium of ectomycorrhizal plants. New Phytologist 145: 301–309.

Pérez-Moreno, J. and D.J. Read, 2001a. Nutrient transfer from soil nematodes to plants: a direct pathway provided by the mycorrhizal mycelial network. Plant, Cell and Environment 24: 1219–1226.

Pérez-Moreno, J. and D.J. Read, 2001b. Exploitation of pollen by mycorrhizal mycelial systems with special reference to nutrient recycling in boreal forests. Proceedings of the Royal Society of London B 268: 1329–1335.

Perret, X., C. Staehelin, and W.J. Broughton, 2000. Molecular basis of symbiotic promiscuity. Microbiology and Molecular Biology Review 64: 180–201.

Perry, R.N., 1996. Chemoreception in plant parasitic nematodes. Annual Review of Phytopathology 34: 181–199.

Perry, R.N., 1997. Plant signals in nematode hatching and attraction. In: C. Fenoll, F. Grundler, and S. Ohl, eds. Celullar and Molecular Aspects of Plant–Nematode Interactions. Kluwer Academic Publishers, Dordrecht, pp. 38–50.

Rasmann, S., T.G. Kollner, J. Degenhardt, I. Hiltpold, S. Toepfer, U. Kuhlmann, J. Gershenzon, and T.C.J. Turlings, 2005. Recruitment of entomopathogenic nematodes by insect-damaged maize roots. Nature 434: 732–737.

Reinhart, K.O., A. Royo, W.H. van der Putten, and K. Clay, 2005. Soil feedback and pathogen activity in *Prunus serotina* throughout its native range. Journal of Ecology 93: 890–898.

Reitz, M., K. Rudolph, I. Schroder, S. Hoffmann-Hergarten, J. Hallmann, and R.A. Sikora, 2000. Lipopolysaccharides of *Rhizobium etli* strain G12 act in potato roots as an inducing agent of systemic resistance to infection by the cyst nematode *Globodera pallida*. Applied and Environmental Microbiology 66: 3515–3518.

Reynolds, H.L., A. Packer, J.D. Bever, and K. Clay, 2003. Grassroots ecology: plant–microbe–soil interactions as drivers of plant community structure and dynamics. Ecology 84: 2281–2291.

Richardson, D.M., N. Allsopp, C.M. D'Antonio, and S.J. Milton, 2000. Plant invasions—the role of mutualisms. Biological Reviews 75: 65–93.

Rodríguez-Echeverría, S., M.A. Pérez-Fernández, S. Vlaar, and T.M. Finan, 2003. Analysis of the legume-rhizobia symbiosis in shrubs from central western Spain. Journal of Applied Microbiology 95: 1367–1374.

Rodriguez-Kabana, R., 1991. Biological control of plant-parasitic nematodes. Nematropica 21: 111–122.
Roncadori, R.W., 1997. Interactions between arbuscular mycorrhizas and plant parasitic nematodes in agro-ecosystems. In: A.C. Gange and V.K. Brown, eds. Multitrophic Interactions in Terrestrial Systems. The 36th Symposium of the British Ecological Society. Blackwell Science, Oxford.
Sanders, I.R., J.P. Clapp, and A. Wiemken, 1996. The genetic diversity of arbuscular mycorrhizal fungi in natural ecosystems—A key to understanding the ecology and functioning of the mycorrhizal symbiosis. New Phytologist 133: 123–134.
Sawada, H., L.D. Kuykendall, and J.M. Young, 2003. Changing concepts in the systematics of bacterial nitrogen-fixing legume symbionts. The Journal of General and Applied Microbiology 49: 155–179.
Schnee, C., T. Kollner, M. Held, T. Turlings, J. Gershenzon, and J. Degenhardt, 2006. The products of a single maize sesquiterpene synthase form a volatile defense signal that attracts natural enemies of maize herbivores. Proceedings of the National Academy of Sciences of the United States of America 103: 1129–1134.
Scholl, E.H., J.L. Thorne, J.P. McCarter, and D.M. Bird, 2003. Horizontally transferred genes in plant-parasitic nematodes: a high-throughput genomic approach. Genome Biology 4: R39.
Schüßler, A., D. Schwarzott, and C. Walker, 2001. A new fungal phylum, the *Glomeromycota*: phylogeny and evolution. Mycological Research 105: 1413–1421.
Siddiqui, Z.A. and I. Mahmood, 1996. Biological control of plant parasitic nematodes by fungi: A review. Bioresource Technology 58: 229–239.
Simard, S.W., M.D. Jones, and D.M. Durall, 2002. Carbon and nutrient fluxes within and between mycorrhizal plants. In: M.G.A. van der Heijden and I.R. Sanders, eds. Mycorrhizal Ecology. Springer-Verlag, Berlin Heidelberg, pp. 33–74.
Smith, G.S. and D.T. Kaplan, 1988. Influence of mycorrhizal fungus, phosphorus, and burrowing nematode interactions on growth of rough lemon citrus seedlings. Journal of Nematology 20: 539–544.
Smith, S.E. and D.J. Read, 1997. Mycorrhizal Symbiosis. Academic Press, London.
Stirling, G.R., 1991. Biological Control of Plant Parasitic Nematodes: Progress, Problems and Prospects. CB International, Wallinford.
Sturhan, D., 1988. New host and geographical records of nematode-parasitic bacteria of the *Pasteuria penetrans* group. Nematologica 34: 350–356.
Sturz, A.V. and J. Kimpinski, 2004. Endoroot bacteria derived from marigolds (*Tagetes* spp.) can decrease soil population densities of root-lesion nematodes in the potato root zone. Plant and Soil 262: 241–249.
Turnbull, M.H., R. Goodall, and G.R. Stewart, 1995. The impact of mycorrhizal colonization upon nitrogen source utilization and metabolism in seedlings of *Eucalyptus maculata* Hook. Plant, Cell and Environment 18: 1386–1394.
Turnbull, M.H., S. Schmidt, P.D. Erskine, S. Richards, and G.R. Stewart, 1996. Root adaptation and nitrogen source acquisition in natural ecosystems. Tree Physiology 16: 941–948.
Twomey, U., P. Warrior, B.R. Kerry, and R.N. Perry, 2000. Effects of the biological nematicide, DiTera (R), on hatching of *Globodera rostochiensis* and *G. pallida*. Nematology 2: 355–362.
Ulenbroek, J. and J. Brijloo, 1958. Isolation and structure of a nematicidal principle occurring in *Tagetes* roots. Recueil des travaux chimiques des Pays-Bas 77: 1004–1009.
Ulrich, A. and I. Zaspel, 2000. Phylogenetic diversity of rhizobial strains nodulating *Robinia pseudoacacia* L. Microbiology 146: 2997–3005.
van der Heijden, M.G.A., J.N. Klironomos, M. Ursic, P. Moutoglis, R. Streitwolf-Engel, T. Boller, A. Wiemken, and I.R. Sanders, 1998. Mycorrhizal fungal diversity determines plant biodiversity, ecosystem variability and productivity. Nature 396: 69–72.
Van der Putten, W.H., 2003. Plant defense belowground and spatiotemporal processes in natural vegetation. Ecology 84: 2269–2280.
Van der Putten, W.H. and B.A.M. Peters, 1997. How soil-borne pathogens may affect plant competition. Ecology 78: 1785–1795.
Van der Putten, W.H., P.W.T. Maas, W.J.M. Van Gulik, and H. Brinkman, 1990. Characterization of soil organisms involved in the degeneration of *Ammophila arenaria*. Soil Biology and Biochemistry 22: 845–852.

Van der Putten, W.H., C. Vandijk, and B.A.M. Peters, 1993. Plant-specific soil-borne diseases contribute to succession in foredune vegetation. Nature 362: 53–56.

Van der Putten, W.H. and C.D. Van der Stoel, 1998. Plant parasitic nematodes and spatio-temporal variation in natural vegetation. Applied Soil Ecology 10: 253–262.

van Tol, R., A.T.C. van der Sommen, M.I.C. Boff, J. van Bezooijen, M.W. Sabelis, and P.H. Smits, 2001. Plants protect their roots by alerting the enemies of grubs. Ecology Letters 4: 292–294.

Villenave, C. and P. Cadet, 1998. Interactions of *Helicotylenchus dihystera, Pratylenchus pseudopratensis*, and *Tylenchorhynchus gladiolatus* on two plants from the Soudano-Sahelian zone of West Africa. Nematropica 28: 31–39.

Warrior, P., L. Rehberger, M. Beach, P. Grau, G. Kirfman, and J. Conley, 1999. Commercial development and introduction of Ditera TM, a new nematicide. Pesticide Science 55: 343–389.

Weston, L.A. and S.O. Duke, 2003. Weed and crop allelopathy. Critical Reviews in Plant Sciences 22: 367–389.

Whipps, J.M. and K.G. Davies, 2000. Success in biological control of plant pathogens and nematodes by microorganisms. In: G. Gurr and S. Wratten, eds. Biological Control: Measures of Success, Kluwer Academic Publishers, pp. 231–269.

Wills, R.T. and J. Kinnear, 1993. Threats to the biota of the Stirling Range. In: C. Thompson, G.P. Hall, and G.R. Friend, eds. Mountains of Mystery—A Natural History of the Stirling Range. Department of Conservation and Land Management, Como, Western Australia, pp. 135–141.

Xavier, L.J.C. and J.J. Germida, 2003. Selective interactions between arbuscular mycorrhizal fungi and *Rhizobium leguminosarum* bv. *viceae* enhance pea yield and nutrition. Biology and Fertility of Soils 37: 261–267.

Yeates, G.W., 1987. Nematode feeding and activity—the importance of development stages. Biology and Fertility of Soils 3: 143–146.

Yeates, G.W., T. Bongers, R.G.M. Degoede, D.W. Freckman, and S.S. Georgieva, 1993. Feeding habits in soil nematode families and genera—an outline for soil ecologists. Journal of Nematology 25: 315–331.

Young, I. and K. Ritz, 2005. The habitat of soil microbes. In: R.D. Bardgett, M. Usher, and D. Hopkins, eds. Biological Diversity and Function in Soils. Cambridge University Press, Cambridge, pp. 31–43.

Zacheo, G., 1993. Introduction. In: M.W. Khan, ed. Nematode Interactions. Chapman & Hall, London, pp. 1–25.

Zentmyer, G.A., 1980. *Phytophthora cinnamomi* and the Diseases It Causes. American Phytopathological Society, St Paul, MN.

20 Resistance to Air Pollutants: From Cell to Community

Jeremy Barnes, Alan Davison, Luis Balaguer, and Esteban Manrique-Reol

CONTENTS

The race is not to the swift, nor the battle to the strong . . . but time and chance happeneth to them all.

—Ecclesiastes 9:11

INTRODUCTION

The generation of energy by the burning of fossil fuels, all manner of industrial processes, the biodegradation of wastes, and some farming operations lead to the release of wide range of contaminants into the air. Most have little or no discernible effect on the environment, because the resulting concentrations in the atmosphere are well below levels known to be toxic or because they are not toxic to biological systems. Others attain levels that are known to threaten human health and to damage both fauna and flora. The situation is not new because there have been air pollution problems of one kind or another since fire was first used and metals were first smelted, but the unbridled expansion of industry in many parts of the world over the past century has resulted in problems on an unprecedented scale, with impacts extending from the local or regional to global level.

Although air pollution can take various forms (i.e., dusts, smoke, fumes, aerosols, or mist), this chapter focuses on resistance and adaptation to the most common gaseous pollutants. Stringent control measures have resulted in a steady decline in the emissions of several pollutants in developed regions (e.g., sulfur dioxide [SO_2]); however, ground-level concentrations of some of the most potent gases (e.g., ozone [O_3]) continue to increase (Penkett 1988, Boubel et al. 1994, Stockwell et al. 1997). Locally, ground-level concentrations of some pollutants may be high enough to result in severe foliar injury under conditions favoring accumulation in the atmosphere (i.e., periods of high solar radiation, favorable temperatures, or temperature inversions), whereas potentially damaging concentrations of others (e.g., O_3) maybe generated at a considerable distance from the source (Bell 1984, Boubel et al. 1994, Krupa 1996). Long distance transport is usually favored by the high levels of irradiance and stable atmospheric conditions associated with slow-moving high-pressure systems in the northern hemisphere. Under such conditions, there is poor dispersal of polluted air masses, and pollutant concentrations, although typically lower than those experienced near to point sources, may be high enough to result in subtle changes in plant physiology, growth, and community composition. Such effects are not necessarily associated with the appearance of typical visible symptoms of injury, but are more common and just as debilitating (Wolfenden and Mansfield 1991, Davison and Barnes 1992, 1998).

It would seem that all plants possess the encoded capability for the perception, signaling, and response to air pollutants; however, differential expression under the influence of genetic and environmental factors can result in constitutive and inducible differences between the reaction norms of plants within and between populations to the same air pollution insult. In this chapter, we discuss aspects related to the ecotoxicology of airborne pollutants. We begin by reviewing what is known about the mechanisms underlying differential resistance to the most common gaseous pollutants, and then attempt to scale-up from responses at the cellular level to those affecting resistance at the plant, population, and community level. A generic model is used to provide a conceptual framework within which to discuss the mechanisms underlying differential resistance. Where appropriate, we have elected to focus on plant responses to O_3. Not only because the authors are more familiar with the literature relating to this pollutant than any other, but also because O_3 is now recognized to be one of the most potent and widespread toxic agents to which vegetation is exposed in the field (Davison and Barnes 1992, Kärenlampi and Skärby 1996, Fuhrer et al. 1997, Davison and Barnes 1998). Moreover, increasing concentrations of the pollutants pose a growing threat to vegetation in many regions (Penkett 1988, Stockwell et al. 1997), and exciting advances have recently been made in our understanding of the mechanisms underlying the genetic basis of resistance to O_3 (Kangasjärvi et al. 1994, Schraudner et al. 1994, Alscher et al. 1997, Pell et al. 1997, Schraudner et al. 1997).

CELLULAR LEVEL

The processes controlling differential resistance to pollutants are considered within the framework of a conceptual model (Figure 20.1), discussed first by Ariens et al. (1976) and later by Tingey et al. (Tingey and Taylor 1981, Hogsett et al. 1988, Tingey and Andersen 1991), where resistance* is envisaged to be governed by constitutive and inducible differences in a complex sequence of events that either reduces pollutant penetration to the target (i.e., avoidance mechanisms) or enhances the ability of plant tissues to withstand the pollutant and its products once it has penetrated to the target (i.e., tolerance mechanisms). However, various feedbacks can also influence plant response, especially in relation to pollutant detoxification and the repair of injury. These processes are initially dependent on the constitutive resources available, whereas subsequent responses may be governed by the regulation of gene expression, post-translational modification of enzymes (e.g., phosphorylation/dephosphorylation), and the synthesis of secondary defense-related metabolites. If these responses are not sufficient

* Herein defined after Roose et al. (1982) as the "relative ability of a genotype to maintain normal growth and remain free from injury in a polluted environment. A trait that is quantitative, rather than qualitative, as resistance need not be complete."

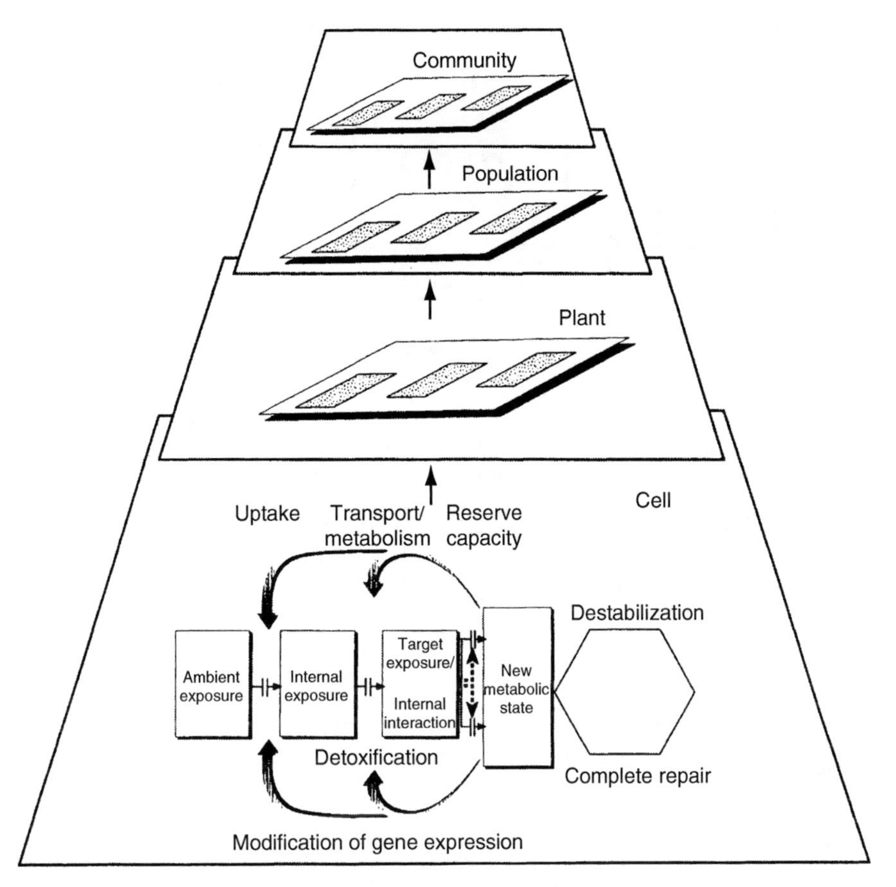

FIGURE 20.1 Conceptual model showing the processes that govern the sensitivity of plants to gaseous air pollutants. Pollutant levels that exceed the capacity of avoidance/tolerance mechanisms will result in cellular destabilization—an effect that underpins changes in the performance of the individual and shifts in the genetic composition of both populations and communities.

to prevent damage at the cellular level, then there will be destabilization and injury that will be reflected in downstream consequences at the level of the individual, population, and community. The mechanisms conferring resistance may be independent (i.e., pollutant-specific) or broadly based (i.e., a number of different pollutants trigger the same coordinated defense reaction); broadly based responses can result in cross-resistance to several pollutants (and possibly to a number of other environmental stresses), whereas there is growing evidence that cross-tolerance may be restricted to pollutants (and other stresses) that provoke similar insult on the same target (Schraudner et al. 1997, Barnes and Wellburn 1998).

Uptake

The dose of a pollutant absorbed by plant tissues plays a key role in determining effects on metabolism and physiology, and in the description and quantification of dose–response relationships (Taylor et al. 1998, Runeckles 1992). Uptake is predominantly controlled by rates of foliar gas exchange, with conventional approaches focusing on the importance of the cuticle and stomata in controlling the rate at which pollutants diffuse into individual leaves (Mansfield and Freer-Smith 1984). However, it is important to recognize that factors operating at different scales of resolution influence the rate of uptake, in addition to those operating at the level of the individual leaf. At a higher scale of resolution (i.e., scaling-up),

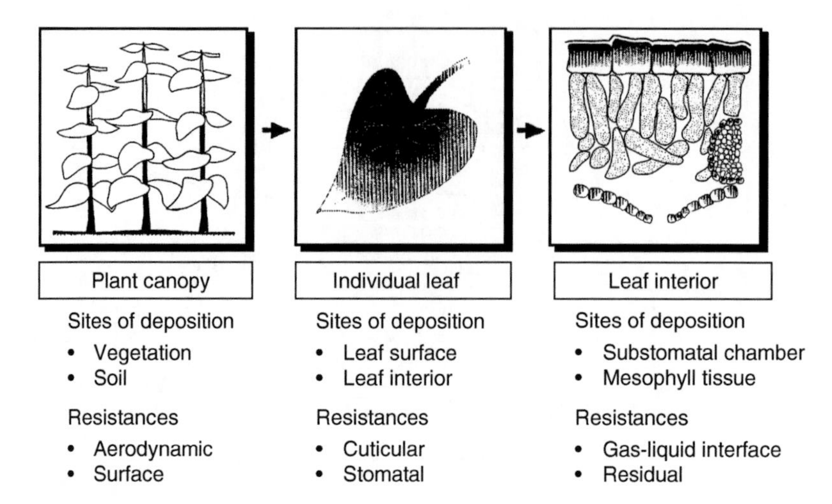

FIGURE 20.2 The different scales of resolution that need to be considered in determining rates of pollutant uptake. (Redrawn from Tanaka, K., Furusawa, I., Kondo, N., and Tanaka, K., *J. Plant Cell Physiol.*, 29, 743, 1988. With permission.)

aerodynamic factors influence uptake at the canopy level, whereas at a finer scale of resolution (i.e., scaling-down) physicochemical factors determine uptake at the interface between the plant and its external surroundings (Figure 20.2).

The size, density, and shape of the canopy all have a pronounced effect on the concentration of a pollutant to which individual leaves are exposed. Relatively few studies have addressed this issue, but it has been shown that the concentration of O_3 (and other pollutants) declines as one passes down through the canopy to soil level (Bennett and Hill 1973). As a result, leaves within a dense canopy tend to be exposed to lower concentrations than those on the surface or at the edges. However, because the air movement within a dense canopy is reduced, concentrations within it tend to be less prone to short-term fluctuations (Runeckles 1992). Other features of the leaves (e.g., leaf thickness, Rubisco content, ratio of mesophyll cell surface area to projected surface area, etc.) and of the environment (e.g., reduced levels of irradiance, increased temperature, higher humidity, etc.) also differ within the canopy in relation to those on the outside. This may strongly influence the uptake and effects of pollutants one leaves at different positions within the canopy and is an important consideration when attempting to extrapolate to the field from laboratory-based studies, identify the magnitude of the response of different species in a mixture, and establish the impact of pollutant at different developmental stages, since certain species or particular growth stages (e.g., seedlings) may be protected from exposure to potentially damaging pollutant concentrations by other elements of the canopy.

At the leaf-level, the flux of the pollutant to the leaf interior (J) is a function of the concentration gradient between the atmosphere (i.e., the concentration of the pollutant in the surrounding air, C_a) and the leaf interior (i.e., the concentration of the pollutant in the intercellular air spaces, C_i), and the sum of the physical, chemical, and biological resistances (ΣR) to diffusion from source to sink. Mathematically, this is generally expressed in a form analogous with Ohm's law, where:

$$J = (C_a - C_i)\Sigma R.$$

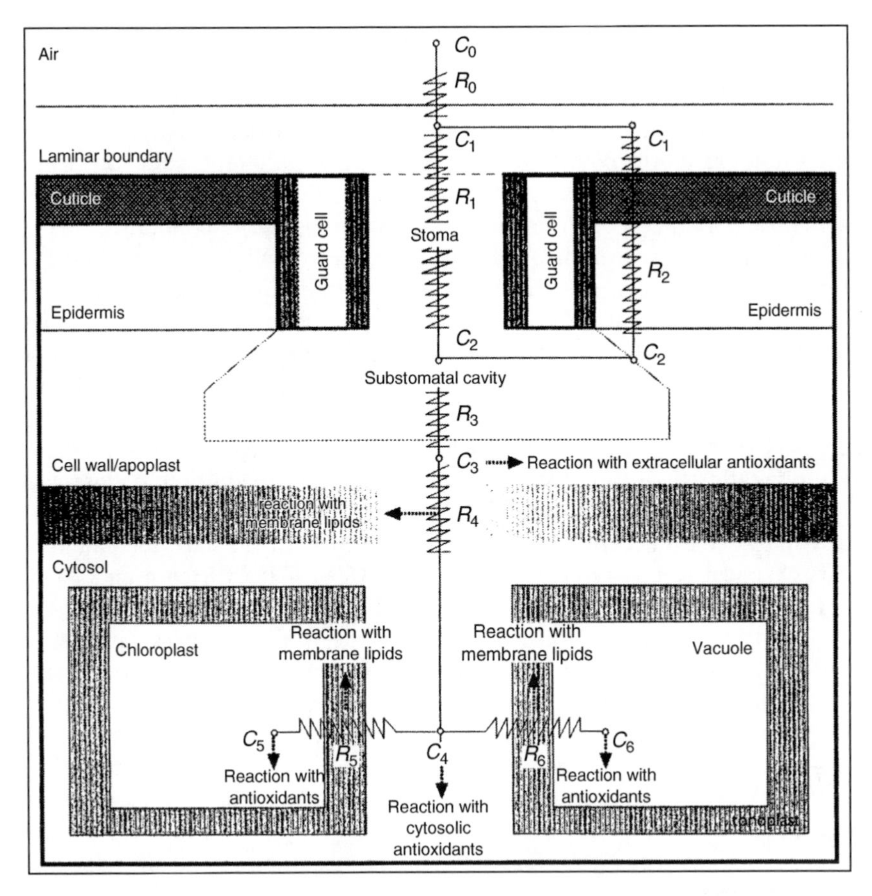

FIGURE 20.3 Resistance analog model indicating the network of physical, chemical, and biological impedances influencing the rate of diffusion of gaseous pollutants from the external atmosphere to the target.

The various impedances (ΣR) are generally visualized as a network of resistances to gas flow (Figure 20.3), using Gaastra's (1959) fundamental principles governing the tortuous route taken by effluxing water vapor molecules. The most important resistances at the leaf level are recognized to be those governing the movement of the pollutant into the leaf interior (i.e., stomatal resistance, r_1, and cuticular resistance, r_2), the rate of deposition on the hydrated surface of mesophyll cells, and the extent of sorption and reaction on the surface of the cuticle and in the substomatal cavities. These resistances are under genetic and environmental control, as well as being influenced by the physicochemical characteristics of the gas in question.

The rate of diffusion of gaseous pollutants through the cuticular membrane is commonly several orders of magnitude lower than that through the stomata (Lendzian 1984) and may be considered negligible for some reactive gases, such as O_3 (Kersteins and Lendzian 1989). Hence, the dose of the pollutant absorbed at the leaf level is predominantly controlled by factors determining stomatal conductance (i.e., stomatal aperture and frequency). Some plants are known to exhibit intrinsically higher stomatal conductance than others (Körner 1994), and, in general, these tend to be more susceptible to damage (Reich 1987, Becker et al. 1989, Darrall 1989). However, the response of stomata to external stimuli (e.g., irradiance, vapor pressure deficit (VPD), soil moisture content, the presence of pollutants in the atmosphere, atmospheric CO_2 concentrations, etc.) can strongly influence the rate of pollutant uptake through effects on stomatal aperture (Mansfield and Freer-Smith 1984, Darrall 1989, Wolfenden and Mansfield 1991, Wolfenden et al. 1992, Heath and Taylor 1997). Differences in stomatal conductance between sensitive and resistant individuals are rarely sufficient to result in complete exclusion of the pollutant from the leaf interior; therefore, it is generally concluded that mechanisms resulting in the avoidance of pollutant uptake are not the only factors determining resistance to airborne pollutants.

Physical leaf characteristics such as leaf thickness, mesophyll cell surface area, internal air space volume, cell wall thickness, and the volume of the aqueous matrix of the cell wall influence the eventual concentration of the pollutant and its dissolution products in the apoplast. Plants show considerable variation in all of these attributes. Mesophyll cell surface area: projected leaf area, for example, ranges from typical values of between 10 and 40 for mesophytes, but may be as high as 70 for some xerophytes (Nobel and Walker 1985, Pfanz 1987). There may also be systematic differences in anatomy along altitudinal gradients, which can contribute to differential resistance (Körner et al. 1989). In addition, pollutant molecules in the intercellular space or substomatal cavity are partitioned across the gas-to-liquid interface at a rate determined by the solubility of the gas in the extracellular fluid (Nobel 1974) and its chemical reactivity in the liquid phase (Heath 1988, Heath and Taylor 1997), factors influenced by temperature and possibly radiation (Barnes et al. 1996, Cape 1997). Differences in physicochemical properties between the most common gaseous pollutants result in substantial differences in their solubility in water and rates of diffusion in air (Nobel 1974), factors that, independent of other considerations, result in substantial differences in the rate at which individual gaseous pollutants are taken up (Runeckles 1992, Taylor et al. 1998). Leaf surface characteristics (such as surface wetness, was composition, micromorphology, etc.) can also influence the extent of sorption onto foliar surfaces (Wellburn et al. 1997), whereas reactions with other gases in the boundary layer or in the substomatal cavities may constitute a significant sink for some pollutants, for example, O_3 (Hewitt and Terry 1992, Salter and Hewitt 1992).

In most instances, difficult and time-consuming measurements of physical leaf characteristics are not undertaken, so it has become common practice to express rates of pollutant uptake on the basis of the flux to the leaf interior (i.e., that impinging on the mesophyll cell surface). This represents what is often termed the absorbed or effective dose of the pollutant, and can be readily estimated using Fickian diffusion principles from knowledge of boundary layer, stomatal and cuticular conductances, correcting for differences in the diffusivities between water vapor and the pollutant of interest. Hence, the flux of O_3 to the leaf interior (J_{O_3}) may be described as

$$J_{O_3} = g_b + 0.612\, g_{H_2O}(O_a - O_i),$$

where g_b is the turbulent boundary layer conductance, 0.612 is the difference in the binary diffusivities of water vapor and ozone in air (Nobel 1983), g_{H_2O} represents the stomatal conductance to water vapor, and O_a and O_i represent the concentrations of O_3 in the external atmosphere and in the intercellular spaces, respectively. No correction needs to be made for uptake through the cuticle in this case, since the cuticle is considered to represent a virtually impermeable barrier to O_3 (Kersteins and Lendzian 1989), whereas measurements of the intercellular O_3 concentration (O_i) suggest it is close to zero (Laisk et al. 1989). On a cautionary note, it is important to emphasize two points. First, such calculations represent the gross flux to the leaf interior—for some gases where the plant may act as a source as well as a sink (e.g., H_2S, NH_3 NO), correction for effluxing gas molecules is required to enable estimates of the net flux of the pollutant. Second, fluxes determined in the above manner take no account of differences in physical leaf characteristics or internal resistances influencing the rate at which the pollutant (or its products) is delivered to the eventual target. Potentially more informative models are available that enable estimates of the extent of penetration to the plasmalemma and beyond (Chameides 1989, Ramge et al. 1992, Plöchl et al. 1993), but these have rarely been used due to the uncertainties surrounding the complex solution chemistry of certain gases (e.g., O_3) and the relative importance of scavenging/transformation mation in the cell wall region.

Metabolism

Once the pollutant has penetrated as far as the mesophyll cell surface, it may be metabolized, sequestered, or excreted. Recent attention has focused on the rate at which pollutants (and their reactive products, including a variety of free radical species) are immobilized and detoxified at either the first barrier encountered after entry into the leaf (i.e., in the apoplast) or subsequently, after penetration into the cell proper. In any consideration of the processes underlying pollutant detoxification, it is important to recognize that dissolution in the apoplast can result in different types of stress on cellular constituents. Many gases (e.g., HF, NO_x, NH_3, and SO_2) induce acidification of subcellular compartments, whereas others (e.g., O_3, PAN, NO_x) result in oxidative stress, and some (e.g., SO_2) produce both (Malhotra and Khan 1984, Hippeli and Elstner 1996, Mudd 1996). The extent of damage resulting from the former is related to the ability of cells to buffer the increase in acidity or to excrete protons to the external media (Slovik 1996, Burkhardt and Drechsel 1997), whereas the influence of the latter is dependent on the efficiency of endogenous antioxidant systems that scavenge free radical and active oxygen species before they can react with cellular constituents (Kangasjärvi et al. 1994, Luwe and Heber 1995, Alscher et al. 1997).

Detoxification systems capable of protecting sensitive targets from the oxidative stress imposed by pollutants and their derivatives are common in plants, as in animals, and are subject to strict genetic control. Most attention has focused on those systems located in the various intracellular compartments (in chloroplasts, mitochondria, cytosol, and peroxisomes); however, similar systems are intimately associated with the plasmalemma and the cell wall (Figure 20.4). In recent years, the latter has attracted increased attention since there is growing evidence that some pollutants (e.g., O_3, SO_2, NO_2) and their dissolution products may be scavenged and detoxified/transformed at the mesophyll cell surface (i.e., in the apoplast). The aqueous matrix of the cell wall is now recognized to contain significant quantities of ascorbic acid (vitamin C) and polyamines as well as isoforms of Cu/Zn superoxide dismutase (SOD), ascorbate peroxidase (APX), and nonspecific peroxidases (GPODs) (Polle and Rennenberg 1993, Luwe and Heber 1995, Ogawa et al. 1996, Dietz 1996), which are known to function as antioxidants (Polle and Rennenberg 1993, Kangasjärvi et al. 1994, Alscher et al. 1997, Polle 1997). Research is still at an early stage and many questions remain to be answered, but preliminary model calculations based on the scavenging of O_3 (rather than its dissolution products) by apoplastic ascorbic acid indicate that detoxification processes operating in the apoplast may be sufficient to provide at least limited protection against O_3 (Chameides 1989, Polle and Rennenberg 1993, Lyons et al. 1999), a finding supported by several independent lines of evidence (Lyons et al. 1998). Although the relative importance of scavenging in the cell wall region remains to be established, there is a growing opinion that factors such as the extracellular concentration of ascorbic acid may play a central role in determining resistance to O_3 (Conklin et al. 1996, Lyons et al. 1998), as well as influencing the impacts of SO_2 (Dietz 1996) and NO_2 (Ramge et al. 1992).

Pollutants and their reactive products that breach the extracellular defense, or are produced from the reaction with plasmalemma constituents, must be scavenged by intracellular detoxification systems if damage is to be averted. There is, for example, strong evidence linking SO_2 tolerance with the activity of intracellular enzymes such as catalase (CAT),

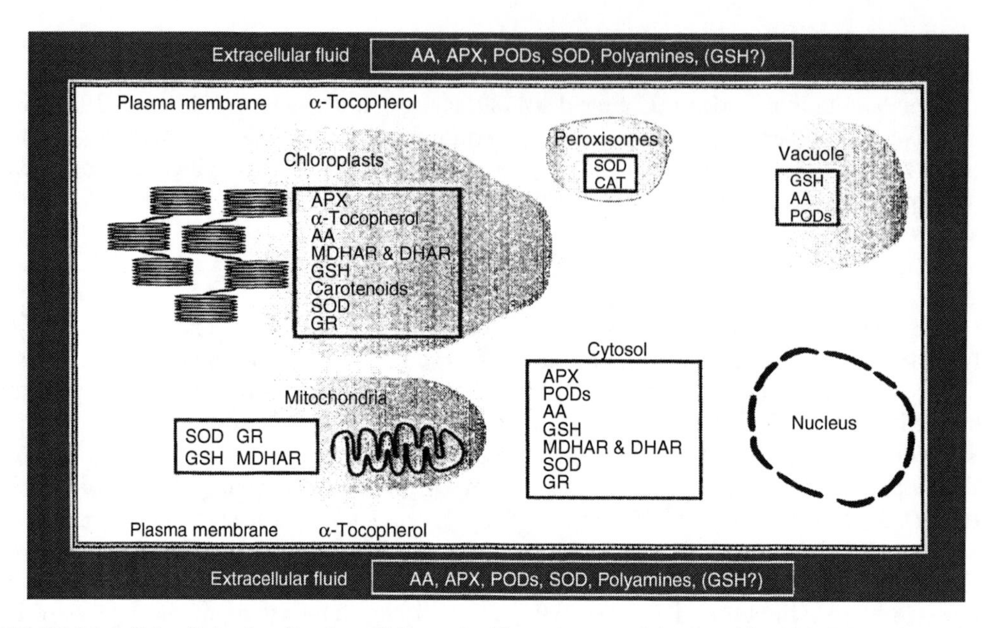

FIGURE 20.4 Subcellular localization of the antioxidant systems. AA, ascorbic acid; GSH, glutathione; APX, ascorbate peroxidase EC. 1.11.1.11; MDHAR, monodehydroascorbate radical reductase EC. 1.1.5.4; DHAR, dehydroascorbate radical reductase EC. 1.8.5.1; GR, glutathione reductase EC 1.6.4.2; SOD, superoxide dismutase EC. 1.15.1.1; CAT, catalase EC. 1.11.1.6; PODs, nonspecific peroxidase (sometimes referred to as guaiacol peroxidase) EC. 1.11.1.7.

superoxide dismutase (SOD), glutathione synthetase (GS), glutathione reductase (GR), glutathione peroxidase (GPX), and glutathione transferase (GST) (Tanaka et al. 1988, Ranieri et al. 1992, Lea et al. 1998), as well as with levels of glutathione (GSH) and ascorbic acid (ASC) (Madamanchi and Alscher 1994). Furthermore, recent work has drawn attention to the importance of the subcellular localization of these systems in relation to the protection afforded against different pollutants. For example, SO_2 tolerance in transgenic plants engineered to overexpress GR in different cellular compartments indicates that it is the cytoplasmic activity of this enzyme, rather than that of the plastidic forms, which is important in determining SO_2 tolerance (Aono et al. 1993, Broadbent 1995, Lea et al. 1998). There is also considerable evidence linking components of the cellular antioxidant system (ASC, GSH, polyamines, α-tocopherol, carotenoids, SOD, ascorbate peroxidase [APX], GR, and CAT) with O_3 tolerance (Heath 1988, Kangasjärvi et al. 1994, Hippeli and Elstner 1996, Mudd 1996, Alscher et al. 1997, Pell et al. 1997, Heath and Taylor 1997). However, the degree of protection afforded by these intracellular systems in relation to that achieved by scavenging in the cell wall region (the primary site of O_3 action) remains poorly understood (Lyons et al. 1998), and it is interesting to note that transgenic plants overproducing particular antioxidant enzymes in different intracellular compartments rarely display enhanced O_3 resistance (Pitcher et al. 1991, Van Camp et al. 1994, Pitcher and Zilinskas 1996, Torsethaugen et al. 1997).

In addition, toxicity thresholds are influenced by the efficiency of metabolic processes resulting in the utilization, sequestration, and excretion of pollutants. The plasmalemma, for example, is known to constitute an important obstacle impeding the penetration of pollutants (and their products) to their intracellular sites of action (Herschbach et al. 1995). Some pollutants (e.g., O_3) react readily with membrane constituents (Mudd 1996, Heath and Taylor 1997). The uptake of others (i.e., SO_2-derived sulfate [SO_4^{2-}], sulfite [SO_3^{2-}], and bisulfite [HSO_3^-] is limited by the activity of membrane-bound carriers and transformation processes

(Pfanz et al. 1990). Indeed, research on the inheritance of SO_2 resistance in *Cucumis sativus* L. suggests that differences in intrinsic membrane properties may contribute to variations in SO_2 resistance (Bressan et al. 1981). Once inside the cell, the products of some (e.g., SO_4^{2-}, NO_3^-, and NH_4^+) may be metabolized via the usual channels, sequestered for later use, stored indefinitely, or volatilized. For example, SO_2-derived SO_4^{2-} can either be metabolized to yield elevated levels of water-soluble nonprotein thiols (such as cysteine, γ-glutamylcysteine, and glutathione), which can be degraded at a later date to provide reduced S to support new growth (De Kok 1990), or it can be sequestered on a semipermanent basis in the vacuole (presuming there is sufficient available energy and H^+ ions to facilitate its transport across the tonoplast) (Cram 1990, Kaiser et al. 1989, Slovik 1996). In contrast, SO_3^- may be photoreduced and, after volatilization, be reemitted, mainly as H_2S. This pathway was originally considered to form a possible pathway for the detoxification of SO_2 (Rennenberg 1984), but current opinion suggests that the contribution of such emissions to the detoxification of environmentally relevant SO_2 concentrations may be negligible (Stuhlen and De Kok 1990).

Gene Expression

The photoautotrophic habit adopted by plants has resulted in the evolution of a sophisticated battery of mechanisms that renders them, with exception of all but a few microbes, the most adaptable of all multicellular organisms on the planet (Smith 1990). This flexibility includes the capacity to sense and react to the presence of airborne pollutants, as to other environmental stimuli, in a manner directed at sustaining survival to reproduction, a goal that may or may not be achieved, depending on the extent of metabolic flexibility and the degree of stress imposed at the cellular level.

Exposure to pollutants and other oxidative stresses induces changes in the expression of defense-related genes, posttranslational modification of enzymes (e.g., phosphorylation/dephosphorylation), and the synthesis of secondary metabolites—resulting in increases in the threshold for damage, that is, acclimation (Kangasjärvi et al. 1994, Schraudner et al. 1994). Recent work suggests that the pattern of changes induced by some pollutants (e.g., O_3) reflects on orchestrated series of events, triggered by disparate oxidative syndromes, which resembles the hypersensitive response provoked by pathogen attack (Alscher et al. 1997, Pell et al. 1997, Schraudner et al. 1997). This raises the question of whether the patterns of defense-related gene expression triggered by one pollutant are similar to those induced by another, whether the same signal transduction pathways are involved, and whether the similarity in response results in enhanced tolerance to a range of pollutants, and possibly other oxidative stresses. Opinions differ. However, based on the fact that particular genotypes, specific transformants, and plants treated with specific protectants (e.g., EDU) may be sensitive to one pollutant but not to another (Barnes and Wellbum 1998 and references therein), whereas the direction and the extent of the responses of the same species to different pollutant combinations vary in individual genotypes grown under common conditions (Bender and Weigel 1992), we take the view that despite the common responses observed at the level of gene expression, specific mechanisms must underlie tolerance to different air pollutants (i.e., tolerance to one pollutant is independent of that to another), as the action and subcellular localization of the stresses imposed by pollutants differ. This conclusion is supported by observations that differences in stomatal behavior/conductance (i.e., avoidance mechanisms) commonly govern the similarity in response to different pollutants (Winner et al. 1991). It is also interesting to note that some pollutants (e.g., SO_2 and NO_2) are much less effective than others (e.g., O_3) in eliciting changes in antioxidant gene expression (Schraudner et al. 1994, Schraudner et al. 1997), whereas other stresses (such as wounding, necrotizing pathogens, or elevated levels of UV-B radiation), which are known to elicit strong and rapid changes in defense-related gene expression, have been shown to reduce the extent of O_3 injury (Yalpani et al. 1994, Rao et al. 1996, Övar et al. 1997).

PLANT LEVEL

Where pollutant uptake exceeds the capacity of the detoxification/repair systems to prevent damage, there may be a host of adverse consequences on plant physiology resulting, ultimately, in the death of plant tissues. The oxidative stress imposed by O_3, for instance, is reflected in a decline in the photosynthetic capacity of individual leaves (Kangasjärvi et al. 1994, Pell et al. 1997), increased rates of maintenance respiration (Darrall 1989, Wellburn et al. 1997), enhanced retention of fixed carbon in leaves (Cooley and Manning 1987, Balaguer et al. 1995), and accelerated rates of leaf senescence (Alscher et al. 1997, Pell et al. 1997)—effects that are reflected in reduced growth and reproductive potential. Under some circumstances, the pollutant can induce localized cell death, resulting in typical visible symptoms of foliar injury. However, what aspect of performance should be used as an indication of resistance in an ecological context? In the case of crops, this is relatively straightforward since the impact of the pollutant on yield and marketable product is clearly the important feature. However, it is more difficult for natural vegetation. Because annual or monocarpic species must produce seeds to survive, seed output is an obvious criterion to use. However, what should be used to rank the performance of perennial species? Many iteroparous perennials live for decades or even centuries (Harberd 1960, Harper 1977). Yet, more often than not, assessments of resistance have been based on the degree of visible foliar damage or impacts on plant growth rate relative to that of controls, with little consideration or understanding of whether these features are important in an ecological context (Davison and Barnes 1998). It has, for example, become common practice to rank species in terms of susceptibility on the basis of visible symptoms of injury. This tends to lead to the intuitive conclusion that the affected species must suffer from some ecological disadvantage in the field, and conversely that unmarked species do not. This may be a serious misconception. First, the expression of symptoms is affected by many factors, including soil water deficit, vapor pressure deficit, photon flux density, temperature (Balls et al. 1996), and possibly UV-B radiation (Thalmiar et al. 1995). Second, there is usually little relation between relative sensitivity in terms of visible symptoms and effects on growth or seed production (Heagle 1979, Fernandez-Bayon et al. 1992, Bergmann et al. 1995); some taxa show highly significant effects of O_3 on growth but no visible symptoms, whereas others exhibit extensive visible symptoms but no effects on growth (Davison and Barnes 1998 and references therein). Third, and possibly most importantly, species that show injury are not necessarily debilitated in competition with other species that do not. This is clearly demonstrated in the work of Chappelka and colleagues (Chappelka et al. 1997, Barbo et al. 1998) on communities containing sensitive species such as blackberry (*Rubus cuneifolius*) and tall milkweed (*Asclepias exaltata*), shown to proliferate in field studies, despite the development of extensive visible symptoms of O_3 injury (Figure 20.5). Observations such as these led Davison and Barnes (1998) to conclude that "visible symptoms are best regarded as evidence of a biochemical response to ozone. They do not necessarily indicate sensitivity in terms of growth reduction and they are not evidence of an ecological impact."

Many other assessments of resistance of wild species have been based on the ratio of harvest weight in treated plants to that of controls, whereas others have used the ratio of mean relative growth rate. Most have used aboveground weight, and few have measured root weight or parameters of ecological importance such as specific root length or root length/leaf area ratio (Pell et al. 1997). This is important because the choice of measure can influence both the apparent magnitude of the ozone response and the relative ranking (Macnair 1991,

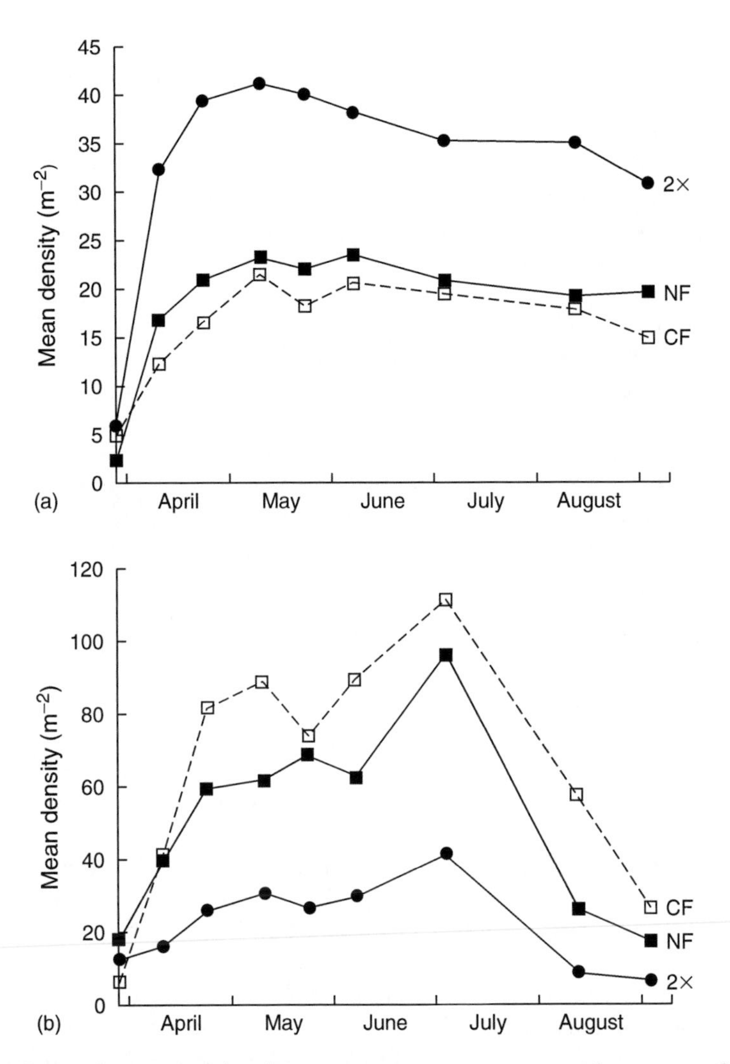

FIGURE 20.5 Effects of open-top chamber ozone exposure on two understory species in an early successional forest community; changes in mean density of (a) blackberry (*Rubus cuneifolius*), a species that develops extensive visible symptoms of O_3 injury and has therefore often been classified as sensitive, and (b) bahia grass (*Paspalum notatum*), a species that does not show visible injury and has therefore often been classified as resistant. Treatments: NF, nonfiltered ambient air; CF, charcoal-filtered air; 2×, 2× ambient O_3. (From Davison, A.W. and Barnes, J.D., *New Phytol.*, 139, 135, 1998. With permission; based on original data presented by Barbo, D.N., Chappelka, A.H., and Stolte, K.W., *Proc. 8th Bienn. South. Silvicult. Res. Counc.*, Auburn, Alabama, 1–3, 1994.)

Ashmore and Davison 1996). The difficulties are apparent in the hypothetical example given by Davison and Barnes (1998), where straightforward comparisons of the effects of O_3 on the growth of a fast-growing ruderal and a slower-growing stress tolerator (Grime 1979) would yield erroneous results. Attempts to find broad relationships between resistance and adaptive or ecological characters have proved disappointing. Preliminary studies conducted by Reiling and Davison (1992) on 32 species revealed a weak negative relationship between the effect of O_3 and plant relative growth rate in clean air, implying that approximately 30% of the variation between taxa was related to differences in inherent growth rate. However, in a more comprehensive study of 43 species, in which one of the principal aims was to investigate relationships between O_3 sensitivity and CSR strategy (Grime 1979), O_3 resistance was found to be significantly correlated with only one other trait, mycorrhizal status, and no relation existed between O_3 sensitivity and R in clean air or plant growth strategy (Grime et al. 1997). In contrast, SO_2 resistance was significantly correlated with 12 other traits. However, conclusions drawn from such studies are difficult to interpret because the effects on growth may be influenced by many factors, including seed characteristics and seed provenance; some experiments have used seeds collected in the field, whereas others have used commercial seeds or a mixture of sources (Davison and Barnes 1998). Hence, maternal effects caused by the parental environment (Roach and Wulff 1987) may contribute to some of the differences in ranking reported for the same species, for example, *Phleum alpinum* (Mortensen 1994, Pleijel and Danielsson 1997). There may also be substantial intraspecific variation in the response of different genotypes within a population to the same air pollution insult (see Section "Population Level").

Much attention has focused on the relative partitioning of dry matter between the root and the shoot of crop plants, and the observed impacts of specific pollutants (e.g., O_3) are often interpreted as universal (Övar et al. 1997). However, experiments with a range of wild species show that the situation is much more complicated, and subtle effects on allocation that are probably of greater ecological significance than changes in mass are common. In the legume *Lotus corniculatus*, Warwick and Taylor (1995) found that O_3 had no effect on allometric root/shoot growth, but caused a large reduction in specific root length, and there are other cases in which decreased allocation to the root has been found to be associated with compensatory changes in thickness, so length is unaffected (Taylor and Ferris 1996). Some species (e.g., *Arrhenatherum elatius*, *Rumexacetosa*) may even show an increased allocation to the root when exposed to ozone (Reiling and Davison 1992); others such as clover show the greatest decrease in allocation not to the roots, but to the storage and overwintering organs, that is, the stolons (Wilbourn et al. 1995, Fuhrer 1997). One of the most instructive studies of pollutant impacts on resource allocation in wild species was performed by Bergmann et al. (1995, 1996). They exposed 17 herbaceous species from seedling stage to flowering to two O_3 regimes with different dynamics: CF + 70 nL L^{-1} per 8 h and CF + 60% ambient +30 nL L^{-1}. Responses varied with exposure regime, and the weight of some species was reduced to about 60% of controls, but the most striking differences were in resource allocation (Figure 20.6). Most showed a proportionate change between shoot mass and reproductive effort (bottom left quadrant of Figure 20.6), but two species (*Chenopodium album* and *Matricaria discoidea*) showed a greater vegetative shoot weight and reduced reproductive allocation. Conversely, *Papaver dubium* and *Trifolium arvense* exhibited reduced shoot mass sand increased allocation to seed/flowers. Such shifts in resource allocation may help to explain why O_3 is sometimes found to stimulate growth and highlight the need for greater understanding of the control of resource allocation in species that have different reproductive and survival strategies.

Relative rankings of resistance may also be biased by growth stage/developmental status, since plants do not appear to be equally sensitive to pollutants at all stages in their life cycle (Davison and Barnes 1998). Our own work (Lyons and Barnes 1998) on *Plantago major*, for example, shows that seedlings are much more sensitive to O_3 than juvenile or mature plants; O_3-induced declines in accumulated biomass appeared to be almost entirely due to effects on seedling relative growth rate in this species, whereas seed production is most affected during the early stages of flowering (Figure 20.7; Davison and Barnes 1998). Compensatory changes in growth and morphology may also limit the impacts of prolonged exposure to O_3 (Lyons and Barnes 1998 and references therein). In the absence of such effects, it is conceivable that the impacts of O_3 (and other pollutants) would be considerably greater than they are.

The impacts of pollutants in the field may also be modified by a multitude of factors, including management practices, soil water deficit, mineral nutrition, nutrition,

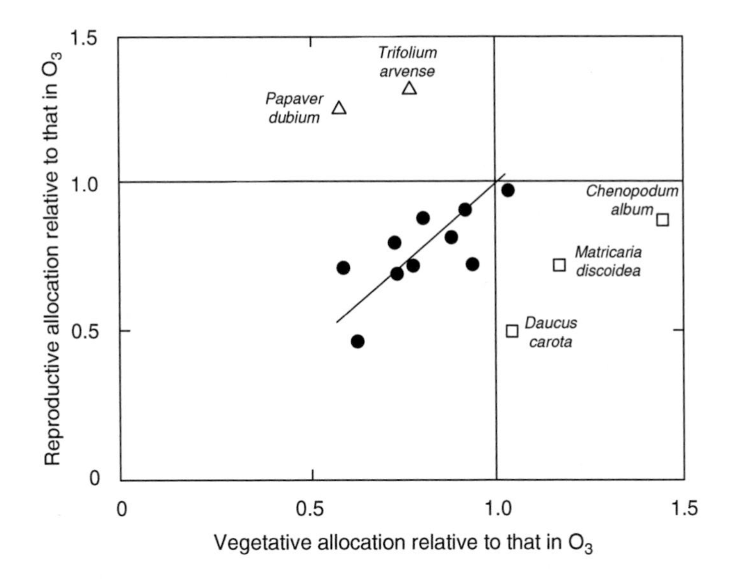

FIGURE 20.6 Effects of O_3 on the relative allocation of herbaceous species exposed from seedling to flowering to two O_3 regimes; charcoal-filtered air plus 70 nL L^{-1} O_3 8 h day^{-1} or charcoal-filtered air plus 60% ambient O_3 plus 30 nL L^{-1}. Changes in allocation expressed as the ratio of that in O_3 to that in CF air. (From Davison, A.W. and Barnes, J.D., *New Phytol.*, 139, 135, 1998. With permission; based on the data collected by Bergmann, E., Bender, J., and Weigel, H.-J. *Water, Air Soil Pollut.*, 85, 1437, 1995; Bergmann, E., Bender, J., and Weigel, H.-J., *Exceedance of Critical Loads and Levels*, M. Knoflacher, R. Schneider, and G. Soja, eds, Federal Ministry for Environment, Youth and Family, Vienna, 1996.)

other pollutants, frost, disease, and herbivory (Davison and Barnes 1992, Barnes et al. 1996, Wellburn et al. 1997, Barnes and Wellburn 1998, Davison and Barnes 1998). Work on crops, for example, indicates that the impacts of O_3 are commonly reduced under conditions in which soil water deficit results in a decline in stomatal conductance and hence pollutant uptake (Tingey and Taylor 1981, Darrall 1989, Tingey and Andersen 1991, Wolfenden et al. 1992, Wellburn et al. 1997). However, field observations relating to possible pollutant–water stress interactions in wild species are rare and anecdotal. Showman (1991) reported that visible oxidant injury in wild species in Ohio and Indiana was virtually absent in a year when it was dry and ozone was high, but widespread in a year when it was not as dry and ozone was lower. Similarly, Davison and Barnes (1998) drew attention to the fact that in heavily polluted regions of southern Europe, visible symptoms of oxidant injury are common in irrigated crops, and there are effects on yield; however, in nonirrigated areas subjected to severe summer drought, there are virtually no records of symptoms in wild species.

POPULATION LEVEL

There is growing evidence that pollutants, like other novel stresses imposed by human activities, can bring about changes in the genetic composition of populations (Roose et al. 1982, Taylor and Pitelka 1992). The phenomenon was first revealed through the work of Bell et al. (1991) on the evolution of SO_2 resistance in grassland species in industrialized regions of the United Kingdom. However, convincing experimental evidence of evolution of resistance to regional-scale pollutants (such as O_3) has only recently been reported. In a way, this is rather surprising, since the requirements to drive the evolution of O_3 resistance in wild

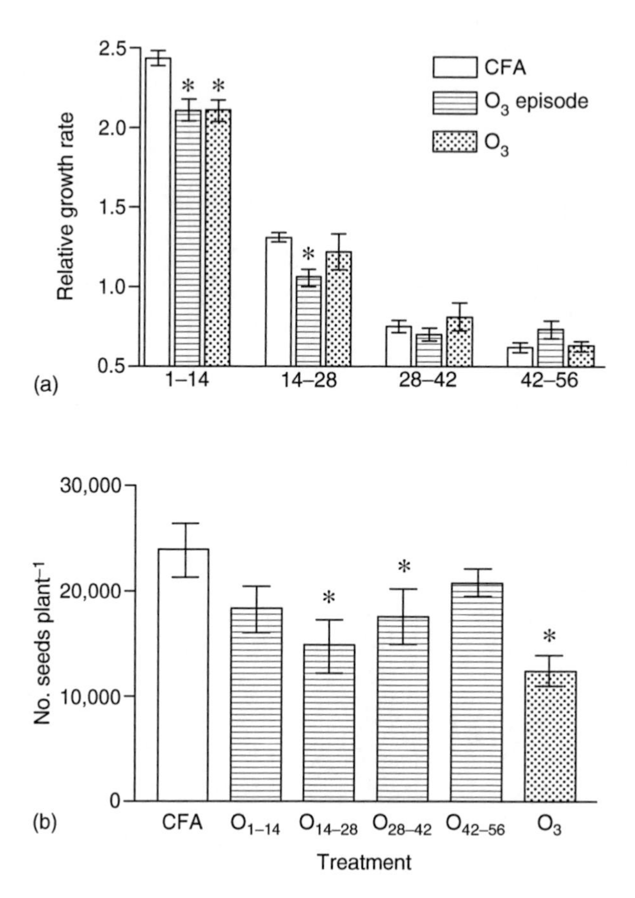

FIGURE 20.7 Effects of O_3 on (a) growth and (b) seed production in *Plantago major* Valsain exposed in duplicate controlled environment chambers to charcoal/Purafil-filtered air (CFA), O_3 (CFA plus 70 nL L^{-1} O_3 7 h day^{-1}), or 14 day episodes of O_3 (i.e., windows) administered for 1–14 days (O_{1-14}), 14–28 days (O_{14-28}), 28–42 days (O_{28-42}), or 42–56 days (O_{42-56}). Asterisks indicate significant ($P = 0.05$) differences from plants maintained in CFA. (From Lyons, T.M. and Barnes, J.D., *New Phytol.*, 138, 83, 1998. With permission.)

species have been recognized for many years (Lyons et al. 1997, Davison and Barnes 1998). Our own artificial selection studies (Whitfield et al. 1997) on resistant and sensitive populations of *P. major* indicate the potential for evolution of resistance/sensitivity to O_3 within a matter of only a few generations in this short-lived species. Based on the effects of a 2 week exposure to 70 nL L^{-1} O_3 for 7 h day^{-1} on rosette diameter, selection from an initially sensitive population led to a line with significantly enhanced O_3 resistance, although it was possible to select a line with greater sensitivity than the original population. Conversely, selection from an initially resistant population led to a line with increased sensitivity, but not to a line with enhanced resistance (Figure 20.8). Subsequent experiments have shown that the resistance of the selected lines is maintained, and differences are reflected in contrasting effects on growth and seed production.

The earliest suggestion that ambient levels of O_3 maybe high enough to drive the selection of resistant genotypes in the filed was provided by Dunn (1959), who worked on *Lupinus bicolor* in the Los Angeles basin. He attributed differences in the performance of populations of this species to oxidant smog and commented that "the stress was so severe that some

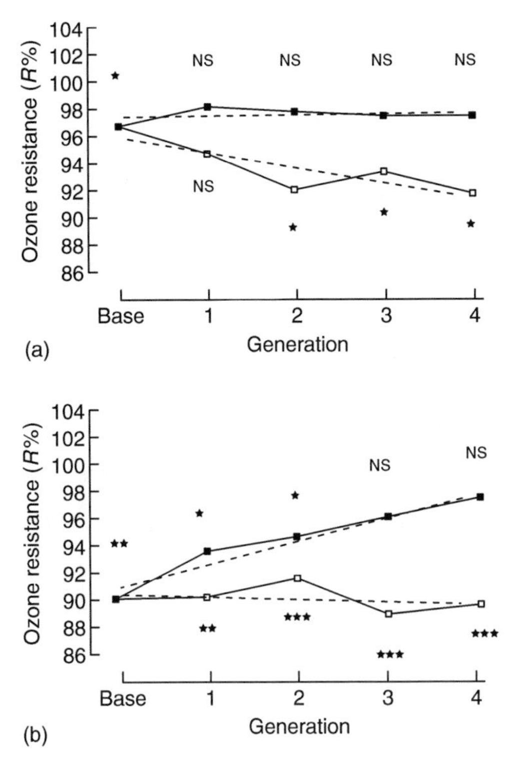

FIGURE 20.8 Change in O_3 resistance over four generations in lines selected for sensitivity (▫) and resistance (▪) from two population of *Plantago major* based on effects on rosette diameter. Data presented indicate effects on plant relative growth rate (R); $R\% = [R_{O_3}/R_{CF}] \times 100$. Plants were exposed in controlled-environment chambers to charcoal/Purafil-filtered air or O_3 (CFA plus 70 nl L^{-1} O_3 7 h^{-1}). Dashed lines represent linear regression fits for selections. Data for each generation in each selected line were subjected to ANOVA. Probabilities ($^*P < 0.05$; $^{**}P < 0.01$; $^{***}P < 0.001$) indicate the significance of O_3 effects on each generation. (From Whitfield, C.P., Davison, A.W., and Ashenden, T.W., *New Phytol.*, 137, 645, 1997. With permission.)

populations failed to set seed"—an effect expected to contribute to rapid evolution. It is only through recent work (reviewed by Macnair 1993, Davison and Barnes 1998) on *Populus tremuloides*, *Trifolium repens*, and *P. major* that convincing experimental evidence has been provided to support the evolution of resistance to O_3 in the field. Our own studies on *P. major* represent the only case in which heritable change in resistance has been shown to occur in the field over a period of time when O_3 levels increased (Reiling and Davison 1992, 1995). However, the crucial question in all of the studies conducted to date is whether the observed differences in resistance between populations have arisen in response to O_3 or to other correlated environmental factors (Bell et al. 1991). It was suggested by Roose et al. (1982) that because traits affecting sensitivity to air pollutants may simultaneously reflect adaptations to other natural stresses and vice versa, resistance might be indirect. Our own work on 41 European populations of *P. major* indicates that O_3 resistance is significantly correlated with the C_3 concentration at or near the site of collection (Figure 20.9), and similar findings have been reported by Berrang et al. for *Populus tremuloides* originating from parts of the United States with different O_3 climates (Davison and Barnes 1998). However, in some cases, O_3 resistance has also been shown to correlate with other variables (Reiling and Davison 1992). Although the available evidence is consistent with the evolution of O_3 resistance,

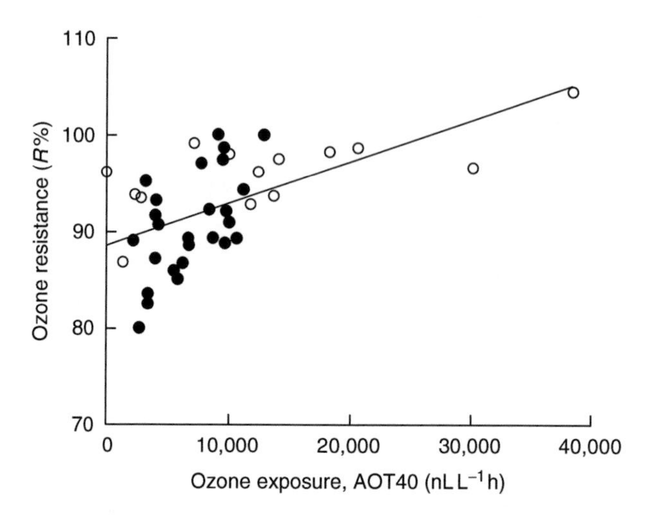

FIGURE 20.9 Ozone resistance of seed-grown *Plantago major* populations plotted against the O_3 exposure (based on the accumulated O_3 exposure above a 40 nL L^{-1} threshold, that is, AOT40) at the collection sites. Resistance determined as the mean relative growth rate in O_3 (70 nL L^{-1} O_3 7 h day^{-1}) as a percentage of that in charcoal/Purafil-filtered air. ●, UK populations; o continental European populations. Regression 88.7 + 0.00043(AOT40) $r = 0.538$, $P < 0.0001$. (From Davison, A.W. and Barnes J.D., *New Phytol.*, 139, 135, 1998. With permission; based on data presented by Reiling, K. and Davison, A.W., *New Phytol.*, 120, 29, 1992 [●] and Lyons, T.M., Barnes, J.D., and Davison, A.W., *New Phytol.*, 136, 503, 1997 [o].)

correlations do not definitively prove a cause–effect relationship. Therefore, it has been necessary to try to eliminate other possibilities. This has been achieved using partial correlations to remove the effect of climatic differences between collection sites. In some cases, this has reduced the significance of regressions consistent with the evolution of O_3 resistance (Berrang et al. 1991, Reiling and Davison 1992); in other studies it has made little difference to the significance of regressions (Lyons et al. 1991). Although the experimental data are generally consistent with the evolution of resistance to O_3, the work graphically illustrates the difficulties in interpreting the observed spatial variability in pollution resistance.

Based on the assertion of Bell et al. (1991) that "if populations are evolving it should be possible to demonstrate a change in resistance over time, as with evolution to other novel stresses," Davison and Reiling (1995) compared the ozone resistance of *P. major* populations grown from seed collected from the same sites over a 6 year period in which ambient O_3 concentrations increased. They demonstrated that two populations increased in resistance, and Wolff and Morgan-Richards (unpublished data 1997) have recently proven (using random amplified polymorphic DNA primers [RAPDs]) that the later populations are subsets of the earlier ones. This is consistent with directional in situ selection rather than a catastrophic loss and replacement of the populations by migration. However, difficulties in interpretation remain, since one of the reasons that O_3 levels increased was because it was sunnier and warmer. This probably led to a greater incidence of soil moisture deficit and photoinhibition, but no records are available for the collection sites. Hence, the possibility that other factors may have contributed to the evolution of O_3 resistance cannot be dismissed. It is also important to recognize that the evolution of resistance to O_3 may be associated with costs or a loss of fitness in unpolluted environments, as with other cases of directional selection motivated by novel stresses (Roose et al. 1982, Macnair 1993, Reiling and Davison 1995), but little is known about the nature of these costs with respect to O_3 resistance (Davison and Barnes 1998). Furthermore, the rate of evolution of resistance would be expected to be influenced by many factors, including the mode of reproduction, the form of sexual reproduction (determining the degree of inbreeding), the dynamics of gene flow within and between populations, generation time, the presence of seed banks, the extent of the loss of fitness induced by the pollutant, and the timing of selection in relation to the plants' life cycle (Roose et al. 1982, Taylor and Pitelka 1992, Macnair 1993, Lyons and Barnes 1998). Thus, even in areas exposed to potentially damaging pollutant concentrations for long periods, it is possible to find species that persistently show typical visible symptoms of injury (Davison and Barnes 1998).

COMMUNITY LEVEL

Studies on isolated taxa indicate that there is wide variation in resistance between species to gaseous pollutants. This is exemplified in our own screening of the O_3 sensitivity of 30 species using seed collected from central England, using a standard O_3 exposure of 70 nL L^{-1} for 7 h day^{-1} over 2 weeks (Reiling and Davison 1992). The minimum detectable effect of O_3 on growth rate in this test was approximately 5%, and species exhibited a range of sensitivities (Table 20.1); the most sensitive species were affected (in terms of effects on plant relative growth rate) to a similar extent as some of the most sensitive crop species (e.g., tobacco Bel-W3). These and other studies (Fuhrer et al. 1997) indicate that herbaceous species exhibit wide-ranging sensitivity to O_3. However, there maybe as much variation within species as between species (see Section "Population Level"), and there is no guarantee that controlled trials are reminiscent of responses in the field, where a range of additional factors must be considered. Consequently, such studies contribute little other than indicating the range of potential responses to gaseous pollutants.

Where pollutants emanate from a point source, it is possible to determine effects on biodiversity by standard ecological and multivariate methods, especially where records can be repeated over time (Musselman et al. 1992). There is, for example, a wealth of literature documenting the effects of acidifying air pollutants on epiphytic lichen communities (Nimis et al. 1991). For regional pollutants such as O_3 it is more difficult, because the pollutant does not usually show sharp gradients within defined boundaries. Consequently, measurable effects on the structure of plant communities may be restricted to a few special cases. Westman (1979, 1985), for example, showed that percentage cover and species richness were strongly influenced by O_3 (and other oxidants) in Californian coastal sage scrub. Preston (1985) reported similar effects for SO_2. Interestingly, ordination approaches suggested a greater effect at sites with low foliar cover than at high ones, but Preston did not comment on this. The same types of studies are likely to prove unfruitful in less impacted areas, because of the lack of sharp gradients in oxidant concentrations and the difficulty in locating appropriate control sites. Consequently, progress in understanding the impacts of regional pollutants (such as O_3) on plant communities has, and will probably continue, to depend on experimental exposures using open-top chambers (OTCs) and relatively simple species mixtures. To date, majority of research have focused on herbaceous plant communities, especially seminatural grasslands (Ashmore and Davison 1996, Kärenlampi and Skärby 1996, Fuhrer 1997, Fuhrer et al. 1997, Davison and Barnes 1998). Consequently, there is very little known about the responses of the wide range of vegetation types that are found across Europe and the United States, although there is no reason to believe that these will necessarily respond the same as some of the simple mixtures that have been studied experimentally (Fuhrer et al. 1997).

Investigations on simple grass/clover mixtures indicate that the effects of O_3 depend on the relative sensitivity of the competing species (Fuhrer 1997) and on management practices (Fuhrer et al. 1997, Davison and Barnes 1998). Because red clover (*Trifolium pretense*) and timothy (*Phleum pretense*) are about equally sensitive to ozone, both are equally affected in

TABLE 20.1
Impact of O_3 on Growth and Root/Shoot Dry Matter Partitioning in 32 Taxa[a]

	R week^{-1}			$K = R_{root}/R_{shoot}$		
	Control	+O_3	*R*%	Control	+O_3	*K*%
Arrhenatherum elatius	1.96	1.85	−6	0.84	1.01	+20***
Avena fatua	1.97	2.00	+2	0.87	0.80	−8
Brachypodium pinnatum	1.58	1.48	−6	0.75	0.81	+8*
Bromus erectus	1.93	1.92	0	1.05	1.18	−12*
Bromus sterilis	1.77	1.73	−2	1.02	1.08	+6
Cerastium fontanum	2.18	1.96	−10***	0.67	0.65	−3
Chenopodium album	1.99	1.97	−1	0.62	0.53	−14***
Deschampsia flexuosa	1.81	1.73	−4	0.95	0.96	+1
Desmazeria rigida	1.21	1.10	−9	1.23	0.75	−39**
Epilobium hirsutum	1.61	1.58	−2	0.83	0.62	−25**
Festuca ovina	1.35	1.27	−6	0.58	0.60	+4
Holcus lanatus	1.64	1.57	−4*	0.93	0.91	−2
Hordeum murinum	1.84	1.74	−5*	0.83	0.79	−5
Koeleria macrantha	1.73	1.63	−6***	0.77	0.42	−45*
Lolium perenne Talbot	2.03	1.98	−2	0.65	0.88	+35*
Nicotiana tabacum Bel-W3	2.27	1.91	−16***	0.98	0.89	−9
Pisum sativum	1.95	1.80	−8***	1.07	1.24	+16
Conquest						
Plantago coronopus	2.28	1.98	−13***	0.76	0.79	+4
Plantago lanceolata	2.17	1.98	−9**	1.10	0.72	−34**
Plantago major[b]						
1	2.46	1.88	−24***	0.95	0.86	−9***
2	2.36	1.81	−23**	0.90	0.83	−8**
Plantago major Athens	1.79	1.77	−1	0.90	0.96	+7
Plantago maritime	1.76	1.66	−6	0.87	0.95	+9
Plantago media	1.29	1.33	+3	1.16	1.25	+8
Poa annua	1.56	1.45	−7*	1.03	0.92	−11*
Poa trivialis	1.34	1.29	−4	0.95	0.89	−6
Rumex acetosa	1.72	1.71	−1	0.63	0.85	+35*
Rumex acetosella	1.71	1.54	−10*	1.06	1.08	+2
Rumex obtusifolius	2.16	1.97	−9**	0.87	0.83	−5
Teucrium scorodomia 0221	1.74	1.56	−10**	0.70	0.70	0
Teucrium scorodonia 0223	1.74	1.57	−10**	0.84	0.70	−17**
Urtica dioica	2.59	2.29	−12***	1.54	1.35	−12***

Source: From Reiling, K. and Davison, A.W., *New Phytol.*, 120, 29, 1992.

[a] Plants were raised in duplicate controlled environment chambers ventilated with charcoal/Purafil-filtered air (control; <5 nL L^{-1} O_3) or O_3 (70 nL L^{-1} O_3 for 7 h day^{-1} for 2 weeks). *R*, mean plant relative growth rate; *K*, allometric root/shoot coefficient; *R*%, the % change in *R*; *K*%, the % change in *K*. Asterisks denote probability of difference between control and fumigated plants: *0.05, **0.01, ***0.001. With the exception of pea (*Pisum sativum* L. cv. Conquest, supplied by Batchelors foods) and tobacoo (*Nicotiana tabacum* L. cv. Bel-W3, supplied by IPO, Wageningen), which were included for comparative purposes, all seeds were supplied by the UCPE seed bank, Sheffield University, UK.

[b] Because of limited space in the fumigation chambers, only five to six taxa could be grown and tested at a time. To ensure reproducibility, one species (*Plantago major*) was tested twice—once at the beginning (1) and once at the end (2) of the series of experiments.

competition and the ratio in biomass is unaffected. In contrast, because white clover (*T. repens*) is more sensitive than its usual companion grasses, it tends to decline in competition. The primary effect is on the stolons; if O_3 concentrations decrease then plants may recover, but there can be lasting effects on stolon density (Wilbourn 1991). Figure 20.10

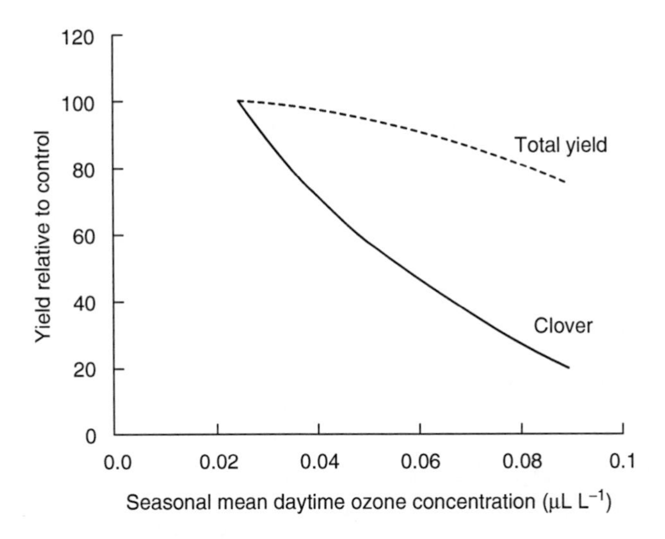

FIGURE 20.10 Effects of O_3 (seasonal daytime mean) on relative yield and clove content of managed grassland. (Redrawn from Fuhrer, J., Ecological Advances and Environmental Impact Assessment, Gulf Publishing Company, Houston, 1997. With permission; based on the regression of four datasets.)

shows the recent collation of data by Fuhrer (1997) from white clover experiments performed in the United States and Europe. This reveals two important points: (1) there is good agreement in dose–response relationships between experimental studies, despite difference in varieties, exposure techniques, and climate and (2) total forage yield tends to be much less affected than that of the sensitive component (in this case, clover).

The few OTC studies conducted on communities other than grass/clover mixtures indicate that ambient levels of SO_2 and O_3 may already be high enough to modify species composition in parts of Europe and the United States Working on O_3, Ashmore and Ainsworth (1995) found little effect on the total biomass of mixtures of two grasses and forbs sown as seeds in pots of unamended acid soil, but the forb component declined with increasing exposure. Similar changes in forbs (*Campanula rotundifolia*, *Leontodon hispidus*, *Lotus corniculatus*, *Sanguisorba minor*) were reported by Ashmore et al. (1995) in a simulated calcareous grassland mixture. Bearing in mind the simplified nature of the community and the absence of interacting factors such as water deficit, these data provide a tentative indication that species composition might be affected in parts of central and northern Europe in high-O_3 years. Where pollutant concentrations are variable from year to year, long-term effects would be expected to depend on the magnitude of the changes in years with high pollutant concentrations and the capacity of the community to recover in between. The strongest experimental evidence that ambient levels of O_3 have a significant ecological impact comes from work conducted in the United States. Duchelle et al. (1983) determined the effects of ambient O_3 on the productivity of natural vegetation in a high meadow in the Shenandoah National Park. Over 3 years, aboveground biomass was increased by filtration and the cumulative dry weights for charcoalfiltered (CF), nonfiltered (NF), and ambient air (AA) treatments were significantly different at 1.38, 1.09, and 0.89 kg m^{-2}, respectively, but data relating to effects on species richness were not collected. Working in the same area, Barbo et al. (1994, 1998) exposed an early successional forest community to AA, CF, NF, and 2 × AA. They found changes in species performance, canopy structure, species richness, and diversity index consistent with the view that oxidants have resulted in a shift in vegetation dominance in some heavily polluted regions. Comparably large effects have not been reported in Europe, but Evans and Ashmore (1992) showed that filtration affected the forb component of a seminatural grassland in a year when O_3 concentrations were relatively high (17% of days with maximum hourly average >60 ppb). In contrast, there is much evidence that ambient levels of SO_2 are high enough to change community structure in some regions, and several excellent reviews are available on the subject (Kozlowski 1985, Bell et al. 1991, Armentano and Bennett 1992, Taylor and Pitelka 1992).

CONCLUSIONS

In this chapter, we have attempted to evaluate some of the features that underlie the observed variation in the resistance of plants to the most common gaseous air pollutants. However, many key questions remain to be answered. Comparative analysis at the molecular level is beginning to reveal evidence of similar defense-related responses to a range of stresses, and the use of mutants and genetic transformation techniques, already of importance in our improved understanding of specific genetic sequences, should eventually allow the dissection of the molecular mechanisms underlying resistance, as well as yielding approaches that may be used to manipulate the resistance of crop plants.

We have chosen to focus on O_3 since this is one of the most potent gases to which plants are regularly exposed in the filed, and it is likely to constitute a continuing threat to vegetation for the foreseeable future. Although there is growing literature documenting the effects of this and other gaseous pollutants on wild species, most studies have measured a small number of response variables, commonly under controlled or semicontrolled conditions, with little regard to their ecological significance. This work demonstrates which aspects of growth and reproduction are sensitive to pollutants and the potential range of responses that exist both within and between species, but it is of limited value in assessing how pollutants affect the fitness of plants in communities in the field where a range of additional factors must be taken into consideration. The few studies that have been performed on simple plant communities in the filed indicate that there may be winners and losers at the individual, population, and community levels. These differential responses may result in the most fit genotypes predominating in future generations in polluted regions, and there can be changes in the genetic structure of populations and dominance relationships within individual communities. In many cases, however, it is difficult to ascribe directional selection to individual pollutants, as other factors of the physical and biological environments interact in a variety of ways that may collectively influence the direction and extent of selection. Even in circumstances in which the pollutant may not be the principal factor underlying evolutionary changes in population structure, the existence of resistant populations in polluted regions is of considerable ecological significance. A major challenge is to devise experiments in the future in which the impacts of gaseous pollutants on natural ecosystems can be assessed; to use the words of Smith (1990): "this will require the marriage—or at least the co-habitation—of specialists in different disciplines (e.g. molecular biologists, biochemists, ecologists & physiologists), at present not the most congenial of bed-fellows."

ACKNOWLEDGMENTS

The authors acknowledge support from the Royal Society, the Spanish Ministry of Science and Culture, and Spanish Interministerial Commission for Science and Technology, the Natural Environment Research Council, the European Union, the Swales Foundation (administered by Newcastle University), and the UK Overseas Development Agency. The chapter was written during J.B.'s tenure as a Royal Society Research Fellow.

REFERENCES

Alscher, R.G., J.L. Donahue, and C.L. Cramer, 1997. Reactive oxygen species and antioxidants: Relationships in green cells. Physiologia Plantarum 100: 224–233.

Aono, M., A. Kubo, H. Saji, K. Tanaka, and N. Kondo, 1993. Enhanced tolerance to photoxidative stress of transgenic *Nicotiana tabacum* with high chloroplastic glutathione reductase activity. Journal of Plant Cell Physiology 34: 129–135.

Ariens, E.L., A.M. Simonis, and J. Offermerier, 1976. Introduction to General Toxicology. Academic Press, New York.

Armentano, T.V. and J.P. Bennett, 1992. Air pollution effects on the diversity and structure of communities. In: J. Barker and D.T. Tingey, eds. Air Pollution Effects on Biodiversity. Van Nostrand Reinhold, New York, pp. 159–176.

Ashmore, M.R. and N. Ainsworth, 1995. The effects of cutting on the species composition of artificial grassland communities. Functional Ecology 9: 708–712.

Ashmore, M.R. and A.W. Davison, 1996. Towards a critical level of ozone for natural vegetation. In: L. Kärenlampi, and L. Skärby, eds. Critical Levels for Ozone in Europe: Testing and finalizing the Concepts. UN-ECE Workshop Report. University of Kuopio, Kuopio, pp. 58–71.

Ashmore, M.R., R.H. Thwaites, N. Ainsworth, D.A. Cousins, S.A. Power, and A.J. Morton, 1995. Effects of ozone on calcareous grassland communities. Water, Air and Soil Pollution 85: 1527–1532.

Balaguer, L., J.D. Barnes, A. Panicucci, and A.M. Borland, 1995. Production and utilization of assimilated in what (*Triticum aestivum* L.) leaves exposed to elevated CO_2 and/or O_3. New Phytologist 129: 557–568.

Balls, G.R., D. Palmer-Brown, and G.E. Sanders, 1996. Investigating microclimatic influence on ozone injury in clover (*Trifolium subterraneum*) using artificial neural networks. New Phytologist 132: 271–280.

Barbo, D.N., A.H. Chappelka, and K.W. Stolte, 1994. Ozone effects on productivity and diversity of an early successional forest community. Proceedings of the Eighth Biennial Southern Silvicultural Research Council, Auburn, Alabama, November 1–3, pp. 291–298.

Barbo, D.N., A.H. Chappelka, G.L. Somers, M.S. Miller-Goodman, and K.W. Stolte, 1998. Diversity of an early successional plant community as influenced by ozone. New Phytologist 138: 653–662.

Barnes, J.D. and A.R. Wellburn, 1998. Air pollutant combinations. In: L.J. de Kok, and I. Stuhlen, eds. Responses of Plant Metabolism to Air Pollution and Global Change. Backhuys, Leiden, pp. 147–164.

Barnes, J.D., M.R. Hull, and A.W. Davison, 1996. Impacts of air pollutants and rising CO_2 in wintertime. In: M. Yunus, and M. Iqbal, eds. Plant Response to Air Pollution. Wiley, London, pp. 35–166.

Becker, K., M. Saurer, A. Egger, and J. Fuhrer, 1989. Sensitivity of white clover to ambient ozone in Switzerland. New Phytologist 112: 235–243.

Bell, J.N.B., 1984. Air pollution problems in western Europe. In: M.J. Koziol, and F.R. Whatley, eds. Gaseous Air Pollutants and Plant metabolism. Butterworths, London, pp. 3–24.

Bell, J.N.B., M.R. Ashmore, and G.B. Wilson, 1991. Ecological genetic and chemical modification of the atmosphere. In: G.E. Taylor, L.F. Pitelka, and M.T. Clegg, eds. Ecological Genetics and Air Pollution. Springer-Verlag, New York, pp. 33–59.

Bender, J. and H.-J. Weigel, 1992. Crop responses to mixtures of air pollutants. In: H.J. Jäger, M. Unsworth, L. De Temmerman, and P. Mathy, eds. Effects of Air Pollution on Agricultural Crops in Europe. CEC Air Pollution Research Report No. 46. CEC, Brussels, pp. 445–454.

Bennett, J.H. and A.C. Hill, 1973. Absorption of gaseous air pollutants by a standardized plant canopy. Journal of Air Pollution Control and Waste Management 23: 203–206.

Bergmann, E., J. Bender, and H.-J. Weigel, 1995. Growth responses and foliar sensitivities of native herbaceous species to ozone exposure. Water, Air and Soil Pollution 85: 1437–1442.

Bergmann, E., J. Bender, and H. Weigel, 1996. Effects of chronic ozone stress on growth and reproduction capacity of native herbaceous plants. In: M. knoflacher., R. Schneider, and G. Soja, eds. Exceedance of Critical Loads and Levels. Federal Ministry for Environment, Youth and Family, Vienna, pp. 177–185.

Berrang, P., D.F. Karnosky, and J.P. Bennett, 1991. Natural selection for ozone tolerance in *Populus tremuloides*: An evaluation of nationwide rends. Canadina Journal of Forest Research 21: 1091–1097.
Boubel. R.W., D.L. Fox, D.B. Turner, and A.C. Stern, 1994. Fundamentals of Air pollution, 3rd edn. Academic Press, London, pp. 19–60.
Bressan, R.A., L. LeCurreux, L.G. Wilson, P. Filner, and L.R. Baker, 1981. Inheritance of resistance to sulphur dioxide in cucumber. Horticultural Science 16: 332–333.
Broadbent, P., G.P. Creissen, B. Kular, A.R. Wellburn, and P.M. Mullineaux, 1995. Oxidative stress responses in transgenic tobacco containing altered levels of glutathione reductase activity. Plant Journal 8: 247–255.
Burkhardt, J. and P. Drechsel, 1997. The synergism between SO_2 oxidation and manganese leaching on spruce needles—a chamber experiment. Environmental Pollution 95: 1–11.
Cape, J.N., 1997. Photochemical oxidants—what else is in the atmosphere besides ozone? Phyton 37: 45–58.
Chameides, W.L., 1989. The chemistry of ozone deposition to plant leaves: role of ascorbic acid. Environmental Science and Technology 23: 595–600.
Chappelka, A.H., J. Renfro, G. Somers, and B. Nash, 1997. Evaluation of ozone injury on foliage of black cherry (*Prunus serotina*) and tall milkweed (*Asclepias exaltata*) in the Great Smoky Mountains National Park. Environmental Pollution 95: 13–18.
Conklin, P.L., E.H. Williams, and R.L. Last, 1996. Environmental stress sensitivity of an ascorbic acid-deficient *Arabidopsis* mutant. Proceedings of the National Academy of Sciences USA 93: 9970–9974.
Cooley, D.R. and W.J. Manning, 1987. The impact of ozone on assimilate partitioning in plants: A review. Environmental Pollution 47: 95–113.
Cram, W.J., 1990. Uptake and transport of sulphate. In: H. Rennenberg, C. Brunold, L.J. De Kok, and I. Stuhlen, eds. Sulphur Nutrition and Assimilation in Higher Plants. SPB Academic Publishing, The Hague, pp. 3–11.
Darrall, N.M., 1989. The effects of air pollutants on physiological processes in plants. Plant, Cell and Environment 12: 1–30.
Davison, A.W. and J.D. Barnes, 1992. Patterns of air pollution: The use of the Critical Loads concept as a basis for abatement strategy. In: M.D. Newsom, ed. Managing the Human Impact on the Natural Environment: Patterns and Processes. Bellhaven, New York, pp. 109–129.
Davison, A.W. and J.D. Barnes, 1998. Effects of ozone on wild plants. New Phytologist 139: 135–151.
Davison, A.W. and K. Reiling, 1995. A rapid change in ozone resistance in *Plantago major* after summers with high ozone concentrations. New Phytologist 131: 227–343.
De Kok, L.J., 1990. Sulphur metabolism in plants exposed to atmospheric sulfur. In: H. Rennenberg, C. Brunold, L.J. De Kok, and I. Stuhlen, eds. Sulphur Nutrition and Assimilation in Higher Plants. SPB Academic Publishing, The Hague, pp. 111–130.
Dietz, K.-J., 1996. Functions and responses of the leaf apoplast under stress. Progress in Botany 58: 221–254.
Duchelle, S.F., J.M. Skelly, T.L. Sharick, B. Chevone, Y.S. Yang, and J.E. Nielessen, 1983. Effects of ozone on the productivity of natural vegetation in a high meadow of the Shenandoah National park of Virginia. Journal Environmental Management 17: 299–308.
Dunn, D.B., 1959. Some effects of air pollution on *Lupinus* in the Los Angeles area. Ecology 40: 621–625.
Evans, P. and M.R. Ashmore, 1992. The effects of ambient air on a semi-natural grassland community. Agriculture, Ecosystems and Environment 38: 91–97.
Fernandez-Bayon, J.M., J.D. Barnes, J.H. Ollerenshaw, and A.W. Davison, 1992. Physiological effects of ozone on cultivars of watermelon (*Citrullus lanatus*) and muskmelon (*Cucumis melo*) widely grown in Spain. Environmental Pollution 81: 199–206.
Fuhrer, J., 1997. Ozone sensitivity of managed pastures. In: P.N. Chereminisoff, ed. Ecological Advances and Environmental Impact Assessment. Advances in Environmental Control Technology Series. Gulf Publishing Company, Houston, pp. 681–706.
Fuhrer, J., L. Skärby, and M.R. Ashmore, 1997. Critical levels for ozone effects on vegetation in Europe. Environmental Pollution 97: 91–106.

Gaastra, P., 1959. Photosynthesis of crop plants. Mededelingen van de Landbouwhogeschool et Wageningen 59: 1–68.

Grime, J.P., 1979. Plant Strategies and Vegetation Processes. Wiley, London.

Grime, J.P., K. Thompson, R. Hunt, J.G. Hodgson, J.H.C. Cornelissen, I.H. Rorison, G.A.F. Hendry, T.W. Ashenden, A.P. Askew, S.R. Band, R.E. Booth, C.C. Bossard, B.D. Campbell, J.E.L. Cooper, A.W. Davison, P.L. Gupta, W. Hall, D.W. Hand, M.A. Hannah, S.H. Hillier, D.J. Hodkinson, A. Jalili, Z. Liu, J.M.L. Mackey, N. Matthews, M.A. Mowforth, A.M. Neal, R.J. Reader, K. Reiling, W. Ross-Fraser, R.E. Spencer, F. Sutton, D.E. Tasker, P.C. Thorpe, and J. Whitehouse, 1997. Integrated screening validates primary axis of specialization in plants. Oikos 79: 259–281.

Harberd, D.J., 1960. Observations on population structure and longevity in *Festuca rubra* L. New Phytologist 60: 202–206.

Harper, J.L., 1977. Population Biology of Plants. Academic Press, London.

Heagle, A.S., 1979. Ranking of soybean cultivars for resistance to ozone using different ozone doses and response measures. Environmental Pollution 19: 1–10.

Heath, R.L., 1988. Biochemical mechanisms of pollutant stress. In: W.W. Heck, O.C. Taylor, and D.T. Tingey, eds. Assessment of Crop Loss from Air Pollutants. Elsevier, London, pp. 259–286.

Heath, R.L. and G.E. Taylor, 1997. Physiological processes and plant responses to zone exposure. In: H. Sandermann, A.R. Wellburn, and R.L. Heath, eds. Forest Decline and Ozone: A Comparison of Controlled Chamber and Field Experiments. Springer-Verlag, Berlin, pp. 317–368.

Herschbach, C., L.J. De Kok, and H. Rennenberg, 1995. Net uptake of sulfate and its transport to the shoot in spinach plants fumigated with H_2S or SO_2: Does atmospheric sulphur affect the "inter-organ" regulation of sulfur nutrition? Botanica Acta 108: 41–46.

Hewitt, N. and G. Terry, 1992. Understanding ozone-plant chemistry. Environmental Science and Technology 26: 1890–1891.

Hippeli, S. and E.F. Elstner, 1996. Mechanisms of oxygen activation during plant stress: biochemical effects of air pollutants. Journal of Plant Physiology 148: 249–257.

Hogsett, W.E., D.T. Tingey, and E.H. Lee, 1988. Ozone exposure indices: concepts for development and evaluation of their use. In: W.W. Heck, O.C. Taylor, and D.T. Tingey, eds. Assessment of Crop Loss from Air Pollutants. Elsevier, London, pp. 107–138.

Kaiser, G., E. Martinoia, G. Schroppel-Meier, and U. Heber, 1989. Active transport of sulphate into the vacuole of plant cells provides halotolerance and can detoxify SO_2. Journal of Plant Physiology 133: 756–763.

Kangasjärvi, J., J. Talvinen, M. Utriainen, and R. Karjalainen, 1994. Plant defence systems induced by zone. Plant Cell and Environment 17: 783–794.

Kärenlampi, L., and L. Skärby, 1996. Critical levels for ozone in Europe: testing and finalizing the concepts UN-ECE Workshop Report. University of Kuopio, Kuopio, pp. 1–355.

Kersteins, G. and K.J. Lendzian, 1989. Interactions between ozone and plant cuticles. I. Ozone deposition and permeability. New Phytologist 112: 13–19.

Körner, C., 1994. Leaf diffusive conductance in the major vegetation types of the globe. In: E.-D. Schulze and M.M. Caldwell, eds. Ecophysiology of Photosynthesis. Springer-Verlag, Berlin, pp. 463–490.

Körner, C.H., M. Neumayer, S.P. Menendez-Riedl, and A. Smeets-Schiel, 1989. Functional morphology of mountain plants. Flora 182: 353–383.

Kozlowski, T.T., 1985. SO_2 effects on plant community structure. In: W.E. Winner, H.A. Mooney, and R.A. Goldstein, eds. Sulphur Dioxide and Vegetation. Stanford University Press, Stanford, pp. 431–453.

Krupa, S.V., 1996. The role of atmospheric chemistry in the assessment of crop growth and productivity. In: M. Yunus and M. Iqbal, eds. Plant Response to Air Pollution. Wiley, London, pp. 35–73.

Laisk, A., O. Kull, and H. Moldau, 1989. Ozone concentration in leaf intercellular air spaces is close to zero. Journal of Plant Physiology 90: 1163–1167.

Lea, P.J., F.A.M. Wellburn, A.R. Wellburn, G.P. Creissen, and P.M. Mullineaux, 1998. Use of transgenic plants in the assessment of responses to atmospheric pollutants. In: L.J. de Kok and I. Stuhlen, eds. Responses of Plant Metabolism to Air Pollution and Global Change. Backhuys, Leiden, pp. 241–250.

Lendzian, K.J., 1984. Permeability of plant cuticles to gaseous air pollutants. In: M.J. Koziol, and F.R. Whatley, eds. Gaseous Air Pollutants and Plant Metabolism.Butterworth, London, pp. 77–81.
Luwe, M. and U. Heber, 1995. Ozone detoxification in the apoplasm and symplasm of spinach, bean and beech leaves at ambient and elevated concentrations of ozone in air. Planta 197: 448–455.
Lyons, T.M. and J.D. Barnes, 1998. Influence of plant age on ozone resistance in *Plantago major*. New Phytologist 138: 83–89.
Lyons, T.M., J.D. Barnes, and A.W. Davison, 1997. Relationships between ozone resistance and climate in European populations of *Plantago major*. New Phytologist 136: 503–510.
Lyons, T.M., M. Plöchl, E. Turcsanyi, and J.D. Barnes, 1998. Extracellular ascorbate: A protective screen against ozone? In: S.B. Agrawal, S. Madhoolika, M. Agrawal, and D.T. Krizek, eds. Environmental Pollution and Plant Responses. CRC Press/Lewis Publication, New York.
Lyons, T., J.H. Ollerenshaw, and J.D. Barnes, 1999. Impacts of ozones on *Plantago major*: apoplastic and symplastic antioxidant status. New Phytologist 141: 253–263.
Macnair, M.R., 1991. Genetics of the resistance of plants to pollutants. In: G.E. Taylor, L.F. Pitelka, and M.T. Clegg, eds. Ecological Genetics and Air Pollution. Springer-Verlag, New York, pp. 127–136.
Macnair, M.R., 1993. The genetics of metal tolerance in vascular plants. New Phytologist 124: 541–559.
Madamanchi, N.R. and R.G. Alscher, 1994. Metabolic bases to differences in sensitivity of two pea cultivars to sulphur dioxide. Journal of Plant Physiology 97: 88–93.
Malhotra, S.S. and A.A. Khan, 1984. Biochemical and physiological impact of major pollutants. In: M. Treshow, ed. Air Pollution and Plant Life. Wiley, London, pp. 123–146.
Mansfield, T.A. and P.H. Freer-Smith, 1984. The role of stomata in resistance mechanisms. In: M.J. Koziol and F.R. Whatley, eds. Gaseous Air Pollutants and Plant Metabolism. Butterworth, London, pp. 131–146.
Mortensen, L., 1994. Further studies on the effects of ozone concentration on growth of subalpine plant species. Norwegian Journal of Agricultural Science 8: 91–97.
Mudd, J.B., 1996. Biochemical basis for the toxicity of ozone. In: M. Yunus and M. Iqbal, eds. Plant Response to Air Pollution. Wiley, London, pp. 267–283.
Musselman, R.C., D.G. Fox, C.G. Shaw III, and W.H. Moir, 1992. Monitoring atmospheric effects on biological diversity. In: J. Barker and D.T. Tingey, eds. Air Pollution Effects on Biodiversity. Van Nostrand Reinhold, New York, pp. 52–71.
Nimis, P.L., G. Lazzarin, and D. Gasparo, 1991. Lichens as bioindicators of air pollution by SO_2 in the Veneto Region (NE Italy). Studia Geobotanica 11: 3–76.
Nobel, P.S., 1974. Introduction to Biophysical Plant Physiology, 2nd edn. Freeman, San Francisco, CA.
Nobel, P.S., 1983. Biochemical Plant Physiology and Ecology. Freeman, New York.
Nobel, P.S. and D.B. Walker, 1985. Structure of leaf photosynthetic tissue. In: J. Barber and N.R. Baker, eds. Photosynthetic Mechanisms and the Environment. Elsevier, New York, pp. 501–536.
Ogawa, K., S. Kanematsu, and K. Asada, 1996. Intra- and extra-cellular localization of "cytosolic" CuZn-superoxide dismutase in spinach leaf and hypocotyls. Journal of Plant Cell Physiology 37: 790–799.
Örvar, L.B., J. McPherson, and B.E. Ellis, 1997. Pre-activating wounding response in tobacco prior to high-level ozone exposure prevents Necrotic injury. Plant Journal 11: 203–212.
Pell, E.J., C.D. Schlagnhaufer, and R.N. Arteca, 1997. Ozone-induced oxidative stress: Mechanisms of action and reaction. Physiologia Plantarum 100: 264–273.
Penkett, S.A., 1988. Indications and causes of ozone increase in the troposphere. In: F.S. Rowland and I.S.A. Isaksen, eds. The Changing Atmosphere. Wiley, London, pp. 91–103.
Pfanz, H., 1987. Uptake and Distribution of Sulphur Dioxide in Plant Cells and Plant Organelles. Effects on Metabolism. Masters Thesis, Maximillian University, Wurzburg, 136 pp.
Pfanz, H., K.-J. Dietz, I. Weinerth, and B. Oppmann, 1990. Detoxification of sulphur dioxide by apoplastic peroxidase. In: C.H. Brunold, L.J. De Kok, and I. Stuhlen, eds. Sulphur Nutrition and Assimilation in Higher Plants. SPB Academic, The Hague, pp. 229–233.
Pitcher, L.H. and B.A. Zilinskas, 1996. Overexpression of copper/zinc superoxide dismutase in the cytosol of transgenic tobacco confers partial resistance to ozone-induced foliar necrosis. Journal of Plant Physiology 110: 583–588.

Pitcher, L.H., E. Brennan, A. Hurley, P. Dunsmuir, J.M. Tepperman, and B.A. Zilinskas, 1991. Over production of petunia chloroplastic copper/zinc superoxide dismutase does not confer ozone tolerance in transgenic tobacco. Journal of Plant Physiology 97: 452–455.

Pleijel, T.H. and H. Danielsson, 1997. Growth of 27 herbs and grasses in relation to ozone exposure and plant strategy. New Phytologist 135: 361–367.

Plöchl, M., P. Ramge, F.-W. Badeck, and G.H. Kohlmaier, 1993. Modeling of deposition and detoxification of ozone in plant leaves: a contribution to subproject BIATEX. In: P.M. Borrell, ed. Proceedings of the EUROTRAC Symposium '92. SPB Academic, The Hague, pp. 748–752.

Polle, A., 1997. Defense against photoxidative damage in plants. In: K. Scandalios, ed. Oxidative Stress and the Molecular Biology of Antioxidant Defenses. Laboratory Press, Cold Spring Harbor, pp. 623–665.

Polle, A. and H. Rennenberg, 1993. Significance of antioxidants in plant adaptation to environmental stress. In: L. Fowden, T.A. Mansfield, and I. Stoddart, eds. Plant Adaptation to Environmental Stress. Chapman and Hall, London, pp. 263–273.

Preston, K.P., 1985. Effects of sulphur dioxide pollution on a California sage scrub community. Environmental Pollution 51: 179–195.

Ramge, P., F.-W. Badeck, M. Plöchl, and G.H. Kohlmaier, 1992. Apoplastic antioxidants as decisive elimination factors within the uptake process of nitrogen dioxide into leaf tissues. New Phytologist 125: 771–785.

Ranieri, A., M. Durante, A. Volterrani, G. Lorenzini, and G.F. Soldatini, 1992. Effects of low SO_2 levels on superoxide dismutase and peroxidase isoenzymes in two different wheat cultivars. Biochem Physiol Pflanzen 188: 67–71.

Rao, M.V., G. Paliyath, and D.P. Ormrod, 1996. Ultraviolet-B and ozone-induced biochemical changes in antioxidant enzymes of *Arabidopsis thaliana*. Journal of Plant Physiology 110: 125–136.

Reich, P.B., 1987. Quantifying plant response to ozone: A unifying theory. Tree Physiology 3: 63–91.

Reiling, K. and A.W. Davison, 1992. The response of native, herbaceous species to ozone; growth and fluorescence screening. New Phytologist 120: 29–37.

Reiling, K. and A.W. Davison, 1995. Effects of ozone on stomatal conductance and photosynthesis in populations of *Plantago major* L. New Phytologist 129: 587–594.

Rennenberg, H., 1984. The fate of the excess sulphur in higher plants. Annual Review of Plant Physiology 35: 121–153.

Roach, D.A. and R. Wulff, 1987. Maternal effects in plants. Annual Review of Ecological Systems 18: 209–235.

Roose, M.L., A.D. Bradshaw, and T.M. Roberts, 1982. Evolution of resistance to gaseous air pollution. In: M.H. Unsworth and D.P. Ormrod, eds. Effects of Gaseous Air Pollutants in Agriculture and Horticulture. Butterworth, London, pp. 379–409.

Runeckles, V.C., 1992. Uptake of ozone by vegetation. In: A.S. Lefohn, ed. Surface Level Ozone Exposures and their Effects on Vegetation. Lewis, Chelsea, pp. 157–188.

Salter, L. and N. Hewitt, 1992. Ozone-hydrocarbon interactions in plants. Phytochemistry 31: 4045–4050.

Schraudner, M., U. Graf, C. Langebartels, and H. Sandermann, 1994. Ambient ozone can induce plant defence reactions in tobacco. Proceedings of the Royal Society of Edinburgh 102B: 55–61.

Schraudner, M., C. Langebartels, and H. Sandermann, 1997. Changes in the biochemical status of plant cells induced by the environmental pollutant ozone. Physiologia Plantarum 100: 274–280.

Showman, R.E., 1991. A comparison of ozone injury to vegetation during moist and drought years. Journal of Environmental Management Association 41: 63–64.

Slovik, S., 1996. Chronic SO_2^- and NO_{x^-} pollution interferes with the K^+ and Mg^{2+} budget of Norway spruce trees. Journal of Plant Physiology 148: 276–286.

Smith, H., 1990. Signal perception, differential expression within multigene families and the molecular basis of phenotypic plasticity. Plant, Cell and Environment 13: 585–594.

Stockwell, W.R., G. Kramm, H.-E. Scheel, V.A. Mohnen, and W. Seiler, 1997. In: H. Sandermann, A.R. Wellburn, and R.L. Heath, eds. Ozone Formation, Destruction, and Exposure in Europe and the United States. Forest Decline and Ozone: A Comparison of Controlled chamber and Field Experiments. Springer-Verlag, Berlin, pp. 1–38.

Stuhlen, I. and L.J. De Kok, 1990. Whole plant regulation of sulfur metabolism a theoretical approach and comparison with current ideas on regulation of nitrogen metabolism. In: H. Rennenberg,

C. Brunold, L.J. De Kok, and I. Stuhlen, eds. Sulphur Nutrition and Assimilation in Higher Plants. SPB Academic Publishing, The Hague, pp. 71–91.
Tanaka, K., I. Furusawa, N. Kondo, and K. Tanaka, 1988. SO_2 tolerance of tobacco plants regenerated from paraquat-tolerant callus. Journal of Plant Cell Physiology 29: 743–746.
Taylor, G. and R. Ferris, 1996. Influence of air pollution on root physiology and growth. In: M. Yunus and M. Iqbal, eds. Plant Response to Air Pollution. Wiley, London, pp. 375–394.
Taylor, G.E. and L.F. Pitelka, 1992. Genetic diversity of plant populations and the role of air pollution. In: J. Barker and D.T. Tingey, eds. Air Pollution Effects on Biodiversity. Van Nostrand Reinhold, New York, pp. 111–130.
Taylor, G.E., P.J. Hanson Jr., and D.B. Baldochi, 1998. Pollutant deposition to individual leaves and plant canopies: sites of regulation and relationship to injury. In: W.W. Heck, O.C. Taylor, and D.T. Tingey, eds. Assessment of Crop Loss from Air Pollutants. Elsevier, London, pp. 227–257.
Thalmair, M., G. Bauw, S. Thie, T. Dohring, C. Langebartels, and H. Sandermann, 1995. Ozone and ultraviolet-B effects on the defence-related proteins beta-1,3-glucanase and chitinase in tobacco. Journal of Plant Physiology 148: 222–228.
Tingey, D.T. and C.P. Andersen, 1991. The physiological basis of differential plant sensitivity to changes in atmospheric quality. In: G.E. Taylor, L.F. Pitelka, and M.T. Clegg, eds. Ecological Genetics and Air Pollution. Springer-Verlag, Berlin, pp. 209–236.
Tingey, D.T. and G.E. Taylor, 1981. Variation in plant response to ozone: a conceptual model of physiological events. In: M. Unsworth, and D.P. Ormrod, eds. Effects of Gaseous Air Pollutants in Agriculture and Horticulture. Butterworth, London, pp. 113–138.
Torsethaugen, G., L.H. Pitcher, B.A. Zilinskas, and E.J. Pell, 1997. Overproduction of ascorbate peroxidase in the tobacco chloroplast does not provide protection against ozone. Journal of Plant Physiology 114: 529–537.
Van Camp, W., H. Willekens, C. Bowler, M. Van Montagu, D. Inzé, P. Reupold-Popp, H. Sandermann, and C. Langebartels, 1994. Elevated levels of superoxide dismutase protect transgenic plants against ozone damage. Biotechnology 12: 165–168.
Warwick, K.R. and G. Taylor, 1995. Contrasting effects of tropospheric ozone on five native herbs which coexist in calcareous grassland. Global Change Biology 1: 143–151.
Wellburn, A.R., J.D. Barnes, P.W. Lucas, A.R. McLeod, and T.A. Mansfield, 1997. Controlled O_3 exposures and field observations of O_3 effects in the UK. In: H. Sandermann, A.R. Wellburn, and R.L. Heath, eds. Forest Decline and Ozone: A Comparison of Controlled Chamber and Field Experiments. Springer-Verlag, Berlin, pp. 201–248.
Westman, W.E., 1979. Oxidant effects on California coastal sage scrub. Science 205: 1001–1003.
Westman, W.E., 1985. Air pollution injury to coastal sage scrub in the Santa Monica Mountains, Southern California. Water, Air and Soil Pollution 26: 19–41.
Whitfield, C.P., A.W. Davison, and T.W. Ashenden, 1997. Artificial selection and heritability of ozone resistance in two populations of *Plantago* major L. New Phytologist 137: 645–655.
Wilbourn, S., 1991. The Effects of Ozone on Grass-Clover Mixtures. Ph.D. Thesis, University of Newcastle Upon Tyne, U.K.
Wilbourn, S., A.W. Davison, and J.H. Ollerenshaw, 1995. The use of an unenclosed field fumigation system to determine the effects of ozone on a grass-clover mixture. New Phytologist 129: 23–32.
Winner, W.E., J.S. Coleman, C. Gillespie, H.A. Mooney, and E.J. Pell, 1991. Consequences of evolving resistance to air pollution. In: G.E. Taylor, L.F. Pitelka, and M.T. Clegg, eds. Ecological Genetics and Air Pollution. Springer-Verlag, Berlin, pp. 107–202.
Wolfenden, J. and T.A. Mansfield, 1991. Physiological disturbances in plants caused by air pollutants. Proceedings of the Royal Society Edinburgh 97B: 117–138.
Wolfenden, J., P.A. Wookey, P.W. Lucas, and T.A. Mansfield, 1992. Action of pollutants individually and in combination. In: J. Barker and D.T. Tingey, eds. Air Pollution Effects on Biodiversity. Van Nostrand Reinhold, New York, pp. 72–92.
Yalpani, N., A.J. Enyedi, J. León, and I. Raskin, 1994. Ultraviolet light and ozone stimulate accumulation of salicylic acid, pathogenesis-related proteins and virus resistance in tobacco. Planta 193: 372–376.

21 Canopy Photosynthesis Modeling

Wolfram Beyschlag and Ronald J. Ryel

CONTENTS

INTRODUCTION

Photosynthesis models are an important development for estimating gas-flux rates of plants. These models have been used to estimate fluxes from the level of the single leaf to community carbon fluxes across the globe. Questions addressed with photosynthesis models involve differences in carbon exchange among plant communities, plant community responses to climate change, and basic ecological concepts concerning resource acquisition, competition for light, and effects of stress.

Primary productivity of a whole plant or canopy is the cumulative carbon gain from all photosynthetically active organs. Because of differences in age, physiology, and exposure to microclimatic conditions, organs are not equally productive, and measurements of individual elements generally do not represent the behavior of the whole plant or canopy. However, by accounting for structural and microclimatic differences between foliage elements, photosynthesis models can provide integrated estimates of photosynthesis and water vapor exchange for the whole canopy.

In this chapter, a class of photosynthesis models is presented that scale up from single-leaf estimates to the whole plant and entire canopies. A general description, perspectives on use, and recent developments are presented as well as relevant details for model development. Model formulations described here range from simple for homogeneous single-species canopies, to complex for diverse multispecies canopies, and are suitable for addressing a range of ecological questions. This presentation is broken into two parts; the Section "Model Overview and Perspectives" is designed to introduce the reader to the general approach used in constructing canopy photosynthesis models, whereas the Section "Model Development" contains sufficient mathematical detail to aid in model use and development.

MODEL OVERVIEW AND PERSPECTIVES

Whole-plant and canopy photosynthesis models contain integrated submodels for (1) single-leaf photosynthesis and (2) the linkage between foliage elements and the physical environment (Figure 21.1). Whole-plant or canopy photosynthesis is modeled by dividing the canopy structure into subregions of similar foliage characteristics (density, orientation, and physiological properties). Microclimatic conditions affecting foliage elements within these subunits are determined at regular intervals, and photosynthesis rates are then calculated for defined foliage classes within each subunit. Whole-plant or canopy rates are calculated as sums of rates for canopy subunits weighted by foliage density. Canopy photosynthesis models often include three general assumptions: (1) the photosynthetic activity of a plant organ relates to its maximum photosynthetic capacity, and reflects age, phenology, acclimation, and physiological condition of the plant; (2) the photosynthetic rate of individual foliage elements depends on interactions with microclimatic conditions (e.g., intercepted radiation, leaf temperature, CO_2 partial pressure within the leaf); and (3) microclimatic conditions within plant canopies result from interactions of macroclimate above the canopy, structural and physical properties of canopy elements, and relative position of foliage elements. An additional assumption is often included: (4) the photosynthesis rate of each foliage element is independent of the rates of the other elements.

SINGLE-LEAF PHOTOSYNTHESIS AND CONDUCTANCE MODELS

Single-leaf photosynthesis is the basic unit of canopy photosynthesis models, and carbon assimilation for this basic unit depends on photosynthetic characteristics and surrounding microclimatic conditions. These models are of two basic types: (1) empirical models, where mathematical relationships are formulated between measured variables and (2) mechanistic models, where mathematical formulations are more closely linked to the physiology of

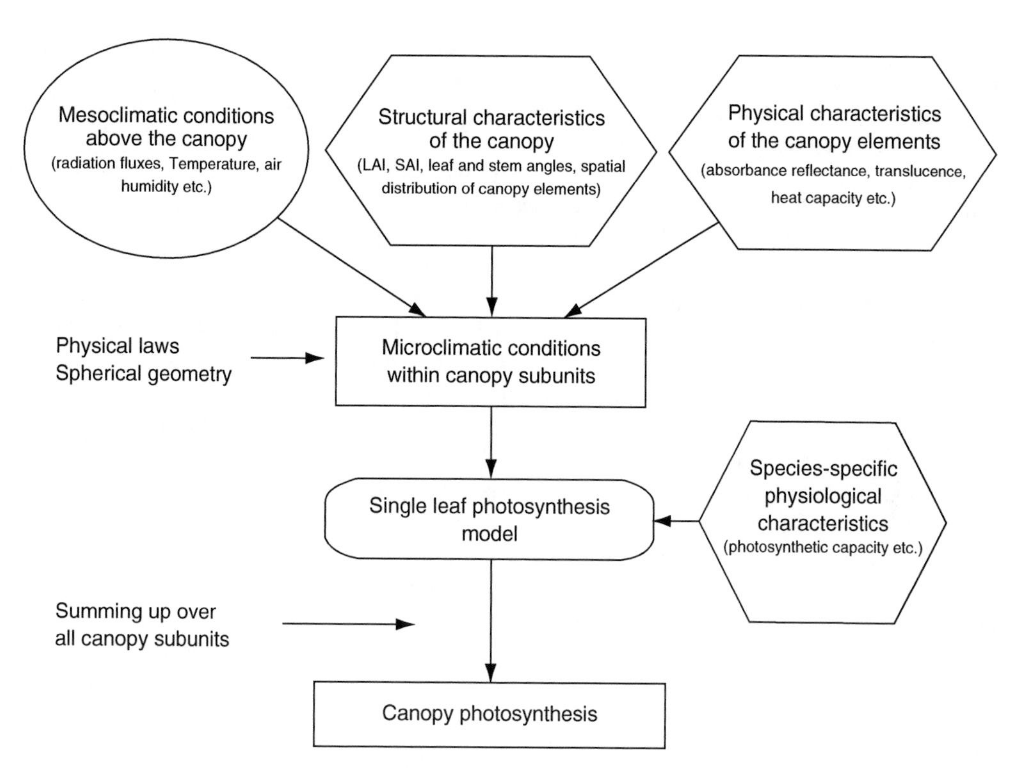

FIGURE 21.1 Flow diagram of a typical canopy photosynthesis model. Input parameters are in oval and hexagonal boxes, the latter representing parameters provided for each canopy subsection. Variables calculated by the model are shown within rectangular boxes.

photosynthesis. Since mechanistic models are more closely linked to physiological processes, they are more suitable for addressing how changes in environmental conditions affect photosynthesis (e.g., Harley et al. 1992, Tenhunen et al. 1994). Because parameterization is often simpler, however, empirical models are often used when questions concern canopy structure or light competition (e.g., Ryel et al. 1990, 1994). Models exist for both C_3 and C_4 metabolic pathways of single-leaf photosynthesis. Development details for mechanistic and empirical models for single-leaf C_3 and C_4 photosynthesis are contained in the Section "Model Development".

Important companions to models for single-leaf photosynthesis are models for stomatal conductance. Since CO_2 is supplied to the mesophyll through the stomata, and water vapor exits the leaf through these same pores, stomatal diffusive conductance is an essential component in modeling single-leaf photosynthesis. Although many factors affect stomatal conductance (including irradiance, temperature, air humidity, CO_2 partial pressure, plant water status, and endogenic rhythms), controlling mechanisms of stomatal regulation are not fully understood. As a result, truly mechanistic models do not exist for stomatal conductance. The existing empirical models can be divided into uncoupled models where stomatal conductance is calculated as a function of external environmental conditions, and coupled models where the rate of leaf photosynthesis is linked to conductance. Models for stomatal conductance are usually linked with formulations that account for foliage boundary-layer diffusional effects (see Gates 1980, Nobel 1983, Schuepp 1993).

Whole-Plant/Canopy Models

Whole-plant/canopy photosynthesis is accomplished by simultaneously calculating single-leaf fluxes at multiple locations within the plant canopy. Since microclimatic differences in light intensity, temperature, humidity, and CO_2 concentration result in variable rates of photosynthesis and transpiration throughout the canopy, describing the interaction between canopy structure and its environment is essential to providing realistic predictions. Of the microclimatic variables affecting gas-flux rates, variability in incident light intensity is usually responsible for much of the heterogeneity in rates of net photosynthesis and transpiration within the canopy. This occurs primarily because of the strong light and temperature dependence of photosynthesis and stomatal conductance, but also because of the effect of vapor-pressure deficit on transpiration as manifested by radiation-induced increases in leaf temperature. Other factors that add to variability of gas-flux rates within the canopy include photosynthetic characteristics that vary with depth in the canopy (Beyschlag et al. 1990, Niinemets 1997, Drouet and Bonhomme 2004) and turbulence in the canopy, which can significantly alter temperature and humidity gradients.

Foliage intercepts both longwave (>3000 nm) and shortwave (400–3000 nm) radiation. The portion of the shortwave spectrum where absorption by chlorophyll *a* and *b* is high is often referred to as photosynthetically active photon flux (PFD), and may vary from full sunlight at the top of the canopy to less than 1% of full sunlight deep within the canopy (Pearcy and Sims 1994). Shortwave radiation (including PFD) incident on foliage is the sum of three fluxes: direct solar beam, diffuse radiation from the sky, and diffuse radiation reflected and transmitted by other foliage elements (Baldocchi and Collineau 1994). Position of the sun and cloud cover affect fluxes of direct solar beam radiation, and both solar altitude and azimuth are important in relationship to foliage. Solar direct beam flux depends on latitude, date, time of day, and orientation of the foliage elements. Diffuse radiation from the sky emanates from the hemisphere of the sky, and may be relatively constant across the hemisphere with clear or uniformly overcast skies (but also see Spitters et al. 1986). Reflection and transmission of direct beam and sky diffuse radiation within the canopy constitutes leaf diffuse radiation, with flux as a function of the proximity and optical properties (transmittance and reflectance) of adjacent foliage.

Absorbed shortwave (I_S) and longwave (I_L) radiation affect the leaf energy balance, and in conjunction with convection, leaf transpiration, and leaf longwave emittance, affect leaf temperature. Longwave radiation emanates to the leaf surface from the sky, soil surface, and from surrounding foliage, and fluxes are related to the temperature and emissivity of the radiation surfaces. Convective heat transfer (C_l) between the leaf and the surrounding air varies with air and leaf temperatures, and wind speed across the leaf surface. Leaf transpiration rate affects latent heat loss (H_l) from the leaf. Leaf temperature results from a balance of energy gains and losses, which may be written as

$$I_S a_S + I_L a_L = C_l + H_l + L_l, \tag{21.1}$$

where a_S and a_L are the fractions of intercepted shortwave and longwave radiation, respectively, and L_l is longwave radiation emittance from the leaf surface. Formulations for convective and latent heat transfer and leaf emittance may be found in Norman (1979) and Gates (1980). Energy balance routines to calculate leaf temperature require iterative calculation procedures when linked to stomatal conductance, and resulting model formulations are generally more complex (Caldwell et al. 1986, Ryel and Beyschlag 1995). However, when leaves are small or narrow in stature, the assumption that leaf and air temperature are identical is often made (Ryel et al. 1990, 1993, Wang and Jarvis 1990).

Uniform Monotypic Plant Stands

Single-species plant communities with relatively homogeneous foliage distributions are modeled with the simplest canopy photosynthesis models. Generally, this model structure is

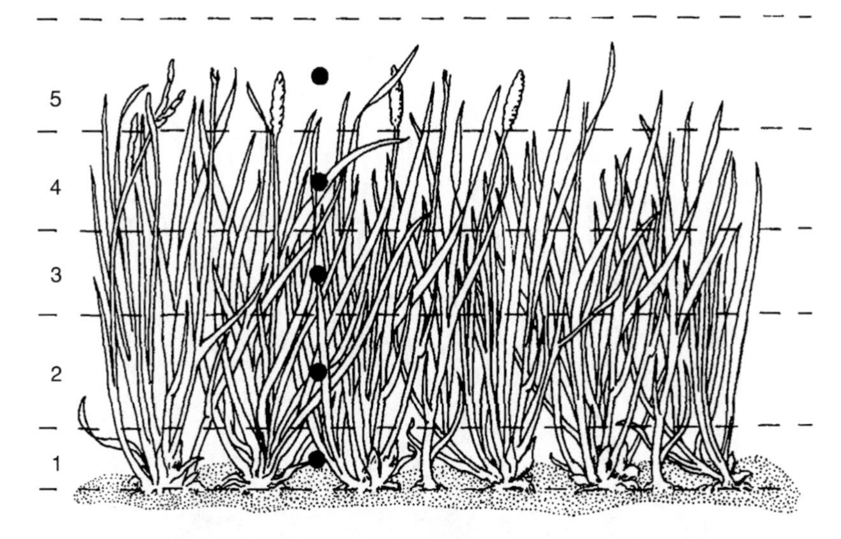

FIGURE 21.2 Uniform single-species grass canopy subdivided into five layers. Foliage density and orientation are assumed similar within each layer. Calculations for light interception and photosynthesis are conducted for the points as shown within the layers.

limited to grass (e.g., lawns, pastures) or crop canopies, but may also include forest canopies with relatively uniform tree cover. Models for these plant stands divide the canopy into layers of approximately uniform foliage density and orientation (Figure 21.2), and interception of radiation and photosynthesis is calculated for points located within the center of these layers. These models are considered one dimensional since foliage is assumed to be horizontally uniform, and differences in radiation interception occur only in the vertical dimension. Details for modeling light relations within uniform monotypic plant stands are contained in the Section "Model Development".

Uniform Multispecies Plant Canopies

Canopies with relatively homogeneous mixtures of two or more species (e.g., grasslands and crop/weed mixtures) can be modeled as simple extensions of the model for uniform monotypic plant stands. Relatively uniformly distributed foliage is assumed for all species, but the vertical distribution may vary by species. A simple situation arises when one species overtops another (e.g., Tenhunen et al. 1994), but the typical canopy has foliage elements mixed within canopy layers (Figure 21.3) (e.g., Ryel et al. 1990, Beyschlag et al. 1992). As with the uniform multispecies plant canopy model, intercepted radiation and photosynthesis is calculated for points located within the center of layers defined by uniformity in foliage density and orientation for each species. Details for this model type are contained in the Section "Model Development".

Inhomogeneous Canopies

Plant canopies with clumped or discontinuous vegetation cannot be realistically represented with models that only vary foliage density vertically. Within these canopies, foliage interception of light is affected by neighboring vegetation that may not be positioned at uniform distances or compass direction. If gaps occur within the canopy, light may reach plants from the sides, and not simply above the foliage. With this complexity, a three-dimensional light-interception model is necessary to represent these canopies. The common approach to

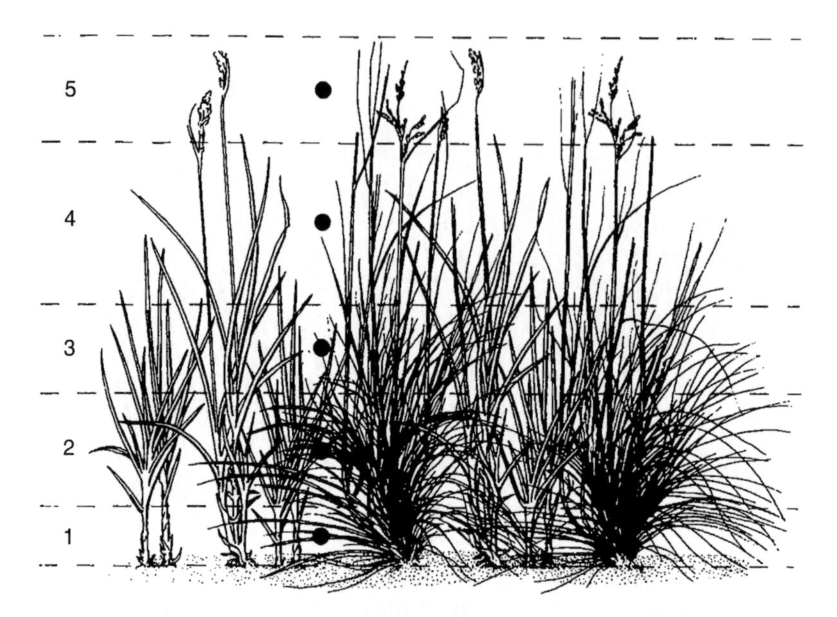

FIGURE 21.3 Two-species uniform canopy subdivided into five layers. Foliage density and orientation are assumed similar by species within each layer. Calculations for light interception and photosynthesis are conducted for the points as shown within the layers.

modeling heterogeneous canopies is to fit individual plants or clumps of vegetation with suitable three-dimensional geometric shapes, including cubes (Fukai and Loomis 1976), cones (Oker-Blom and Kellomäki 1982a,b, Kuuluvainen and Pukkala 1987, Oker-Blom et al. 1989), ellipses (Charles-Edwards and Thornley 1973, Mann et al. 1979, Norman and Welles 1983, Wang and Jarvis 1990), and cylinders (Brown and Pandolfo 1969, Ryel et al. 1993). Cescatti (1997) developed a model structure allowing for radial heterogeneity within individual tree crowns. Regions within these shapes are assumed to have relatively similar foliage density and orientation, and light interception and photosynthesis are calculated at points within these subregions (Figure 21.4).

Big-Leaf Models

In contrast to models for inhomogeneous canopies, big-leaf models simplify rather than increase canopy structural complexity. In these models (e.g., Sellers et al. 1992, Amthor 1994), properties of the whole canopy are reduced to that of a single leaf, and modified equations for single-leaf net photosynthesis and conductance are used for calculating whole-canopy (big-leaf) flux rates. These models have the advantage of fewer parameters, greatly reduced complexity of development, and substantially less time required for model simulation. Big-leaf models are often used when flux rates are modeled for several vegetation communities at the landscape scale (Kull and Jarvis 1995).

Although attractive because of their simplicity, big-leaf models have serious limitations. Parameters for big-leaf models cannot be directly measured, and simple arithmetic means of parameters for individual leaves are inadequate because most functions involving light transmission and gas fluxes are nonlinear (Leuning et al. 1995, Jarvis 1995, de Pury and Farquhar 1997). McNaughton (1994) illustrates this problem by showing that the average canopy conductance preserving whole-canopy transpiration flux differed from the conductance necessary to preserve whole-canopy CO_2 assimilation. Despite these problems, big-leaf models may be suitable if the big leaf is separated into sunlit and shaded elements (de Pury

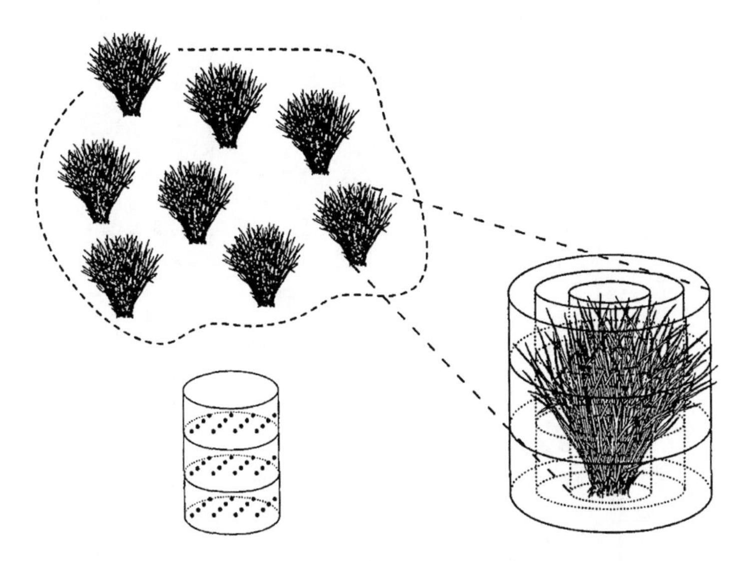

FIGURE 21.4 Individual plant represented as a series of concentric cylinders subdivided into layers as used by the model of Ryel et al. (1993). Foliage density and orientation are assumed similar within each layer for an individual plant. Individual plants can be grouped to form a multi-individual canopy with calculations conducted for each member or representative members. Light attenuation for an individual plant would be affected by neighboring plants when such a canopy is defined. The matrix of points indicates locations where light interception and photosynthesis are calculated.

and Farquhar 1997, Wang 2000, Dai et al. 2004) or calculated fluxes are calibrated to outputs from more detailed canopy models across an appropriate range of meteorological conditions, or to whole-canopy flux measurements (Fan et al. 1995, Raulier et al. 1999).

Examples

Canopy photosynthesis models have been used to address a wide variety of topics including basic plant ecophysiology (Barnes et al. 1990, Beyschlag et al. 1990, Ryel et al. 1993), environmental change (Ryel et al. 1990, Reynolds et al. 1992), and crop management (Grace et al. 1987a,b). Model outputs can also provide carbon-gain inputs to allocation and growth models (Johnson and Thornley 1985, Charles-Edwards et al. 1986, Reynolds et al. 1987, Buwalda 1991, Webb 1991). Two examples of model use are briefly discussed below.

Roadside Grasses

The neophytic grass *Puccinellia distans* has recently invaded roadsides in central Europe, which are dominated by the highly competitive grass *Elymus repens*. Beyschlag et al. (1992) showed that *P. distans* could coexist in garden plots with the highly competitive grass *E. repens* when regular mowing reduced the competitive advantage of *E. repens* for light. Ryel et al. (1996) conducted in situ experiments along roadsides where mowed and unmowed portions of the same roadway were compared. A multispecies, homogeneous canopy photosynthesis model was used to estimate reductions in net photosynthesis for *P. distans* due to the presence of *E. repens* within the mowed and unmowed plots. Simulations indicated that little difference in net photosynthesis occurred between mowed and unmowed plots (Figure 21.5), eliminating mowing as the primary factor contributing to this coexistence. Subsequent experiments indicated that shallow soil depth was the primary factor contributing to coexistence (Beyschlag et al. 1996).

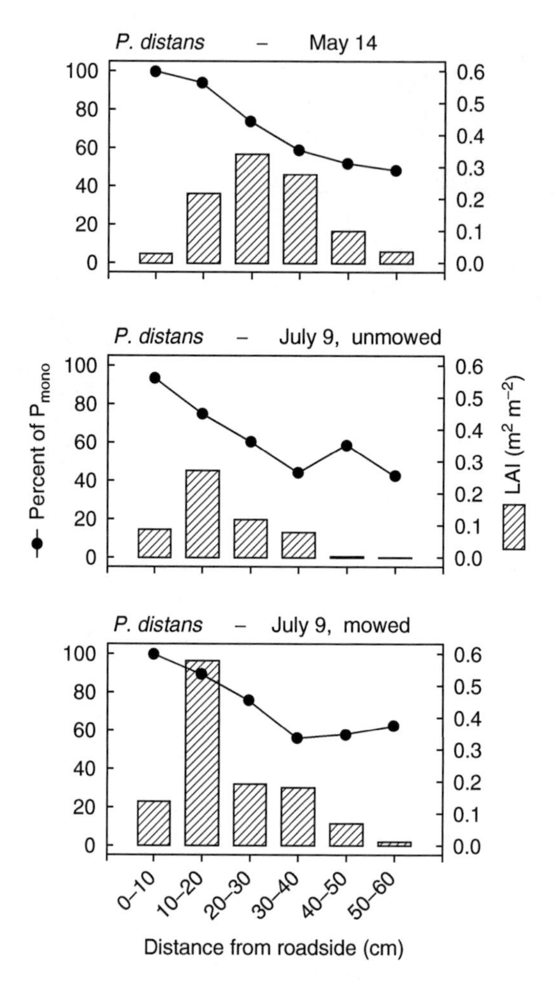

FIGURE 21.5 Model calculations of relative photosynthesis rates (lines) for *Puccinellia distans* for roadside canopies at six distances from the road edge at the beginning of the mowing experiment in May (*upper*) and for unmowed (*center*) and mowed (*lower*) plots in July. Relative photosynthesis rates are expressed as a percent of the rate with *Elymus repens* removed from the canopy. Total foliage area of *P. distans* is also shown (bars). (From Ryel, R.J., Beyschlag, W., Heindl, B., and Ullmann, I., *Bot. Acta*, 109, 441, 1996. With permission.)

Effects of Needle Loss on Spruce Photosynthesis

Needle loss in conifers is a prevalent symptom of forest decline. Beyschlag et al. (1994) assessed the effect of needle loss on whole-plant photosynthesis for forests of young *Picea abies*. A photosynthesis model for inhomogeneous canopies was used to evaluate light interception and net photosynthesis in these canopies. Simulation results indicated that in sparse canopies, needle loss resulted in significant reductions in whole-plant net photosynthesis. However, in more dense canopies, little reduction occurred (Figure 21.6) as in canopies without needle loss, shaded foliage contributed little to whole-plant net photosynthesis.

Future Directions

Canopy photosynthesis models are one method of effectively estimating whole-canopy gas fluxes (Ruimy et al. 1995), and provide a link between measurable single-leaf photosynthesis

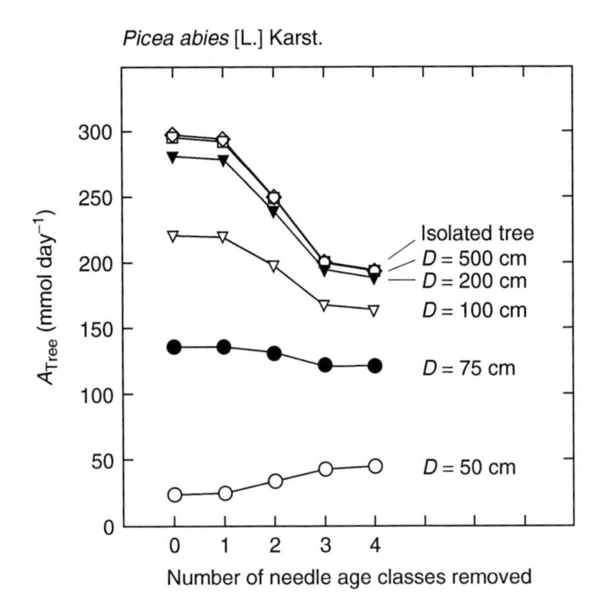

FIGURE 21.6 Model calculations of absolute changes in daily overall tree photosynthesis of experimental 7 year old spruce tree as a function of the number of needle age classes removed and the stand density of a simulated canopy. Simulated stands were created with trees equally spaced (D = distance center to center) as indicated. (From Beyschlag, W., Ryel, R.J., and Dietsch, C., *Trees Struct. Funct.*, 9, 51, 1994. With permission.)

and rates of photosynthesis in whole plants or canopies. Development of these models is an ongoing endeavor, with both increasing complexity and simplification characteristic of new advances. Research objectives will play an important role in influencing the direction of future model developments.

Model complexity will be increased through the addition of other phenomena and more realistic structural design. Additional phenomena may include stomatal patchiness (e.g., Pospisilová and Šantrucek 1994, Eckstein et al. 1996), sunflecks (Pearcy and Pfitsch 1994), penumbra (Oker-Blom 1985, Ryel et al. 2001), leaf clumping (Baldocchi et al. 2002, Cescatti and Zorer 2003), leaf flutter (Roden and Pearcy 1993, Roden 2003), photoinhibition (Werner et al. 2001), and nonsteady-state stomatal dynamics (Pearcy et al. 1997). Macro- and microclimate linkages with the canopy may also be improved, with better representations of air turbulence and concentration gradients, particularly using large-eddy simulation of airflow within canopies (Dwyer et al. 1997). Plant structural complexity may also increase as indicated by the developments of Cescatti (1997). Modular format of model structure allowing for relatively easy replacement of components with new routines enhance the increase in model complexity (Reynolds et al. 1987).

Canopy photosynthesis models may also be reduced in complexity particularly with research directed at landscape level fluxes. Approaches similar to the big-leaf models may be important, but necessitate dealing with problems inherent with model structure and parameterization (Medlyn et al. 2003). Canopy-level models of photosynthesis based on remote sensing data may also see more development in the future (Ustin et al. 1993, Field et al. 1994, Running et al. 2000, Ahl et al. 2004).

Canopy photosynthesis models will be increasingly linked to growth models (e.g., Charles-Edwards et al. 1986, Webb 1991, Hoffmann 1995) as growth models become further developed (e.g., Conner and Fereres 1999). Development of growth models require better knowledge of linkages between carbon gain and structural changes within the canopy, particularly as affected by light climate. Other linkages may occur between canopy photosynthesis models and cellular automation models (Wolfram 1983) to address plant succession and the formation of stable vegetation patterns (Hogeweg et al. 1985, van Tongeren and Prentice 1986, Czaran and Bartha 1989, Silvertown et al. 1992).

MODEL DEVELOPMENT

In this section, we review model development for both single-leaf photosynthesis and light attenuation by foliage. Models described in this section have been selected because they have been successfully used to address a broad range of ecological questions, and because they characterize the formulations within this class of models. The level of detail provided is sufficient for the reader to gain a basic understanding of the modeling process and to aid in development of similar models. The section concludes with a brief discussion of model parameterization and validation.

C_3 SINGLE-LEAF PHOTOSYNTHESIS

C_3 photosynthesis is the most widespread metabolic pathway for plant carbon assimilation. Because of this, substantially more effort has been focused on developing models for C_3 photosynthesis than for C_4 photosynthesis. Portions of C_3 photosynthesis models are also contained within C_4 photosynthesis models.

Mechanistic

The model developed by Farquhar et al. (1980) and Farquhar and von Caemmerer (1982) for C_3 single-leaf photosynthesis is the most commonly used mechanistic model. Their original model assumed limitations in photosynthesis by the activity of the CO_2-binding enzyme RuBP-1,5-carboxidismutase (Rubisco) and by the RuBP regeneration capacity of the Calvin cycle as mediated by electron transport. Sharkey (1985) added an inorganic phosphate limitation to photosynthesis that was incorporated into models by Sage (1990) and Harley et al. (1992). This presentation follows the model development of Harley et al. (1992), which is suitable for ecological applications.

With 0.5 mol of CO_2 released in the cycle for photorespiratory carbon oxidation for each mol of O_2 reduced, net photosynthesis (A) may be expressed as

$$A = V_C - 0.5 \cdot V_O - R_d = V_C\left(1 - \frac{0.5 \cdot O}{\tau \cdot C_i}\right) - R_d, \tag{21.2}$$

where V_C and V_O are carboxylation and oxygenation rates at Rubisco, C_i and O are the partial pressures of CO_2 and O_2 in the intercellular air space, R_d is CO_2 evolution rate in the light excluding photorespiration, and τ is the specificity factor for Rubisco (Jordan and Ogren 1984).

As discussed earlier, the rate of carboxylation (V_C) is limited by three factors and is set as the minimum of

$$V_C = \min\{W_C, W_j, W_p\}, \tag{21.3}$$

where W_C is the carboxylation rate limited by the quantity, activation state, and kinetic properties of Rubisco, W_j is the carboxylation rate limited by the rate of RuBP regeneration in the Calvin cycle, and W_p is the carboxylation rate limited by available inorganic phosphate. A becomes

$$A = \left(1 - \frac{0.5 \cdot O}{\tau \cdot C_i}\right) \cdot \min\left\{W_C, W_j, W_p\right\} - R_d. \tag{21.4}$$

Michaelis–Menten kinetics are assumed for W_C with competitive inhibition by O_2, and W_C is written

$$W_C = \frac{V_{C_{max}} \cdot C_i}{C_i + K_C(1 + O/K_O)}, \tag{21.5}$$

where $V_{C_{max}}$ is the maximum carboxylation rate, with K_C and K_O the Michaelis constants for carboxylation and oxygenation, respectively.

W_j is assumed proportional to the electron transport rate (J) and is written as

$$W_j = \frac{J \cdot C_i}{4 \cdot (C_i + O/\tau)} \tag{21.6}$$

with the additional assumption that sufficient ATP and NADPH are generated by four electrons to regenerate RuBP in the Calvin cycle (Farquhar and von Caemmerer 1982). J is a function of incident photosynthetic photon flux density (PFD) and was formulated empirically by Harley et al. (1992) with the equation of Smith (1937) as

$$J = \frac{\alpha \cdot I}{\left(1 + \frac{\alpha^2\ I^2}{J_{max}^{\ 2}}\right)^{1/2}}, \tag{21.7}$$

where α is quantum efficiency, I is incident PFD, and J_{max} is the light-saturated rate of electron transport.

Carboxylation rate as limited by phosphate (W_p) is expressed as

$$W_p = 3 \cdot \text{TPU} + \frac{V_O}{2} = 3 \cdot \text{TPU} + \frac{V_C \cdot 0.5 \cdot O}{C_i \cdot \tau}, \tag{21.8}$$

where TPU is the rate at which phosphate is released during starch and sucrose production by triose-phosphate utilization.

The temperature dependencies of factors K_c, K_O, R_d, and τ were expressed empirically by Harley et al. (1992) as

$$K_c,\ K_O,\ R_d, \text{ and } \tau = \exp[c - \Delta H_a/(R \cdot T_K)], \tag{21.9}$$

where c is a scaling factor, ΔH_a is the energy of activation, R is the gas constant, and T_K is leaf temperature.

J_{max} and $V_{C_{max}}$ are also temperature dependent and Harley et al. (1992) used activation and deactivation energies based on Johnson et al. (1942)

$$J_{max} \text{ and } V_{C_{max}} = \frac{\exp[c - \Delta H_a/(R \cdot T_K)]}{1 + \exp[(\Delta S \cdot T_K - \Delta H_d)/(R \cdot T_K)]}, \tag{21.10}$$

where ΔH_d is the deactivation energy and ΔS is an entropy term.

Empirical

The model of Thornley and Johnson (1990) has been widely used and gives good fits to measured data. Gross photosynthesis (P) is expressed as

$$P = \frac{\alpha \cdot I_l \cdot P_m}{\alpha \cdot I_l + P_m}, \tag{21.11}$$

where α is the quantum efficiency, I_l is incident PFD, and P_m is the maximum gross photosynthesis rate at saturating PFD. A factor (θ) for resistance between the CO_2 source and site of photosynthesis, determined from fitting measured data, can be added to obtain

$$P = \frac{1}{2\theta}\left\{\alpha \cdot I_l + P_m - \left[(\alpha \cdot I_l + P_m)^2 - 4\theta \cdot \alpha \cdot I_l \cdot P_m\right]^{1/2}\right\} \tag{21.12}$$

for $0 < \theta < 1$. Johnson et al. (1989) contains a typical application of this model.

Another useful empirical model was developed by Tenhunen et al. (1987), which uses formulations from Smith (1937) for the light and CO_2 dependency of net photosynthesis. Equation 21.7 is used for the light dependency of A, and a similar formulation is used for the CO_2 dependency. This equation is also used for the light dependency of carboxylation efficiency. Equation 21.10 is used for the temperature dependence of the maximum capacity of photosynthesis.

C_4 Single-Leaf Photosynthesis

C_4 photosynthesis involves a CO_2 concentrating mechanism coupled with the C_3 photosynthesis cycle. In the concentrating process, phosphoenolpyruvate (PEP) is carboxylated in the mesophyll cells, transferred to the bundle sheath cells, and decarboxylated before entering the C_3 cycle (Peisker and Henderson 1992). An initial submodel calculates the pool of inorganic carbon in the bundle sheath cells.

Mechanistic

A simple C_4 photosynthesis model is that of Collatz et al. (1992) who assumed that the carboxylation catalyzed by PEP carboxylase is linearly related to CO_2 concentration of the internal mesophyll air space. The model does not consider light dependencies of PEP carboxylase activity and other activation processes (Leegood et al. 1989). Chen et al. (1994) proposed a more complex model with the C_4 cycle controlled by PEP carboxylase. The rate of this cycle (V_4) is described by Michaelis–Menten kinetics and related to mesophyll CO_2 concentration by

$$V_4 = \frac{V_{4m} \cdot C_m}{C_m + k_p}, \tag{21.13}$$

where C_m is mesophyll CO_2 concentration and k_p is a rate constant. The maximum reaction velocity (V_{4m}) is related to incident PFD (I_p) by

$$V_{4m} = \frac{\alpha_p \cdot l_p}{\left(1 + \alpha_p^2 \cdot l_p^2 / V_{pm}{}^2\right)^{1/2}} \tag{21.14}$$

with α_p a fitted parameter and V_{pm} the potential maximum activity of PEP carboxylase. The C_3 and C_4 cycles are linked as

$$V_4 = V_b + A_n, \tag{21.15}$$

where V_b is the diffusion flux of CO_2 between the bundle sheath and the mesophyll and A_n (net photosynthesis) is the net CO_2 exchange rate between the atmosphere and the mesophyll intercellular air space. A_n is calculated using Equation 21.2 with C_i replaced with the CO_2 concentration in the bundle sheath cells. Equation 21.13 and Equation 21.15 are solved iteratively to obtain A_n by balancing the CO_2 and O_2 concentrations in the bundle sheath cells (see Chen et al. 1994 for details).

Empirical

Two approaches to empirical C_4 photosynthesis model are discussed here. Dougherty et al. (1994) used the minimum of photosynthetic capacities limited by light (A_l) and by intercellular CO_2 (A_2). A_l is expressed with a nonrectangular hyperbola as

$$A_l = \frac{A_m + \alpha I^2\sqrt{A_m^2 - 2A_m\alpha I(2\beta - 1) + \alpha^2 I^2}}{2\beta}, \tag{21.16}$$

where α is quantum efficiency, β is an empirical shape parameter, I is incident PFD, and A_m is maximum photosynthetic capacity. A_2 is expressed as

$$A_2 = A_m \frac{c_i}{c_i + 1/E_c}, \tag{21.17}$$

where E_c is an empirical index of leaf CO_2 efficiency and c_i is intracellular CO_2. The temperature dependence of A_m uses Equation 21.10.

Thornley and Johnson (1990) describe two empirical formulations of C_4 photosynthesis. In one formulation, energy necessary for pumping CO_2 into the bundle sheath is assumed independent of that available to the bundle sheath for C_3 photosynthesis and photorespiration. In the second model, a common supply of energy to both mesophyll and bundle sheath is assumed.

Stomatal Conductance

Models for stomatal diffusive conductance are either coupled or uncoupled with leaf photosynthesis rates, and are coupled with various environmental factors. Presently all models are empirical in design. Coupled stomatal models are recommended for addressing questions more physiological in nature.

Coupled Models

A simple, but effective coupled model for stomatal conductance was developed by Ball et al. (1987). Stomatal conductance (g_s) is related linearly to net photosynthesis (A) and relative humidity (h_s), and expressed as

$$g_s = k \cdot A \cdot (h_s/c_s), \tag{21.18}$$

with c_s the mole fraction of CO_2 at the leaf surface. Although g_s does not respond directly to net photosynthesis, relative humidity, or CO_2 at the leaf surface, this model often corresponds well to measured data.

This model has been modified by Leuning (1995) to be consistent with the findings of Mott and Parkhurst (1991) that stomata respond to the rate of transpiration. Interactions between gas molecules leaving and entering stomata (Leuning 1983) are also considered, and g_s is expressed as

$$g_s = g_o + a_1 \cdot A/[(c_s - \Gamma) \cdot (1 + D_s/D_o)] \tag{21.19}$$

with a_1 and D_o the empirical coefficients, Γ the CO_2 compensation point, and g_o the stomatal conductance when A approaches zero; D_s and c_s are humidity deficit and CO_2 concentration at the leaf surface, respectively. Tuzet et al. (2003) further link stomatal conductance to leaf water potential, which in turn is linked to soil water potential in the rooting zone.

Uncoupled Models

In uncoupled models of stomatal conductance, factors considered to influence g_s include incident PFD, leaf temperature, water vapor mole fraction difference, air humidity, leaf water potential, and soil water potential (Jarvis 1976, Whitehead et al. 1981, Caldwell et al. 1986, Jones and Higgs 1989, Beyschlag et al. 1990, Lloyd 1991, Ryel et al. 1993, 2002). Regression equations often relate g_s to one or more of these environmental factors. Uncoupled conductance models can be easier to parameterize than coupled models, and often give good correspondence to measured data.

Whole-Plant and Canopy Models

Models for light attenuation through canopies and interception by foliage are linked with single-leaf photosynthesis models and form the basis of whole-plant and canopy photosynthesis models. Although other environmental factors affecting photosynthesis throughout the canopy can be included in model developments, modeling of canopy light relations is the most important aspect of the canopy photosynthesis models. In this section, we illustrate the approach commonly used to model light attenuation through plant canopies. Models for attenuation and interception of direct beam, sky diffuse, and leaf reflected and transmitted diffuse within canopies with varying complexity are presented, with incident PFD calculated for both sunlit and shaded leaves.

Uniform Monotypic Plant Canopies

Models for single-species plant communities with relatively homogeneous distribution of foliage are constructed by dividing the canopy into layers of relatively homogeneous foliage density and orientation, and interception of radiation is calculated for points at the center of these layers (Figure 21.2). A random distribution of foliage elements is typically assumed in these models (but see, e.g., Caldwell et al. 1986). The thickness of layers is determined by the foliage density distribution in the canopy, and ideally layers have leaf-area index less than $0.5\,m^2$ foliage area m^{-2} ground area to facilitate attenuation of leaf-diffuse (Norman 1979). Foliage areas of leaves and stems (or branches) are often separated as they typically have different optical properties, photosynthetic rates, and orientations. Leaf inclination, which is the angle of the major axis of the foliage element from horizontal, and azimuth angle, which is the directional alignment, comprise the components of foliage orientation. Leaf inclination is most simply defined as a constant for a layer, but can have a defined distribution (Norman 1979, Campbell 1986). Azimuth orientation of foliage elements may be considered random or nonrandom in distribution (Lemeur 1973, Caldwell et al. 1986). The model development presented here considers constant inclination and random azimuth orientation of foliage in each layer, which is applicable to most uniform canopies.

With randomly distributed foliage within a layer, the penetration of direct beam PFD declines exponentially with passage through increasing amounts of foliage. The relative area or fraction of sunlit foliage (P_i) for the sample point within in layer i (bottom layer = 1, top layer = n) is

$$P_i = \exp\left(\sum_{m=i}^{n} (L_m\ Kl_m + S_m\ Ks_m) l_m\right), \tag{21.20}$$

where L_i and S_i are leaf and stem densities (m^2 m^{-3}) in layer i, respectively, and l_i is the path length (m) of PFD through layer i. Kl_i and Ks_i are the light extinction coefficients for leaves and stems, respectively, and for fixed leaf inclination (α_i) and random azimuth can be calculated (Duncan et al. 1967) for inclination of sun above horizon (β) as

$$K_i = \cos\alpha_i\ \sin\beta \tag{21.21}$$

for α_i less than β, and otherwise as

$$K_i = \frac{2}{\pi}\sin\alpha_i\ \cos\beta\ \sin\theta_i + \left(1 - \frac{\theta_i}{90}\right)\cos\alpha_i\ \sin\beta, \tag{21.22}$$

where θ_i is the angle (0–90°) that satisfies $\cos\theta_i = \cot\alpha_i\ \tan\beta$.

The flux of direct beam PFD does not decline with attenuation through foliage (Figure 21.7), but the flux incident on foliage is a function of the orientation of the leaf surface relative to the sun. Incident flux of PFD can hit either the upper or the lower surface of the leaf surface, and the flux for the upper leaf surface can be expressed (Burt and Luther 1979) as

$$Qu_i = B(\cos\alpha_i\ \sin\beta - \sin\alpha_i\ \cos\beta\ \cos\delta_i) \tag{21.23}$$

and for the lower surface as

$$Ql_i = B(\cos(180 - \alpha_i)\sin\beta - \sin(180 - \alpha_i)\cos\beta\cos\delta_i), \tag{21.24}$$

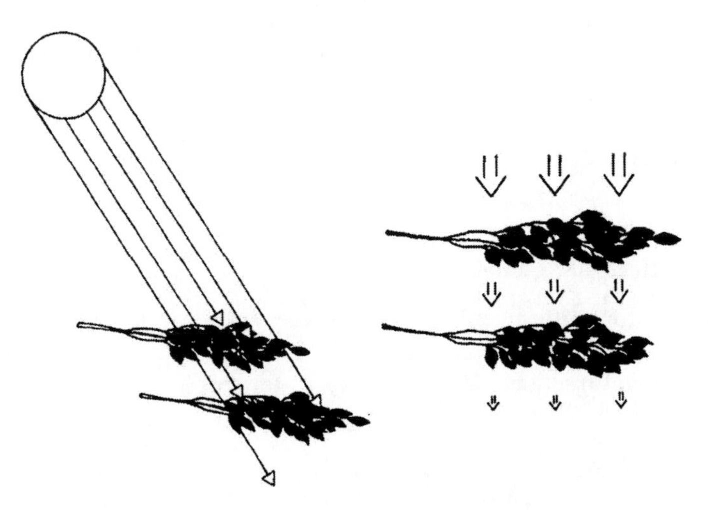

FIGURE 21.7 Attenuation of direct beam (*left*) and sky diffuse (*right*) radiation through foliage. The fraction of foliage illuminated by direct beam sunlight declines with interception by foliage, but the radiation flux does not change in the sunlit portion. In contrast, radiation flux for sky diffuse (and for scattered diffuse) declines with passage through foliage, but all foliage receives similar flux.

where B is the PFD flux on a surface normal to the solar beam and δ is the azimuth angle of the major axis of the leaf surface relative to the sun.

Attenuation of sky-diffuse PFD through the canopy is considered from the midpoint of concentric bands of equal width or area (skybands) that divide the hemisphere of sky above the canopy. Unlike direct beam PFD flux, the flux of sky diffuse PFD declines with passage through increasing amounts of foliage (Figure 21.7), and can be expressed as

$$D_{i,w} = D_{\text{sky}} A_w \exp\left(\sum_{m=i}^{n} (L_m\ Kl_m + S_m\ Ks_m) l_m\right), \tag{21.25}$$

where w is the skyband (e.g., $w = 1, 2, \ldots, 9$ for inclination bands $= 0°–10°, 10°–20°, \ldots, 80°–90°$), D_{sky} is the sky diffuse PFD flux on a horizontal surface, and A_w (view factor, see Duncan et al. 1967 for formulations) is the fractional portion of the sky hemisphere within band w.

Diffuse radiation reflected and transmitted by foliage is difficult to accurately portray in canopy models. Although leaf optical properties, foliage density, and characteristics of adjacent foliage contribute to scattering, the process of scattering is complex. Simplistic approaches have been proposed with one approach to assume that radiation striking the upper foliage surface reflects upward or transmits downward, with the opposite occurring for radiation striking the lower surface. Using this approach (see Norman 1979, Ryel et al. 1990), downward PFD flux from the midpoint of layer i would be

$$\begin{aligned} Td_i = {} & t_i L_i (Qu_i P_i + Du_i) + r_i L_i (Ql_i P_i + Dl_i) + Tu_{i-1} r_i L_i + Td_{i+1} L_i \\ & + Td_{i+1}(1 - L_i - S_i), \end{aligned} \tag{21.26}$$

whereas the upward flux would be

$$\begin{aligned} TU_i = {} & r_i L_i (Qu_i P_i + Du_i) + t_i L_i (Ql_i P_i + Dl_i) + Td_{i+1} r_i L_i + Tu_{i-1} L_i \\ & + Tu_{i-1}(1 - L_i - S_i), \end{aligned} \tag{21.27}$$

where r_i and t_i are leaf reflectance and transmittance for PFD, respectively.

Sunlit leaves have incident PFD fluxes from direct beam, sky diffuse, and reflected and transmitted diffuse. The total PFD incident on both sides of a sunlit leaf in layer i is

$$I_i = Qu_i + Ql_i + Du_i + Dl_i + Tu_{i-1} + Td_{i+1}. \tag{21.28}$$

Incident fluxes for shaded leaves are calculated similarly, but without direct beam components Qu_i and Ql_i.

Uniform Multispecies Plant Canopies

The model structure for uniform multispecies canopies is a generalization of the single-species models. Equations describing interception of PFD can be extended from those for monotypic plant stands. The sunlit fraction of foliage in layer i for species x is simply

$$P_{x,i} = \exp\left(\sum_{m=i}^{n}\left(\sum_{y=x}^{N}(L_{y,m} Kl_{y,m} + S_{y,m} Ks_{y,m})\right) l_m\right), \tag{21.29}$$

where N is the number of species in the canopy and corresponds to Equation 21.20. Equation 21.25 through Equation 21.27 are similarly extended to apply to multiple species (Ryel et al. 1990).

Inhomogeneous Canopies

The cylinder model of Ryel et al. (1993) is used as an example of models for complex canopy structure as it contains the spectrum of complexity found in these models.

Individual plants within the canopy are fitted to concentric cylinders and horizontal layers of relatively uniform foliage density and orientation (Figure 21.4), a process analogous to the layer divisions in uniform canopies. A three-dimensional array of points (Figure 21.4) is used to sample the plant canopy for calculation of light interception and gas exchange. Consistent calculations of flux rates throughout the course of a day may require 1000 or more points within an individual plant crown (Ryel et al. 1993, Falge et al. 1997). Values for points within the canopy are appropriately weighted, averaged, and summed to generate whole-plant flux rates. Simulations may be conducted for all plants within the canopy (e.g., Grace et al. 1987b), for individuals representing plants of similar structure and stature (e.g., Falge et al. 1997), or for one individual when all plants are assumed to be similar in structure and relatively uniformly distributed (e.g., Beyschlag et al. 1994, Ryel et al. 1994). Whole canopy rates can be calculated as the sum of individual plants and expressed per ground area if desired for comparisons between vegetation communities or plots.

Many of the model equations are analogous to the homogeneous canopy model and are simply generalizations of those formulations. The fraction of sunlit foliage (analogous to Equation 21.20) is calculated for each sample point k in plant x as

$$P_{x,k} = \exp\left(\sum_{y=1}^{N_P}\left(\sum_{i=1}^{n_{y,l}}\sum_{m=1}^{n_{y,c}}\left(L_{y,i,m}\,Kl_{y,i,m} + S_{y,i,m}\,Ks_{y,i,m}\right) l_{y,i,m}\right)\right), \qquad (21.30)$$

where N_P is the total number of plants in the canopy, and $n_{y,l}$ and $n_{y,c}$ are the number of layers and subcylinders in plant individual y, respectively. The flux of sky diffuse PFD is calculated for both sky bands, and azimuth directions to account for differential placement of neighboring plants. Sky diffuse flux (analogous equation to Equation 21.25) is calculated as

$$D_{x,p,w,a} = D_{\text{sky}}A_{w,a}\exp\left(\sum_{y=1}^{N_P}\left(\sum_{i=1}^{n_l}\sum_{m=1}^{n_c}\left(L_{y,i,m}\,Kl_{y,i,m} + S_{y,i,m}\,Ks_{y,i,m}\right) l_{y,i,m}\right)\right), \qquad (21.31)$$

where $A_{w,a}$ is the view factor including azimuth orientation (a) of skyband subsections.

As with the uniform canopy model, reflected and transmitted diffuse within diverse canopies is difficult to model. One approach is to calculate average reflected and transmitted diffuse for horizontal layers of points within the canopy using formulations similar to Equation 21.26 and Equation 21.27 (Norman and Welles 1983, Ryel et al. 1993). Cescatti (1997) used a similar approach, but used a weighted average for the flux at each point within a layer.

PARAMETERIZATION

A lengthy treatise of data-collection methods is too voluminous for this chapter, but an overview is provided. The reader is referred to the cited literature for further details. Estimating model parameters is often challenging and time consuming because of the difficulty involved in measuring flux rates and structural parameters. Since parameter estimates can affect model outputs and study conclusions, parameter estimation must be done carefully.

Assessing model sensitivity to parameter estimates (Forrester 1961, Steinhorst et al. 1978) should be conducted to evaluate the robustness of model outputs.

Model Parameters for Single-Leaf Photosynthesis

Parameters for single-leaf photosynthesis and stomatal models are often derived from gas exchange measurements (e.g., Harley et al. 1992, Falge et al. 1997). For the model of Farquhar et al. (1980), the initial slope of the A versus c_i relationship characterizes the activity status of Rubisco, and A at saturating light and CO_2 relates to the maximum RuBP regeneration rate of the Calvin cycle (von Caemmerer and Farquhar 1981). The initial slope of the relationship A versus incident PFD at saturating CO_2 is an estimate of the maximum quantum-use efficiency of the light reaction of photosynthesis, and can also be measured with chlorophyll fluorescence (Schreiber et al. 1994). Biochemical parameters are often obtained from the literature, but some may vary considerably within or among species (Evans and Seemann 1984, Keys 1986).

Canopy Structural Parameters

Both destructive and nondestructive methods are used to estimate density and distribution of foliage. Destructive methods often involve harvesting foliage from portions of the plant canopy (e.g., Caldwell et al. 1986, Beyschlag et al. 1994) with foliage area estimated with analytical devices such as the leaf-area meter. Within canopy light measurements are the most commonly used nondestructive methods for estimating canopy structure (Norman and Campbell 1989), and are often combined with inverted light extinction equations to estimate foliage area in many types of canopies (Lang et al. 1985, Perry et al. 1988, Walker et al. 1988). This estimation procedure has been automated with the LAI-2000 plant canopy analyzer (LI-COR, Lincoln, NE, USA) that efficiently calculates foliage area and average leaf inclination despite some limitations (Gower and Norman 1991, Deblonde et al. 1994, Stenberg et al. 1994). High-contrast fish-eye photography has also been used to estimate foliage area (Anderson 1971, Bonhomme and Chartier 1972). An additional nondestructive within-canopy method, the linear probe or inclined-point-quadrat method uses the frequency of contacts of foliage obtained from the repeated movement of a rod oriented at a fixed angle through the canopy (Warren Wilson 1960, Caldwell et al. 1983). Estimates of leaf area from outside the canopy are most commonly conducted using the normalized difference vegetation index (NDVI). This nondestructive method using remotely sensed images of visible and near infrared reflectances obtained from aircraft or satellite (Field 1991, Wang et al. 2005) is effective at the community level where more detailed measurements are time consuming. Ground truthing using the above methods, however, is necessary to calibrate the NDVI to actual LAI.

Foliage orientation can be measured in situ (Caldwell et al. 1986), with the LAI-2000 analyzer or by linear probe (Warren Wilson 1960, 1963). Leaf reflectance and transmittance of PFD, total shortwave and longwave radiation are often measured with a spectroradiometer (Brunner and Eller 1977) or obtained from suitable literature values (e.g., Sinclair and Thomas 1970, Ross 1975).

Model Validation

Model validation is necessary to assess whether model simulations mimic the dynamics of the modeled system. Model validation is conducted by comparing model output with independently measured variables (Forrester 1961, Innis 1974) to evaluate model performance. Comparisons are often made graphically, but statistical evaluation methods have been proposed (Kitanidis and Bras 1980). Simulations extrapolating beyond the range of validation run the risk of producing faulty results, but are often performed to investigate system dynamics in response to hypothetical conditions. However, validations are conducted under a range of conditions to increase the confidence of extrapolations using the model.

Diurnal course measurements of single-leaf net photosynthesis, transpiration, intercellular CO_2, and stomatal conductance for H_2O are usually compared with simulated results for photosynthesis and conductance models (Figure 21.8; see also Tenhunen et al. 1987, Ryel et al. 1993). Validation of light interception models are conducted by comparing model outputs to light sensor measurements taken within the canopy (Figure 21.9; see also Norman and Welles 1983, Beyschlag et al. 1994). Model predictions of whole-plant or canopy gas fluxes can be compared with several types of field measurements for model validation including whole-plant gas exchange measured within large gas-exchange cuvettes (Figure 21.10); sap-flow measurements (Cermak et al. 1973, Ishida et al. 1991, Falge et al. 2000) of a suitable sample of plants within a canopy (Dye and Olbrich 1993), and eddy-correlation measurements (Baldocchi et al. 1986, Verma 1990, Dugas et al. 1991) of gases within and above the canopy (Baldocchi and Harley 1995, Aber et al. 1996).

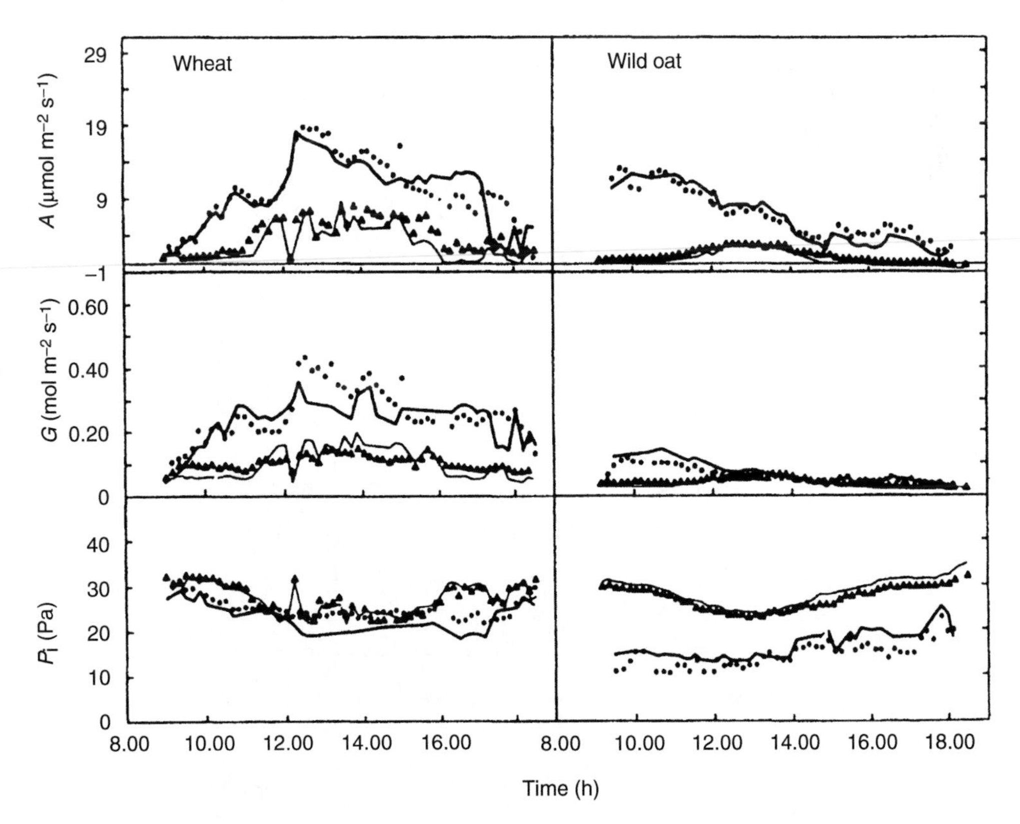

FIGURE 21.8 Model validation conducted for single-leaf photosynthesis model of Tenhunen et al. (1987) for wheat (*Triticum aestirum*) and wild oat (*Avena fatua*). Model output for the course of a single day was compared with data measured by gas exchange for net photosynthesis (A), stomatal conductance (G), and intercellular CO_2 (P_i). (From Beyschlag, W., Barnes, P.W., Ryel, R.J., Caldwell, M.M., and Flint, S.D., *Oecologia*, 82, 374, 1990. With permission.)

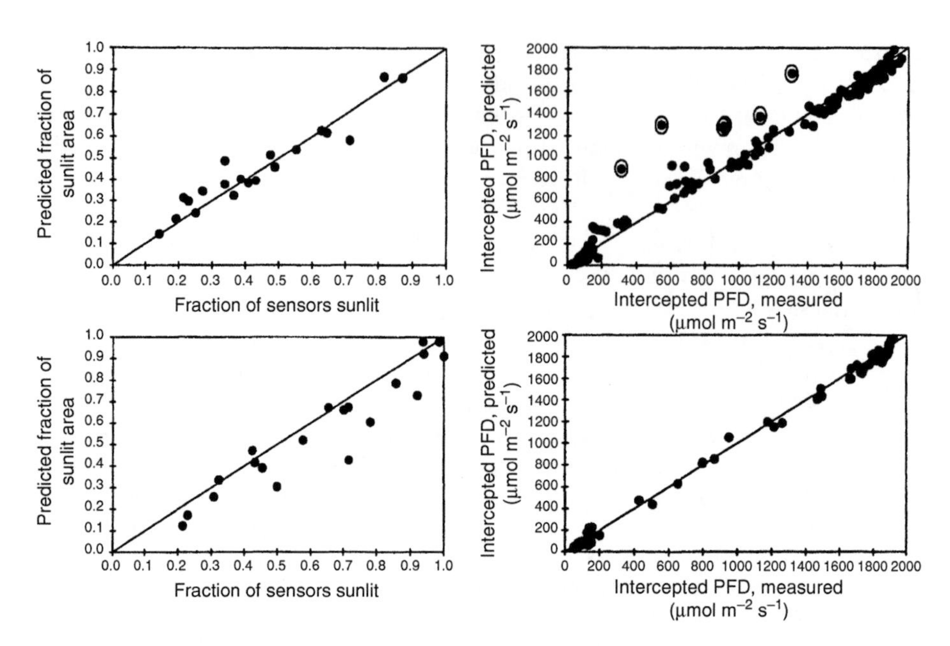

FIGURE 21.9 Model validation for light flux in uniform canopy as attenuated through mixtures of wheat and wild oat (*upper left*) and through monocultures of wheat or wild oat (*lower left*). Measured data were collected with photodiodes mounted on a long stick inserted into the canopy at defined canopy depths. Fluxes below 400 μmol m^2 s^{-1} are for shaded leaves and sensors. The fraction of foliage sunlit was also compared between model and measured data for mixtures (*upper right*) and monocultures (*lower right*). Measured flux data was bimodal in nature and allowed for estimating the portion of fully sunlit sensors. Data points from measurements made deep in the canopy where individual photodiodes may not have been fully sunlit are circled. (From Ryel, R.J., Barnes, P.W., Beyschlag, W., Caldwell, M.M., and Flint, S.D., *Oecologia,* 82, 304, 1990. With permission.)

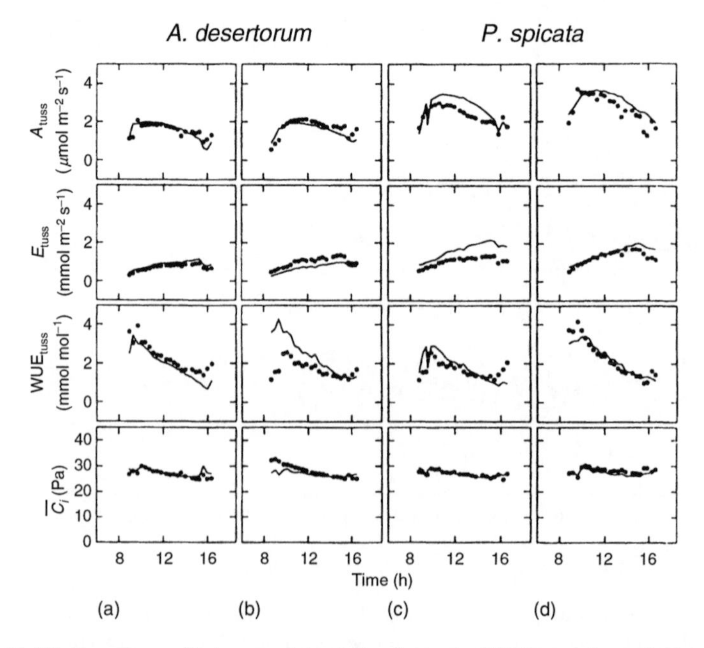

FIGURE 21.10 Validation for cylinder model of Ryel et al. (1993) conducted for whole-plant net photosynthesis (A), transpiration (E), water-use efficiency (WUE), and average intercellular CO_2 concentration (C_i) for two tussocks of crested wheat grass (*Agropyron desertorum* a,b), and two of blue-bunch wheat grass (*Pseudoroegneria spicata* c,d). Measured data were from whole-plant gas-exchange cuvettes. (From Ryel, R.J., Beyschlag, W., and Caldwell, M.M., *Functional Ecology*, 7, 115, 1993. With permission.)

ACKNOWLEDGMENTS

Part of this work has been funded by the Deutsche Forschungsgemeinschaft, Bonn (SFB 251, University of Würzburg) and the Utah Agricultural Experiment Station. We would also like to thank Mr. H. Weigel, Bielefeld for the artwork in Figure 21.2 through Figure 21.4 and Figure 21.7.

REFERENCES

Aber, D.A., P.B. Reich, and M.L. Goulden, 1996. Extrapolating leaf CO_2 exchange to the canopy: A generalized model of forest photosynthesis compared with measurements by eddy correlation. Oecologia 106: 257–267.

Ahl, D.E., S.T. Gower, D.S. Mackay, S.N. Burrows, J.M. Norman, and G.R. Diak, 2004. Heterogeneity of light use efficiency in a northern Wisconsin forest: implications for modeling net primary production with remote sensing. Remote Sensing of Environment 93: 168–178.

Amthor, J.S., 1994. Scaling CO_2-photosynthesis relationships from the leaf to the canopy. Photosynthesis Research 39: 321–350.

Anderson, M.C., 1971. Radiation and crop structure. In: Z. Sestak, J. Catsky, and P.G. Jarvis, eds. Plant Photosynthetic Production. Manual of Methods. W. Junk Publishers, The Hague, Pages 412–466.

Baldocchi, D.D. and S. Collineau, 1994. The physical nature of solar radiation in heterogeneous canopies: spatial and temporal attributes. In: M.M. Caldwell and R.W. Pearcy, eds. Exploitation of Environmental Heterogeneity by Plants. Academic Press, San Diego, CA, pp. 21–71.

Baldocchi, D.D. and P.C. Harley, 1995. Scaling carbon dioxide and water vapour exchange form leaf to canopy in a deciduous forest. II. Model testing and application. Plant, Cell and Environment 18: 1157–1173.

Baldocchi, D.D., S.B. Verma, D.R. Matt, and D.E. Anderson, 1986. Eddy-correlation measurements of carbon dioxide efflux from the floor of a deciduous forest. Journal of Applied Ecology 23: 967–976.

Baldocchi, D.D., K.B. Wilson, and L.H., Gu, 2002. How the environment, canopy structure and canopy physiological functioning influence carbon, water and energy fluxes of a temperate broad-leaved deciduous forest—an assessment with the biophysical model CANOAK. Tree Physiology 22: 1065–1077.

Ball, J.T., I.E. Woodrow, and J.A. Berry, 1987. A model predicting stomatal conductance and its contribution to the control of photosynthesis under different environmental conditions. In: J. Biggens, ed. Progress in Photosynthesis Research, Vol. IV. Martinus Nijhoff Publishers, Dordrecht, pp. 221–224.

Barnes, P.B., W. Beyschlag, R.J. Ryel, S.D. Flint, and M.M. Caldwell, 1990. Plant competition for light analyzed with a multispecies canopy model. III. Influence of canopy structure in mixtures and monocultures of wheat and wild oat. Oecologia, 82: 560–566.

Beyschlag, W., P.W. Barnes, R.J. Ryel, M.M. Caldwell, and S.D. Flint, 1990. Plant competition for light analyzed with a multispecies canopy model. II. Influence of photosynthetic characteristics on mixtures of wheat and wild oat. Oecologia 82: 374–380.

Beyschlag, W., R.J. Ryel, and I. Ullmann, 1992. Experimental and modeling studies of competition for light in roadside grasses. Botanica Acta 105: 285–291.

Beyschlag, W., R.J. Ryel, and M.M. Caldwell, 1994. Photosynthesis of vascular plants: Assessing canopy photosynthesis by means of simulation models. In: E.-D. Schulze, and M.M. Caldwell, eds. Ecophysiology of Photosynthesis. Ecological Studies Vol. 100. Springer Verlag, Berlin, Heidelberg, New York, pp. 409–430.

Beyschlag, W., R.J. Ryel, and C. Dietsch, 1994. Shedding of older needle age classes does not necessarily reduce photosynthetic primary production of Norway spruce. Analysis with a three-dimensional canopy photosynthesis model. Trees Structure and Function 9: 51–59.

Beyschlag, W., R.J. Ryel, I. Ullmann, and J. Eckstein, 1996. Experimental studies on the competitive balance between two Central European roadside grasses with different growth forms.

2. Controlled experiments on the influence of soil depth, salinity and allelopathy. Botanica Acta 109: 449–455.
Bonhomme, R. and P. Chartier, 1972. The interpretation and automatic measurement of hemispherical photographs to obtain sunlit foliage area and gap frequency. Israel Journal of Agricultural Research 22: 53–61.
Brown, P.S. and J.P. Pandolfo, 1969. An equivalent obstacle model for the computation of radiative flux in obstructed layers. Agricultural Meteorology 6: 407–421.
Brunner, U. and B.M. Eller, 1977. Spectral properties of juvenile and adult leaves of *Piper betle* and their ecological significance. Physiologia Plantarum 41: 22–24.
Burt, J.E. and F.M. Luther, 1979. Effect of receiver orientation on erythema dose. Photochemistry and Photobiology 29: 85–91.
Buwalda, J.G., 1991. A mathematical model of carbon acquisition and utilization by kiwifruit vines. Ecological Modelling 57: 43–64.
Caldwell, M.M., G.W. Harris, and R.S. Dzurec, 1983. A fiber optic point quadrat system for improved accuracy in vegetation sampling. Oecologia 59: 417–418.
Caldwell, M.M., H.P. Meister, J.D. Tenhunen, and O.L. Lange, 1986. Canopy structure, light microclimate and leaf gas exchange of *Quercus coccifera* L. in a Portuguese macchia: measurements in different canopy layers and simulations with a canopy model. Trees Structure and Function 1: 25–41.
Campbell, G., 1986. Extinction coefficients for radiation in plant canopies calculated using an ellipsoidal inclination angle distribution. Agricultural and Forest Meteorology 36: 317–321.
Cermak, J., M. Deml, and M. Penka, 1973. A new method of sap flow rate determination in trees. Biologia Plantarum 15: 171–178.
Cescatti, A., 1997. Modelling the radiative transfer in discontinuous canopies of asymmetric crowns. I. Model structure and algorithms. Ecological Modelling 101: 263–274.
Cescatti, A. and R. Zorer, 2003. Structural acclimation and radiation regime of silver fir (Abies alba Mill.) shoots along a light gradient. Plant Cell and Environment 26: 429–442.
Charles-Edwards, D.A., and J.H.M. Thornley, 1973. Light interception by an isolated plant. A simple model. Annals of Botany 37: 919–928.
Charles-Edwards, D.A., D. Doley, and G.M. Rimmington, 1986. Modelling Plant Growth and Development. Academic Press, Sydney.
Chen, D.-X., M.B. Coughenour, A.K. Knapp, and C.E. Owensby, 1994. Mathematical simulation of C_4 grass photosynthesis in ambient and elevated CO_2. Ecological Modelling 73: 63–80.
Collatz, G.J., M. Ribas-Carbo, and J.A. Berry, 1992. Coupled photosynthesis-stomatal conductance model for leaves of C_4 plants. Australian Journal of Plant Physiology 19: 519–538.
Connor, D.J. and E. Fereres, 1999. A dynamic model of crop growth and partitioning of biomass. Field Crops Research 63: 139–157.
Czaran, T. and S. Bartha, 1989. The effect of spatial pattern of community dynamics: A comparison of simulated and field data. Vegetatio 83: 229–239.
Dai, Y.J., R.E. Dickinson, and Y.P. Wang, 2004. A two-big-leaf model for canopy temperature, photosynthesis, and stomatal conductance. Journal of Climate 17: 2281–2299.
Deblonde, G., M. Penner, and A. Royer, 1994. Measuring leaf area index with the LI-COR LAI-2000 in pine stands. Ecology 75: 1507–1511.
Dougherty, R.L., J.A. Bradford, P.I. Coyne, and P.L. Sims, 1994. Applying an empirical model of stomatal conductance to three C-4 grasses. Agricultural and Forest Meteorology 67: 269–290.
Drouet, J.L. and R. Bonhomme, 2004. Effect of 3D nitrogen, dry mass per area and local irradiance on canopy photosynthesis within leaves of contrasted heterogeneous maize crops. Annals of Botany 93: 699–710.
Dugas, W.A., L.J. Fritschen, L.W. Gay, A.A. Held, A.D. Matthias, D.C. Reicosky, P. Steduto, and J.L. Steiner, 1991. Bowen ratio, eddy correlation, and portable chamber measurements of sensible and latent heat flux over irrigated spring wheat. Agricultural and Forest Meteorology 56: 1–20.
Duncan, W.C., R.S. Loomis, W.A. Williams, and R. Hanau, 1967. A model for simulating photosynthesis in plant communities. Hilgardia 38: 181–205.
Dwyer, M.J., E.G. Patton, and R.H. Shaw, 1997. Turbulent kinetic energy budgets from a large-eddy simulation of airflow above and within a forest canopy. Boundary Layer Meteorology 84: 23–43.

Dye, P.J. and B.W. Olbrich, 1993. Estimating transpiration from 6-year-old *Eucalyptus grandis* trees: Development of a canopy conductance model and comparison with independent sap flux measurements. Plant, Cell and Environment 16: 45–53.

Eckstein, J., W. Beyschlag, K.A. Mott, and R.J. Ryel, 1996. Changes in photon flux can induce stomatal patchiness. Plant, Cell and Environment 19: 1066–1074.

Evans, J.R. and J.R. Seemann, 1984. Differences between wheat genotypes in specific activity of ribulose-1,5-bisphosphate carboxylase and the relationship to photosynthesis. Plant Physiology 74: 759–765.

Falge, E., W. Graber, R. Siegwolf, and J.D. Tenhunen, 1995. A model of the gas exchange response of *Picea abies* to habitat conditions. Trees Structure and Function 10: 277–287.

Falge, E., R.J. Ryel, M. Alsheimer, and J.D. Tenhunen, 1997. Effects of stand structure and physiology on forest gas exchange: A simulation study for Norway spruce. Trees Structure and Function 11: 436–448.

Falge, E., J.D. Tenhunen, R. Ryel, M. Alsheimer, and B. Kostner, 2000. Modelling age- and density-related gas exchange of *Picea abies* canopies in the Fichtelgebirge, Germany. Annals of Forest Science 57: 229–243.

Fan, S.-M., M.L. Goulden, J.W. Munger, B.C. Daube, P.S. Bakwin, S.C. Wofsy, J.S. Amthor, D.R. Fitzjarrald, K.E. Moore, and T.R. Moore, 1995. Environmental controls on the photosynthesis and respiration of a boreal lichen woodland: A growing season of whole-ecosystem exchange measurements by eddy correlation. Oecologia 102: 443–452.

Farqhuar, G.D. and S. von Caemmerer, 1982. Modelling of photosynthetic response to environmental conditions. In: O.L. Lange, P.S. Nobel, C.B. Osmond, and H. Ziegler, eds. Encyclopedia of Plant Physiology, New Series, vol. 12B. Springer Verlag, Berlin, Heidelberg, New York, pp. 549–587.

Farquhar, G.D., S. von Caemmerer, and J.A. Berry, 1980. A biochemical model of photosynthetic CO_2 assimilation in leaves of C_3 species. Planta 149: 78–90.

Field, C.B., 1991. Ecological scaling of carbon gain to stress and resource availability. In: H.A. Mooney, W.E. Winner, and E.J. Pell, eds. Integrated Responses of Plants to Stress. Academic Press, San Diego, CA, pp. 35–65.

Field, C.B., J.A. Gamon, and J. Peñelas, 1994. Remote sensing of terrestrial photosynthesis. In: E.-D. Schulze and M.M. Caldwell, eds. Ecophysiology of Photosynthesis. Ecological Studies Vol. 100. Springer Verlag, Berlin, Heidelberg, New York, pp. 511–527.

Forrester, J.W., 1961. Industrial Dynamics. MIT Press, Cambridge, MA.

Fukai, S. and R.S. Loomis, 1976. Leaf display and light environments in row-planted cotton communities. Agricultural Meteorology 17: 353–379.

Gates, D.M., 1980. Biophysical Ecology. Springer Verlag, Berlin, Heidelberg, New York.

Gower, S.T. and J.M. Norman, 1991. Rapid estimation of leaf area index in conifer and broad-leaf plantations. Ecology 72: 1896–1900.

Grace, J.C., P.G. Jarvis, and J.M. Norman, 1987a. Modelling the interception of solar radiant energy in intensively managed stands. New Zealand Journal of Forestry Science 17: 193–209.

Grace, J.C., D.A. Rook, and P.M. Lane, 1987b. Modelling canopy photosynthesis in *Pinus radiata* stands. New Zealand Journal of Forestry Science 17: 210–228.

Harley, P.C., J.D. Tenhunen, and O.L. Lange, 1986. Use of an analytical model to study limitations on net photosynthesis in *Arbutus unedo* under field conditions. Oecologia 70: 393–401.

Harley, P.C., R.B. Thomas, J.F. Reynolds, and B.R. Strain, 1992. Modelling photosynthesis of cotton grown in elevated CO_2. Plant, Cell and Environment 15: 271–282.

Hoffmann, F., 1995. FAGUS, a model for growth and development of beech. Ecological Modelling 83: 327–348.

Hogeweg, P., B. Hesper, C.P. van Schail, and W.G. Beeftink,1985. Patterns in vegetation succession, an ecomorphological study. In J. White, ed. The Population Structure of Vegetation. Dr. W. Junk Publishers, Dordrecht, pp. 637–666.

Innis, G.S., 1974. A spiral approach to ecosystem simulation, I. In G.S. Innis and R.V. O'Neill, eds. Systems Analysis of Ecosystems. International Cooperative Publishing House, Fairland, ML, pp. 211–386.

Ishida, T., G.S. Campbell, and C. Calissendorff, 1991. Improved heat balance method for determining sap flow rate. Agricultural and Forest Meteorology 56: 35–48.

Jarvis, P.G., 1976. The interpretation of the variations in leaf water potential and stomatal conductance found in canopies in the field. Philosophical Transactions of the Royal Society of London B 273: 593–610.

Jarvis, P.G., 1995. Scaling processes and problems. Plant, Cell and Environment 18: 1079–1089.

Johnson I.R. and J.H.M. Thornley, 1985. Dynamic model of the response of a vegetative Grass crop to light temperature and nitrogen. Plant, Cell and Environment 8: 485–499.

Johnson, F., H. Eyring, and R. Williams, 1942. The nature of enzyme inhibitions in bacterial luminescence; sulfanilamide, urethane, temperature, and pressure. Journal of Cell Comparative Physiology 20: 247–268.

Johnson I.R., Parsons, A.J., and M.M. Ludlow, 1989. Modelling photosynthesis in monocultures and mixtures. Australian Journal of Plant Physiology 16: 501–516.

Jones, H.G. and K.H. Higgs, 1989. Empirical models of the conductance of leaves in apple orchards. Plant, Cell and Environment 12: 301–308.

Jordan, D.B. and W.L. Ogren, 1984. The CO_2/O_2 specificity of ribulose 1,5-bisphosphate carboxylase/oxygenase: Dependence on ribulose-bisphosphate concentration, pH and temperature. Planta 161: 308–313.

Keys, A.J., 1986. Rubisco: Its role in photorespiration. Proceedings of the Royal society of London B 313: 325–336.

Kitanidis, P.K. and R.L. Bras, 1980. Real-time forecasting with a conceptual hydrologic model. 2. Applications and results. Water Resource Research 16: 1034–1044.

Kull, O. and P.G. Jarvis, 1995. The role of nitrogen in a simple scheme to scale up photosynthesis from leaf to canopy. Plant, Cell and Environment 18: 1174–1182.

Kuuluvainen, T. and T. Pukkala, 1987. Effect of crown shape and tree distribution on the spatial distribution of shade. Agricultural and Forest Meteorology 40: 215–231.

Lang, A.R.G., Y. Xiang, and J.M. Norman, 1985. Crop structure and the penetration of direct sunlight. Agricultural and Forest Meteorology 35: 83–101.

Leegood, R.C., M.D. Adcock, and H.D. Doncaster, 1989. Analysis of the control of photosynthesis in C_4 plants by changes in light and carbon dioxide. Philosophical Transactions of the Royal Society of London B 323: 339–355.

Lemeur, R., 1973. A method for simulating the direct solar radiation regime of sunflower, Jerusalem artichoke, corn and soybeans using actual stand structural data. Agricultural Meteorology 12: 229–247.

Leuning, R., 1983. Transport of gases into leaves. Plant, Cell and Environment 6: 181–194.

Leuning, R., 1995. A critical appraisal of a combined stomatal-photosynthesis model for C_3 plants. Plant, Cell and Environment 18: 339–355.

Leuning, R., F.M. Kelliher, D.G.G. dePury, and E.-D. Schulze, 1995. Leaf nitrogen, photosynthesis, conductance and transpiration: scaling from leaves to canopies. Plant, Cell and Environment 18: 1183–1200.

Lloyd, J., 1991. Modelling stomatal responses to environment in *Macadamia integrifolia*. Australian Journal of Plant Physiology 18: 649–660.

Mann, J.E., G.L. Curry, and P.J.H. Sharp, 1979. Light penetration by isolated plants. Agricultural Meteorology 20: 205–214.

McNaughton, K.G., 1994. Effective stomatal and boundary-layer resistances of heterogeneous surfaces. Plant, Cell and Environment 17: 1061–1068.

Medlyn, B., D. Barrett, J. Landsberg, P. Sands, and R. Clement, 2003. Conversion of canopy intercepted radiation to photosynthate: Review of modelling approaches for regional scales. Functional Plant Biology 30: 153–169.

Mott, K.A. and D.F. Parkhurst, 1991. Stomatal responses to humidity in air and helox. Plant, Cell and Environment 14: 509–515.

Niinemets, Ü., 1997. Distribution patterns of foliar carbon and nitrogen as affected by tree dimensions and relative light conditions in the canopy of *Picea abies*. Trees Structure and Function 11: 144–154.

Nobel, P., 1983. Biophysical Plant Physiology and Ecology. W.H. Freeman and Company, New York.

Norman, J.M., 1979. Modeling the complete crop canopy. In: B.J. Barfield and J.F. Gerber, eds. Modification of the Aerial Environment of Crops. American Society of Agricultural Engineers, St. Joseph, MI, pp. 249–277.

Norman, J.M. and J.M. Welles, 1983. Radiative transfer in an array of canopies. Agronomy Journal 75: 481–488.

Norman, J.M. and G.S. Campbell, 1989. Canopy structure. In: R.W. Pearcy, J.R. Ehleringer, H.A. Mooney, and P.W. Rundel, eds. Plant Physiological Ecology. Field Methods and Instrumentation. Chapman and Hall, London, New York, pp. 301–325.

Oker-Blom, P., 1985. The influence of penumbra on the distribution of direct solar radiation in a canopy of scots pine. Photosynthetica 19: 312–317.

Oker-Blom, P. and S. Kellomäki, 1982a. Effect of angular distribution of foliage on light absorption and photosynthesis in the plant canopy: theoretical computations. Agricultural Meteorology 26: 105–116.

Oker-Blom, P. and S. Kellomäki, 1982b. Theoretical computations on the role of crown shape in the absorption of light by forest trees. Mathematical Biosciences 59: 291–311.

Oker-Blom, P., T. Pukkala, and T. Kuuluvainen, 1989. Relationship between radiation interception and photosynthesis in forest canopies: effect of stand structure and latitude. Ecological Modelling 49: 73–87.

Pearcy, R.W., L.J. Gross, and D. He, 1997. An improved dynamic model of photosynthesis for estimation of carbon gain in sunfleck light regimes. Plant, Cell and Environment 20: 411–424.

Pearcy, R.W. and W.A. Pfitsch, 1994. The consequences of sunflecks for photosynthesis and growth of forest understory plants. In: E.-D. Schulze and M.M. Caldwell, eds. Ecophysiology of Photo synthesis. Ecological Studies Vol. 100. Springer Verlag, Berlin, Heidelberg, New York, pp. 343–359.

Pearcy, R.W. and D.A. Sims, 1994. Photosynthetic acclimation to changing light environments: Scaling from the leaf to the whole plant. In: M.M. Caldwell and R.W. Pearcy, eds. Exploitation of Environmental Heterogeneity by Plants. Academic Press, San Diego, CA, pp. 145–174.

Peisker, M. and S.A. Henderson, 1992. Carbon: terrestrial C_4 plants. Plant, Cell and Environment 15: 987–1004.

Perry, S.G., A.B. Fraser, D.W. Thompson, and J.M. Norman, 1988. Indirect sensing of plant canopy structure with simple radiation measurements. Agricultural and Forest Meteorology 42: 255–278.

Pospisilová, J. and J. Šantrucek, 1994. Stomatal patchiness. Biologia Plantarum. 36: 421–453.

de Pury, D.G.G. and G.D. Farquhar, 1997. Simple scaling of photosynthesis from leaves to canopies without the errors of big-leaf models. Plant, Cell and Environment 20: 537–557.

Raulier, F., P.Y. Bernier, and C.H. Ung, 1999. Canopy photosynthesis of sugar maple (*Acer saccharum*): Comparing big-leaf and multilayer extrapolations of leaf-level measurements. Tree Physiology 19: 407–420.

Reynolds, J.F., B. Acock, R.L. Dougherty, and J.D. Tenhunen, 1987. A modular structure for plant growth simulation models. In: J.S. Pereira. and J.J. Landsberg, eds. Biomass Production by Fast Growing Trees. Kluwer Academic Publishers, Dordrecht, pp. 123–34.

Reynolds, J.F., J.L. Chen, P.C. Harley, D.W. Hilbert, R.L. Dougherty, and J.D. Tenhunen, 1992. Modeling the effects of elevated CO_2 on plants: Extrapolating leaf response to a canopy. Agricultural and Forest Meteorology 61: 69–94.

Roden, J.S., 2003. Modeling the light interception and carbon gain of individual fluttering aspen (Populus tremuloides Michx) leaves. Trees Structure and Function 17: 117–126.

Roden, J.S. and R.W. Pearcy, 1993. The effect of flutter on the temperature of poplar leaves and its implications for carbon gain. Plant, Cell and Environment 16: 571–577.

Ross, J., 1975. Radiative transfer in plant communities. In: J.W. Monteith, ed. Vegetation and the Atmosphere. Academic Press, London, New York, San Francisco, pp. 13–55.

Ruimy, A., P.G. Jarvis, D.D. Baldocchi, and B. Saugier, 1995. CO_2 fluxes over plant canopies and solar radiation: A review. Advances in Ecological Research 26: 1–68.

Running, S.W., P.E. Thornton, R. Nemani, and J.M. Glassy, 2000. Global terrestrial gross and net primary productivity from the Earth Observing System. In: O. Sala, R. Jackson, and H. Mooney, eds. Methods in Ecosystem Science. Springer Verlag, New York, pp. 44–57.

Ryel, R.J. and W. Beyschlag, 1995. Benefits associated with steep foliage orientation in two tussock grasses of the American Intermountain West. A look at water-use-efficiency and photoinhibition. Flora 190: 251–260.

Ryel, R.J., P.W. Barnes, W. Beyschlag, M.M. Caldwell, and S.D. Flint, 1990. Plant competition for light analyzed with a multispecies canopy model. I. Model development and influence of enhanced UV-B conditions on photosynthesis in mixed wheat and wild oat canopies. Oecologia 82: 304–310.

Ryel, R.J., W. Beyschlag, and M.M. Caldwell, 1993. Foliage orientation and carbon gain in two tussock grasses as assessed with a new whole-plant gas-exchange model. Functional Ecology 7: 115–124.

Ryel, R.J., W. Beyschlag, and M.M. Caldwell, 1994. Light field heterogeneity among tussock grasses: Theoretical considerations of light harvesting and seedling establishment in tussocks and uniform tiller distributions. Oecologia 98: 241–246.

Ryel, R.J., W. Beyschlag, B. Heindl, and I. Ullmann, 1996. Experimental studies of the competitive balance between two Central European roadside grass with different growth forms. 1. Field experiments on the effects of mowing and maximum leaf temperatures on competitive ability. Botanica Acta 109: 441–448.

Ryel, R.J, E. Falge, U. Joss, R. Geyer, and J.D. Tenhunen, 2001. Penumbral and foliage distribution effects on *Pinus sylvestris* canopy gas exchange. Theoretical and Applied Climatology 68: 109–124.

Ryel, R.J., M.M. Caldwell, C.K. Yoder, D. Or, and A.J. Leffler. 2002. Hydraulic redistribution in a stand of *Artemisia tridentata*: evaluation of benefits to transpiration assessed with a simulation model. Oecologia 130: 173–184.

Sage, R.F. 1990. A model describing the regulation of ribulose-1,5-bisphosphate carboxylase, electron transport, and triose-phosphate use in response to light intensity and CO_2 in C_3 plants. Plant Physiology 94: 1728–1734.

Schreiber, U., W. Bilger, and C. Neubauer. 1994. Chlorophyll fluorescence as a nonintrusive indicator for rapid assessment if *in vivo* photosynthesis. In: Schulze, E.-D., and M.M. Caldwell, eds. Ecophysiology of Photosynthesis. Ecological Studies Vol. 100. Springer Verlag, Berlin, Heidelberg, New York, pp. 49–70.

Schuepp, P.H., 1993. Leaf boundary layers. Tansley Review No. 59. New Phytologist 125: 477–507.

Sellers, P.J., J.A. Berry, G.J. Collatz, C.B. Field, and F.G. Hall. 1992. Canopy reflectance, photosynthesis and transpiration. III. A reanalysis using improved leaf models and a new canopy integration scheme. Remote Sensing Environment 42: 187–216.

Sharkey, T.D. 1985. Photosynthesis in intact leaves of C_3 plants: physics, physiology and rate limitations. The Botanical Review 51: 53–105.

Silvertown, J., S. Holtier, J. Johnson, and P. Dale, 1992. Cellular automaton models of interspecific competition for space—the effect of pattern on process. Journal of Ecology 80: 527–534.

Sinclair, R. and B.A. Thomas, 1970. Optical properties of leaves of some species in arid south Australia. Australian Journal of Botany 18: 261–273.

Smith, E., 1937. The influence of light and carbon dioxide on photosynthesis. General Physiology 20: 807–830.

Spitters, C.J.T., H.A.J.M. Toussaint, and J. Goudriaan, 1986. Separating the diffuse and direct component of global radiation and its implications for modeling canopy photosynthesis. Part I. Components of incoming radiation. Agricultural and Forest Meteorology 38: 217–229.

Steinhorst, R.K., H.W. Hunt, G.S. Innis, and K.P. Haydock, 1978. Sensitivity analyses of the ELM model. In: Innis, G.S., ed. Grassland Simulation Model. Ecological Studies Vol. 26. Springer Verlag, Berlin, Heidelberg, New York, pp. 231–255.

Stenberg, P., S. Linder, H. Smolander, and J. Flowerellis, 1994. Performance of the LAI-2000 plant canopy analyzer in estimating leaf area Index of some Scots pine stands. Tree Physiology 14: 981–995.

Tenhunen, J.D., P.C. Harley, W. Beyschlag, and O.L. Lange, 1987. A model of net photosynthesis for leaves of the sclerophyll *Quercus coccifera*. In: J.D. Tenhunen, F. Catarino, O.L. Lange, and W.C. Oechel, eds. Plant Response to Stress—Functional Analysis in Mediterranean Ecosystems. Springer Verlag, Berlin, Heidelberg, New York, pp. 339–354.

Tenhunen, J.D., R.A. Siegwolf, and S.F. Oberbauer, 1994. Effects of phenology, physiology and gradient in community composition, structure and microclimate on tundra ecosystem CO_2 exchange. In: E.-D. Schulze and M.M. Caldwell, eds. Ecophysiology of Photosynthesis. Ecological Studies Vol. 100. Springer-Verlag, Berlin, Heidelberg, New York, pp. 431–460.

Thornley, J.H.M., and I.R. Johnson, 1990. Plant and Crop Modelling. A Mathematical Approach to Plant and Crop Physiology. Clarendon Press, Oxford.

Tongeren, O. van and I.C. Prentice, 1986. A spatial simulation model for vegetation dynamics. Vegetation 65: 163–173.

Tuzet, A., A. Perrier, and R. Leuning, 2003. A coupled model of stomatal conductance, photosynthesis and transpiration. Plant Cell and Environment 26: 1097–1116.

Ustin, S.L., M.O. Smith, J.B. Adams, 1993. Remote sensing of ecological processes: A strategy for developing and testing ecological models using spectral misture analysis. In: J.R. Ehleringer and C.B. Field, eds. Scaling Physiological Processes: Leaf to Globe. Academic Press, San Diego, CA, pp. 339–357.

Verma, S.B., 1990. Micrometeorological methods for measuring surface fluxes of mass and energy. In: N.S. Goel and J.M. Norman, eds. Instrumentation for Studying Vegetation Canopies for Remote Sensing in Optical and Thermal Infrared Regions. Remote Sensing Reviews Vol. 5. Harwood Academic Publishers, Chur, pp. 99–115.

Von Caemmerer, S. and G.D. Farquhar, 1981. Some relationships between the biochemistry of photosynthesis and the gas exchange of leaves. Planta, 153: 376–387.

Walker, G.K., R.E. Blackshaw, and J. Dekker, 1988. Leaf area and competition for light between plant species using direct sunlight transmission. Weed Technology 2: 159–165.

Wang, Y.P., 2000. A refinement to the two-leaf model for calculating canopy photosynthesis. Agricultural and Forest Meteorology 101: 143–150.

Wang, Y.P. and P.G. Jarvis. 1990. Description and validation of an array model—MAESTRO. Agricultural and Forest Meteorology 51: 257–280.

Wang, Q., S. Adiku, J. Tenhunen, and A. Granier, 2005. On the relationship of NDVI with leaf area index in a deciduous forest site. Remote Sensing of Environment 94: 244–255.

Warren Wilson, J., 1960. Inclined point quadrats. New Phytologist 58: 1–8.

Warren Wilson, J., 1963. Estimation of foliage denseness and foliage angle by inclined point quadrats. Australian Journal of Botany 11: 95–105.

Webb, W.L. 1991. Atmospheric CO_2, climate change, and tree growth: a process model I. Model structure. Ecological Modelling 56: 81–107.

Werner C., R.J. Ryel, O. Correia, and W. Beyschlag, 2001. Effects of photoinhibition on whole-plant carbon gain assessed with a photoinhibition model. Plant, Cell and Environment 24: 27–40.

Whitehead, D., D.V.W. Okali, and F.E. Fasehun, 1981. Stomatal response to environmental variables in two tropical rainforest species during the dry season in Nigeria. Journal of Applied Ecology 18: 581–587.

Wolfram, S., 1983. Statistical mechanics of cellular automata. Reviews of Modern Physics 55: 601–643.

22 Ecological Applications of Remote Sensing at Multiple Scales

John A. Gamon, Hong-Lie Qiu, and Arturo Sanchez-Azofeifa

CONTENTS

"I'm ruler," said Yertle, "of all that I see.
But I don't see enough. That's the trouble with me.
With this stone for a throne, I look down on my pond
But I cannot look down on the places beyond.
This throne that I sit on, is too, too low down.
It ought to be higher!" he said with a frown.
"If I could sit high, how much greater I'd be!
What a king! I'd be ruler of all that I see!"

—Geisel (1950)

INTRODUCTION

Like Yertle, we are restrained by our senses, which often limit our view to a particular favorite or accessible geographic location. Consequently, much ecological research has traditionally focused on specific organisms, populations, and communities defined by geographic region, and ecology is largely a collection of case histories in search of unifying principles. However, the growing human pressures on the planet's resources are altering ecosystem function at regional to global scales and placing a new urgency on studies that integrate information across spatial and temporal scales. Ecologists are now faced with the challenge of measuring and understanding ecological processes at these multiple scales. Answering this challenge necessarily involves the use of remote sensing, which has the unique ability to provide an objective, synoptic view of the earth and its atmosphere.

In this chapter, remote sensing is primarily defined as the measurement of electromagnetic radiation (reflectance, fluorescence, or longwave emission) with noninvasive sampling (Figure 22.1). Remote implies measurement from a distance (e.g., from an airborne or satellite platform), and many of the most spectacular examples have been from these great distances. However, the fundamental tools and principles of radiation measurement can be applied at almost any scale. Indeed, one of the strengths of optical sampling is that, unlike many other measurement methods, it is eminently scaleable—it can be applied at many spatial scales (from a leaf to the entire globe). It can also be used to examine how information changes with scale, lending insight into processes controlling the interaction between ecosystems and radiation at different scales (Wessman 1992, Ustin et al. 1993, Foody and Curran 1994, Quattrochi and Goodchild 1997). This ability to bridge scales allows us to extend our otherwise limited perception to new domains and to reach a new understanding of complex ecological phenomena. Remote sensing has additional virtues, including the ability to sample nondestructively and without direct contact, thus avoiding the common problem of disturbing or destroying the object of measurement. Because digital remote sensing provides a consistent

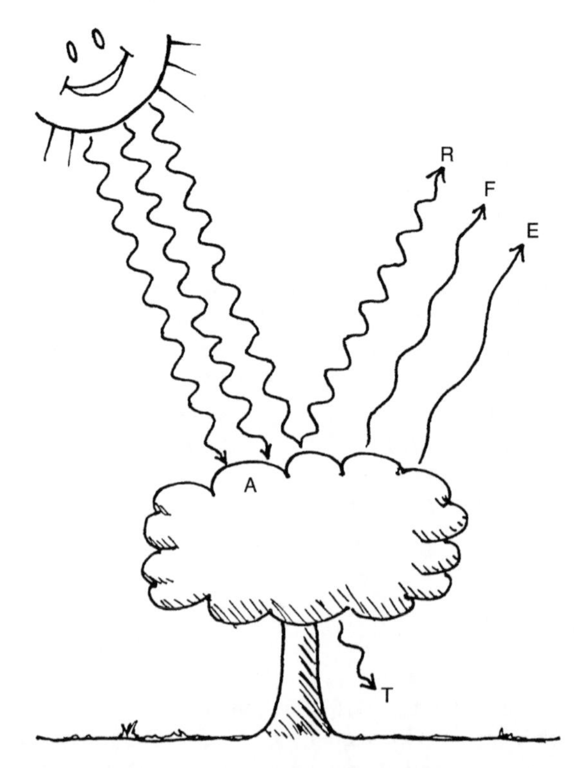

FIGURE 22.1 Schematic illustrating radiation sampled with passive remote sensing. Solar radiation can be absorbed (A), transmitted (T), or reflected (R). Absorbed radiation can be released rapidly as fluorescence (F) or more slowly as long-wave, thermal emission (E). Remotely positioned sensors can infer information about surface features by detecting patterns of reflected, transmitted, fluoresced, or emitted radiation. In a real landscape, multiple scattering and signals from multiple scene components (not illustrated) significantly influence the pattern of radiation detected.

data format, it provides a degree of objectivity that is lacking in many other methods of data collection.

Remote sensing per se is not new. Properly positioned, the eye is a powerful (if subjective) remote-sensing device. Aerial photography has been used for many decades for mapping and reconnaissance. Continued advances in digital and optical technology are providing ever more powerful tools for collecting and analyzing remotely sensed data, and a vast array of sensors are now available for use by the ecological community. These technological advances are redefining the questions that can be addressed by ecologists and are contributing to paradigm shifts in both the concepts and methodology of ecology. Much of what is currently new about remote sensing is the way in which investigators are finding innovative ways of using these tools, often in combination with other methods, to address ecological questions at multiple scales.

The rapid pace of advance in this field, combined with the uncertain future of many current and planned sensors, makes a comprehensive review a difficult, if not impossible, task. Remote sensing is now used to infer a dizzying array of earth surface and atmospheric properties and processes, including surface temperature, moisture, topography, albedo, mineral composition, vegetation cover and type, vegetation dynamics and land use change, atmospheric composition, irradiance, and surface-atmosphere fluxes. Some of these (e.g., surface-atmosphere fluxes) require the incorporation of remotely sensed measurements with models, whereas others (e.g., surface temperature and vegetation cover) can be assessed more directly. Clearly, a full survey of these applications is well beyond the scope of this chapter. For a more thorough coverage of remote sensing applications, the interested reader might pursue current information on the Internet or refer to any of a number of references and reviews on the topic (Lillesand and Kiefer 1987, Hobbs and Mooney 1990, Richards 1993, Ustin et al. 1993, 2004, Foody and Curran 1994, Danson and Plummer 1995, Gholz et al. 1997, Kasischke et al. 1997, Quattrochi and Goodchild 1997, Sabins 1997). Instead of a comprehensive survey, we have chosen to present a few specific applications that illustrate the range, power, and potential of remote sensing for addressing ecological questions at several levels of inquiry. We selected a particular focus on terrestrial vegetation, and the associated stocks and fluxes of carbon. This topic is of critical concern today because human activities are perturbing vegetation cover, composition, and the carbon cycle in significant and measurable ways (Vitousek 1994, Amthor 1995, Houghton 1995). A central goal of global ecology is to clearly define vegetation-atmosphere fluxes in the context of these perturbations.

The chapter begins with a short presentation of a few remote sensing fundamentals. It then illustrates recent examples of remote sensing as a mapping tool, perhaps the most obvious and traditional of remote sensing applications. Newer extensions of this mapping capability are now explicitly considering spectral and temporal aspects of landscapes, and linking this information to process models. We also examine some examples of the application of remote sensing to models of terrestrial carbon flux and consider how new developments offer to improve our understanding of carbon stocks and fluxes. The chapter ends with suggestions for improving the application of remote sensing within the ecological community. The goal is to encourage ecologists to continue exploring innovative ways of applying this powerful and exciting technology.

FUNDAMENTALS

SENSORS

Remote sensing typically involves noncontact measurement of electromagnetic radiation and the inference of patterns and processes from these measurements. Sensors can be classified as imaging and nonimaging instruments (Table 22.1). Imaging sensors view a given ground area (scene) with a characteristic array of detectors or pixels, each covering a specific ground area (instantaneous field of view or IFOV). Sensors can be further divided into passive or active sensors, both of which have advantages and disadvantages. Passive sensors rely on external (typically solar) energy sources, and generally sample during daylight (Figure 22.1). A notable exception to daytime application is the use of passive sensors for monitoring nighttime energy use (Elvidge et al. 1997). Passive sensors can, in principle, sample any wavelength of radiation, provided the energy level is high enough for an adequate signal-to-noise ratio. However, in practice, a given detector is usually restricted to a specific wavelength range. For example, silicon photodiodes are generally sensitive to the near-UV to near-IR range (approximately 350–1100 nm; Pearcy 1989), which corresponds to the spectral region of strongest solar irradiance (Nobel 1991). Other sensors augment silicon detectors with other detectors to achieve a wider spectral range, often extending into the short-wave infrared (up to 2500 nm, Table 22.1).

Atmospheric absorption, primarily by water vapor and other atmospheric gases in specific infrared wavelengths, and atmospheric scattering, particularly by aerosols in the blue and near-UV wavelengths, reduce the usable spectral region when sampling from aircraft or satellite. Because passive detectors are directly dependent on the solar source, they work poorly in some circumstances (e.g., low light or cloudy conditions)—a serious limitation in certain parts of the world, including many polar or tropical regions. By contrast, active sensors can operate independently of light conditions because they sample the return of a signal originating from the instrument. Laser altimeters (light detecting and ranging, or LIDAR) and active microwave (radar) sensors provide good examples of active sensors (Table 22.1). Radar sensors offer the additional advantage of cloud penetration, so are independent of light or weather conditions, and are now emerging as useful sensors for a variety of ecological, hydrological, and topographic studies (Hess and Melack 1994, Kasischke et al. 1997, Smith 1997). LIDAR sensors are now widely applied for detailed studies of forest structure and biomass (Lefsky et al. 2002).

Concept of Scale

Issues of scale must be carefully considered when applying remote sensing to biological or ecological processes. Scale in remote sensing has several dimensions, including the spatial, spectral, and temporal dimensions, each of which provides a rich source of ecologically relevant information (Figure 22.2 and Figure 22.3). Because ecological processes occur over definable time frames and geographic regions, often with characteristic spectral signatures, detectability of a given process varies with the temporal, spatial, and spectral scales of measurement. Although the principles of detection operate similarly at all scales, the actual information content changes with scale, and consequently the methods or tools of interpretation often have to change to match the question at hand. For a variety of physical and biological reasons, these three dimensions tend to be interdependent. At coarser geographic scales (large pixel sizes, grain sizes, or IFOV), processes tend to be detectable over larger time frames, whereas many processes at fine spatial scale (small pixel sizes, grain sizes, or IFOV) are best detected over short time periods. Many fine-scale properties and processes can only be resolved with narrow spectral bands (Figure 22.3). For example, changes in certain key biochemical processes involving photosynthetic regulation of leaves via the xanthophyll cycle can be best detected over seconds to minutes only with narrow spectral bands (Gamon and Surfus 1999); these processes may simply not be relevant or detectable, or the additional presence of other signals may confound their interpretation as the sampling scale changes (Barton and North 2001). By contrast, changes in green vegetation cover over large regions can be readily detected using broad

TABLE 22.1
Selected Sensors Useful for Ecological Applications

Platform	Sensor	Status	No. of Bands	Approximate Pixel Size (m)	Wavelength Range	Source
Satellite	AVHRR	In orbit	5	1100	0.58–12.5 μm	NOAA (USA)
	Landsat Multispectral Scanner (MSS)		4 vis-NIR	79	0.5–1.1 μm	EOSAT, Lanham, MD
	Landsat Thematic Mapper (TM)	In orbit	6 vis-NIR	30	0.45–2.35 μm	EOSAT, Lanham, MD
			1 TIR	120	10.40–12.50 μm	
	Landsat Enhanced Thematic Mapper Plus (ETM+)	In orbit	6 vis-NIR	30	0.45–1.75 μm	USGS, Reston, VA
			1 TIR	60	10.40–12.50 μm	
			1 Pan.	15	0.52–0.90 μm	
	SPOT 1–3		3 vis-NIR	20	0.48–0.71 μm	SPOT Corporation
			1 Pan.	10	0.50–0.73 μm	
	SPOT 4		4 vis-MIR	20	0.50–1.75 μm	SPOT Corporation
			1 Mono.	10	0.61–0.68 μm	
	SPOT 5	In orbit	3 vis-NIR	10	0.50–0.89 μm	SPOT Corporation
			1 MIR	20	1.58–1.75 μm	
			1 Pan.	2.5 or 5.0	0.48–0.71 μm	
	Moderate resolution Imaging spectrometer (MODIS)	In orbit	36	250–1000	0.4–14.4	NASA (USA)
	Hyperion	In orbit	220	30	0.4–2.5 μm	NASA (USA)
Space Shuttle	Shuttle Imaging Radar (SIR)	Flew 1994		Variable	3.0 cm 6.0 cm 24.0 cm	NASA (USA)
Airborne	AVIRIS	Operational	224	20	400–2500 nm	NASA (USA)
	CASI (spectral mode)	Commercially available				ITERS, Calgary, Alberta (Canada)
	Scanning Lidar Imager of Canopies by Echo Recovery (SLICER, laser alimeter)	Operational	1	Variable	1.06 μm	NASA (USA)
Personal spectrometers (diode array, nonimaging)	FieldSpec FR	Commercially available	1512–1582	N/A	350–2500 nm	Analytical Spectral Devices, Inc., Boulder, CO
	S2000	Commercially available	2048	N/A	200–1100 nm	Ocean Optics, Dunedin, FL
	UniSpec (VIS/NIR model)	Commercially available	256	N/A	300–1100 nm	PP Systems, Amesbury MA

Sources: Lillesand, T.M. and Kiefer, R.W. in *Remote Sensing and Image Interpretation*, 2nd edn, Wiley, New York, 1987; Richards, J.A. in *Remote Sensing Digital Image Analysis: An Introduction*, Springer-Verlag, Berlin, 1993; Jensen, J.R. in *Introductory Digital Image Processing: A Remote Sensing Perspective*, 2nd edn, Prentice Hall, Upper Saddle River, New Jersey, 1996; Sabins, F.F. in *Remote Sensing: Principles and Interpretation*, 3rd edn, W.H. Freeman and Company, New York, 1997; vendor literature for companies listed under "source".

Note: Except for SLICER and the personal spectrometers, all sensors are imaging instruments.

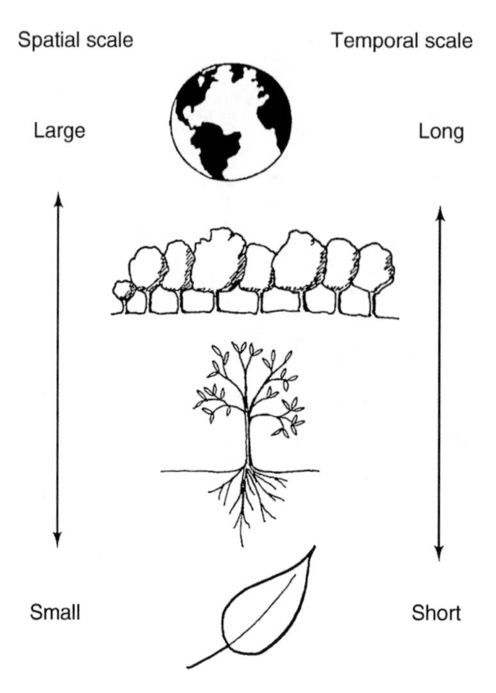

FIGURE 22.2 Interdependent spatial and temporal scales covered by remote sensing. At coarser geographic scales (large pixel sizes, grain sizes, or IFOV), processes tend to be detectable over larger time frames, whereas many processes at fine spatial scale (small pixel sizes, grain sizes, or IFOV) can often be detected over short time periods.

FIGURE 22.3 (See color insert following page 684.) True-color AVIRIS image for a region of Ventura County in southern California showing the Pacific Ocean (*bottom*), the western edge of the Santa Monica Mountains (*center right*), and developed and agricultural areas (*top left*). The different colored bands receding into the background illustrate the spectral dimension (Z dimension), in this case 224 different spectral bands (wavelengths), ranging from near-ultraviolet bands (foreground) to the infrared bands (background). Much of the information content of this image is present in this spectral dimension.

spectral bands, even at larger spatial scales. Sensor look angle and radiation polarization provide additional sampling dimensions that may be relevant in specific cases (Barnsley 1994). For example, certain structural features of vegetation are sensitive to polarization of optical radiation (Vanderbilt et al. 1990) or microwave radiation (Hess and Melack 1994).

Vegetation Indices

In terrestrial systems, vegetated and nonvegetated areas can be readily distinguished based on their contrasting reflectance patterns in the red and near-infrared (NIR) spectral regions (Figure 22.4). This contrast results from differential pigment absorption in the red and NIR wavebands, and illustrates the radiation requirements for photosynthesis. Photosynthetic pigments (primarily chlorophylls and carotenoids) absorb most effectively in the visible region (400–700 nm), in which energy is most abundant and strong enough to drive electron transport, yet weak enough to avoid excessive damage to biological molecules. By contrast, there is insufficient energy in the NIR to drive photosynthesis; consequently, photosynthetic pigments cannot use or absorb these wavelengths, and vegetation canopies effectively scatter (reflect and transmit) most NIR radiation.

Numerous vegetation indices have been developed that characterize the contrasting reflectance on either side of the red edge at 700 nm, including the normalized difference vegetation index (NDVI), the simple ratio (SR), the soil-adjusted vegetation index and its derivatives (SAVI, TSAVI, SARVI), the greenness vegetation index (GVI), the perpendicular vegetation index (PVI), and the enhanced vegetation index (EVI) (Perry and Lautenschlager 1984, Baret and Guyot 1991, Wiegand et al. 1991, Huete et al. 1997, 2002). Other indices are based on the slope or inflection point of the red edge (Curran et al. 1990, Johnson et al. 1994, Gitelson et al. 1996). Essentially, all of these indices collapse the full spectrum into a single, readily usable value that scales with green canopy cover, absorbed radiation, leaf area index

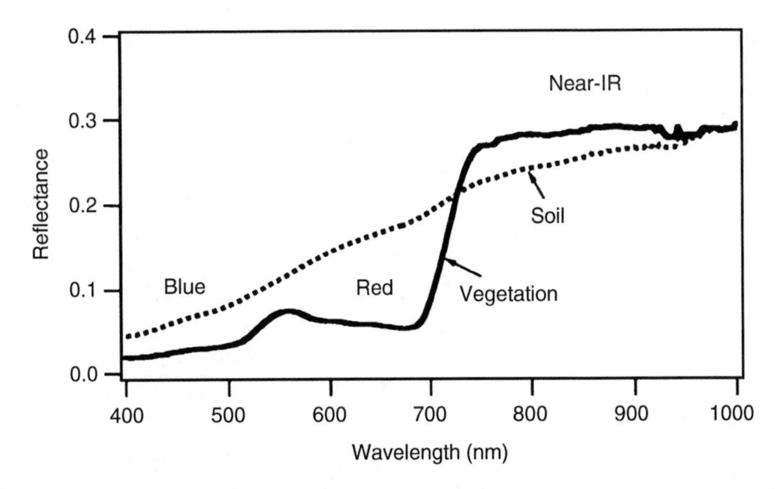

FIGURE 22.4 Typical reflectance spectra for green vegetation and bare soil in the visible (400–700 nm) to NIR (>700 nm) region. Chlorophyll absorbs well in the visible (particularly the blue and red regions), but not in the infrared, resulting in the characteristic vegetation red edge at 700 nm. This red edge feature is generally missing in nonphotosynthetic surfaces. The contrasting red and NIR reflectance is the basis for many vegetation indices that depict relative amounts of green vegetation cover. Note also the slight water absorption feature (dip) between 900 and 1000 nm, which can serve as an indicator of vegetation water content.

(LAI), or related measures of vegetation abundance. Of all these indices, the most commonly used index is undoubtedly the NDVI:

$$NDVI = \frac{R_{NIR} - R_{RED}}{R_{NIR} + R_{RED}}, \tag{22.1}$$

where R_{NIR} is the reflectance in the NIR region and R_{RED} is the reflectance (or radiance) in the red region (Figure 22.4).

Each of these vegetation indices suffers from certain well-discussed limitations (Holben 1986, Myneni and Asrar 1994, Myneni and Williams 1994, Sellers et al. 1996b). Many of these limitations arise because reflectance of any spectral region is necessarily influenced by multiple factors that can confound a simple interpretation of a given reflectance signature. Complex landscapes contain many components that can have multiple, overlapping effects on any single index, and the concept of a pure index becomes elusive, particularly at large spatial scales, where many landscape components contribute to the measured signal. Additional errors arise from atmospheric effects (absorption or scattering), variation in solar angle, and sensor calibration drifts. Added to this is the fact that each sensor has unique spectral characteristics (spectral response and bandwidth). For example, the red and NIR bands vary in width and position across sensors, meaning that, in reality, many NDVI formulations exist, leading to problems when comparing data across sensors. Consequently, it is often necessary to correct indices for these confounding factors, and this is now a common practice with global NDVI datasets (Goward et al. 1994, James and Kalluri 1994, Los et al. 1994, Sellers et al. 1994, 1996b, Townshend 1994). Despite these flaws, all of these vegetation greenness indices provide some measure of light absorption by green vegetation, and thus can be used to distinguish relative levels of energy capture for photosynthetic processes. This fundamental relationship is the basis for most models of photosynthetic carbon uptake driven by remote sensing, as further discussed later.

Some of the most potent and spectacular applications of remote sensing in ecology are derived from the ability to detect spectral features associated with specific, biologically important compounds (Table 22.2). Some of these compounds, notably photosynthetic pigments (chlorophylls and carotenoids) and photoprotective pigments (carotenoids and anthocyanins), are positioned for capturing solar energy; consequently, they are well-suited for detection from above. Because photosynthetic pigments control light absorption for photosynthetic carbon uptake, their detection can be linked to photosynthetic production models.

NDVI provides a good example of an operational index that is influenced by both vegetation structure and chlorophyll content (Figure 22.4), thus providing a powerful measure of vegetation greenness and radiation absorption. Over terrestrial regions, global NDVI dynamics are significantly correlated with seasonal and latitudinal variation of atmospheric CO_2 content (Tucker et al. 1986, Fung et al. 1987), demonstrating a strong influence of terrestrial photosynthesis on atmospheric CO_2 levels. NDVI is also strongly linked to primary production at continental scales (Goward et al. 1985). This realization of strong links between NDVI, CO_2 fluxes, and primary production, along with the increasing availability of standardized global NDVI datasets, has spurred tremendous progress in global studies of carbon flux, as further discussed later.

New Opportunities with Hyperspectral Sensors

The advent of hyperspectral sensors—those with many, narrow, adjacent bands—is now allowing exploration of many additional biologically active compounds with spectral

TABLE 22.2
Examples of Biologically Important Compounds Detectable with Spectral Reflectance

Compounds	Function	References
Chlorophylls	Photosynthesis (PAR absorption)	Gitelson and Merzlyak (1994, 1996, 1997), Gitelson et al. (1996), Carter (1994), Johnson et al. (1994), Peñuelas et al. (1994, 1995a), Gamon et al. (1995), Yoder and Pettigrew-Crosby (1995)
Carotenoids (carotenes and xanthophylls)	Photosynthesis (PAR absorption) and protection (photoprotection, antioxidants)	Gamon et al. (1990, 1992, 1997), Peñuelas et al. (1994, 1995a, 1997a), Filella et al. (1996), Gamon and Surfus (1999), Gitelson et al. (2002), Sims and Gamon (2002)
Flavonoids (anthocyanins)	Protection (antioxidants, pathogen deterrent, sunscreen)	Gould et al. (1995), Coley and Barone (1996), Gamon and Surfus (1999), Gitelson et al. (2001)
Water	Essential for structural support and most metabolic processes	Bull (1991), Carter (1991), Peñuelas et al. (1993, 1994, 1997b), Gao and Goetz (1995), Roberts et al. (1997), Zhang et al. (1997), Gamon et al. (1998), Ustin et al. (1998), Serrano et al. (2000), Sims and Gamon (2003), Claudio et al. (2006)
Lignin	Plant cell wall structure, resists decomposition	Peterson et al. (1998), Wessman et al. (1988), Johnson et al. (1994), Gastellu-Etchegorry et al. (1995)
Nitrogen-containing compounds (e.g., proteins and pigments)	Needed for many metabolic processes	Peterson et al. (1988), Peñuelas et al. (1994), Filella et al. (1995) Gamon et al. (1995), Gastellu-Etchegorry et al. (1995), Johnson et al. (1994), Yoder and Pettigrew-Crosby (1995), Asner and Vitousek (2005)
Cellulose	Plant cell wall structure	Gastellu-Etchegorry et al. (1995), Martin and Aber (1997)

reflectance. Operational methods for reliably quantifying many of these compounds are now available (Table 22.2), and some have been discussed in recent reviews (Curran 1989, Wessman 1990, Peñuelas and Filella 1998). A number of pigments, notably carotenoids (accessory photosynthetic pigments; Young and Britton 1993) and anthocyanins (phenolic pigments; Strack 1997), may serve as indicators of stress or leaf senescence in many cases. Additionally, distinct absorption features for water vapor and liquid water are now remotely detectable (Gao and Goetz 1990, Green et al. 1991, 1993, Roberts et al. 1997, Gamon et al. 1998, Ustin et al. 1998, Serrano et al. 2000, Sims and Gamon 2003). Water vapor features are now used to remove confounding atmospheric effects in the derivation of apparent surface reflectance from uncorrected radiance signals (Green et al. 1991, 1993, Roberts et al. 1997). Liquid water absorption features can provide useful indices of LAI and vegetation cover (Roberts et al. 1997, Gamon et al. 1998, Ustin et al. 1998), canopy moisture status (Penuelas et al. 1993, 1997b, Zhang et al. 1997, Gamon et al. 1998, Ustin et al. 1998), and evapotranspiration (Claudio et al. 2006).

In addition to pigments or water status, absorption features of other biochemically relevant compounds, including lignin or nitrogen, can be detected with hyperspectral sensing, leading to new applications in the detection and mapping of vegetation functional state. Remote detection of lignin or nitrogen content has sometimes provided inputs for models of ecosystem production and N cycling, and successional change (Wessman et al. 1988,

Aber et al. 1990, Matson et al. 1994, Martin and Aber 1997, Peterson et al. 1988, Asner and Vitousek 2005; however, see cautions by Curran 1989, Curran and Kupiec 1995, Grossman et al. 1996).

The emergence of hyperspectral sensors is also spurring the application of new analytical procedures that promise to add insight into ecological questions. In contrast to simple indices, which are typically limited to information from one or two broad spectral bands, these methods often take advantage of the additional information present in the shape of narrow-band spectra, and are often able to resolve subtle features not detectable with broad-band indices. These alternative approaches include derivative spectra (Demetriades-Shah et al. 1990), continuum removal (Clark and Roush 1984), hierarchical foreground or background analysis (Pinzon et al. 1998), and a variety of other feature fitting methods (e.g., Gao and Goetz 1994, 1995). Another promising example is provided by spectral mixture analysis, which models a spectrum as a mix of component endmember spectra (Roberts et al. 1993, 1997, Ustin et al. 1993). These endmembers can be spectra of pure components (e.g., species, canopy components, minerals, or soil types; Roberts et al. 1993, 1998), or can be spectra derived from recognizable landscape components present in the image itself (Tompkins et al. 1997). Because spectral mixture analysis uses the additional leverage of multiple spectral bands (instead of the few bands used in most spectral indices), it can be less sensitive to confounding factors, assuming all of the appropriate spectral endmembers are properly included in the model. Because it unmixes a spectrum to fundamental components (e.g., areas of bare ground and vegetation), it can be applied to situations in which components are smaller than the spatial resolution of the detector (pixel size). On the other hand, spectral mixture analysis is computationally intensive, and the resulting products can be highly dependent on the particular endmembers selected (Roberts et al. 1998), making quantitative interpretations difficult. Although many challenges to interpretation and validation of hyperspectral data remain, the continued emergence of hyperspectral sensors, combined with improved computational capabilities, undoubtedly leads to increased refinement and application of new analytical methods applicable to ecological studies.

REMOTE SENSING AS A FUNCTIONAL MAPPING TOOL

The traditional strength of remote sensing lies in its ability to make objective maps of regions much larger than can be sampled from the ground. When image features are linked to objects or events on the surface, these maps allow us to greatly extend our view of ecologically significant patterns and processes. Historically, aerial photography has provided a useful tool for local- to regional-scale mapping, and this application has been well described elsewhere (Knipling 1969, Yost and Wenderoth 1969, Salerno 1976, Lillesand and Kiefer 1987). In contrast to the high spatial resolution (fine detail) of aerial photography, many current imaging sensors operate at coarser spatial resolution (typically, a pixel size ranging from several meters to a thousand meters in diameter), and this resolution typically degrades with sampling distance (Figure 22.2). However, current imaging spectrometers offer added benefits of a digital data format and supplementary spectral information that can be thought of as multiple data layers for a given image (Figure 22.3). Multitemporal coverage, now provided by many satellite and some aircraft sensors (Table 22.1), adds an additional dimension that is particularly useful in examining time-dependent processes, including phenological, successional, and land-use change (Justice et al. 1985, Malingreau et al. 1989, Hobbs 1990, Hall et al. 1991, DeFries and Townshend 1994a,b). Exploration of spectral and temporal information in a spatial context offers powerful new ways of distinguishing features and functional states. This combination of spatial, temporal, and spectral dimensions in a structured, multidimensional data volume can readily reveal ecologically significant patterns and processes.

The uniform data format provides potent inputs for digital mapping and modeling tools (e.g., geographic information systems; Jensen 1996) that are now revolutionizing the field of ecology. Landscape ecology (Turner and Gardner 1991) and global ecology (Gates 1993, Solomon and Shugart 1993, Walker and Steffen 1996) provide examples of new fields emerging from this synthesis of pattern and process.

A simple illustration of the power of combined spectral, temporal, and spatial information can be illustrated by imagery from the Santa Monica Mountains region of southern California. Ongoing research in this area is seeking improved ways to map vegetation patterns and landscape processes, in part to improve models of fire behavior and assist resource manager in this increasingly urbanized and disturbed region. Remote sensing is providing maps of changing vegetation cover and land usage not attainable from ground surveys due to the mountainous terrain, complex land ownership, and varying successional states and community composition caused by frequent fire, agriculture, land development, and other disturbances (Figure 22.3). NDVI images, derived from reflectance in two bands, the red and NIR (Figure 22.4), readily depict relative levels of green vegetation cover across this landscape. However, a single overpass during spring, the peak biomass period for this landscape, fails to clearly distinguish the several native vegetation types (Figure 22.5a), which for this landscape include chaparral, coastal sage scrub, grassland, southern oak woodland, and riparian woodland (Raven et al. 1986). In spring, these vegetation types have similar reflectance spectra (Figure 22.5a) and are not readily distinguished by NDVI. However, these vegetation types have contrasting seasonal patterns of green leaf display that can be readily captured by multitemporal NDVI sampling (Gamon et al. 1995). This contrasting phenology is readily revealed in the second image at the end of the summer drought (fall 1994), illustrating a general decline in NDVI (darkening across the image) relative to the spring image, with distinct patches emerging that reveal different vegetation types due to the seasonal contrasts in canopy greenness (Figure 22.6), allowing NDVI to more fully separate these vegetation types (compare Figure 22.5a and Figure 22.5b). In this way, the addition of multitemporal sampling can reveal contrasting seasonal trajectories of canopy greenness that are not evident in a single overpass.

In this landscape, key vegetation types can also be readily separated by more fully using the rich spectral dimension present in hyperspectral imagery. With spectral mixture analysis, which provides a tool for separating spectrally distinct landscape components, an individual vegetation type (e.g., coastal sage scrub) can be readily depicted in an endmember fraction map, in which varying intensity represents various quantities of that vegetation type (Figure 22.5c). Elaborations of spectral mixture analysis that employ many spectral endmembers suggest that further separation of vegetation types into dominant species may be possible (Roberts et al. 1998). Spectral mixture analysis is also able to depict relative levels of green or dead biomass associated with vegetation composition, varying seasonal or successional states, or contrasting disturbance regimes (Gamon et al. 1993, Wessman et al. 1997, Roberts et al. 1998). These improvements in the ability to distinguish vegetation types and functional states are possible with a new generation of narrow-band imaging spectrometers now available (Table 22.1). The ability to accurately depict dominant species or functional vegetation types undoubtedly improves ecosystem models that require spatially explicit vegetation maps as inputs.

At the global scale, multitemporal, broad-band satellite imagery capturing seasonal variation in NDVI has often been used to develop improved global vegetation classifications (DeFries and Townshend 1994a,b, DeFries et al. 1995, Sellers et al. 1996b, Nemani and Running 1997). NDVI dynamics detected from satellite are providing valuable insights into changing vegetation activity at global and decadal scales. For example, multiyear observations of AVHRR-derived NDVI for the continent of Africa have documented long-term vegetation dynamics in the Sahel and have demonstrated cyclical patterns of Sahel vegetation

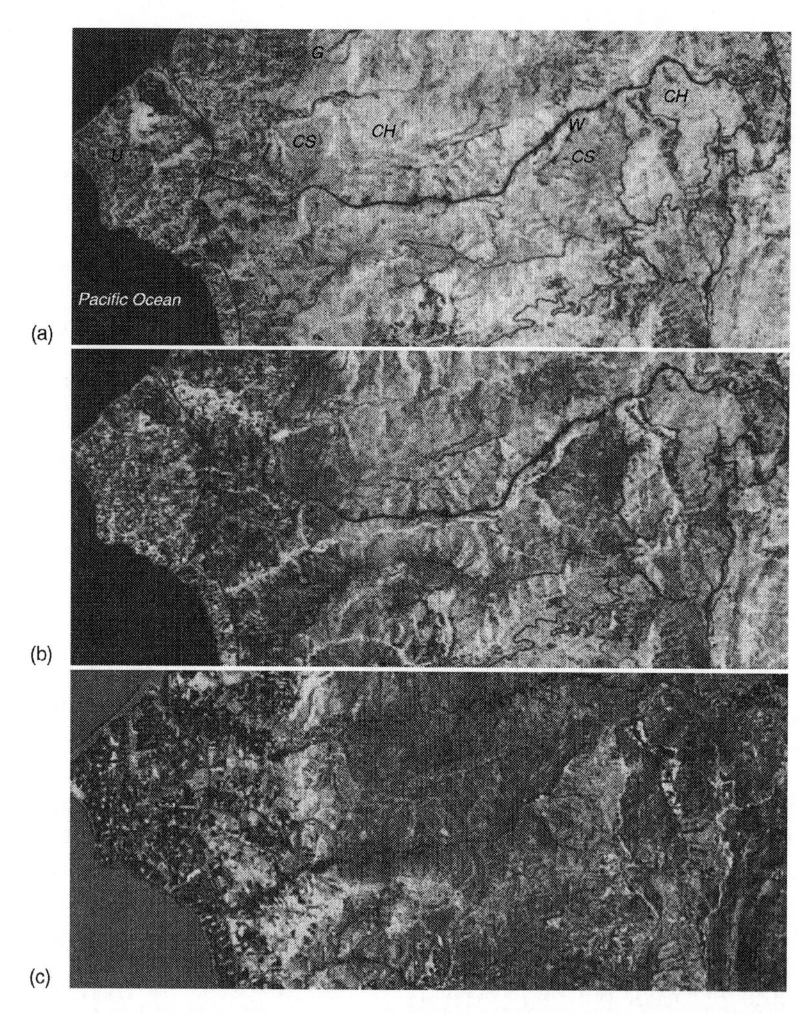

FIGURE 22.5 Images of the Pt. Dume region of the Santa Monica Mountains of southern California. (a) The NDVI in spring 1995; (b) the NDVI in fall 1994 (brighter areas indicate higher NDVI values, signifying more green vegetation). Different native vegetation types, which include chaparral, coastal sage scrub, annual grassland, and riparian areas, are difficult to separate in the spring (a), but are readily distinguished in fall (b) due to the contrasting seasonal patterns of these vegetation types. Vegetation types can also be readily separated with spectral mixture analysis, which models a landscape in terms of endmember fractions, where each endmember represents a spectral type (vegetation type in this case). (c) An example of an endmember fraction image for coastal sage scrub (brighter areas indicate areas with higher coastal sage scrub content). Images derived from NASA's AVIRIS sensor (see Table 22.1). In these images, north is to the right.

(Tucker et al. 1991). This study is particularly notable because it counters the often-stated view that desertification in this region is an irreversible and inevitable process. However, newer findings incorporating rain-use efficiency into this analysis suggest that the impacts of livestock grazing are indeed causing long-term degradation of the region's ecosystems (Hein and De Ridder 2006).

Long-term satellite records are also adding insights into biospheric responses to climate change in northern latitudes. Recent satellite NDVI evidence suggests that early fingerprints of global warming are now detectable by the increased vegetation activity in northern latitudes (Myneni et al. 1997). One possible conclusion from these observations might be that the northern regions are becoming more productive. However, most field studies indicate

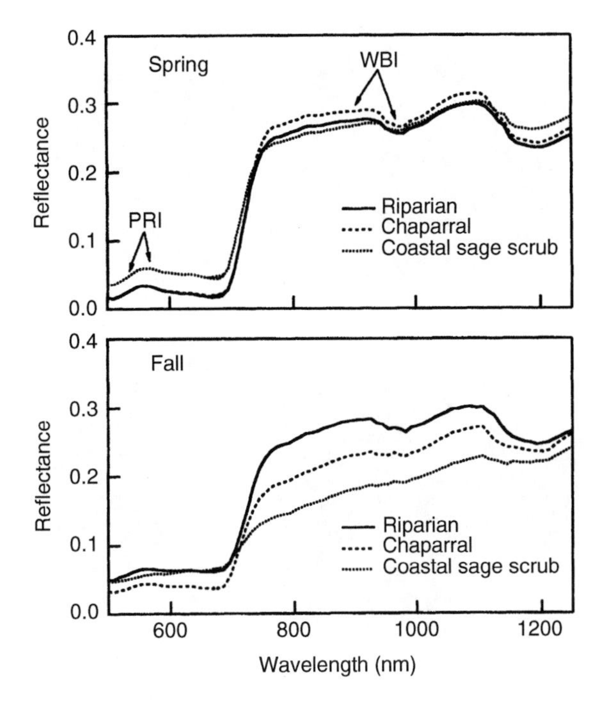

FIGURE 22.6 Spectra for representative types depicted in the AVIRIS image shown in Figure 22.5. Because vegetation types are spectrally similar in spring, they are hard to distinguish from a single overpass during that season. However, the contrasting spectral patterns emerging by fall allow ready separation of distinct vegetation types. The relaxation of the slope in reflectance at the red edge (700 nm) indicates a drop in green LAI as summer drought progresses. Note how the water absorption bands (near 970 and 1200 nm), which are clearly visible in the spring, become less apparent in the fall spectra, indicating a progressive drying of the landscape. Arrows indicate wavebands used for the PRI and the water band index (WBI). (Adapted from Gamon, J.A., Lee, L.-F., Qiu, H.-L., Davis, S., Roberts, D.A., and Ustin, S.L., Summaries of the Seventh Annual JPL Earth Science Workshop, Pasadena, California, 1998.)

that the enhanced respiratory carbon loss from thawing northern ecosystems more than offset any increase in vegetation productivity (Oechel et al. 1993, 2000). Although the exact interpretation of these kinds of large-scale satellite data are subject to question due to the difficulties of direct validation, they clearly indicate that remote sensing plays an increasingly critical role in elucidating regional vegetation change in response to climate change and other perturbations. Clearly, to fully interpret these changing patterns, improved validation at finer scales are needed to supplement these emerging satellite tools. From this, we conclude that, to address complex ecological questions, remote sensing is at its best when applied with ancillary information obtained at a range of scales.

LINKING REMOTE SENSING TO PHOTOSYNTHETIC PRODUCTION

Satellite remote sensing now provides the frequent global coverage of surface reflectance needed to drive models of global net primary productivity (Figure 22.7). The MODIS sensor and the informatics system that makes the data freely available to the scientific community (EOSDIS—NASA's Earth Observing System Data and Information System) are revolutionizing our ability to view changing patterns of global carbon exchange (Running et al. 2004). However, as further discussed later, there are many issues of validation that still need to be addressed. Multiscale remote sensing necessarily plays an important role in this validation.

In the past two decades, remote sensing has emerged as an essential tool for providing the data fields that drive spatially explicit models of photosynthesis and net primary production

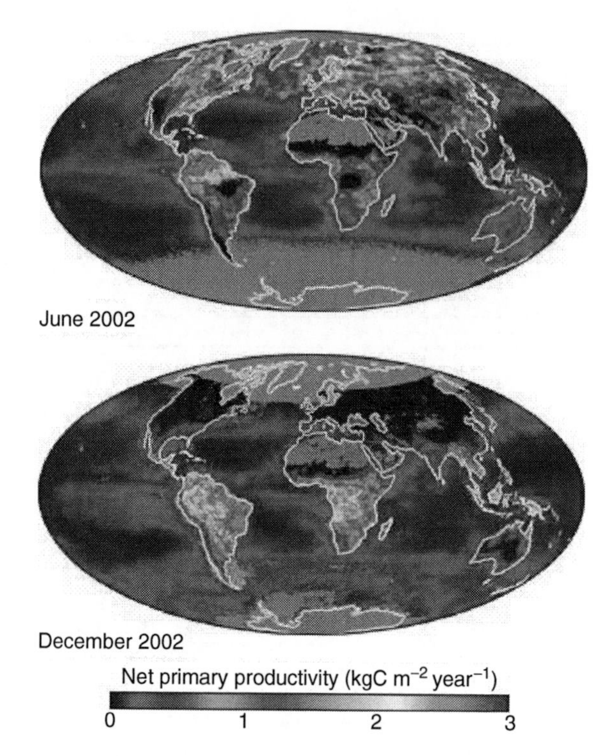

FIGURE 22.7 (See color insert following page 684.) Global NPP estimates for the months of June and December, 2002, derived from the MODIS satellite sensor. (From NASA's Earth Observatory, http://earthobservatory.nasa.gov/)

(NPP), and a wide variety of model formulations now exist. Most of these are variations of the light-use efficiency model. In its basic form, the light-use efficiency model for terrestrial vegetation states that the photosynthetic rate is a function of the absorption of photosynthetically active radiation (APAR) and the efficiency with which that absorbed radiation gets converted to fixed (organic) carbon:

$$\text{Photosynthetic rate} = \text{Efficiency} \times \text{APAR}. \tag{22.2}$$

In this case, the photosynthetic rate is the instantaneous photosynthetic rate. Although not explicitly discussed here, this equation is most appropriately applied to the gross photosynthetic rate, ignoring respiratory losses. To derive net photosynthesis, respiratory processes (photorespiration and mitochondrial respiration) can be added as a separate term in the model. Alternatively, the effects of respiration could be incorporated into the efficiency term.

If integrated over time (typically a growing season) and space (typically canopies, stands, or regions), Equation 22.2 is often expressed as:

$$\text{Primary productivity} = \varepsilon \times \Sigma\text{APAR}, \tag{22.3}$$

where primary productivity is usually expressed as NPP, often estimated by aboveground biomass accumulated in a growing season, ΣAPAR is the annual integral of radiation absorbed by vegetation, and ε represents the efficiency with which absorbed radiation is converted to biomass (Monteith 1977). Again, the respiratory contribution of nonphotosynthetic organs (e.g., stems, branches, and roots) can be incorporated into ε, or more explicitly included as a separate respiratory term or coefficient (Prince and Goward 1995, Ruimy et al. 1996, Landsberg et al. 1997).

Equation 22.2 and Equation 22.3 can be viewed as a simple conceptual model, or can be further elaborated as a more explicit, mechanistic model, in which case, the addition of distinct terms for autotrophic (and possibly heterotrophic) respiration would be appropriate. Similarly, simplistic versions of this model might represent the efficiency (ε) term as a fixed coefficient, whereas more mechanistic versions of this model formulation might treat efficiency as a variable that is continually affected by environmental factors. The APAR and maximum efficiency (ε) values determine the maximum attainable photosynthetic rate, sometimes called the potential photosynthetic rate. Environmental or physiological factors that reduce photosynthetic rate through a variety of mechanisms impact either APAR or light-use efficiency, or both. For example, water stress, temperature extremes, and nutrient limitations can all reduce effective LAI and thus absorbed radiation (APAR). These same conditions can reduce light-use efficiency through downregulation of photosynthetic processes involving stomatal closure, enzyme inactivation, and altered light energy distribution involving photoprotective mechanisms (Björkman and Demmig-Adams 1994, Gamon et al. 1997, 2001; Figure 22.8).

The beauty of the light-use efficiency model is that it can be readily applied at several temporal and spatial scales and can be readily linked to remote sensing, which is also scaleable, allowing application and validation at multiple spatial and temporal levels. The remote sensing link is usually provided by the APAR term, which can be further defined as the product of irradiance of photosynthetically active radiation (PAR) and the fraction of the irradiance that is absorbed by photosynthetic (i.e., green) canopy elements (F_{APAR}).

$$\mathrm{APAR} = F_{\mathrm{APAR}} \times \mathrm{PAR}. \tag{22.4}$$

Both PAR irradiance (Frouin and Pinker 1995) and F_{APAR} can be determined from remote sensing. F_{APAR} can be derived from NDVI or other vegetation indices, and this relationship has been well tested, both with theoretical studies (Kumar and Monteith 1981, Asrar et al. 1984, Sellers 1987, Prince 1991) and empirical measurements at many spatial and

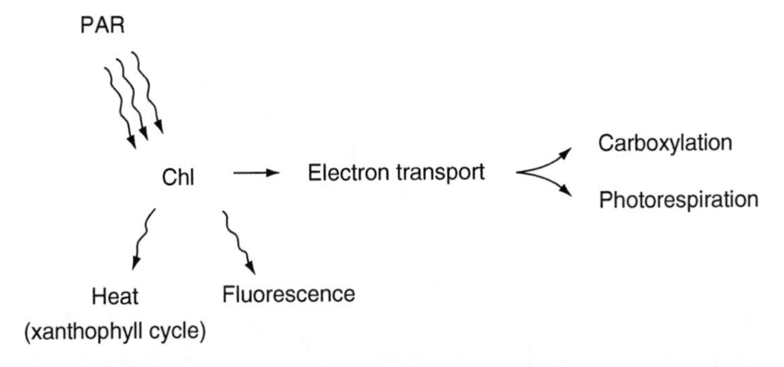

FIGURE 22.8 Schematic depicting possible fates of PAR absorbed by photosynthetic pigments (Chl). Absorbed energy can be used to drive photosynthetic electron transport and carbon uptake via carboxylation. Under conditions of reduced photosynthetic light-use efficiency, the photosynthetic system downregulates, and an increased proportion of absorbed energy is dissipated by fluorescence or heat production. The operation of the xanthophyll cycle is linked to this heat dissipation (Pfündel and Bilger 1994, Demmig-Adams and Adams 1996, Demmig-Adams et al. 1996). Both xanthopyll pigment conversion and chlorophyll fluorescence provide useful means for optically detecting reductions in photosynthetic efficiency. The relative levels of photosynthetic pigments (chlorophylls and carotenoids) provide additional indicators of photosynthetic activity.

temporal scales (Daughtry et al. 1983, Goward et al. 1985, Bartlett et al. 1990, Steinmetz et al. 1990, Gamon et al. 1995, Joel et al. 1997, Sims et al. 2006). Due to the strong links between vegetation indices and radiation absorbed by photosynthetic canopy elements, a number of models explicitly or implicitly use some form of Equation 22.2 and Equation 22.3 to derive photosynthetic fluxes or primary production. These models vary widely in complexity and detail, but can be roughly divided as follows: (1) models that assume a uniform efficiency (fixed coefficient) for all vegetation types (Heimann and Keeling 1989, Myneni et al. 1995) and (2) models that allow efficiency to vary, either according to biome type (Ruimy et al. 1994) or according to dynamic environmental conditions, notably temperature and water availability (Potter et al. 1993, Field et al. 1995, Prince and Goward 1995, Running et al. 2004).

The assumption of a constant efficiency has been carefully examined at many scales using a combination of approaches that include remote sensing, modeling, biomass harvesting, APAR sampling, and flux measurements (Bartlett et al. 1990, Gamon et al. 1993, 1995, Running and Hunt 1993, Runyon et al. 1994, Valentini et al. 1995, Joel et al. 1997, Landsberg et al. 1997). The general conclusion has been that the assumption of a constant efficiency may work well in certain cases, but not in others. For example, in certain annual grasslands, well-managed crops, and possibly deciduous forests, where physiological activity closely tracks canopy greenness and light absorption, it is often possible to model photosynthesis or NPP accurately from APAR alone by assuming a constant efficiency (Monteith 1977, Russell et al. 1989, Gamon et al. 1993). By contrast, in water- or nutrient-stressed canopies (Joel et al. 1997, Landsberg et al. 1997) or in evergreens exposed to periodically unfavorable environmental conditions (Running and Nemani 1988, Hunt and Running 1992, Running and Hunt 1993, Runyon et al. 1994, Gamon et al. 1995), it is inappropriate to assume a constant efficiency, particularly over short periods (hours to months). In these cases, photosynthetic downregulation can be significant, making it difficult to accurately predict fluxes from APAR alone. Because photosynthetic downregulation and the associated reductions in light-use efficiency can exert a significant feedback on atmospheric processes (Sellers et al. 1996a), efficiency should be considered a variable in surface-atmosphere flux models.

There are a number of approaches to defining a variable light-use efficiency term in terrestrial photosynthetic or production models. Some models incorporate different values for efficiency according to biome type (Ruimy et al. 1994) and others link efficiency to temporally varying environmental conditions and physiological status (Potter et al. 1993, Prince and Goward 1995, Running et al. 2004). In their current formulation, most of these models suffer from the fact that some of the information needed to run or validate these models is not readily available at the same scale as the remotely sensed inputs; consequently, some critical model values must be assumed or derived indirectly from measurements often made at inappropriate scales (Hall et al. 1995, Sellers et al. 1995). For example, many models derive the efficiency value from weather station inputs, which are not available at the necessary density for all regions of the Earth (Running et al. 2004). The challenge is to develop sensors and algorithms that provide all the necessary model inputs at the appropriate scale, without reference to external assumptions or excessive model tuning.

The wide variety of current and emerging sensors (Table 22.1) provides a range of tools for evaluating radiation-use efficiency. Some of these methods directly assess light regulation at the level of the photosynthetic reaction center. For example, for terrestrial vegetation, both chlorophyll fluorescence (at approximately 685 and 735 nm) and reflectance at 531 nm provide indicators of fundamental photoregulatory processes linked to carboxylation (Figure 22.8 and Figure 22.9). The fluorescence index $\Delta F/Fm'$ (Genty et al. 1989) is a widely used measure of photosystem II radiation-use efficiency. However, this index is best applied at very close range (mm to cm), in part because of a requirement for saturating light

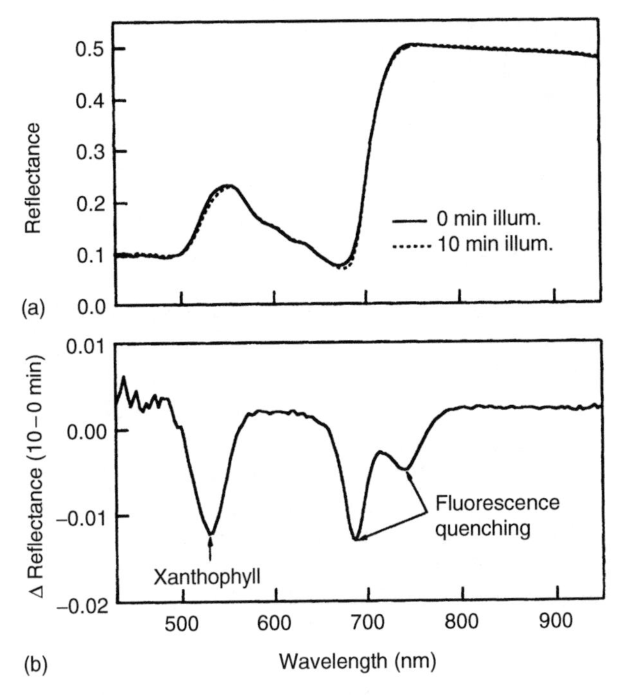

FIGURE 22.9 (a) Reflectance of a single Douglas fir (*Pseudotsuga menziesii*) needle sampled immediately on illumination (0 min illum.) and 10 minutes after illumination (10 min illum.) with irradiance equivalent to full sun. (b) When plotted as a difference spectrum (10 min minus 0 min), subtle changes in apparent reflectance appear that can be attributed to xanthophylls pigment conversion (feature near 531 nm) and chlorophyll fluorescence quenching (double feature near 685 and 735 nm). Optical indices of xanthophylls pigment activity and chlorophyll fluorescence can be used to monitor changing photosynthetic light-use efficiency (Gamon et al. 1997). This Douglas fir needle spectrum was sampled from the Wind River Canopy Crane with a leaf reflectometer. (From Gamon and Surfus, *New Phytol.*, 143, 105, 1999. With permission.)

pulses (Bolhar-Nordenkampf et al. 1989). New algorithms and advances in fluorescence technology, including laser-induced fluorescence, are beginning to relax this limitation (Günther et al. 1994). However, at this time, quantitative applications of chlorophyll fluorescence to efficiency estimation are most easily applied at a very close range (e.g., leaf scales).

In contrast to fluorescence, spectral reflectance is now applied at several spatial scales to monitor the activity of xanthophyll cycle pigments. One common expression of xanthophyll cycle pigment activity is the photochemical reflectance index (PRI):

$$\mathrm{PRI} = \frac{(R_{531} - R_{\mathrm{REF}})}{(R_{531} + R_{\mathrm{REF}})},$$

where R_{531} represents reflectance at 531 nm (the xanthophyll cycle band) and R_{REF} represents reflectance at a reference wavelength (typically 570 nm, Peñuelas et al. 1995b, Gamon et al. 1997, 2001). This index requires a narrow-band detector with bandwidths of approximately 10 nm full-width half maximum (FWHM) or finer, which includes the CASI, AVIRIS sensors, and any of the personal spectrometers listed in Table 22.1.

Because the xanthophyll cycle pigments function in photosynthetic light regulation (Pfündel and Bilger 1994, Demmig-Adams and Adams 1996, Demmig-Adams et al. 1996), and because interconversion of these pigments is detectable with spectral reflectance (Gamon et al. 1990), PRI provides a measure of radiation-use efficiency at the level of the fundamental photosynthetic light reactions. Because these reactions are closely linked to photosynthetic carbon uptake, they can also provide a near-direct index of photosynthetic light-use efficiency (Gamon et al. 1992, 1997, 2001, Penuelas et al. 1995b, 1997a, Filella et al. 1996).

Recently, the advent of new satellite sensors with suitable bands has enabled space-borne measurement of PRI, with some promising results. Hyperion, a prototype hyper-spectral satellite sensor, has allowed calculation of PRI over limited terrestrial regions (Asner et al. 2004). The MODIS sensor has a band at 530 nm and one at 550 nm that were designed for ocean color, so were not originally designed for terrestrial sampling. These bands which approximate the wavelengths used for leaf- and canopy-scale PRI measurements are now evaluated over terrestrial regions to explore the utility of satellite-based PRI measurements. Recent findings indicate that these satellite PRI measurements track seasonal patterns of photosynthetic light-use efficiency remarkably well, and manage to capture periods of reduced photosynthetic activity due to drought (Asner et al. 2004, Rahman et al. 2004). These studies suggest the derivation of terrestrial carbon fluxes entirely from remote sensing remains a reasonable goal. However, further validation is needed to understand whether light-use efficiency models developed at finer spatial scales and tested over short time periods properly capture the same processes over larger time spans and spatial scales. Consequently, it remains unclear whether satellite PRI measurements are really tracking physiological regulatory processes linked to radiation-use efficiency, or whether the significant relationships between satellite PRI measurements and radiation-use efficiency are simply fortuitous. Thus, the ability to develop predictive models using this index remains a central challenge of terrestrial light-use efficiency models. Addressing this challenge necessarily requires experimental approaches and multiscale sampling that is often not done with satellite studies alone.

Models of marine photosynthesis or primary production differ from terrestrial models in detail, but are remarkably similar in concept to those defined earlier (for a recent review, see Behrenfeld and Falkowski 1997). Satellite-driven production models generally derive surface chlorophyll concentration from ocean color measurements, and typically include a term for the maximum photosynthetic rate for a give temperature, scaled by a term (or terms) that accounts for depth-resolved distribution of light and phytoplankton. For example, one version of these models describes marine primary production as:

$$\mathrm{NPP} = C_{\mathrm{sat}} \times Z_{\mathrm{eu}} \times f(\mathrm{PAR}) \times P^{\mathrm{b}}{}_{\mathrm{opt}}(T),$$

where C_{sat} is the phytoplankton chlorophyll concentration (similar to APAR), Z_{eu} refers to the depth to which positive NPP occurs, f(PAR) is the fraction of this depth where photosynthesis is light-saturated (analogous to an efficiency term), and $P^{\mathrm{b}}{}_{\mathrm{opt}}(T)$ refers to the maximum carbon fixation rate as a function of temperature (Field et al. 1998).

Due to the different albedos over land and water, and due to different research histories in terrestrial and marine science communities, satellite sensors are generally designed separately for ocean and terrestrial sampling. With very few exceptions (e.g., Field et al. 1998, Behrenfeld et al. 2001), ocean and terrestrial productivity patterns have been generally reported separately, with very little consideration for how they may influence each other or both be influenced by larger patterns. The advent of the SeaWiFS and MODIS sensors with their global NPP products (Figure 22.7) now provides a basis for a more uniform earth-system approach to understanding primary productivity. For example, the impacts of periodic climate events (such as El Nino-Southern Oscillation) that affect both ocean and terrestrial ecosystems can be more readily understood with the advent of truly global productivity products (Behrenfeld et al. 2001).

FUTURE RECOMMENDATIONS

Although remote sensing has considerable power and promise, there remains substantial barriers to the use of remote sensing for ecological studies. These limitations are both technical and cultural. On the technical side, many processes and properties of interest to ecologists are currently inaccessible or poorly described by remote sensing. For example, remote sensing is particularly useful for depicting the production and energy input end of ecosystem fluxes, but it does not readily depict many important processes associated with allocation, respiration, or belowground activities. Similarly, largely due to the coarse spatial and spectral scales of most satellite sensors, they cannot readily describe species composition, diversity, distribution, and individual demographic variables at the scales traditionally studied by population ecologists. Because remote sensing alone is insufficient to fully characterize many critical ecological states and processes, it is likely that examination of critical ecological question will continue to require multiple approaches.

Because not all variables can be directly derived from remote sensing, we must often evaluate whether it is possible to do without information that was previously considered essential. Furthermore, because remote sensing measures differently from most field sampling methods, it often forces us to rethink how we structure our concepts and models. One example is the concept of LAI, which is generally thought to be an important parameter for terrestrial photosynthesis models (Running and Hunt 1993, Sellers et al. 1996b). This view derives, in part, from the fact that leaf-level gas exchange is typically expressed on a leaf area basis (Field et al. 1989). Furthermore, leaf area is readily estimated for single canopies or vegetation stands, and ecologists, foresters, and agronomists now have a wide array of tools and procedures for determining leaf area (Norman and Campbell 1989, Daughtry 1990, Welles 1990, Chen and Cihlar 1995). Many remote sensing studies attempt to translate vegetation indices into LAI, as if they were necessarily closely related, and this often requires varying calibrations and assumptions for different species or vegetation types (Sellers et al. 1996b). However, as described earlier, remote vegetation indices actual sample light absorption, or effective leaf are index, which incorporates clumping at the leaf and branch scales and can differ substantially from LAI determined by traditional methods (Chen 1996). Consequently, when using remote sensing, we must reconsider the functional significance and relevance of LAI as it has traditionally been applied. One solution is to define a new parameter, *effective* LAI, which is more closely linked to remote vegetation indices, light absorption, and photosynthetic production at the ecosystem level, precluding the need for individual species calibrations.

Similarly, ecologists have long considered the relative importance of species versus functional types in ecosystem function, and this topic has received considerable attention recently as new experimental evidence continues to emerge linking functional categories to ecosystem performance (Grime 1997, Hooper and Vitousek 1997, Tilman et al. 1997, Wardle et al. 1997). Although species continue to be a fundamental unit for many branches of ecology, recent studies suggest that the concept of functional types will gradually emerge as a more accessible unit of ecosystem function (Chapin 1993, Schulze and Mooney 1994, Chapin et al. 1996, 1997, Lavorel et al. 1997, Smith et al. 1997). In many cases, remote sensing may be more able to depict functional types than species, and it is likely that the use of remote sensing further spurs the development of the concept and application of functional types.

A related issue is the concept of biodiversity, here defined as species diversity (analogous to alpha diversity, Whittaker 1972). Most satellite sensors are simply too coarse spectrally and spatially to distinguish individual organisms, an essential prerequisite to counting individuals. Furthermore, remote sensing is generally limited to the canopy surface visible from above, and may miss the diversity associated with understory or belowground vegetation. However, recent work, often using hyperspectral field sensors or high-resolution (small pixel) sensors, have demonstrated a remarkable ability to distinguish different functional types and, in some cases, species (Roberts et al. 1998). These methods can become particularly powerful when time series are employed to distinguish contrasting phenological patterns (Kalacska et al. in press). So although classical measures of species biodiversity may remain elusive with current remote sensing tools, it is time to define a new concept of optical diversity that can be directly determined from remote sensing. Given the synoptic and rapid sampling capabilities of many spectrometers, and the urgent need to identify biodiversity hotspots and identify areas for conservation priority, the application of new remote sensing tools to biodiversity assessment is emerging as a critical research need.

Clearly, remote sensing presents a number of challenges to traditional views and practices in ecology, and this leads to the issue of culture, and how we perform science. The scientific method, with experimental treatments, controls, and independent replicates, is not easily followed with the tools of remote sensing, particularly at the larger scales. One solution to this dilemma lies in the use of natural experiments, in which temporal or spatial dimensions can be substituted for individual treatments. Another approach lies in the linking of remotely sensed parameters to process models that can be run as a series of virtual experiments, which is a common practice in global modeling (Sellers et al. 1996a). In either case, remote sensing frequently requires that we redefine what is considered an acceptable level of evidence, and forces us to continue to reevaluate the process of science itself, leading to new paradigms.

Reluctance to adopt remote sensing among the ecological community is often due to the fuzzy nature of the data derived from great distances; we are often more comfortable with the exactness of readily measurable or directly observable quantities that match the scales best perceived by our senses. However, fine-scale observations may not be directly relevant to critical, large-scale processes. Moreover, there is great benefit in examining patterns and processes across scales. This concern on how information transcends (or fails to transcend) scales can yield valuable insight into complex ecological issues, and remote sensing will undoubtedly continue to provide a principle tool for exploring the influence of scale on these processes (Wessman 1992, Ehleringer and Field 1993, Quattrochi and Goodchild 1997). It is likely that ecologists gain greater comfort with remote sensing as they gain familiarity with issues of scaling. The redefinition of traditional ecological questions to more closely match scales appropriate for remote sensing undoubtedly leads to further progress in ecology.

Historically, remote sensing has been an engineering-driven field disproportionately influenced by military, intelligence, and commercial extractive (e.g., mining) industries; consequently, many of the instruments of remote sensing have been designed in ignorance of the needs of ecology. As a community, ecologists have inherited the tools designed for other purposes and have been struggling to adapt these tools to new uses. This tradition has been changing, most notably with the advent of NASA's Earth Observation System and with new biospheric sensors such as the MODIS sensor (Running et al. 2004; Table 22.1, Figure 22.7). There is a great need for ecologists to continue to influence the design and application of remote sensing tools so that the instrumentation more closely reflects the evolving needs of the ecological community. This can most effectively be done if the larger scientific community works across disciplinary boundaries, and if ecologists gain familiarity with the view from above.

ACKNOWLEDGMENTS

Thanks to D. Roberts and M. Gardner for atmospherically corrected AVIRIS imagery shown in Figure 22.5. The spectra in Figure 22.9 were collected with assistance from personnel at the Wind River Canopy Crane Research Facility located within the T.T. Munger Research

Natural Area in Washington State, USA. This facility is a cooperative scientific venture among the University of Washington, the USFS Pacific Northwest Research Station, and the USFS Gifford Pinchot National Forest. C.J. Fotheringham, S.D. Prince, J. Randerson, D. Roberts, and A. Ruimy provided valuable suggestions in early drafts of the manuscript. Many of the research findings presented here were supported by grants from NASA, NSF, and the US EPA to J.A.G. Additional support was provided in the form of an iCORE Fellowship through the University of Alberta to J.A.G.

REFERENCES

Aber, J.D., C.A. Wessman, D.L. Peterson, J.M. Melillo, and J.H. Fownes, 1990. Remote sensing of litter and soil organic matter decomposition in forest ecosystems. In: R.J. Hobbs and H.A. Mooney, eds. Remote Sensing of Biosphere Functioning. Springer-Verlag, NY, pp. 87–103.

Amthor, J.S., 1995. Terrestrial higher-plant response to increasing atmospheric [CO_2] in relation to the global carbon cycle. Global Change Biology 1: 243–274.

Asner, G.P. and P. Vitousek, 2005. Remote analysis of biological invasion and biogeochemical change. Proceedings of the National Academy of Sciences of the United States of America 102(12): 4383–4386.

Asner, G.P., D. Nepstad, G. Cardinot, and D. Ray, 2004. Drought stress and carbon uptake in an Amazon forest measured with spaceborne imaging spectroscopy. Proceedings of the National Academy of Sciences 101(16): 6039–6044.

Asrar, G., M. Fuchs, E.T. Kanemasu, and J.L. Hatfield, 1984. Estimating absorbed photosynthetic radiation and leaf area index from spectral reflectance in wheat. Agronomy Journal 76: 300–306.

Baret, F. and G. Guyot, 1991. Potentials and limits of vegetation indices for LAI and APAR assessment. Remote Sensing of Environment. 35: 161–173.

Barnsley, M.J., 1994. Environmental monitoring using multiple-view-angle (MVA) remotely-sensed data. In: G. Foody and P. Curran, eds. Environmental Remote Sensing from Regional to Global Scales. Wiley, Chichester, pp. 181–201.

Bartlett, D.S., G.H. Whiting, and J.M. Hartman, 1990. Use of vegetation indices to estimate intercepted solar radiation and net carbon dioxide exchange of a grass canopy. Remote Sensing of Environment 30: 115–128.

Barton, C.V.M. and P.R.J. North, 2001. Remote sensing of canopy light use efficiency using the photochemical reflectance index. Model and sensitivity analysis. Remote Sensing of Environment 78: 264–273.

Behrenfeld, M.J. and P.G. Falkowski, 1997. A consumer's guide to phytoplankton primary productivity models. Limnology and Oceanography 42(7): 1479–1491.

Behrenfeld, M.J., J.T. Randerson, C.R. McClain, G.C. Feldman, S.O. Los, C.J. Tucker, P.G. Falkowski, C.B. Field, R. Frouin, W.E. Esaias, D.D. Kolber, and N.H. Pollack, 2001. Biospheric primary production during an ENSO transition. Science 291(5513): 2594–2597.

Björkman, O. and B. Demmig-Adams, 1994. Regulation of photosynthetic light energy capture, conversion, and dissipation in leaves of higher plants. In: E.-D. Schulze and M.M. Caldwell, eds. Ecophysiology of Photosynthesis. Springer-Verlag, Berlin, pp. 17–47.

Bolhar-Nordenkampf, H.R., S.P. Long, N.R. Baker, G. Oquist, U. Schreiber, and E.G. Lechner, 1989. Chlorophyll fluorescence as a probe of the photosynthetic competence of leaves in the field: a review of current instrumentation. Functional Ecology 3: 497–514.

Bull, C.R., 1991. Wavelength selection for near-infrared reflectance moisture meters. Journal of Agricultural Engineering Research 49: 113–125.

Carter, G.A., 1991. Primary and secondary effects of water concentration on the spectral reflectance of leaves. American Journal of Botany 78: 916–924.

Carter, G.A., 1994. Ratios of leaf reflectances in narrow wavebands as indicators of plant stress. International Journal of Remote Sensing 15: 697–703.

Chapin, F.S., III, 1993. Functional role of growth forms in ecosystem and global processes. In: J.R. Ehleringer and C.B. Field, eds. Scaling Physiological Processes: Leaf to Globe. Academic Press, San Diego, CA, pp. 287–312.

Chapin, F.S., III, R.H. Reynolds, C.M. D'Antonio, and V.M. Eckhart, 1996. The functional role of species in terrestrial ecosystems. In: B. Walker and W. Steffen, eds. Global Change and Terrestrial Ecosystems. Cambridge University Press, Cambridge, pp. 403–428.

Chapin, F.S., III, B.H. Walker, R.J. Hobbs, D.U. Hooper, J.H. Lawton, O.E. Sala, and D. Tilman, 1997. Biotic control over the functioning of ecosystems. Science 277: 500–504.

Chen, J.M., 1996. Optically-based methods for measuring seasonal variation of leaf area index in boreal conifer stands. Agricultural and Forest Meteorology 80: 135–163.

Chen, J.M. and J. Cihlar, 1995. Plant canopy gap-size analysis theory for improving optical measurements of leaf area index. Applied Optics 34: 6211–6222.

Clark, R.N. and T.L. Roush, 1984. Reflectance spectroscopy: quantitative analysis techniques for remote sensing applications. Journal of Geophysical Research 89: 6329–6340.

Claudio, H.C., J.A. Gamon, Y. Cheng, D. Fuentes, A.F. Rahman, H.-L. Qiu, D.A. Sims, H. Luo, and W.C. Oechel, 2006. Monitoring drought effects on vegetation water content and fluxes in chaparral with the 970 nm water band index. Remote Sensing of Environment 103: 304–311.

Coley, P.D. and J.A. Barone, 1996. Herbivory and plant defenses in tropical forests. Annual Review of Ecology and Systematics 27: 305–335.

Curran, P.J., 1989. Remote sensing of foliar chemistry. Remote Sensing of Environment 30: 271–278.

Curran, P.J. and J.A. Kupiec, 1995. Imaging spectrometry: a new tool for ecology. In: F.M. Danson and S.E. Plummer, eds. Advances in Environmental Remote Sensing. Wiley, Chichester, pp. 71–88.

Curran, P.J., J.L. Dungan, and H.L. Gholz, 1990. Exploring the relationship between reflectance red edge and chlorophyll content in slash pine. Tree Physiology 7: 33–48.

Danson, F.M. and S.E. Plummer, eds., 1995. Advances in Environmental Remote Sensing. Wiley, Chichester.

Daughtry, C.S.T., 1990. Direct measurements of canopy structure. Remote Sensing Reviews 5: 45–60.

Daughtry, C.S.T., K.P. Gallo, and M.E. Bauer, 1983. Spectral estimates of solar radiation intercepted by corn canopies. Agronomy Journal 75: 527–531.

DeFries, R.S. and J.R.G. Townshend, 1994a. NDVI-derived land cover classification at a global scale. International Journal of Remote Sensing 15: 3567–3586.

DeFries, R.S. and J.R.G. Townshend, 1994b. Global land cover: comparison of ground-based data sets to classifications with AVHRR data. In: G. Foody and P. Curran, eds. Environmental Remote Sensing from Regional to Global Scales. Wiley, Chichester, pp. 84–110.

DeFries, R.S., C.B. Field, I. Fung, C.O. Justice, S. Los, P.A. Matson, E. Matthews, H.A. Mooney, C.S. Potter, K. Prentice, P.J. Sellers, J.R.G. Townshend, C.J. Tucker. S.L. Ustin, and P.M. Vitousek, 1995. Mapping the land surface for global atmosphere-biosphere models: toward continuous distributions of vegetation's functional properties. Journal of Geophysical Research–Atmospheres 100: 867–882.

Demetriades-Shah, T.H., M.D. Steven, and J.A. Clark, 1990. High resolution derivative spectra in remote sensing. Remote Sensing of Environment 33: 55–64.

Demmig-Adams, B. and W.W. Adams III, 1996. The role of xanthophyll cycle carotenoids in the protection of photosynthesis. Trends in Plant Science 1: 21–26.

Demmig-Adams, B., A.M. Gilmore, and W.W. Adams III, 1996. In vivo functions of carotenoids in higher plants. FASEB Journal 10: 403–412.

Ehleringer, J.R. and C.B. Field, eds., 1993. Scaling Physiological Processes: Leaf to Globe. Academic Press, San Diego, CA.

Elvidge, C., K. Baugh, V. Hobson, E. Kihn, H. Kroehl, E. Davis, and D. Cocero, 1997. Satellite inventory of human settlements using nocturnal radiation emissions: a contribution for the global toolchest. Global Change Biology 3(5): 387–395.

Field, C.B., J.T. Ball, and J.A. Berry, 1989. Photosynthesis: principles and field techniques. In: R.W. Pearcy, J. Ehleringer, H.A. Mooney, and P.W. Rundel, eds. Plant Physiological Ecology: Field Methods and Instrumentation. Chapman and Hall, London, pp. 209–253.

Field, C.B., J.T. Randerson, and C.M. Malmstrom, 1995. Global net primary production: combining ecology and remote sensing. Remote Sensing of Environment 51: 74–88.

Field, C.B., M.J. Behrenfeld, J.T. Randerson, and P. Falkowski, 1998. Primary production of the biosphere: Integrating terrestrial and oceanic components. Science 281: 237–240.

Filella, I., L. Serrano, J. Serra, and J. Peñuelas, 1995. Evaluating wheat nitrogen status with canopy reflectance indices and discriminant analysis. Crop Science 35: 1400–1405.

Filella, I., T. Amaro, J.L. Araus, and J. Peñuelas, 1996. Relationship between photosynthetic radiation-use efficiency of barley canopies and the photochemical reflectance index (PRI). Physiologia Plantarum 96: 211–216.

Foody, G. and P. Curran, eds., 1994. Environmental Remote Sensing from Regional to Global Scales. Wiley, Chichester.

Frouin, R. and R.T. Pinker, 1995. Estimating photosynthetically active radiation (PAR) at the earth's surface from satellite observations. Remote Sensing of Environment 51: 98–107.

Fung, I.Y., C.J. Tucker, and K.C. Prentice, 1987. Application of advanced very high resolution radiometer vegetation index to study atmosphere-biosphere exchange of CO_2. Journal of Geophysical Research 92: 2999–3015.

Gamon, J.A. and J.S. Surfus, 1999. Assessing leaf pigment content and activity with a reflectometer. New Phytologist 143: 105–117.

Gamon, J.A., C.B. Field, W. Bilger, O. Björkman, A.L. Fredeen, and J. Peñuelas, 1990. Remote sensing of the xanthophyll cycle and chlorophyll fluorescence in sunflower leaves and canopies. Oecologia 85: 1–7.

Gamon, J.A., J. Peñuelas, and C.B. Field, 1992. A narrow-waveband spectral index that tracks diurnal changes in photosynthetic efficiency. Remote Sensing of Environment 41: 35–44.

Gamon, J.A., C.B. Field, D.A. Roberts, S.L. Ustin, and R. Valentini, 1993. Functional patterns in an annual grassland during an AVIRIS overflight. Remote Sensing of Environment 44: 239–253.

Gamon, J.A., C.B. Field, M.L. Goulden, K.L. Griffin, A.E. Hartley, G. Joel, J. Peñuelas, and R. Valentini, 1995. Relationships between NDVI, canopy structure, and photosynthesis in three Californian vegetation types. Ecological Applications 5: 28–41.

Gamon, J.A., L. Serrano, and J.S. Surfus, 1997. The photochemical reflectance index: an optical indicator of photosynthetic radiation use efficiency across species, functional types, and nutrient levels. Oecologia 112: 492–501.

Gamon, J.A., L.-F. Lee, H.-L. Qiu, S. Davis, D.A. Roberts, and S.L. Ustin, 1998. A multi-scale sampling strategy for detecting physiologically significant signals in AVIRIS imagery. In: Summaries of the Seventh Annual JPL Earth Science Workshop, January 12–16, 1988, Pasadena, CA.

Gamon, J.A., C.B. Field, A.L. Fredeen, and S. Thayer, 2001. Assessing photosynthetic downregulation in sunflower stands with an optically-based mode. Photosynthesis Research 67: 113–125.

Gao, B.-C. and A.F.H. Goetz, 1990. Column atmospheric water vapor and vegetation liquid water retrievals from airborne image spectrometer data. Journal of Geophysical Research 95: 3549–3564.

Gao, B.-C. and A.F.H. Goetz, 1994. Extraction of dry leaf spectral features from reflectance spectra of green vegetation. Remote Sensing of Environment 47: 369–374.

Gao, B.-C. and A.F.H. Goetz, 1995. Retrieval of equivalent water thickness and information related to biochemical components of vegetation canopies from AVIRIS data. Remote Sensing of Environment 52: 155–162.

Gastellu-Etchegorry, J.P., F. Zagolski, E. Mougin, G. Marty, and G. Giordano, 1995. An assessment of canopy chemistry with AVIRIS—a case study in the Landes Forest, south-west France. International Journal of Remote Sensing 16: 487–501.

Gates, D.M., 1993. Climate Change and its Biological Consequences. Sinauer Associates, Sunderland.

Geisel, T.S., 1950. Yertle the Turtle, and Other Stories. Random House, New York.

Genty, B., J.-M. Briantais, and N.R. Baker, 1989. The relationship between the quantum yield of photosynthetic electron transport and quenching of chlorophyll fluorescence. Biochimica et Biophysica Acta 990: 87–92.

Gholz, H.L., K. Nakane, and H. Shimoda, eds., 1997. The Use of Remote Sensing in the Modeling of Forest Productivity. Kluwer Academic Publishers, Dordrecht.

Gitelson, A. and M.N. Merzlyak, 1994. Spectral reflectance changes associated with autumn senescence of *Aesculus hippocastanum* L. and *Acer platanoides* L. leaves: spectral features and relation to chlorophyll estimation. Journal of Plant Physiology 143: 286–292.

Gitelson, A. and M.N. Merzlyak, 1996. Signature analysis of leaf reflectance spectra: algorithm development for remote sensing of chlorophyll. Journal of Plant Physiology 148: 494–500.

Gitelson, A.A. and M.N. Merzlyak, 1997. Remote estimation of chlorophyll content in higher plant leaves. International Journal of Remote Sensing 18(12): 2691–2697.

Gitelson, A., M.N. Merzlyak, and H.K. Lichtenthaler, 1996. Detection of red edge position and chlorophyll content by reflectance measurements near 700 nm. Journal of Plant Physiology 148: 501–508.

Gitelson, A.A., M.N. Merzlyak, and O.B. Chivkunova, 2001. Optical properties and nondestructive estimation of anthocyanin content in plant leaves. Photochemistry and Photobiology 74(1): 38–45.

Gitelson, A.A., Y. Zur, O.B. Chivkunova, and M.N. Merzlyak, 2002. Assessing carotenoid content in plant leaves with reflectance spectroscopy. Photochemistry and Photobiology 75(3): 272–281.

Gould, K.S., D.N. Kuhn, D.W. Lee, and S.F. Oberbauer, 1995. Why leaves are sometimes red. Nature 378: 241–242.

Goward, S.N., C.J. Tucker, and D.G. Dye, 1985. North American vegetation patterns observed with the NOAA-7 advanced very high resolution radiometer. Vegetatio 64: 3–14.

Goward, S.N., S. Turner, D.G. Dye, and S. Liang, 1994. The University of Maryland improved Global Vegetation Index product. International Journal of Remote Sensing 15: 3365–3395.

Green, R.O., J.E. Conel, J.S. Margolis, C.J. Brugge, and G.L. Hoover, 1991. An inversion algorithm for retrieval of atmospheric and liquid water absorption from AVIRIS radiance with compensation for atmospheric scattering. Proceedings of the Third AVIRIS Workshop, May 20–21, 1991, JPL 91–28: 51–61.

Green, R.O., J.E. Conel, and D.A. Roberts, 1993. Estimation of aerosol optical depth, pressure elevation, water vapor and calculation of apparent surface reflectance from radiance measured by the Airborne Visible-Infrared Imaging Spectrometer (AVIRIS) using MODTRAN2. SPIE Proceedings 1937: 2–12.

Grime, J.P., 1997. Biodiversity and ecosystem function: the debate deepens. Science 277: 1260–1261.

Grossman, Y.L., S.L. Ustin, S. Jacquemoud, and E.W. Sanderson, 1996. Critique of stepwise multiple linear regression for the extraction of leaf biochemistry information from leaf reflectance data. Remote Sensing of Environment 56: 182–193.

Günther, K.P., H.-G. Dahn, and W. Lüdeker, 1994. Remote sensing vegetation status by laser-induced fluorescence. Remote Sensing of Environment 47: 10–17.

Hall, F.G., D.B. Botkin, D.E. Strebel, K.D. Woods, and S.J. Goetz, 1991. Large-scale patterns of forest succession as determined by remote sensing. Ecology 72: 628–640.

Hall, F.G., J.R. Townshend, and E.T. Engman, 1995. Status of some remote sensing algorithms for estimation of land surface state parameters. Remote Sensing of Environment 51: 138–156.

Heimann, M. and C.D. Keeling, 1989. A three-dimensional model of atmospheric CO_2 transport based on observed winds: 2. Model description and simulated tracer experiments. In: D.H. Peterson, ed. Aspects of Climate Variability in the Pacific and the Western Americas. AGU Monograph, 55. American Geophysical Union, Washington, D.C., pp. 237–275.

Hein, L. and N. De Ridder, 2006. Desertification in the Sahel: a reinterpretation. Global Change Biology 12(5): 751–758.

Hess, L.L. and J.M. Melack, 1994. Mapping wetland hydrology and vegetation with synthetic aperture radar. International Journal of Ecology and Environmental Sciences 20: 197–205.

Hobbs, R.J., 1990. Remote sensing of spatial and temporal dynamics of vegetation. In: R.J. Hobbs and H.A. Mooney, eds. Remote Sensing of Biosphere Functioning. Springer-Verlag, NY, pp. 203–219.

Hobbs, R.J. and H.A. Mooney, eds., 1990. Remote Sensing of Biosphere Functioning. Springer-Verlag, NY.

Holben, B.N., 1986. Characteristics of maximum value composite images from temporal AVHRR data. International Journal of Remote Sensing 7: 1417–1434.

Hooper, D.U. and P.M. Vitousek, 1997. The effects of plant composition and diversity on ecosystem processes. Science 277: 1302–1305.

Houghton, R.A., 1995. Land-use change and the carbon cycle. Global Change Biology 1: 275–287.

Huete, A.R., H.Q. Liu, K. Batchily, and W. van Leeuwen, 1997. A comparison of vegetation indices over a global set of TM images for EOS-MODIS. Remote Sensing of Environment 59: 440–451.

Huete, A., K. Didan, T. Miura, E.P. Rodriquez, X. Gao, and L.G. Ferreira, 2002. Overview of the radiometric and biophysical performance of the MODIS vegetation indices. Remote Sensing of Environment 83: 195–213.

Hunt, E.R. and S.W. Running, 1992. Simulated dry matter yield for aspen and spruce stands in the North American boreal forest. Canadian Journal of Remote Sensing 18: 126–133.

James, M.E. and S.N.V. Kalluri, 1994. The Pathfinder AVHRR land data set: an improved coarse resolution data set for terrestrial monitoring. International Journal of Remote Sensing 15: 3347–3363.

Jensen, J.R., 1996. Introductory Digital Image Processing: A Remote Sensing Perspective, 2nd edn. Prentice Hall, Upper Saddle River, NJ.

Joel, G., J.A. Gamon, and C.B. Field, 1997. Production efficiency in sunflower: the role of water and nitrogen stress. Remote Sensing of Environment 62: 176–188.

Johnson, L.F., C.A. Hlavka, and D.L. Peterson, 1994. Multivariate analysis of AVIRIS data for canopy biochemical estimation along the Oregon Transect. Remote Sensing of Environment 47: 216–230.

Justice, C.O., J.R.G. Townshend, B.N. Holben, and C.J. Tucker, 1985. Analysis of the phenology of global vegetation using meteorological satellite data. International Journal of Remote Sensing 6: 1271–1318.

Kalacska, M., G.A. Sanchez-Azofeifa, B. Rivard, T. Caelli, H.P. White, and J.C. Calvo-Alvarado. Ecological Fingerprinting of ecosystem succession: estimating secondary tropical dry forest structure and diversity using imaging spectroscopy. Remote Sensing of Environment (in press).

Kasischke, E.S., J.M. Melack, and M.C. Dobson, 1997. The use of imaging radars for ecological applications—a review. Remote Sensing of Environment 59: 141–156.

Knipling, E.B., 1969. Leaf reflectance and image formation on color infrared film. In: P.L. Johnson, ed. Remote Sensing in Ecology. University of Georgia Press, Athens, GA, pp. 17–29.

Kumar, M. and J.L. Monteith, 1981. Remote sensing of crop growth. In: H. Smith, ed. Plants and the Daylight Spectrum. Academic Press, London, pp. 133–144.

Landsberg, J.J., S.D. Prince, P.G. Jarvis, R.E. McMurtrie, R. Luxmoore, and B.E. Medlyn, 1997. Energy conversion and use in forests: an analysis of forest production in terms of radiation utilisation efficiency (ε). In: H.L. Gholz, K. Nakane, and H. Shimoda, eds. The Use of Remote Sensing in the Modeling of Forest Productivity. Kluwer Academic Publishers, Dordrecht, pp. 273–298.

Lavorel, S., S. McIntyre, J. Landsberg, and T.D.A. Forbes, 1997. Plant functional classifications: from general groups to specific groups based on response to disturbance. Trends in Ecology and Evolution 12: 474–478.

Lefsky, M.A., W.B. Cohen, G.G. Parker, and D.J. Harding, 2002. Lidar remote sensing for ecosystem studies. BioScience 52(1): 19–30.

Lillesand, T.M. and R.W. Kiefer, 1987. Remote Sensing and Image Interpretation, 2nd edn. Wiley, NY.

Los, S.O., C.O. Justice, and C.J. Tucker, 1994. A global 1° by 1° NDVI data set for climate studies derived from the GIMMS continental NDVI data. International Journal of Remote Sensing 15: 3493–3518.

Malingreau, J.P., C.J. Tucker, and N. Laporte, 1989. AVHRR for monitoring global tropical deforestation. International Journal of Remote Sensing 10: 855–867.

Martin, M.E. and J.D. Aber, 1997. High spectral resolution remote sensing of forest canopy lignin, nitrogen, and ecosystem processes. Ecological Applications 72: 431–443.

Matson, P.A., L. Johnson, C. Billow, J. Miller, and R. Pu, 1994. Seasonal patterns and remote spectral estimation of canopy chemistry across the Oregon Transect. Ecological Applications 4: 280–298.

Monteith, J.L., 1977. Climate and the efficiency of crop production in Britain. Philosophical Transactions of the Royal Society of London 281: 277–294.

Myneni, R.B. and G. Asrar, 1994. Atmospheric effects and spectral vegetation indices. Remote Sensing of Environment 47: 390–402.

Myneni, R.B. and D.L. Williams, 1994. On the relationship between FAPAR and NDVI. Remote Sensing of Environment 49: 200–211.

Myneni, R.B., S.O. Los, and G. Asrar, 1995. Potential gross primary productivity of terrestrial vegetation from 1982–1990. Geophysical Research Letters 22: 2617–2620.

Myneni, R.B., C.D. Keeling, C.J. Tucker, G. Asrar, and R.R. Nemani, 1997. Increased plant growth in the northern high latitudes from 1981–1991. Nature 386: 698–702.

Nemani, R. and S. Running, 1997. Land cover characterization using multitemporal red, near-IR, and thermal-IR data from NOAA/AVHRR. Ecological Applications 7: 79–90.

Nobel, P.S., 1991. Physicochemical and Environmental Plant Physiology. Academic Press, San Diego, CA.

Norman, J.M. and G.S. Campbell, 1989. Canopy structure. In: R.W. Pearcy, J. Ehleringer, H.A. Mooney, and P.W. Rundel, eds. Plant Physiological Ecology: Field Methods and Instrumentation. Chapman and Hall, London, pp. 301–325.

Oechel, W.C., S.J. Hastings, G. Vourlitis, M. Jenkins, G. Riechers, and N. Grulke, 1993. Recent change of Arctic tundra ecosystems from a net carbon dioxide sink to a source. Nature 361: 520–523.

Oechel, W.C., G.L. Vourlitis, S.J. Hastings, R.C. Zulueta, L. Hinzman, and D. Kane, 2000. Acclimation of ecosystem CO2 exchange in the Alaskan Arctic in response to decadal climate warming. Nature 406(6799): 978–81.

Pearcy, R.W., 1989. Radiation and light measurements. In: R.W. Pearcy, J. Ehleringer, H.A. Mooney, and P.W. Rundel, eds. Plant Physiological Ecology: Field Methods and Instrumentation. Chapman and Hall, London, pp. 97–116.

Peñuelas, J. and I. Filella, 1998. Visible and near-infrared reflectance techniques for diagnosing plant physiological status. Trends in Plant Science 3: 151–156.

Peñuelas, J., I. Filella, C. Biel, L. Serrano, and R. Save, 1993. The reflectance at the 950–970 nm region as an indicator of plant water status. International Journal of Remote Sensing 14: 1887–1905.

Peñuelas, J., J.A. Gamon, A.L. Fredeen, J. Merino, and C.B. Field, 1994. Reflectance indices associated with physiological changes in nitrogen- and water-limited sunflower leaves. Remote Sensing of Environment 48: 135–146.

Peñuelas, J., F. Baret, and I. Filella, 1995a. Semi-empirical indices to assess carotenoids/chlorophyll a ratio from leaf spectral reflectance. Photosynthetica 31: 221–230.

Peñuelas, J., I. Filella, and J.A. Gamon, 1995b. Assessment of photosynthetic radiation-use efficiency with spectral reflectance. New Phytologist 131: 291–296.

Peñuelas, J., J. Llusia, J. Piñol, and I. Filella, 1997a. Photochemical reflectance index and leaf photosynthetic radiation-use efficiency assessment in Mediterranean trees. International Journal of Remote Sensing 18: 2863–2868.

Peñuelas, J., J. Piñol, R. Ogaya, and I. Filella, 1997b. Estimation of plant water concentration by the reflectance Water Index WI (R900/R970). International Journal of Remote Sensing 18: 2869–2875.

Perry, C.R. and L.F. Laughtenschlager, 1984. Functional equivalence of spectral vegetation indices. Remote Sensing of Environment 14: 169–182.

Peterson, D.L., J.D. Aber, P.A. Matson, D.H. Card, N. Swanberg, C. Wessman, and M. Spanner, 1988. Remote sensing of forest canopy and leaf biochemical content. Remote Sensing of Environment 24: 85–108.

Pfündel, E. and W. Bilger, 1994. Regulation and possible function of the violaxanthin cycle. Photosynthesis Research 42: 89–109.

Pinzon, J.E., S.L. Ustin, C.M. Castenada, and M.O. Smith, 1998. Investigation of leaf biochemistry by hierarchical foreground/background analysis. IEEE Trans. Geoscience and Remote Sensing 36: 1913–1927.

Potter, C.S., J.T. Randerson, C.B. Field, P.A. Matson, P.M. Vitousek, H.A. Mooney, and S.A. Klooster, 1993. Terrestrial ecosystem production: a process model based on global satellite and surface data. Global Biogeochemical Cycles 7: 811–841.

Prince, S.D., 1991. A model of regional primary production for use with coarse-resolution satellite data. International Journal of Remote Sensing 12: 1313–1330.

Prince, S.D. and S.N. Goward, 1995. Global primary production: a remote sensing approach. Journal of Biogeography 22: 815–835.

Quattrochi, D.A. and M.A. Goodchild, eds., 1997. Scale in Remote Sensing and GIS. CRC Press, Boca Raton, FL.

Rahman A.F., V.D. Cordova, J.A. Gamon, H.P. Schmid, and D.A. Sims, 2004. Potential of MODIS ocean bands for estimating CO_2 flux from terrestrial vegetation: A novel approach. Geophysical Research Letters, 31, L10503, doi:10.1029/2004GL019778.
Raven, P.H., H.J. Thompson, and B.A. Prigge, 1986. Flora of the Santa Monica Mountains, California, 2nd edn. Southern California Botanists, Special Publication No. 2. University of California, Los Angeles.
Richards, J.A., 1993. Remote Sensing Digital Image Analysis: An Introduction. Springer-Verlag, Berlin.
Roberts, D.A., J.B. Adams, and M.O. Smith, 1993. Discriminating green vegetation, non-photosynthetic vegetation and soils in AVIRIS data. Remote Sensing of Environment 44: 255–270.
Roberts, D.A., R.O. Green, and J.B. Adams, 1997. Temporal and spatial patterns in vegetation and atmospheric properties from AVIRIS. Remote Sensing of Environment 62: 223–240.
Roberts, D.A., M. Gardner, R. Church, S. Ustin, G. Scheer, and R.O. Green, 1998. Mapping chaparral in the Santa Monica Mountains using multiple endmember spectral mixture models. Remote Sensing of Environment 65: 267–279.
Ruimy, A., B. Saugier, and G. Dedieu, 1994. Methodology for the estimation of terrestrial primary production from remotely sensed data. Journal of Geophysical Research 99: 5263–5283.
Ruimy, A., G. Dedieu, and B. Saugier, 1996. TURC: a diagnostic model of continental gross primary productivity and net primary productivity. Global Biogeochemical Cycles 10: 269–285.
Running, S.W. and E.R. Hunt Jr., 1993. Generalization of a forest ecosystem process model for other biomes, BIOME-BGC, and an application for global-scale models. In: J.R. Ehleringer and C.B. Field, eds. Scaling Physiological Processes: Leaf to Globe. Academic Press, San Diego, CA, pp. 141–158.
Running, S.W. and R.R. Nemani, 1988. Relating seasonal patterns of the AVHRR vegetation index to simulated photosynthesis and transpiration of forests in different climates. Remote Sensing of Environment 24: 347–367.
Running, S.W., R.R. Nemani, F.A. Heinsch, M. Zhao, M. Reeves, and H. Hashimoto, 2004. A continuous satellite-derived measure of global primary production. BioScience 54(6): 547–560.
Runyon, J., R.H. Waring, S.N. Goward, and J.M. Welles, 1994. Environmental limits on net primary production and light-use efficiency across the Oregon Transect. Ecological Applications 4: 226–237.
Russell, G., P.G. Jarvis, and J.L. Monteith, 1989. Absorption of radiation by canopies and stand growth. In: G. Russell, B. Marshall, and P.G. Jarvis, eds. Plant Canopies: Their Growth, Form, and Function. Cambridge University Press, Cambridge, pp. 21–39.
Sabins, F.F., 1997. Remote Sensing: Principles and Interpretation, 3rd edn. W.H. Freeman and Company, NY.
Salerno, A.E., 1976. Aerospace photography. In: E. Schanda, ed. Remote Sensing for Environmental Sciences. Springer-Verlag, Berlin, pp. 11–83.
Schulze, E.-D. and H.A. Mooney, eds., 1994. Biodiversity and Ecosystem Function. Springer-Verlag, Berlin.
Sellers, P.J., 1987. Canopy reflectance, photosynthesis, and transpiration. II. The role of biophysics in the linearity of their interdependence. Remote Sensing of Environment 21: 143–183.
Sellers, P.J., C.J. Tucker, G.J. Collatz, S.O. Los, C.O. Justice, D.A. Dazlich, and D.A. Randall, 1994. A global 1° by 1° NDVI data set for climate studies. Part 2: the generation of global fields of terrestrial biophysical parameters from the NDVI. International Journal of Remote Sensing 15: 3519–3545.
Sellers, P.J., B.W. Meeson, F.G. Hall, G. Asrar, R.E. Murphy, R.A. Schiffer, F.P. Bretherton, R.E. Dickinson, R.G. Ellingson, C.B. Field, K.F. Huemmrich, C.O. Justice, J.M. Melack, N.T. Roulet, D.S. Schimel, and P.D. Try, 1995. Remote sensing of the land surface for studies of global change: models—algorithms—experiments. Remote Sensing of Environment 51: 3–26.
Sellers, P.J., L. Bounoua, G.J. Collatz, D.A. Randall, D.A. Dazlich, S.O. Los, J.A. Berry, I. Fung, C.J. Tucker, C.B. Field, and T.G. Jensen, 1996a. Comparison of radiative and physiological effects of doubled atmospheric CO_2 on climate. Science 271: 1402–1406.
Sellers, P.J., S.O. Los, C.J. Tucker, C.O. Justice, D.A. Dazlich, G.J. Collatz, and D.A. Randall, 1996b. A revised land surface parameterization (SiB2) for atmospheric GCMs. Part II: the generation

of global fields of terrestrial biophysical parameters from satellite data. Journal of Climate 9: 706–737.
Serrano, L., S.L. Ustin, D.A. Roberts, J.A. Gamon, and J. Penuelas, 2000. Deriving water content of chaparral vegetation from AVIRIS data. Remote Sensing of Environment 74: 570–581.
Sims, D.A. and J.A. Gamon, 2002. Relationships between leaf pigment content and spectral reflectance across a wide range of species, leaf structures and developmental stages. Remote Sensing of Environment 81: 337–354.
Sims, D.A. and J.A. Gamon, 2003. Estimation of vegetation water content and photosynthetic tissue area from spectral reflectance: a comparison of indices based on liquid water and chlorophyll absorption. Remote Sensing of Environment 84: 526–537.
Sims, D.A., H. Luo, S. Hastings, W.C. Oechel, A.F. Rahman, and J.A. Gamon, 2006. Parallel adjustments in vegetation greenness and ecosystem CO_2 exchange in response to drought in a Southern California chaparral ecosystem. Remote Sensing of Environment 103: 289–303.
Smith, L.C., 1997. Satellite remote sensing of river inundation area, stage, and discharge: a review. Hydrological Processes 11: 1427–1440.
Smith, T.M., H.H. Shugart, and F.I. Woodward, eds., 1997. Plant Functional Types: Their Relevance to Ecosystem Properties and Global Change. Cambridge University Press, Cambridge.
Solomon, A.M. and H.H. Shugart, eds., 1993. Vegetation Dynamics and Global Change. Chapman and Hall, NY.
Steinmetz, S., M. Guerif, R. Delecolle, and F. Baret, 1990. Spectral estimates of the absorbed photosynthetically active radiation and light-use efficiency of a winter wheat crop subjected to nitrogen and water deficiencies. International Journal of Remote Sensing 11: 1797–1808.
Strack, D., 1997. Phenolic metabolism. In: P.M. Dey and J.B. Harbourne, eds. Plant Biochemistry. Academic Press, San Diego, CA, pp. 387–416.
Tilman, D., J. Knops, D. Wedin, P. Reich, M. Ritchie, and E. Siemann, 1997. Optimization of endmembers for spectral mixture analysis. Science 277: 1300–1302.
Tompkins, S., J.F. Mustard, C.M. Pieters, and D.W. Forsyth, 1997. Optimization of endmembers for spectral mixture analysis. Remote Sensing of Environment 59: 472–489.
Townshend, J.R.G., 1994. Global data sets for land applications from the Advanced Very High Resolution Radiometer: an introduction. International Journal of Remote Sensing 15: 3319–3332.
Tucker, C.J., I.Y. Fung, C.D. Keeling, and R.H. Gammon, 1986. Relationship between atmospheric CO_2 variations and a satellite-derived vegetation index. Nature 319: 195–199.
Tucker, C.J., H.E. Dregne, and W.W. Newcomb, 1991. Expansion and contraction of the Sahara Desert from 1980 to 1990. Science 253: 299–301.
Turner, M.G., and R.H. Gardner, eds., 1991. Quantitative Methods in Landscape Ecology. Springer-Verlag, NY.
Ustin, S.L., M.O. Smith, and J.B. Adams, 1993. Remote Sensing of ecological processes: a strategy for developing and testing ecological models using spectral mixture analysis. In: J.R. Ehleringer and C.B. Field, eds. Scaling of Physiological Processes: Leaf to Globe. Academic Press, San Diego, CA, pp. 339–357.
Ustin, S.L., D.A. Roberts, J.E. Pinzon, S. Jacquemoud, G. Scheer, C.M. Castenada, and A. Palacios, 1998. Estimating canopy water content of chaparral shrubs using optical methods. Remote Sensing of Environment 65: 280–291.
Ustin, S.L., D.A. Roberts, J.A. Gamon, G.P. Asner, and R.O. Green, 2004. Using imaging spectroscopy to study ecosystem processes and properties. BioScience 54(6): 523–533.
Valentini, R., J.A. Gamon, and C.B. Field, 1995. Ecosystem gas exchange in a California grassland: seasonal patterns and implications for scaling. Ecology 76: 1940–1952.
Vanderbilt, V.C., L. Grant, and S.L. Ustin, 1990. Polarization of light by vegetation. In: J. Ross and R.B. Myneni, eds. Photon–Vegetation Interactions: Applications in Optical Remote Sensing and Plant Ecology. Springer-Verlag, Berlin, pp. 194–228.
Vitousek, P.M., 1994. Beyond global warming: ecology and global change. Ecology 75: 1861–1876.
Walker, B. and W. Steffen, eds., 1996. Global Change and Terrestrial Ecosystems. Cambridge University Press, Cambridge.

Wardle, D.A., O. Zackrisson, G. Hörnberg, and C. Gallet, 1997. The influence of island areas on ecosystem properties. Science 277: 1296–1299.

Welles, J.M., 1990. Some indirect methods of estimating canopy structure. Remote Sensing Reviews 5: 31–43.

Wessman, C.A., 1990. Evaluation of canopy biochemistry. In: R.J. Hobbs and H.A. Mooney, eds. Remote Sensing of Biosphere Functioning. Springer-Verlag, NY, pp. 135–156.

Wessman, C.A., 1992. Spatial scales and global change: bridging the gap from plots to GCM grid cells. Annual Review of Ecology and Systematics 23: 175–200.

Wessman, C.A., J. Aber, D. Peterson, and J. Melillo, 1988. Remote sensing of canopy chemistry and nitrogen cycling in temperate forest ecosystems. Nature 335: 154–156.

Wessman, C.A., C.A. Bateson, and T.L. Benning, 1997. Detecting fire and grazing patterns in tallgrass prairie using spectral mixture analysis. Ecological Applications 7: 493–511.

Whittaker, R.H., 1972. Evolution and measurement of species diversity. Taxon 21: 213–251.

Wiegand, C.L., A.J. Richardson, D.E. Escobar, and A.H. Gerbermann, 1991. Vegetation indices in crop assessments. Remote Sensing of Environment 35: 105–119.

Yoder, B.J. and R.E. Pettigrew-Crosby, 1995. Predicting nitrogen and chlorophyll content and concentration from reflectance spectra (400–2500 nm) at leaf and canopy scales. Remote Sensing of Environment 53: 199–211.

Yost, E. and S. Wenderoth, 1969. Ecological applications of multispectral color aerial photography. In: P.L. Johnson, ed. Remote Sensing in Ecology. University of Georgia Press, Athens, GA, pp. 46–62.

Young, A. and G. Britton, eds., 1993. Carotenoids in Photosynthesis. Chapman and Hall, London.

Zhang, M., S.L. Ustin, E. Rejmankova, and E.W. Sanderson, 1997. Remote sensing of salt marshes: potential for monitoring. Ecological Applications 7: 1039–1053.

FIGURE 22.3 True-color AVIRIS image for a region of Ventura County in southern California showing the Pacific Ocean (*bottom*), the western edge of the Santa Monica Mountains (*center right*) and developed and agricultural areas (*top left*). The different colored bands receding into the background illustrate the spectral dimension (Z dimension), in this case 224 different spectral bands (wavelengths), ranging from near-ultraviolet bands (foreground) to the infrared bands (background). Much of the information content of this image is present in this spectral dimension.

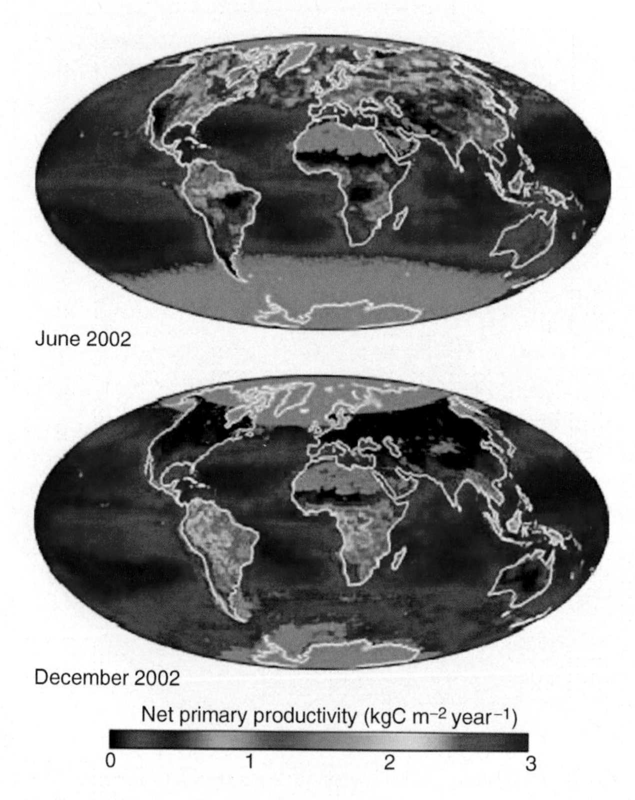

FIGURE 22.7 Global NPP estimates for the months of June and December, 2002, derived from the MODIS satellite sensor. (From NASA's Earth Observatory, http://earthobservatory.nasa.gov/.)

23 Generalization in Functional Plant Ecology: The Species-Sampling Problem, Plant Ecology Strategy Schemes, and Phylogeny*

Mark Westoby

CONTENTS

INTRODUCTION

This chapter is about how we generalize across species, and how we choose species for study.

If a process is universal, it should not matter in which species we study it. An expectation that the processes under study should be universal is implicit in much physiological or developmental genetic work. Correspondingly a single species is studied, chosen for experimental convenience, perhaps tobacco, or *Arabidopsis*. In functional ecology, on the other hand, most generalizations are conditional and comparative. Functional ecology aims to work out how things operate differently in pioneer versus shade-tolerant species, on low-nutrient versus high-nutrient soils, among monocots versus dicots, in rare compared with widespread species.

Functional ecology makes progress through the struggle to generalize, to understand what is similar and what is different about species and situations. Progress is manifested at two levels in publication. In primary journal publications reporting fresh data, comparisons may be between as few as two species, up to tens of species, depending which traits were

* This chapter has not been updated since the first edition of the Handbook.

studied and how difficult and time-consuming it was to characterize them. However, later on, in literature reviews or meta-analyses, numbers of such primary studies are gathered together and generalizations are sought. It is really at the stage of compiling literature reviews that knowledge can be said to become consolidated and reliable. At this stage evidence may be available about tens to hundreds of species (e.g., edited review books about ecophysiology such as Lange et al 1984, Lambers et al 1989, Roy and Garnier 1994, Schulze and Caldwell 1994, Mulkey et al 1996). Similarly many hundreds of field experiments have now been conducted on the major interactions between species such as competition, herbivory, and predation (Connell 1983, Crawley 1983, Schoener 1983, Sih et al 1985, Price et al 1986, Hairston 1989, Goldberg and Barton 1992, Gurevitch et al 1992, Wilson and Agnew 1992, Goldberg 1996). Generalizing across the many primary studies has now become an outstanding problem for ecology.

This chapter argues in the following sequence. First, the reader is briefly reminded of the rules for inferring generalizations from samples. These rules are familiar to ecologists in contexts such as vegetation sampling, but curiously, they have been almost completely ignored in the context of selecting species for study. Some problems in applying the generalization rules to species selection are acknowledged, and the distinction between papers reporting primary data collection and subsequent review and meta-analysis papers becomes important here. Then to arrive at a conditional generalization, species have to be categorized in some way. The three main types of categorization are habitat, ecological strategy (traits of the species itself), and phylogeny. For primary data collection, anyone could use their own basis for categorizing species, but during subsequent meta-analyses, only widely-adopted categorization schemes can come into play. Hence, achieving consensus on at least some categorization schemes should be an important part of the research agenda of comparative functional ecology. A new ecological strategy scheme is proposed for this purpose. Finally phylogeny: species with recent common ancestors are more likely to have similar traits. A brief outline is provided of the recent debate about whether phylogeny should be regarded as an alternative explanation to present-day functionality. More constructively, phylogeny provides a potential tool for choosing study species to arrive more efficiently at generalizations, and to place better-defined boundaries around the generalizations.

INFERENCE RULES FOR GENERALIZING FROM A SAMPLE

The rules for inferring a generalization from a sample have been well appreciated by ecologists for many years, in the context for example of describing vegetation by means of quadrat sampling. The rules are

- Quadrats are placed at random, in other words each possible location within some category should have an equal chance of being sampled.
- Accordingly, the scope and boundaries of the category itself need to be defined. Is it a particular patch of forest that is described, or is it all patches of that forest-type within a continent? As well as conceptual decisions about the scope of the category, there usually need to be some exclusion rules—perhaps sites are not sampled on cliffs because of the practical inconvenience, or in the middle of forest tracks because these are not thought interesting.
- There needs to be some replication of the samples, to give a sense of the range of variation. I will not digress to discuss different indices of the range of variation; biometry texts provide recipes for calculating these and discuss their merits and assumptions. The point is only that all recipes require replication, because the logic by which we generalize about a category requires some indication of the range of variation within the category.

These rules come into play during the design of a sampling procedure. Generalization rules forge the link between the sampling procedure and the scope of the generalizations that can subsequently be inferred. The design of the study is shaped by contemplating what conclusions one might be seeking to draw. The decision what category to sample, together with the exclusion rules, draws boundaries on the conclusions that can be drawn. The legitimacy of the conclusions depends on the equal-chance-of-being-sampled principle, and the replication allows the strength of the conclusions to be assessed.

APPLICATION OF THE INFERENCE RULES TO GENERALIZATION ACROSS SPECIES—AN ISSUE THAT HAS BEEN UNDERESTIMATED

In contexts such as vegetation sampling and manipulative field experiments, the inference rules have been thoroughly assimilated into the practices of ecological researchers. All the more remarkable, then, that choice of species to study still seems not to be perceived as a sampling problem where the rules for generalization need to be applied. Some researchers may quite legitimately argue that they have chosen particular species because they are interested just in them, they have no interest in generalizing. Alternatively they may believe the mechanism they are studying is universal, so it does not matter in which species they study it. However, many studies of the functional ecology or ecophysiology of plants seek to interpret their results by references to categories of plants—shade-tolerant versus pioneer, low versus high altitude, annual versus perennial, inhabitants of low-nutrient versus more fertile soils, rare and endangered versus widespread, and constitutively slow-growing versus capable of rapid growth—in other words, the species are to be interpreted as samples from some category. Nevertheless, these studies rarely address explicitly the representativeness and replication criteria. If the inference rules were followed, the methodology of these papers would first define categories of species—listing all the low-nutrient species in NSW, for example—then indicate exclusion rules that had to be applied—for example, only species available from seed merchants could be considered, and among those, some had to be eliminated because they could not be persuaded to germinate in reasonable numbers within a few days of each other—and then within those boundaries, would confirm that the species studied were chosen at random. However, in reality, most papers in comparative ecophysiology hardly comment at all on the choice of species, and it is quite common for two categories to be compared by means of two species, with no replication. Imagine if a manuscript were submitted in which two vegetation types were compared by means of one quadrat in each, placed at the location the investigator thought most representative or convenient?—it would get short shrift from referees and editors. However, many papers that have been successfully published and are respectfully cited in comparative ecophysiology do exactly this with regard to choice of species.

How have these differences in research culture, in expectations about sampling, come about? Probably one factor has been the physiologist's expectation that truly interesting processes occur reliably in whatever species is studied. Surely the other main factor must have been the sheer difficulty and laboriousness of some types of measurement. Remember also that for an investigator seeking to dissect mechanisms, there are always strong incentives to measure more processes and at more frequent intervals within one species, rather than to extend to further species. Consider investigations of complete carbon budgets for example, including dissecting root respiration into growth and maintenance components. Within the framework of a Ph.D., it is unreasonable to expect more than a couple of species to be studied. Then since there are too few species for replication within categories, it must seem unimportant to apply the equal-chance rule within each category. There are also strong reasons to use species where seeds are available and are known to germinate readily, which can lead to a few species used repeatedly by successive investigators. So in summary, the tradition of using few species and those not chosen according to explicit rules of sampling is entirely understandable, within ecophysiology and indeed in other areas such as genetics and demography of rare species.

Nevertheless, subsequent literature review or meta-analysis has to face the issue: is species coverage in the primary research reports representative, and of what categories? As more and more primary research reports accumulate, this issue is becoming more pressing. One way of defining the problem in species selection is to say that species should be selected to make better generalizations possible to the person who comes along 5–10 years after and writes the literature review.

How might this come about? At one extreme could be imagined a grand overall species-selection design, decided by a well-meaning dictator or committee. Each lab worldwide would be then instructed which species to select. I feel confident this is not going to happen (and indeed should not be allowed to, because the slackening of competitive energy would surely outweigh the benefits of coordination). At the other extreme, and far more likely, is that primary research publications move toward discussing their species selection explicitly, in relation to the inference rules. Explicit discussion could be very valuable. Even when there can be little replication, and when many exclusion rules have been required so that there is only a small choice of species, compilers of literature reviews will nevertheless be better placed if the primary studies have spelled out their species-selection criteria explicitly. Related to explicit discussion, authors should be encouraged to make available as much background information as possible about their species. By background, I mean information that is not referred to in the paper for the purpose of proving a conclusion, but that might subsequently be useful to reviewers or meta-analysts seeking to investigate different questions. The limited page-space in journals need no longer discourage authors and editors from reporting background information about traits and habitats of the study species, since such material can now be placed at websites.

CRITERIA ON WHICH TO COMPARE SPECIES

If species-sampling is one side of a coin, the other side is categories into which species are grouped for comparison. Criteria on which species might be compared fall under three main headings:

- Habitat
- Attributes of the species themselves—ecological strategies
- Phylogeny (or in practice, usually taxonomy)

The distinction between primary research reports and reviews or meta-analyses is important here also. Individual research groups can and do categorize species according to whatever criteria they think meaningful for the question in hand. However, at the stage when knowledge is consolidated across many primary reports, the reviewer can use only categories that were adopted in common by all the primary reports, or alternatively that are simple enough for species to be attributed to them by the reviewer. As a consequence, most reviews group species into rather simple categories: herbaceous versus woody, deciduous versus evergreen, temperate versus tropical and so forth. For categories to be used at review that captured a subtler degree of difference between species, consensus on the categorization scheme would need to be achieved in advance. Developing wider consensus on more expressive categorization schemes is thus a major priority for improved generalization in functional ecology.

HABITAT

I will comment only very briefly on habitat descriptors. Rainfall and temperature are not, nowadays, too much of a problem to attribute to plant species. Maps or climate-interpolating software can produce estimates of yearly means or various seasonal patterns or extremes, for a given location where a species is known to occur. Shading under canopies used often to be described in qualitative categories, but measurements on an absolute scale of PAR have become more typical over the last couple of decades, and have been important in clarifying seedling performance in the shade.

Soil nutrients remain an unresolved problem. Many papers have compared species from infertile versus fertile soils, within particular landscapes. However, synthesis across these papers is very difficult because they do not share common measurements that describe soil fertility of the sites where the species is successful. There is some reason to believe that relatively fertile soils in Australia might fall into an infertile category in northwest Europe, for instance, and this complicates any attempt to relate Australian to European studies. Certainly there are many complexities in measuring soil fertility, but still, it would be a step forward if even one or two lowest-common-denominator measures could be agreed.

TRAITS OF THE SPECIES THEMSELVES—ECOLOGICAL STRATEGY SCHEMES

The literature on plant ecological strategy schemes can be summarized into three main strands of thinking. One strand categorizes species by reference to distribution (realized niche) on one or more gradients, for example, Dyksterhuis (1949) for grazing, Noble and Slatyer (1980) for time after disturbance, Ellenberg (1988) for soil and other habitat features. A second strand categorizes species according to physiognomy (e.g., Raunkiaer 1934, Dansereau 1951, Mueller-Dombois and Ellenberg 1974, Box 1981, Sarmiento and Monasterio 1983, Barkman 1988, Orshan 1989, Prentice et al 1992) and has been active especially within plant geography. In a third strand axes or categories are named according to concepts (as distinct from naming them according to traits or realized niche). Examples include the *r-K* spectrum (Cody 1966, MacArthur and Wilson 1967) and several schemes that have developed this spectrum into a three-cornered arrangement (Greenslade 1972, 1983, Grime 1974, Whittaker 1975, Southwood 1977). The three-cornered schemes add a category of opportunities where the physical environment permits only slow acquisition of resources. This situation is called stress in Grime's CSR triangle (1974, 1979, Grime et al 1988), the best developed three-cornered scheme for plants. The CSR triangle has two dimensions, the *C–S* axis reflecting adaptation to opportunities for rapid growth versus continuing enforcement of slow growth (Competitors to Stress-tolerators), the *R*-axis reflecting adaptation to disturbance (Ruderals).

Among conceptual strategy schemes the CSR triangle is the most widely cited in textbooks (e.g., Cockburn 1991, Colinvaux 1993, Begon et al 1996, Crawley 1996, Ingrouille 1992), reflecting wide acceptance that exploiting opportunities for fast versus slow growth, and coping with disturbance, are two of the most important forces shaping the ecologies of plants within landscapes. Yet papers in functional ecology do not routinely report CSR axis scores for the species studied, for the reason that there is no explicit quantitative protocol for scoring a species from anywhere worldwide (see qualitative and partly subjective keys in Grime 1984, Grime et al 1988). In other words, the CSR scheme is widely cited for conceptual discussion but not widely adopted for practical comparisons. The only general-purpose scheme that has been widely adopted is Raunkiaer's life-form categorization (1907, English translation 1934), based on the location of the buds where regrowth arises after the unfavorable season of the year. Raunkiaer life-forms are very easily attributed for most species (that is why the scheme is widely adopted), but the scheme conveys only a modest amount of information about differences between species.

In summary, then, those existing schemes that can easily be applied worldwide capture very few of the differences between species, especially with regard to how they exploit different opportunities within multispecies vegetation and between different sites in a landscape. On the other hand, schemes that seek to be more expressive about these differences between species are not so designed that species anywhere can readily be categorized. Consequently schemes such as the CSR triangle have not been able to be used to group species worldwide during literature reviews or meta-analyses. In this context, I have recently (Westoby 1998) proposed a new LHS (leaf-height-seed) scheme (Box 23.1) designed to express at least some of the differences between species addressed by the CSR triangle and related schemes, while using axes readily measured on the plant itself, and therefore offering the potential for worldwide comparison.

BOX 23.1
Proposed LHS Plant Ecology Strategy Scheme

The LHS scheme (Westoby 1998) would consist of three axes:

- Typical specific leaf area (SLA) (of mature leaves, developed in full light, or the fullest light the species naturally grows in)
- Typical height of the canopy of the species at maturity
- Typical seed mass

The strategy of a species would be characterized in the scheme by a position in a 3D volume. Each dimension is known to vary widely between species at any given level of the other two, thus the volume occupied by present-day species extends considerably in all three dimensions. Each of these traits is correlated with a number of others, but they have not been chosen only as conveniently measured indicators. Rather it is believed that they themselves are fundamental trade-offs controlling plant strategies. They are fundamental because it is ineluctable that a species cannot both deploy a large light-capturing area per gram and also build strongly reinforced leaves that may have long-lives; cannot support leaves high above the ground without incurring the expense of a tall stem; cannot produce large, heavily provisioned seeds without producing fewer of them per gram of reproductive effort.

As would be expected for traits of such ecological importance, plants have some capacity to shift trait values in response to the circumstances they find themselves in. In other words, none of SLA, height at maturity, or seed mass are absolute constants within species. Nevertheless, variation between species is much greater than within species, and many previous authors have seen no insuperable difficulty in recording characteristic species values for comparative purposes (e.g., for height at maturity Hubbell and Foster 1986, Grime et al 1988, Keddy 1989, Bugmann 1996, Chapin et al 1996). All three axes would be log-scaled, reflecting the fact that the difference between 30 and 31 m (to take canopy height at maturity as an example) is not nearly so important as the difference between 30 and 130 cm.

SPECIFIC LEAF AREA

SLA is the light-catching area deployed per unit of previously photosynthesized dry mass allocated to the purpose. SLA is like an expected rate of return on investment. High SLA permits (given favorable growth conditions) a shorter payback time on a gram of dry matter invested in a leaf (Poorter 1994). At first glance it might appear that a low rate of return on investment would not be evolutionarily competitive, but low SLA species achieve greater leaf life span (Reich et al 1992, 1997),

BOX 23.1 (continued)
Proposed LHS Plant Ecology Strategy Scheme

through extra structural strength and sometimes through allocation to tannins, phenols, or other defensive compounds. Therefore light capture per gram invested can be at least as great in a low-SLA species when considered through the whole life of the investment. Reich et al (1997) have shown across six biomes that SLA is closely correlated with mass-based net photosynthetic capacity and mass-based leaf N as well as leaf life span. Higher leaf water content and reduced lamina depth can both contribute to higher SLA (Witkowski and Lamont 1991, Garnier and Laurent 1994, Cunningham et al 1999). Grime et al (1997) found SLA to be among the major contributors to the primary axis of specialization they identified by ordination of 67 traits among 43 species, corresponding to the *C–S* axis of the CSR scheme.

Potential relative growth rate RGR, measured on exponentially growing seedlings given plentiful water and nutrients, has been seen as an indicator of responsiveness to favorable conditions (e.g., Grime and Hunt 1975, Leps et al 1982, Loehle 1988, Poorter 1989, Reich et al 1992, Aerts and van der Peijl 1993, Chapin et al 1993, van der Werf et al 1993, Turner 1994). Because potRGR is made up of net assimilation rate x leaf fraction x SLA, variation in SLA necessarily influences potRGR. Indeed, in most comparative studies SLA has been the largest of the three sources of variation in potRGR (Poorter 1989, Poorter and Remkes 1990, Poorter and Lambers 1991, Lambers and Poorter 1992, Reich et al 1992, Garnier and Freijsen 1994, Saverimuttu and Westoby 1996, Cornelissen et al 1996, Grime et al 1997, Hunt and Cornelissen 1997, Poorter and van der Werf 1998). High SLA species can have strategies associated with rapid production of new leaf during early life. Faster turnover of plant parts permits also a more flexible response to the spatial patchiness of light and soil resources (Grime 1994b). On the other hand, species with low SLA and long-lived leaves can eventually accumulate a much greater mass of leaf and capture a great deal of light in that way; and the long mean residence time of nutrients made possible by leaf longevity permits a progressively larger share of nitrogen pools to be sequestered (Aerts and van der Pijl 1993).

CANOPY HEIGHT AT MATURITY

Height obviously conditions how plants make a living, in different ways depending on vegetation dynamics. In some vegetation types a characteristic vertical profile of leaf area and light attenuation persists over time, through the turnover of individual plants. Species with canopies at different depths in this profile are operating at different light incomes, heat loads, wind speeds, humidities, and with different capital costs for supporting leaves and lifting water to the leaves. In other vegetation types disturbances, or the death of large individual trees, destroy canopy cover and daylight becomes available near the ground. The successional process that ensues can be understood as a race upward for the light. Because light descends from above, the leading species at a given time have a considerable advantage. In this race, unlike a standard athletic contest, there is not a single winner determined after a fixed distance. Rather any species that is among the leaders at some stage during the race is a winner, in that being among the leaders for a reasonable period permits a sufficient carbon profit to be accumulated for the species to ensure it runs also in subsequent races. The entry in subsequent races may occur via vegetative regeneration, via a stored seed bank, or via dispersal to other locations, but the prerequisite for any of these is sufficient carbon accumulation at some stage during vegetative growth. Races are restarted when a new disturbance destroys the accumulated stem height. The duration of an individual race can be measured in years, or ideally in units of biomass accumulation, calibrating intervals between disturbances to the productivity of a site. However, within a race-series with some typical race duration, one finds successful growth strategies that have been designed by natural selection to be among the leaders early in a race, and other successful strategies that join the leaders at various later stages. Species that achieve most of their lifetime photosynthesis with leaves deployed at

(*continued*)

BOX 23.1 (continued)
Proposed LHS Plant Ecology Strategy Scheme

10–50 cm have different stem tissue properties from those designed for 1–5 m, and those in turn are different from species that achieve 30–40 m. The canopy height that species have been designed by natural selection to achieve is the simplest measure of this spectrum of strategies.

SEED MASS

Seed mass variation expresses a species' chance of successfully dispersing a seed into an establishment opportunity, from a given area of ground already occupied by a species. Seed mass is also quite a good indicator of a cotyledon-stage seedling's ability to survive various hazards.

Species with smaller seed mass can produce more seeds from within a given reproductive effort, and seed mass therefore is the best easy predictor of seed output per square meter of canopy cover. It might be thought that distance of dispersal would be the major influence on a species' chance of dispersing a seed to a forest gap or another establishment opportunity. However, dispersal distances have not proved tidily related to dispersal morphology, to seed mass, or to any other plant attribute (reviewed in Hughes et al 1994). Among unassisted species, larger seeds do not travel as far from a given height of release, but on the other hand larger seeds tend to have wings, arils, and so on or to be released from a greater height. Similarly among wind assisted species, larger seeds tend to have larger wings or longer pappuses. Because reduced dispersal associated with larger seed mass tends to be counteracted by extra investment in dispersal-assisting structures, or sometimes by being released from a taller plant, the net effect is that dispersal distance is not tidily related to any of these attributes. Seed mass (as a surrogate for seed output per ground area occupied) is the best predictor, for the present, of the chance that an occupied site will disperse a propagule to an establishment opportunity.

Species with larger seed mass have been shown experimentally to survive better under a variety of different seedling hazards (tabulated in Westoby et al 1996), including drought, removal of cotyledons, and dense shade below the compensation point. The tendency to survive longer applies only during cotyledon phase, whereas seed reserves are deployed into the fabric of the seedling (Saverimuttu and Westoby 1996). Capacity to continue growth into later seedling life under a low-light level is determined more by canopy architecture and leaf properties (Kitajima 1994). It seems likely that tolerance of seedling hazards is endowed not by seed mass as such, but by a tendency for larger seeds to retain more metabolic reserves uncommitted to the fabric of the seedling over a longer period, and therefore available to support respiration when in carbon deficit (Westoby et al 1996).

LHS SCHEME IN RELATION TO GRIME'S CSR TRIANGLE

Where each axis of the CSR scheme implies a complex of plant traits (e.g., Grime et al 1997), the LHS scheme has axes defined by single quantitative traits. The benefit of the LHS scheme's simple protocol for positioning a species outweighs any loss of information content in the LHS axes compared with the CSR axes, for the purpose of facilitating worldwide comparisons of species.

The CSR scheme has been made triangular rather than rectangular because the most stressful and most frequently disturbed corner is said not to be occupied (Grime et al 1988), or because ineluctable trade-offs are said to prevent a species from getting highly adapted to more than one of the three primary strategies *C*, *S*, or *R* (Grime 1994a). The idea that a whole quadrant is missing due to the combination of high stress and high disturbance has been criticized (Grubb 1985) and experiments with crossed gradients of fertility and disturbance (Campbell and Grime 1992, Burke and Grime 1996) have not produced wholly unoccupied space at the low-fertility high-disturbance corner. The LHS scheme avoids prejudging the question whether any particular corner of the LHS volume is not viable.

BOX 23.1 (continued)
Proposed LHS Plant Ecology Strategy Scheme

Another difficulty in the CSR scheme is the ruderality axis. Adaptation to disturbance might in principle include adaptations for surviving individual disturbances, together with adaptations for completing life history within a short interval between disturbances, together with adaptations for dispersing through space or time to freshly disturbed locations. Grubb (1985) criticized the CSR scheme for not distinguishing continuing from episodic disturbance. According to Grime (Grime et al 1988, Grime and Hillier 1992) the scheme is for adults not juveniles: a given adult strategy can occur in combination with several different juvenile strategies, which has the effect of separating out dispersal and seed bank strategies from the main CSR categorization of a species. The LHS scheme disentangles these disparate elements to some extent. The canopy height at maturity axis reflects adaptation to the interval between disturbances (calibrated in units of height growth rather than time). The seed mass axis (more exactly its inverse, seed number per mass allocated to seed production) reflects the potential for dispersal to freshly disturbed locations. Adaptations for continuing the lineage through particular types of disturbance (e.g., lignotubers for resprouting after fire, soil seed banks with a light requirement for germination following soil turnover, basal tillering in graminoids for grazing tolerance) have deliberately been left outside the LHS scheme, since they do not lend themselves to any simple generalization.

In summary of Box 23.1, the LHS scheme captures a substantial part of the same spectra of strategy variation as the CSR scheme, while resolving some difficulties with it. SLA variation (the L dimension) is crucial to the CS axis (Grime et al 1988, 1997), which is to leaf longevity, mean residence time of nutrients, soil nutrient adaptation, and potential RGR. Canopy height at maturity (the H dimension) is arguably the most central single trait that needs to be adjusted to the duration of the growth opportunity between disturbances (*R*-axis); it is also treated by Grime et al (1988) as a significant predictor of *C* versus *S* strategy. The LHS scheme does not prejudge what parts of the LHS volume will be occupied, compared with the CSR triangle, which decides a priori that the high-*S*-high-*R* quadrant is not a viable strategy. By separating out seed mass (*S* dimension) as a distinct axis, it expresses something about dispersal to new growth opportunities, independently of what is expressed by canopy height about the duration of the growth opportunity between disturbances. Seed mass also expresses some significant differences between species about seedling establishment. Most importantly, the LHS axes chosen require little enough effort to estimate that experimentalists may be willing to report them for their species with a view to subsequent meta-analysis by others, even though they have no immediate use for the data themselves.

PHYLOGENY

Regrettably, phylogenetic relatedness is often interpreted as an alternative reason (sometimes called phylogenetic constraint) why species should be similar (Hodgson and Mackey 1986, Kelly and Purvis 1993, Harvey 1996, Silvertown and Dodd 1996). Common ancestry or phylogeny is seen as a source of confounding or error that requires controlling for; in competition with explanations that invoke natural selection or functionality continuing into the present day.

This competing explanation approach is incorrect, with regard to explaining present-day ecological function. Phylogenetic niche conservatism is commonplace; hence, species can often have similar trait combinations *both* because they are phylogenetically related, *and* because they are subject to similar continuing forces of natural selection. The issues of interpretation have been debated elsewhere, for example in a Forum in *Journal of Ecology*

(Ackerly and Donoghue 1995, Harvey et al 1995a,b, Rees 1995, Westoby et al 1995a–c, 1996, 1998, Price 1997) and are summarized in Box 23.2 to Box 23.4.

Precisely because functionally important traits are sometimes phylogenetically conservative, phylogeny can and should be seen not as a source of confounding, a technical difficulty

BOX 23.2
Frequently Asked Questions (FAQs) about Phylogeny and Functional Ecology

FAQ1: When related species tend to be similar (e.g., seed mass more similar within than between genera), should this be attributed to phylogenetic constraint?

Answer to FAQ1: No. The term constraint clearly implies that the trait has been under directional selection toward different values, but nevertheless has failed to respond to selection. (Remaining unchanged due to absence of directional selection, or due to continuing convergent selection, cannot usefully be called constraint, or inertia, an alternative sometimes seen.) There are two reasons why similarity of species with a common ancestor should not be regarded as positive evidence for constraint or failure to respond to directional selection.

First, given that differences between species, genera, or families are under consideration, the hypothesized constraint needs to have applied over millions, perhaps tens of millions of years. Thus features of genetic architecture that might be measured in a present-day population and might restrict response to selection over tens of generations, such as low heritability or genetic correlations between traits, could only account for constraint in this context if the low heritability or the genetic correlations survived a million years of mutation and genetic rearrangement under directional selection. This would be sufficiently surprising that it certainly should not be accepted as a null hypothesis, especially not for quantitative traits such as seed mass. Rather it is a decidedly strong biological hypothesis: a definite mechanism for the constraint should be proposed, and means sought to test it.

Second, there are alternative well-established mechanisms through which species could tend to maintain similar traits over time after diverging from a common ancestor. Therefore correlation of a trait with phylogeny cannot be regarded as evidence for constraint rather than for continuing functionality. Phylogenetic niche conservatism is a process whereby because ancestors have a particular constellation of traits, their descendants tend to be most successful using similar ecological opportunities, and so natural selection tends to maintain the same traits among most if not all descendant lineages. Niche conservatism is at least as likely a cause of similarity among related species as constraint—more likely, for quantitative traits—and explicitly invokes ecological functionality continuing into the present day.

Sometimes one will see the term "phylogenetic effect" used to refer to the tendency for phylogenetically related species to have similar traits. This term is defensible provided it means "effect" only in a purely statistical sense, a label for variation correlated with phylogeny. However, the temptation seems strong to see a phylogenetic effect as somehow an alternative causal interpretation to ecological functionality, and this is wrong. The term phylogenetic effect was better eschewed (Westoby et al 1995c). If constraint is inferred, a specific mechanism should be proposed. If not, one might refer to phylogenetic conservatism, identifying the pattern in the outcome without hinting at any particular mechanism.

FAQ2: Is it true that phylogenetically related species are nonindependent as evidence for present-day ecological function?

Answer to FAQ2: Yes in part, but mainly no: actually formulating the question around the term independence is not helpful (see Box 23.3). The grain of truth in this idea is that a correlation across present-day species between traits X and Y might be caused through a cross-correlation with Z rather than reflecting a direct functional relationship between X and Y. Related species are more likely to have similar values for Z. However, the argument usually connected to the claim about nonindependence is that radiations, separate divergences on the phylogenetic tree, *are*

BOX 23.2 (continued)
Frequently Asked Questions (FAQs) about Phylogeny and Functional Ecology

independent events, and therefore that a test for correlated divergence deconfounds the $X–Y$ correlation to a large extent (Harvey et al 1995a) from third-variable influences (see Box 23.4 for further discussion of the sense in which correlated-change analysis deconfounds or partials out third variables). This implication that analyzing divergences rather than present-day species is an improved, phylogenetically corrected method for assessing ecological function is unsafe for two reasons.

First, the problem of cross-correlation with third variables is not confined to related species. Hence analyzing for correlations in divergence rather than correlations across present-day species does not overcome the well-known problems of inferring causation from correlation.

Second, just because an $X–Y$ correlation is cross-correlated with Z, this does not necessarily mean Z is the true cause. It remains just as likely that the true mechanism runs from X to Y, and Z is a secondary correlate, so far as anyone can tell from the correlation pattern alone. In such cross-correlated situations, it is not conducive to sensible interpretation to deconfound $X–Y$ from any influence of Z without at the same time looking at the raw $X–Y$ correlation. This is all the more so when the third, fourth, and so on variables from which $X–Y$ is deconfounded are not explicitly identified, but rather are an aggregate of all variables that have been conservative down the phylogenetic tree.

In summary, evolution by natural selection has given rise to cross-correlated patterns of traits among present-day species. Selection for ecological functionality has inherently been confounded with phylogeny during the history of evolution, and statistical corrections are not capable of converting that inherently confounded history into the ideal experiment in which phylogeny is orthogonally crossed with present-day function. In this situation the credibility of a hypothesis connecting traits to ecological functions cannot be judged according to the pattern of correlation and cross-correlation alone, but must rest also on whether the physiological or morphological mechanism is convincing, and the outcomes are well tested in field experiments. While everyone should be aware that correlation cannot prove causation, it is important to remember also that disappearance of correlation after correction or partialling does not *disprove* causation.

The qualification "as evidence for present-day ecological function" in FAQ2 is important. If the issues under study were to do with the historical process of evolutionary divergence, then naturally data about present-day species should be transformed by hanging on the phylogenetic tree (Grafen 1989) to give rise to inferred data about the radiations.

FAQ3: Is it obligatory to correct for effects of phylogeny?

Answer to FAQ3: No (when concerned with present-day function). Although advocates of phylogenetic correction or correlated-divergence analysis (Kelly and Purvis 1993, Rees 1993, Harvey 1996, Silvertown and Dodd 1996) have taken the view that cross-species correlation analysis has been superseded, correlated-divergence analysis cannot be considered obligatory because: (a) Tests for correlated evolutionary divergence (phylogenetic correction procedures) do *not* reliably control for all potentially confounding third variables, see FAQ2, and (b) Phylogenetic correction *does* remove from consideration correlations that have been phylogenetically conservative, many of which may also reflect present-day function (phylogenetic niche conservatism), see also FAQ2. A trait can perfectly well be functional, but have arisen in only one or a few separate radiations, so that a correlated-divergence analysis would never show statistical significance. Conversely, a trait can be repeatedly correlated with an ecological outcome across many radiations or phylogenetically independent contrasts, but nevertheless not be the true cause.

to be overcome, but more positively, as a basis on which to select species for study. Through better species selection we may hope to arrive at generalizations more efficiently and with better-defined boundaries on the generalization. People concerned with methods of phylogenetic analysis have mostly been using datasets already in existence, and have not as yet paid

BOX 23.3
Meanings of Independence and Adaptation

The debate over phylogenetic correction has (like most debates) gotten issues of how we obtain reliable knowledge mixed together with issues of semantics. Certain key words need comment:

Independence: Arguments for phylogenetic correction typically begin from the formulation by Felsenstein (1985), where species do not represent independent data points, on the grounds that related species will have similarities by reason of common ancestry. To claim that species lack independence purely because they have similarities cannot be justified. If correlation with another trait were sufficient to vitiate independence, either two species that both occurred in Europe could not be considered independent, or two species that both had alternate leaves. Carried to its logical conclusion, evidence could never be found for anything, because some correlate could always be found that would be regarded as vitiating the independence.

In general, independence is not an absolute property, but makes sense only in the context of a particular model of causation. The issue is whether two species represent separate items of evidence for that causation process. The model connected with the Felsenstein formulation of nonindependence focuses on the process of change in a trait. The present-day trait value is viewed as caused by the past process of change, rather than the process of change being caused by an attraction toward the present-day trait value, an attraction arising from ecological functionality. The claim that species are not independent items of evidence, rather the change along each phylogenetic branch is an independent item, makes sense only in the context of this particular model of the generating process. Price (1997) gives an example of a model where the evolutionary process positions species in trait space according to the present-day ecological context, and shows formally that a better test of that process is obtained by considering each species an independent case than by considering each radiation an independent case.

Adaptation: A sector of the scientific community wishes to reserve adaptation to refer only to the natural selection under which a trait first emerged, excluding natural selection that may be maintaining it in the present day (Gould and Vrba 1982, Harvey and Pagel 1991). Although others continue with a broader usage of adaptation that can refer also to ecological functionality in the present day (see Williams 1992, Reeve and Sherman 1993 for balanced discussion), advocates of phylogenetic correction have chosen to insist that tests for adaptation must exclude trait-maintenance (Harvey et al 1995a). In practical effect, this definition of adaptation insists that questions about the emergence of traits are legitimate, whereas questions about trait maintenance are not.

Under these circumstances the word adaptation is best avoided, for the present. Throughout this chapter traits have been referred to as functional, or those that have ecological significance, to avoid getting sidetracked by this issue of the definition of adaptation.

much attention to species-selection designs. However, for experimentalists who collect new data, sufficient effort is involved for each species that it is worth thinking carefully about how that effort should be allocated.

Species-selection design, like any other aspect of design, depends on the question under study. It is important to be careful about the exact formulation of questions invoking phylogeny. A range of different question formulations and corresponding species-selection designs are discussed in Westoby et al (1998), where a more general overview is provided.

A traditional idea is that one should compare species within a genus rather than more distantly related species, for the reason that other unmeasured attributes are less likely to vary in such a comparison. This idea is actually not a very good compromise. On the one hand, it is

BOX 23.4
Relationship between Correlated Divergence Analysis (Phylogenetic Correction) and Partialling Out the Cross-Correlation with a Third Variable

According to people who believe correlated-divergence analysis should be obligatory (e.g., Harvey 1996), one of its major benefits is in deconfounding an X–Y correlation, partialling out potential influences of whatever third traits $Z1$, $Z2$, and so on may be phylogenetically conservative. Westoby et al (1995a) described phylogenetic correction as extracting variation in this sense and discarding it from consideration as potentially related to ecological function, but in response Harvey et al (1995a) asserted that correlated-change procedures should not be regarded as extracting any component from the cross-species dataset. What, then, are the similarities and differences between correlated-divergence analyses and partial correlation analysis?

Correlated-divergence analysis transforms a species x traits data table by hanging it on the tree (Grafen 1989), producing a new dataset where each row is a radiation or node in the phylogenetic tree, and each column is a measure of divergence in a trait at the radiation in question. In the simplest case, the measure of divergence would simply be the difference in the trait between the two species descended from a branch-point. (There are various complications where three or more branches descend from a node, or where branch-lengths are not assumed equal, but the essential logic is the same for these more complicated cases.) The question is whether divergence in Y is correlated with divergence in X, tested by fitting a regression through the origin to the data-points derived one from each radiation.

Thus correlated-divergence analysis has similarities to a paired design, such as if one set out to study 1000 biology students, and paired each one with a humanities student, matched for age, gender, and University. Then if one wished to analyze for a relationship between biology versus humanities and attending live drama, the number of plays attended during the preceding year for each biology student would be subtracted from the number attended for the corresponding humanities student, and one would test whether the difference in plays attended was significantly different from zero. In correlated-divergence analysis, species are similarly matched into pairs, using the criterion of common ancestry, which has the effect also of pairing them according to any number of phylogenetically conservative traits. Depending on the species-selection design, pairs may be deliberately contrasted on some attribute, for example, soil habitat, or may simply be random species descended from a branch-point in the phylogenetic tree. In any event, the point of subtracting trait values between pair-members is to remove from consideration trait-variation associated with matters for which pairs have been matched, such as age, gender, and University. Although this has advantages for some questions, looking only at the differences also has distinct disadvantages. Suppose there was some tendency for humanities students to attend more live drama, but the tendency of females rather than males to attend live drama was much stronger—this second fact, putting the first in perspective, would be rendered invisible by the pairing and subtraction process. Further, if one then drew the conclusion that the average humanities student attended more plays than the average biology student, this might be quite wrong if a greater proportion of biology students were female.

not a safe means of controlling for the influence of third variables. Other unmeasured variables are quite capable of varying within genera as well as between genera. On the other hand, there is no way to tell how far a generalization from a within-genus study extends to other lineages. The within-genus study sacrifices all power to assess generality across lineages, without decisively controlling for third variables.

To assess the consistency of a pattern across lineages, typical designs are based on phylogenetically independent contrasts or PICs. A phylogenetic contrast, or radiation, is a branch-point in the phylogeny and the set of branches descending from it (Felsenstein 1985, Grafen 1989, Harvey and Pagel 1991). In the simplest case it is a pair of species descended from a common ancestor. (Using pairs maximizes the number of PICs in the design relative to the number of species required.) The independence refers to the set of contrasts within a particular study being independent of each other, representing separate divergences or radiations in the phylogenetic tree. Each PIC then provides one replicate for testing whether a divergence in attribute X has consistently been associated with a divergence in Y across separate evolutionary divergences.

Suppose the phylogenetic tree adopted is simply the existing taxonomy, and the aim is to select say 20 PICs from a pool of candidate species. A simple rule for obtaining further PICs is to include new genera in preference to more than two species within a genus, new families in preference to more than two genera within a family, new orders in preference to more than two families within an order, and so forth. The effect of this rule is to spread sampling out across the phylogenetic tree, so any PIC-based design will have some degree of breadth of coverage of different lineages. Nevertheless, unless there are a very large number of PICs some lineages may go unrepresented, and nothing in the simple rule described allocates equal representation to different major branches. To achieve these design aims one might spread PICs through the phylogenetic tree more systematically, for example by treating as blocks major branches of the angiosperm tree such as rosids, asterids, and palaeoherbs. No study known to me has implemented such a design as yet.

PIC-based designs are good for assessing consistency of a relationship across many lineages, but have countervailing disadvantages (several complications of using PICs are discussed in more detail in Westoby et al 1998). Probably the most important is that they will usually not satisfy the equal-chance-of-being selected rule in the species selected from a particular habitat or strategy. Suppose, for example, we wish to contrast species from infertile soils with species from more fertile soils (e.g., Cunningham et al 1999, Wright and Westoby 1999). For each species chosen from an infertile soil, a related species will be sought on fertile soil to form a phylogenetically independent contrast. This means that from the list of all species on fertile soil, species are more likely to be chosen if they belong to genera or families that are also present on infertile soil. The species chosen according to a PIC-based design will not give a fair representation of the overall shift in the frequency distribution of (say) species leaf sizes between habitats. Specifically, they will tend to underestimate the contribution from families and orders that are present in one habitat but not the other.

One can select PICs branching across higher as well as lower taxonomic levels, so in principle it might be possible to select species in such a way that they both constituted a set of PICs between two habitats, and also were proportionately representative of the phylogenetic species-mixture occurring within each habitat, but such a design has never been attempted to my knowledge. This is on the premise that a PIC between (say) orders within a superorder is constructed by selecting one species at random from each order. In some designs such a PIC can also be constructed by estimating each order's trait value from species within that order that have also been used to build PICs between families, genera, or species. However, this is only possible when the species have been randomly sampled from the phylogeny within each order, not when the PICs have deliberately been contrasted for (say) leaf size, or soil habitat (Westoby et al 1998). No doubt species-selection on the basis of phylogeny is an area where many further developments and improvements can be expected over the next few years.

CONCLUSION

The current situation in functional plant ecology is that quite a large number of detailed field experiments and ecophysiological studies on one or a few species have accumulated, more than have been satisfactorily digested, interpreted, and generalized. Emerging wide-area applied problems, notably global change of climate and land use, are creating urgent demand for plant functional type classifications that might permit worldwide generalizations (Steffen et al 1992, Körner 1993, Woodward and Cramer 1996, Smith et al 1997). The gradual accumulation of comparative information in electronic databases is reaching critical mass, allowing patterns to have their generality quantified much more widely and quickly than a decade ago. Together, these trends mean that generalization across species and the associated topic of ecological strategy schemes are becoming keys to research progress in functional plant ecology over the next 10–20 years.

In this context the selection of species for study is an issue deserving closer attention than it has received up to the present. The maxim is to be explicit. This means describing explicitly the boundaries on categories of species that are to be compared in any given study. Ideally one would then select replicate species at random within those categories. This is a counsel of perfection that will be hard to meet in practice, but again, authors should be encouraged to think about and list explicitly whatever exclusion rules they have found it necessary to use, that prevented them from choosing at random from the whole list of species within a particular category. The work of subsequent literature review and generalization must surely become more rigorous and powerful once reviewers have available to them a clearer knowledge of what sort of species have been studied and what sorts have been avoided.

REFERENCES

Ackerly, D.D. and M.J. Donoghue, 1995. Phylogeny and ecology reconsidered. Journal of Ecology 83: 730–732.

Aerts, R. and M.J. van der Peijl, 1993. A simple model to explain the dominance of low-productive perennials in nutrient-poor habitats. Oikos 66: 144–147.

Barkman, J.J., 1988. New systems of plant growth forms and phenological plant types. In: M.J.A. Werger, P.J.M. van der Aart, H.J. During, and J.T.A. Verhoeven, eds. Plant Form and Vegetation Structure. Adaptation, Plasticity and Relation to Herbivory. SPB Academic Publishing, The Hague, pp. 9–44.

Begon, M., J.L. Harper, and C.R. Townsend, 1996. Ecology, 3rd edn. Blackwell Science, Oxford.

Box, E.O., 1981. Macroclimate and Plant Forms. Dr W. Junk, The Hague.

Bugmann, H., 1996. Functional types of trees in temperate and boreal forests: classification and testing. Journal of Vegetation Science 7: 359–370.

Burke, M.J.W. and J.P. Grime, 1996. An experimental study of plant community invasibility. Ecology 77: 776–790.

Campbell, B.D. and J.P. Grime, 1992. An experimental test of plant strategy theory. Ecology 73: 15–29.

Chapin F.S. III, K. Autumn, and F. Pugnaire, 1993. Evolution of suites of traits in relation to environmental stress. American Naturalist 139: 1293–1304.

Chapin, F.S. III, M. Bret-Harte, M. Syndonia, S.E. Hobbie, and H. Zhong, 1996. Plant functional types as predictors of transient responses of arctic vegetation to global change. Journal of Vegetation Science 7: 347–358.

Cockburn, A., 1991. An Introduction to Evolutionary Ecology. Blackwell Science, Oxford.

Cody, M.L., 1966. A general theory of clutch size. Evolution 20: 174–184.

Colinvaux, P., 1993. Ecology 2. Wiley, New York.

Connell, J.H., 1983. On the prevalence and relative importance of interspecific competition: evidence from field experiments. American Naturalist 122: 661–696.

Cornelissen, J.H.C., P. Castro Diez, and R. Hunt, 1996. Seedling growth, allocation and leaf attributes in a wide range of woody plant species and types. Journal of Ecology 84: 755–765.

Crawley, M.J., 1983. *Herbivory*. Blackwell Scientific Publications, Oxford.

Crawley, M.J. (ed.) 1996. Plant Ecology, 2nd edn. Blackwell Scientific, Oxford.

Cunningham, S.A., B. Summerhayes, and M. Westoby, 1999. Evolutionary divergences in leaf structure and chemistry, comparing rainfall and soil nutrient gradients. Ecological Monographs 69: 569–588.

Dansereau, P., 1951. Description and recording of vegetation upon a structural basis. Ecology 32: 172–229.

Dyksterhuis, E.J., 1949. Condition and management of rangeland based on quantitative ecology. Journal of Range Management 2: 104–115.

Ellenberg, H., 1988. Vegetation of Central Europe, 4th edn. Cambridge University Press, New York.

Felsenstein, J., 1985. Phylogenies and the comparative method. American Naturalist 125: 1–15.

Garnier, E. and A.H.J. Freijsen, 1994. On ecological inference from laboratory experiments conducted under optimum conditions. In: J. Roy and E. Garnier, eds. A Whole-Plant Perspective on Carbon-Nitrogen Interactions. SPB Publishing, The Hague, pp. 267–292.

Garnier, E. and G. Laurent, 1994. Leaf anatomy, specific mass and water content in congeneric annual and perennial grasses. New Phytologist 128: 725–736.

Goldberg, D.E., 1996. Competitive ability: definitions, contingency and correlated traits. Philosophical Transactions of the Royal Society B 351: 1377–1385.

Goldberg, D.E. and A.M. Barton, 1992. Patterns and consequences of interspecific competition in natural communities: A review of field experiments with plants. American Naturalist 139: 771–801.

Gould, S.J. and E. Vrba, 1982. Exaptation—a missing term in the science of form. Paleobiology, 8: 4–15.

Grafen, A., 1989. The phylogenetic regression. Philosophical Transactions of the Royal Society of London, Series B 205: 581–598.

Greenslade, P.J.M., 1972. Evolution in the Staphylinid genus *Priochirus* (Coleoptera). Evolution 26: 203–220.

Greenslade, P.J.M., 1983. Adversity selection and the habitat templet. American Naturalist 122: 352–365.

Grime, J.P., 1974. Vegetation classification by reference to strategies. Nature 250: 26–31.

Grime, J.P., 1977. Evidence for the existence of three primary strategies in plants and its relevance to ecological and evolutionary theory. American Naturalist 111: 1169–1194.

Grime, J.P., 1979. Plant Strategies and Vegetation Processes. Wiley, Chichester.

Grime, J.P., 1984. The ecology of species, families and communities of the contemporary British flora. New Phytologist 98: 15–33.

Grime, J.P., 1994a. Defining the scope and testing the validity of CSR theory: A response to Midgley, Laurie and Le Maitre. Bulletin of the South African Institute of Ecologists 13: 4–7.

Grime, J.P., 1994b. The role of plasticity in exploiting environmental heterogeneity. In: M.M. Caldwell and R. Pearcy, eds. Exploitation of Environmental Heterogeneity in Plants. Academic Press, San Diego, CA, pp. 1–18.

Grime, J.P. and S.H. Hillier, 1992. The contribution of seedling regeneration to the structure and dynamics of plant communities and larger units of landscape. In: M. Fenner, ed. Seeds: The Ecology of Regeneration in Plant Communities. CAB International, Wallingford, UK, pp. 349–364.

Grime, J.P. and R. Hunt, 1975. Relative growth rate: its range and adaptive significance in a local flora. Journal of Ecology 63: 393–342.

Grime, J.P., J.G. Hodgson, and R. Hunt, 1988. Comparative Plant Ecology. Unwin-Hyman, London.

Grime, J.P. and 33 others, 1997. Integrated screening validates a primary axis of specialization in plants. Oikos 79: 259–281.

Grubb, P.J., 1985. Plant populations and vegetation in relation to habitat, disturbance and competition: Problems of generalization. In J. White, ed. The Population Structure of Vegetation. Junk, Dordrecht, pp. 595–621.

Gurevitch, J., L.L. Morrow, A. Wallace, and J.S. Walsh, 1992. A meta-analysis of field experiments on competition. American Naturalist 140: 539–572.

Hairston, N.G., Sr. 1989. Ecological Experiments: Purpose, Design and Execution. Cambridge University Press, Cambridge.

Harvey, P.H., 1996. Phylogenies for ecologists. Journal of Animal Ecology 65: 255–263.

Harvey, P.H. and M.D. Pagel, 1991. The Comparative Method in Evolutionary Biology. Oxford University Press, Oxford.

Harvey, P.H, A.F. Read, and S. Nee, 1995a. Why ecologists need to be phylogenetically challenged. Journal of Ecology 83: 535–536.

Harvey, P.H, A.F. Read, and S. Nee, 1995b. Further remarks on the role of phylogeny in comparative ecology. Journal of Ecology 83: 735–736.

Hubbell, S.P. and R.B. Foster, 1986. Commonness and rarity in a tropical forest: implications for tropical tree diversity. In: M.E. Soule, ed. Conservation Biology. Sinauer, Sunderland, MA, pp. 205–231.

Hughes, L., M. Dunlop, K. French, M. Leishman, B. Rice, L. Rodgerson, and M. Westoby, 1994. Predicting dispersal spectra: A minimal set of hypotheses based on plant attributes. Journal of Ecology 82: 933–950.

Hunt, R. and J.H.C. Cornelissen, 1997. Physiology, allocation and growth rate: a reexamination of the Tilman model. American Naturalist 150: 122–130.

Ingrouille, M., 1992. Diversity and Evolution of Land Plants. Chapman and Hall, London.

Keddy, P.A., 1989. Competition. Chapman and Hall, London.

Kelly, C.K. and A. Purvis, 1993. Seed size and establishment conditions in tropical trees: On the use of taxonomic relatedness in determining ecological patterns. Oecologia 94: 356–360.

Kitajima, K., 1994. Relative importance of photosynthetic traits and allocation pattern as correlates of seedling shade tolerance of 13 tropical trees. Oecologia 98: 419–428.

Körner, Ch., 1993. Scaling from species to vegetation: the usefulness of functional groups. In E.-D. Schulze and H.A. Mooney, eds. Biodiversity and Ecosystem Function. Springer-Verlag, Berlin, pp. 117–140.

Lambers, H. and H. Poorter, 1992. Inherent variation in growth rate between higher plants: a search for physiological causes and ecological consequences. Advances in Ecological Research 23: 188–261.

Lambers, H., H. Konings, M.L. Cambridge, and T.L. Pons, eds, 1989. Causes and Consequences of Variation in Growth Rate and Productivity of Higher Plants. SPB Academic Publishing, The Hague.

Lange, O.L., P.S. Nobel, C.B. Osmond, and H. Ziegler, eds, 1984. Encyclopaedia of Plant Physiology, New Series, Vol 12B. Physiological Plant Ecology II. Water Relations and Carbon Assimilation. Springer-Verlag, NY.

Leps, J., J. Osborna-Kosinova, and K. Rejmanek, 1982. Community stability, complexity and species life-history strategies. Vegetatio 50: 53–63.

Loehle, C., 1988. Tree life history strategies: The role of defenses. Canadian Journal of Forest Research 18: 209–222.

MacArthur, R.H. and E.O. Wilson, 1967. The Theory of Island Biogeography. Princeton University Press, Princeton, NJ.

Mueller-Dombois, D. and H. Ellenberg, 1974. Aims and Methods of Vegetation Ecology. Wiley, New York.

Mulkey, S.S., R.L. Chazdon, and A.P. Smith, eds, 1996. Tropical Forest Plant Ecophysiology. Chapman and Hall, London.

Noble, I.R. and R.O. Slatyer, 1980. The use of vital attributes to predict successional changes in plant communities subject to recurrent disturbances. Vegetatio 43: 5–21.

Orshan, G., 1989. Plant Pheno-Morphological Studies in Mediterranean Ecosystems. Geobotany Volume 12, Junk, The Hague.

Poorter, H., 1989. Interspecific variation in relative growth rate: On ecological causes and physiological consequences. In: H. Lambers et al., eds. Causes and Consequences of Variation in Growth Rate and Productivity in Higher Plants. SPB Academic Publishing, The Hague, pp. 45–68.

Poorter, H., 1994. Construction costs and payback time of biomass: A whole plant perspective. In: J. Roy and E. Garnier, eds. A Whole-Plant Perspective on Carbon-Nitrogen Interactions. SPB Publishing, The Hague, pp. 111–127.

Poorter, H. and H. Lambers, 1991. Is interspecific variation in relative growth rate positively correlated with biomass allocation to the leaves? American Naturalist 138: 1264–1268.

Poorter, H. and C. Remkes, 1990. Leaf area ratio and net assimilation rate of 24 wild species differing in relative growth rates. Oecologia 83: 553–559.
Poorter, H. and A. van der Werf, 1998. Is inherent variation in RGR determined by LAR at low irradiance and by NAR at high irradiance? A review of herbaceous species. In: Lambers, H. Poorter, H, and Van Vuaren M.M.I., eds. Inherent Variation in Plant Growth. Physiological Mechanisms and Ecological Consequences. Backhuys, Leiden, The Netherlands, pp. 309–336.
Prentice, I.C., W. Cramer, S.P. Harrison, R. Leemans, R.A. Monserud, and A.A. Solomon, 1992. A global biome model based on plant physiology and dominance, soil properties and climate. Journal of Biogeography 19: 117–134.
Price, T., 1997. Correlated evolution and independent contrasts. Philosophical Transactions of the Royal Society London B 352: 519–529.
Price, P.W., M. Westoby, B. Rice, P.R. Atsatt, R.S. Fritz, J.N. Thompson and K. Mobley, 1986. Parasite mediation in ecological interactions. Annual Reviews of Ecology and Systematics 17: 487–505.
Raunkiaer, C., 1934. The Life Forms of Plants and Statistical Plant Geography. Clarendon Press, Oxford.
Rees, M., 1993. Trade-offs among dispersal strategies in British plants. Nature 366: 150–152.
Rees, M., 1995. EC-PC comparative analyses? Journal of Ecology 83: 891–892.
Reeve, H.K. and P.W. Sherman, 1993. Adaptation and the goals of evolutionary research. Quarterly Review of Biology 68: 1–32.
Reich, P.B., M.B. Walters, and D.S. Ellsworth, 1992. Leaf life-span in relation to leaf, plant and stand characteristics among diverse ecosystems. Ecological Monographs 62: 365–392.
Reich, P.B., M.B. Walters, and D.S Ellsworth, 1997. From tropics to tundra: Global convergence in plant functioning. Proceedings of the National Academy of Science 94: 13730–13734.
Roy, J. and E. Garnier, eds, 1994. A Whole-Plant Perspective on Carbon-Nitrogen Interactions. SPB Publishing, The Hague.
Sarmiento, G. and M. Monasterio, 1983. Life forms and phenology. In: F. Bourliere, ed. Ecosystems of the World 13: Tropical Savannas. Elsevier, Amsterdam, pp. 79–108.
Saverimuttu, T. and M. Westoby, 1996. Seedling longevity under deep shade in relation to seed size. Journal of Ecology 84: 681–689.
Schoener, T.W., 1983. Field experiments on interspecific competition. American Naturalist 122: 240–285.
Schulze, E.-D. and M.M. Caldwell, eds, 1994. Ecophysiology of Photosynthesis. Springer-Verlag, Berlin.
Sih, A., P. Crowley, M. McPeek, J. Petranka, and K. Strohmeier, 1985. Predation, competition and prey communities. Annual Review of Ecology and Systematics 16: 269–311.
Silvertown, J. and M. Dodd, 1996. Comparing plants and connecting traits. Philosophical Transactions of the Royal Society of London Series B 351: 1233–1239.
Smith, T.M., H.H. Shugart, and F.I. Woodward, eds, 1997. Plant Functional Types: their relevance to ecosystem properties and global change. Cambridge University Press, Cambridge.
Southwood, T.R.E., 1977. Habitat, the templet for ecological strategies. Journal of Animal Ecology 46: 337–365.
Steffen, W.L., B.H. Walker, J.S. Ingram, and G.W. Koch, eds, 1992. Global Change and Terrestrial Ecosystems: The Operational Plan. International Geosphere-Biosphere Program, IGBP Report No. 21, Stockholm.
Turner, I.M., 1994. A quantitative analysis of leaf form in woody plants from the world's major broadleaved forest types. Journal of Biogeography 21: 413–419.
van der Werf, A., M. van Nuenen, A.J. Visser, and H. Lambers, 1993. Contribution of physiological and morphological plant traits to a species' competitive ability at high and low nitrogen supply. Oecologia 94: 434–440.
Westoby, M., 1998. A leaf-height-seed (LHS) plant ecology strategy scheme. Plant and Soil 199: 213–227.
Westoby, M., M.R. Leishman, and J.M. Lord, 1995a. On misinterpreting the "phylogenetic correction." Journal of Ecology 83: 531–534.

Westoby, M., M.R. Leishman, and J.M. Lord, 1995b. Further remarks on phylogenetic correction. Journal of Ecology 83: 727–730.

Westoby, M., M.R. Leishman, and J.M. Lord, 1995c. Issues of interpretation after relating comparative datasets to phylogeny. Journal of Ecology, 83: 892–893.

Westoby, M., M.R. Leishman, and J.M. Lord, 1996. Comparative ecology of seed size and seed dispersal. Philosophical Transactions of the Royal Society B 351: 1309–1318.

Westoby, M., S. Cunningham, C.M. Fonseca, J.M. Overton, and I.J. Wright, 1998. Phylogeny and variation in light capture deployed per unit investment in leaves: designs for selecting study species with a view to generalizing. In: H. Lambers et al., ed. Variation in Growth Rate and Productivity of Higher Plants. Backhuys Publishers, Leiden, The Netherlands, pp. 539–566.

Whittaker, R.H., 1975. Communities and Ecosystems. Macmillan, New York.

Williams, G.C., 1992. Natural Selection: Domains, Levels, and Challenges. Oxford University Press, Oxford.

Wilson, J.B. and A.D.Q. Agnew, 1992. Positive-feedback switches in plant communities. Advances in Ecological Research 23: 263–336.

Witkowski, E.T.F. and B.B. Lamont, 1991. Leaf specific mass confounds leaf density and thickness. Oecologia 88: 486–493.

Woodward, F.I. and W. Cramer, 1996. Plant functional types and climatic changes: introduction. Journal of Vegetation Science 7: 306–308.

Wright, I.J. and M. Westoby, 1999. Differences in seedling growth behaviour among species: trait correlations across species, and trait shifts along nutrient compared to rain gradients. Journal of Ecology 87: 85–97.

Index

B

C

D

E

F

G

H

M

N

O

P

Q

R

T

U

V

W

X

Y

Z